dynamics of fluids in porous media

Jacob Bear
Department of Civil Engineering
Technion—Israel Institute of Technology, Haifa

DOVER PUBLICATIONS, INC.
New York

This Dover edition, first published in 1988, is an unabridged, corrected republication of the work first published by the American Elsevier Publishing Company, Inc., New York, 1972, in its Environmental Science Series.

Library of Congress Cataloging-in-Publication Data

Bear, Jacob.
Dynamics of fluids in porous media / by Jacob Bear.
p. cm.
Reprint. Originally published: New York: American Elsevier Pub. Co., 1972. Originally published in series: Environmental science series (New York, 1972–). With corrections.
Bibliography: p.
Includes index.
ISBN-13: 978-0-486-65675-5
ISBN-10: 0-486-65675-6
1. Fluid dynamics. 2. Groundwater flow. I. Title.
TA357.B38 1988
620.1'064—dc19 87-34940
CIP

Manufactured in the United States by LSC Communications
4500057068
www.doverpublications.com

This book is dedicated to
my dear friend, Colonel S. Alton,
who fell defending his country
in the Six Days' War (June 5–11, 1967).

Contents

CHAPTER 5

CHAPTER 8

CHAPTER 9

CHAPTER 10

CHAPTER 11

Jacob Bear

Jacob Bear is one of the world's foremost hydrologists. Presently, he has 67 publications to his credit and is Professor of Hydrology and Deputy Vice President at the Technion-Israel Institute of Technology in Haifa, where he has also been head of both the Hydraulics Laboratory and the Water Resources Research Center. In addition to his years at the Technion, Dr. Bear has taught and developed courses in hydrology at the Massachusetts Institute of Technology, the University of California at Berkeley, the University of Wisconsin, Princeton University, and New Mexico Institute of Mines and Technology. He has also been a planning engineer, a hydrologist, and a consultant on hydrology to Water Planning for Israel Ltd., the Ministry of Agriculture of Israel, and, most recently, Yalon-Balasha, consultants and engineers on large scale development projects in Iran and the Benue Plateau in Nigeria.

A native of Israel, Dr. Bear graduated with honors from the Technion in 1953, where he also received his Dipl. Eng. and M.Sc. degrees in 1954 and 1957 respectively. In 1960 he obtained his Ph.D at the University of California at Berkeley. His studies abroad include scholarships at the Hague and the University of California at Berkeley, and a three-month U.N. scholarship for a study tour in the Netherlands, France and the U.S.A.

Active in many related fields, Dr. Bear's memberships include the Israel Association of Geodesy and Geophysics, the International Association for Hydraulic Research, the National Committee of the Hydrological Decade, the editorial board of the *Israel Journal of Earth Sciences,* and Israel's Water Planning Committee. He has been chairman of the Hydrology Section of Israel Association of Geodesy and Geophysics, the National Committee of the Hydrological Decade, and the Section of Flow through Porous Media of the International Association for Hydraulic Research. A speaker at numerous international conferences, Dr. Bear's lectures include discussions on intercepting fresh water above the interface in a coastal aquifer, the optimal yield of an aquifer, hydrodynamic dispersion in porous media, immiscible displacement in fractured porous media, and hydrologic education in Israel.

Preface

This book is an attempt to present, in an ordered manner, the theory of dynamics (actually, also of statics) of fluids in porous media, as applicable to many disciplines of science and engineering. For some years I have taught courses on flow through porous media, and have treated this subject as a part of other courses, such as ground water hydrology, while at the Technion—Israel Institute of Technology, at M.I.T. where I spent my sabbatical leave (1966–7), and at several other institutions. I have felt the lack of a suitable textbook on this subject. Ideally, such a text should start from first principles of fluid mechanics and mechanics of continua, should show the passage from the microscopic to the macroscopic level of treatment, should emphasize the special features of porous media, establish the macroscopic theory and then show how it is applied to cases of practical interest.

It is rather surprising that in spite of its importance in many fields of practical interest, such as petroleum engineering, ground water hydrology, agricultural engineering and soil mechanics, so small number of treatises is available on fluids in porous media. This circumstance is even more surprising in view of the vast amount of literature published on the subject in a number of scientific and engineering journals. Although dynamics of fluids in porous media could become an interesting interdisciplinary course serving several departments, I believe that the relatively small number of courses offered by universities on the subject is due in part to lack of a suitable textbook. To overcome this lack I prepared notes for my own classes, which I present here in the form of a book, hoping that it will serve others in a similar situation.

The book is designed primarily for advanced undergraduate students and for graduates in fields such as ground water hydrology, soil mechanics, soil physics, drainage and irrigation engineering, sanitary engineering, petroleum engineering and chemical engineering, where flow through porous media plays a fundamental role. The book, I hope, will also serve the needs of scientists and engineers already active in these fields, who require a sound theoretical basis for their work. The emphasis in this book is on understanding the microscopic phenomena occurring in porous media and on their macroscopic description. The reader is led to grasp the meanings of the various parameters and coefficients appearing in the macroscopic descriptions of problems of flow through porous media, and their actual determination, as well as the limitations and approximations inherent in their description. In each case, the objective is to achieve a clear formulation of the flow problem considered and a complete mathematical statement of it in terms of partial differential equations and a set of initial and boundary conditions. Once a flow problem is stated properly in mathematical terms, three methods of solution are possible in principle: analytic solution, numerical solution aided by high speed digital computers and solution by means of laboratory models and analogs. All three tools are described in this book. Typical examples of analytic solutions are scattered throughout the book, but no attempt is made to present a collection of a large number of solved problems. The principles of the numerical method of solution are presented, and a detailed description is given of laboratory models and analogs, their scaling and applications.

Mathematics is employed extensively and the reader is expected to have a good background in advanced engineering mathematics, including such subjects as vector analysis, Cartesian tensor analysis, partial differential equations and elements of the theory of functions.

No attempt is made to give a complete citation of all published literature or to indicate the first author on a particular subject. References selected for citation are those I think represent a more important point of view, are more appropriate from the educational point of view or are more readily available for the average reader.

Obviously a single book, even of this size, cannot include everything related to the subject treated. Although we consider porous media in general, the discussion is limited to media with relatively large pores, thus excluding clays and media with micropores or colloidal-size particles. Similarly, chemical and electrochemical surface phenomena are excluded. The discussion is restricted to Newtonian fluids.

With these objectives and limitations in mind, the book starts with examples of two important porous media: the ground water aquifer and the oil reservoir. An attempt is made to define porous media, and the continuum approach is introduced as a tool for treating phenomena in porous media. This requires the definition of a "representative elementary volume" based on the definition of porosity. Chapter 2 includes a summary of some important fluid and porous media properties. In chapter 3, the concepts of pressure and piezometric head are introduced. Chapter 4 starts with the definition of velocities and fluxes in a fluid continuum. Then the equations of conservation of mass, momentum and energy in a fluid continuum are presented, and using a porous medium conceptual model these equations are averaged to obtain the basic equations that describe flow through porous media: the equations of volume and mass conservation, including the equation of mass conservation of a species in solution (also called the equation of hydrodynamic dispersion), and the motion equation for the general case of an anisotropic medium and inhomogeneous fluid. Although the basic equations of motion and of mass conservation are developed from first principles in chapter 4, chapters 5 and 6 return to these topics, discussing them from a different point of view, perhaps more suitable for the reader who is less versed in fluid mechanics. Chapter 5 presents the equation of motion, starting from its original one-dimensional form (as suggested by Darcy on the basis of experiments), and extending it to three-dimensional flow, compressible fluids and anisotropic media. This chapter also contains a review of theoretical derivations of Darcy's law. My objectives in presenting this and similar reviews is to indicate research methods, such as the use of conceptual and statistical models. A section on the motion equation at high Reynolds numbers is also included.

In chapter 6, the control volume approach is introduced as a general tool for developing mass conservation equations. Special attention is devoted to deformable media. Also included in this chapter is the stream function and its relationship to the piezometric head. Once the continuity or mass conservation equations have been established, the next natural step is to consider the initial and boundary conditions. These are discussed in detail in chapter 7. Special attention is given to the phreatic surface boundary condition and to its description in the hodograph plane. The second part of this chapter contains a discussion on various analytic and numerical solution techniques.

Upon reaching this point, the reader should be able to state a problem of flow through porous media in terms of an appropriate partial differential equation and a set of initial and boundary conditions. He should also know the major methods of solution (analog solutions are discussed in chapter 11).

Chapter 8 deals with the problem of flow in unconfined aquifers. This is a problem often encountered in ground water hydrology and in drainage. The Dupuit assumptions are explained and employed to derive the continuity equations for unconfined flow. The hodograph method, as a tool for solving two-dimensional, steady phreatic flow problems, is discussed in detail with many examples. Several linearization techniques and solutions of the nonlinear equation of unconfined flow are also presented in this chapter.

In chapter 9 the discussion, hitherto confined to single-phase flow, is extended to polyphase flow in porous media, a topic of special interest in petroleum engineering. Starting from the fundamental concepts of saturation, capillary pressure and relative permeability, the motion and continuity equations are established. The case of unsaturated flow as treated by soil physicists is presented as a special case of flow of immiscible fluids, where one of the fluids—the air—is stationary and at constant pressure. Special cases of interest, dealing with infiltration into soils, are considered in more detail. A new concept is introduced: that of an abrupt interface as an approximation replacing the actual transition zone that occurs between two fluids, whether miscible or immiscible. A detailed discussion is presented on the coastal interface, of great interest to ground water hydrologists.

Chapter 10 deals with hydrodynamic dispersion. Again, although the fundamental equation is developed from first principles in chapter 4, a review of several other theories leading to this equation is presented. Special attention is given to the coefficient of dispersion and its relationship to matrix and flow characteristics. A section on heat and mass transfer completes the discussion on hydrodynamic dispersion.

Chapter 11 presents the use of models and analogs, both as research tools and as tools for solving boundary value problems. Following the presentation of a general method for deriving analog scales, a detailed description is given of the sand box model, the electric analogs of various types, the Hele–Shaw analogs and the membrane analog. Recommendations for application are indicated in each case.

In brief, this is the subject matter I have chosen to cover in this book. I have made an effort to present the information in such a way as to require a minimum of supplementary material, except for those who wish to dig more deeply into the subject. A large number of problems and exercises is included in this book.

I should like to express my appreciation to the many individuals who, through their comments and criticism, have contributed to the completion of this book. Special thanks are due to Dr. Y. Bachmat, Dr. C. Braester, Mr. E. A. Hefez and E. Goldshlager, for the help they have given me in reading, discussing and constructively criticizing the draft. Thanks are also due to the Department of Civil Engineering at M.I.T., and especially to Professor C. L. Miller, head of the department, Professor A. T. Ippen and Professor D. R. F. Harleman, who made it possible for me to write a large part of this book while spending a most fruitful year as a visiting professor at M.I.T.

The heaviest burden involved in writing this book was borne by my wife, Siona, who had to put up with the many inconveniences that are unavoidable when one is engaged in writing a book. For her constant encouragement to me throughout the various stages of writing, my hearty gratitude.

I realize that an attempt to represent a systematic account of a theory, such as I have made here, is bound to have defects. I will accept with gratitude all readers' suggestions directed toward the improvement of this book.

Haifa, Israel
1972

Jacob Bear

CHAPTER 1

Introduction

1.1 Aquifers, Ground Water and Oil Reservoirs

Flow through porous media is a topic encountered in many branches of engineering and science, e.g., ground water hydrology, reservoir engineering, soil science, soil mechanics and chemical engineering. Although our objective in this book was to present only the fundamental aspects of this topic, common to all these scientific and applied fields, we thought it appropriate to begin by presenting examples of porous media, and fluids in them, as encountered in practice. The aquifer, which is the porous medium domain treated by the ground water hydrologist, and the oil reservoir, which is the porous medium domain treated by the reservoir engineer, will serve as typical examples for this purpose. Following is a brief description of these domains and the fluids present in them.

1.1.1 Definitions

An *aquifer* (a ground water basin) is a geologic formation, or a stratum, that (a) contains water, and (b) permits significant amounts of water to move through it under ordinary field conditions. Todd (1959) traces the term aquifer to its Latin origin: *aqua*, meaning water, and *-fer* from *ferre*, meaning to bear.

In contradistinction, an *aquiclude* is a formation that may contain water (even in appreciable amounts), but is incapable of transmitting significant quantities under ordinary field conditions. A clay layer is an example. For all practical purposes, an *aquiclude* is considered an *impervious formation*.

An *aquitard* is a semipervious geologic formation transmitting water at a very slow rate as compared to the aquifer. However, over a large (horizontal) area it may permit the passage of large amounts of water between adjacent aquifers, which it separates from each other. It is often referred to as a *leaky formation*. An *aquifuge* is an impervious formation that neither contains nor transmits water.

Ground water is a term used to denote all waters found beneath the ground surface. However, the ground water hydrologist, who is primarily concerned with the water contained in the zone of saturation (par. 1.1.2), uses the term *ground water* to denote water in this zone. In drainage of agricultural lands, or agronomy, the term *ground water* is used also to denote the water in the partially saturated layers above the water table. In this book we shall use the term *ground water* mainly in the sense employed by the ground water hydrologist.

That portion of rock not occupied by solid matter is the *void space* (also *pore space*, *pores*, *interstices* and *fissures*). This space contains water and/or air. Only *connected*

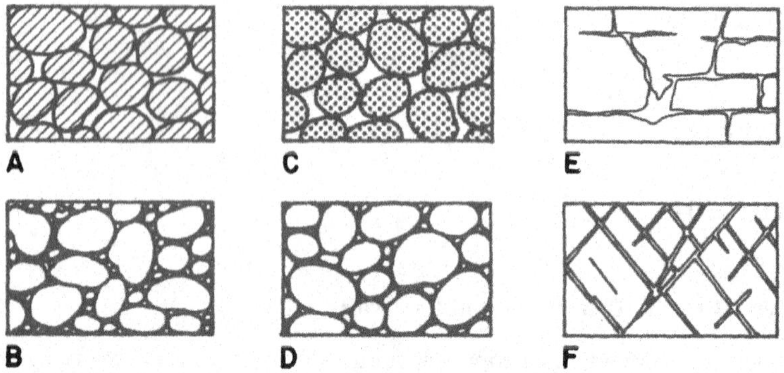

FIG. 1.1.1. Diagram showing several types of Rock Interstices. A. Well-sorted sedimentary deposit having high porosity; B. Poorly sorted sedimentary deposit having low porosity; C. Well-sorted sedimentary deposit consisting of pebbles that are themselves porous, so that the deposit as a whole has a very high porosity; D. Well-sorted sedimentary deposit whose porosity has been diminished by the deposition of mineral matter in the interstices; E. Rock rendered porous by solution; F. Rock rendered porous by fracturing. (After Meinzer, 1942.)

interstices can act as elementary conduits within the formation. Figure 1.1.1 (after Meinzer 1942) shows several types of rock interstices. Interstices may range in size from huge limestone caverns to minute subcapillary openings in which water is held primarily by adhesive force. The interstices of a rock can be grouped in two classes: original interstices, mainly in sedimentary and igneous rocks, created by geologic processes at the time the rock was formed, and secondary interstices, mainly in the form of fissures, joints and solution passages, developed after the rock was formed.

1.1.2 The Moisture Distribution in a Vertical Profile

Subsurface water may be divided vertically into zones depending on the relative proportion of the pore space occupied by water: a *zone of saturation*, in which all pores are completely filled with water, and an overlying *zone of aeration*, in which the pores contain both gases (mainly air and water vapor) and water.

Figure 1.1.2 shows a schematic distribution of subsurface water. Water (e.g., from precipitation and/or irrigation) infiltrates the ground surface, moves downward, primarily under the influence of gravity, and accumulates, filling all the interconnected interstices of the rock formation above some impervious stratum. A *zone of saturation* is thus formed above this impervious bedrock. The saturated zone (fig. 1.1.2) is bounded above by a *water table*, or *phreatic surface*. This is a surface on which the pressure is atmospheric. It is revealed by the water level in a well penetrating the aquifer, in which the flow is essentially horizontal (sec. 6.6). Actually, saturation extends a certain distance above the water table, depending on the type of soil (par. 9.4.2). Wells, springs and streams are fed by water from the zone of saturation.

The *zone of aeration* extends from the water table to the ground surface. It usually

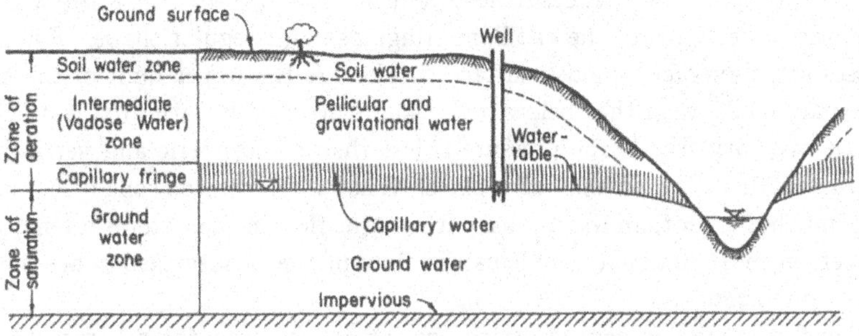

FIG. 1.1.2. The distribution of subsurface water.

consists of three subzones: the *soil water zone* (or *belt of soil water*), the *intermediate zone* (or *vadose water zone*) and the *capillary zone* (or *capillary fringe*).

The soil water zone is adjacent to the ground surface and extends downward through the root zone. The moisture distribution in this zone is affected by conditions at the ground surface: seasonal and diurnal fluctuations of precipitation, irrigation, air temperature and air humidity, and by the presence of a shallow water table. Water in this zone moves downward during infiltration (e.g., from precipitation, flooding of the ground surface or irrigation; sec. 9.4), and upward by evaporation and plant transpiration. Temporarily, during a short period of excessive infiltration, the soil in this zone may be almost completely saturated (*gravitational water*).

After an extended period of drainage without a supply of water at the soil surface, the amount of moisture remaining in the soil is called *field capacity* (par. 9.4.3). Below field capacity, the soil contains capillary water, in the form of continuous films around the soil particles, held by surface tension. This water is moved by capillary action and is available to plants. Below a moisture content—called the *hygroscopic coefficient* (maximum moisture that an initially dry soil will adsorb when brought in contact with an atmosphere of 50% relative humidity at 20°C)—the water in the soil is called *hygroscopic water*. Water is then unavailable to plants as it forms thin films of moisture adhering strongly to the surface of soil particles (fig. 9.4.2).

The intermediate zone extends from the lower edge of the soil water zone to the upper limit of the capillary zone. It does not exist when the water table is too high, in which case the capillary fringe may extend into the soil water zone, or even to the ground surface. Nonmoving, or *pellicular*, water in the intermediate zone is held in place by hygroscopic and capillary forces. Temporarily, water moves downward through this zone as gravitational water.

The capillary fringe extends upward from the water table. Its thickness depends on the soil properties and on the uniformity of pore sizes. The capillary rise ranges from practically nothing in coarse materials to as much as $2 \div 3$ m and more in fine materials (e.g., clay). A detailed discussion on the capillary rise is given in paragraph 9.4.2. Within the capillary zone there is usually a gradual decrease in moisture content with height above the water table. Just above the water table, the pores are practically saturated. Moving higher, only the smaller connected pores contain water. Still higher, only the smallest connected pores are still filled with water. Hence, the upper limit of the capillary fringe has an irregular shape. For practical purposes some average, smooth surface is taken as the upper limit of the capillary fringe (par. 9.4.2), such that below it the soil is practically saturated (say, $> 75\%$).

In the capillary fringe, the pressure is less than atmospheric and vertical as well as horizontal flow of water may take place. When the saturated zone below the water table is much thicker than the capillary fringe, the flow in the latter is often neglected. However, in most drainage problems the flow in the unsaturated zones may be of primary importance.

Obviously, numerous complications are introduced into the schematic moisture distribution described here by the great variability in pore sizes, the presence of permeability layers and by the temporary movement of infiltrating water.

1.1.3 Classification of Aquifers

The following brief review of some geological formations that serve as aquifers is based on a work by Thomas (1952).

Most aquifers consist of unconsolidated or partly consolidated gravel and sand. They are located in abandoned or buried valleys, in plains and in intermontane valleys. Some are of a limited area; others may extend over large areas. Their thickness may also vary from several meters to several hundred meters.

Sandstone and conglomerate are the consolidated equivalent of sand and gravel. In these rocks the individual particles have been cemented together, thus reducing permeability.

Limestone formations, varying widely in thickness, density, porosity and permeability, serve as important aquifers in many parts of the world, especially when sizeable proportions of the original rock have been dissolved and removed. Openings in limestone may range from microscopic original pores to large fractures and caverns forming subterranean channels. By dissolving the rock along fractures and fissures the water tends to enlarge them, thus increasing permeability with time. Ultimately, a limestone terrane develops into a karst region. Its macroscopic behavior (i.e., on a large scale) is probably similar to that of a sand and gravel aquifer; on a smaller scale, this similarity is questionable (sec. 1.2). Figure 1.1.1 e and f show examples of rocks rendered porous by solution and fracturing.

Volcanic rocks may form permeable aquifers. Basalt flows are very permeable. The pore space of a basalt aquifer may not be as large as that of loose sands and gravels, but the permeability, owing to the cavernous character of the openings, may be many times greater. Most shallow intrusive rocks in the form of sills, dikes or plugs are low in permeability, and many of them are impervious enough to serve as barriers to ground water flow.

Crystalline and metamorphic rocks are relatively impervious and constitute poor aquifers. When such rocks occur near the ground surface, some permeability may develop by weathering and fracturing.

Clay and coarser materials mixed with clay, although in general having a high porosity, are relatively impervious owing to the small size of their pores.

Aquifers may be regarded as underground storage reservoirs that are replenished *naturally* by precipitation and influent streams, or through wells and other *artificial* recharge methods. Water leaves the aquifer naturally through springs or effluent streams and artificially through pumping wells.

The thickness and other vertical dimensions of an aquifer are usually much smaller than the horizontal lengths involved. Therefore, throughout this book, all drawings describing flow in aquifers are *highly distorted*. The reader should not be misled by the distorted scales of such figures.

Aquifers may be classed as unconfined or confined depending upon the presence or absence of a water table.

A *confined aquifer* (fig. 1.1.3), also known as a *pressure aquifer*, is one bounded above and below by impervious formations. In a well penetrating such an aquifer, the water level will rise above the base of the confining formation; it may or may not reach the ground surface. A properly constructed *observation well* (or a piezometer) has a relatively short screened section (yet not too short with respect to the size of the openings; see sec. **1.3**) such that it indicates the *piezometric head* (sec. **3.3**) at a specific point (say, the center of the screen). The water levels in a number of observation wells tapping a certain aquifer define an imaginary surface called the *piezometric*

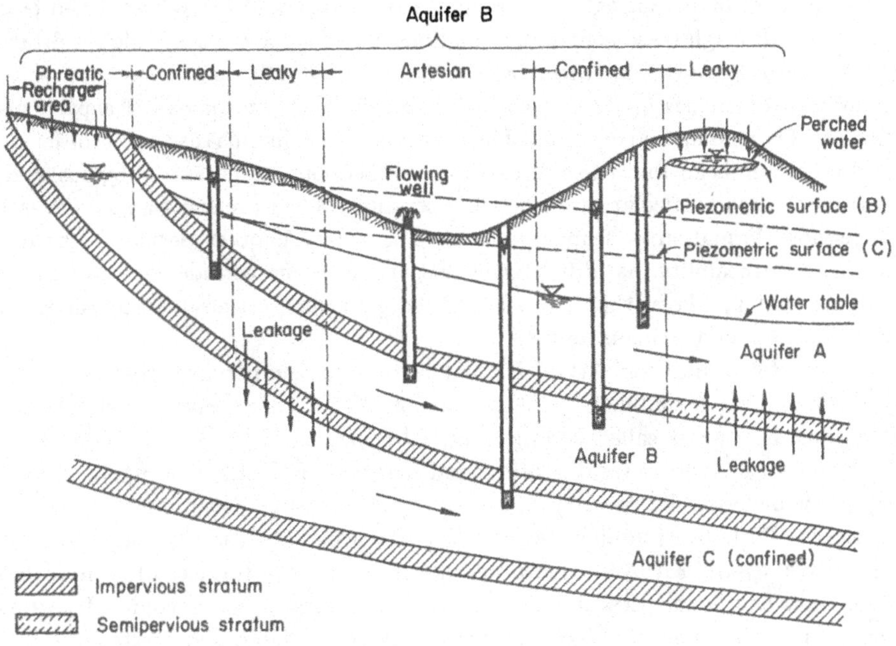

FIG. 1.1.3. Types of aquifers.

surface (or *isopiestic surface*). When the flow in the aquifer is essentially horizontal, such that equipotential surfaces are vertical, the depth of the piezometer opening is immaterial; otherwise, a different piezometric surface is obtained for piezometers that have openings at different elevations. Fortunately, *except in the neighborhood of outlets such as partially penetrating wells or springs*, the *flow in aquifers is essentially horizontal*.

An *artesian aquifer* is a confined aquifer (or a portion of it) where the elevations of the piezometric surface (say, corresponding to the base of the confining layer) are above ground surface. A well in such an aquifer will flow freely without pumping (*artesian* or *flowing well*). Sometimes the term *artesian* is used to denote a confined aquifer.

Water enters a confined aquifer through an area between confining strata that rise to the ground surface, or where an impervious stratum ends underground, rendering the aquifer unconfined. The region supplying water to a confined aquifer is called a *recharge area*.

A *phreatic aquifer* (also called *unconfined aquifer* or *water table aquifer*) is one with a water table (*phreatic surface*) serving as its upper boundary. Actually, above the phreatic surface is a capillary fringe, often neglected in ground water studies. A phreatic aquifer is recharged from the ground surface above it, except where impervious layers of limited horizontal area exist between the phreatic surface and the ground surface.

Aquifers, whether confined or unconfined, that can lose or gain water through either or both of the formations bounding them above and below, are called *leaky aquifers*. Although these bounding formations may have a relatively high resistance to the flow of water through them, over the large (horizontal) areas of contact involved significant quantities of water may leak through them into or out of a particular aquifer. The amount and direction of leakage is governed in each case by the difference in piezometric head that exists across the semipervious formation. Obviously, the decision in each particular case whether a certain stratum overlying an aquifer is an impervious formation, a semipervious one, or simply another pervious formation having a permeability that differs from that of the aquifer considered, is not a clear-cut one. Often, a layer that is considered semipervious (or leaky) is thin relative to the thickness of the main aquifer.

A phreatic aquifer (or part of it) that rests on a semipervious layer is a *leaky phreatic aquifer*. A confined aquifer (or part of it) that has at least one semipervious confining stratum is called a *leaky confined aquifer*.

Figure 1.1.3 shows several aquifers and observation wells. The upper phreatic aquifer is underlain by two confined ones. In the recharge area, aquifer B becomes phreatic. Portions of aquifers A, B and C are leaky, with the direction and rate of leakage determined by the elevation of the piezometric surfaces of each of these aquifers. The boundaries between the various confined and unconfined portions may vary with time as a result of changes in water table and piezometric head elevations. A special case of a phreatic aquifer is the *perched aquifer* (fig. 1.1.3) that occurs wherever an impervious (or relatively impervious) layer of limited horizontal area is located between the water table of a phreatic aquifer and the ground surface. Another ground water body is then built above this impervious layer. Clay or loam lenses in sedimentary deposits have shallow perched aquifers above them. Sometimes these aquifers exist only a relatively short time as they drain to the underlying phreatic aquifer.

1.1.4 Properties of Aquifers

The general properties of an aquifer to transmit, store and yield water are further defined numerically through a number of aquifer parameters. A detailed analysis of these parameters is given throughout this book. Here we shall present a brief general description of some of them in order to supplement the definition of aquifers given above.

The *hydraulic conductivity* indicates the ability of the aquifer material to conduct water through it under hydraulic gradients. It is a combined property of the porous medium and the fluid flowing through it (sec. 5.5). When the flow in the aquifer is essentially horizontal, the *aquifer transmissivity* indicates the ability of the aquifer to transmit water through its entire thickness. It is the product of the hydraulic conductivity and the thickness of the aquifer (sec. 6.4).

The *storativity of an aquifer* (sometimes called the *coefficient of storage*) indicates the relationship between the changes in the quantity of water stored in an aquifer and the corresponding changes in the elevations of the piezometric surface (or the water table in an unconfined aquifer).

The storativity of a confined aquifer is defined as that volume of water released from (or added to) a vertical column of aquifer of unit horizontal cross-section, per unit of decline (or rise) of the piezometric head (secs. 6.3 and 6.4). Figure 1.1.4a illustrates this concept. The storativity of a confined aquifer is caused by the compressibility of the water and the elastic properties of the aquifer (or of the porous matrix) as a whole; the elasticity of the solid grains, particles, etc., is usually neglected. A detailed discussion is given in chapter 6.

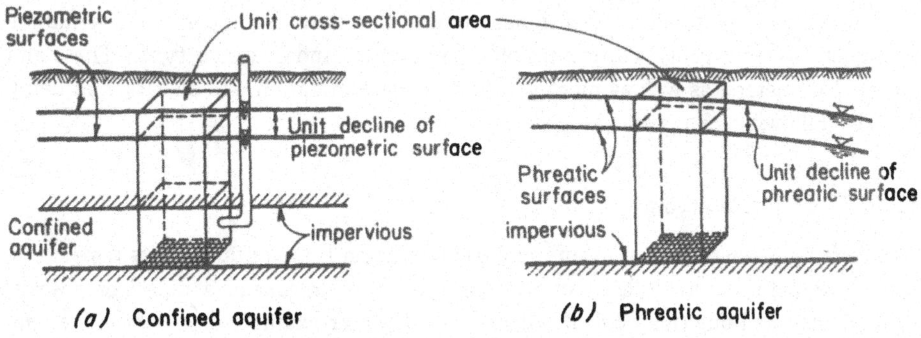

FIG. 1.1.4. Illustrative sketches for defining storativity.

In a phreatic aquifer, the definition of storativity given above remains essentially unchanged, except that the decline is in the phreatic surface. However, a different mechanism causes the variation in the quantity of water stored in a column of aquifer. In the case of a phreatic aquifer, water is actually drained out of the pore space, and air is substituted as the water table drops. However, not all water contained in the pore space is removed by gravity drainage (say, toward a depression in the ground water table caused by a pumping well). A certain amount of water is held in place against gravity in the interstices between grains under molecular forces and surface tension. Hence, the storativity of a phreatic aquifer is less than the porosity by a factor called *specific retention* (the ratio of water retained against gravity to the bulk volume of a soil sample). Reflecting this phenomenon, the storativity of a phreatic aquifer is often referred to as *specific yield*. The term *effective porosity* is also often used in this context. However, one should be careful not to confuse this usage of the term with that effective porosity referring to flow through a porous medium (sec. 5.1). A detailed discussion of specific yield and specific retention is presented in section 9.4.3.

The *elastic storativity* resulting from compressibility of aquifer and water is much smaller than the specific yield. As an illustration, the storativity of most confined aquifers falls in the range between 10^{-3} and 10^{-5}, whereas the specific yield of most alluvial aquifers falls in the range between 10 and 25%. This indicates that for the same volume of withdrawal (or recharge), changes in piezometric surface elevations are much larger in a confined than in a phreatic aquifer.

In defining storativity for a confined aquifer it is assumed that no time lag is involved and that water is released immediately upon change in head. However, especially in fine-grained materials, there may occur an appreciable time lag due to low hydraulic conductivity restricting movement of water out of storage. This is also true for a phreatic aquifer where the dewatering process takes time.

A parameter characterizing a leaky aquifer is the *resistance* of the semiconfining layer. It is defined by the ratio of the thickness of this layer to its hydraulic conductivity. As this value becomes larger, the leakage through this layer diminishes. Another parameter, the *leakage factor*, is the root of the product of the transmissivity of the aquifer and the resistance of the semipervious layer.

The various parameters mentioned above are used as guides in determining whether a certain geological formation is an aquifer, and of which type. These and other parameters, as well as methods for their determination, are discussed in detail in the following chapters.

1.1.5 The Oil Reservoir

The *oil or gas reservoir* is a porous geologic formation that contains in its pore space, in addition to water, at least one hydrocarbon (oil or gas) in a liquid or gaseous phase. Throughout this book it will serve as another example of a porous medium of practical interest to which one may apply the theory of dynamics of fluids in porous media. Although this is by no means a text on reservoir or petroleum engineering, it seems appropriate, as in paragraph 1.1.1 in connection with the aquifer, to present some background comments related to the reservoir and to the production of oil and gas from it. These comments are based mainly on Muskat (1949) and Amyx et al. (1960).

Although several theories exist on the origin of hydrocarbons, scientists currently tend to accept the *organic theory*, according to which hydrocarbons evolve as the decomposition products of organic material (vegetable and animal) from organisms that lived during early geological ages. The beds, rich in organic material from which hydrocarbons originated, are called *source beds*. In these beds, the hydrocarbons are created in the form of a large number of tiny bubbles surrounded by the water that fills the pore space. From the source beds, the hydrocarbon droplets migrate to the *reservoir rock* where they accumulate in quantities of possible commercial interest.

The primary forces causing migration of hydrocarbons are buoyancy and capillarity. As oil and gas are lighter than the water surrounding them in the rock's pore space, they will in general have an upward flow component (sec. 9.8). The buoyant forces lead to the separation of the gas, oil and water bodies in the pore space of the reservoir rock. To enable accumulation, the upward migration of hydrocarbons to higher beds in the stratigraphic sequence must be restricted by a blanket of impervious (or nearly impervious) material that serves as an upper boundary of the hydrocarbon-bearing rock. Thus, either a natural barrier that causes an abrupt change in the flow direction, or a trap that prevents outflow under prevailing conditions, must exist for hydrocarbon accumulation to take place. Clays, shales and general argillaceous rocks, such as sandy shales and marls, constitute the most common oil-confining strata. In a way, an oil reservoir is similar to a confined aquifer (par. 1.1.3).

Sometimes the sealing blanket is a water-saturated fine grained stratum with a much lower permeability than that of the reservoir rock itself. In this case, such a stratum serves as a barrier to upward seepage through capillary interfacial flow resistance at the point of contact between strata of different pore-size distributions (par. 9.2.5).

Figures 1.1.5 and 1.1.6 show several examples of elementary reservoir traps (after Wilhelm 1945) by a sectional view and by a view of contours of the structural environment. Some reservoirs are complex and result from a combination of two or more of the elementary trap features. One should recognize that a trap is a necessary but not a sufficient condition for hydrocarbon accumulation.

Most producing oil reservoirs are made of sandstone, limestone and dolomite formations. Occasionally other types of rocks also prove productive.

Within the oil reservoir, gravitational forces cause less dense fluids to seek higher positions in the trap. Capillary forces tend to cause the wetting fluid to rise into the interstices containing nonwetting fluid, thus counteracting the effect of gravity in segregating the fluids. In general, water is the wetting fluid with respect to oil and gas, while oil is the wetting fluid with respect to gas.

In fig. 1.1.7 typical hydrocarbon distributions in reservoirs are shown under equilibrium conditions, with transition zones between water and oil and between

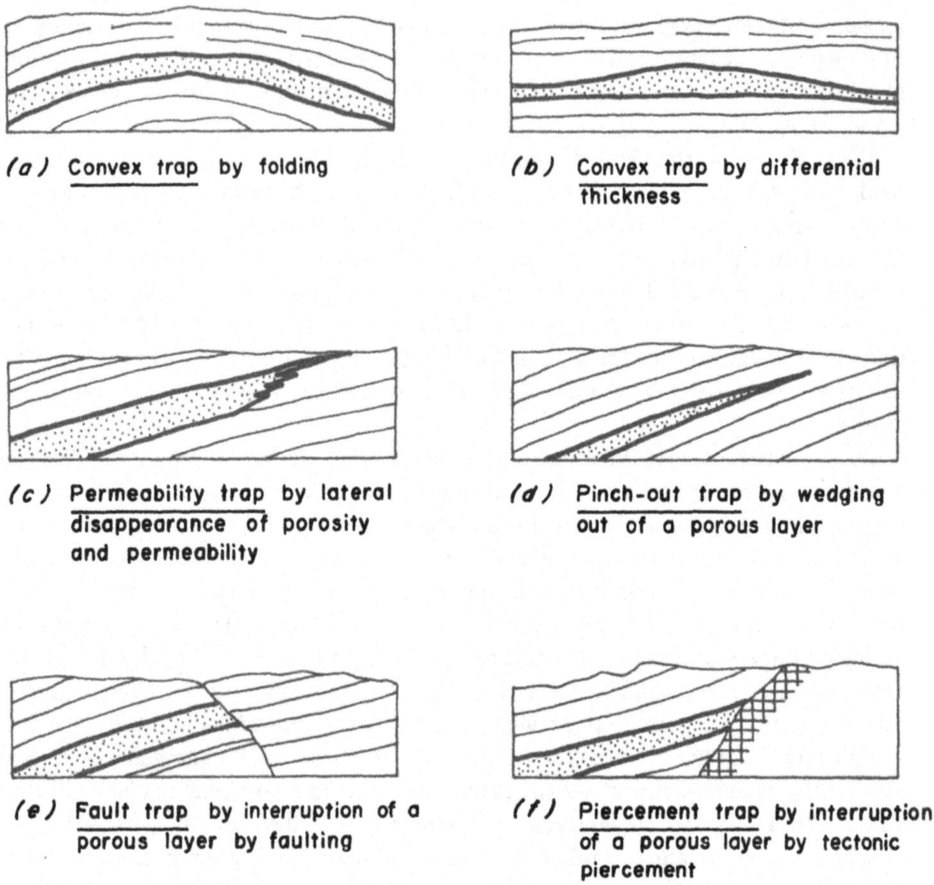

FIG. 1.1.5. Elementary traps in sectional view (Wilhelm, 1945).

oil and gas (par. 9.2.3). Under certain conditions of pressure and temperature in the trap, no gas is present. The oil zone contains small amounts of water (connate water; par. 9.2.4). The fraction of pore space occupied by water increases with depth in the oil–water transition zone so that the base of this transition zone is defined by a completely water-saturated pore space.

An oil (or gas) well field is comprised of a group of wells penetrating one or more subsurface reservoirs and producing oil (or gas) from these reservoirs. Several sources of energy exist that cause the hydrocarbons (oil or gas) in the reservoir to move toward the wells and to rise to the surface. The major types of energy available for oil and gas production (Muskat 1949) are: (a) energy of compression of oil, gas (in gas caps), and water within the reservoir; (b) gravitational energy of oil in the upper parts of the formation as compared with that at greater depths; (c) energy of compression and solution of the gas dissolved in the oil (and to some extent in the water); and (d) the energy of compression in the water in reservoirs (and aquifers) contiguous to and intercommunicating with the oil-bearing rock.

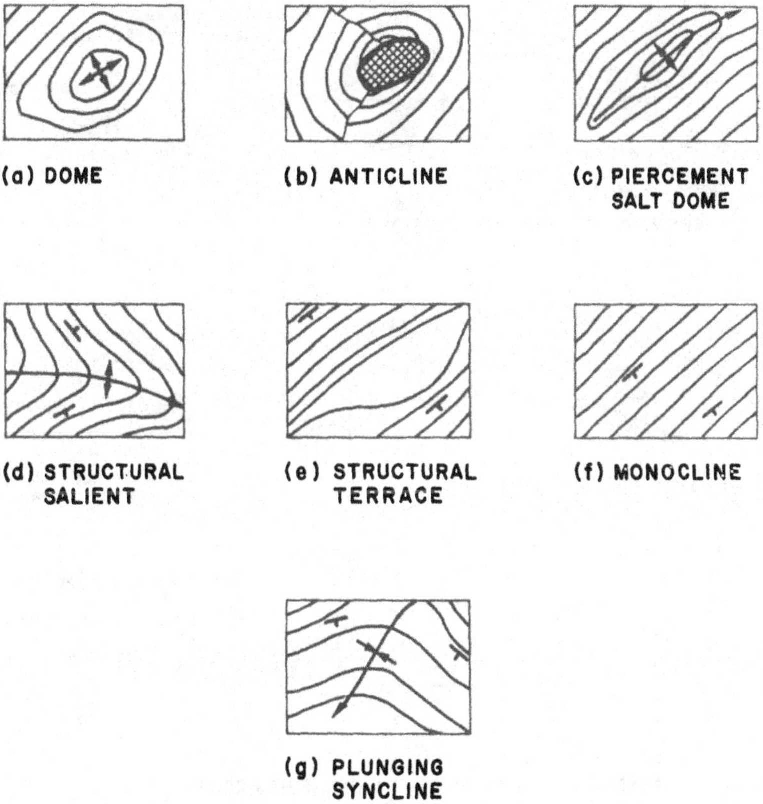

FIG. 1.1.6. Contours of structural environment (Wilhelm, 1945).

Additional minor forms of energy that may control the reservoir performance should be mentioned for the sake of completeness. These are the differential energy of the internal surfaces of the porous rock for the different fluid phases, and the compression of the rock itself (Muskat 1949, p. 365).

As the wells provide outlets for these forms of energy, the energy is expended by the action of forces or pressures exerted in the direction of lower energy levels. These forces serve to overcome the resistance of the rock to the fluid's flow toward the wells.

Among the four major energy sources, the first is less significant than the others.

Some reservoirs are closed, so that the associated volume of water is quite small and no gas cap or associated active water are present. In this case the energy available to displace the hydrocarbons toward the wells is solely that resulting from liberation of gas from solution in the reservoir oil, with subsequent expansion and expulsion of the oil. This is the third type of energy source listed above. The reservoir in this case is referred to as a *solution-gas drive reservoir*, or a *depletion-drive reservoir*. This type of drive is characterized by rapid pressure decline and low efficiency of oil

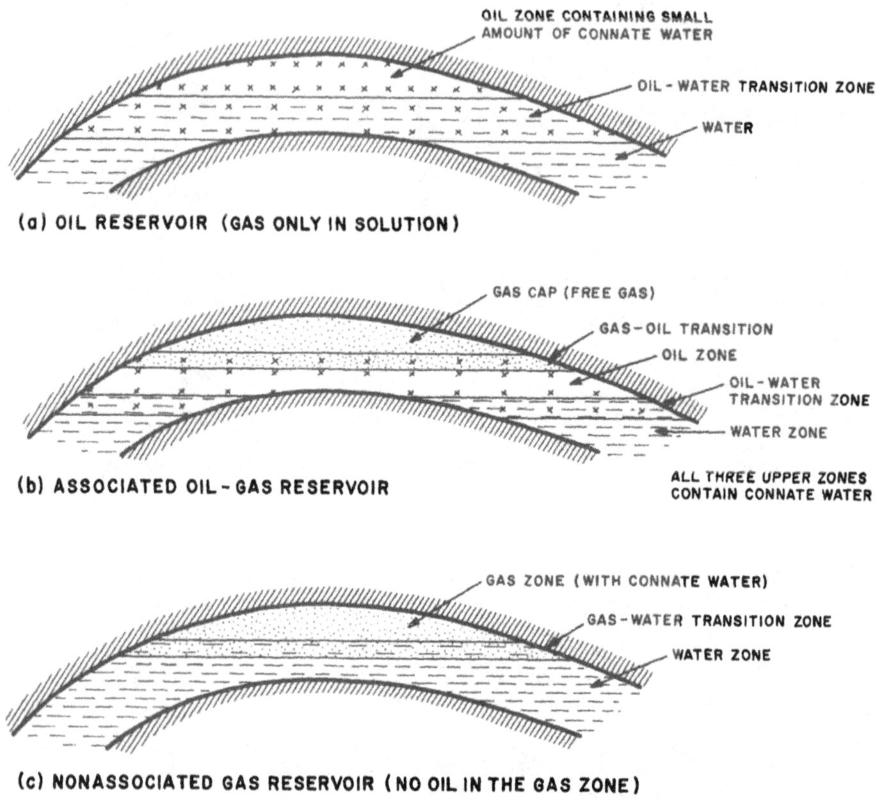

FIG. 1.1.7. Typical hydrocarbon distributions in reservoirs (Amyx et al., 1960).

recovery. It usually results in recovery of only a small fraction of the total oil volume present in the formation.

When the source of energy in a reservoir is an expanding free gas cap, but no associated active water body is present, the driving mechanism is referred to as a *gas-cap-drive* (or sometimes *gas-cap-expansion* or *external gas drive*). This is basically a displacement-type drive, the gas displacing the oil ahead as it expands because of pressure reduction.

The second type of energy source listed above, gravitational, does not, in general, play an important role as a driving mechanism until the reservoir becomes substantially depleted. In some high-relief reservoirs, where the producing wells are located structurally low, oil recovery by *gravitational segregation* may be rather substantial.

A petroleum reservoir associated with aquifers so active that little or no pressure drop occurs in the petroleum reservoir by the withdrawal of hydrocarbon fluids is referred to as a *water-drive reservoir*. In the water-drive process, water from surrounding aquifers enters the reservoir almost as fast as the hydrocarbon fluid is withdrawn, thus preventing any substantial decline in pressure. This displacement-type drive is by far the most efficient *natural* reservoir driving mechanism. The efficiency of water displacement is usually greater than that of gas displacement, regardless of the wetting characteristics of the reservoir rock. When the reservoir formation has a steep dip, the oil–water interface will be of limited area and will provide an *edgewater* boundary. We then have an *edgewater drive* (e.g., fig. 9.5.8*f*). For gently sloping formations, the water–oil interface may underlie an appreciable part of the oil zone. Wells producing above such a water body are subject to a *bottom-water-drive* (fig. 9.7.10) if the water is mobile and is permitted to invade the oil reservoir at a rate sufficient to replace the withdrawn fluid.

Most petroleum reservoirs are subject to one or more drives, either simultaneously, or at various times during their productive life. Such reservoirs are referred to as *combination-drive reservoirs*.

In addition to the natural drives described above, artificial supplementation of natural energy by fluid injection (through wells) may be effected to improve the production efficiency of hydrocarbons from reservoirs. The fluid injection may involve the return of gas, water (*water flooding*) or gas and water. Such operations are often termed *pressure maintenance operations*. A more general term, which includes such techniques as injection of miscible (with oil) fluids, thermal recovery methods (involving the injection of hot water or steam into the reservoir) and internal combustion methods, is *secondary recovery operations*. The basic difference between secondary recovery and pressure maintenance operations is that the initial condition for the former is a state of virtually complete depletion of the reservoir pressure or the original natural oil-expulsion energy, whereas in the latter fluid injection is undertaken during the primary oil producing processes before such states are reached.

Details of the various flooding techniques are given in all texts on reservoir engineering. The theory underlying water flooding and gas injection is considered in sections 9.3 and 9.5 and that underlying miscible flooding techniques is presented in chapter 10.

1.2 The Porous Medium

Two examples of practical interest—the aquifer encountered in ground water hydrology, and the oil reservoir encountered in petroleum engineering—were introduced in section 1.1. In both cases, flow takes place through a *porous medium*. Actually, by introducing the term *porous medium*, and by considering *flow of fluids through a porous medium*, we have made a basic and important step in our way of thinking. We have introduced the concept of a *continuum*, which is common to most branches of physics.

The continuum approach is discussed in detail in the following section. Here we shall attempt to define, or at least to describe in relative terms, a porous medium. Examples of porous materials are numerous. Soil, porous or fissured rocks, ceramics, fibrous aggregates, filter paper, sand filters and a loaf of bread are just a few. Somewhat less obvious examples, but still part of this group, are large geologic formations of karstic limestone, where the open passages (such as solution channels or caverns) may be of substantial size and far apart. All of these materials have some characteristics in common that permit them to be grouped and classified as porous media.

Initially, we may attempt to describe a porous medium as a "solid with holes." Obviously, a hollow metal cylinder would not normally be classed as a porous medium, nor would a solid block with isolated holes or pores, since we seek to define a porous medium in connection with flow through the medium, and not, for example, in connection with thermal insulation. We might try to improve our definition by stipulating that the pores are *interconnected*, with at least several *continuous paths* from one side of the medium to the other, and by somehow specifying a better distribution (in either a *regular or random manner*) of holes and paths over the entire porous medium domain. Although we are now describing an acceptable porous medium model, the medium described lacks the possibility of exchange of fluid between adjacent paths, and especially those aspects related to three-dimensional flow, which requires a spatial distribution of channels (or paths) within the domain.

Summarizing these preliminary remarks, and looking at figure 1.1.1, we may try to define a porous medium as (Bear, Zaslavsky and Irmay 1968):

(a) a portion of space occupied by *heterogeneous* or *multiphase* matter. At least one of the phases comprising this matter is not *solid*. They may be gaseous and/or liquid phases. The solid phase is called the *solid matrix*. That space within the porous medium domain that is not part of the solid matrix is referred to as *void space* (or *pore space*).

(b) The solid phase should be distributed throughout the porous medium within the domain occupied by a porous medium; solid must be present inside each representative elementary volume (par. 1.3.2). An essential characteristic of a porous medium is that the *specific surface* (sec. 2.6) of the solid matrix is *relatively high*. In many respects, this characteristic dictates the behavior of fluids in porous media. Another basic feature of a porous medium is that the various openings comprising the void space are *relatively narrow*.

(c) At least some of the pores comprising the void space should be interconnected. The interconnected pore space is sometimes termed the *effective pore space*. As far as flow through porous media is concerned, *unconnected pores* may be considered as part of the solid matrix. Certain portions of the interconnected pore space may, in fact, also be ineffective as far as flow through the medium is concerned. For example, pores may be *dead-end pores* (or *blind pores*), i.e., pores or channels with only a narrow single connection to the interconnected pore space, so that almost no flow occurs through them. Another way to define this porous medium characteristic is by requiring that any two points within the effective pore space may be connected by a curve that lies completely within it. Moreover, except for special cases, any two such points may be connected by many curves with an arbitrary maximal distance between any two of them. For a finite porous medium domain, this maximal distance is dictated by the domain's dimensions.

The features described above can scarcely be called a definition as they involve several terms, such as "relatively small," "more or less evenly distributed," and "relatively narrow," that are comparative, rather than absolute. However, in combination, these terms convey to the reader something of the nature of a porous medium. Rideal (1958) also tries to indicate the various characteristics of porous materials, emphasizing the difficulty in arriving at an exact definition still sufficiently general to be applied to the wide variety of porous media. To some of the features described, numerical values may be assigned. To others, mainly those related to the geometry of the solid surfaces, no such values may be assigned. In fact, it is this difficulty in defining the geometry of the solid surfaces, which act as boundaries to the flow in the void space, that forces us to introduce the continuum approach as a tool for handling phenomena in porous media.

1.3 The Continuum Approach to Porous Media

1.3.1 The Molecular and Microscopic Levels

The main purpose of section 1.2 is to demonstrate the hopelessness of any attempt to describe in an exact manner the geometry of the internal solid surfaces that bound the flow domain inside a porous medium. Directing our attention to the fluid or fluids contained in the void space, and trying to describe phenomena associated with them, such as motion, mass transport, etc., the same difficulties are encountered.

First, the concept of the fluid itself requires some further elaboration. Actually, fluids are composed of a large number of molecules (overlooking the existence of a submolecular structure) that move about, colliding with each other and with the solid walls of the container in which they are placed. By employing theories of classical mechanics, we could fully describe a given system of molecules: e.g., given their initial positions in space and their momenta, we could predict their future positions. However, despite the apparent simplicity of this approach, it is exceedingly difficult to solve the problem of the motion of even three molecules (assuming that we know *all* the forces, which is also doubtful). With the advent of high speed digital computers, the many body problem can be attacked, in principle, numerically. It is still impossible, however, to determine the motion of 10^{23} molecules in one gram mole of gas. In addition, because the number of molecules is so large, their initial positions and momenta cannot actually be determined, for example, by observation.

It is the embarrassingly large number of equations that ultimately provides a way out, at least under certain conditions. Instead of treating the problems, say of fluid motion, at the *molecular level* or *viewpoint* described above, we may adopt a different approach, statistical in nature, to derive information regarding the motion of a system composed of many molecules. By the *statistical approach* we mean one in which the results of an analysis or an experiment are presented only in statistical form. This means that we can determine the average value of successive measurements, but we cannot predict with certainty the outcome of a single measurement in the future. In this context, an "experiment" means, for example, the position of a certain molecule at a certain time. *Statistical mechanics* is an analytical science by which statistical properties of the motion of a very large number of molecules (or of particles in general) may be inferred from laws governing the motion of individual molecules. Many texts are available on this subject, and the reader is referred to them (e.g., Landau and Lifschitz 1958).

When the purpose of abandoning the molecular level of treatment is the description of phenomena as a fluid continuum, the statistical approach is referred to as the macroscopic approach. Our ultimate objective is to handle phenomena in porous media. We shall need for that purpose a still higher, or coarser, level of treatment, reached by averaging phenomena in the fluid continuum filling the void space. We choose, therefore, to refer to the fluid continuum level as the *microscopic* one. At the microscopic level, we overlook the actual or molecular structure of a fluid and regard it as a continuum.

Essential to the treatment of fluids as continua is the concept of a *particle*. A particle is an ensemble of many molecules contained in a small volume. Its size is much larger than the mean free path of a single molecule. It should, however, be sufficiently small as compared to the considered fluid domain that by averaging fluid and flow properties over the molecules included in it, meaningful values, i.e., values relevant to the description of bulk fluid properties, will be obtained. These values are then related to some centroid (par. 4.1.1) of the particle. Then, at every point in the domain occupied by a fluid, we have a particle possessing definite dynamic and kinematic properties.

Associated with the question of particle size, or of the elementary volume that should be considered as a point—a *physical* or *material point*—within the fluid continuum, is the definition of the fluid's *density* or *specific mass*.

Density is the ratio between the mass Δm of an amount of matter and the volume ΔU occupied by it. If we consider a mathematical point and wish to assign to it a density value, ρ, such that this value will represent the density of a volume of fluid for which this point is a mass centroid, we have to decide on the volume to use.

Following Prandtl and Tietjens (1934), let us consider a point P in the fluid, and let Δm_i denote the fluid mass in a sufficiently large volume, ΔU_i, for which P is a centroid. The average density, ρ_i, of the fluid in ΔU_i is $\rho_i = \Delta m_i / \Delta U_i$. Obviously, if ΔU_i is too large, say of the order of magnitude of the entire field of flow, it is meaningless to assign the value ρ_i to the point P, i.e., to represent the ratio $\Delta m_i / \Delta U_i$ for the fluid in the vicinity of P. This is especially true when the fluid is inhomogeneous. To determine how small ΔU_i should be in order for ρ_i to represent the fluid in the neighborhood of P, we gradually reduce ΔU_i around P, determining the ratio $\Delta m_i / \Delta U_i$ for a sequence of volumes ΔU_i: $\Delta U_1 > \Delta U_2 > \Delta U_3 \cdots$. The results of these computations are shown in figure 1.3.1. If we start from a sufficiently large ΔU_i, gradual changes may be observed in ρ_i if the fluid is inhomogeneous. Fluctuations around $\rho_i = \rho_i(\Delta U_i)$ diminish as ΔU_i becomes smaller. Then as ΔU_i converges on P, a range of practically no changes in ρ_i with changes in ΔU_i is observed. However, as ΔU_i is made smaller and the number of molecules in it becomes smaller too, a range is reached below a certain volume ΔU_0 where the number of molecules in ΔU_i is so small that any further reduction of ΔU_i appreciably affects the ratio $\Delta m_i / \Delta U_i$. This happens when the characteristic length dimension of ΔU_i becomes of the order of magnitude of the average distance λ between the molecules (mean free path of molecules). As $\Delta U_i \to 0$ very wild fluctuations in the ratio $\Delta m_i/\Delta U_i$ are observed, and it is meaningless to use ρ_i as a definition for the density of the fluid at P. Hence the fluid's density at P is defined as:

$$\rho(P) = \lim_{\Delta U_i \to \Delta U_0} \rho_i = \lim_{\Delta U_i \to \Delta U_0} (\Delta m_i/\Delta U_i). \tag{1.3.1}$$

The characteristic volume ΔU_0 is called the ***physical point*** (or ***material point***) of the fluid at the *mathematical point* P. The volume ΔU_0 may now be identified with the *volume of a particle* at P. By the procedure just described, a material made of a collection of molecules in a vacuum is replaced by a continuum filling the entire space. We obtain a fictitious smooth medium (instead of the molecules), called fluid, for each point of which a continuous function of space ρ is defined. For any two close points P and P':

$$\rho(P) = \lim_{P' \to P} \rho(P').$$

In an inhomogeneous fluid, it is interesting to define a characteristic length L:

$$L = \rho/(\partial\rho/\partial l) \tag{1.3.2}$$

or lengths, say, in the direction of the three coordinates:

$$L_x = \rho/(\partial\rho/\partial x); \quad L_y = \rho/(\partial\rho/\partial y); \qquad L_z = \rho/(\partial\rho/\partial z)$$

that characterize the macroscopic changes in ρ in the field of flow; it indicates how rapid these changes in ρ are. The volume L^3 (or $L_x L_y L_z$) may be used as the upper limit for the range of ΔU_i for which ρ_i is independent of ΔU_i (fig. 1.3.1). One should note that in (1.3.2):

$$\frac{\partial\rho}{\partial l} = \lim_{\Delta l \to \Delta l_0} \frac{\rho[l + (\Delta l/2)] - \rho[l - (\Delta l/2)]}{\Delta l}$$

where $\lambda < \Delta l_0 < L$, is a characteristic length of the fluid at P, with $(\Delta l_0)^3$ of the order of ΔU_0.

Obviously, when the fluid is a gas at very low pressure, λ may be very large and Δl_0 and ΔU_0 may be as well.

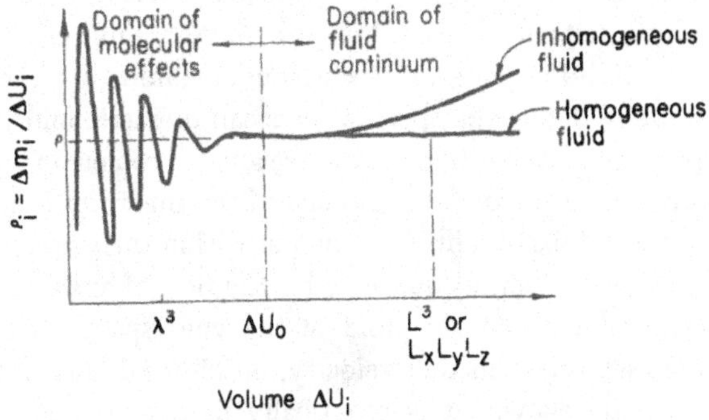

FIG. 1.3.1. Definition of fluid density.

Knudsen (1934) defines a dimensionless number $K_n = \lambda/L$ (called a *Knudsen* number). When $K_n < 0.01$, the flowing fluid may be considered as a continuum to which macroscopic considerations (e.g., in the form of partial differential equations) are applicable. This then sets the limits for Δl_0 or ΔU_0. When $K_n \sim 1$ we have a *slip-flow regime* (par. 5.3.3) and when $K_n > 1$ we have *Knudsen flow* or "*free molecular flow.*"

A comment seems appropriate regarding time involved in the process described above, because of the continuous rapid motion of the molecules in and out of any volume around P. One should consider figure 1.3.1 as representing what happens at a specified (or mathematical) instant of time around point P, or should regard each point of the curve as an average of several observations taken over an interval of time, Δt_0, around the considered instant of time. To determine Δt_0 we must go through arguments similar to those involved in the determination of ΔU_0. The interval of time should not be too large (as we should then lose information regarding temporal variations of ρ, if such variations take place), i.e., it should be smaller than a characteristic time, T, defined by $T = \rho/(\partial\rho/\partial t)$. On the other hand, it should not be shorter than a *mean free time* (Landau and Lifschitz 1958) of a molecule (an interval of time during which each molecule collides once on the average with another molecule). The value of Δt_0 may be obtained from a curve similar to that shown in figure 1.3.1, but with Δt_i as abscissa and $\Delta m_i/\Delta t_i$ as ordinate.

Many other physical phenomena in fluids, observed through their macroscopic manifestations, are the outcome of perpetual molecular motion. Among these we have *mass transport by molecular diffusion*, *heat transfer* and *momentum transfer*, which manifests itself in the form of internal friction or viscosity. In each of these cases, because we are unable to treat transfer phenomena on a molecular level, we average the transfer produced by the individual molecules and pass to a higher level—that of a fluid continuum, referred to in the present text as the microscopic level. In order to describe the various transfer phenomena at this higher level, *transfer coefficients* are needed. They are molecular diffusivity, thermal diffusivity, kinematic viscosity, etc. Occasionally, to understand phenomena, or the meaning of the microscopic parameters described and their relation to basic molecular properties, it may prove instructive to revert back to the molecular point of view.

To conclude, our discussion has led us from the molecular level to the microscopic level of treating physical phenomena. We now have a fluid continuum enclosed by solid surfaces—the solid surfaces of the porous medium. At each point of this fluid continuum we may define the specific physical, dynamic and kinematic properties of the fluid particle. Can we, however, solve a flow problem in a porous medium at this level? In principle we have at our disposal the theory of fluid mechanics, so that we may derive the details of a fluid's behavior within the void space. For example, we may use the Navier–Stokes equations for the flow of a viscous fluid to determine the velocity distribution of the fluid in the void space, satisfying specified boundary conditions, say, of vanishing velocity, on all fluid–solid interfaces. However, as we have already shown, it is practically impossible, except in especially simple cases, such as a medium composed of straight capillary tubes, to describe in any exact mathematical manner the complicated geometry of the solid surfaces that bound the flow domain within the porous solid matrix. Moreover, it is often difficult to define the boundary conditions themselves. Consequently, any solution by means of this approach is precluded. In view of the discussion presented above, the obvious way to circumvent these difficulties is to pass to a coarser level of averaging—to the *macroscopic level.* This is again a *continuum approach,* but on a higher level.

The following paragraph deals with the macroscopic approach to dynamics of fluids in porous media.

1.3.2 Porosity and Representative Elementary Volume

In paragraph 1.3.1 we have seen that essential to the concept of a continuum is the particle, or the physical point, or the representative volume over which an average is performed. This is also true in the passage from the microscopic to the macroscopic level. Our task now is, therefore, to determine the size of the representative porous medium volume around a point P within it. From the discussion in the previous paragraph we know that this volume should be much smaller than the size of the entire flow domain, as otherwise the resulting average cannot represent what happens at P. On the other hand, it must be enough larger than the size of a single pore that it includes a sufficient number of pores to permit the meaningful statistical average required in the continuum concept.

When the medium is inhomogeneous, say, with porosity varying in space, the upper limit of the length dimension of the representative volume should be a characteristic length that indicates, as in (1.3.2), the rate at which changes in porosity take place. The lower limit is related to the size of the pores or grains.

We shall now define *volumetric porosity* and the *representative elementary volume* (REV) *associated with it,* following the same procedure as that described in paragraph 1.3.1 for the definition of density. We choose to define the REV through the concept of porosity, which seems to be the basic porous matrix property. Porosity is, in a sense, the equivalent of density (mass per unit volume) discussed in paragraph 1.3.1. This might be more easily visualized if we defined the REV through the concept of the *solid's bulk density* (*unit weight of solids* = mass of solid per unit volume of medium).

Let P be a mathematical point inside the domain occupied by the porous medium. Consider a volume ΔU_i (say, having the shape of a sphere) much larger than a single pore or grain, for which P is the centroid. For this volume we may determine the ratio:

$$n_i \equiv n_i(\Delta U_i) = (\Delta U_v)_i/\Delta U_i \tag{1.3.3}$$

where $(\Delta U_v)_i$ is the volume of void space within ΔU_i. Repeating the same procedure, a sequence of values $n_i(\Delta U_i)$, $i = 1, 2, 3, \ldots$ may be obtained by gradually shrinking the size of ΔU_i around P as a centroid: $\Delta U_1 > \Delta U_2 > \Delta U_3 \cdots$.

For large values of ΔU_i, the ratio n_i may undergo gradual changes as ΔU_i is reduced, especially when the considered domain is inhomogeneous (e.g., layers of

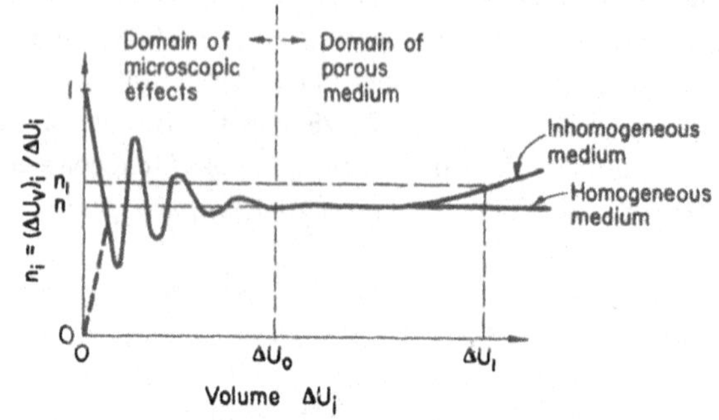

FIG. 1.3.2. Definition of porosity and representative elementary volume.

soil). Below a certain value of ΔU_i, depending on the distance of P from boundaries of inhomogeneity, these changes or fluctuations tend to decay, leaving only small-amplitude fluctuations that are due to the random distribution of pore sizes in the neighborhood of P. However, below a certain value ΔU_0 we suddenly observe large fluctuations in the ratio n_i. This happens as the dimensions of ΔU_i approach those of a single pore. Finally, as $\Delta U_i \to 0$, converging on the mathematical point P, n_i will become either one or zero, depending on whether P is inside a pore or inside the solid matrix of the medium. Figure 1.3.2 shows the relationship between n_i and ΔU_i.

The medium's *volumetric porosity* $n(P)$ at point P is defined as the limit of the ratio n_i as $\Delta U_i \to \Delta U_0$:

$$n(P) = \lim_{\Delta U_i \to \Delta U_0} n_i\{\Delta U_i(P)\} = \lim_{\Delta U_i \to \Delta U_0} \frac{(\Delta U_v)_i(P)}{\Delta U_i}. \tag{1.3.4}$$

For values of $\Delta U_i < \Delta U_0$, we must consider the actual presence of pores and solid particles; in this range there is no single value that can represent the porosity at P. The volume ΔU_0 is therefore the *representative elementary volume* (REV) or the *physical* (or *material*) point of the porous medium at the mathematical point P.

The limiting process in (1.3.4) is sometimes called the *extrapolated limit* (Hubbert 1956). Obviously, the limit $\Delta U_i \to 0$ is meaningless. From the definition of the REV it follows that its dimensions are such that the effect of adding or subtracting one or several pores has no significant influence on the value of n.

We shall assume that both ΔU_0 and ΔU_v vary smoothly in the vicinity of P. Then:

$$n(P) = \lim_{P' \to P} n(P') \tag{1.3.5}$$

which means that n is a continuous function of the position of P within the porous medium.

Thus, by introducing the concept of porosity and the definition of REV, we have replaced the actual medium by a fictitious continuum in which we may assign values of any property (whether of the medium or of the fluids filling the void space) to any mathematical point in it. Sometimes one must define REVs of the medium on the basis of parameters other than porosity.

1.3.3 Areal and Linear Porosities

We may define *areal porosity* (n_A) and *linear porosity* (n_L) at a point within the domain occupied by a porous medium by using a procedure similar to that described above for volumetric porosity.

Type 1, areal porosity. Consider a point P in a porous medium domain and a plane, whose normal is in the direction of the unit vector **1j**, passing through it. Let $(\Delta A_j)_i$, $i = 1, 2, 3, \ldots$ denote a sequence of areas with centroids at P, such that $(\Delta A_j)_1 > (\Delta A_j)_2 > (\Delta A_j)_3 \cdots$. As the planes in question pass through both the solid matrix and the void space, each area $(\Delta A_j)_i$ will contain an "area of solids" $(\Delta A_{sj})_i$ and an "area of voids" $(\Delta A_{vj})_i$. As for the volumetric porosity (par. 1.3.2), we first define a sequence of ratios:

$$(n_{Aj})_i = (\Delta A_{vj})_i/(\Delta A_j)_i$$

and plot them as a function of the areas $(\Delta A_j)_i$. A curve similar to that shown in figure 1.3.2 is obtained. Using this curve, we define the smallest area $(\Delta A_j)_0$ for which no large fluctuations occur as the *representative elementary area* (REA) of the medium facing the direction **1j** at P. Then the *directional* areal porosity at P for the area facing the direction **1j** is obtained by a limiting process similar to (1.3.4):

$$n_{Aj}(P) = \lim_{(\Delta Aj)_i \to (\Delta Aj)_0} (n_{Aj})_i = \lim_{(\Delta Aj)_i \to (\Delta Aj)_0} \frac{(\Delta A_{vj})_i}{(\Delta A_j)_i}. \tag{1.3.6}$$

As in (1.3.5), we shall also assume:

$$n_{Aj}(P) = \lim_{P' \to P} n_{Aj}(P'). \tag{1.3.7}$$

To determine the relationship between n_{Aj} and n at some point P of the porous medium, we consider an REV having the shape of a cylinder whose centroid is at P. The cylinder's axis is in the direction **1j**. Its normal cross-section is $(\Delta A_j)_0$ and its height is $L_j = \Delta U_0/(\Delta A_j)_0$ in the direction **1j**. The volume of voids within this cylinder is:

$$n(P)\Delta U_0 = \int_{s(P)-Lj/2}^{s(P)+Lj/2} \Delta A_{vj}|_{s'}\, ds' = \int_{s(P)-Lj/2}^{s(P)+Lj/2} [n_{Aj}(s')](\Delta A_j)_0\, ds'$$

$$= (\Delta A_j)_0 \int_{s(P)-Lj/2}^{s(P)+Lj/2} n_{Aj}(s')\, ds' = [(\Delta A_j)_0]\overline{n_{Aj}}L_j = \overline{n_{Aj}}\Delta U_0;$$

$$\overline{n_{Aj}} = \frac{1}{L_j} \int_{s(P)-Lj/2}^{s(P)+Lj/2} n_{Aj}(s)\, ds,$$

where $\overline{n_{Aj}}$ is the average value of n_{Aj} along the cylinder's axis and s is the length measured along this axis.

If instead of the cylinder we take an REV of any shape, we shall obtain:

$$n(P)\Delta U_0 = \int_{(L_j)} [n_{Aj}(s)](\Delta A_j)_0(s)\, ds.$$

Since $(\Delta A_j)_0(s)$ is always positive and $n_{Aj}(s)$ is continuous and independent of $(\Delta A_j)_0$ we may write:

$$\int_{(L_j)} [n_{Aj}(s)](\Delta A_j)_0(s)\, ds = \overline{n_{Aj}} \int_{(L_j)} (\Delta A_j)_0(s)\, ds = \overline{n_{Aj}}\Delta U_0.$$

Where L_j is the length of the segment inside the REV of a line passing through P in the direction $\mathbf{1j}$, and $\overline{(\Delta A_j)_0}$ is the average value of $(\Delta A_j)_0$:

$$\Delta U_0 = \int_{(L_j)} (\Delta A_j)_0(s)\, ds; \qquad \overline{(\Delta A_j)_0} = \frac{1}{L_j}\int_{(L_j)} (\Delta A_j)_0(s)\, ds.$$

In both cases the final result is:

$$n(P) = \overline{n_{Aj}}(P). \tag{1.3.8}$$

Stated verbally, the volumetric porosity (or simply the porosity) of a porous medium at a point is equal to the average value of the directional areal porosity at that point. However, since $n(P)$ is independent of the direction of $\mathbf{1j}$, it follows that $\overline{n_{Aj}}(P)$ must also be independent of direction, and it is sufficient to define an average areal porosity (or simply areal porosity) $\overline{n_A}$ (or n_A).

Type 2, linear porosity. In a manner similar to that described for areal porosity, it is possible to define for a point P in the porous medium a *representative elementary length* (REL) and a corresponding *linear porosity*. It can then be shown that the average value of the linear porosity is equal to the areal porosity, and hence also to the volumetric porosity.

1.3.4 Velocity and Specific Discharge

At a point P in a porous medium domain through which flow takes place, let $(\Delta A_j)_0$ denote an REA facing the direction $\mathbf{1j}$. Actually, flow takes place only through that portion of $(\Delta A_j)_0$ not occupied by solids. Fluid particles passing through $(\Delta A_j)_0$ have velocity vectors that differ from each other both in direction and in magnitude. We refer to these velocities (of fluid particles within the pores) as *local* or *microscopic velocities*. In order to obtain the total discharge through $(\Delta A_j)_0$, we must consider only velocity components V_j in the direction $\mathbf{1j}$. Hence:

$$Q_{j0} = \int_{(\Delta A_{vj})_0} V_j\, dA_{vj}; \qquad V_j = \mathbf{V}\cdot\mathbf{1j}. \tag{1.3.9}$$

We now define the *specific discharge* q_{j0} through $(\Delta A_j)_0$ in the direction $\mathbf{1j}$ by:

$$q_{j0} = Q_{j0}/(\Delta A_j)_0$$

i.e., we divide the discharge by the *total* area of the REA. Then, assuming that flow takes place through the entire pore space:

$$q_{j0} = \frac{1}{(\Delta A_j)_0} \int_{(\Delta A_{vj})_0} V_j dA_{vj} = \frac{1}{(\Delta A_j)_0} (\Delta A_{vj})_0 \frac{1}{(\Delta A_{vj})_0} \int_{(\Delta A_{vj})_0} V_j dA_{vj} = n_{Aj} \langle\langle V_j \rangle\rangle \tag{1.3.10}$$

where:

$$\langle\langle V_j \rangle\rangle = \frac{1}{(\Delta A_{vj})_0} \int_{(\Delta A_{vj})_0} V_j \, dA_{vj}$$

is the average value of V_j taken over $(\Delta A_{vj})_0$. In (1.3.10), $\langle\langle V_j \rangle\rangle$ is the component in the direction $\mathbf{1j}$ of the average velocity vector, and q_{j0} is the component in the direction $\mathbf{1j}$ of the specific discharge vector through the REA. If we consider another cross-section through the medium, parallel to the first but at some small distance from it, the value of n_{Aj} may vary appreciably. Moreover, according to (1.3.10), q_{j0} at P still depends on the direction of $\mathbf{1j}$.

To circumvent this difficulty, we perform a second average over the REAs facing the direction $\mathbf{1j}$ of an REV centered at P:

$$\bar{q}_j \equiv q_j = \frac{1}{L_j} \int_{(L_j)} q_{j0}(s) \, ds = \frac{1}{L_j} \int_{(L_j)} \left[\frac{1}{(\Delta A_j)_0} \int_{(\Delta A_{vj})_0} V_j \, dA_{vj} \right] ds = \frac{1}{\Delta U_0} \int_{(\Delta U_v)_0} V_j dU_v = n \bar{V}_j \tag{1.3.11}$$

where:

$$\bar{V}_j = \frac{1}{(\Delta U_v)_0} \int_{(\Delta U_v)_0} V_j \, dU_v; \qquad n = \frac{(\Delta U_v)_0}{\Delta U_0} \quad \text{and} \quad \overline{\langle\langle V_j \rangle\rangle} = \bar{V}_j.$$

Equation (1.3.11) relates the component in the direction $\mathbf{1j}$ of the average specific discharge vector (discharge per unit area of porous medium) to the component in the direction $\mathbf{1j}$ of the average velocity vector—both averages taken over the REV around P. Therefore, omitting the average symbol:

$$\mathbf{q} = n\mathbf{V} \tag{1.3.12}$$

where boldface type denotes vector. The relationship (1.3.12) is sometimes called *Dupuit–Forchheimer's equation*.

By combining (1.3.10) with (1.3.11), we obtain:

$$q_j \equiv \bar{q}_j = \frac{1}{L_j} \int_{(L_j)} \left[\frac{1}{(\Delta A_j)_0(s)} \int_{(\Delta A_{vj})_0(s)} V_j dA_{vj} \right] ds. \tag{1.3.13}$$

Had we assumed, for the sake of simplicity, that the REV had the shape of a cylinder of constant cross-section, $(\Delta A_j)_0 = \Delta U_0/L_j$, equation (1.3.11) could have been written as:

$$q_j = \frac{1}{L_j} \int_{(L_j)} n_{\Delta j} \langle\langle V_j \rangle\rangle \, ds \qquad \text{or:} \qquad q_j = \frac{1}{\Delta U_0} \int_{(\Delta U_v)_0} V_j \, dU_v \qquad (1.3.14)$$

or, since $V_j = 0$ for any point in the volume $[\Delta U_0 - (\Delta U_v)_0]$, we obtain:

$$q_j \equiv \bar{q}_j = \frac{1}{\Delta U_0} \int_{(\Delta U_0)} V_j \, dU$$

or, in vector form:

$$\mathbf{q} \equiv \bar{\mathbf{q}} = \frac{1}{\Delta U_0} \int_{(\Delta U_0)} \mathbf{V} \, dU. \qquad (1.3.15)$$

In this last form, the interpretation of the specific discharge as an average parameter becomes obvious and requires no further explanation.

We have thus satisfied the requirements that (a) components of the specific discharge vector are continuous functions of position in the vicinity of any point within the medium, and (b) the absolute value $|\bar{q}|$ of the average specific discharge vector $\mathbf{q} (\equiv \bar{\mathbf{q}})$ is independent of the direction and the areal porosity of a single cross-section taken at the considered point. It may therefore be considered a continuum (or macroscopic) property of the flow in the porous medium.

Sometimes temporal fluctuation may take place in the local velocity. Following the approach described for smoothing out such fluctuations, so that $\mathbf{q}$ also becomes a continuous function of time, we must perform another averaging operation (Irmay, in Bear, Zaslavsky and Irmay 1968):

$$\mathbf{q} = \frac{1}{\Delta t_0} \int_{(\Delta t_0)} \mathbf{q}' \, dt = \frac{1}{\Delta t_0} \int_{(\Delta t_0)} dt \int_{(\Delta U_0)} \mathbf{V} \, (dU/\Delta U_0) \qquad (1.3.16)$$

where $\mathbf{q}'$ is $\mathbf{q}$ of (1.3.15). The time interval Δt_0 in (1.3.16) is the ***representative elementary time*** (abbreviated as RET) of the flow in the porous medium.

The time averaging in (1.3.16) follows also from the discussion in paragraph 1.3.1 where we considered the rapid motion of molecules. By introducing the time average in (1.3.16), we obtain "smooth" temporal variations of $\mathbf{q}$ at a physical point in space at a "physical" instant of time.

1.3.5 Concluding Remarks

In the present text we shall adopt the continuum approach. Accordingly, the actual multiphase porous medium is replaced by a fictitious continuum: a structureless substance, to any point of which we can assign kinematic and dynamic variables and parameters that are continuous functions of the spatial coordinates of the point and of time. Sometimes we shall replace the multiphase porous medium by a number of overlapping continua. Each of these represents one phase (solid, fluid 1, fluid 2, etc.), and fills the entire porous medium domain. To every point in space we may assign properties of any of these continua. Interactions may take place between these continua.

The variables and the parameters of the various fictitious continua, averaged over an REV, enable us to describe flow, and other phenomena within a porous medium domain, by means of *partial differential equations.* Such equations describe what happens at every physical point in space and at every physical instant of time.

In applying the continuum approach to the dynamics of fluids in porous media we shall feel the need to introduce macroscopic medium parameters or coefficients to accommodate the observed phenomena and to enable us to make the passage from the microscopic to the macroscopic, continuum, level. One such parameter is porosity as defined in paragraph 1.3.2. Other parameters will be permeability, the dispersivity, etc. To a certain extent, all such parameters may be called *parameters of ignorance*; they are introduced because of our inability to solve the problem on the microscopic level. In principle, it is possible to calculate these coefficients from information supplied from the lower (e.g., microscopic or molecular) levels of treatment. In practice, however, this is not possible and they must be deduced from *actual experiments* in which the various phenomena related to these parameters are observed. In spite of the unavoidable use of experiments for the determination of coefficients, certain advantages are gained by introducing simplified conceptual models of the medium, and of the various phenomena taking place in them, and by analyzing such models in an exact analytical or statistical manner (par. 4.5.1).

There is more in the continuum approach than the enormous simplification achieved in presenting and solving problems of flow in porous media. Even if we could describe and solve such problems at the microscopic level (i.e., obtain as solution a description of what happens at points within a single pore), such solutions would be of no practical value. In fact, there would be no easy way even to verify these solutions as no instruments are available for measuring values of dependent variables at the microscopic level.

From the discussion above it follows that a relationship exists between the size of the REV of a porous medium and the measuring instrument. Any instrument introduced to measure macroscopic properties has a "window" of a certain size through which it is in contact with the porous medium. A piezometer open at its bottom is an example. Such an instrument "sees," or averages, a certain property over a certain porous medium volume around it. Since the macroscopic parameters (both coefficients and dependent variables) have the meaning of average values over an REV around the considered point, a measured quantity will be useful only if it has the same meaning, i.e., if the measuring device also averages over the same REV.

The last observation may be demonstrated by considering the incompressible Navier–Stokes equations for flow of a liquid continuum in a gravity field. Neglecting inertial terms, these take the form:

$$\text{grad}(p + \gamma z) = \mu \nabla^2 \mathbf{V} \tag{1.3.17}$$

$$\text{div}\,\mathbf{V} = 0 \tag{1.3.18}$$

where p is pressure, z is elevation, and $\nabla^2\mathbf{V} \equiv (\nabla^2 V_x)\mathbf{1x} + (\nabla^2 V_y)\mathbf{1y} + (\nabla^2 V_z)\mathbf{1z}$. For *very slow motion*, by taking the divergence of both sides of (1.3.17), and noting that the operations div and ∇^2 on the right-hand side may be performed in a reverse order, we obtain:

$$\operatorname{div}\operatorname{grad}(p + \gamma z) = \operatorname{div}(\mu \nabla^2\mathbf{V}). \tag{1.3.19}$$

In view of (1.3.18), after some vector manipulations, equation (1.3.19) reduces to (Schlichting 1955):

$$\nabla^2 p = 0 \quad \text{or} \quad \nabla^2 \varphi = 0; \qquad \varphi = z + p/\gamma. \tag{1.3.20}$$

This means that in a fluid continuum filling the void space of a porous medium domain, in which the motion is indeed very slow, the pressure $p = p(x, y, z)$ or the piezometric head $\varphi = \varphi(x, y, z)$ are potential functions.

Nothing will be changed if the medium is anisotropic. However, in paragraph 6.2.2 it will be shown that the continuity equation for an incompressible fluid in a medium which is homogeneous but anisotropic, is:

$$K_x \frac{\partial^2 p}{\partial x^2} + K_y \frac{\partial^2 p}{\partial y^2} + K_z \frac{\partial^2 p}{\partial z^2} = 0$$

or:

$$K_x \frac{\partial^2 \varphi}{\partial x^2} + K_y \frac{\partial^2 \varphi}{\partial y^2} + K_z \frac{\partial^2 \varphi}{\partial z^2} = 0 \tag{1.3.21}$$

which indicates that p or φ are not potential functions. The apparent contradiction becomes clear if we distinguish between the local, or microscopic, pressure and piezometric head satisfying (1.3.20) and the average, or macroscopic, pressure and piezometric head satisfying (1.3.21).

Exercises

1.1 Karstic aquifers, with large solution channels and caverns are usually regarded in ground water hydrology as ordinary porous media. Under what conditions is this approach justified?

1.2 The hydraulic conductivity of a medium is given by $K = K_0(1 + ax)$, where a and K_0 are constants. What is the length L_x characterizing this heterogeneous medium? Give an interpretation of L_x.

CHAPTER 2

Fluid and Porous Matrix Properties

The following sections include definitions and discussions of fluid and porous matrix properties that appear in the theory of dynamics of fluids in porous matrices. Whereas only a brief discussion is presented on fluid properties, a somewhat more elaborate discussion is presented on matrix properties that are of special interest in connection with dynamics of fluids in porous media.

2.1 Fluid Density

2.1.1 Definitions

Fluid density (ρ) is defined as the mass of the fluid per unit volume. In general, it varies with pressure (p) and temperature (T) according to relations called *equations of state*:

$$\rho = \rho(p, T) \qquad \text{or} \qquad f(\rho, p, T) = 0. \tag{2.1.1}$$

The dimensions of ρ are ML^{-3} in the physical (M, L, T) system of dimensions and FT^2L^{-4} in the technical (F, L, T) system. Special cases of (2.1.1) are presented below.

The equation of state is usually rather complicated. However, in special cases it has simple forms. For a *perfect gas* the equation of state takes the form:

$$\rho = p/RT; \qquad \partial\rho/\partial p = \rho/p; \qquad \beta = (1/\rho)\, \partial\rho/\partial p = 1/p \tag{2.1.2}$$

where T *is taken constant* (*isothermal conditions*), β is the gas compressibility, and R is the gas constant (universal gas constant divided by the molecular weight of the gas) related to the specific heats c_p and c_v at constant pressure and constant volume, respectively, by:

$$R = c_v(c_p/c_v - 1). \tag{2.1.3}$$

For *real gases*, equation (2.1.2) is replaced by:

$$\rho = p/Z(p, T)RT \tag{2.1.4}$$

where $Z(p, T)$, called the *compressibility—or Z—factor*, expresses the deviation from the perfect gas law. For an isothermal process Z is a function of p only. For a perfect gas, $Z = 1$.

A single-component system can exist in three possible phases: solid, liquid and gas, each having a different equation of state.

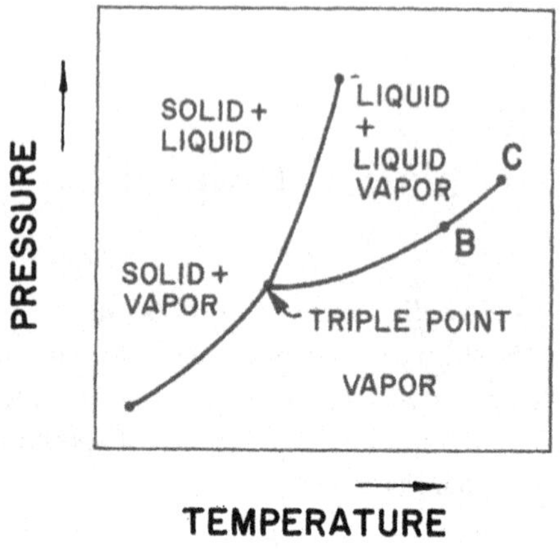

FIG. 2.1.1. Schematic pressure—temperature diagram of a single phase component.

A means of illustrating the behavior of a single-phase component is the pressure–temperature diagram shown in figure 2.1.1. The three curves shown divide the p–T plane into three regions, each corresponding to a single phase. Each curve gives values of p and T at which the two adjacent phases can coexist in equilibrium. The intersection of these curves is called the *triple point*. At the conditions of p and T specified by the pressure–temperature along the portion BC of the curve of figure 2.1.1, liquid and vapor coexist in equilibrium; otherwise only one phase exists. Above the curve, the substance in question exists only as a liquid. Below the curve it exists as a vapor.

Point C, where the p–T curve terminates, is called the *critical point* of the system. For a single-component system the critical point is defined as the highest value of pressure and temperature at which two phases of a particular fluid (i.e., liquid and gas or vapor) can still coexist. For a multicomponent system a different definition is required (e.g., Amyx et al. 1960).

Another means of illustrating the phase changes of a single-component system is the p–v–T diagram shown schematically in figure 2.1.2. Point C is the critical point. The dashed curve encloses the region where the two phases coexist. The solid lines are curves of constant temperatures (isotherms). Two definitions are needed to understand figure 2.1.2. A *bubble point* is defined (for a single-component system) as that state in which the substance is entirely in the liquid phase, and any slight reduction in pressure (or increase in volume) at the substance's fixed temperature produces a vapor phase; similarly, at a fixed pressure and volume, a slight increase in temperature produces a vapor phase. A *dew point* is defined (for a single-component system) as that state in which the substance is entirely in the vapor phase, and any slight increase in pressure (or reduction in volume) produces a liquid phase at constant temperature; similarly, at fixed pressure and volume, a slight reduction in temperature produces a liquid phase.

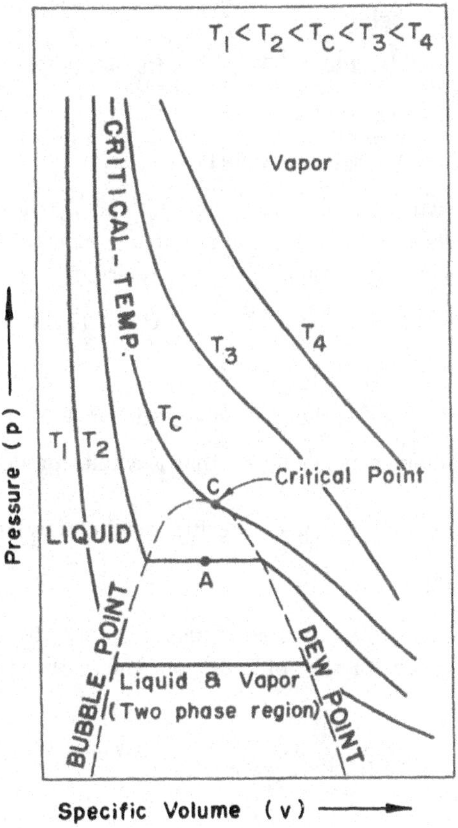

FIG. 2.1.2. Schematic pressure-volume-temperature diagram for a single component fluid.

A related property is the *specific weight* (γ) defined as the weight of fluid per unit volume:

$$\gamma = \rho g \tag{2.1.5}$$

where g is the gravity acceleration. The dimensions of γ are $ML^{-2}T^{-2}$ in the physical system of dimensions, and FL^{-3} in the technical system.

A third, dimensionless, property is the *specific gravity* or *relative density* (δ) of a fluid, defined as the ratio of the density of a fluid to that of pure water at 4°C. (Engineers, using the Fahrenheit scale, sometimes use 60°F as a standard temperature.) For a gas, the specific gravity is defined as the ratio of its density to that of hydrogen or air at some stated temperature.

In the c.g.s. system of units, density is measured in g/cm^3; specific weight is measured in dynes/cm^3. In the English engineers' system, density and specific weight are measured in $lb_f\ \sec^2/ft^4$, and in lb_f/ft^3, respectively.

The specific gravity of liquids is usually measured by some form of hydrometer that has its special scale. The most common scales are the *Baume' scale* and the

API scale. For liquids lighter than water:

$$\delta_{60/60} = 140/(130 + °\text{Be}'); \qquad °\text{Be}' = (140/\delta) - 130 \tag{2.1.6}$$

and for liquids heavier than water:

$$\delta_{60/60} = 145/(145 - °\text{Be}') \tag{2.1.7}$$

where $\delta_{60/60}$ is the specific gravity of the liquid at 60°F relative to water at 60°F.

The American Petroleum Institute has adopted a slightly different hydrometer for oils lighter than water for which the scale, referred to as the *API scale,* is:

$$\delta_{60/60} = 141.5/(131.5 + °\text{API}); \qquad °\text{API} = (141.5/\delta) - 131.5. \tag{2.1.8}$$

For any temperature, T, we have:

$$\delta(T) = \delta_{60}/[1 + a(T - 60)]; \qquad a = \exp(0.0106 \times °\text{API} - 8.05). \tag{2.1.9}$$

The density of an inhomogeneous liquid that is a solution depends on the solute's concentration.

Figure 2.1.3 shows the variation of specific weight of water with temperature.

2.1.2 Mixture of Fluids

The weight of a component of a mixture of fluids is equal to the product of the molecular weight and the mole fraction of that component in the mixture:

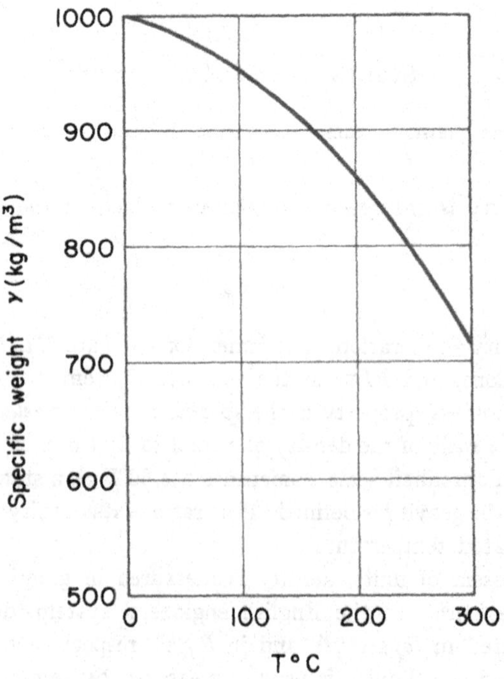

FIG. 2.1.3. Variation of specific weight of water with temperature.

$$W_i = x_i M_i; \qquad W = \sum_{(i)} W_i \tag{2.1.10}$$

where W is the total weight, W_i is the weight of the ith component of the mixture, x_i is the mole fraction of the ith component in the mixture, and M_i is the molecular weight of the ith component.

The volume of a component in a mixture (U_i) is the product of the weight of that component and its specific volume (v_i) at the prevailing conditions of pressure and temperature. The total volume (U) and specific weight (γ) of the mixture are, therefore:

$$U = \sum_{(i)} U_i = \sum_{(i)} W_i v_i = \sum_{(i)} x_i M_i v_i; \qquad \gamma = W/U = \sum_{(i)} x_i M_i / \sum_{(i)} x_i M_i v_i. \tag{2.1.11}$$

2.1.3 Measurement of Density

The most commonly used method for determining the density of a liquid is the *hydrometer method* (fig. 2.1.4). The hydrometer can be used for density determination at atmospheric pressure (fig. 2.1.4) or at any other pressure in a pressure cylinder.

The *bicapillary pycnometer* (fig. 2.1.5) is a tool for accurate determination of density. The density of the liquid sample drawn into the pycnometer is determined from its volume and weight.

The specific gravity of gases can be determined by weighing on an analytical balance. In this method the weight of a glass bulb fitted with inlet and outlet stopcocks is determined on an analytical balance. Then the weight is determined for the bulb filled with the gas and finally for the bulb filled with air. Instead of determining the weights directly, the buoyancy force acting on a sealed glass bulb

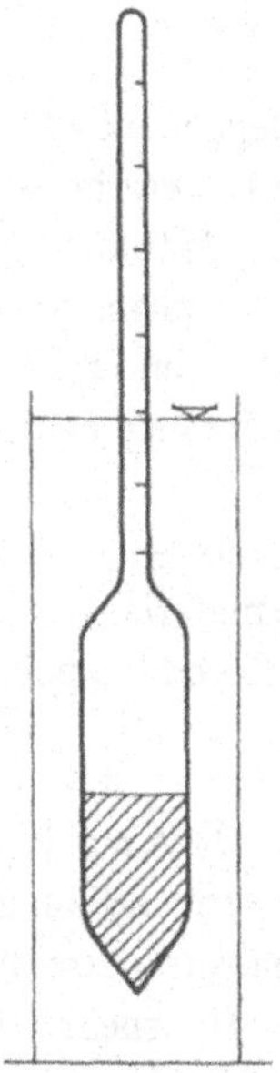

FIG. 2.1.4. The hydrometer.

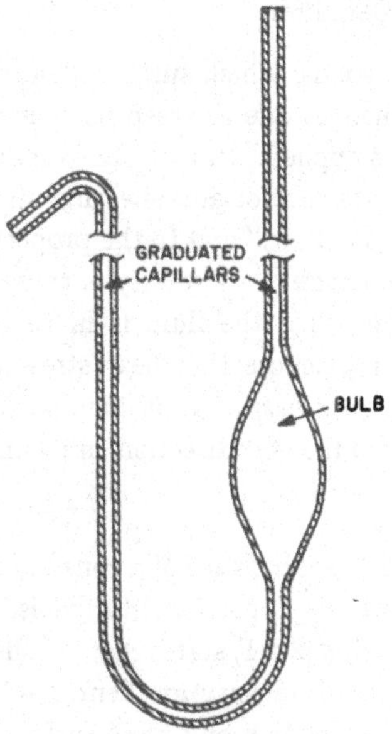

FIG. 2.1.5. Bicapillary pycnometer.

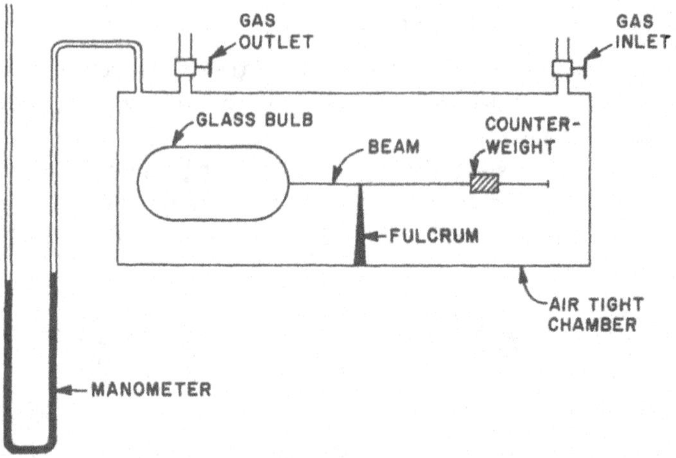

FIG. 2.1.6. Determining relative density of gas by the buoyancy method (Burcik, 1961).

suspended in the gas environment can be measured. The instrument is shown schematically in figure 2.1.6. The gas pressure is adjusted so as to balance the beam.

Another method for determining gas density is based on the effusion of the gas through a small orifice. The method is based on *Graham's law*, which states that the rate of effusion of a gas through an orifice is inversely proportional to the square root of the density of the gas.

2.2 Fluid Viscosity

2.2.1 Definition

Unlike solids, which suffer a deformation or strain independent of time (except for slow changes due to creep phenomena), fluids continue to deform as long as a shear stress is applied. In fact, fluids may be defined as materials that continue to deform in the presence of any shearing stress. We refer to this continuous deformation of a fluid as "flow," and to the property by virtue of which a fluid resists any deformation as *viscosity*. Viscosity is thus a measure of the reluctance of the fluid to yield to shear when the fluid is in motion.

Let τ_{yx} denote the shear stress exerted in the *direction* $+x$ on a fluid surface whose *outer normal is in the direction* $+y$ (when the material at greater y exerts a shear in the $+x$ direction on the material at lesser y). Then for a point P (fig. 2.2.1):

$$\tau_{yx} = \mu\, \partial u/\partial y. \tag{2.2.1}$$

For the special case of a constant velocity gradient, $\partial u/\partial y$ is replaced by u_0/b. The constant of proportionality, μ, is called the *dynamic viscosity* of the fluid.

Equation (2.2.1) states that the shear force per unit area (shear stress) is proportional to the local velocity gradient. When τ_{yx} is defined as the shear stress exerted in the $+x$ direction on a fluid surface of constant y by the fluid in the region of lesser y, equation (2.2.1) becomes:

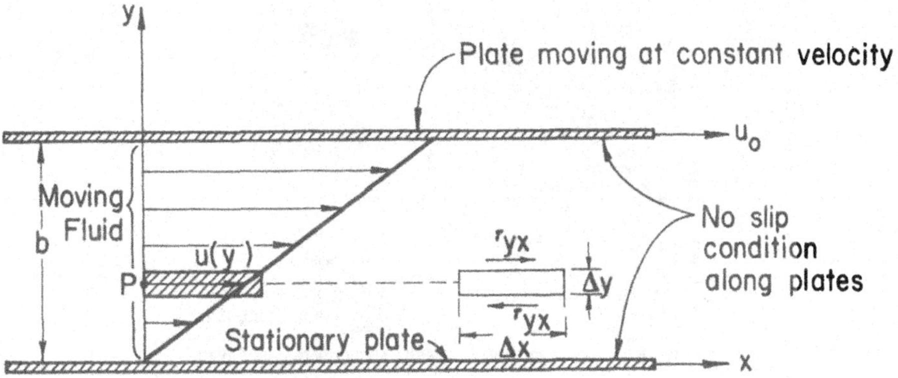

FIG. 2.2.1. Flow between two parallel plates to illustrate viscosity.

$$\tau_{yx} = -\mu \, \partial u / \partial y. \tag{2.2.2}$$

Another interpretation of τ_{yx}, as defined by (2.2.2), is that it gives the flow of x-momentum across any plane of constant y (carried by molecules that cross this plane in the $+y$ direction). Equation (2.2.2) is known as *Newton's law of viscosity.* Fluids whose behavior obeys this law are called *Newtonian fluids.*

2.2.2 Non-Newtonian Fluids

All gases and most simple liquids are *Newtonian fluids.* Fluids that do not obey (2.2.2) are called *non-Newtonian fluids.* Among the non-Newtonian fluids we may mention:

(a) *Bingham plastic,* which exhibits a yield stress at zero shear rate, followed by a straight-line relationship between shear stress and shear rate. This means that a certain stress (shear stress) must be exceeded for flow to begin.

(b) *Pseudoplastic fluid* is characterized by a progressively decreasing slope of shear stress versus shear rate (i.e., μ decreases with increasing rate of shear).

(c) In *dilatant fluid* the apparent viscosity (slope of curve) increases with increasing shear rate.

These three examples (fig. 2.2.2a) are of time-independent non-Newtonian fluids. Some fluids are more complex in that their apparent viscosity depends not only on the shear rate, but also on the time the shear rate has been applied. There are two general classes of such fluids (fig. 2.2.2b):

(a) *thixotropic fluids,* whose apparent viscosity depends on time of shearing as well as on the shear rate; as the fluid is sheared from a state of rest, it breaks down on a molecular scale, but then the structural reformation will increase with time. If allowed to rest, the fluid builds up slowly, and eventually regains its original consistency.

(b) In *rheopectic fluids* the molecular structure is formed by shear and the behavior is opposite to that of thixotropy.

Only Newtonian fluids are considered in this book.

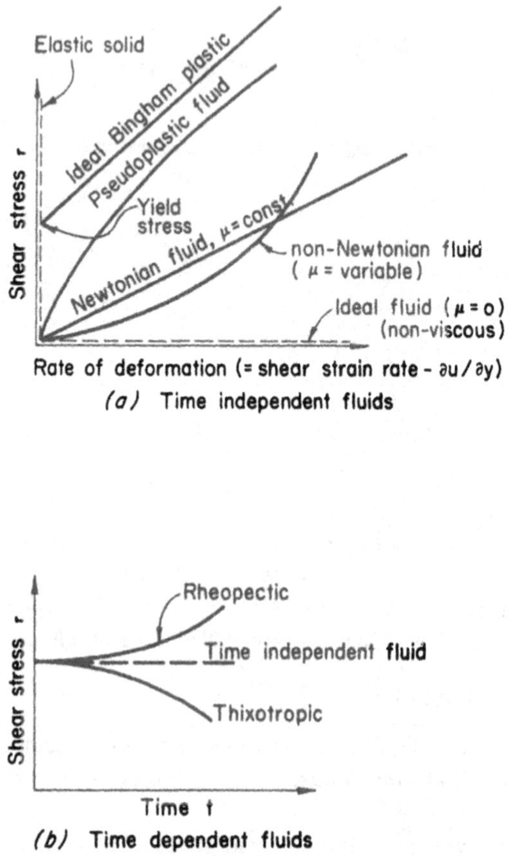

FIG. 2.2.2. Behavior of non-Newtonian fluids.

2.2.3 Units

The dimensions of dynamic viscosity are: $ML^{-1}T^{-1}$ or $FL^{-2}T$.

The unit of dynamic viscosity in the c.g.s. system is dyne sec/cm^2 = (g cm/sec^2) · (sec/cm^2) = g cm^{-1} sec^{-1}. This unit is called the *poise*. Another useful unit is the *centipoise* (1 cp) = 0.01 poise. Water at 20°C has a dynamic viscosity of 1.0 centipoise.

Kinematic, or molecular kinematic, viscosity, ν, is the ratio μ/ρ appearing frequently in fluid dynamics. Its dimensions (only kinematic) are L^2T^{-1} in both MLT and FLT systems. In the c.g.s. system the unit of kinematic viscosity is 1 cm^2/sec, called 1 *stoke*, with the *centistoke* as a commonly used unit.

Sometimes other units, depending on instruments used for measuring viscosity, are used (see below).

2.2.4 Effect of Pressure and Temperature

The viscosity of fluids varies with pressure and temperature. For most fluids the viscosity shows a rather pronounced variation with temperature, but is relatively insensitive to pressure until rather high pressures have been attained. For gases at twice the critical temperature, variations of viscosity with temperature are quite small until pressures of the order of the critical pressure have been reached. The viscosity of liquids usually rises with pressure at constant temperature. Water is an exception to this rule; its viscosity decreases with pressure at constant temperature. For most cases of practical interest, however, the effect of pressure on the viscosity of liquids can be ignored. For example, for petroleum oils we assume that an increase in pressure of 200 psi produces a viscosity change equivalent to that produced by a 1°F decrease in temperature (Popovich and Hering 1959).

Changes in temperature cause opposite variations in the viscosity of gases and liquids: a decrease in temperature causes the viscosity of a gas at low density to decrease while it causes the viscosity of a liquid to rise. For gases at high pressures, however, an increase in temperature causes a decrease in viscosity. These differences in behavior stem from the basic mechanism of molecular momentum exchange that gives rise to viscosity. Table 2.2.1 gives typical values of viscosity of some liquids.

Table 2.2.1

Dynamic Viscosity of Some Liquids (in centipoise)

	Temperature				
Liquid	0°C	10°C	20°C	40°C	70°C
Water	1.787	1.310	1.002	0.653	0.407
Benzene	0.902	0.759	0.649	0.492	0.351
Chloroform	0.700	0.626	0.564	0.465	—
Ethyl alcohol	0.177	0.145	0.119	0.827	0.504
Methyl alcohol	0.813	0.686	0.591	0.450	—
Ether	0.286	0.258	0.234	0.197	—

2.2.5 Measurement of Viscosity

Viscosity of liquids is determined by instruments called viscosimeters or viscometers. The viscosimeters for liquids are the *Ostwald viscosimeter* (fig. 2.2.3) and other capillary tube viscosimeters. In these viscosimeters, the viscosity is deduced from the comparison of the times required for a given volume of the tested liquid and of a reference liquid to flow through a given capillary tube under specified initial head conditions. When so desired, the temperature of the liquid can be kept constant by immersing the instrument in a temperature-controlled water bath.

The *Saybolt viscosimeter* is also a capillary tube viscosimeter in which the kinematic viscosity is determined from the time required for a known volume of liquid to drip out of a container through a known capillary tube. The (kinematic) viscosity thus obtained is then expressed in units called *Saybolt seconds*.

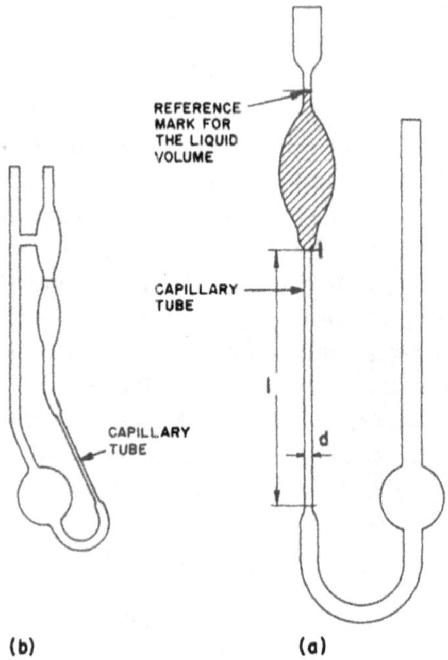

FIG. 2.2.3. Ostwald viscosimeters.

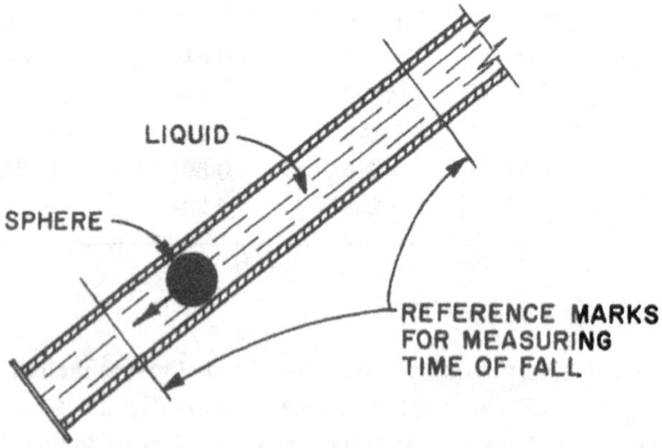

FIG. 2.2.4. Schematic diagram of the falling ball viscosimeter.

Another instrument commonly used for determining viscosity of a liquid is the *falling (or rolling) ball viscosimeter* (fig. 2.2.4), which is based on Stoke's law for a sphere falling in a fluid under the effect of gravity. A polished steel ball is dropped into a glass tube of a somewhat larger diameter containing the tested liquid, and the time required for the ball to fall at constant velocity through a specified distance between reference marks is recorded. The rolling ball forces fluid through the crescent between the ball and the tube wall. The rolling ball viscosimeter will give good results as long as the fluid flow in the tube remains in the laminar range. In certain instruments of this type both pressure and temperature may be controlled.

Other often used viscosimeters are of the *rotational or Couette* type consisting of two concentric cylinders, with the space between containing the liquid whose viscosity is to be measured. Either the outer cylinder or the inner one is rotated at a constant speed, and the rotational deflection of the other cylinder (accomplished against a calibrated spring) becomes a measure of the liquid's viscosity. Some instruments are based on the damping of oscillations of a solid sphere or disc in the tested liquid.

The viscosity of a gas can be derived from the recorded pressure drop between the ends of a capillary tube through which the gas is allowed to flow. An example of such an instrument is the *Rankine viscosimeter* in which the pressure gradient for the gas flowing through the capillary tube can be very small.

2.3 Fluid Compressibility

Compressibility (β) is the measure of volume (and related density) changes when a substance is subjected to changes in normal pressures or tensions. It is defined for *isothermal conditions* by:

$$\beta = -\frac{1}{U}\frac{dU}{dp} = \frac{1}{\rho}\frac{d\rho}{dp} \tag{2.3.1}$$

where U is the volume of a given mass of substance, p is pressure, ρ is density, and the minus sign indicates a decrease in volume as pressure increases. The reciprocal of compressibility is known as the *modulus of elasticity* E:

$$E = -\frac{dp}{dU/U} = \frac{dp}{d\rho/\rho}. \tag{2.3.2}$$

For a homogeneous fluid, incompressibility therefore means $d\rho/dp = 0$, i.e., $\rho =$ const.

One should note that in (2.3.1) the derivative dU/dp is the material derivative (par. 4.1.4) and not the local one $\partial U/\partial p$. This means that β in (2.3.1) is defined with respect to a constant mass of fluid.

Neglecting changes in solute concentration, the density of a fluid depends on pressure and temperature; therefore, compressibility under isothermal conditions is often defined as:

$$\beta = (\partial\rho/\partial p)/\rho; \qquad T = \text{const.}$$

We also have a *coefficient of expansion* (or *isobaric thermal expansion*) defined at pressure p as:

$$\beta_p = -(\partial\rho/\partial T)/\rho; \qquad p = \text{const.}$$

However, in these two definitions the partial derivative is used only to emphasize that in defining β, T is maintained constant, while in defining β_p, p is maintained constant.

For β independent of pressure, we obtain from (2.3.1):

$$U = U_0 \exp[-\beta(p - p_0)] \quad \text{or} \quad \rho = \rho_0 \exp[\beta(p - p_0)] \tag{2.3.3}$$

where U_0 and ρ_0 are, respectively, the volume and density at the reference pressure p_0. For small values of $(p - p_0)$ the *equation of state* (2.3.3) may be approximated by the first terms of its series expansion:

$$\rho = \rho_0[1 + \beta(p - p_0)]. \tag{2.3.4}$$

The equation of state (2.3.3) describes rather well the behavior of most liquids, though the presence of large quantities of dissolved gases causes deviations.

For an ideal gas under isothermal conditions, the equation of state is the Boyle–Mariotte law:

$$pU = (m/M)RT, \quad \text{or} \quad \rho \equiv m/U = (M/RT)p; \tag{2.3.5}$$

where m is the mass of gas of volume U, M is the molecular weight of the gas, R is the gas constant and T is the absolute temperature. From (2.3.5) we obtain:

$$d\rho/dp = M/RT; \qquad d\rho/dp = \rho/P. \tag{2.3.6}$$

By comparing (2.3.6) with (2.3.1) we see that in this case (T = const) $\beta = 1/p$, i.e., the compressibility, β, is the reciprocal of the gas pressure, and no more a constant. In a similar way we obtain $\beta_p = 1/T$.

For real gases, the deviation from (2.3.5) is taken into account by introducing the empirically determined Z-factor (par. 2.1.1).

Muskat (1937) suggests the following general equation of state:

$$\rho = \rho_0(p/p_0)^m \exp[\beta(p - p_0)] \tag{2.3.7}$$

where for incompressible liquids: $m = 0$, $\beta = 0$ and for compressible liquids: $m = 0$, $\beta \neq 0$, and (2.3.7) reduces to (2.3.3); for gases in an isothermal process: $\beta = 0$, $m = 1$, and (2.3.7) reduces to (2.3.6); for gases in an adiabatic process: $\beta = 0$, $m = c_v/c_p$, where c_v and c_p are, respectively, the specific heats at constant volume and constant pressure.

2.4 Statistical Description of Porous Media

For various purposes some description of the geometric properties of the porous matrix is required. However, owing to its complexity, the geometry of the solid cannot be defined by stating the equations that describe the surface bounding it. Instead, certain macroscopic (or average) geometric properties, such as porosity (sec. 2.5) and specific surface (sec. 2.6), are defined, determined experimentally and employed as parameters describing or actually reflecting the geometry of a porous matrix.

Another possible approach for achieving a more detailed description of a porous matrix is based on the introduction of some statistical characterization of the matrix. Some relatively simple examples of statistical descriptions of porous media are presented in this section.

2.4.1 Particle-size Distribution

Granular materials such as soils are often described by their *particle-size distribution*. This immediately raises the question of what is meant by *particle size*, as, except for a sphere or a cube, the size of a particle cannot be uniquely defined by a single linear dimension. In general, a particle size depends in each case on the length dimension that is measured (or determined indirectly) and on the method of measurement. The two main methods for determining particle sizes and particle-size distributions are *sieve analysis*, for particles larger than approximately 0.06 mm, and *hydrometer analysis*, for smaller particles. In sieve analysis, the tested granular material is shaken on a sieve with square openings of specified size, so that the "size" of a particle is based on the side dimension of a square hole in a screen. However, as particles are seldom spherical or of any regular shape that would permit one to determine whether or not they would slip through a sieve hole, this arbitrary definition of size actually means only that we have measured some dimension of a particle that will permit it to slip through a square hole. In hydrometer analysis, the size of a particle is the diameter of a sphere that settles in water at the same velocity as the particle. Other methods include direct measurement by microscope, electron microscope and electrical counting of particles passing through an aperture of known diameter. Some indirect methods based on statistics are mentioned later in this section.

Figure 2.4.1a shows some typical particle- or grain-size distribution curves obtained by plotting the percentage by weight passing through each sieve in a sieve analysis. A porous medium composed entirely, or almost entirely, of particles of a particular size is called *uniform*. If the grains are of different sizes, the granular porous medium is said to be *well graded*. Some commonly used classifications of soils are given in figure 2.4.1b. The mixed names given to some soils in figure 2.4.1a indicate that these are well graded soils; the first adjective always indicates the dominant fraction. These curves are also used to determine various "effective" or average grain sizes. For example, often a grain diameter greater than that of 10% of the particles by weight is called the *effective grain size* (or *Hazen's effective grain size* denoted by d_{10}). Using d_{10} and d_{60}, uniformity, or *Hazen's effective grain size coefficient* C_u, is defined by:

$$C_u = d_{60}/d_{10}. \tag{2.4.1}$$

A well graded soil has a lower C_u. A soil having a uniformity coefficient smaller than 2 is considered uniform. Effective grain size is often related to the soil's permeability (par. 5.1.1).

Another coefficient mentioned in the literature is the *coefficient of gradation* C_g:

$$C_g = (d_{30})^2/d_{60}d_{10}. \tag{2.4.2}$$

Various characteristic grain diameters (d_{10}, d_{50}, d_{60}, d_{85}) are used in the design of granular filters for various purposes. In drainage of agricultural lands, a filter or gravel envelope is used to prevent the entrance of fine sand and silt into the drains.

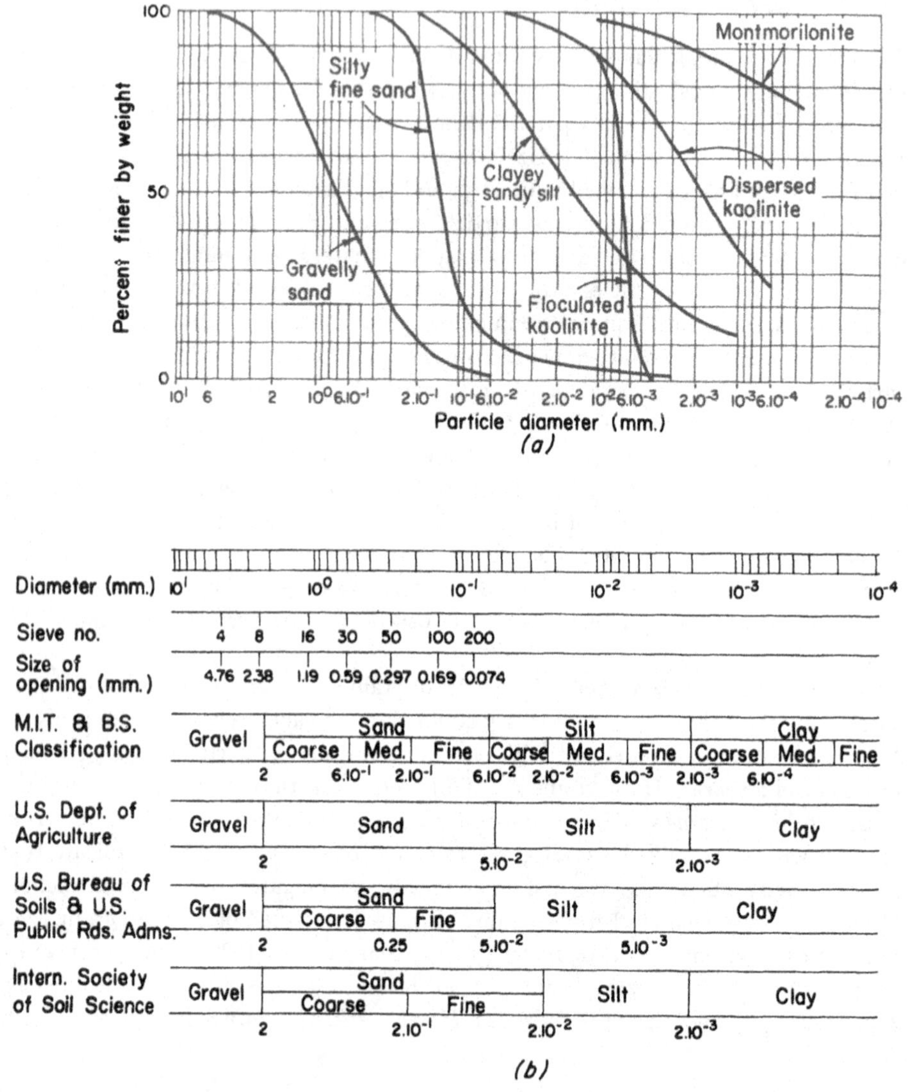

FIG. 2.4.1. Typical grain size distribution curves for various soils with major used classifications.

For example, according to the criterion proposed by the U.S. Bureau of Reclamation, the design of a filter is based on:

$$d_{50} \text{ of filter}/d_{50} \text{ for base material} = 5 \text{ to } 10, \text{ for uniform material}$$

$$= 12 \text{ to } 28, \text{ for graded material.}$$

Figure 2.4.2 shows the soil triangle for the basic soil textural classes used by the U.S. Soil Conservation Service.

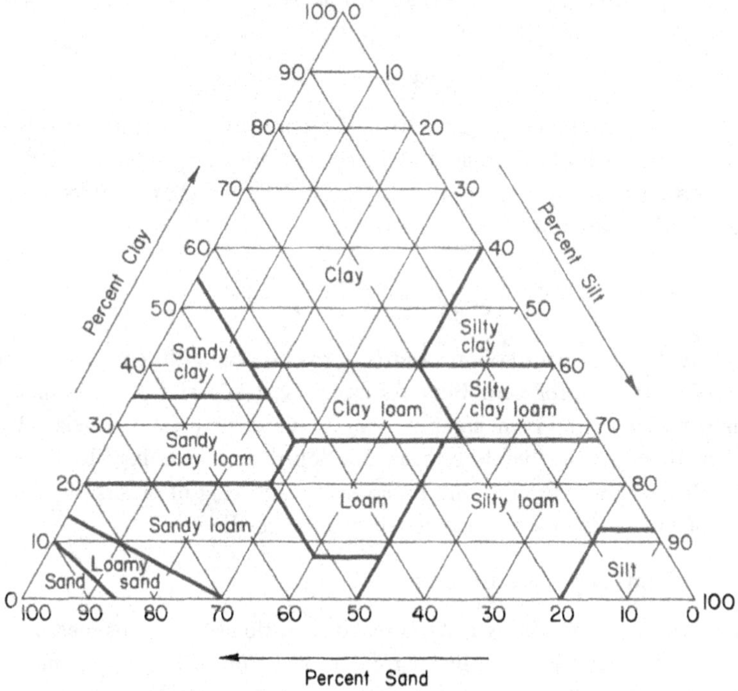

FIG. 2.4.2. Soil triangle of the basic soil textural classes (U.S. Soil Conservation Service).

2.4.2 Pore-size Distribution

There is obviously no need to elaborate on the fact that a detailed description of the pore space is impossible and that, as with respect to solid grains in a granular material, only a statistical description of one kind or another is possible. The determination of a *pore-size distribution* is of special importance in consolidated materials for which a grain-size distribution cannot be obtained.

One way to define a "pore size" is by defining the *pore diameter* δ at a point within the pore space of a porous medium as the diameter of the largest sphere that contains this point while still remaining entirely within the pore space. Thus by attaching a pore diameter to each point of the pore space, a pore-size distribution may be defined by determining what fraction, α, of the total pore volume, U_v, has a pore diameter between δ and $\delta + d\delta$ (Scheidegger 1960). For this distribution:

$$\int_0^\infty \alpha(\delta)\, d\delta = 1. \tag{2.4.3}$$

Because no single length dimension can be employed to describe the size or the geometric shape of the pores, it is sometimes convenient to visualize the holes as short circular capillary tubes. The pressure required to force a nonwetting liquid (e.g., mercury) into the pore spaces can then be related to the diameter, δ, of the pore. For example, from (9.2.13) we obtain:

$$\delta = 4\sigma \cos\theta / p_c \tag{2.4.4}$$

where δ is some equivalent diameter of the pore into which a nonwetting fluid is just entering, θ is contact angle and p_c is the capillary pressure required to force the nonwetting fluid into the pore. Ritter and Drake (1945) introduced the pore-size distribution function $D(\delta)$ defined as:

$$D(\delta) = \frac{2p_c}{\delta} \frac{d(U_v - U)}{dp_c} \tag{2.4.5}$$

where U_v is the total pore volume and U is the volume of injected nonwetting fluid.

In general, at least for consolidated materials, it is easier to determine the grain-size distribution of a given sample than to determine its pore-size distribution. Several methods have, therefore, been developed for obtaining the latter from the former. Most of these theories are based on the packing of grains or on a statistical analysis of cut-sections of consolidated porous media.

2.4.3 Other Statistical Descriptions

Several methods for obtaining a statistical description of a porous medium have been based on the introduction of a linear random function of the porous medium. Fara and Scheidegger's (1961) procedure may serve as an example. Let an arbitrary line be drawn through the porous medium. In practice this is done on an enlarged photomicrograph of the porous medium. Points on this line are defined by giving their arc length, s, from an arbitrary selected origin. Certain segments of this line pass through the void space, while others pass through the solid matrix. Fara and Scheidegger introduce a function $f(s)$ whose value is defined as equal to $+1$ if the point corresponding to s is in the void space, and otherwise to -1. The function $f(s)$ defined in this way is a random function that describes the porous medium. As long as the medium is homogeneous and isotropic, the statistical properties of $f(s)$ do not depend on the position or direction of the line. Figure 2.4.3 shows an example

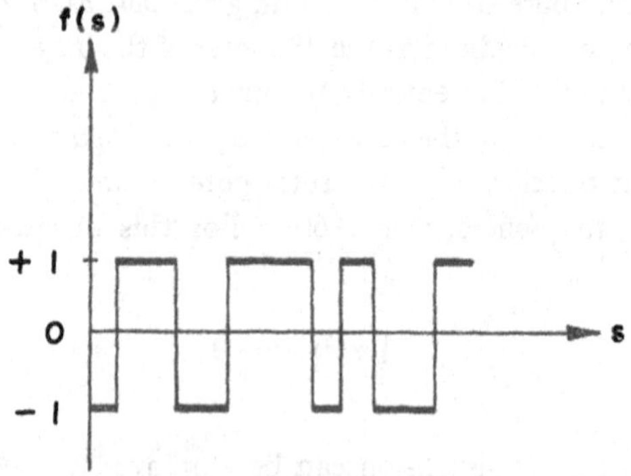

FIG. 2.4.3. Example of the function $f(s)$.

of a function $f(s)$. The average $\bar{f}$ of f, defined by:

$$\bar{f} = \lim_{s \to \infty} \left[\int_{-s}^{+s} f(s')\, ds' \Big/ \int_{-s}^{+s} ds' \right] \tag{2.4.6}$$

is related to the medium's porosity, n, by:

$$\bar{f} = 2n - 1. \tag{2.4.7}$$

As another statistical way of describing the properties of $f(s)$, Fara and Scheidegger (1961) suggest the spectrum $Z(k)$ of $f(s)$:

$$Z(k) = \frac{1}{2\pi} \int_{-\infty}^{+\infty} \frac{1 - \exp(iks)}{is} f(s)\, ds. \tag{2.4.8}$$

Prager (1961a) employs a function $g(\mathbf{r})$, where $\mathbf{r}$ is the position vector, such that $g(\mathbf{r}) = 1$ if $\mathbf{r}$ is in the void space and $g(\mathbf{r}) = 0$ if $\mathbf{r}$ is in the solid space. He then obtains information about the shape of the void space from the average of the product $g(\mathbf{r})g(\mathbf{r} + \boldsymbol{\rho})$, keeping $\boldsymbol{\rho}$ fixed. In spite of the random nature of g, the two-point average $S(\boldsymbol{\rho})$ is a perfectly regular function of the relative position, $\boldsymbol{\rho}$, of the two points concerned; it represents the probability that a line segment having the length and direction of the vector $\boldsymbol{\rho}$ will, when thrown at random on an appropriate cross-section of the medium, land with both its ends in the void space. At $\rho = 0$, $S = n$; at $\rho = \infty$, $S = n^2$. The initial slope of S plotted against ρ is $-\frac{1}{4}$ times the pore surface area, and the integral from 0 to ∞ of $(S - n^2)$ with respect to ρ, is a characteristic length of the medium. He also derives other medium characteristics from a three-point average.

Haring and Greenkorn (1970) present a random network model of a porous medium with nonuniform pores. The model pore space is represented by a large number of randomly oriented, straight cylindrical pores. They achieve non-uniformity by assigning two-parameter distributions to pore radius and pore length. Their statistical derivations result in expressions for bulk model parameters such as capillary pressure, permeability and longitudinal and transversal dispersivities, which are consistent with known empirical behavior of porous media.

2.5 Porosity

2.5.1 Porosity and Effective Porosity

Porosity (n), or *volumetric porosity*, a macroscopic porous medium property, is defined as the ratio of volume of the void space (U_v) to the bulk volume (U_b) of a porous medium (say, of a rock sample):

$$n = U_v/U_b = (U_b - U_s)/U_b \tag{2.5.1}$$

where U_s is the volume of solids within U_b. Usually the porosity, a dimensionless quantity, is expressed in percentages.

The definitions of *areal porosity* and *linear porosity* were presented in paragraph 1.3.3. When U_b in (2.5.1) is the total void space, regardless of whether pores are interconnected or not, the porosity is referred to as ***absolute or total porosity***. However, from the standpoint of flow through the porous medium, only *interconnected pores* are of interest. Hence the concept of *effective porosity*, n_e, defined as the ratio of the interconnected (or effective) *pore volume*, $(U_v)_e$, to the total volume of the medium, is introduced:

$$n_e = (U_v)_e/U_b, \qquad (U_v)_e + (U_v)_{ne} = U_v \tag{2.5.2}$$

where $(U_v)_{ne}$ is the noneffective volume (volume of noninterconnected pores). Unless specifically defined otherwise, porosity in this book means this kind of effective porosity.

Another type of pores, which seem to belong to the class of interconnected pores but contribute very little to the flow, are the *dead-end pores* (fig. 2.5.1) or *stagnant pockets* (Coats and Smith 1964). Owing to their geometry (e.g., with a constricted opening), the fluid in such pores is practically stagnant. In certain mechanisms of flow (i.e., diffusion and dispersion phenomena, sec. 10.4), or in the relationship between specific discharge and average velocity, it is important to take into consideration the effects of dead-end pores.

Various investigators report that in fine-textured porous media (e.g., clays) there are indications of an immobile or highly viscous water layer (e.g., $< 0.5\ \mu$) on the particle surfaces that makes the effective porosity (with respect to flow through the porous medium) much smaller than the measured one. On the basis of experimental evidence they also conclude that the viscosity of the water increases as the particle surface is approached (e.g., Philip 1961). This phenomenon may lead to still another definition of effective porosity.

The term effective porosity, denoted by n_e, is also frequently used (as in this text) to denote the *specific yield* (or drainable porosity) discussed in paragraph 7.1.4 and section 9.4.

Sometimes, two types of pores are encountered in a porous medium: *original*

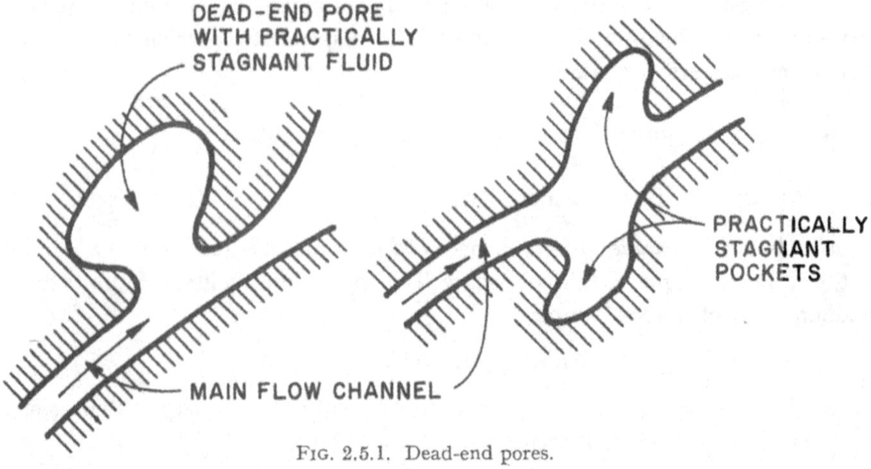

FIG. 2.5.1. Dead-end pores.

porosity, which develops in the process of deposition that forms the rock, and *induced porosity*, added at a later stage by geological and chemical processes (e.g., fissures and solution channels).

The void ratio (e) is defined as the ratio between the volume of voids and the volume of solids:

$$e = \frac{U_v}{U_s} = \frac{n}{1-n}; \quad n = \frac{e}{1+e}; \quad U_v = \frac{e}{1+e} U_b; \quad U_s = \frac{1}{1+e} U_b. \qquad (2.5.3)$$

2.5.2 Porosity, Structure and Packing

Many porous media are made of a mixture of discrete large and small grains or particles that are either loose (as dune sand) or held together by compression and cementing material (e.g., sandstone). The quantity of finer particles has a marked effect on the porosity. The porosity of consolidated materials depends mainly on the degree of cementation. The porosity of unconsolidated materials depends on the packing of the grains, their shape, arrangement and size distribution.

A starting point in understanding the manner in which porosity depends on structure and on the mode of packing is to consider simple models, such as regular packings of uniform spheres or rods. Figure 2.5.2 shows typical examples of such media and their porosities. Graton and Fraser (1935) analyzed the porosity of various packing arrangements of uniform spheres. The least compact arrangement of uniform spheres is that of cubical packing (fig. 2.5.2a) with a porosity of 47.6%. The most compact packing of uniform spheres is the rhombohedral one (where each sphere is tangential to 12 neighboring spheres) where the porosity is 25.96%. Intermediate arrays have porosities falling between these limits. In these and other cases of spheres of equal size, the porosity is independent of the radius of the spheres.

The particle-size distribution may appreciably affect the resulting porosity, as small particles may occupy pores formed between the large particles, thus reducing porosity. Hence, other parameters being equal, poorly sorted sediments will have a considerably lower porosity than well sorted ones (fig. 2.5.3).

Other factors that affect porosity are compaction, consolidation and cementation. Because compacting forces vary with depth, porosity will also vary with depth, especially in clays and shales. Krumbein and Sloss (1951) indicate a reduction in sandstone porosity from 52 to 41% and in shale porosity from 60 to 6% as depth increases from 0 to 6000 ft. Athy (1930) expressed the change in porosity with depth as $n = n_0 \exp(-\alpha d)$ where α is a coefficient and d is depth below ground surface. Most of the pore reduction is due to the inelastic, hence irreversible, effects of intergranular movement.

Consolidated sedimentary rocks are derived from initially unconsolidated grains (e.g., sand) that have undergone significant cementation at areas of grain contact. As the pore space is filled with cementing material, great reduction in porosity may take place.

Table 2.5.1 gives typical porosity values of various materials.

Table 2.5.1

Typical Porosity Values of Natural Sedimentary Materials[a]

Sedimentary Material	Porosity Value (percent)	Sedimentary Material	Porosity Value (percent)
Peat soil	60–80	Fine-to-medium mixed sand	30–35
Soils	50–60	Gravel	30–40
Clay	45–55	Gravel and sand	30–35
Silt	40–50	Sandstone	10–20
Medium-to-coarse mixed sand	35–40	Shale	1–10
Uniform sand	30–40	Limestone	1–10

[a] After Todd 1959.

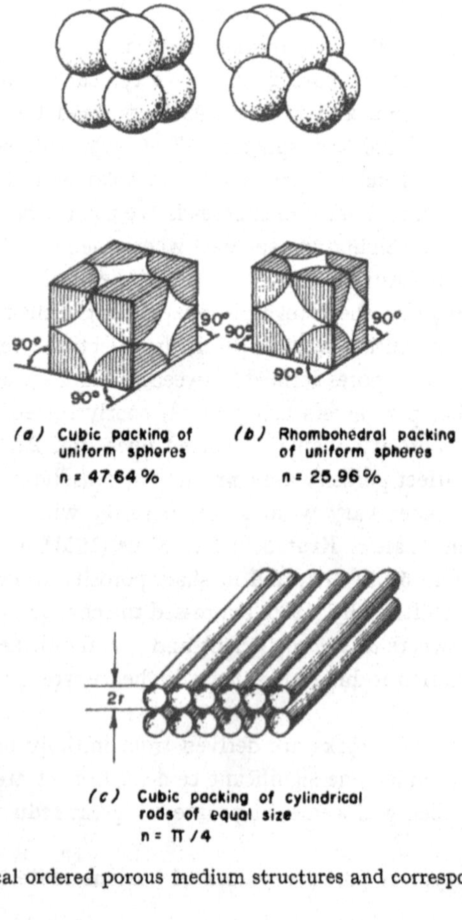

Fig. 2.5.2. Typical ordered porous medium structures and corresponding porosities.

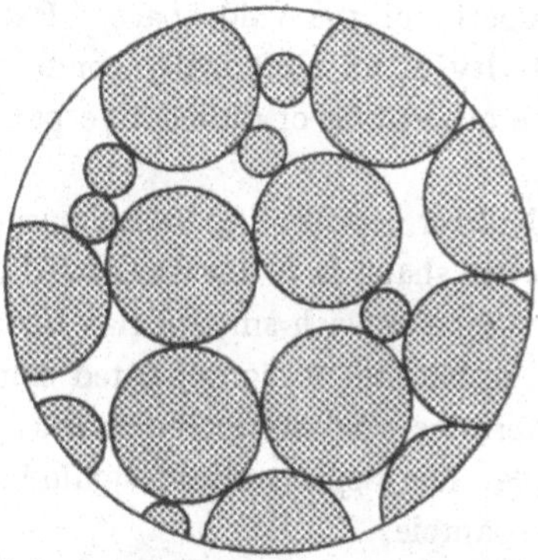

(a) WELL SORTED MATERIAL n = ~32%

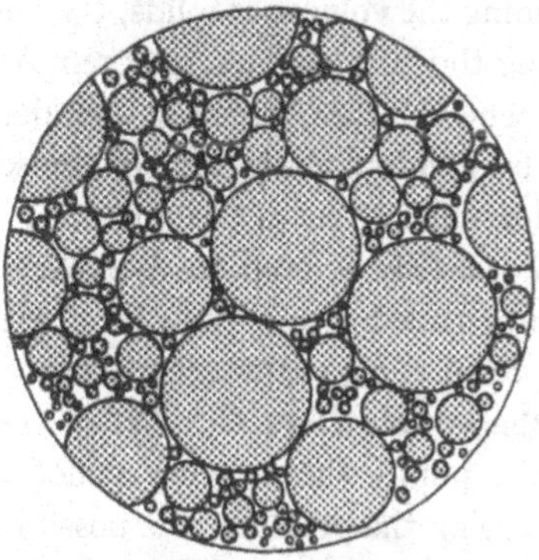

(b) POORLY SORTED MATERIAL n = ~17%

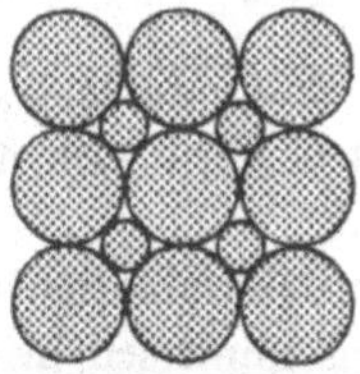

(c) CUBIC ARRANGEMENT OF SPHERICAL GRAINS OF TWO SIZES n = ~12.5%

FIG. 2.5.3. Effect of sorting on porosity.

2.5.3 Porosity Measurement

A great many methods have been developed for determining porosity, mainly of consolidated rocks (encountered in oil reservoirs). From the definition of porosity it is obvious that common to all methods is the need to determine two of the three volumes: the total or bulk volume of the sample (U_b), its pore volume (U_v) and the volume of its solid matrix (U_s). The various methods based on such volume determinations, called ***direct methods***, differ from each other in the way these volumes are determined. Other methods are available, called ***indirect methods***, based on the measurement of some property of the void space. Examples of such properties are the electrical conductivity of an electrically conducting fluid filling the void space of the sample, or the absorption of radioactive particles by a fluid filling the void space of the sample.

The simplest direct method for determining bulk volume of a consolidated sample with a well defined geometric shape is to measure its dimensions. The method is applicable only to cylindrical cores with smooth flat surfaces and 90° angles. The usual procedure, however, applicable to consolidated samples of both regular and irregular shapes, is to determine volumetrically or gravimetrically the volume of fluid displaced by the sample. The displaced fluid should be prevented from penetrating the pore space of the sample.

Among the direct methods are included:

(a) a method for determining the volume of solids, U_s (and from it, U_v), by crushing the sample after determining the bulk volume, thus removing all pores (including the noninterconnected ones!). The volume of solids is then determined by fluid displacement in a pycnometer, or by dividing the weight of the solids by the solid's specific weight (2.65–2.67 g/cm³ for natural rock materials).

(b) *The mercury injection method* is used to determine U_b and U_v. The tested sample is placed in a chamber filled to a certain level with mercury, with a known volume of air at a known pressure (e.g., atmospheric pressure) above it. The volume of mercury displaced by the sample gives U_b. When the pressure in the mercury is increased by a volumetric pump, the mercury penetrates the pore space of the sample. By gradually increasing the pressure it is possible to determine, in addition to the total pore volume U_v, the entire capillary pressure curve (par. 9.2.6). In general, the method is not suitable for low permeability samples as very high pressures are required.

(c) Another direct method for determining U_b involves observing the change in weight of a pycnometer when it is filled with mercury and when it is filled with mercury and the tested sample.

(d) A method for gravimetric determination of U_b involves observing the loss in weight of a sample when it is immersed in a fluid.

(e) A compression chamber may be used for determining U_s. Several instruments are available, all based on the Boyle–Mariotte gas law. Basically, two chambers are connected by a valve (fig. 2.5.4). The tested sample is placed in the chamber of volume U_1. The pressure in this chamber is p_1. The second chamber (volume U_2), initially evacuated of gas, is connected to the first one by opening the valve between them, thus permitting the gas to expand isothermically. If the final pressure is p_2, we have:

$$(U_1 - U_s)p_1 = (U_1 - U_s + U_2)p_2$$

or:

$$U_s = U_1 - \frac{p_2}{p_1 - p_2} U_2. \tag{2.5.4}$$

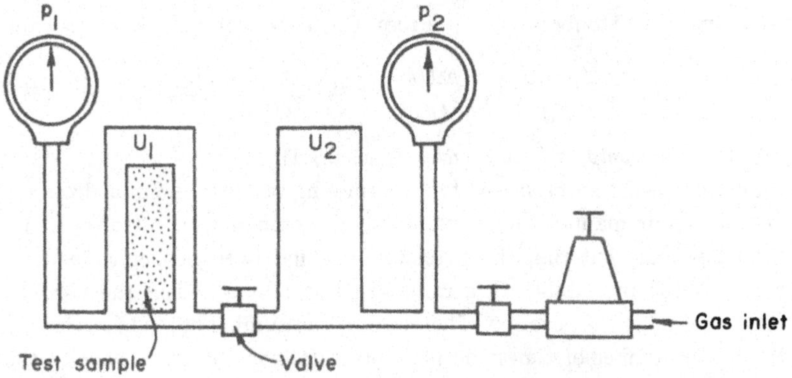

FIG. 2.5.4. Porosimeter based on Boyle-Mariotte's law.

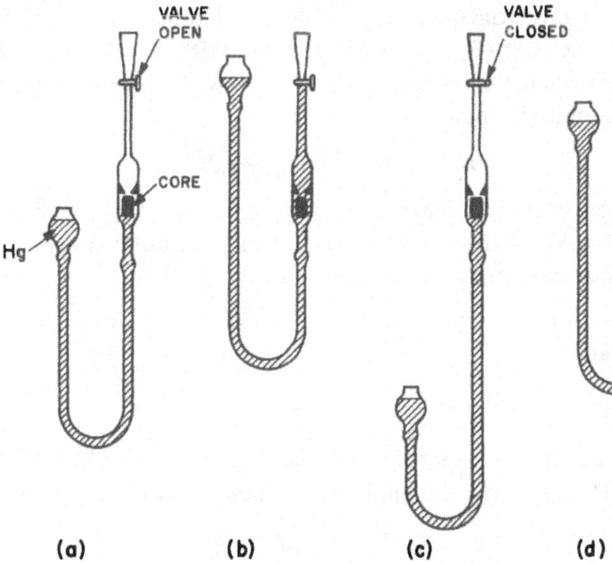

FIG. 2.5.5. Schematic diagram of the Washburn-Bunting porosimeter (after Monicard, 1965).

(f) In the Washburn–Bunting porosimeter, the pore volume is determined by measuring the volume of air (at atmospheric pressure) extracted from the pore space of the sample by creating a partial vacuum in the porosimeter (fig. 2.5.5). The partial volume is obtained by manipulating the mercury reservoir.

(g) The *gas expansion method*, also based on Boyle–Mariotte's gas law, is perhaps the most widely used method for determining porosity. The test is usually carried out at a constant temperature. The sample is placed in a chamber of volume U_1, into which gas is injected. The gas pressure at this stage is p_1. The gas is then allowed to expand into another chamber of volume U_2, where the initial pressure (i.e., before being connected to the first chamber) is p_0. The ultimate pressure in the two connected chambers is p_2. From Boyle–Mariotte's law we obtain:

$$\frac{p_1(U_1-U_s)}{Z(p_1)}+\frac{p_0U_2}{Z(p_0)}=\frac{p_2(U_1+U_2-U_s)}{Z(p_2)} \tag{2.5.5}$$

where $Z(p)$ is the *compressibility factor* (par. 2.1.1).

(h) Employing a *statistical method* for determining porosity, a pin is dropped many times in a random manner on an enlarged photomicrograph of a section of the consolidated porous material, the porosity of which is to be determined. It can be shown that the probability of a random point falling within the pore space of this section is equal to the porosity. Therefore, as the number of pin tosses is increased, the ratio of the number of times the pin's point falls in the pore space to the total number of tosses approaches the value of the porosity (e.g., Chalkley et al. 1949).

An example of an *indirect method* is one based on the absorption of radioactive particles by a fluid (liquid or gas) saturating the sample. Norel (1967) developed a method based on this principle. The incident radiation is provided by a source of radioactive Prometeum ($P_m{}^{147}$), which together with a Samarium target emits α particles. The absorbing fluid is Xenon, an inert gas with a high absorption capacity. The porosity is obtained from:

$$n=(\ln N_1-\ln N_2)/\mu_{Xe}L \tag{2.5.6}$$

where μ_{Xe} is the coefficient of linear absorption of Xenon, L is the thickness of the sample, and N_1 and N_2 are, respectively, total counts of radiation with a dry sample and with a sample saturated with Xenon.

2.6 Specific Surface

2.6.1 Definitions

The *specific surface* (M) of a porous material is defined as the total interstitial surface area of the pores (A_s) per unit bulk volume (U_b) of the porous medium:

$$M=A_s/U_b; \qquad [M]=L^{-1}. \tag{2.6.1}$$

For example, the specific surface of a porous material made of identical spheres of radius R in a cubical packing is $M=4\pi R^2/(2R)^3=\pi/2R$. It thus becomes obvious that fine materials will exhibit a much greater specific surface than will coarse materials. Some fine porous materials contain an enormous surface area per unit volume. For example the specific area of sandstone may be of the order of 1500 cm^2/cm^3. Carman (1938) gives the range of $1.5\times10^2-2.2\times10^2$ cm^{-1} for the specific surface of sand.

If a granular medium consists of a mixture of m fractions, each made up of N_i identical spheres of radius r_i, the total area, A_s, and volume, U_s, of the solid spheres are:

$$A_s=\sum_{i=1}^{m}4\pi r_i^2N_i; \qquad U_s=\sum_{i=1}^{m}\tfrac{4}{3}\pi r_i^3N_i=(1-n)U_b.$$

Then:

$$M = \sum_{i=1}^{m} 4\pi r_i^2 N_i \bigg/ \left[\sum_{i=1}^{m} \tfrac{4}{3}\pi r_i^3 N_i/(1-n) \right] = 3(1-n) \sum_{i=1}^{m} f_i/r_i = 3(1-n)/\bar{r} \qquad (2.6.2)$$

where f_i is the volume fraction of the ith fraction ($= \frac{4}{3}\pi r_i^3 N_i/U_s$) and $\bar{r}$ is the harmonic mean radius.

Sometimes the term "specific" is used to indicate "per unit volume of solid material." Then, denoting the latter specific area by M_s, we obtain:

$$M = A_s/U_b = A_s(1-n)/U_s = (1-n)M_s. \qquad (2.6.3)$$

For spheres of uniform radius r:

$$M_s = 4\pi r^2 / \tfrac{4}{3}\pi r^3 = 3/r. \qquad (2.6.4)$$

We may also define a specific area M_v with respect to a unit pore volume:

$$M = A_s/U_b = A_s n/U_s = nM_v; \qquad M_v/M_s = (1-n)/n \qquad (2.6.5)$$

or with respect to a unit weight of the material (e.g., for sandstone we may have 0.5–5.0 m^2/g whereas for shale we may have 100 m^2/g).

Thus, specific area of a porous material is affected by porosity, by mode of packing, by the grain size and by the shape of the grains. For example, disc-shaped particles will exhibit a much larger specific area than will spherical ones.

Specific surface plays an important role in cases that involve absorption of materials from the fluid flowing through the medium. It is also important in the design of filter columns, reactor columns and ion exchange columns.

2.6.2 Measurement of Specific Surface

Several methods of surface area determination are described below. It is important to realize that the results obtained by different methods differ, as in each method we measure a *different surface*. Depending on the size of the irregularities we wish to take into account, different surface areas are obtained. In each case the definition of surface area and the method of its measurement should be chosen according to the particular problem on hand.

Obviously the specific surface of natural porous media can be determined only by indirect or statistical methods. Some of them are described briefly below.

Type 1, the statistical method developed by Chalkey et al. (1949). This method, described in paragraph 2.5.3(h), can be employed for determining the specific surface of a consolidated porous material. A needle of length l is dropped at random a great many times on an enlarged photomicrograph of a section of the porous material. A count is kept of the number of times (α) the pin's end points fall within the void space and of the number of times (β) the pin intersects the perimeter of pores. The specific surface is then found from:

$$M = 4n\beta m/l\alpha \qquad (2.6.6)$$

where n is porosity. This method is considered one of the best available. Many other matrix properties can be derived with it.

Type 2, adsorption methods. These are based on the adsorption of gas or a vapor by a solid surface. The solid's surface area is determined from the quantity of gas adsorbed on it, assuming that the gas covers the entire surface of the solid with a uniform monomolecular film.

Type 3, the heat of wetting method. Here, the surface area is calculated from the heat of wetting liberated when the dry medium is immersed in a liquid (Wenzel 1938).

Type 4, a method based on fluid flow. Developed by Kozeny (1927) and modified by later workers (see detailed discussion in par. 5.10.3), this method suggested a relationship between the permeability of a medium and its specific area. Using this relationship one can obtain the specific area by conducting experiments leading to the determination of the medium's permeability. Brooks and Purcell (1952) compare specific area derived by Kozeny's equation and by other methods.

2.7 Matrix and Medium Compressibility

Under natural conditions, a porous medium volume at some depth in a ground water aquifer or in an oil reservoir is subjected to an internal stress or hydrostatic pressure (p) of the fluid saturating the medium, and to an external stress (σ) exerted by the formation in which the particular volume is embedded. The effect of overburden load is included in this stress. Only $1-d$ consolidation is considered here.

The coefficient of bulk compressibility α'_b is defined for a saturated porous medium as the fractional change in the bulk volume of the porous medium with a unit change in σ:

$$\alpha'_b = -\frac{1}{U_b}\frac{dU_b}{d\sigma}\bigg|_{p=\text{const}} \tag{2.7.1}$$

where U_b is the volume of a fixed mass of the porous medium. Because generally, in practice, σ remains virtually constant, another coefficient of bulk compressibility (α_b) is often defined with respect to a unit change in p:

$$\alpha_b = -\frac{1}{U_b}\frac{dU_b}{dp}\bigg|_{\sigma=\text{const}} \tag{2.7.2}$$

Note that a third coefficient of bulk compressibility (α) with respect to σ' is defined below. Geerstma (1957) proposed two more kinds of compressibility in addition to α_b: (a) *rock* (or *solid*) *matrix compressibility* (α_s), which is the fractional change in volume of the solid matrix (U_s) with a unit change in pressure; and (b) *pore compressibility* (α_p; designated by Hall (1953) as *effective rock compressibility*), defined as the fractional change in pore volume (U_p) with unit change in pressure:

$$\alpha_s = -\frac{1}{U_s}\frac{dU_s}{dp}\bigg|_{\sigma=\text{const}} \qquad \alpha_p = -\frac{1}{U_p}\frac{dU_p}{dp}\bigg|_{\sigma=\text{const}} \tag{2.7.3}$$

We therefore have:

$$U_b = U_s + U_p; \quad dU_b/dp = dU_s/dp + dU_p/dp$$

$$U_s = (1-n)U_b; \quad U_p = nU_b; \quad \alpha_b = (1-n)\alpha_s + n\alpha_p \tag{2.7.4}$$

at $\sigma = \text{const}$. When $(1-n)\alpha_s \ll \alpha_b$, we have $\alpha_b = n\alpha_p$. Typical values of α_p are of the order of $10^{-5} \div 10^{-6}$ reciprocal psi.

We thus see that various definitions exist for bulk, matrix and rock compressibilities (e.g., $\alpha'_s = -(1/U_s)\, dU_s/d\sigma|_{p=\text{const}}$ and $\alpha'_p = -(1/U_p)\, dU_p/d\sigma|_{p=\text{const}}$); each definition requires an appropriate experimental technique. Geerstma (1957), as did several others, developed a theory and performed experiments of rock compressibility that explain rock deformation under oil reservoir conditions. For example, when saturating a rock sample with a fluid, immersing it in a pressure vessel containing the saturating fluid and imposing a hydrostatic pressure on the fluid, he could observe changes in U_s (volume of solids in the sample) and calculate α_s. A diagram of necessary equipment is shown in figure 2.7.1 (Hall 1953). The amount of liquid "squeezed" out of the core is indicated by the liquid rise in a calibrated small-bore tube.

In order to obtain the effect of external pressure alone, the pressure in the fluid saturating the core is maintained constant (say, atmospheric). Thus, only a hydrostatic external pressure can be investigated by this apparatus. By an appropriate hydraulic pressure system, the saturated core can be subjected to a constant or a variable internal pressure and overburden or external pressure.

For extreme compaction pressures, all materials show some irreversible change in porosity due to distortion and crushing of matrix elements (e.g., grains).

In the discussion thus far no account has been taken of the stresses arising within the solid matrix itself.

Figure 2.7.2a shows a typical vertical cross-section through a confined compressible aquifer of thickness b at time t. Figure 2.7.2b shows the details at the upper impervious confining boundary of the aquifer. To simplify the discussion, we shall consider a granular noncohesive porous matrix with grain sizes such that molecular and interparticle forces may be neglected. Any horizontal plane of area A passing through

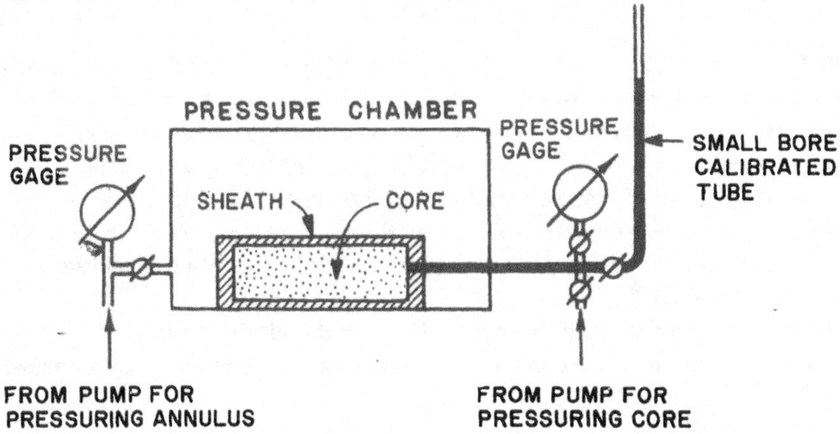

Fig. 2.7.1. Schematic representation of a rock-compressibility device (after Hall, 1953).

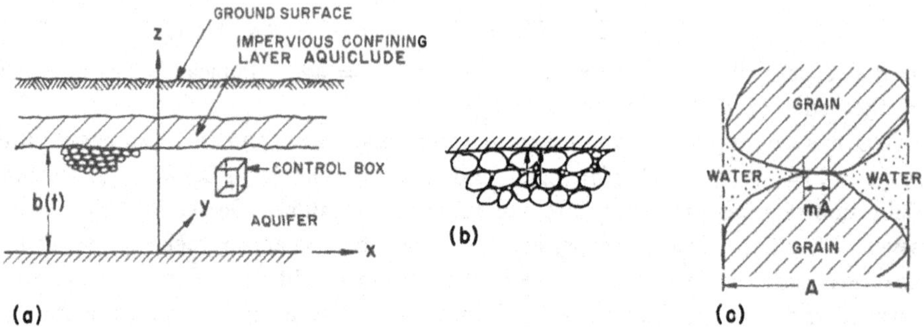

FIG. 2.7.2. A confined compressible aquifer.

an aquifer, whether confined or phreatic, will intersect both solid grains and void spaces. According to Terzaghi (1925, 1943 and 1960), the total load of soil and water (and actually, also that resulting from atmospheric pressure, par. 6.3.3) above this plane is balanced by *interparticle stresses* in the porous matrix (e.g., soil) and by the pressure in the water. Taking into account the contact areas of solid and of void space intersected by the horizontal plane, we may write:

$$\sigma = p(1 - m) + \sigma_s\, m \tag{2.7.5}$$

where σ is the total stress over the total area A, mA is the area of solid–solid contact, $(1 - m)A$ is the area of water–solid contact (fig. 2.7.2c), p is the pressure in the water and σ_s is the stress in the solid (i.e., force per unit *solid area*). The actual value of m is small ($m \ll 1$); however, σ_s is very high and probably equals the yield stress of the solid at the surface. Terzaghi called the product $m\sigma_s = \sigma'$ the *effective stress* of the solid matrix. In the water $(1 - m)p \approx p$, so that (2.7.5) becomes:

$$\sigma = \sigma' + p. \tag{2.7.6}$$

Thus the effective stress (σ') is equal to the difference between the total stress on the bulk porous medium sample (σ) and the pressure (p, sometimes called *neutral stress*). In (2.7.6), σ, σ' and p of course have their usual macroscopic meaning. This equation, first introduced by Terzaghi (1925, 1943 and 1960), and further analyzed by Skempton (1961), is fundamental in soil mechanics. The idea leading to it is called the *effective stress concept*. A similar theory, which also leads to the concept of effective stress, was developed by Gersevanov (1933; cited in Scheidegger 1960) and Gassmann (1951).

Changes in porosity result from variations in the effective stress σ', which, by (2.7.6), depends on the stress σ caused by the external load, and on the water pressure:

$$d\sigma = d\sigma' + dp. \tag{2.7.7}$$

In (2.7.6) a positive pressure ($p > 0$) means compression. Similarly, σ and σ' are taken as positive in (2.7.6) when they are compressive stresses, i.e., directed into the surface upon which they act (as is the case in most soil mechanics problems). In general, however, a positive-stress component means tension. Then, with $p > 0$ still meaning compression, we have $\sigma = \sigma' - p$ as in (6.3.62).

When dealing with three-dimensional consolidation, we must take into account the fact that stress is a second-order symmetrical tensor. Hence (2.7.6) is written in terms of stress components in the form:

$$\sigma_{ij} = \sigma'_{ij} + p\delta_{ij}, \qquad i, j = 1, 2, 3. \tag{2.7.8}$$

The Kronecker symbol, δ_{ij}, is used in the second term on the right-hand side of (2.7.8) because the pressure in the pore space is isotropic. Experience has shown that in soils deformation occurs as an integrated result of the usually irreversible microscopic movements of many irregular particles under stress patterns set up by the applied load and the random geometry of the individual grain-to-grain contact (Scott **1963**). After removing the cause of motion, each particle will find itself under a new stress system that, while requiring small local adjustments that may take some time, will not return the particles to their original positions. Changes in position will continue until a new equilibrium is reached. Hence we have inelastic deformation. This also means that the soil's response to stresses depends on the loading history. This is especially so in granular unconsolidated materials. In rocks or consolidated materials we have an approximately elastic deformation. On the other hand, in clays, which are highly plastic, the time lag may be of considerable importance; the assumption of immediate response may lead to erroneous results. Another important observation is that since it takes some time for the porosity, n, or void ratio, e, to change, the effective stress, σ', remains at first unchanged and then varies gradually. This means that any increment, σ, produces an immediate response in the form of a pressure rise in the water above the equilibrium that existed initially. Then this pressure is reduced as σ' increases and e decreases as drainage of water takes place. This is the process of *consolidation*. If the porous medium in question is a soil stratum, an aquifer or an oil reservoir, the vertical direction is usually a principal direction of $\underset{=}{\boldsymbol{\sigma}}$.

Limiting ourselves to one-dimensional (in the vertical direction) stress, and to deformation in the same direction, figure 2.7.3 shows the relationship between the void ratio, $e = n/(1 - n)$, and changes in effective stress in a granular soil. Initially, large deformations take place as a result of grain movements and readjustments. At higher vertical stresses these movements decrease. Eventually, most of the deformation results from the elastic compressibility of the solid grains themselves. The curve AB shows inelastic recovery after the external load is removed. At higher stresses the behavior of the porous matrix is more elastic. At very high stresses, e.g., such as occur in deep oil reservoirs, the slope steepens as the grains break at the points of contact (Roberts and de Souza **1958**). Such materials are beyond the scope of the present book. A detailed discussion on consolidation of clays and cohesive soils is given by Scott (**1963**).

Obviously, no analytic expression can be obtained for $n = n(\sigma')$. However, for

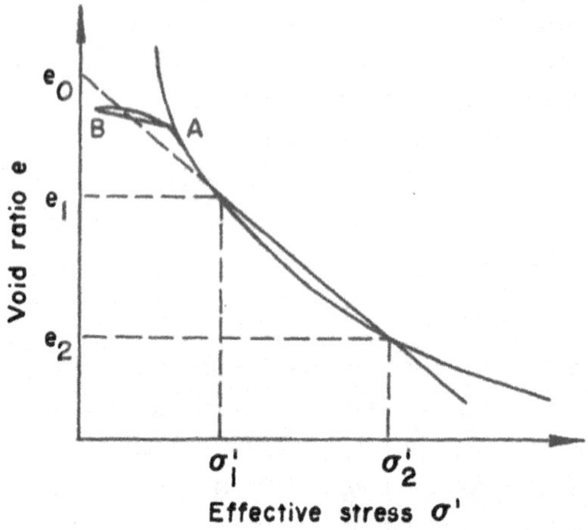

FIG. 2.7.3. Relationship between e and σ' for granular soil (after Roberts and de Souza, 1958).

small changes (as from σ'_1 to σ'_2 in fig. 2.7.3), we may assume a linear relationship:

$$e = e_0 - \alpha^* \sigma'; \qquad \partial e/\partial \sigma' = -\alpha^* \tag{2.7.9}$$

where we extend the straight line so that $e = e_0$ for $\sigma' = 0$, and α^* is a coefficient expressing soil's (not solid's) compressibility. For cohesive soils, the expression corresponding to (2.7.9) is:

$$e = e_0 - C_c \log(\sigma'/\sigma'_0) \tag{2.7.10}$$

where C_c is called the *compression index.*

As the deformation of the solid matrix is caused by the effective stress defined above, it is possible to define additional coefficients of compressibility, this time with respect to a unit change in effective stress. For example:

$$\alpha' = -\frac{1}{U_b}\frac{dU_b}{d\sigma'}. \tag{2.7.11}$$

Finally, one should note that in problems of flow through compressible media, two motions actually take place simultaneously: that of the fluid and that of the solid matrix. This problem is discussed in detail in paragraphs 6.3.1 and 6.3.2.

Exercises

2.1 Determine (a) Hazen's effective grain size coefficient C_u and (b) the coefficient of gradation C_g, for the soils whose particle size distribution is shown in fig. 2.4.1a.

2.2 A pore size distribution function $D(r_i)$, similar to (2.4.5) is defined by:

$$D(r_i) = \frac{p_c}{r_i}\frac{dS_{nw}}{dp_c}$$

where $D(r_i)$ gives the fraction of the pore volume contributed by pores with radii between r_i and $r_i + dr$. Show how this function can be obtained from (2.4.4).

2.3 Use the following definitions: $W_s = \gamma_s U_s$ = weight of solids in U, $\gamma_w = W_w/U_v$, $\delta_s = \gamma_s/\gamma_w = W_s/\gamma_w U_s$, $w = W_w/W_s$ = water content (by weight), $S = U_w/U_v$ = degree of saturation, $\gamma_t = (W_s + W_w)/U$, $\gamma^* = \gamma_t - \gamma_w$ = buoyant unit weight of soil, to show that:

(a) $Se = \delta_s w$

(b) $\gamma_t = [(1 + w)/(1 + e)]\gamma_s$

(c) for $S = 1$, $\gamma^* = [(\delta_s - 1)/(1 + e)]\gamma_w$

(d) for $S \neq 1$, $W_w = S\gamma_w eW_s/\gamma_s$.

2.4 For a porous medium made of spheres of constant diameter packed in a cubical arrangement, determine:

(a) the volumetric porosity,

(b) the range of variation of areal porosity.

2.5 Show that the definitions of α_b, α_p and α_s satisfy the relationship

$$\alpha_b = (1 - n)\alpha_s + n\alpha_p.$$

CHAPTER 3

Pressure and Piezometric Head

Only a brief discussion of the concepts of pressure, stress, piezometric head, etc., is given here. For additional information the reader is referred to standard texts on fluid mechanics.

The discussion is related first to a *liquid continuum*. By averaging point values over the fluid contained in the void space of a representative elementary volume (REV) of the porous medium, macroscopic average values are obtained and assigned to the centroids of the REV.

3.1 Stress at a Point

In the study of a *fluid continuum*, as in the study of continua in general, we distinguish between two types of forces: *body forces* (e.g., gravity and centrifugal forces), which reach into the medium and act throughout the volume without any physical contact, and *surface forces*, which include all forces exerted on a boundary by its surroundings through direct contact. The external forces acting on any volume of liquid give rise to a condition of internal *stress* (force per unit area).

To study the surface forces, we examine a small, yet finite, portion, $\delta\mathbf{A}$, of the boundary surface around a point P of a body (fig. 3.1.1). The resultant force $\delta\mathbf{F}$, acting over the area $\delta\mathbf{A}$, whose outward normal is in the direction of $\mathbf{1n}$, can be resolved into components normal (direction indicated by the unit vector $\mathbf{1n}$) and tangential (direction indicated by the unit vector $\mathbf{1s}$) to the area. These are indicated in figure 3.1.2 by δF_n and δF_s, respectively. The *normal stress* σ_{nn} and the *shear stress* σ_{ns} at P are now defined by the following limiting processes:

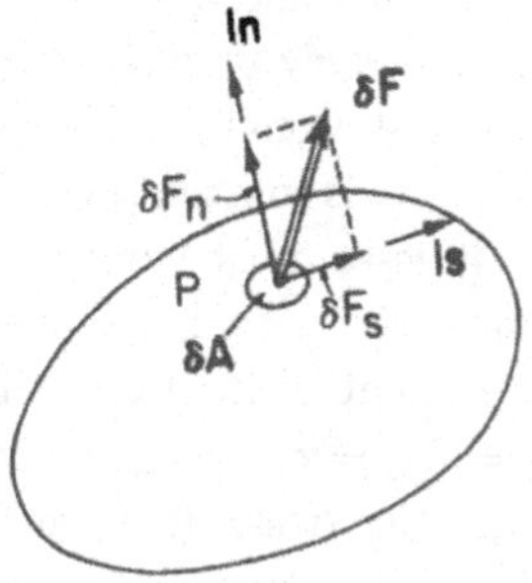

FIG. 3.1.1. Resolution of a surface force.

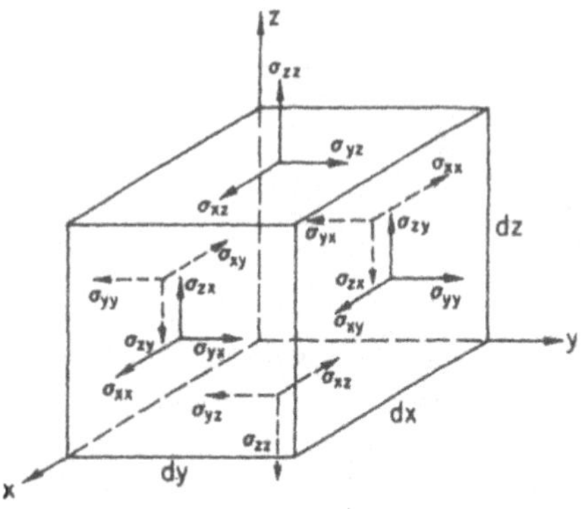

FIG. 3.1.2. Stresses on an element of fluid continuum.

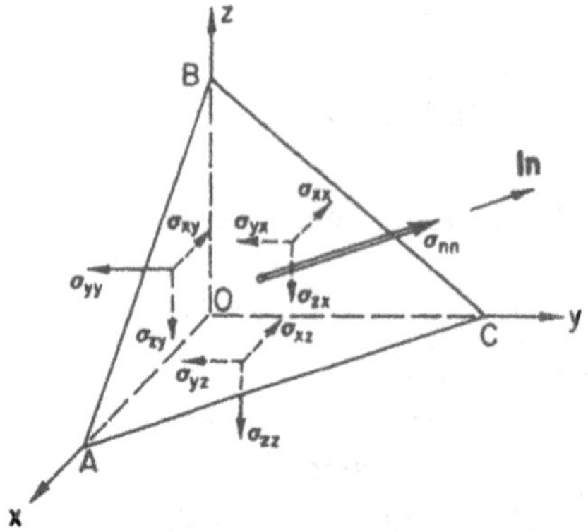

FIG. 3.1.3. An infinitesimal tetrahedron.

$$\sigma_{nn} = \lim_{\delta A \to 0} \frac{\delta F_n}{\delta A} = \frac{dF_n}{dA}; \qquad \sigma_{ns} = \lim_{\delta A \to 0} \frac{\delta F_s}{\delta A} = \frac{dF_s}{dA} \tag{3.1.1}$$

where $\delta A \to 0$ is to be interpreted as the *extrapolated limit* defined in section 1.3. Since most fluids cannot withstand appreciable tensile normal stress (directed out from the body) we shall restrict the discussion henceforth to compressive normal stresses (as is usually done in fluid mechanics). This means that σ_{nn} is a compressive stress.

In the section that follows, we shall employ a right Cartesian coordinate system xyz with z vertically upward. A double-subscript scheme will be used to identify the stresses: the second subscript will indicate the direction of the normal to the plane associated with the stress, while the first subscript will denote the direction of the stress itself. For example, σ_{yx} is the value of a shear stress acting in the direction parallel to y on a plane whose outward normal is in a direction parallel to the direction of $+x$. If both directions are positive (i.e., parallel to $+x$ and to $+y$), we shall say that the stress component is positive. Otherwise, it is negative. The various stresses are shown in figure 3.1.2. The normal stresses are σ_{xx}, σ_{yy} and σ_{zz}. The shear stresses are σ_{xy}, σ_{yx}, σ_{xz}, σ_{zx}, σ_{yz} and σ_{zy}.

Consider first the case of a *stationary or uniformly moving fluid*. Since a fluid cannot withstand a shear stress, a stationary fluid must be completely free of shear stress. In a uniform flow, the velocity is constant everywhere, and hence, by Newton's viscosity law (2.2.2), all shear stresses must vanish. Assuming that the only body force is that of gravity, we can perform force balances in the x, y and z directions for an infinitesimal prismatic element of fluid and then let the size of the element shrink to zero. We obtain $\sigma_{xx} = \sigma_{yy} = \sigma_{zz} = \sigma_{nn}$, which means that for a stationary or for a uniformly flowing fluid, the stress at a particular point is independent of direction, and is hence a scalar. This quantity is identical to the negative of the *thermodynamic pressure* defined in the kinetic theory of gases (i.e., arising from the perfectly elastic collision of fluid molecules). It is negative because σ_{nn} is a compressive stress.

In the general case of a moving viscous fluid, all stresses shown in figure 3.1.2 are present. In order to determine the stress at a particular point, it is convenient to examine an infinitesimal tetrahedron of fluid as shown in figure 3.1.3.

Using Newton's law of motion in the direction $\mathbf{1n}$ (normal to the inclined surface) and dropping (in the limiting process) the gravity and inertial terms, we may express σ_{nn} in terms of the nine stresses shown in figure 3.1.2. We obtain:

$$\sigma_{nn} = \sigma_{xx}\alpha_{nx}^2 + \sigma_{xy}\alpha_{nx}\alpha_{ny} + \sigma_{xz}\alpha_{nx}\alpha_{nz} + \sigma_{yy}\alpha_{ny}^2 + \sigma_{yz}\alpha_{ny}\alpha_{nz} + \sigma_{yx}\alpha_{ny}\alpha_{nx} + \sigma_{zz}\alpha_{nz}^2 + \sigma_{zx}\alpha_{nz}\alpha_{nx} + \sigma_{zy}\alpha_{nz}\alpha_{ny} \qquad (3.1.2)$$

where α_{nx}, α_{ny} and α_{nz} are the direction cosines of $\mathbf{1n}$ with respect to the unit vectors $\mathbf{1x}$, $\mathbf{1y}$ and $\mathbf{1z}$, respectively.

In a similar manner, we may determine two orthogonal components of shear stress on ABC in terms of the same nine stress components.

Since the inclination of the surface ABC in figure 3.1.3 is arbitrary, stresses on all planes, i.e., for any set of direction cosines, may be determined in terms of the nine components on the orthogonal reference planes. This means that (3.1.2) may be considered as an expression that transforms the nine components defining the stress at a point in the xyz coordinate system into components in any other coordinate system (say x', y', z' or n, s_1, s_2). A physical quantity that is transformed in this manner is a *second-order tensor*. Therefore the stress at a point is a *second-order tensor*. It is customary to write the nine components of the stress tensor $\boldsymbol{\sigma}$ in the form:

$$\begin{pmatrix} \sigma_{xx} & \sigma_{xy} & \sigma_{xz} \\ \sigma_{yx} & \sigma_{yy} & \sigma_{yz} \\ \sigma_{zx} & \sigma_{zy} & \sigma_{zz} \end{pmatrix}$$

recognizing that equal subscripts denote normal stress and different ones denote shear stress. A typical (scalar) component can also be denoted by σ_{ij} or $\sigma_{x_ix_j}$ where $x_1 = x$, $x_2 = y$, $x_3 = z$. It can be shown (see any text on fluid mechanics) that in ordinary fluids the stress tensor is symmetric, i.e.,

$$\sigma_{ij} = \sigma_{ji}. \qquad (3.1.3)$$

There is a symmetry about the diagonal of normal stresses. Thus the stress at a point is described by nine scalar components, of which six are different from each other.

A characteristic property of a second-order tensor is that the sum of components $(\sigma_{11} + \sigma_{22} + \sigma_{33})$ along the diagonal (the *first invariant*) is independent of the orientation of the coordinate axes. For the stress tensor considered here, one third of this quantity is often called the *bulk stress*:

$$\bar{\sigma} = \tfrac{1}{3}(\sigma_{xx} + \sigma_{yy} + \sigma_{zz}). \qquad (3.1.4)$$

Since $\bar{\sigma}$ has no directional properties, it is a scalar quantity. For a nonviscous fluid, all normal stresses at a point are equal; consequently, each of them must be equal to the bulk stress. Furthermore, it may generally be shown that for such a fluid (as well as for a viscous Newtonian fluid if we invoke the Stokes assumption $\lambda + \frac{2}{3}\mu = 0$, where λ is the second coefficient of viscosity), the magnitude of the bulk stress is equal to the thermodynamic pressure:

$$-\bar{\sigma} = p \tag{3.1.5}$$

where the minus sign results from the fact that in general negative normal stresses occur in a fluid. Although (3.1.5) has been shown to be a good approximation for a perfect gas, experience shows that it may be used with confidence also for most liquids (or incompressible Newtonian fluids). Hence, for a fluid at rest, or in the absence of shear stresses, the stress is described as hydrostatic, and is expressed by:

$$\begin{pmatrix} -p & 0 & 0 \\ 0 & -p & 0 \\ 0 & 0 & -p \end{pmatrix}$$

in which p is called the *hydrostatic pressure.*

The stress tensor σ_{ij} (actually this is a typical component which is used as a symbol for the entire tensor as $i, j = 1, 2, 3$) may always be written as a sum:

$$\sigma_{ij} = -p\delta_{ij} + P_{ij} \tag{3.1.6}$$

where δ_{ij} is the Kronecker δ, and P_{ij} is called the *viscous stress tensor.* In this case, the mean of the three normal stresses is given by:

$$\bar{\sigma} = -p + \bar{P}; \qquad \bar{P} = \tfrac{1}{3}(P_{11} + P_{22} + P_{33}). \tag{3.1.7}$$

In the case of hydrostatic stress, P_{ij} vanishes. A *perfect fluid* is one for which P_{ij} vanishes identically.

The discussion until now has applied to a liquid continuum. When the fluid fills the pore space of a porous medium, the various quantities considered above (e.g., pressure and stress component) must be averaged over an REV of the medium (par. 1.3.2), as only such average quantities are measurable. Thus, the average pressure $\bar{p}$ is defined by:

$$\bar{p} = \frac{1}{(\Delta U_0)_v} \int_{(\Delta U_0)_v} p \, dU_v. \tag{3.1.8}$$

This pressure is sometimes referred to as *pore pressure*, although in the present book we shall refer to it simply as pressure. This means that in measuring pressure in the medium, the instrument should be large enough to "see" the volume over which such an average is taken.

Unless otherwise stated, all pressures in this text are *gage pressures*. Gage pressure and absolute pressure are interrelated by: Gage pressure = absolute pressure − local atmospheric pressure.

3.2 Hydrostatic Pressure Distribution

From hydrostatics we know that for a liquid continuum at rest or in uniform flow in a gravity field, pressure variations obey:

$$-\frac{\partial p}{\partial z} = \gamma; \qquad \frac{\partial p}{\partial x} = \frac{\partial p}{\partial y} = 0. \tag{3.2.1}$$

In a *homogeneous liquid*, the density (ρ) is a constant. Hence, for two points of elevations z_1 and z_2, such that $z_2 - z_1 = d > 0$, we obtain:

$$z + p/\gamma = \text{const} \quad \text{or:} \quad p_2 - p_1 = \gamma(z_1 - z_2) \tag{3.2.2}$$

which is the equation for ***hydrostatic pressure distribution***. From (3.2.2) it follows that p decreases with elevation. If the gage pressure $p_2 = 0$ (i.e., $=$ atmospheric pressure) at a free surface of elevation z_2, we have $p \equiv p_1 = \gamma d$.

In a saturated porous medium (3.2.1) and (3.2.2) are still valid, with variables having their averages taken over representative elementary volumes around points in the medium.

Figure 3.2.1 shows a hydrostatic pressure distribution below a water table (taken as $p = 0$) for a homogeneous fluid. The region above the water table is considered in section 9.4.

In a *compressible (still single-component) fluid*, the density (ρ) varies with pressure and temperature. The appropriate equation of state should be introduced in (3.2.1) before integrating. For example, for a perfect gas under isothermal conditions we have $\gamma = g\rho = Cp$, $C =$ a constant. Then:

$$\int_{p_1}^{p} \frac{dp}{p} = -\int_{z_1}^{z} C\,dz; \quad \ln\frac{p}{p_1} = C(z_1 - z) = \frac{g\rho_1}{p_1}(z_1 - z), \qquad p = p_1 \exp\left[-\frac{\gamma_1}{p_1}(z - z_1)\right]. \tag{3.2.3}$$

3.3 Piezometric Head

The quotient p/γ in (3.2.2) is called the ***pressure head***. It represents the ***pressure energy*** (or ***flow work***) ***per unit weight*** of fluid. It is the net work done by a unit weight

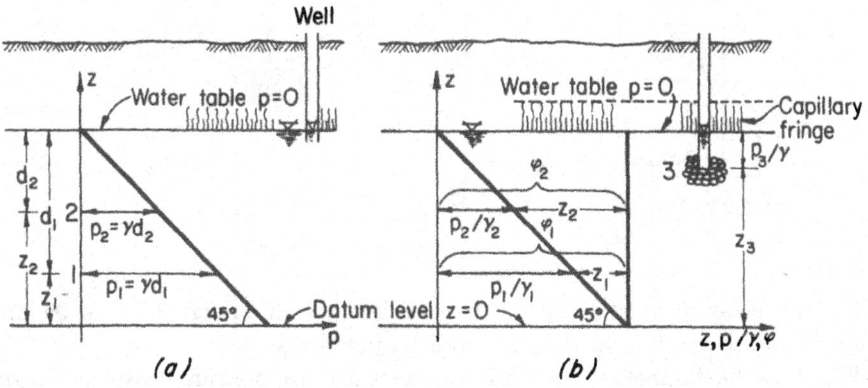

FIG. 3.2.1. Hydrostatic pressure distribution and piezometric head (fluid at rest).

of incompressible fluid against the pressure difference existing along its flow. For a compressible fluid under isothermal conditions the pressure head is defined by:

$$\int_{p_0}^{p} \frac{dp}{g\rho(p)} \quad \text{(sec. 5.9)}.$$

In (3.2.2), z represents *the elevation head*, i.e., the potential energy per unit weight of fluid. Sometimes the pressure energy and the potential energy are referred to a unit mass; the corresponding expressions are then p/ρ and gz.

The sum of the pressure head and the elevation head is called the *piezometric head* (φ):

$$\varphi = z + p/\gamma. \tag{3.3.1}$$

It is indicated, for each point within a fluid, or as an average φ for each point within a saturated porous medium domain, by the elevation of fluid in a pipe of a sufficiently large diameter to avoid capillary effects within it, and whose opening is such that the flow is presumed to be moving undisturbed by it. We shall refer to this pipe as a *piezometer* (fig. 3.2.2).

Since z is measured from some arbitrarily chosen *datum level*, the piezometric head (dim L) is also measured from the same datum level; one should always indicate the datum level when speaking about the piezometric head.

In a liquid at rest the piezometric head is constant everywhere. If a water table (phreatic surface) is present, its elevation indicates the piezometric head (fig. 3.2.2). In a moving fluid, φ varies as a function of space and time. However, for a fluid in uniform flow, the pressure distribution is still hydrostatic. Figure 3.2.2 shows the pressure head, elevation head and piezometric head in uniform flow through a horizontal confined aquifer.

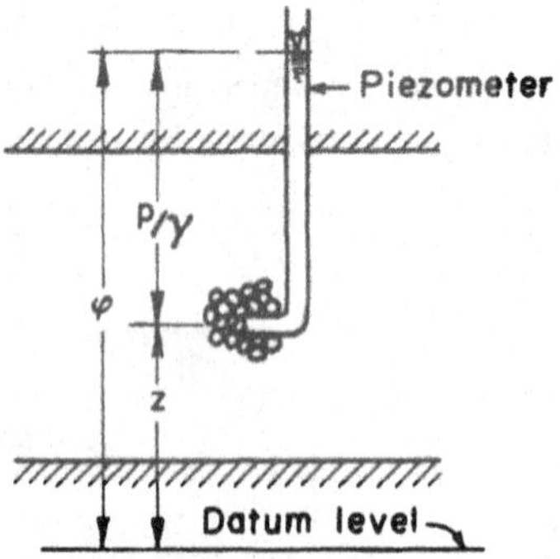

FIG. 3.2.2. Piezometric head.

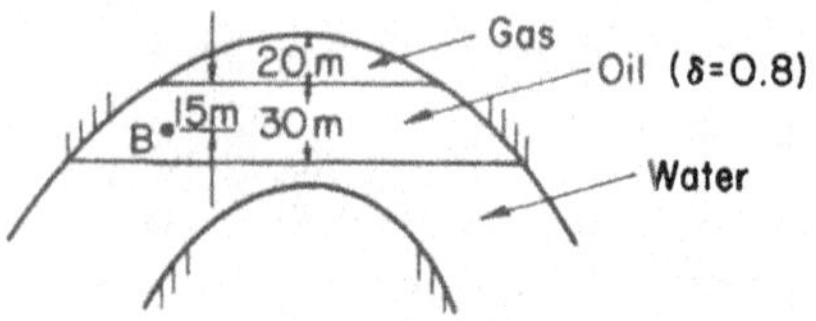

FIG. 3.E.1.

Exercises

3.1 The pressure at the top of the anticline shown in fig. 3.E.1 is 90 atm. To what level will the oil rise in a well terminated at B?

3.2 Use the flownet of fig. 6.6.1 to determine the pressure distribution along the dam's base.

CHAPTER 4

The Fundamental Fluid Transport Equations in Porous Media

In this chapter, the fundamental equations describing the transport of a fluid in a porous medium are developed. These are partial differential equations which describe dynamic and kinematic relationships between fluid, medium and flow parameters at a point within a considered flow domain. The description is a macroscopic one in the sense explained in sec. 1.3. Accordingly, in the derived equations both the fluids and the porous matrix are considered as continua, each filling the entire space.

4.1 Particles, Velocities and Fluxes in a Fluid Continuum

4.1.1 Definitions of Particles and Velocities

As in mechanics of rigid bodies, the temporal rate of change in position of a fluid particle is its velocity. However, since a fluid undergoes a continuous deformation as it moves, it is essential to express the velocity at its various points. The velocity vector $\mathbf{V}$ at any point of a fluid continuum is, therefore, written as the *extrapolated limit* (par. 1.3.1) approached by the ratio between the displacement vector, $\delta\mathbf{s}$, of a particle along its path, and the corresponding increment of time, δt, as the latter approaches zero:

$$\mathbf{V} = \lim_{\delta t \to 0} \frac{\delta \mathbf{s}}{\delta t}. \tag{4.1.1}$$

Obviously, "point," "particle" and the limit $\delta t \to 0$, have here the sense considered in section 1.3.

In what follows, the components of the velocity vector $\mathbf{V}$ in any coordinate system will be denoted by V_i ($i = 1, 2, 3$ in a three-dimensional space), or sometimes by V_x, V_y, V_z, in an x, y, z Cartesian coordinate system fixed in space. The limit operation (4.1.1) must be performed separately for each of the velocity components. In general, we have:

$$V_i = f(x_1, x_2, x_3, t) \equiv f(\mathbf{x}, t).$$

We must still define a fluid particle and its path to which the definition of velocity given above refers. A *fluid particle* is defined as an ensemble of molecules included in a certain volume. This volume is associated with the REV of a fluid continuum defined in paragraph 1.3.1. However, one should realize that the size of the REV is not a single constant value. When we speak of an REV we actually mean a volume that may vary within a certain range, the limits of which are determined on one end by the point (fig. 1.3.1), where wild fluctuations occur in the ratio of mass to volume, and on the other hand by some characteristic (macroscopic) length of interest (such as the external boundaries of the flow domain, the distance between two measuring points or the distance along which variations in density may be assumed to be linear). Thus the size of the fluid particle should be much larger than the mean free path of a single molecule, yet it should be sufficiently small so that when the density is determined for it, a meaningful point value is obtained. This value is then related to the *centroid of the particle.*

Let us first consider a homogeneous, single-species fluid. The individual molecules are in continuous natural motion. If we label a cloud of initially close molecules, say those within an REV, after a certain time interval, even if the fluid as a whole is stationary, those labeled particles will spread out and occupy a larger volume around the centroid of the initial cloud. The flux of labeled molecules is governed in this case by molecular diffusion.

The limitations of size of an REV as described above mean that, after a certain time interval, a "new" particle has to be defined for the same centroid so that it has the same number of molecules (or mass) as the original one. If the homogeneous fluid considered is in motion, the flux of the labeled particles is with respect to the moving fluid. Hence, the molecules comprising a moving particle must be relabeled whenever they occupy a volume that is too large. Each time, the end of the path of a "former" particle becomes the centroid of the new particle. By following this procedure, a continuous *path of a fluid particle* is obtained. We speak of following the path of *a particle,* although the molecules comprising this particle are continuously changing.

Let us next consider a *heterogeneous* (*multicomponent, multispecies*) *fluid,* which is a solution, the density of which may vary as a result of variations in the concentration of the various species comprising it. For such fluids the definitions given above for particle, velocity, etc., are insufficient. Instead, let us focus our attention on a certain species α of a fluid (*solution, system*) composed of a mixture of N chemical species. One advantage of using the continuum approach is that any number of different continua may occupy the same portion of space at the same time (*interpenetrating continua*; Rose 1966). Also at the same point in space we may have any number of "particles" at the same time, as long as they are of different species, and/or referring to a different property of the α-species.

In a volume dU of space occupied by a multispecies fluid system, let dm_α and dm denote the instantaneous masses of the α species, and of the fluid system, respectively. We may then define a mass density ρ_α (in the sense of sec. 1.3) of the α species as the mass of the α species per unit volume of fluid (solution):

$$\rho_\alpha = dm_\alpha/dU; \qquad \sum_{\alpha=1}^{N} \rho_\alpha = \sum_{\alpha=1}^{N} (dm_\alpha/dU) = \left(\sum_{\alpha=1}^{N} dm_\alpha\right) \bigg/ dU = dm/dU = \rho \qquad (4.1.2)$$

where ρ is the density of the fluid system.

In general, the velocity distribution and pathlines of α particles will be different from those of the fluid system. At any (mathematical) point of the fluid system, a velocity vector must be defined for each of the system's components as well as for the fluid system as a whole.

Sometimes use is made of the definition of *mass fraction* ω_α of species α defined as mass of species α per unit mass of solution:

$$\omega_\alpha = \rho_\alpha/\rho; \qquad \sum_{\alpha=1}^{N} \omega_\alpha = 1. \tag{4.1.3}$$

The velocity $\mathbf{V}_\alpha$ (with respect to a fixed coordinate system) of the α-species at a point P is the statistical average velocity within dU of the individual molecules of the α species, i.e., the sum of velocities divided by the number of molecules. The velocity $\mathbf{V}_\alpha$ is thus a *microscopic level* parameter derived from the velocities at the *molecular level.* The *mass average velocity* $\mathbf{V}^*$ (also called velocity of material point of the fluid system) is then defined (Bird, Stewart and Lightfoot 1960) by:

$$\mathbf{V}^* = \left(\sum_{\alpha=1}^{N} \rho_\alpha \mathbf{V}_\alpha\right) \Big/ \sum_{\alpha=1}^{N} \rho_\alpha = \left(\sum_{\alpha=1}^{N} \rho_\alpha \mathbf{V}_\alpha\right) \Big/ \rho = \sum_{\alpha=1}^{N} \mathbf{V}_\alpha \omega_\alpha. \tag{4.1.4}$$

In (4.1.4), $\rho_\alpha \mathbf{V}_\alpha$ is the *mass flux,* i.e., mass passing through a unit area per unit time, placed perpendicularly to the flow, of the α species with respect to a fixed coordinate system. Obviously, the sum $\sum_{(\alpha)} \rho_\alpha \mathbf{V}_\alpha$ over all α species should be equal to the mass flux of the system represented by $\rho\mathbf{V}^*$. A perfect *pitot tube* placed in the flow of a nonhomogeneous liquid measures $\mathbf{V}^*$. Actually the interpretation of $\mathbf{V}^*$ is that it represents the *momentum per unit mass* of the flowing liquid, since $\rho\mathbf{V}^*$ represents momentum per unit volume. Equation (4.1.4) then says that the momentum per unit volume (of solution) of the fluid system is equal to the sum of momenta of the individual species.

The *volume-average velocity* $\mathbf{V}'$ is defined for a multicomponent system by:

$$\mathbf{V}' = \sum_{(\alpha)} \rho_\alpha u_\alpha \mathbf{V}_\alpha \tag{4.1.5}$$

where u_α is the *partial specific volume* defined by: $u_\alpha = \partial U/\partial m_\alpha$, with

$$\sum_{\alpha=1}^{N} (\partial U/\partial m_\alpha)(dm_\alpha/dU) = 1 = \sum_{\alpha=1}^{N} \rho_\alpha u_\alpha.$$

The volume averaged velocity $\mathbf{V}'$ is used mainly when the volume of the solution changes with concentration.

In a homogeneous incompressible single-species fluid ($\alpha = 1$ only, and $\rho_\alpha = \rho =$ const) $\mathbf{V}_\alpha \equiv \mathbf{V}' \equiv \mathbf{V}^*$. In general, these velocities differ both in direction and in magnitude.

Using the same line of thought, the concept of a particle and its velocity may be extended to other properties of the fluid as a whole, or of a component α of it, in addition to the fluid's mass, or the mass of the α-species discussed above. In the preceding discussion, the fluid's *mass particle* was considered. However, as there are other properties associated with the fluid's molecules, such as momentum, kinetic energy, etc., we may also define particles composed of fixed amounts of these other properties. For example, we may define a particle of kinetic energy of the fluid at a point in a fluid continuum. In general, the mass average velocity $\mathbf{V}^*$ and the kinetic energy average velocity $\mathbf{V}_{ke}$ are different at every point in the space. In a multicomponent system, we may also define particles possessing a certain property in one of the species, and follow them along their path in the flow domain.

An *extensive property* of a substance is one that depends on the mass of the substance to which this property is referred. Volume, mass, energy, momentum and kinetic energy are examples of extensive properties. The property itself may be a scalar property, a vectorial property, or a tensorial property of any rank. We shall denote by G the amount of such property in a fluid system. We may also consider an extensive property of a species (say, the α-species) of a multicomponent system. When a particle of G or G_α is followed in a flow domain, its instantaneous velocity is denoted by $\mathbf{V}_G$ or $\mathbf{V}_{G\alpha}$. We may also speak of an *extensive entity* that depends on the volume to which it is referred.

With each extensive property G (or G_α) we may associate an *intensive property* γ (or γ_α), which is the *amount* of that property *per unit mass* of the fluid system (or of the α-species). We may also introduce a density g (or g_α) of G (or G_α), defined as the amount of entity G (or of G_α) per unit volume of the fluid system. Thus, for a given volume U, we have:

$$G_\alpha(x, t) = \int_{[U(\mathbf{x})]} \rho_\alpha(\mathbf{x}', t)\gamma_\alpha(\mathbf{x}', t)\, dU(\mathbf{x}') = \int_{[U(\mathbf{x})]} g_\alpha(\mathbf{x}', t)\, dU(\mathbf{x}') \tag{4.1.5a}$$

where $\mathbf{x}$ and $\mathbf{x}'$ denote the coordinates of the centroid of U, and of some arbitrary point within U, both with respect to a fixed coordinate system.

With these concepts, we may extend the definition of an average velocity and define a velocity $\mathbf{V}_G$ of an extensive property G of a multicomponent system by:

$$\mathbf{V}_G = \sum_{(\alpha)} (g_\alpha \mathbf{V}_{G_\alpha}) / \sum_{(\alpha)} g_\alpha \tag{4.1.6}$$

where $\mathbf{V}_{G_\alpha}$ is the velocity of propagation of G_α (compare with (4.1.4) and (4.1.5)).

4.1.2 Diffusive Velocities and Fluxes

Each of the fluid particles, defined in paragraph 4.1.1 with respect to properties of the fluid system as a whole, or to properties of individual species in a multicomponent fluid, has its own pathline in the flow domain and its own velocity along this path. Obviously a particle possesses a pathline only if its identity is preserved during a finite period of time. This is true for particles of conservative properties. All the properties considered here are associated with (and carried by) the fluid's matter itself, namely by its molecules that are in a continuous kinetic motion giving rise to *molecular diffusion*.

Most molecular diffusion laws depend primarily upon the assumption of a linear homogeneous relation between *diffusive mass fluxes* and the driving forces producing mass movement. The *diffusive velocities* $\hat{\mathbf{V}}'_\alpha$, and $\hat{\mathbf{V}}^*_\alpha$ of the α-species (mass) particle with respect to the volume average velocity $\mathbf{V}'$ and with respect to the mass average velocity $\mathbf{V}^*$, respectively, are:

$$\hat{\mathbf{V}}'_\alpha = \mathbf{V}_\alpha - \mathbf{V}'; \qquad \hat{\mathbf{V}}^*_\alpha = \mathbf{V}_\alpha - \mathbf{V}^*. \tag{4.1.7}$$

Other diffusive velocities

$$\hat{\mathbf{V}}^*_G = \mathbf{V}_G - \mathbf{V}^*; \qquad \hat{\mathbf{V}}^*_{G_\alpha} = \mathbf{V}_{G_\alpha} - \mathbf{V}^*; \qquad \hat{\mathbf{V}}'_G = \mathbf{V}_G - \mathbf{V}'; \text{ etc.} \tag{4.1.8}$$

may also be defined. Similarly, we define *diffusive property fluxes* of the particles with respect to either $\mathbf{V}'$, or $\mathbf{V}^*$. For example, we have the diffusive mass fluxes of the α-species, $\mathbf{J}^*_\alpha$ and $\mathbf{J}'_\alpha$, with respect to the mass average velocity and the volume average velocity, respectively:

$$\mathbf{J}^*_\alpha = \rho_\alpha \hat{\mathbf{V}}^*_\alpha = \rho_\alpha(\mathbf{V}_\alpha - \mathbf{V}^*); \qquad \sum_{\alpha=1}^{N} \mathbf{J}^*_\alpha = 0; \qquad \mathbf{J}'_\alpha = \rho_\alpha \hat{\mathbf{V}}'_\alpha = \rho_\alpha(\mathbf{V}_\alpha - \mathbf{V}'). \tag{4.1.9}$$

We may extend this concept to any other density and define *diffusive fluxes* such as:

$$\begin{aligned} \mathbf{J}^*_{G_\alpha} &= g_\alpha \hat{\mathbf{V}}^*_{G_\alpha} = g_\alpha(\mathbf{V}_{G_\alpha} - \mathbf{V}^*); \qquad & \mathbf{J}^*_G &= g\hat{\mathbf{V}}^*_G = g(\mathbf{V}_G - \mathbf{V}^*) \\ \mathbf{J}'_{G_\alpha} &= g_\alpha \hat{\mathbf{V}}'_{G_\alpha} = g_\alpha(\mathbf{V}_{G_\alpha} - \mathbf{V}'); \qquad & \mathbf{J}'_G &= g\hat{\mathbf{V}}'_G = g(\mathbf{V}_G - \mathbf{V}'). \end{aligned} \tag{4.1.10}$$

In section 4.3 we shall consider laws relating fluxes $\mathbf{J}^*_{G_\alpha}$ and $\mathbf{J}'_{G_\alpha}$ to driving forces. Here let us consider the special case where $g_\alpha = \rho_\alpha$, i.e., the case of the diffusive mass fluxes $\mathbf{J}^*_\alpha$ and $\mathbf{J}'_\alpha$.

Under conditions for which forced pressure and thermal diffusion are neglected, Lightfoot and Cussler (1965) suggest the following equation as the most general linear relation between mass movement and concentration gradients in an N-species system, neglecting all quadratic and higher powers of concentration gradients:

$$\mathbf{J}^{ac}_\alpha = C_\alpha(\mathbf{V}_\alpha - \mathbf{V}^a) = -\sum_{\beta=1}^{N} D^{ac}_{\alpha\beta} \nabla C_\beta \tag{4.1.11}$$

where: C_β is the β concentration, $\mathbf{V}^a$ is the reference velocity ($\mathbf{V}^*$, or $\mathbf{V}'$), and the superscripts ac in the flux $\mathbf{J}^{ac}_\alpha$ and in the molecular diffusivity $D^{ac}_{\alpha\beta}$ indicate that the magnitude of the latter depends upon the choice of C_β and $\mathbf{V}^a$.

In *binary systems* (i.e., with two species only), Fick's law (for $\nabla p = \nabla T = 0$) takes the following forms for the species α.

(a) Diffusive mass flux with respect to mass-average velocity ($g_\alpha = \rho_\alpha$; $\mathbf{V}_{G_\alpha} = \mathbf{V}_\alpha$):

$$\mathbf{J}^*_\alpha = \rho_\alpha(\mathbf{V}_\alpha - \mathbf{V}^*) = -\rho D_{\alpha\beta} \nabla \omega_\alpha. \tag{4.1.12}$$

in which $D_{\alpha\beta}$ is the *binary diffusivity*.

(b) Diffusive volumetric flux with respect to volume-average velocity, assuming no volume changes upon mixing ($g_\alpha = v_\alpha$ = volume fraction of species α, $\mathbf{V}_{G_\alpha} = \mathbf{V}_{v_\alpha}$):

$$\mathbf{J}'_{v_\alpha} = v_\alpha(\mathbf{V}_{v_\alpha} - \mathbf{V}') = -D_{\alpha\beta} \nabla v_\alpha. \tag{4.1.13}$$

(c) Diffusive mass flux with respect to volume-average velocity ($g_\alpha = \rho_\alpha$; $\mathbf{V}_{G_\alpha} = \mathbf{V}_\alpha$):

$$\mathbf{J}'_\alpha = \rho_\alpha(\mathbf{V}_\alpha - \mathbf{V}') = -D_{\alpha\beta}\nabla\rho_\alpha. \tag{4.1.14}$$

By eliminating $\rho_\alpha\mathbf{V}_\alpha$ from (4.1.12) and (4.1.14), we obtain

$$\rho_\alpha(\mathbf{V}^* - \mathbf{V}') = -D_{\alpha\beta}(\nabla\rho_\alpha - \rho\nabla\omega_\alpha) = -\omega_\alpha D_{\alpha\beta}\nabla\rho \tag{4.1.15}$$

or, for the fluid as a whole:

$$\rho(\mathbf{V}^* - \mathbf{V}') = -D_{\alpha\beta}\nabla\rho. \tag{4.1.16}$$

4.1.3 The Eulerian and Lagrangian Points of View

There are two methods for describing motion in a fluid system. The first, in which the history of individual particles is described, is called the *Lagrangian method*, while the second, which focuses attention on fixed points of space, is called the *Eulerian method*. Prandtl (1952) indicates that both were used by Euler.

In the Lagrangian method, the coordinates of a moving particle (e.g., the center of mass of a mass particle) are represented as functions of time. To distinguish among the various particles, we label each of them by the coordinates (called *fluid*, *material* or *Lagrangian coordinates*) of the particle's position at some initial time t_0 (say, $t_0 = 0$). These coordinates, $\boldsymbol{\xi}(\xi, \eta, \zeta)$, are sometimes also referred to as *convected coordinates*. The position at any later time, t, of a particle initially at point $\boldsymbol{\xi}(\xi, \eta, \zeta)$ is given, in a Cartesian spatial (or Eulerian) system, by its three coordinates:

$$x = x(\xi, \eta, \zeta, t); \quad y = y(\xi, \eta, \zeta, t); \quad z = z(\xi, \eta, \zeta, t). \tag{4.1.17}$$

We say that the position $\mathbf{x}$ of the particle is a function of t and of its initial position $\boldsymbol{\xi}$. In a compact form we write:

$$\mathbf{x} = \mathbf{x}(\boldsymbol{\xi}, t), \quad \text{or} \quad x_i = x_i(\boldsymbol{\xi}, t); \quad i = 1, 2, 3 \tag{4.1.18}$$

where:

$$\boldsymbol{\xi} \equiv (\xi, \eta, \zeta); \quad \mathbf{x} \equiv (x_1, x_2, x_3) \equiv (x, y, z); \quad \mathbf{x}(\boldsymbol{\xi}, 0) = \boldsymbol{\xi}.$$

It is assumed that the motion is continuous, i.e., the two functions $\mathbf{x}$ and $\boldsymbol{\xi}$ are continuously differentiable up to order three and single valued, i.e., two particles once distinct remain so for all time, so that (4.1.18) may be inverted to give the initial position of the particle $\boldsymbol{\xi}(\xi, \eta, \zeta)$ knowing its position $\mathbf{x}$ at time t:

$$\xi = \xi(\mathbf{x}, t); \qquad \eta = \eta(\mathbf{x}, t); \qquad \zeta = \zeta(\mathbf{x}, t). \tag{4.1.19}$$

Aris (1962) discusses the meaning of the continuity assumptions and shows that a necessary and sufficient condition for the existence of inverse functions is that the *Jacobian* J should not vanish, where J is defined by the determinant:

$$J \equiv \frac{\partial(x, y, z)}{\partial(\xi, \eta, \zeta)} = \begin{vmatrix} \partial x/\partial\xi & \partial x/\partial\eta & \partial x/\partial\zeta \\ \partial y/\partial\xi & \partial y/\partial\eta & \partial y/\partial\zeta \\ \partial z/\partial\xi & \partial z/\partial\eta & \partial z/\partial\zeta \end{vmatrix}. \tag{4.1.20}$$

A physical interpretation of J is given in paragraph 4.1.4.

For a given $\boldsymbol{\xi}$, equations (4.1.18) may be thought of as the parametric equations of the *pathline* of a particle that at $t = 0$ started from the point (ξ, η, ζ).

When a multicomponent system is considered, the mass particles of species α are labeled by $\boldsymbol{\xi}_\alpha(\xi_\alpha, \eta_\alpha, \zeta_\alpha)$. Then, the instantaneous velocity $\mathbf{V}_\alpha$ of a species α particle (having $\boldsymbol{\xi}_\alpha$ as its initial position) is given symbolically by:

$$\mathbf{V}_\alpha(\boldsymbol{\xi}_\alpha, t) = \left.\frac{\partial \mathbf{x}}{\partial t}\right|_{\boldsymbol{\xi}_\alpha = \text{const}} \tag{4.1.21}$$

It is thus the rate of change of position $\mathbf{x}$ of the particle $\boldsymbol{\xi}_\alpha$; it may be used for a particle of any entity, and not only for a mass particle. In a single-species fluid, the subscript α may be dropped. In the Lagrangian approach, x, y, z—the *spatial coordinates of particles*—are the *dependent* variables. In (4.1.21) we have three equations for the velocity components. The solution of these equations is given by (4.1.17), where ξ, η, ζ are constants of integration equal to the values of x, y, z, respectively, at $t = 0$.

In the Eulerian approach we investigate what happens at specific points that are fixed in space within the field of flow as different particles pass through them in the course of time. Accordingly, a complete description of the flow involves an *instantaneous* picture of the velocities at *all* points in the field. In unsteady flow, this instantaneous picture changes with time. The Eulerian velocity field, $\mathbf{V}$, is, therefore, given in a Cartesian coordinate system by the components:

$$V_x = V_x(x, y, z, t); \qquad V_y = V_y(x, y, z, t); \qquad V_z = V_z(x, y, z, t). \tag{4.1.22}$$

Here, the *velocities* are the *dependent variables*, whereas x, y, z, t are the independent ones; the coordinates (x, y, z) identify a fixed point in the flow domain, relative to the fixed frame of reference. In general, the Eulerian approach is more convenient and is more frequently used. However, in certain cases, as when dealing with material front movements, the Lagrangian approach is more useful.

4.1.4 The Substantial Derivative

To indicate partial differentiation with respect to time of any fluid property (which may be a tensor of any rank), following a particular particle of that property in the Lagrangian approach, and to avoid having to emphasize this fact by adding the subscript $\boldsymbol{\xi} = \text{const}$ or $\boldsymbol{\xi}_\alpha = \text{const}$ as in (4.1.21), the notation $D(\;)/Dt$ is used. It is then called a *material* (or *substantial*) derivative. The terms *hydrodynamic* and *total derivative* are also found in the literature.

Consider a property B_α of a G_α particle. For example, we may consider the velocity $\mathbf{V}^*$, the density ρ or the position coordinates (x, y, z), etc., of a fluid mass particle. The particle in question has an identity label $\boldsymbol{\xi}_\alpha = \text{const}$, made up of the particle's coordinates at $t = 0$.

Employing the Lagrangian formulation for following particles, we wish to determine the temporal rate of change of B_α, associated with this particular particle of fixed identity, as the particle is displaced in a flow domain. The particle's changing position is given by the parametric equations: $x_i = x_i(\xi_\alpha, t)$ and its velocity by $\partial x_i(\xi_\alpha, t)/\partial t|_{\xi_\alpha=\text{const}} = Dx_i/Dt \equiv (V_{G_\alpha})_i$.

Since here $B_\alpha = B_\alpha(x, y, z, t)|_{\xi_\alpha=\text{const}}$, using the chain rule of partial differentiation, we obtain:

$$\frac{DB_\alpha}{Dt} = \frac{\partial B_\alpha}{\partial x}\frac{\partial x}{\partial t}\bigg|_{\xi_\alpha=\text{const}} + \frac{\partial B_\alpha}{\partial y}\frac{\partial y}{\partial t}\bigg|_{\xi_\alpha=\text{const}} + \frac{\partial B_\alpha}{\partial z}\frac{\partial z}{\partial t}\bigg|_{\xi_\alpha=\text{const}} + \frac{\partial B_\alpha}{\partial t}. \tag{4.1.23}$$

We therefore have:

$$\frac{DB_\alpha}{Dt} = \frac{\partial B_\alpha}{\partial t} + (V_{G_\alpha})_x\frac{\partial B_\alpha}{\partial x} + (V_{G_\alpha})_y\frac{\partial B_\alpha}{\partial y} + (V_{G_\alpha})_z\frac{\partial B_\alpha}{\partial z}. \tag{4.1.24}$$

In vector notation (4.1.24) may be written as:

$$DB_\alpha/Dt = \partial B_\alpha/\partial t + \mathbf{V}_{G_\alpha}\cdot \operatorname{grad} B_\alpha \equiv \partial B_\alpha/\partial t + (\mathbf{V}_{G_\alpha}\cdot \nabla)B_\alpha. \tag{4.1.25}$$

This may also be written as:

$$\frac{DB_\alpha}{Dt} = \frac{\partial B_\alpha}{\partial t} + \operatorname{div}(\mathbf{V}_{G_\alpha}B_\alpha) - B_\alpha \operatorname{div} \mathbf{V}_{G_\alpha}. \tag{4.1.26}$$

The right-hand side of (4.1.25) is the sum of two terms. The first, $\partial B_\alpha/\partial t$ is called the *local derivative*; it gives (in unsteady flow) the temporal variations at a fixed point in space. The second term, called the *convective derivative*, represents the change in B_α caused by the convection of the particle under consideration from one location to a second location with a different value of B_α. For example, if B_α is the density ρ (or ρ_α) of a fluid mass particle (or a species α mass particle) whose velocity is $\mathbf{V}^*$ (or $\mathbf{V}_\alpha$), we obtain from (4.1.25):

$$D\rho/Dt = \partial\rho/\partial t + \mathbf{V}^*\cdot \operatorname{grad}\rho \quad \text{or:} \quad D\rho_\alpha/Dt = \partial\rho_\alpha/\partial t + \mathbf{V}_\alpha\cdot \operatorname{grad}\rho_\alpha. \tag{4.1.27}$$

As a second example, let $B_\alpha = \mathbf{V}^*$ of a fluid mass particle. We obtain:

$$D\mathbf{V}^*/Dt \equiv \mathbf{a} = \partial\mathbf{V}^*/\partial t + (\mathbf{V}^*\cdot \operatorname{grad})\mathbf{V}^*; \qquad \mathbf{a} = \text{acceleration} \tag{4.1.28}$$

which, in terms of components in a Cartesian coordinate system, may be written as: $a_1 = \partial V^*_1/\partial t + V^*_1\,\partial V^*_1/\partial x + V^*_2\,\partial V^*_1/\partial y + V^*_3\,\partial V^*_1/\partial z$, etc.

In all these examples, the material derivative of the considered property is the temporal rate of change of the property from the point of view of an observer moving with the particle.

A ***material*** (*or* ***fluid***) ***surface*** is a surface always composed of the same particles. The equation describing this surface is:

$$F(x, y, z, t) = 0. \tag{4.1.29}$$

For this case we obtain from (4.1.25):

$$DF/Dt \equiv \partial F/\partial t + (\mathbf{V}^*\cdot \nabla)F = 0. \tag{4.1.30}$$

In certain cases, B_α is such that:

$$\partial B_\alpha/\partial t + \operatorname{div}(\mathbf{V}_{G_\alpha} B_\alpha) = 0. \tag{4.1.31}$$

(See interpretation in sec. 4.2.) By combining (4.1.26) with (4.1.31), we obtain:

$$\operatorname{div} \mathbf{V}_{G_\alpha} = -(1/B_\alpha)\, DB_\alpha/Dt. \tag{4.1.32}$$

If we consider *a constant amount of property*, i.e., $G_\alpha = \text{const} = B_\alpha\, \delta U$ where δU is a volume of fluid, we obtain from (4.1.32):

$$\operatorname{div} \mathbf{V}_{G_\alpha} = -\frac{1}{(G_\alpha/\delta U)} \frac{D(G_\alpha/\delta U)}{Dt} = \frac{1}{\delta U} \frac{D\,\delta U}{Dt}. \tag{4.1.33}$$

From (4.1.33) it follows that the *divergence of the velocity field* $\mathbf{V}_{G_\alpha}$ may be interpreted as the relative rate of growth of the volume of fluid occupied by a fixed amount G_α as it advances at that velocity. It is also the *dilatation* (or *cubical dilatation*) defined as the rate of change of the volume occupied by $G_\alpha = \text{const}$ per unit volume.

Aris (1962) develops (4.1.33) for the special case of a fluid mass particle and $\mathbf{V}_{G_\alpha} = \mathbf{V}^*$ by considering an elementary volume $\delta U_0 = d\xi_1\, d\xi_2\, d\xi_3$ which deforms into $\delta U = dx_1\, dx_2\, dx_3$. He obtains:

$$\delta U_0 = d\xi_1\, d\xi_2\, d\xi_3; \qquad \delta U = J\, d\xi_1\, d\xi_2\, d\xi_3 = J\, \delta U_0 \tag{4.1.34}$$

where the Jacobian J of the transformation of variables expressed by (4.1.20) is:

$$J = \partial(x_1, x_2, x_3)/\partial(\xi_1, \xi_2, \xi_3). \tag{4.1.35}$$

This follows from $x_i = x_i(\xi_1, \xi_2, \xi_3)$; $dx_i = (\partial x_i/\partial \xi_j)\, d\xi_j$. For an incompressible fluid $\delta U = \delta U_0$, $J = 1$. By inserting (4.1.34) into (4.1.33), keeping in mind that the original volume does not vary, we obtain:

$$DJ/Dt = [\partial(\mathbf{V}_{G_\alpha})_1/\partial x_1 + \partial(\mathbf{V}_{G_\alpha})_2/\partial x_2 + \partial(\mathbf{V}_{G_\alpha})_3/\partial x_3]J \tag{4.1.36}$$

or:

$$\operatorname{div} \mathbf{V}_{G_\alpha} = \frac{1}{J} \frac{DJ}{Dt}. \tag{4.1.37}$$

For the special case of $\mathbf{V}_{G_\alpha} \equiv \mathbf{V}^*$, and $B_\alpha \equiv \rho$, we have from (4.1.32):

$$\operatorname{div} \mathbf{V}^* = -\frac{1}{\rho} \frac{D\rho}{Dt}. \tag{4.1.38}$$

For $\rho = \text{const}$, the volume of a fixed mass is constant; hence $\operatorname{div} \mathbf{V}^* = 0$, which is the *equation of incompressibility of a homogeneous fluid.*

From (4.1.33), since $G_\alpha/\delta U = g_\alpha$, we also have:

$$\operatorname{div} \mathbf{V}_{G_\alpha} = -(1/g_\alpha) Dg_\alpha/Dt \tag{4.1.39}$$

which, when combined with (4.1.37), yields:

$$g_\alpha\, DJ/Dt + J\, Dg_\alpha/Dt \equiv D(g_\alpha J)/Dt = 0 \tag{4.1.40}$$

which may be regarded as the Lagrangian statement of conservation.

4.2 The General Conservation Principle

In the present section, we are still considering a flowing fluid continuum. To generalize the discussion, we shall consider a multicomponent fluid such that, following the continuum approach, each component in itself is a continuum filling the entire space.

Consider a certain initial amount G_α of an extensive fluid property (e.g., mass), contained at time t within a portion of space of volume U and enclosed by a surface S (fig. 4.2.1). If we consider what happens to the initial amount G_α as it moves, the volume occupied by it generally changes continuously so that $U = U(\mathbf{x}, t)$ (Lagrangian approach). The (temporal) rate of change of G_α is given by DG_α/Dt. Our objective is to express this Lagrangian change by means of the Eulerian approach, considering the volume U as a control volume fixed in the xyz frame of reference, rather than a material volume.

During a time interval from t to $t' = t + \Delta t$, the material volume (system) moves and deforms. Its position and shape at $t + \Delta t$ are shown in dashed lines in figure 4.2.1. The system at time t (volume U and surface S) and the system at time $t + \Delta t$ (volume U' and surface S') are composed of three regions: $U = U_1 + U_2$, $U' = U_2 + U_3$. The region U_2 is common to U and U'. The temporal rate of change of G_α for the moving system (Lagrange point of view) is given by:

$$\left.\frac{DG_\alpha}{Dt}\right|_{\text{system}} = \lim_{\Delta t \to 0}\left\{\frac{[(G_\alpha)_2 + (G_\alpha)_3]_{t+\Delta t} - [(G_\alpha)_1 + (G_\alpha)_2]_t}{\Delta t}\right\} \tag{4.2.1}$$

where

$$(G_\alpha)_i = \int_{(U_i)} \rho_\alpha \gamma_\alpha \, dU = \int_{(U_i)} g_\alpha \, dU, \qquad i = 1, 2, 3.$$

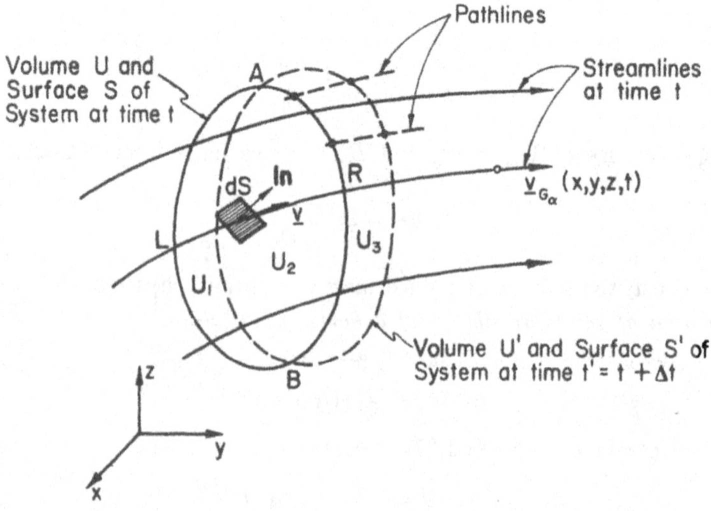

FIG. 4.2.1. Nomenclature for deriving the transport theorem.

Equation (4.2.1) may be written as:

$$\left.\frac{DG_\alpha}{Dt}\right|_{\text{system}} = \lim_{\Delta t \to 0}\left\{\frac{[(G_\alpha)_2]_{t+\Delta t} - [(G_\alpha)_2]_t}{\Delta t}\right\} + \lim_{\Delta t \to 0}\frac{[(G_\alpha)_3]_{t+\Delta t}}{\Delta t} - \lim_{\Delta t \to 0}\frac{[(G_\alpha)_1]_t}{\Delta t}. \tag{4.2.2}$$

As $t \to 0$, the volume U_2 approaches U, i.e., that of the system at $t = 0$, so that in the limit, the first term on the right-hand side of (4.2.2) becomes:

$$\lim_{\Delta t \to 0}\left\{\frac{[(G_\alpha)_2]_{t+\Delta t} - [(G_\alpha)_2]_t}{\Delta t}\right\} = \frac{\partial}{\partial t}(G_\alpha) = \frac{\partial}{\partial t}\int_{(U)} \rho_\alpha \gamma_\alpha \, dU = \int_{(U)} \frac{\partial}{\partial t}(\rho_\alpha \gamma_\alpha)\, dU \tag{4.2.3}$$

where, as $t \to 0$, $U_2 \to U$.

From the definition of $(G_\alpha)_3$ we have (see fig. 4.2.1)

$$\frac{[(G_\alpha)_3]}{\Delta t} \cong \frac{1}{\Delta t}\int_{(ARB)} \rho_\alpha \gamma_\alpha \mathbf{V}_{G_\alpha}(t)|_{ARB} \cdot \mathbf{dS}\, \Delta t.$$

Hence, the integrand of the second integral on the right-hand side of (4.2.2) represents approximately the average during Δt of the rate of efflux of the property in question that has crossed through the portion of surface ARB common to U_2 and U_3. In the limit as $\Delta t \to 0$, this ratio $(G_\alpha)_3/\Delta t$ becomes the exact rate of efflux of G_α through the surface ARB of the control volume U. Similarly, the last limit on the right-hand side of (4.2.2) gives the exact rate of influx of G_α into U through the surface ALB common to U_1 and U_2. Together, the last two terms on the right-hand side of (4.2.2) give the net efflux of G_α across the entire surface S of U at time t. Using the outward normal unit vector to an areal element of the surface S, this efflux can be expressed by:

$$\int_{(S)} g_\alpha \mathbf{V}_{G_\alpha} \cdot \mathbf{dS}.$$

Thus (4.2.2) becomes:

$$\left.\frac{DG_\alpha}{Dt}\right|_{\text{system}} = \frac{\partial}{\partial t}\int_{(U)} g_\alpha \, dU + \int_{(S)} g_\alpha \mathbf{V}_{G_\alpha} \cdot \mathbf{dS}. \tag{4.2.4}$$

Since the control volume U in (4.2.4) remains fixed and saturated, the differentiation and integration may be interchanged.

On the left-hand side of (4.2.4), we have the change in G_α as the system moves. On the right-hand side, the volume U at time t and its surface S are fixed in space, and the velocity $\mathbf{V}_{G_\alpha}$ is with respect to a fixed frame of reference. *Instantaneously* the control volume coincides with the moving system. Therefore, the time rate of change of G_α as the system moves is observed from the control volume U. The objective stated above has thus been achieved in the form of (4.2.4).

Another approach for deriving (4.2.4), attributable to Reynolds (Aris 1962), may be of interest. Using the notation of paragraph 4.1.4 (x_i coordinates with respect to the fixed frame of reference and ξ_i are the material coordinates), we start from (4.1.5a) written as:

$$G_\alpha(t) = \int_{U(t)} g_\alpha(\mathbf{x}, t)\, dU \tag{4.2.5}$$

where $G_\alpha(t)$ is the amount of G_α instantaneously contained in $U(t)$. Since this integral is over the varying volume $U(t)$, we cannot interchange integration and differentiation when taking the material derivative DG_α/Dt. However, if we transform (4.1.5a) from the $\mathbf{x}$ coordinate system to the material coordinates $\boldsymbol{\xi}$ in which $U = U_0$ at $t = t_0$, $\mathbf{x} = \mathbf{x}(\boldsymbol{\xi}, t)$, and the two systems are related to each other by (4.1.34) and (4.1.35), we obtain:

$$\begin{aligned}
\frac{DG_\alpha}{Dt} &= \frac{d}{dt}\int_{[U(t)]} g_\alpha(\mathbf{x}, t)\, dU = \frac{d}{dt}\int_{(U_0)} g_\alpha\,[\mathbf{x}(\boldsymbol{\xi}, t), t]\, dU_0 \\
&= \int_{(U_0)} \left(\frac{dg_\alpha}{dt} J + g_\alpha \frac{dJ}{dt}\right) dU_0 = \int_{(U_0)} \left(\frac{dg_\alpha}{dt} + g_\alpha \operatorname{div} \mathbf{V}_{G_\alpha}\right) J\, dU_0 \\
&= \int_{[U(t)]} \left(\frac{dg_\alpha}{dt} + g_\alpha \operatorname{div} \mathbf{V}_{G_\alpha}\right) dU = \int_{[U(t)]} \frac{\partial g_\alpha}{\partial t}\, dU + \int_{[U(t)]} \operatorname{div}(g_\alpha \mathbf{V}_{G_\alpha})\, dU.
\end{aligned} \tag{4.2.6}$$

By applying Gauss' theorem, this becomes identical to (4.2.4) where $U = U(t)$ and $S = S(t)$ have instantaneously constant values. Equation (4.2.4) is also called *Reynold's transport theorem*. Our next objective is to obtain a general conservation principle valid for any extensive fluid property (Prager 1961; De Groot and Mazur 1963; Scipio 1967).

In the absence of sources and sinks within U, changes in G_α can result only from changes in the net outflow through the surface S. Hence:

$$\int_{(U)} \frac{\partial g_\alpha}{\partial t}\, dU + \int_{(S)} g_\alpha \mathbf{V}_{G_\alpha} \cdot d\mathbf{S} = 0, \qquad \int_{(U)} \frac{\partial g_\alpha}{\partial t}\, dU + \int_{(U)} \operatorname{div}(g_\alpha \mathbf{V}_{G_\alpha})\, dU = 0. \tag{4.2.7}$$

The last equation is identical to (4.2.4) if $DG_\alpha/Dt = 0$.

If production of the property takes place within U at a (temporal) rate I_α per unit volume as time proceeds, as a result of internal (e.g., chemical) processes, this may also cause changes in G_α. For this case (4.2.7) becomes (De Groot and Mazur 1962):

$$\int_{(U)} \frac{\partial g_\alpha}{\partial t}\, dU + \int_{(S)} g_\alpha \mathbf{V}_{G_\alpha} \cdot d\mathbf{S} = \int_{(U)} I_\alpha\, dU. \tag{4.2.8}$$

One can easily recognize that (4.2.8) is written from the Eulerian point of view. It states that the rate of increase of G_α within U is equal to the sum of the rate at which the considered property of the species crosses into U through the surface S and the rate at which this property is produced within U. Applying Gauss' theorem to the surface integral in (4.2.8) it is possible to rewrite this equation in the form:

$$\int\limits_{(U)}\left[\frac{\partial g_\alpha}{\partial t} + \operatorname{div}(g_\alpha \mathbf{V}_{G_\alpha}) - I_\alpha\right] dU = 0. \tag{4.2.9}$$

Since the volume U is arbitrary, the integrand must vanish for (4.2.9) to be valid for all volumes. Therefore:

$$\partial g_\alpha/\partial t + \operatorname{div}(g_\alpha \mathbf{V}_{G_\alpha}) = I_\alpha. \tag{4.2.10}$$

Equation (4.2.10) is the general conservation principle of a property of the species α.

4.3 Equations of Mass, Momentum and Energy Conservation in a Fluid Continuum

The general conservation equation developed in section 4.2 will now be applied to various extensive fluid properties.

4.3.1 Mass Conservation of a Species

For a species α of a multicomponent fluid system, we insert $g_\alpha = \rho_\alpha$ and $\mathbf{V}_{G_\alpha} = \mathbf{V}_\alpha$ in (4.2.10), and obtain:

$$\partial \rho_\alpha/\partial t + \operatorname{div}(\rho_\alpha \mathbf{V}_\alpha) = I_\alpha \tag{4.3.1}$$

where I_α (dim: $M_\alpha L^{-3} T^{-1}$) is the rate at which mass of the species α is produced per unit volume of the system by chemical reactions. Equation (4.3.1) is the *equation of continuity (or of mass conservation) for species* α as written from the Eulerian point of view (par. 4.1.3).

By combining (4.3.1) with (4.1.24) for a material derivative, we obtain:

$$D\rho_\alpha/Dt + \rho_\alpha \operatorname{div} \mathbf{V}_\alpha = I_\alpha \tag{4.3.2}$$

where $D\rho_\alpha/Dt = (\partial \rho_\alpha/\partial t)|_{\xi_\alpha = \text{const}}$. For $I_\alpha = 0$ (4.3.2) becomes:

$$D\rho_\alpha/Dt + \rho_\alpha \operatorname{div} \mathbf{V}_\alpha = 0. \tag{4.3.3}$$

Introducing the specific volume (volume per unit mass) $v_\alpha = 1/\rho_\alpha$, we may rewrite (4.3.3) as:

$$D(1/v_\alpha)/Dt + (1/v_\alpha) \operatorname{div} \mathbf{V}_\alpha = 0; \qquad \operatorname{div} \mathbf{V}_\alpha = (1/v_\alpha) Dv_\alpha/Dt \tag{4.3.4}$$

which leads to the interpretation of div $\mathbf{V}_\alpha$ as the relative rate of growth of the volume of fluid occupied by a unit mass of species α as it moves at the velocity $\mathbf{V}_\alpha$ (comparable with (4.1.33) and the interpretation that follows).

Now let ρ_α be the property B_α of a fluid mass particle. Then, with $B_\alpha = \rho_\alpha$ and $\mathbf{V}_{G_\alpha} = \mathbf{V}^*$ (4.1.25) becomes:

$$D^*\rho_\alpha/Dt = \partial \rho_\alpha/\partial t + \mathbf{V}^* \cdot \operatorname{grad} \rho_\alpha. \tag{4.3.5}$$

Or, using the definition of a material derivative:

$$D^*\rho_\alpha/Dt + \rho_\alpha \operatorname{div} \mathbf{V}_\alpha + (\mathbf{V}_\alpha - \mathbf{V}^*) \cdot \operatorname{grad} \rho_\alpha = I_\alpha. \tag{4.3.6}$$

By introducing the definition (4.1.9) for $\mathbf{J}^*$ in (4.3.1), and combining the resulting equation with (4.3.5), we obtain:

$$\partial\rho_\alpha/\partial t + \operatorname{div}(\rho_\alpha \mathbf{V}^*) + \operatorname{div} \mathbf{J}^*_\alpha = I_\alpha; \qquad D^*\rho_\alpha/Dt + \rho_\alpha \operatorname{div} \mathbf{V}^* + \operatorname{div} \mathbf{J}^*_\alpha = I_\alpha \quad (4.3.7)$$

(compare with (4.3.2) and note the difference between $D(\ )/Dt$ and $D^*(\ )/Dt$). In the absence of sources and sinks for the mass of the species α, $I_\alpha = 0$ and (4.3.7) becomes:

$$D^*\rho_\alpha/Dt + \rho_\alpha \operatorname{div} \mathbf{V}^* + \operatorname{div} \mathbf{J}^*_\alpha = 0. \quad (4.3.8)$$

In (4.3.7), $\rho_\alpha \mathbf{V}^*$ is the convective component of the mass flux of the α species, $\mathbf{J}^*_\alpha = \rho_\alpha \hat{\mathbf{V}}^*_\alpha$ is the conductive (or diffusive) component of the same flux. In terms of the volume averaged velocity $\mathbf{V}'$, we have:

$$\partial\rho_\alpha/\partial t = -\operatorname{div}(\rho_\alpha \mathbf{V}' + \mathbf{J}'_\alpha) + I_\alpha. \quad (4.3.9)$$

If our system is a *binary* one, we have $\alpha = 1, 2$ only. Expressing $\mathbf{J}^*_\alpha$ in (4.3.7) by (4.1.12), and $\mathbf{J}'_\alpha$ in (4.3.9) by (4.1.14), we obtain:

$$\partial\rho_\alpha/\partial t - \operatorname{div}[\rho D_{\alpha\beta} \operatorname{grad} \rho_\alpha/\rho] + \operatorname{div}(\rho_\alpha \mathbf{V}^*) = I_\alpha \quad (4.3.10)$$

and:

$$\partial\rho_\alpha/\partial t - \operatorname{div}(D_{\alpha\beta} \operatorname{grad} \rho_\alpha) + \operatorname{div}(\rho_\alpha \mathbf{V}') = I_\alpha. \quad (4.3.11)$$

For a stationary fluid, $\mathbf{V}' = 0$, and if we also have $I_\alpha = 0$ (4.3.11) reduces to *Fick's second law of diffusion* (for a binary system), or simply, the *diffusion equation*:

$$\partial\rho_\alpha/\partial t = \operatorname{div}(D_{\alpha\beta} \operatorname{grad} \rho_\alpha). \quad (4.3.12)$$

For $D_{\alpha\beta} = \text{const}$ (4.3.12) becomes:

$$\partial\rho_\alpha/\partial t = D_{\alpha\beta} \nabla^2 \rho_\alpha. \quad (4.3.13)$$

4.3.2 Mass Conservation of a Fluid System

By summing (4.3.1) or (4.3.7) over all the species α of a multicomponent system, we obtain:

$$\partial\rho/\partial t + \operatorname{div}(\rho \mathbf{V}^*) = 0 \quad (4.3.14)$$

which is the *equation of continuity (or mass conservation) for the fluid*; it expresses the conservation of mass in the spatial—Eulerian—description.

The equation of mass conservation (4.3.14) may also be obtained by inserting $g_\alpha = \rho$ and $\mathbf{V}_{G_\alpha} = \mathbf{V}^*$ in (4.2.10) with $I_\alpha = 0$ (as mass is not produced).

Equation (4.3.14) can also be written as:

$$\partial\rho/\partial t + \rho \operatorname{div} \mathbf{V}^* + \mathbf{V}^* \cdot \operatorname{grad} \rho = 0; \qquad D\rho/Dt + \rho \operatorname{div} \mathbf{V}^* = 0 \quad (4.3.15)$$

where $D\rho/Dt = \partial\rho/\partial t + \mathbf{V}^* \cdot \operatorname{grad} \rho$. With (4.1.40) we have:

$$D(\rho J)/Dt = 0 \quad (4.3.16)$$

which expresses the *mass conservation of the fluid in terms of the material coordinates.*

For an incompressible homogeneous fluid, i.e., for a fluid of constant mass density ρ, equation (4.3.14) reduces to:

$$\operatorname{div} \mathbf{V}^* = 0. \tag{4.3.17}$$

Consider now the property *specific volume*, v, of a moving fluid. By inserting $g_\alpha = \rho v = 1$, and $\mathbf{V}_{G_\alpha} = \mathbf{V}'$ (volume average velocity), $I_\alpha = 0$ (i.e., no "production" of volume as the fluid is assumed incompressible), in (4.2.10), we obtain:

$$\operatorname{div} \mathbf{V}' = 0. \tag{4.3.18}$$

This is the *equation of volume conservation* of a multicomponent fluid. In view of (4.1.16), (4.3.18) becomes for a binary system:

$$\operatorname{div} \mathbf{V}' = \operatorname{div} \mathbf{V}^* + \operatorname{div}[(D_{\alpha\beta}/\rho) \operatorname{grad} \rho] = 0; \qquad \operatorname{div}[\mathbf{V}^* + (D_{\alpha\beta}/\rho) \operatorname{grad} \rho] = 0. \tag{4.3.19}$$

When $\rho\mathbf{V}^* \gg D_{\alpha\beta} \operatorname{grad} \rho$ (i.e., $\mathbf{V}^*$ is such that mass transport by convection is much larger than mass transport by molecular diffusion), we have:

$$\operatorname{div} \mathbf{V}^* \simeq 0. \tag{4.3.20}$$

4.3.3 Conservation of Linear Momentum of a Species α

The momentum, or linear momentum, of a mass fluid contained in a volume U is given by $\int_{(U)} \rho\mathbf{V}^* \, dU$. This is an example of an extensive property G_α as defined in section 4.2.

If in (4.2.10) we insert $g_\alpha = \rho_\alpha \mathbf{V}_\alpha$ (momentum of species α per unit volume of mixture), $\mathbf{I}_\alpha = \mathbf{I}_{m\alpha}$ and $\mathbf{V}_{G_\alpha} = \mathbf{V}_{m\alpha}$, we obtain:

$$\partial(\rho_\alpha \mathbf{V}_\alpha)/\partial t + \operatorname{div}(\rho_\alpha \mathbf{V}_\alpha \mathbf{V}_{m\alpha}) = \mathbf{I}_{m\alpha} \tag{4.3.21}$$

where $\mathbf{I}_{m\alpha}$ is the rate of production of the momentum density $\rho_\alpha \mathbf{V}_\alpha$, $\mathbf{V}_{m\alpha}$ is the velocity at which the momentum of the species α propagates (velocity of α species momentum particles) in flow domain, and $\mathbf{V}_\alpha \mathbf{V}_{m\alpha}$ is the dyadic product: $\mathbf{V}_\alpha \mathbf{V}_{m\alpha} = V_{\alpha i} V_{m\alpha j} \mathbf{1x}_i \mathbf{1x}_j$, $i, j = 1, 2, 3$. In (4.3.21):

$\underline{\mathbf{J}}_{m\alpha} = \rho_\alpha \mathbf{V}_\alpha \mathbf{V}_{m\alpha} =$ flux of momentum of species α with respect to a fixed frame of reference

$\underline{\mathbf{J}}^*_{m\alpha} = \rho_\alpha \mathbf{V}_\alpha \hat{\mathbf{V}}^*_{m\alpha} = \rho_\alpha \mathbf{V}_\alpha (\mathbf{V}_{m\alpha} - \mathbf{V}^*) =$ flux of momentum of species α with respect to the mass average velocity

where $\underline{\mathbf{J}}_{m\alpha}$ and $\underline{\mathbf{J}}^*_{m\alpha}$ are second-rank tensors and $\mathbf{V}_\alpha \mathbf{V}_{m\alpha}$ and $\mathbf{V}_\alpha \mathbf{V}^*$ are dyadic products (different from the scalar products $\mathbf{V}_\alpha \cdot \mathbf{V}_{m\alpha}$ and $\mathbf{V}_\alpha \cdot \mathbf{V}^*$). Hence:

$$\underline{\mathbf{J}}_{m\alpha} = \rho_\alpha \mathbf{V}_\alpha \mathbf{V}^* + \underline{\mathbf{J}}^*_{m\alpha}. \tag{4.3.22}$$

By inserting this expression in (4.3.21), we obtain:

$$\frac{\partial(\rho_\alpha \mathbf{V}_\alpha)}{\partial t} + \operatorname{div}(\rho_\alpha \mathbf{V}_\alpha \mathbf{V}^*) + \operatorname{div}(\underline{\mathbf{J}}^*_{m\alpha}) = \mathbf{I}_{m\alpha}. \tag{4.3.23}$$

Truesdell (1960) presents a detailed discussion of (4.3.23) and interprets the meaning of the various terms appearing in it. As our main interest is the conservation of the linear momentum of the fluid as a whole (as a basis for the derivation of the motion equation), we shall not pursue this topic any further.

4.3.4 Conservation of Linear Momentum of a Fluid System

Here we insert in (4.2.10): $g_\alpha = \rho\mathbf{V}^*$ and $\mathbf{V}_{G\alpha} = \mathbf{V}_m =$ velocity of momentum particles, and $\mathbf{I}_\alpha = \sum_{(\alpha)}\rho_\alpha\mathbf{F}_\alpha$ where $\mathbf{F}_\alpha$ is the external force per unit mass of species α acting on species α particles. The last expression is the consequence of Newton's second law of mechanics, which states that the rate of production of momentum density equals the resultant of all external forces. We obtain:

$$\partial(\rho\mathbf{V}^*)/\partial t + \operatorname{div}(\rho\mathbf{V}^*\mathbf{V}_m) = \sum_{(\alpha)} \rho_\alpha\mathbf{F}_\alpha \qquad (4.3.24)$$

where $\mathbf{V}^*\mathbf{V}_m$ is a dyadic product of $\mathbf{V}^*$ and $\mathbf{V}_m$. In (4.3.24), let:

$\underline{\mathbf{J}}_m = \rho\mathbf{V}^*\mathbf{V}_m =$ momentum flux with respect to a fixed frame of reference,

$\underline{\mathbf{J}}^*_m = \rho\mathbf{V}^*\hat{\mathbf{V}}^*_m = \rho\mathbf{V}^*(\mathbf{V}_m - \mathbf{V}^*) =$ momentum flux with respect to $\mathbf{V}^*$. Hence:

$$\underline{\mathbf{J}}_m = \underline{\mathbf{J}}^*_m + \rho\mathbf{V}^*\mathbf{V}^*. \qquad (4.3.25)$$

Both $\underline{\mathbf{J}}_m$ and $\underline{\mathbf{J}}^*_m$ are second-rank tensors and $\mathbf{V}^*\mathbf{V}^*$ and $\mathbf{V}^*\mathbf{V}_m$ are dyadic products. By combining (4.3.24) and (4.3.25), we obtain:

$$\frac{\partial(\rho\mathbf{V}^*)}{\partial t} + \operatorname{div}(\rho\mathbf{V}^*\mathbf{V}^*) + \operatorname{div}\underline{\mathbf{J}}^*_m = \sum_{(\alpha)} \rho_\alpha\mathbf{F}_\alpha. \qquad (4.3.26)$$

Like mass (e.g., equation (4.3.9)), momentum is also transported in two ways, simultaneously: by the bulk fluid motion ($\rho\mathbf{V}^*\mathbf{V}^*$) and as a diffusive flux by the molecular motion ($\underline{\mathbf{J}}^*_m$). The latter is equal to the negative of the *pressure tensor*, or the second-rank symmetrical *stress tensor* $\underline{\boldsymbol{\sigma}}$ (components σ_{ij}; sec. 3.1):

$$\underline{\mathbf{J}}^*_m = -\underline{\boldsymbol{\sigma}}. \qquad (4.3.27)$$

Usually $\underline{\boldsymbol{\sigma}}$ is split into two parts: $-p\underline{\boldsymbol{\delta}} =$ the contribution of the pressure p, and $\underline{\mathbf{P}} =$ the contribution of the viscous forces (*viscous stress tensor* or *shear stress tensor*). We have, by (3.1.6):

$$\sigma_{ij} = -p\delta_{ij} + P_{ij} \qquad (4.3.28)$$

where $\underline{\boldsymbol{\delta}}$ (components δ_{ij}) is the Kronecker $\underline{\boldsymbol{\delta}}$, and P_{ij} is interpreted as the flux in the jth direction of the ith component of the momentum. We thus have:

$$\underline{\mathbf{J}}_m = \rho\mathbf{V}^*\mathbf{V}^* + p\underline{\boldsymbol{\delta}} - \underline{\mathbf{P}}. \qquad (4.3.29)$$

In a perfect fluid, or also in a fluid at rest, $P_{ij} = 0$, so that the stresses within it are purely isotropic ($\sigma_{ij} = -p_0\delta_{ij}$; $p_0 =$ hydrostatic pressure).

The momentum conservation equation (4.3.26) may now be rewritten as:

$$\frac{\partial(\rho \mathbf{V}^*)}{\partial t} + \operatorname{div}(\rho \mathbf{V}^*\mathbf{V}^*) + \operatorname{div}(p\underline{\boldsymbol{\delta}}) - \operatorname{div} \underline{\mathbf{P}} = \sum_{(\alpha)} \rho_\alpha \mathbf{F}_\alpha \tag{4.3.30}$$

or:

$$\frac{\partial(\rho V^*_i)}{\partial t} + \frac{\partial}{\partial x_j}(\rho \mathbf{V}^*_i\mathbf{V}^*_j) + \frac{\partial p}{\partial x_i} - \frac{\partial P_{ij}}{\partial x_j} = \sum_{(\alpha)} \rho_\alpha (F_\alpha)_i \tag{4.3.31}$$

where the summation convention is applicable.

Equation (4.3.30) can also be brought to the form:

$$\rho\, D\mathbf{V}^*/Dt = -\operatorname{grad} p + \operatorname{div} \underline{\mathbf{P}} + \sum_{(\alpha)} \rho_\alpha \mathbf{F}_\alpha; \qquad \rho\, D\mathbf{V}^*/Dt - \operatorname{div} \underline{\boldsymbol{\sigma}} = \sum_{(\alpha)} \rho_\alpha \mathbf{F}_\alpha \tag{4.3.32}$$

where $\rho\, D\mathbf{V}^*/Dt$ = mass per unit volume times acceleration; $-\operatorname{grad} p$ = pressure force per unit volume; $\operatorname{div} \underline{\mathbf{P}}$ = viscous force per unit volume; $\sum_{(\alpha)}\rho_\alpha \mathbf{F}_\alpha$ = gravitational force per unit volume.

The equation of momentum conservation of the fluid (4.3.32) is also called Cauchy's *equation of motion* of the fluid. Actually this is equivalent to Newton's second law of motion for continuous media. In this form, it states that the product of mass and acceleration is equal to the sum of pressure forces, viscous forces and external body forces, all terms referring to a unit volume of fluid.

In the absence of viscous forces (4.3.32) becomes:

$$\rho \frac{D\mathbf{V}^*}{Dt} = -\operatorname{grad} p + \sum_{(\alpha)} \rho_\alpha \mathbf{F}_\alpha \tag{4.3.33}$$

which is known as *Euler's equation of motion.*

The body force $\sum_{(\alpha)}\rho_\alpha \mathbf{F}_\alpha$ appearing in the motion equation may be due to gravity. Then:

$$\sum_{(\alpha)} \rho_\alpha \mathbf{F}_\alpha = \rho \mathbf{g}. \tag{4.3.34}$$

As a further step in the development of the motion equation, $\underline{\mathbf{P}}$ (related to the internal friction of the fluid) and $\operatorname{div} \underline{\mathbf{P}}$ (viscous force) are expressed in terms of velocity gradients and fluid properties. For example, for a *Newtonian fluid,* this leads to the *Navier–Stokes equations* (see Bird, Stewart and Lightfoot 1960, or any text on fluid mechanics). Written in Cartesian coordinates x_i ($i = 1, 2, 3$), they are:

$$\begin{aligned}
\rho \frac{DV^*_1}{Dt} &\equiv \rho\left(\frac{\partial V^*_1}{\partial t} + V^*_1 \frac{\partial V^*_1}{\partial x_1} + V^*_2 \frac{\partial V^*_1}{\partial x_2} + V^*_3 \frac{\partial V^*_1}{\partial x_3}\right)\\
&= \sum_{(\alpha)} \rho_\alpha (F_\alpha)_1 - \frac{\partial p}{\partial x_1} + \frac{\mu}{3}\frac{\partial}{\partial x_1}(\operatorname{div} \mathbf{V}^*) + \mu \nabla^2 V^*_1 + 2\frac{\partial V^*_1}{\partial x_1}\frac{\partial \mu}{\partial x_1}\\
&\quad - \tfrac{2}{3} \operatorname{div} \mathbf{V}^* \frac{\partial \mu}{\partial x_1} + \left(\frac{\partial V^*_2}{\partial x_1} + \frac{\partial V^*_1}{\partial x_2}\right)\frac{\partial \mu}{\partial x_2} + \left(\frac{\partial V^*_1}{\partial x_3} + \frac{\partial V^*_3}{\partial x_1}\right)\frac{\partial \mu}{\partial x_3}
\end{aligned} \tag{4.3.35}$$

and similar equations for $\rho\, DV^*_2/Dt$ and $\rho\, DV^*_3/Dt$; μ is the fluid viscosity. When both ρ and μ are constant (4.3.17) is valid, and the Navier–Stokes equations can be written symbolically as:

$$\rho \frac{D\mathbf{V}^*}{Dt} = \sum_{(\alpha)} \rho_\alpha \mathbf{F}_\alpha - \operatorname{grad} p + \mu \nabla^2 \mathbf{V}^*. \tag{4.3.36}$$

In all the equations above $\sum_{(\alpha)} \rho_\alpha \mathbf{F}_\alpha$ stands for the resultant body force per unit volume acting on the fluid. In the usual case (of the earth's gravity field) this force is expressed by $\rho \mathbf{g}$.

In the more general case, when the force is a conservative one that can be derived from a potential:

$$\sum_{(\alpha)} \rho_\alpha \mathbf{F}_\alpha = \rho \mathbf{F}; \qquad \mathbf{F} = -\operatorname{grad} \Omega,$$

we may rewrite in (4.3.32), and in subsequent equations, the sum $(-\rho^{-1} \operatorname{grad} p + \rho^{-1} \sum_{(\alpha)} \rho_\alpha \mathbf{F}_\alpha)$ as $-(\rho^{-1} \operatorname{grad} p + \operatorname{grad} \Omega)$. For $\rho = \text{const}$ this sum becomes $-\operatorname{grad}(\Omega + p/\rho)$. Another case of special interest is that of compressible fluids where ρ is a function of pressure only; $\rho = \rho(p)$. Then, since

$$\frac{1}{\rho} \nabla p = \nabla \int_{p_0}^{p} \frac{dp}{\rho(p)},$$

we may replace the above sum by: $\nabla\{\Omega + \int_{p_0}^{p} [dp/\rho(p)]\}$. In the case of gravity forces $\Omega = gz$, we obtain $\nabla\{gz + \int_{p_0}^{p} [dp/\rho(p)]\}$.

In principle, given the Navier–Stokes equations (three equations), a mass conservation equation, two equations of state for ρ and μ and appropriate conditions on the external boundaries of a flow domain, the velocity and pressure distribution in the flow domain may be derived.

For a fluid that is a multicomponent system, the equations of motion can be obtained by summing equations corresponding to the individual components.

4.4 Constitutive Assumption and Coupled Processes

4.4.1 General Considerations

In regions where variables change sufficiently smoothly, the various conservation principles discussed in section 4.3, applicable to all continua, are expressed in the form of partial differential *field equations*. However, as they contain no information regarding the properties (e.g., mechanical properties), or the intrinsic excitation–response relationship of the particular continua under consideration, they form an undetermined system, insufficient to yield specific answers unless further equations are supplied. This statement becomes more obvious when the number of unknowns is compared with the number of field equations expressing the various conservation principles. This fact should already be clear from the basic conservation statement (4.2.10), which is a single equation in at least four variables (g_α and three components of $\mathbf{V}_{G_\alpha}$), depending on the tensorial rank of g_α.

The additional information is contained in what is known as *constitutive assumptions* or *constitutive equations* (also known as *rheological equations of state*). These equations define the intrinsic response, which depends on the internal constitution of the particular materials considered. Usually constitutive assumptions take the form of relationships between fluxes (in the broader sense) and driving forces. Among the fluxes we may list those of mass, momentum and energy of various kinds. In (4.3.27) we have shown the identity between the momentum flux $-\mathbf{J}^*_m$ and the stress tensor $\underline{\boldsymbol{\sigma}}$. Hence, the linear relationship between stress and strain in an *elastic solid* is an example of a constitutive assumption. The *Newtonian fluid*, the *Stokesian fluid*, the perfectly *plastic material*, etc., each has its own stress–strain relationship (constitutive assumption). Other examples of constitutive assumptions are:

(a) that the flux of thermal energy is a linear function of the temperature gradient;

(b) that the flux (by molecular diffusion) of the mass of a species in a binary mixture is linearly proportional to the gradient of the species' density.

One should note at this point that we often prefer to refer to these relationships as assumptions, rather than equations, although outwardly the assumptions take the form of equations. We do so to emphasize that these equations define the *assumed* behavior of ideal continua that are *mathematical models* of particular classes of continua encountered in nature. Although mathematically a constitutive assumption presented in the form of an equation is a definition, it is arrived at by physical experience, perhaps fortified by experimental evidence. Because of their dependence on observations and experimental evidence, the constitutive equations are often referred to as *phenomenological equations* (or laws). However, it is rarely, if ever, possible to determine all the basic equations of a theory by physical experience alone. Every theory abstracts and simplifies the natural phenomena it is intended to describe. One of the more powerful approaches for abstracting and simplifying natural phenomena is the use of *conceptual models*. In each case, the simplification is carried to the point where the model is still amenable to mathematical treatment, yet is not so simple as to miss those features of the studied phenomena it is intended to describe (see sec. 4.5 for a detailed discussion on conceptual models). Once a conceptual model has been established, basic principles of physics and mathematical procedures are applied to it, leading to a theory that describes the investigated phenomena. Here phenomena include also the excitation-response relationships of natural materials for which ideal materials serve as models.

Almost invariably, when a constitutive equation is derived from an analysis of a conceptual model it includes a coefficient that must be determined experimentally. There is no way to derive such a coefficient from a theoretical analysis alone. This point becomes obvious if we recognize that a constitutive assumption is part of the continuum level of approach, whereas the real phenomena described by it are at the molecular level (par. 1.3.1). From the continuum viewpoint, the internal constitution of bodies (responsible for the fact that the same external forces cause different responses in different bodies) must be understood in the microscopic (i.e., of the fluid or solid continuum) sense. Thus, the coefficient appearing in a constitutive assumption should be regarded as an "ignorance factor" behind which is hidden our inability to describe the phenomenon at the molecular level at which it really occurs. Molecular diffusion and heat and momentum transfers may serve as examples of phenomena occurring at the molecular level. At their microscopic fluid continuum level, the descriptions require such coefficients as molecular diffusivity, thermal conductivity and viscosity. Although some models are more refined than others and may lead to a better understanding of a coefficient appearing in a constitutive assumption, or its internal structure and dependence upon certain, more elementary, medium parameters, there will always remain unknown numerical constants, or constitutive functions, that must eventually be determined from experiments.

The simplest constitutive assumptions take the form of a linear relationship between a flux and a driving force, with a coefficient that is independent of the magnitude of both the latter and the former. However, this should be considered only as a first approximation, valid for all practical purposes (within our measuring capabilities) for what is known as *low intensity processes*, where the gradients that define the driving forces are relatively small. In general, the coefficients vary and are not constant. In our attempt to employ simplified models, rather than relate the flux to a nonlinear expression in the driving force (e.g., containing higher powers of the latter), we sometimes prefer to relate it to the first power of the driving force, incorporating the nonlinearity in the coefficient itself.

4.4.2 Principles to be Used in Forming Constitutive Equations

In order to formulate a definite constitutive equation on the basis of assembled experimental data, Truesdell (1961, see also Truesdell and Toupin 1960) suggests the following list of mathematical principles to be used as guidelines.

Principle 1, consistency. Any constitutive equation must be compatible with the general principles of balance of mass, momentum and energy.

Principle 2, coordinate invariance. Constitutive assumptions must be stated as equations that hold in all inertial coordinate systems, at any fixed time, as we do not expect physical phenomena to be affected by the particular coordinate system chosen arbitrarily for their description. This means that constitutive equations must be framed in tensor language to ensure that they express the same relationship in all coordinate systems.

Principle 3, just setting. When combined, constitutive equations and balance (of mass, momentum, energy, etc.) equations related to the same set of variables, should yield a compatible set of equations leading to a unique and stable solution corresponding to physically meaningful initial and boundary conditions.

Principle 4, dimensional invariance. Constitutive equations depend on the number of dimensionally invariant moduli or material constants. The dimensionally independent moduli or material constants (such as viscosity, etc.) on which the response of the material may depend should be specified in a constitutive equation in a way that is consistent with the classical Buckingham π-theorem (sec. 11.2). Obviously, as a consequence, any constitutive equation should also be dimensionally homogeneous.

Principle 5, material indifference. This is the most important and most frequently used criterion for formulating constitutive equations. The principle states that *the (intrinsic) response of a material is independent of the observer*; i.e., the response must be seen to be the same by all observers, for otherwise it would not be intrinsic to the material.

Principle 6, material invariance (material isomorphism). Constitutive equations invariant with respect to a group of transformations of the material coordinates, a subgroup of the full orthogonal group, are said to have a material symmetry characterized by the subgroup. If there exists no preferred direction with respect to a certain constitutive material property, we say that the material is *isotropic* with respect to that property. Otherwise the material is said to be *aeolotropic* or *anisotropic*. A material that is isotropic with respect to one property need not be isotropic with respect to other properties.

Principle 7, equipresence. In the most general physical situations, mass, momentum and energy (and electromagnetism, when it is considered) are present simultaneously. In classical theories, however, variables describing these phenomena are divided somewhat arbitrarily into so-called "classes," assuming that members of one class do not affect members of other classes. For example, we associate stress with rate of strain, and heat flux with temperature gradient, but we do not cross-link them. Many physical phenomena, however, involve interaction among members of different classes, so that the separation into classes is quite arbitrary, resulting only from the gradual discovery of individual phenomena, and/or development of instrumentation for measuring energies and fluxes. Independent variables present in one constitutive equation should be present in all. Examples of interactions are mass transfer caused by temperature gradient (thermal-diffusion effect) and energy flux caused by concentration gradients (diffusion-thermo effect).

4.4.3 Coupled Processes

There exist a large number of constitutive equations describing irreversible processes in the form of linear relationships between fluxes and driving forces. For example:

Fourier's law relating heat flow to temperature gradient;
Fick's law relating flow of matter of a species in a multicomponent system to its concentration gradient;
Ohm's law relating electrical current to electrical potential gradient;
Newton's law (for a Newtonian fluid) relating shearing force to velocity gradient, etc.

In all these cases we see a simple, linear dependence of a flow on some conjugated force. However, this simple relationship does not always hold. As early as **1801**, Rouss (Katchalsky and Curran **1965**) carried out an experiment that showed that the application of an electromotive force in a porous medium may produce not only a flow of an electric charge, but also a nonconjugated flow of fluid. On the other hand, the application of hydrostatic pressure was found to produce a nonconjugated electric current. Several years later Seebeck showed the *thermoelectric phenomena* (a temperature gradient in a bimetallic system establishes a gradient of electric potential) and Peltier showed that the passage of electric current through the system caused an isothermal flow of heat. We may summarize these phenomena by stating that there may exist *coupling* between *forces* of one type and *flows* of another type. For sufficiently slow processes, any flow may depend in a direct and linear manner not only on its conjugated force, but also on nonconjugated forces that are present. Although the thermodynamic study of coupled phenomena was pioneered by Kerlin in 1853, it has recently become, though in a modified form, one of the essential features of modern thermodynamics (Katchalsky and Curran 1965).

The above-mentioned *cross-transport phenomena*, or *coupled processes*, verify the principle of equipresence mentioned in the preceding paragraph. In addition to the phenomena already mentioned, we also have: *Soret effect* (or phenomenon of *thermal diffusion*)—mass flux caused by temperature gradient (in addition to mass flux caused by concentration gradient), and *Dufour effect*— the reciprocal of the Soret effect—heat flux caused by concentration gradient, in addition to the heat flux caused by temperature gradient.

In the same class of cross effects is the contribution to the component of flow in one direction (say, to q_i) by a component of the (conjugated) force (J_j) in another direction in an anisotropic medium. In this case, each component of the driving force is considered as a driving force in itself, and each component of the flux as the sum effect of the three driving forces (components of the actual driving force).

Nonequilibrium thermodynamics is restricted primarily to the study of (assumed) linear phenomena as described above, although most phenomena are probably nonlinear. Very little of a sufficiently general nature is known outside the domain of linear relationships between forces and fluxes. However, this is not a very serious restriction since the linear approximation seems to describe rather well a wide range of slow processes of practical interest occurring when the system considered is not too far from a state of equilibrium.

Mathematically the cross effects are described by the addition of new terms to the constitutive equations discussed above. For example, for thermal diffusion, assuming a linear relationship as in the basic constitutive equation of molecular diffusion, a term proportional to the temperature gradient is added to the right-hand side of Fick's law. The new law:

$$\mathbf{J}^*_\alpha = -D_{11}\nabla C_\alpha - D_{12}\nabla T \tag{4.4.1}$$

states that the mass flux of the α-species is caused both by a concentration gradient, ∇C_α (ordinary molecular diffusion), and by a temperature gradient, ∇T. Equation (4.4.1), as do all similar linear expressions, defines certain coefficients: D_{11} = molecular diffusivity, and D_{12} = Soret coefficient.

Onsager (1931), and later Casimir (1945), developed a theory, referred to in the present text as *Onsager's theory*, upon which a systematic, macroscopic *thermodynamics of irreversible processes* can be based. In this theory, the set of equations expressing the relationships between fluxes and forces is written in the form:

$$\begin{aligned} \mathbf{J}_1 &= L_{11}\mathbf{X}_1 + L_{12}\mathbf{X}_2 + L_{13}\mathbf{X}_3 + \cdots + L_{1n}\mathbf{X}_n \\ \mathbf{J}_2 &= L_{21}\mathbf{X}_1 + L_{22}\mathbf{X}_2 + L_{23}\mathbf{X}_3 + \cdots + L_{2n}\mathbf{X}_n \\ &\vdots \\ \mathbf{J}_n &= L_{n1}\mathbf{X}_1 + L_{n2}\mathbf{X}_2 + L_{n3}\mathbf{X}_3 + \cdots + L_{nn}\mathbf{X}_n \end{aligned} \tag{4.4.2}$$

or, in abbreviated form:

$$\mathbf{J}_\alpha = \sum_{\beta=1}^{n} L_{\alpha\beta}\mathbf{X}_\beta; \qquad \alpha, \beta = 1, 2, \ldots, n \tag{4.4.3}$$

where the $\mathbf{J}_\alpha$s are fluxes the $\mathbf{X}_\alpha$s are thermodynamic forces, and the coefficients $L_{\alpha\beta}$ are independent of both the fluxes and the forces. Although we have used the vector symbol for the fluxes and the forces, it should be emphasized at this point that in the general case not all fluxes and forces must be vectors; they may be tensors of any rank. These "forces" have nothing in common with forces in the Newtonian sense: they cause irreversible phenomena such as heat transfer, mass transfer, chemical reaction, etc. We may then speak of a *generalized flux* $\mathbf{J}$ and *generalized force* $\mathbf{X}$. Thus, if n flows, or processes, $\mathbf{J}_1, \mathbf{J}_2, \ldots, \mathbf{J}_n$ simultaneously take place in a system, we may assign a conjugate force, $\mathbf{X}_1, \mathbf{X}_2, \ldots, \mathbf{X}_n$ to each flow, and obtain the fluxes from (4.4.2).

The coefficients $L_{\alpha\beta}$ appearing in (4.4.2) are called *phenomenological coefficients* (or *cross coefficients*). Here "phenomenological" is used to mean that the coefficients must be experimentally determined. Central to the theory of coupled processes is *Onsager's fundamental theorem* (or *Onsager's law*) that, provided a proper choice is made of the fluxes, $\mathbf{J}_\alpha$, and the forces, $\mathbf{X}_\alpha$, the *matrix of phenomenological coefficients*, $L_{\alpha\beta}$, appearing in (4.4.3) is symmetric, i.e.,

$$L_{\alpha\beta} = L_{\beta\alpha}; \qquad \alpha \neq \beta. \tag{4.4.4}$$

The reader is referred to texts on thermodynamics of irreversible processes (e.g., De Groot 1963) for an explanation of what is meant by a proper choice of forces and fluxes. These identities are called *Onsager's reciprocal relations*. They express a relationship between any pair of reciprocal phenomena (e.g., thermal diffusion and Dufour effect) arising from mutual interference between simultaneously occurring irreversible processes (e.g., heat conduction and molecular diffusion). The laws of Fourier, Fick, Ohm and others, express a linear relationship between a flux and its conjugated force, the "straight" coefficient of proportionality being $L_{\alpha\alpha}$ of (4.4.2). All the "straight" coefficients appear on the diagonal of the matrix of coefficients $L_{\alpha\beta}$. However, the equations (4.4.2) also state that any flow, $\mathbf{J}_\alpha$, may be driven by forces $\mathbf{X}_1, \mathbf{X}_2, \ldots, \mathbf{X}_n$ if the cross coefficients $L_{\alpha\beta}$ ($\alpha \neq \beta$) differ from zero. An alternative form of (4.4.3) is:

$$\mathbf{X}_\alpha = \sum_{\beta=1}^{n} R_{\alpha\beta}\mathbf{J}_\beta, \qquad \alpha, \beta = 1, 2, \ldots, n \tag{4.4.5}$$

where the matrix of coefficients $R_{\alpha\beta}$ is the inverse of the matrix $L_{\alpha\beta}$:

$$R_{\alpha\beta} = \text{cofactor of } L_{\alpha\beta}/|L| \tag{4.4.6}$$

where $|L|$ is the determinant of the matrix of the coefficients $L_{\alpha\beta}$. While the coefficients $L_{\alpha\beta}$ express flows per unit force, and hence have the characteristics of generalized conductivities or mobilities, the coefficients $R_{\alpha\beta}$ express force per unit flow and hence represent generalized resistivities.

In general, fluxes and forces may have different tensorial characters (some are scalars, others are vectors and still others are second-rank tensors) whose components transform in different ways under rotations and reflections of the Cartesian coordinate system. This fact is acknowledged in the *Curie–Prigogine symmetry principle* (Curie 1908). According to this principle, in *isotropic systems*, entities whose tensorial characters differ by an odd integer cannot interact. Put another way, the combination $\sum_{(\beta)} L_{\alpha\beta}\mathbf{X}_\beta$ is possible only in a case when the thermodynamic forces $\mathbf{X}_\beta$ are tensors of the same order, or where the difference in order is even. From Curie's principle, it therefore follows that we must consider separately scalar, vector and tensor quantities, and must write an equation of the form (4.4.3) for each tensorial type.

In an isotropic system, the scalar coefficients $L_{\alpha\beta}$ ($\alpha, \beta = 1, 2, \ldots, n$) appearing in each of the equations (4.4.3) form a *square matrix* (having n rows and n columns). This is a symmetrical matrix where $L_{\alpha\beta} = L_{\beta\alpha}$. Very often one finds in the literature that $L_{\alpha\beta}$ is referred to as a second-rank tensor in the n-dimensional space. This is incorrect, however. Although the matrix $L_{\alpha\beta}$ has the outward appearance of a tensor, from the latter we also require that its components obey certain transformation rules when axes of the space are changed. In fact the n-dimensional space of $L_{\alpha\beta}$ is what is known as a *vector space* (or an affine vector space) in which, unlike the *metric space*, no metric exists (i.e., we cannot define in such a space a distance between two points). In fact the situation is ever more complicated than the usual case when the fluxes and forces are not scalars. For example, if they are vectors, let us write (4.4.3) as a matrix product:

$$\begin{pmatrix} \mathbf{J}_1 \\ \mathbf{J}_2 \\ \mathbf{J}_3 \\ \vdots \\ \mathbf{J}_n \end{pmatrix} = \begin{pmatrix} L_{11} & L_{12} & \cdots & L_{1n} \\ L_{21} & L_{22} & \cdots & L_{2n} \\ L_{31} & L_{32} & \cdots & L_{3n} \\ \vdots & \vdots & & \vdots \\ L_{n1} & L_{n2} & & L_{nn} \end{pmatrix} \cdot \begin{pmatrix} \mathbf{X}_1 \\ \mathbf{X}_2 \\ \mathbf{X}_3 \\ \vdots \\ \mathbf{X}_n \end{pmatrix}. \tag{4.4.7}$$

We may refer to the matrix $(\mathbf{J}_\alpha)$ as a vector in the α-vector-space and to $(\mathbf{X}_\beta)$ as another vector in this space. However, each component of this vector is in itself a vector. The matrix $(L_{\alpha\beta})$ in this space relates these two vectors to each other. No absolute value of other vectors can be defined since there is no common measure for the different components of any of these vectors.

Consider now a *nonisotropic system*. For the case of a single flux, $\mathbf{J}$, produced by a single force, $\mathbf{X}$, equation (4.4.3) becomes:

$$\mathbf{J} = \underline{\mathbf{L}}\mathbf{X} \tag{4.4.8}$$

or, in indicial notation, employing the summation convention:

$$J_i = L_{ij}X_j, \qquad i, j = 1, 2, 3. \tag{4.4.9}$$

In this case the i-space is a *metric* space. J_i and X_j are components of the vectors $\mathbf{J}$ and $\mathbf{X}$, respectively, in this space. The coefficients L_{ij} are components of a second-rank symmetrical *tensor*. Here the term tensor is justified. However, if we extend the discussion to several coupled phenomena, in an anisotropic system, we obtain:

$$\mathbf{J}_\alpha = \underline{\mathbf{L}}_{\alpha\beta}\mathbf{X}_\beta. \tag{4.4.10}$$

For $\mathbf{J}_\alpha$ and $\mathbf{X}_\beta$, which are vectors, we may write for a three-dimensional Cartesian space:

$$J_{\alpha i} = L_{\alpha\beta ij}X_{\beta j}; \qquad \alpha, \beta = 1, 2, \ldots, n; \qquad i, j = 1, 2, 3. \tag{4.4.11}$$

In this case we *cannot* refer to $L_{\alpha\beta ij}$ as a fourth-rank tensor (although it is often loosely referred to as such), as it is a matrix that belongs to a *mixed space* that is affine with respect to the α coordinates and metric with respect to the i-coordinates. $L_{\alpha\beta ij}$ is symmetric separately in $\alpha\beta$ and in ij:

$$L_{\alpha\beta ij} = L_{\beta\alpha ij} = L_{\beta\alpha ji} = L_{\alpha\beta ji}. \tag{4.4.12}$$

Thus, the phenomenological coefficients in the anisotropic case form a square, n-dimensional matrix $\underline{\mathbf{L}}_{\alpha\beta}$; each of its components is a second-rank symmetrical tensor. Similar considerations may be applied to cases where $\mathbf{J}$ and $\mathbf{X}$ are of other tensorial ranks.

The absolute magnitude of the coupling coefficients ($\alpha \neq \beta$) is restricted by the magnitude of the straight coefficients ($\alpha = \beta$). By considering entropy production $s(= \sum_{(\alpha)} \mathbf{J}_\alpha \mathbf{X}_\alpha)$, which should be nonnegative (De Groot 1963), it can be shown that in the case of two vector flows and two conjugate vector forces in an isotropic system (where the $L_{\alpha\beta}$s are scalars) we obtain:

$$s = \mathbf{J}_1 \cdot \mathbf{X}_1 + \mathbf{J}_2 \cdot \mathbf{X}_2 = L_{11}(\mathbf{X}_1)^2 + (L_{12} + L_{21})\mathbf{X}_1 \cdot \mathbf{X}_2 + L_{22}(\mathbf{X}_2)^2 \geqslant 0. \tag{4.4.13}$$

Since either $\mathbf{X}_1$ or $\mathbf{X}_2$ may be made to vanish, we have the requirements that:

$$L_{11}(\mathbf{X}_1)^2 \geqslant 0; \qquad L_{22}(\mathbf{X}_2)^2 \geqslant 0 \tag{4.4.14}$$

so that both straight coefficients, L_{11} and L_{22}, must be positive. Furthermore, the quadratic form (4.4.13) will remain positive-definite only if:

$$\begin{vmatrix} L_{11} & L_{12} \\ L_{21} & L_{22} \end{vmatrix} = L_{11}L_{22} - L_{12}L_{21} \geqslant 0. \tag{4.4.15}$$

This relationship serves as a restriction on the possible magnitudes of the coupling coefficients L_{12} and L_{21}. Using Onsager's reciprocal relationship, this condition becomes:

$$L_{11}L_{22} \geqslant L^2{}_{12}. \tag{4.4.16}$$

For any number of fluxes and forces in an isotropic system, with $s \geqslant 0$ and $L_{ii} \geqslant 0$, we obtain:

$$L_{\alpha\alpha}L_{\beta\beta} \geqslant \tfrac{1}{4}(L_{\alpha\beta} + L_{\beta\alpha})^2. \tag{4.4.17}$$

In a single flux case in a nonisotropic system (4.4.17) may still be used; however, α and β stand for the coordinate directions: $\alpha, \beta = 1, 2, 3$.

4.5 A Porous Medium Model

So far, the discussion in the present chapter has been related to a fluid continuum filling the void space. However, as was explained in section 1.3, it is unpractical to treat transport phenomena in porous media by referring only to the fluid continuum filling the void space, owing to our ignorance of the detailed geometry of the fluid–solid interfaces. Instead, the continuum approach was suggested. According to this approach the actual, multiphase, porous medium is replaced by a fictitious continuum, to any mathematical point of which are assigned values of the several variables and parameters. These are *average values* obtained by averaging any particular variable over the representative elementary volume (REV) of the medium around the point considered. A problem arises concerning how these averaging procedures should be carried out.

In the following paragraphs a certain *statistical model* of flow through porous media is proposed and analyzed, leading to the basic (macroscopic) equations governing the flow of fluids in porous media. The model and its analysis are presented here without any general introductory remarks on statistical models as tools for obtaining macroscopic descriptions of investigated phenomena. Such remarks are postponed until the discussion on theories of hydrodynamic dispersion in section 10.3. The reader is referred to the first part of paragraph 10.3.2 for general comments on statistical models.

4.5.1 The Conceptual Model Approach

One of the most powerful tools in investigating phenomena in complex systems is the *conceptual model* employed in all branches of physics. According to this approach a complicated physical phenomenon, or system, the mathematical treatment of which is practically impossible, is replaced by some fictitious, simpler phenomenon or system, that *is* amenable to mathematical treatment. By analyzing the fictitious phenomenon, or system results are obtained, often in the form of laws or mathematical relationships among various parameters of the investigated phenomenon. Various parameters of the conceptual model also appear in these laws, and are then related to the parameters of the real phenomenon. An example is the *kinetic theory of gases*, where gas molecules are visualized as balls colliding with each other; instead of considering the actual forces acting among these molecules, we introduce a statistical picture of molecules moving at random, colliding with each other according to some assumed rule. This fictitious model is sufficiently simple and can be treated mathematically.

It is difficult to define a conceptual model, or briefly, a model (not to be confused with "models" discussed in chap. 11, which are small-scale reproductions of prototypes). Here the term is used to denote a simplified way of visualizing a phenomenon that cannot be directly observed microscopically, but for which macroscopic excitations and responses can be observed and measured. A model may be a representation of either an object, such as visualizing a porous medium as a system of interconnected capillary tubes, or of a process, such as visualizing the movement of a tracer in flow through porous media as a three-dimensional random flight of tracer particles.

The kinetic theory of gases is based on a model in which each gas molecule is visualized as a rigid, noninteracting tiny sphere. Each such sphere is assumed to move in a straight line, occasionally colliding with another molecule it meets at random. By analyzing these motions and collisions, a relationship is obtained between the macroscopic pressure in the gas and the kinetic energy of the gas molecules. However, when compared with experimental results, it is found that this *conceptual model* of an ideal gas predicts the *observed* behavior of a real gas with increasing accuracy the lower the number of molecules per unit volume at a given temperature. This simplified model fails to provide additional information unless the model is further refined, or additional assumptions are introduced.

Other examples of conceptual models that lead from the molecular level to the description of phenomena at the fluid continuum level, and to appropriate fluid parameters, are those dealing with the viscosity of a fluid, its thermal conductivity and its molecular diffusivity (see, for example, Bird, Lightfoot and Stuart 1960).

In all these examples, the procedure is composed of three steps. In the first step, the complicated system is represented by a simplified conceptual model. The model should be made simple enough to be amenable to mathematical treatment, yet should incorporate as many as possible of the features and factors of the investigated phenomenon, without becoming too complicated. If, in order to make the model amenable to mathematical treatment we must dispose of too many essential features of the investigated phenomenon, we had better choose another model. Once the model is chosen, the second step is to analyze the model by available theoretical tools, and to derive mathematical relationships that describe the investigated phenomenon. These relationships show how the various active variables (fluxes, forces, etc.) depend on each other. They also show which factors have, *according to the chosen model*, no influence on the investigated problem. The only way to test the validity of laws derived in this way is to perform controlled experiments in the laboratory (or to observe phenomena in nature). Such controlled experiments, which comprise the third step of this approach, will test the validity of the derived relationships among the variables. No theory developed by this approach can be accepted without first being verified by experiments. This also means that all macroscopic parameters and coefficients appearing in equations resulting from model analysis should be measurable quantities. In certain cases the experiments will indicate the range over which the developed theory is sufficiently accurate for practical purposes.

Relationships or laws derived from models (e.g., Fick's law or Fourier's law) always contain parameters or coefficients made up of various elementary properties and parameters of the model. In general, such relationships also contain numerical coefficients. However, in most cases, even if we trust the relationship between a gross coefficient and its elementary components, the *numerical coefficient must be determined experimentally*. This stems from the fact that the model is chosen so that it simulates features and behavior of the prototype, but not to the extent of numerical coefficients. With these thoughts in mind, a question sometimes arises as to why we bother with the model in the first place, since in any case we must eventually go back to the laboratory to determine the required coefficients. The answer is that by employing the conceptual model approach we gain an understanding of the investigated phenomenon and the role of the various factors that affect it. We also gain an insight into the internal structure of the various coefficients appearing in the equations that describe the investigated phenomenon. All this information is needed for planning the laboratory experiments.

Throughout this book many examples are given of the use of the conceptual model approach. Some of the models are physical, while others are mathematical or statistical in nature (e.g., a model visualizing diffusion as the random walk of a drunkard). In the next paragraph, and in the sections that follow, a conceptual model of a porous medium and of an averaging procedure is presented, the analysis of which leads to Darcy's law and to the laws of conservation of the total fluid mass and of a species in solution. In section 5.10, a review is given of several models leading to Darcy's law. In section 9.3, porous medium models are presented that are capable of simulating capillary pressure and two-phase flows. Models of hydrodynamic dispersion are presented in section 10.3. In all these cases, the model is presented as an attempt to simulate, and thus to explain, phenomena observed in nature or in the laboratory. Sometimes several models are equally successful in explaining the relationship between observed excitations and responses. However, we must emphasize again that the proof of the validity of a model, and the only way to determine coefficients, is always the experiment.

4.5.2 A Model of Flow through a Porous Medium

In order to employ the conceptual model approach for the derivation of a theory of flow through porous media, we must first introduce a simplified porous medium model that will be amenable to a mathematical treatment and that will incorporate the main features of a porous medium, as described in section 1.2. As our objective here is to analyze flow through the pore space, we must supplement the porous medium model by statements, in the form of a set of assumptions and constraints, regarding the fluid and the flow regime in the model.

One of the more essential features of a porous medium, in connection with the flow of a fluid through it, is that it *restricts the transport of the fluid to well defined channels*. Because of the immediate presence of the walls of the solid matrix, the velocity of a fluid particle at a point in the void space is essentially in the direction parallel to the walls, and not normal to them. Accordingly we propose to employ a model that has this property as one of its essential features. We shall adopt here the porous medium model proposed by Bear and Bachmat (1966, 1967).

Bear and Bachmat (1966, 1967) visualize the void space of a porous medium as composed of a spatial network of interconnected random passages (*channels* or *tubes*) of varying length, cross-section and orientation, and *junctions*, where channels meet; at least three channels meet at each junction. The main difference between a channel and a junction is that for a channel, because of its elongated shape, an axis may be defined, whereas a junction has no definite direction in space. The channels and the junctions have a more or less uniform spatial distribution. The solid matrix is assumed to be rigid and noninteracting with the fluid filling the void space. For the laminar flow considered in the analysis, each channel of the model defines in space a streamtube in which the pattern of streamlines is fixed, although the direction of flow along them may be reversed. In a junction, there is no fixed pattern of streamlines. It is assumed that the volume of a junction is much smaller than that of a channel. This and other constraints (e.g., in the form of assumptions) that are incorporated in the proposed model may be removed later in the process of refining the model, which should be considered just one example of many possible models. Other models are presented in several places in the present book.

The fluid completely saturating the porous medium model is a single-phase binary system containing two components: a solvent and a solute (tracer). The fluid is assumed to be chemically inactive, viscous, Newtonian, of a constant temperature and inhomogeneous, i.e., its density and viscosity may vary from point to point as a result of variations in solute concentration. However, as a first approximation it is assumed that the fluid is incompressible, i.e., the changes in volume due to changes in solute concentration may be neglected. This assumption is practically valid for low solute concentrations.

The flow regime is assumed to be *laminar*. The active forces are those due to pressure, gravity and shear resulting from the fluid's viscosity. It is also assumed that in our model the fluid loses energy only during passage through the narrow channels and not while passing from one channel to the next through a junction. Thus, the network of channels connected to each other by junctions produces average gradients of pressure, density, viscosity and solute concentration in any elementary volume that includes a sufficiently large number of channels and junctions. These average gradients are practically independent of the geometric shape of a single channel within the elementary volume. On the other hand, local deviations from the average (assumed to be much smaller than the average itself) at points within the void space depend strongly on the local geometry of the solid matrix.

4.5.3 Frames of Reference

Two orthogonal Cartesian coordinate systems will be used (fig. 4.5.1a): (a) x_1, x_2, x_3 fixed in space (abbreviated as x_i, $i = 1, 2, 3$); and (b) a right-handed Cartesian coordinate system formed at a point M on a channel's axis by the three base vectors: $\mathbf{1\lambda}_1$ in the direction of the *tangent* to the curve at M, $\mathbf{1\lambda}_2$ in the direction of the *principal normal* to the curve, and $\mathbf{1\lambda}_3$ in the direction of the *binormal*. This coordinate system is called the *triad* or *trihedral* at the point. The equation of an axis of a channel of length L is (fig. 4.5.1a):

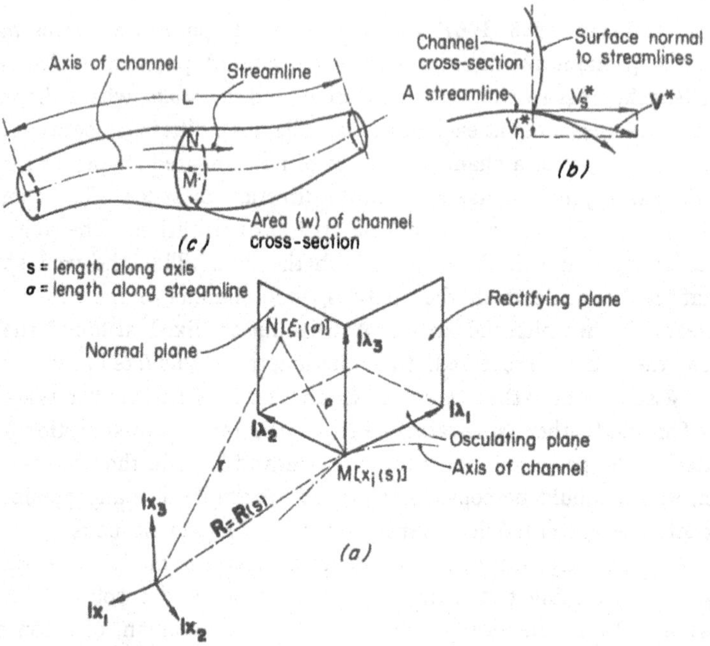

FIG. 4.5.1. Model of a channel.

$$\mathbf{R} = \mathbf{R}(s) \quad \text{or:} \quad x_i = x_i(s); \quad 0 \leqslant s \leqslant L \tag{4.5.1}$$

where s denotes the length measured along the axis of the channel. The unit vector $\mathbf{1\lambda}_1$ is tangent to the axis and shows the direction in which s increases.

The three unit vectors, $\mathbf{1\lambda}_1$, $\mathbf{1\lambda}_2$ and $\mathbf{1\lambda}_3$, are related to each other by the *Frenet–Serret* formulas (e.g., Spiegel 1959).

For a point N located on a plane normal to the channel's axis at point M of the latter, the *local coordinates* ξ_i in the fixed coordinate system x_i are:

$$\xi_i = x_i(s) + \lambda_j \cos(\mathbf{1\lambda}_j, \mathbf{1x}_i); \qquad i = 1, 2, 3, \quad j = 2, 3 \tag{4.5.2}$$

with summation on j. In a vectorial form, equation (4.5.2) can be written as:

$$\mathbf{r} = \mathbf{R}(s) + \boldsymbol{\rho} \tag{4.5.3}$$

where $\boldsymbol{\rho}$ is the radius vector of N in the $\lambda_2\lambda_3$ plane.

Each point N belongs to a streamline passing through it. Denoting the length measured along this streamline from the entrance to the channel (where both $s = 0$ and $\sigma = 0$) by σ, the coordinates ξ_i of N can be expressed in the form:

$$\xi_i = \xi_i[\sigma(s)]. \tag{4.5.4}$$

The relationship between σ and s is unique; for a given s (i.e., a given channel cross-section) there is a unique value of σ for any point N on the corresponding cross-section. The (local) ξ_i coordinates are related to the x_i coordinates by:

$$\frac{d\xi_i}{ds} = (1+\alpha)\frac{dx_i(s)}{ds} + \beta\frac{d^2x_i(s)}{ds^2} + \gamma\frac{d^3x_i(s)}{ds^3} \tag{4.5.5}$$

where:

$$\alpha = -(\rho_1/\rho)(d\lambda_3/d\sigma)(d\sigma/ds);$$

$$\beta = (\rho/\rho_1)[\lambda_3[\sigma(s)] + \rho_1(d\lambda_2/d\sigma)(d\sigma/ds)] + (d\rho/ds)[\lambda_2[\sigma(s)] - \rho_1(d\lambda_3/d\sigma)(d\sigma/ds)];$$

$$\gamma = \rho[\lambda_2[\sigma(s)] - \rho_1(d\lambda_3/d\sigma)(d\sigma/ds)];$$

ρ and ρ_1 are the radii of curvature and of torsion, respectively, of the axis at the considered point:

$$\rho = 1/|d^2\mathbf{R}/ds^2|; \qquad \rho_1 = -(1/\rho^2)[(d\mathbf{R}/ds)\cdot(d^2\mathbf{R}/ds^2)\times(d^3\mathbf{R}/ds^3)].$$

4.5.4 An Averaging Procedure

The various conservation equations derived in section **4.3** are valid for any point in the fluid continuum filling the void space of a porous medium. They are also valid for any point in the void space of the porous medium model described in paragraph 4.5.3. However, as explained in section 1.3, we are interested only in average values of the various flow parameters appearing in these equations, where the averages are taken over the ***representative elementary volume*** (REV) of the porous medium. An average $\bar{g}(P)$ at point P of some intensive fluid property g (amount of property per unit volume of fluid) can be defined by:

$$\bar{g}(P) = \frac{1}{(\Delta U_0)_v}\int_{[(\Delta U_0)_v]_P} g(\mathbf{x}')\,dU_v(\mathbf{x}') \tag{4.5.6}$$

where $[(\Delta U_0)_v]_P$ is the volume of void space within an REV that is centered at point P. However, this averaging procedure does not manifest a certain intrinsic property of the porous medium that is also an essential feature of the porous medium model described above. Underlying the choice of the model presented above was the observation that the presence of channels dictates a priori that flow can take place only in the directions of the channels; not every possible direction in space can be a flow direction. For example, no flow can take place in the direction of the normal to the solid walls of a channel. Hence it is proposed to carry out the averaging process in two steps. In the first step the averaging will be performed over a cross-section of a channel, assigning the resulting average to the centroid of the channel's cross-section, which is, by definition, a point on the channel's axis. Once this step has been carried out for all cross-sections of the porous medium model within the considered REV, only the flow along the network of channel axes will have to be considered. In the second step, these point values will be averaged over all channel axes inside the REV.

Let $g(\sigma)$ be some scalar variable inside the void space of the porous medium model. The average value $\langle g(s)\rangle$ over a channel cross-section is given by:

$$\langle g(s)\rangle = \frac{1}{w(s)} \int_{[w(s)]} g(\sigma)\, dw(\sigma) \tag{4.5.7}$$

or:

$$\langle g(s)\rangle = \frac{1}{w(s)} \int_{[w(s)]} g(s; \lambda_2, \lambda_3)\, d\lambda_2\, d\lambda_3; \qquad d\lambda_2\, d\lambda_3 = dw \tag{4.5.8}$$

where $w(s)$ is the area of the considered cross-section, and λ_2, λ_3 are coordinates of a point on this cross-section. If g is differentiable in the void space, then:

$$\langle \operatorname{grad} g\rangle = \left\langle \frac{\partial g}{\partial s}\right\rangle \mathbf{1\lambda}_1 + \left\langle \frac{\partial g}{\partial \lambda_2}\right\rangle \mathbf{1\lambda}_2 + \left\langle \frac{\partial g}{\partial \lambda_3}\right\rangle \mathbf{1\lambda}_3. \tag{4.5.9}$$

Equation (4.5.9) can be applied to any vector field in the void space. For example, for velocity we have:

$$\langle \mathbf{V}\rangle = \langle V_s\rangle \mathbf{1\lambda}_1 + \langle V_{\lambda_2}\rangle \mathbf{1\lambda}_2 + \langle V_{\lambda_3}\rangle \mathbf{1\lambda}_3. \tag{4.5.10}$$

Since:

$$\mathbf{V} = V\mathbf{1\sigma}; \qquad V = |\mathbf{V}| \tag{4.5.11}$$

where $\mathbf{1\sigma}$ is a unit vector in the direction of the tangent of the streamline at a point on the cross-section, we also have:

$$V_s = \mathbf{V}\cdot\mathbf{1\lambda}_1 = V\mathbf{1\sigma}\cdot\mathbf{1\lambda}_1 = V\frac{d\xi_i}{d\sigma}\frac{dx_i(s)}{ds}. \tag{4.5.12}$$

By combining (4.5.12) with (4.5.5), and assuming that the curvature and tortuosity of the channel's axis are small with respect to changes in some characteristic length of the channel's cross-section, we obtain:

$$V_s \approx V/(d\sigma/ds). \tag{4.5.13}$$

If variations of g on the circumference of a channel's cross-section are such that $|g - \tilde{g}| \ll \tilde{g}$, where $\tilde{g}$ is the average value of g taken over the circumference, we have (Bear and Bachmat 1966, 1967):

$$\frac{\partial\langle g\rangle}{\partial s} \cong \left\langle \frac{\partial g}{\partial s}\right\rangle + (\tilde{g} - \langle g\rangle)\frac{\partial w}{w\,\partial s}. \tag{4.5.14}$$

This expression becomes an exact one for a circular cross-section. For $w =$ const (or approximately so), or when $\tilde{g} \approx \langle g\rangle$, we obtain:

$$\partial\langle g\rangle/\partial s \cong \langle \partial g/\partial s\rangle. \tag{4.5.15}$$

The transport equations obtained as a result of the first averaging step (i.e., describing what happens at points on a channel's cross section) will be called *local transport equations*.

In the second step of the averaging process we average over the void space of the

REV of volume $n\,\Delta U_0 (= \Delta U_{0,\text{chan}} + \Delta U_{0,\text{junct}})$. If G denotes some extensive fluid property associated with the intensive property g, then:

$$G = \int_{(n\Delta U_0)} g\,dU_v = \int_{(\Delta U_{\text{chan}})} g\,dU_{\text{chan}} + \int_{(\Delta U_{\text{junct}})} g\,dU_{\text{junct}}. \tag{4.5.16}$$

Neglecting ΔU_{junct} (following our porous medium model), we obtain:

$$G = \int_{(\Delta U_{\text{chan}})} g\,dU_{\text{chan}} = \sum_{(L_i)} \int_0^{L_i} \langle g \rangle w(s)\,ds. \tag{4.5.17}$$

The average $\bar{g}$ of g over the void space of REV is expressed for a point P in the porous medium domain by:

$$\bar{g}(P) = \frac{1}{(n\,\Delta U_0)_P} \int_{(n\Delta U_0)_P} g\,dU_v. \tag{4.5.18}$$

In this way $\bar{g}$ is defined for every point in the porous medium domain in accordance with the continuum approach (sec. 1.3). We assume that for the properties g considered here, $\bar{g}$ is differentiable up to second order.

From (4.5.17) and (4.5.18) it follows that when the junctions' volume is neglected, we have:

$$\overline{\langle g \rangle} = \bar{g}. \tag{4.5.19}$$

The value of g at a point within the void space may be represented as the sum of the average value of the REV and a local deviation:

$$g = \bar{g} + \mathring{g}; \qquad \overline{\mathring{g}} = 0. \tag{4.5.20}$$

For products of variables or parameters, e.g., g and f, we have:

$$\overline{gf} = \bar{g}\bar{f} + \overline{\mathring{g}\mathring{f}}; \qquad \overline{g^2} = \bar{g}^2 + \overline{\mathring{g}^2}. \tag{4.5.21}$$

In (4.5.21), $\bar{g}$ is the average, $\overline{\mathring{g}^2}$ is the variance and $\overline{\mathring{g}\mathring{f}}$ is the covariance. It is important to note that $\overline{\mathring{g}^2}$ measures the dispersion of g around $\bar{g}$.

The actual execution of the averaging to obtain $\bar{g}$ and $\overline{\mathring{g}^2}$ is impossible without information on the exact distribution of the values of g within the void space of the REV. Instead, we shall treat g as a *random variable* within the void space of the REV. Moreover, since we are interested only in the measurable mean values themselves, there is no need to specify a particular distribution function. All that is required is to make some general assumptions regarding the moments of correlation among the variables represented by g and f. For example, if we assume that g and f are independent of each other, then $\overline{\mathring{g}\mathring{f}} = 0$. We shall also make the assumption that the average of three deviations $\overline{\mathring{g}\mathring{f}\mathring{c}}$ appearing in $\overline{gfc} = \bar{g}\bar{f}\bar{c} + \bar{g}\overline{\mathring{f}\mathring{c}} + \bar{f}\overline{\mathring{g}\mathring{c}} + \bar{c}\overline{\mathring{g}\mathring{f}} + \overline{\mathring{c}\mathring{g}\mathring{f}}$ is negligible, i.e., $\overline{\mathring{g}\mathring{f}\mathring{c}} \approx 0$.

From the definition of the average, and of the porous medium model with rigid solid boundaries (i.e., $n = \text{const}$) it follows that:

$$\overline{(\partial g/\partial t)} = \partial \bar{g}/\partial t. \tag{4.5.22}$$

In analogy to (4.5.14) or (4.5.15), we may assume:

$$\overline{(\partial g/\partial \xi_j)} = \partial \bar{g}/\partial x_j + (\bar{g} - \bar{\bar{g}})\partial \ln(n\,\Delta U_0)/\partial x_j. \tag{4.5.23}$$

When $\bar{g} = \bar{\bar{g}}$, we have:

$$\overline{(\partial g/\partial \xi_j)} = \partial \bar{g}/\partial x_j \tag{4.5.24}$$

respectively, where $\bar{\bar{g}}$ is the average of g taken over the solid surface bounding the void space of the REV. The approximation of $\bar{g} = \bar{\bar{g}}$ is well justified for $g = \rho, \mu, p$ and ρ_α. However, for the velocity that vanishes at the solid walls, so that $\bar{\bar{V}} = 0$, we have:

$$\overline{\left(\frac{\partial V}{\partial \xi_j}\right)} = \frac{\partial \bar{V}}{\partial x_j} + \bar{V}\frac{\partial \ln(n\Delta U_0)}{\partial x_j} = \frac{\partial \bar{V}}{\partial x_j} + \bar{V}\frac{\partial \ln(\Delta U_{0v})}{\partial x_j}. \tag{4.5.25}$$

Only in a homogeneous medium where $\Delta U_{0v} = (n\,\Delta U_0) = \text{const}$, equation (4.5.25) reduces to:

$$\overline{\left(\frac{\partial V}{\partial \xi_j}\right)} = \frac{\partial \bar{V}}{\partial x_j}. \tag{4.5.26}$$

Equation (4.5.26) may also be used for an inhomogeneous medium if we assume that ΔU_{0v} is practically constant. In what follows we shall employ mainly (4.5.26). The use of (4.5.25) in developing the fundamental transport equations is presented by Bear and Bachmat (1966, 1967).

4.6 Equations of Volume and Mass Conservation

The development presented in this section and in the subsequent one follows that presented by Bear and Bachmat (1966, 1967).

4.6.1 Equation of Volume Conservation

Let us define an elementary volume U_0 of a channel between two cross-sections w_1 at s_1 and w_2 at $s_2 = s_1 + \Delta s$:

$$U_0 = \int_{s_1}^{s_2} w(s)\,ds. \tag{4.6.1}$$

Equation (4.3.18) may be written for the volume U_0 as:

$$\int_{(U_0)} \operatorname{div} \mathbf{V}'\,dU = 0 \tag{4.6.2}$$

or, in view of the Gauss divergence theorem, and noting that at the solid wall $\mathbf{V}' = 0$:

$$\int_{(U_0)} \operatorname{div} \mathbf{V}' \, dU = \int_{(w_1+w_2+S)} \mathbf{V}' \cdot d\mathbf{S} = \int_{(w_1+w_2)} \mathbf{V}' \cdot d\mathbf{S} = [\langle V'_s \rangle w]_{s=s_2} - [\langle V'_s \rangle w]_{s=s_1} = 0 \tag{4.6.3}$$

where S is the surface area of the channel between the two cross-sections and V'_s is the projection of the volume-averaged velocity $\mathbf{V}'$ on the direction of the channel's axis. Dividing (4.6.3) by U_0 and passing to the limit as $\Delta s \to 0$ (and $w(s_2) \to w(s_1) = w$), we obtain:

$$\partial(\langle V'_s \rangle w)/\partial s = 0 \tag{4.6.4}$$

or:

$$w\, \partial\langle V'_s \rangle/\partial s + \langle V'_s \rangle\, \partial w/\partial s = 0; \qquad \partial\langle V'_s \rangle/\partial s + (\langle V'_s \rangle/w)\, \partial w/\partial s = 0. \tag{4.6.5}$$

From (4.5.14), with $g \equiv V'_s$ and with $\tilde{g} = \tilde{V}'_s = 0$, we obtain:

$$\partial\langle V'_s \rangle/\partial s = \langle \partial V'_s/\partial s \rangle - \langle V'_s \rangle\, \partial w/w\, \partial s. \tag{4.6.6}$$

Hence, by combining (4.6.5) and (4.6.6), we obtain:

$$\langle \partial V'_s/\partial s \rangle = 0 \tag{4.6.7}$$

which is the *local* (i.e., averaged over a cross-section) *equation of fluid volume conservation.*

From (4.5.23) written for $g \equiv V'_j$, we obtain:

$$\overline{(\partial V'_j/\partial \xi_j)} = \partial \bar{V}'_j/\partial x_j + (\bar{V}'_j/n\, \Delta U_0)\, \partial(n \Delta U_0)/\partial x_j. \tag{4.6.8}$$

Since by (4.3.18) $\partial V'_j/\partial \xi_j \equiv 0$, and assuming $\Delta U_0 \approx$ const, we obtain from (4.6.8):

$$\partial \bar{V}'_j/\partial x_j + (\bar{V}'_j/n)\, \partial n/\partial x_j = 0 \tag{4.6.9}$$

or:

$$\partial(n \bar{V}'_j)/\partial x_j = 0 \tag{4.6.10}$$

as the *volume conservation equation* for an incompressible fluid in a porous medium. In view of (1.3.11), where $\mathbf{V}$ has the meaning of a volumetric velocity $\mathbf{V}'$, we obtain from (4.6.10):

$$\partial q_j/\partial x_j = 0; \qquad \operatorname{div} \mathbf{q} = 0 \tag{4.6.11}$$

as the *volume conservation equation.* For a homogeneous medium, $n =$ const, we also have:

$$\partial \bar{V}'_j/\partial x_j = 0; \qquad \operatorname{div} \mathbf{V}' = 0. \tag{4.6.12}$$

In (4.6.8) through (4.6.12), as in what follows, the summation convention will be employed.

4.6.2 Equation of Mass Conservation of a Species in Solution

By writing (4.3.1) for the elementary volume U_0, we obtain:

$$\int_{s_1}^{s_2} \left\langle \frac{\partial \rho_\alpha}{\partial t} \right\rangle w\, ds = [\langle \rho_\alpha V_{\alpha s} \rangle w]_1 - [\langle \rho_\alpha V_{\alpha s} \rangle w]_2 - \int_{(S)} \rho_\alpha \mathbf{V}_\alpha \cdot \mathbf{dS} + \int_{s_1}^{s_2} \langle I_\alpha \rangle w\, ds \qquad (4.6.13)$$

where $V_{\alpha s} = \mathbf{V}_\alpha \cdot \mathbf{1\lambda}_1$ is the projection of $\mathbf{V}_\alpha$ on the axis of the channel considered. In (4.6.13):

$$\rho_\alpha \mathbf{V}_\alpha \cdot \mathbf{dS} = -\rho_\alpha V_{\alpha s}\, dw + \rho_\alpha V_{\alpha n}\, dP\, ds$$

where dP is an element of the length of a channel's cross-section's boundary. In the absence of phenomena such as adsorption or dissolution, which cause a transfer of species α solute through the solid envelope (S) of U_0, $V_{\alpha n} = 0$. When also $I_\alpha = 0$, by dividing (4.6.13) by U_0 and passing to the limit $\Delta s \to 0$, we obtain:

$$\left\langle \frac{\partial \rho_\alpha}{\partial t} \right\rangle = -\frac{1}{w} \frac{\partial [\langle \rho_\alpha V_{\alpha s} \rangle w]}{\partial s} + \frac{1}{w} \overline{\rho_\alpha V_{\alpha s}} \frac{dw}{ds}. \qquad (4.6.14)$$

In view of (4.5.14) with $g \equiv \rho_\alpha V_{\alpha s}$ and $\tilde{g} \equiv \overline{\rho_\alpha V_{\alpha s}} = 0$, we obtain from (4.6.14):

$$\langle \partial \rho_\alpha / \partial t \rangle = -\langle \partial (\rho_\alpha V_{\alpha s}) / \partial s \rangle. \qquad (4.6.15)$$

From (4.1.14) and (4.5.13), we obtain:

$$\rho_\alpha V_{\alpha s} = \rho_\alpha [V_{\alpha s} - V'_s] + \rho_\alpha V'_s = -D_{\alpha\beta} \frac{\partial \rho_\alpha}{\partial \sigma} \left(\frac{d\sigma}{ds} \right) + \rho_\alpha V' \left(\frac{d\sigma}{ds} \right)^{-1}. \qquad (4.6.16)$$

By introducing (4.6.16) into (4.6.15) and passing to the fixed Cartesian coordinates, x_i, we obtain:

$$\left\langle \frac{\partial \rho_\alpha}{\partial t} \right\rangle = \left\langle \frac{\partial}{\partial \xi_i} \left(D_{\alpha\beta} \frac{\partial \rho_\alpha}{\partial \xi_j} \frac{d\xi_j}{d\sigma} \frac{d\sigma}{ds} \right) \frac{d\xi_i}{d\sigma} \frac{d\sigma}{ds} \right\rangle - \left\langle \frac{\partial}{\partial \xi_i} \left[\rho_\alpha V' \left(\frac{d\sigma}{ds} \right)^{-1} \right] \cdot \frac{d\xi_i}{d\sigma} \frac{d\sigma}{ds} \right\rangle \qquad (4.6.17)$$

or:

$$\left\langle \frac{\partial \rho_\alpha}{\partial t} \right\rangle = \left\langle \frac{\partial}{\partial \xi_i} \left(D_{\alpha\beta} T^*_{ij} \frac{\partial \rho_\alpha}{\partial \xi_j} \right) \right\rangle - \left\langle \frac{\partial}{\partial \xi_i} (\rho_\alpha V'_i) \right\rangle \qquad (4.6.18)$$

where

$$V'_i = V(d\xi_i/d\sigma), \qquad T^*_{ij} = T_{ij}(d\sigma/ds)^2, \quad \text{and} \quad T_{ij} = (d\xi_i/d\sigma)(d\xi_j/d\sigma).$$

Equation (4.6.18) is the *local equation of mass conservation of a species in solution.* We shall return below to the detailed interpretation of T_{ij} and T^*_{ij}. At this point it is sufficient to note that $(d\xi_i/d\sigma)(d\xi_j/d\sigma)$ represents a matrix whose nine elements (for $i, j = 1, 2, 3$) are products of cosines of angles between the direction of a streamline at a point and the coordinate axis.

By averaging (4.6.18) over the void space of an REV, and assuming $\Delta U_0 = \text{const}$, we obtain:

$$\frac{\partial \bar{\rho}_\alpha}{\partial t} = \frac{\partial}{\partial x_i}\left(D_{\alpha\beta}\overline{T^*_{ij}}\frac{\partial \bar{\rho}_\alpha}{\partial x_j}\right) - \frac{\partial}{\partial x_i}\overline{(\rho_\alpha V'_i)} - \frac{\overline{\rho_\alpha V'_i}}{n}\frac{\partial n}{\partial x_i}. \tag{4.6.19}$$

The last term on the right-hand side of (4.6.19) vanishes for a homogeneous medium where $n = \text{const.}$

The expression $\rho_\alpha V'_i$ represents the instantaneous mass flux of the α species carried by the liquid. This flux can be separated into two parts:

$$\rho_\alpha V'_i = \rho_\alpha \bar{V}'_i + \rho_\alpha \mathring{V}'_i$$

where $\rho_\alpha \bar{V}'_i$ is the flux carried by the average fluid motion through the REV, and $\rho_\alpha \mathring{V}'_i$ is the *dispersive flux* resulting from the velocity fluctuations. It is important to note that the *dispersive flux* considered here is different from the *diffusive flux* appearing inside the brackets of the first term on the right-hand side of (4.6.19). The dispersive flux of the α-species expresses the rate at which mass is transported because of the velocity variations in the void space of the REV, where the solute's density distribution may be taken instantaneously as stationary, i.e., $\partial\rho_\alpha/\partial t = 0$. Under such conditions we obtain from (4.3.1) with $I_\alpha = 0$:

$$\rho_\alpha\, \partial V_{\alpha j}/\partial \xi_j + V_{\alpha j}(\partial\rho_\alpha/\partial \xi_j) = 0;$$

$$\rho_\alpha = -(\mathbf{V}_\alpha \cdot \operatorname{grad} \rho_\alpha)(\operatorname{div} \mathbf{V}_\alpha)^{-1} \tag{4.6.20}$$

where the interpretation of $\mathbf{V}_\alpha$ is given by (4.3.4).

For the average flux we have:

$$\overline{\rho_\alpha V'_i} = \bar{\rho}_\alpha \overline{V'_i} + \overline{\mathring{\rho}_\alpha \mathring{V}'_i}; \qquad \overline{\rho_\alpha \mathring{V}'_i} \equiv \overline{\mathring{\rho}_\alpha \mathring{V}'_i}. \tag{4.6.21}$$

Assuming (a) that there is no correlation between the local velocity of a liquid particle at a mathematical point inside the void space and the concentration gradient, i.e., $\overline{\mathring{V}'_j(\partial\mathring{\rho}_\alpha/\partial \xi_j)} = 0$, and (b) that the mean of the product of three or more deviations is zero, we obtain from (4.6.20), following Bear and Bachmat (1965, 1967):

$$\overline{\mathring{\rho}_\alpha \mathring{V}'_i} \equiv \overline{\rho_\alpha \mathring{V}'_i} = -\overline{V_{\alpha j}\frac{\partial \rho_\alpha}{\partial \xi_j}(\operatorname{div}\mathbf{V}_\alpha)^{-1}\mathring{V}'_i}$$

$$\approx -\overline{(\operatorname{div}\mathbf{V}_\alpha)^{-1}}\frac{\partial \bar{\rho}_\alpha}{\partial x_j}\overline{V_{\alpha j}\mathring{V}'_i} - \overline{V_{\alpha j}}\frac{\partial \bar{\rho}_\alpha}{\partial x_j}\overline{(\operatorname{div}\mathbf{V}_\alpha)^{-1}\mathring{V}'_i}. \tag{4.6.22}$$

Expressing $\mathbf{V}_\alpha$ by (4.1.14):

$$V_{\alpha i} = (V_{\alpha i} - V'_i) + V'_i = -\frac{D_{\alpha\beta}}{\rho_\alpha}\frac{\partial \rho_\alpha}{\partial \xi_i} + V'_i \tag{4.6.23}$$

and neglecting the effect of the second term on the right-hand side of (4.6.22), Bear and Bachmat (1965, 1967) obtain:

$$\overline{\rho_\alpha \mathring{V}'_i} \approx -\overline{\mathring{V}'_i \mathring{V}'_j (\operatorname{div}\mathbf{V}_\alpha)^{-1}}\frac{\partial \bar{\rho}_\alpha}{\partial x_j} = -D_{ij}\frac{\partial \bar{\rho}_\alpha}{\partial x_j} \tag{4.6.24}$$

where D_{ij}, to be considered in detail below, is defined by:

$$D_{ij} = \overline{\mathring{V}'_i \mathring{V}'_j}\,\overline{(\operatorname{div} \mathbf{V}_\alpha)^{-1}}. \tag{4.6.25}$$

By inserting (4.6.20) through (4.6.25) in (4.6.19), and assuming $n = \text{const}$, we obtain:

$$\partial \bar{\rho}_\alpha / \partial t = \partial[(D_{ij} + D_{\alpha\beta}\overline{T^*_{ij}})\partial \bar{\rho}_\alpha / \partial x_j]/\partial x_i - \partial(\bar{\rho}_\alpha \overline{V'_i})/\partial x_i. \tag{4.6.26}$$

If $n \neq \text{const}$, the term $-\ (\rho_\alpha \overline{V'_i}/n)\ \partial n/\partial x_i$ should be added on the right-hand side of (4.6.26). In view of (4.6.12) for an incompressible fluid, equation (4.6.26) becomes:

$$\partial \bar{\rho}_\alpha / \partial t = \partial[(D_{ij} + D_{\alpha\beta}\overline{T^*_{ij}})\partial \bar{\rho}_\alpha / \partial x_j]/\partial x_i - \bar{V}'_i\, \partial \bar{\rho}_\alpha / \partial x_i \tag{4.6.27}$$

or, in vector and tensor notation:

$$\partial \bar{\rho}_\alpha / \partial t = \operatorname{div}[(\underline{\mathbf{D}} + D_{\alpha\beta}\overline{\underline{\mathbf{T}}^*}) \operatorname{grad} \bar{\rho}_\alpha] - \bar{\mathbf{V}}' \operatorname{grad} \bar{\rho}_\alpha. \tag{4.6.28}$$

In (4.6.28), $\underline{\mathbf{D}}$ (a second-rank tensor relating the dispersive flux vector to grad $\bar{\rho}_\alpha$; dim L^2T^{-1}) is the *coefficient of mechanical dispersion*, $D_{\alpha\beta}\overline{\underline{\mathbf{T}}^*} \equiv \overline{\underline{\mathbf{D}}^*_{\alpha\beta}}$ (dim L^2T^{-1}) is the *coefficient of molecular diffusion in a porous medium*, and $\overline{\underline{\mathbf{T}}^*}$ (second-rank tensor; dimensionless) is the medium's *tortuosity*. $\underline{\mathbf{D}}$ is discussed in detail in paragraph 10.4.1; $\overline{\underline{\mathbf{T}}^*}$ is discussed in detail in section 4.8. Equations (4.6.26) through (4.6.28) may also be written in terms of the specific discharge $\mathbf{q}$ ($= \bar{\mathbf{V}}'/n$).

Equation (4.6.28) is the *equation of mass conservation of an α-species*, averaged over an REV of a porous medium. It is also called the *equation of hydrodynamic dispersion* (chap. 10).

4.6.3 Equation of Mass Conservation

Following the procedure outlined in paragraph 4.6.2, we obtain from (4.3.14) and (4.1.16):

$$\left\langle \frac{\partial \rho}{\partial t} \right\rangle = \left\langle -\frac{\partial}{\partial s}(\rho V^*_s) \right\rangle = \left\langle -\frac{\partial}{\partial s}(\rho V'_s) + D_{\alpha\beta}\frac{\partial \rho}{\partial s} \right\rangle. \tag{4.6.29}$$

Then, in a manner similar to that leading to (4.6.18), we obtain from (4.6.29):

$$\left\langle \frac{\partial \rho}{\partial t} \right\rangle = \left\langle \frac{\partial}{\partial \xi_i}\left[D_{\alpha\beta} T^*_{ij} \frac{\partial \rho}{\partial \xi_j}\right] - \frac{\partial}{\partial \xi_i}(\rho V'_i) \right\rangle \tag{4.6.30}$$

which is the *mass conservation equation for a liquid*, averaged over a channel's cross-section.

From (4.1.16) we obtain:

$$\langle \rho(V'_i - V^*_i) \rangle = \left\langle -D_{\alpha\beta} T^*_{ij} \frac{\partial \rho}{\partial \xi_j} \right\rangle. \tag{4.6.31}$$

This expression can also be obtained from (4.6.29) and (4.6.30). Whereas (4.1.16) governs molecular diffusion in a fluid continuum, equation (4.6.31) governs the flow in an elementary porous medium channel. It can easily be verified that (4.6.30) can be obtained from (4.6.18) by summing all values of α.

By averaging (4.6.30) over the entire void space of an REV, subject to the various assumptions listed in the previous paragraph (including $n = \text{const}$), or directly by summing (4.6.26) for all values of α, we obtain:

$$\partial\bar{\rho}/\partial t = \partial[(D_{ij} + D_{\alpha\beta}\overline{T^*_{ij}})\partial\bar{\rho}/\partial x_j]/\partial x_i - \partial(\bar{\rho}\bar{V}'_i)/\partial x_i \tag{4.6.32}$$

which is the *equation of mass conservation of an incompressible inhomogeneous fluid,* averaged over an REV of a porous medium. If $n \neq \text{const}$, the term $-\overline{(\rho V'_i/n)}\,\partial n/\partial x_i$ should be added on the right-hand side of (4.6.32). In (4.6.32), we have:

$$\overline{\mathring{\rho}\mathring{V}'_i} = -D_{ij}\,\partial\bar{\rho}/\partial x_j \tag{4.6.33}$$

where D_{ij} defined by (4.6.33) is related to the liquid system as a whole and may differ from the D_{ij} defined by (4.6.25) for the α component alone.

We may also start from the following equations of mass conservation, averaged over a channel's cross-section:

$$\langle\partial\rho/\partial t\rangle = -\langle\partial(\rho V^*_i)/\partial\xi_i\rangle. \tag{4.6.34}$$

When averaged over an REV, subject to the assumption of $n = \text{const}$, equation (4.6.34) yields:

$$\partial\bar{\rho}/\partial t = -\partial(\overline{\rho V^*_i})/\partial x_i \tag{4.6.35}$$

and:

$$\partial\bar{\rho}/\partial t = \partial\left(D^*_{ij}\frac{\partial\bar{\rho}}{\partial x_j}\right)\Big/\partial x_i - \partial(\bar{\rho}\overline{V^*_i})/\partial x_i \tag{4.6.36}$$

where the dispersive flux due to fluctuations in the mass averaged velocity is:

$$\overline{\mathring{\rho}\mathring{V}^*_i} = -D^*_{ij}\,\partial\bar{\rho}/\partial x_j. \tag{4.6.37}$$

By comparing (4.6.32) with (4.6.36), neglecting the difference between the dispersive fluxes due to fluctuations in the volume-averaged velocity and those due to fluctuations in the mass-averaged velocity (with respect to the difference in fluxes resulting from the averaged velocities themselves), we obtain:

$$\rho\,(\overline{V^*_i} - \overline{V'_i}) = -D_{\alpha\beta}\overline{T^*_{ij}}\,\partial\bar{\rho}/\partial x_j. \tag{4.6.38}$$

The same result can be obtained directly by averaging (4.1.16) following the two-step averaging procedure employed here, and neglecting the same difference in dispersive fluxes.

The mass conservation equation (4.6.36) may also be obtained by summing (4.6.26) for all values of α. One should note that $\sum \mathbf{J}^*_\alpha = 0$ whereas $\sum \mathbf{J}'_\alpha = -D_{\alpha\beta}\nabla\rho$, where $\mathbf{J}^*_\alpha$ and $\mathbf{J}'_\alpha$ are defined by (4.1.12) and (4.1.14), respectively.

In the absence of motion, $D_{ij} = 0$, $\bar{\mathbf{V}}' = 0$ and (4.6.32) becomes the mass conservation equation describing molecular diffusion in a porous medium:

$$\partial\bar{\rho}/\partial t = \partial(D_{\alpha\beta}\overline{T^*_{ij}}\,\partial\bar{\rho}/\partial x_j)/\partial x_i. \tag{4.6.39}$$

For $\bar{\rho}\overline{\mathbf{V}^*} \gg \overline{\mathring{\rho}\mathring{\mathbf{V}}^*}$, i.e., the mass flux carried by the average flow is much larger than the mass flux resulting from velocity fluctuations, we obtain from (4.6.35):

$$\partial\bar{\rho}/\partial t = -\,\partial(\bar{\rho}\overline{V^*}_i)/\partial x_i. \tag{4.6.40}$$

4.7 Equation of Motion

The starting point in this case is the equation of conservation of linear momentum of a fluid system (4.3.30):

$$\partial(\rho\mathbf{V}^*)/\partial t + \operatorname{div}(\rho\mathbf{V}^*\mathbf{V}^*) + \operatorname{div}(p\underline{\boldsymbol{\delta}}) - \operatorname{div}\underline{\mathbf{P}} = \sum_{(\alpha)} \rho_\alpha \mathbf{f}_\alpha \tag{4.7.1}$$

where the body force $\sum_{(\alpha)}\rho_\alpha\mathbf{f}_\alpha$ is due to gravity:

$$\sum_{(\alpha)} \rho_\alpha\mathbf{f}_\alpha = \rho\mathbf{g} = -\,\rho g\mathbf{1z}. \tag{4.7.2}$$

In view of the mass conservation equation (4.3.14), (4.7.1) may also be written as:

$$\rho\,\partial\mathbf{V}^*/\partial t + (\rho\mathbf{V}^*\cdot\operatorname{grad})\mathbf{V}^* + \operatorname{div}(p\underline{\boldsymbol{\delta}}) - \operatorname{div}\underline{\mathbf{P}} = \sum_{(\alpha)}\rho_\alpha\mathbf{f}_\alpha. \tag{4.7.3}$$

In these equations $\operatorname{div}\underline{\mathbf{P}}$ ($\equiv \partial P_{ij}/\partial x_j$) represents the density of the force due to the fluid's viscosity, which resists the motion. It is a viscous force per unit volume of fluid. Let us denote this force by the symbol $\mathbf{R}$.

Written for the velocity component V^*_s at a point inside an elementary channel, (4.7.3) becomes:

$$\rho\,\partial V^*_s/\partial t = -\,(\rho\mathbf{V}^*\cdot\operatorname{grad})V^*_s - \rho g\,\partial z/\partial s - \partial p/\partial s + R_s. \tag{4.7.4}$$

At this point in the development of the motion equation we introduce (following Bear and Bachmat 1966, 1967) an *assumption* (equivalent, in a way, to the *constitutive assumptions* discussed in sec. 4.3) that the force resisting the motion of a particle at a point inside an elementary channel is proportional to the particle's mass-averaged velocity at that point, and is acting in a direction opposite to that of the local velocity vector:

$$\mathbf{R} = -\,(\mu/B)\mathbf{V}^* \tag{4.7.5}$$

where B (dim L^2) may be called *conductance of a channel* at the considered point. For example, in laminar flow (Poiseuille's law) at a point at a distance r from the axis of a straight circular tube of radius r_0, $B = (r^2_0 - r^2)/4$. Equation (4.7.5) may be derived by dimensional analysis. B is a function of the shape of the channel's cross-section, and of the location of the considered point on it.

From (4.7.4) and (4.7.5) we obtain:

$$\rho\,\partial V^*_s/\partial t + \rho\mathbf{V}^*\cdot\operatorname{grad}V^*_s = -\,\rho g\,\partial z/\partial s - \partial p/\partial s - \mu V^*_s/B. \tag{4.7.6}$$

We now introduce the approximation:

$$\mathbf{V}^*\cdot\operatorname{grad}V^*_s \approx 0, \tag{4.7.7}$$

i.e., we neglect the convective acceleration in (4.7.6), assuming it to be much smaller than $\mu V^*_s/\rho B$. Equation (4.7.6) then reduces to:

$$\rho\, \partial V^*_s/\partial t = -\rho g\, \partial z/\partial s - \partial p/\partial s - \mu V^*_s/B. \tag{4.7.8}$$

Multiplying both sides of (4.7.8) by $B\, d\sigma/ds$, we obtain, in view of (4.5.13):

$$B\rho\, \partial V^*/\partial t + \mu V^* = -B(\partial p/\partial s + \rho g\, \partial z/\partial s)(\partial\sigma/\partial s). \tag{4.7.9}$$

Passing to the fixed coordinate system x_i, by multiplying by $d\xi_i/d\sigma$ and averaging over the cross-section, we obtain:

$$\left\langle V^*_i + \frac{B\rho}{\mu}\frac{\partial V^*_i}{\partial t}\right\rangle = -\left\langle \frac{BT^*_{ij}}{\mu}\left(\frac{\partial p}{\partial \xi_j} + \rho g\frac{\partial z}{\partial \xi_j}\right)\right\rangle. \tag{4.7.10}$$

By averaging (4.7.10) over the REV of the porous medium model, we obtain:

$$\overline{V^*_i} + \overline{B\left(\frac{\rho}{\mu}\right)\frac{\partial V^*_i}{\partial t}} = -\overline{BT^*_{ij}}\left[\overline{\frac{1}{\mu}\frac{\partial p}{\partial x_j}} + \overline{\left(\frac{\rho}{\mu}\right)}g\frac{\partial z}{\partial x_j}\right]. \tag{4.7.11}$$

In deriving (4.7.11), we have assumed that:

(a) no correlation exists between the liquid's properties at a given point and properties of the medium at that point. This means, for example, that $\overline{\mathring{B}(\mathring{\rho/\mu})} = 0$.

(b) No correlation exists between the instantaneous values of liquid properties at a given point and the temporal rate of change of the velocity at that point (i.e., $\overline{(\mathring{\rho/\mu})(\partial \mathring{V}^*_i/\partial t)} = 0$;

(c) no correlation exists between the pressure gradient in the liquid at a given point and the direction of the streamline at that point or the position of the considered point on the cross-section. This means that $\overline{B\mathring{T}^*_{ij}(\partial \mathring{p}/\partial \xi_j)} = 0$.

For the low solute concentrations considered here, and with $|\mathring{\rho}/\bar{\rho}| \ll 1$, $|\mathring{\mu}/\bar{\mu}| \ll 1$, we may assume (Bear and Bachmat 1966, 1967) that $\overline{(\rho/\mu)} \cong \bar{\rho}/\bar{\mu}$ and $\overline{(1/\mu)} \cong 1/\bar{\mu}$. By inserting these approximations in (4.7.11), we obtain:

$$\overline{V^*_i} + \frac{\bar{B}\bar{\rho}}{\bar{\mu}}\frac{\partial \overline{V^*_i}}{\partial t} = -\frac{k_{ij}}{n\bar{\mu}}\left(\frac{\partial \bar{p}}{\partial x_j} + \bar{\rho} g\frac{\partial z}{\partial x_j}\right) \tag{4.7.12}$$

where we have denoted:

$$k_{ij} = n\overline{BT^*_{ij}};\qquad T^*_{ij} = \left(\frac{d\sigma}{ds}\right)^2 T_{ij} = \left(\frac{d\sigma}{ds}\right)^2 \frac{d\xi_i}{d\sigma}\frac{d\xi_j}{d\sigma}. \tag{4.7.13}$$

Equation (4.7.12) is the *average equation of motion for an inhomogeneous fluid in laminar flow through a porous medium.*

The same equation in terms of the volumetric average velocity $\bar{\mathbf{V}}'$ is:

$$\overline{V'_i} - D_{\alpha\beta}\overline{T^*_{ij}}\,\overline{\frac{1}{\rho}\frac{\partial \rho}{\partial x_j}} + \frac{\bar{B}\bar{\rho}}{\bar{\mu}}\frac{\partial}{\partial t}\left[\overline{V'_i} - D_{\alpha\beta}\overline{T^*_{ij}}\,\overline{\frac{1}{\rho}\frac{\partial \rho}{\partial x_j}}\right] = \frac{-k_{ij}}{n\bar{\mu}}\left(\frac{\partial \bar{p}}{\partial x_j} + \bar{\rho} g\frac{\partial z}{\partial x_j}\right). \tag{4.7.14}$$

Assuming that the flux due to molecular diffusion resulting from a density gradient is much smaller than the convective flux, i.e., $|D_{\alpha\beta}\underline{\mathbf{T}}^*\,\mathrm{grad}\,\bar{\rho}| \ll |\bar{\rho}\bar{\mathbf{V}}^*|$, we obtain, in view of (4.6.38), an equation of motion identical to (4.7.12), in which $\mathbf{V}^*$ is replaced by $\mathbf{V}'$. By multiplying (4.7.12) by n, we obtain:

$$q_i + (\bar{B}/\bar{\nu})\, \partial q_i/\partial t = -\,(k_{ij}/\bar{\mu})(\partial \bar{p}/\partial x_j + \bar{\rho} g\, \partial z/\partial x_j) \tag{4.7.15}$$

where $\bar{\nu}$ is the average kinematic viscosity ($\simeq \bar{\mu}/\bar{\rho}$) and $\mathbf{q}$ is the specific discharge. Thus, the motion equation derived in the form of (4.7.12) or (4.7.15) is valid for an inhomogeneous fluid in laminar flow through an anisotropic porous medium. The coefficient k_{ij}—the medium's permeability—will be discussed in detail below.

The second term on the left-hand side of the motion equation (4.7.12) expresses the local acceleration. For flows in which the local inertial forces can be neglected with respect to the viscous (resistance) forces, equations (4.7.12) and (4.7.15) become:

$$\overline{V^*}_i = -\,(k_{ij}/n\bar{\mu})(\partial \bar{p}/\partial x_j + \bar{\rho} g\, \partial z/\partial x_j) \tag{4.7.16}$$

$$q_i = -\,(k_{ij}/\bar{\mu})(\partial \bar{p}/\partial x_j + \bar{\rho} g\, \partial z/\partial x_j). \tag{4.7.17}$$

Several authors (e.g., Polubarinova-Kochina 1952, 1965) have indeed shown that the local acceleration term in the motion equation tends very rapidly (e.g., within a fraction of a second) to zero after the onset of flow. Hence, one is justified in deleting this term from the equation of motion. The equations of motion (4.7.16) and (4.7.17) have been derived by assuming that the inertial forces (both convective and local) are much smaller than the viscous force. Prandtl and Tietjens (1934) refer to such a motion as *creeping motion*. Such a motion is characterized by a low *Reynolds number* that expresses the ratio between the inertial force and the viscous (frictional) force.

It is also possible to follow steps similar to those described above and to derive a motion equation in which the inertial terms remain (Bachmat 1965).

A discussion on the parameters $\underset{\sim}{\mathbf{k}}$ and $\underset{\sim}{\overline{\mathbf{T}^*}}$ appearing in this paragraph is presented in the following section.

For a homogeneous incompressible fluid, $\bar{\rho} = \text{const}$, $\bar{\mu} = \text{const}$ and we may write the motion equation (4.7.17) in terms of the piezometric head $\bar{\varphi} = z + \bar{p}/\bar{\gamma}$:

$$q_i = -\,(k_{ij}\,\bar{\gamma}/\bar{\mu})\; \partial \bar{\varphi}/\partial x_j. \tag{4.7.18}$$

A detailed discussion of this equation—which is the extension of Darcy's experimental law to three-dimensional flow in anisotropic media—is given in chapter 5.

4.8 Tortuosity and Permeability

4.8.1 Relationship Between Tortuosity and Permeability

In (4.6.18) we defined T_{ij} and T^*_{ij}. For a porous medium model composed of straight capillary tubes, $d\sigma/ds = 1$ and $T^*_{ij} \equiv T_{ij}$. In what follows, we shall, for the sake of simplicity, focus the discussion on T_{ij}, although the discussion can easily be extended to T^*_{ij}.

Let us now consider in detail the interpretation of T_{ij}, following Bear and Bachmat (1966, 1967). This dimensionless parameter is an *operator* that transforms the components F_i (in the coordinate system x_i) of any external driving force (e.g., ∇p, $-g\mathbf{1z}$, ∇C) acting on a small fluid volume at a point inside the void space, into the components $F^{(\sigma)}{}_j$ of the force's projection on the direction of the streamline at that point. The transformation is carried out in two steps. First the given force is projected on the direction of the streamline, i.e., $\mathbf{F} \to \mathbf{F} \cdot \mathbf{1\sigma}$ or $\mathbf{F} \to F \cos(\mathbf{F}, \mathbf{1\sigma})$. In terms of components in the x_i system, this can be written as:

$$F_i \to F_i \, d\xi_i/d\sigma. \tag{4.8.1}$$

Then, the components (in the x_i system) of this projection are obtained:

$$F \cos(\mathbf{F}, \mathbf{1\sigma}) \to F \cos(\mathbf{F}, \mathbf{1\sigma}) \cos(\mathbf{1\sigma}, \mathbf{1x}_j).$$

Or, in terms of components in the x_i system:

$$F_i \frac{d\xi_i}{d\sigma} \to F_i \frac{d\xi_i}{d\sigma}\frac{d\xi_j}{d\sigma} \equiv F_i T_{ij} \equiv F^{(\sigma)}{}_j. \tag{4.8.2}$$

By this definition:

$$F_i T_{ij} = T_{ji} F_i = F^{(\sigma)}{}_j \tag{4.8.3}$$

which means that T_{ij} is a *symmetrical, second-rank matrix.*

By averaging (4.8.2) over an REV, assuming that the components F_i are statistically independent of T_{ij}, we obtain:

$$\overline{F^{(\sigma)}{}_j} = \overline{T_{ji}}\,\overline{F_i} = \overline{F_i}\,\overline{T_{ij}}; \qquad \overline{T_{ij}} = \overline{T_{ji}}. \tag{4.8.4}$$

$\overline{T_{ij}}$ is a *nonrandom porous medium operator* (property) that transforms the average components of an external force acting at a physical point of a porous medium into the average components (in the x_i system) of its projections along the streamlines. Here we shall assume that the determinant of $\overline{T_{ij}}$ does not vanish so that the inverse matrix of $\overline{T_{ij}}$ exists.

In the special case where $\overline{\mathbf{F}^{(\sigma)}}$ is colinear with $\overline{\mathbf{F}}$, the transformation $\overline{T_{ij}}$ causes a change only in the magnitude of $\overline{\mathbf{F}}$:

$$\overline{F^{(\sigma)}{}_j} = \overline{T}\,\overline{F_j}; \qquad \overline{T_{ij}} = \tfrac{1}{3}\overline{T_{kk}}\delta_{ij} = \overline{T}\delta_{ij} \tag{4.8.5}$$

where δ_{ij} is the Kronecker delta.

A direction in the porous medium for which (4.8.5) is valid is called a ***principal direction*** of the transformation $\overline{T_{ij}}$; $\overline{\mathbf{F}}$ and $\overline{T}$ corresponding to this direction are the ***eigenvector*** and the ***eigenvalue*** of the transformation, respectively. A porous medium is said to be ***isotropic*** at a given point, with respect to the transformation $\overline{T_{ij}}$, if (4.8.5) is valid for every direction of $\overline{\mathbf{F}}$ acting at that point. In this case, every direction in space is a principal direction. In all other cases, the porous medium is ***anisotropic*** with respect to $\overline{T_{ij}}$.

In the general theory of linear transformations it is shown that any symmetrical transformation, such as $\overline{T_{ij}}$, always possesses at least one group of three eigenvectors, which are perpendicular to each other and may be used as a base, and three corresponding eigenvalues. The directions of these three eigenvectors are the principal directions of the anisotropic medium with respect to the transformation considered. If $\boldsymbol{\sigma}^{(1)}$, $\boldsymbol{\sigma}^{(2)}$ and $\boldsymbol{\sigma}^{(3)}$ are three such base vectors, then we obtain for (4.8.4):

$$\overline{F^{(\sigma)}}_1 = \overline{T^{(1)}}\,\overline{F}_1; \qquad \overline{F^{(\sigma)}}_2 = \overline{T^{(2)}}\,\overline{F}_2; \qquad \overline{F^{(\sigma)}}_3 = \overline{T^{(3)}}\,\overline{F}_3 \tag{4.8.6}$$

with $\overline{T_{ij}}$ in this system taking the form (Prager 1961):

$$\overline{T_{ij}} = \overline{T^{(1)}}\sigma^{(1)}{}_i\sigma^{(1)}{}_j + \overline{T^{(2)}}\sigma^{(2)}{}_i\sigma^{(2)}{}_j + \overline{T^{(3)}}\sigma^{(3)}{}_i\sigma^{(3)}{}_j \tag{4.8.7}$$

or, using the definition of $\overline{T_{ij}}$ given in (4.6.18):

$$\begin{aligned}\overline{T_{ij}} &= \overline{\left(\frac{d\xi_1}{d\sigma}\right)^2}\frac{dx_i}{dx_1}\frac{dx_j}{dx_1} + \overline{\left(\frac{d\xi_2}{d\sigma}\right)^2}\frac{dx_i}{dx_2}\frac{dx_j}{dx_2} + \overline{\left(\frac{d\xi_3}{d\sigma}\right)^2}\frac{dx_i}{dx_3}\frac{dx_j}{dx_3} \\ &= \overline{T^{(1)}}\delta_{i1}\delta_{j1} + \overline{T^{(2)}}\delta_{i2}\delta_{j2} + \overline{T^{(3)}}\delta_{i3}\delta_{j3}.\end{aligned} \tag{4.8.8}$$

This can also be written in matrix form:

$$\overline{T_{ij}} = \begin{pmatrix} \overline{T^{(1)}} & 0 & 0 \\ 0 & \overline{T^{(2)}} & 0 \\ 0 & 0 & \overline{T^{(3)}} \end{pmatrix}. \tag{4.8.9}$$

In the general case of anisotropy, there exists one and only one triplet $\boldsymbol{\sigma}^{(1)}$, $\boldsymbol{\sigma}^{(2)}$ and $\boldsymbol{\sigma}^{(3)}$. For an isotropic medium, $\overline{T^{(1)}} = \overline{T^{(2)}} = \overline{T^{(3)}} = \overline{T}$, so that

$$\overline{T_{ij}} = \overline{T}\delta_{ij} = \begin{pmatrix} \overline{T} & 0 & 0 \\ 0 & \overline{T} & 0 \\ 0 & 0 & \overline{T} \end{pmatrix} = \overline{T}\begin{pmatrix} 1 & 0 & 0 \\ 0 & 1 & 0 \\ 0 & 0 & 1 \end{pmatrix}. \tag{4.8.10}$$

In this case any three orthogonal directions are principal directions of the transformation $\overline{T_{ij}}$.

The local transformation $\overline{T_{ij}}$ is a *second-rank symmetric tensor*. This follows from the observation that when the coordinate system, x_i, is transformed into another one, x'_i, we have

$$T_{i'j'} = \frac{\partial x_{i'}}{\partial x_k}\frac{\partial x_{j'}}{\partial x_l}T_{kl} \tag{4.8.11}$$

which is the law of transformation of second-rank tensors. Hence, the average $\overline{T_{ij}}$ is also a second-rank symmetrical tensor. In general curvilinear coordinates y^i, we have:

$$T_{ij} = \frac{dy^i}{d\sigma}\frac{dy^j}{d\sigma}; \qquad T^{i'j'} = \frac{\partial y^{i'}}{\partial y^k}\frac{\partial y^{j'}}{\partial y^l}T^{kl}, \tag{4.8.12}$$

i.e., T^{ij} are the contravariant components of a second-order tensor.

The coefficient $\overline{nBT^*{}_{ij}} = k_{ij}$ in (4.7.12) is the *medium's permeability*. This is an average (or macroscopic) medium property that measures the ability of the porous medium to transmit fluid through it. The discussion presented on T_{ij} and T^*_{ij} above is also applicable to BT^*_{ij} and its tensorial rank, as $(d\sigma/ds)^2$ is invariant at a point. One may refer to BT^*_{ij} as the *directional conductivity* of the channel at some point in it (Bear and Bachmat 1966, 1967). For BT^*_{ij} we have:

$$\overline{BT^*_{ij}} = \bar{B}\overline{T^*_{ij}} + \overline{\mathring{B}\mathring{T}^*_{ij}}. \tag{4.8.13}$$

Assuming that no correlation exists between the conductance B at a point on a channel's axis and the direction of the streamline at that point, i.e., $\overline{\mathring{B}\mathring{T}^*_{ij}} = 0$, then the principal directions of $\overline{T^*_{ij}}$ and $\overline{BT^*_{ij}}$ coincide. With this assumption we also have:

$$\overline{BT^*_{ij}} = \bar{B}\overline{T^*_{ij}} \tag{4.8.14}$$

where $\bar{B}$ (dim L^2) is the *average medium conductance.*

When the coordinate axes coincide with the principal directions of k_{ij}, we have:

$$k_{ij} = \begin{cases} k_{ij} & \text{for} \quad i = j \\ 0 & \text{for} \quad i \neq j. \end{cases} \tag{4.8.15}$$

If the medium is isotropic with respect to k_{ij}, the motion equation (4.7.17) becomes:

$$q_i = -\frac{k\bar{\gamma}}{\bar{\mu}}\left(\frac{1}{\bar{\gamma}}\frac{\partial p}{\partial x_j} + \frac{\partial z}{\partial x_j}\right); \qquad \bar{\gamma} = \bar{\rho}g \tag{4.8.16}$$

where

$$k \equiv \tfrac{1}{3}k_{ii} = k_{11} = k_{22} = k_{33}.$$

For a homogeneous fluid (e.g., in ground water hydrology), it is convenient to work with the *hydraulic conductivity* K (sec. 5.5):

$$\underset{=}{\mathbf{K}} = \underset{=}{\mathbf{k}}\bar{\gamma}/\bar{\mu}; \qquad K_{ij} = k_{ij}\bar{\gamma}/\bar{\mu} \tag{4.8.17}$$

which is a combined property of the medium and the fluid. Because $\bar{\gamma}/\bar{\mu}$ is invariant, $\underset{=}{\mathbf{K}}$ is also a second-rank symmetrical tensor.

Thus, by the analysis of the porous medium model suggested by Bear and Bachmat (1966, 1967), the permeability of a porous medium is shown to depend on three, more elementary, porous medium properties: porosity n, average medium conductance $\bar{B}$ and average parameter $\overline{T^*_{ij}}$. $\bar{B}$ is related to the cross-sections of the elementary channels through which the flow takes place. Although its dimensions are those of length squared, one should recognize that, at least as far as the present porous medium model is concerned, the characteristic length here is one characterizing the cross-section of a channel and not its length. In paragraph 10.4.2, a characteristic channel length $\bar{L}$ is shown to be another elementary medium parameter that affects the medium's dispersivity.

The average (or macroscopic) medium parameter $\overline{T_{ij}}$ (or $\overline{T^*_{ij}}$) may be interpreted as the *medium's tortuosity*. One should recall that the discussion here is restricted to the case of *laminar flow of a Newtonian fluid* in a porous medium, visualized as a network of interconnected narrow channels through which flow takes place. In such a model, every channel is a stream tube fixed in space; the streamlines in it have a fixed geometry. To understand the reason for using the term tortuosity for $\overline{T^*_{ij}}$, let us begin by referring to Carman (1937) who proposed a *tortuosity factor* when studying the relationship among the average velocity in a tortuous capillary tube, the piezometric head difference between the ends of the tube, and the straight line distance between its ends. Carman (1937) introduces the effect of tortuosity in two ways:

(a) as it affects the velocity: let the direction of the straight line of length L, connecting the two ends of a tortuous tube of length L_e, be defined as the direction s, and the projection on the direction s of the average velocity V in the tube (i.e., in the direction tangential to the tube's axis) be u_s. Even if $|V|$ is constant, the component u_s varies. Carman (1937) suggests that if $\bar{V}$ is the magnitude of the average tangential velocity, the mean value $\bar{u}_s$ of u_s is given by:

$$\bar{u}_s = \bar{V}(L/L_e) \tag{4.8.18}$$

(b) as it affects the driving force: if $|\overline{\nabla\varphi}|_s$ is the absolute value of the component in the direction s of the mean hydraulic gradient, which acts as the driving force in the porous medium, then Carman (1937) suggests that:

$$|\overline{\nabla\varphi}|_s = \frac{\Delta\varphi}{L_e}\frac{L_e}{L}. \tag{4.8.19}$$

Then, starting from the extension of Poiseuille's law to flow in a noncircular tube:

$$V = \frac{R^2\gamma}{m\mu}\frac{\Delta\varphi}{L_e} \tag{4.8.20}$$

where R is the hydraulic radius of the tube and m is a numerical coefficient (shape factor) accounting for the noncircular shape of the tube, Carman (1937) obtains:

$$\bar{u}_s = \frac{\overline{R^2}\gamma(L/L_e)^2}{m\mu}|\overline{\nabla\varphi}|_s = \frac{(\overline{R^2}\gamma/m)T}{\mu}|\overline{\nabla\varphi}|_s \tag{4.8.21}$$

where $T = (L/L_e)^2 < 1$ is called the *tortuosity of the porous medium*. Recognizing that (L/L_e) in (4.8.21) stands for the average of the local cosine of the angle between the direction of the tube's axis and the direction s, the reason for referring to $\overline{T^*_{ij}} = \overline{(d\xi_i/ds)(d\xi_j/ds)}$ as the medium's tortuosity becomes obvious. Actually, in (4.8.21), T stands for the average value of T, and L_e for the average length of the streamlines in the void space of a porous medium between two parallel planes a distance L apart in the direction s. $\overline{T^*_{ij}}$, however, is an extension of the tortuosity concept to the more general case of three-dimensional flow and a medium that may be anisotropic. In fact, from the definition of permeability $k_{ij} = n\overline{BT^*_{ij}}$ (for $\overline{\mathring{B}\mathring{T}^*_{ij}} = 0$), it follows that the anisotropy is actually due to anisotropy in the nature of the tortuosity $\overline{T^*_{ij}}$. In the general case of $k_{ij} = n\overline{BT^*_{ij}}$, we must also consider the spatial distribution of B. To demonstrate this observation, consider a single capillary tube in the direction denoted by the unit vector $\mathbf{1s}$ in a field with average gradient $\mathbf{J} = -\nabla\varphi$. The vector $\mathbf{J}$ has three components J_i $(i = 1, 2, 3)$ along the three Cartesian axes. Since flow can take place *only along the tube*, each of the three components J_i contributes to the flow in the tube through its projections J_{si} along the tube's axis. Together, the driving force is $J_s = J_i \cos(\mathbf{1s}, \mathbf{1x}_i)$, where the summation convention is employed. We assume that the average velocity in the tube is governed by Poiseuille's law in the form $V_s = BJ_s$, where B is the tube's conductance. In order to obtain the components V_j of $\mathbf{V}$ in the direction of the axes x_j, we write:

$$V_j = V_s \cos(\mathbf{1s}, \mathbf{1x}_j),$$

so that

$$V_j = BJ_s \cos(\mathbf{1s}, \mathbf{1x}_j) = BJ_i \cos(\mathbf{1s}, \mathbf{1x}_i) \cos(\mathbf{1s}, \mathbf{1x}_j) \tag{4.8.22}$$

where $V = |\mathbf{V}|$, $J = |\mathbf{J}|$. Passing to the average over a large number of tubes in various directions, and introducing $\mathbf{q} = n\overline{\mathbf{V}}$, we obtain:

$$q_j = n\overline{BT^*_{ij}}J_i; \qquad T^*_{ij} = \cos(\mathbf{1s}, \mathbf{1x}_i) \cos(\mathbf{1s}, \mathbf{1x}_j). \tag{4.8.23}$$

By using this simple model, it was thus shown again that tortuosity is a *second-rank symmetrical tensor.*

Actually, some authors (e.g., Carman 1937) prefer to define tortuosity as $(L_e/L)^2 > 1$. This is only a matter of definition. On the other hand, the definition of tortuosity presented by some authors as L_e/L (ratio of flow-path length to sample length; Collins 1961) is a mistake. This last erroneous definition of tortuosity arises from failure to recognize effect (a) discussed above, and taking into account only effect (b) (Kozeny 1927; see discussion on the Kozeny and the Kozeny–Carman formulas in sec. 5.10). The tortuosity factor also appears in some of the equations proposed for flows at large Reynolds numbers (see sec. 5.11).

Estimates on the numerical value of the tortuosity factor $(L/L_e)^2$ are given by several authors. Irmay (in chap. 5 of Bear, Zaslavsky and Irmay 1968) mentions the value 0.4 for Carman's tortuosity factor L/L_e. However, if $(L/L_e)^2 = 0.4$, then $(L/L_e) = \sqrt{0.4} \approx 2/\pi$. The value $T = (L/L_e)^2 = 0.4$ was obtained by introducing $c_0 = \frac{1}{2}$ in the Kozeny–Carman equation (5.10.19) for permeability:

$$k = c_0 T n^3/M_s^2(1-n)^2 = Tn^3/2M_s^2(1-n)^2 \tag{4.8.24}$$

where n = porosity and M_s is the specific surface area (per unit volume of solid material), and comparing it with Carman's semiempirical formula (5.10.20):

$$k = n^3/5M_s^2(1-n)^2. \tag{4.8.25}$$

Carman (1937) mentions the empirical value $L/L_e = 1/\sqrt{2} = 0.71$, which is close to the values mentioned above. Other values mentioned in the literature for L/L_e vary in the range 0.56 to 0.8.

For an isotropic medium the tortuosity tensor discussed above reduces to a single scalar $\overline{T^*}$. It is shown in paragraph 10.4.3 (and by Bear and Bachmat 1966) that in this case $\overline{T} = \overline{T_{11}} = \overline{T_{22}} = \frac{1}{3}$, and $\overline{T_{12}} = \overline{T_{21}} = 0$. To obtain $\overline{T^*}$ we must multiply $\overline{T}$ by $\overline{(d\sigma/ds)^2} > 1$, which introduces the effect of divergence of streamlines in a porous medium (e.g., an unconsolidated medium) that cannot be visualized as made up of capillary tubes of constant cross-section. If we assume that the angle θ between a channel axis and a streamline varies between $\theta = 0°$ and $\theta = 90°$, such that $\theta = 45°$ can be chosen as a representative value, we obtain:

$$\overline{(d\sigma/ds)^2} = \overline{\sec^2\theta} = \sec^2 45° = 2.$$

Then:

$$\overline{T^*} = 2\overline{T} = \tfrac{2}{3} \tag{4.8.26}$$

can be used as an estimate of the tortuosity of unconsolidated media. Saffman (1960; see par. 10.3.2) also mentions the value $\frac{2}{3}$ as the ratio between the coefficient of longitudinal dispersion (D_L) and molecular diffusion (D_d) in a saturated medium at rest. In the next paragraph it will be shown that this ratio is equal to the tortuosity $\overline{T^*}$.

4.8.2 Tortuosity and other Transport Coefficients

The discussion in this paragraph is limited to transport phenomena which take place *only* in a single-phase fluid saturating the porous medium; no transport takes place through the solid phase. Also there is no interaction (e.g., in the form of ion-exchange, adsorption, etc.) between constituents of the fluid and the solid surfaces of the porous matrix.

For a homogeneous liquid ($\bar{\rho}$ = const) (4.7.18) becomes:

$$\mathbf{q} = -(\underline{\mathbf{k}}\bar{\rho}g/\bar{\mu}) \operatorname{grad} \bar{\varphi}; \qquad \bar{\varphi} = z + \bar{p}/\bar{\rho}g. \tag{4.8.27}$$

For molecular diffusion of a homogeneous liquid ($\bar{\rho}$ = const) at rest ($\mathbf{V}^* = 0$), we obtain from (4.1.12):

$$\mathbf{J}^*_\alpha = -D_d \operatorname{grad} \bar{\rho}_\alpha.$$

For the same liquid, the average molecular diffusion flux vector per unit area of voids is:

$$\bar{\mathbf{J}}^*_\alpha = -\underline{\mathbf{D}}^*_d \operatorname{grad} \bar{\rho}_\alpha \tag{4.8.28}$$

or, per unit area of porous medium:

$$\mathbf{q} = n\bar{\mathbf{J}}^* = -nD_d\overline{\underline{\mathbf{T}}^*} \cdot \operatorname{grad} \bar{\rho}_\alpha = -n\underline{\mathbf{D}}^*_d \cdot \operatorname{grad} \bar{\rho}_\alpha \tag{4.8.29}$$

where $\underline{\mathbf{D}}^*_d = \overline{\underline{\mathbf{T}}^*} D_d$ is the *coefficient of molecular diffusion in a porous medium.* It is sometimes better to refer to the product $n\underline{\mathbf{D}}^*_d$ as the coefficient of molecular diffusion in a porous medium. Equation (4.8.28) or (4.8.29) can be used to determine $\underline{\mathbf{T}}^*$.

Equations similar to (4.8.27) also describe transport phenomena in other branches of physics. For example, Ohm's law in electricity is:

$$\mathbf{i} = -K_E \operatorname{grad} E \tag{4.8.30}$$

where K_E is electrical conductivity, E is the electrical potential and $\mathbf{i}$ is the electric flux vector. If these transport phenomena take place in a porous medium (e.g., an electrolyte fully saturating an electrically *nonconducting porous medium*), considerations similar to those leading to (4.8.29) will yield:

$$\mathbf{q}_i = -nK_E \overline{\underline{\mathbf{T}}^*} \cdot \operatorname{grad} \bar{E} \tag{4.8.31}$$

where $\mathbf{q}_i$ is the average current flux per unit area of medium (voids + solid). Thus $n\overline{T^*}\lambda$ may be used to define any conductive property of a porous medium, where λ is the conductive property in the bulk (i.e., in the absence of a porous matrix). Examples are thermal conductivity, or electrical conductivity of the fluid saturating the medium, where the *porous matrix is assumed nonconductive*, and the flux is taken *per unit area of porous medium.* Equation (4.8.31) also permits an experimental determination of $\overline{\underline{\mathbf{T}}^*}$ (Klinkenberg 1941), by comparing the electrical resistivity of the porous material saturated by an electrolyte to the bulk resistivity of the same electrolyte. In reservoir engineering this ratio is often called the *formation factor.* Here it is equal to $1/n\overline{T^*}$. Additional comments on the formation factor and its relationship to tortuosity are given in the following paragraph.

4.8.3 Formation Factor and Resistivity Index (Amyx et al. 1960) in Reservoir Engineering

The *formation factor* F is one of several electrical properties of fluid-saturated porous media often used in studies of flow through porous media and in the interpretation of electric logs. For example, electrical resistivity is used widely as a measure of brine saturation in rocks.

As defined by Archie (1942), the formation factor F is the ratio of the resistivity (ρ_0) of the porous material (resistance of a uniform cube having sides of 1 unit length, to a one-dimensional electric current that enters through one face and leaves through an opposite face), saturated with an electrolyte, to the bulk resistivity (ρ_w) of the same electrolyte:

$$F = \rho_0/\rho_w. \tag{4.8.32}$$

A NaCl solution at a concentration higher than 10 gr/liter is usually used.

Consider a parallelepiped block of NaCl solution of length L and cross-sectional area A. Its resistance to flow along L is given by:

$$R_w = \rho_w L/A. \tag{4.8.33}$$

If we have, instead, a porous rock with the same external dimensions, and the solids assumed nonconductive, the resistance to the flow is R_0:

$$R_0 = \rho_0 L/A. \tag{4.8.34}$$

Because of the presence of the solid matrix, only part (A_e) of A is available for current flow. Also, the average length of current flow is increased from L to L_e (average length that an ion must traverse along its way from the inflow to the outflow face). Hence:

$$R_0 = \rho_w \frac{L_e}{A_e} = \rho_0 \frac{L}{A}; \qquad F = \frac{\rho_0}{\rho_w} = \frac{L_e/L}{A_e/A} = \frac{1/\sqrt{\overline{T^*}}}{A_e/A} \tag{4.8.35}$$

where $\overline{T^*}$ is the tortuosity of the porous medium.

If the void space contains a nonconductive hydrocarbon as well as water (chap. 9), the cross-sectional area A'_e ($< A_e$) available to the current flow is still smaller. Also, the actual length of pathlines is larger (L'_e instead of L_e). Hence, the resistance R_t of a parallelepiped block of porous medium is:

$$R_t = \rho_w \frac{L'_e}{A'_e} = \rho_t \frac{L}{A}; \qquad \rho_w = \rho_0 \frac{L}{A}\frac{A_e}{L_e}. \tag{4.8.36}$$

This is then used to define another electrical property of the porous rock—the resistivity index I:

$$I = \rho_t/\rho_0 = (A_e/A'_e)/(L_e/L'_e). \tag{4.8.37}$$

Thus both F and I are shown to be functions of effective path length and effective cross-sectional area. In order to relate these quantities to more elementary parameters of the porous medium, e.g., porosity n and tortuosity $\overline{T^*}$, idealized models of the porous medium must be introduced.

Figure 4.8.1a shows a model suggested by Wyllie and Spangler (1952). The cross-sectional areas of the pore openings vary along their length, but in such a manner that the sum of the areas of the pores is constant. L_e represents the average path

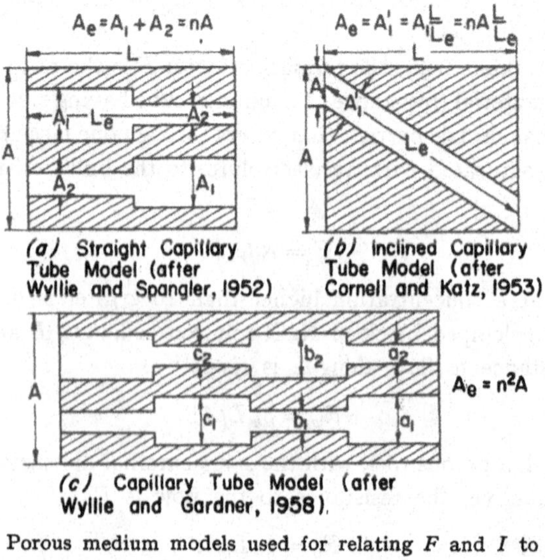

FIG. 4.8.1. Porous medium models used for relating F and I to n and T^*.

length through the pores. We obtain from (4.8.35):

$$F = \frac{L_e/L}{n} \quad \text{or} \quad F = \frac{1}{n\sqrt{\overline{T^*}}}; \qquad \overline{T^*} = \left(\frac{L}{L_e}\right)^2. \tag{4.8.38}$$

Obviously, the discrepancy between the expression for F here and at the end of the previous paragraph stems from the fact that here the differences between the direction of the average flow and that of the local velocities are not recognized.

If hydrocarbons fill part of the pore space so that S_w is the saturation of water (wetting phase; see chap. 9), then:

$$A'_e = nS_wA; \qquad I = (L'_e/L)/S_w. \tag{4.8.39}$$

In a second model (fig. 4.8.1b) suggested by Cornell and Katz (1953), the effective length L_e is greater than L. Here we do take into account the fact that the actual flow is in a direction different from that of L. We obtain (fig. 4.8.1b):

$$F = (L_e/L)^2/n = 1/n\overline{T^*}$$

which is identical to the relationship proposed at the end of the previous paragraph. For a medium partly saturated by a nonconductive hydrocarbon, we obtain:

$$A'_e = A''_1 = A_1L/L'_e = nS_w AL/L'_e; \qquad I = (L'_e/L_e)^2/S_w. \tag{4.8.40}$$

A third model, suggested by Wyllie and Gardner (1958), is shown in figure 4.8.1c. This model is discussed in detail in paragraph 9.3.2, where it is used for the derivation of an expression for relative permeability. For this model:

$$F = (L_e/L)/n^2 = 1/n^2\sqrt{\overline{T^*}}; \qquad \text{for} \quad L_e = L, \quad F = 1/n^2 \tag{4.8.41}$$

$$A'_e = (nS_w)^2A, \qquad L'_e = L_e = L; \qquad I = 1/S^2_w. \tag{4.8.42}$$

From this discussion it follows that the formation factor can be expressed in the following form:

$$F = C(\overline{T^*})n^{-m} \tag{4.8.43}$$

where $C(\overline{T^*})$ is some function of the tortuosity $\overline{T^*}$ and m is a function of the number of reductions in pore opening sizes or closed-off channels. The value of m has been shown above to vary in the range from 1 to 2. As $\overline{T^*} < 1$, it is suggested that C should be 1 or smaller (Amyx et al. 1960).

Both F and I depend, in one way or another, on the tortuosity $L/L_e = \sqrt{\overline{T^*}}$. However, as L_e cannot be measured directly, one must rely on empirical correlations based on laboratory measurements. For example, Winsauer et al. (1952) determine the ratio L/L_e by measuring the transit time of ions flowing through a saturated rock under an electric potential difference. By correlating their data with the separately measured product, Fn, they obtain:

$$\left(\frac{L_e}{L}\right)^{5/3} = Fn \quad \text{or} \quad F = \frac{1}{n(\overline{T^*})^{5/6}} \tag{4.8.44}$$

which is not much different from the expression $F = 1/n\overline{T^*}$ given at the end of the previous paragraph. The deviation from the theory is an indication of the deviation of the simplified model, on which the theory was based, from the actual pore system.

Archie (1942) correlates observed formation factors with porosity and permeability, and concludes that:

$$F = n^{-m} \tag{4.8.45}$$

where m is the *cementation factor*. According to Archie (1942), m varies from $1.8 \div 2.0$ for consolidated sandstone, to 1.3 for unconsolidated sand. For the resistivity of a partially saturated medium, Archie suggests:

$$I = \rho_t/\rho_0 = (S_w)^{-m_1} \tag{4.8.46}$$

where m_1 is a constant that he calls the *saturation exponent* ($m_1 \approx 2$ for clean sand). Williams (1950) suggests a value of $m_1 = 2.7$ for consolidated sand. Values of m_1 in the range 2–2.7 are mentioned by various authors.

Among the theoretical expressions for F presented in the literature we may mention that of Maxwell (1881), who studied the classical problem of flow of electricity through a two-phase region of uniform spheres imbedded in a fluid continuum. His result for a suspension of nonconducting spheres is:

$$F = (3 - n)/2n. \tag{4.8.47}$$

Maxwell's work is based on a solution of Laplace's equation for steady flow and on simple Ohm–Fourier type laws for the flux-potential relationships. The observed resistivity (reciprocal of conductivity) of the solid–fluid mixture is derived from the solution of these equations for each phase. It is interesting to note that (for the dilute system considered) F is independent of the resistivity of the fluid and the size of the spheres (as long as they are nonconducting).

A detailed discussion on tortuosity and the formation factor is also given by Carman (1956).

Obviously the electric conductivity of a heterogeneous solid–fluid system, where both solid and fluid are electrically conductive, is much more complicated and cannot be expected to be adequately represented as a combination of two resistances in parallel. Various researchers have proposed models based on a combination of resistances both in parallel and in series. Among the works on the formation factor when both the solid and the fluid are electrically conductive we may mention those of Wyllie and Rose (1950), Wyllie and Spangler (1952), Wyllie and Southwick (1954), Hill and Milburn (1956), de Witte (1955, 1957) and Waxman and Smits (1968). Obviously, once we have flow (in this case, of an electric current) both in the solid phase and in the fluid filling the void space, *the concept of tortuosity is no longer applicable* (as it is a purely geometrical property of the void space alone—visualized as an interconnected network of channels).

In addition to ionic conduction through the liquid (electrolyte), and electronic conduction through the solid phase (when the latter is conductive), there exists a third mode of electric conduction that further complicates the picture, i.e., *surface conduction*. This is a special form of ionic transport of electricity that takes place at the solid–liquid interface by means of exchange mechanisms. A discussion of this conduction, explained by the theory of the electrical double layer at the solid–liquid interface, is presented by Pfannkuch (1969). Under certain conditions, surface conduction can be an appreciable part of the total conduction of an electrolyte in a porous medium.

The formation factor F has been treated here in some detail because of its importance in reservoir engineering, and also because it is sometimes proposed as a first step in the determination of a medium's tortuosity. Tortuosity is introduced here as one of the fundamental geometrical porous medium properties. The three fundamental properties n, $\bar{B}$ and $\underline{\overline{\mathbf{T}^*}}$ make up the medium's permeability $\underline{\mathbf{k}}$ in the laminar flow regime considered here.

The various formulas presented above for the relationship among F, $\overline{T^*}$ and n seem sufficient to indicate the difficulties inherent in this approach. Another difficulty stems from the very definition of the ratio L/L_e. When L_e is defined as the average path of fluid particles through the porous medium, two interpretations of this average are possible. We may average the actual lengths of pathlines—disregarding the fact that fluid particles move along these pathlines at different velocities, following the velocity distribution across the elementary passages. In this case, tortuosity is a *purely geometrical concept*. Such average pathlines may be obtained (conceptually) by averaging the pathlines of all particles passing at a certain instant through a given porous medium cross-section. On the other hand, the average may be obtained by averaging lengths of pathlines of all particles passing through the area of a given medium cross-section during a given period of time. In this case tortuosity becomes a *kinematical property* as the velocity distribution, affected by the shape of the elementary channels and by the slip or no-slip boundary condition at the fluid–solid interface, will affect the resulting average pathline. Obviously, the method by which the tortuosity is determined experimentally also determines the type of tortuosity obtained. It may be of interest to note that tortuosity as obtained by the average procedure given in (10.4.29) is geometrical tortuosity, as directions of streamlines are not weighted according to the velocity distribution. It is as if in each channel we have piston-like flow at an average velocity $\langle V_s \rangle$ (obtained by the first averaging step) in the direction s of the channel's axis. The velocity distribution is taken care of by the factor B. This means that the tortuosity thus obtained is applicable also to other piston-like flows (e.g., molecular diffusion, electricity, etc.); obviously we refer only to flows taking place exclusively in the channel system, and not through the solid.

Several authors also extend the concept of tortuosity to the simultaneous flow of immiscible fluids (chap. 9) and to unsaturated flow (sec. 9.4). In this case, tortuosity must be defined for each fluid and it becomes a function of the saturation of that fluid. Just as an example, we may mention the empirical result derived by Burdine (1952):

$$T|_{S_w=1.0}/T|_{S_w} = \left[\frac{(S_w - S_{w0})}{(1 - S_{w0})}\right]^2 \tag{4.8.48}$$

where S_w is the wetting fluid saturation, and S_{w0} is the irreducible wetting fluid saturation.

Exercises

4.1 Use the following definitions:

ρ_α = mass of species α per unit volume of solution,
M_α = molecular weight of α species,
$\mathbf{V}_\alpha$ = velocity of α species with respect to fixed coordinates,
x_α = molar fraction $= (\rho_\alpha/M_\alpha)/\sum_{\beta=1}^{N}(\rho_\beta/M_\beta)$,

to determine a molar average velocity $\mathbf{V}''$.

4.2 With the definition $B = \sum_{(\alpha)} \omega_\alpha B_\alpha$, show that:

$$\sum_{(\alpha)}\left(\omega_\alpha \frac{DB_\alpha}{Dt}\right) = \frac{D^*B}{Dt} + \frac{1}{\rho}\sum_{(\alpha)} \operatorname{div}(\rho_\alpha B_\alpha \hat{\mathbf{V}}^*) - \sum_{(\alpha)} \frac{B_\alpha}{\rho}\left[\frac{\partial B_\alpha}{\partial t} + \operatorname{div}(\rho_\alpha \mathbf{V}_\alpha)\right]$$

where B_α is a property of the α species mass particles.

4.3 The specific volume of a species α is defined by $v_\alpha = 1/\rho_\alpha$. Show that

$$\operatorname{div} \mathbf{V}_\alpha = \frac{1}{v_\alpha}\frac{Dv_\alpha}{Dt}$$

Give an interpretation to this expression.

CHAPTER 5

The Equation of Motion of a Homogeneous Fluid

Although the basic equation governing fluid motion through porous media was developed in chapter 4 from first principles, it might be interesting to review briefly the elementary approach leading to this equation in the case of a homogeneous fluid, starting from the experiments of Darcy in 1856.

To some readers this simplified presentation is sufficient, as it gives the required equations of motion without going into their detailed, sometimes mathematically difficult, derivation (as was done in chap. 4). This is especially true in the case of practical engineers and/or ground water hydrologists who are interested only in the flow of a homogeneous fluid—namely water.

In the present chapter, all variables and parameters have meaning in a porous medium domain regarded as a continuum.

5.1 The Experimental Law of Darcy

In 1856, Henry Darcy investigated the flow of water in vertical homogeneous sand filters in connection with the fountains of the city of Dijon, France. Figure 5.1.1 shows the experimental set-up he employed (Darcy 1856). From his experiments, Darcy concluded that the rate of flow (volume per unit time) Q is (a) proportional to the constant cross-sectional area A, (b) proportional to $(h_1 - h_2)$ and (c) inversely

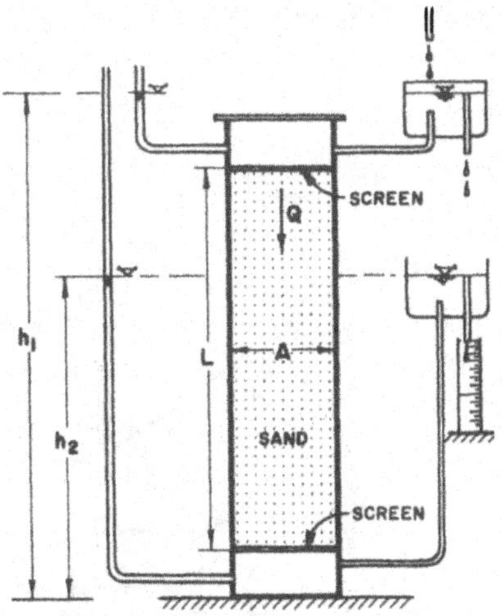

FIG. 5.1.1. Darcy's experiment.

proportional to the length L (as the symbols are defined in fig. 5.1.1). When combined, these conclusions give the famous Darcy formula:

$$Q = KA(h_1 - h_2)/L \tag{5.1.1}$$

where K is a coefficient of proportionality to be discussed below. The lengths h_1 and h_2 are measured with respect to some arbitrary (horizontal) datum level.

One can easily recognize here that h is the *piezometric head* and $h_1 - h_2$ is the difference in piezometric head across the filter of length L. As the piezometric head describes (in terms of head of water) the sum of pressure and potential energies of the fluid per unit weight, $(h_1 - h_2)/L$ is to be interpreted as *hydraulic gradient*. Denoting this gradient by J $(= (h_1 - h_2)/L)$ and defining the specific discharge, q, as discharge per unit cross-sectional area normal to the direction of flow ($q = Q/A$), we obtain:

$$q = KJ; \qquad J = (h_1 - h_2)/L \tag{5.1.2}$$

as another form of Darcy's formula (or Darcy's law).

Figure 5.1.2 shows how Darcy's law (5.1.1) may be extended to flow through an inclined homogeneous porous medium column:

$$Q = KA(\varphi_1 - \varphi_2)/L; \qquad q = K(\varphi_1 - \varphi_2)/L = KJ; \qquad \varphi_i = z + p_i/\gamma. \tag{5.1.3}$$

The *energy loss* $\Delta\varphi = \varphi_1 - \varphi_2$ is due to friction in the flow through the narrow tortuous paths of the porous medium. Actually, the total mechanical energy of the fluid includes a kinetic energy term. However, since, in general, changes in the piezometric head are much larger than changes in the kinetic energy head along the flow path, the kinetic energy is neglected when considering the head loss *along* the flow.

It is important to note that (5.1.3) states that the flow takes place from a higher piezometric head to a lower one and not from a higher pressure to a lower pressure. As shown in figure 5.1.2, $p_1/\gamma < p_2/\gamma$, i.e., the flow is in the direction of *increasing* pressure and *decreasing* head. It is *only* in the special case of horizontal flow, i.e.,

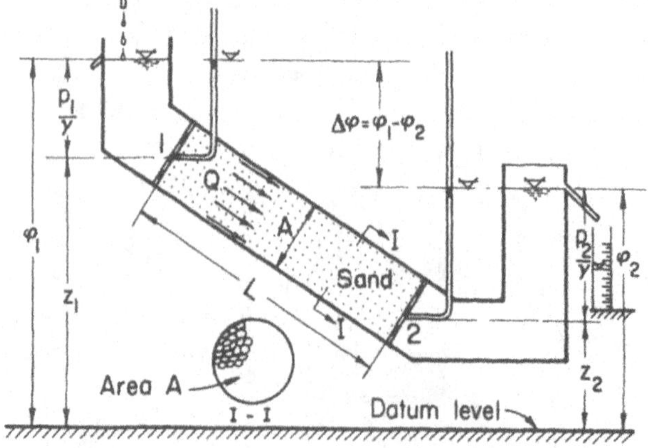

FIG. 5.1.2. Seepage through an inclined sand filter.

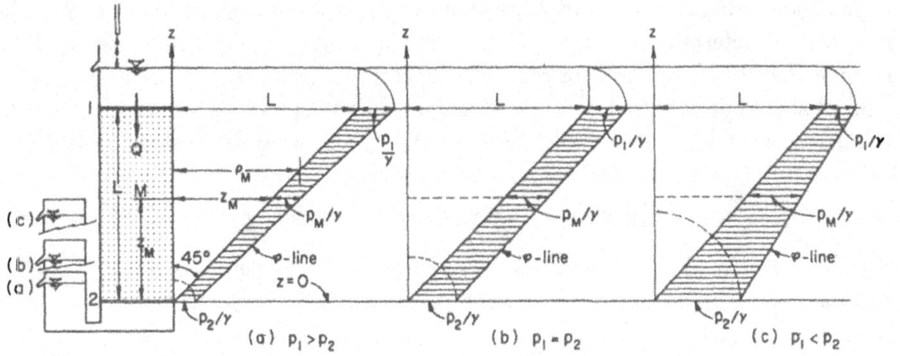

FIG. 5.1.3. Pressure variations in vertical downward flow through a homogeneous medium.

$z_1 = z_2$, that we may write:

$$Q = KA(p_1 - p_2)/\gamma L. \tag{5.1.4}$$

In the vertical flow shown in figure 5.1.1, we have:

$$Q = KA\,[(p_1 - p_2)/\gamma L + 1]. \tag{5.1.5}$$

Figure 5.1.3 shows several cases of vertical downward flow where (a) $p_1 > p_2$, (b) $p = \text{const}$ and (c) $p_1 < p_2$. The different cases are obtained by controlling the water elevation in the downstream constant-level reservoir. Case (b) is of special interest as it involves flow with the pressure maintained constant everywhere along the sand column.

The difference in piezometric head $\Delta\varphi = \varphi_1 - \varphi_2$ (i.e., in the energy of the water per unit weight, which may also be expressed as a difference in energy per unit mass, $\Delta\varphi' = \Delta(gz + p/\rho)$, ρ being the density of the fluid), is called the *driving head*. The hydraulic gradient $J = (\varphi_1 - \varphi_2)/L$ (or, $J' = gJ = g(\varphi_1 - \varphi_2)/L$) is the *driving force* here, causing the fluid to move toward the point of lower energy.

Actually, flow takes place through only part of the cross-sectional area A of the column in figure 5.1.2, the remaining part being occupied by the solid matrix of the porous medium. Since it was shown (par. 1.3.3) that the average areal porosity is equal to the volumetric porosity, n, the *average velocity*, V, through the column is:

$$V = Q/nA = q/n. \tag{5.1.6}$$

Sometimes, even in the flow of a single homogeneous fluid, part of the fluid in the pore space is immobile (or practically so). This may occur when the flow takes place in a fine-textured medium where adhesion (the attraction to the solid surface of the porous matrix of the molecules of fluid adjacent to it) is important or when the porous matrix includes a large number of dead-end pores. In this case one may define an *effective porosity with respect to the flow through the medium* n'_e $(< n)$ such that:

$$V = q/n'_e. \tag{5.1.7}$$

One should distinguish clearly between the specific discharge, q (to be used, for example, to determine volumes of fluid passing through given surfaces), and the average velocity (to be used, for example, for front, or particle, movements). Neither concept should be confused with the actual, or microscopic, local velocity of the fluid at microscopic points inside the pore space. Considering dimensions, one should note that q, V and K have the same dimensions:

$$[q] = L/T; \qquad [V] = L/T; \qquad [K] = L/T. \tag{5.1.8}$$

The coefficient of proportionality, K, appearing in Darcy's law (5.1.1), is called *hydraulic conductivity*, and is discussed in section 5.5.

5.2 Generalization of Darcy's Law

5.2.1 Isotropic Medium

The experimentally derived form of Darcy's law (for a homogeneous incompressible fluid) was limited to one-dimensional flow. When the flow is three-dimensional, the obvious formal generalization of Darcy's law, say in the form (5.1.3), is:

$$\mathbf{q} = K\mathbf{J} = -K \operatorname{grad} \varphi \tag{5.2.1}$$

where $\mathbf{q}$ is the *specific flux vector* with components q_x, q_y, q_z in the directions of the Cartesian x, y, z coordinates, respectively, and $\mathbf{J} = -\operatorname{grad} \varphi$ is the *hydraulic gradient* with components $J_x = -\partial\varphi/\partial x$, $J_y = -\partial\varphi/\partial y$, $J_z = -\partial\varphi/\partial z$, in the x, y, z directions, respectively. When the flow takes place through a homogeneous isotropic medium, so that as in section 5.1, the coefficient K is a constant scalar, we may write (5.2.1) as three equations:

$$q_x = KJ_x = -K\,\partial\varphi/\partial x; \qquad q_y = KJ_y = -K\,\partial\varphi/\partial y; \qquad q_z = KJ_z = -K\,\partial\varphi/\partial z. \tag{5.2.2}$$

Or, for flow in any direction given by the unit vector $\mathbf{1s}$:

$$q_s = -K\,\partial\varphi/\partial s. \tag{5.2.3}$$

Equations (5.2.2) and (5.2.3) remain valid for three-dimensional flow through inhomogeneous isotropic media where $K = K(x, y, z)$.

Although we have started this section by suggesting (5.2.1) as a formal generalization of the *equation of motion* to three-dimensional flow, there exist a large number of theories that support this extension. One such derivation is presented in detail in section 4.7 leading to (4.7.18). Several other derivations are presented in section 5.10.

In an isotropic medium, K is a scalar and the vectors $\mathbf{q}$ and $\operatorname{grad} \varphi$ are colinear. From (5.2.1) and the definition of $\operatorname{grad} \varphi$ it also follows that the vector $\mathbf{q}$ is normal everywhere to the equipotential surfaces $\varphi = \text{const}$.

Darcy's law (5.2.1) is sometimes written in the form:

$$\mathbf{q} = -\operatorname{grad}(K\varphi) = -\operatorname{grad} \Phi; \qquad \Phi = K\varphi \tag{5.2.4}$$

where Φ is called the *velocity potential* (following the terminology used in hydrodynamics, where the velocity is derived from a potential: $\mathbf{V} = -\operatorname{grad}\Phi$). Actually, in flow through porous media we should refer to Φ as the *specific discharge potential.* Scheidegger (1960) discusses the two forms—(5.2.1) and (5.2.4)—and concludes that both are possible. Hubbert (1940) also discusses this problem and uses thermodynamic reasoning to show that (5.2.4) is incorrect. However, his discussion, which is based on the microscopic flow pattern, cannot readily be extended to inhomogeneous media.

Consider an inhomogeneous isotropic medium where $K = K(x, y, z)$. Then from (5.2.4) it follows that:

$$\mathbf{q} = -K \operatorname{grad}\varphi - \varphi \operatorname{grad} K. \tag{5.2.5}$$

This means that for a domain where $\varphi = \text{const}$ everywhere we have flow due to the fact that K varies. This is obviously impossible as we cannot have flow with no source of energy. Another proof for the correctness of (5.2.1) is given by Jones (1962). Hence (5.2.4) is an erroneous way of writing Darcy's law, except for *a homogeneous isotropic* medium, in which case (5.2.1) and (5.2.4) become identical.

In general, it is recommended that Darcy's law be written in terms of the piezometric head φ. One should note that both Φ and φ are defined except for an arbitrary constant.

Other forms of (5.2.1) are:

$$\mathbf{q} = -\frac{k}{\mu}\left[\frac{\partial p}{\partial x}\mathbf{1x} + \frac{\partial p}{\partial y}\mathbf{1y} + \left(\frac{\partial p}{\partial z} + \rho g\right)\mathbf{1z}\right]$$

$$= -\frac{k}{\mu}(\operatorname{grad} p + \rho g \mathbf{1z}) = -\frac{k}{\mu}(\operatorname{grad} p - \rho \mathbf{g}) \tag{5.2.6}$$

where $K = k\rho g/\mu$ and $\mathbf{g} = -g\mathbf{1z}$ is the vector of gravity acceleration directed downward. Hubbert (1940) defined a potential $\varphi' = p/\rho + gz$ that is the energy per unit mass of the fluid. Then:

$$\mathbf{q} = -(k\rho/\mu) \operatorname{grad}\varphi'. \tag{5.2.7}$$

5.2.2 Anisotropic Medium

A medium is said to be homogeneous with respect to a certain property if that property is *independent of position* within the medium. Otherwise the medium is said to be *heterogeneous.* A medium is said to be *isotropic* with respect to a certain property if that property is *independent of direction* within the medium. If at a point within the medium a property of the medium, e.g., permeability or thermal conductivity, varies with direction, the medium is said to be *anisotropic* (or *aleotropic*) at the considered point with respect to that property. The main property of interest to us in this book is permeability. Hence, unless otherwise specified, in this book, isotropy or anisotropy is with respect to the medium's permeability.

In natural materials, anisotropy is encountered in soils and in geological formations that serve as reservoirs or aquifers. Sediments are commonly deposited in such a way that their permeability in one direction (usually the horizontal one, unless later tilting of the formation occurs) is greater than in other directions. In most stratified materials the resistance to the flow is smaller (i.e., permeability is greater) along the planes of deposition than across them. Piersol et al. (1940) mention ratios of horizontal to vertical permeabilities of sandstone of 1.5÷3. Muskat (1937, p. 111) lists 65 pairs of sand samples, more than two-thirds of which had a larger permeability in the direction parallel to the bedding plane than normal to it. The quotient of the two values ranged from 1 to 42.

Stratified soils are usually anisotropic. The stratification may result from the shape of the particles. For example, plate-shaped particles (e.g., mica) will generally be oriented with the flat side down. Both sedimentation and pressure of overlying material cause flat particles to be oriented with their longest dimensions parallel to the plane on which they settle. Later, this produces flow channels parallel to the bedding plane, differing from those oriented normal to this plane, and the medium becomes anisotropic. Alternating layers of different texture also give rise to anisotropy (see par. 5.8.3). However, in order for a stratified formation of this kind to be qualified as an *anisotropic homogeneous* medium, the thickness of the individual layers should be much smaller than lengths of interest. There is no use in attempting to determine the permeability of such a formation from a core whose size is smaller than the thickness of a single stratum. In many aquifers, fractures produce very high permeability in the direction along the fractures, whereas the permeability of the rock in the direction normal to the fractures is much smaller. In carbonate rocks, solution of the rock takes place by means of the flowing water. This produces solution channels that develop mainly in the direction of the flow; the rock becomes anisotropic, with a very high permeability in the general direction of these channels. In many soils (e.g., loess), vertical joints, root holes and animal burrows give rise to anisotropy in permeability, with vertical permeability being greater than horizontal. In some soils structural fissures may develop more readily in some directions than in others, and the soil will exhibit anisotropy.

As Darcy's law, say in the form of (5.2.2), presents a single linear relationship between the components of the specific discharge and those of the hydraulic gradient, it would be natural to extend it to anisotropic media by writing the most general linear relationship between components of flux ($\mathbf{q}$) and those of hydraulic gradient ($\mathbf{J} = -\operatorname{grad}\varphi$):

$$\begin{aligned} q_x &= K_{xx}J_x + K_{xy}J_y + K_{xz}J_z \\ q_y &= K_{yx}J_x + K_{yy}J_y + K_{yz}J_z \\ q_z &= K_{zx}J_x + K_{zy}J_y + K_{zz}J_z \end{aligned} \tag{5.2.8}$$

where x, y, z are Cartesian coordinates and $K_{xx}, K_{xy}, \ldots, K_{zz}$ are nine constant coefficients. A constant is not included in these general linear relationships because we want $\mathbf{q} = 0$ for $\mathbf{J} = 0$. Anisotropic permeability is discussed in detail in section 5.6.

Although we have written Darcy's law for anisotropic media as a heuristic extension of the law for isotropic media, theories exist that lead directly to Darcy's law for anisotropic media. One such theory is presented in section 4.7; other theories are presented in section 5.10.

The discussion on anisotropic permeability is continued in section 5.6, where details are given on directional permeability, principal directions, etc.

Several other cases that require special forms of the equation of motion are employed throughout the present book.

(a) *Compressible homogeneous fluids* are discussed in section 5.9. The motion equation is (5.9.2).

(b) *Inhomogeneous fluids* are discussed in section 4.7. The motion equation is (4.7.17).

(c) *Multiphase flows*, where each phase fills only part of the void space, are discussed in detail in paragraph 9.3.1. The case of unsaturated flow is discussed in paragraph 9.4.4.

(d) *Flow at large Reynolds numbers* is discussed in section 5.11.

5.3 Deviations From Darcy's Law

5.3.1 The Upper Limit

As the specific discharge increases, Darcy's law, which specifies a linear relationship between the specific discharge, $\mathbf{q}$, and the hydraulic gradient, $\mathbf{J}$, has been shown by many investigators to be invalid. Figure 5.11.1 shows typical experimental results leading to this conclusion. A definition of a range of validity of Darcy's law seems, therefore, appropriate.

In flow through conduits, the *Reynolds number* (Re), a dimensionless number expressing the ratio of inertial to viscous forces, is used as a criterion to distinguish between laminar flow occurring at low velocities and turbulent flow (see any text on fluid mechanics). The critical Re between laminar and turbulent flow in pipes is around 2100. By analogy, a Reynolds number is defined also for flow through porous media:

$$\mathrm{Re} = qd/\nu \tag{5.3.1}$$

where d is some length dimension of the porous matrix, and ν is the kinematic viscosity of the fluid. Although, by analogy to the Reynolds number for pipes, d should be a length dimension representing the elementary channels of the porous medium, it is customary (probably because of the relative ease of determining it) to employ some representative dimension of the grains for d (in an unconsolidated porous medium). Often the mean grain diameter is taken as the length dimension, d, in (5.3.1). Sometimes d_{10} is used, i.e., the grain size that exceeds the size (diameter) of 10% of the material by weight. The term d_{50} is also mentioned in the literature as a representative grain diameter. Collins (1961) suggests $d = (k/n)^{1/2}$, where k is permeability, and n is porosity, as the representative length dimension to be used in the Reynolds number. Ward (1964) uses $k^{1/2}$ as the representative length d.

In practically all cases, *Darcy's law is valid as long as the Reynolds number based on average grain diameter does not exceed some value between* 1 *and* 10.

Continuing the analogy with flow in conduits, flow through porous media has often been expressed as a relationship between some friction factor and Re. The one most commonly used is the Darcy–Weisbach friction factor f:

$$f = 2g\,dJ/V^2 = 2d(\Delta\varphi/L)(\gamma/\rho V^2); \qquad J = (\Delta\varphi/L) \tag{5.3.2}$$

obtained from the Darcy–Weisbach formula: $\Delta\varphi/L = (f/d)V^2/2g$. In (5.3.2) $\Delta\varphi$ is the head drop over a length L in the direction of the flow, and d is the diameter of the pipe. In flow through porous media d is taken as the same representative length as that used in (5.3.1) and V is replaced by q. Fanning uses another definition for the friction factor:

$$(\Delta\varphi/L) = (f/R)V^2/2g; \qquad f = 2gRJ/V^2 = 2R(\Delta\varphi/L)(\gamma/\rho V^2) \tag{5.3.3}$$

$$\text{or:} \quad f = \tfrac{1}{2}d(\Delta\varphi/L)(\gamma/\rho V^2) \tag{5.3.4}$$

where R is *hydraulic radius* of the pipe $(d/4)$.

When the results of experiments of flow through porous media are plotted as a relationship between the Fanning friction factor and the Reynolds number, a curve is obtained, as shown in figure 5.3.1. The straight line portion of this curve can be expressed by $f = c/\mathrm{Re}$ where c is some constant. This relationship can be then written as:

$$\frac{d}{2}(\Delta\varphi/L)(\gamma/\rho q^2) = c\nu/qd; \qquad q = KJ \tag{5.3.5}$$

where $K = d^2g/2c\nu$ and $J = \Delta\varphi/L$. This is nothing but Darcy's linear relationship

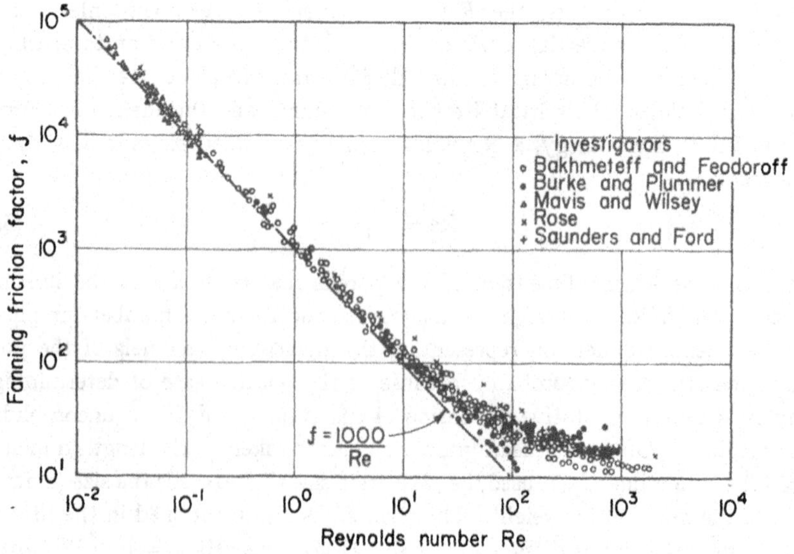

FIG. 5.3.1. Relationship between Fanning's friction factor and Reynolds number for flow through granular porous media (Rose, 1945).

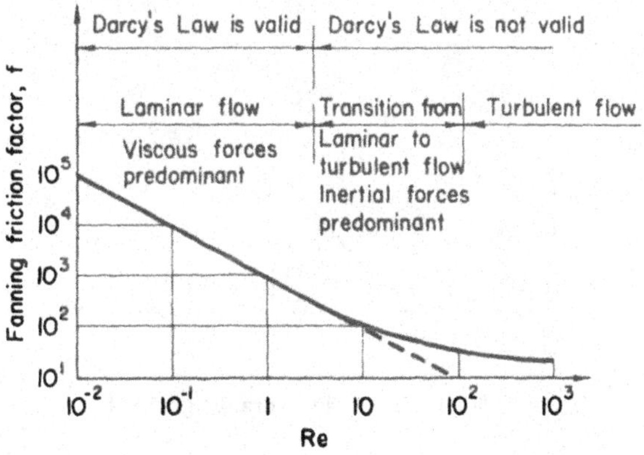

FIG. 5.3.2. Schematic classification of flow through porous media.

between q and J. Many investigators suggest mathematical and graphic correlations to fit the experimental curve (e.g., Lindquist 1933).

As Re (i.e., q for constant d and ν) increases, we observe a deviation from the linear relationship. However, one should be careful not to identify this deviation with a transition from laminar flow, for which Darcy's law is valid, to turbulent flow. A detailed discussion on flow at large Reynolds numbers is given in section 5.11. Let us present here only the three regions that may be distinguished in flow through porous media (fig. 5.3.2).

(a) At low Reynolds numbers (i.e., at low velocity for a constant d and ν), we have a region where the flow is laminar, viscous forces are predominant, and the linear Darcy law is valid. The upper limit of this range is at a value of Re between 1 and 10.

(b) As Re increases, we observe a transition zone. At the lower end of this zone we have the passage from the laminar regime, where viscous forces are predominant, to another laminar regime where inertial forces govern the flow. At the upper end of the transition zone we have a gradual passage to turbulent flow. Some authors suggest Re 100 for the upper limit of the laminar flow regime, which is often referred to as the *nonlinear laminar flow regime.*

(c) At high Re we have turbulent flow.

Darcy's law is valid only in region (a).

5.3.2 The Lower Limit

Some authors discuss a lower limit of applicability for Darcy's law of saturated flow through porous media. Irmay (in chap. 5 of Bear et al. 1968) mentions the existence of a *minimum (or initial) gradient* J_0 below which there is very little flow. For clay soils, J_0 may exceed 30. Irmay attributes this phenomenon to the rheological non-Newtonian behavior of the water. With this concept Darcy's law becomes (fig. 5.3.3):

$$\mathbf{q} = 0 \quad \text{for} \quad J \leqslant J_0; \qquad \mathbf{q} = K\mathbf{J}(J - J_0)/J \quad \text{for} \quad J \geqslant J_0; \qquad J = |\mathbf{J}|. \quad (5.3.6)$$

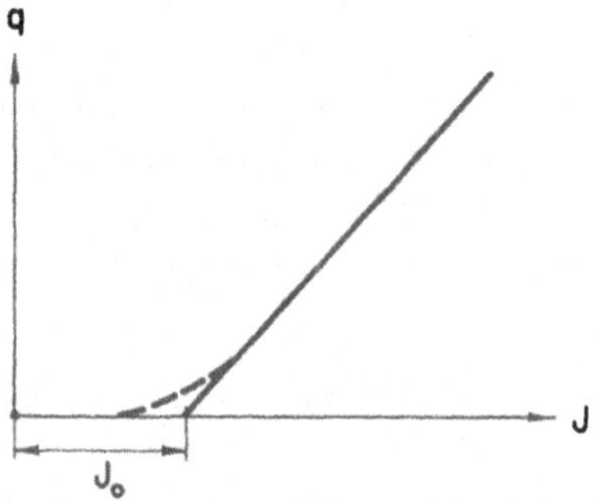

FIG. 5.3.3. Threshold hydraulic gradient.

Some authors attribute J_0 to the fact that the *streaming potential* generated by the flow, particularly in fine-grained soils, can produce small countercurrents along the pore walls in a direction opposite that of the main flow. This, at least in fine-grained cohesive soils, may result in no flow before the *threshold gradient* J_0 is exceeded.

Swartzendruber (1962) reviews some previous works in which reference is made to non-Darcy behavior in liquid-saturated porous media, and analyzes velocity-gradient data in order to verify a modified flow equation. Based on experimental data, he suggests the following equation for one-dimensional flow:

$$q = 0 \qquad \text{for} \quad 0 \leqslant J < J_0$$
$$q = B[J - \exp\{-CJ\}] \quad \text{for} \quad J \geqslant J_0 \tag{5.3.7}$$

which has a zero slope at $J = 0$. Following earlier works (e.g., Von Engelhardt and Tunn 1955; Low 1961) he attributes (5.3.7) to non-Newtonian (shear-rate dependent) liquid viscosity caused by clay–water interaction. Engelhardt and Tunn (1955) also recognize an alternative mechanism consisting of adsorbed immobile water layers on the particle surfaces, with Newtonian conditions prevailing elsewhere. A detailed review on non-Darcian flow of water in soils (laminar region) is also given by Kutilek (1969) and by Bolt and Groenvelt (1969).

5.3.3 The Slip Phenomenon

Another case in which deviations from Darcy's law have been observed is that of gas flow at low pressure. This case is of interest because the flow of air is often used as a means for determining permeability. Various authors report that air permeability is higher than liquid permeability in the same porous medium. It is known from fluid mechanics that when the diameter of a capillary tube approaches the mean free path of the molecules, gases flow through it more rapidly than would be predicted by Poiseuille's law. The same is true for flow of gas at low pressure through porous media; the flow is faster than would be predicted by Darcy's law. Contrary to laminar flow theory (on which Darcy's law is based), where we assume zero fluid velocity at the solid wall with shear taking place in the fluid, here there is no intimate contact of the fluid molecules with the solid wall and the velocity at the solid–fluid interface has a finite value, not necessarily zero. Thus, whenever the mean free path of the gas molecules approaches the dimensions of the flow conduit, the individual gas molecules are in motion at this interface and contribute an additional flux. This phenomenon is called *slip phenomenon*, or *Klinkenberg effect* (or *slip flow*, *Knudsen flow*, *free molecular flow*; par. 1.3.1). The slip phenomenon also occurs in gas–liquid flow systems where effective permeability to the gas (par. 10.3.2) is affected by it.

Klinkenberg (1941), using a glass capillary tube as a model, derives the following expression for the permeability of a gas:

$$k_g = k_l(1 + 4c\lambda/r) = k_l(1 + b/p) \tag{5.3.8}$$

where k_g is the permeability to gas, k_l is the permeability to liquid or high density gas, λ is the mean free path of the gas molecules under the mean pressure p at which k_g is determined, $c \sim 1$ is a proportionality factor, r is the radius of the capillary tube in Klinkenberg's model, and b is a constant for a gas–solid system that depends on the mean free path of the gas and the size of the openings in the porous medium. Since the openings are a function of the permeability, k_l, b is also a function of k_l. Examples for b suggested by the American Petroleum Institute are: $b = 3$, 0.5 and 0.1 at $k_l = 0.03$, 0.05 and 200 millidarcies, respectively (a straight line on a log–log paper).

Scheidegger (1960) reviews several other works leading to formulas for k_g.

5.4 Rotational and Irrotational Motion

5.4.1 The Potential and Pseudopotential

In the previous sections, the specific discharge was related to the observable quantity φ (piezometric head). By observing φ at different points in a field of flow, we assume the existence of simple surfaces in space on which the piezometric head φ is a constant. In order to establish the presence of such surfaces, $\varphi = \text{const}$, we examine the equation of such a surface:

$$\operatorname{grad} \varphi \cdot d\mathbf{s} = 0 \tag{5.4.1}$$

which expresses the fact that any displacement vector in the surface must be perpendicular to the vector grad φ.

Following Morse and Feshbach (1953), let us first consider the equation of this surface in the form

$$\mathbf{F} \cdot d\mathbf{s} = F_x\,dx + F_y\,dy + F_z\,dz = 0 \tag{5.4.2}$$

where $\mathbf{F} = \mathbf{F}(x, y, z)$ is some vector field (component F_x, F_y, F_z). Equation (5.4.2) is integrable under certain conditions. If there exists a scalar function $G = G(x, y, z)$ such that

$$F_x = a\,\partial G/\partial x; \qquad F_y = a\,\partial G/\partial y; \qquad F_z = a\,\partial G/\partial z \tag{5.4.3}$$

then the equation for the surface considered may be integrated to yield $G(x, y, z) =$

const, and we obtain a family of such surfaces by varying the value of the constant. The quantity a, which may be function of x, y, z, is called an integrating factor. To derive the criterion for whether an equation for the surfaces exists in an integral form, let us start by assuming that a function G as defined above indeed exists, and then consider the expression:

$$F_x(\partial F_z/\partial y - \partial F_y/\partial z) + F_y(\partial F_x/\partial z - \partial F_z/\partial x) + F_z(\partial F_y/\partial x - \partial F_x/\partial y) \equiv \mathbf{F} \cdot \operatorname{curl} \mathbf{F} = 0. \tag{5.4.4}$$

By inserting (5.4.3) into (5.4.4), we obtain:

$$a\frac{\partial G}{\partial x}\left[\frac{\partial}{\partial y}\left(a\frac{\partial G}{\partial z}\right) - \frac{\partial}{\partial z}\left(a\frac{\partial G}{\partial y}\right)\right] + a\frac{\partial G}{\partial y}\left[\frac{\partial}{\partial z}\left(a\frac{\partial G}{\partial x}\right) - \frac{\partial}{\partial x}\left(a\frac{\partial G}{\partial z}\right)\right]$$
$$+ a\frac{\partial G}{\partial z}\left[\frac{\partial}{\partial x}\left(a\frac{\partial G}{\partial y}\right) - \frac{\partial}{\partial y}\left(a\frac{\partial G}{\partial x}\right)\right] = 0. \tag{5.4.5}$$

This means that if a function G (continuous and with continuous first and second derivatives) exists, the expression (5.4.4) vanishes identically. If a vector:

$$\mathbf{H} = (\partial F_z/\partial y - \partial F_y/\partial z)\mathbf{1x} + (\partial F_x/\partial z - \partial F_z/\partial x)\mathbf{1y} + (\partial F_y/\partial x - \partial F_x/\partial y)\mathbf{1z} \equiv \operatorname{curl} \mathbf{F} \tag{5.4.6}$$

exists, it follows from (5.4.4), written as $\mathbf{F} \cdot \mathbf{H} = 0$, that it is perpendicular at every point to $\mathbf{F}$. Therefore it is possible to obtain the equation for the normal surfaces in an integral form as $G(x, y, z) =$ const. The scalar function G is called a *pseudo-potential.* In certain special cases, a is a constant and may be set equal to -1. In these cases the scalar function G is called the *potential function* for vector field $\mathbf{F}$, and the surfaces $G =$ const are called *equipotential surfaces.*

From this discussion it follows that the condition for the existence of surfaces $G(x, y, z) =$ const normal to $\mathbf{F}$ is (5.4.4) (i.e., $\mathbf{F} \cdot \operatorname{curl} \mathbf{F} = 0$). Conversely, if (5.4.4) is valid, surfaces normal to $\mathbf{F}$ exist.

With this discussion in mind, we may now identify G with the piezometric head φ, and may refer to it by the mathematical term *pseudopotential.* Replacing a in (5.4.3) with the hydraulic conductivity $K = K(x, y, z)$, we may identify $\mathbf{F}$ with the specific discharge $\mathbf{q}$. Then:

$$\mathbf{q} = -K(x, y, z) \operatorname{grad} \varphi. \tag{5.4.7}$$

For the homogeneous case ($K =$ const), equation (5.4.7) may be written as:

$$\mathbf{q} = -\operatorname{grad}(K\varphi) = -\operatorname{grad} \Phi, \qquad \Phi = K\varphi. \tag{5.4.8}$$

A flow governed by (5.4.8) is called *potential flow.* Then we may identify G with Φ and refer to it by the mathematical term *potential*; Φ is then the specific discharge potential (sec. 5.2). Very often both φ and Φ are referred to just as "potentials," although in the mathematical sense, φ is not a potential. It should be emphasized again that the form (5.4.8) and the use of the velocity potential Φ is *restricted only to homogeneous isotropic porous media.*

5.4.2 Irrotational Flow

When written in terms of the specific discharge vector $\mathbf{q}$ (5.4.4) becomes

$$q \cdot \text{curl}\, \mathbf{q} = 0 \tag{5.4.9}$$

where:

$$\text{curl}\, \mathbf{q} = \nabla x \mathbf{q} = \begin{vmatrix} \mathbf{1x} & \mathbf{1y} & \mathbf{1z} \\ \partial/\partial x & \partial/\partial y & \partial/\partial z \\ q_x & q_y & q_z \end{vmatrix}$$

$$= \mathbf{1x}\left(\frac{\partial q_z}{\partial y} - \frac{\partial q_y}{\partial z}\right) + \mathbf{1y}\left(\frac{\partial q_x}{\partial z} - \frac{\partial q_z}{\partial x}\right) + \mathbf{1z}\left(\frac{\partial q_y}{\partial x} - \frac{\partial q_x}{\partial y}\right). \tag{5.4.10}$$

Equation (5.4.9) is satisfied either in the trivial case of $\mathbf{q} = 0$ or when

$$\text{curl}\, \mathbf{q} = 0. \tag{5.4.11}$$

The vector $\mathbf{w} = \frac{1}{2}$ curl $\mathbf{q}$ is called the *vorticity vector*. A flow satisfying (5.4.11) is called an *irrotational flow*. We have seen in paragraph 5.4.1 that a vector field satisfying (5.4.11) in a simply connected region always has a single-valued potential Φ or a pseudopotential φ. This also follows from the identity curl (grad Φ) $\equiv 0$. Thus, when a velocity potential exists (and this is possible only in homogeneous isotropic domains), the two terms, *potential flow* and *irrotational flow*, are synonyms. From (5.4.11), we have for an irrotational field of flow:

$$\frac{\partial q_z}{\partial y} = \frac{\partial q_y}{\partial z}; \qquad \frac{\partial q_x}{\partial z} = \frac{\partial q_z}{\partial x}; \qquad \frac{\partial q_y}{\partial x} = \frac{\partial q_x}{\partial y} \tag{5.4.12}$$

simultaneously and for all points within the flow domain.

It is interesting to note (Zaslavsky 1962) that in two-dimensional flow, since $q_z = 0$ and $\partial q_x/\partial z = 0$, $\partial q_y/\partial z = 0$, only $(\text{curl}\, \mathbf{q})_z \neq 0$ so that (5.4.9) is always satisfied. A family of equipotentials always exists, and although the medium is inhomogeneous (yet isotropic), the streamlines are everywhere normal to it.

An irrotational flow is sometimes called a *circulation-free flow*, where the *circulation* Γ is defined as the line integral, at a given time, t, of the tangential velocity component about any closed contour, C:

$$\Gamma = \oint_{(C)} \mathbf{q} \cdot d\mathbf{s} = \iint_{(A)} \text{curl}\, \mathbf{q} \cdot d\mathbf{A} \tag{5.4.13}$$

where a positive direction of $\mathbf{1s}$ corresponds to a counterclockwise direction around the surface A enclosed by C. When curl $\mathbf{q} \equiv 0$, we also have $\Gamma = 0$.

An important conclusion from the fact that $\Gamma = 0$ is that the line integral of $\mathbf{q} \cdot d\mathbf{s}$ from a fixed point $P_0(x_0, y_0, z_0)$ to any point $P(x, y, z)$ in a flow domain where $\mathbf{q} = -\text{grad}\, \Phi$, is independent of the path connecting these points. Hubbert (1956) (sec. 5.9) used this fact to determine the potential (actually a pseudopotential) φ for the flow of a compressible fluid.

Thus the irrotational vector field $\mathbf{q}$ is characterized by the following three equivalent properties:

$$\mathbf{q} = -\operatorname{grad}\Phi, \qquad \Phi = \Phi(x, y, z) \text{ is a scalar point function}$$

$$\int_{(C)} \mathbf{q} \cdot d\mathbf{s} \equiv 0 \qquad \text{for any closed curve } C \text{ which does not contain a singularity}$$

$$\operatorname{curl}\mathbf{q} \equiv 0 \qquad \text{for every point.}$$

The term *lamellar* is also applied to an irrotational field ($\operatorname{curl}\mathbf{q} = 0$) where the vector is equal to the gradient of a scalar field. The vector field $\mathbf{q}$ in the more general case of flow in a nonhomogeneous medium, where $\mathbf{q} = -K(x, y, z)\operatorname{grad}\varphi$ and $\mathbf{q} \cdot \operatorname{curl}\mathbf{q} = 0$, is called a *complex lamellar field*. The condition required for a vector field to be "complex lamellar" is that the two vectors, $\mathbf{q}$ and $\operatorname{curl}\mathbf{q}$, should be orthogonal to each other at any point. This is less restrictive than the requirement $\operatorname{curl}\mathbf{q} = 0$.

For the isotropic inhomogeneous porous medium, we have:

$$\begin{aligned}\operatorname{curl}\mathbf{q} &= \operatorname{curl}(-K\operatorname{grad}\varphi) = -K\operatorname{curl}\operatorname{grad}\varphi - \operatorname{grad}K \times \operatorname{grad}\varphi \\ &= -\operatorname{grad}K \times \operatorname{grad}\varphi = \operatorname{grad}\varphi \times \operatorname{grad}K = -\mathbf{q} \times \operatorname{grad}\ln K \quad (5.4.14)\end{aligned}$$

which shows that the flow is *rotational* except for the special case where the vectors $\operatorname{grad}K$ and $\operatorname{grad}\varphi$ (or $\mathbf{q}$ and $\operatorname{grad}\ln K$) are parallel everywhere.

One should recall that here $\mathbf{q}$ has the sense of an average taken over the flow within the elementary pores. We must emphasize this point because the flow inside each of the pores is viscous and hence rotational in character. Jacob (1950) suggests that by the averaging procedure the rotations occurring within the individual pores balance out, so that the average flow is irrotational and a velocity potential exists. In an inhomogeneous medium, in spite of averaging, the flow is rotational.

The flow is also rotational in the case of a homogeneous anisotropic medium where $\mathbf{q} = -\underline{\mathbf{K}}\operatorname{grad}\varphi$. In this case we do not have an integrating factor (par. 5.4.1) a, such that $q_x = -a\,\partial\varphi/\partial x$, $q_y = -a\,\partial\varphi/\partial y$, $q_z = -a\,\partial\varphi/\partial z$.

In spite of the discussion above, we shall often refer to φ as a potential even when the medium is nonhomogeneous and/or anisotropic. The use of this term is too entrenched in everyday use to be changed.

5.5 Hydraulic Conductivity of Isotropic Media

5.5.1 Hydraulic Conductivity and Permeability

The coefficient of proportionality K, appearing in the various forms of Darcy's law discussed in sections 5.1 and 5.2, is called *hydraulic conductivity*. Sometimes the term "coefficient of permeability" is also used. In an isotropic medium it may be defined, using (5.2.1), as the specific discharge per unit hydraulic gradient. It is a scalar (dimensions L/T) that expresses the ease with which a fluid is transported through a porous matrix. It is, therefore, a coefficient that depends on both matrix and fluid properties. The relevant fluid properties are density, ρ, and viscosity, μ, or, in the combined form of kinematic viscosity, ν. The relevant solid matrix properties are mainly grain- (or pore-) size distribution, shape of grains (or pores), tortuosity, specific surface and porosity. From analytic derivations of Darcy's law (sec. 5.10), or from a dimensional analysis, it follows that we may express the hydraulic conductivity K as (Nutting 1930):

$$K = k\gamma/\mu = kg/\nu \tag{5.5.1}$$

where k (dim L^2)—called the *permeability* (or *intrinsic permeability*) of the porous matrix—depends solely on properties of the solid matrix, and γ/μ represents the influence of the fluid's properties. With (5.5.1), Darcy's law (5.2.4) may be written as:

$$q_i = -K\,\partial\varphi/\partial x_i = -(k\rho g/\mu)\,\partial\varphi/\partial x_i = -(k\rho g/\mu)\,\partial(p/\rho g + z)/\partial x_i; \qquad i = 1, 2, 3. \tag{5.5.2}$$

Then, if $\rho =$ const, we have:

$$q_i = -(k/\mu)\,\partial(p + \rho g z)/\partial x_i; \qquad i = 1, 2, 3 \tag{5.5.3}$$

or, for horizontal flow:

$$q_i = -(k/\mu)\,\partial p/\partial x_i; \qquad i = 1, 2. \tag{5.5.4}$$

Various formulas are described in the literature, relating k to the various properties of the solid matrix. Some of these formulas are *purely* empirical, as, for example, the relationship shown in figure 5.5.1 where k (in cm^2) is related to a mean (or "effective") grain diameter d (in microns) by:

$$k = 0.617 \times 10^{-11} d^2. \tag{5.5.5}$$

Another example is the Fair and Hatch (1933) formula developed from dimensional

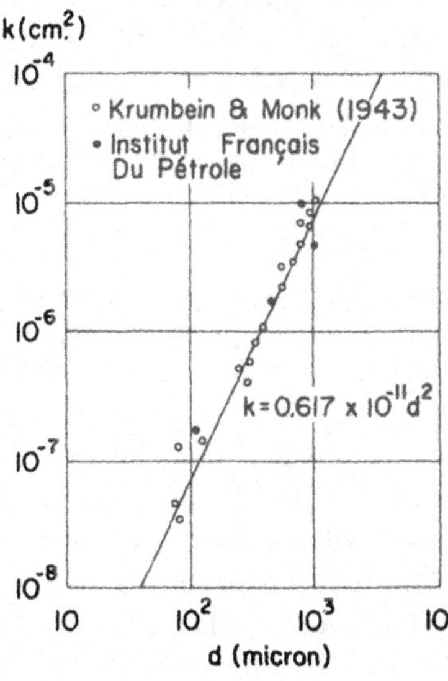

FIG. 5.5.1. Variation of intrinsic permeability (k) with grain diameter (d).

considerations and verified experimentally:

$$k = \frac{1}{m}\left[\frac{(1-n)^2}{n^3}\left(\frac{\alpha}{100}\sum\frac{P}{d_m}\right)^2\right]^{-1} \tag{5.5.6}$$

where m is a *packing factor*, found experimentally to be about 5, α is a sand-shape factor, varying from 6.0 for spherical grains to 7.7 for angular ones, P is the percentage of sand held between adjacent sieves, and d_m is the geometric mean rated sizes of the adjacent sieves.

Other formulas are *purely theoretical*, obtained from the theoretical derivation of Darcy's law. An example of such relationship is (4.7.13), in which permeability is related to porosity, to tortuosity and to conductance of the elementary matrix channels. Several other examples are given in section 5.10 where theoretical derivations of Darcy's law are presented. The better known ones are the Kozeny equation (5.10.15) and the Kozeny–Carman equation (5.10.18). In the latter equation we have an empirical coefficient. This is typical of a third class of formulas for permeability—the *semiempirical* ones, where the theoretical analysis, usually using some conceptual model, leads to a relationship between k and various matrix parameters; however, the numerical coefficients must be determined experimentally for each particular porous matrix (or group of similar porous matrices).

The various formulas mentioned have the general form:

$$k = f_1(s)f_2(n)d^2 \tag{5.5.7}$$

where s is a dimensionless parameter (or parameters) that expresses the effect of the shape of the grains (or pores), $f_1(s)$ is called *shape factor*, $f_2(n)$ is called the porosity factor and d is the *effective* (or *mean*) *diameter* of the grains. Usually d is taken as the harmonic mean diameter as defined by (2.6.2), or as d_{10}—i.e., a diameter such that 10% (by weight) of the porous matrix consists of grains smaller than it. Other definitions for d are also used. Examples for $f_1(s)$ and $f_2(n)$ may be obtained by comparing (5.5.7) with the various formulas for k given in section 5.10.

Often, as in (5.5.2), the product $f_1(s)f_2(n)$ appears as a single dimensionless coefficient, C, in the relationship between k and d. We then have:

$$k = Cd^2; \qquad K = Cd^2\rho g/\mu. \tag{5.5.8}$$

Harleman et al. (1963) report a fairly good agreement between experimental values of k and those derived from an equation similar to (5.5.8) with $C = 6.54 \times 10^{-4}$ (where d is in cm and k in cm^2). They obtained this value from the relationship for C appearing in (5.10.31).

When k varies in space, i.e., $k = k(x, y, z)$, we refer to the porous medium as being *inhomogeneous* or *heterogeneous*. When, at some point, k of saturated flow varies with direction, we refer to the medium at that point as *anisotropic* (sec. 5.6). Spatial and temporal variations in the fluid's temperature, or in the concentration of solids dissolved in it, will, by affecting ρ and μ, cause K to vary both in space and in time.

Under certain conditions k may also vary with time. Such variation may be caused by external loads that produce stresses that change the structure and texture of the porous matrix. Subsidence and consolidation are phenomena associated with changes in permeability. Among other factors producing changes in k of saturated flow, we may mention the effect of solution of the solid matrix (which may, over prolonged times, produce large channels and cavities), and the swelling of clay. When k is determined for a given core that contains argillaceous material, drying of the core may shrink the clay, especially bentonite, causing the permeability to air of the dried core to be higher than would be the case when determined with water. Fresh water in a core, as contrasted with salt water, may cause the clay to swell, thereby reducing k.

Under certain conditions there may be biological activity in the medium producing a growth that tends to clog the matrix, thus reducing k with time. In laboratory work additives (e.g., formaldehyde) are used to prevent such clogging. When water containing fines is introduced into the soil (e.g., in artificial replenishment by surface water containing suspended load), another clogging process occurs.

Permeability of unsaturated flow is discussed in detail in chapter 9, particularly in section 9.4.

5.5.2 Units and Examples

Various units are used in practice for hydraulic conductivity, K (dim L/T). Hydrologists prefer the unit m/d (meters per day). Soil scientists often use cm/sec. In the United States, as in many countries using the English system of units, two other units are commonly employed by hydrologists. One is *laboratory* or *standard hydraulic conductivity* defined, using (5.1.3), as the total discharge (Q) of water at 60°F, expressed in gallons per day, through a porous medium cross-sectional area (A), expressed in ft², under a hydraulic gradient ($(\varphi_1 - \varphi_2)/L$) of 1 ft per ft. With this definition, the units of K are gal/day ft². In a similar way, again using (5.1.3), a *field*, or *aquifer hydraulic conductivity*, is defined as the discharge of water at field temperature, through a cross-sectional area of an aquifer 1 ft thick and 1 mile wide under a hydraulic gradient of 1 ft/mile. The unit is the same as for the laboratory K. Following are some conversions among these units:

$$1 \text{ US gal/day ft}^2 = 4.72 \times 10^{-5} \text{ cm/sec} = 4.08 \times 10^{-2} \text{ m/d}.$$

Permeability, k (dim L^2), is measured in the metric system in cm² or in m². In the English system, the unit is ft². For water at 20°C we have the conversion:

$$K = 1 \text{ cm/sec is equivalent to } k = 1.02 \times 10^{-5} \text{ cm}^2.$$

Reservoir engineers often use the unit *darcy* based on (5.5.4) written as: $k = (Q/A)\mu/(\Delta p/\Delta x)$. Accordingly, the darcy is defined by:

$$1 \text{ darcy} = \frac{[1(\text{cm}^3/\text{sec})/\text{cm}^2]\ 1 \text{ centipoise}}{1 \text{ atmosphere/cm}}. \tag{5.5.11}$$

Thus a porous medium is said to have a permeability of one darcy if a single-phase fluid of 1 centipoise viscosity that completely fills the void space of the medium will flow through it at a rate of 1 cm³/sec per cm² cross-sectional area under a pressure

Table 5.5.1

Typical Values of Hydraulic Conductivity and Permeability

<table>
<tr><td>$-\log_{10}$ · K(cm/s)</td><td>−2</td><td>−1</td><td>0</td><td>1</td><td>2</td><td>3</td><td>4</td><td>5</td><td>6</td><td>7</td><td>8</td><td>9</td><td>10</td><td>11</td></tr>
<tr><td>Permeability</td><td colspan="4">Pervious</td><td colspan="4">Semipervious</td><td colspan="6">Impervious</td></tr>
<tr><td>Aquifer</td><td colspan="5">Good</td><td colspan="4">Poor</td><td colspan="5">None</td></tr>
<tr><td rowspan="2">Soils</td><td colspan="2">Clean gravel</td><td colspan="3">Clean sand or sand and gravel</td><td colspan="4">Very fine sand, silt, loess, loam, solonetz</td><td colspan="5"></td></tr>
<tr><td colspan="4"></td><td colspan="2">Peat</td><td colspan="3">Stratified clay</td><td colspan="5">Unweathered clay</td></tr>
<tr><td>Rocks</td><td colspan="4"></td><td colspan="3">Oil rocks</td><td colspan="2">Sandstone</td><td colspan="2">Good limestone, dolomite</td><td colspan="3">Breccia, granite</td></tr>
<tr><td>$-\log_{10}$ · k(cm²)</td><td>3</td><td>4</td><td>5</td><td>6</td><td>7</td><td>8</td><td>9</td><td>10</td><td>11</td><td>12</td><td>13</td><td>14</td><td>15</td><td>16</td></tr>
<tr><td>$\log_{10}k$(md)</td><td>8</td><td>7</td><td>6</td><td>5</td><td>4</td><td>3</td><td>2</td><td>1</td><td>0</td><td>−1</td><td>−2</td><td>−3</td><td>−4</td><td>−5</td></tr>
</table>

or equivalent hydraulic gradient of 1 atmosphere per cm.

In (5.5.11), 1 centipoise $= 10^{-2}$ poise $= 10^{-2}$ dyne sec/cm^2 and 1 atmosphere $=$ 1.0132×10^6 dynes/cm^2. The conversion from the darcy to the area units is given by:

$$1 \text{ darcy} = 9.8697 \times 10^{-9}\,\text{cm}^2 = 1.062 \times 10^{-11}\,\text{ft}^2$$
$$= 9.613 \times 10^{-4}\,\text{cm/sec (for water at 20°C)}$$
$$= 1.4156 \times 10^{-2}\,\text{US gal/min ft}^2 \text{ (for water at 20°C)}.$$

In many cases the darcy is a rather large unit so that the millidarcy (md) (10^{-3} darcy) is used.

Typical examples of permeabilities of oil reservoir are summarized by Katz et al. (1959).

Table 5.5.1 (Irmay in Bear, Zaslavsky and Irmay 1968) gives a summary of hydraulic conductivity and permeability. In this table, following the United States Bureau of Reclamation, K is expressed in units of hydraulic conductivity class:

$$K_c = -\log_{10}K(\text{cm/sec}).$$

5.6 Anisotropic Permeability

In sections 4.8 and 5.2, the permeability, $\underline{\mathbf{k}}$, and the hydraulic conductivity, $\underline{\mathbf{K}}$, of a porous medium were shown to be second-rank tensors. Some brief comments on second-rank tensors are incorporated in paragraph 5.6.1. The reader is referred to such books as Morse and Feshbach (1953), Spain (1956), Spiegel (1959), Prager (1961), and Aris (1962) for additional information on this subject.

Although the discussion below is restricted to the x, y, z (or x_i, $i = 1, 2, 3$) Cartesian

coordinate system, it can easily be extended to curvilinear coordinate systems.
The porous medium is assumed to be homogeneous.

5.6.1 The Principal Directions

The relationship between the specific discharge $\mathbf{q}(q_1, q_2, q_3)$ and the gradient $\mathbf{J}(J_1, J_2, J_3) \equiv -\operatorname{grad}\varphi$ in the general case of an anisotropic medium, can be written as (5.2.8) or in the more compact forms:

$$\mathbf{q} = \underline{\mathbf{K}} \cdot \mathbf{J} \quad \text{or} \quad q_i = K_{ij}J_j \quad (i, j = 1, 2, 3, \quad \text{or} \quad x, y, z, \quad \text{or} \quad x_1, y_1, z_1). \quad (5.6.1)$$

In (5.6.1), as elsewhere in this book unless otherwise stated, *Einstein's summation convention* (or the *double-index summation convention*) is implied. According to this convention, in any product of terms, a suffix (subscript or superscript) repeated twice (and only twice) is held to be summed over its range of values (usually 1, 2, 3). For example, the second equation in (5.6.1) should be understood to stand for the three equations:

$$\begin{aligned} q_1 &= K_{11}J_1 + K_{12}J_2 + K_{13}J_3 \\ q_2 &= K_{21}J_1 + K_{22}J_2 + K_{23}J_3 \\ q_3 &= K_{31}J_1 + K_{32}J_2 + K_{33}J_3. \end{aligned} \quad (5.6.2)$$

The *nine* components K_{ij} in a three-dimensional space, or *four* components in a two-dimensional one, define the hydraulic conductivity tensor. They are often written in the compact matrix forms:

$$\underline{\mathbf{K}} = \begin{bmatrix} K_{11} & K_{12} & K_{13} \\ K_{21} & K_{22} & K_{23} \\ K_{31} & K_{32} & K_{33} \end{bmatrix} \quad \text{or} \quad \underline{\mathbf{K}} = \begin{bmatrix} K_{11} & K_{12} \\ K_{21} & K_{22} \end{bmatrix}. \quad (5.6.3)$$

Because $\underline{\mathbf{K}}$ is a symmetrical tensor (i.e., $K_{ij} = K_{ji}$), we have only *six different components* (or three in a two-dimensional field). Another form of writing the three equations of (5.6.2) is as a single matrix equation:

$$\begin{bmatrix} q_1 \\ q_2 \\ q_3 \end{bmatrix} = \begin{bmatrix} K_{11} & K_{12} & K_{13} \\ K_{21} & K_{22} & K_{23} \\ K_{31} & K_{32} & K_{33} \end{bmatrix} \cdot \begin{bmatrix} J_1 \\ J_2 \\ J_3 \end{bmatrix}. \quad (5.6.4)$$

The mixed component $K_{x_ix_j} (\equiv K_{ij})$ may be interpreted as the coefficient that, when multiplied by the component J_{x_j} of the hydraulic gradient $\mathbf{J}$, gives the contribution of the latter to the specific discharge q_{x_i} in the x_i direction. The flow q_{x_i} is the sum of specific discharges caused by $J_{x_1}, J_{x_2}, J_{x_3}$.

At this point, the reader is reminded of some properties of second-rank tensors.

(a) Given the components K_{ij} of a tensor $\underline{\mathbf{K}}$ in an x_i coordinate system, its components K'_{pq} in an x'_p system, obtained from x_i by rotation, are given by:

$$K'_{pq} = K_{ij}\alpha_{pi}\alpha_{qj}; \qquad i, j, p, q = 1, 2, 3 \quad (5.6.5)$$

where α_{mn} is the direction cosine between the axes x'_m and x_n: $\alpha_{mn} = \cos(\mathbf{1x}_m, \mathbf{1x}_n)$. In fact, to establish that a given entity is a second-rank tensor, we must show that its components transform according to (5.6.5).

(b) Another way to establish that a given entity is a second-rank tensor is the *quotient law* that states that if K_{ij} are nine quantities, and **B** and **C** are vectors (with components B_i and C_i, respectively), such that the B_i's are completely independent of the K_{ij}'s, and $C_i = K_{ij}B_j$, then the K_{ij}'s are components of a second-rank tensor.

(c) When $\det K \neq 0$, where $\det K$ denotes the determinant of the 3×3 matrix made up of the nine components K_{ij} of a second-rank tensor, it is possible to find the components of the *reciprocal* or *conjugate tensor* $\underset{-}{\mathbf{W}}$ ($\equiv \underset{-}{\mathbf{K}}^{-1}$) of $\underset{-}{\mathbf{K}}$ by

$$W_{ij} \equiv (\underset{-}{\mathbf{K}}^{-1})_{ij} = \frac{\text{cofactor of } K_{ij}}{\det K} \tag{5.6.6}$$

where the *cofactor* (or *minor*) of K_{ij} is the determinant obtained from $\det K$ by deleting the ith row and the jth column, and associating the sign $(-1)^{i+j}$ to this determinant.

The direction in space specified by the unit vector **1u** (with component u_i) is called a principal direction of the *tensor* $\underset{-}{\mathbf{K}}$ (component K_{ij}) if the associated vector $K_{ij}u_i$ is parallel to **1u**, that is, if this vector can be written in the form Ku_j, where K is a scalar. For a principal direction u_i of a symmetrical tensor K_{ij}, we then have the relationship $K_{ij}u_i - Ku_j = 0$ or:

$$(K_{ij} - K\delta_{ij})u_i = 0; \qquad \delta_{ij}\begin{cases} = 1 & \text{for } i = j \\ = 0 & \text{for } i \neq j \end{cases} \tag{5.6.7}$$

where δ_{ij} is the *Kronecker* δ. This relation represents a system of three linear homogeneous equations for the components of the unit vector u_i. Since the trivial solution $u_i = 0$ is not compatible with the condition $\mathbf{1u} \cdot \mathbf{1u} = 1$, the determinant of the coefficients in (5.6.7) must vanish. Indicating this determinant by a typical element, we have:

$$|K_{ij} - K\delta_{ij}| = 0. \tag{5.6.8}$$

Development of (5.6.8) gives a cubic equation for K called the *characteristic equation* of the symmetrical tensor K_{ij}:

$$K^3 - I_1K^2 - I_2K - I_3 = 0 \tag{5.6.9}$$

where I_1, I_2, and I_3 are scalars independent of the coordinate system. They are the *basic invariants* of the symmetrical tensor K ($I_1 = K_{ii} = K_{11} + K_{22} + K_{33}$ = sum of elements of the main diagonal of the matrix of K_{ij} = *trace* of K; $I_2 = \frac{1}{2}(K_{ii}K_{jj} - K_{ij}K_{ji})$ = sum of minors of these elements; $I_3 = (K_{ij}K_{jk}K_{ki} - 3K_{ij}K_{ji}K_{kk} + K_{ii}K_{jj}K_{kk})/6 = \det K$ = determinant of the matrix of K_{ij}).

The roots $K^{(1)}$, $K^{(2)}$ and $K^{(3)}$ of the characteristic equation are called *principal values* (characteristic values or *eigenvalues*) of the symmetrical tensor K_{ij}; they can be shown to be real numbers. With each of them we associate a unit vector,

$\mathbf{1u}^{(1)}$, $\mathbf{1u}^{(2)}$ and $\mathbf{1u}^{(3)}$, respectively; these are mutually orthogonal if $K^{(1)}$, $K^{(2)}$ and $K^{(3)}$ are distinct. These unit vectors indicate three principal directions in space corresponding to the principal values $K^{(1)}$, $K^{(2)}$ and $K^{(3)}$, respectively. The components of the tensor with the distinct principal values $K^{(1)}$, $K^{(2)}$ and $K^{(3)}$, and the corresponding orthogonal principal directions $\mathbf{1u}^{(1)}$, $\mathbf{1u}^{(2)}$ and $\mathbf{1u}^{(3)}$, can be written in the form:

$$K_{ij} = K^{(1)}u^{(1)}{}_i u^{(1)}{}_j + K^{(2)}u^{(2)}{}_i u^{(2)}{}_j + K^{(3)}u^{(3)}{}_i u^{(3)}{}_j. \tag{5.6.10}$$

Using $\mathbf{1u}^{(1)}$, $\mathbf{1u}^{(2)}$ and $\mathbf{1u}^{(3)}$ as base vectors of an orthogonal coordinate system x'_1, x'_2, x'_3, in which the components of the tensor $\underline{\mathbf{K}}$ are K'_{ij}, we obtain $K'_{11} = K^{(1)}$, $K'_{22} = K^{(2)}$, $K'_{33} = K^{(3)}$, $K'_{12} = K'_{13} = K'_{23} = 0$, so that the matrix of tensor K_{ij} of (5.6.10) has the diagonal form:

$$K'_{ij} = \begin{bmatrix} K^{(1)} & 0 & 0 \\ 0 & K^{(2)} & 0 \\ 0 & 0 & K^{(3)} \end{bmatrix}. \tag{5.6.11}$$

In this form, the tensor is called a *diagonal tensor*. Conversely, three orthogonal axes are called principal axes of a symmetrical tensor of the second order if, for these axes, the matrix of this tensor assumes a diagonal form. When the principal values are distinct, there exists only a single system of principal axes.

If $K^{(2)} = K^{(3)} \neq K^{(1)}$, the matrix has the diagonal form (5.6.11), but with $K'_{22} = K'_{33} = K^{(2)} = K^{(3)}$. However, once the x' axis is given the directions of $\mathbf{1u}^{(1)}$, this diagonal form is independent of the choice of y' and z' axes.

Finally, if all three principal values are equal, any unit vector obtained by a linear combination of $u^{(1)}{}_i$, $u^{(2)}{}_i$ and $u^{(3)}{}_i$ indicates a principal direction. In other words, *any* direction of space is a principal direction and any system of rectangular axes is a system of principal axes. The tensor $\underline{\mathbf{K}}'$, with $K^{(1)} = K^{(2)} = K^{(3)} = K$, then becomes:

$$\underline{\mathbf{K}}' = \begin{bmatrix} K & 0 & 0 \\ 0 & K & 0 \\ 0 & 0 & K \end{bmatrix} = K \begin{bmatrix} 1 & 0 & 0 \\ 0 & 1 & 0 \\ 0 & 0 & 1 \end{bmatrix} \quad \text{or:} \quad K'_{ij} = K\delta_{ij}. \tag{5.6.12}$$

(d) Given the components K_{ij} of a tensor $\underline{\mathbf{K}}$ in an x_i $(i = 1, 2, 3)$ coordinate system, we can always find three principal directions x'_i $(i = 1, 2, 3)$ such that in it the tensor $\underline{\mathbf{K}}'$ (components K'_{ij}) will have a diagonal form, i.e., $K'_{ij} = 0$ for $i \neq j$. In fact, using this condition together with (5.6.5), it is possible to obtain expressions relating the K_{ij}s to K'_{ij}s. For the two-dimensional case, these relationships are:

$$K'_{11} = \frac{K_{11} + K_{22}}{2} + \frac{K_{11} - K_{22}}{2} \cos 2\theta + K_{12} \sin 2\theta$$

$$K'_{22} = \frac{K_{11} + K_{22}}{2} - \frac{K_{11} - K_{22}}{2} \cos 2\theta - K_{12} \sin 2\theta$$

$$K'_{12} = K'_{21} = 0 \tag{5.6.13}$$

where θ is the angle between x_i and x'_i.

It can readily be verified that there are two values of θ, differing by 90°, for which $K'_{12} = K'_{21} = 0$. These are solutions of:

$$\tan 2\theta = 2K_{12}/(K_{11} - K_{22}). \tag{5.6.14}$$

By differentiation it can be shown that in these two directions K'_{11} has in one case the largest value, and in the other case the smallest one. Following our definition above, these directions are the principal directions, and the values of K'_{11} and K'_{22} in these directions are the principal values of the tensor $\underline{\mathbf{K}}'$:

$$\underline{\mathbf{K}}' \equiv \begin{bmatrix} K'_{11} & 0 \\ 0 & K'_{22} \end{bmatrix} \tag{5.6.15}$$

where now x'_1 and x'_2 are the principal directions.

(e) Given the principal axes x'_1, x'_2, with the tensorial property $\underline{\mathbf{K}}$ defined by (5.6.15), the components of this tensor in any other coordinate system x_1, x_2 making an angle θ with the x'_1, x'_2 system are given by:

$$K_{11} = \frac{K'_{11} + K'_{22}}{2} + \frac{K'_{11} - K'_{22}}{2}\cos 2\theta; \qquad K_{12} = -\frac{K'_{11} - K'_{22}}{2}\sin 2\theta;$$

$$K_{22} = \frac{K'_{11} + K'_{22}}{2} - \frac{K'_{11} - K'_{22}}{2}\cos 2\theta. \tag{5.6.16}$$

(f) A most convenient way of representing these relationships (and deriving angles and components) in a graphic form is by using *Mohr's circle* (fig. 5.6.1) (Terzaghi 1943).

The reason for presenting this brief summary here is that the entire discussion is obviously applicable to the second-rank tensors $\underline{\mathbf{k}}$ and $\underline{\mathbf{K}}$.

It is also possible to express the relationship between $\mathbf{q}$ and $\mathbf{J}$ by the *hydraulic resistivity tensor* $\underline{\mathbf{W}}$. This is a symmetrical, second-rank tensor, the conjugate tensor (denoted as $\underline{\mathbf{K}}^{-1}$) of $\underline{\mathbf{K}}$. Its nine components W_{ij} are obtained from those of $\underline{\mathbf{K}}$ by (5.6.6), as $W_{ij} \equiv (\underline{\mathbf{K}}^{-1})_{ij}$. In detail:

$$W_{11} = (K_{22}K_{33} - K_{23}K_{32})/\det K; \qquad W_{12} = -(K_{21}K_{33} - K_{23}K_{31})/\det K$$

$$W_{21} = -(K_{12}K_{33} - K_{13}K_{32})/\det K; \qquad W_{22} = (K_{11}K_{33} - K_{13}K_{31})/\det K \quad \text{etc.} \tag{5.6.17}$$

When x_1, x_2, x_3 are *principal directions*, $\det K = K_{11}K_{22}K_{33}$, $K_{ij} = 0$ for $i \neq j$, i.e., $K_{12} = K_{23} = K_{13} = \cdots = 0$, and $W_{11} = 1/K_{11}$, $W_{22} = 1/K_{22}$, $W_{33} = 1/K_{33}$, $W_{12} = W_{13} = W_{23} = \cdots = 0$.

Using $\underline{\mathbf{W}}$, the relationship between $\mathbf{q}$ and $\mathbf{J}$ becomes:

$$\mathbf{J} = \underline{\mathbf{W}} \cdot \mathbf{q}; \qquad J_i = W_{ij}q_j, \qquad i, j = 1, 2, 3. \tag{5.6.18}$$

Being a second-rank tensor, the transformation of $\underline{\mathbf{K}}$ (components $K_{x_1x_1}$, $K_{x_1x_2}$, etc.) from one coordinate system (say, x_i) into $\underline{\mathbf{K}}'$ (components $K_{x'_1x'_1}$, $K_{x'_1x'_2}$, etc.)

in another coordinate system (say, x'_i), is given by (5.6.5). If the x_is coincide with the principal directions of hydraulic conductivity, we have:

$$K_{x'_i x'_j} = K_{x_1x_1}\alpha_{i1}\alpha_{j1} + K_{x_2x_2}\alpha_{i2}\alpha_{j2} + K_{x_3x_3}\alpha_{i3}\alpha_{j3}. \tag{5.6.19}$$

For the two-dimensional case (say, $x_1 \equiv x$, $x_2 \equiv y$), we have:

$$K_{x'x'} = \frac{K_{xx} + K_{yy}}{2} + \frac{K_{xx} - K_{yy}}{2}\cos 2\theta + K_{xy}\sin 2\theta$$

$$K_{x'y'} = K_{y'x'} = -\frac{K_{xx} - K_{yy}}{2}\sin 2\theta + K_{xy}\cos 2\theta$$

$$K_{y'y'} = \frac{K_{xx} + K_{yy}}{2} - \frac{K_{xx} - K_{yy}}{2}\cos 2\theta - K_{xy}\sin 2\theta. \tag{5.6.20}$$

The directions x, y and x', y' and the angle θ are shown in figure 5.6.1a. One should note that the symbol $\underline{\mathbf{K}}'$ stands for the hydraulic conductivity tensor when written in the $x'y'$ coordinate system to emphasize that, by referring its components to another coordinate system, the property itself—the hydraulic conductivity—does not change.

If x and y are the orthogonal principal directions of hydraulic conductivity, so that $K_{xy} = K_{yx} = 0$, (5.6.20) becomes:

$$\begin{matrix} K_{x'x'} \\ K_{y'y'} \end{matrix} = \frac{K_{xx} + K_{yy}}{2} \pm \frac{K_{xx} - K_{yy}}{2}\cos 2\theta; \qquad K_{x'y'} = -\frac{K_{xx} - K_{yy}}{2}\sin 2\theta. \tag{5.6.21}$$

If, however, x', y' are principal directions but x, y are not, the angle θ may be obtained from K_{xx}, K_{yy} and K_{zz} by (5.6.14). Then:

$$\begin{matrix} K_{x'} \equiv K_{x'x'} \\ K_{y'} \equiv K_{y'y'} \end{matrix} = \frac{K_{xx} + K_{yy}}{2} \pm \left[\left(\frac{K_{xx} - K_{yy}}{2}\right)^2 + K_{xy}^2\right]^{1/2}. \tag{5.6.22}$$

In figure 5.6.1b Mohr's circle is used to determine the hydraulic conductivity in any coordinate system, x', y', given that x and y (with known components $K_{xx} \equiv K_x$ and $K_{yy} \equiv K_y$) are the principal directions.

Figure 5.6.1c shows how Mohr's circle is used to determine principal directions and principal hydraulic conductivities, given the hydraulic conductivity in any x, y coordinate system. Figure 5.6.1d shows how $\underline{\mathbf{K}}$ is transformed from one coordinate system to another one making an angle, θ, with the first. One should note that in (5.6.21), $K_{x'y'}$ may become negative.

In certain porous media, we have two principal hydraulic conductivities that are equal, e.g., $K_x = K_y \neq K_z$, where one subscript is used to indicate that x, y, z are principal directions. In such cases, any direction in the xy plane is a principal direction. Such media are said to have *cross* or *transverse anisotropy* (also *axi-symmetric anisotropy*). Media made of a large number of thin layers of alternating permeability exhibit this property.

In an *isotropic medium* $K_x = K_y = K_z = K$ and every direction is a principal direction; K_{ij} may then be written as (5.6.12).

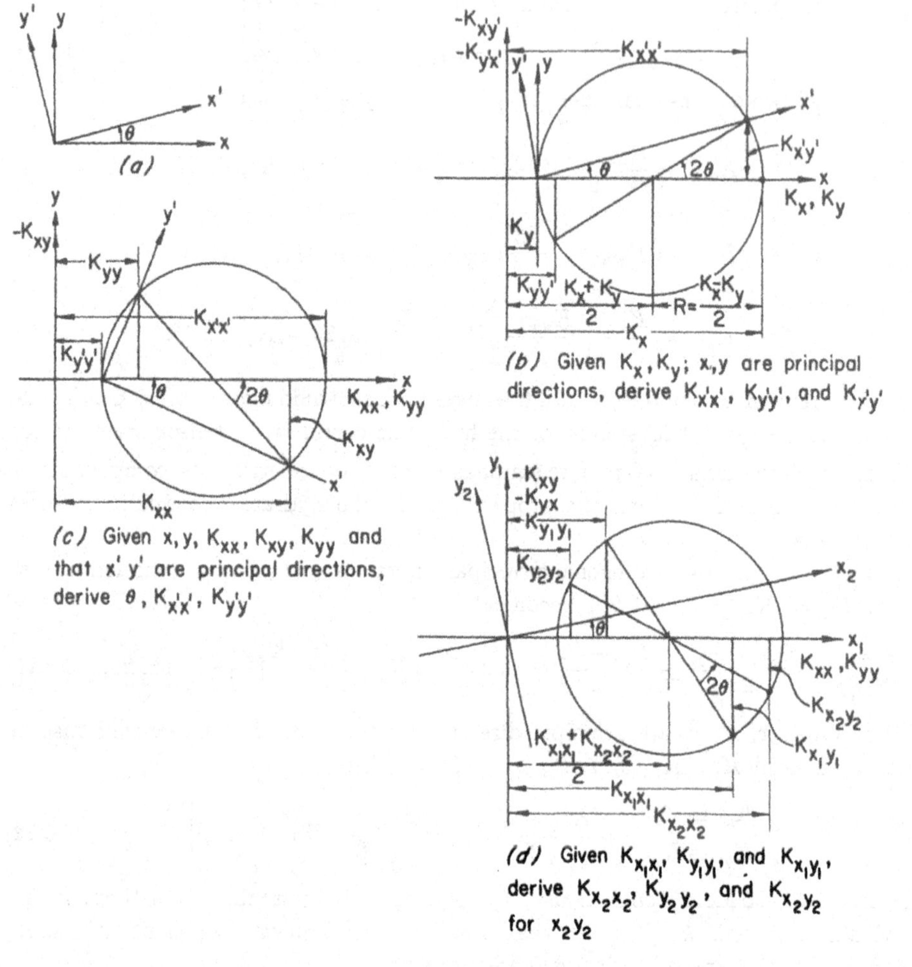
(a)
(b) Given Kx, Ky; x,y are principal directions, derive Kx'x', Ky'y', and Kx'y'
(c) Given x, y, Kxx, Kxy, Kyy and that x' y' are principal directions, derive θ, Kx'x', Ky'y'
(d) Given Kx1x1, Ky1y1, and Kx1y1, derive Kx2x2, Ky2y2, and Kx2y2 for x2y2

FIG. 5.6.1. The use of Mohr's circle for hydraulic conductivity in two dimensions.

Once the principal values and principal axes (say x, y, z) have been determined, the relationships among the components of the vectors $\mathbf{q}$ and $\mathbf{J}$ take the form:

$$q_x = K_x J_x; \qquad q_y = K_y J_y; \qquad q_z = K_z J_z. \tag{5.6.23}$$

"Two-dimensional flow" has already been mentioned several times in the present chapter and will be referred to many times throughout the remaining chapters of this book. By two-dimensional flow, say in the x_1x_2 plane, we mean a flow such that $\partial(\ )/\partial x_3 \equiv 0$ everywhere, where ( ) stands for any property of the fluid (pressure, velocity, etc.). In flow through an isotropic three-dimensional medium this requirement immediately leads to the conclusion that both J_{x_3} and q_{x_3} must vanish everywhere. However, in general, when the medium is anisotropic, having $J_{x_3} \equiv 0$ does not necessarily mean that $q_{x_3} \equiv 0$, because in this case we still have contributions to q_{x_3} from J_{x_1} and J_{x_2}:

$$q_{x_3} = K_{31}J_{x_1} + K_{32}J_{x_2}. \tag{5.6.24}$$

Hence two-dimensional flow is possible when $K_{31} = K_{32} = K_{13} = K_{23} = 0$ or, in general, when $K_{31}J_{x_1} + K_{32}J_{x_2} = 0$. When x_1, x_2 and x_3 are principal directions, then having one component of **J** vanish is sufficient to produce two-dimensional flow in planes normal to the direction of that component. Finally, one should note that the various components of K must obey the relationships (4.4.15) through (4.4.17).

5.6.2 Directional Permeability

From (5.6.1) it follows that in an anisotropic medium the vectors **q** and **J** are non-colinear except when they are in the direction of one of the principal axes. This means that the directions of the streamlines (sec. 6.5) do not coincide with those of the normals to the equipotentials. The angle θ between the vectors **q** (components: q_x, q_y, q_z) and **J** (components: J_x, J_y, J_z) is given by:

$$\cos\theta = \mathbf{q}\cdot\mathbf{J}/qJ; \qquad q \equiv |\mathbf{q}|, \qquad J \equiv |\mathbf{J}|. \tag{5.6.25}$$

When x, y, z are the principal directions of hydraulic conductivity, we have:

$$\cos\theta = (K_xJ_x^2 + K_yJ_y^2 + K_zJ_z^2)/qJ. \tag{5.6.26}$$

Consider the following two cases.

Case 1, directional hydraulic conductivity in the direction of the flow. Following the definition of hydraulic conductivity in Darcy's law for an isotropic medium, let us define hydraulic conductivity as the ratio between the specific discharge at a point and the component of the gradient in the direction of **q**. We shall refer to it as a *directional hydraulic conductivity* (or if only the medium property is considered, as *directional permeability*) *in the direction of the flow.* Denoting this hydraulic conductivity by K_q, we may write (fig. 5.6.2a):

$$K_q = \frac{q}{J\cos\theta}; \qquad q = K_q(J\cos\theta). \tag{5.6.27}$$

From (5.6.25) and (5.6.26) it follows that:

$$K_q = q/J\cos\theta = q^2/(\mathbf{q}\cdot\mathbf{J}) = q^2/(\underline{\mathbf{K}}\mathbf{J})\cdot\mathbf{J} = \frac{K_x^2J_x^2 + K_y^2J_y^2 + K_z^2J_z^2}{K_xJ_x^2 + K_yJ_y^2 + K_zJ_z^2}. \tag{5.6.28}$$

Let β_1, β_2, β_3 denote the angles between the direction of **q** and the three principal axes, x, y, z, respectively. Then:

$$q_x = K_xJ_x = q\cos\beta_1, \qquad q_y = K_yJ_y = q\cos\beta_2, \qquad q_z = K_zJ_z = q\cos\beta_3. \tag{5.6.29}$$

By combining (5.6.28) and (5.6.29) we obtain:

$$\frac{1}{K_q} = \frac{\cos^2\beta_1}{K_x} + \frac{\cos^2\beta_2}{K_y} + \frac{\cos^2\beta_3}{K_z}. \tag{5.6.30}$$

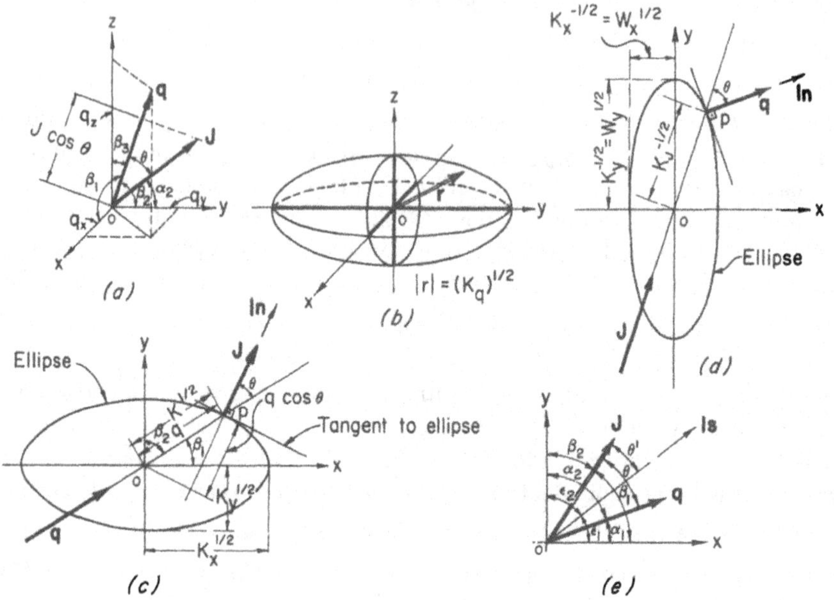

FIG. 5.6.2. Ellipsoids and ellipses of directional hydraulic conductivity.

In the xyz coordinate system, a radius vector $\mathbf{r}$, which is colinear with $\mathbf{q}$, has the components: $x = r \cos \beta_1$; $y = r \cos \beta_2$; $z = r \cos \beta_3$. Hence, (5.6.30) becomes:

$$\frac{r^2}{K_q} = \frac{x^2}{K_x} + \frac{y^2}{K_y} + \frac{z^2}{K_z}. \tag{5.6.31}$$

By drawing a segment of length $r = (K_q)^{1/2}$ in the direction of $\mathbf{q}$, we obtain from (5.6.31):

$$\frac{x^2}{K_x} + \frac{y^2}{K_y} + \frac{z^2}{K_z} = 1 \tag{5.6.32}$$

which is the canonical equation of an *ellipsoid* (fig. 5.6.2b) in the x, y, z coordinate system. Its semiaxes are: $K_x^{1/2}$, $K_y^{1/2}$ and $K_z^{1/2}$ in the x, y and z directions, respectively. In two dimensions, we have the ellipse $x^2/K_x + y^2/K_y = 1$ shown in figure 5.6.2c. The ellipsoid (or ellipse in two dimensions) thus gives directional permeability in the direction of the flow. Knowing β_1, β_2 and β_3, we may determine K_q from them or from (5.6.30). Given $\mathbf{q}$ (in magnitude and direction) and K_x, K_y and K_z, we may determine K_q from (5.6.28).

Case 2, directional hydraulic conductivity in the direction of the gradient. Suppose that we know the direction of the hydraulic gradient $\mathbf{J}$. We may then define hydraulic conductivity by the ratio between the component of the specific discharge $\mathbf{q}$ in the direction of the gradient and the gradient itself. This would then be the *directional hydraulic conductivity* (K_J) *in the direction of the gradient.* Thus:

$$K_J = q \cos \theta / J; \qquad q \cos \theta = K_J J \tag{5.6.33}$$

with:

$$J_x = J \cos \alpha_1, \qquad J_y = J \cos \alpha_2, \qquad J_z = J \cos \alpha_3,$$

where $\alpha_1, \alpha_2, \alpha_3$ are the angles between the direction of **J** and x, y, z respectively. We obtain from (5.6.26) and (5.6.33):

$$K_J = \frac{1}{J^2}(K_x J_x^2 + K_y J_y^2 + K_z J_z^2). \tag{5.6.34}$$

With a radius vector r such that $x = r \cos \alpha_1$, $y = r \cos \alpha_2$, $z = r \cos \alpha_3$, (5.6.34) becomes:

$$\frac{r^2}{1/K_J} = \frac{x^2}{1/K_x} + \frac{y^2}{1/K_y} + \frac{z^2}{1/K_z}. \tag{5.6.35}$$

By drawing $r = (K_J)^{-1/2}$ in the direction of **J**, we obtain a canonical equation of an ellipsoid:

$$\frac{x^2}{1/K_x} + \frac{y^2}{1/K_y} + \frac{z^2}{1/K_z} = 1. \tag{5.6.36}$$

However, in this case the semiaxes are $K_x^{-1/2}$, $K_y^{-1/2}$, $K_z^{-1/2}$. In two dimensions the equation is $x^2/(1/K_x) + y^2/(1/K_y) = 1$, and we have the ellipse shown in figure 5.6.2d. Since in the principal axes of permeability, $W_x = 1/K_x$, $W_y = 1/K_y$, $W_z = 1/K_z$, equation (5.6.36) may be written as:

$$\frac{x^2}{W_x} + \frac{y^2}{W_y} + \frac{z^2}{W_z} = 1 \tag{5.6.37}$$

so that the major semiaxes are also $W_x^{1/2}$, $W_y^{1/2}$, $W_z^{1/2}$. The length of any radius vector will give the value of $K_J^{-1/2} \equiv W_J^{1/2}$.

We thus have two definitions of directional hydraulic conductivity. They are distinct and should not be interchanged. Whenever the direction of the specific discharge **q** is given in a medium with known K_x, K_y and K_z, directional conductivity in that direction is given by (5.6.30). On the other hand, given **J**, directional conductivity in that direction is determined by (5.6.34).

Consider a particular direction (specified by α_1, α_2, α_3) in the porous material. For that direction we determine K_q by (5.6.30) and K_J by (5.6.34) in *two different flows*. From these two equations it follows that:

$$\begin{aligned} \frac{K_J}{K_q} &= \left(\frac{\cos^2\alpha_1}{K_x} + \frac{\cos^2\alpha_2}{K_y} + \frac{\cos^2\alpha_3}{K_z}\right)(K_x \cos^2\alpha_1 + K_y \cos^2\alpha_2 + K_z \cos^2\alpha_3) \\ &= 1 + \frac{(K_x - K_y)^2}{K_x K_y} \cos^2\alpha_1 \cos^2\alpha_2 + \frac{(K_y - K_z)^2}{K_y K_z} \cos^2\alpha_2 \cos^2\alpha_3 \\ &\quad + \frac{(K_x - K_z)^2}{K_x K_z} \cos^2\alpha_1 \cos^2\alpha_3 \geqslant 1. \end{aligned} \tag{5.6.38}$$

On the other hand, for a *particular flow*, we may determine both K_q and K_J, i.e., *in two different directions*, in the soil. Then, from (5.6.28) and (5.6.33), it follows that

$$K_J/K_q = \cos^2\theta \leqslant 1. \tag{5.6.39}$$

Marcus and Evenson (1961) give graphs for the ratio K_J/K_q of (5.6.38) for the two-dimensional case where:

$$K_J/K_q = 1 + [(K_x - K_y)^2/4K_xK_y] \sin^2 2\alpha_1. \tag{5.6.40}$$

To understand the significance of (5.6.38), we should associate with case 1 the flow in a narrow and relatively long sand-packed column. The sand is anisotropic with known principal conductivities K_1, K_2 and K_3 ($\equiv K_x$, K_y and K_z), but unknown principal directions (with respect, say, to the column's axis). In this case, the column's geometry virtually compels the liquid to move in a certain direction, namely in that of the column's axis, and (5.6.30) is applicable. On the other hand, case 2 corresponds to a very wide and relatively thin sample between two parallel equipotential plane boundaries. The gradient is practically parallel to the normal to both planes. The fluid will move in the direction of least resistance. Here we know the direction of the gradient and hence (5.6.34) is applicable. Thus K_q is measured when the flow is forced in a certain direction, whereas K_J is measured when the fluid is free to choose the path of least resistance, and consequently, highest conductivity (Marcus and Evenson 1961). It becomes obvious that knowledge of both $\underline{\mathbf{K}}$ (e.g., by knowing the principal values K_1, K_2 and K_3 and the principal directions) and the boundary conditions is required to determine the flow pattern. This conclusion has an important bearing on the determination of hydraulic conductivity by *permeameters* (see sec. 5.7) in anisotropic media. Most cases have intermediate width–length ratios and fall between the extreme cases 1 and 2, discussed above.

The two cases considered above may be combined in the following way. Consider a direction $\mathbf{1s}$ defined by angles ε_1, ε_2 and ε_3 with respect to the axes x, y, z. The directions of $\mathbf{J}$ and $\mathbf{q}$ are defined by the angles (α_1, α_2, α_3) and (β_1, β_2, β_3), respectively. Figure 5.6.2e shows this situation for a two-dimensional flow. Then:

$$q_s = \mathbf{q} \cdot \mathbf{1s} = K_s J_s = K_s J \cos\theta' \tag{5.6.41}$$

where θ' is the angle between $\mathbf{J}$ and $\mathbf{1s}$. Since

$$\cos\theta' = \cos\alpha_1 \cos\varepsilon_1 + \cos\alpha_2 \cos\varepsilon_2 + \cos\alpha_3 \cos\varepsilon_3,$$

we obtain

$$K_s = \frac{K_x \cos\alpha_1 \cos\varepsilon_1 + K_y \cos\alpha_2 \cos\varepsilon_2 + K_z \cos\alpha_3 \cos\varepsilon_3}{\cos\alpha_1 \cos\varepsilon_1 + \cos\alpha_2 \cos\varepsilon_2 + \cos\alpha_3 \cos\varepsilon_3} \tag{5.6.42}$$

which expresses K_s for the direction $\mathbf{1s}$ in the flow in terms of K_x, K_y, K_z and the directions of $\mathbf{1s}$ and $\mathbf{J}$. In two dimensions (fig. 5.6.2e):

$$K_s = (K_x \cos\alpha_1 \cos\varepsilon_1 + K_y \sin\alpha_1 \sin\varepsilon_1)/\cos(\alpha_1 - \varepsilon_1). \tag{5.6.43}$$

Case 1 above corresponds to $(\varepsilon_1, \varepsilon_2, \varepsilon_3) \equiv (\beta_1, \beta_2, \beta_3)$; case 2 corresponds to $(\varepsilon_1, \varepsilon_2, \varepsilon_3) \equiv (\alpha_1, \alpha_2, \alpha_3)$.

From $q_s = q \cos \theta''$, where θ'' is the angle between the directions of **q** and **1s**, we may also obtain:

$$K_s = \frac{\cos \beta_1 \cos \varepsilon_1 + \cos \beta_2 \cos \varepsilon_2 + \cos \beta_3 \cos \varepsilon_3}{(\cos \beta_1 \cos \alpha_1)/K_1 + (\cos \beta_2 \cos \alpha_2)/K_2 + (\cos \beta_3 \cos \alpha_3)/K_3} \tag{5.6.44}$$

which gives K_s in terms of K_x, K_y, K_z and the directions of **q** and **1s**.

The ellipsoid (5.6.32) may be used to determine the direction of **J**, given **q**. Consider the surface $F = \text{const}$, where:

$$2F = x^2/K_x + y^2/K_y + z^2/K_z. \tag{5.6.45}$$

The unit vector normal to this surface at some point, $P(x, y, z)$, on it is:

$$\mathbf{1n} = \nabla F/|\nabla F| = [(x/K_x)\mathbf{1x} + (y/K_y)\mathbf{1y} + (z/K_z)\mathbf{1z})]/|\nabla F|. \tag{5.6.46}$$

The direction OP is also the direction of the specific discharge vector **q**. On the other hand, the gradient is given by:

$$\mathbf{J} = J_x\mathbf{1x} + J_y\mathbf{1y} + J_z\mathbf{1z} = (q_x/K_x)\mathbf{1x} + (q_y/K_y)\mathbf{1y} + (q_z/K_z)\mathbf{1z}. \tag{5.6.47}$$

Hence the direction of **J** is that of the normal to the ellipsoid at P. The two-dimensional case is shown in figure 5.6.2c. To determine the magnitude of **J** consider another ellipsoid:

$$Jq \cos \theta = (\mathbf{q} \cdot \mathbf{J}) = q_xJ_x + q_yJ_y + q_zJ_z = (q_x^2/K_x) + (q_y^2/K_y) + (q_z^2/K_z) = 2G. \tag{5.6.48}$$

This is an ellipsoid in a coordinate system $q_xq_yq_z$. The axes are the same as those of the ellipsoid represented by (5.6.32). If we use the same ellipsoid (i.e., the same scales), OP will represent $|\mathbf{q}|$, $2G = 2F = 1$ and $J = 1/q \cos \theta$. Figure 5.6.2c gives the two-dimensional case.

In a similar manner, one may obtain the direction and magnitude of **q** given, **J**, from the ellipsoid (5.6.36) or (5.6.37). The **q** is again in the direction of the normal to the ellipsoid at P, and $q = 2F/J \cos \theta$. In both cases, we may determine q or J from $2G$:

$$qJ \cos \theta = q_x^2/K_x + q_y^2/K_y + q_z^2/K_z = K_xJ_x^2 + K_yJ_y^2 + K_zJ_z^2 = 2G \tag{5.6.49}$$

without referring to the figures. The directional permeabilities are obtained from:

$$K_q = q^2/2G, \qquad K_J = 2G/J^2, \qquad K_qK_J = (q/J)^2.$$

The discussion above can be extended to x, y, z that are not principal direction by using:

$$qJ \cos \theta = (\mathbf{q} \cdot \mathbf{J}) = K_{xx}J_x^2 + K_{yy}J_y^2 + K_{zz}J_z^2 + 2K_{xy}J_xJ_y + 2K_{xz}J_xJ_z + 2K_{yz}J_yJ_z$$
$$= 2G = K_JJ^2 \tag{5.6.50}$$

which describes the ellipsoid for K_J, and

$$qJ\cos\theta = (\mathbf{q}\cdot\mathbf{J}) = W_{xx}q_x^2 + W_{yy}q_y^2 + W_{zz}q_z^2 + 2(W_{xy}q_xq_y + W_{xz}q_xq_z + W_{yz}q_yq_z)$$
$$= 2G = q^2/K_q \qquad (5.6.51)$$

which describes the ellipsoid for K_q.

5.7 Measurement of Hydraulic Conductivity

5.7.1 General

Darcy's law, in fact, any law describing a physical phenomenon, is meaningless unless we can determine the numerical value of the coefficient or coefficients appearing in it. Here it is the hydraulic conductivity K (or the intrinsic permeability k) of a porous medium.

The following principles underlie the experimental determination, in field or laboratory, of coefficients appearing in the macroscopic description of phenomena.

(a) We assume a flow pattern (geometry, boundaries, etc.) such that (i) we can derive an analytical solution describing it and (ii) it can be produced in an experiment (in the laboratory or in the field). The mathematical solution relates the dependent variables (usually head, pressure, velocity, discharge, etc.) to the independent ones (time and position in space) and to the various coefficients (e.g., of the medium or of the fluid).

(b) An experiment is performed reproducing the chosen flow pattern. All measurable quantities appearing in the analytical solution are determined.

(c) The various coefficients are computed by inserting these quantities into the analytical solution.

These steps are applicable to both field and laboratory experiments. Sometimes uncontrolled natural phenomena in the field (e.g., the decaying discharge of a spring or the fluctuations of a water table) may be used as experiments from which coefficients of the natural environment (e.g., the aquifer) may be derived.

Hydraulic conductivity may be derived by both field and laboratory methods. Field pumping tests in wells are often used for determining aquifer coefficients such as storativity and transmissivity (KD). When the aquifer's thickness (D) is known, these tests also yield K. Various drainage formulas (e.g., those describing the two-dimensional flow in a parallel-drain system) are often used by drainage engineers. Another method often used by drainage engineers for determining hydraulic conductivity *in situ* is the *auger-hole method* (e.g., Luthin 1966).

In the laboratory, the hydraulic conductivity (or the permeability) is determined by means of an instrument called a *permeameter*. In a permeameter, the flow pattern is one of steady or unsteady one-dimensional flow through a small cylindrical porous medium sample. In general the samples are disturbed, except when they are obtained by means of special techniques and instruments designed to produce undisturbed samples. It is very difficult to obtain undisturbed samples of unconsolidated materials.

To avoid clogging the tested sample with entrapped air, bacterial growth (especially in prolonged experiments) or fines, filtered water, de-aired and sterilized (e.g., by adding Formaldehyde), is recommended.

Only saturated flow is considered in the present section. The determination of relative permeability is discussed in paragraph 9.3.7.

5.7.2 The Constant Head Permeameter

In this instrument, the one most commonly used for determining K, the flow is one-dimensional and steady. The porous medium sample, usually in the shape of a right circular cylinder of length L and cross-sectional area A (fig. 5.7.1), is placed between two porous plates that provide almost no resistance to the flow through them. A constant head difference, $\Delta\varphi$, is applied across the tested sample, producing a steady flow at a rate Q. When an incompressible fluid (a liquid) is used, hydraulic conductivity and permeability are determined from Darcy's law:

$$K = k\gamma/\mu = QL/A\,\Delta\varphi. \tag{5.7.1}$$

To obtain more reliable results, several tests are performed under different heads, $\Delta\varphi$. If a gas is employed for determining permeability, both the gas compressibility and the Klinkenberg effect (par. 5.3.3) must be taken into account.

The gas permeability of an ideal gas, k_g, may be determined from the solution of the one-dimensional gas flow equation (par. 6.2.3), or the one-dimensional motion equation:

$$\rho Q = -\rho A(k/\mu)\,\partial p/\partial x \tag{5.7.2}$$

where ρQ is the gas mass flux in the horizontal x direction, μ is the gas viscosity (assumed constant) and A is the cross-section of the core.

By integrating from $x = 0$, $p = p_1$ to $x = L$, $p = p_2$:

$$Q_b L = (k_g A/2\mu p_b)(p_1^2 - p_2^2); \qquad k_g = 2\mu p_b Q_b L/A(p_1^2 - p_2^2) = \mu p_b Q_b L/A(p_1 - p_2)\bar{p} \tag{5.7.3}$$

where $\bar{p} = (p_1 + p_2)/2$ and Q_b and p_b are some base values of discharge and corresponding pressure (say atmospheric). To obtain the liquid permeability k_l from k_g, we employ (5.3.8) with $p = \bar{p}$ and b corresponding to $\bar{p}$.

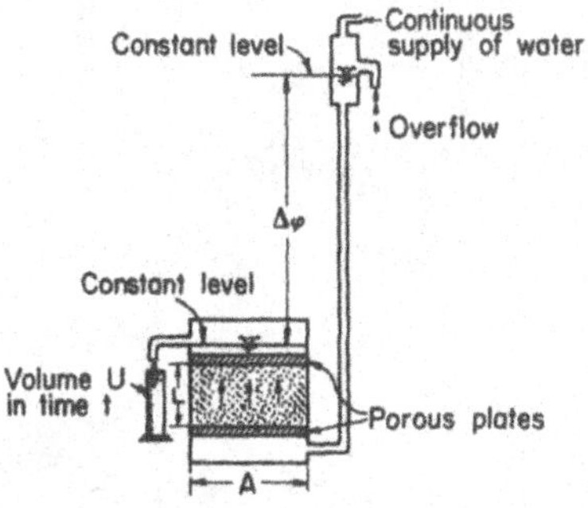

FIG. 5.7.1. Schematic diagram of constant head permeameter.

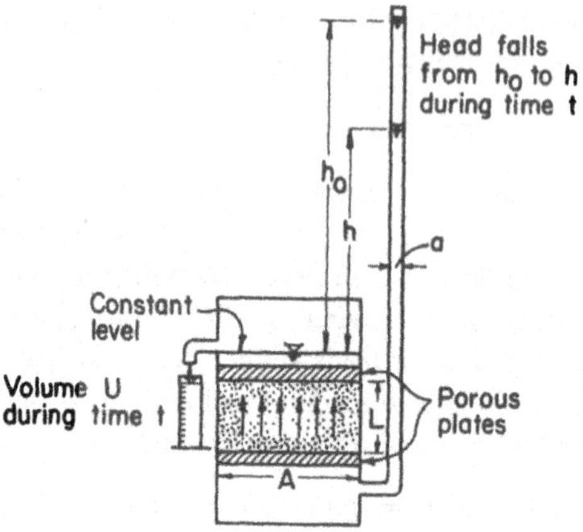

FIG. 5.7.2. Schematic diagram of falling head permeameter.

5.7.3 The Falling Head Permeameter

In a permeameter of this kind (fig. 5.7.2), used when a high head is desired, head and discharge vary during the test. With the nomenclature shown in figure 5.7.2, we have:

$$a = dU/dt = KAh/L; \qquad dU = a\,dh; \qquad dh/h = (KA/aL)\,dt;$$

$$K = (al/At)\ln(h_0/h). \tag{5.7.4}$$

Another type of permeameter with a varying head is the one shown in figure 5.7.3. This permeameter, called the *nondischarging permeameter*, is especially suitable for nonconsolidated media under very low heads. Hydraulic conductivity is obtained from:

$$K = (AL/2at)\ln(h_0/h). \tag{5.7.5}$$

5.7.4 Determining Anisotropic Hydraulic Conductivity

The hydraulic conductivity tensor contains six different components (sec. 5.6). We

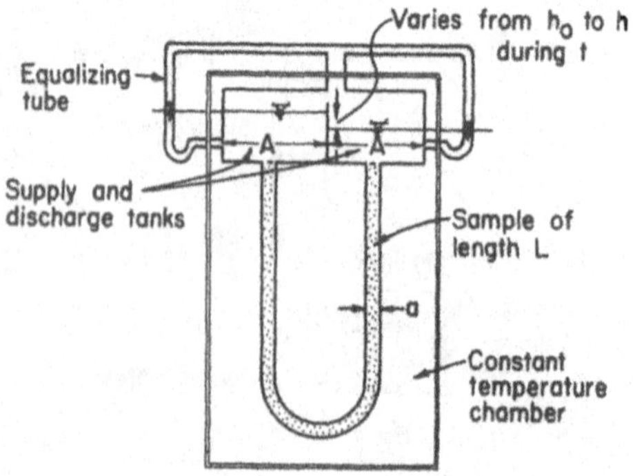

FIG. 5.7.3. Schematic diagram of nondischarging permeameter.

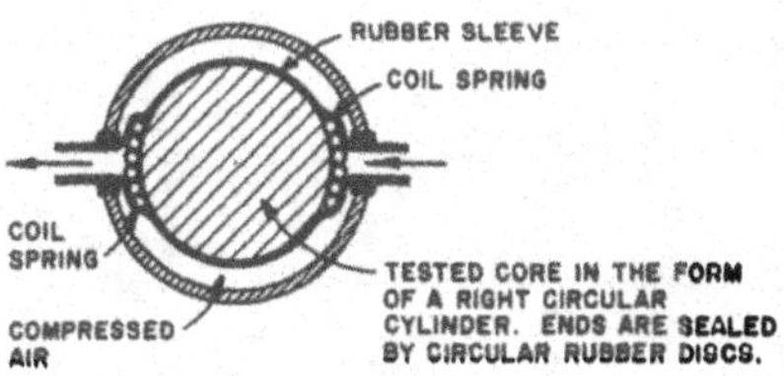

FIG. 5.7.4. Schematic section through apparatus for determining transverse permeability.

therefore need six independent experiments in order to derive them. When the three major axes are known, we need only three independent experiments. In the case of transversal- (or axisymmetrical) anisotropy, and given the directions of the major axes, we need only two experiments (to determine the two principal hydraulic conductivities).

In the latter case it is possible, for example, to cut small circular plugs along the core's axis (assumed to be a principal direction) and to determine their permeability by a linear flow experiment. Then another test is performed on a long core section employing transverse flow; figure 5.7.4 shows an example of an arrangement for producing steady plane gas flow transverse to the core's axis. The compressed air prevents bypass flow around the core.

Following the steps described in paragraph 5.7.1, we seek a solution for the flow pattern taking place in the cross-section normal to the core's axis. $\varphi = \varphi(x, y)$, or $\varphi = \varphi(r, \theta)$, satisfies $\nabla^2\varphi = 0$. Such a solution is presented, for example, by Collins (1961). From this solution k_t can be derived.

Thus from one axial experiment and one transversal experiment, we obtain the two principal permeabilities (assuming, of course, that these two directions are indeed principal directions).

Rühl and Schmidt (1957) review anisotropy of permeability. Measurement of anisotropic permeability is also described by Johnson and Hughes (1948), de Boodt and Kirkham (1953), Maasland and Kirkham (1955), Hutta and Griffiths (1955), and others.

5.8 Layered Porous Media

5.8.1 Flow Normal and Parallel to the Medium Layers

Porous media making up aquifers and oil reservoirs are seldom homogeneous with respect to permeability. However, when a rock system is composed of distinct layers, the average permeability of the system can be determined for some simple flow cases. Because in the present section we consider only the flow of a homogeneous liquid, we shall refer to the hydraulic conductivity, K, of the layer, rather than to its permeability.

Consider the flow of a homogeneous fluid (ρ, μ = const) parallel to the combination of N layers of porous rock (fig. 5.8.1). This situation where ***flow is parallel to the layers*** may occur when at the boundaries of the flow domain (say points (1) and (2))

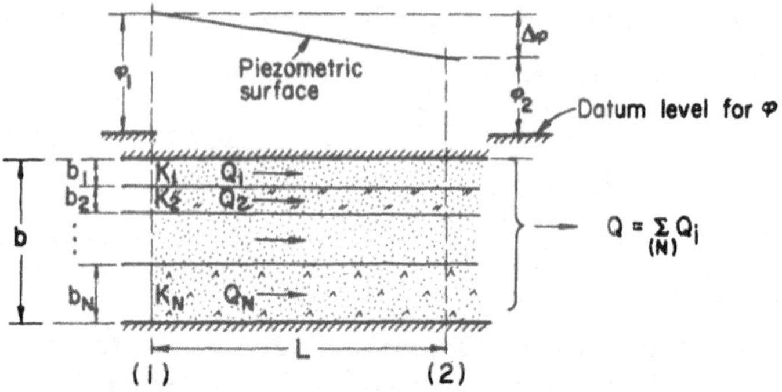

FIG. 5.8.1. Flow parallel to layers in a confined aquifer.

we have equipotentials normal to the layers as boundary conditions. Let K_i, b_i and Q_i be, respectively, the hydraulic conductivity, the thickness and the discharge per unit width of aquifer (measured normal to the cross-section given in fig. 5.8.1) in the ith layer. Then the total discharge, Q, is the sum of the individual discharge rates Q_i, each expressed by Darcy's law:

$$Q = \sum_{i=1}^{N} Q_i; \qquad b = \sum_{i=1}^{N} b_i; \qquad Q_i = K_i b_i \frac{\Delta\varphi}{L} \tag{5.8.1}$$

whereas the gradient $J = \Delta\varphi/L$ is maintained constant in the different layers (i.e., equipotentials are everywhere normal to the layers). Hence:

$$Q = \sum_{i=1}^{N} Q_i = \sum_{i=1}^{N} K_i b_i \frac{\Delta\varphi}{L} = \frac{\Delta\varphi}{L} \cdot \sum_{i=1}^{N} T_i \tag{5.8.2}$$

where $T_i = K_i b_i$ is the *transmissivity* of the ith layer.

If we wish to consider an equivalent hydraulic conductivity $\bar{K}^P$ (where the superscript P is added to indicate that this is the equivalent $\bar{K}$ for flow parallel to the layers), such that the same discharge rate, Q, will be conducted through the same aquifer of thickness, b, under the same gradient, $\Delta\varphi/L$, we have:

$$Q = \bar{K}^P b \frac{\Delta\varphi}{L} = \bar{T}^P \frac{\Delta\varphi}{L}; \qquad \bar{T}^P = \bar{K}^P b = \text{aquifer transmissivity.} \tag{5.8.3}$$

By comparing (5.8.2) with (5.8.3), we obtain:

$$\bar{K}^P = \sum_{i=1}^{N} K_i b_i \Big/ \sum_{i=1}^{N} b_i \quad \text{or:} \quad \bar{T}^P = \sum_{i=1}^{N} T_i. \tag{5.8.4}$$

If instead of layers we have a continuous variation of hydraulic conductivity (say, in the vertical z direction), such that $K = K(z)$, the total discharge, Q, parallel to the stratification, through a thickness, b, is:

$$dQ = K(z) \frac{\Delta\varphi}{\Delta L} dz; \qquad \frac{\Delta\varphi}{\Delta L} = \text{const}; \qquad Q = \frac{\Delta\varphi}{\Delta L} \int_0^b K(z)\, dz. \tag{5.8.5}$$

Hence:

$$\bar{K}^p = \frac{1}{b}\int_0^b K(z)\,dz; \qquad \bar{K}^p b \equiv \bar{T}^p = \int_0^b K(z)\,dz. \tag{5.8.6}$$

As a second simple case, consider flow in a direction perpendicular to the layers (fig. 5.8.2). Here the discharge, Q, remains unchanged whereas the total drop of head, $\Delta\varphi$, is the sum of the individual $\Delta\varphi_i$s:

$$\Delta\varphi = \sum_{i=1}^{N} \Delta\varphi_i; \qquad L = \sum_{i=1}^{N} L_i.$$

Hence:

$$Q = K_i b \frac{\Delta\varphi_i}{L_i}; \qquad \Delta\varphi_i = \frac{L_i Q}{K_i b}; \qquad \Delta\varphi = \sum_{i=1}^{N} \Delta\varphi_i = \frac{Q}{b}\sum_{i=1}^{N} \frac{L_i}{K_i}. \tag{5.8.7}$$

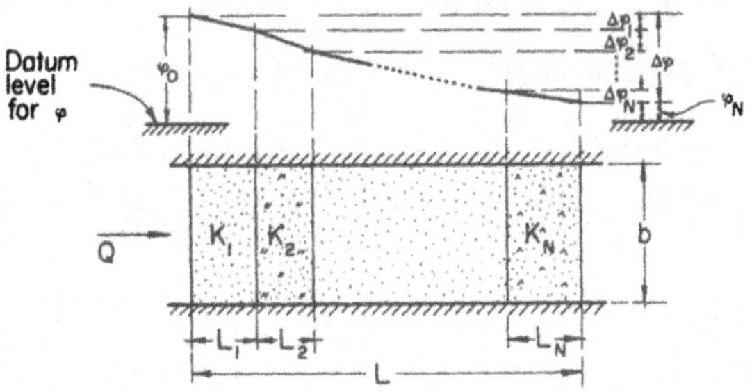

(a) Flow normal to layers in a confined aquifer.

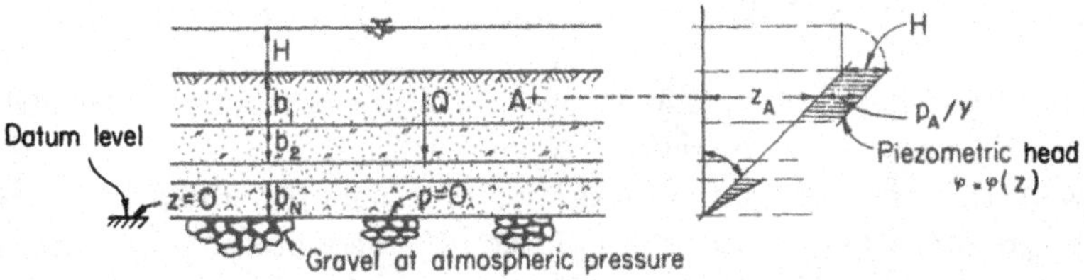

(b) Vertical flow through stratified soil. Water escapes into atmospheric pressure.

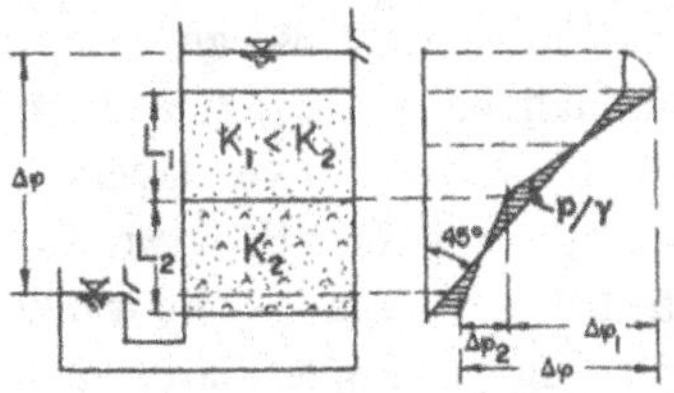

(c) Vertical (saturated) flow through stratified soil with a zone of negative pressure.

FIG. 5.8.2. Flow normal to layers.

Again, if we consider an equivalent homogeneous hydraulic conductivity $\bar{K}^N$, such that the same discharge, Q, will be conducted through the length, L, we obtain

$$Q = \bar{K}^N b\, \Delta\varphi/L \qquad \text{and hence:} \qquad L/\bar{K}^N = \sum_{i=1}^{N} (L_i/K_i). \tag{5.8.8}$$

It is of interest to note that if one of the K_is is zero (an impervious layer), $\bar{K}^N = 0$, i.e., the whole system is rendered impervious. However, it also follows from (5.8.8), as well as from (5.8.11) below, that the flow is determined by the resistances of the layers (i.e., by the layer of highest resistance) and not by their K_is and L_is separately.

If instead of layers we have a continuous variation of K (say, in the horizontal x direction), it can easily be shown that:

$$\frac{L}{\bar{K}^N} = \int_0^L \frac{dx}{K(x)}. \tag{5.8.9}$$

If we write Darcy's law (5.1.3) in the form:

$$Q = KA\, \Delta\varphi/L = \Delta\varphi/R; \qquad R = L/KA \tag{5.8.10}$$

where R is the resistance to the flow of a block of porous medium of length L (in the direction of the flow) and cross-sectional area A we may rewrite (5.8.4) for flow parallel to layers ($A_i \equiv b_i$) as:

$$1/\bar{R}^P = \sum_{i=1}^{N} (1/R_i); \qquad R_i = L/K_i b_i; \qquad \bar{R}^P = L/\bar{K}^P b \tag{5.8.11}$$

and (5.8.8) for flow normal to the layers as:

$$\bar{R}^N = \sum_{i=1}^{N} R_i; \qquad R_i = L_i/K_i b; \qquad \bar{R}^N = L/\bar{K}^N b. \tag{5.8.12}$$

One can easily recognize in (5.8.11) and (5.8.12) the laws for combining electrical resistances in parallel and in series.

Sometimes, in the passage from one layer to a more permeable one, a region of negative pressure may take place (fig. 5.8.2c). If these pressures are not too large and no air entry is possible, the flow may still remain a saturated one (for other cases, see discussion in sec. 9.4).

From (5.8.4) and (5.8.8) it follows that $\bar{K}^P > \bar{K}^N$, that is, the hydraulic conductivity is greater in the direction of the stratification. This can easily be shown by first considering two layers and then gradually increasing the number of layers while computing $\bar{K}^P$ and $\bar{K}^N$ of n layers as if only two layers are present: one composed of $(n - 1)$ layers, and the other of the nth layer.

Several investigators have studied the equivalent hydraulic conductivity of series-parallel models of porous media. For example, Cardwell and Parsons (1945) studied an ordered arrangement of blocks of soil in the form of a chessboard, with the black and white squares representing soil blocks of either hydraulic conductivity K_1 or K_2. They showed that the average hydraulic conductivity for such a system

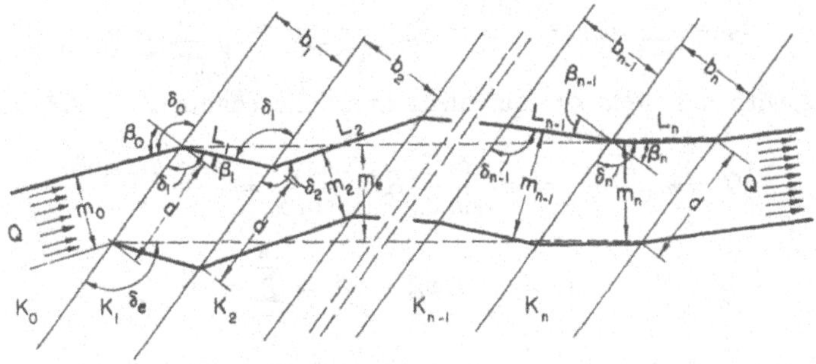

FIG. 5.8.3. Arbitrarily directed flow through parallel layered soil (after Marcus and Evenson, 1961).

always lies between the arithmetic- and harmonic-mean conductivities, expressed by (5.8.5) and (5.8.8), respectively. Marshal (1962) also presented an average conductivity for a series-parallel model.

5.8.2 Equivalent Hydraulic Conductivity of Arbitrarily Directed Flow

The analysis in this paragraph is based on the discussion of streamline refraction at a plane of discontinuity in hydraulic conductivity, as presented in paragraph 7.1.10.

Consider the sequence of parallel (homogeneous, isotropic) layers of hydraulic conductivity, K_i, and thickness, b_i $(i = 0, 1, \ldots, n-1, n, \ldots)$, shown in figure 5.8.3. At each interface between different hydraulic conductivities, (7.1.63) is valid. Written in terms of the angles δ_m, (7.1.63) becomes:

$$K_m/K_{m+1} = \tan\delta_{m+1}/\tan\delta_m. \tag{5.8.13}$$

Following Vreedenburgh (1937; see also Maasland 1957), who derives an expression for a system of two layers, Marcus and Evenson (1961) derive an expression for an equivalent hydraulic conductivity of n layers, not necessarily of equal thickness (fig. 5.8.3).

From figure 5.8.3 it follows that:

$$d = m_0/\sin\delta_0 = m_1/\sin\delta_1 = \cdots \quad \text{or:} \quad m_i = m_0 \sin\delta_i/\sin\delta_0.$$

Also:

$$L_1 = b_1/\sin\delta_1; \qquad L_2 = b_2/\sin\delta_2; \qquad \ldots L_n = b_n/\sin\delta_n. \tag{5.8.14}$$

The flow between the two streamlines in the ith layer is given by:

$$Q = K_i m_i (\Delta\varphi/\Delta L)_i = K_i m_i \Delta\varphi_i/\Delta L_i; \qquad \Delta\varphi_i = Q\,\Delta L_i/K_i m_i \tag{5.8.15}$$

or, in view of (5.8.13) and (5.8.14)

$$\Delta\varphi_i = Q(b_i \sin\delta_0/K_i m_0 \sin^2\delta_i). \tag{5.8.16}$$

And the total loss of head through n layers:

$$\Delta\varphi = \sum_{i=1}^{n} \Delta\varphi_i = \frac{Q \sin \delta_0}{m_0} \sum_{i=1}^{n} \frac{b_i}{K_i \sin^2\delta_i} = \frac{Q}{d} \sum_{i=1}^{n} \frac{b_i}{K_i \sin^2\delta_i}. \tag{5.8.17}$$

Comparing with a system of equivalent hydraulic conductivity $\bar{K}^\delta$, where:

$$Q = m_e \bar{K}^\delta \frac{\Delta\varphi}{L_e} = m_0 \frac{\sin \delta_e}{\sin \delta_0} \bar{K}^\delta \frac{\Delta\varphi}{b_e/\sin \delta_e} = \frac{d\bar{K}^\delta \Delta\varphi \sin^2\delta_e}{b_e}$$

$$\Delta\varphi = \sum_{i=1}^{n} \Delta\varphi_i; \qquad b_e = \sum_{i=1}^{n} b_i$$

we obtain:

$$\frac{\sum_{i=1}^n b_i}{\bar{K}^\delta \sin^2\delta_e} = \sum_{i=1}^{n} \frac{b_i}{K_i \sin^2\delta_i} \quad \text{or:} \quad \bar{K}^\delta = \frac{\sum_{i=1}^n b_i}{\sin^2\delta_e} \Bigg/ \sum_{i=1}^{n} \frac{b_i}{K_i \sin^2\delta_i}. \tag{5.8.18}$$

Equation (5.8.18) gives the equivalent sought for hydraulic conductivity. Since $1/\sin^2\delta_e = 1 + \cot^2\delta_e$ and from (7.1.63) we have $K_0/\bar{K}^\delta = \cot \delta_0/\cot \delta_e$, we obtain

$$1/\sin^2\delta_e = 1 + (\bar{K}^\delta/K_0)^2 \cot^2\delta_0. \tag{5.8.19}$$

Substituting (5.8.19) in (5.8.18) and eliminating the unknown angle δ_e, yields:

$$\frac{\bar{K}^\delta}{1 + (\bar{K}^\delta/K_0)^2 \cot^2\delta_0} = \sum_{i=1}^{n} b_i \Bigg/ \sum_{i=1}^{n} \frac{b_i}{K_i \sin^2\delta_i}. \tag{5.8.20}$$

5.8.3 A Layered Medium as an Equivalent Anisotropic Medium

Following the work of Vreedenburgh (1937) and Maasland (1957), who considered a two-layered system, Marcus and Evenston (1961) derive a relationship between the equivalent hydraulic conductivities $\bar{K}^\delta$, $\bar{K}^N$ and $\bar{K}^P$.

From the geometry of figure 5.8.3 and (5.8.14) it follows that:

$$\sum_{i=1}^{n} b_i \cot \delta_i = \sum_{i=1}^{n} L_i \cos \delta_i = \cot \delta_e \sum_{i=1}^{n} b_i. \tag{5.8.21}$$

From (5.8.13) we have:

$$K_i/K_m = \cot \delta_i/\cot \delta_m. \tag{5.8.22}$$

By inserting (5.8.21) in (5.8.22) we obtain:

$$\sum_{i=1}^{n} b_i(K_i/K_m) \cot \delta_m = \cot \delta_e \sum_{i=1}^{n} b_i \quad \text{or:} \quad \cot \delta_m = K_m \cot \delta_e \frac{\sum_{i=1}^n b_i}{\sum_{i=1}^n K_i b_i}. \tag{5.8.23}$$

Rewriting (5.8.18), we obtain:

$$\bar{K}^\delta \sum_{i=1}^{n} \frac{b_i}{K_i \sin^2\delta_i} = \frac{\sum_{i=1}^n b_i}{\sin^2\delta_e} \quad \text{or:} \quad \bar{K}^\delta \sum_{i=1}^{n} (b_i/K_i)(1 + \cot^2\delta_i) = \frac{\sum_{i=1}^n b_i}{\sin^2\delta_e}. \tag{5.8.24}$$

Hence:

$$\bar{K}^\delta \sum_{i=1}^{n} (b_i/K_i) + \bar{K}^\delta \sum_{m=1}^{n} (b_m/K_m) \cot^2\delta_m = \sum_{i=1}^{n} b_i/\sin^2\delta_e$$

or, with (5.8.23):

$$\bar{K}^{\delta} \sum_{i=1}^{n} \frac{b_i}{K_i} + \bar{K}^{\delta} \sum_{m=1}^{n} b_m K_m \frac{(\sum_{i=1}^{n} b_i)^2}{(\sum_{i=1}^{n} K_i b_i)^2} \cdot \cot^2\delta_e = \frac{\sum_{i=1}^{n} b_i}{\sin^2\delta_e}. \tag{5.8.25}$$

Now, from (5.8.4) and (5.8.7), (5.8.25) becomes:

$$1/\bar{K}^{\delta} = \sin^2\delta_e/\bar{K}^N + \cos^2\delta_e/\bar{K}^P. \tag{5.8.26}$$

This last equation is identical to (5.6.30)—written for two dimensions, with $\bar{K}^P \equiv K_x$, $\bar{K}^N \equiv K_y$—which describes the directional hydraulic conductivity in the direction of the specific discharge given here by the angle δ_e. This shows that a layered soil is equivalent in its macroscopic, or average, behavior to an anisotropic soil with principal hydraulic conductivities $\bar{K}^P$ and $\bar{K}^N$. However, for this statement to be valid, we must require that the individual layers be thin with respect to the over-all dimensions of the flow domain; otherwise the concept of average or equivalent hydraulic conductivity at a point is meaningless. Accordingly, if we consider the general case (of any combination of layers) shown in figure 5.8.3, $\bar{K}^N$, $\bar{K}^P$ and hence $\bar{K}^{\delta}$ (for a given direction δ_e) will vary from point to point in the medium. We then have an equivalent inhomogeneous anisotropic medium. A case of special interest occurs when we have alternating layers (K_1, b_1), (K_2, b_2), (K_1, b_1), (K_2, b_2),... etc. Then we have a medium that behaves, on the average, as would an equivalent homogeneous anisotropic medium (fig. 5.8.4).

5.8.4 Girinskii's Potential

We have emphasized in paragraph 5.2.1 that only for an isotropic homogeneous medium may we use a specific discharge potential, Φ, such that $q_x = -\partial\Phi/\partial x$ and

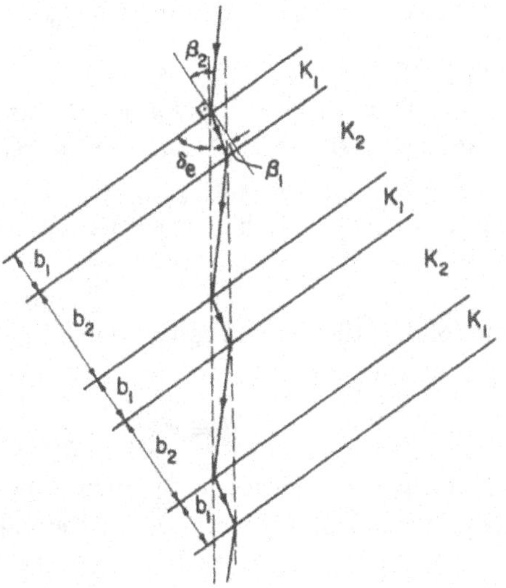

FIG. 5.8.4. Arbitrarily directed flow through a system of regularly alternating homogeneous isotropic layers.

$q_z = -\partial\Phi/\partial z$. Girinskii (1946; see also Polubarinova-Kochina 1952 and 1962, and Aravin and Numerov 1953 and 1965) introduced another potential, called *Girinskii's potential*, Φ^g, applicable only to a horizontally stratified formation where $K = K(z)$, such that the discharge will be derived by taking its derivative.

Consider first an approximately horizontal flow in a stratified phreatic aquifer with a horizontal bottom at $z = 0$. Using Dupuit assumptions (sec. 8.1) for this aquifer (fig. 8.1.1b), we have:

$$q'_x(x, y, t) = -\int_0^{h(x,y,t)} K(z) \frac{\partial h(x, y, t)}{\partial x} dz = -\frac{\partial h(x, y, t)}{\partial x} \int_0^h K(z)\, dz$$

$$q'_y(x, y, t) = -\int_0^{h(x,y,t)} K(z) \frac{\partial h(x, y, t)}{\partial y} dz = -\frac{\partial h(x, y, t)}{\partial y} \int_0^h K(z)\, dz \qquad (5.8.27)$$

where $\mathbf{q}'$ is the specific discharge per unit width through the entire thickness of flow. Girinskii defined his *potential for a phreatic aquifer* as:

$$\Phi^g = \int_0^h (h - z)K(z)\, dz; \qquad h = h(x, y, t). \qquad (5.8.28)$$

Using Leibnitz' rule of differentiation, we obtain:

$$\frac{\partial \Phi^g}{\partial x} = \frac{\partial}{\partial x} \int_0^h (h - z)K(z)\, dz = \int_0^{h(x,y,t)} \frac{\partial h}{\partial x} K(z)\, dz + \frac{\partial h}{\partial x} [(h - z)K(z)]_{z=h}$$

$$= \int_0^h \frac{\partial h}{\partial x} K(z)\, dz = -q'_x. \qquad (5.8.29)$$

Similarly, $\partial\Phi^g/\partial y = -q'_y$. Thus Girinskii's potential, Φ^g, as defined by (5.8.28), is *a specific discharge potential for a stratified phreatic aquifer with essentially horizontal flow*.

For the special case $K = \text{const}$ (i.e., homogeneous aquifer), Φ^g becomes:

$$\Phi^g = K\, h^2/2. \qquad (5.8.30)$$

From this expression it follows that φ' introduced in (8.1.4) is nothing but Φ^g/K. The continuity equation in this case is:

$$\partial q'_x/\partial x + \partial q'_y/\partial y = 0 \quad \text{or:} \quad \nabla^2\Phi^g \equiv \partial^2\Phi^g/\partial x^2 + \partial^2\Phi^g/\partial y^2 = 0. \qquad (5.8.31)$$

This means that $\Phi^g = \Phi^g(x, y, t)$ is a harmonic function of position in the xy plane. This means that all analytical and analog methods of potential theory are applicable also to Φ^g.

Girinskii also extends his potential to a horizontally stratified confined aquifer of constant thickness, b, where the flow is horizontal. In this case:

$$\Phi^g = \int_0^b [\varphi(x, y, t) - z]K(z)\,dz = \int_0^b \frac{p}{\gamma}K(z)\,dz; \qquad \frac{\partial \Phi^g}{\partial x} = \int_0^b \frac{\partial \varphi(x, y, t)}{\partial x} K(z)\,dz = -q'_x \tag{5.8.32}$$

and a similar expression for q'_y. Here also Φ^g is a harmonic function. For $K = \text{const}$,

$$\Phi^g = Kb(\varphi - b/2). \tag{5.8.33}$$

The potential Φ^g can also be extended to a phreatic aquifer with $\varphi = \varphi(x, y, z)$, without employing the Dupuit assumptions, by replacing b in (5.8.32) with $h(x, y)$.

As Φ^g is a potential function, it is possible to define a conjugate function—Girinskii's stream function Ψ^g—to it such that $\Delta\Psi^g$ will give the discharge between two streamlines (through the entire thickness of the phreatic or confined aquifer).

It can be shown that Ψ^g also satisfies the Laplace equation.

5.9 Compressible Fluids

In sections 4.7 and 5.1, the equation of motion for a homogeneous incompressible fluid was presented in terms of the piezometric head, $\varphi = z + p/\rho g$ (sum of pressure and potential energies per unit weight). Hubbert (1940) extended the concept of piezometric head to include compressible fluids where $\rho = \rho(p)$, introducing a potential defined by:

$$g\varphi^* = gz + \int_{p_0}^{p} \frac{dp}{\rho(p)} \quad \text{or:} \quad g\varphi^* = g\int_{z_0}^{z} dz + \int_{p_0}^{p} \frac{dp}{\rho(p)}. \tag{5.9.1}$$

The corresponding equation of motion for an isotropic medium becomes:

$$\mathbf{q} = -(k\rho g/\mu)\nabla\varphi^*. \tag{5.9.2}$$

For an anisotropic medium, k must be replaced by the second-rank tensor $\underline{\mathbf{k}}$.

To derive (5.9.1) Hubbert (1940) starts from the potential of a fluid, defined as its mechanical energy per unit mass, and from the fact that energy is a relative quantity measurable by the amount of work required to effect any given transformation from some arbitrary initial (reference) state (elevation z_0, pressure p_0, density ρ_0 and velocity V_0) to a specified final state (z, p, ρ, V). Accordingly, the potential of a fluid at a specified point is the work required to transform a unit mass of fluid from an arbitrary chosen reference state, to the point under consideration. Taking $z = 0$, $p = p_0$, $\rho = \rho_0$ (or the specific volume $v_0 = 1/\rho_0$) and $V = 0$ as the initial state, Hubbert (1940) computes the amount of work required to transform a unit mass of fluid from this reference state to some arbitrary state described by z, ρ, (or $v = 1/\rho$), p and V:

$$g\varphi^* = gz - p_0 v_0 + \int_{v}^{v_0} p\,dv + pv + V^2/2. \tag{5.9.3}$$

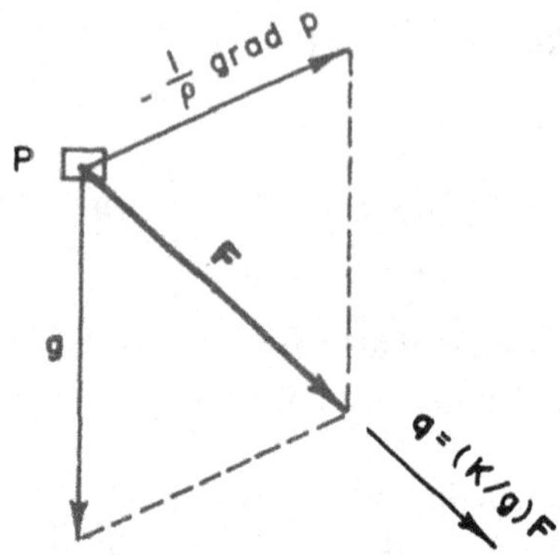

FIG. 5.9.1. Forces acting on a unit mass of fluid.

Since $\int p\, dv = \int d(pv) - \int v\, dp$, he obtains from (5.9.3) for an incompressible fluid $(v = v_0)$:

$$g\varphi^* = gz + (p - p_0)v + V^2/2 = gz + (p - p_0)/\rho + V^2/2 \tag{5.9.4}$$

and for a compressible fluid:

$$g\varphi^* = gz + \int_{p_0}^{p} \frac{dp}{\rho} + \frac{V^2}{2}. \tag{5.9.5}$$

By neglecting the kinetic energy per unit mass, $V^2/2$, we obtain (5.9.1). Equation (5.9.5) is the most general form of the potential. In order for the φ^* potential, as defined by (5.9.5), to be unique, it must be stipulated that *the density of the fluid ρ is a function of the pressure only*; otherwise the integral in (5.9.5) will be indeterminate. This condition is generally satisfied in the case of liquids; for gases it is satisfied under isothermal conditions and under adiabatic conditions. The main difference in the potential between a compressible fluid (say, a gas), and an incompressible one, is that in the former additional work must be done to compress the fluid from its volume at the reference state to that of the final state.

Hubbert (1940, 1956) also considers the driving forces acting on a unit mass of fluid flowing in a porous medium. The total driving force, **F** (acting, say, on a volume element of fluid contained in the porous medium domain), is the sum of the net force due to pressure and of the force due to gravity (body force). Figure 5.9.1 shows these forces (per unit mass) acting on a fluid particle of density ρ, at some point, P, in the flow domain: the force per unit mass due to pressure is given by $-(1/\rho)$ grad p; the force per unit mass due to gravity is given by **g** (vector; positive when directed downward). When combined, we have:

$$\mathbf{F} = \mathbf{g} + (-1/\rho \operatorname{grad} p) = -[\operatorname{grad}(gz) + 1/\rho \operatorname{grad} p]. \tag{5.9.6}$$

Assuming that the specific discharge **q** is proportional to the force **F** as expressed by (5.9.6), and in its direction, we obtain:

$$\mathbf{q} = (K/g)\mathbf{F} = -(K/g)(g \operatorname{grad} z + 1/\rho \operatorname{grad} p) \tag{5.9.7}$$

where K is hydraulic conductivity. Compare with (4.7.17).

The question arises whether or not this flow field, $\rho = \rho(p)$, is a potential one (so that $\mathbf{q}$ may be derived by taking the gradient of some (scalar) potential function φ^*). The same question might be asked in the case of $\rho = \rho(C)$, where C is the concentration of some dissolved solute. A criterion for whether $\mathbf{F}$ of (5.9.7) has a potential $\varphi^* = \varphi^*(x, y, z)$, so that it may be written in the form: $\mathbf{F} = -g \operatorname{grad} \varphi^*$ (g is simply a constant), is whether or not the field is *irrotational* (sec. 5.4). This means that we require that curl $\mathbf{F} = 0$. From (5.9.7), we have:

$$\begin{aligned} \operatorname{curl} \mathbf{F} = \operatorname{curl}[-g \operatorname{grad} z - (1/\rho \operatorname{grad} p)] &= -\operatorname{grad}(1/\rho) \operatorname{grad} p \\ &= \operatorname{grad} p \times \operatorname{grad}(1/\rho). \end{aligned} \tag{5.9.8}$$

This is a vector- or cross-product of two vectors: grad p and grad$(1/\rho)$. Therefore, we have curl $\mathbf{F} = 0$ if (a) grad $p = 0$, i.e., the case of constant pressure, (b) grad$(1/\rho) = 0$, corresponding to $\rho =$ const, or (c) if the two vectors are colinear, i.e., the case where surfaces of constant density coincide with isobaric (of constant pressure) surfaces (barotropic fluid). This third case corresponds to the case where *ρ is a function of p only.* We may, therefore, conclude that we have a potential (or irrotational field of flow) such that we may write (5.9.2) only when $\rho = \rho(p)$ (and not, for example, when $\rho = \rho(p, T^0)$ where T^0 is temperature). In the general case of $\rho = \rho(p, C, T^0)$ the flow is not a potential one. Some comments on this general case are given by DeWiest (1969).

To derive an expression for φ^*, Hubbert (1956) considers the work performed by the force $\mathbf{F}$ in the field of flow between two points P and P_0. In a potential field (sec. 5.4) this work must be independent of the path chosen between these points. Hence:

$$\varphi^*(P) - \varphi^*(P_0) = -1/g \int_{p_0}^{p} \mathbf{F} \cdot d\mathbf{s} = \int_{0}^{z} dz + \int_{p_0}^{p} \frac{dp}{g\rho(p)} = z + \int_{p_0}^{p} \frac{dp}{g\rho(p)} \tag{5.9.9}$$

where we introduce $z(P_0) = 0$, $p(P_0) = p_0$. Often we use $p_0 = 0$. In (5.9.9), the integral from p_0 to p may be taken along any path, s. This concludes the justification of (5.9.2).

5.10 Derivations of Darcy's Law

In paragraph 5.1.1 Darcy's law was presented (as originally presented by Darcy) as an empirical relationship based on experiments of steady flow in a vertical column of homogeneous sand. The various extensions of the law, e.g., to three-dimensional flow, to unsteady flow, etc., were first suggested as heuristic generalizations, and were then justified a posteriori by numerous planned experiments as well as by their success in predicting the flow in porous media systems of practical interest.

Nevertheless, many attempts have been made to derive Darcy's law, or, more generally, the motion equations in porous media, from the basic principles underlying the theory of hydrodynamics. In most cases the *conceptual model approach* discussed in detail in paragraph 4.5.1 was employed.

The present section includes a brief description of several examples of the derivation of Darcy's law for saturated flow using conceptual models. These examples are in addition to the detailed development presented in section 4.7. Examples related to flow at large Re are presented in paragraph 5.11.2, and those related to relative permeability in paragraph 9.3.2. Detailed reviews are given by Scheidegger (1960) and Irmay (1964d).

5.10.1 Capillary Tube Models

Probably the simplest models from which Darcy's law may be derived are those made of capillary tubes in one arrangement or another. The starting point in all these models is Hagen–Poisseuille's law governing the steady flow through a single, straight circular capillary tube of diameter δ, oriented in the direction $\mathbf{1s}$:

$$Q_s = -\frac{\pi\delta^4\rho g}{128\mu}\frac{\partial\varphi}{\partial s} \quad \text{and:} \quad V_s = \frac{Q_s}{\pi\delta^2/4} = -\frac{\delta^2}{32}\frac{\rho g}{\mu}\frac{\partial\varphi}{\partial s} \tag{5.10.1}$$

where s is length measured along the tube, Q_s is total discharge and V_s is average velocity in the tube. The analogy between (5.10.1) and Darcy's law is obvious. In (5.10.1), $\delta^2/32$ is analogous to the medium permeability, k. Obviously, any development based on Hagen–Poisseuille's law will eventually lead to a linear relationship between the velocity and the piezometric head gradient. The difference among the various models is only in the relationships they yield among k and properties of various media.

Figure 5.10.1a shows a bundle of such parallel tubes, all of the same diameter,

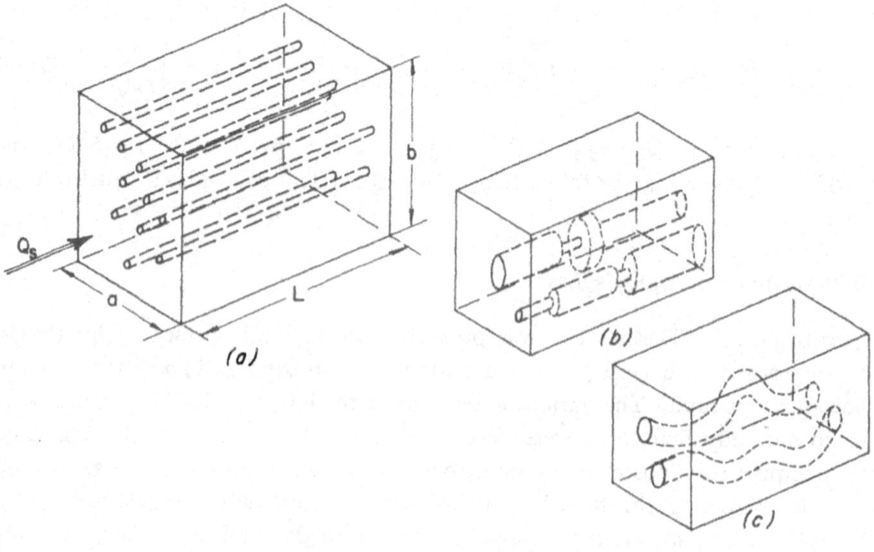

FIG. 5.10.1. Capillary tube models.

δ, embedded in a solid. If there are N such tubes per unit area of cross-section of the model (normal to the direction of flow), then the specific discharge through the porous block is:

$$q_s \equiv Q_s/ab = -N\frac{\pi\delta^4\rho g}{128\mu}\frac{d\varphi}{ds}. \qquad (5.10.2)$$

Since the porosity (n) of this model is given by $n = N(\pi\delta^2/4)L/L = N\pi\delta^2/4$ we obtain from (5.10.2):

$$q_s = -(k\rho g/\mu)\,\partial\varphi/\partial s; \qquad k = n\delta^2/32 \qquad (5.10.3)$$

which is again analogous to Darcy's law, with k related both to n and to δ; these may be considered as representing some average pore diameter. Obviously here, as in all other models, the numerical coefficient 1/32 is meaningless so far as an actual porous medium is concerned; it is commonly replaced by some arbitrary coefficient that must be determined experimentally. Some authors suggest replacing this factor with another one that represents tortuosity.

The next step in making the model more like an actual porous medium is to assume that the diameter of the capillary tubes is not uniform. In such a case (5.10.2) becomes:

$$q_s = -\sum_{i=1}^{m} N_i\frac{\pi\delta^4_i}{128}\frac{\rho g}{\mu}\frac{\partial\varphi}{\partial s} = -k\frac{\rho g}{\mu}\frac{\partial\varphi}{\partial s}; \qquad k = \sum_{i=1}^{m} N_i\frac{\pi\delta^4_i}{128} \qquad (5.10.4)$$

where N_i is the number of tubes of diameter δ_i per unit of cross-sectional area of medium.

A drawback of the models discussed so far is that they give permeability only in one direction. To overcome this difficulty the model is usually modified so that only one-third of the capillary tubes are placed in each of three mutually orthogonal directions. This will lead to a permeability lower than that defined by (5.10.3):

$$k = n\delta^2/96. \qquad (5.10.5)$$

Scheidegger (1953, 1960) suggested an improvement of the "bundle of capillaries" model by relating the average diameter of the bundle, δ, to the pore-size distribution $\alpha(\delta)$ of the medium as defined in section 2.8. According to this definition, $\alpha(\delta)\,d\delta$ represents the fraction of the pore volume made up of pores whose diameter is between δ and $\delta + d\delta$. The volume taken up by capillaries of diameter between δ and $\delta + d\delta$, parallel to any of the mutually orthogonal directions x_i in a piece of model of length Δx_i and unit cross-sectional area normal to $\mathbf{1x}_i$, is $\frac{1}{3}n\,\Delta x_i \cdot \alpha(\delta)\,d\delta$. Since the frontal area of these capillaries is $\frac{1}{3}n\alpha(\delta)\,d\delta$, the specific discharge is:

$$q_{x_i} = \tfrac{1}{3}n\int_0^\infty V_{x_i}(\delta)\alpha(\delta)\,d\delta = -\frac{1}{96}\frac{\rho g}{\mu}n\frac{\partial\varphi}{\partial x_i}\int_0^\infty \delta^2\alpha(\delta)\,d\delta \qquad (5.10.6)$$

where $V_{x_i}(\delta)$ is the average velocity in capillaries of the diameter δ given by (5.10.1). By comparison with Darcy's law, we obtain:

$$k = \frac{n}{96} \int_0^\infty \delta^2 \alpha(\delta)\, d\delta. \tag{5.10.7}$$

Scheidegger (1953, 1960) also considers the case where the capillary tubes are tortuous and of a variable diameter (e.g., as obtained by combining the models shown in figures 5.10.1b and c). He suggests the following expression for k:

$$k = (n/96T^2) \Big/ \left[\int_0^\infty \delta^2 \alpha(\delta)\, d\delta \right]^2 \int_0^\infty [\alpha(\delta)/\delta^6]\, d\delta \tag{5.10.8}$$

where T is his "tortuosity factor" (see par. 4.8.2), defined as the ratio of the length of the actual flow channel to the length of the porous medium.

Other models based on capillary tubes (e.g., Fatt's (1956) network model) are described in the literature (see, for example, the review by Scheidegger 1960).

5.10.2 Fissure Models

Several authors (e.g., Irmay 1955) employ narrow capillary fissures (or slits of constant width) as a model representing a porous medium. A fractured rock would probably be the porous medium closest to such a model (fig. 5.10.2a).

The starting point for analysis of such a model is the solution of Navier–Stokes' equation for average velocity, V, in a single fissure of constant width, b, bounded by two parallel impervious planes (fig. 5.10.2b):

$$V = (b^2/12)(\rho g/\mu) J \tag{5.10.9}$$

where J is the hydraulic gradient (details are given in par. 11.4.3). If we consider a porous medium made of a large number of parallel fissures of this kind (fig. 5.10.2c), we obtain:

$$q = Vb/(a+b) = nV = n\frac{b^2}{12}\frac{\rho g}{\mu} J; \quad k = nb^2/12 = [n^3/(1-n)^2]\, a^2/12. \tag{5.10.10}$$

In this way k is related either to a "pore size" b or to a "grain size" a. The coefficient 1/12 serves here as the shape factor $f(s)$ discussed in paragraph 5.5.1.

Irmay (1955) considers a model composed of three sets of such mutually orthogonal fissures. Figure 5.10.2a shows a cross-section through the model. By analyzing the flow through the model with $b_1 = b_2 = b$ and $a_1 = a_2 = a$, and assuming no loss of head at the junctions, Irmay (1955) derives the permeability, k, of a fissured porous medium:

$$q = \frac{(a+b)^2 - a^2}{(a+b)^2} g \frac{b^2}{12\nu} \frac{a+b}{a} J \tag{5.10.11}$$

$$k = \frac{(a+b)^2 - a^2}{(a+b)^2} \frac{a+b}{a} \frac{b^2}{12} = (1 - m^{1/3})^3 (1 + m^{1/3}) a^2/12m \tag{5.10.12}$$

where $m = 1 - n = a^3/(a+b)^3$.

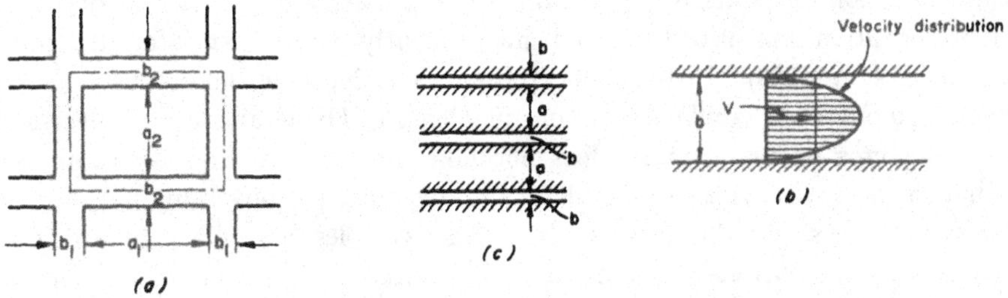

FIG. 5.10.2. A capillary fissure model.

Fissure models were also analyzed by Snow (1965), Parsons (1965), and by others. By considering networks of fissures they obtained expressions for anisotropic permeability.

5.10.3 Hydraulic Radius Models

In flow through pipes and open channels the concept of *hydraulic radius* R, defined as the ratio between area of cross-section and wetted perimeter, is often used. For example, in a circular pipe of radius r, $R = r/2$. Another definition for R is the ratio of volume of a conduit filled with fluid to its wetted surface. This last definition, combined with a visualization of the porous medium as a network of interconnected channels or passages, leads to the following relationship between R and the specific surface (M) defined by (2.6.1): $R = n/M$. Thus, R represents an equivalent (or average) hydraulic radius of the exceedingly complicated flow channels.

Returning now to Poisseuille's equation for flow in a pipe, and replacing δ ($= 2r$) by $4R$, we obtain:

$$V_s = -\frac{R^2}{2}\frac{\rho g}{\mu}\frac{\partial\varphi}{\partial s}\begin{cases}\text{from (5.10.3):} & k = nR^2/2\\ \text{from (5.10.5):} & k = nR^2/6.\end{cases} \qquad (5.10.13)$$

A more general expression for k, based on the hydraulic radius, would be

$$k = f'_1(s)f'_2(n)R^2 \qquad (5.10.14)$$

where, as in paragraph 5.5.1, $f'_1(s)$ is a dimensionless *shape factor* and $f'_2(n)$ is a function of the porosity (*porosity factor*). Most investigators are concerned with relating R in (5.10.14) to some measurable property of the porous matrix (e.g., some mean grain diameter) and determining the function $f(n)$. Other authors (e.g., Wyllie and Spangler 1952), trying to compensate for the noncircular shape of the conduits in an actual porous medium, suggest that the factor 2 in the denominator be generalized and replaced by a *shape factor*. Wyllie and Spangler (1952) propose values of this factor in the range 2.5–3.0.

One of the most widely accepted derivations of permeability and its relationship to porous medium properties is the one proposed by Kozeny (1927) and later modified by Carman (1937, 1956). In his book, Carman (1956) mentions that Kozeny actually developed his theory without knowing of an earlier work by Blake (1922), who arrived at an expression for permeability similar to the one derived by Kozeny (1927). The Kozeny equation was also developed independently a few years after Kozeny by Fair and Hatch (1933). In the theory proposed by Kozeny, the porous medium is treated as a bundle of capillary tubes of equal length. These tubes are not necessarily of circular cross-section, yet he does not take into account velocity components normal to the tubes' axes as a result of the diverging or converging nature of the flow in the tubes. In this respect his model resembles a capillary tube model.

By solving the Navier–Stokes equations simultaneously for all channels passing through a cross-section normal to the flow in the porous medium, he obtains the equations for motion and permeability in the forms:

$$q = -(c_0 n^3/\mu M^2)\,\mathrm{grad}\,p; \qquad k = c_0 n^3/M^2 \tag{5.10.15}$$

where c_0 is a numerical coefficient called *Kozeny's constant*; it varies slightly according to the geometrical form of the individual channels in the model ($c_0 = 0.5$ for a circle, $c_0 = 0.562$ for a square, $c_0 = 0.597$ for an equilateral triangle and $c_0 = 0.667$ for a strip). The expression for k in (5.10.15) is called *Kozeny's equation*.

With the hydraulic radius $R = n/M$, the k's in (5.10.13) become

$$k = n^3/2M^2 \quad \text{and} \quad k = n^3/6M^2. \tag{5.10.16}$$

If we express the specific surface with respect to a unit volume of solid, rather than with respect to a unit volume of porous medium (M_s in par. 2.6.1), we obtain from (5.10.15):

$$k = c_0[n^3/(1-n)^2]/M_s^2. \tag{5.10.17}$$

Carman (1937) used this form of Kozeny's equation with $c_0 = \frac{1}{5}$:

$$k = [n^3/(1-n)^2]/5M_s^2. \tag{5.10.18}$$

Carman's values for c_0, which he claims give the best agreement with experiments, are $\frac{1}{2}$ and $\frac{1}{6}$ as in (5.10.16). The relationship (5.10.18) is known as the *Kozeny–Carman equation*. In this equation the porosity factor $f(n)$ is expressed by $n^3/(1-n)^2$. It is also possible to define some mean particle size by $d_m = 6/M_s$. Then (5.10.18) becomes:

$$k = \frac{d_m^2}{180}\frac{n^3}{(1-n)^2}.$$

Among other modifications of Kozeny's equation mentioned in the literature it is of interest to mention at this point the introduction of *tortuosity*, discussed in detail in section 4.8. From (4.8.21) in which the tortuosity T is already included and where $q = n\bar{u}_s$, we obtain the following modified forms of (5.10.15) and (5.10.18):

$$k = c_0 T n^3/M^2 \quad \text{or} \quad k = c_0 T[n^3/(1-n)^2]/M_s^2 \tag{5.10.19}$$

where $T = (L/L_e)^2 < 1$. Sometimes c_0T is combined into a single coefficient, c. Carman (1938, 1956) reported that $c = c_0T = \frac{1}{5}$ in (5.10.19) fits well with experimental data. Hence:

$$k = n^3/5M^2 \quad \text{or} \quad k = [n^3/(1-n)^2]/5M_s^2. \tag{5.10.20}$$

With $L/L_e = 1/\sqrt{2}$, $T = (L/L_e)^2 = \frac{1}{2}$, $c_0 = 1/2.5 = 0.4$, which Carman (1956) mentions as plausible for noncircular sections. The coefficient $c = c_0 T$ is also often called Kozeny's constant.

Several authors (e.g., Sullivan and Hertel 1942) prefer to write (5.10.17) in terms of some average (or equivalent) grain diameter, d, related to the specific surface, M_s, of the medium:

$$k = f(n)f(s)d^2 \tag{5.10.21}$$

where the porosity factor is expressed by $n^3/(1-n)^2$ and $f(s)$ is a shape factor such that $f(s)d^2 = c_0/M_s^2$. For spheres of constant diameter, d, $M_s = 6/d$.

5.10.4 Resistance to Flow Models

A fluid moving relative to a solid boundary exerts a force on the boundary. This force is caused by two factors. The first is shear stresses, due to viscosity and velocity gradients at the boundary surface which give rise to forces tangential to the surface. The second is pressure variations along the surface that cause forces normal to the surface. The vector sum of the normal and tangential surface forces, integrated over the entire surface of the considered solid body, gives a resultant force. The component of this force in the direction of the relative velocity, $\mathbf{V}$, past the body is called *drag* $\mathbf{D}$. The component normal to the velocity is called a *lift* or *lateral force*. The drag is the sum of the frictional (D_f) and pressure (D_p) components. The following expressions define useful drag coefficients (see any text on fluid mechanics):

$$D_f = C_f \frac{\rho V^2}{2} A_f; \qquad D_p = C_p \frac{\rho V^2}{2} A_p; \qquad D = C_D \frac{\rho V^2}{2} A \tag{5.10.22}$$

where A_f and A_p are suitably chosen reference areas, C_D is the *total drag coefficient* and A is the frontal area normal to V. Analogous coefficients are usually defined for the lift.

Resistance to flow (or drag) models have been employed by various investigators to derive the permeability of a porous medium.

The simplest approach is to use the drag on a single sphere of diameter d as an analogy to the drag on the particles of a porous medium:

$$D = 3\pi\, d\mu V \tag{5.10.23}$$

(i.e., proportional to the first power of the velocity). Equation (5.10.23) may also be written as:

$$D = C_D \rho \frac{V^2}{2} A = \frac{24}{\mathrm{Re}} \rho \frac{V^2}{2} \frac{\pi d^2}{4}; \qquad C_D = \frac{24}{\mathrm{Re}}; \qquad \mathrm{Re} = \frac{\rho V d}{\mu}. \tag{5.10.24}$$

This is *Stokes' equations for drag* and for drag coefficient. Experiments show that the expression for C_D in (5.10.24) is valid for $\mathrm{Re} < 1$. Above this value, inertia can no longer be neglected, and Stokes' solution does not apply. Oseen (1910) suggests for $\mathrm{Re} < 2$ (laminar nonlinear flow):

$$C_D = 24/\mathrm{Re} + 4.5 \tag{5.10.25}$$

and for $1000 < \mathrm{Re} < 10{,}000$ (i.e., in turbulent flow) $C_D = 0.4$.

If in flow through porous media we replace V by q, and the shear stress D/A by $-\gamma\, dJ$, (in analogy, for example, to the drag force in uniform flow in an open channel given by γRS, where R is the hydraulic radius and S is the slope of the channel), we obtain from (5.10.24):

$$-\gamma\, dJ = \frac{24\mu}{\rho q d}\rho\frac{q^2}{2} \quad \text{or} \quad q = -\frac{1}{12}d^2\frac{\gamma}{\mu}J = -k\frac{\gamma}{\mu}J \tag{5.10.26}$$

which is Darcy's law. If the same changes are introduced in (5.10.25), we obtain Dupuit–Forchheimer's equation (5.11.1).

Obviously this is a very crude model, as the effect of neighboring particles is not taken into account. Also, the simple relationships between the shear stresses τ and J are questionable. Rumer and Drinker (1966) and Rumer (1969) also start from the resistance of a single sphere, but write it as:

$$D = \lambda\mu\, dV_s \tag{5.10.27}$$

where λ is a coefficient that takes into account the effect of neighboring spheres (for a single sphere in an infinite fluid $\lambda = 3\pi$), V_s is the local average velocity of flow around the particle in the s-direction (q_s/n) and d is diameter of the sphere. In a porous medium the coefficient λ depends on the streamline configuration around the sphere; it is therefore a function of the geometry of the pore space.

If we have N such spheres in a representative elementary volume of a porous medium that has the form of a cylinder of length ds in the direction of the flow and cross-section dA: $N = (1-n)\, dA\, ds/\beta d^3$; $\beta = \pi/6$. The total force D_t is:

$$D_t = ND. \tag{5.10.28}$$

For particles other than spheres $\beta \neq \pi/6$.

Summing the forces acting on the element (fig. 5.10.3) leads to:

$$pn\, dA - \left(p + \frac{\partial p}{\partial s}ds\right)n\, dA - (\rho g n\, dA\, ds)\,\partial z/\partial s - D_t = 0$$

$$-\frac{\partial\varphi}{\partial s} - \frac{D_t}{\rho g n\, dA\, ds} = 0; \qquad \varphi = z + p/\rho g \tag{5.10.29}$$

where p is understood to be the average pressure. By inserting the value of D_t from (5.10.28), we obtain for $q_s = nV_s$:

$$-\frac{\partial\varphi}{\partial s} = \frac{\lambda(1-n)\mu}{\beta d^2 n\rho g}V_s; \qquad q_s = -\frac{\beta n^2}{\lambda(1-n)}d^2\frac{\rho g}{\mu}\frac{\partial\varphi}{\partial s} \tag{5.10.30}$$

or:

$$q_s = -(Cd^2\rho g/\mu)\,\partial\varphi/\partial s; \qquad C = k/d^2 = \beta n^2/[\lambda(1-n)] \tag{5.10.31}$$

where C is a coefficient depending on the geometry of the pore system. The product

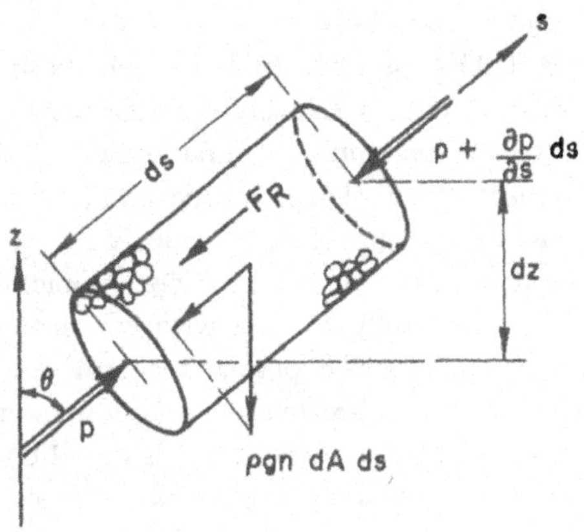

FIG. 5.10.3. Diagram of forces acting on the fluid within a cylindrical element of the porous field (Rumer and Drinker, 1966).

Cd^2 is recognized here as the (intrinsic) permeability, k, discussed in paragraph 5.5.1. When ρ varies, we cannot introduce the piezometric head, φ, and we obtain:

$$q_s = -(k/\mu)(\partial p/\partial s + \rho g\, \partial z/\partial s). \tag{5.10.32}$$

Harleman et al. (1963) verified experimentally the relationship for k included in (5.10.31).

Iberall's (1950) model consists of randomly distributed circular cylindrical fibers of the same diameter. The drag force, D', per unit length of a single fiber surrounded by similar fibers, all oriented in the direction of the flow with moderate separations, is given by:

$$D' = 4\pi\mu V \tag{5.10.33}$$

where V is the velocity of the fluid at a distance from the fiber (equivalent to the average velocity in a porous medium). Assuming that: (a) there are N filaments of unit length in a unit volume; (b) $N/3$ filaments are arrayed in each; and (c) the separation between fibers and the length of individual fibers are both large compared to the diameters of the fibers (i.e., the fibers are characterized by large porosity). The total drag force (per unit volume) on fibers having the direction of the flow can be equated to the pressure drop per unit length:

$$\Delta p/L = (4\pi N/3)\mu V. \tag{5.10.34}$$

Introducing the drag force on a cylinder perpendicular to the stream, his analysis leads to:

$$k = \frac{3}{16}\frac{n}{1-n}d^2\frac{2-\ln \mathrm{Re}}{4-\ln \mathrm{Re}}; \qquad \mathrm{Re} = \frac{\rho V d}{\mu}. \tag{5.10.35}$$

An outcome of (5.10.35) is that k is a function of Re (i.e., of q). This dependence on q (although the variations caused by it are relatively small) is introduced in this drag theory in a way completely different from the accepted approach of associating this dependence with the inertial forces.

Happel and Brenner (1965) present an extensive work on flow relative to assemblages of particles (this reference includes a summary of many earlier works by the same authors, written together or individually). This is an extension or generalization of the works described above. Happel and Brenner start by considering the force on a single particle, assuming no interaction with neighboring particles (dilute system); then proceed to consider particle interaction, expressed by the parameter λ in (5.10.27), and finally to deal with concentrated systems.

When solid particles comprising the porous medium are distributed in space in some repetitive manner (e.g., a square array in two-dimensional flow) they employ a model which they call a *cell model*. This model is based on the assumption that a repetitive, three-dimensional assemblage of solid particles may be regarded as consisting of a number of identical unit cells, each of which contains a single solid particle surrounded by a fluid envelope containing a volume of fluid sufficient to make the porosity of the cell identical to that of the entire assemblage. The outstanding feature of Happel and Brenner's (1965) approach is that they obtain Darcy's law with detailed expressions for k, including the numerical coefficients.

Finally, they consider flow through porous media, in which the arrangement and shape of particles vary in a manner that makes an exact geometrical specification of a boundary value problem, as in the cell model, impossible. Instead, they establish the plausibility of Darcy's law (i.e., the relationship between head drop and fluid velocity) for the case of a viscous incompressible fluid at low Reynolds numbers in an isotropic medium on the basis of similarity.

As other authors do, employing the same approach (e.g., Hubbert 1956), they assume that the inertial terms may be neglected, and start from the creeping motion equation in the form:

$$\nabla p = \mu \nabla^2 \mathbf{V}. \tag{5.10.36}$$

By using the modified inspectional analysis method described in detail in paragraph 11.2.7, they obtain from (5.10.37):

$$p_r/l_r = \mu_r V_r/l^2{}_r; \quad \text{or} \quad pl/\mu V = c = \text{const.} \tag{5.10.37}$$

Thus the constancy of the dimensionless group $pl/\mu V$ characterizes all solutions of the creeping motion equation (5.10.36). From this result it follows that, in general, all solutions of the creeping motion equation will correspond to:

$$V = (l^2/c)(\Delta p/\mu l) \tag{5.10.38}$$

where p is the average pressure, V is the average velocity used as a characteristic velocity and l^2/c corresponds to Darcy's permeability. Happel and Brenner extend this result to three-dimensional flow in porous media, in which the resistance to the flow will be experienced in the same direction as the flow itself (i.e., in isotropic media). When adding the effect of gravity as well, they obtain:

$$\mathbf{q} = -(k/\mu)\nabla(p + \rho g z). \tag{5.10.39}$$

5.10.5 Statistical Models

Although some of the models discussed above seem satisfactory in terms of the objective of yielding Darcy's law, they may be considered unsatisfactory in that they involve a high degree of simplification of actual porous media and the flow through them. This is especially so because in most cases, to permit a theoretical or mathematical treatment, e.g., in the form of a solution of the Navier–Stokes equations, an *ordered* porous medium model is chosen, whereas actual porous media are highly disordered. Obviously, the choice of a simplified model stems from the fact that it is hopeless to try to solve the Navier–Stokes equations for disordered media. In order to introduce disorder into our considerations, methods of *statistical mechanics* must be applied. A detailed discussion of the statistical approach and of some statistical models is given in paragraphs 4.5.9 and 10.3.2, and need not be repeated here.

As an example of the application of a statistical model, consider Scheidegger's (1954, 1960) work, discussed in detail in paragraph 10.3.2. There it leads to (10.3.44) as a macroscopic description of the spreading of tracer particles in flow through porous media. The fourth of the assumptions on which this derivation is based may be expressed by Poiseuille's law written in the form:

$$\mathbf{V} = \boldsymbol{\xi}/\Delta t = -[(B/\mu)\,\mathrm{grad}\,p]\cos\theta; \qquad |\boldsymbol{\xi}| = \xi \tag{5.10.40}$$

where $\boldsymbol{\xi}$ is the elementary displacement (vector), Δt is the duration of an elementary displacement, $\mathbf{V}$ is the velocity of the particle, B is the conductance of the elementary channel at the point through which the particle moves, $-\mathrm{grad}\,p$ is the external (uniform and steady) driving force and θ is the angle between the direction of $\boldsymbol{\xi}$ and that of $-\mathrm{grad}\,p$.

Let each elementary displacement be decomposed into two components: $\xi\cos\theta$ in the direction of the external force and $\xi\sin\theta$ normal to it. Then, in an *isotropic medium*, because of the equal probabilities of movement in all direction, the normal components will be eliminated by the process of averaging over the ensemble, and the average displacement will be colinear with $-\mathrm{grad}\,p$. We obtain:

$$\bar{\mathbf{V}} = -(\overline{B\cos^2\theta}/\mu)\,\mathrm{grad}\,p$$

$$\overline{\xi\sin\theta} = 0, \qquad \overline{\xi\cos\theta} = -\overline{(B/\mu)\,\mathrm{grad}\,p\,\cos^2\theta\,\Delta t} = \bar{\mathbf{V}}\,\Delta t \tag{5.10.41}$$

where $\bar{\mathbf{V}}$ is the average displacement velocity in the porous medium. When no correlation exists between B and $\cos^2\theta$, we have $\overline{B\cos^2\theta} = \bar{\mathbf{B}}\,\overline{\cos^2\theta}$. One may recognize in $\overline{\cos^2\theta}$ the tortuosity tensor $\underline{\mathbf{T}}$ (sec. 4.8) for an isotropic medium.

By introducing $\overline{B\cos^2\theta} = k$ (medium's permeability), Scheidegger (1954, 1960) obtains:

$$\mathbf{V} = \mathbf{q}/n = -(k/\mu n)\,\mathrm{grad}\,p. \tag{5.10.42}$$

Thus, in Scheidegger's statistical model, the randomness is in the movement of the fluid particles, whereas in the statistical model considered in sections 4.5 and 4.7 the randomness is attributed to the porous medium. Scheidegger (1954, 1960) considers only isotropic media and therefore he can split the probability density, p, defined by (10.3.44), into a product of three identical probabilities. His work may easily be extended to anisotropic media by stating that the probability distribution governing the individual displacements $\boldsymbol{\xi}$ is a function that has no special symmetry properties.

An interesting development is suggested by de Josselin de Jong (1969), using a model of interconnected channels as shown in figure 5.10.4 (see also par. 10.3.2). Flow is caused by the average head gradient, $\nabla\varphi$, which makes an angle α with the $+x$ axis. In a channel making an angle θ with the $+x$ axis, the gradient and the discharge are $|\nabla\varphi| \cos(\theta - \alpha)$ and $q(\theta) = \lambda(\theta)|\nabla\varphi| \cos(\theta - \alpha)$, respectively; $\lambda(\theta)$ is the conductivity of the channel in the direction θ. De Josselin de Jong (1969) assumes that the probability $P(\theta; \theta + d\theta)$ of a particle choosing a direction between θ and $\theta + d\theta$ is proportional to the total fluid discharge Q in that direction:

$$P(\theta; \theta + d\theta) = n(\theta)q(\theta)\, d\theta/Q \tag{5.10.43}$$

where $n(\theta)\, d\theta$ is the number of channels having an angle between θ and $\theta + d\theta$. The total number of channels under consideration is:

$$N = \int_{-\pi/2}^{+\pi/2} n(\theta)\, d\theta$$

where the integral covers only half the circle. Thus the total discharge through N channels is given by:

$$Q = \int_{-\pi/2+\alpha}^{+\pi/2+\alpha} n(\theta)q(\theta)\, d\theta = |\nabla\varphi| \int_{-\pi/2+\alpha}^{+\pi/2+\alpha} n(\theta)\lambda(\theta) \cos(\theta - \alpha)\, d\theta. \tag{5.10.44}$$

The integration is carried only over the range $-\pi/2 + \alpha < \theta < \pi/2 + \alpha$ because only channels in these directions carry fluid away from the junction points.

With l the length of a channel, and $\cos\theta$ and $\sin\theta$ its projections in the x and

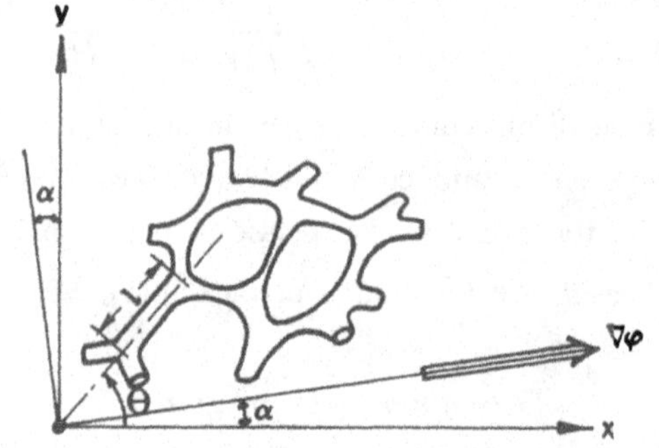

FIG. 5.10.4. De Josselin de Jong's model of interconnected channels (de Josselin de Jong, 1958).

y directions, respectively, he then computes the mean values $\bar{x} \equiv \overline{l \cos \theta}$ and $\bar{y} \equiv \overline{l \sin \theta}$:

$$\bar{x} = \frac{1}{Q} \int_{-\pi/2+\alpha}^{\pi/2+\alpha} n(\theta) l \cos \theta \, q(\theta) \, d\theta; \qquad \bar{y} = \frac{1}{Q} \int_{-\pi/2+\alpha}^{+\pi/2+\alpha} n(\theta) l \sin \theta \, q(\theta) \, d\theta. \quad (5.10.45)$$

The mean residence time ($\bar{t}$) of a particle in a channel is given by:

$$\bar{t} = \overline{lA(\theta)/q(\theta)} = \frac{1}{Q} \int_{-\pi/2+\alpha}^{\pi/2+\alpha} n(\theta) l A(\theta) \, d\theta \quad (5.10.46)$$

where $A(\theta)$ is the cross-sectional area of the channels $n(\theta)\, d\theta$.

The general case of anisotropy is obtained by assigning to the $n(\theta)\, d\theta$ channels an arbitrary distribution with respect to θ of the combined conductivities $n(\theta)\lambda(\theta)\, d\theta$ and the combined cross-sectional area $n(\theta)A(\theta)\, d\theta$. Every possible distribution can be expressed by a Fourier series. Let these be:

$$n(\theta)\lambda(\theta)/N \equiv \sum_{n=0}^{\infty} A_{2n} \cos 2n(\theta - \alpha_{2n})$$

$$n(\theta)A(\theta)/N \equiv \sum_{n=0}^{\infty} B_{2n} \cos 2n(\theta - \beta_{2n}) \quad (5.10.47)$$

where only the even terms appear.

By performing the integrations in (5.10.45) and (5.10.46), de Josselin de Jong obtains:

$$\bar{V}_x = \frac{\bar{x}}{\bar{t}} = \left(\frac{A_0}{2B_0} + \frac{A_2}{4B_0} \cos 2\alpha_2\right) \frac{\partial \varphi}{\partial x} + \frac{A_2}{4B_0} \sin 2\alpha_2 \frac{\partial \varphi}{\partial y}$$

$$\bar{V}_y = \frac{\bar{y}}{\bar{t}} = \frac{A_2}{4B_0} \sin 2\alpha_2 \frac{\partial \varphi}{\partial x} + \left(\frac{A_0}{4B_0} + \frac{A_2}{4B_0} \cos 2\alpha_2\right) \frac{\partial \varphi}{\partial y}. \quad (5.10.48)$$

Thus the four coefficients appearing on the right-hand sides of (5.10.48) are components of the second-rank symmetrical tensor of hydraulic conductivity divided by the porosity.

5.10.6 Averaging the Navier–Stokes Equations

Many researchers obtain the motion equation by averaging, in one form or another, the Navier–Stokes equations over some representative elementary volume of the porous medium. An example of this approach is presented in section 4.7 where the motion equation is derived by averaging the Navier–Stokes equations, in which the inertial terms are neglected, using the porous medium model described in paragraph 4.5.2 and the averaging procedure described in paragraph 4.5.4. In this case the porous medium model is composed of a spatial network of interconnected channels.

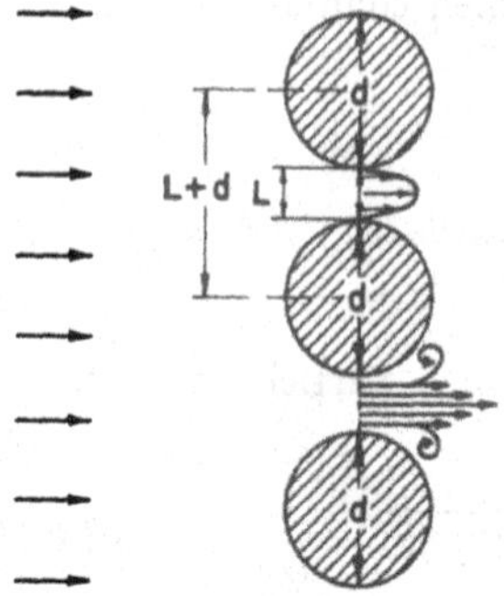

FIG. 5.10.5. Irmay's (1958) flow model.

A similar work is presented by Whitaker (1966), who averages the Navier–Stokes equations, without the inertial terms, over a representative volume of a porous medium, and obtains the motion equation for the general case of an anisotropic medium. The same approach is employed by others, including Hubbert (1956) and Prager (1961a). Raats (1965) bases his development of the macroscopic equations of conservation of mass, momentum (i.e., the motion equation) and energy on the theory of mixtures.

Finally, as an example of an averaging carried out over a specific porous medium model that does not neglect the inertial terms in the Navier–Stokes equations, consider the work of Irmay (1958). Irmay starts from the Navier–Stokes equations written as:

$$g\,\partial E/\partial x = \tfrac{1}{2}\,\partial(V_y^2 + V_z^2 - V_x^2)/\partial x + \partial(V_x V_y)/\partial y + \partial(V_x V_z)/\partial z - \partial V_x/\partial t + \nu(\partial^2 V_x/\partial x^2 + \partial^2 V_x/\partial y^2 + \partial^2 V_x/\partial z^2) \quad (5.10.49)$$

and similar equations for the y and z directions. The total energy of flow per unit weight of fluid is:

$$E = \varphi + V^2/2g = z + p/\rho g + V^2/2g.$$

Irmay (1958) then computes the space average (denoted by an overscore) of (5.10.49) for a model made of spheres of diameter d (see fig. 5.10.5), representing a homogeneous isotropic model of porosity $n = L/(L + d)$. By reasons of homogeneity and isotropy, $\bar{V}_y = 0$, $\bar{V}_z = 0$. Assuming that no correlation exists among the velocity components V_x, V_y and V_z, we have:

$$\overline{V_z V_y} = 0, \qquad \overline{V_x V_z} = 0, \qquad \partial(\overline{V_x V_y})/\partial_y = 0, \qquad \partial(\overline{V_x V_z})/\partial_z = 0.$$

For an incompressible homogeneous fluid, the continuity equation (par. 4.3.2) is: $\partial V_x/\partial x + \partial V_y/\partial y + \partial V_z/\partial z = 0$ and hence,

$$\partial V_x/\partial x = -\,\partial V_y/\partial Y - \partial V_z/\partial z \quad \text{and} \quad \partial^2 V_x/\partial x^2 = -\,\partial^2 V_y/\partial x\,\partial y - \partial^2 V_z/\partial x\,\partial z.$$

From these considerations Irmay (1958) obtains:

$$\overline{\partial^2 V_x/\partial x^2} = -\,\overline{\partial^2 V_y/\partial x\,\partial y} - \overline{\partial^2 V_z/\partial x\,\partial z} = 0. \quad (5.10.50)$$

The specific discharge and its time derivation are given by $\bar{V}_x = q/n$ and $\partial \bar{V}_x/\partial t = (\partial q/\partial t)/n$.

As the fluid adheres to the walls, we have $V_x = 0$. Between the grains we have a certain distribution of V_x values, with a maximum value somewhere between adjacent grains. For this distribution $\partial^2 V_x/\partial y^2 < 0$, $\partial^2 V_x/\partial z^2 < 0$. By averaging, we obtain:

$$\partial^2 \bar{V}_x/\partial y^2 + \partial^2 \bar{V}_x/\partial z^2 = -\,[\beta(1-n)^2/n^3 d^2]q \tag{5.10.51}$$

where β is a numerical shape factor that depends on the shape of the grains but not on the porosity or the diameter. Since div $\mathbf{V} = 0$, we have:

$$\tfrac{1}{2}\,\partial(V_x^2)/\partial x = V_x\,\partial V_x/\partial x = -\,V_x\,\partial V_y/\partial y - V_x\,\partial V_z/\partial z \quad \text{or} \quad \tfrac{1}{2}\,\partial\overline{(V_x^2)}/\partial x = 0. \tag{5.10.52}$$

At the vena contracta between the grains, $V_y = 0$, $V_z = 0$. Hence, in the converging zone $\partial(V_y^2 + V_z^2)/\partial x < 0$. Because of the large velocities under consideration, there is often separation of flow in the diverging zone, where again: $\partial(V_y^2 + V_z^2)/\partial x < 0$. In the zone of separation we have a milder divergence. This decreases the effective quadratic terms $V_x\,\partial V_x/\partial x$, $V_y\,\partial V_x/\partial y$ and $V_z\,\partial V_x/\partial z$ of divergent flow. Irmay, therefore, estimates:

$$\overline{\partial(V_y^2 + V_z^2)/\partial x} = -\,\alpha \bar{V}_x^2/L = -\,[\alpha(1-n)/n^3 d]q^2 \tag{5.10.53}$$

where α is a numerical shape factor. Finally, he obtains:

$$\bar{J}_x = -\,\partial\bar{E}/\partial x = Wq + bq^2 + c\,\partial q/\partial t \tag{5.10.54}$$

where:

$$W = [\beta(1-n)^2/n^3](\nu/gd^2); \qquad b = \tfrac{1}{2}[\alpha(1-n)/n^3](1/gd); \qquad c = 1/ng.$$

Similar expressions are obtained for $\bar{J}_y$ and $\bar{J}_z$. From (5.10.54) Irmay concludes that: (a) at low values of Re:

$$\bar{J}_x = Wq_x \quad \text{or} \quad q_x = K\bar{J}_x; \qquad K = 1/W = (gd^2/\beta\nu)\,[n^3/(1-n)^2],$$

which is Darcy's law with Kozeny's expression for K; W is hydraulic resistivity (par. 5.6.1); (b) at higher values of Re and steady flow (5.10.54) becomes Forchheimer's empirical formula (5.11.1) (The coefficient, b, depends on n and on the average grain diameter, d, but is independent of viscosity.); (c) when flow is unsteady, the term $c\,\partial q/\partial t$ of (5.10.54) should remain. However, Polubarinova-Kochina (1952) shows that except for the very onset of motion (perhaps during a fraction of a second), this additional term tends very rapidly to zero.

5.10.7 Ferrandon's Model

Ferrandon's (1948) model is presented here as one of the models leading to understanding of anisotropy in permeability.

Ferrandon (1948) considers an area, A, at a point, P, in the porous medium domain (fig. 5.10.6), with a unit vector $\mathbf{1n}$ (components n_i in an x_i Cartesian coordinate system) normal to it. Let $\mathbf{1m}$ be a unit vector (components m_i in the x_i coordinate

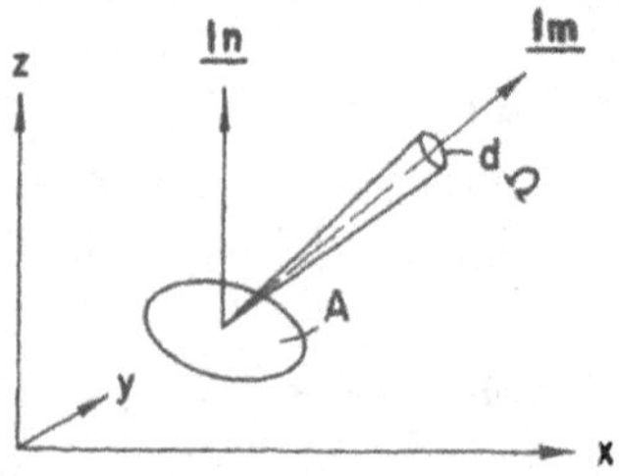

FIG. 5.10.6. Nomenclature for Ferrandon's (1948) model.

system) in an arbitrary direction, and $d\Omega$ an infinitesimal solid angle whose axis is in the direction **1m** and whose vertex is at P on A. Ferrandon assumes that the specific discharge contribution, dq_n, to flow in the direction **1n** through area A from elementary flow tubes parallel to the direction **1m**, and whose combined cross-sectional area is $c\,d\Omega$, is proportional to the component of the gradient of the piezometric head φ:

$$d\mathbf{q_m} = -\frac{k\gamma}{\mu}\,\mathbf{1m}\,\frac{\partial\varphi}{\partial m}\,c\,d\Omega; \qquad dq_n = \mathbf{1n}\cdot d\mathbf{q_m} = -\frac{k\gamma}{\mu}\,\mathbf{1n}\cdot\mathbf{1m}\,\frac{\partial\varphi}{\partial m}\,c\,d\Omega \quad (5.10.55)$$

where c and k are assumed functions of the components m_i.

Since $\partial\varphi/\partial m = m_j\,\partial\varphi/\partial x_j$, the summation convention being applied, we obtain:

$$dq_n = -\frac{k\gamma}{\mu}\,n_i m_i m_j\,\frac{\partial\varphi}{\partial x_j}\,c\,d\Omega. \quad (5.10.56)$$

In order to obtain the total specific discharge through A we must integrate dq_n over half-space, with respect to A, on the side of **1n** of the area A. Since the total solid area in such case is 2π, we obtain:

$$q_n = -n_i\,\frac{\gamma}{\mu}\,\frac{\partial\varphi}{\partial x_j}\int_0^{2\pi} kcm_i m_j\,d\Omega = -n_i\,\frac{\gamma}{\mu}\,k_{ij}\,\frac{\partial\varphi}{\partial x_j}$$

or:

$$k_{ij} = \int_0^{2\pi} kcm_i m_j\,d\Omega; \qquad q_j = -\frac{k_{ij}\gamma}{\mu}\,\frac{\partial\varphi}{\partial x_j} \quad (5.10.57)$$

which is Darcy's law for three-dimensional flow in an anisotropic medium.

5.11 Flow at Large Reynolds Numbers

5.11.1 The Phenomenon

The discussion in sections 5.1 through 5.10 is limited to laminar flow through porous media at low Reynolds numbers (say Re $< 1 \div 10$) for which a proportionality

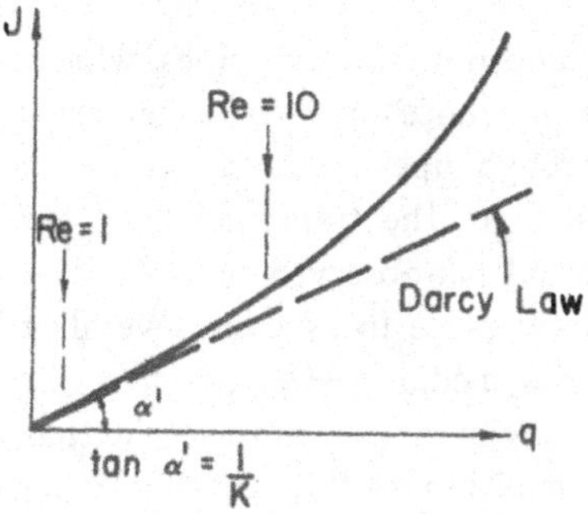

FIG. 5.11.1. Schematic curve representing experimental relationship between flux and hydraulic gradient.

relationship exists between the flux **q** and the hydraulic gradient **J**.

As Re increases beyond this range, the ratio J/q, say, in one-dimensional flow, gradually also increases. Figure 5.11.1 gives a schematic representation of results of a typical one-dimensional experiment in which the rate of flow is gradually increased. The deviation from Darcy's law is clearly seen in this figure. The motion equation in the range above that to which Darcy's (linear) law is applicable (par. 5.3.1) is discussed in this section.

Reviews of the subject are presented, among others, by Scheidegger (1960) and Kirkham (1967a). Except for the work of Bachmat (1967) there is no published work known to the author on flow at large Re in anisotropic media.

5.11.2 Turbulence, Inertial Forces and Separation

Forchheimer (1901) is probably the first to suggest a nonlinear relationship between J and q at large Reynolds numbers. His one-dimensional motion equation takes the form:

$$J = Wq + bq^2 = (W + bq)q \qquad (5.11.1)$$

where W and b are constants (fig. 5.11.2). At first sight, the second power seems to indicate that Forchheimer attributes nonlinearity to the appearance of turbulence, as the head gradient in turbulent flow through rough pipes is also proportional to the square of the velocity. However, there seem to be several basic differences

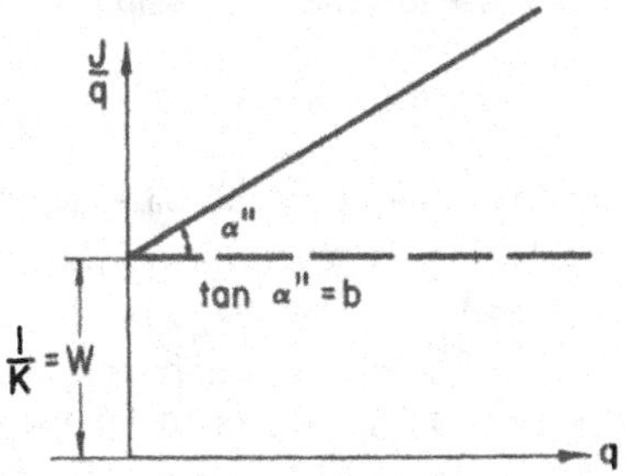

FIG. 5.11.2. Graphical representation of (5.11.1).

between the case on hand and flow through pipes, which make the attempt to base the explanation on the appearance of turbulence rather questionable. The first is that in turbulent flow through pipes the linear term Wq of (5.11.1) does not appear. Secondly, in flow through pipes the transition from laminar to turbulent flow is rather sharp and not gradual (as observed in fig. 5.3.1). Finally, the critical Re at which the transition takes place in pipes is several orders of magnitude higher than in flow through porous media. Attempts to explain these discrepancies by the inhomogeneity of the medium (for example, actual turbulence starts at the larger pores while in the smaller ones flow is still laminar) also fail when they are analyzed thoroughly (e.g., Scheidegger 1960).

Schneebeli (1955), Dudgeon (1966), Wright (1968) and others perform experiments to determine the Reynolds number at which turbulence starts. Most of them observe that the onset of turbulence occurs at Re in the range 60 to 150. These results indicate that the deviation from Darcy's law (which occurs at Re $= 1 \div 10$) cannot be attributed to turbulence. Instead, Lindquist (1933), Schneebeli (1955), Hubbert (1956), Scheidegger (1960) and many others, propose attributing the deviation from Darcy's law to *inertial forces*, which at low Reynolds numbers are negligible in comparison with viscous forces. The inertial forces are proportional to the square of the velocity and are independent of the viscosity. This is probably a better explanation for the second power in (5.11.1). Carman (1937), Schneebeli (1955) and Scheidegger (1960) emphasize that the Hagen–Poisseuille formula is not valid for tortuous tubes, as in such tubes the inertial forces do not vanish as they do in straight ones, and we have the relationship between the pressure drop (Δp) and the discharge (Q) in the general form:

$$\Delta p = a_1 Q + a_2 Q^2; \qquad a_1, a_2 = \text{constants}. \tag{5.11.2}$$

Another model often used for flow through porous media is that of drag forces (e.g., Rumer and Drinker 1966 and Rumer 1969). From the discussion of this model in paragraph 5.10.4 it follows that using inertial forces to explain the second power (5.11.1) is more appropriate. At low Reynolds number, Stokes' law—analogous to Darcy's law—states that the drag resistance acting on a solid body moving in a fluid is directly proportional to the velocity of the moving body with respect to the fluid. At higher Re—i.e., in the nonlinear laminar flow regime—Rumer and Drinker (1966 and Rumer 1969) make use of the expression:

$$D = C_D \alpha d^2 (\rho V_s^2/2) \tag{5.11.3}$$

in which αd^2 is the cross-sectional area of a solid particle and C_D is the "drag coefficient" which for incompressible flow is generally considered to be a function of the Reynolds number of the flow and also the shape of the particle. For flow in the Stokes range, $C_D = (2n\lambda)/(\alpha \text{ Re})$, where λ is a factor which introduces the effect of neighboring spheres, so that (5.11.3) and (5.10.27) become identical, leading to (5.10.30) for linear laminar flow. Re is defined by: Re $= q_s d/\nu$. Making use of the more general expression (5.11.3), the general force balance equation (5.10.31) becomes:

$$-\frac{\partial \varphi}{\partial s} = \frac{\alpha(1-n)}{2\beta n^3 d}\frac{C_D}{g}q_s^2 = \frac{\alpha}{2n^2}\left(\frac{1-n}{\lambda\beta}\right)^{1/2}\frac{1}{g\sqrt{k}}C_D q_s^2 \tag{5.11.4}$$

where k is defined by (5.10.31). In (5.11.4) βd^3 is the volume of a solid particle. The drag coefficient, C_D, may be related to a dimensionless friction factor, f_k, defined (Ward 1964) as:

$$f_k = -\left(\frac{g\sqrt{k}}{q_s^2}\right)\frac{\partial \varphi}{\partial s} = \left(\frac{\alpha}{2n^2}\right)\left[\frac{1-n}{\lambda\beta}\right]^{1/2} C_D \tag{5.11.5}$$

which depends upon both the pore system geometry and the Reynolds number. For linear laminar flow, $C_D = (2n\lambda)/(\alpha \text{ Re})$ so that (5.11.5) leads to:

$$f_k = 1/\text{Re}k; \qquad \text{Re}k = q_s\sqrt{k}/\nu \tag{5.11.6}$$

and Darcy's law is obtained by combining (5.11.5) with (5.11.6). If f_k is expressed by the more general expression

$$f_k = 1/\text{Re}k + C_1(\text{Re}/\text{Re}k) \tag{5.11.7}$$

(Goldstein 1938), we obtain (Sunada 1965):

$$-\frac{\partial \varphi}{\partial s} = \frac{\nu}{gk}q_s + \frac{C_1 d}{gk}q_s^2. \tag{5.11.8}$$

In this last form, the motion equations include both a linear term and a quadratic one.

As another example leading to a nonlinear motion equation, consider the porous medium model proposed and analyzed by Blick (1966). This model (fig. 5.11.3) consists of a bundle of parallel capillary tubes with orifice plates spaced along each tube, separated by a distance equal to the mean pore diameter. If the mean pore diameter is δ, then the number of orifices per unit length is δ^{-1}. The diameter of the orifice is assumed to be smaller than the mean pore diameter.

The following assumptions are made for the momentum balance on the model element. The medium is rigid; the fluid is homogeneous and Newtonian, filling the pore space fully; no adsorbed phase is present; the flow is one-dimensional. By applying the momentum theory to the control volume in figure 5.11.3b, and dividing by the volume $\pi\delta^2 \Delta x/4$, Blick (1966) obtains:

$$-\partial p/\partial x = 4\tau/\delta + C_D\rho V^2/2\delta + \partial(\rho V^2)/\partial x + \partial(\rho V)/\partial t - \rho F_B = 0 \tag{5.11.9}$$

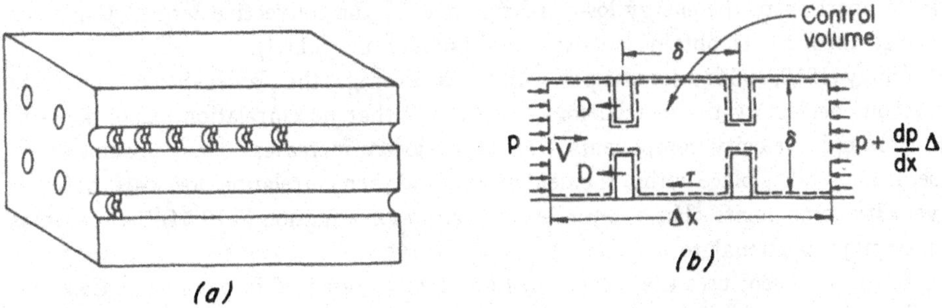

FIG. 5.11.3. Blick's capillary orifice model (Blick, 1966).

where V is velocity, τ is shear stress at the wall, C_D is drag coefficient of orifice plate, and F_B is body force per unit mass. For steady flow and for a body force due to gravity only (5.11.9) becomes:

$$-\partial p/\partial x = 4\tau/\delta + C_D\rho V^2/2\delta + \rho V\,\partial V/\partial x - \rho g\cos(\mathbf{1x}, \mathbf{g}). \qquad (5.11.10)$$

The shear stress can be expressed as:

$$\tau = C_f \tfrac{1}{2}\rho V^2 \qquad (5.11.11)$$

or:

$$\tau = C_f \frac{1}{2}\frac{V\mu}{\delta}\,\mathrm{Re} \qquad (5.11.12)$$

with (5.11.12) and $q = nV$, equation (5.11.10) becomes:

$$-\partial p/\partial x = (2C_f\,\mathrm{Re}/\delta^2 n\mu)q + (C_D\rho/2n^2\delta)q^2 + (\rho q/n^2)\,\partial q/\partial x - \rho g\cos(\mathbf{1x}, \mathbf{g}). \quad (5.11.13)$$

For horizontal flow, the last term on the right-hand side of (5.11.13) vanishes. For gas flow under isothermal conditions, Blick (1966) obtains:

$$-\frac{\partial p}{\partial x} = \frac{2C_f\,\mathrm{Re}}{n\delta^2}\mu q + \frac{C_D\rho}{2\delta n^2}q^2. \qquad (5.11.14)$$

Thus, the procedure produces a theoretical expression for the viscous resistance (reciprocal of permeability) and for the inertial resistance coefficient. The former produces head loss along the tube while the latter produces local losses proportional to the second power of the velocity. If we consider only the linear term in (5.11.1), then, by comparison, we have $1/W = k\rho g/\mu$ and hence: $k = (1/2C_f\,\mathrm{Re})n\delta^2$. Since the skin friction coefficient for laminar flow in a capillary is given by the Hagen–Poisseuille law as $C_f = 16/\mathrm{Re}$, we obtain $k = n\delta^2/32$, which is the same as (5.10.3).

Various investigators derive the motion equation by averaging the Navier–Stokes equations (see par. 5.10.6). Hubbert (1956), by averaging the Navier–Stokes equations, neglecting the inertial forces, obtains Darcy's linear law. He therefore concludes that the deviation from Darcy's law occurring at high Reynolds number is due to the inertial forces. Ahmed (1967) also averages the Navier–Stokes equations, without neglecting the inertial terms. During the development leading to the equation of motion he shows that as energy losses are due to turbulence, they may be neglected in comparison to the energy losses represented by the convective terms. Using the Gauss theorem he obtains an expression similar to (5.11.1).

Irmay (1958, 1964d; see also par. 5.10.6) also averages the Navier–Stokes equations without neglecting the inertial forces, assuming that no correlation exists between microscopic velocity components. As the velocity increases, Irmay assumes that separation takes place within the field of flow and then correlation does exist between velocity components. His analysis leads to a motion equation (5.10.54) that contains a term proportional to q^2.

Bachmat (1965) uses the porous medium model presented in paragraph 4.5.2 and follows the procedure of averaging the Navier–Stokes equations over the pore volume of this model as described in section 4.7, but without neglecting the convective term $\mathbf{V} \cdot \text{grad}\, \mathbf{V}$.

In the three works mentioned here—those of Irmay (1958, 1964d), Ahmad (1967) and Bachmat (1965)—the nonlinear term is obtained by averaging the Navier–Stokes equations, taking the inertial terms into account. However, the last-named investigators did not examine the conditions under which the volume or areal integral (over the representative elementary volume) appearing in the averaging procedure attains negative, positive or zero values. Irmay did prove that when separation occurs the average of the term $\mathbf{V} \cdot \text{grad}\, \mathbf{V}$ has a negative value (see details in par. 5.10.6). It seems, however, that Irmay's averaging was not performed over a sufficiently large volume. It appears that in all cases of steady *uniform flow* in a homogeneous medium, the average over a sufficiently large volume will lead to a zero value of the above term (i.e., $\overline{\partial(V_i V_j)/\partial x_i} = \partial(\overline{V_i V_j})/\partial x_i = 0$). This conclusion is also valid for the analyses carried out by Ahmad and Bachmat.

Thus, rather than assume fixed streamlines, one may base the explanation of the deviation from Darcy's law on the assumption that at high Re streamlines change, possibly with separation. The concept of a critical Reynolds number at which the deviation from Darcy's law occurs is thus associated with the explanation that attributes the deviation to the onset of flow separation.

If we attribute the deviation from Darcy's law to inertial forces only, no critical Re should exist. One may then define a critical Re (as does Ward 1964) as that value of Re where deviation starts to be felt, according to some agreement as to which deviation this refers to. Scheidegger (1960) also mentions a critical Reynolds number—two to three orders of magnitude higher than that at which turbulence begins. Dudgeon (1966) distinguishes on the basis of experiments several subregions of high Re.

Chauveteau and Thirriot (1967) perform experiments in several two-dimensional models that permit visual observation of streamlines and streaklines. In a typical model, which has the shape of a conduit that diverges and converges alternately, they observe streaklines. For $\text{Re} < 2$, the flow obeys Darcy's law and the streamlines remain fixed. As Re increases, streamlines start to shift and fixed eddies begin to appear in the diverging areas of the model. They become larger as Re increases. At $\text{Re} = 75$ turbulence appears and starts to spread out as Re increases. Turbulence covers some 50% of the flow domain at $\text{Re} = 115$ and 100% of it at $\text{Re} = 180$. The deviation from Darcy's law is observed at $\text{Re} = 2$–3. Thus the deviation from Darcy's law as the velocity increases is associated with gradual shifting of streamlines due to the curvature of the microscopic solid walls of the pore space.

To summarize, in the various explanations presented above for deviation from Darcy's law, inertial forces play the primary role. Yet, according to one explanation, they produce turbulence. According to another explanation these forces are always present, but they gradually become predominant (with respect to viscous forces) at large Re. Finally, one may attempt to explain the deviation by the separation of the flow caused at large Re by the inertial forces at microscopic points of the pore space where flow diverges or is curved. Most experiments indicate that actual turbulence occurs at Re values at least one order of magnitude higher than the Re at which deviation from Darcy's law is observed.

5.11.3 Some Examples of Proposed Nonlinear Motion Equations

As a result of various theoretical or experimental investigations, some of them described in the preceding paragraph, a large number of nonlinear motion equations appear in the literature. Scheidegger (1960) presents a review of many of them. They may be divided into three groups. In group I, the coefficients are not related to any specific fluid and medium properties. Group II contains expressions with coefficients more or less related to fluid and medium properties, and including unspecified numerical parameters. The coefficients appearing in the equations belonging to group III are similar in nature to those of group II, but the numerical parameters are exactly specified.

Some examples follow of nonlinear motion equations from each of these groups. Except for (5.11.25), all equations refer to an isotropic porous medium.

Group I

Forchheimer (1901) $$J = Wq + bq^2, \qquad W, b = \text{const} \tag{5.11.15}$$

Forchheimer (1930) $$J = Wq + bq^m, \qquad 1.6 \leqslant m \leqslant 2 \tag{5.11.16}$$

Forchheimer (1901) $$J = Wq + bq^2 + cq^3 \tag{5.11.17}$$

where the last term is added to make the equation fit experimental results.

Polubarinova-Kochina (1952) $$J = Wq + bq^2 + c\, \partial q/\partial t \tag{5.11.18}$$

White (1935) $$\Delta p/\Delta x = aq^{1.8} \tag{5.11.19}$$

based on experiments with air flow. Other authors report powers of 2.23 and 2.49.

Group II

Scheidegger (1960) $$J = C_1 \frac{\nu T^2}{gn} q + C_2 \frac{T^3}{gn^2} q^2 \tag{5.11.20}$$

where C_1 and C_2 are coefficients depending on the grain-size distribution, and T is tortuosity. This equation is obtained from a porous medium model consisting of capillaries in series. Ergun and Orning (1949), following Kozeny and Carman (see Scheidegger 1960) obtained

$$\begin{aligned} J &= \frac{5\alpha(1-n)^2 \nu M_s^2}{gn^3} q + \frac{\beta(1-n)M_s}{8gn^3} q^2 \\ &= 180\alpha \frac{(1-n)^2\nu}{gn^3d^2} q + \frac{3\beta(1-n)}{4gn^3d} q^2 \end{aligned} \tag{5.11.21}$$

where α and β are shape factors, and M_s is specific area per unit volume of solid, $d = 6/M_s$.

Burke and Plummer (1928, see Scheidegger 1960)

$$J = [K_0(1-n)/gn^3d^2]q^2 \tag{5.11.22}$$

where K_0 is constant. This equation probably corresponds only to flow at very large Re.

Rumer and Drinker (1966; par. 5.11.2)

$$J = (\alpha C_D/n^2)\sqrt{(1-n)/\lambda\beta k}\, q^2/2g \tag{5.11.23}$$

where α and β are defined volume and surface-shape factors, λ is a factor that takes into account the effect of neighboring spheres, C_D is a drag coefficient of a particle (depending on Re) and k is the medium's permeability.

Irmay (1958, 1964d)

$$J = \alpha \frac{\nu(1-n)^2}{gd^2(n-n_0)^3} q + \frac{\beta(1-n)}{gd(n-n_0)^3} q^2 + \frac{1}{g(n-n_0)} \frac{\partial q}{\partial t} \tag{5.11.24}$$

where α and β are constant shape factors and n_0 is the "ineffective" porosity.

Bachmat (1965)

$$\underline{\mathbf{k}} \cdot \mathbf{J} = \frac{\nu}{g}\left(1 + \frac{q\beta}{n\nu}\right)\mathbf{q} \tag{5.11.25}$$

which is the only known equation for three-dimensional flow in an anisotropic medium. In (5.11.25) $\underline{\mathbf{k}}$ is the permeability tensor, given by (4.7.14), $\beta = \overline{B\,\mathrm{div}(-\mathbf{1}l)}$ is a geometrical coefficient, and $\mathbf{1}l$ is a unit vector denoting the orientation of a pathline of a fluid at a point.

Blick (1966; par. 5.11.2)

$$J = \left(\frac{32\nu}{gnD^2}\right)q + \left(\frac{C_D}{2Dgn^2}\right)q^2 \tag{5.11.26}$$

where $D = 3(1-n)/n$ is the diameter of the capillary, $C_D = (1 - h^2/D^2)(D^4/h^4 - 1)/C_0^2$, h is the diameter of the narrow opening (fig. 5.11.3b) and C_0 is the coefficient of the opening.

Ahmad (1967)

$$J = (\mu/\gamma k)q + (1/g\sqrt{Ck})q^2 \tag{5.11.27}$$

where C is the coefficient appearing in $k = Cd^2$.

Group III

Ergun (1952)

$$J = 150 \frac{(1-n)^2\nu}{n^3gd^2} q + 1.75 \frac{(1-n)}{n^3gd} q^2 \tag{5.11.28}$$

Schneebeli (1955)

$$J = 1100 \frac{\nu}{gd^2} q + \frac{12}{gd} q^2 \tag{5.11.29}$$

for spheres. This equation is proposed for Re > 2.

Carman (1937)

$$J = 180 \frac{(1-n)^2 \nu}{n^3 g d^2} q + 2.87 \frac{\nu^{0.1}(1-n)^{1.1}}{n^3 g d^{1.1}} q^{1.9} \qquad (5.11.30)$$

obtained from experiments.

Ward (1964)

$$J = \frac{\nu}{gk} q + \frac{0.55}{g\sqrt{k}} q^2; \qquad (5.11.31)$$

Irmay (1964d)

$$J = 180 \frac{(1-n)^2 \nu}{g n^3 d^2} q + 0.6 \frac{1-n}{g n^3 d} q^2; \qquad \mathrm{Re} > 1 \div 10. \qquad (5.11.32)$$

5.12 Seepage Forces and Stresses

5.12.1 The Forces

In saturated flow through a porous medium a force is exerted by the fluid on the solid matrix. These forces have an important effect in many engineering problems. Following Terzaghi (1943) and Taylor (1948), three forces, per unit volume of porous medium, act on the solid particles of the porous medium through which seepage occurs (Rizenkampf 1938).

Force 1, the weight of the solids $\mathbf{F}_1$ *acting in a downward direction*

$$\mathbf{F}_1 = -\gamma_s(1-n)\mathbf{1z} \equiv -\gamma_s(1-n)\nabla z = -\gamma_s \nabla z/(1+e) \qquad (5.12.1)$$

where γ_s is the specific weight of the dry solid particles, n is porosity and e is the void ratio.

Force 2, the buoyancy (or uplift) force $\mathbf{F}_2$ *(equal to the resultant of the liquid pressure acting on the solid particles).* Since the force per unit volume of solid due to water pressure is equal to $-\nabla p$, we have:

$$\mathbf{F}_2 = -(1-n)\nabla p. \qquad (5.12.2)$$

Force 3, the drag (or shear) $\mathbf{F}_3$ *of the fluid at the solid–fluid interface, per unit volume of porous medium.* The drag is equal to:

$$\mathbf{F}_3 = n\gamma_f \mathbf{J} = -n\gamma_f \nabla \varphi \qquad (5.12.3)$$

where γ_f is the fluid's specific weight. $\mathbf{F}_3$ is sometimes called the *seepage force.* For an isotropic medium $\mathbf{F}_3 = (n\gamma_f/K)\mathbf{q}$. Being in the direction of $\mathbf{J}$, the drag (shear) in an isotropic medium is always normal to the equipotential surfaces (φ = const). However, in anisotropic media, the direction of this drag force does not coincide with that of the streamlines. One should recall that the available energy per unit weight of fluid, φ, is used up in driving the fluid through the medium. This energy is dissipated as viscous friction at the solid–fluid interfaces and, as always when energy is dissipated through friction, a force, or drag, is exerted on the solid matrix in the direction of the motion. Another way of expressing the same idea is by saying that the porous matrix offers a resistance to the flow of the fluid through it. According to Darcy's law this force, per unit mass of fluid, is proportional to the specific discharge, and in a direction opposite to it. Hence the force exerted on the solid matrix by the fluid, per unit mass of fluid, is equal to $+(g/K)\mathbf{q} = (g/K)(K\mathbf{J}) = g\mathbf{J}$ (see sec. 5.9). When written as force per unit volume of porous medium we obtain $\mathbf{F}_3$ of (5.12.3).

The resultant force per unit volume of porous medium acting on the solid matrix is obtained by adding the above three forces:

$$\begin{aligned}\mathbf{F} &= \mathbf{F}_1 + \mathbf{F}_2 + \mathbf{F}_3 = -\gamma_s(1-n)\nabla z - (1-n)\nabla p + n\gamma_f\mathbf{J}\\ &= -(1-n)\nabla(p+\gamma_s z) + n\gamma_f\mathbf{J} = -\gamma_s(1-n)\nabla z - (1-n)\nabla p - n\gamma_f\nabla(z + p/\gamma_f)\\ &= -\gamma_{\text{total}}\nabla z - \nabla p = -(\gamma_f + \gamma^*)\nabla z - \nabla p = \gamma_f\mathbf{J} - \gamma^*\nabla z \qquad (5.12.4)\end{aligned}$$

where γ_{total} is the specific weight of the porous medium (fluid included) and $\gamma^* = (\gamma_s - \gamma_f)(1-n)$ is the *specific submerged weight of the solid matrix.*

Figure 5.12.1 gives a graphic representation of the forces involved in (5.12.4). For the sake of simplicity it is assumed that all forces act in the vertical xz plane. The triangle ACE represents Darcy's law and the definition of **J**. From the diagram we also see that the total force **F** can either be obtained as a sum of the *submerged weight* $(-\gamma^*\mathbf{1z})$ and the *seepage force* $\gamma_f\mathbf{J}$ (which is equal to the drag per unit volume of fluid) or as the sum of the total weight $(-\gamma_t\mathbf{1z})$ and the force $(-\nabla p)$ exerted by the pressure alone. The latter is sometimes called *boundary neutral force.* The resultant force **F** may be decomposed into three components in a Cartesian coordinate system:

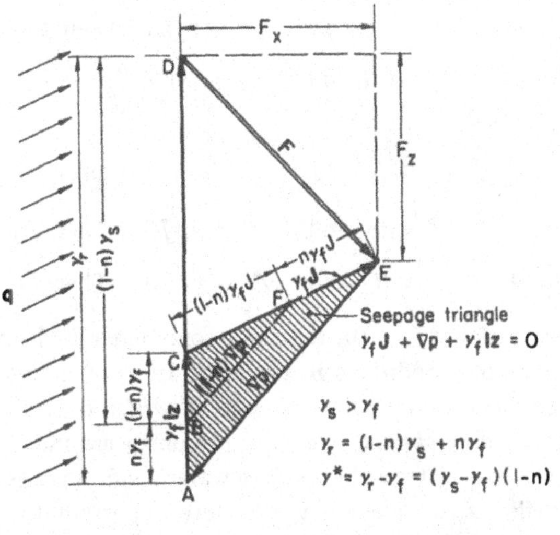

FIG. 5.12.1. Forces polygon in flow through porous media (Irmay, 1964c).

$$F_x = -(1-n)\,\partial p/\partial x + (n\gamma_f/K)q_x = -(1-n)\,\partial p/\partial x - n\gamma_f\,\partial\varphi/\partial x = -\,\partial p/\partial x = \gamma_f J_x$$
$$F_y = -(1-n)\,\partial p/\partial y + (n\gamma_f/K)q_y = -(1-n)\,\partial p/\partial y - n\gamma_f\,\partial\varphi/\partial y = -\,\partial p/\partial y = \gamma_f J_y$$
$$F_z = -\gamma_s(1-n) - (1-n)\,\partial p/\partial z + (n\gamma_f/K)q_z = -\gamma_s(1-n)$$
$$-(1-n)\,\partial p/\partial z - n\gamma_f\,\partial\varphi/\partial z = -\gamma^* - \gamma_f\,\partial\varphi/\partial z = -\gamma^* + \gamma_f J_z \qquad (5.12.5)$$

where the expressions with components of **q** apply only to isotropic media. In the case of anisotropic media we must introduce the appropriate relationship between **q** and **J**. When no flow occurs, $\mathbf{J} = 0$. In order to obtain a downward-directed force we must have $-\gamma^* + \gamma_f J_z > 0$; $\gamma^* > 0$ when $\gamma_s > \gamma_f$ and $\gamma^* < 0$ where $\gamma_s < \gamma_f$ (e.g., when the solid is ice).

From (5.12.4) we see that we may define a *force potential* ϕ such that the force **F** is derived from it:

$$\phi = \gamma_s(1-n)z + (1-n)p + n\gamma_f\varphi = \gamma_{\text{total}}z + p = (\gamma_f + \gamma^*)z + p$$
$$\mathbf{F} = -\operatorname{grad}\phi \qquad (5.12.6)$$

where ϕ is defined except for a constant (Polubarinova–Kochina 1952, 1962). When $\nabla^2\varphi = 0$ (and therefore $\nabla^2 p = 0$) we also have $\nabla^2\phi = 0$. **F** and ϕ are useful in the analysis of stresses in deformable soils (sec. 6.3).

5.12.2 Piping and Quicksand

The component F_z may be zero, positive (upward) or negative (downward). In upward flow, $J_z > 0$. In one-dimensional flow in isotropic media we may distinguish several cases of interest.

Upward flow:

(a) $J_z > 0$, $\gamma^* > 0$ and $\gamma^* < \gamma_f J_z$. Then $F_z > 0$.

(b) $J_z > 0$, $\gamma^* > 0$ and $\gamma^* > \gamma_f J_z$. Then $F_z < 0$.

(c) $J_z > 0$, $\gamma^* < 0$ (i.e., $\gamma_s < \gamma_f$). Then $F_z > 0$.

Downward flow:

(d) $J_z < 0$, $\gamma^* > 0$ (i.e., $\gamma_s > \gamma_f$). Then $F_z < 0$.

(e) $J_z < 0$, $\gamma^* < 0$ and $|\gamma^*| > \gamma_f|J_z|$. Then $F_z > 0$.

(f) $J_z < 0$, $\gamma^* < 0$ and $|\gamma^*| < \gamma_f|J_z|$. Then $F_z < 0$.

Whenever in upward flow ($J_z > 0$), the force component F_z is directed downward ($F_z < 0$), we have stable conditions in the soil. In cases a and c, however, F_z is directed upward and a situation known as *quick condition* occurs. Sand under quick condition is often called *quicksand*. Quick conditions occur more readily at the exit of water from the soil to the atmosphere or to a water pond. In the design of structures underneath which flow takes place it is therefore important to determine the *exit gradient* at the soil surface, taking care that it remains below some critical value. Thus, quick conditions occur when in case a of upward flow:

$$F_z \geqslant 0; \qquad J_z \geqslant \gamma^*/\gamma_f; \qquad J_z > 0; \qquad \gamma^* > 0. \tag{5.12.7}$$

The gradient corresponding to $F_z = 0$ is called the *critical gradient* J_{cr}

$$J_{cr} = \gamma^*/\gamma_f. \tag{5.12.8}$$

Since the ratio γ^*/γ_f is usually close to unity (e.g., for $n = 0.35$, $J_{cr} = 0.72$; for $n = 0.2$, $J_{cr} = 1.12$), the critical gradient is usually taken as equal to one. Thus a saturated sand becomes quick if upward flow at a gradient close to unity, or greater, occurs at its surface.

The quick condition is also frequently termed a condition of *fluidized bed.*

Under quick conditions, the strength of the soil becomes zero. As the shear strength of cohesionless soils is directly proportional to the *effective stress* (sec. 2.7) in the soil, a quick condition is one where the shear strength of the soil is zero owing to zero effective stress. Hence it is possible to determine the critical conditions by considering the effective stress in the soil.

In soil, stresses are caused by the weight of the soil itself (solid matrix + water), as well as by any external load applied to the soil. When the ground surface is horizontal and the soil is homogeneous, or at least homogeneous in the horizontal direction, the vertical stresses at a depth D are obtained from:

$$\sigma = \int_0^D \gamma_{\text{total}}(z)\, dz. \tag{5.12.9}$$

One should note that when the phreatic table is present at some depth below the ground surface, γ_{total} above the phreatic surface is the specific weight of the dry soil, whereas γ_{total} below this surface is the specific weight of the combined solid–water system. When an external load is present at the ground surface, its effect at D should also be taken into account.

Consider a point B in the sand layer of thickness D shown in figure 5.12.2. The sand is covered by a pond of water ($\gamma_f \equiv \gamma_w$) of depth H. In all three cases, the stress at point B is given by:

$$\sigma_B = H\gamma_w + D\gamma_t; \qquad \gamma_t = (1-n)\gamma_s + n\gamma_w \equiv \gamma_{\text{total}}. \tag{5.12.10}$$

It is caused by the total weight of the water and the sand above B. Let us now determine the *effective stress* ($\sigma' = \sigma - p$) in the three cases.

Case 1, no flow, $q = 0$. The pressure at B is: $p_B = (H + D)\gamma_w$. Hence, the effective stress at B is:

$$\sigma'_B = \sigma_B - p_B = H\gamma_w + D\gamma_t - (H+D)\gamma_w = D(\gamma_t - \gamma_w) = D\gamma^*;$$
$$\gamma^* = \gamma_t - \gamma_w = (1-n)(\gamma_s - \gamma_w). \tag{5.12.11}$$

Case 2, downward flow, $q_z < 0$

$$p_B = (H + D - \Delta\varphi)\gamma_w, \qquad \sigma'_B = D(\gamma_t - \gamma_w) + \gamma_w \Delta\varphi = D\gamma^* + \gamma_w \Delta\varphi. \tag{5.12.12}$$

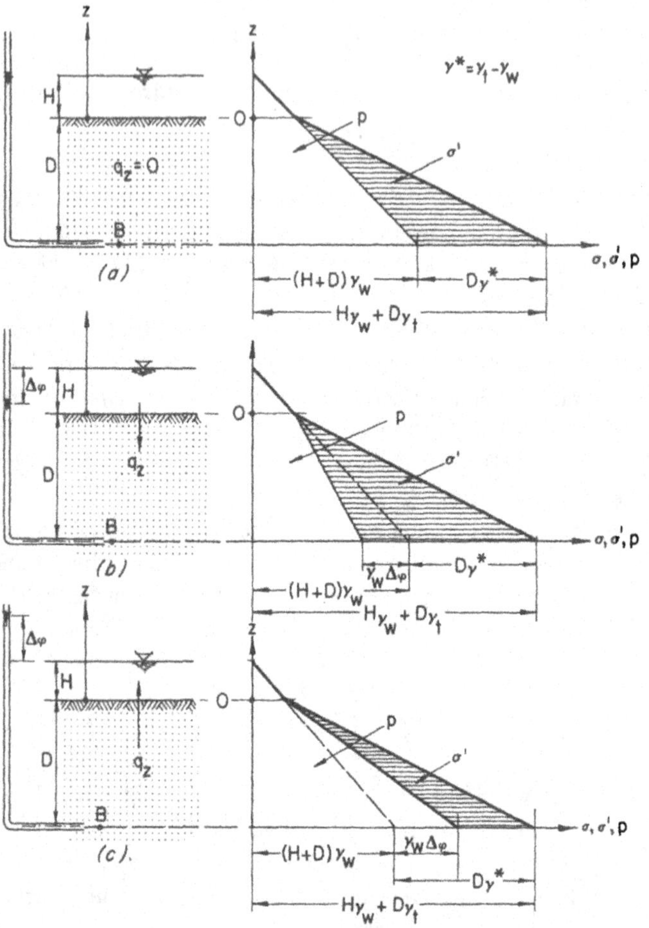

FIG. 5.12.2. Distribution of vertical effective stress in soil.

Case 3, upward flow, $q_z > 0$

$$p_B = (H + D + \Delta\varphi)\gamma_w, \qquad \sigma'_B = D(\gamma_t - \gamma_w) - \gamma_w \Delta\varphi = D\gamma^* - \gamma_w \Delta\varphi. \quad (5.12.13)$$

The distribution of σ' for the three cases is shown in figure 5.12.2. In downward flow, as $\Delta\varphi$ increases, σ' also increases and the soil is stable. In upward flow, we may reach a critical value $(\Delta\varphi)_{cr}$ of $\Delta\varphi$ (corresponding to a certain vertically upward specific discharge) such that $\sigma' = D\gamma^* - \gamma_w(\Delta\varphi)_{cr} = 0$:

$$(\Delta\varphi)_{cr} = D\gamma^*/\gamma_w. \quad (5.12.14)$$

Since $\Delta\varphi/D = J_z$, we have the critical gradient:

$$(J_z)_{cr} = \gamma^*/\gamma_w, \quad (5.12.15)$$

i.e., the ratio of the submerged specific weight to the specific weight of water.

Equation (5.12.15) is identical to (5.12.8). Whenever an external load exists, it is taken into account in the computation of σ. Both (5.12.8) and (5.12.15) hold for an unloaded soil.

In designing hydraulic structures, a gradient of no more than $(J_z)_{cr}/\alpha$, where $\alpha = 4 \div 5$ (safety factor) is actually permitted. Quick conditions may be prevented if an external load is applied to hold the sand in place.

The total force acting on the grains in unconsolidated media may loosen particles, especially fines, from their fixed position in the soil structure. When the force acts in the direction of the flow, these particles may be removed permanently from their place, leaving behind a local cavity, and may eventually be washed out of the soil causing the phenomenon of *piping*. Returning now to the list of six cases mentioned above, we see that quick conditions or piping will occur in cases a and c in upward flow and in case e in downward flow.

In a general three-dimensional flow pattern, piping will occur wherever the total force **F** acting on the solid particles makes an angle smaller than 90° with the specific discharge vector. This means:

$$\mathbf{F} \cdot \mathbf{q} < 0; \qquad F_1 q_1 + F_2 q_2 + F_3 q_3 < 0, \quad \text{or} \quad \mathbf{F} \cdot (\underline{\mathbf{K}} \cdot \mathbf{J}) < 0. \qquad (5.12.16)$$

In view of (5.12.5), and assuming x, y, z to be principal directions of the medium's hydraulic conductivity, equation (5.12.16) becomes:

$$K_x \gamma_f (J_x)^2 + K_y \gamma_f (J_y)^2 + K_z (\gamma_f J_z - \gamma^*) J_z < 0 \qquad (5.12.17)$$

or:

$$\gamma_f [K_x (J_x)^2 + K_y (J_y)^2 + K_z (J_z)^2] - K_z J_z \gamma^* < 0.$$

Another way of obtaining a general criterion for piping is to determine the conditions for $\sigma' = 0$.

Exercises

5.1 Water at 10°C is flowing through the sand column shown in fig. 5.1.1 ($L = 120$ cm, $A = 200$ cm^2, $h_1 - h_2 = 120$ cm). The sand in the column has a porosity $n = 0.36$ and hydraulic conductivity $K = 20$ m/d. Determine:

(a) The rate of flow (Q) and the specific discharge (q).

(b) The average velocity of flow in the column (V).

5.2 Using a radioactive tracer, the average velocity of flow in an aquifer was found to be $V = 0{,}75$ m/d. The slope of the piezometric surface at that point was $J = 0.002$. Determine the hydraulic conductivity (K) of the aquifer if the aquifer's porosity is $n = 0.2$.

5.3 The slope of the piezometric surface at a distance $r = 5$m from the center of a well pumping $Q = 200$ m^3/h from a confined aquifer of constant thickness (D) is $J = 0.2$. (a) Determine the transmissivity $T = KD$ of the aquifer. (b) How will the slope be affected if pumpage continues at the same rate while the water's temperature varies?

5.4 Draw the piezometric head and pressure head distributions along the column of homogeneous sand of length $L = 100$ cm shown in Fig. 5.1.1.

Case (a)	$h_1 = 115$ cm,	$h_2 = 15$ cm
Case (b)	$h_1 = 125$ cm,	$h_2 = 15$ cm
Case (c)	$h_1 = 115$ cm,	$h_2 = 25$ cm

Draw a conclusion from the resulting pressure distributions.

5.5 Given a homogeneous isotropic confined aquifer of constant thickness $D = 20$ m, porosity $n = 0.2$ and hydraulic conductivity $K = 15$ m/d, in which uniform flow takes place. Two observation wells, at A and B located $L = 1200$ m apart in the direction of the flow indicate piezometric heads of $\varphi_B = +3$ m and $\varphi_A = +5.4$ m above mean sea level. The sand's average grain diameter is $d = 1$ mm. The water's kinematic viscosity is $\nu = 0.013$ cm²/s.

(a) Is Darcy's law applicable?

(b) Determine the average flow velocity (V) in the pores.

5.6 Given a sloping oil reservoir of thickness $D = 100$ ft. and permeability $k = 0.1$ darcy in which a uniform flow of oil takes place. At point A the elevation is $+200$ ft. (above mean sea level) and the pressure is 3200 psi. At point B, located upslope of A in the direction of the flow, the elevation is $+250$ ft. The distance from A to B is 1 mile. The discharge in the reservoir is 5 bbl/d/(ft normal to the flow). Determine the pressure at B if the oil has the following properties: $\mu = 1.2$ cp, $\delta = 0.77$. Note: 1 bbl $= 5614.58$ ft³.

5.7 Given a confined aquifer with a thickness that varies linearly. Two observation wells, at A and B, a distance L apart along a streamline, show piezometric heads of φ_A and φ_B, respectively. The aquifer's thickness at A and B is D_A and D_B $(< D_A)$, respectively.

(a) Determine the discharge Q in the aquifer assuming that the flow is everywhere in the direction of the aquifer's axis (midway between the two impervious confining planes).

(b) Draw the piezometric line for $\varphi_A > \varphi_B$ and $\varphi_A < \varphi_B$.

5.8 Given three observation wells in a homogeneous isotropic aquifer:

Well	A	B	C
Coordinate x	0	300 m	0
Coordinate y	0	0	200 m
Piezometric head	+ 10 m	+ 11.5 m	+ 8.4 m

Determine the specific discharge $\mathbf{q}$ if the hydraulic conductivity K is known.

5.9 Let an aquifer be inhomogeneous with $K = ax + by + c$. Derive the partial differential equation for $\varphi = \varphi(x, y)$ in the aquifer.

5.10 A fully penetrating well is pumping at a constant rate $Q = 600\ \text{m}^3/\text{h}$ from a confined sandy aquifer of thickness $D = 50$ m and average grain diameter of 0.1 cm. What is the domain around the well for which Darcy's law is applicable?

5.11 To determine the liquid permeability (k_l) of a core, a test is carried out with air (k_g) at an average pressure (p) of 1.5 atm. Determine k_l if $k_g = 0.1$ md and b of Eq. (5.3.8) is related to k_l by $b = 0.77 k_l^{-0.39}$.

5.12 A porous medium has a permeability of 1 darcy.
(a) What is the medium's permeability in cm²?
(b) What is the hydraulic conductivity of this medium for water ($\nu_w = 0.013$ cm²/s) and for oil ($\nu_0 = 1.8$ cm²/s)?

5.13 Let the observation wells of Ex. 5.8 be located in a homogeneous anisotropic aquifer with

$$[K_{ij}] = \begin{bmatrix} 20 & 6 \\ 6 & 10 \end{bmatrix}; \qquad K \text{ in m/d.}$$

Determine $\mathbf{q}$. Note: The flow is two-dimensional in the xy-plane.

5.14 In two-dimensional flow, the hydraulic conductivity is given in the xy coordinate system by

$$[K_{ij}] = \begin{bmatrix} K_{xx} & K_{xy} \\ K_{yx} & K_{yy} \end{bmatrix} = \begin{bmatrix} 9 & -4 \\ -4 & 3 \end{bmatrix}; \qquad K \text{ in m/d.}$$

(a) If $\mathbf{J} = 5^0/_{00}$ makes an angle $\beta_1 = 30°$ with the $+x$ axis, determine $\mathbf{q}$.
(b) What are the principal directions of the anisotropic porous medium?

5.15 Let the x and y axes of Ex. 5.13 be rotated by 30° in a clockwise direction. Determine the components of $\mathbf{K}$ in the new, $x'y'$, coordinate system.

5.16 Let $K_x = 36$ m/d and $K_y = 9$ m/d be the principal values of the hydraulic conductivity of an anisotropic soil in the x and y directions, respectively.
(a) If the gradient $\mathbf{J}$ makes an angle $\beta_1 = 45°$ with the $+x$ axis (fig. 5.6.2e), determine graphically the directional hydraulic conductivity K_J in the direction of $\mathbf{J}$ and the direction of the specific discharge $\mathbf{q}$ (angle with the $+x$ axis).
(b) If in the same soil, $\mathbf{q}$ is in a direction that makes an angle $\alpha_1 = 45°$ with the $+x$ axis, determine the directional hydraulic conductivity K_q in the direction of $\mathbf{q}$ and the direction of $\mathbf{J}$ (β_1 with the $+x$ axis).

(c) Compare the values of K_q and K_J for a direction in the soil making an angle ε_1 with the $+x$ axis.

5.17 For the porous medium of Ex. 5.13, determine the principal directions, and write down the components of **K** in the principal directions.

5.18 Given a specific discharge **q** and a gradient **J** in two-dimensional flow in the xy-plane:

$$q_x = 0.010 \text{ m/d}; \qquad q_y = 0.005 \text{ m/d}, \qquad J_x = 1^0/_{00}, \qquad J_y = 2^0/_{00}.$$

(a) Determine **K** if x and y are principal directions.

(b) Determine the hydraulic conductivity in the direction of the flux (K_q).

(c) Determine the hydraulic conductivity in the direction of the gradient (K_J).

(d) Determine the hydraulic conductivity in the direction which makes an angle $\alpha = 30°$ with the $+x$ axis.

5.19 Let the two reservoirs in the nondischarging permeameter shown in fig. 5.7.3 have different cross-sectional areas A_1 and A_2. Develop an expression for the hydraulic conductivity K.

5.20 A test was performed in order to determine the permeability of a core of length $L = 5$ cm and cross-sectional area $A = \pi$ cm^2, using air ($\mu = 0.0172$ cp) under isothermal conditions. With pressures of $p_1 = 2.0$ atm. and $p_2 = 1.5$ atm., at the inflow and outflow faces of the core, respectively, the air discharge determined at atmospheric pressure was $Q = 12.4$ cm^3/s. Determine the air permeability k.

5.21 Prove (5.8.9).

5.22 Let the aquifer of Ex. 5.5 be inhomogeneous, with $K = (0.04x + 20)$ m/d. Points A and B are at $x = 0$ and 1.2 km, respectively. Determine the specific discharge in the aquifer (q) and the equivalent hydraulic conductivity ($\bar{K}$) of the soil between A and B.

5.23 Draw the pressure distribution along the stratified sand column shown in fig. 5.8.2c and determine the pressure at B (a) for $K_2 = 6K_1 = 60$ m/d, (b) for $K_1 = 2K_2 = 40$ m/d. In both cases $L_1 = 60$ cm. $L_2 = 40$ cm, depth of water above A is 20 cm, depth of water above C is 10 cm.

5.24 Let the aquifer of fig. 5.8.2a be composed of only two layers $L_1 = 700$ m, $L_2 = 400$ m, $K_1 = 20$ m/d, $K_2 = 10$ m/d. Determine Q for $\Delta\varphi = 1.8$ m and $b = 50$ m.

5.25 The stratified soil of Fig. 5.8.1 is made of N_1 layers of sand of thickness b_1 and hydraulic conductivity K_1, each, and N_2 layers of sandy-clay of thickness b_2 and hydraulic conductivity K_2 each. Determine the equivalent horizontal hydraulic conductivity (K_H) of the layered soil.

5.26 Let the layered soil of fig. 5.8.2c be composed of the same combination of layers as in Ex. 5.25. Determine the equivalent vertical hydraulic conductivity of the layered soil (K_V).

5.27 It is sometimes said that the flow normal to the layers in a layered soil as shown in fig. 5.8.2a is determined by the layer of lowest hydraulic conductivity. Is this statement correct? Use the definition of layer resistance $C_i = L_i/K_i$ and discuss.

For the case $N = 3$, $K_1 = K_3 \neq K_2$, determine the relationship between Q/Q_0 and the ratios K_1/K_2 and L_2/L_1, where Q_0 is the discharge under the same head difference, but with $L_2 = 0$. The total length L remains always unchanged.

5.28 A well of radius r_w is located at the center of a confined circular aquifer of radius r_e and constant thickness b. The hydraulic conductivity varies: $K = K_1$ for $r_w < r < r_1$, $K = K_2$ for $r_1 < r < r_e$. Determine the well's discharge Q_w if the piezometric heads φ_w and φ_e are maintained constant at $r = r_w$ and $r = r_e$, respectively.

5.29 Water enters the confined aquifer shown in fig. 5.E.1 at A and leaves at C in the form of a spring.

(a) Determine the piezometric head at an observation well located at B.

(b) Under what conditions will the well at B be a flowing (artesian) well?

5.30 Determine Girinskii's potential Φ^g for horizontal flow in a two layered confined aquifer: K_1 for $0 < z < b_1$, K_2 for $b_1 < z < (b_1 + b_2)$. Note that $\varphi = \varphi(x, y, t)$.

5.31 Determine Girinskii's potential Φ^g for phreatic flow in a two layered unconfined aquifer: K_1 for $0 < z < b_1$ and K_2 for $b_1 < z < (b_1 + b_2)$.

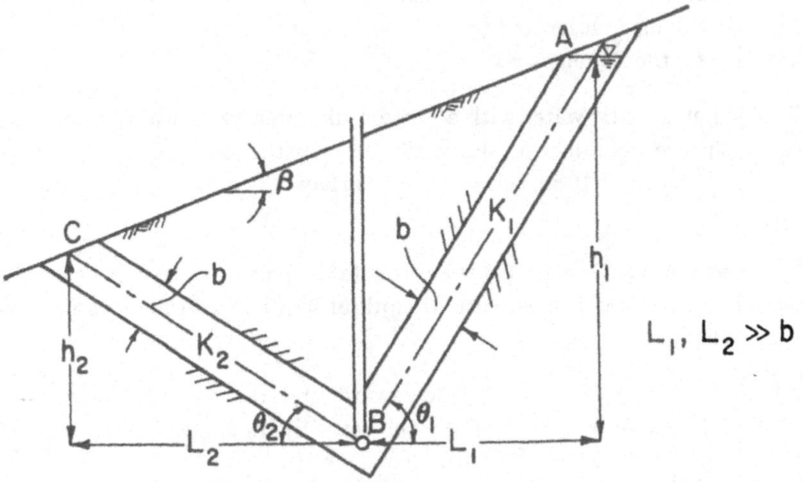

FIG. 5.E.1.

(a) Assume everywhere $0 < h < b_1$.
(b) Assume everywhere $b_1 < h < (b_1 + b_2)$.

5.32 a) Show that a potential Φ defined by:

$$\Phi = \int_0^{h(x,y)} (\varphi - z)K(z)\, dz, \qquad \varphi = \varphi(x, y, z) = z + p/\gamma$$

satisfies the Laplace equation in steady flow without accretion.

b) Show that this potential reduces to Girinskii's potential when the Dupuit assumptions are employed.

c) Assume $K(z) = K_0(1 + Az)$. If $q'_x \neq 0$, $q'_y = 0$, where $\mathbf{q}'$ is the discharge per unit width through an aquifer, determine q'_x if the water table elevations h_1 and h_2 above a horizontal bottom are known at two points located a distance L apart.

5.33 Determine the pressure distribution for steady isothermal flow of gas in a horizontal sand column of length L and cross-sectional area A. The pressures p_0 and p_L are maintained constant at $x = 0$ and $x = L$.

5.34 A well of radius r_w is located at the center of a circular gas reservoir of radius r_e.

(a) Determine the well's mass discharge $M = \rho_{at}Q_{at}$ (isothermal conditions) if the pressures p_w and p_e are maintained constant at $r = r_w$ and $r = r_e$, respectively.
(b) Determine the pressure distribution $p = p(r)$.

5.35 Let a liquid flow through the sand column of Ex. 5.33 such that (5.11.1) is applicable with $W = \nu/gk$, $b = 0.55/g\sqrt{k}$, $k = d^2/360$, $d = 1$ mm, $\nu = 1$ cs. Determine the discharge for $\Delta p = 1$ atm. and $L = 100$ cm.

5.36 Repeat Ex. 5.35 assuming that the cross-sectional area of the column varies linearly along L according to $A(x) = A_0 + ax$ and the flow is everywhere in the direction of the column's axis.

5.37 Assuming static water with a horizontal water table which coincides with the ground surface, determine (a) the vertical stress in the soil, σ, and (b) the effective stress, σ', at a depth of 10 m below ground surface. The soil is homogeneous with $\gamma_t = 2.1$ t/m³.

5.38 Repeat Ex. 5.37 when (a) the water table is lowered to 2 m below ground surface, and (b) water is ponded with a depth of 2 m above ground surface, but no flow takes place.

5.39 Repeat Ex. 5.38 when (a) downward and (b) upward flow takes place at a specific discharge of $q_z = 1$ m/d. The soil's hydraulic conductivity is 5 m/d.

5.40 For what value of upward q_z will quick conditions occur in the case of Ex. 5.39(b)?

CHAPTER 6

Continuity and Conservation Equations for a Homogeneous Fluid

A *homogeneous fluid* here is a single-species fluid, or a homogeneous mixture. If the fluid is compressible, its density depends on pressure only. We exclude the dependence of density on solute concentration. A comment on the conservation equation for an inhomogeneous incompressible fluid is given in section 6.7.

The equation of motion considered in the previous chapter involves two dependent variables: the specific discharge vector, and the pressure (or piezometric head). Altogether we have four unknowns (three components of $\mathbf{q}$ and φ or p) and only three equations, one for each component of $\mathbf{q}$. To obtain a complete description of the flow we must employ an additional basic law, namely that of conservation of matter, expressed in the form of a continuity equation. If ρ and μ also vary, equations of state, such as $\rho = \rho(p)$, must also be introduced.

In chapter 4 the mass conservation equation for flow in porous media was derived from basic principles of fluid mechanics, as equation (4.6.40). To simplify the discussion, it was assumed there that the medium was stable, inert and nondeformable, and that the liquid was incompressible. The objective in the present chapter is to relax some of these restrictions. We shall also change our approach somewhat by considering mass conservation for an REV of a specified shape, rather than of an arbitrary shape as in chapter 4. This approach—the control volume approach—will be advantageous to those readers who are less familiar with some of the mathematics involved in dealing with the general conservation statement. It will be shown to be of particular importance in those cases where we wish to employ the approximate "hydraulic approach" of incorporating a complicated boundary condition (such as a phreatic surface (sec. 7.1) or a leaky boundary) in the differential equation stated in a space of $(n - 1)$ dimensions, instead of the original n (3 or 2) dimensions.

6.1 The Control Volume

In the Lagrangian approach (par. 4.1.3), the analysis is carried out with respect to a specified mass of fluid. This mass—the particle—though changing in shape, position or other properties as it moves, always remains unchanged.

A second possibility, employing Euler's approach, is to focus our attention on a definite volume, fixed in space, called the *control volume* (or *control box*). The shape of the control volume is arbitrary. Its boundaries are called *control surfaces*;

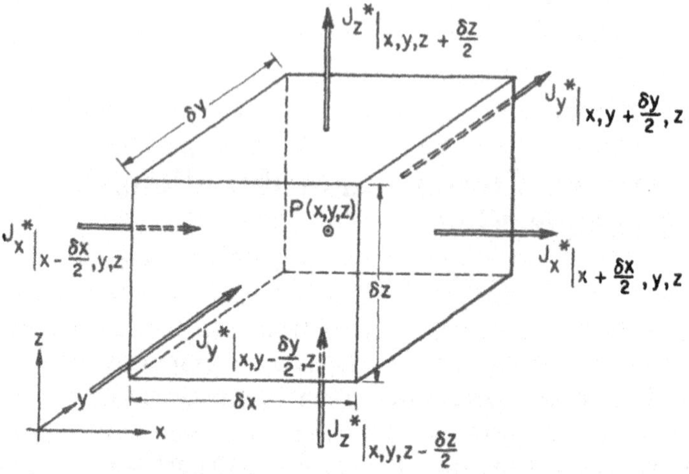

FIG. 6.2.1. Mass conservation for a control volume.

they always form a closed surface in space. The amount and identity of matter in the control volume may change with time, but the *shape and position of this volume remain fixed*. Again this volume may be small, i.e., an REV in the sense of section 1.3, or much larger, but still finite, depending on the problem investigated.

To facilitate discussion, we shall give the control volume a definite shape according to the particular coordinate system employed (e.g., a parallelepiped box in Cartesian coordinates). Yet, this shape does not affect the resulting equation (compare with (4.2.9)).

6.2 Mass Conservation in a Nondeformable Porous Matrix

6.2.1 The Basic Continuity Equation

In this section, the *equation of mass conservation*, or the *continuity equation*, is obtained by considering a small control volume in Cartesian coordinates x, y, z (or x_1, x_2, x_3). If the medium is anisotropic, we shall assume, unless otherwise specified, that x, y, z are the principal directions of permeability with k_x, k_y, k_z the principal values.

Consider a control volume of dimensions δx, δy, δz, parallel, respectively, to the x, y, z coordinates around a point $P(x, y, z)$ in the porous medium domain. Let the vector $\mathbf{J}^*$, with components J^*_x, J^*_y, J^*_z in the x, y, z directions, respectively, denote the mass flux (mass per unit area per unit time) of a fluid of density ρ. Referring to figure 6.2.1, the excess of inflow over outflow during a short time interval, δt, through the surfaces of the control volume that are perpendicular to the x direction, may be expressed by the difference:

$$[J^*_x|_{x-(\delta x/2),y,z} - J^*_x|_{x+(\delta x/2),y,z}]\,\delta y\,\delta z\,\delta t$$

or, approximately, by developing in a Taylor series around P, neglecting terms of order $(\delta x)^2$ and higher, as:

$$-(\partial J^*_x/\partial x)\,\delta x\,\delta y\,\delta z\,\delta t.$$

Repeating this procedure for the other two directions and adding, we obtain the excess of inflow over outflow through the control volume surfaces:

$$-[\partial J^*_x/\partial x + \partial J^*_y/\partial y + \partial J^*_z/\partial z]\,\delta x\,\delta y\,\delta z\,\delta t.$$

By the principle of mass conservation, this must be equal to the change of mass within the control volume during δt given by $[\partial(n\rho\,\Delta U_0)/\partial t]\,\delta t$, where $\Delta U_0 = \delta x\,\delta y\,\delta z =$ constant is the volume of the control box. Hence, we obtain:

$$-[\partial J^*_x/\partial x + \partial J^*_y/\partial y + \partial J^*_z/\partial z] = \partial(\rho n)/\partial t \quad \text{or} \quad \operatorname{div} \mathbf{J}^* + \partial(\rho n)/\partial t = 0. \tag{6.2.1}$$

The mass flux $\mathbf{J}^*$ may now be expressed by:

$$\mathbf{J}^* = \rho\mathbf{q} = \rho n\mathbf{V}^* \tag{6.2.2}$$

where ρ and $\mathbf{q}$ denote average values; the dispersion (chap. 10) caused by fluctuations in velocity is neglected here. Also, because the fluid is homogeneous, and molecular diffusion is neglected, $\mathbf{V}^* \equiv \mathbf{V}$. From (6.2.1) we now obtain:

$$\partial(\rho q_x)/\partial x + \partial(\rho q_y)/\partial y + \partial(\rho q_z)/\partial z + \partial(\rho n)/\partial t = 0; \qquad \operatorname{div}(\rho\mathbf{q}) + \partial(\rho n)/\partial t = 0. \tag{6.2.3}$$

For a *nondeformable medium*, $n =$ constant and (6.2.3) becomes:

$$\operatorname{div}(\rho\mathbf{q}) + n\,\partial\rho/\partial t = 0. \tag{6.2.4}$$

This can also be expressed as:

$$\rho \operatorname{div} \mathbf{q} + \mathbf{q}\cdot\operatorname{grad}\rho + n\,\partial\rho/\partial t = 0. \tag{6.2.5}$$

In many practical problems $\mathbf{q}\cdot\operatorname{grad}\rho \ll n\,\partial\rho/\partial t$ (i.e., spatial variations in ρ are much smaller than the local temporal ones). Then (6.2.5) becomes:

$$\rho \operatorname{div} \mathbf{q} + n\,\partial\rho/\partial t = 0. \tag{6.2.6}$$

Another conclusion that may be derived from (6.2.5) is that when the fluid is incompressible (i.e., $\operatorname{div} \mathbf{q} = 0$, yet the fluid is such that ρ varies because of variations in solute concentrations in it), steady flow (i.e., $\partial\rho/\partial t = 0$) can occur only when $\mathbf{q}\cdot\nabla\rho = 0$, that is, streamlines are everywhere tangent to surfaces $\rho =$ constant. Mixing due to hydrodynamic dispersion (chap. 10) is neglected.

We may now introduce piezometric head $\varphi^* = \varphi^*(x, y, z, t)$ or pressure $p = p(x, y, z, t)$ as independent variables instead of $\mathbf{q}$. For the general case of an inhomogeneous anisotropic medium, i.e., $k_x = k_x(x, y, z)$, $k_y = k_y(x, y, z)$, $k_z = k_z(x, y, z)$, where x, y, z are everywhere the principal directions, and the fluid is compressible, $\rho = \rho(p)$, we use (5.9.1) and (5.9.2) to obtain from (6.2.4):

$$\operatorname{div}\left(\rho^2 g \frac{\underline{\mathbf{k}}}{\mu} \operatorname{grad}\varphi^*\right) = n\frac{\partial\rho}{\partial t} \tag{6.2.7}$$

$$\frac{\partial}{\partial x}\left(\rho^2\frac{gk_x}{\mu}\frac{\partial\varphi^*}{\partial x}\right) + \frac{\partial}{\partial y}\left(\rho^2\frac{gk_y}{\mu}\frac{\partial\varphi^*}{\partial y}\right) + \frac{\partial}{\partial z}\left(\rho^2\frac{gk_z}{\mu}\frac{\partial\varphi^*}{\partial z}\right) = n\frac{\partial\rho}{\partial t} \tag{6.2.8}$$

or:

$$\frac{\partial}{\partial x_i}\left(\rho^2 g \frac{k_{ij}}{\mu}\frac{\partial \varphi^*}{\partial x_j}\right) = n\frac{\partial \rho}{\partial t} \tag{6.2.9}$$

where $k_{ij} = 0$ for $i \neq j$, $k_{11} = k_x$, $k_{22} = k_y$, $k_{33} = k_z$, $x_1 = x$, $x_2 = y$, $x_3 = z$. Equations (6.2.7) and (6.2.9) are also applicable to the general case of anisotropy where the x_i axes are not in the principal direction, and $\mathbf{k}$ is the second-rank symmetrical tensor. Since: $g \operatorname{grad} \varphi^* = g\mathbf{1z} + (1/\rho)\operatorname{grad} p$ (sec. 5.9), we may rewrite (6.2.8) in the form:

$$\frac{\partial}{\partial x}\left(\frac{k_x\rho}{\mu}\frac{\partial p}{\partial x}\right) + \frac{\partial}{\partial y}\left(\frac{k_y\rho}{\mu}\frac{\partial p}{\partial y}\right) + \frac{\partial}{\partial z}\left(\frac{k_z\rho}{\mu}\left[\frac{\partial p}{\partial z} + \rho g\right]\right) = n\frac{\partial \rho}{\partial t}. \tag{6.2.10}$$

In many practical problems, especially in gas flow in reservoirs, the term ρg expressing the gravity effect in (6.2.10) is much smaller than the pressure gradient $\partial p/\partial z$ and is, therefore, neglected. For similar reasons, the term $g\,\partial\rho/\partial z$ is sometimes omitted from (6.2.13) below.

For a *homogeneous anisotropic medium,* we set $k_x = C_1$, $k_y = C_2$, $k_z = C_3$ in (6.2.7) through (6.2.10), where C_1, C_2 and C_3 are constants. If the medium is *homogeneous and isotropic,* equation (6.2.7) becomes:

$$k\left[\frac{\partial}{\partial x}\left(\frac{g\rho^2}{\mu}\frac{\partial \varphi^*}{\partial x}\right) + \frac{\partial}{\partial y}\left(\frac{g\rho^2}{\mu}\frac{\partial \varphi^*}{\partial y}\right) + \frac{\partial}{\partial z}\left(\frac{g\rho^2}{\mu}\frac{\partial \varphi^*}{\partial z}\right)\right] = n\frac{\partial \rho}{\partial t}. \tag{6.2.11}$$

Assuming also that the hydraulic conductivity $K = k\rho g/\mu$ is *approximately a constant,* we obtain:

$$K\left[\frac{\partial}{\partial x}\left(\rho\frac{\partial \varphi^*}{\partial x}\right) + \frac{\partial}{\partial y}\left(\rho\frac{\partial \varphi^*}{\partial y}\right) + \frac{\partial}{\partial z}\left(\rho\frac{\partial \varphi^*}{\partial z}\right)\right] = n\frac{\partial \rho}{\partial t} \tag{6.2.12}$$

or:

$$K\left(\frac{\partial^2 p}{\partial x^2} + \frac{\partial^2 p}{\partial y^2} + \frac{\partial^2 p}{\partial z^2} + g\frac{\partial \rho}{\partial z}\right) = ng\frac{\partial \rho}{\partial t}. \tag{6.2.13}$$

The various continuity equations in this paragraph must be supplemented by an equation of state $\rho = \rho(p)$.

For *steady flow* we must introduce $\partial\rho/\partial t = 0$ in (6.2.4) through (6.2.13).

6.2.2 Continuity Equation for an Incompressible Fluid

If the fluid is incompressible, $\rho =$ const, and in a nondeforming medium we obtain:

$$\partial q_x/\partial x + \partial q_y/\partial y + \partial q_z/\partial z = 0; \qquad \operatorname{div} \mathbf{q} = 0. \tag{6.2.14}$$

A vector field $\mathbf{a}$ (here $\mathbf{q}$) for which $\operatorname{div} \mathbf{a} = 0$ is called a *solenoidal vector field* When we also have $\mathbf{a} = \operatorname{grad} \phi$, the vector field $\mathbf{a}$ is called a *Laplacian field.* We shall see below that in the case at hand, φ satisfies the Laplace equation. In flow through porous media we have a Laplacian field *only* when the medium is isotropic and homogeneous.

For an incompressible fluid, $\rho = \text{const}$, $\mu = \text{const}$, we may introduce $\varphi = z + p/\gamma$, and $\underline{\mathbf{K}} = \underline{\mathbf{k}}\rho g/\mu$.

We then obtain from (6.2.10):

$$\frac{\partial}{\partial x}\left(K_x \frac{\partial \varphi}{\partial x}\right) + \frac{\partial}{\partial y}\left(K_y \frac{\partial \varphi}{\partial y}\right) + \frac{\partial}{\partial z}\left(K_z \frac{\partial \varphi}{\partial z}\right) = 0 \qquad (6.2.15)$$

$$\operatorname{div}(\underline{\mathbf{K}} \cdot \operatorname{grad} \varphi) = 0; \qquad \frac{\partial}{\partial x_i}\left(K_{ij} \frac{\partial \varphi}{\partial x_j}\right) = 0. \qquad (6.2.16)$$

For an *isotropic inhomogeneous medium*, equation (6.2.16) becomes:

$$\operatorname{div}(K \cdot \operatorname{grad} \varphi) = 0; \qquad K = K(x, y, z)$$

$$K \operatorname{div} \operatorname{grad} \varphi + \operatorname{grad} K \cdot \operatorname{grad} \varphi = 0$$

$$\nabla^2 \varphi + \operatorname{grad} \varphi \cdot \operatorname{grad}(\ln K) = 0. \qquad (6.2.17)$$

In the special case $\nabla\varphi \perp \nabla K$, φ satisfies the Laplace equation.

Two special cases of inhomogeneity exist (Georgitza 1969):

(a) the Helmholtz *inhomogeneous medium*, where K satisfies Helmholtz' equation:

$$\nabla^2 K^{1/2} + \alpha^2 K^{1/2} = 0; \qquad \alpha = \text{a real constant}; \qquad K = K(x, y, z). \qquad (6.2.18)$$

Then, by introducing a new variable φ':

$$\varphi'(x, y, z, t) = \varphi(x, y, z, t) K^{1/2} \qquad (6.2.19)$$

we have from (6.2.17):

$$\nabla^2 \varphi' + \alpha^2 \varphi' = 0 \qquad (6.2.20)$$

which is the *Helmholtz equation* for φ'; in general it is easier to solve (6.2.20) than (6.2.17);

(b) the *harmonically inhomogeneous medium*, where K satisfies

$$\nabla^2 K^{1/2} = 0. \qquad (6.2.21)$$

In this case φ' satisfies the Laplace equation:

$$\nabla^2 \varphi' = \varphi \nabla^2 K^{1/2} + K^{-1/2}(K \nabla^2 \varphi + \nabla K \cdot \nabla \varphi) \equiv 0. \qquad (6.2.22)$$

If the *fluid is incompressible* and the *matrix is homogeneous*, we obtain from (6.2.10) and from (6.2.16), respectively:

$$K_x \frac{\partial^2 p}{\partial x^2} + K_y \frac{\partial^2 p}{\partial y^2} + K_z \frac{\partial^2 p}{\partial z^2} = 0; \qquad K_x \frac{\partial^2 \varphi}{\partial x^2} + K_y \frac{\partial^2 \varphi}{\partial y^2} + K_z \frac{\partial^2 \varphi}{\partial z^2} = 0. \qquad (6.2.23)$$

Finally, if the *fluid is incompressible* and the *matrix is homogeneous and isotropic*, we obtain:

$$\nabla^2 \varphi \equiv \frac{\partial^2 \varphi}{\partial x^2} + \frac{\partial^2 \varphi}{\partial y^2} + \frac{\partial^2 \varphi}{\partial z^2} = 0; \qquad \operatorname{div}(\operatorname{grad} \varphi) = 0 \qquad (6.2.24)$$

$$\nabla^2 p \equiv \frac{\partial^2 p}{\partial x^2} + \frac{\partial^2 p}{\partial y^2} + \frac{\partial^2 p}{\partial z^2} = 0; \qquad \text{div}(\text{grad}\, p) = 0 \tag{6.2.25}$$

where the symbol ∇^2 is the Laplacian operator.

Equation (6.2.24), which describes the potential distribution $\varphi = \varphi(x, y, z, t)$, or equation (6.2.25), which describes the pressure distribution $p = p(x, y, z, t)$ in a field of flow of an incompressible fluid in a homogeneous, isotropic, nondeformable matrix, is called the *Laplace equation.* It occurs in many problems of flow through porous media, in ground water flow and oil reservoir engineering.

It is interesting to note that (6.2.24) and (6.2.25) do not contain K. This means that when K is not included in the boundary conditions, the distribution of φ is purely geometrical. Once $\varphi = \varphi(x, y, z, t)$ is obtained, the discharge Q itself depends on K. In other words, Q/K can be derived from $\varphi = \varphi(x, y, z, t)$ with no additional information regarding K.

Equations (6.2.14) through (6.2.25) apply to steady flow as well as to nonsteady flow of an incompressible fluid. In the latter case, the variations in time are introduced by time-dependent boundary conditions.

If the flow is two dimensional in the horizontal xy plane, all terms in (6.2.17) through (6.2.25) involving first or second derivatives with respect to z vanish.

6.2.3 Continuity Equation for a Compressible Fluid

By introducing (2.1.4) into (6.2.10) and neglecting the gravity term, we obtain for isothermal gas flow:

$$\frac{\partial}{\partial x}\left(\frac{k_x}{2\mu}\frac{1}{Z(p)}\frac{\partial p^2}{\partial x}\right) + \frac{\partial}{\partial y}\left(\frac{k_y}{2\mu}\frac{1}{Z(p)}\frac{\partial p^2}{\partial y}\right) + \frac{\partial}{\partial z}\left(\frac{k_z}{2\mu}\frac{1}{Z(p)}\frac{\partial p^2}{\partial z}\right) = n\frac{\partial}{\partial t}\left(\frac{p}{Z(p)}\right). \tag{6.2.26}$$

Because $Z(p)$ is available only in graphic form (6.2.26) can only be solved numerically. For an ideal gas, we insert $Z = 1$ in (6.2.26). If the medium is homogeneous and isotropic, we then obtain, with μ assumed constant:

$$\nabla^2 p^2 = (n\mu/kp)\, \partial p^2/\partial t. \tag{6.2.27}$$

For a steady two-dimensional flow in the xy plane of a gas in a reservoir with a homogeneous anisotropic matrix, assuming also that $\mu = \text{const}$, we obtain:

$$k_x\, \partial^2 p^2/\partial x^2 + k_y\, \partial^2 p^2/\partial y^2 = 0 \tag{6.2.28}$$

and for an isotropic matrix:

$$\nabla^2_{x,y} p^2 \equiv \partial^2 p^2/\partial x^2 + \partial^2 p^2/\partial y^2 = 0, \qquad \nabla^2_{x,y} \equiv \partial^2/\partial x^2 + \partial^2/\partial y^2. \tag{6.2.29}$$

For a slightly compressible liquid, under isothermal flow conditions, we use (2.3.3) from which we obtain:

$$\beta = (1/\rho)\, \partial\rho/\partial p = \text{const}; \qquad \partial\rho/\partial p = \rho\beta$$

$$\partial\rho/\partial x = \rho\beta\, \partial p/\partial x, \text{ etc.}; \quad \text{or} \quad \text{grad}\, \rho = \rho\beta\, \text{grad}\, p; \qquad \partial\rho/\partial t = \rho\beta\, \partial p/\partial t. \tag{6.2.30}$$

These expressions may now be introduced into any of the continuity equations. For example, for a homogeneous isotropic matrix, assuming $\mu =$ const, we obtain from (6.2.10) and (6.2.30):

$$\frac{k}{\mu}\nabla^2 p + \frac{\beta k}{\mu}\left[\left(\frac{\partial p}{\partial x}\right)^2 + \left(\frac{\partial p}{\partial y}\right)^2 + \left(\frac{\partial p}{\partial z}\right)^2\right] + \frac{2kg\rho\beta}{\mu}\frac{\partial p}{\partial z} = n\beta\frac{\partial p}{\partial t}$$

$$\frac{k}{\mu}\nabla^2 p + \frac{\beta k}{\mu}(\nabla p)^2 + \frac{2kg\rho\beta}{\mu}\frac{\partial p}{\partial z} = n\beta\frac{\partial p}{\partial t}. \tag{6.2.31}$$

In many cases, the gravitational term (the last on the right-hand side of (6.2.31)) is relatively small and may be neglected. A further modification of this equation is sometimes employed when it is assumed that:

$$\nabla^2 p \equiv \frac{\partial^2 p}{\partial x^2} + \frac{\partial^2 p}{\partial y^2} + \frac{\partial^2 p}{\partial z^2} \gg \beta\left[\left(\frac{\partial p}{\partial x}\right)^2 + \left(\frac{\partial p}{\partial y}\right)^2 + \left(\frac{\partial p}{\partial z}\right)^2\right].$$

Then (6.2.31) becomes:

$$\nabla^2 p = (n\beta\mu/k)\, \partial p/\partial t \tag{6.2.32}$$

or, in terms of ρ:

$$\nabla^2 \rho = (n\beta\mu/k)\, \partial \rho/\partial t \tag{6.2.33}$$

where we have neglected $2\rho g\beta\, \partial\rho/\partial z$ on the left-hand side of (6.2.31). These equations are often used in reservoir engineering.

If we wish to express the continuity equation in terms of φ^*, we notice that with the equation of state (2.3.3), we have:

$$\varphi^* = z + \int_{p_0}^{p}\frac{dp}{g\rho(p)} = z + \int_{p_0}^{p}\frac{dp}{g\rho_0 \exp[\beta(p - p_0)]} = z + \frac{1}{g\rho_0\beta} - \frac{1}{g\rho\beta} \tag{6.2.34}$$

$$g\frac{\partial\varphi^*}{\partial x} = \frac{1}{\beta\rho^2}\frac{\partial\rho}{\partial x} = \frac{1}{\rho}\frac{\partial p}{\partial x}; \qquad g\frac{\partial\varphi^*}{\partial y} = \frac{1}{\rho}\frac{\partial p}{\partial y}; \qquad g\frac{\partial\varphi^*}{\partial z} = g + \frac{1}{\rho^2\beta}\frac{\partial\rho}{\partial z} = g + \frac{1}{\rho}\frac{\partial p}{\partial z}$$

$$g\frac{\partial\varphi^*}{\partial t} = \frac{1}{\rho}\frac{\partial p}{\partial t} = \frac{1}{\beta\rho^2}\frac{\partial\rho}{\partial t}; \qquad \nabla\varphi^* = \nabla z + \frac{1}{g\beta\rho^2}\nabla\rho = \mathbf{1z} + \frac{1}{\rho g}\nabla p. \tag{6.2.35}$$

By inserting (6.2.35) and (6.2.30) into (6.2.11) for the isotropic, homogeneous medium and $\mu =$ const, we obtain:

$$\frac{k}{\mu}\nabla^2\varphi^* + \frac{2k\rho g\beta}{\mu}\left[\left(\frac{\partial\varphi^*}{\partial x}\right)^2 + \left(\frac{\partial\varphi^*}{\partial y}\right)^2 + \left(\frac{\partial\varphi^*}{\partial z}\right)^2\right] - \frac{2kg\rho\beta}{\mu}\frac{\partial\varphi^*}{\partial z} = n\beta\frac{\partial\varphi^*}{\partial t}$$

$$\frac{k}{\mu}\nabla^2\varphi^* + \frac{2k\rho g\beta}{\mu}(\nabla\varphi^*)^2 - \frac{2kg\rho\beta}{\mu}\frac{\partial\varphi^*}{\partial z} = n\beta\frac{\partial\varphi^*}{\partial t}. \tag{6.2.36}$$

Usually, the quadratic term on the left-hand side is relatively small and may be neglected. Introducing hydraulic conductivity, $K = k\rho g/\mu$, we then obtain:

$$K\nabla^2\varphi^* - 2Kg\rho\beta\frac{\partial\varphi^*}{\partial z} = n\rho\beta g\frac{\partial\varphi^*}{\partial t}. \qquad (6.2.37)$$

In general, hydraulic conductivity K for a compressible fluid is a function of ρ. By repeating the above procedure, but assuming that although the fluid is compressible, the hydraulic conductivity $k\rho g/\mu$ of the medium is a constant, one obtains:

$$K\,\nabla^2\varphi^* - K\rho g\beta\,\partial\varphi^*/\partial z = n\rho\beta g\,\partial\varphi^*/\partial t. \qquad (6.2.38)$$

Sometimes (Jacob 1950), equation (2.3.4) is taken as the equation of state. Then, equation (6.2.30) becomes:

$$\operatorname{grad}\rho = \rho_0\beta\operatorname{grad}p, \qquad \partial\rho/\partial t = \rho_0\beta\,\partial p/\partial t$$

and:

$$\varphi^* = z + \int_{p_0}^{p}\frac{dp}{g\rho_0(1+\beta p)} = z + \frac{1}{\beta g\rho_0}\ln(1+\beta p) + \text{const}, \qquad \frac{\partial\varphi^*}{\partial t} = \frac{1}{\rho g}\frac{\partial p}{\partial t};$$

$$\nabla\varphi^* = \nabla z + \frac{1}{\rho g}\nabla p. \qquad (6.2.39)$$

We then obtain for $K = \text{const}$:

$$K\,\nabla^2\varphi^* + K\rho_0\beta g(\nabla\varphi^*)^2 - K\rho_0\beta g\,\partial\varphi^*/\partial z = n\rho_0\beta g\,\partial\varphi^*/\partial t \qquad (6.2.40)$$

which, when the quadratic term is neglected, is similar to (6.2.38), except that ρ_0 here takes the place of ρ in (6.2.38).

6.3 Mass Conservation in a Consolidating Medium

The process of transient fluid flow through a porous medium that deforms in time is called *consolidation*. The problem of mass conservation in a deforming medium is a rather complicated one. In ground water hydrology it is encountered when pumping takes place in a confined, completely saturated, aquifer. We assume in such a case that the aquifer as a whole behaves as an elastic body, though we neglect the compressibility of the individual solid grains themselves. The problem is also encountered in soil mechanics (consolidation of soil strata under structures), in problems of soil subsidence, etc.

As in section 6.2, here again the medium is assumed to be completely saturated by a single-phase, compressible or incompressible, homogeneous liquid. However, it will be assumed that the porous matrix may undergo deformation during flow as a result of changes in effective stresses (sec. 2.7) within it. These may be caused by changes in external loading and/or by changes in the water pressure during flow (e.g., by pumping, or drainage).

In general we may say that under such conditions both the l quid and the solid are in motion. If we consider the control volume of figure 6.2.1, we must also take into account the variation in the mass of solids inside the imaginary *fixed* control box that retains its shape, size and position during flow.

Because of the complexity of the problem, the discussion below is carried out in several steps with different degrees of approximation. Our starting point is (6.2.3), which may be rewritten for a *deformable* (or consolidating) porous medium in the form:

$$\frac{\partial(\rho q_x)}{\partial x} + \frac{\partial(\rho q_y)}{\partial y} + \frac{\partial(\rho q_z)}{\partial z} + n\frac{\partial \rho}{\partial t} + \rho\frac{\partial n}{\partial t} = 0. \tag{6.3.1}$$

6.3.1 Vertical Compressibility Only

The discussion in section 2.7 is applicable to the case considered in this paragraph. From (2.7.7) and (2.7.9) we obtain:

$$\partial e/\partial t = -\alpha^* \partial\sigma'/\partial t = -\alpha^*(\partial\sigma/\partial t - \partial p/\partial t) \tag{6.3.2}$$

$$\frac{\partial e}{\partial t} \equiv \frac{\partial}{\partial t}\left(\frac{n}{1-n}\right) = \frac{1}{(1-n)^2}\frac{\partial n}{\partial t} = -\alpha^*\left(\frac{\partial\sigma}{\partial t} - \frac{\partial p}{\partial t}\right). \tag{6.3.3}$$

In writing (6.3.3) we have already made an *assumption*, namely that $e = e(\sigma')$ only. This is similar to the assumption inherent in (6.2.30), as the rigorous definition of β (like that of other coefficients of compressibility) is $\beta = (1/\rho)\, d\rho/dp$, from which we should have obtained $d\rho/dt = \beta\rho\, dp/dt$. Assuming, as was done in (6.2.30), that $\rho = \rho(p)$ only, we obtain $\partial\rho/\partial t = \beta\, \partial p/\partial t$ and $\nabla\rho = \beta\, \nabla p$. Since a compressibility experiment is always carried out with a *fixed mass* of the compressed substance, the coefficient of compressibility should have been defined in terms of the total derivative.

If, as in section 6.2, we introduce $\partial\rho/\partial t = \beta\rho\, \partial p/\partial t$ (i.e., assuming $\rho = \rho(p)$) together with (6.3.3) into (6.3.1), we obtain:

$$\frac{\partial(\rho q_x)}{\partial x} + \frac{\partial(\rho q_y)}{\partial y} + \frac{\partial(\rho q_z)}{\partial z} = -\beta\rho n\frac{\partial p}{\partial t} + \alpha^*\rho(1-n)^2\left(\frac{\partial\sigma}{\partial t} - \frac{\partial p}{\partial t}\right). \tag{6.3.4}$$

In most ground water flow problems, the load remains unchanged, i.e., $\partial\sigma/\partial t = 0$. Then:

$$\frac{\partial(\rho q_x)}{\partial x} + \frac{\partial(\rho q_y)}{\partial y} + \frac{\partial(\rho q_z)}{\partial z} = -\rho[n\beta + (1-n)^2\alpha^*]\frac{\partial p}{\partial t}. \tag{6.3.5}$$

Taking a somewhat different approach, let U_b be the bulk volume of some porous medium element (not a control box), with U_s and U_w the volumes of solids and water (or voids), respectively, within it:

$$U_b = U_s + U_w; \qquad U_s = (1-n)U_b = \frac{1}{1+e}U_b. \tag{6.3.6}$$

Still considering only vertical soil consolidation, we define the *soil coefficient of compressibility* α by:

$$\alpha = -\frac{1}{U_b}\frac{\partial U_b}{\partial\sigma'}. \tag{6.3.7}$$

Note that the partial differentiation here, as in (6.3.3), denotes that the bulk volume, while undergoing deformation, *essentially remains stationary*. This should be considered an approximation. In (6.3.24) we shall relax this restriction. From (6.3.7) we obtain:

$$\alpha \frac{\partial \sigma'}{\partial t} = - \frac{1}{U_b} \frac{\partial U_b}{\partial t}. \tag{6.3.8}$$

But since we assume that the volume of solid remains unchanged:

$$U_s = (1 - n)U_b = \text{const}; \qquad \frac{\partial U_s}{\partial t} = 0 = - U_b \frac{\partial n}{\partial t} + (1 - n) \frac{\partial U_b}{\partial t}$$

$$\frac{\partial U_b}{\partial t} = \frac{U_b}{1 - n} \frac{\partial n}{\partial t}; \qquad \frac{\partial n}{\partial t} = - (1 - n)\alpha \frac{\partial \sigma'}{\partial t} = - (1 - n)\alpha \left(\frac{\partial \sigma}{\partial t} - \frac{\partial p}{\partial t} \right). \tag{6.3.9}$$

Hence, we obtain:

$$\frac{\partial(\rho q_x)}{\partial x} + \frac{\partial(\rho q_y)}{\partial y} + \frac{\partial(\rho q_z)}{\partial z} = - \rho\beta n \frac{\partial p}{\partial t} + \rho\alpha(1 - n) \left(\frac{\partial \sigma}{\partial t} - \frac{\partial p}{\partial t} \right) \tag{6.3.10}$$

and for $\partial\sigma/\partial t = 0$:

$$\frac{\partial(\rho q_x)}{\partial x} + \frac{\partial(\rho q_y)}{\partial y} + \frac{\partial(\rho q_z)}{\partial z} = - \rho [\beta n + \alpha(1 - n)] \frac{\partial p}{\partial t}. \tag{6.3.11}$$

It can easily be verified that (6.3.4) and (6.3.10) are identical since: $\alpha^* = (1 + e)\alpha = \alpha/(1 - n)$. Since the left-hand side of (6.3.11) expresses the *excess of mass outflow over mass inflow per unit volume of medium and per unit time*, the right-hand side may be interpreted as the *mass* of fluid released from the unit volume of the medium during the same unit of time. Accordingly we may define a *specific mass storativity* S^*_{0p} *related to pressure changes* by:

$$S^*_{0p} = \rho [\beta n + \alpha(1 - n)]. \tag{6.3.12}$$

It is the mass of fluid released (or added) to a unit volume of porous medium per unit decline (or increase) of pressure. If we use (6.2.35) to replace the pressure by the piezometric head φ^*, we obtain:

$$S^*_{0\varphi^*} = \rho^2 g [\beta n + \alpha(1 - n)] \tag{6.3.13}$$

where $S^*_{0\varphi^*}$ is the *specific mass storativity related to potential changes*. As n approaches unity (i.e., we have a fluid bulk with no medium), equation (6.3.11) becomes identical to (4.3.14). If we neglect medium compressibility ($\alpha = 0$) we obtain the equations derived in section 6.2.

Our next step is to express $\mathbf{q}$ and p in (6.3.11) in terms of the measurable quantity φ^*, and to introduce the fact that the solid matrix is not stationary. In doing so, the first observation is that Darcy's law, as discussed in chapter 5, describes flow *with respect to the solid matrix* (Biot 1956). For the sake of illustrating the problems involved, consider a loose granular matrix. Since in this case we have a deformable matrix where solid grains and any bulk volume associated with it change their position with respect to a fixed coordinate system, at a velocity $\mathbf{V}_s$ we have:

$$\mathbf{V} = \mathbf{V}_r + \mathbf{V}_s; \qquad \mathbf{q} = \mathbf{q}_r + n\mathbf{V}_s \tag{6.3.14}$$

where $\mathbf{q}$ is the specific discharge with respect to the fixed coordinates, $\mathbf{q}_r$ is the specific discharge with respect to the moving solid particles and n is the instantaneous porosity. Here $\mathbf{V}_s$, an average over a sufficiently large number of solid grains, is the velocity of the bulk volume U_b associated with the moving grains.

With (6.3.14), the basic fluid mass conservation equation (6.2.3) for a control volume becomes:

$$\operatorname{div}[\rho(\mathbf{q}_r + n\mathbf{V}_s)] + \rho\, \partial n/\partial t + n\, \partial\rho/\partial t = 0. \tag{6.3.15}$$

Let us now repeat the analysis from section 6.2, this time for the flow of solids through the control box. If ρ_s is the density of the solid grains, we have:

$$\operatorname{div}[\rho_s(1-n)\mathbf{V}_s] + \partial[\rho_s(1-n)]/\partial t = 0. \tag{6.3.16}$$

At this point we introduce the assumption, valid for all practical purposes, that although the medium as a whole undergoes deformation (reflected in the change of porosity), the solid grains themselves are rigid, i.e., $\rho_s = \text{const}$. Hence:

$$\operatorname{div}[(1-n)\mathbf{V}_s] + \partial(1-n)/\partial t = 0 \tag{6.3.17}$$

or:

$$\operatorname{div}[(1-n)\mathbf{V}_s] = (1-n)\operatorname{div}\mathbf{V}_s - \mathbf{V}_s \cdot \operatorname{grad} n = \partial n/\partial t. \tag{6.3.18}$$

From (6.3.15) and (6.3.18) we obtain:

$$\operatorname{div}(\rho\mathbf{q}_r) + \rho\operatorname{div}\mathbf{V}_s + n\mathbf{V}_s \cdot \operatorname{grad}\rho + n(\partial\rho/\partial t) = 0. \tag{6.3.19}$$

For a homogeneous incompressible fluid (6.3.19) becomes:

$$\operatorname{div}\mathbf{q}_r + \operatorname{div}\mathbf{V}_s = 0. \tag{6.3.20}$$

From (6.2.30), i.e., $\rho = \rho(p)$ only, and (6.3.19), we obtain:

$$\operatorname{div}(\rho\mathbf{q}_r) + \rho\operatorname{div}\mathbf{V}_s = -n\rho\beta(\partial p/\partial t + \mathbf{V}_s \cdot \operatorname{grad} p) = -n\rho\beta\, d_s p/dt \tag{6.3.21}$$

where $d_s(\ )/dt = \partial(\ )/\partial t + \mathbf{V}_s \cdot \operatorname{grad}(\ )$, i.e., the Lagrangian total derivative concept *with respect to moving solids.*

From (4.1.33) and the interpretation of divergence that follows, it is obvious that if we consider a fixed mass of solid grains moving at a velocity $\mathbf{V}_s$, the bulk volume U_b may undergo deformation so that:

$$\operatorname{div}\mathbf{V}_s = \frac{1}{\delta U_b}\frac{d_s(\delta U_b)}{dt}. \tag{6.3.22}$$

Since the volume of solids remains constant:

$$\delta U_s = (1-n)\,\delta U_b = \text{const},$$

we have:

$$\frac{d_s(\delta U_s)}{dt} = 0 = (1-n)\frac{d_s(\delta U_b)}{dt} - \delta U_b \frac{d_s n}{dt}; \qquad \frac{1}{\delta U_b}\frac{d_s(\delta U_b)}{dt} = \frac{1}{1-n}\frac{d_s n}{dt}. \tag{6.3.23}$$

Hence:

$$\operatorname{div} \mathbf{V}_s = \frac{1}{\delta U_b}\frac{d_s(\delta U_b)}{dt} = \frac{1}{(1-n)}\frac{d_s n}{dt} = -\alpha' \frac{d_s\sigma'}{dt} = -\alpha' \frac{d_s(\sigma - p)}{dt} \tag{6.3.24}$$

where:

$$\alpha' = -\frac{1}{\delta U_b}\frac{d_s(\delta U_b)}{d\sigma'}$$

is the coefficient of compressibility, taking into account the fact that the solids move and that α' is defined for a *fixed mass of moving solids*. One should recall that this is still the Hookian expression for one-dimensional compression (see sec. 2.7).

Inserting (6.3.24) into (6.3.21), we obtain for $d_s\sigma/dt = 0$:

$$\operatorname{div}(\rho \mathbf{q}_r) + \rho(\alpha' + n\beta)\, d_s p/dt = 0 \tag{6.3.25}$$

which is the *partial differential equation of mass conservation in a deformable medium*. If $d_s\sigma/dt \neq 0$ (e.g., by external loading), equation (6.3.25) becomes:

$$\operatorname{div}(\rho \mathbf{q}_r) + \rho n\beta\, d_s p/dt - \rho\alpha'(d_s\sigma/dt - d_s p/dt) = 0. \tag{6.3.26}$$

Since in general $\mathbf{V}_s \cdot \operatorname{grad} p \ll \partial p/\partial t$ (or when we assume $\mathbf{V}_s \approx 0$), we have $\mathbf{q}_r \to \mathbf{q}$, $d_s(\ )/dt \to \partial(\ )/\partial t$ so that (6.3.25) reduces to:

$$\operatorname{div}(\rho \mathbf{q}) + \rho(\alpha' + n\beta)\, \partial p/\partial t = 0. \tag{6.3.27}$$

Our next step is to express $\mathbf{q}_r$ in (6.3.25) in terms of the potential φ^* (for a compressible fluid). As a *first approximation* we may follow De Wiest (1966) and assume that:

$$\mathbf{q}_r = -K \operatorname{grad} \varphi^*. \tag{6.3.28}$$

The $\operatorname{div}(\rho\mathbf{q}_r)$ is the same as $\operatorname{div}(\rho\mathbf{q})$ discussed in section 6.2. For example, from (6.2.37), recognizing all the approximate assumptions involved, we obtain for (6.3.25):

$$K\nabla^2\varphi^* - 2K\rho g\beta\, \partial\varphi^*/\partial z - (\alpha' + n\beta)(d_s\varphi^*/dt - V_s)\rho g = 0 \tag{6.3.29}$$

where V_s is the vertical component of $\mathbf{V}_s$ and:

$$d_s p/dt = (d_s\varphi^*/dt - V_s)\rho g; \qquad d_s\varphi^*/dt = \partial\varphi^*/\partial t + V_s\, \partial\varphi^*/\partial z. \tag{6.3.30}$$

Actually, if we wish to be rigorous, we should take into account the fact that K in (6.3.29) depends on n, e.g., in the form of (5.10.17), which varies during deformation.

In (6.3.30) we still have V_s, which may be interpreted as the velocity of settlement or distortion of solids.

Let us now refer the flow to a coordinate system (ξ, η, ζ) moving at a velocity V_s (in the vertical direction):

$$\xi = x; \qquad \eta = y; \qquad \zeta = z - V_s t; \qquad \tau = t.$$

Since: $d_s(\ )/dt = -V_s\, \partial(\ )/\partial\zeta + \partial(\ )/\partial\tau$, we obtain from (6.3.29):

$$K \nabla^2_{\xi,\eta,\zeta}\varphi^* - 2K\rho g\beta\, \partial\varphi^*/\partial\zeta - (\alpha' + n\beta)\, \partial p/\partial\tau = 0. \tag{6.3.31}$$

Finally, with (6.2.35), this becomes:

$$K \nabla^2_{\xi,\eta,\zeta}\varphi^* - 2K\rho g\beta\, \partial\varphi^*/\partial\zeta = S_s\, \partial\varphi^*/\partial\tau; \qquad S_s = (\alpha' + n\beta)\rho g \tag{6.3.32}$$

where S_s defines a specific volumetric storativity.

Cooper (1966) relates flow to a moving coordinate system, which, since we assumed motion only in the z direction, will be x, y, z' ($z' = z - V_s t$). He then obtains in this system (denoted by the prime):

$$K \nabla'^2\varphi^* - 2K\rho g\beta\, \partial\varphi^*/\partial z' = \rho g(\alpha' + n\beta)\, \partial\varphi^*/\partial t = S_s\, \partial\varphi^*/\partial t \tag{6.3.33}$$

in which the relative specific discharge is expressed by $(q_r)_z = -K\, \partial\varphi^*/\partial z'$. Equation (6.3.33) does not include V_s. If we neglect the second term on the left-hand side of (6.3.33), we obtain:

$$K \nabla'^2\varphi^* = \rho g(\alpha' + n\beta)\, \partial\varphi^*/\partial t = S_s\, \partial\varphi^*/\partial t. \tag{6.3.34}$$

Cooper (1966) defends the use of $\partial\varphi^*/\partial z'$ rather than $\partial\varphi^*/\partial z$ in $(q_r)_z$, by suggesting that since the resistance to the flow depends on the number of solid grains traversed by the flow (Rumer and Drinker (1966) and Rumer (1969) derive Darcy's law by this approach), then when $\Delta z'$ is considered, the number of grains traversed remains unchanged, thus justifying the assumption of a constant K.

A few historical remarks should be introduced here. Jacob (1940, 1950), probably the first to consider flow in an elastic aquifer, considers deformation only in a vertical direction, but neglects to recognize that solids are also moving. He obtains:

$$\operatorname{div}(\rho q) = -\rho(\alpha' + n\beta)\, \partial p/\partial t. \tag{6.3.35}$$

However, in his derivation he makes an error by assuming that the control volume itself undergoes deformation, i.e., he expresses the change of mass of water within the box as $\partial(\rho n\, \Delta x\, \Delta y\, \Delta z)/\partial t = \Delta x\, \Delta y\, \partial(n\rho\, \Delta z)/\partial t = \Delta x\, \Delta y\, \Delta z\, \partial(\rho n)/\partial t + \Delta x\, \Delta y \rho n \cdot \partial(\Delta z)/\partial t$. The error is in this last term. He then introduces Darcy's law in the form $\mathbf{q} = -K \operatorname{grad} \varphi$, assuming that K is a constant (i.e., neglecting its dependence on ρ), and obtains for an isotropic homogeneous medium:

$$K\rho \nabla^2\varphi + K\rho^2\beta g[(\partial\varphi/\partial x)^2 + (\partial\varphi/\partial y)^2 + (\partial\varphi/\partial z)^2 - \partial\varphi/\partial z] = \rho(\alpha' + n\beta)\, \partial p/\partial t. \tag{6.3.36}$$

Jacob (1950) suggests that in most ground water problems, the second term on the left-hand side of (6.3.36) is small and may be neglected.

Irmay (in Bear, Zaslavsky and Irmay 1968) carries out the control box analysis obtaining:

$$\operatorname{div}(\rho\mathbf{q}) = -\rho^2 g[\alpha'(1 - n) + \beta n]\, \partial\varphi/\partial t \tag{6.3.37}$$

where $g\rho\, \partial\varphi/\partial t \cong \partial p/\partial t$ and defines a *specific volumetric storativity*:

$$S_0 = \rho g[\alpha'(1 - n) + \beta n] \tag{6.3.38}$$

as the volume of water released from (or added to) a unit soil volume per unit decline (or rise) of head. Equation (6.3.38) is similar in form to (6.3.13) except for an additional ρ appearing in the latter. In both cases, $\mathbf{q}$ is with respect to the fixed coordinates. As written, equation (6.3.37) involves no approximation or error. It is only when we wish to express $\mathbf{q}$ by φ^* that we must consider the movement of the solid particles.

De Wiest (1966a) also obtains (6.3.37). He then expresses $\mathbf{q}$ as $\mathbf{q} = -(k\rho g/\mu) \nabla\varphi^*$, recognizing that this is an approximation in view of the fact that Darcy's law gives the specific discharge with respect to the moving grains. He obtains:

$$K \nabla^2\varphi^* - 2\rho g K \,\partial\varphi^*/\partial z = S_0 \,\partial\varphi^*/\partial t \tag{6.3.39}$$

where the specific storativity S_0 is defined by (6.3.38). By assuming that $K = k\rho g/n =$ const and that $\mathbf{q} \cdot \nabla p \ll \rho \nabla \mathbf{q}$ and $\varphi^* \approx \varphi$, we obtain from (6.3.37) and (6.3.38):

$$K \nabla^2\varphi = S_0 \,\partial\varphi/\partial t; \qquad \mathscr{D} \nabla^2\varphi = \partial\varphi/\partial t \tag{6.3.40}$$

where $\mathscr{D} = K/S_0$; $\mathscr{D}$(dim L^2/T) is sometimes called the diffusivity of the porous medium.

6.3.2 Extensions to Three Phases and to Three-Dimensional Consolidation

Following Biot (1941, 1956), de Josselin de Jong et al. (1963) and, mainly, Verruijt (1965, 1969), let us extend the discussion of the previous paragraphs to the case where the porous medium is composed of three phases: a solid (denoted by a subscript s), a liquid (subscript l) and a gas (subscript g). However, we shall assume that the medium is almost completely saturated so that the total amount of gas in it is small. We shall also assume that horizontal deformations do not vanish so that we have the general case of three-dimensional consolidation.

Let ρ_s, ρ_l and ρ_g denote the densities of the solid, the liquid and the gas, respectively. The specific mass discharge $\mathbf{J}$ is related to the specific (volume) discharge $\mathbf{q}$ by:

$$\mathbf{J}_s = \rho_s \mathbf{q}_s; \qquad \mathbf{J}_l = \rho_l \mathbf{q}_l; \qquad \mathbf{J}_g = \rho_g \mathbf{q}_g. \tag{6.3.41}$$

A bulk volume U_b of the porous medium contains a volume $U_s = (1 - n)U_b \equiv n_s U_b$ of solid. If the degree of saturation is S_r, the volume of liquid is $U_l = nS_r U_b \equiv n_l U_b$ and the volume of gas is $U_g = (1 - S_r)nU_b \equiv n_g U_b$.

We may now apply the equation of mass conservation (6.2.3) to each of three components:

$$\operatorname{div} \mathbf{J}_s + \partial(n_s\rho_s)/\partial t = 0; \qquad \operatorname{div} \mathbf{J}_s + \partial[(1 - n)\rho_s]/\partial t = 0 \tag{6.3.42}$$

$$\operatorname{div} \mathbf{J}_l + \partial(n_l\rho_l)/\partial t = 0; \qquad \operatorname{div} \mathbf{J}_l + \partial(nS_r\rho_l)/\partial t = 0 \tag{6.3.43}$$

$$\operatorname{div} \mathbf{J}_g + \partial(n_g\rho_g)/\partial t = 0; \qquad \operatorname{div} \mathbf{J}_g + \partial[(1 - S_r)n\rho_g]/\partial t = 0. \tag{6.3.44}$$

Darcy's law expresses the specific discharge relative to the moving solids. Following Verruijt (1965, 1969) we shall postulate that for the homogeneous isotropic medium under consideration:

$$\mathbf{V}_l - \mathbf{V}_s = -\frac{K}{n_l} \operatorname{grad} \varphi^* = -\frac{K}{S_r n} \operatorname{grad} \varphi^* \tag{6.3.45}$$

where φ^* is defined in section 5.9, $\rho(p)$ stands for $\rho_l(p)$ and $\mathbf{V}_l$ and $\mathbf{V}_s$ are the velocities of the liquid and the solid, respectively, with respect to a fixed coordinate system. The hydraulic conductivity K is assumed to be a constant, in spite of its dependence on S_r, ρ_l, and n, which vary. An extension to anisotropic media is also possible. Equation (6.3.45) can also be used to define $\mathbf{q}_r$, the relative specific discharge with respect to the solid grains (par. 6.3.1):

$$\mathbf{q}_r = (\mathbf{V}_l - \mathbf{V}_s)S_r n = -K \operatorname{grad} \varphi^*. \tag{6.3.46}$$

As stated above, Verruijt assumes that the medium is almost completely saturated, i.e., $(1 - S_r) \ll 1$, and that $\mathbf{V}_l \approx \mathbf{V}_g$, i.e., the relative velocity of the gas with respect to the liquid is zero.

The equations of state of the three phases are:

$$\rho_s = \text{const}; \qquad \rho_l = \rho_{0l} \exp(\beta p); \qquad \rho_g = \rho_{0g}(p/p_{0g}) \tag{6.3.47}$$

where ρ_{0l} and ρ_{0g} are reference densities. Scott (1963) suggests that for soil grains: $\rho_s = \rho_{0s}[1 + \beta_s p]$.

By inserting these expressions in (6.3.42) through (6.3.44), we obtain: for the solid:

$$\partial n/\partial t + \mathbf{V}_s \cdot \operatorname{grad} n - (1 - n) \operatorname{div} \mathbf{V}_s = 0; \qquad \operatorname{div}[(1 - n)\mathbf{V}_s] + \partial(1 - n)/\partial t = 0 \tag{6.3.48}$$

for the liquid:

$$\operatorname{div}(S_r n \rho_l \mathbf{V}_l) + \partial(S_r n \rho_l)/\partial t = 0 \tag{6.3.49}$$

for the gas:

$$\operatorname{div}[(1 - S_r)n\rho_g \mathbf{V}_g] + \partial[(1 - S_r)n\rho_g]/\partial t = 0$$

or, since $\mathbf{V}_g \approx \mathbf{V}_l$:

$$\mathbf{V}_l \cdot \operatorname{grad} \ln[(1 - S_r)n\rho_g] + \operatorname{div} \mathbf{V}_l + \partial[\ln(1 - S_r)]/\partial t + \partial \ln p/\partial t + \partial \ln n/\partial t = 0. \tag{6.3.50}$$

By eliminating $\partial S_r/\partial t$ from (6.3.49) and (6.3.50), we obtain:

$$\partial \ln n/\partial t + (1 - S_r + S_r p\beta)\, \partial \ln p/\partial t + \mathbf{V}_l\, [\operatorname{grad} \ln n + (1 - S_r + S_r p\beta) \operatorname{grad} \ln p] + \operatorname{div} \mathbf{V}_l = 0. \tag{6.3.51}$$

By taking the divergence of (6.3.45), we obtain:

$$\operatorname{div}[n_l(\mathbf{V}_l - \mathbf{V}_s)] = -(K/\rho_l g)[\nabla^2 p - \beta(\operatorname{grad} p)^2]; \qquad n_l = S_r n. \tag{6.3.52}$$

By eliminating $\operatorname{div} \mathbf{V}_l$ from (6.3.51) and (6.3.52), and expressing $\partial n/\partial t$ by (6.3.48), we obtain:

$$\begin{aligned}(K/\rho_l g)\, \nabla^2 p = {} & S_r \operatorname{div} \mathbf{V}_s - S_r n(1 - S_r + S_r p\beta)\, \partial \ln p/\partial t \\ & + n\mathbf{V}_l\{S_r(1 - S_r + S_r p\beta) \operatorname{grad} \ln p - \operatorname{grad} S_r\} \\ & + n\mathbf{V}_s \cdot \operatorname{grad} S_r + (K/\rho_l g)\beta(\operatorname{grad} p)^2. \end{aligned} \tag{6.3.53}$$

Recalling that $\delta = 1 - S_r \ll 1$ and that in ground water flow we also have $p\beta \ll 1$, we may neglect terms of order δ and $p\beta$ with respect to terms of order 1. If, following Verruijt (1965, 1969), we also disregard the scalar products on the right-hand side of (6.3.52) as terms of second order, we obtain:

$$\operatorname{div} \mathbf{q}_r \cong S_r n(\operatorname{div} \mathbf{V}_l - \operatorname{div} \mathbf{V}_s) = -K \nabla^2 \varphi^* = -\frac{K}{\rho g} \nabla^2 p \tag{6.3.54}$$

and hence (6.3.53) is replaced by:

$$(K/\rho_l g) \nabla^2 p = \operatorname{div} \mathbf{V}_s + n\beta' \, \partial p/\partial t \tag{6.3.55}$$

where $\beta' = \beta + (1 - S_r)/p$ and $1 - S_r \ll 1$. For small deviations of p from some initial value, β' becomes a constant. The term $(1 - S_r)/p$ introduces the effect of the gas.

By starting from (6.3.21), i.e., neglecting the presence of gas, with $\mathbf{q}_r = n(\mathbf{V}_l - \mathbf{V}_s)$, and $\rho = \rho_l$, we obtain:

$$(K/\rho g) \nabla^2 p = \operatorname{div} \mathbf{V}_s + n\beta \, \partial p/\partial t. \tag{6.3.56}$$

Equation (6.3.56) is the same as (6.3.55), except for the difference between β and β'.

Equation (6.3.55) relates pressure, p, to the velocity of solid displacement, $\mathbf{V}_s$; p and $\mathbf{V}_s$ are the *two independent variables.*

From the definition of *dilatation*, ε (*volume strain, dilatational strain*), in section 4.1, we have:

$$\varepsilon = \operatorname{div} \mathbf{u}; \qquad \mathbf{V}_s = \partial \mathbf{u}/\partial t; \qquad \operatorname{div} \mathbf{V}_s = \partial \varepsilon/\partial t. \tag{6.3.57}$$

In (6.3.57) $\mathbf{u}$ is the displacement vector of the solid material. We therefore have:

$$(K/\rho g) \nabla^2 p = \partial \varepsilon/\partial t + n\beta' \, \partial p/\partial t. \tag{6.3.58}$$

Under the conditions considered in the present section, equation (6.3.58) governs the flow; it takes the place of the conservation equations considered in the previous sections. Since it contains two variables, p and ε, a second equation is needed to describe the flow completely. This equation is derived by considering the *elastic behavior of the solid matrix.*

Verruijt's (1965, 1969) analysis of the elastic behavior of the isotropic, perfectly elastic solid matrix results in the following set of equations:

$$\varepsilon = \operatorname{div} \mathbf{u} \tag{6.3.59}$$

$$\mu \nabla^2 u_{x_i} + (\lambda + \mu) \, \partial \varepsilon/\partial x_i = \partial \sigma/\partial x_i; \qquad i = 1, 2, 3 \tag{6.3.60}$$

$$(K/\rho g) \nabla^2 \sigma = \partial \varepsilon/\partial t + n\beta' \, \partial \sigma/\partial t \tag{6.3.61}$$

where the stress $\underline{\boldsymbol{\sigma}}$ (components σ_{ij}) in the porous matrix is related to the effective stress $\underline{\boldsymbol{\sigma}}'$ (components σ'_{ij}) and to the pressure p ($p > 0$ for compression) by:

$$\sigma_{ij} = \sigma'_{ij} - p\delta_{ij}; \qquad i, j = 1, 2, 3 \tag{6.3.62}$$

$\varepsilon = \varepsilon_{xx} + \varepsilon_{yy} + \varepsilon_{zz}$ is the incremental volume strain (sum of diagonal components of the strain tensor $\underline{\boldsymbol{\varepsilon}}$), $\lambda = \nu E/(1 + \nu)(1 - 2\nu)$, E is Young's modulus, ν is Poisson's ratio and $\mu = E/[2(1 - \nu)]$. The reader who wishes to dig deeper into the detailed development carried out by Verruijt (1965, 1969) should consult texts on theory of elasticity (e.g., Sokolnikoff 1964) or soil mechanics (e.g., Scott 1963).

Equations (6.3.59), (6.3.60) and (6.3.61) form a set of five equations for the five independent variables: $\sigma, \varepsilon, u_x, u_y, u_z$. By differentiating the first equation of (6.3.60) with respect to x, the second with respect to y and the third with respect to z, and adding the results, we obtain:

$$(\lambda + 2\mu)\, \nabla^2\varepsilon = \nabla^2\sigma. \tag{6.3.63}$$

Thus (6.3.61) and (6.3.63) form a set of two equations from which ε and σ may be derived. These equations were first derived by Biot (1941).

By integrating (6.3.63), we obtain:

$$(\lambda + 2\mu)\varepsilon = \sigma + f(x, y, z, t); \qquad \nabla^2 f = 0. \tag{6.3.64}$$

When $f \equiv 0$, equation (6.3.61) becomes (Verruijt 1965, 1969):

$$(K/\rho_l g)\, \nabla^2\sigma = (\alpha + n\beta')\, \partial\sigma/\partial t \tag{6.3.65}$$

which is similar to some of the equations derived in the previous section. In this form it was originally derived by Terzaghi (1923). Verruijt (1965, 1969) gives some examples in which problems of flow in leaky aquifers are solved by solving (6.3.61) and (6.3.63).

Thus, assuming the validity of Darcy's law and the generalized Hooke law, a theory was developed for flow in a consolidating medium, without actually defining a storage coefficient. The differences between the results here and in the previous section stem mainly from the different assumptions (especially those related to the relative motion and to the constancy of K) in the various derivations.

In spite of the difficulties in defining a storage coefficient and relating it to the elastic properties of the medium and the liquid, it seems that for practical purposes, especially in the field of ground water hydrology, the introduction of such a definition would be advantageous (sec. 6.4). Since in practical cases the coefficients must, in any event, be determined by actual field experiments (such as pumping tests, etc.), the exact dependence of storativity on the compressibility of the medium and the soil need not enter into the definition itself.

A theory of flow in compressible media was also presented by Scheidegger (1960).

6.3.3 Barometric Efficiency of Aquifers

As explained in section 2.7, the total load of soil and water, and that resulting from atmospheric pressure above any horizontal plane passing through an aquifer, is balanced by the effective (interparticle) stresses and by the pressure in the water. Since part of this external load results from the atmospheric pressure that is transmitted to the aquifer, changes in this pressure cause changes in water pressure in the aquifer. Such pressure changes produce fluctuations in the water table in observation wells that penetrate the aquifer (Jacob 1940; Parker and Stringfield 1950; and Tuinzaad 1954). This phenomenon is observed in wells tapping confined aquifers

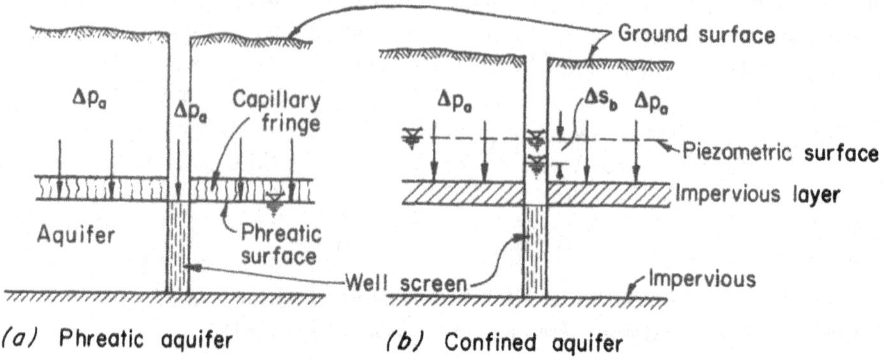

FIG. 6.3.1. Effect of atmospheric pressure variations on water level in observation wells.

and is ordinarily not observed in wells tapping phreatic aquifers (fig. 6.3.1). The reason for this difference is that in a phreatic aquifer, any change Δp_a in atmospheric pressure is transmitted directly to the entire water table in the aquifer through the zone of aeration (fig. 1.1.3), and simultaneously to the water table in a well tapping this aquifer. In a confined aquifer, any change Δp_a is transmitted to the water in the aquifer only (indirectly) through the confining layer.

Following Jacob (1940) and the discussion in section 2.7 and paragraph 6.3.1, i.e., considering vertical compressibility only, we obtain from (2.7.7):

$$dp_a = d\sigma' + dp \tag{6.3.66}$$

where p is the pore pressure in the water and $d\sigma'$ is the change in effective stress produced by dp_a. In general, $p_a = f(t)$.

In a well tapping a confined aquifer (fig. 6.3.1b), the increase Δp_a forces water from the well into the aquifer. This movement, indicated by a drawdown in the well by an amount Δs_b, will continue until it is balanced by the counteracting pressure change in the water. Thus:

$$dp_a = \gamma\, ds_b + dp \tag{6.3.67}$$

where γ is the specific weight of the water. Equation (6.3.67) expresses an equilibrium, thus assuming that the volume of water leaving (or entering) the well as a result of atmospheric pressure changes is negligible.

The *barometric efficiency BE* is defined as the ratio of water level change ds_b to the atmospheric pressure change, expressed in terms of water height:

$$BE = ds_b/(dp_a/\gamma). \tag{6.3.68}$$

Values of BE in the range 0.25–0.75 have been observed. In order to relate the barometric efficiency to aquifer and water properties, we combine (6.3.66) through (6.3.68), obtaining:

$$BE = ds_b/(dp_a/\gamma) = (dp_a - dp)/dp_a = d\sigma'/(d\sigma' + dp) = 1/(1 + dp/d\sigma'). \tag{6.3.69}$$

Following a procedure similar to that outlined in paragraph 6.3.1, i.e., considering only vertical compressibility of aquifer and incompressible solid grains, for a given volume of solids, U_s, we obtain:

$$U_s = \text{const} = U_b - U_w = (1 - n)U_b$$

$$\frac{\partial U_s}{\partial \sigma'} = 0 = \frac{\partial}{\partial \sigma'}(1 - n)U_b = -U_b \frac{\partial n}{\partial \sigma'} + (1 - n)\frac{\partial U_b}{\partial \sigma'}$$

$$\partial \sigma'/\partial p = \beta n/\alpha \tag{6.3.70}$$

where

$$\beta = -(1/U_w)\,\partial U_w/\partial p; \qquad \alpha = -(1/U_b)\,\partial U_b/\partial \sigma'.$$

Hence, equation (6.3.69) becomes:

$$BE = \frac{\beta n}{\alpha + \beta n} < 1. \tag{6.3.71}$$

As $BE > 0$ this means that for $dp_a > 0$ we have $ds_b > 0$ (i.e., water level is dropping).

It is also possible to relate the specific storativity S_0 of (6.3.38) to BE:

$$S_0 = \gamma[\beta n + \alpha(1 - n)] = \alpha\gamma[1/(1 - BE) - n]. \tag{6.3.72}$$

The barometric efficiency, which is a constant for any given aquifer, is a measure of the aquifer's rigidity.

6.4 Continuity Equations for Flow in Confined and Leaky Aquifers

6.4.1 The Horizontal Flow Approximation

In sections 6.2 and 6.3 the control volume approach was used to develop the mass conservation equations for flow through porous media. In the present section, we shall focus our attention on the special case of ground water in an aquifer (sec. 1.1). Both confined and leaky aquifers will be considered. Flow in phreatic aquifers is considered only in part (see also chap. 8).

The main characteristic of flow in an aquifer is that it is *essentially horizontal*, or that it may be approximated as such. This is true, for example, in confined horizontal homogeneous and isotropic aquifers with fully penetrating wells. It is still a good approximation when the thickness of the aquifer varies, but in such a way that the variations are much smaller than the thickness itself. The approximation fails in regions where the flow has a vertical component as, for example, in the vicinity of a partially penetrating well, or an outlet in the form of a spring. Similarly, flow in a leaky aquifer (sec. 1.1) always has a vertical component, yet in view of the fact that the thickness of the aquifer is much less than the horizontal lengths involved, the assumption of essentially horizontal flow may be considered a good approximation.

Therefore, under certain conditions, instead of considering the flow as three dimensional, with $\varphi = \varphi(x, y, z, t)$, we may treat the problem in terms of an average head, $\bar{\varphi} = \bar{\varphi}(x, y, t)$, where the average is taken along a vertical line extending from the bottom to the top of the aquifer.

We shall make use of this assumption of essentially horizontal flow in the discussion of the Dupuit assumptions in connection with unconfined flow (sec. 8.1).

A second important feature of the flow in confined and leaky aquifers is that although the water is compressible, we neglect the variations in ρ and assume that we have essentially a homogeneous fluid of constant ρ. We may then apply the conservation statement to the volume of the fluid and assume that K, and hence transmissivity $T = Kb$, is independent of time. We may also use φ defined by (3.3.1) rather than φ^* of (5.9.1).

6.4.2 Flow in a Confined Aquifer

For the purpose of the discussion in this section, the variations in time of aquifer thickness b are neglected, as they are assumed much smaller than b itself. The following aquifer parameters will be used.

The *aquifer's transmissivity*, T, is defined in essentially horizontal flow as the discharge of water through the entire thickness of the aquifer per unit horizontal length of aquifer perpendicular to the direction of the flow and per unit hydraulic gradient. It is equal to the product of average (over a vertical line) hydraulic conductivity and the aquifer's thickness:

$$T = \bar{K}b; \qquad \bar{K} = \frac{1}{b}\int_0^b K(z)\,dz. \tag{6.4.1}$$

In an inhomogeneous aquifer, $T = T(x, y)$ because either b or $\bar{K}$, or both, may be functions of x and y.

The *aquifer's storativity* S (dimensionless) is defined (par. 1.1.4) as the volume of water (ΔU_w) released from (or added to) storage in the aquifer per unit horizontal area of aquifer and per unit decline (or rise) of the average (over the vertical) piezometric head in the aquifer:

$$S = \Delta U_w / \Delta A\, \Delta\bar{\varphi}. \tag{6.4.2}$$

The storativity is the outcome of the elastic properties of the medium and the water. However, in defining it by (6.4.2) we do not specify the exact nature of this dependence as was studied in section 6.3. In an inhomogeneous aquifer S may vary with location, but we assume that it undergoes no changes in time. Defining the *specific storativity* (or the *medium's storativity*) S_s (dim L^{-1}) as the volume of water released from (or added to) storage in the aquifer per unit (bulk) volume of aquifer and per unit decline (or rise) of head, we have:

$$S = S_s b. \tag{6.4.3}$$

In (6.4.3) S_s is assumed independent of z. In (6.3.38), S_s is denoted by S_0.

With these definitions in mind, consider the essentially horizontal flow (i.e., $\bar{\varphi} \approx \varphi$) in the confined aquifer shown in figure 6.4.1. Writing the continuity of flow for the control volume $b\,\Delta x\,\Delta y$, we obtain, using (6.4.2):

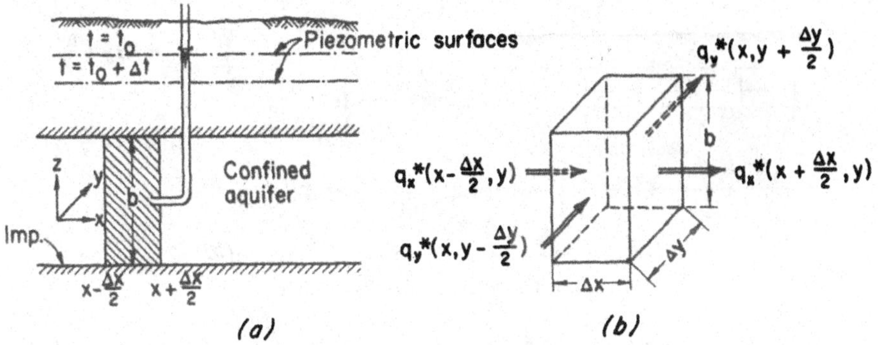

FIG. 6.4.1. Flow in a confined aquifer.

$$\Delta t \left\{\Delta y \left[q^*_x\left(x - \frac{\Delta x}{2}, y\right) - q^*_x\left(x + \frac{\Delta x}{2}, y\right)\right]\right.$$
$$\left. + \Delta x\left[q^*_y\left(x, y - \frac{\Delta y}{2}\right) - q^*_y\left(x, y + \frac{\Delta y}{2}\right)\right]\right\}$$
$$= S\,\Delta x\,\Delta y\,[\varphi(t + \Delta t) - \varphi(t)]$$

where $\mathbf{q}^* = -T \operatorname{grad} \varphi$ is the discharge per unit width of aquifer.

Dividing by $\Delta x\, \Delta y\, \Delta t$, and passing to the limits as $\Delta x, \Delta y, \Delta t \to 0$, we obtain:

$$-(\partial q^*_x/\partial x + \partial q^*_y/\partial y) = S\, \partial\varphi/\partial t$$
$$\operatorname{div}(T \operatorname{grad} \varphi) = S\, \partial\varphi/\partial t. \tag{6.4.4}$$

In an anisotropic aquifer, T is a second-rank tensor. If x and y are principal directions (with principal values T_x and T_y, respectively), equation (6.4.4) becomes:

$$\frac{\partial}{\partial x}\left(T_x \frac{\partial \varphi}{\partial x}\right) + \frac{\partial}{\partial y}\left(T_y \frac{\partial \varphi}{\partial y}\right) = S \frac{\partial \varphi}{\partial t}. \tag{6.4.5}$$

If the medium is isotropic, we obtain:

$$\frac{\partial}{\partial x}\left(T \frac{\partial \varphi}{\partial x}\right) + \frac{\partial}{\partial y}\left(T \frac{\partial \varphi}{\partial y}\right) = S \frac{\partial \varphi}{\partial t}$$
$$T \nabla^2 \varphi + \operatorname{grad} \varphi \cdot \operatorname{grad} T = S\, \partial\varphi/\partial t. \tag{6.4.6}$$

For a homogeneous isotropic aquifer, $T =$ constant, and (6.4.6) becomes:

$$T \nabla^2 \varphi \equiv T(\partial^2\varphi/\partial x^2 + \partial^2\varphi/\partial y^2) = S\, \partial\varphi/\partial t. \tag{6.4.7}$$

By comparing (6.4.7) with the equations developed in paragraph 6.3.2, we can see which terms have been neglected here as a result of the assumptions stated at the beginning of this section. For example, by making the following changes in (6.3.29): $\mathbf{V}_s = 0$ (no movement of solid matrix), $S = \rho g(\alpha + n\beta)b$, $T = Kb$, $\varphi^* \to \varphi$ and $\partial\varphi^*/\partial z \to 0$, we obtain (6.4.7). The approach taken here is a less rigorous one, yet is sufficiently accurate for all practical purposes.

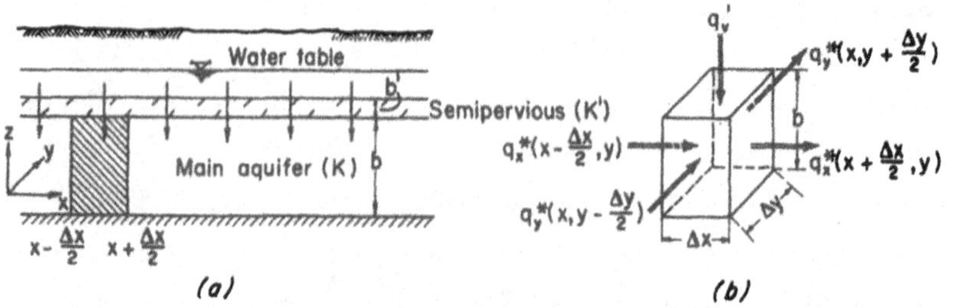

FIG. 6.4.2. Flow in a leaky aquifer.

By comparing (6.4.7) with (6.3.40), sometimes an *aquifer diffusivity* is defined by $\mathscr{D} = T/S$ (dim L^2/T).

6.4.3 Flow in a Leaky Aquifer

A leaky aquifer was defined in paragraph 1.1.2. We shall use the nomenclature of figure 6.4.2, in which we have a main aquifer and a semipervious layer ($K \gg K'$). Again we assume that the flow in the main aquifer is essentially horizontal and therefore, in view of the law of streamline refraction (7.1.63), the flow in the semipervious layer is essentially vertical. The semipervious layer is overlain by another pervious formation in which the piezometric head distribution may or may not be affected by the flow in the main aquifer. Other types of leaky aquifers are also possible (e.g., Hantush 1964).

Conservation of volume (since we neglect the small variations in water density) leads to:

$$\Delta t \left\{ \Delta y \left[q^*_x\left(x - \frac{\Delta x}{2}, y\right) - q^*_x\left(x + \frac{\Delta x}{2}, y\right) \right] \right.$$
$$\left. + \Delta x \left[q^*_y\left(x, y - \frac{\Delta y}{2}\right) - q^*_y\left(x, y + \frac{\Delta y}{2}\right) \right] + q'_v \Delta x \Delta y \right\}$$
$$= S \Delta x \Delta y [\varphi(t + \Delta t) - \varphi(t)].$$

Going to the limit as $\Delta x, \Delta y, \Delta t \to 0$, and with $q'_v = -K'(\partial\varphi'/\partial z)$, where φ' is the piezometric head distribution in the semipervious layer, we obtain for a homogeneous isotropic main aquifer:

$$T(\partial^2\varphi/\partial x^2 + \partial^2\varphi/\partial y^2) + K'(\partial\varphi'/\partial z) = S\, \partial\varphi/\partial t. \tag{6.4.8}$$

In (6.4.8), $\partial\varphi'/\partial z$ is taken at the interface between the two layers.

Since in the semipervious layer we have assumed vertical flow only, we have:

$$K'(\partial^2\varphi'/\partial z^2) = S'\, \partial\varphi'/\partial t \tag{6.4.9}$$

where S' is the storativity in the semipervious layer. Obviously, at the interface between the two layers we have $\varphi = \varphi'$. Often, as in figure 6.4.2a, we have another aquifer, above the semipervious layer, where the potential is φ_0. We may then express the vertical leakage by $q'_v = K'(\varphi_0 - \varphi)/b'$. Equation (6.4.8) then becomes:

$$T\,\nabla^2_{xy}\varphi + (\varphi_0 - \varphi)/\sigma' = S\,\partial\varphi/\partial t \tag{6.4.10}$$

where $\sigma' = b'/K'$ (dim T) is the resistance of the semipervious layer. Rewriting (6.4.10) as:

$$\nabla^2_{xy}\varphi + (\varphi_0 - \varphi)/\lambda^2 = (S/T)\,\partial\varphi/\partial t \tag{6.4.11}$$

we obtain another leaky aquifer parameter $\lambda = (T\sigma')^{1/2}$, called *leakage factor*, that determines the areal distribution of the leakage.

It is of interest to note another way of deriving (6.4.11). We start from the three-dimensional continuity equation for $\varphi = \varphi(x, y, z, t)$ in a homogeneous aquifer:

$$\partial^2\varphi/\partial x^2 + \partial^2\varphi/\partial y^2 + \partial^2\varphi/\partial z^2 = (S_s/K)\,\partial\varphi/\partial t \tag{6.4.12}$$

where S_s is specific storativity, and integrate it with respect to z over the constant height b of the aquifer. Since:

$$\int_0^b \frac{\partial^2\varphi}{\partial x^2}\,dz = \frac{\partial^2}{\partial x^2}\int_0^b \varphi(x, y, z, t)\,dz = b\,\frac{\partial^2\tilde{\varphi}(x, y, t)}{\partial x^2}$$

$$\int_0^b \frac{\partial^2\varphi}{\partial y^2}\,dz = \frac{\partial^2}{\partial y^2}\int_0^b \varphi(x, y, z, t)\,dz = b\,\frac{\partial^2\tilde{\varphi}(x, y, t)}{\partial y^2}$$

$$\int_0^b \frac{\partial^2\varphi}{\partial z^2}\,dz = \left(\frac{\partial\varphi}{\partial z}\right)_{z=b} - \left(\frac{\partial\varphi}{\partial z}\right)_{z=0}; \qquad \int_0^b \left(\frac{\partial\varphi}{\partial t}\right) dz = b\,\frac{\partial\tilde{\varphi}}{\partial t}$$

and assuming $\partial\varphi/\partial z = 0$ at $z = 0$, and $K\partial\varphi/\partial z = (\varphi_0 - \tilde{\varphi})/\sigma'$ at $z = b$, we obtain:

$$b\left(\frac{\partial^2\tilde{\varphi}}{\partial x^2} + \frac{\partial^2\tilde{\varphi}}{\partial y^2}\right) + \frac{(\varphi_0 - \tilde{\varphi})}{\sigma' K} = \left(\frac{S_s b}{K}\right)\frac{\partial\tilde{\varphi}}{\partial t}$$

where:

$$\tilde{\varphi}(x, y, t) = \frac{1}{b}\int_{z=0}^{b} \varphi(x, y, z, t)\,dz$$

is the average value of φ over b. Hence, with $T = Kb$ and $S = S_s b$, we obtain:

$$T\left(\frac{\partial^2\tilde{\varphi}}{\partial x^2} + \frac{\partial^2\tilde{\varphi}}{\partial y^2}\right) + \frac{\varphi_0 - \tilde{\varphi}}{\sigma'} = S\,\frac{\partial\tilde{\varphi}}{\partial t} \tag{6.4.13}$$

which shows that φ in (6.4.11) should be interpreted as $\tilde{\varphi}$. A more general discussion of this subject is presented in the following paragraph.

In this way, with relatively simple assumptions and definitions of S and T, we may derive equations governing the flow in ground water aquifers.

Before leaving this subject, let us consider the error introduced by assuming horizontal flow in the aquifer in the case where the aquifer is anisotropic.

We start from the definition

$$\bar{\varphi}(x, y) = \frac{1}{b}\int_0^b \varphi(x, y, z)\, dz,$$

and integrate it by parts. We obtain:

$$\bar{\varphi}(x, y) = \frac{1}{b} z\varphi(x, y, z)\Big|_{z=0}^{z=b} - \frac{1}{b}\int_0^b z\frac{\partial\varphi(x, y, z)}{\partial z}\, dz = \varphi(x, y, b) + \frac{1}{K_z b}\int_0^b zq_z\, dz.$$

Letting $\varphi_0 > \varphi(x, y, b)$, we have $q_z < 0$, with $q_z = 0$ for $z = 0$ and $q_z = [\varphi_0 - \varphi(x, y, b)]/\sigma'$ for $z = b$. Hence:

$$-\frac{b}{2K_z}\frac{\varphi_0 - \varphi(x, y, b)}{\sigma'} < [\bar{\varphi} - \varphi(x, y, b)] < 0$$

or:

$$0 < \frac{\bar{\varphi} - \varphi(x, y, b)}{\varphi(x, y, b) - \varphi_0} < \frac{b}{2K_z\sigma'} = \frac{1}{2}\left(\frac{b}{b'}\right)\left(\frac{K'_z}{K_z}\right) = \frac{1}{2}\left(\frac{K_x}{K_z}\right)\left(\frac{b}{\lambda}\right)^2 \ll 1 \qquad (6.4.14)$$

as in general $K_x > K_z$ and $b \ll \lambda$.

6.4.4 Averaging the Exact Equations over a Vertical Line

Continuity for an elementary control box in an aquifer (in the absence of sources or sinks inside the aquifer) is given by:

$$-\left(\frac{\partial q_x}{\partial x} + \frac{\partial q_y}{\partial y} + \frac{\partial q_z}{\partial z}\right) = S_s\frac{\partial\varphi}{\partial t} \qquad (6.4.15)$$

where S_s is specific storativity. We shall use the nomenclature of figure 6.4.3, where $-N'$ and N'' are rates of accretion in the (vertical) directions as indicated. We

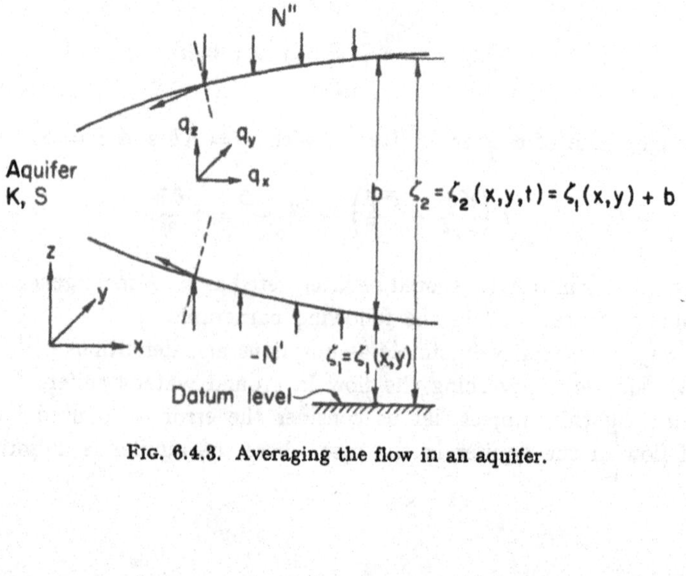

FIG. 6.4.3. Averaging the flow in an aquifer.

shall generalize the discussion, following Hantush (1964), by assuming that the upper boundary is a phreatic surface with accretion so that $\zeta_2 = \zeta_2(x, y, t)$.

Along a boundary with accretion (par. 7.1.6) which moves at a velocity $\mathbf{V}_B = \mathbf{q}_B/n_e$, where n_e is the effective porosity, we have:

$$\mathbf{q} - \mathbf{N} = \mathbf{q}_B \tag{6.4.16}$$

where $\mathbf{N}$ is positive if it is directed downward. At $\zeta_2 = \zeta_2(x, y, t)$ for the vertical component $(V_{B2})_z$ of $\mathbf{V}_{B2}$, we have:

$$(V_{B2})_z = \frac{d\zeta_2}{dt} = \frac{\partial\zeta_2}{\partial x}\frac{dx}{dt} + \frac{\partial\zeta_2}{\partial y}\frac{dy}{dt} + \frac{\partial\zeta_2}{\partial t} = \frac{\partial\zeta_2}{\partial x}\frac{q_x}{n_e} + \frac{\partial\zeta_2}{\partial y}\frac{q_y}{n_e} + \frac{\partial\zeta_2}{\partial t} \tag{6.4.17}$$

and at $\zeta_1 = \zeta_1(x, y)$, we have for the vertical component $(V_{B1})_z$ of $\mathbf{V}_{B1}$:

$$(V_{B1})_z = \frac{d\zeta_1}{dt} = \frac{\partial\zeta_1}{\partial x}\frac{dx}{dt} + \frac{\partial\zeta_1}{\partial y}\frac{dy}{dt}. \tag{6.4.18}$$

Hence, along the upper and the lower boundaries, the specific discharges are:

$$\begin{aligned} q_z|_{\zeta_2} &= -N'' + q_x\,\partial\zeta_2/\partial x + q_y\,\partial\zeta_2/\partial y + n_e\,\partial\zeta_2/\partial t \\ q_z|_{\zeta_1} &= -N' + q_x\,\partial\zeta_1/\partial x + q_y\,\partial\zeta_1/\partial y. \end{aligned} \tag{6.4.19}$$

Following Hantush (1964), we now multiply all the terms appearing in (6.4.15) by dz and integrate with respect to z from $\zeta_1(x, y)$ to $\zeta_2(x, y, t)$ using the Leibnitz rule of differentiation. We obtain:

$$\int_{\zeta_1}^{\zeta_2} (\partial q_x/\partial x)\,dz + \int_{\zeta_1}^{\zeta_2} (\partial q_y/\partial_y)\,dz + \int_{\zeta_1}^{\zeta_2} (\partial q_z/\partial z)\,dz = -S_s \int_{\zeta_1}^{\zeta_2} (\partial\varphi/\partial t)\,dz$$

$$\begin{aligned} \int_{\zeta_1}^{\zeta_2} (\partial q_z/\partial z)\,dz = q_z|_{\zeta_2} - q_z|_{\zeta_1} &= -N'' + q_x|_{\zeta_2}\,\partial\zeta_2/\partial x + q_y|_{\zeta_2}\,\partial\zeta_2/\partial y + n_e\,\partial\zeta_2/\partial t \\ &\quad + N' - q_x|_{\zeta_1}\,\partial\zeta_1/\partial x - q_y|_{\zeta_1}\,\partial\zeta_1/\partial y \end{aligned}$$

$$\int_{\zeta_1}^{\zeta_2} (\partial\varphi/\partial t)\,dz = \frac{\partial}{\partial t}\int_{\zeta_1}^{\zeta_2} \varphi\,dz - \varphi|_{\zeta_2}\frac{\partial\zeta_2}{\partial t}$$

$$\int_{\zeta_1}^{\zeta_2} \frac{\partial q_x}{\partial x}\,dz = \frac{\partial}{\partial x}\int_{\zeta_1}^{\zeta_2} q_x\,dz - q_x|_{\zeta_2}\frac{\partial\zeta_2}{\partial x} + q_x|_{\zeta_1}\frac{\partial\zeta_1}{\partial x}$$

and a similar expression for the derivative $\partial q_y/\partial y$. Hence:

$$\frac{\partial}{\partial x}\left[\int_{\zeta_1}^{\zeta_2} q_x\,dz\right] + \frac{\partial}{\partial y}\left[\int_{\zeta_1}^{\zeta_2} q_y\,dz\right] - (N'' - N') + n_e\frac{\partial\zeta_2}{\partial t} = -S_s\left\{\frac{\partial}{\partial t}\int_{\zeta_1}^{\zeta_2} \varphi\,dz - \varphi|_{\zeta_2}\frac{\partial\zeta_2}{\partial t}\right\}. \tag{6.4.20}$$

By introducing the average head $\bar{\varphi}$ taken over $b = \zeta_2 - \zeta_1$

$$\bar{\varphi}(x, y, t) = \frac{1}{\zeta_2 - \zeta_1} \int_{\zeta_1(x,y)}^{\zeta_2(x,y,t)} \varphi(x, y, z')\, dz'; \qquad b = \zeta_2 - \zeta_1 \tag{6.4.21}$$

we obtain from (6.4.20) for a homogeneous aquifer ($K = \text{const}$)

$$\frac{\partial}{\partial x}\left[\frac{\partial(b\bar{\varphi})}{\partial x} - \varphi(x, y, \zeta_2, t)\frac{\partial\zeta_2}{\partial x} + \varphi(x, y, \zeta_1, t)\frac{\partial\zeta_1}{\partial x}\right]$$

$$+ \frac{\partial}{\partial y}\left[\frac{\partial(b\bar{\varphi})}{\partial y} - \varphi(x, y, \zeta_2, t)\frac{\partial\zeta_2}{\partial y} + \varphi(x, y, \zeta_1, t)\frac{\partial\zeta_1}{\partial y}\right] + (N'' - N')/K - \frac{n_e}{K}\frac{\partial\zeta_2}{\partial t}$$

$$= \frac{S_s}{K}\left[\frac{\partial(b\bar{\varphi})}{\partial t} - \varphi(x, y, \zeta_2, t)\frac{\partial\zeta_2}{\partial t}\right]. \tag{6.4.22}$$

Depending on the nature of the variations of ζ_1, ζ_2, N' and N'', equation (6.4.22) may be simplified considerably. Equations similar to (6.4.22) can also be derived for $K \neq \text{const}$.

As an example, consider a homogeneous tilted leaky aquifer of constant thickness b. The aquifer is underlain by an impervious formation and overlain by a semipervious formation such that $N'' = K'(\partial\varphi'/\partial z)|_{z=\zeta_2}$, where $\varphi'(x, y, z, t)$ is the head in the semipervious layer. The two bounding surfaces of the aquifer at $z = \zeta_1(x, y)$ and $z = \zeta_2(x, y)$ are tilted with respect to the horizontal by a small angle ε, so that:

$$\zeta_1 = \varepsilon x, \qquad \zeta_2 = \varepsilon x + b.$$

Under these conditions (6.4.22) becomes:

$$\partial^2\bar{\varphi}/\partial x^2 + \partial^2\bar{\varphi}/\partial y^2 - \frac{\varepsilon}{b}\partial(\varphi|_{\zeta_2} - \varphi|_{\zeta_1})/\partial x + (K'/Kb)(\partial\varphi'/\partial z)|_{\zeta_2} = (S_s/K)\,\partial\bar{\varphi}/\partial t. \tag{6.4.23}$$

In general, the quantity $\partial(\varphi|_{\zeta_2} - \varphi|_{\zeta_1})/\partial x$ is small (Hantush 1964). When multiplied by ε (< 0.01) it becomes insignificantly small and may be neglected. Then (6.4.23) becomes:

$$\partial^2\bar{\varphi}/\partial x^2 + \partial^2\bar{\varphi}/\partial y^2 + (K'/Kb)(\partial\varphi'/\partial z)|_{\zeta_2} = (S_s/K)\,\partial\bar{\varphi}/\partial t \tag{6.4.24}$$

independent of the tilt, provided that the tilt is sufficiently small (compare with (6.4.8)).

Let the leakage through the semipervious layer be assumed to be such that:

$$\partial\varphi'/\partial z|_{\zeta_2} = (\varphi_0 - \varphi|_{\zeta_2})/b' \simeq (\varphi_0 - \bar{\varphi})/b'$$

where b' is the thickness of the semipervious layer and φ_0 is the head above this layer. Then (6.4.24) becomes (6.4.13).

Following De Wiest (1966a), consider the case of a confined aquifer ($N' = N'' = 0$), where $\zeta_1 = \text{const}$, $\zeta_2 = \zeta_1 + b(t)$, $n_e = n$. This means that we wish to consider the effect of the elastic consolidation on the thickness b. Equation (6.4.22) becomes:

$$b\left(\frac{\partial^2\bar{\varphi}}{\partial x^2} + \frac{\partial^2\bar{\varphi}}{\partial y^2}\right) - \frac{n}{K}\frac{\partial b}{\partial t} = \frac{S_s}{K}\left[\frac{\partial(b\bar{\varphi})}{\partial t} - \varphi|_{\zeta_2}\frac{\partial b}{\partial t}\right] \simeq \frac{S_s}{K}b\frac{\partial\bar{\varphi}}{\partial t}. \tag{6.4.25}$$

We now express S_s by $S^*_{0\varphi*}$ of (6.3.13): $S_s = S^*_{0\varphi*}/\rho = \rho g[\beta n + \alpha(1-n)]$ and $\partial b/\partial t$ by: $\partial b/\partial t = \alpha b\, \partial p/\partial t = \alpha b \rho g\, \partial\bar{\varphi}/\partial t$. Then (6.4.25) becomes:

$$\partial^2\bar{\varphi}/\partial x^2 + \partial^2\bar{\varphi}/\partial y^2 = (S/T)\, \partial\bar{\varphi}/\partial t; \qquad S = \rho g b(\alpha + n\beta). \tag{6.4.26}$$

6.4.5 The Boltzmann Transformation

Consider the equation:

$$(S/T)\, \partial s/\partial t = \partial^2 s/\partial x^2 \tag{6.4.27}$$

obtained from (6.4.7) by introducing the drawdown $s = \varphi_0 - \varphi$ and considering flow in only one dimension. Its solution usually takes the form $s = s(x, t)$, or $F(s, x, t) = 0$ where $\partial F/\partial s \neq 0$. Other possible forms are $x = x(s, t)$, $t = t(s, x)$.

Sometimes it may be of interest to consider a solution $x = x(s, t)$, i.e., the location of a front of a given drawdown propagating in the flow domain. In order to derive such expressions we must first transform (6.4.27) into an equation with x as the dependent variable.

Following King and Corey (1962), we obtain from $x = x(s, t)$ and $s = s(x, t)$:

$$dx = \frac{\partial x}{\partial s} ds + \frac{\partial x}{\partial t} dt = \frac{\partial x}{\partial s}\left[\frac{\partial s}{\partial x} dx + \frac{\partial s}{\partial t} dt\right] + \frac{\partial x}{\partial t} dt$$

or:

$$\partial s/\partial t = -(\partial s/\partial x)(\partial x/\partial t). \tag{6.4.28}$$

After some additional mathematical manipulations we also obtain:

$$\partial^2 s/\partial x^2 = -(\partial^2 x/\partial s^2)/(\partial x/\partial s)^3. \tag{6.4.29}$$

By substituting (6.4.28) and (6.4.29) in (6.4.27), we obtain:

$$\frac{S}{T}\frac{\partial x}{\partial t} = \frac{\partial^2 x}{\partial s^2} \Big/ \left(\frac{\partial x}{\partial s}\right)^2 \tag{6.4.30}$$

as a partial differential equation for $x = x(s, t)$. At this point we assume that the method of separation of variables is applicable so that:

$$x = \sigma(s)\tau(t). \tag{6.4.31}$$

Taking the necessary derivatives of (6.4.31) and inserting in (6.4.30), we obtain:

$$\frac{S}{T}\tau\frac{\partial \tau}{\partial t} = \frac{1}{\sigma}\frac{\partial^2 \sigma}{\partial s^2} \Big/ \left(\frac{\partial \sigma}{\partial s}\right)^2. \tag{6.4.32}$$

Hence:

$$\frac{S}{T}\tau\frac{\partial \tau}{\partial t} = \frac{1}{\sigma}\frac{\partial^2 \sigma}{\partial s^2}\left(\frac{\partial \sigma}{\partial s}\right)^2 = k = \text{const}$$

which is equivalent to the two ordinary equations:

$$\tau\frac{d\tau}{dt} = k\, T/S; \qquad \frac{1}{\sigma}\frac{d^2\sigma/ds^2}{(d\sigma/ds)^2} = k. \tag{6.4.33}$$

The solution of the first equation in (6.4.33) is:

$$\tau = \sqrt{2kTt/S + C_1}; \qquad C_1 = \text{const.} \tag{6.4.34}$$

From the second equation of (6.4.33) we obtain:

$$\sigma = [d^2\sigma/ds^2]/[k(d\sigma/ds)^2]. \tag{6.4.35}$$

This means that given a value of s, and having solved (6.4.35), we obtain a unique value for $\sigma(s)$, constant for the entire flow domain. Hence, the solution for x takes the form:

$$x = \sigma(s)\sqrt{2kTt/S + C_1}.$$

If the advancing front starts from the source of disturbance (a well or a boundary) such that at $x = 0$, $t = 0$ for the given s, we obtain $C_1 = 0$. Hence:

$$x = \sigma(s)\sqrt{\frac{2kTt}{S'}}; \qquad \frac{x^2}{t} = \frac{\sigma^2(s)S}{2KT} = \lambda^2(s), \qquad \lambda(s) = \sigma(s)\sqrt{\frac{S}{2T}} \tag{6.4.36}$$

where $\lambda(s)$ is a single-valued function of s only.

Thus, a given drawdown s will propagate in the flow domain such that x^2/t equals a constant which depends on s and on the medium's properties S and T.

The development presented above suggests that:

$$\lambda(s) = xt^{-1/2} \tag{6.4.37}$$

be considered as a new single variable of the problem on hand. With this new variable (6.4.27) becomes:

$$\frac{T}{S}\frac{d^2s}{d\lambda^2} + \frac{\lambda}{2}\frac{ds}{d\lambda} = 0. \tag{6.4.38}$$

The solution $\lambda = \lambda(s)$ of (6.4.38), which may be presumed to be a unique one, can be transformed back into $s = s(x, t)$ by means of (6.4.37).

The transformation $\lambda(s) = xt^{-1/2}$ is known as the *Boltzmann transformation* (Boltzmann 1894; Crank 1956; Carslaw and Jaeger 1959), originally suggested by the fact that many solutions of problems with constant diffusivity (here T/S) are functions of $(xt^{-1/2})$ only. The usefulness of this transformation is limited by the fact that initial and boundary conditions must also be expressible in terms of λ only.

For $\sqrt{T/S} =$ const, an elementary solution of (6.4.38) is

$$\lambda = A\,\text{erfc}(\lambda/2\sqrt{T/S}) = A\,\text{erfc}(x/2\sqrt{Tt/S}) = A\,\text{erfc}\,u^{1/2}; \quad u = Sx^2/4Tt. \tag{6.4.39}$$

Another approach to the derivation of the Boltzmann transformation is mentioned in paragraph 9.4.6.

6.5 Stream Functions

In chapter 5 the description of flow through porous media was based on the concept of the potential or the piezometric head, recognizing that it is the difference in piezometric head that causes flow. The piezometric head is a measurable quantity determined by using piezometers or observation wells. The motion equation (e.g., in the form of Darcy's law), relating a flux to the head gradient, was accepted as an experimental law or was derived analytically by introducing a certain constitutive assumption and then verified experimentally.

Instead of observing head variations in an experiment and relating the flux to them, one may start by observing streamlines in an experiment of steady flow through a porous medium domain, and then relate the flux to these observations (Zaslavsky 1962). The streamline, as well as the pathline and streakline, is to be understood as an average concept.

In practical terms, it is impossible to label a single fluid particle (say in an experiment of flow through a porous medium) and observe its motion. Instead we label a group of particles occupying a small neighborhood, or we continuously inject a tracer into a point in a steadily moving fluid. In laminar flow, in spite of hydrodynamic dispersion (chap. 10), and in the case of a continuous injection, in spite of the lateral dispersion, it is possible to define the average path of the particles and to use it in describing the flow.

It will be shown in the following paragraphs that a stream function may be derived by assuming the existence of pathlines and streamlines. Then the flux components may be expressed in terms of this stream function (Zaslavsky 1962).

6.5.1 Pathlines, Streamlines, Streaklines and Fronts

A *pathline* of a fluid particle is the locus of its positions in space as time passes. It is thus the *trajectory* of a particle of *fixed identity*. It is a Lagrangian concept (par. 4.1.3). The pathline is described by the solutions of the three parametric equations:

$$dx/V_x(x, y, z, t) = dy/V_y(x, y, z, t) = dz/V_z(x, y, z, t) = dt \tag{6.5.1}$$

(par. 4.1.3).

The solution of (6.5.1) can also be written in the parametric form:

$$\lambda = \lambda(x, y, z, t); \qquad \chi = \chi(x, y, z, t), \qquad \omega = \omega(x, y, z, t)$$

where $\omega =$ const, $\lambda =$ const and $\chi =$ const describe surfaces in space. Each set of three such surfaces defines (at their intersection) the position of a particle in space at time t. Together they describe the pathlines of the different particles. In order for ω, λ and χ to be functionally independent, we require that the Jacobian J satisfies:

$$J \equiv \frac{\partial(\lambda, \chi, \omega)}{\partial(x, y, z)} \neq 0 \quad \text{or} \quad J \equiv \nabla\omega \cdot \nabla\lambda \times \nabla\chi \neq 0 \tag{6.5.2}$$

(see (4.1.20)).

Yih (1957) and Nelson (1963) also present these solutions for pathlines.

In flow through porous media, V_x, V_y, V_z are components of the average velocity vector, with $V_x = q_x/n$, $V_y = q_y/n$, $V_z = q_z/n$.

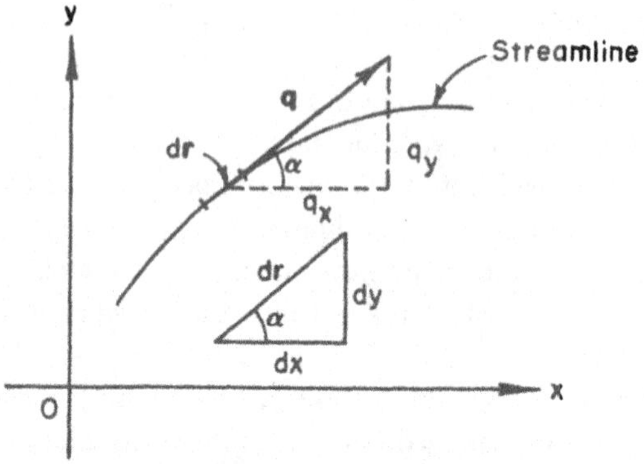

FIG. 6.5.1. Streamline and specific discharge in plane flow.

At any instant of time, there is at every point in a flow domain a velocity (or specific discharge) vector with a definite direction. The *instantaneous* curves that are at every point tangent to the direction of the velocity at that point are called *streamlines* of the flow. The mathematical expression defining a streamline is therefore $\mathbf{q} \times d\mathbf{r} = 0$, or:

$$\frac{dx}{q_x(x, y, z, t_0)} = \frac{dy}{q_y(x, y, z, t_0)} = \frac{dz}{q_z(x, y, z, t_0)} \tag{6.5.3}$$

(fig. 6.5.1) where t_0 indicates a certain time, and $d\mathbf{r}$ is an element of arc along a streamline. Equation (6.5.3) is valid for both isotropic and anisotropic media. For a flow described by (5.6.23), i.e., where x, y, z are the principal directions of permeability, equation (6.5.3) becomes:

$$dx/(K_x\, \partial\varphi/\partial x) = dy/(K_y\, \partial\varphi/\partial y) = dz/(K_z\, \partial\varphi/\partial z). \tag{6.5.4}$$

For an isotropic medium, $K_x = K_y = K_z = K$ in (6.5.4).

Had we started from postulating the presence of an equipotential surface $\varphi =$ const, we could consider an elementary displacement $d\mathbf{s}$ in the direction normal to this surface. Then:

$$\text{grad}\, \varphi \times d\mathbf{s} = 0; \qquad \frac{dx}{\partial\varphi/\partial x} = \frac{dy}{\partial\varphi/\partial y} = \frac{dz}{\partial\varphi/\partial z} \tag{6.5.5}$$

which define curves in space normal to the equipotential surfaces. These are the streamlines. Multiplying (6.5.4), written for an isotropic medium, by K, we obtain (6.5.5). Thus, in an isotropic medium, streamlines are perpendicular to the equipotential surfaces. Since the differential equations (6.5.4) define what happens at a point, we may have $K = K(x, y, z)$, i.e., a nonhomogeneous medium. These two independent approaches to the definition of streamlines are possible; both are based on observations of field phenomena.

In *steady flow*, i.e., one in which flow characteristics remain invariant with time, streamlines and pathlines coincide. In *unsteady flow* they may be distinct. In unsteady flow we can speak only of an instantaneous picture of the streamlines, as the picture varies continuously.

A *streakline* represents the locus of points occupied, or to be occupied, by fluid particles, all of which pass through a certain fixed point in space at some previous time. It is most easily visualized as a dye stream in water or a smoke filament in air. In unsteady flow, streaklines differ from streamlines. In steady flow they coincide.

A *front* (*fluid surface* or *material surface*) is a surface (or curve in two-dimensional flow) that always consists of the same particles. Let the particles forming a continuous surface within the flow domain be labeled by a tracer at some initial time $t = 0$. It is then possible to determine the subsequent positions of this surface by tracing the paths of the individual fluid particles comprising it (chap. 9). The equation of a front may be written as $F(x, y, z, t) = 0$, and (4.1.30) is applicable.

The *interface* is a special case of a front—it separates two immiscible liquids. Prandtl and Tietjens (1934) proved a theorem, attributable to Lagrange, that an interface (including an impervious boundary that is a fluid–solid interface) always consists of the same particles (see sec. 7.1).

6.5.2 The Stream Function in Two-Dimensional Flow

In two-dimensional flow in the xy plane (6.5.3) becomes (fig. 6.5.1):

$$dx/q_x = dy/q_y, \quad \text{or} \quad q_y\,dx - q_x\,dy = 0. \tag{6.5.6}$$

The solution of (6.5.6) is

$$\Psi = \Psi(x, y) = \text{const} \tag{6.5.7}$$

which describes the instantaneous geometry of the streamlines. The condition for (6.5.6) to be an exact differential of some function $\Psi = \Psi(x, y)$ is $\partial q_x/\partial x + \partial q_y/\partial y = 0$, which is nothing but the continuity equation in this case: div $\mathbf{q} = 0$. Since div $\mathbf{q} = 0$ describes flow of an incompressible fluid in a nondeformable medium (i.e., $\partial(\rho n)/\partial t = 0$) the stream function Ψ as defined here is valid *only* for such a case. From (6.5.6), written as $dy/dx = q_y/q_x = f(x, y)$, it follows that it defines for any point in the xy plane an angle, $\alpha = \tan^{-1} f(x, y)$, which the tangent to the integral curve (6.5.7) makes with the $+x$ axis. Equation (6.5.7) describes a family of curves, for various values of the constant, all of which have the prescribed direction in the plane.

Since Ψ is an exact differential, along any streamline we have:

$$d\Psi = \frac{\partial \Psi}{\partial x}\,dx + \frac{\partial \Psi}{\partial y}\,dy = q_y\,dx - q_x\,dy = 0 \tag{6.5.8}$$

from which we may obtain the relationships for the specific discharge components:

$$q_x = -\,\partial\Psi/\partial y; \qquad q_y = \partial\Psi/\partial x. \tag{6.5.9}$$

The function $\Psi = \Psi(x, y)$, which is *constant along streamlines* (or $d\Psi = 0$), is called the *stream function* of two-dimensional flow (dim L^2/T). It is sometimes called the *Lagrange stream function*, as it was introduced by Lagrange as he sought

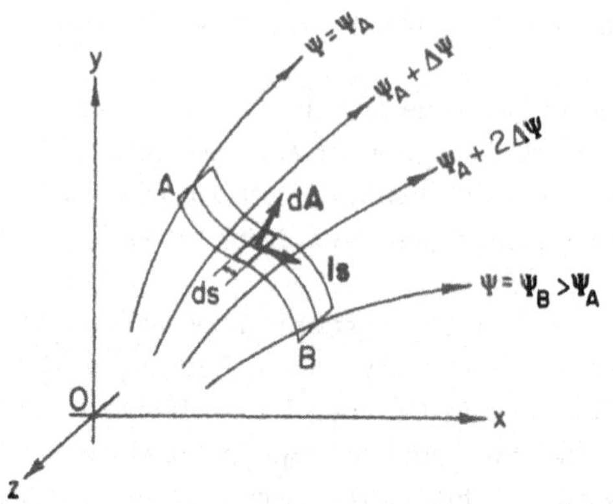

FIG. 6.5.2. The relationship between the stream function and the discharge.

a solution to the differential equation of a streamline (6.5.3) in two-dimensional flow.

The physical interpretation of the *generating function* Ψ (as it is used to generate the specific discharge) may be obtained as follows. Figure 6.5.2 shows some streamlines labelled $\Psi_A, \Psi_A + \Delta\Psi, \Psi_A + 2\,\Delta\Psi$, etc. The direction of increasing Ψ (or grad Ψ) is chosen such that grad Ψ is obtained from grad Φ by a counterclockwise rotation. Let $d\mathbf{A}$ be an element of area of unit width (in the z direction) and length ds, where ds is the length of an element of arc of an arbitrary path connecting points A and B (on the streamlines Ψ_A and Ψ_B, respectively). The direction of $d\mathbf{A}$ is perpendicular to $d\mathbf{s}$ (where $d\mathbf{A} = dA\,\mathbf{1n}$, $d\mathbf{s} = ds\mathbf{1s}$ and $\mathbf{1n}$ is obtained from $\mathbf{1s}$ by a counterclockwise rotation). We therefore have: $d\mathbf{A} = \mathbf{1z} \times d\mathbf{s}$ (i.e., $\mathbf{1s}$, $\mathbf{1n}$ and $\mathbf{1z}$ form a right-handed rectangular system). Consider the integral:

$$Q_{AB} = \int_A^B \mathbf{q} \cdot d\mathbf{A} = \int_A^B \mathbf{q}(\mathbf{1z} \times d\mathbf{s}) = -\int_A^B q_x\,dy - q_y\,dx = -\int_A^B d\Psi = \Psi_A - \Psi_B. \qquad (6.5.10)$$

Thus the total discharge between two streamlines (actually two stream surfaces and the planes $z = 0$, $z = 1$) is given by the difference between the values of the stream functions corresponding to these lines. The dimensions of Ψ are volume per unit time per unit width. It is interesting to note that the *path* chosen for the integration between A and B is *immaterial*, as the result of the integration in (6.5.10) depends only on the boundary points A and B. This means that Ψ may be regarded as an indeterminate function to the extent of an additive constant. According to the sign convention employed here, $\Psi_A < \Psi_B$. An impervious boundary of a flow domain, with the flow always tangential to it, invariably coincides with a streamline. One can always start labeling the streamlines from an impervious boundary.

6.5.3 The Stream Functions in Three-Dimensional Flow

The solution of the two equations in (6.5.3) that describe the geometry of streamlines in space takes the form of two independent relationships:

$$\lambda = \lambda(x, y, z) = \text{const}$$

$$\chi = \chi(x, y, z) = \text{const}. \tag{6.5.11}$$

These represent two families of surfaces whose intersections are the streamlines. Along each streamline, both χ and λ are constant. Since the streamlines are imbedded in these surfaces, the latter are called *stream surfaces*. The functions λ and χ are called *stream functions of three-dimensional flow*. In mathematics, the streamline thus obtained is called the *characteristic curve* of (6.5.3).

The equation of a streamline may, accordingly, be written as:

$$\Psi[\lambda(x, y, z), \chi(x, y, z)] = 0. \tag{6.5.12}$$

Since on each of the stream surfaces, the corresponding stream function has a constant value, and the flow is tangential to this surface, we have:

$$q_n = \mathbf{q} \cdot \mathbf{1n} = 0 \tag{6.5.13}$$

for each of these surfaces. In view of (7.1.3), this becomes:

$$q_x \frac{\partial \lambda}{\partial x} + q_y \frac{\partial \lambda}{\partial y} + q_z \frac{\partial \lambda}{\partial z} = 0; \qquad q_x \frac{\partial \chi}{\partial x} + q_y \frac{\partial \chi}{\partial y} + q_z \frac{\partial \chi}{\partial z} = 0. \tag{6.5.14}$$

These two equations contain the same three specific discharge components q_x, q_y, q_z. Any one of the latter can be eliminated by a simultaneous solution to yield:

$$q_x : q_y : q_z = \left(\frac{\partial \lambda}{\partial y}\frac{\partial \chi}{\partial z} - \frac{\partial \lambda}{\partial z}\frac{\partial \chi}{\partial y}\right) : \left(\frac{\partial \lambda}{\partial z}\frac{\partial \chi}{\partial x} - \frac{\partial \lambda}{\partial x}\frac{\partial \chi}{\partial z}\right) : \left(\frac{\partial \lambda}{\partial x}\frac{\partial \chi}{\partial y} - \frac{\partial \lambda}{\partial y}\frac{\partial \chi}{\partial x}\right) \tag{6.5.15}$$

which may be expressed alternatively by introducing a function $a = a(x, y, z)$:

$$\left.\begin{aligned} q_x &= a\left(\frac{\partial \lambda}{\partial y}\frac{\partial \chi}{\partial z} - \frac{\partial \lambda}{\partial z}\frac{\partial \chi}{\partial y}\right) \\ q_y &= a\left(\frac{\partial \lambda}{\partial z}\frac{\partial \chi}{\partial x} - \frac{\partial \lambda}{\partial x}\frac{\partial \chi}{\partial z}\right) \\ q_z &= a\left(\frac{\partial \lambda}{\partial x}\frac{\partial \chi}{\partial y} - \frac{\partial \lambda}{\partial y}\frac{\partial \chi}{\partial x}\right) \end{aligned}\right\} \quad \text{or} \quad \mathbf{q} = a(x, y, z)\, \text{grad}\, \lambda \times \text{grad}\, \chi. \tag{6.5.16}$$

For incompressible flow, since $\mathbf{q}$ satisfies div $\mathbf{q} = 0$ (par. 6.2.2), a can be a function of χ and λ, only. It can be shown that there is no loss of generality in taking $a \equiv 1$. Yih (1957), in presenting the theory given above, shows that for steady flow of a compressible fluid $a = 1/\rho$.

For two-dimensional flow, say in the xy plane, planes $z = \text{const}$ play the role of stream surfaces. With $\lambda = z$, we obtain:

$$q_x = -(\partial \chi/\partial y); \qquad q_y = (\partial \chi/\partial x) \tag{6.5.17}$$

which are the same equations as (6.5.9) when $\chi \equiv \Psi$ (the Lagrange stream function).

A physical interpretation of the stream functions in three-dimensional flow may be obtained as follows. Figure 6.5.3 shows two pairs of stream surfaces: χ, $\chi + \Delta\chi$, λ, $\lambda + \Delta\lambda$. The directions of $\mathbf{q}$, $\nabla\lambda$ and $\nabla\chi$ are chosen so that they form a right-handed coordinate system. The four stream surfaces form a stream tube in space of cross-sectional area dS (in a surface normal to the streamlines). The discharge through this tube is given by:

$$Q = -\iint_{(dS)} \mathbf{q}\cdot d\mathbf{S}; \qquad dS = d\nu\, d\mu \sin\theta;$$

$$|q| = |\operatorname{grad}\chi|\,|\operatorname{grad}\lambda| \sin\theta; \qquad \operatorname{grad}\chi = \frac{d\chi}{d\nu}\frac{1}{\sin\theta}, \qquad \operatorname{grad}\lambda = \frac{d\lambda}{d\mu}\frac{1}{\sin\theta}$$

$$Q = -\iint_{(dS)} \frac{d\chi}{d\nu}\frac{1}{\sin\theta}\frac{d\lambda}{d\mu}\frac{1}{\sin\theta}\sin^2\theta\, d\nu\, d\mu = -\int_{\lambda}^{\lambda+\Delta\lambda}\int_{\chi}^{\chi+\Delta\chi} d\chi\, d\lambda$$

$$= -\Delta\chi\,\Delta\lambda = -(\chi_2 - \chi_1)(\lambda_2 - \lambda_1) \tag{6.5.18}$$

where $d\mathbf{S} = \mathbf{1n}\, dS$, and $\mathbf{1n}$ is in the direction of the streamlines. Thus, equation (6.5.18) is a generalization of the two-dimensional expression (6.5.10) derived for Ψ.

The pair of constants in (6.5.11) must be interrelated by an arbitrary function to obtain a complete solution of (6.5.3). This is done by using the boundary conditions of the flow domain. For example, an impervious boundary is a stream surface, as the flow is everywhere tangential to it.

Axially symmetric flow is one in which the flow takes place in a series of planes, all of which pass through a common line—the *axis of symmetry*—and where the flow

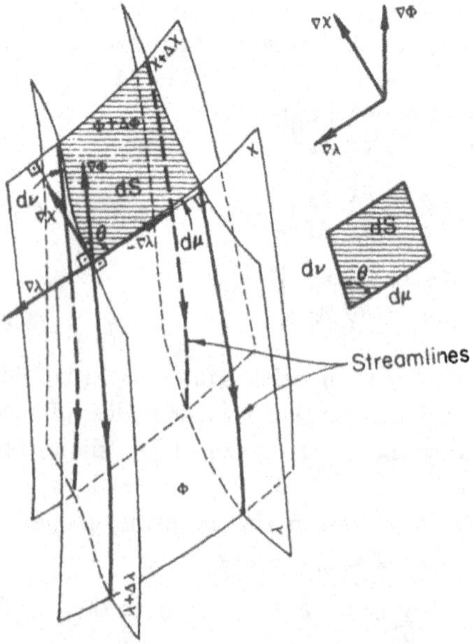

FIG. 6.5.3. Streamlines and stream surfaces in three-dimensional flow in a homogeneous isotropic medium.

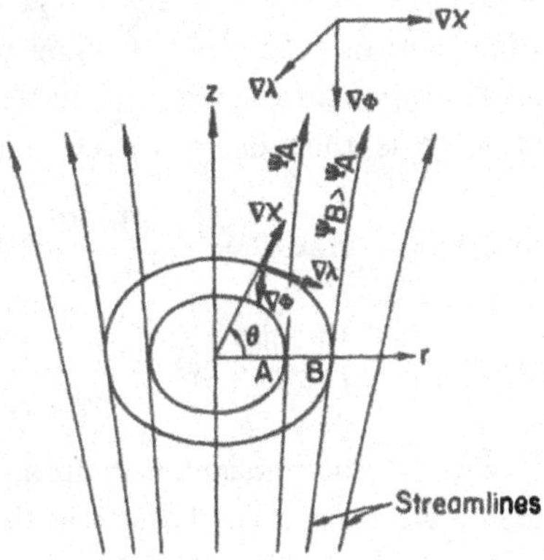

FIG. 6.5.4. Stokes' stream function in axially symmetric flow.

pattern is identical in each of these planes. Thus, these meridional planes may be taken as one of the stream surfaces expressed by (6.5.11). By setting $\lambda = -\theta$ in (6.5.16) written in cylindrical coordinates (r, θ, z) we obtain for $a \equiv 1$:

$$\mathbf{q} = \begin{vmatrix} \mathbf{1r} & \mathbf{1\theta} & \mathbf{1z} \\ \dfrac{\partial\lambda}{\partial r} & \dfrac{\partial\lambda}{r\,\partial\theta} & \dfrac{\partial\lambda}{\partial z} \\ \dfrac{\partial\chi}{\partial r} & \dfrac{1}{r}\dfrac{\partial\chi}{\partial\theta} & \dfrac{\partial\chi}{\partial z} \end{vmatrix} = \begin{vmatrix} \mathbf{1r} & \mathbf{1\theta} & \mathbf{1z} \\ 0 & -\dfrac{1}{r} & 0 \\ \dfrac{\partial\chi}{\partial r} & \dfrac{1}{r}\dfrac{\partial\chi}{\partial\theta} & \dfrac{\partial\chi}{\partial z} \end{vmatrix} \qquad (6.5.19)$$

from which:

$$q_r = -\frac{1}{r}\frac{\partial\chi}{\partial z}, \qquad q_\theta = 0, \qquad q_z = \frac{1}{r}\frac{\partial\chi}{\partial r}. \qquad (6.5.20)$$

The stream function χ here (dim L^3/T) is identical to *Stokes' stream function* for axially symmetric flow. The latter is also often denoted by Ψ. Originally, Stokes derived this stream function for axially symmetric flow of an incompressible fluid in a manner similar to the development of the Lagrange stream function for two-dimensional plane flow, namely, by considering the discharge passing through an arbitrary curve connecting two points in the plane. This approach is given in all hydrodynamic textbooks (e.g., Rouse 1959) and is not repeated here.

With the nomenclature of figure 6.5.4, the discharge through the ring between the two cylindrical stream surfaces (denoted by Ψ) passing through points A and B is:

$$Q_{AB} = \int_A^B \mathbf{q} \cdot d\mathbf{A} = \int_A^B \mathbf{q} \cdot 2\pi r\, dr\, \mathbf{1z} = 2\pi \int_A^B \frac{1}{r}\frac{\partial\Psi}{\partial r} r\, dr = 2\pi(\Psi_B - \Psi_A). \quad (6.5.21)$$

It thus has the same physical interpretation as the stream functions considered above.

In the two-dimensional case (homogeneous medium), as well as in the axisymmetrical case, we have one component of the specific discharge vanishing throughout the flow domain: $q_z = 0$ and $q_\theta = 0$, respectively. Zaslavsky (1962) extended these considerations to three-dimensional orthogonal curvilinear coordinates, y^i, in which one of the specific discharge components, say $q_3 \equiv 0$ in the entire flow, while $q_1 \neq 0$, $q_2 \neq 0$. A stream function Ψ is then defined such that:

$$\mathbf{q} \equiv q_1 \mathbf{1y}^1 + q_2 \mathbf{1y}^2 = -\operatorname{grad} \Phi = -\left(\frac{1}{h_1}\frac{\partial \Phi}{\partial y^1}\mathbf{1y}^1 + \frac{1}{h_2}\frac{\partial \Phi}{\partial y^2}\mathbf{1y}^2\right)$$

$$= \mathbf{1y}^3 \times \operatorname{grad} \Psi = -\frac{1}{h_2 h_3}\frac{\partial \Psi}{\partial y^2}\mathbf{1y}^1 + \frac{1}{h_1 h_3}\frac{\partial \Psi}{\partial y^1}\mathbf{1y}^2 \tag{6.5.22}$$

where h_i are the scale factors of the considered coordinate system (see some text on curvilinear coordinates). He then shows that the discharge Q between points 1 and 2 along a streamline $\Psi = \text{const}$ in the y^3 coordinate surface is given by:

$$dQ = \mathbf{q} \cdot \mathbf{1n}\, dA = dy^3\left(\frac{\partial \Psi}{\partial y^1} dy^1 + \frac{\partial \Psi}{\partial y^2} dy^2\right) = d\Psi\, dy^3$$

$$\mathbf{1n}\, dA = h_3\, dy^3 [h_2\, dy^2 \mathbf{1y}^1 - h_1\, dy^1 \mathbf{1y}^2] \tag{6.5.23}$$

which, upon integrating twice, along the lines connecting points 1 and 2 and along a length $(b_2 - b_1)$ in the y^3 direction, yields:

$$Q = (b_2 - b_1)(\Psi_2 - \Psi_1)$$

where Ψ is independent of the y^3 coordinate.

Thus, through a unified approach, the stream functions of Lagrange and Stokes have been shown to be special cases of the general definition of the stream functions in three-dimensional flow. The presentation here follows that of R. von Mises (1908), Prazil (1913), Lagally (1927), Yih (1957), Zaslavsky (1962) and Nelson (1963). Robertson (1965) comments that an indication of this approach is mentioned by Euler as early as 1755, and by Jacobi and Clebsch around 1850.

6.5.4 The Partial Differential Equations for the Lagrange and Stokes Stream Functions

Consider two-dimensional flow in an isotropic homogeneous medium in the xy plane. Since $\mathbf{q} = -\operatorname{grad} \Phi$ $(\Phi = K\varphi)$ and the flow is irrotational (potential), we have $\operatorname{curl} \mathbf{q} = 0$ (par. 5.4.2). By inserting (6.5.9) in $(\operatorname{curl} \mathbf{q})_z = 0$, we obtain:

$$\nabla^2 \Psi \equiv \partial^2 \Psi / \partial x^2 + \partial^2 \Psi / \partial y^2 = 0 \tag{6.5.24}$$

which is the partial differential equation for Ψ. Note that in this case, no equation can be obtained from $\operatorname{div} \mathbf{q} = 0$. Thus, Ψ in the present flow also satisfies the Laplace equation, and hence may be considered as the specific discharge potential corresponding to a second flow having the curves $\Phi = \text{const}$ as streamlines.

In the case of an inhomogeneous (yet isotropic) medium, $K = K(x, y, z)$, we have pseudopotential flow, with:

$$\mathbf{q} = -K(x, y, z) \operatorname{grad} \varphi; \qquad \mathbf{J} = -\operatorname{grad} \varphi = \mathbf{q}/K. \tag{6.5.25}$$

Hence:

$$\operatorname{curl} \mathbf{J} = \operatorname{curl}(\mathbf{q}/K) = 0$$

$$(\operatorname{curl} \mathbf{q}/K)_z = \frac{\partial}{\partial x}\left(\frac{q_y}{K}\right) - \frac{\partial}{\partial y}\left(\frac{q_x}{K}\right) = 0.$$

By inserting (6.5.9) in the last equation, we obtain:

$$\frac{\partial}{\partial x}\left(\frac{1}{K}\frac{\partial \Psi}{\partial x}\right) + \frac{\partial}{\partial y}\left(\frac{1}{K}\frac{\partial \Psi}{\partial y}\right) = 0$$

$$\frac{1}{K}\nabla^2\Psi - \frac{1}{K^2}\left(\frac{\partial K}{\partial x}\frac{\partial \Psi}{\partial x} + \frac{\partial K}{\partial y}\frac{\partial \Psi}{\partial y}\right) = 0$$

$$K\,\nabla^2\Psi - \operatorname{grad} K \cdot \operatorname{grad} \Psi = 0, \qquad \nabla^2\Psi - \operatorname{grad} \ln K \cdot \operatorname{grad} \Psi = 0. \tag{6.5.26}$$

When $\nabla K \perp \nabla\Psi$, Ψ satisfies $\nabla^2\Psi = 0$. In the case of a Helmholtz inhomogeneous medium considered in paragraph 6.2.2, equation (6.5.26) may be simplified by introducing the transformation:

$$\Psi' = \Psi K^{-1/2}. \tag{6.5.27}$$

We then obtain:

$$\nabla^2\Psi' - \frac{1}{K^{-1/2}}(\nabla^2 K^{-1/2})\Psi' = 0. \tag{6.5.28}$$

If $K^{-1/2}$ is a solution of:

$$\nabla^2 K^{-1/2} \pm g(x, y)K^{-1/2} = 0 \tag{6.5.29}$$

then Ψ' satisfies:

$$\nabla^2\Psi' \pm g(x, y)\Psi' = 0. \tag{6.5.30}$$

If the medium is also anisotropic ($K_x \neq K_y$), let us use (5.6.18):

$$\mathbf{J} = -\operatorname{grad}\varphi = \underline{\mathbf{W}} \cdot \mathbf{q},$$

with x and y as principal directions. Then:

$$\mathbf{J} = W_x q_x \mathbf{1x} + W_y q_y \mathbf{1y}$$

where $W_x \equiv W_{xx}$, $W_y \equiv W_{yy}$. Then, from $\operatorname{curl} \mathbf{J} = \operatorname{curl}(-\operatorname{grad}\varphi) = 0$, we obtain:

$$\frac{\partial}{\partial x}(W_y q_y) - \frac{\partial}{\partial y}(W_x q_x) = \frac{\partial}{\partial x}\left(W_y\frac{\partial \Psi}{\partial x}\right) + \frac{\partial}{\partial y}\left(W_x\frac{\partial \Psi}{\partial y}\right)$$

$$= \frac{\partial}{\partial x}\left(\frac{1}{K_y}\frac{\partial \Psi}{\partial x}\right) + \frac{1}{\partial y}\left(\frac{1}{K_x}\frac{\partial \Psi}{\partial y}\right) = 0$$

$$K_x\frac{\partial^2\Psi}{\partial x^2} + K_y\frac{\partial^2\Psi}{\partial y^2} - \left(K_x\frac{\partial \ln K_y}{\partial x}\frac{\partial \Psi}{\partial x} + K_y\frac{\partial \ln K_x}{\partial y}\frac{\partial \Psi}{\partial y}\right) = 0. \tag{6.5.31}$$

For a homogeneous anisotropic medium, this becomes:

$$K_x\, \partial^2\Psi/\partial x^2 + K_y\, \partial^2\Psi/\partial y^2 = 0. \tag{6.5.32}$$

For axially symmetric flow (in the z direction), we obtain from curl $\mathbf{q} = 0$ with:

$$q_1 = q_r, \qquad q_2 = q_\theta, \qquad q_3 = q_z; \qquad q_\theta = 0, \qquad \partial(\ )/\partial\theta = 0$$

$$\partial q_r/\partial z - \partial q_z/\partial r = 0. \tag{6.5.33}$$

And by inserting (6.5.20), where $\chi = \Psi$ is the Stokes stream function, we obtain:

$$\frac{\partial^2\Psi}{\partial r^2} - \frac{1}{r}\frac{\partial\Psi}{\partial r} + \frac{\partial^2\Psi}{\partial z^2} = 0; \qquad \nabla^2{}_{r,\theta,z}\Psi - \frac{2}{r}\frac{\partial\Psi}{\partial r} = 0. \tag{6.5.34}$$

It is important to notice that (6.5.34) is basically different from

$$\nabla^2{}_{r,\theta,z}\Phi \equiv \frac{\partial^2\Phi}{\partial r^2} + \frac{1}{r}\frac{\partial\Phi}{\partial r} + \frac{\partial^2\Phi}{\partial z^2}$$

in the sign preceding the term $(1/r)\, \partial\Psi/\partial r$ so that $\nabla^2\Psi \neq 0$ in this case. Therefore, it is impossible to combine Φ and Ψ in this case of axially symmetric flow to form the complex function $\Phi + i\Psi$ of the variable $\rho + iz$ as in the two-dimensional case (sec. 7.8).

In the case of a nonhomogeneous medium, the flow is pseudopotential, and the equation for the stream function (denoted by Ψ), derived from curl $\mathbf{J} = 0$, with $\partial(\ )/\partial\theta = 0$, becomes:

$$\partial J_r/\partial z - \partial J_z/\partial r = 0; \qquad \frac{\partial}{\partial z}\left(\frac{q_r}{K}\right) - \frac{\partial}{\partial r}\left(\frac{q_z}{K}\right) = 0 \tag{6.5.35}$$

$$\nabla^2{}_{r,\theta,z}\Psi - \frac{2}{r}\frac{\partial\Psi}{\partial r} - \frac{1}{K}\left(\frac{\partial\Psi}{\partial r}\frac{\partial K}{\partial r} + \frac{\partial\Psi}{\partial z}\frac{\partial K}{\partial z}\right) = 0. \tag{6.5.36}$$

If the medium is also anisotropic: $K_r \neq K_z$, we obtain:

$$K_r\frac{\partial^2\Psi}{\partial r^2} - K_r\frac{1}{r}\frac{\partial\Psi}{\partial r} + K_z\frac{\partial^2\Psi}{\partial z^2} - \left(\frac{K_z}{K_r}\frac{\partial K_r}{\partial z}\frac{\partial\Psi}{\partial z} + \frac{K_r}{K_z}\frac{\partial K_z}{\partial z}\frac{\partial\Psi}{\partial z}\right) = 0. \tag{6.5.37}$$

For a homogeneous anisotropic medium, we obtain:

$$K_r\left(\frac{\partial^2\Psi}{\partial r^2} - \frac{1}{r}\frac{\partial\Psi}{\partial r}\right) + K_z\frac{\partial^2\Psi}{\partial z^2} = 0. \tag{6.5.38}$$

In three-dimensional flow it is possible to start from (6.5.16) and derive two simultaneous partial differential equations for λ and χ from curl $\mathbf{q} = 0$, or from curl $\mathbf{J} = 0$.

It is also possible to use $\mathbf{q}$ from (6.5.22) in order to obtain a partial differential equation for Ψ in a homogeneous medium in terms of the special y^i coordinates from curl $\mathbf{q} = 0$ (Zaslavsky 1962):

$$\frac{\partial}{\partial y^1}\left(\frac{h_2}{h_1 h_3}\frac{\partial\Psi}{\partial y^1}\right) + \frac{\partial}{\partial y^2}\left(\frac{h_1}{h_2 h_3}\frac{\partial\Psi}{\partial y^2}\right) = 0. \tag{6.5.39}$$

For a nonhomogeneous medium we insert $\mathbf{q}$ from (6.5.22) into curl$(\mathbf{q}/K) = 0$.

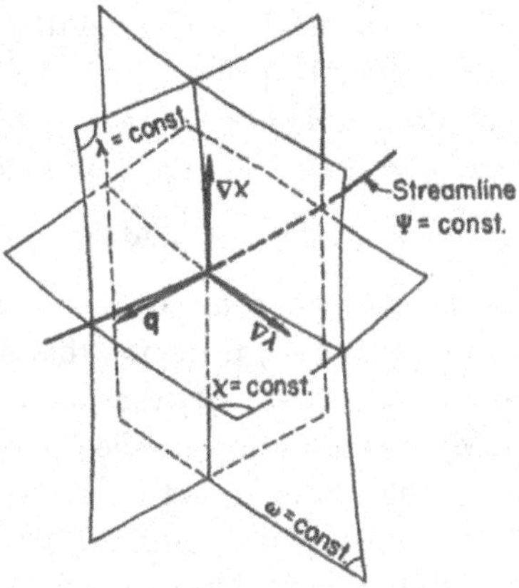

FIG. 6.5.5. Natural coordinates for three-dimensional flow. λ = const. and χ = const. are stream surfaces.

The partial differential equation describing flow of an incompressible inhomogeneous fluid is developed in section 6.7.

6.5.5 The Relationships Between the Potential and the Stream Functions

From $\mathbf{q} = \nabla\lambda \times \nabla\chi$ it follows that $\mathbf{q}$, λ and χ form a right-hand orthogonal coordinate system. The three coordinate surfaces are λ = const, χ = const, and a third surface, ω = const, which is everywhere normal to the $\mathbf{q}$ (fig. 6.5.5). $\mathbf{1\omega}$, $\mathbf{1\lambda}$ and $\mathbf{1\chi}$ are thus the *natural coordinates* of the flow constructed at every point of the streamline.

In two-dimensional flow, it is sometimes convenient to choose the two mutually orthogonal families of curves Φ = const and Ψ = const as coordinate curves. We then refer to these coordinates as *natural*, or *intrinsic coordinates*. Starting from $\mathbf{q} = -\operatorname{grad}\Phi$, we find that $1/h_1 = 1/h_2 = q$. In the natural coordinates:

$$\nabla^2(\ ) = q^2[\partial^2(\ )/\partial\Phi^2 + \partial^2(\ )/\partial\Psi^2] \tag{6.5.40}$$

$$\nabla^2(\ ) = q\left[\frac{\partial}{\partial s_\Phi}\left(\frac{1}{q}\frac{\partial(\ )}{\partial s_\Phi}\right) + \frac{\partial}{\partial s_\Psi}\left(\frac{1}{q}\frac{\partial(\ )}{\partial s_\Psi}\right)\right] \tag{6.5.41}$$

where s_Φ and s_Ψ are lengths measured along the coordinate axes Φ and Ψ, respectively.

In three-dimensional flow, the natural coordinates at a point on a streamline are (a) in the direction $\mathbf{1T}$ tangential to the streamline, (b) in the direction $\mathbf{1N}$ (the direction of $d(\mathbf{1T})/ds$), called the *principal normal*, and (c) in the direction $\mathbf{1B}$, such that $\mathbf{1B} = \mathbf{1T} \times \mathbf{1N}$, called the *binormal*. It is sometimes convenient to write and solve the Laplace equation in these coordinates.

Denoting the stream function maintained constant along a streamline by Ψ = const, where $\Psi = \Psi(\lambda, \chi)$, we have (par. 4.1.4) along the streamline:

$$D\Psi/Dt \equiv (\mathbf{q}/n)\cdot\operatorname{grad}\Psi = 0 \tag{6.5.42}$$

where $\partial\Psi/\partial t$ was deleted because the flow is a steady one. From (6.5.42) it follows that the vectors $\mathbf{q}$ and grad Ψ (which are normal to the surfaces $\Psi = \text{const}$, hence $\omega = \text{const}$) are perpendicular to each other. This statement is independent of the type of relationship we have between $\mathbf{q}$ and φ.

Consider now the case of flow in a *homogeneous isotropic domain*, where $\mathbf{q} = -\operatorname{grad}\Phi$, $\Phi = K\varphi$. From $\mathbf{q} = \operatorname{grad}\lambda \times \operatorname{grad}\chi$ it follows:

$$-\operatorname{grad}\Phi = \operatorname{grad}\lambda \times \operatorname{grad}\chi; \qquad \operatorname{grad}\Phi = \operatorname{grad}\chi \times \operatorname{grad}\lambda$$

so that grad Φ, grad χ and grad λ now form an orthogonal curvilinear system of coordinates at a point. Also from (6.5.42) it follows that since $\operatorname{grad}\Phi \cdot \operatorname{grad}\Psi = 0$, the *streamline is everywhere normal to the equipotential surfaces* $\Phi = \text{const}$. In an isotropic medium, surfaces $\Phi = \text{const}$ are identified with surfaces $\omega = \text{const}$.

In two-dimensional flow in the xy plane, we have identified the direction of $\mathbf{1z}$ with $\mathbf{1\lambda}$ (par. 6.5.3), so that the xy plane coincides with a $\lambda = \text{const}$ plane, and $\Psi \equiv \chi$. From $\operatorname{grad}\Phi = \operatorname{grad}\Psi \times \operatorname{grad} z$, we obtain:

$$\frac{\partial\Phi}{\partial x} = \frac{\partial\Psi}{\partial y}; \qquad \frac{\partial\Phi}{\partial y} = -\frac{\partial\Psi}{\partial x} \tag{6.5.43}$$

which are the well known *Cauchy–Riemann conditions for two-dimensional flow in a homogeneous isotropic medium.* From (6.5.43), or also directly from (6.5.42), it follows that everywhere in a homogeneous isotropic medium the curves $\Phi = \text{const}$ are orthogonal to $\Psi = \text{const}$ (fig. 6.6.2). This is actually the substance of the Cauchy–Riemann conditions. We may also write the equations:

$$q_x = -(\partial\Phi/\partial x) = -(\partial\Psi/\partial y); \qquad q_y = -(\partial\Phi/\partial y) = (\partial\Psi/\partial x) \tag{6.5.44}$$

which, again, yield the same conditions.

When the discussion is extended to *inhomogeneous, yet isotropic, media,* we have: $\mathbf{q} = -K\operatorname{grad}\varphi$, and (6.5.42) becomes $(K/n)\operatorname{grad}\varphi \cdot \operatorname{grad}\Psi = 0$. This again indicates that at every point *streamlines are normal to surfaces* $\varphi = \textit{const}$. For the two-dimensional case, we have $K\operatorname{grad}\varphi = \operatorname{grad}\Psi \times \operatorname{grad} z$ from which we obtain:

$$K\frac{\partial\varphi}{\partial x} = \frac{\partial\Psi}{\partial y}; \qquad K\frac{\partial\varphi}{\partial y} = -\frac{\partial\Psi}{\partial x}. \tag{6.5.45}$$

It may also be derived from

$$q_x = -K(\partial\varphi/\partial x) = -(\partial\Psi/\partial y);$$

$$q_y = -K(\partial\varphi/\partial y) = \partial\Psi/\partial x.$$

In a *homogeneous isotropic medium*, another stream function ψ is often defined, such that $\psi = \Psi/K$, and in (6.5.10) $Q_{AB}/K = \psi_B - \psi_A$. Then the Cauchy–Riemann conditions for the orthogonal families of curves φ, ψ, become:

$$\partial\varphi/\partial x = \partial\psi/\partial y; \qquad \partial\varphi/\partial y = -\partial\psi/\partial x. \tag{6.5.46}$$

Figure 6.6.2 shows the orthogonality of grad φ and grad ψ.

When the medium is *anisotropic*, with x, y, z as principal directions, we obtain:

$$K_x\frac{\partial\varphi}{\partial x} = \frac{\partial\Psi}{\partial y} = -q_x; \qquad K_y\frac{\partial\varphi}{\partial y} = -\frac{\partial\Psi}{\partial x} = -q_y. \tag{6.5.47}$$

Here,

$$\operatorname{grad}\varphi\cdot\operatorname{grad}\Psi\neq 0 \quad \left(\text{as: } K_x\frac{\partial\varphi}{\partial x}\frac{\partial\Psi}{\partial x}+K_y\frac{\partial\varphi}{\partial y}\frac{\partial\Psi}{\partial y}=0\right)$$

so that the equipotential curves $\varphi=$ const *are not* orthogonal to the streamlines $\Psi=$ const. However, q_x and q_y may still be derived from Ψ by (6.5.9).

In axisymmetric flow, we have for the anisotropic case:

$$q_r=-K_r\frac{\partial\varphi}{\partial r}=-\frac{1}{r}\frac{\partial\Psi}{\partial z}; \qquad q_z=-K_z\frac{\partial\varphi}{\partial z}=\frac{1}{r}\frac{\partial\Psi}{\partial r} \tag{6.5.48}$$

where Ψ is the Stokes stream function. In the homogeneous isotropic case, we may also use the velocity potential $\Phi=K\varphi$ in (6.5.48).

With $\mathbf{q}$ defined by (6.5.22), the relationships between φ and Ψ for the isotropic inhomogeneous case are:

$$\frac{\partial\varphi}{\partial y^1}=\frac{h_1}{h_2h_3}\frac{1}{K}\frac{\partial\Psi}{\partial y^2}; \qquad \frac{\partial\varphi}{\partial y^2}=-\frac{h_2}{h_1h_3}\frac{1}{K}\frac{\partial\Psi}{\partial y^1}. \tag{6.5.49}$$

Equations (6.5.49) may be combined to yield $-K\,\nabla\varphi\cdot\nabla\Psi=0$, i.e., curves $\varphi=$ const are normal to curves $\Psi=$ const.

As a conclusion of the above discussion, it is always preferable to use the piezometric head φ, which is a measurable quantity. The specific discharge potential Φ is a concept applicable only to homogeneous isotropic media. The basic stream function, on the other hand, is Ψ. For homogeneous isotropic media we may also use the stream function ψ.

It is of interest to note that in non-Darcy flow in a two-dimensional homogeneous isotropic porous medium, where the motion equation is (par. 5.11.2):

$$\mathbf{J}=(W+bq)\mathbf{q} \quad \text{or:} \quad \begin{aligned}-\partial\varphi/\partial x&=F(q)q_x\\ -\partial\varphi/\partial y&=F(q)q_y.\end{aligned} \tag{6.5.50}$$

In view of (6.5.9):

$$-(\partial\varphi/\partial x)=F(q)(-\partial\Psi/\partial y); \qquad -(\partial\varphi/\partial y)=F(q)(\partial\Psi/\partial x)$$

and hence:

$$\frac{\partial\varphi/\partial x}{\partial\varphi/\partial y}=-\frac{\partial\Psi/\partial y}{\partial\Psi/\partial x}; \qquad \frac{\partial\varphi}{\partial x}\frac{\partial\Psi}{\partial x}+\frac{\partial\varphi}{\partial y}\frac{\partial\Psi}{\partial y}=0; \qquad \nabla\varphi\cdot\nabla\Psi=0. \tag{6.5.51}$$

This means that also in this case, at every point, equipotential lines $\varphi=$ const are orthogonal to streamlines $\Psi=$ const. As from (6.5.50) it follows that $\mathbf{J}$ $(=-\operatorname{grad}\varphi)$ and $\mathbf{q}$ are colinear, this conclusion is identical to the one derived from (6.5.42).

6.5.6 Solving Problems in the φ–ψ Plane

For certain boundary conditions, it will be more advantageous to solve a problem of plane flow using the φ–ψ network as independent variables (intrinsic coordinates mentioned in par. 6.5.5) instead of x, y or r, θ.

From (6.5.47), by using the laws of implicit function theory regarding change of variables, we obtain for an *inhomogeneous, anisotropic* medium with x, y as principal directions:

$$\frac{1}{K_x}\frac{\partial x}{\partial \varphi} = \frac{\partial y}{\partial \Psi}; \qquad K_y \frac{\partial x}{\partial \Psi} = -\frac{\partial y}{\partial \varphi}. \tag{6.5.52}$$

By differentiating the first equation with respect to φ and the second one with respect to Ψ, or vice versa, and adding, we obtain:

$$\frac{\partial}{\partial \varphi}\left(\frac{1}{K_x}\frac{\partial x}{\partial \varphi}\right) + \frac{\partial}{\partial \Psi}\left(K_y \frac{\partial x}{\partial \Psi}\right) = 0; \qquad \frac{\partial}{\partial \varphi}\left(\frac{1}{K_y}\frac{\partial y}{\partial \varphi}\right) + \frac{\partial}{\partial \Psi}\left(K_x \frac{\partial y}{\partial \Psi}\right) = 0. \tag{6.5.53}$$

For a *homogeneous, anisotropic* medium, we obtain:

$$\frac{1}{K_x}\frac{\partial^2 x}{\partial \varphi^2} + K_y \frac{\partial^2 x}{\partial \Psi^2} = 0; \qquad \frac{1}{K_y}\frac{\partial^2 y}{\partial \varphi^2} + K_x \frac{\partial^2 y}{\partial \Psi^2} = 0. \tag{6.5.54}$$

For a *homogeneous, isotropic* medium, with $K_x = K_y = K$, $\Psi = K\psi$ and $\Phi = K\varphi$:

$$\partial^2 x/\partial \varphi^2 + \partial^2 x/\partial \psi^2 = 0 \quad \text{or} \quad \partial^2 x/\partial \Phi^2 + \partial^2 x/\partial \Psi^2 = 0 \tag{6.5.55}$$

$$\partial^2 y/\partial \varphi^2 + \partial^2 y/\partial \psi^2 = 0 \quad \text{or} \quad \partial^2 y/\partial \Phi^2 + \partial^2 y/\partial \Psi^2 = 0. \tag{6.5.56}$$

Obviously, boundary conditions must be specified in terms of

$$x = x(\varphi, \psi), \qquad y = y(\varphi, \psi), \qquad \partial x/\partial \varphi = f(\varphi, \psi), \quad \text{etc.}$$

Once the solution $x = x(\varphi, \Psi)$ has been derived, it is possible to obtain $y = y(\varphi, \Psi)$ either by integrating the second of (6.5.52) along equipotential curves $\varphi = \text{const}$:

$$y = \int_{(\varphi)} \frac{1}{K_x}\frac{\partial x}{\partial \varphi}\, d\Psi \tag{6.5.57}$$

or by integrating the first of (6.5.52) along streamlines $\Psi = \text{const}$:

$$y = \int_{(\Psi)} K_y \frac{\partial x}{\partial \Psi}\, d\varphi. \tag{6.5.58}$$

The corresponding equations for a homogeneous isotropic medium are:

$$y = \int_{(\varphi)} (\partial x/\partial \varphi)\, d\psi; \qquad y = \int_{(\psi)} (\partial x/\partial \psi)\, d\varphi. \tag{6.5.59}$$

6.6 Flow Nets and Ground Water Contour Maps

6.6.1 The φ–ψ Flow Net

The solution of problems of flow through porous media is usually given in the form of the piezometric head distributions $\varphi = \varphi(x, y, z)$ for steady flow, or $\varphi = \varphi(x, y, z, t)$ for unsteady flow. From these distributions we may define *equipotential* (*or isopiestic*) *surfaces* $\varphi = \varphi(x, y, z) =$ const in three-dimensional flow, or equipotential curves $\varphi = \varphi(x, y) =$ const in two-dimensional flow (say, in the xy plane). In unsteady flow, these surfaces or curves change with time, but at any given instant we have a definite family of surfaces or curves.

As explained in section 6.5, the statement of a problem of flow through a porous medium can be also made, and its solution presented, in terms of the stream function. In three-dimensional flow we have two stream functions λ and χ (par. 6.5.3), with surfaces $\lambda = \lambda(x, y, z) =$ const, being families of stream surfaces. In two-dimensional flow, in the xy plane, for example, we have the stream function Ψ (or $\psi = \Psi/K$ in an isotropic medium), with curves $\Psi =$ const (or $\psi =$ const) as streamlines.

In two-dimensional flow, a plot of equipotentials and streamlines is called a *flow net*. Examples of flow nets are given in figures 7.10.2, 7.10.3, 7.10.4, and 7.10.5. In all these cases the flow net is obtained analytically. In section 7.10 graphic techniques for deriving the flow net are explained. The use of models and analogs of various types for deriving the flow net is explained in chapter 11. Of special interest is the electric analog of the electrolytic tank type (par. 11.5.1).

The relationships between the functions φ and ψ, or Φ and Ψ are discussed in detail in paragraph 6.5.5. In an isotropic homogeneous medium we may use either the pair φ, ψ or the pair Φ, Ψ. In each case the *two families of curves* $\varphi =$ *const and* $\psi =$ *const* (or $\Phi =$ const and $\Psi =$ const) *are mutually orthogonal*. When the medium is not isotropic or is inhomogeneous, only φ and Ψ can be employed. In an isotropic medium, according to (6.5.49), the family of curves $\varphi =$ const is orthogonal to the family of curves $\Psi =$ const. When the medium is homogeneous but anisotropic, it is always possible to employ the technique described in section 7.4 first to transform the given domain into an equivalent isotropic one, to obtain the flow net in this transformed domain and then to transform the flow net back to the original domain. Equation (7.4.26) is applicable here. The result is a flow net in which *streamlines are not orthogonal* to equipotentials. Obviously, this results from the fact that in anisotropic media $\mathbf{q}$ and $\mathbf{J}$ are not colinear (sec. 5.6). Figure 6.6.1 shows a non-orthogonal flow net in an anisotropic medium; numbers on equipotentials are ratios φ/H in percents, while numbers on streamlines give Ψ/Q in percents.

Figure 6.6.2a shows a portion of a flow net with two streamtubes in a homogeneous isotropic medium. It is customary to draw the flow net such that the difference $\Delta\varphi$ between any two adjacent equipotentials is constant. The difference $\Delta\psi$ between any two adjacent streamlines is also constant (equal to the discharge $\Delta Q/K$ through the streamtube). Because streamlines behave as impervious boundaries of the streamtube:

$$\Delta Q = K\,\Delta n_1(\Delta\varphi/\Delta s_1) = K\,\Delta n_2(\Delta\varphi/\Delta s_2)$$

where ΔQ is the discharge per unit thickness (dim L^2T^{-1}). Hence:

$$\Delta n_1/\Delta s_1 = \Delta n_2/\Delta s_2, \tag{6.6.1}$$

i.e., in a homogeneous medium, the ratio $\Delta n/\Delta s$ of the sides of the rectangles must remain constant throughout the flow net.

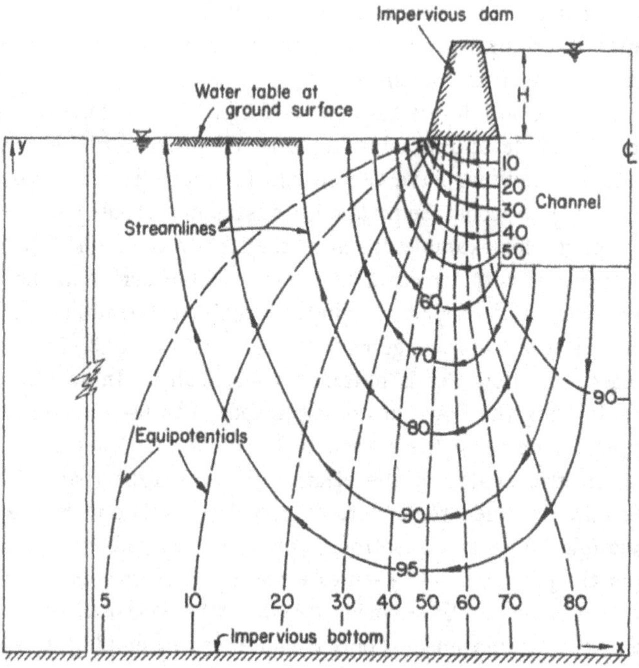

FIG. 6.6.1. Flow-net in an anisotropic medium (Todd and Bear, 1959).

When the medium is inhomogeneous, we have:

$$K_1\,\Delta n_1(\Delta\varphi/\Delta s_1) = K_2\,\Delta n_2(\Delta\varphi/\Delta s_2); \qquad K_1/(\Delta s_1/\Delta n_1) = K_2/(\Delta s_2/\Delta n_2) \quad (6.6.2)$$

and the ratio between the sides of the curvilinear rectangles varies. If in a flow net $\Delta n \simeq$ const (i.e., streamlines are nearly parallel) we have $K_1/K_2 = \Delta s_1/\Delta s_2$ or $K_1/K_2 = (\Delta\varphi_2/\Delta s_2)/(\Delta\varphi_1/\Delta s_1)$. This means that the hydraulic conductivity is inversely proportional to the hydraulic gradient. Therefore, in such a case, the region of wide contour spacing corresponds to a higher hydraulic conductivity, whereas contours will be crowded in regions of lower hydraulic conductivity.

In many cases of homogeneous media, however, the flow net is drawn so that approximate curvilinear squares are formed (fig. 6.6.2b). In this case $\Delta s_i = \Delta n_i$ everywhere. This is done only because in some graphic methods it is easier to draw curvilinear squares than to draw curvilinear rectangles. In the case of a flow net made up of squares:

$$\Delta Q = K\,\Delta\varphi. \quad (6.6.3)$$

In certain cases (e.g., when the electrolytic tank analog is used) the flow net is drawn such that we have m streamtubes, each carrying the same discharge ΔQ, and n equal drops in piezometric head from the highest φ $(= \varphi_{max})$ to the lowest φ $(= \varphi_{min})$ in the flow domain. Both m and n are integers. Then:

$$\Delta\varphi = (\varphi_{\max} - \varphi_{\min})/n; \qquad \Delta Q = Q_{\text{total}}/m.$$

$$Q_{\text{total}} = m\,\Delta Q = mK\,\Delta n\,\frac{\Delta\varphi}{\Delta s} = mK\,\Delta n\,\frac{\varphi_{\max} - \varphi_{\min}}{n\,\Delta s} = \frac{m}{n}\,\frac{\Delta n}{\Delta s}\,K(\varphi_{\max} - \varphi_{\min}) \tag{6.6.4}$$

or:

$$\frac{\Delta n}{\Delta s} = \frac{n}{m}\,\frac{Q_{\text{total}}}{K(\varphi_{\max} - \varphi_{\min})}. \tag{6.6.5}$$

Again, the ratio $\Delta n/\Delta s$ (in general $\neq 1$) remains constant throughout the flow domain. Sometimes n and m (not necessarily integers) are chosen such that squares with $\Delta s = \Delta n$ are obtained. Then:

$$Q_{\text{total}} = \frac{m}{n}\,K(\varphi_{\max} - \varphi_{\min}). \tag{6.6.6}$$

In a zoned porous medium, where K varies abruptly from K_1 to K_2 along a specified boundary, we have refraction of both equipotentials and streamlines across this boundary (par. 7.1.10). In plane flow through a porous medium of constant thickness, K should be replaced by the transmissivity $T = Kb$, with Q referring to the discharge through the entire thickness.

It is important to emphasize that when a homogeneous isotropic flow domain has boundaries on which the boundary conditions are given in terms of φ, and the flow is steady, described by the Laplace equation, the flow net is independent of the hydraulic conductivity. It depends only on the geometry of the flow domain. This is also true when one of the boundaries is a phreatic surface.

6.6.2 The Ground Water Contour Map

We shall use here the term ground water flow (or aquifer flow) in the sense discussed in section 6.4, i.e., a flow that is essentially horizontal. In general, the only information that we have about what happens in the aquifer is the piezometric heads observed and recorded at observation wells. In a phreatic aquifer these heads give the elevation

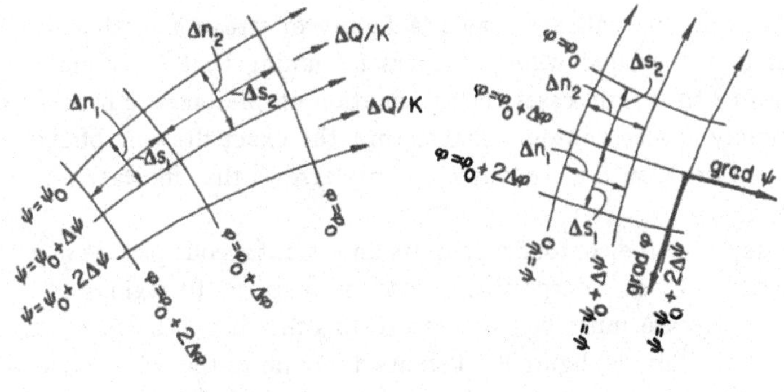

(a) Δn ≠ Δs *(b)* Δn = Δs

FIG. 6.6.2. A portion of a flow-net.

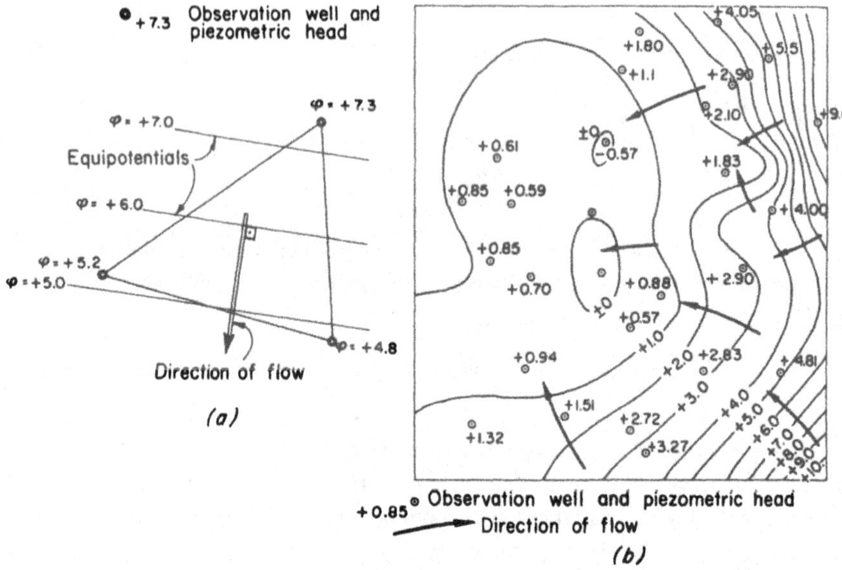

FIG. 6.6.3. Ground water contour map with arrows indicating directions of flow.

of the water table at the observation wells (neglecting the capillary fringe and employing the Dupuit assumption (chap. 8)). Except for special cases, such as when impervious faults are known to be present in the aquifer, we may assume that the piezometric head is a continuous function of the plane coordinates. Hence, using the information on values of φ at discrete points (observation wells), smooth equipotential curves φ = const are drawn, usually by linear interpolation, over the flow domain enclosed between the observation wells. Care must be taken to use only the information from wells tapping the aquifer being considered. An example of drawing equipotential curves, also called *ground water contours*, is given in figures 6.6.3a and 6.6.3b. The contours give the shape of the piezometric surface (or phreatic surface in a water table aquifer).

Once the equipotentials are drawn, a family of streamlines that are everywhere orthogonal to the ground water contours (assuming that the aquifer is isotropic) can be drawn. In certain cases the information on piezometric heads is insufficient to draw closely spaced contours that permit the exact drawing of streamlines. In such cases we draw on the contour map only arrows that indicate direction of flow (fig. 6.6.3b).

In unsteady flow the contour map gives an instantaneous picture of what happens in the aquifer. When interpreting a contour map (as to regions of recharge and discharge, etc.), care must be taken to distinguish between steady and unsteady situations. For example, figure 6.6.4 shows a mound in the water table of a phreatic aquifer. If this map describes steady flow, there *must be* a source of water inside the area enclosed by the + 5 contour curve. If, however, the flow is unsteady and this is only an instantaneous picture that changes with time, the water leaving the area

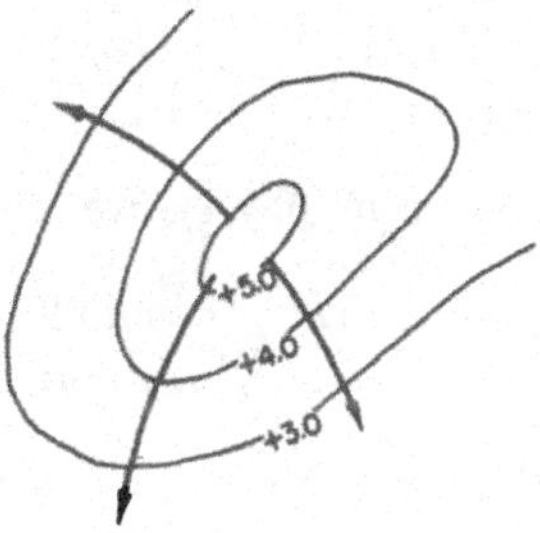

FIG. 6.6.4. A mound in the phreatic surface.

along the streamlines is taken from storage within the aquifer itself, producing a ubiquitous drop in the water table.

In reservoir engineering contours are usually drawn of pressure rather than of piezometric head.

6.7 The Partial Differential Equations Describing Flow of an Inhomogeneous Incompressible Fluid in Terms of Ψ

This section is a continuation of the discussion presented in paragraph 6.5.4. In the present section we consider isothermal flow of an incompressible fluid whose density varies as a result of changes in the concentration of some solute dissolved in it. From the discussion presented at the beginning of section 6.5, it follows that the concept of the stream function Ψ is a general one, applicable also to flow of inhomogeneous fluids. It would therefore be advantageous to express the required partial differential equations in terms of this function. The problem of flow of an inhomogeneous fluid is discussed in detail in chapter 10.

6.7.1 Two-Dimensional Flow

Consider flow in a two-dimensional vertical xz plane. The medium is assumed homogeneous and isotropic ($k =$ constant). Since we allow both ρ and μ to vary, the specific discharge is given by (4.7.17). With the definition of the stream function Ψ, and the relationship (6.5.43), we obtain:

$$q_x = -\frac{k}{\mu}\frac{\partial p}{\partial x} = -\frac{\partial \Psi}{\partial z}$$

$$q_z = -\frac{k}{\mu}\frac{\partial p}{\partial z} - \frac{k\rho g}{\mu} = +\frac{\partial \Psi}{\partial x}. \tag{6.7.1}$$

By cross-differentiation of (6.7.1) we obtain:

$$\frac{\partial q_x}{\partial z} - \frac{\partial q_z}{\partial x} = -\frac{\partial(k/\mu)}{\partial z}\frac{\partial p}{\partial x} - \frac{k}{\mu}\frac{\partial^2 p}{\partial x\,\partial z} + \frac{\partial(k/\mu)}{\partial x}\frac{\partial p}{\partial z} + \frac{k}{\mu}\frac{\partial^2 p}{\partial x\,\partial z} + \frac{\partial(k\rho g/\mu)}{\partial x}$$

$$= -\left(\frac{\partial^2 \Psi}{\partial z^2} + \frac{\partial^2 \Psi}{\partial x^2}\right) \tag{6.7.2}$$

or:

$$\frac{\partial^2\Psi}{\partial x^2}+\frac{\partial^2\Psi}{\partial z^2}=\frac{\partial(k/\mu)}{\partial z}\frac{\partial p}{\partial x}-\frac{\partial(k/\mu)}{\partial x}\frac{\partial p}{\partial z}-\frac{\partial}{\partial x}\left(\frac{k\rho g}{\mu}\right). \tag{6.7.3}$$

By introducing $\partial p/\partial x$ and $\partial p/\partial z$ from (6.7.1), we obtain:

$$\nabla^2_{xz}\Psi=-\left(\frac{\partial\ln\mu}{\partial z}\frac{\partial\Psi}{\partial z}+\frac{\partial\ln\mu}{\partial x}\frac{\partial\Psi}{\partial x}\right)-\frac{kg}{\mu}\frac{\partial\rho}{\partial x}. \tag{6.7.4}$$

From (5.4.11) it follows that:

$$\partial q_x/\partial z-\partial q_z/\partial x\equiv-\nabla^2_{xz}\Psi=(\operatorname{curl}\mathbf{q})_y, \tag{6.7.5}$$

i.e., the y component of curl $\mathbf{q}$. This means that in the case of an inhomogeneous fluid, the flow is *rotational*. Actually this conclusion is not new, as for an inhomogeneous fluid, we could not find a potential Φ such that $\mathbf{q}=-\operatorname{grad}\Phi$. Here we have also found the value of the vorticity of the field; it is expressed by the right-hand side of (6.7.4). The first term on the right-hand side of (6.7.4) gives the contribution of viscosity variations to the vorticity, whereas the second one gives the contribution of the density gradient. If $\rho=\rho(x,y,z)$ is known, one can determine everywhere the value of $(\operatorname{curl}\mathbf{q})_y$.

For the special case of a fluid where $\mu=$ const yet ρ varies, we obtain from (6.7.4):

$$\nabla^2_{xz}\Psi=-(kg/\mu)\,\partial\rho/\partial x. \tag{6.7.6}$$

If $\rho=\rho(x,y,z,t)$ is known, equation (6.7.6) is the Poisson equation for Ψ whose solution is known from potential theory:

$$\Psi(x,y,z,t)=-\frac{kg}{2\pi\mu}\iint\limits_{(A)}\frac{\partial\rho(\xi,t)}{\partial\xi}\ln[(x-\xi)^2+(z-\zeta)^2]^{1/2}\,d\xi\,d\zeta \tag{6.7.7}$$

where $[(x-\xi)^2+(x-\zeta)^2]^{1/2}$ is distance from the point being considered to the point where the vortex of strength $-(kg/\mu)\,\partial\rho/\partial x$ is present. Integration must be effected over all vortices, i.e., over the entire area A where $\partial\rho/\partial x\neq0$. The resulting Ψ is regular in the entire field (de Josselin de Jong 1960, 1969a).

From the condition of continuity div $\mathbf{q}=0$ and (6.7.1), we obtain for $\mu=$ const:

$$\nabla^2{}_{xz}p=-\partial(\rho g)/\partial z \tag{6.7.8}$$

(also valid for $\nabla^2{}_{xyz}p$ in three-dimensional flow).

By cross differentiation and summation of the Cauchy–Riemann condition (6.5.43), together with (6.7.6), we obtain:

$$\partial^2\Phi/\partial x\,\partial z-\partial^2\Phi/\partial z\,\partial x=\nabla^2_{xz}\Psi=-(kg/\mu)\,\partial\rho/\partial x\neq0\quad\text{if}\quad\rho\neq\text{const}. \tag{6.7.9}$$

This shows that Φ *is not single valued* and therefore has no particular physical significance in the flow domain (actually we already know that no potential exists for an inhomogeneous fluid). Instead, de Josselin de Jong (1960, 1969a) suggests for two-dimensional flow in a homogeneous isotropic field an *inhomogeneous fluid potential* θ:

$$\theta = \frac{k}{\mu}(p + \rho g z). \tag{6.7.10}$$

This means that for a *constant k and* μ,

$$\nabla\theta = \frac{k}{\mu}(\nabla p + \rho g\,\nabla z) + (kgz/\mu)\,\nabla\rho = -\,\mathbf{q} + (kgz/\mu)\,\nabla\rho;$$

$$\nabla^2\theta = -\,\nabla\cdot\mathbf{q} + (kg/\mu)\nabla\cdot(z\,\nabla\rho)$$

or, since $\nabla\cdot\mathbf{q} = 0$,

$$\nabla^2\theta = (kg/\mu)\nabla\cdot(z\,\nabla\rho) = (kg/\mu)\,[z\,\nabla^2\rho + \partial\rho/\partial z]. \tag{6.7.11}$$

This is again a Poisson equation showing singularities in θ of magnitude $(kg/\mu)\nabla\cdot(z\,\nabla\rho)$. Noting that $\partial^2\theta/\partial x\,\partial z = \partial^2\theta/\partial z\,\partial x$, we may conclude that the function θ is a single-valued one in the field of flow. The solution of (6.7.11) is given by:

$$\theta = -\frac{kg}{2\pi\mu}\iint_{(A)} \nabla\cdot[z\,\nabla\rho]\ln r\cdot dA. \tag{6.7.12}$$

The discussion given here was used by de Josselin de Jong as the basis for a singularity distribution method for analyzing multiple fluid flow through porous media (par. 9.5.5).

In all the cases discussed above, it is assumed that the functions $\rho = \rho(x, y, z, t)$ and $\mu = \mu(x, y, z, t)$ are known. Actually, in order to know them we must solve simultaneously the hydrodynamic dispersion equation taking into account a gradual variation of ρ (chap. 10).

If we wish to extend the discussion to an homogeneous *anisotropic medium*, the stream function Ψ satisfies (6.5.47). Hence we start from:

$$q_x = -\frac{k_x}{\mu}\frac{\partial p}{\partial x} = -\frac{\partial\Psi}{\partial z}; \qquad q_z = -\frac{k_z}{\mu}\frac{\partial p}{\partial z} - \frac{k_z\rho g}{\mu} = +\frac{\partial\Psi}{\partial x}. \tag{6.7.13}$$

Repeating the procedure described above for an isotropic medium, we obtain:

$$k_x\frac{\partial^2\Psi}{\partial x^2} + k_z\frac{\partial^2\Psi}{\partial z^2} = -\left(k_x\frac{\partial\ln\mu}{\partial x}\frac{\partial\Psi}{\partial x} + k_z\frac{\partial\ln\mu}{\partial x}\frac{\partial\Psi}{\partial z}\right) - \frac{k_x k_z g}{\mu}\frac{\partial\rho}{\partial x}. \tag{6.7.14}$$

6.7.2 Axisymmetric Flow

From (6.5.20), we obtain for the Stokes stream function χ:

$$q_r = -\frac{1}{r}\frac{\partial\chi}{\partial z} = -\frac{k}{\mu}\frac{\partial p}{\partial r}; \qquad q_z = \frac{1}{r}\frac{\partial\chi}{\partial r} = -\frac{k}{\mu}\left(\frac{\partial p}{\partial z} + \rho g\right). \tag{6.7.15}$$

By eliminating p from (6.7.15) we obtain:

$$\frac{\partial}{\partial z}\left(\frac{\mu}{kr}\frac{\partial\chi}{\partial z}\right) + \frac{\partial}{\partial r}\left(\frac{\mu}{kr}\frac{\partial\chi}{\partial r}\right) = -\frac{\partial(\rho g)}{\partial r} \tag{6.7.16}$$

for k, $\mu =$ const, this becomes:

$$\frac{1}{r}\frac{\partial^2 \chi}{\partial z^2} - \frac{1}{r^2}\frac{\partial \chi}{\partial r} + \frac{1}{r}\frac{\partial^2 \chi}{\partial r^2} = -\frac{kg}{\mu}\frac{\partial \rho}{\partial r}. \tag{6.7.17}$$

Exercises

6.1 Rewrite (6.2.15) (a) in polar coordinates for an isotropic medium, (b) in cylindrical coordinates for a medium with axisymmetrical anisotropy.

6.2 Prove that

$$\operatorname{div} \mathbf{V}^* = -\frac{1}{\rho n}\frac{d(\rho n)}{dt}.$$

Hint: Start from (6.2.3).

6.3 Derive the continuity equation for flow of an incompressible fluid in a homogeneous anisotropic rigid porous medium, where:

$$q_i + \frac{B}{\nu}\frac{\partial q_i}{\partial t} = -\frac{k_{ij}\gamma}{\mu}\frac{\partial \varphi}{\partial x_j}.$$

6.4 Develop the continuity equation in terms of the pressure $p = p(x, y, z, t)$ for flow of a slightly compressible liquid in a nonhomogeneous isotropic medium. The equation of state (isothermal conditions) is

$$\rho = \rho_0[1 + \beta(p - p_0)].$$

6.5 Rewrite (6.2.9) in detail (a) when the cartesian coordinates x_i are principal directions, (b) when they are not principal directions, and (c) when the medium is homogeneous and isotropic.

6.6 Prove Helmholtz' equation (6.2.20).

6.7 Develop the continuity equation for flow of an incompressible fluid in a nonhomogeneous, nondeformable anisotropic porous medium, (a) when x, y, z are principal directions, and (b) when they are not.

6.8 Prove that

$$K\nabla^2\varphi^* - K\rho g\beta\, \partial\varphi^*/\partial z = n\beta\rho g\, \partial\varphi^*/\partial t$$

assuming that although the fluid is compressible, the hydraulic conductivity $K = k\rho g/\mu$ is a constant.

6.9 The water level in a well in a confined aquifer rises by 0.008′ in response to a barometric pressure drop of 0.0145′ of water. Determine the barometric efficiency BE of the aquifer.

6.10 Show that (6.3.15) can also be written as

$$\operatorname{div}(\rho \mathbf{q}_r) + n\, d_s\rho/dt + \rho\, d_s n/dt + \rho n \operatorname{div} \mathbf{V}_s = 0$$

where

$$d_s(\ )/dt = \partial(\ )/\partial t + \mathbf{V}_s \cdot \operatorname{grad}(\ ).$$

6.11 Rewrite (6.4.7) in polar coordinates.

6.12 Show that the total flow $\mathbf{Q}(x, y)$ through an inhomogeneous aquifer of variable thickness $b = b(x, y)$ and hydraulic conductivity $K = K(x, y, z)$, assuming that the flow is essentially horizontal, is expressed by

$$\mathbf{Q} = -T(x, y)\,\nabla\varphi; \qquad T = \int_0^{b(x,y)} K(x, y, z)\, dz = b\bar{K}.$$

6.13 Show that the total flow $\mathbf{Q}(x, y)$ through an inhomogeneous confined aquifer (with horizontal bottom) of variable thickness $b = b(x, y)$ and variable hydraulic conductivity $K = K(x, y)$ is expressed by

$$\mathbf{Q} = -T(x, y)\,\nabla\tilde{\varphi} - [\tilde{\varphi} - \varphi(x, y, b)]K\nabla b,$$

where

$$T = K(x, y)b \quad \text{and} \quad b\tilde{\varphi} = \int_0^{b(x,y)} \varphi(x, y, z)\, dz.$$

Rewrite this expression for equipotentials which are practically vertical.

6.14 Let the flow to a well in a phreatic aquifer be essentially horizontal with $q_r = -K\,\partial h/\partial r$, $h = h(r, t)$ being the height of the water table above a horizontal impervious bottom. Derive the continuity equation for $h = h(r, t)$ in the vicinity of the well.

6.15 Use (6.4.22) to derive the equation of continuity for flow in a confined aquifer, with T and S denoting aquifer transmissivity and storativity, respectively.

6.16 A fully penetrating well taps the upper of two leaky aquifers described by the profile shown in fig. 6.E.1. Develop the equations describing steady flow in the aquifers. Assume no drawdown in the phreatic aquifer (i.e., the phreatic surface remains horizontal and at a constant elevation).

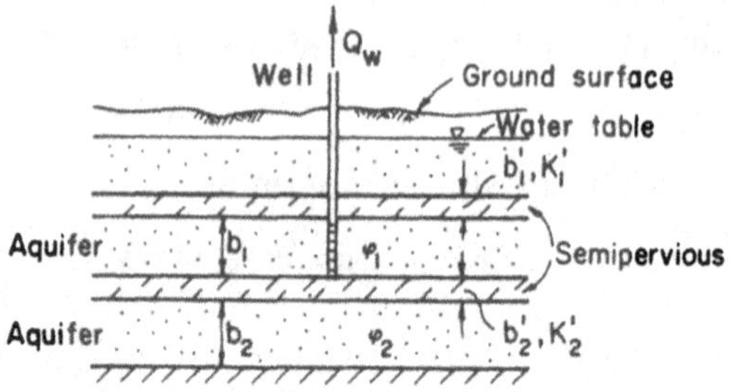

FIG. 6.E.1.

6.17 Develop the continuity equation for unsteady flow in a leaky confined aquifer, bounded both from above and below by semi-pervious layers (b', K'; b'', K''), (a) when we are interested in a regional flow problem, and (b) when we are interested in flow to a single well in the main aquifer (b, K).

Note. Assume that the flow in the aquifer is essentially horizontal and denote by φ' and φ'' the potentials in the aquifers above and below the considered aquifer.

6.18 A streamfunction in two-dimensional flow is defined by

$$\Psi = x^3 - y^3,$$

show that the flow could not be a potential flow.

6.19 What is the partial differential equation for Ψ if the medium is (a) inhomogeneous with $K = ax$, but isotropic. (b) Homogeneous, but anisotropic with $K_x = ay$, $K_y = by$?

6.20 The stream function of a two-dimensional irrotational flow is defined by

$$\Psi = b(x^2 - y^2).$$

(a) Determine $\mathbf{q}$ at point (1,1).

(b) Find an expression for the potential function Φ when the aquifer is isotropic and homogeneous.

6.21 Can a flow exist with Ψ of Ex. 6.20 in a nonhomogeneous field where $K = ax$?

CHAPTER 7

Solving Boundary and Initial Value Problems

The various partial differential equations derived in chapters 4 and 6, as well as those to be derived in chapters 8, 9 and 10, describe what happens at a point within a domain in which a certain physical phenomenon occurs. In most cases the independent variables are (x, y, z) or (x, y, z, t). The dependent variables for which a solution is sought are piezometric head, pressure, saturation, specific discharge, solute concentrations, temperature (in nonisothermal flows), etc. When a problem is described simultaneously by a number of dependent variables, the same number of equations is needed for a complete solution. Each such equation describes interrelationships inherent in all phenomena belonging to the same class, where a class of phenomena includes all phenomena in which the same mechanism, or process, is active, e.g., mass transfer, heat conduction, momentum transfer. In a way, we may refer to the partial differential equations as *mathematical models* of the actual phenomena.

Being a general description of a *class of phenomena*, it is obvious that the partial differential equation itself does not contain any information concerning the specific values of quantities characterizing a specific case of a phenomenon. Therefore, any partial differential equation has an infinite number of possible solutions, each of which corresponds to a particular case of the phenomenon.

To obtain from this multitude of possible solutions one particular solution corresponding to a certain specific problem of interest, it is necessary to provide supplementary information that is not contained in the partial differential equation. The supplementary information that, together with the partial differential equation, defines an individual problem should include specifications of: (a) the geometry of the domain in which the phenomenon being considered takes place, (b) all physical coefficients and parameters that affect the phenomenon considered (e.g., medium and fluid parameters), (c) initial conditions that describe the initial state of the system considered, and (d) the interaction of the system under consideration with surrounding systems, i.e., conditions on the boundaries of the domain in question.

The present chapter includes a discussion of the initial and boundary conditions of flow of fluids through porous media and some comments on analytical and numerical solution methods. For the details of these methods, especially the standard methods of solving partial differential equations not included here, the reader is referred to texts on mathematical physics and numerical methods. The use of models and analogs as tools for solving problems is discussed in chapter 11. The discussion will be confined to the following three equations:

the Laplace equation, $\nabla^2\varphi = 0$

the Poisson equation, $\nabla^2_{xy}\varphi + N/T = 0$ and

the heat conduction equation, $T\,\nabla^2_{xy}\varphi = S\,\partial\varphi/\partial t.$

7.1 Initial and Boundary Conditions

In solving a specific physical problem, such as the flow of a liquid through a specified porous medium domain, it is necessary to choose from the infinite number of possible solutions only that *one* that satisfies certain additional conditions imposed by the physical situation *at the boundaries* of the considered domain. These are called *boundary conditions*, and the problem is referred to as a *boundary value problem*. If the problem is one in which the dependent variables are also time dependent (i.e., unsteady flow), the boundary conditions must be specified for all times $t \geqslant 0$. In addition, certain conditions, called *initial conditions*, must be satisfied at all points of the domain being considered at the particular instant of time at which the physical process begins. The problem is then referred to as an *initial value problem*.

Boundary and initial conditions are usually obtained from field observations or from controlled experiments. As such, they always involve certain errors that may affect the solution of the problem. This effect is considered in section 7.2, where the stability of the solution is discussed. In other cases, mainly in those involving the prediction of a system's future behavior under conditions that do not necessarily exist at present, the boundary conditions are a mathematical statement of certain hypotheses, often based on past experience. We shall see below (sec. 7.2) that certain restrictions must be imposed on boundary conditions before they can be accepted.

In general, any specification of boundary conditions for the second-order partial differential equations considered here should include: (a) the geometric shape of the boundary, and (b) a statement of how the dependent variable, for example φ (or p or Ψ), and/or its derivatives, vary on the boundary.

Flow in both two- and three-dimensional porous medium domains is considered here. Let the domain be denoted by D. In three dimensions it is bounded by a surface S; in two dimensions by a curve C (fig. 7.1.1). In certain cases, the flow domain may be imagined to extend to infinity. We then speak of an *unbounded domain*. However, we shall see (sec. 7.2) that we must then impose an additional restriction on the value of the dependent variable as the boundary at infinity is approached.

In an anisotropic medium the directions x, y, z will be principal directions of the anisotropy.

For three-dimensional domains, the fixed boundary surface S (fig. 7.1.1a) may be described mathematically (in a Cartesian coordinate system x, y, z) by an equation whose general form is:

$$F(x, y, z) = 0. \tag{7.1.1}$$

In two dimensions (x, y), the equation of C (fig. 7.1.1b) is:

$$F(x, y) = 0. \tag{7.1.2}$$

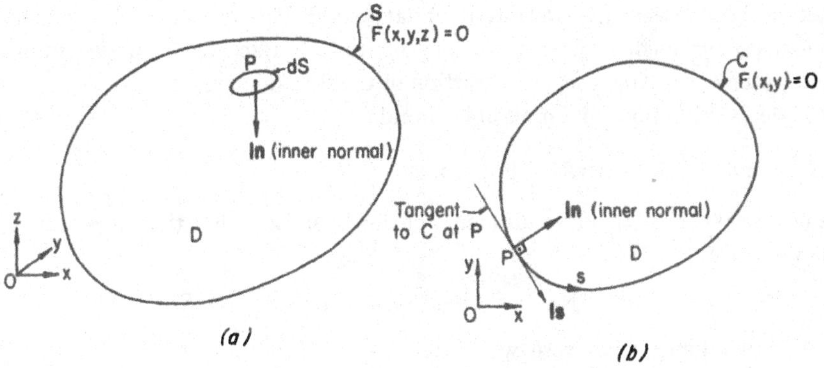

FIG. 7.1.1. Two- and three-dimensional flow domains.

Let **1n** denote a unit vector perpendicular to the boundary (i.e., normal to a plane that is tangential to it) at point P and directed inward (inner normal, fig. 7.1.1a). This unit vector can then be expressed in terms of the unit vectors **1x, 1y, 1z** (or **1x, 1y,** in two dimensions) in the directions x, y, z (or x, y), respectively. Accordingly, for S:

$$|\nabla F|\mathbf{1n} = \nabla F = \frac{\partial F}{\partial x}\mathbf{1x} + \frac{\partial F}{\partial y}\mathbf{1y} + \frac{\partial F}{\partial z}\mathbf{1z}; \qquad |\nabla F|^2 = \left(\frac{\partial F}{\partial x}\right)^2 + \left(\frac{\partial F}{\partial y}\right)^2 + \left(\frac{\partial F}{\partial z}\right)^2 \tag{7.1.3}$$

for C:

$$|\nabla F|\mathbf{1n} = \nabla F = \frac{\partial F}{\partial x}\mathbf{1x} + \frac{\partial F}{\partial y}\mathbf{1y}; \qquad |\nabla F|^2 = \left(\frac{\partial F}{\partial x}\right)^2 + \left(\frac{\partial F}{\partial y}\right)^2. \tag{7.1.4}$$

Denoting the angles that **1n** makes with the $+\,x$, $+\,y$ and $+\,z$ directions by α_{nx}, α_{ny} and α_{nz}, respectively, we have:

$$\cos\alpha_{nx} = (\partial F/\partial x)/|\nabla F|; \qquad \cos\alpha_{ny} = (\partial F/\partial y)/|\nabla F|; \qquad \cos\alpha_{nz} = (\partial F/\partial z)/|\nabla F| \tag{7.1.5}$$

and similar expressions in two dimensions.

When an impervious boundary or an assumed abrupt interface (steady as well as nonsteady) between two fluids is being considered, we shall make use of the important *theorem attributable to Lagrange* (Prandtl and Tietjens **1934**). According to this theorem "the boundary surface is always composed of the same particles." Prandtl and Tietjens (**1934**) indicate that in this form the theorem is too general and certain restrictions should be added. It is obviously true for smooth boundaries of infinite area. In paragraph **4.1.3** we referred to such a surface as a *material surface*.

In the two-dimensional case (fig. **7.1.1b**), we may define at every point P of the boundary a unit vector **1s** in the direction of the tangent to the boundary at that point, such that **1s** is obtained from **1n** by a rotation of 90° in a *clockwise direction*. This s, n Cartesian coordinate system is called the *intrinsic coordinate system* located on the considered boundary surface.

Let us now review the various boundary conditions often encountered in flow through porous media. The conditions on the interface between two immiscible liquids, each occupying a separate portion of the flow domain, are given in chapter 9, which deals with flow of immiscible liquids.

7.1.1 Boundary of Prescribed Potential

The potential is prescribed for all points of this boundary. For the three-dimensional case we write:

$$\varphi = \varphi(x, y, z), \quad \text{or} \quad \varphi = \varphi(x, y, z, t) \text{ on } S. \tag{7.1.6}$$

For the two-dimensional case we write:

$$\varphi = \varphi(x, y), \quad \text{or} \quad \varphi = \varphi(x, y, t), \quad \text{or} \quad \varphi = \varphi(s), \quad \text{or} \quad \varphi = \varphi(s, t) \text{ on } C. \tag{7.1.7}$$

A boundary of this kind occurs whenever the flow domain is adjacent to a body of a liquid continuum. At every point on such a boundary, since pressure p and elevation z are the same when the point is approached from both sides, the piezometric head $\varphi = z + p/\gamma$ is the same as that in the liquid at the point adjacent to it. In ground water flow, this occurs at the interface between a saturated porous medium and a river, lake or sea.

A special case of (7.1.6) and (7.1.7) is:

$$\varphi = \varphi_0 = \text{const on } S \text{ or } C. \tag{7.1.8}$$

In this case the boundary is an *equipotential surface* (or *curve* in two-dimensional flow). For example, the piezometric head is the same at all points of a body of stationary liquid with a horizontal free surface. This piezometric head is, therefore, also the boundary condition at points on the interface between a porous medium and this body of liquid. Segments AB and MN of figure 7.1.2 are boundaries of this type. In (7.1.8), the piezometric head may vary with time, i.e., $\varphi = \varphi_0(t)$.

In certain cases, especially in reservoir engineering, it is customary to prescribe the distribution of the pressure p on the boundary, rather than that of the potential φ. This is necessary, for example, when a solution is sought for equations such as (6.2.25) or (6.2.27).

When the boundary is an *equipotential surface*, the vector $\mathbf{J} = -\nabla\varphi$ is perpendicular

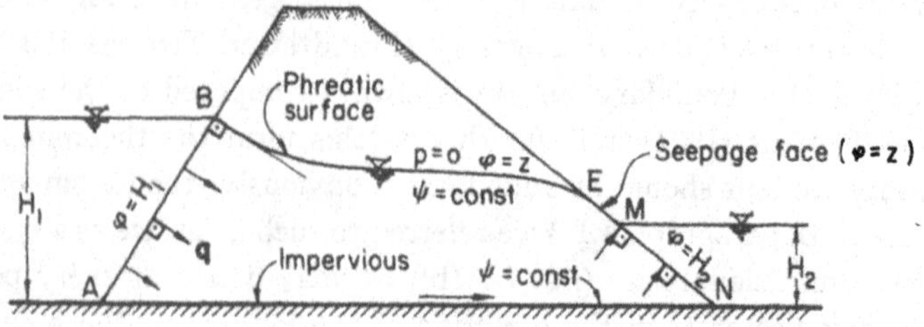

FIG. 7.1.2. Boundaries and boundary conditions in an earth embankment (Steady flow, isotropic medium).

to it, and therefore, parallel to **1n**. We then have:

$$\mathbf{J} \times \mathbf{1n} = 0; \qquad \nabla\varphi \times \mathbf{1n} = 0. \tag{7.1.9}$$

With (7.1.5), we then have for points along the boundary:

$$(\partial\varphi/\partial x)/\cos\alpha_{nx} = (\partial\varphi/\partial y)/\cos\alpha_{ny} = (\partial\varphi/\partial z)/\cos\alpha_{nz} \tag{7.1.10}$$

or:

$$(\partial\varphi/\partial x)/\cos\alpha_{nx} = (\partial\varphi/\partial y)/\cos\alpha_{ny} \tag{7.1.11}$$

for two-dimensional flow. We may use (6.5.46) to rewrite (7.1.11) in terms of ψ for an isotropic medium or (6.5.47) to rewrite it in terms of Ψ for an anisotropic medium.

In the theory of partial differential equations, a problem in which only this type of boundary condition is encountered is called a *Dirichlet, or a first, boundary value problem*. We speak, for example, of the Dirichlet problem for the Laplace equation.

7.1.2 Boundary of Prescribed Flux

Along a boundary of this type, the flux normal to the boundary surface (or curve in two-dimensional flow) is prescribed for all points of the boundary as a function of position (and of time in unsteady flow):

$$q_n = \mathbf{q} \cdot \mathbf{1n} = q_n(x, y, z); \quad \text{or} \quad q_n = q_n(x, y, z, t) \tag{7.1.12}$$

where q_n is the component of $\mathbf{q}$ normal to the boundary, and **1n** is in the direction of the inner normal. *In an isotropic medium,* this boundary condition may be expressed in terms of the potential φ:

$$\nabla\varphi \cdot \mathbf{1n} \equiv \partial\varphi/\partial n = f(x, y, z, t) \text{ on } S \tag{7.1.13}$$

where $f(x, y, z, t)$ is a known function for all points on the boundary, and n is the distance measured along **1n**.

A special case is the *impervious boundary* (AN in fig. 7.1.2) where the flux vanishes everywhere on the boundary. Thus:

$$\mathbf{q} \cdot \mathbf{1n} = q_n = 0, \qquad \text{for anisotropic media}$$

$$\nabla\varphi \cdot \mathbf{1n} = \partial\varphi/\partial n = 0, \quad \text{for isotropic media.} \tag{7.1.14}$$

With the geometry of the boundary described by (7.1.1), and $\mathbf{1n} = \nabla F/|\nabla F|$, (7.1.14) becomes:

$$\underline{\mathbf{K}}\, \nabla\varphi \cdot \nabla F \equiv K_x \frac{\partial\varphi}{\partial x}\frac{\partial F}{\partial x} + K_y \frac{\partial\varphi}{\partial y}\frac{\partial F}{\partial y} + K_z \frac{\partial\varphi}{\partial z}\frac{\partial F}{\partial z} = 0, \quad \text{for anisotropic media}$$

$$\nabla\varphi \cdot \nabla F \equiv \frac{\partial\varphi}{\partial x}\frac{\partial F}{\partial x} + \frac{\partial\varphi}{\partial y}\frac{\partial F}{\partial y} + \frac{\partial\varphi}{\partial z}\frac{\partial F}{\partial z} = 0, \quad \text{for isotropic media.} \tag{7.1.15}$$

In two-dimensional flow, an impervious boundary is also a streamline. We have in this case:

$$\Psi = \text{const} \quad \text{or} \quad \Psi = \Psi(t) \text{ on } C$$

$$d\Psi/ds = 0, \qquad \nabla\Psi \times \mathbf{1n} = 0 \text{ on } C \tag{7.1.16}$$

where s is distance measured along the boundary. From

$$\frac{d\Psi}{ds} = \frac{\partial\Psi}{\partial x}\frac{dx}{ds} + \frac{\partial\Psi}{\partial y}\frac{dy}{ds} = 0,$$

and with β_{sx}, β_{sy} the angles which the vector $\mathbf{1s}$ makes with the $+x$ and $+y$ axes, respectively ($dx/ds = \cos\beta_{sx}$, $dy/ds = \cos\beta_{sy} = \sin\beta_{sx}$), we obtain:

$$(\partial\Psi/\partial x)\cos\beta_{sx} + (\partial\Psi/\partial y)\cos\beta_{sy} = 0. \tag{7.1.17}$$

Using the Cauchy–Riemann relationships (6.5.45) for an isotropic medium, we obtain:

$$-(\partial\varphi/\partial y)\cos\beta_{sx} + (\partial\varphi/\partial x)\cos\beta_{sy} = 0; \qquad \partial\varphi/\partial y = (\partial\varphi/\partial x)\tan\beta_{sx}. \tag{7.1.18}$$

For an anisotropic medium, we employ (6.5.47) and obtain:

$$-K_y(\partial\varphi/\partial y)\cos\beta_{sx} + K_x(\partial\varphi/\partial x)\cos\beta_{sy} = 0; \qquad \partial\varphi/\partial y = (\partial\varphi/\partial x)(K_x/K_y)\tan\beta_{sx}. \tag{7.1.19}$$

In the theory of partial differential equations, a problem having this type of boundary condition of prescribed flux (in terms of φ) is called the *Neumann, or the second, boundary value problem.* In two-dimensional flow this is equivalent to the first boundary value problem in terms of Ψ.

A third, or Cauchy boundary value problem, occurs when both the potential and its normal derivative are prescribed on the boundary in the combined form:

$$\frac{\partial\varphi}{\partial n} + \lambda(x, y, z)\varphi = f(x, y, z); \quad \text{or} \quad \frac{\partial\varphi}{\partial n} + \lambda(s)\varphi = f(s) \tag{7.1.20}$$

where λ and f are known functions. Such a condition is enountered in problems of flow through porous media with semipervious boundary.

In general, we have a *mixed boundary value problem* in which the Dirichlet conditions apply over a part of the boundary and the Neumann conditions apply over the remaining part.

The Dirichlet, Neumann and Cauchy types of problems also occur in other branches of physics (e.g., heat conduction and electric flow). However, certain other conditions, which are actually special cases of the first two types, often occur in flow through porous media because of the effect of gravity and the presence of capillary phenomena. The following paragraphs deal with these boundary conditions.

7.1.3 The Steady Free (or Phreatic) Surface without Accretion

The *phreatic surface* (*BE* in fig. 7.1.2) was defined and discussed in paragraph 1.1.2 and section 3.2. In this book, the term *free surface* is used to denote that surface on which the pressure is atmospheric (assuming that this pressure exists everywhere in the air phase filling the void space), arbitrarily taken as $p = 0$. Actually, in a real situation, a *capillary fringe* (par. 9.4.2) may be located above, on the side, or below a free surface. We shall reserve the term *phreatic surface* (derived from the

Greek word *phrear-atos*, meaning a well or water table) to denote the surface of $p = 0$, encountered in ground water hydrology, where the capillary fringe is located *above* it. In this paragraph, we neglect the presence of moisture above or outside the free surface (e.g., in the capillary fringe), and assume that the free surface is an abrupt interface between air and water in the void space. Although this kind of boundary occurs in the general case of two-phase flow, assuming an abrupt interface between them (chap. 9), the boundary conditions on a phreatic surface will be considered here in detail because of the importance of this surface in many ground water and drainage problems.

The location and geometric shape of the free surface are unknown. In fact they constitute part of the required solution. A boundary of this type is sometimes called a *floating boundary*. Following the usual procedure, we must specify for the boundary being considered (a) its geometry in the form of (7.1.1) and (b) the conditions to be satisfied along it.

As the pressure at all points of the free surface is taken as $p = 0$, we have from $\varphi(x, y, z) = z + p/\gamma$:

$$\varphi(x, y, z) = z \quad \text{or} \quad \varphi(x, y, z) - z = 0. \tag{7.1.21}$$

This is thus a boundary of prescribed potential. As (7.1.21) gives for any time t a relationship between the coordinates of points of the free surface, it may be considered as the equation describing its geometry. Thus:

$$F(x, y, z) \equiv \varphi(x, y, z) - z = 0 \text{ on } S. \tag{7.1.22}$$

The difficulty, however, stems from the fact that the distribution $\varphi(x, y, z)$, and hence $F(x, y, z)$, is unknown before the problem is solved.

The free surface, being a steady interface between water and air, behaves as an impervious boundary for which (7.1.14) holds. However, we must express this condition in term of φ alone. For the case of an anisotropic medium, we therefore obtain from (7.1.15) and (7.1.22):

$$F(x, y, z) \equiv \varphi(x, y, z) - z = 0 \text{ on } S$$

$$\nabla F = (\partial\varphi/\partial x)\mathbf{1x} + (\partial\varphi/\partial y)\mathbf{1y} + (\partial\varphi/\partial z)\mathbf{1z} - \mathbf{1z} = \nabla\varphi - \mathbf{1z}. \tag{7.1.23}$$

Thus:

$$\mathbf{q} \cdot \mathbf{1n} = 0 = -K_x\left(\frac{\partial\varphi}{\partial x}\right)^2 - K_y\left(\frac{\partial\varphi}{\partial y}\right)^2 - K_z\left(\frac{\partial\varphi}{\partial z}\right)^2 + K_z\frac{\partial\varphi}{\partial z}$$

or:

$$K_x\left(\frac{\partial\varphi}{\partial x}\right)^2 + K_y\left(\frac{\partial\varphi}{\partial y}\right)^2 + K_z\left(\frac{\partial\varphi}{\partial z}\right)^2 - K_z\frac{\partial\varphi}{\partial z} = 0 \text{ on } S. \tag{7.1.24}$$

This is the boundary condition along a steady free (or phreatic) surface without accretion in an anisotropic medium. The nonlinear nature of this boundary condition (and some of the following ones) is obvious.

In an isotropic medium, $K_x = K_y = K_z$, and (7.1.24) becomes:

$$|\nabla\varphi|^2 - \partial\varphi/\partial z = \left(\frac{\partial\varphi}{\partial x}\right)^2 + \left(\frac{\partial\varphi}{\partial y}\right)^2 + \left(\frac{\partial\varphi}{\partial z}\right)^2 - \left(\frac{\partial\varphi}{\partial z}\right) = 0 \text{ on } S. \tag{7.1.25}$$

In two-dimensional flow, in the vertical xz plane, $(\partial\varphi/\partial y) = 0$ in (7.1.24) and (7.1.25).

Another way of obtaining this boundary condition in two-dimensional flow is by using the fact that, in both isotropic and anisotropic media, the steady free surface without accretion is a streamline. Therefore, the condition along it becomes:

$$\Psi = \text{const on } C.$$

From $d\Psi/ds = 0$ and $d\varphi/ds = dz/ds = \cos\beta_{sz}$, with $dx/ds = \cos\beta_{sx}$, and using the Cauchy–Riemann relationship (6.5.45) for an anisotropic medium, or the relationship (6.5.46) for an isotropic medium, we obtain (7.1.24) or (7.1.25) in two dimensions, respectively.

Finally, the boundary condition on a free surface may be obtained by noting that (a) it is a material, or fluid surface, and (b) that all particles belonging to this surface have a certain property (B of par. 4.1.4) that remains unchanged. Here, this property is that the pressure remains unchanged: $p = 0$. To avoid repetition, we shall employ this approach in the following paragraph. The steady flow case may then be obtained as a special case.

It is of interest to note that (7.1.25) is independent of K. The equation of the steady free surface in an isotropic medium is a purely geometric statement; it depends only on the geometry of the flow domain. The same observation was made with respect to (6.2.25) where the distribution of φ was shown to be independent of K. It is only the specific discharge $\mathbf{q}$ that depends on K.

Since along the free surface $\varphi = z$, and $d\varphi/ds = dz/ds = \cos\beta_{sz} = \sin\beta_{sx}$, we have for an isotropic medium:

$$q_s = -K\,\partial\varphi/\partial s = -K\sin\beta_{sx}. \tag{7.1.26}$$

This gives the specific discharge tangential to the free surface. The maximum possible value of q_s is when $s \equiv z$. Then $\mathbf{q} = -q\mathbf{1z}$, or $q = K$, and the flow is vertically downward.

7.1.4 The Unsteady Free (or Phreatic) Surface Without Accretion

The capillary fringe (par. 9.4.2) above (or outside) the free surface is again neglected. The unsteady free surface without accretion is a *material or fluid surface* always consisting of the same fluid particles. As mentioned in paragraph 4.1.3, this is strictly true only when this surface is of infinite area. Since this surface is moving, it can be represented, in a fashion similar to (7.1.1), by an equation whose general form is:

$$F(x, y, z, t) = 0. \tag{7.1.27}$$

From (4.1.30) it follows that:

$$\partial F/\partial t + \mathbf{V} \cdot \operatorname{grad} F = 0 \tag{7.1.28}$$

where $\mathbf{V}$ is the velocity of particles belonging to the free surface. This velocity may be related to the specific discharge $\mathbf{q}$ at the free surface by:

$$\mathbf{q} = n_e \mathbf{V}$$

where n_e has the meaning of an *effective porosity with respect to flow.* When the free surface is steady, and particles move only "in" the surface, n_e may be assumed to be equal to the porosity n. The same is true when a moving free surface always separates a region of complete saturation from a region of zero saturation (i.e., with no liquid in the voids). In reality, such a situation never occurs. When a free surface advances into a region of dry porous medium, $n = n_e$. If this region is not completely dry, but contains a certain moisture content c_r retained in the soil, say from a previous recession of the free surface, then $n_e = n - c_r$. If the free surface is receding, evacuating a completely saturated region, the drainage is never complete as a certain amount of water always remains in the pores above or outside a receding free surface, owing to surface phenomena. This means that c_r varies with time, after a sufficiently large time approaching a certain value c_0, so that n_e approaches $n - c_0$. Sometimes c_0 is called *stagnant or ineffective porosity* (volume of water retained per unit volume of soil, Irmay 1956). Moreover, the volume of water evacuated from the void space (per unit volume of porous medium) varies with time as the free surface declines, since the drainage process is, in general, a slow one. Bear, Zaslavsky and Irmay (1968) refer to the ratio of instantaneous volume of water removed from the soil per unit horizontal area and unit drop of a phreatic surface as an *instantaneous dewatering coefficient* (*drainage coefficient* or *aerated porosity*) m. Always, $m < (n - c_0)$.

The conclusion from the discussion here is that n_e may vary with time and with the initial and final water contents (c_i and c_f, respectively). Irmay (in Bear, Zaslavsky and Irmay 1968) suggests:

$$n_e = |c_i - c_f|. \tag{7.1.29}$$

Equation (7.1.28) can now be rewritten as:

$$\partial F/\partial t + (\mathbf{q}/n_e) \cdot \operatorname{grad} F = 0. \tag{7.1.30}$$

With F defined by (7.1.23), we have:

$$(\partial\varphi/\partial t) + (1/n_e)\mathbf{q} \cdot (\nabla\varphi - \mathbf{1z}) = 0. \tag{7.1.31}$$

For an isotropic medium, $\mathbf{q} = -K\nabla\varphi$, and (7.1.31) becomes:

$$(\partial\varphi/\partial t) - (K/n_e)[(\nabla\varphi)^2 - \nabla\varphi \cdot \mathbf{1z}] = 0$$

$$\frac{\partial\varphi}{\partial t} - \frac{K}{n_e}\left[\left(\frac{\partial\varphi}{\partial x}\right)^2 + \left(\frac{\partial\varphi}{\partial y}\right)^2 + \left(\frac{\partial\varphi}{\partial z}\right)^2 - \frac{\partial\varphi}{\partial z}\right] = 0. \tag{7.1.32}$$

For anisotropic media, we obtain from (7.1.31):

$$\frac{\partial\varphi}{\partial t} - \frac{1}{n_e}\left[K_x\left(\frac{\partial\varphi}{\partial x}\right)^2 + K_y\left(\frac{\partial\varphi}{\partial y}\right)^2 + K_z\left(\frac{\partial\varphi}{\partial z}\right)^2 - K_z\left(\frac{\partial\varphi}{\partial z}\right)\right] = 0. \tag{7.1.33}$$

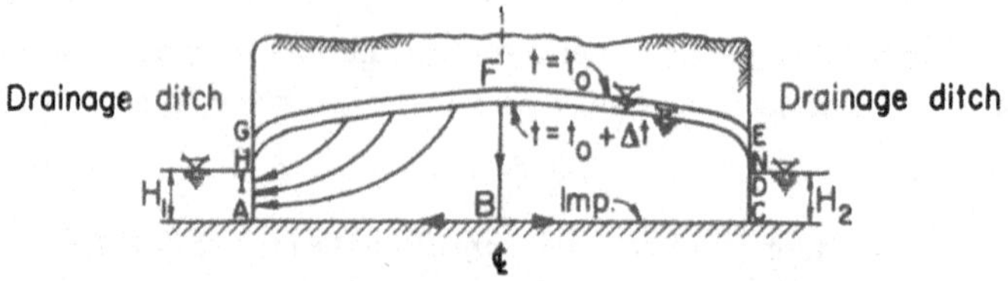

FIG. 7.1.3. Instantaneous picture of streamlines in the case of a receding phreatic surface between two drainage ditches, after cessation of irrigation.

Equations (7.1.32) and (7.1.33) are the boundary conditions (in terms of φ).

One may also regard the free surface as an *iso-B_α surface* (Irmay in Bear, Zaslavsky and Irmay 1968), where pressure is the property maintained constant. With $B_\alpha = p$, and $\mathbf{V}_{G\alpha} = \mathbf{q}/n_e$ (4.1.25) becomes

$$\frac{\partial p}{\partial t} + \frac{1}{n_e}\mathbf{q}\cdot \operatorname{grad} p = 0. \tag{7.1.34}$$

Since $\partial p/\partial t = \gamma\, \partial\varphi/\partial t$ and $\nabla p = \gamma(\nabla\varphi - \mathbf{1z})$, we can easily obtain (7.1.32) or (7.1.33) from (7.1.34).

Unlike the stationary free surface without accretion, the moving free surface is not a stream surface (or streamline, in two dimensions). The instantaneous streamlines are not perpendicular to the phreatic surface. Figure 7.1.3 shows an instantaneous picture of a receding phreatic surface between two drainage ditches.

The boundary condition along a steady phreatic surface can be obtained from the nonsteady one by introducing $\partial\varphi/\partial t = 0$ in (7.1.32) and (7.1.33).

In many cases these nonlinear boundary conditions are linearized to the form:

$$(\partial\varphi/\partial t) + (K_z/n_e)\, \partial\varphi/\partial z = 0 \tag{7.1.35}$$

by neglecting the quadratic terms. It can be shown (par. 8.4.5 and Dagan 1964) that this condition is the first approximation of (7.1.33).

7.1.5 The Steady Free (or Phreatic) Surface with Accretion

Accretion (N) is the rate (volume per unit horizontal area per unit time) at which liquid is added to an aquifer, or a saturated region of porous medium, at its free surface boundary. Negative accretion occurs when liquid is withdrawn from a liquid body through its free surface. The best examples would be in the case of an aquifer with a phreatic surface: accretion is the natural replenishment, i.e., that portion of the precipitation that actually infiltrates the soil surface and reaches the water table. Negative accretion is evaporation (or evapotranspiration) from the water table. In ground water flow we shall refer, therefore, to N as positive when it is directed vertically downward. Segment EFE of figure 7.1.4 is a steady phreatic surface with accretion by natural replenishment (or irrigation) in a drainage problem.

By applying the requirement of continuity at a steady phreatic surface, we obtain:

$$\mathbf{N}\cdot\mathbf{1n} = q_n \quad \text{or} \quad (\mathbf{N} - \mathbf{q})\cdot\mathbf{1n} = 0; \qquad \mathbf{N} \equiv -N\mathbf{1z}. \tag{7.1.36}$$

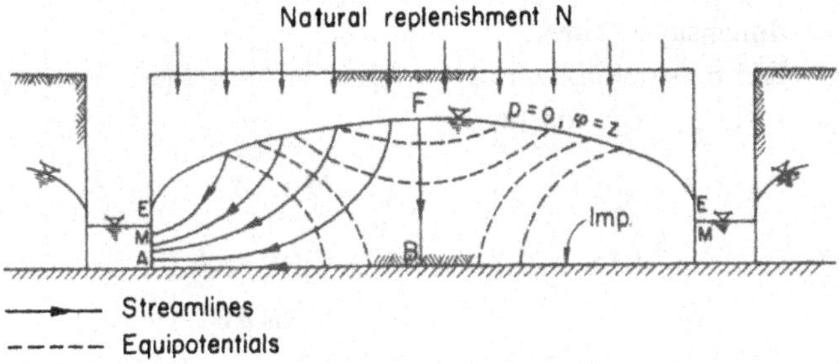

FIG. 7.1.4. Steady phreatic surface with accretion.

Following the analysis presented in paragraphs 7.1.3 and 7.1.4, we obtain from (7.1.23) and (7.1.36):

$$(\mathbf{N} - \mathbf{q}) \cdot \nabla F = 0; \qquad (\mathbf{N} - \mathbf{q}) \cdot (\nabla \varphi - \mathbf{1z}) = 0 \tag{7.1.37}$$

from which we obtain:

$$K_x \left(\frac{\partial \varphi}{\partial x}\right)^2 + K_y \left(\frac{\partial \varphi}{\partial y}\right)^2 + K_z \left(\frac{\partial \varphi}{\partial z} - \frac{K_z + N}{2K_z}\right)^2 = \left(\frac{K_z - N}{2K_z}\right)^2 K_z. \tag{7.1.38}$$

This is the required boundary condition. For isotropic media, we introduce $K_x = K_y = K_z$ in (7.1.38). For two-dimensional flow in the vertical xz plane, the boundary condition can be obtained either by setting $\partial \varphi / \partial y = 0$ in (7.1.38) or by noting that because of continuity we have along the phreatic surface with $N =$ const (fig. 7.1.5):

$$q_z = -N = \partial \Psi / \partial x; \qquad \Psi = -Nx + \text{const} \tag{7.1.39}$$

(Polubarinova-Kochina 1952, 1962). Hence, with:

$$d\Psi/ds = -N\, dx/ds = (\partial \Psi / \partial x)\, dx/ds + (\partial \Psi/\partial z)\, dz/ds \tag{7.1.40}$$

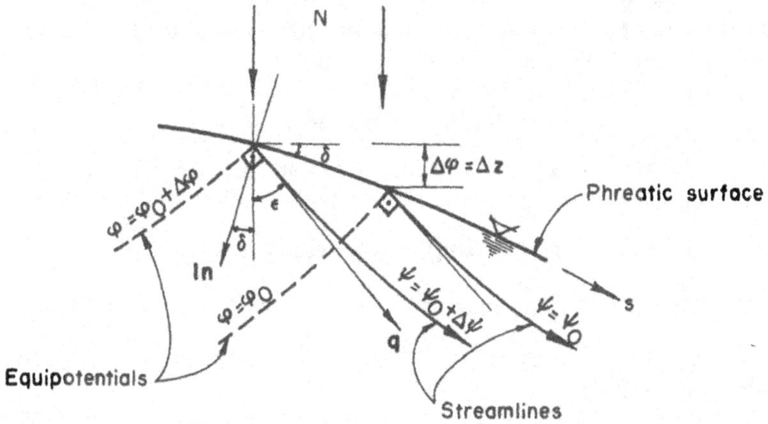

FIG. 7.1.5. Streamlines intersecting a steady phreatic surface with accretion (Isotropic medium).

and using the same procedure as outlined in paragraph 7.1.3, we obtain (7.1.38) written for two-dimensional flow.

With angles ε and δ as defined on figure 7.1.5, we have the following relationship at the phreatic surface. By continuity:

$$N \cos \delta = q \cos(\varepsilon + \delta); \qquad q = N \cos \delta / \cos(\varepsilon + \delta)$$

$$q_x = - K_x \, \partial\varphi/\partial x = - q \sin \varepsilon = - N \cos \delta \sin \varepsilon / \cos(\varepsilon + \delta)$$

$$q_z = - K_z \, \partial\varphi/\partial z = - q \cos \varepsilon = - N \cos \delta \cos \varepsilon / \cos(\varepsilon + \delta).$$

Since $\varphi = z$, we obtain:

$$d\varphi/ds = (\partial\varphi/\partial x)\, dx/ds + (\partial\varphi/\partial z)\, dz/ds = (\partial\varphi/\partial x) \cos \beta_{sx} + (\partial\varphi/\partial z) \cos \beta_{sz} = \cos \beta_{sz}. \tag{7.1.41}$$

Then, with $\beta_{sx} = \delta$, we obtain:

$$\tan \delta = \frac{\partial\varphi}{\partial x} + \frac{\partial\varphi}{\partial z} \tan \delta = \frac{N}{K_x} \frac{\cos \delta \sin \varepsilon}{\cos(\varepsilon + \delta)} + \frac{N}{K_z} \frac{\cos \delta \cos \varepsilon}{\cos(\varepsilon + \delta)} \tan \delta$$

$$\tan \delta = \left(\frac{N}{K_x} \frac{\cos \delta \sin \varepsilon}{\cos(\varepsilon + \delta)} + \frac{N}{K_z} \frac{\sin \delta \cos \varepsilon}{\cos(\varepsilon + \delta)} \right) \tag{7.1.42}$$

from which one can determine the direction ε of the streamline. For the isotropic case, $K_x = K_z = K$:

$$\tan \delta = \frac{N}{K} \tan(\varepsilon + \delta); \qquad \varepsilon = \tan^{-1}\left(\frac{K \tan \delta}{N}\right) - \delta. \tag{7.1.43}$$

7.1.6 The Unsteady Free (or Phreatic) Surface with Accretion

The unsteady free surface with accretion is also a surface on which the pressure p is maintained constant. We may therefore use (4.1.25) with $B_\alpha \equiv p$, and $\mathbf{V}_{B_\alpha}$ the velocity of propagation of the free surface on which the property $B_\alpha \equiv p = 0$ is maintained constant. This velocity is in a direction normal to the moving surface.

Continuity requires that at a moving interface:

$$(\mathbf{q} - \mathbf{N}) \cdot \mathbf{1n} = n_e \mathbf{V}_{B_\alpha} \cdot \mathbf{1n} = n_e V_{B_\alpha} \tag{7.1.44}$$

where n_e is effective porosity defined in paragraph 7.1.5. With (7.1.44), equation (4.1.25) becomes:

$$\partial p/\partial t + (1/n_e)(\mathbf{q} - \mathbf{N}) \cdot \operatorname{grad} p = 0. \tag{7.1.45}$$

Since on the free surface,

$$\varphi = z, \qquad \partial p/\partial t = \gamma \, \partial\varphi/\partial t, \qquad \nabla p = \gamma(\nabla\varphi - \mathbf{1z}), \qquad \mathbf{N} \equiv - \mathbf{1z} N,$$

we obtain from (7.1.45)

$$\frac{\partial\varphi}{\partial t} - \frac{1}{n_e}\left[K_x \left(\frac{\partial\varphi}{\partial x}\right)^2 + K_y \left(\frac{\partial\varphi}{\partial y}\right)^2 + K_z \left(\frac{\partial\varphi}{\partial z}\right)^2 - \frac{\partial\varphi}{\partial z}(K_z + N) + N\right] = 0. \tag{7.1.46}$$

Equation (7.1.46) may also be obtained by using (7.1.28) with $F \equiv \varphi(x, y, z, t) - z = 0$, and $\mathbf{V} = (\mathbf{q} - \mathbf{N})/n_e$.

Equation (7.1.46) may also be linearized (Dagan 1964, and sec. 8.4) to the form:

$$n_e \, \partial\varphi/\partial t + (K_z + N) \, \partial\varphi/\partial z - N = 0. \tag{7.1.47}$$

7.1.7 Boundary of Saturated Zone (or of Capillary Fringe)

As explained in paragraph 9.4.2, the capillary fringe is an approximate concept according to which we assume that a completely saturated zone exists beyond the phreatic or free surface. The pressure in this zone is negative (i.e., less than atmospheric), varying from $p = 0$ to the critical capillary pressure $p = -p_c$ at the inner side of the surface that serves as the external boundary of the capillary fringe (figure 7.1.6). On the outer side of this boundary we *assume* the presence of air only, at atmospheric pressure. For this reason, the boundary of saturation is sometimes also referred to as a free surface. Flow takes place within the capillary fringe and streamlines cross the phreatic surface in both directions. Even in steady flow, the phreatic surface is no longer a stream surface. As in the case of the free surface, the position and shape of this boundary are unknown a priori.

When the flow is steady and no accretion occurs along it, the boundary of the saturated zone is a stream surface. The two conditions on it are: (a) $p = -p_c =$ const, i.e., $\varphi = z - p_c/\gamma$, and (b) $q_n = 0$. These are the same conditions as those corresponding to the free surface (par. 7.1.3). The boundary conditions (7.1.24)

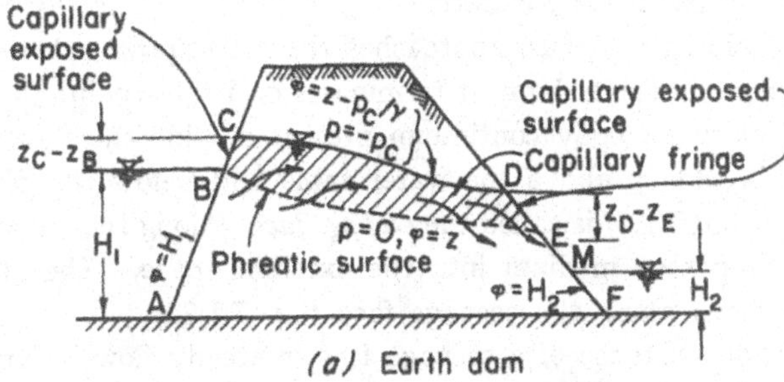

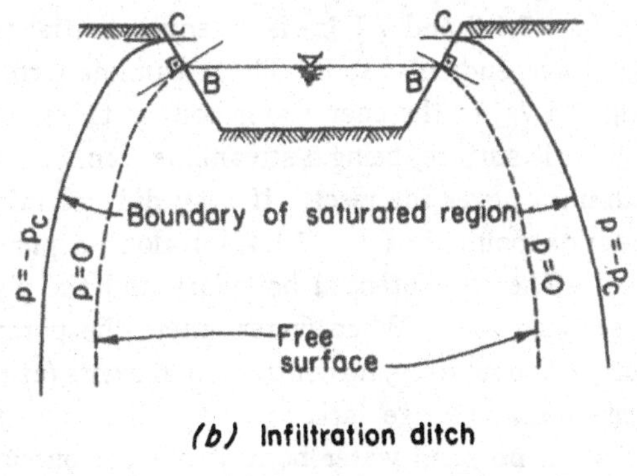

FIG. 7.1.6. Boundary of saturated zone.

and (7.1.25) are, therefore, also applicable here.

A similar analysis for the cases of steady flow with accretion and of unsteady flow with or without accretion will lead to the conclusion that in all cases, the free surface conditions are also applicable to the boundary of the saturated zone.

7.1.8 The Seepage Face

From (7.1.24) it follows that along a steady phreatic surface, since:

$K_x(\partial\varphi/\partial x)^2 + K_y(\partial\varphi/\partial y)^2 + K_z(\partial\varphi/\partial z)^2 \geqslant 0, \quad K_z\,\partial\varphi/\partial z \geqslant 0$ or $q_z = -K_z\,\partial\varphi/\partial z \leqslant 0.$

Similarly, from (7.1.38) it follows that since $K_x, K_y, K_z \geqslant 0$ and $K_x(\partial\varphi/\partial x)^2 + K_y(\partial\varphi/\partial y)^2 \geqslant 0$, we have:

$$\left(\frac{K_z - N}{2K_z}\right)^2 - \left(\frac{\partial\varphi}{\partial z} - \frac{K_z + N}{2K_z}\right)^2 \geqslant 0; \qquad \left(\frac{\partial\varphi}{\partial z} - \frac{N}{K_z}\right)\left(1 - \frac{\partial\varphi}{\partial z}\right) \geqslant 0$$

from which, with $N > 0$ (by our definition), we have:

$$0 \leqslant \frac{N}{K_z} \leqslant \frac{\partial\varphi}{\partial z} \leqslant 1; \qquad 0 \geqslant -N \geqslant q_z = -K_z\frac{\partial\varphi}{\partial z} \geqslant -K_z. \tag{7.1.48}$$

Or, the vertical component of flow along a phreatic surface is always directed downward; the phreatic surface always has a downward slope in the direction of the flow.

This should have been obvious from $\varphi = z$ on the phreatic surface where φ decreases in the direction of the flow. The same considerations can be shown to apply to the boundary of saturation.

As the steady phreatic surface approaches the *downstream external boundary* of a flow domain (e.g., an earth dam), it terminates on it at a point that is above the water table of the water-body continuum present outside the flow domain. The segment of the boundary above the water table and below the phreatic surface (EM in figs. 7.1.2 and 7.1.4) is called the *seepage face*. Along the seepage face, water emerges from the porous medium into the external space. The emerging water usually trickles down along the seepage face (fig. 7.1.7a).

The necessity for existence of a seepage face in steady flow is further indicated by the following observation. The interface between the flow domain and the open water body (MN in figs. 7.1.2 and 7.1.7a) is an equipotential surface. In isotropic media, streamlines are perpendicular to it. The streamline terminating at the water table (point M of fig. 7.1.7a) of the open water body is therefore also perpendicular to it. The steady phreatic surface, being a streamline, cannot terminate at the same point, as two streamlines never intersect. If they did, an infinite velocity would occur at the intersection point, but by (7.1.48) q along a phreatic surface cannot exceed K_z downward. The equipotential boundary and the phreatic surface must be separated by a seepage face. When water enters the porous medium through an upstream equipotential boundary, no contradiction exists (at point B of fig. 7.1.2), and there is no need for a seepage face.

The same is true when no open water body is present outside the flow domain, but the phreatic surface approaches an impervious boundary (fig. 7.1.7b). The

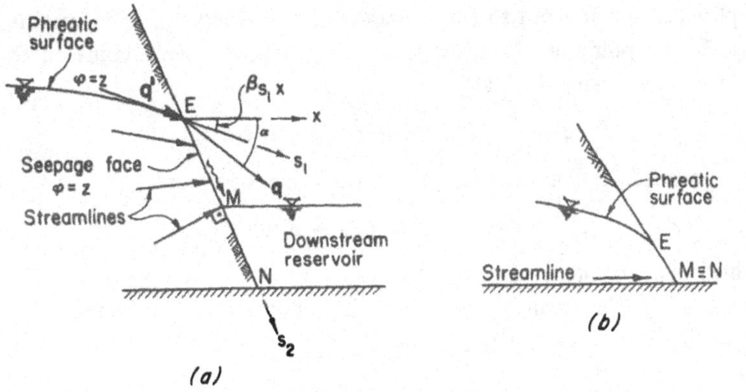

FIG. 7.1.7. Seepage face (Isotropic medium).

impervious boundary, being a streamline, cannot intersect the phreatic surface. A seepage face must separate them.

The seepage face is part of the boundary of the phreatic flow domain. Its geometry is generally known, *except* for its upper limit, which is also lying on the (a priori) unknown phreatic surface. The location of this point is, therefore, part of the required solution.

The seepage face is exposed to the atmospheric pressure (neglecting the thin layer of water flowing above it). Actually, in order for water to emerge from the porous medium domain, the pressure just inside the boundary should be somewhat higher than atmospheric. With $p = 0$ along the seepage face, we have the boundary conditions:

$$\varphi = z \text{ on } S \text{ or } C$$

and:

$$\frac{d\varphi}{ds} = \frac{dz}{ds} = \cos \beta_{sz} = \frac{\partial \varphi}{\partial x} \cos \beta_{sx} + \frac{\partial \varphi}{\partial z} \cos \beta_{sz} \text{ on } C \qquad (7.1.49)$$

as in the case of a phreatic surface. However, whereas in the latter case the entire shape and position of the surface are unknown, here only the upper limiting point (E in fig. 7.1.2) is unknown. Also, unlike the steady phreatic surface, the seepage face is not a streamline, as streamlines terminate along it.

Since the seepage face $F(x, y, z) = 0$ is an equipressure surface, we have along it:

$$\nabla p \times \nabla n = 0; \qquad \nabla(\varphi - z) \times \nabla F = 0.$$

In two-dimensional flow in the vertical xz plane (7.1.41) is valid. When the seepage face is a straight line, β_{sx} and β_{sz} remain constant. Equations (7.1.41) may be expressed in terms of q_x, q_z or in terms of Ψ.

To show that the steady or nonsteady free surface is tangent to the seepage face at its limiting point, consider the two-dimensional case shown in figure 7.1.7a; the contact is at point E. The unit vectors $\mathbf{1s}_1$ and $\mathbf{1s}_2$ are in the directions of the tangent to the phreatic surface and to the seepage face, respectively. The specific discharge at point E as a point on the seepage face is $\mathbf{q}$. At the same point on the phreatic surface we have, using (7.1.41),

$$\mathbf{q}' = -(K\,d\varphi/ds_1)\mathbf{1s}_1 = -K\sin\beta_{s_1x}\mathbf{1s}_1.$$

The components of $\mathbf{q}$ and of $\mathbf{q}'$ in the direction of s_1 are:

$$q\cos(\alpha - \beta_{s_1x}) = -K\sin\beta_{s_1x}. \tag{7.1.50}$$

The component of $\mathbf{q}'$ in the direction of $\mathbf{1s}_2$ is: $-K\,\partial\varphi/\partial s_2 = -K\cos\beta_{s_2z} = -K\sin\beta_{s_2x}$. By equating components in the $\mathbf{1s}_2$ direction we have:

$$q\cos(\beta_{s_2x} - \alpha) = -K\sin\beta_{s_2x}.$$

Dividing (7.1.50) by the last equation, and rearranging terms, we obtain:

$$\cos\alpha\sin(\beta_{s_1x} - \beta_{s_2x}) = 0.$$

Hence, either $\alpha = 90°$ (see fig. 7.3.20c) or $\beta_{s_1x} = \beta_{s_2x}$, which means that the phreatic surface is tangent to the seepage face. For the steady phreatic surface, which is also a streamline, $\mathbf{q}$ is in the direction of $\mathbf{1s}_1$ and the analysis may be simplified.

In paragraph 7.3.4 we shall reconsider this problem by mapping the boundaries in the hodograph plane.

7.1.9 Capillary Exposed Faces

When a capillary fringe is present above the phreatic surface (fig. 7.1.6a), or beyond a free surface (fig. 7.1.6b), the boundary of the saturated region intersects the downstream external boundaries of the porous medium at a curve (D in the two-dimensional flow in fig. 7.1.6a) that is at some distance above the seepage face. Similarly, it intersects the upstream boundary of the porous medium domain at a curve (C in figs. 7.1.6a and 7.1.6b) that is at some distance above the upstream equipotential boundary that terminates at the upstream water table (point B). The segments BC and DE (fig. 7.1.6) are thus boundaries of the flow domain. They are called *capillary exposed boundaries* (Bear, Zaslavsky and Irmay 1968). Their geometry is known (as it coincides with the external boundaries of the flow domain), except for their upper points (or curves) C and D, which also belong to the a priori unknown boundary of the saturated region. In unsteady flow, the position of the upper limit of the capillary exposed surface varies with time.

A similar situation exists under confined conditions (segments BC in fig. 7.1.8).

When no accretion takes place, the capillary exposed boundaries are streamlines—as no flow can take place across them. The boundary conditions of an impervious boundary are, therefore, applicable. At the upper limits of the capillary exposed boundaries (points C and D, fig. 7.1.6a) the pressure is $-p_c$. Along the segment BC, the pressure varies between zero at B and $-p_c$ at C. Along the downstream segment, DE, the pressure varies from $-p_c$ at D to zero at E. Because of the head losses along these segments:

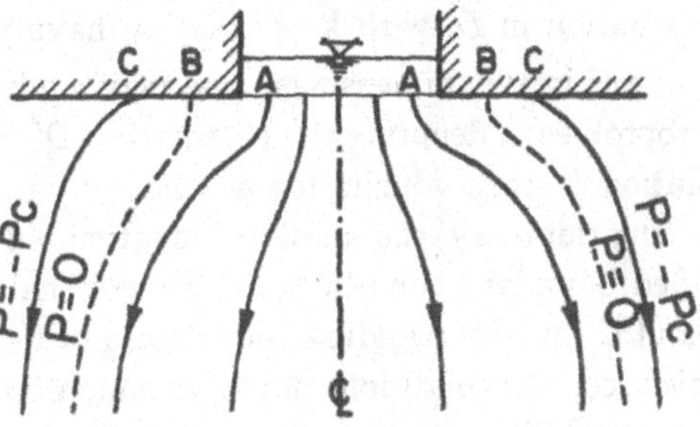

FIG. 7.1.8. Capillary exposed surfaces (AC) under a confining layer.

$$(z_C - z_B) < p_c/\gamma < (z_D - z_B). \tag{7.1.51}$$

When accretion takes place along the capillary exposed boundaries in two-dimensional flow in the vertical xz plane, equations (7.1.39) and (7.1.40) are applicable. From these equations we obtain for an anisotropic medium:

$$-N = -K_z(\partial\varphi/\partial z) + K_x(\partial\varphi/\partial x)\tan\beta_{sx}. \tag{7.1.52}$$

7.1.10 Discontinuity in Permeability

Figure 7.1.9 shows a boundary (S) of this type. It divides the three-dimensional flow domain D into two subdomains, each having a different permeability: $\mathbf{k}'$ in subdomain D', and $\mathbf{k}''$ in subdomain D''. Instead of solving for φ, satisfying the

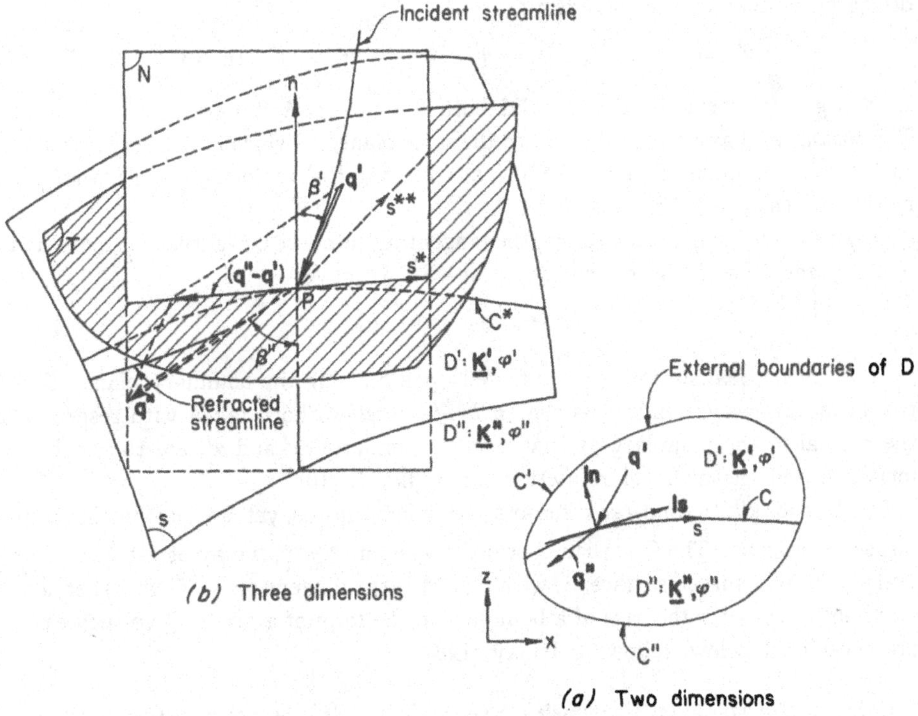

FIG. 7.1.9. Boundary between regions of different permeability.

partial differential equation in D, with $\underline{\mathbf{k}} = \underline{\mathbf{k}}(x, y, z)$ having a discontinuity along the surface S (or curve C in two dimensions), one may equivalently decompose the problem into two subproblems, denoting the potential in D' by φ', and in D'' by φ''. A simultaneous solution is then sought for φ' and φ'', satisfying the considered partial differential equation, say the Laplace equation, in their respective subdomains, and specified boundary conditions on the external boundaries S' (for φ') and S'' for φ'' (C' and C'' in the two-dimensional case of fig. 7.1.9a). In addition, φ' and φ'' must satisfy certain conditions on the common boundary S—the surface of discontinuity in permeability. As two variables in two regions are involved, two conditions must be specified along this boundary: one for φ' and one for φ'' (or two conditions, each in terms of both φ' and φ'').

Since the elevation z and the pressure p are the same when a point on the boundary is approached from both sides, we have:

$$z' = z'', \qquad p' = p'', \qquad \varphi' = \varphi'' \text{ for all points on } S. \tag{7.1.53}$$

In two-dimensional flow, since both φ' and φ'' are functions of the length s measured along the boundary curve C (fig. 7.1.9a), we also have:

$$\partial\varphi'/\partial s = \partial\varphi''/\partial s \text{ on } C. \tag{7.1.54}$$

The second condition on S or C is obtained from the requirement of continuity of flux across the boundary. This is expressed through the components of specific discharge normal to the boundary:

$$q'_n = q''_n; \qquad \mathbf{q}' \cdot \mathbf{1n} = \mathbf{q}'' \cdot \mathbf{1n}; \qquad (\mathbf{q}'' - \mathbf{q}') \cdot \mathbf{1n} = 0. \tag{7.1.55}$$

In this general form (7.1.55) is valid both for two- and three-dimensional flows. This means (a) that $\mathbf{q}'$, $\mathbf{q}''$ and $\mathbf{1n}$ are in the same plane, N, which is also perpendicular to the tangent plane T (fig. 7.1.9b), and (b) that the vector $(\mathbf{q}'' - \mathbf{q}')$, which is normal to $\mathbf{1n}$, is in the plane T.

Denoting by s^* the distance along the tangent to the curve C^* (curve of intersection of the plane N with the boundary surface S), we have:

$$\partial\varphi'/\partial s^* = \partial\varphi''/\partial s^* \text{ on } C^*. \tag{7.1.56}$$

Thus, an *incident streamline* is refracted upon crossing the boundary plane. β' is the *angle of incidence* and β'' is the *angle of refraction* (both taken with respect to the normal to the boundary at that point). Similarly, α' and α'' are the angle of incidence and the angle of refraction for $\mathbf{J}$ (fig. 7.1.10).

Let $\underline{\mathbf{k}}'$ and $\underline{\mathbf{k}}''$ (second-rank tensors) be homogeneous yet anisotropic in their respective regions. The Cartesian coordinate system s^*, s^{**}, n may be established at every point of a curved surface, S, with s^*, s^{**} in the tangent plane T. It is therefore sufficient to consider the case of a boundary in the form of a plane. Two such cases are considered below, following Irmay (1964).

Case 1, media have axially symmetric permeabilities with x, y, z as principal directions. Let both media have axially symmetric permeabilities, and let x, y, z be the principal

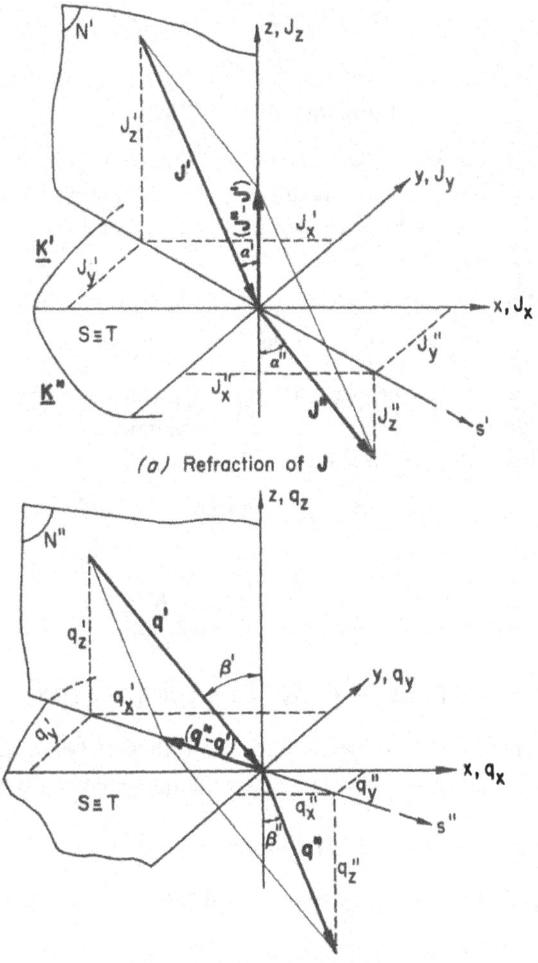

(*b*) Refraction of **q**

FIG. 7.1.10. Refraction at a plane boundary between two permeabilities.

directions of both **k**′ and **k**″ (fig. 7.1.10). In such case $k'_x = k'_y \neq k'_z$; $k''_x = k''_y \neq k''_z$. The plane $\bar{S}$ coincides with the xy plane. From (7.1.53) it follows that:

$$\partial\varphi'/\partial x = \partial\varphi''/\partial x; \qquad J'_x = J''_x; \qquad (\mathbf{J}' - \mathbf{J}'')\cdot\mathbf{1x} = 0$$

$$\partial\varphi'/\partial y = \partial\varphi''/\partial y; \qquad J'_y = J''_y; \qquad (\mathbf{J}' - \mathbf{J}'')\cdot\mathbf{1y} = 0. \qquad (7.1.57)$$

This means that the vectors' difference $(\mathbf{J}' - \mathbf{J}'')$ is normal to both $\mathbf{1x}$ and $\mathbf{1y}$ and hence in the direction of $\mathbf{1z}$ (fig. 7.1.10a). Also, $\mathbf{J}'$, $\mathbf{J}''$ and $\mathbf{1z}$ are in the same plane, N', that intersects the xy plane along s'. Since:

$$\tan\alpha' = J'_s/J'_z; \qquad \tan\alpha'' = J''_s/J''_z; \qquad J'_s = J''_s$$

we have:

$$\tan\alpha'/\tan\alpha'' = J''_z/J'_z. \tag{7.1.58}$$

But from (7.1.55), $K'_z J'_z = K''_z J''_z$. Hence:

$$\tan\alpha'/\tan\alpha'' = K'_z/K''_z. \tag{7.1.59}$$

This is the *law of refraction of the gradient* **J**. Equation (7.1.59) is also applicable to two-dimensional flow (in the $s'z$ plane). One can then derive from it the refraction of the equipotential lines (which are normal to **J**). Only the permeability normal to the boundary appears in (7.1.59).

In figure 7.1.10b,

$$q'_z = q''_z; \qquad (\mathbf{q}'' - \mathbf{q}') \cdot \mathbf{1z} = 0 \tag{7.1.60}$$

so that q', q'' and $\mathbf{1z}$ are in one plane ($s''z$ plane), which is perpendicular to the xy plane. Also, the vector $(\mathbf{q}'' - \mathbf{q}')$ must be in a plane that is normal to $\mathbf{1z}$ (i.e., in the xy plane). From figure 7.1.10b:

$$\tan\beta' = q'_s/q'_z; \qquad \tan\beta'' = q''_s/q''_z \tag{7.1.61}$$

but since

$$q'_z = q''_z \quad \text{and} \quad q'_{s''} = K'_{s''}J'_{s''}, \qquad q''_{s''} = K''_{s''}J''_{s''}; \qquad J'_{s''} = J''_{s''}$$

we obtain:

$$\tan\beta'/\tan\beta'' = K'_s/K''_s = K'_x/K''_x = k'_x/k''_x. \tag{7.1.62}$$

This *law of refraction of streamlines* is also applicable to two-dimensional flow (in the $s''z$ plane). Note that it depends only on the permeability parallel to the boundary plane.

When both media are isotropic (7.1.62) becomes:

$$\tan\beta'/\tan\beta'' = K'/K''. \tag{7.1.63}$$

From (7.1.63) it follows that:

(a) when $K' \gg K''$, $\beta' \gg \beta''$ and the refracted streamline approaches the normal to the boundary on passing from one layer to another, less pervious than the first; (b) when $K'' \gg K'$, $\beta'' \gg \beta'$, and the refracted streamline becomes practically horizontal on passing from a less pervious (e.g., semipervious) to a more pervious layer. This justifies the assumption of practically horizontal flow in a leaky, relatively thin, aquifer.

Figure 7.1.11a shows a more pervious layer between two less pervious ones. Figure 7.1.11b shows a less pervious layer between two more pervious ones.

When the contrast between the two permeabilities is very high, e.g., $K''/K' \to \infty$ in figure 7.1.11b, especially when we are interested in the flow in the K' layer, the interfaces between the layers behave as approximately equipotential boundaries.

Figure 7.1.12 shows the refraction of equipotentials and streamlines.

Case 2, anisotropic media with x, y, z in arbitrary directions. Consider the general

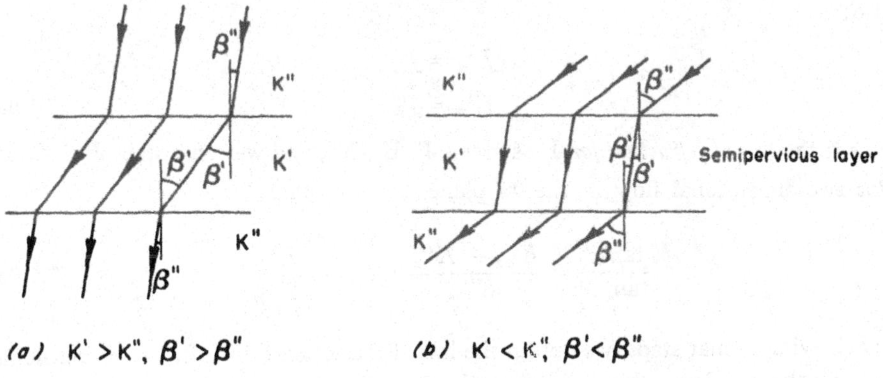

FIG. 7.1.11. Refraction of streamlines at the interface between two permeabilities.

case where the axes x, y, z do not coincide with any of the principal directions of $\underline{\mathbf{k}}'$ or $\underline{\mathbf{k}}''$. Equations (7.1.55) and (7.1.57), with n replaced by z, and figure 7.1.10 are still valid. From figure 7.1.10 we also have:

$$J'_y/J'_x = J''_y/J''_x; \qquad q'_y/q'_x = q''_y/q''_x. \tag{7.1.64}$$

Using (7.1.55) in the form $q'_z = q''_z$, we obtain:

$$K'_{zx}J'_x + K'_{zy}J'_y + K'_{zz}J'_z = K''_{zx}J''_x + K''_{zy}J''_y + K''_{zz}J''_z$$

or:

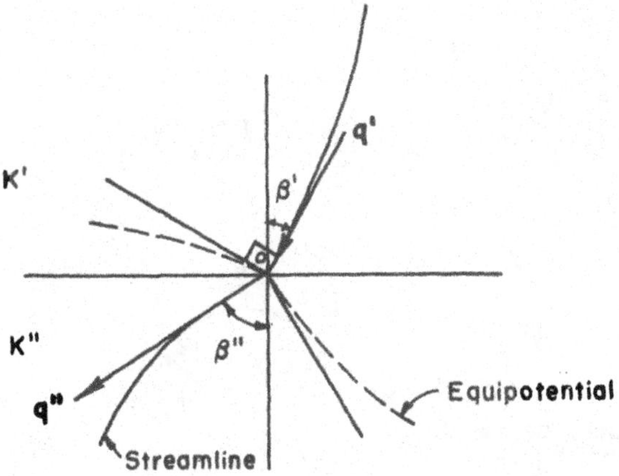

FIG. 7.1.12. Refraction of an equipotential line and a streamline at the interface between two isotropic media.

$$J''_z = [(K'_{zx} - K''_{zx})J'_x + (K'_{zy} - K''_{zy})J'_y + K'_{zz}J'_z]/K''_{zz}$$

also:

$$J''_y = J'_y$$
$$J''_x = J'_x. \tag{7.1.65}$$

This is the *law of refraction for* **J**, as given **J**′, **K**′, **K**″, and we can obtain **J**″ from it. For two-dimensional flow in the xz plane, this becomes:

$$\frac{\tan\alpha'}{\tan\alpha''} = \frac{K'_{zx} - K''_{zx}}{K''_{zz}} \tan\alpha' + \frac{K'_{zz}}{K''_{zz}}. \tag{7.1.66}$$

Following similar steps, we obtain the law of refraction for **q**. In two-dimensional flow in the xz plane, this law takes the form:

$$\frac{\cot\beta''}{\cot\beta'} = \frac{W''_{xz} - W'_{xz}}{W'_{xx}} \cot\beta'' + \frac{W''_{xx}}{W'_{xx}} \tag{7.1.67}$$

where **W** is the hydraulic resistivity.

A special kind of streamline, the refraction of which is sometimes sought at a discontinuity of permeability, is the phreatic surface or the interface (sec. 9.7). Figure 7.1.13 gives the nomenclature for the refraction of the phreatic surface. For the sake of simplicity, let the two media be homogeneous and isotropic and the flow be steady. The following conditions are satisfied at point P:

$$q_{na} = q_{nb}; \qquad \partial\varphi_a/\partial s = \partial\varphi_b/\partial s \quad \text{or} \quad q_{sa}/K_a = q_{sb}/K_b. \tag{7.1.68}$$

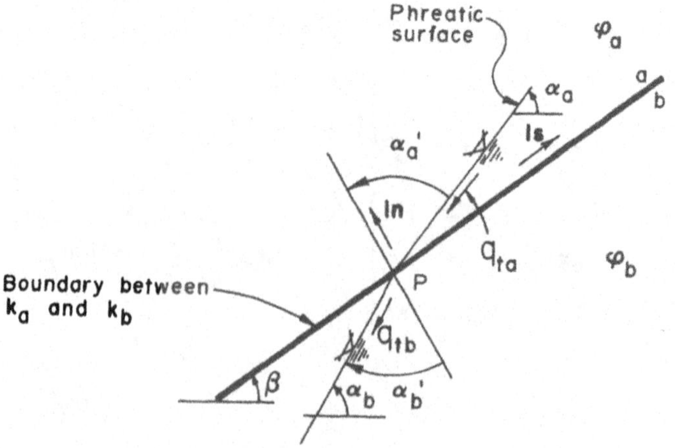

FIG. 7.1.13. Refraction of a phreatic surface at the interface between two isotropic media.

From (7.1.68) we obtain the known refraction conditions:

$$\frac{q_{sa}/q_{na}}{K_a} = \frac{q_{sb}/q_{nb}}{K_b} \quad \text{or} \quad \frac{\tan \alpha'_a}{K_a} = \frac{\tan \alpha'_b}{K_b}. \tag{7.1.69}$$

However, in this case we have additional conditions:

$$q_{ta} = -K_a \sin \alpha_a; \qquad q_{tb} = -K_b \sin \alpha_b$$

with:

$$\begin{aligned} q_{ta} \cos \alpha'_a = q_{na} &= -K_a \sin \alpha_a \cos \alpha'_a \\ q_{tb} \cos \alpha'_b = q_{nb} &= -K_b \sin \alpha_b \cos \alpha'_b \\ q_{ta} \sin \alpha'_a = q_{sa} &= -K_a \sin \alpha_a \sin \alpha'_a \\ q_{tb} \sin \alpha'_b = q_{sb} &= -K_b \sin \alpha_b \sin \alpha'_b. \end{aligned} \tag{7.1.70}$$

By combining (7.1.68) with (7.1.70), we obtain:

$$\begin{aligned} K_a \sin \alpha_a \cos \alpha'_a &= K_b \sin \alpha_b \cos \alpha'_b \\ \sin \alpha_a \sin \alpha'_a &= \sin \alpha_b \sin \alpha'_b \end{aligned} \tag{7.1.71}$$

or:

$$\begin{aligned} K_a \sin \alpha_a \sin (\beta - \alpha_a) &= K_b \sin \alpha_b \sin(\beta - \alpha_b) \\ \sin \alpha_a \cos(\beta - \alpha_a) &= \sin \alpha_b \cos(\beta - \alpha_b). \end{aligned} \tag{7.1.72}$$

Thus, for a given β (7.1.72) relates α_a to α_b. Possible solutions are:

(a) $\alpha_a = 0$, $\alpha_b = 0$, independent of β. This case corresponds to a horizontal phreatic surface (at both sides of P) that is a stagnation point.

(b) $\alpha_a = \alpha_b = \beta$. This case corresponds to a phreatic surface that is tangential to the discontinuity on both sides of P.

(c) $\qquad 0 \leqslant \beta \leqslant \pi/2; \qquad \alpha_a, \alpha_b < \beta;$

$$\begin{aligned} \operatorname{tg} \alpha_a &= \frac{K_a + K_b}{2K_a} \operatorname{tg} \beta + \left[\left(\frac{K_a + K_b}{2K_a} \operatorname{tg} \beta\right)^2 + \frac{K_b}{K_a}\right]^{1/2} \\ \operatorname{tg} \alpha_b &= \frac{K_a + K_b}{2K_b} \operatorname{tg} \beta + \left[\left(\frac{K_a + K_b}{2K_b} \operatorname{tg} \beta\right)^2 + \frac{K_a}{K_b}\right]^{1/2} \end{aligned} \tag{7.1.73}$$

(d) $\qquad \pi/2 \leqslant \beta \leqslant \pi; \qquad \alpha_a, \alpha_b < (\beta - \pi/2);$

α_a and α_b given by (7.1.73).

In conclusion, unlike the general case of a refracting streamline, a unique solution is obtained in each case for α_a and α_b. A similar analysis can also be carried out for a stationary interface (sec. 9.7).

7.1.11 A Note on Anisotropic Media

In certain cases, it is inconvenient to arrange for the principal directions of permeability x', y', z' to coincide with the horizontal–vertical x, y, z coordinate system. Three approaches are then possible.

(a) Treat the problem in the x, y, z coordinate system. This means that the partial differential equation describing the flow must be given in these coordinates and all boundary conditions must also be specified in terms of the x, y, z system.

It is sometimes easier to write the boundary conditions first in terms of x', y', z' and then to transform them into the x, y, z system by transforming both $\bar{\mathbf{K}}$ and the coordinates. In this process, the angles between the x, y, z and the x', y', z' axes must be determined (sec. 5.6).

(b) Treat the problem in the x', y', z' system. There is no difficulty in writing the boundary conditions in this system. In fact, some of the boundary conditions given in this section are written for the general anisotropic case. One should, however, pay special attention to the free surface boundary condition. There, the condition $\varphi = z$ becomes:

$$\varphi = x' \cos(\mathbf{1x}', \mathbf{1z}) + y' \cos(\mathbf{1y}', \mathbf{1z}) + z' \cos(\mathbf{1z}', \mathbf{1z}) \tag{7.1.74}$$

where $\mathbf{1x}'$, $\mathbf{1y}'$, $\mathbf{1z}'$ are unit vectors in the x', y', z' directions.

For example, the boundary condition on a steady phreatic surface in the vertical x', z' plane with permeabilities $K_{x'}, K_{z'}$, becomes:

$$K_{x'}\left[\left(\frac{\partial\varphi}{\partial x'}\right)^2 - \frac{\partial\varphi}{\partial x'}\sin\alpha\right] + K_{z'}\left[\left(\frac{\partial\varphi}{\partial z'}\right)^2 - \frac{\partial\varphi}{\partial z'}\cos\alpha\right] = 0 \tag{7.1.75}$$

where $\alpha \equiv (\mathbf{1x}', \mathbf{1x})$.

(c) Use the procedure discussed in section 7.4 to transform the problem into one in an equivalent isotropic domain and solve it in that domain.

7.1.12 Boundary Conditions in Terms of Pressure or Density

In reservoir engineering, where in most cases the fluids are compressible with $\rho = \rho(p)$, and in chemical engineering, the equations to be solved are given in terms of pressures, so that boundary conditions must also be specified in terms of pressure. The various boundary conditions can be derived by following the steps described in the previous paragraphs. For specific discharge we must use:

$$\mathbf{q} = -(k/\mu)(\nabla p + \rho g \mathbf{1z}). \tag{7.1.76}$$

7.2 A Well Posed Problem

In chapter 6, the various partial differential equations describing flow through porous media were developed. In the first section of this chapter we have discussed in detail the boundary and initial conditions encountered in problems of flow through porous media. With this information, any flow problem can now be stated mathematically:

(a) by defining the flow domain by giving the equation (or equations) of the surface (or surfaces) bounding it; part of the boundary may be at infinity; sometimes the boundary is unknown a priori;

(b) by determining the dependent variable (or variables) by means of which the flow is described; usually one of the variables will be pressure $p = p(x, y, z, t)$ or piezometric head $\varphi = \varphi(x, y, z, t)$;

(c) by specifying the partial differential equations to be satisfied by the dependent variables at all points within the flow domain;

(d) by specifying the boundary conditions that must be satisfied by the dependent variables at all points of the boundary;

(e) by prescribing the initial conditions when time is one of the independent variables.

The nature of the initial and boundary conditions (presented in sec. 7.1) is motivated by the physical reality of the flow problem considered, although it is not always obvious which boundary condition should be applied in a given situation. The physicist or the engineer usually derives the boundary and initial conditions by stating the physical reality in mathematical terms. Mathematicians often seek also a motivation from a purely mathematical point of view. In this book we have been following the physical–engineering approach.

A mathematical problem that corresponds to a physical reality should satisfy the following basic requirements (Courant and Hilbert 1962):

(a) The solution must *exist* (existence).

(b) The solution must be *uniquely determined* (uniqueness).

(c) The solution should *depend continuously on the data* (stability).

The first requirement states simply that a solution does, in fact, exist. The second requirement stipulates completeness of the problem—leeway or ambiguity should be excluded unless inherent in the physical situation. The third requirement means that a variation of the given data (e.g., boundary and initial conditions) in a sufficiently small range leads to an arbitrary small change in the solution. This requirement is also valid for approximate (e.g., numerical) solutions. We require that a small error in satisfying the equation be reflected by only a small deviation of the approximate solution from the true one. If small errors in the data do not result in correspondingly small errors in the solution, we should be inclined to think that our mathematical model of the physical phenomenon is badly formulated.

Any problem that satisfies these three requirements is called a *well* (or *properly*) *posed problem*.

In mathematical texts on the solution of partial differential equations (e.g., Garabedian 1964), before an actual solution is undertaken a rigorous analysis is usually carried out to prove *existence*, *uniqueness* and *stability* of the solution to various boundary value problems. In the present text we shall assume implicitly, without proof, that all mathematically stated problems are well posed. This follows from the fact that our problems are not mere mathematical exercises, where it is not always obvious that they are well posed, but are attempts to describe actual physical phenomena. It should be noted that certain problems mentioned in this book (e.g., those with a phreatic boundary condition) do not yet have proofs in the mathematical literature that they are indeed well posed. Obviously this does not exclude the possibility that mathematically stated problems describing physical phenomena are not well posed because of mistakes, overdetermination or simply because of ignorance as to the appropriate boundary conditions to be specified in certain situations.

7.3 Description of Boundaries in the Hodograph Plane

In this section, the hodograph plane is introduced only as a tool for describing the conditions along the boundaries of a domain of steady flow in a porous medium. In section 8.3, this description will be used as part of the *hodograph method* employed for solving problems of steady flow through porous media where a free surface is part of the boundary of the flow domain. Here, the description in the hodograph plane of the flow domain, together with its boundaries, should be regarded as just a tool by means of which one obtains a better understanding of the boundary conditions, specific discharge components, stagnation points, directions of flow, etc., without actually solving the problem. One can also use this tool to check whether or not one's assumptions on boundary conditions and on the kinematics of flow at the boundaries are compatible, since whenever these elements of the problem are indeed compatible, the representation of the boundaries in the hodograph plane forms a closed boundary (although portions of it may be at infinity).

The main advantage of using a hodograph plane is that whereas the geometric shape and position of the phreatic surface (or the interface) in the physical plane are unknown a priori, they are *completely specified* in the hodograph plane.

Although the literal meaning of the term *hodograph* is a "graph of velocities," we shall use it here to mean a graph of specific discharge.

The elements of the hodograph theory were presented by Helmholtz and Kirchhoff in the 19th century when they studied discontinuous motion in classical hydrodynamics. Prandtl and Tietjens (1934), who attribute the term "hodograph" to Hamilton, and many other texts in hydrodynamics describe the hodograph method mainly in connection with the treatment of two-dimensional steady free surface flows. Hamel (1934) developed the theory of the hodograph method in detail for the study of flow through porous media in connection with free surfaces and seepage surfaces. Dachler (1936) introduced the hodograph method to ground water flows. Muskat (1937), who describes the method, also mentions Davison (1932) as one who applied the method to the problem of an earth dam with vertical walls. Vedernikov (1934), Polubarinova-Kochina (1952, 1962), Aravin and Numerov (1953), Harr (1962) and Bear and Dagan (1962) describe and apply the method to various problems. Henry (1959), Bear and Dagan (1962a, 1964) and de Josselin de Jong (1964, 1965), among many others, apply the method to the solution of salt water–fresh water interface problems in ground water hydrology. Kidder (1956), Van Quy (1963) and others apply the method to problems encountered in reservoir engineering.

7.3.1 The Hodograph Plane

Let Γ be some curve in a two-dimensional field of flow (fig. 7.3.1a) in the xy plane—*the physical plane*. A certain specific discharge vector, $\mathbf{q}$, with components q_x and q_y in the x and y directions, respectively, exists at every point along this curve ($\mathbf{q}_A$ at A, $\mathbf{q}_B$ at B, etc.). Consider now another plane with a Cartesian coordinate system q_x, q_y parallel to x and y, respectively; its origin is at $0'$ (fig. 7.3.1b). From $0'$ we now draw the vectors $\mathbf{q}_A$, $\mathbf{q}_B$, $\mathbf{q}_C$, etc. The end points (A', B', C', etc.) of these vectors

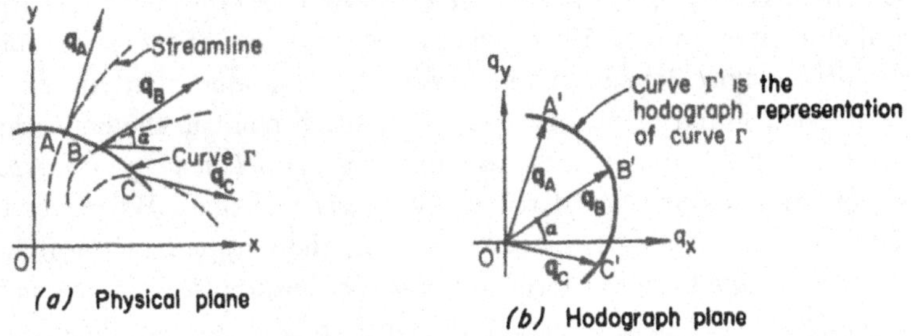

FIG. 7.3.1. Hodograph representation of a curve.

form a curve, Γ', which describes the specific discharge (in magnitude and in direction) at the corresponding points A, B, C, etc. along the curve Γ in the physical plane. We say that curve Γ' is the *hodograph* (or *hodograph representation*) of Γ. We may also regard this as a *mapping* procedure. Points (and curves) in the physical plane are mapped onto corresponding points (and curves) in the hodograph plane.

Figure 7.3.2 shows the hodograph of a curve, Γ, which is a streamline in the physical plane (and in steady flow is also a pathline or a trajectory of a fluid particle). The hodograph of an *isotach* (or *isovel*), i.e., a line along which the absolute value of the specific discharge $|\mathbf{q}|$ is a constant, is a circle centered at $0'$ and with radius $|\mathbf{q}|$ (dashed circle in fig. 7.3.2b).

The boundaries of a flow domain are usually composed of several straight or curved segments along each of which certain conditions regarding the distribution of the potentials or fluxes are given. Knowing these boundary conditions, it is possible to obtain the hodographs of the various boundary segments. The points at which adjacent segments meet each other are mapped onto the hodograph plane as the points at which the hodographs of these segments intersect each other. In most cases, although it is possible to obtain the hodograph of a boundary segment, it is not possible to identify points along it as being hodograph representations of particular points in the physical plane. However, we shall see both here and in section 8.3 that this is not necessary.

When the entire boundary of the flow domain (including segments at infinity)

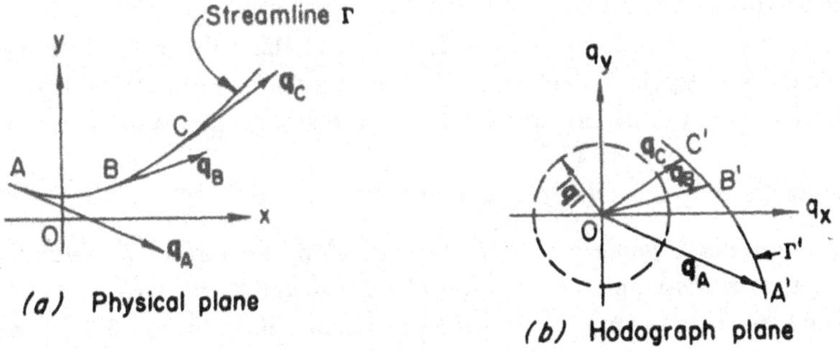

FIG. 7.3.2. Hodograph representation of a streamline.

is represented in the hodograph plane, it encloses a certain portion of this plane. If, in the flow domain, no specific discharge vector occurs at more than one point, each point of the flow domain is represented by a unique point in the enclosed portion of the hodograph plane, and conversely, each point in the hodograph plane represents one, and only one, point in the physical plane (*one-to-one correspondence*). Only cases of this kind are considered in the present section. Whenever the same specific discharge occurs at more than one point, the mapping of the flow domain in the hodograph plane takes the form of a *Riemann surface* (see example in sec. 9.6).

Although the analysis in the present section is related to two-dimensional flow, it can easily be extended to three-dimensional flow. We shall then have a *hodograph space* with a Cartesian coordinate system q_x, q_y, q_z parallel to x, y, z directions, respectively (Bear, Zaslavsky and Irmay 1968). Boundary surfaces in the physical plane will be represented by surfaces in this space. The extension is straightforward and requires no further explanation. Obviously, a graphic representation becomes cumbersome.

Instead of mapping the flow domain onto the hodograph plane (or space, in three-dimensional flow), one may consider another plane—the *gradient plane* (Bear, Zaslavsky and Irmay 1968)—with Cartesian coordinates J_x and J_y, parallel to x and y, respectively, onto which the flow domain may be mapped by considering the gradient vector $\mathbf{J}$ at each point. In three-dimensional flow this will be a *gradient space* with coordinates $J_x, J_y, J_z\,(= -\,\partial\varphi/\partial z)$. In isotropic media, since $\mathbf{q} = K\mathbf{J}$, the vectors $\mathbf{J}$ and $\mathbf{q}$ are parallel and differ only in absolute magnitude. In homogeneous isotropic media, the ratio $|\mathbf{q}|/|\mathbf{J}| = K =$ constant in the entire flow domain.

The following discussion will be limited mainly to two-dimensional flow in the vertical xy plane. We use $+\,y$ here as an axis directed vertically upward because of the special meaning that will be assigned to z in sections 7.8 and 8.3 (namely the complex number $z = x + iy$). The medium will be assumed homogeneous. The hodographs are given for an isotropic medium, although the mathematical expressions for the case of an anisotropic medium are also given. Unless otherwise specified, the horizontal and vertical directions will then be assumed to be principal directions. The corresponding hodographs may easily be obtained from these expressions. In general, a separate discussion of anisotropic media is not necessary, because one can always transform a problem given in an anisotropic medium into one in an isotropic medium (sec. 7.4).

Correctly drawn, if the flow domain is always to the right of the boundary when the latter is traversed in a clockwise direction, its mapping in the hodograph plane is to the left side of the boundary when traversed in the same direction.

7.3.2 Boundaries in the Hodograph Plane

Type 1, impervious boundary. At each point along an *impervious boundary*, or a boundary that is a streamline, equation (7.1.17) is valid. By writing this equation for two-dimensional flow in the vertical xy plane, and introducing (6.5.10), we obtain

$$q_y = q_x \tan \beta_{sx}. \tag{7.3.1}$$

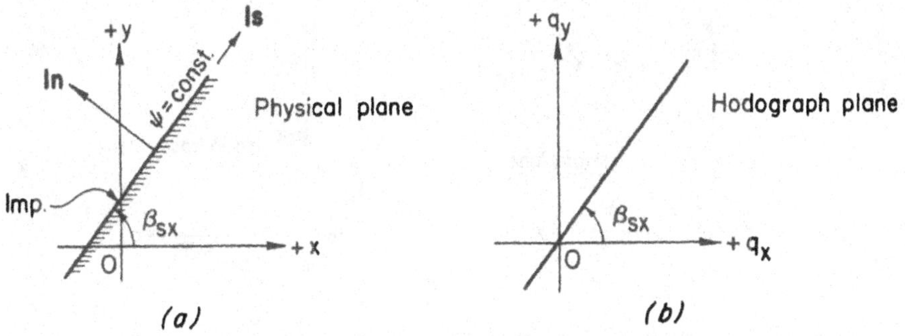

FIG. 7.3.3. Hodograph of an impervious boundary.

In general, β_{sx} varies from point to point along a curved impervious boundary (and so do the components q_x, q_y) in an unknown manner so that the mapping of points along the boundary in the physical xy plane onto the hodograph q_xq_y plane is not possible. However, when the impervious boundary takes the form of a *straight line* in the direction **1s**, making an angle β_{sx} with the $+\,x$ axis, its representation in the hodograph plane (or, simply, its hodograph) is a straight line passing through the origin and making the same angle β_{sx} with the $+\,q_x$ axis. In other words, the hodograph of a straight impervious boundary is a parallel straight line passing through the origin in the hodograph plane (fig. 7.3.3). Actually this should be obvious since flow is tangent to an impervious boundary.

One should recall that this does not mean we can identify specific points on this line as being the mapping of specific points on the considered boundary. All we know is that the various points along the boundary are mapped onto an as yet unknown segment of this line.

Equation (7.3.1), and hence also figure 7.3.3, is valid for isotropic as well as anisotropic media. However, if we wish to describe this boundary in the gradient plane (7.3.1) becomes;

$$J_y = J_x(K_x/K_y)\tan\beta_{sx} \tag{7.3.2}$$

and the hodograph (in the J_xJ_y plane) is no more parallel to the boundary in the physical plane.

Type 2, equipotential boundary. At each point along an *equipotential boundary* in an isotropic medium, we obtain from (7.1.11):

$$q_y = q_x \tan\alpha_{nx} = -\,q_x \operatorname{cotg}\beta_{sx} = q_x\tan(\beta_{sx} - 90^\circ). \tag{7.3.3}$$

This means that when the equipotential boundary is a straight line making a constant angle β_{sx} with the $+\,x$ axis, its hodograph is a straight line through the origin, making an angle $(\beta_{sx} - 90^\circ)$ with the $+\,q_x$ axis, i.e., normal to the boundary in the physical plane (fig. 7.3.4).

In an anisotropic medium, we obtain from (7.1.11):

$$q_y = q_x(K_y/K_x)\tan(\beta_{sx} - 90^\circ) \tag{7.3.4}$$

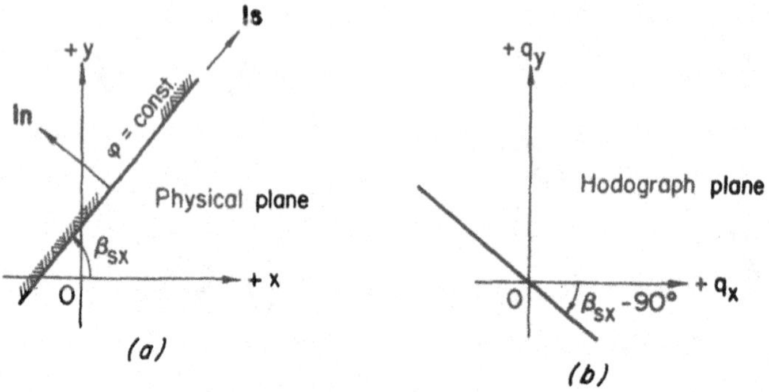

FIG. 7.3.4. Hodograph of an equipotential boundary (Isotropic medium).

which is also a straight line through the origin making an angle α with the $+x$ axis, such that $\alpha = \tan^{-1}[(K_y/K_x)\tan(\beta_{sx} - 90°)]$. However, in this case:

$$J_y = J_x \tan(\beta_{sx} - 90°) \tag{7.3.5}$$

i.e., the boundary is represented in the gradient plane by a straight line through the origin, perpendicular to the boundary in the physical plane.

Type 3, seepage face. For points along a *seepage face*, from (7.1.49) (with y replacing z), we obtain:

$$q_y = q_x(K_y/K_x)\tan(\beta_{sx} - 90°) - K_y. \tag{7.3.6}$$

For the isotropic case, this becomes:

$$q_y = q_x \tan(\beta_{sx} - 90°) - K. \tag{7.3.7}$$

The hodograph of a seepage face in the form of a straight line making an angle β_{sx} with the $+x$ axis in an isotropic medium is a straight line in the hodograph plane, passing through the point $(0, -K)$, and making an angle of $(\beta_{sx} - 90°)$ with the $+q_x$ axis. It is therefore perpendicular to the seepage face in the physical plane (fig. 7.3.5).

For the gradient plane, both for isotropic and anisotropic media, (7.3.6) becomes:

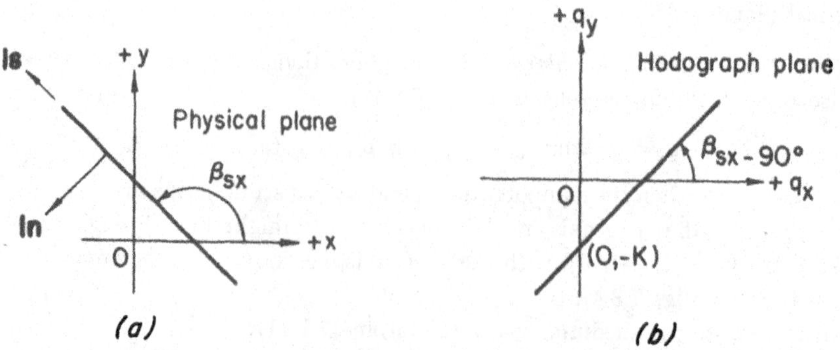

FIG. 7.3.5. Hodograph of a seepage face (Isotropic medium).

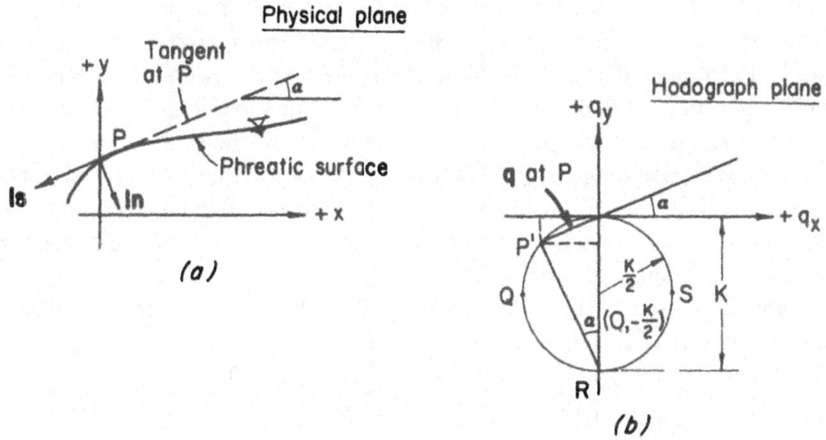

FIG. 7.3.6. Hodograph of a phreatic surface (Isotropic medium).

$$J_y = J_x \tan(\beta_{sx} - 90°) - 1. \tag{7.3.8}$$

Type 4, steady phreatic surface without accretion. For points along a *steady phreatic surface without accretion,* we have from (7.1.24):

$$\frac{q_x^2}{K_x} + \frac{q_y^2}{K_y} + q_y = 0; \qquad \frac{q_x^2}{[(K_xK_y)^{1/2}/2]^2} + \frac{(q_y + K_y/2)^2}{(K_y/2)^2} = 1. \tag{7.3.9}$$

For an isotropic medium, this becomes:

$$q_x^2 + (q_y + K/2)^2 = (K/2)^2. \tag{7.3.10}$$

The hodograph of a steady free surface in an isotropic porous medium is, therefore, a circle centered at $(0, -K/2)$, with radius $K/2$ (fig. 7.3.6).

Let P be a point on the phreatic surface at which the tangent to the latter makes an angle α with the $+x$ axis. This is also the direction of $\mathbf{q}$ at P. According to the hodograph transformation, P is mapped onto P' of the hodograph plane (fig. 7.3.6b). Unlike the previous boundaries, here it is possible to identify the corresponding points in both planes. The specific discharge q_s is therefore $q_s = -K \sin \alpha$ (compare with (7.1.26)).

Three observations based on figure 7.3.6 are: (a) the highest possible value of $-q_y$ along a phreatic surface is $-q_y = K$ (point R). At that point we have $q_x = 0$ and $q_y = -K$. This means that in the limiting case (point R), $\mathbf{q}$ is directed vertically downward. (b) The lowest value is $-q_y = 0$, where also $q_x = 0$, $q_y = 0$. Hence, q_y is never positive (i.e., the water table always slopes down, or is horizontal as a limiting case). (c) The highest and lowest possible values of q_x are $q_x = +K/2$ and $q_x = -K/2$, respectively. Then $q_y = -K/2$ (points Q and S).

Type 5, unsteady phreatic surface without accretion. Irmay (in chap. 6 of Bear, Zaslavsky and Irmay 1968, following Morel-Seytoux 1961) uses (7.1.32) written as:

$$q_x^2 + q_y^2 + (q_z + K/2)^2 = Kn_e\, \partial\varphi/\partial t + K^2/4 \tag{7.3.11}$$

to introduce an *instantaneous hodograph representation* of a phreatic surface flow without accretion in an isotropic medium. This is an equation of a sphere in a q_x, q_y, q_z Cartesian coordinate system (*hodograph space*) with $R = (Kn_e\,\partial\varphi/\partial t + K^2/4)^{1/2}$ and center at $(0, 0, -K/2)$. However, one must be careful in this interpretation of (7.3.11) as R also depends on the coordinates of the considered point on the phreatic surface (since $\varphi = \varphi(x, y, z, t)$). This means that the interpretation of (7.3.11) is that, for *every point* on the phreatic surface and for every instant, the hodograph representation is a point on a circle, the radius R of the circle varying from point to point and with time.

Several conclusions may be drawn from (7.3.11). (a) The extreme values of q_z are attained when $q_x = q_y = 0$. In general

$$-(Kn_e\,\partial\varphi/\partial t + K^2/4)^{1/2} - K/2 < q_z < +(Kn_e\,\partial\varphi/\partial t + K^2/4)^{1/2} - K/2. \quad (7.3.12)$$

When $\partial\varphi/\partial t > 0$, both positive and negative values of q_z are possible. When $q_z = 0$, q_x and q_y may have different positive, or negative, values. However, the extreme values are attained when $q_x = q_y$, $q_z = -K/2$. (b) When $\partial\varphi/\partial t < 0$ (7.3.12) is still valid, but always $q_z < 0$. (c) Conclusions (a) and (b) above are also valid for two-dimensional flow.

These conclusions may easily be derived from figure 7.3.7, which gives the hodograph representation of (7.3.11) written for the xy plane.

One should recall that in the discussion above, $\partial\varphi/\partial t$ varies with time. A similar analysis may be carried out for the anisotropic case. The instantaneous hodograph representation is an ellipsoid in three-dimensional flow and an ellipse in two-dimensional flow.

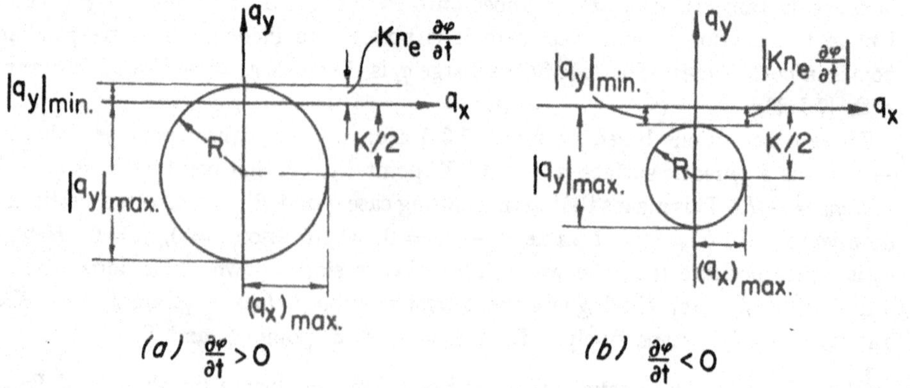

FIG. 7.3.7. Instantaneous representation of a phreatic surface in unsteady flow (after Irmay in Bear, Zaslavsky and Irmay, 1968).

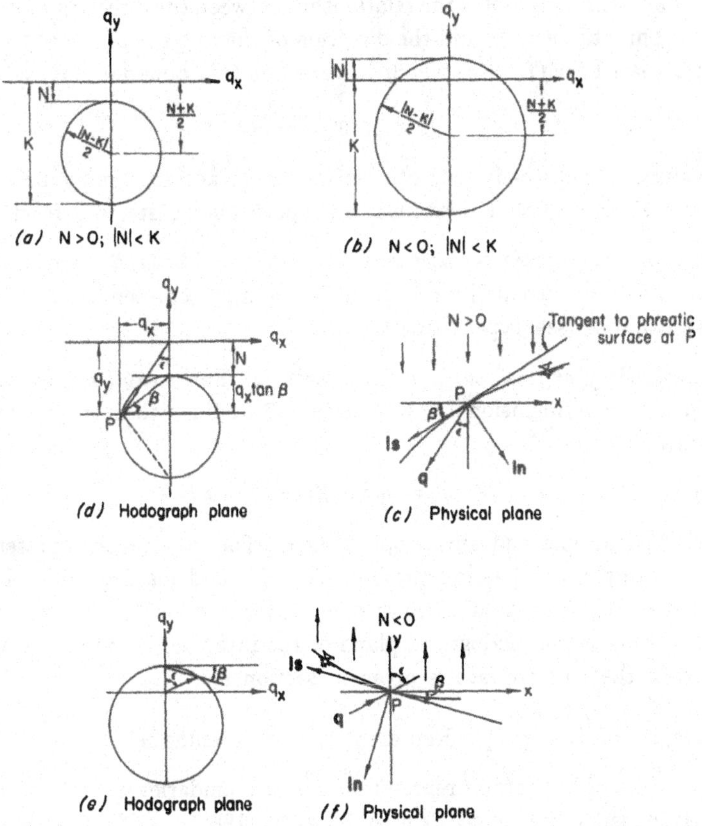

FIG. 7.3.8. Hodograph of free surface with accretion (Isotropic medium).

Type 6, steady phreatic surface with accretion. For a *steady phreatic surface with accretion* in the vertical *xy* plane, we have from (7.1.38):

$$\frac{q_x^2}{[((N-K_y)/2)(K_x/K_y)^{1/2}]^2}+\frac{[q_y+((N+K_y)/2)]^2}{((N-K_y)/2)^2}=1. \qquad (7.3.13)$$

The hodograph representation is, therefore, an ellipse centered at $(0, -(N+K_y)/2)$. For the isotropic case, this becomes:

$$q_x^2+[q_y+(N+K)/2]^2=[(N-K)/2]^2. \qquad (7.3.14)$$

This is a circle in the hodograph plane, with center at $(0, -(N+K)/2)$ and radius $|N-K|/2$. Two possibilities, according to whether $N>0$ (i.e., infiltration), or $N<0$ (i.e., evaporation), are shown on figure 7.3.8.

From (7.3.14) or from figure 7.3.8, several conclusions may be drawn. (a) For $N \geqslant 0$, $q_y \leqslant 0$. The smallest value of q_y is $q_y=-K$; the highest value is $q_y=-N$. (b) For $N \leqslant 0$, q_y may be positive or negative. Its highest value is $q_y=|N|$ and smallest value is $q_y=-K$.

Figures 7.3.8c through f show the relationship between the direction of $\mathbf{q}$ at a point P along the phreatic surface, and the direction of the tangent to the phreatic surface at P. It is based on (7.1.40), which is valid for this boundary, rewritten as:

$$q_y = q_x \tan \beta - N.$$

The analysis for unsteady phreatic surface with accretion may be carried out in a manner similar to that described for type 5 above, starting from (7.1.46).

Type 7, capillary exposed surfaces without accretion. The capillary exposed surfaces without accretion are streamlines. Their hodograph representation is, therefore, that of streamlines (see type 1 above).

Type 8, capillary exposed surface with accretion. For a *capillary exposed surface with accretion* in two-dimensional flow, equation (7.1.52) is valid. It may be rewritten in the form:

$$q_y = q_x \tan \beta_{sx} - N \tag{7.3.15}$$

valid for both isotropic and anisotropic media. The hodograph representation is, therefore, a straight line passing through $(0, -N)$ and making an angle β_{sx} with the $+q_x$ axis (i.e., it is parallel to the boundary).

The description in the hodograph plane of the interface between two immiscible liquids and of the outflow face is given in section 9.6.

7.3.3 Examples of Hodograph Representation of Boundaries

Several examples of hodograph representations of boundaries in cases of unconfined flow are given below, following Bear and Dagan (1962). Additional examples are given by Harr (1962) and by Polubarinova-Kochina (1952, 1962). In all cases we shall assume that the porous medium is isotropic and that the flow is steady. At certain points subscripts A, B, L, R are used to indicate that the point is approached from above, from below, from the left side, or from the right side, respectively.

In the first few problems, a detailed explanation is given of how to arrive at the required hodograph. Then, as the reader becomes more familiar with the construction of the hodograph of the boundary, the examples are given with only brief comments.

Example 1, an earth dam with parallel vertical walls (fig. 7.3.9). The flow domain is $ABCDEA$. Its boundaries are: an equipotential line, AB, a phreatic surface, BC, a seepage face, CD, an equipotential line, DE, and an impervious line, EA. Following the discussion in paragraph 7.3.2, these are now represented in figure 7.3.9b: BA, AE and ED_B are represented by straight lines through the origin making an angle $\beta_{sx} = 0$ with the $+q_x$ axis; the phreatic surface is represented by a circle of a radius $R = K/2$, centered at $(0, -K/2)$; the seepage face CD_A is a straight line through $(0, -K/2)$ normal to the seepage face in the physical field. The hodograph representation of points that are common to two boundary segments are at the intersection of their hodograph representation. In this way, points A', B', C', D'_A, D'_B and E' are obtained. One should note here that whereas the flow domain is

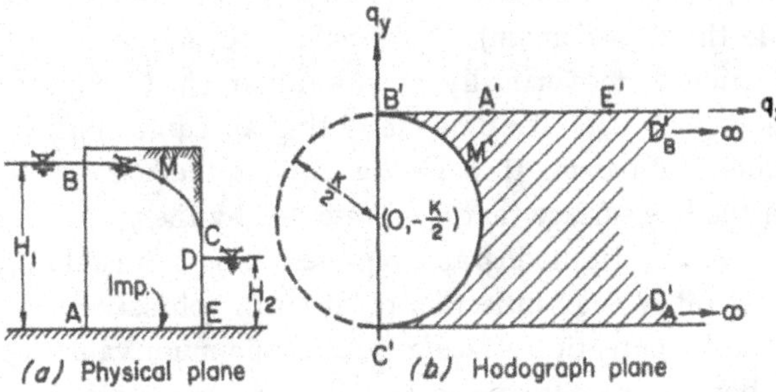

FIG. 7.3.9. Hodograph of boundaries for example 1.

on the right side of the boundary traveling from A to B, etc., it is on the left side of the boundary traveling from A' to B', etc.

Example 2, an earth dam above a pervious stratum (fig. 7.3.10). At A we have $q_x = \varepsilon \to 0$ (i.e., a very small positive value). AI is a streamline ($q_x > 0$, $q_y = 0$) along which q_x increases, reaching some maximum value at M. Then it decreases again to $q_x \to \varepsilon \to 0$. This yields the *cut* $A'M'I'$ in the hodograph plane. Here, as in the other examples, the cut is a device that helps us not to violate the one-to-one correspondence between points in the z and in the hodograph planes; in the latter plane, points just above and points below the cut are regarded as distinct. At B: $q_x = 0$, $q_y = -\varepsilon$; at H: $q_x = 0$, $q_y = +\varepsilon$. BC is a horizontal equipotential line.

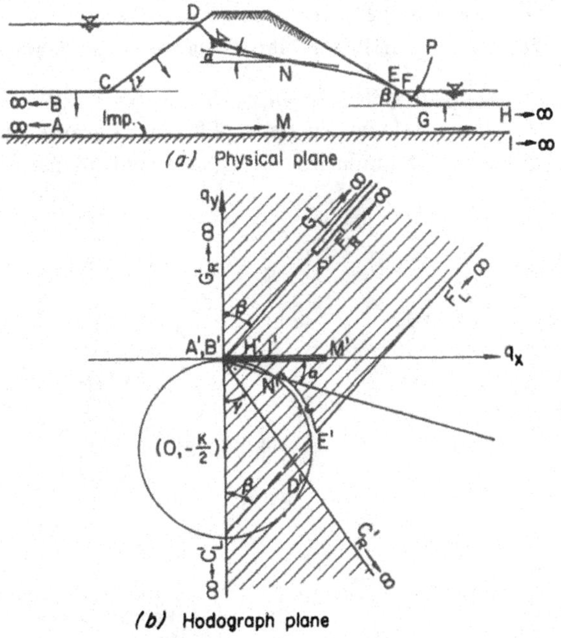

FIG. 7.3.10. Hodograph of boundaries for example 2.

It is therefore mapped as a vertical line through the origin. At C, because the angle $DCB > \pi$ (inside the flow domain), the specific discharge is infinite. However, it changes its direction from vertically downward at C_L to a downward direction perpendicular to CD at C_R. The line CD, being an equipotential line, is mapped on the straight line C'_R0 passing through the origin in the hodograph plane. Point D is also a point of the free surface that is represented by the circle of radius $R = K/2$, centered at $(0, -K/2)$. Its hodograph representation is therefore at the point of intersection D'. Point N is an inflection point on the phreatic surface DE. At this point the slope of the phreatic surface attains a minimum value (and so does $|q_x|$). Point E belongs both to the phreatic surface (E_L) and to the seepage face (E_R) whose hodograph representation is $E'F'_L$. We thus obtain the cut $E'N'$. The velocity at F is infinite in the direction normal to the equipotential FG. This leads to F'_R and F'_L at infinity. The segment $E'F'_L$, representing the seepage face, is a straight line normal to the seepage face and passing through $(0, -K)$. Along FG (with infinite velocities in a direction normal to FG at both F_R and G_L), there is a point P of finite velocity. This gives the cut $F'_RP'G'_L$. At G the velocity is again infinite, but changing direction from perpendicular to EF at G_L to perpendicular to GH at G_R. G_RH is a horizontal equipotential. Its representation in the hodograph plane coincides with the q_y axis. At H: $q_y = -\varepsilon$, $q_x = 0$. At I, $q_x = \varepsilon$, $q_y = 0$. This completes the description of the boundary. The physical flow domain is mapped onto the domain $A'B'C'_LC'_RD'N'E'F'_L(F'_R)P'G'_LG'_RH'I'M'A'$.

Example 3, an earth dam with change of slope on its upstream face (Polubarinova-Kochina 1952, 1962). This example is introduced to show what happens when there is an abrupt change of slope on the upstream surface of the dam (fig. 7.3.11). There is no tail water. No further explanations seem necessary.

Example 4, a parallel drain system in an aquifer of finite depth (Bear and Dagan 1962). The physical plane is shown on figure 7.3.12a. Because lines of symmetry are streamlines (impervious boundaries), it is sufficient to analyze the segment

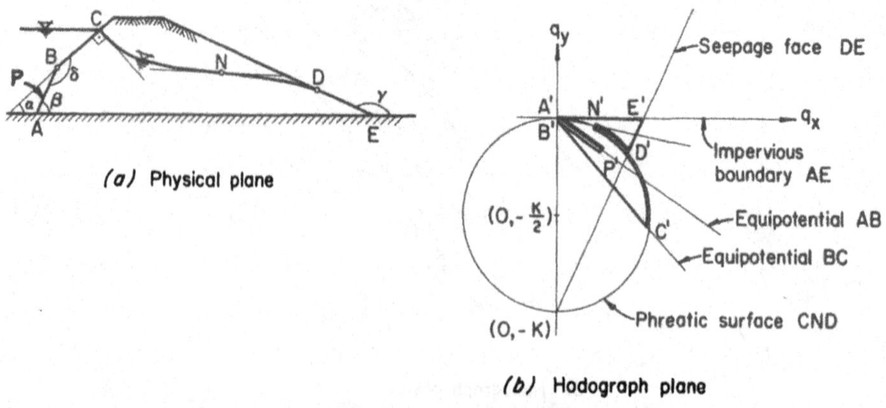

FIG. 7.3.11. Hodograph of boundaries for example 3.

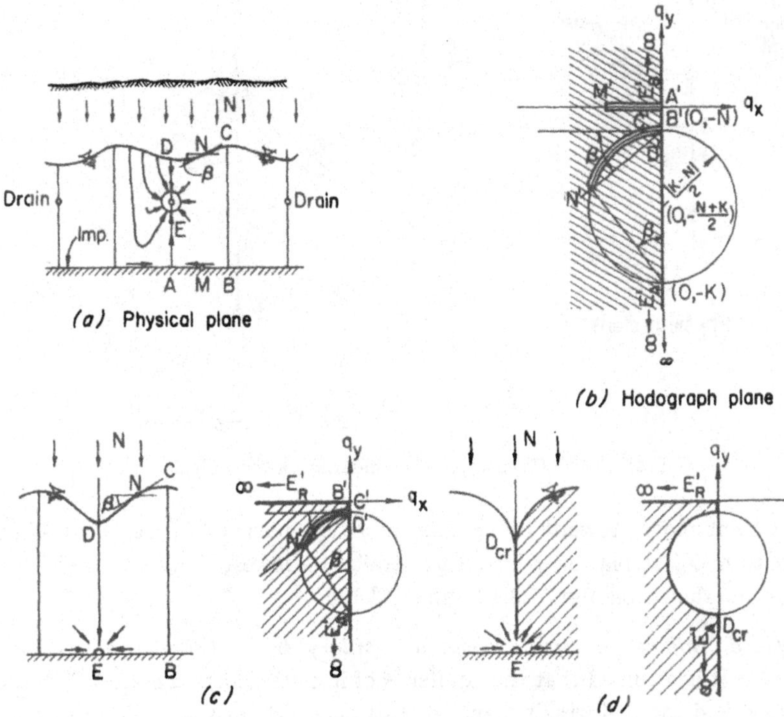

FIG. 7.3.12. Hodograph of boundaries for example 4.

ABCDEA, where E is the drain. The constant natural replenishment is N. We have here a phreatic surface with accretion (DC) as a boundary of the flow domain. A boundary of this kind is represented in the hodograph plane by a circle centered at $(0, -(K+N)/2)$ and with radius $|K-N|/2$.

The following observation guides the drawing of the hodograph representation of the boundaries: the specific discharge is infinite at the vicinity of every point source or sink; its direction varies around the point. Hence at E_A: $q_x = 0$, $q_y = -\infty$; at A_B: $q_x = 0$, $q_y = +\infty$. The specific discharge at a corner point where the boundary changes its direction in an abrupt manner by an angle α is zero when $0 \leq \alpha \leq \pi$, and infinite when $\pi < \alpha \leq 2\pi$. Hence, at B_A: $q_x = 0$, $q_y = -\varepsilon \rightarrow 0$; at B_L: $q_x = -\varepsilon \rightarrow 0$, $q_y = 0$; at A_R: $q_x = -\varepsilon$, $q_y = 0$; at A_A: $q_x = 0$, $q_y = +\varepsilon$; at M: $q_y = 0$ and q_x attains some maximum (absolute) value. Along the phreatic surface $q_x \leq 0$, $q_y < 0$. N is an inflection point along the phreatic surface. At D: $q_x = 0$, $q_y = -N$. At the inflection point, the tangent to the phreatic surface makes an angle β with the $+x$ axis.

When the drain is located on the impervious bottom (point E replacing A), the hodograph representation is shown on figure 7.3.12c. As we increase the discharge of the drain, the elevation of D (in the physical plane) is reduced, with $q_x = 0$. The

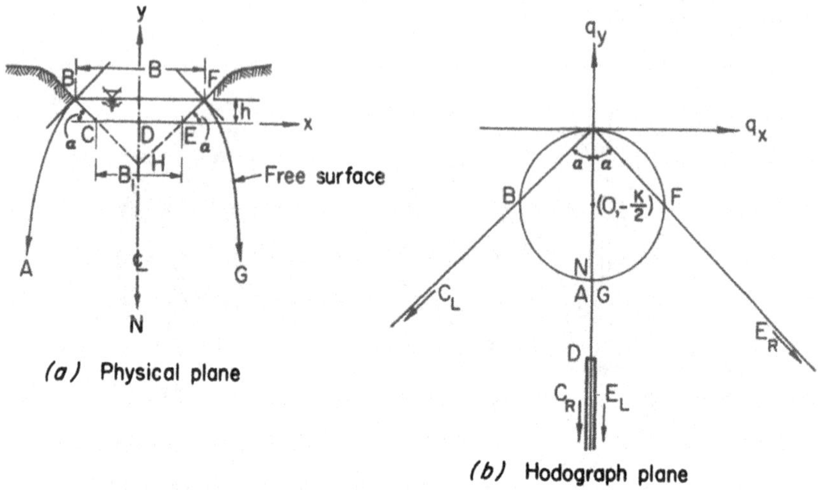

FIG. 7.3.13. Hodograph of boundaries for example 5.

inflection point moves toward D. Finally a critical point D_{cr} is reached, when N coincides with D_{cr} (replacing D), and q_y attains its lowest value of $-K$. These situations are shown on figs. 7.3.12c and 7.3.12d.

Example 5, infiltration ditch without a capillary fringe (Polubarinova-Kochina 1952, 1962). It is assumed that the aquifer is of infinite depth. Figure 7.3.13 shows the physical and the hodograph plane for this case. At G, N and A, $q_x = 0$, $q_y = -K$; at C_R and E_L, $q_x = 0$, $q_y = -\infty$; at C_L and C_R, q is infinite in the directions normal to BC and EF, respectively. At D, $q_x = 0$ and $|q_y|$ becomes minimum. From figure 7.3.13 it is possible to derive q_x and q_y for points B and F. At $y = -\infty$, $q_y = -K$ along the line ANG.

If the flat bottom disappears, i.e., the channel has the shape of the triangle BHF of figure 7.3.13a, the cut in the hodograph plane will disappear.

Example 6, an infiltration ditch with capillary fringe in an aquifer of infinite depth (Bear and Dagan 1962). Here CB and GH are capillary exposed surfaces that are also streamlines. AB and HI are boundaries of the saturated flow domain represented in the hodograph plane by the circle of radius $K/2$ and center at $(0, -K/2)$. At E, $|q_y| > K$ (fig. 7.3.14g, h).

7.3.4 Intersection of Boundaries of Different Types

In each of the examples given in paragraph 7.3.3, we obtained the hodograph representation of a point in the physical plane where boundaries of different types intersect each other, at the intersection of their hodographs. We could then read the flow characteristics at that point from the hodograph. A summary of such points is given in this paragraph.

Several examples are shown in figure 7.3.15 of an *intersection of a phreatic surface without accretion and the water table in an upstream reservoir.*

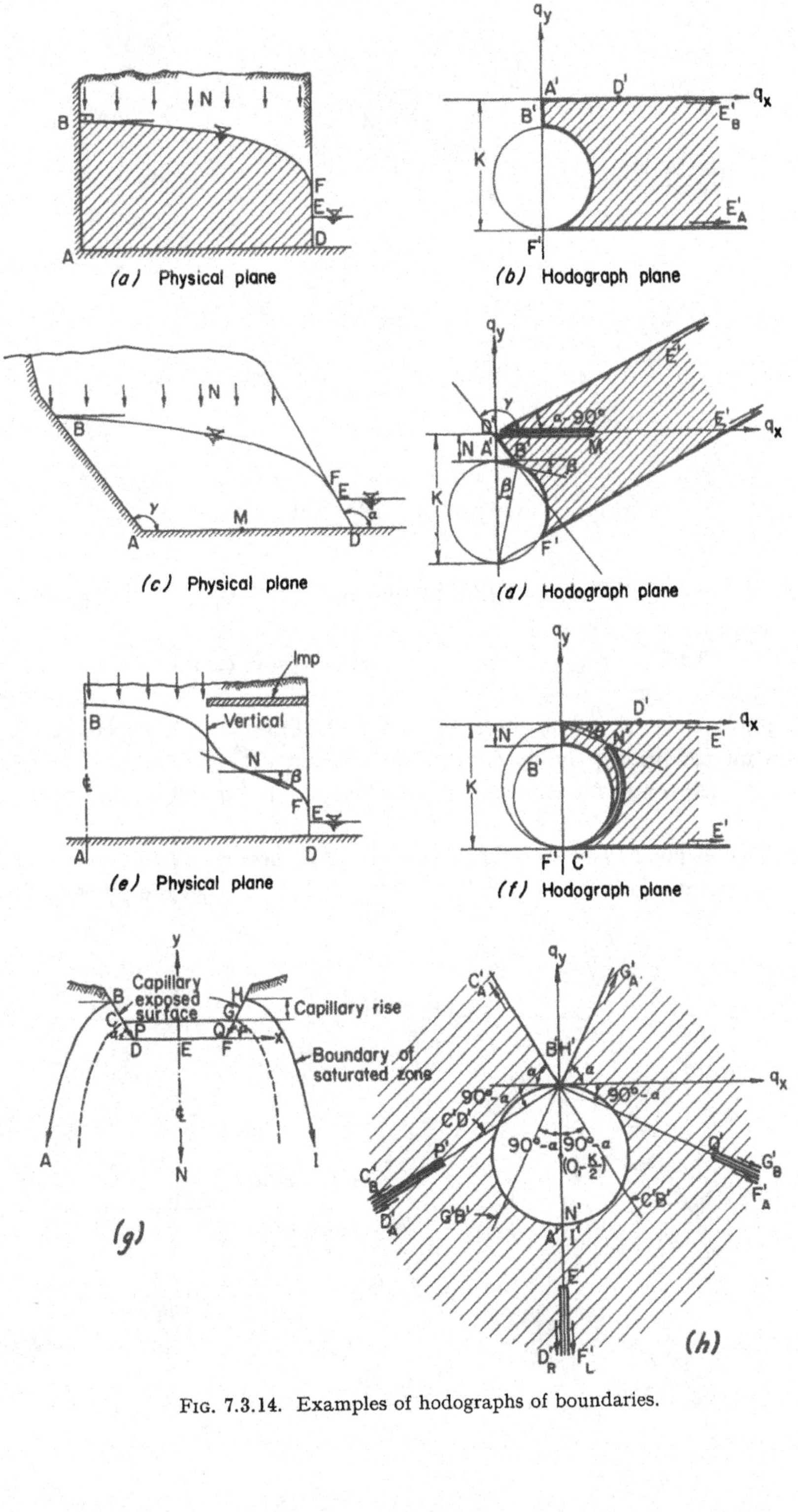

FIG. 7.3.14. Examples of hodographs of boundaries.

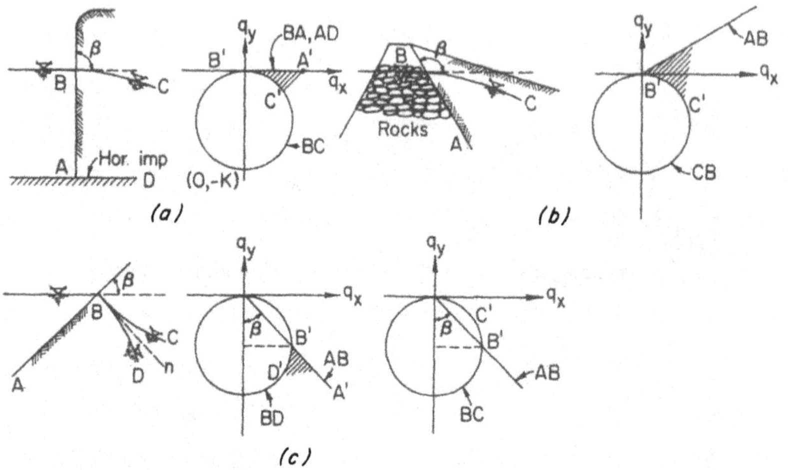

FIG. 7.3.15. Intersection of a phreatic surface and a water table in an upstream reservoir.

In figure 7.3.15a, $\beta = \pi/2$ and B' is at the origin ($q_x = 0$, $q_y = 0$). This situation occurs at point B of figure 7.3.9.

In figure 7.3.15b, $\beta > \pi/2$, B' is at the origin ($q_x = 0$, $q_y = 0$). This case occurs in a rock-filled dam.

In figure 7.3.15c, $\beta < \pi/2$, there are two possibilities for B', depending on the downstream conditions; in both cases we have at B: $q_x = K \cos \beta \sin \beta$, $q_y = -K \cos^2 \beta$. (Also see, for example, points B and F in fig. 7.3.13, and point C in fig. 7.3.11.)

Examples of an *intersection of a phreatic surface without accretion and a seepage face* are shown in figure 7.3.16. In figure 7.3.16c, $\beta < \pi/2$ and the phreatic surface

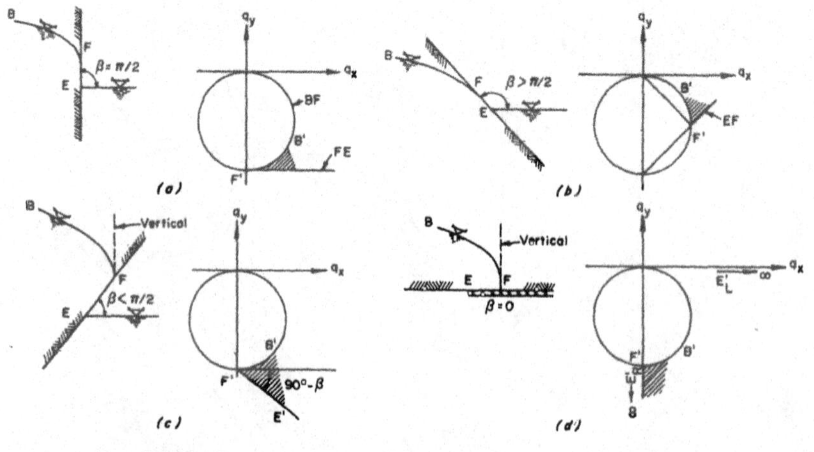

FIG. 7.3.16. Intersection of a phreatic surface and a seepage face.

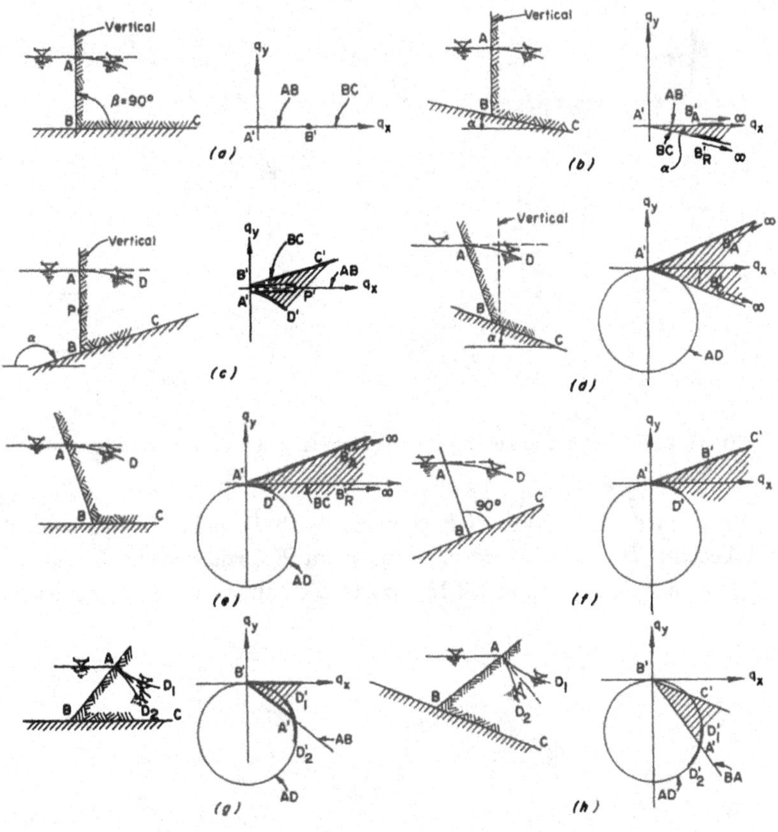

FIG. 7.3.17. Intersection of upstream equipotentials and an impervious boundary.

is tangent to the vertical at F. In figure 7.3.16d, $\beta = 0$, and the phreatic surface is again tangent to the vertical at F.

Examples of an *intersection of an upstream equipotential with an impervious boundary* are shown in figure 7.3.17.

Several examples of the *intersection of an impervious boundary with a downstream equipotential* are shown on figure 7.3.18.

In all these cases, if the angle $A\hat{B}C < 90°$, point B is a stagnation point ($q_x = q_y = 0$); if $A\hat{B}C > 90°$, the specific discharge at B is infinite (direction changes from parallel to AB to normal to BC); if $A\hat{B}C = 90°$, q at B is finite.

Examples of an *intersection of a seepage face with the water table in a downstream reservoir* are: point D in figure 7.3.9, point F in figure 7.3.10, point E in figures 7.3.14a–f. The specific discharge at this point is always infinite. In the hodograph it is the "intersection" of two parallel lines: through the origin normal to the reservoir's boundary (representing the latter), and through the point $(0, -K)$ normal to the seepage face (which is the continuation of the equipotential).

In figure 7.3.19, examples are shown of an *intersection of an impervious boundary*

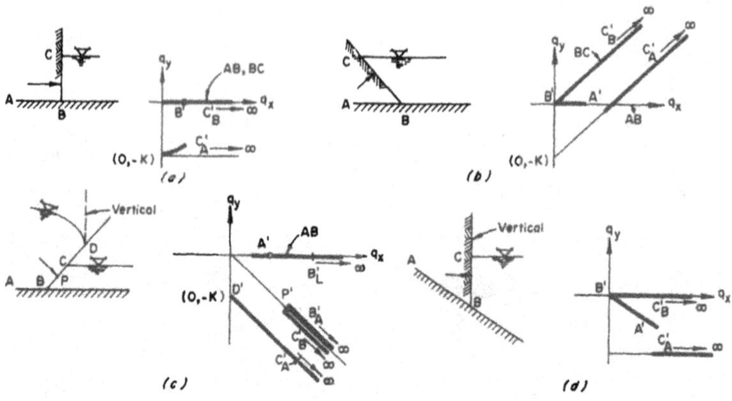

FIG. 7.3.18. Intersection of impervious boundary and downstream reservoir.

with a seepage face. In 7.3.14a, if E coincides with D, point D' will coincide with E'_B and become D'_L ($q_x = +\infty$, $q_y = 0$), point E'_A will become D'_A ($q_x = +\infty$, $q_y = -(N + K)$). If, in figure 7.3.18c, point C disappears (i.e., coincides with B),

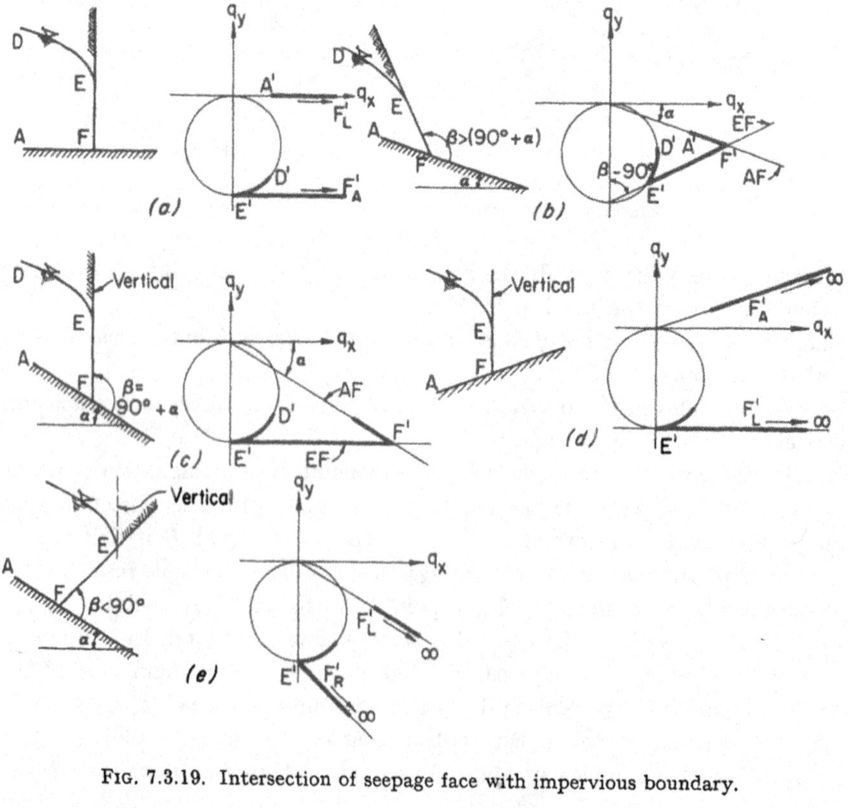

FIG. 7.3.19. Intersection of seepage face with impervious boundary.

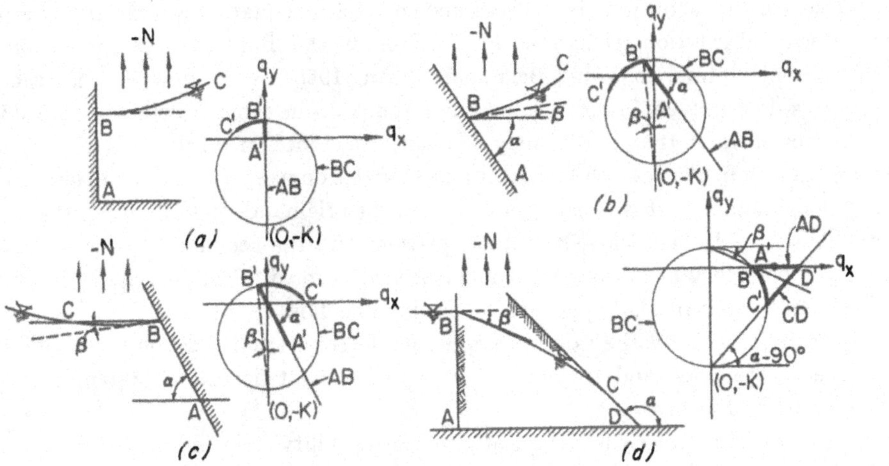

FIG. 7.3.20. Intersection of impervious boundary and phreatic surface with negative accretion.

point C'_A will become B'_R and the cut $C'_B P' B'_A$ will disappear.

The *intersection of a phreatic surface with accretion and an impervious boundary* is illustrated by point B in figures 7.3.14c and 7.3.14e.

Point C in figure 7.3.14e is an example of an *intersection of a phreatic surface with accretion and a phreatic surface without accretion.*

Several examples of an *intersection of a phreatic surface with negative accretion (evaporation) and an impervious boundary* are shown in figure 7.3.20. These are based on figures 7.3.8b, e, f.

Point B of figure 7.3.20d serves as an example of the *intersection of a phreatic surface with negative accretion (evaporation) and an upstream equipotential.*

Intersection of two equipotentials occurs when the slope on an upstream reservoir boundary (point B of fig. 7.3.11), or a downstream one, changes abruptly. With δ as indicated in figure 7.3.11a, we obtain for $\delta < \pi$, point B' at $q_x = 0$, $q_y = 0$; for $\delta > \pi$, point B' will be at infinity with $q_x/q_y = \tan \beta$ for B_B and $q_x/q_y = \tan \alpha$ for B_A.

When the change of slope takes place on a downstream reservoir boundary, if $\delta < 180°$ (the angle again taken inside the flow domain), the velocity at point B (where the slope changes) vanishes; if $\delta > 180°$, it is infinite. Similar considerations may be applied to change of slope along a seepage face, and a capillary exposed surface.

7.4 The Relations Between Solutions of Flow Problems in Isotropic and Anisotropic Media

The continuity equations in chapter 6 were presented, in general, for both isotropic and anisotropic media. Obviously, it is more difficult to solve the equations related to anisotropic media than those related to isotropic ones, especially since the governing equation in the latter case is the well known Laplace equation for which numerous solutions and solution techniques are available in the literature. Our intention in the present section, following Bear and Dagan (1965), is to show that instead of solving a flow problem in a given anisotropic domain, one can use certain relationships by means of which the problem may be transformed into one of flow in an *equivalent isotropic domain*. Once a solution is obtained for this domain (by analytical, numerical or analog methods), it is transformed back to the original domain by using the same relationships. In each case, both the equations and the geometry of the boundary and the boundary conditions on it must be transformed. The analysis is based on the homogeneity of the equations describing the flow.

In paragraph 11.2.7 the same procedure used here is applied to solutions obtained by means of models and analogs. Then the equivalent (isotropic) system becomes the model or the analog system.

The principal directions of anisotropic permeability coincide with the axes of the Cartesian (horizontal–vertical) coordinate system x, y, z. When the principal directions (say, x', y', z') of $\underline{\mathbf{k}}$ do not coincide with the (horizontal–vertical) x, y, z Cartesian system of coordinates, the entire analysis described below is carried out in terms of x', y', z'. However, one should be careful when a phreatic surface is present (see discussion in par. 7.1.11).

7.4.1 The Flow Equations

Consider flow in two porous medium systems: in the given anisotropic flow domain, referred to as the *prototype*, and in an *equivalent* isotropic domain. The flow in both systems is fully specified by a partial differential equation (which describes continuity) and by appropriate initial and boundary conditions. Let subscripts p and e denote variables and parameters in the prototype system and in the equivalent one, respectively.

Neglecting the elastic properties of the medium and the liquid, the continuity equation governing the flow of each of two liquids ($i = 1, 2$) separated by an abrupt interface in a homogeneous porous medium domain is (6.2.23). Written for the p- and e-systems, they are:

$$K^{(i)}{}_{xp}\, \partial^2\varphi^{(i)}{}_p/\partial x_p{}^2 + K^{(i)}{}_{yp}\, \partial^2\varphi^{(i)}{}_p/\partial y_p{}^2 + K^{(i)}{}_{zp}\, \partial^2\varphi^{(i)}{}_p/\partial z_p{}^2 = 0;$$

$$\partial^2\varphi^{(i)}{}_e/\partial x_e{}^2 + \partial^2\varphi^{(i)}{}_e/\partial y_e{}^2 + \partial^2\varphi^{(i)}{}_e/\partial z_e{}^2 = 0 \qquad (7.4.1)$$

where $\varphi^{(i)} = z + p/\gamma^{(i)}$. Equation (7.4.1) and the following equations should be read twice: once for $i = 1$ and once for $i = 2$. When the flow of a single liquid is considered, the superscript i is omitted.

In the presence of specific elastic storativity S_0 (sec. 6.3), the continuity equation is:

$$K^{(i)}{}_{xp}\, \partial^2\varphi^{(i)}{}_p/\partial x_p{}^2 + K^{(i)}{}_{yp}\, \partial^2\varphi^{(i)}{}_p/\partial y_p{}^2 + K^{(i)}{}_{zp}\, \partial^2\varphi^{(i)}{}_p/\partial z_p{}^2 = S^{(i)}{}_{0p}\, \partial\varphi^{(i)}{}_p/\partial t_p \qquad (7.4.2)$$

$$K^{(i)}{}_e(\partial^2\varphi^{(i)}{}_e/\partial x_e{}^2 + \partial^2\varphi^{(i)}{}_e/\partial y_e{}^2 + \partial^2\varphi^{(i)}{}_e/\partial z_e{}^2) = S^{(i)}{}_{0e}\, \partial\varphi^{(i)}{}_e/\partial t_e$$

where

$$K^{(i)}{}_{xp} = k_{xp}\gamma^{(i)}{}_p/\mu^{(i)}{}_p, \qquad S^{(i)}{}_{0p} = [n_p\beta^{(i)}{}_p + (1-n_p)\alpha_p]\gamma^{(i)}{}_p, \quad \text{etc.}$$

By Darcy's law, the elementary discharge δQ in the x direction through an area $\delta y\, \delta z$, is:

$$(\delta Q^{(i)}{}_x)_p = -k_{xp}\frac{\gamma^{(i)}{}_p}{\mu^{(i)}{}_p}\frac{\partial\varphi^{(i)}{}_p}{\partial x_p}\,\delta y_p\,\delta z_p; \qquad (\delta Q^{(i)}{}_x)_e = -k_e\frac{\gamma^{(i)}{}_e}{\mu^{(i)}{}_e}\frac{\partial\varphi^{(i)}{}_e}{\partial x_e}\,\delta y_e\,\delta z_e. \tag{7.4.3}$$

Similar expressions may be written for the y and z directions.

The average velocity in the x direction is:

$$V_{xp} = \frac{\delta x_p}{\delta t_p} = -\frac{k_{xp}}{n_p}\frac{\gamma^{(i)}{}_p}{\mu^{(i)}{}_p}\frac{\partial\varphi^{(i)}{}_p}{\partial x_p}; \qquad V_{xe} = \frac{\delta x_e}{\delta t_e} = -\frac{k_e}{n_e}\frac{\gamma^{(i)}{}_e}{\mu^{(i)}{}_e}\frac{\partial\varphi^{(i)}{}_e}{\partial x_e} \tag{7.4.4}$$

and similar expressions for the y and z directions.

Along a phreatic surface that is part of the boundary of the flow domain, we have:

$$\varphi^{(i)}{}_p = z_p; \qquad \varphi^{(i)}{}_e = z_e. \tag{7.4.5}$$

The boundary condition (7.1.46) for a phreatic surface with accretion is:

$$K^{(1)}{}_{xp}\left(\frac{\partial\varphi^{(1)}{}_p}{\partial x_p}\right)^2 + K^{(1)}{}_{yp}\left(\frac{\partial\varphi^{(1)}{}_p}{\partial y_p}\right)^2 + K^{(1)}{}_{zp}\left[\left(\frac{\partial\varphi^{(1)}{}_p}{\partial z_p}\right)^2 - \frac{\partial\varphi^{(1)}{}_p}{\partial z_p}\right] - N_p\left(\frac{\partial\varphi^{(1)}{}_p}{\partial z_p} - 1\right) = n'_p\frac{\partial\varphi^{(1)}{}_p}{\partial t_p}$$

$$K^{(1)}{}_e\left[\left(\frac{\partial\varphi^{(1)}{}_e}{\partial x_e}\right)^2 + \left(\frac{\partial\varphi^{(1)}{}_e}{\partial y_e}\right)^2 + \left(\frac{\partial\varphi^{(1)}{}_e}{\partial z_e}\right)^2 - \frac{\partial\varphi^{(1)}{}_e}{\partial z_e}\right] - N_e\left(\frac{\partial\varphi^{(1)}{}_e}{\partial z_e} - 1\right) = n'_e\frac{\partial\varphi^{(1)}{}_e}{\partial t_e}. \tag{7.4.6}$$

Along an interface between liquids 1 and 2, neglecting the capillary pressure (sec. 9.5): $p_1 = p_2$. Hence:

$$\gamma^{(2)}{}_p\varphi^{(2)}{}_p - \gamma^{(1)}{}_p\varphi^{(1)}{}_p = (\Delta\gamma)_p z_p; \quad \gamma^{(2)}{}_e\varphi^{(2)}{}_e - \gamma^{(1)}{}_e\varphi^{(1)}{}_e = (\Delta\gamma)_e z_e; \quad \Delta\gamma = \gamma^{(2)} - \gamma^{(1)} \tag{7.4.7}$$

where n' ($< n$) is the effective porosity with respect to phreatic storage.

If the volume of liquid U is of interest:

$$\delta U = Q\,\delta t \quad \text{or} \quad \delta U = n\,\delta x\,\delta y\,\delta z. \tag{7.4.8}$$

Other equations and initial and boundary conditions required to describe the investigated flow may also be listed here.

7.4.2 Relationships Among Parameters in the Two Systems

Let subscript r denote the ratio between the values of corresponding variables or parameters in the two systems:

$$x_r = (\delta x)_e/(\delta x)_p; \qquad y_r = (\delta y)_e/(\delta y)_p; \qquad z_r = (\delta z)_e/(\delta z)_p$$
$$t_r = (\delta t_e)/(\delta t)_p; \qquad Q_r = (\delta Q)_e/(\delta Q)_p; \qquad \varphi_r = (\delta\varphi)_e/(\delta\varphi)_p$$
$$K_{xr} = K_e/K_{xp}; \qquad K_{yr} = K_e/K_{yp}; \qquad K_{zr} = K_e/K_{zp}$$
$$\gamma^{(i)}{}_r = \gamma^{(i)}{}_e/\gamma^{(i)}{}_p \quad \text{etc.} \tag{7.4.9}$$

These ratios may be considered as *scales* between the two systems. Using (7.4.9), we now introduce:

$$(\delta x)_p = (\delta x)_e/x_r; \qquad (\delta y)_p = (\delta y)_e/y_r; \qquad (\delta z)_p = (\delta z)_e/z_r; \qquad (\delta\varphi)_p = (\delta\varphi)_e/\varphi_r; \quad \text{etc.}$$

into all the equations of the p-system. For example, from (7.4.1), we obtain:

$$\frac{k_e x_r^2}{k_{xr}\varphi^{(i)}{}_r}\frac{\partial^2\varphi^{(i)}{}_e}{\partial x_e^2} + \frac{k_e y_r^2}{k_{yr}\varphi^{(i)}{}_r}\frac{\partial^2\varphi^{(i)}{}_e}{\partial y_e^2} + \frac{k_e z_r^2}{k_{zr}\varphi^{(i)}{}_r}\frac{\partial\varphi^{(i)}{}_e}{\partial z_e^2} = 0. \tag{7.4.10}$$

By comparing this equation with the equation for the e-system in (7.4.1), we may conclude that the two equations will become identical if the following relationships exist:

$$\frac{k_e}{\varphi^{(i)}{}_r}\frac{x_r^2}{k_{xr}} = \frac{k_e}{\varphi^{(i)}{}_r}\frac{y_r^2}{k_{yr}} = \frac{k_e}{\varphi^{(i)}{}_r}\frac{z_r^2}{k_{zr}} = C = \text{arbitrary const} \tag{7.4.11}$$

or:

$$x_r^2/k_{xr} = y_r^2/k_{yr} = z_r^2/k_{zr} = C.$$

In (7.4.11) we have *two conditions*. Another form of (7.4.11) is:

$$x_r^2 k_{xp} = y_r^2 k_{yp} = z_r^2 k_{zp} = Ck_e. \tag{7.4.12}$$

It can be easily verified (Samsioë 1931) that by applying the transformations derived from (7.4.12) with $C = 1$:

$$(\delta x)_p = (k_{xp}/k_e)^{1/2}(\delta x)_e, \quad (\delta y)_p = (k_{yp}/k_e)^{1/2}(\delta y)_e, \quad (\delta z)_p = (k_{zp}/k_e)^{1/2}(\delta z)_e \tag{7.4.13}$$

and:

$$\varphi^{(i)}{}_r = 1; \qquad (\delta\varphi^{(i)})_p = (\delta\varphi^{(i)})_e \tag{7.4.14}$$

to the first (prototype) equation of (7.4.1), we obtain the second equation of (7.4.1). This means that if the flow in an anisotropic domain is described by the first equation of (7.4.1), it is always possible to use (7.4.13) and (7.4.14) in order to transform the flow domain into an equivalent isotropic one, in which the flow is described by the second equation of (7.4.1). Obviously, one may introduce any arbitrary permeability as k_e. Of special interest is the choice $k_e = k_{zp}$. Then (7.4.13) becomes:

$$(\delta x)_p = (k_{xp}/k_{zp})^{1/2}(\delta x)_e; \qquad (\delta y)_p = (k_{yp}/k_{zp})^{1/2}(\delta y)_e; \qquad (\delta z)_p = (\delta z)_e. \tag{7.4.15}$$

This means that the equivalent isotropic domain is obtained from the anisotropic prototype (where, in general, $k_{zp} < k_{xp}$, $k_{zp} < k_{yp}$) by leaving vertical lengths unchanged, while horizontal lengths are contracted:

$$(\delta x)_e = (k_{zp}/k_{xp})^{1/2}(\delta x)_p; \qquad (\delta y)_e = (k_{zp}/k_{yp})^{1/2}(\delta y)_p$$

An important conclusion is that whenever the prototype is anisotropic, the equivalent system is a distorted one with $x_r \neq y_r \neq z_r$. This conclusion is essential in the construction of models and analogs (chap. 11).

By introducing (7.4.9) into the remaining prototype equations given in paragraph 7.4.1, and comparing the resulting equations with their corresponding ones in the equivalent system, we obtain the following conditions necessary for the corresponding equations in the two systems to become identical.

From (7.4.2):

$$x_r^2/K^{(i)}{}_{xr}\varphi^{(i)}{}_r = y_r^2/K^{(i)}{}_{yr}\varphi^{(i)}{}_r = z_r^2/K^{(i)}{}_{zr}\varphi^{(i)}{}_r = t_r/S^{(i)}{}_{0r}\varphi^{(i)}{}_r. \tag{7.4.16}$$

From (7.4.3):

$$Q^{(i)}{}_{xr} = k_{xr}\frac{\gamma^{(i)}{}_r}{\mu^{(i)}{}_r}\frac{\varphi^{(i)}{}_r}{x_r}y_r z_r = K^{(i)}{}_{xr}\frac{\varphi^{(i)}{}_r}{x_r}y_r z_r; \quad \text{(also } Q^{(i)}{}_{xr} = Q^{(i)}{}_{yr} = Q^{(i)}{}_{zr} = Q^{(i)}{}_r\text{)}. \tag{7.4.17}$$

From (7.4.4):

$$V_{xr} = \frac{x_r}{t_r} = \frac{k_{xr}}{n_r}\frac{\gamma^{(i)}{}_r}{\mu^{(i)}{}_r}\frac{\varphi^{(i)}{}_r}{x_r} = K^{(i)}{}_{xr}\frac{\varphi^{(i)}{}_r}{n_r x_r}. \tag{7.4.18}$$

From (7.4.5):

$$\varphi^{(i)}{}_r = z_r. \tag{7.4.19}$$

From (7.4.6):

$$K^{(i)}{}_{xr}\left(\frac{\varphi^{(1)}{}_r}{x_r}\right)^2 = K^{(i)}{}_{yr}\left(\frac{\varphi^{(1)}{}_r}{y_r}\right)^2 = K^{(i)}{}_{zr}\left(\frac{\varphi^{(1)}{}_r}{z_r}\right)^2 = K^{(i)}{}_{zr}\frac{\varphi^{(1)}{}_r}{z_r} = N_r$$

$$= N_r\frac{\varphi^{(1)}{}_r}{z_r} = n'_r\frac{\varphi^{(1)}{}_r}{t_r}. \tag{7.4.20}$$

From (7.4.7):

$$\gamma^{(1)}{}_r\varphi^{(1)}{}_r = \gamma^{(2)}{}_r\varphi^{(2)}{}_r = (\Delta\gamma)_r z_r. \tag{7.4.21}$$

One should notice that certain conditions are repeated in these equations. Let us now examine several examples.

Problem 1, flow of a single liquid in a confined domain, neglecting the elasticity of the medium and the liquid. The flow is described by (7.4.1), (7.4.3) and (7.4.4) and $i = 1$. The parameters involved are:

ten in the *p*-system:

$$\delta x)_p; (\delta y)_p; (\delta z)_p; K_{xp}\left(= k_{xp}\frac{\gamma_p}{\mu_p}\right); K_{yp}; K_{zp}; (\delta t)_p; n_p; (\delta Q)_p; (\delta\varphi)_p$$

eight in the *e*-system:

$$(\delta x)_e; (\delta y)_e; (\delta z)_e; K_e; (\delta t)_e; n_e; (\delta Q)_e; (\delta\varphi)_e$$

The four conditions applicable are the two conditions in (7.4.11), (7.4.17) and (7.4.18). No additional information is gained by writing (7.4.17) and (7.4.18) for the other two directions. Hence, out of the eight parameters of the equivalent system, four may be chosen arbitrarily, provided they do not contradict the specified conditions.

Among the various possible groups of four arbitrarily chosen parameters, some are more attractive, especially when the e-system is a model or an analog. For example, consider the choice of a certain $(\delta\varphi)_e$ to correspond to a certain $(\delta\varphi)_p$, i.e., we choose φ_r. We also choose Q_r, t_r and n_r. Then, by (7.4.11), (7.4.17) and (7.4.18), we have:

$$x_r = (t_r\varphi_r/n_r)^{1/2}K^{1/2}{}_{xr} = (t_r\varphi_r/n_r)^{1/3}(Q_r/\varphi_r)^{1/3}(K_{yp}K_{zp}/K^2{}_{xp})^{1/6}$$
$$y_r = (t_r\varphi_r/n_r)^{1/2}K^{1/2}{}_{yr} = (t_r\varphi_r/n_r)^{1/3}(Q_r/\varphi_r)^{1/3}(K_{xp}K_{zp}/K^2{}_{yp})^{1/6}$$
$$z_r = (t_r\varphi_r/n_r)^{1/2}K^{1/2}{}_{zr} = (t_r\varphi_r/n_r)^{1/3}(Q_r/\varphi_r)^{1/3}(K_{xp}K_{yp}/K^2{}_{zp})^{1/6} \qquad (7.4.22)$$

and:

$$K_e = (Q_r/\varphi_r)^{2/3}(t_r\varphi_r/n_r)^{-1/3}(K_{xp}K_{yp}K_{zp})^{1/3}. \qquad (7.4.23)$$

If we choose $n_r = \varphi_r = t_r = Q_r = 1$ (i.e., values of these parameters are identical in both systems), we obtain:

$$K_e = (K_{xp}K_{yp}K_{zp})^{1/3} \qquad (7.4.24)$$

$$x_r = K^{1/2}{}_{xr} = (K_{yp}K_{zp}/K^2{}_{xp})^{1/6}; \qquad y_r = K^{1/2}{}_{yr}, \qquad z_r = K^{1/2}{}_{zr}. \qquad (7.4.25)$$

The relationships (7.4.22) and (7.4.23) follow directly from (7.4.11), (7.4.17) and (7.4.18), which are also valid for other cases. Because t_r and n_r always appear as a group (t_r/n_r), at least one of them must be chosen. Also, from (7.4.22) it follows that only one of the three length ratios may be chosen.

Conditions similar to (7.4.22) and (7.4.23) may also be derived for two-dimensional flow. The number of parameters is reduced to seven, the number of conditions to three (only one of (7.4.11) remains), so that four may be chosen arbitrarily. Again choosing $n_r = \varphi_r = t_r = Q_r = 1$, we obtain:

$$x_r = (K_e/K_{xp})^{1/2} = (K_{yp}/K_{xp})^{1/4}, \qquad y_r = (K_e/K_{yp})^{1/2} = (K_{xp}/K_{yp})^{1/4},$$
$$K_e = (K_{xp}K_{yp})^{1/2}. \qquad (7.4.26)$$

Only at most one of the three length scales may be chosen arbitrarily.

Once we determine a set of transformations, they are used to transform (into the equivalent system) those boundary conditions not included in the basic equations.

Problem 2, flow of a single liquid under confined conditions, when the elastic properties of the liquid and the medium are not neglected. The flow is described by (7.4.2) through (7.4.4). The *nine* parameters of the equivalent system are: $(\delta x)_e$, $(\delta y)_e$, $(\delta z)_e$, K_e, $(\delta t)_e$, n_e, $(\delta Q)_e$, $(\delta\varphi)_e$ and S_{0e}. The five conditions applicable are: (7.4.16)—three conditions, (7.4.17) and (7.4.18). Again, four parameters may be chosen arbitrarily. For example, choosing φ_r, t_r, Q_r, n_r, we obtain x_r, y_r, z_r and K_e from (7.4.22) and (7.4.23). Also from (7.4.16) and (7.4.18), we have:

$$S_{0r} = n_r/\varphi_r. \tag{7.4.27}$$

Often we choose $x_r = (K_{zp}/K_{xp})^{1/2}$, $\varphi_r = Q_r = n_r = 1$. Then:

$$y_r = (K_{zp}/K_{yp})^{1/2}; \qquad z_r = 1, \qquad K_e = (K_{xp}K_{yp})^{1/2}, \qquad S_{0r} = 1,$$
$$t_r = K_{zp}/(K_{xp}K_{yp})^{1/2}. \tag{7.4.28}$$

In two-dimensional flow in the horizontal xy plane, the number of parameters is reduced to eight, the number of conditions to four (only two in (7.4.16)), aad again four parameters may be chosen arbitrarily.

Problem 3, flow of a single liquid with a boundary condition of a free surface with accretion. The equations describing the flow are: (7.4.1), (7.4.3) through (7.4.5) and (7.4.6). The parameters involved are: $(\delta x)_e$, $(\delta y)_e$, $(\delta z)_e$, K_e, $(\delta t)_e$, $(\delta Q)_e$, $(\delta\varphi)_e$, n_e, n'_e and N_e (ten parameters). Seven independent conditions are available: (7.4.11) and (7.4.17) through (7.4.20). Hence, three parameters may be chosen arbitrarily. Choosing φ_r, Q_r and n_r, we obtain:

from (7.4.19): $\quad z_r = \varphi_r$

from (7.4.11): $\quad x_r = \varphi_r(K_{zp}/K_{xp})^{1/2}; \qquad y_r = \varphi_r(K_{zp}/K_{yp})^{1/2}$

from (7.4.17): $\quad K_e = (Q_r/\varphi^2{}_r)(K_{xp}K_{yp})^{1/2}$

from (7.4.18): $\quad t_r = (\varphi^3{}_r n_r/Q_r)(K^2{}_{zp}/K_{xp}K_{yp})^{1/2}$

from (7.4.20): $\quad N_r = K_{zr} = (Q_r/\varphi^2{}_r)(K_{xp}K_{yp}/K^2{}_{zp})^{1/4}; \quad$ and $\quad n'_r = n_r. \qquad (7.4.29)$

This last expression also holds for other possible choices of parameters. For the special case $\varphi_r = Q_r = n_r = 1$, equation (7.4.29) becomes:

$$z_r = 1; \qquad x_r = (K_{zp}/K_{xp})^{1/2}; \qquad y_r = (K_{zp}/K_{yp})^{1/2}; \qquad K_e = (K_{xp}K_{yp})^{1/2};$$
$$t_r = (K^2{}_{zp}/K_{xp}K_{yp})^{1/2}; \qquad N_r = (K_{xp}K_{yp}/K^2{}_{zp})^{1/2}; \qquad n'_r = 1. \tag{7.4.30}$$

Problem 4, flow in a confined domain of two liquids separated by an abrupt interface. Equations (7.4.1), (7.4.3), (7.4.4) and (7.4.7), which describe the flow, lead to seven conditions: (7.4.11), (7.4.17), (7.4.18) and (7.4.21)—two from each of the first three equations and one from the last one. The 13 parameters involved are: $(\delta x)_e$, $(\delta y)_e$, $(\delta z)_e$, k_e, $(\delta t)_e$, $(\delta Q)_e$, $(\delta\varphi^{(1)})_e$, $(\delta\varphi^{(2)})_e$, $\gamma^{(1)}{}_e$, $\gamma^{(2)}{}_e$, $\mu^{(1)}{}_e$, $\mu^{(2)}{}_e$ and n_e. Hence, six parameters may be chosen arbitrarily. However, since from (7.4.17), $\gamma^{(1)}{}_r\varphi^{(1)}{}_r/\mu^{(1)}{}_r = \gamma^{(2)}{}_r\varphi^{(2)}{}_r/\mu^{(2)}{}_r$ and from (7.4.21), $\gamma^{(1)}{}_r\varphi^{(1)}{}_r = \gamma^{(2)}{}_r\varphi^{(2)}{}_r$, we must have:

$$\mu^{(1)}{}_r = \mu^{(2)}{}_r, \quad \text{i.e.,} \quad (\mu_1/\mu_2)_p = (\mu_1/\mu_2)_e. \tag{7.4.31}$$

If we choose: n_r, z_r, k_e, $\gamma^{(1)}{}_r$, $\gamma^{(2)}{}_r$, $\mu^{(1)}{}_r$, we obtain:

$$x_r = z_r(k_{zp}/k_{xp})^{1/2}, \qquad y_r = z_r(k_{zp}/k_{yp})^{1/2}$$
$$\mu^{(2)}{}_r = \mu^{(1)}{}_r; \qquad \varphi^{(1)}{}_r = z_r(\Delta\gamma)_r/\gamma^{(1)}{}_r; \qquad \varphi^{(2)}{}_r = z_r(\Delta\gamma)_r/\gamma^{(2)}{}_r$$
$$t_r = (n_r z_r/k_{zr})\mu^{(1)}{}_r/(\Delta\gamma)_r; \qquad Q_r = z_r{}^2 k_{zr}(\Delta\gamma)_r k_{zp}/\mu_r(k_{xp}k_{yp})^{1/2}. \tag{7.4.32}$$

For the special choice: $n_r = z_r = \gamma^{(1)}{}_r = \gamma^{(2)}{}_r = \mu^{(1)}{}_r = 1$, and $k_e = k_{zp}$, we obtain:

$$x_r = (k_{zp}/k_{xp})^{1/2}, \qquad y_r = (k_{zp}/k_{yp})^{1/2}, \qquad \varphi^{(1)}{}_r = z_r = \mu^{(2)}{}_r = t_r = 1,$$
$$Q_r = (k^2{}_{zp}/k_{xp}k_{yp})^{1/2}. \tag{7.4.33}$$

If the flow is two-dimensional, the number of parameters is 12, the number of conditions is six, and again six parameters may be chosen arbitrarily.

Problem 5, flow with a free surface boundary, of two liquids separated by an abrupt interface. This problem is similar to problem 4, except that we must add (7.4.5) and the condition (7.4.19). This leaves five parameters to be chosen arbitrarily. However, in a phreatic aquifer we must have: $(\Delta\gamma/\gamma^{(1)})_r = 1$.

To summarize, the procedure described above involves the following steps.

(a) Write down *all* the equations (sometimes these are approximate equations) describing the flow in the anisotropic system and in the equivalent isotropic one.

(b) Use the ratios (7.4.9)—with additional ratios when necessary—to rewrite the equation for the prototype in terms of the equivalent system's parameters.

(c) Derive the conditions required to make the two sets of equations identical. Examine, and leave a set which includes only independent conditions.

(d) The difference between the number of parameters (constants and variables) involved in the equivalent system and the number of independent conditions indicates the number of parameters (or their scales) that may be chosen arbitrarily. The choice of parameters is immaterial.

(e) Use the derived scales to convert the anisotropic system (including the geometry of its boundaries) into an equivalent isotropic one, and then again to transform the solution back to the given anisotropic system.

7.4.3 Examples

Example 1. The equation describing *flow of a single liquid in an infinite homogeneous anisotropic confined aquifer* ($T_x \neq T_y$) *of constant thickness, b,* is:

(6.4.5): $$T_x\, \partial^2 s/\partial x^2 + T_y\, \partial^2 s/\partial y^2 = S\, \partial s/\partial t, \qquad Q_x = T_x \delta y\, \partial s/\partial x \tag{7.4.34}$$

where s is the drawdown. Following a procedure similar to that outlined for problem 1 above, and choosing $s_r = Q_r = t_r = S_r = 1$, we obtain the transformations:

$$x_r = (T_{yp}/T_{xp})^{1/4}; \qquad y_r = (T_{xp}/T_{yp})^{1/4}; \qquad T_e = (T_{xp}T_{yp})^{1/2}. \tag{7.4.35}$$

Consider now the case of flow to a single well pumping at a constant rate, Q_w, from this aquifer. The drawdown s in an isotropic aquifer is given by (Theis 1935):

$$s_e = (Q_w/4\pi T_e)W[S_e(x^2{}_e + y^2{}_e)/4t_eT_e]; \qquad W(u) = \int_u^\infty (e^{-x}/x)\, dx \tag{7.4.36}$$

where $W(u)$ is called the *well function.*

We may now employ (7.4.35) in order to obtain the solution for an anisotropic aquifer where the flow is described by (7.4.34): $(Q_w)_e = (Q_w)_p$; $s_e = s_p$, $S_e = S_p$, $t_e = t_p$, $(\delta x)_e = (\delta x)_p(T_{yp}/T_{xp})^{1/4}$, $(\delta y)_e = (\delta y)_p(T_{xp}/T_{yp})^{1/4}$; $T_e = (T_{xp}T_{yp})^{1/2}$. We obtain:

$$s_p = \frac{(Q_w)_p}{4\pi(T_{xp}T_{yp})^{1/2}} W\left[\frac{S_p\{x_p^2(T_{yp}/T_{xp})^{1/2} + y_p^2(T_{xp}/T_{yp})^{1/2}\}}{4t_p(T_{xp}T_{yp})^{1/2}}\right].$$

Or, omitting the subscript p:

$$s = [Q_w/4\pi(T_xT_y)^{1/2}]W[S(x^2T_y + y^2T_x)/4t(T_xT_y)]. \tag{7.4.37}$$

Equation (7.4.37) may be rewritten for the anisotropic domain in the form:

$$s(r, t) = (Q_w/4\pi T_e)W(Sr^2/4tT_q) \tag{7.4.38}$$

where $T_e = (T_xT_y)^{1/2}$ and $T_q = T_x/[\cos^2\theta + (T_x/T_y)\sin^2\theta]$ is the directional transmissivity in the direction of the flow (radial direction); $\theta = \tan^{-1}(y/x)$.

Example 2. Consider the steady flow in a leaky aquifer described for an isotropic medium by introducing $s = \varphi_0 - \tilde{\varphi}$ and $\partial\varphi/\partial t = 0$ in (6.4.13). For an anisotropic aquifer, this equation becomes:

$$T_{xp}\,\partial^2 s_p/\partial x_p^2 + T_{yp}\,\partial^2 s_p/\partial y_p^2 = s_p/\sigma'_p; \qquad Q_{xp} = T_{xp}y_p\,\partial s/\partial x_p. \tag{7.4.39}$$

In an isotropic medium we have:

$$T_e(\partial^2 s_e/\partial x_e^2 + \partial^2 s_e/\partial y_e^2) = s_e/\sigma'_e; \qquad Q_{xe} = T_e y_e\,\partial s/\partial x_e. \tag{7.4.40}$$

From these equations, the following conditions are obtained:

$$x_r^2/T_{xr} = y_r^2/T_{yr} = \sigma'_r; \qquad Q_r = T_{xr}s_r y_r/x_r. \tag{7.4.41}$$

It can be verified that the boundary conditions add no new information. We thus have three conditions for the six parameters: $(\delta x)_e$, $(\delta y)_e$, T_e, $(\delta Q)_e$, σ'_e, $(\delta s)_e$. Three may be chosen arbitrarily. Choosing $Q_r = s_r = \sigma'_r = 1$, we obtain:

$$x_r = (T_{yp}/T_{xp})^{1/4}, \qquad y_r = (T_{xp}/T_{yp})^{1/4}; \qquad T_e = (T_{xp}T_{yp})^{1/2}. \tag{7.4.42}$$

The solution of (7.4.40), i.e., for the isotropic aquifer, is:

$$s_e(x, y) = (Q_{we}/2\pi T_e)K_0[(x_e^2 + y_e^2)/(T_e\sigma'_e)^{1/2}] \tag{7.4.43}$$

where K_0 is the modified Bessel function of the second kind and order zero. Using (7.4.42) this solution can be transformed into the required solutions of (7.4.39):

$$s(x, y) = \frac{Q_w}{2\pi(T_xT_y)^{1/2}} K_0\left[\left(\frac{x^2(T_y/T_x)^{1/2} + y^2(T_x/T_y)^{1/2}}{\sigma'(T_xT_y)^{1/2}}\right)^{1/2}\right]$$

$$= \frac{Q_w}{2\pi(T_xT_y)^{1/2}} K_0\left[\left(\frac{x^2T_y + y^2T_x}{\sigma'T_xT_y}\right)^{1/2}\right] = \frac{Q_w}{2\pi(T_xT_y)^{1/2}} K_0\left[\frac{r'}{\lambda'}\right];$$

$$\lambda'^2 = \sigma'(T_xT_y)^{1/2}; \qquad r'^2 = x^2\,T_y/T_x + y^2\,T_x/T_y \tag{7.4.44}$$

where subscript p was omitted.

7.5 Superposition and Duhamel's Principle

7.5.1 Superposition

The fact that most of the continuity equations developed in chapter 6, and the boundary conditions (sec. 7.1), are *linear* permits us to use a powerful tool—the *principle of superposition*—in solving them with mixed, or inhomogeneous, boundary conditions. By applying this principle, complicated problems may be reduced to simpler ones.

Briefly, the principle of superposition states that if $\varphi_1 = \varphi_1(x, y, z, t)$ and $\varphi_2 = \varphi_2(x, y, z, t)$ are two *particular solutions* of a *homogeneous linear partial differential equation*, $L(\varphi) = 0$, where L represents a linear operator (e.g., in (6.2.24): $L(\ ) \equiv \partial^2(\ )/\partial x^2 + \partial^2(\ )/\partial y^2 + \partial^2(\ )/\partial z^2$), then any linear combination of φ_1 and φ_2:

$$\varphi = C_1\varphi_1 + C_2\varphi_2 \tag{7.5.1}$$

where C_1 and C_2 are arbitrary constants is also a solution of $L(\varphi) = 0$. Or, in general, if $\varphi_i = \varphi_i(x, y, z, t)$, $i = 1, 2, \ldots, n$, are particular solutions of $L(\varphi) = 0$, then:

$$\varphi = \sum_{i=1}^{n} C_i\varphi_i \tag{7.5.2}$$

where the C_i's are constants is also a solution of this equation. In each case the constants are adjusted so that the boundary conditions are also satisfied. For example, if the boundary condition on a surface S is $a_0\varphi + a_1\,\partial\varphi/\partial x + a_2\,\partial\varphi/\partial y + a_3\,\partial\varphi/\partial z = f(x, y, z, t)$, and if we have on S:

$$a_0\varphi_1 + a_1\,\partial\varphi_1/\partial x + a_2\,\partial\varphi_1/\partial y + a_3\,\partial\varphi_1/\partial z = f_1(x, y, z, t),$$

and

$$a_0\varphi_2 + a_1\,\partial\varphi_2/\partial x + a_2\,\partial\varphi_2/\partial y + a_3\,\partial\varphi_2/\partial z = f_2(x, y, z, t),$$

then, according to (7.5.1), we have: $f = C_1f_1 + C_2f_2$. Sometimes the sum in (7.5.2) must be extended to infinity in order to satisfy the boundary conditions.

The solution φ in (7.5.1) and (7.5.2), when the coefficients are determined so that boundary conditions are satisfied, is called the *complete solution of the homogeneous equation*.

Sometimes the equation considered is a *nonhomogeneous* one (i.e., it includes a term that does not involve the dependent variable φ, as, for example, equation (8.4.2)). If $\varphi = \varphi_0$ is any solution (no matter how special) of the nonhomogeneous equation, then any solution of the nonhomogeneous equation can be written in the form:

$$\varphi = \varphi_0 + C_1\varphi_1 + C_2\varphi_2.$$

φ_0 is called the *particular integral* of the nonhomogeneous equation while $C_1\varphi_1 + C_2\varphi_2$, which is the *complete solution* of the associated homogeneous equation, obtained from the given nonhomogeneous one by deleting the term that does not contain φ, is called the *complementary solution* of the nonhomogeneous equation.

Consider the following examples for the use of the principle of superposition. Let the boundary S of a flow domain D be composed of n segments $S^{(1)}, S^{(2)}, \ldots, S^{(n)}$ on each of which we have the boundary condition of a constant potential, i.e., $\varphi = \varphi^{(i)} = \text{const}$ on $S^{(i)}$. The flow in D is governed by the Laplace equation:

$$\partial^2\varphi/\partial x^2 + \partial^2\varphi/\partial y^2 + \partial^2\varphi/\partial z^2 = 0. \tag{7.5.3}$$

Let $\varphi_1 = \varphi_1(x, y, z)$ be a solution of (7.5.3) satisfying the boundary conditions $\varphi_1 = \varphi^{(1)}$ on $S^{(1)}$, $\varphi_1 = 0$ elsewhere, i.e., on $S - S^{(1)}$. Similarly $\varphi_2 = \varphi_2(x, y, z)$ is a solution of (7.5.3) satisfying condition $\varphi_2 = \varphi^{(2)}$ on $S^{(2)}$, $\varphi_2 = 0$ on $S - S^{(2)}$, etc. It can easily be verified that $\varphi = \varphi_1 + \varphi_2 + \cdots + \varphi_n$ is a solution of (7.5.3) that satisfies the given boundary conditions on S.

The principle of superposition means that the presence of one boundary condition does not affect the response produced by the presence of other boundary or initial conditions, and that there are no interactions among the responses produced by the different boundary conditions. Therefore, to analyze the combined effect of a number of boundary conditions (excitations) we may start by solving for the effect of each individual excitation and then combine the results.

It is most important to note that the free surface boundary condition (sec. 7.1) has the form of a *nonlinear* partial differential equation. Hence, the principle of superposition *cannot* be applied to free surface flows.

In ground water hydrology this principle is used to obtain total drawdown caused by a number of wells operating simultaneously in an infinite aquifer and flow to wells superimposed on existing flow fields.

Since the stream function Ψ (in two-dimensional flow or in axisymmetrical flows) satisfies linear partial differential equations such as (6.5.24), (6.5.32) or (6.5.34), the principle of superposition is applicable to it as well.

In all cases where superposition is employed, one should be careful that boundary conditions are satisfied.

7.5.2 Unsteady Flow with Boundary Conditions Independent of Time

Consider the case of unsteady flow described by:

$$\partial^2\varphi/\partial x^2 + \partial^2\varphi/\partial y^2 + \partial^2\varphi/\partial z^2 = (S_s/K)\,\partial\varphi/\partial t \tag{7.5.4}$$

with initial conditions: $\varphi = f(x, y, z)$ in D, at $t = 0$, and boundary conditions:

$$L(\varphi) = g(x, y, z) \quad \text{on } S \text{ at } \quad t \geqslant 0$$

where

$$L(\varphi) = \alpha_0\varphi + a_1(\partial\varphi/\partial x) + a_2(\partial\varphi/\partial y) + a_3(\partial\varphi/\partial z); \qquad a_1, a_2, a_3 = \text{constants}.$$

Let $\varphi^{(1)} = \varphi^{(1)}(x, y, z)$ be a solution of the steady flow problem such that it satisfies:

$$\partial^2\varphi/\partial x^2 + \partial^2\varphi/\partial y^2 + \partial^2\varphi/\partial z^2 = 0 \text{ in } D$$
$$L(\varphi) = g(x, y, z) \text{ on } S, \qquad t \geqslant 0. \tag{7.5.5}$$

Let $\varphi^{(2)} = \varphi^{(2)}(x, y, z, t)$ be a solution of the nonsteady flow described by (7.5.4) with the initial condition $\varphi = f(x, y, z) - g(x, y, z)$ in D, and boundary condition $L(\varphi) = 0$ on S for $t \geqslant 0$. It can easily be verified that $\varphi = \varphi^{(1)} + \varphi^{(2)}$ satisfies both the equation and the initial and boundary conditions of the original problem. The latter has been reduced to two subproblems: one of steady flow with prescribed boundary conditions, and another of unsteady flow with prescribed initial conditions and with homogeneous boundary conditions.

7.5.3 Unsteady Flow with Time-Dependent Boundary Conditions

Superposition is performed here with respect to time, using Duhamel's theorem (1833). According to this theorem, if $\varphi = f(x, y, z, \tau, t)$ represents the potential at point $P(x, y, z)$ of D at time t, satisfying:

$$\partial^2\varphi/\partial x^2 + \partial^2\varphi/\partial y^2 + \partial^2\varphi/\partial z^2 = (S_s/K)\,\partial\varphi/\partial t \tag{7.5.6}$$

with initial conditions $\varphi = 0$ everywhere in D, and boundary conditions $\varphi = g(x, y, z, \tau)$ on S, then the solution $\varphi = \varphi(x, y, z, t)$ of the problem in which the initial conditions are zero, while the boundary condition, $\varphi = g(x, y, z, t)$, is given by:

$$\varphi(x, y, z, t) = \int_0^t \frac{\partial}{\partial t} f(x, y, z, \tau, t - \tau)\, d\tau. \tag{7.5.7}$$

Equation (7.5.7) is also called the *superposition integral*. Another form of this integral is the following.

Let $\varphi = f(x, y, z, t)$ be the potential distribution satisfying (7.5.6) with $\varphi \equiv 0$ initially everywhere in D, while $\varphi = 1$ is maintained on S. Then the solution of the problem when the boundary condition on S is time dependent, i.e., $\varphi = g(t)$, is given by:

$$\varphi(x, y, z, t) = \int_0^t g(\tau) \frac{\partial}{\partial t} f(x, y, z, t - \tau)\, d\tau. \tag{7.5.8}$$

In this form f has the meaning of a *step response*.

Other forms of this integral are also given in the literature (e.g., Cheng 1959).

If we now have a problem where $\varphi = \varphi(x, y, z, t)$ satisfies (7.5.6) with initial conditions $\varphi = f(x, y, z)$ at $t = 0$ in D, and boundary conditions $\varphi = g(x, y, z, t)$ on S, we decompose the problem into two subproblems $\varphi = \varphi^{(1)} + \varphi^{(2)}$, where $\varphi^{(1)}$ satisfies (7.5.6), is equal to $f(x, y, z)$ initially and vanishes everywhere on S, whereas $\varphi^{(2)}$ satisfies (7.5.6), is zero everywhere initially and satisfies $\varphi = g(x, y, z, t)$ on S. The solution $\varphi^{(2)}$ is obtained by using Duhamel's principle as described above.

Actually this decomposition expresses the fact that responses produced by several excitations operating simultaneously are independent. Hence the resulting response may be obtained as an algebraic sum of the individual responses resulting from the various individual excitations operating one at a time.

Duhamel's principle (7.5.7) can be applied to other equations such as (6.4.7) or (6.4.13). The general form of these equations, which describe essentially horizontal flows in aquifers, may be written as:

$$\frac{\partial}{\partial x}\left(T_x \frac{\partial\varphi}{\partial x}\right) + \frac{\partial}{\partial y}\left(T_y \frac{\partial\varphi}{\partial y}\right) + A(x, y, t)\varphi + B(x, y, t) = S\frac{\partial\varphi}{\partial t} \tag{7.5.9}$$

where T_x and T_y may be functions of x and y, but not of t. The general form of the boundary condition (7.1.20) is:

$$C_1 \frac{\partial \varphi}{\partial x} + C_2 \frac{\partial \varphi}{\partial y} + C_3 \varphi = g(x, y, t) \tag{7.5.10}$$

where C_1, C_2 and C_3 are functions of x, y only. The initial condition is $\varphi = f(x, y)$.

Suppose $F(x, y, \tau, t)$ is a solution of the same problem, except that $B(x, y, t)$ and $g(x, y, t)$ are replaced by $B(x, y, \tau)$ and $g(x, y, \tau)$, τ being a constant value of t. Then the solution of the problem stated by (7.5.9) through (7.5.10) is:

$$\varphi(x, y, t) = \frac{\partial}{\partial t} \int_0^t F(x, y, \tau, t - \tau)\, d\tau = f(x, y) + \int_0^t \frac{\partial}{\partial t} F(x, y, \tau, t - \tau)\, d\tau. \tag{7.5.11}$$

7.6 Direct Integration in One-Dimensional Problems

7.6.1 Solution of the One-Dimensional Continuity Equation

In many one-dimensional problems, where $\varphi = \varphi(x)$ only, a solution can be obtained by integrating Darcy's law. For example, for steady uniform flow of an incompressible fluid in a homogeneous, isotropic confined aquifer of constant thickness D, we obtain:

$$Q = - KD\, d\varphi/dx; \qquad d\varphi = -(Q/KD)\, dx \tag{7.6.1}$$

$$\int_{\varphi_0}^{\varphi(x)} d\varphi = -(Q/KD) \int_{x_0}^{x} dx; \qquad \varphi_0 - \varphi(x) = (Q/KD)(x - x_0) \tag{7.6.2}$$

where the assumption Q = const is equivalent to a statement of continuity. Equation (7.6.2) gives the variation of φ in the direction of the flow.

The same solution is obtained by integrating the one-dimensional Laplace equation derived from (6.2.24):

$$\partial^2\varphi/\partial x^2 = 0; \qquad \varphi(x) = Ax + B \tag{7.6.3}$$

where A and B are constants to be determined from the boundary conditions. For example, with $\varphi = \varphi_0$ at $x = x_0$ and $\varphi = \varphi_1$ at $x = x_1$ we obtain: $A = (\varphi_1 - \varphi_0)/(x_1 - x_0)$ and $B = (\varphi_0 x_0 - \varphi_1 x_0)/(x_1 - x_0)$. Hence:

$$\varphi_0 - \varphi(x) = \frac{x - x_0}{x_1 - x_0} (\varphi_1 - \varphi_0). \tag{7.6.4}$$

It is of interest to note that here, as in other cases of flow through a homogeneous isotropic porous medium, that the piezometric head distribution $\varphi(x)$ is independent of K. It is only Q that is determined by K and $\varphi(x)$.

In many cases of practical interest D may vary in the direction of the flow. However, when the thickness D is small with respect to distances involved (in the general direction of the flow), and variations in D are small with respect to D, we assume that the flow is essentially parallel to an average plane passing midway between the two confining layers. Thus, two-dimensional flow in the vertical xz plane, becomes a one-dimensional flow in the direction s. Under such conditions

$$Q = \text{const} = -KD(s)\,d\varphi/ds; \quad d\varphi = -Q\,ds/KD(s); \quad \varphi(s_1) - \varphi(s_2) = (Q/K)\int_{s_1}^{s_2} ds/D(s). \tag{7.6.5}$$

Sometimes the hydraulic conductivity varies in the direction of s. Then, with $K(s)D(s) = T(s) =$ transmissivity of aquifer, equation (7.6.5) becomes:

$$\varphi(s_1) - \varphi(s_2) = Q\int_{s_1}^{s_2} ds/T(s) \tag{7.6.6}$$

or:

$$Q = \bar{T}\,\frac{\varphi(s_1) - \varphi(s_2)}{s_2 - s_1}; \qquad \frac{1}{\bar{T}} = \frac{1}{s_2 - s_1}\int_{s_1}^{s_2} \frac{ds}{T(s)} \tag{7.6.7}$$

where $\bar{T}$ is the average transmissivity of the aquifer between s_1 and s_2. Thus (7.6.7) is nothing but a generalization of (5.8.9). Note that the drop of head here is along s, which does not necessarily coincide with the horizontal distance, say, x.

In the case of $T = T(s)$ considered above, the continuity equation takes the form:

$$\frac{d}{dx}\left[T(s)\,\frac{d\varphi}{ds}\right] = 0. \tag{7.6.8}$$

For steady one-dimensional flow of a compressible fluid in an inhomogeneous medium, the continuity equation is obtained from (6.2.26):

$$\frac{d}{dx}\left[\frac{k_g(x)}{\mu}\,\frac{p}{Z(p)}\,\frac{dp}{dx}\right] = 0 \tag{7.6.9}$$

where k_g is the gas permeability. For $Z(p) = 1.0$, and taking into account the Klinkenberg effect (par. 5.3.3), equation (7.6.9) becomes:

$$\frac{d}{dx}\left[\frac{k(x)p}{\mu}\left(1 + \frac{b}{p}\right)\frac{dp}{dx}\right] = 0; \qquad k_g \equiv k, \quad b = \text{const}. \tag{7.6.10}$$

This equation can be integrated, yielding:

$$\frac{k(x)p}{\mu}\left(1 + \frac{b}{p}\right)\frac{dp}{dx} = A, \qquad A = \text{const}. \tag{7.6.11}$$

Then, integrating again, with boundary conditions of specified pressures at $x = 0$ with $x = L$, yields:

$$\tfrac{1}{2}[p^2(0) - p^2(L)] + b[p(0) - p(L)] = -A\mu\int_0^L \frac{dx}{k(x)} \tag{7.6.12}$$

or, with $\bar{p} = [p(0) + p(L)]/2$ and $\Delta p = p(0) - p(L)$:

$$\bar{p}\,\Delta p + b\,\Delta p = A\mu \int_0^L \frac{dx}{k(x)} = -A\mu\frac{L}{\bar{k}}; \qquad \frac{L}{\bar{k}} = \int_0^L \frac{dx}{k(x)}$$

$$(\bar{k}/\mu)\bar{p}(1 + b/\bar{p})\,\Delta p/L = -A; \qquad A = \bar{q}\bar{p}; \qquad \bar{q} = -(k/\mu)(1 + b/\bar{p})\,\Delta p/L. \quad (7.6.13)$$

Additional examples of solutions obtained by direct integration of the steady flow continuity equation are given in paragraph 8.2.3 (for phreatic flow).

7.6.2 Advance of a Wetting Front

Let water infiltrate the horizontal bottom of a pond of infinite area. The depth of water in the pond is H. The *wetting front* is the surface (here, a horizontal plane) behind which we assume complete saturation. This assumption is an approximation, as, in view of the discussion presented in section 9.4, in this situation we have unsaturated flow with a gradual change in saturation. The wetting front advancing into the dry soil serves as a boundary to the flow domain behind it. We may regard this boundary as a free surface on which $p = 0$, or as a boundary of saturation on which $p = -p_c$. At time t the front is at a distance $\zeta(t)$ below the pond's bottom.

For the flow behind a front advancing in a homogeneous medium, we have:

$$\partial^2\varphi/\partial z^2 = 0; \qquad \varphi = Az + B. \quad (7.6.14)$$

With $\varphi = H(t)$ at $z = 0$ and $\varphi = -\zeta - p_c/\gamma$ at $z = -\zeta$, we obtain:

$$\varphi(z, t) = \left[1 + \frac{H(t) + p_c/\gamma}{\zeta(t)}\right] z + H(t). \quad (7.6.15)$$

The solution (7.6.15) may serve as an example of the statement, made several times in chapter 6, that the Laplace equation does not necessarily mean that flow is steady. The flow in a domain bounded by a phreatic surface or by an advancing wetting front, although unsteady, is also described by the Laplace equation (or by the corresponding equation for inhomogeneous or anisotropic media).

The position of the front advancing at a velocity $V_f(t)$ may be derived from:

$$V_f(t) = \frac{d\zeta(t)}{dt} = \frac{K}{n_e}\frac{d\varphi(z, t)}{dz}\bigg|_{z=-\zeta} \quad (7.6.16)$$

where n_e is the effective porosity (with respect to the advancing front). Hence:

$$\frac{K}{n_e}\,dt = \frac{\zeta\,d\zeta}{\zeta + H(t) + p_c/\gamma}. \quad (7.6.17)$$

For $H(t) = \text{const} = H_0$, we obtain:

$$\frac{Kt}{(H_0 + p_c/\gamma)n_e} = \frac{\zeta}{H_0 + p_c/\gamma} - \ln\left(1 + \frac{\zeta}{H_0 + p_c/\gamma}\right). \quad (7.6.18)$$

Irmay (in Bear, Zaslavsky and Irmay 1968) also presents solutions for the general case of a front advancing along s, for vertical upflow and for horizontal flow (linear and radial). Dicker and Sevian (1965) also discuss this problem.

7.7 The Method of Images

7.7.1 Principles

Although actually a particular application of superposition (par. 7.5.1), the method of images will be considered in greater detail because of its usefulness, especially in cases of sources and sinks (e.g., in the form of wells) near boundaries.

Let $\varphi = \varphi(x, y, t)$ be the solution in the half plane $y > 0$ of the partial differential equation:

$$\frac{\partial^2 \varphi}{\partial x^2} + \frac{\partial^2 \varphi}{\partial y^2} - \frac{\varphi}{\lambda^2} = \frac{S}{T}\frac{\partial \varphi}{\partial t} \tag{7.7.1}$$

satisfying the equipotential boundary conditions:

$$\varphi(x, 0, t) = 0. \tag{7.7.2}$$

Equation (7.7.1) describes unsteady flow in a leaky aquifer. When $\lambda = \infty$ it describes unsteady flow in a confined aquifer. The function $\varphi = \varphi(x, y, t)$, hitherto defined only in the half plane $y > 0$, can be continued in the other half $y < 0$, by defining it in that area as (Hydrologisches Colloquium 1964):

$$\varphi(x, y, t) = -\varphi(x, -y, t). \tag{7.7.3}$$

If y is negative, $-y$ is positive, so that if $\varphi = \varphi(x, y, t)$ is known for positive values of y, by (7.7.3), φ is known both for $y > 0$ and $y < 0$. Thus $\varphi(x, y)$ extended in this way is an odd function of y and the function $\varphi(x, y, t)$ in the half plane $y < 0$ is called an *odd image* of $\varphi(x, y, t)$ in the plane $y > 0$. Since:

$$\frac{\partial \varphi(x, y, t)}{\partial x} = -\frac{\partial \varphi(x, -y, t)}{\partial x}, \qquad \frac{\partial^2 \varphi(x, y, t)}{\partial x^2} = -\frac{\partial^2 \varphi(x, -y, t)}{\partial x^2}$$

$$\frac{\partial \varphi(x, y, t)}{\partial y} = +\frac{\partial \varphi(x, -y, t)}{\partial y}, \qquad \frac{\partial^2 \varphi(x, y, t)}{\partial y^2} = -\frac{\partial^2 \varphi(x, -y, t)}{\partial y^2} \tag{7.7.4}$$

it follows that $\varphi(x, y, t)$ also satisfies (7.7.1) for $y < 0$, and that φ, $\partial\varphi/\partial x$ and $\partial\varphi/\partial y$ are continuous at $y = 0$ (actually φ and $\partial\varphi/\partial x$ are zero along $y = 0$). Moreover, it follows that (7.7.3), which defines $\varphi(x, y, t)$ for $y < 0$, is also valid for $y > 0$. For if we substitute $-y'$ for y in (7.7.3) we obtain:

$$\varphi(x, y', t) = -\varphi(x, y', t)$$

which is valid for $y' > 0$. The same is true for (7.7.4). Finally, from (7.7.4) and Darcy's law, we obtain:

$$q_x(x, y, t) = -q_x(x, -y, t)$$

$$q_y(x, -y, t) = q_y(x, -y, t) \tag{7.7.5}$$

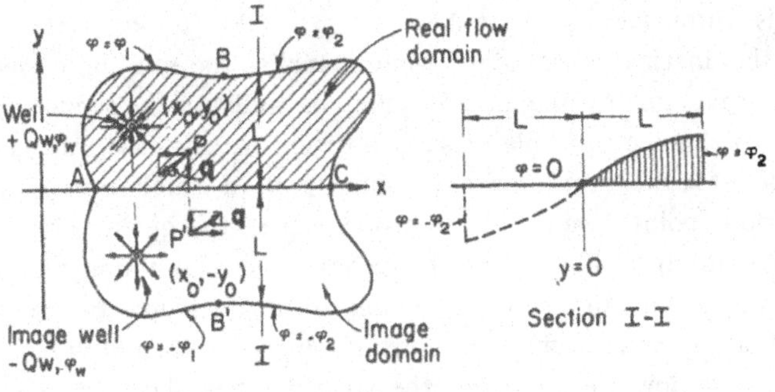

(a) Image domain for φ=0 along y=0

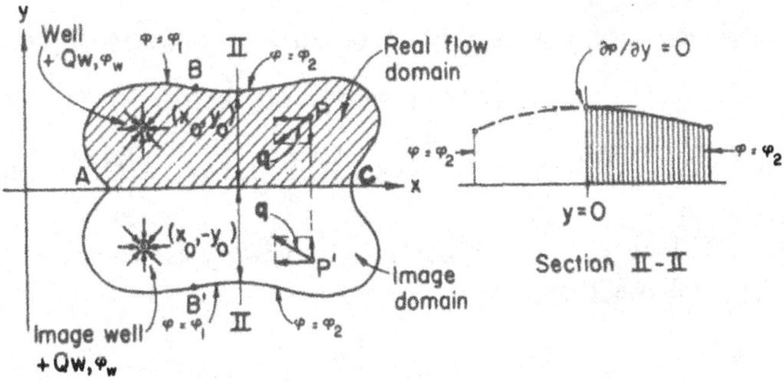

(b) Image domain for ∂φ/∂y=0 along y=0

FIG. 7.7.1. The method of images.

valid for positive and negative values of y. Consequently, the flow patterns in both half planes are images of each other. From (7.7.5) it follows that for $y = 0$, $q_x = 0$, i.e., the flow is normal to the x-axis.

When φ satisfies certain boundary conditions along a specified boundary in the half plane $y > 0$ (say, ABC in fig. 7.7.1a), it follows from (7.7.3) and (7.7.4) that the extended function in the half plane $y < 0$ satisfies analogous boundary conditions on the image in the line $y = 0$ of the boundaries (boundaries $CB'A$ in $y < 0$). For example, when $\varphi = 0$ along $y = 0$, a circular well of radius r_w, discharge Q_w and piezometric head φ_w, with its center at (x_0, y_0), will have an image well at $(x_0, -y_0)$ of radius r_w, discharge $-Q_w$ and piezometric head $-\varphi_w$ (fig. 7.7.1a). Owing to the change of sign of the x component of the specific discharge $\mathbf{q}$, the well's discharge also changes its sign. Figure 7.7.1a shows the boundary and the boundary condition for the case of a finite flow domain.

Thus, instead of solving a problem described by (7.7.1) in the domain $ABCA$ (fig. 7.7.1a), satisfying $\varphi = 0$ along the boundary AC, we may solve the same equation in the domain $ABCB'A$, with appropriate boundary conditions along $CB'A$, and apply the solution only to the real flow domain. The satisfaction of the condition along $y = 0$ is automatically assured.

Obviously, this method is employed when it makes the solution easier. The case of sources and sinks (i.e., recharging and pumping wells) in the semi-infinite domain $y > 0$ is the best known example of such a case. In this case, instead of solving the flow problem for the semi-infinite domain $y > 0$, in which sources and sinks are present at various points, we solve the problem (i.e., the given partial differential equation) in the infinite plane in which we have a double number of sources and sinks, i.e., an image sink to every source and an image source to every sink. The basic idea is that by superposing the flows produced by the individual sources and sinks, we obtain a flow that satisfies the specified boundary conditions.

The method of images is also applicable to the case of the impervious boundary condition $\partial\varphi/\partial n = 0$ along a straight line boundary, e.g., $\partial\varphi/\partial y = 0$ along $y = 0$ in figure 7.7.1b. In this case the function $\varphi(x, y, t)$ in the half plane $y < 0$ is an *even image* of $\varphi(x, y, t)$ in the half plane $y > 0$.

The function $\varphi = \varphi(x, y, t)$ is said to be an *even function* of y:

$$\varphi(x, y, t) = \varphi(x, -y, t) \tag{7.7.6}$$

$$\frac{\partial\varphi(x, y, t)}{\partial x} = \frac{\partial\varphi(x, -y, t)}{\partial x} \quad \text{and} \quad \frac{\partial\varphi(x, y, t)}{\partial y} = -\frac{\partial\varphi(x, -y, t)}{\partial y}. \tag{7.7.7}$$

Since $\partial\varphi/\partial y = 0$ along $y = 0$, then φ, $\partial\varphi/\partial x$ and $\partial\varphi/\partial y$ are continuous along $y = 0$. From (7.7.7) it follows that:

$$\begin{aligned} q_x(x, y, t) &= q_x(x, -y, t) \\ q_y(x, y, t) &= -q_y(x, -y, t). \end{aligned} \tag{7.7.8}$$

Sources and sinks near an impervious straight line boundary are handled in a way similar to that described above for the equipotential boundary, except that in the present case the image of a sink is also a sink having the same strength.

Obviously the straight line boundaries considered above need not coincide with any of the coordinate axes, and the equipotential boundary may be any $\varphi = \varphi_0 =$ const and not necessarily $\varphi = 0$.

It is also possible to apply the method of images to the case where a straight line boundary separates two regions of different transmissivities. Another case that can be handled by this method is an eccentric well in a circular aquifer. When sources and sinks are present in a flow domain, the boundary of which is composed of a number of parallel or intersecting straight lines, multiple images (sometimes infinite in number) are required to satisfy the boundary conditions. Examples of these cases are given in the following paragraph.

The method of images is also applicable to three-dimensional flows, where equipotential and impervious boundaries take the form of planes.

7.7.2 Examples

When point sources or sinks (e.g., pumping and recharging wells in aquifers) are

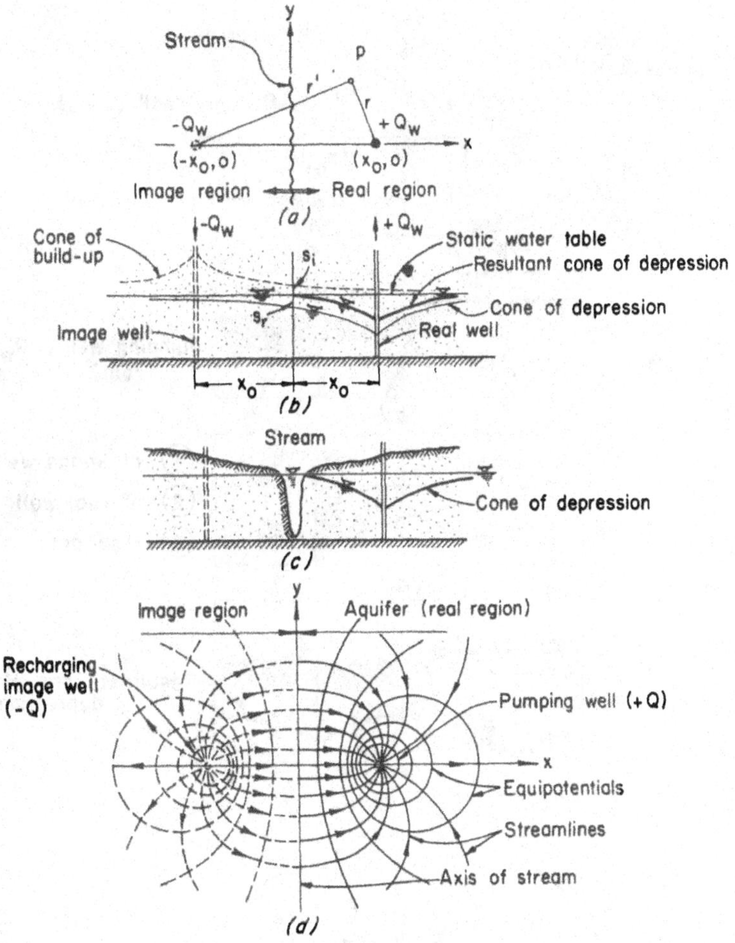

FIG. 7.7.2. The method of images applied to a well near a recharge boundary.

present in a domain bounded by straight line boundaries that are either equipotential or impervious, the domain with its boundaries is replaced by a fictitious domain that is infinite in area. The images are chosen so that the conditions along the boundaries are satisfied. Therefore, the flow pattern sought, produced by the sources and sinks in the real bounded domain, is identical to that obtained for the same area when it is part of the fictitious domain.

Several examples are shown in figures 7.7.2 through 7.7.6. Figure 7.7.2 shows a well near a stream that is an equipotential boundary. Figures 7.7.2b and 7.7.2c also give the physical interpretation of this case. With:

$$s(x, y) = (Q_w/2\pi T) \ln(R/r) \tag{7.7.9}$$

describing the drawdown at a point (x, y) caused by a well pumping at a constant

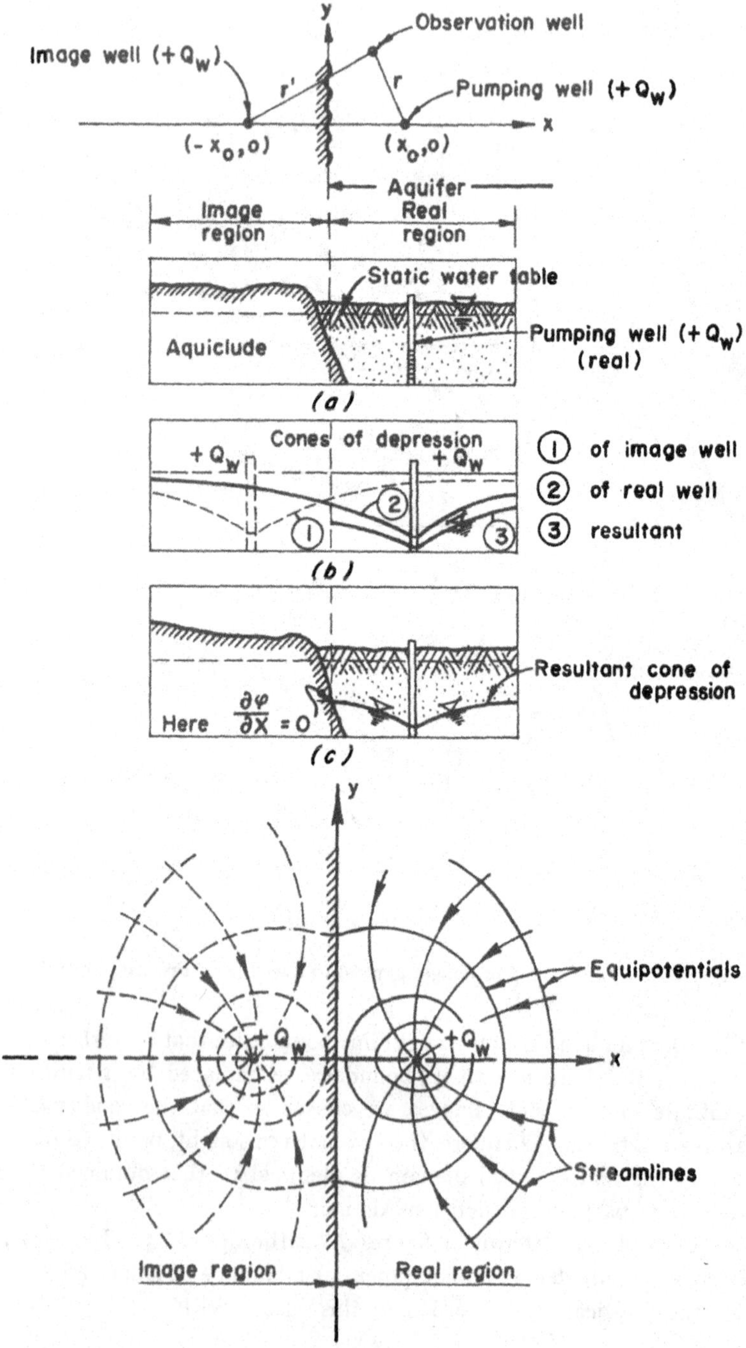

FIG. 7.7.3. The method of images applied to a well near an impervious boundary.

rate Q_w from an *infinite* confined aquifer of transmissivity T and radius of influence R (i.e., the distance at which $s = 0$),

$$\begin{aligned}\varphi_0 - \varphi(x, y) \equiv s(x, y) &= (Q_w/2\pi T) \ln R/r + (-Q_w/2\pi T) \ln R/r' = (Q_w/2\pi T) \ln r'/r \\ &= (Q_w/4\pi T) \ln((x + x_0)^2 + y^2)/((x - x_0)^2 + y^2) \qquad (7.7.10)\end{aligned}$$

(with symbols explained by the figure) gives the drawdown at any observation point (x, y) *in the real domain* $x > 0$. It is obtained by the superposition of the drawdown of the image (pumping) well and the build-up of the image (injection) well, each operating in the fictitious infinite domain. The resulting flow net for the real domain $x > 0$ is shown in figure 7.7.2d. For $r_w \ll x_0$:

$$s(r_w) \approx (Q_w/2\pi T) \ln(2x_0/r_w). \qquad (7.7.11)$$

Figure 7.7.3 shows a well in the semi-infinite domain bounded by an impervious boundary along $x = 0$. In this case the image well is also a pumping well. Hence:

$$\varphi(r) - \varphi_w = \frac{Q_w}{2\pi T} \ln \frac{r}{r_w} + \frac{Q_w}{2\pi T} \ln \frac{r'}{r_w} = \frac{Q_w}{2\pi T} \ln \frac{rr'}{r_w^2}. \qquad (7.7.12)$$

The resulting flow net is shown in figure 7.7.3d.

Figures 7.7.4 and 7.7.5 show several additional cases that require no further

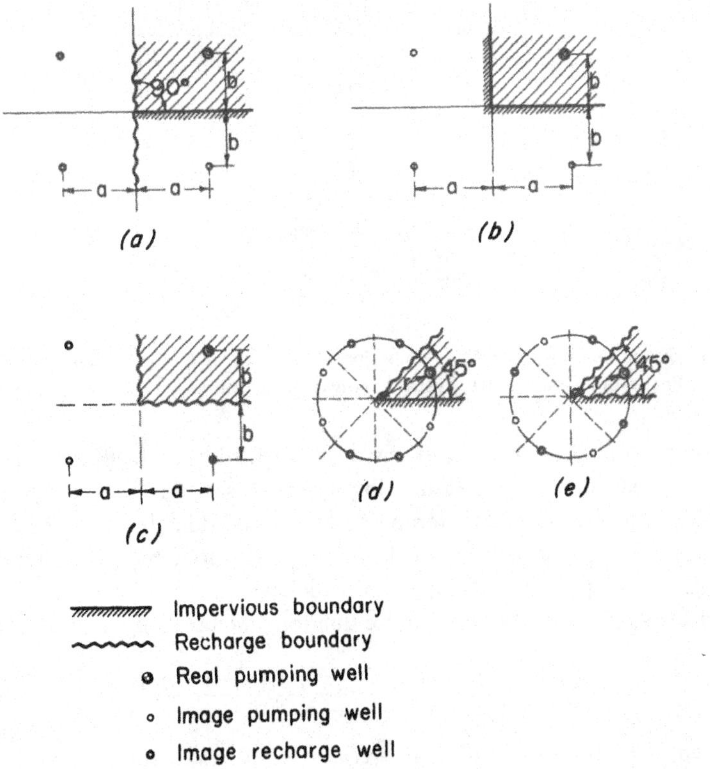

FIG. 7.7.4. Examples of image-well systems for a pumping well near various aquifer boundaries.

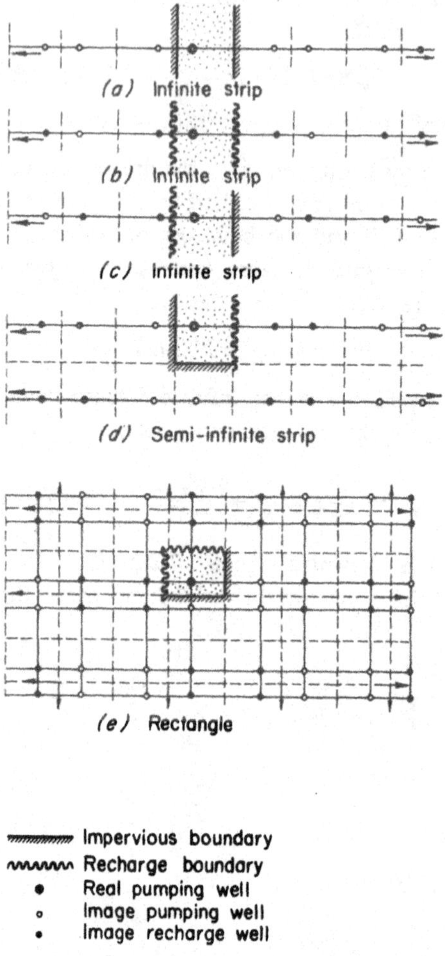

FIG. 7.7.5. Example of image-well systems for a pumping well near various aquifer boundaries (arrows indicate that image well systems continue to infinity).

explanation. In each case it can easily be verified that together, all the sources and sinks produce a potential distribution that satisfies the specified boundary conditions, and the drawdown (or potential) in the domain of interest is obtained by superposition of the individual drawdowns produced by the individual sources and sinks, each operating in the infinite domain.

For example, for a well in an infinite quadrant bounded by a stream (fig. 7.7.4c):

$$s = \frac{Q_w}{4\pi T} \ln \frac{[(x-a)^2 + (y-b)^2][(x+a)^2 + (y+b)^2]}{[(x+a)^2 + (y-b)^2][(x-a)^2 + (y+b)^2]}. \tag{7.7.13}$$

Next consider the example of an *eccentric point sink* (pumping well) at a point $W(x_0, y_0)$ in a circular flow domain of radius r_e; the boundary condition on the

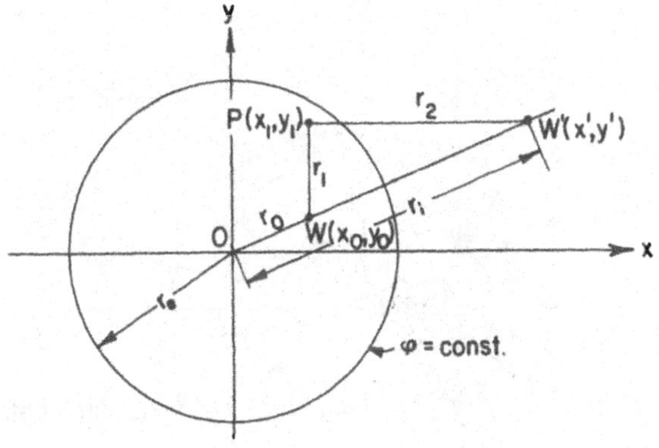

FIG. 7.7.6. Flow to an eccentric well.

circumference of the circle is $\varphi = \text{const} = \varphi_0$ or $s \equiv \varphi_0 - \varphi = 0$ (fig. 7.7.6).

The image in this case is a source (recharge well) located at point $W'(x', y')$, which is the inverse of W with respect to the circle (par. 8.3.2). Equations (8.3.14) and (8.3.15) are applicable (with $\rho = r_e$, $x = x_0$ and $y = y_0$). Hence, with the solution (7.7.9) for a single pumping well in an infinite domain, we obtain by superposition:

$$s = \varphi_0 - \varphi(r) = \frac{Q_w}{2\pi T} \ln \frac{R}{r_1} - \frac{Q_w}{2\pi T} \ln \frac{aR}{r_2} = \frac{Q_w}{2\pi T} \ln \frac{r_2}{a r_1}; \qquad r_i^2 = \frac{r_e^4}{r_0^2} \quad (7.7.14)$$

where aR is the radius of influence of the image well; a well determined by the boundary conditions on the circle. With $M(X, Y)$ denoting a point on the circle we have:

$$X^2 + Y^2 = r_e^2 \quad (7.7.15)$$

and from inserting $s = 0$ in (7.7.14):

$$(X - x')^2 + (Y - y')^2 = a^2[(X - x_0)^2 + (Y - y_0)^2]. \quad (7.7.16)$$

The last two equations will become identical if $a^2 = r_e^2/r_0^2$. Hence, the drawdown equation becomes:

$$s = \frac{Q_w}{2\pi T} \ln \frac{r_2}{r_1} \frac{r_0}{r_e}. \quad (7.7.17)$$

Finally, consider a sink (pumping well) of constant rate Q_w near a discontinuity in aquifer transmissivity (in two-dimensional plane flow): $T = T'$ for $x > 0$, $T = T''$ for $x < 0$ (fig. 7.7.7).

The problem may be stated as follows. Determine: φ' in $x > 0$ and φ'' in $x < 0$, such that $\nabla^2\varphi' = 0$ in $x > 0$ and $\nabla^2\varphi'' = 0$ in $x < 0$ and the boundary conditions on $x = 0$ are:

$$\varphi' = \varphi'', \qquad T' \, \partial\varphi'/\partial x = T'' \, \partial\varphi''/\partial x. \quad (7.7.18)$$

The image well is placed as usual at $(-x_0, 0)$. However, its strength is CQ_w, where C is as yet an unknown constant. By superposition, the drawdown at a point

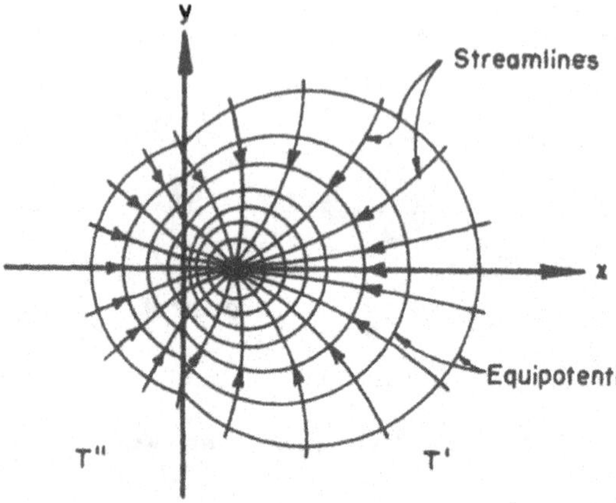

FIG. 7.7.7. Well near a straight line boundary between regions of different transmissivities.

in the region $x > 0$ is given by:

$$s \equiv \varphi_0 - \varphi' = \frac{Q_w}{4\pi T'} \ln \frac{R^2}{(x - x_0)^2 + y^2} + \frac{CQ_w}{4\pi T'} \ln \frac{R^2}{(x + x_0)^2 + y^2}. \quad (7.7.19)$$

For the region $x < 0$, we have:

$$s = \varphi_0 - \varphi'' = \frac{BQ_w}{4\pi T''} \ln \frac{R^2}{(x - x_0)^2 + y^2} \quad (7.7.20)$$

where B is another, yet unknown, constant. By inserting (7.7.19) and (7.7.20) in (7.7.18) we obtain:

$$B = 2/(1 + N); \qquad C = (1 - N)/(1 + N)$$

where $N = T''/T'$. Hence:

$$s = \varphi_0 - \varphi' = \frac{Q_w}{4\pi T'} \ln \frac{R^2}{(x - x_0)^2 + y^2} + \frac{1 - N}{1 + N} \frac{Q_w}{4\pi T'} \ln \frac{R^2}{(x - x_0)^2 + y^2}; \qquad x > 0$$

$$s = \varphi_0 - \varphi'' = \frac{2N}{N + 1} \frac{Q_w}{4\pi T''} \ln \frac{R^2}{(x - x_0)^2 + y^2}; \qquad x < 0.$$

The flow net is shown in figure 7.7.7.

7.8 Methods Based on the Theory of Functions

Unless otherwise stated, the porous medium considered in this section is homogeneous and isotropic. Hence, the specific discharge potential $\Phi = K\varphi$ will be used. Except for a brief review of some concepts, the presentation is based on the assumption that the reader is familiar with the theory of functions. The reader is referred for additional information to a large number of texts available on this subject, or to texts in which this theory is applied to hydrodynamics (e.g., Churchill 1948; Rothe, Pohlhausen and Ollendorf 1951; Milne-Thomson 1960).

7.8.1 Complex Variables and Analytic Functions

A *complex number* α is an expression of the form $\alpha = a + ib$, where a and b are real numbers and i is the imaginary unit satisfying $i^2 = -1$. The numbers a and b are referred to as the *real and imaginary parts of* α, respectively.

Geometrically, it is convenient to represent a complex number $a + ib$ by a point $P(a, b)$ in a diagram called the *Argand diagram* (or the *complex* α*-plane*, or simply the α*-plane*); a and b are the coordinates of the point in a Cartesian coordinate system in this plane. Figure 7.8.1 shows the z-plane where the coordinates are x and y.

It is also possible to define the point $P(x, y)$, or the corresponding complex number $z = x + iy$, in the z- (or xy-) plane in terms of polar coordinates r, θ (fig. 7.8.1):

$$r = |z| = |x + iy| = \sqrt{x^2 + y^2}; \qquad \theta = \tan^{-1}(y/x). \tag{7.8.1}$$

The complex number z in the z-plane is thus represented by the vector OP. $r = |z|$ is called the *absolute value*, or the *modulus* of z; the angle θ is called the *argument* or *amplitude* of z. Since $x = r\cos\theta$, $y = r\sin\theta$ and $\exp(i\theta) = \cos\theta + i\sin\theta$, we may also express z by:

$$z = r(\cos\theta + i\sin\theta); \qquad z = r\exp(i\theta). \tag{7.8.2}$$

The number $\bar{z}$ is called the conjugate of the number $z = x + iy$:

$$\bar{z} = x - iy = r(\cos\theta - i\sin\theta) = r\exp(-i\theta); \qquad |\bar{z}| = |z|. \tag{7.8.3}$$

In the z-plane of figure 7.8.1, $\bar{z}$ is represented by point $P\ (x, -y)$, which is the reflection of P in the x-axis.

A function $f(z)$ is said to be *analytic* (*regular, holomorphic*) in a region R of the complex plane if $f(z)$ has a finite derivative at each point of R, and if $f(z)$ is single-valued in R. It is said to be *analytic at a point* z if z is an interior point of some region where $f(z)$ is analytic. It can be shown that if $f(z)$ is analytic at a point z, then the derivative $f'(z)$ is continuous at z (Hildebrand 1962). This means that if $\zeta = f(z)$,

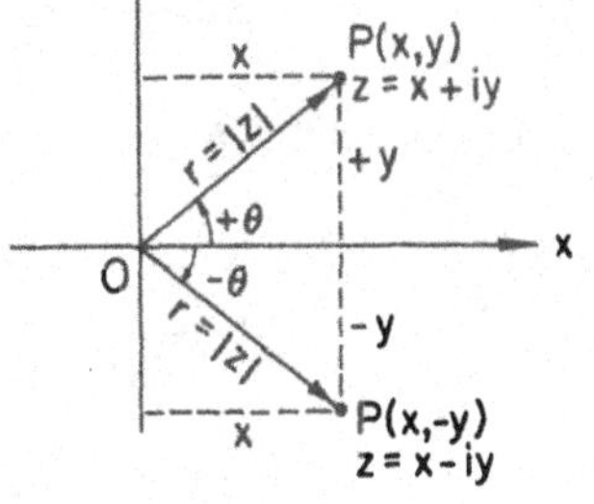

FIG. 7.8.1. The z-plane.

we require that the limit:

$$f'(z) \equiv \frac{d\zeta}{dz} = \lim_{\Delta z \to 0} \frac{f(z + \Delta z) - f(z)}{\Delta z} = \lim_{\Delta z \to 0} \frac{\Delta \zeta}{\Delta z} \tag{7.8.4}$$

exists uniquely as $\Delta z \to 0$ from *any direction in the complex plane.* We shall be especially interested in approaching along a line such that $\Delta y = 0$ (i.e., along a line parallel to the real axis), or such that $\Delta x = 0$, where $\Delta z = \Delta x + i\,\Delta y$.

Consider a function $\zeta = f(z) = u(x, y) + iv(x, y)$ for which (7.8.4) does exist uniquely, independently of the manner in which $\Delta z \to 0$, where $u = u(x, y)$ and $v = v(x, y)$ are the real and imaginary parts of ζ, respectively. Then:

$$\frac{d\zeta}{dz} = \frac{\partial u}{\partial x} + i\frac{\partial v}{\partial x} = -i\frac{\partial u}{\partial y} + \frac{\partial v}{\partial y} \tag{7.8.5}$$

is obtained either by letting $\Delta y \to 0$ first, and then letting $\Delta x \to 0$, or by letting $\Delta x \to 0$ first and then letting $\Delta y \to 0$. Hence the following conditions must be satisfied:

$$\frac{\partial u}{\partial x} = \frac{\partial v}{\partial y}; \qquad \frac{\partial u}{\partial y} = -\frac{\partial v}{\partial x}. \tag{7.8.6}$$

These equations are known as the *Cauchy–Riemann conditions.* The single-valued complex function $\zeta = f(z) = u(x, y) + iv(x, y)$, for which $\partial u/\partial x$, $\partial u/\partial y$, $\partial v/\partial x$ and $\partial v/\partial y$ are continuous in R, is analytic in R *if and only if* the Cauchy–Riemann conditions (7.8.6) are satisfied. Then the derivative $d\zeta/dz$ exists uniquely for any manner of approach to the limit.

When u and v are expressed in polar coordinates r, θ, the Cauchy–Riemann conditions (7.8.6) take the form:

$$\partial u/\partial r = (1/r)\,\partial v/\partial \theta; \qquad (1/r)\,\partial u/\partial \theta = -\,\partial v/\partial r. \tag{7.8.7}$$

Since all partial derivatives are continuous, and the order of differentiation is immaterial, we can obtain from (7.8.6):

$$\begin{aligned} \partial^2 u/\partial x^2 + \partial^2 u/\partial y^2 &= 0; \qquad \nabla^2 u = 0 \\ \partial^2 v/\partial x^2 + \partial^2 v/\partial y^2 &= 0; \qquad \nabla^2 v = 0. \end{aligned} \tag{7.8.8}$$

This means that the real and imaginary parts of an analytic function satisfy the Laplace equation.

With the Cauchy–Riemann conditions (7.8.6), we can determine either $u = u(x, y)$ or $v = v(x, y)$ if the other is known:

$$du = \frac{\partial u}{\partial x}dx + \frac{\partial u}{\partial y}dy = \frac{\partial v}{\partial y}dx - \frac{\partial v}{\partial x}dy; \qquad dv = \frac{\partial v}{\partial x}dx + \frac{\partial v}{\partial y}dy = -\frac{\partial u}{\partial y}dx + \frac{\partial u}{\partial x}dy$$

$$u(x, y) - u(x_0, y_0) = \int_{(x_0,y_0)}^{(x,y)} \left(\frac{\partial v}{\partial y}dx - \frac{\partial v}{\partial x}dy\right);$$

$$v(x, y) - v(x_0, y_0) = \int_{(x_0,y_0)}^{(x,y)} \left(-\frac{\partial u}{\partial y} dx + \frac{\partial u}{\partial x} dy \right). \tag{7.8.9}$$

When $\zeta = f(z) = u(x, y) + iv(x, y)$ is single valued, then to each point z there corresponds one and only one value of ζ. To investigate whether the reverse is true, i.e., to every point (u, v) there corresponds only one point (x, y), we must solve $u = u(x, y)$ and $v = v(x, y)$ for x and y. These equations can be uniquely solved if the *Jacobian*:

$$J = \frac{\partial(u, v)}{\partial(x, y)} = \begin{vmatrix} \partial u/\partial x & \partial u/\partial y \\ \partial v/\partial x & \partial v/\partial y \end{vmatrix}; \qquad J|_{z=z_0} = |f'(z_0)|^2 \tag{7.8.10}$$

defined in paragraph 4.1.3 does not vanish in some region around any point for which ζ is defined.

The two functions $u(x, y)$ and $v(x, y)$, which are respectively the real and imaginary parts of an analytic function, are known as *conjugate functions*. They are also called *conjugate harmonic functions* (as each of them solves the Laplace equation).

If a function $f(z)$ is not analytic at some point z_0, but is analytic at every point in the neighborhood of z_0, then z_0 is said to be a *singular point*, or a singularity of $f(z)$. For example, $f(z) = 1/(z - z_0)$ is analytic at every point except at $z = z_0$, where it is discontinuous; hence $z = z_0$ is a singular point.

Consider the two families of curves in the z-plane, $u(x, y) = C_1$ and $v(x, y) = C_2$, where C_1 and C_2 are constants. The slope of the first family of curves is $dy/dx = -(\partial u/\partial x)/(\partial u/\partial y)$, and that of the second family of curves is $dy/dx = -(\partial v/\partial x)/(\partial v/\partial y)$, which by using the Cauchy–Riemann conditions (7.8.6) becomes $dy/dx = (\partial u/\partial y)/(\partial u/\partial x)$. Hence the two families of curves are *orthogonal* to each other.

Some functions $f(z)$ take on more than one value for each value of z. The functions $z^{1/2}$ and $\ln z$ may serve as examples. The first, $f(z) = z^{1/2} = r^{1/2} \exp(i\theta/2)$ takes on two values, one the negative of the other, depending on the choice of θ. However, if θ is restricted to the range $0 \leqslant \theta \leqslant \pi$, $z^{1/2}$ takes on just one value for each point z of the complex plane. For $\ln z = \ln |z| + i\theta$, where $\theta = \tan^{-1}(y/x)$, we have an infinite number of values of $\ln z$ for each value of z, since for θ we have an infinite number of angles differing by integral multiples of 2π. We may again limit ourselves to $0 \leqslant \theta \leqslant 2\pi$ and speak of this value as the principal value of θ for the logarithm.

A *branch* of a multiple-valued function $f(z)$ is any single-valued analytic function that for each value of z assumes one of the values of $f(z)$. The requirement of analyticity prevents a branch of a function from taking on a random selection of the values of the function. A *branch cut* is a boundary introduced so that the corresponding branch is single valued and analytic throughout the open region (i.e., the region without its boundary) bounded by the cut (Churchill 1948).

In the examples above, a ray $\theta =$ const from the origin is the branch cut. Curves running from the origin to infinity may also serve as branch cuts. Since the cut serves to make θ unique, in certain functions it does not begin at $z = 0$, but at some other point $z = z_0$. Such an origin of a branch cut for a multiple-valued function is called a *branch point* of the function. A branch point is a singularity or a singular point.

7.8.2 The Complex Potential and the Complex Specific Discharge

Following the discussion of $\Phi = \Phi(x, y)$ and $\Psi = \Psi(x, y)$ in paragraphs 6.5.2, 6.5.4 and 6.5.5, where it is shown, from physical consideration, that Φ and Ψ satisfy the Cauchy–Riemann conditions, we may now construct the function:

$$\zeta = f(z) = \Phi(x, y) + i\Psi(x, y). \tag{7.8.11}$$

The function ζ is analytic and we may apply the results of paragraph 7.8.1 to it. With $u = \Phi$ and $v = \Psi$, we obtain from (7.8.5) and (7.8.6):

$$d\zeta/dz = \partial\Phi/\partial x + i\,\partial\Psi/\partial x = \partial\Phi/\partial x - i\,\partial\Phi/\partial y = -q_x + iq_y \tag{7.8.12}$$

or:

$$-d\zeta/dz = q_x - iq_y = q\cos\theta - iq\sin\theta = q\exp(-i\theta) \tag{7.8.13}$$

where q and θ are the absolute magnitude and the direction of $\mathbf{q}$. The complex function $\bar{\omega} = q_x - iq_y$ is referred to as the *conjugate complex specific discharge.* The conjugate of $\bar{\omega}$, $\omega = q_x + iq_y = q\exp(i\theta)$, is the *complex specific discharge.* It can easily be verified that $\bar{\omega}$ is analytic whereas ω is not. We also have:

$$|d\zeta/dz| = (q_x^2 + q_y^2)^{1/2} = q; \qquad q^2 = \omega\bar{\omega}. \tag{7.8.14}$$

Thus, given Φ and Ψ, we may always construct $\zeta = f(z)$ by (7.8.11). Conversely, any analytic function describes a possible two-dimensional flow of an incompressible fluid; its real part describes the potential distribution, whereas its imaginary part describes the stream function.

A point where $q = 0$ is called a *stagnation* point. From (7.8.14) it follows that at such a point $d\zeta/dz = 0$. At a stagnation point both q_x and q_y vanish; hence, also, $\partial\Phi/\partial x = 0$, $\partial\Psi/\partial y = 0$, and streamlines intersect each other or abruptly change direction.

The stagnation point is a singular point (par. 7.8.1). Other types of singular point are (Karman and Biot 1940):

(a) points where the specific discharge q is infinite (converging toward the point or diverging from it). This is a *logarithmic singularity.* Point sources and sinks (par. 7.8.2) are examples of logarithmic singularities. In the vicinity of such a point, $d\zeta/dz$ is indeterminate, and all streamlines intersect each other. Equipotentials crowd together as the point is approached.

(b) The tip of a corner ($\alpha \leqslant \pi$) around which flow takes place is a singular point called a *vortex point.* There the streamlines crowd together, none intersecting each other, while equipotentials intersect each other (par. 7.8.6). The specific discharge at such a point is infinite.

(c) At a *saddle point,* a finite number of streamlines meet each other, the other streamlines bypassing it. Flow inside a corner may serve as an example (par. 7.8.6).

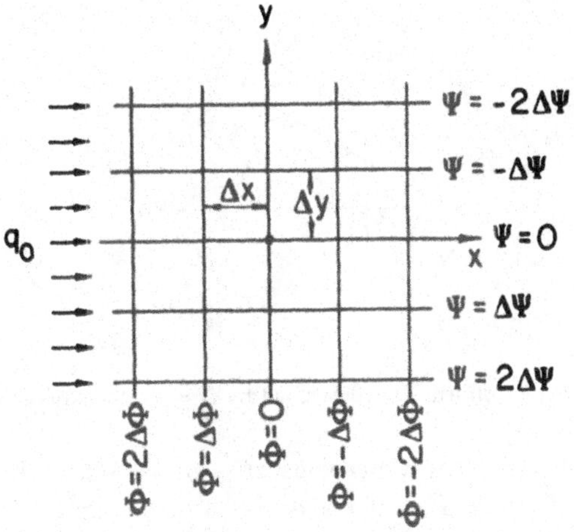

FIG. 7.8.2. Uniform flow described by $\zeta = q_0 z$.

Example 1

$$\zeta = f(z) = -q_0 z = -q_0 x - iq_0 y; \qquad q_0 > 0$$

$$\Phi = -q_0 x; \qquad \Psi = -q_0 y$$

$$q_x = -\frac{\partial\Phi}{\partial x} = q_0; \qquad q_y = -\frac{\partial\Phi}{\partial y} = 0; \qquad -\frac{d\zeta}{dz} = q_0 = q_x - iq_y. \tag{7.8.15}$$

This function describes uniform flow $\mathbf{q} = q_0\mathbf{1x}$ in the x direction in the infinite xy plane (fig. 7.8.2). With $q_0 > 0$, $iq_0 z$ describes a uniform flow $\mathbf{q} = q_y\mathbf{1y}$ in the $+y$ direction.

Example 2

$$\zeta = -q_0 z \exp(-i\alpha) = -q_0(x + iy)(\cos\alpha - i\sin\alpha)$$

$$\Phi = -q_0(x\cos\alpha + y\sin\alpha); \qquad \Psi = -q_0(y\cos\alpha - x\sin\alpha)$$

$$-\frac{d\zeta}{dz} = q_0\exp(-i\alpha) = q_0(\cos\alpha - i\sin\alpha) = q_x - iq_y$$

$$q_x = q_0\cos\alpha; \qquad q_y = q_0\sin\alpha. \tag{7.8.16}$$

With $q_0 > 0$, this is a uniform flow ($q = q_0$) in the infinite xy plane such that $\mathbf{q}$ makes an angle α with the $+x$ axis (fig. 7.8.3). Equipotentials are the straight lines: $y = xq_0 \cot\alpha + C/\sin\alpha$.

Example 3

$$\zeta = -az^2 = -a(x + iy)^2; \qquad a > 0$$

$$\Phi = -a(x^2 - y^2); \qquad \Psi = -2axy = -r^2 a\sin 2\theta$$

$$-\frac{d\zeta}{dz} = 2az = 2a(x + iy) = q_x - iq_y$$

$$q_x = 2ax; \qquad q_y = -2ay. \tag{7.8.17}$$

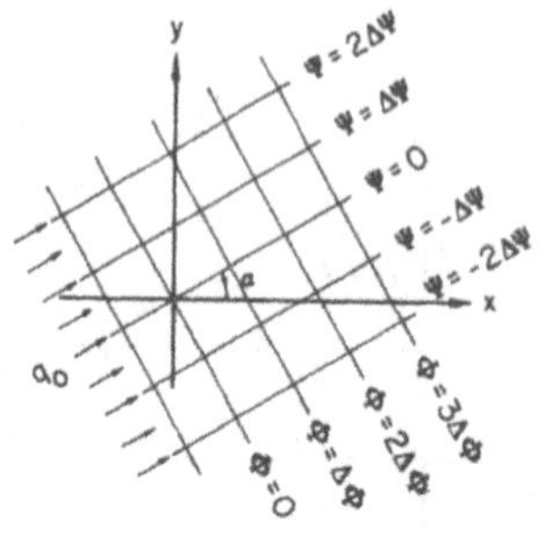

FIG. 7.8.3. Uniform flow described by $\zeta = -q_0 z \exp(-i\alpha)$.

Equipotentials are the family of hyperbolas $x^2 - y^2 = \text{const}$. The streamlines are hyperbolas $xy = \text{const}$, with $\Psi = 0$ for $\theta = 0$ and $\theta = \pi/2$ (fig. 7.8.4). At the origin, $d\zeta/dz = 0$; therefore, it is a singular point that in this case is a stagnation point. When a is purely imaginary, the streamlines and the equipotentials of figure 7.8.4 are interchanged.

From figure 7.8.4 it follows that (7.8.17) describes flow inside a right-angle corner.

Example 4

$$\zeta = -a \sin z = -a(\sin x \cosh y + i \cos x \sinh y); \qquad a > 0$$

$$\Phi = -a \sin x \cosh y; \qquad \Psi = a \cos x \sinh y$$

$$-\frac{d\zeta}{dz} = a \cos z = a(\cos x \cosh y - i \sin x \sinh y) = q_x - iq_y$$

$$q_x = a \cos x \cosh y; \qquad q_y = a \sin x \sinh y$$

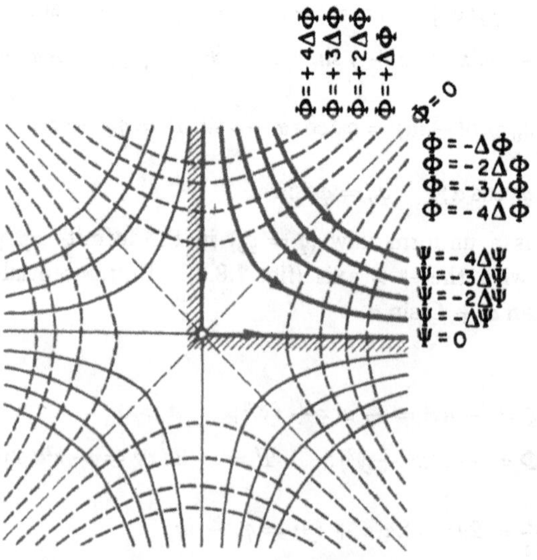

FIG. 7.8.4. Flow in a corner described by $\zeta = -az^2$.

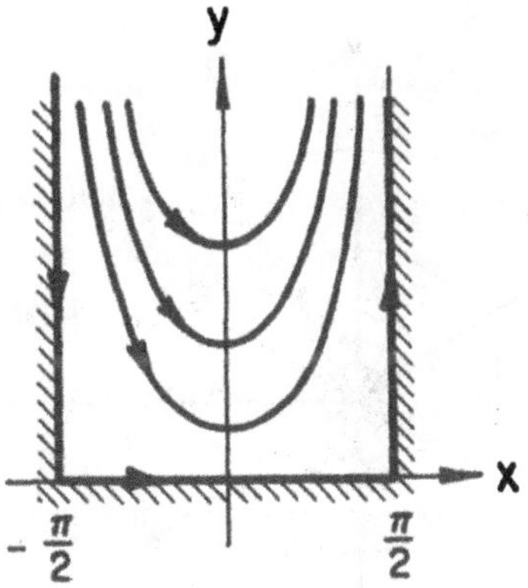

FIG. 7.8.5. Flow in a semi-infinite strip described by $\zeta = -a \sin z$; $a > 0$.

$$q_x = -\frac{\partial \Phi}{\partial x} = a \cos x \cosh y; \qquad \frac{\partial \Phi}{\partial x} = 0 \quad \text{at} \quad x = n\pi \pm \pi/2$$

$$q_y = -\frac{\partial \Phi}{\partial y} = a \sin x \sinh y = 0 \quad \text{at} \quad y = 0. \tag{7.8.18}$$

Hence the complex potential ζ describes flow in a semi-infinite strip bounded by $y \geqslant 0$, $-\pi/2 \leqslant x \leqslant +\pi/2$ (fig. 7.8.5).

Examples involving point sources and sinks are given in paragraph 7.8.3. Examples involving flow inside and around corners are given in paragraph 7.8.6.

Thus, given a function ζ, we can derive the corresponding Φ and Ψ that describe *some* flow. However, the pair of functions Φ and Ψ do not necessarily describe the flow in a *given* flow domain in the z-plane with specified boundary conditions. In fact, Φ and Ψ, or in the combined form ζ, are the functions that comprise the solution sought for any problem. The method of conformal mapping (par. 7.8.4) is a tool to achieve this solution.

7.8.3 Sources and Sinks

Consider the complex potential:

$$\zeta = m \ln z = m(\ln r + i\theta); \qquad \Phi = m \ln r; \qquad \Psi = m\theta \tag{7.8.19}$$

where m (> 0) is real.

The equipotentials are circles $r =$ const. The streamlines are rays $\theta =$ const ($0 < \theta \leqslant 2\pi$). Hence ζ is the complex potential of the flow converging to a *sink* at the origin $z = 0$ (fig. 7.8.6). A pumping well is an example of a sink. The *strength* of the sink, defined as the rate of flow of fluid across a closed curve enclosing the sink, is given by the increase in Ψ corresponding to a closed circuit about the origin, and hence has the value $2\pi m$. If the total discharge of the sink is Q (volume per unit time per unit thickness of the two-dimensional field of flow), then $2\pi m = Q$.

In a similar manner, $\zeta = -m \ln z$ ($m > 0$) represents a *source* of strength $-2\pi m$. An injection well is an example of a source.

For a sink at point z_0 of the xy plane the complex potential is:

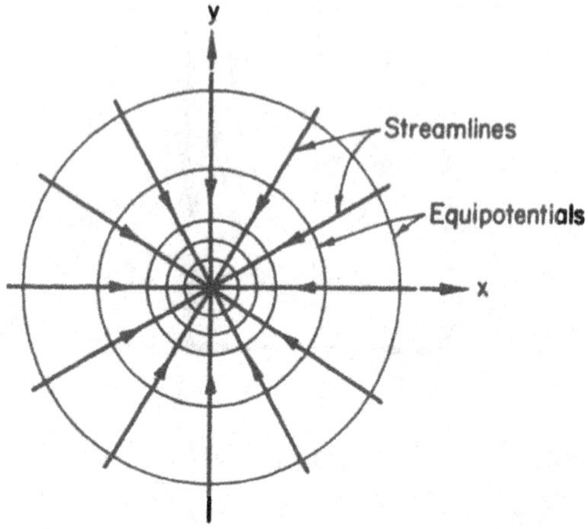

FIG. 7.8.6. Flow to a sink described by $\zeta = m \ln z$; $m > 0$.

$$\zeta = m \ln(z - z_0); \qquad z - z_0 = r_1 \exp(i\theta_1) \tag{7.8.20}$$

where r_1 is the radial distance measured from z_0 and θ_1 is the angle that r_1 makes with the $+ x$-axis (measured in a counterclockwise direction); we have:

$$- d\zeta/dz = - m/(z - z_0) = - (m/r_1)(\cos\theta_1 - i \sin\theta_1) = q_x - iq_y$$

$$q_x = - m \cos\theta_1/r_1; \qquad q_y = - m \sin\theta_1/r_1; \qquad q = m/r_1 = Q/2\pi r_1. \tag{7.8.21}$$

The point z_0 is a singular point (logarithmic singularity) in the infinite z-plane. As $r_1 \to \infty$, $\Phi \to \infty$.

The principle of superposition (sec. 7.5) is applicable to both Φ and Ψ. It is therefore also applicable to ζ. Let us apply this principle to groups of sources and sinks.

Example 1. For a source and a sink of equal strength m located at point $(d, 0)$ and $(- d, 0)$, respectively, in the xy plane (fig. 7.8.7), we have:

$$\zeta = m \ln(z + d) - m \ln(z - d) = m \ln[(z + d)/(z - d)] \tag{7.8.22}$$

where $m = Q/2\pi$ and Q is the rate of flow of the sink or the source. With $z - d = r_1 \exp(i\theta_1)$ and $z + d = r_2 \exp(i\theta_2)$, equation (7.8.22) becomes:

$$\zeta = m \ln(r_2/r_1) + mi(\theta_2 - \theta_1). \tag{7.8.23}$$

Hence:

$$\Phi = m \ln(r_2/r_1) = \frac{m}{2} \ln \frac{(x + d)^2 + y^2}{(x - d)^2 + y^2} \tag{7.8.24}$$

$$\Psi = m(\theta_2 - \theta_1) = m \tan^{-1} \frac{- 2yd}{x^2 + y^2 - d^2}. \tag{7.8.25}$$

Equipotentials $\Phi = c_1$, $\exp(c_1/m) = c = \text{const}$, are:

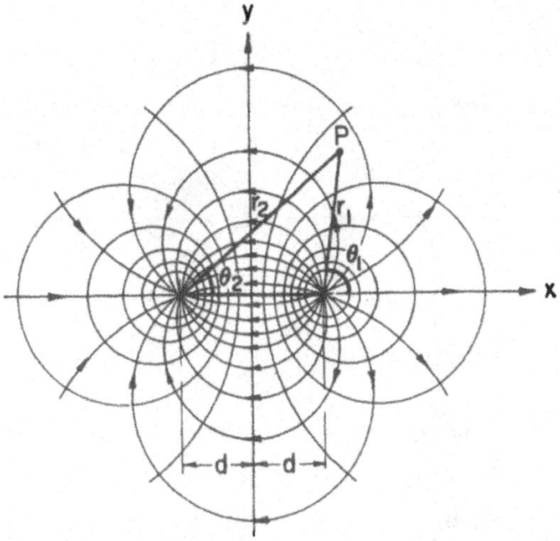

FIG. 7.8.7. A source $(d, 0)$ and a sink $(-d, 0)$ of equal strength m described by $\zeta = m \ln[(z + d)/(z - d)]$, $m > 0$.

$$\left(x + d\frac{1+c}{1-c}\right)^2 + y^2 = \left(\frac{2\sqrt{c}d}{1-c}\right)^2. \tag{7.8.26}$$

This is a family of circles (Apolonios circles) with centers at $(-d(1 + c)/(1 - c), 0)$ and radii $2\sqrt{c}\,d/(1 - c)$. The equipotential $\Phi = 0$, $c = 1$, is the line $x = 0$.

The streamlines $\Psi = \text{const} = c_1$, $\tan \Psi/m = c = \tan(\theta_2 - \theta_1)$ are

$$x^2 + \left(y + \frac{d}{c}\right)^2 = (d\sqrt{1 + 1/c^2})^2. \tag{7.8.27}$$

These are also circles with centers at $(0, -d/c)$, and radii $d\sqrt{1 + 1/c^2}$.

Example 2. Two sources of equal strength m are located at $(d, 0)$ and $(-d, 0)$ in the infinite z-plane. The complex potential is:

$$\zeta = -m \ln(z - d) - m \ln(z + d) = -m \ln(x^2 - y^2 - d^2 + 2ixy) \tag{7.8.28}$$

or:

$$\zeta = -m \ln\{(r_1 r_2) \exp[i(\theta_1 + \theta_2)]\}.$$

Hence:

$$\Phi = -m \ln r_1 r_2 = \frac{m}{2} \ln[(x^2 - y^2 - d^2)^2 + (2xy)^2] \tag{7.8.29}$$

$$\Psi = -m \tan^{-1}[2xy/(x^2 - y^2 - d^2)]. \tag{7.8.30}$$

The streamlines are thus hyperbolas with centers at the origin (fig. 7.8.8). At $z = 0$ we have a stagnation point (saddle-point type).

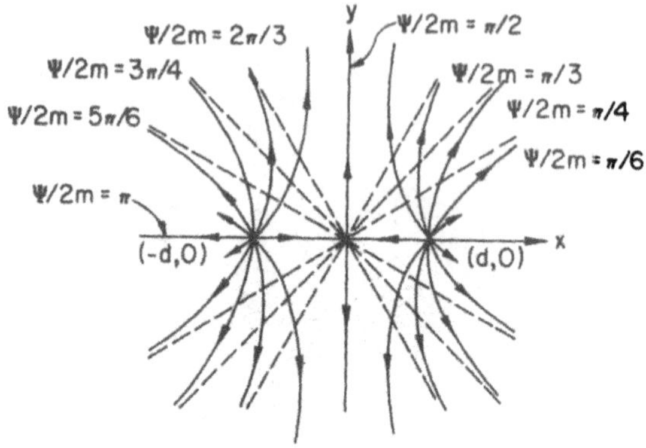

FIG. 7.8.8. Two sources of equal strength described by $\zeta = -m\ln(z^2 - d^2)$; $m>0$.

Example 3. Two of an infinite array of equidistant sinks (spaced a distance d apart) along the x-axis are shown in figure 7.8.9. The complex potential for the infinite array is given by:

$$\zeta = m\ln[\ldots(z+2d)(z+d)z(z-d)(z-2d)\ldots]$$

$$= m\sum_{-\infty}^{+\infty}\ln[z-nd] = -m\ln\sin\frac{\pi z}{d} \tag{7.8.31}$$

$$\Phi = \frac{m}{2}\ln\frac{1}{2}\left(\cosh\frac{2\pi y}{d} - \cos\frac{2\pi x}{d}\right) \tag{7.8.32}$$

$$\Psi = m\tan^{-1}\left[\left(\tanh\frac{\pi y}{d}\right)\Big/\tan\left(\frac{\pi x}{d}\right)\right]. \tag{7.8.33}$$

Stagnation points are located at z, satisfying:

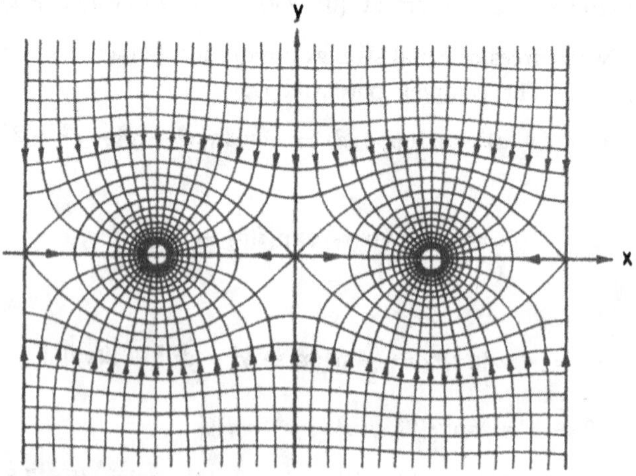

FIG. 7.8.9. An infinite array of sinks described by $\zeta = m\ln\sin(\pi z/d)$.

$$d\zeta/dz = \frac{\pi \cos(\pi z/d)}{d \sin(\pi z/d)} = \frac{\pi}{d} \operatorname{cotg}(\pi z/d) = 0; \qquad \frac{\pi z}{d} = (2n+1)\frac{\pi}{2};$$

$$n = 0, \pm 1, \pm 2, \ldots; \qquad z = (2n+1)\, d/2 \qquad (= d/2, 3\,d/2, \text{ etc.}).$$

The example above may be generalized to any arrangement of n sources and sinks. Let sources and sinks of strength m_i ($m_1, m_2, \ldots, m_n$, positive and negative) be located at points z_i ($z_1, z_2, \ldots, z_n$) of the complex plane. The combined complex potential is given by:

$$\zeta = m_1 \ln(z - z_1) + m_2 \ln(z - z_2) + \cdots = \sum_{i=1}^{n} m_i \ln(z - z_i) \tag{7.8.34}$$

whence we obtain:

$$\Phi = \sum_{i=1}^{n} m_i \ln r_i; \qquad \Psi = \sum_{i=1}^{n} m_i \theta_i \tag{7.8.35}$$

where r_i and θ_i denote the modulus and argument of the complex number $(z - z_i)$.

Equation (7.8.34) may also be extended to the case of a continuous distribution of point sources (or sinks) of constant strength $m\, d\xi$ per length element $d\xi$, uniformly distributed along a segment ($-\,d$ to $+\,d$) of a straight line:

$$\zeta = m \int_{-d}^{+d} \ln(z - \xi)\, d\xi = m[(z + d)\ln(z + d) - (z - d)\ln(z - d) - 2d]. \tag{7.8.36}$$

Example 4. A single sink at the origin in uniform flow ($q_x = -\,q_0 = \text{const}$, $q_y = 0$) in the $-\,x$ direction is shown in figure 7.8.10. By superposition of (7.8.19) and (7.8.15), we obtain:

$$\zeta = q_0 z + m \ln z, \qquad m > 0 \tag{7.8.37}$$

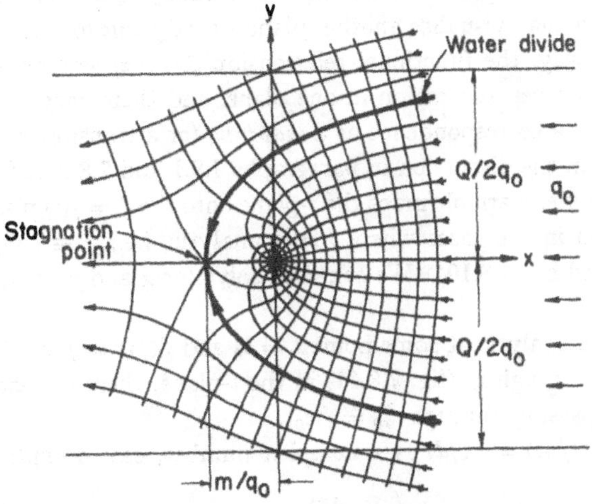

FIG. 7.8.10. A single sink in uniform flow described by $\zeta = q_0 z + m \ln z$; $m > 0$.

$$\Phi = q_0 x + m \ln r = q_0 x + \frac{m}{2} \ln(x^2 + y^2) \tag{7.8.38}$$

$$\Psi = q_0 y + m \tan^{-1}(y/x) = q_0 y + m\theta \tag{7.8.39}$$

$$q_x = -\,\partial\Phi/\partial x = -\,q_0 - mx/(x^2+y^2); \quad q_y = -\,\partial\Phi/\partial y = -\,my/(x^2+y^2). \tag{7.8.40}$$

From (7.8.40) it follows that a stagnation point exists at $y = 0$, $x = -\,m/q_0$. This also follows from the solution of $d\zeta/dz = q_0 + m/z = 0$.

Consider the flow in the upper half plane $y \geqslant 0$. For $x \leqslant 0$ and $y = 0$, $\theta = \pi$. Therefore for this part we have $\Psi = m\pi$. The equation for this streamline is:

$$m\pi = q_0 y + m\theta. \tag{7.8.41}$$

As $\theta \to 0$, $y \to m\pi/q_0$. Since $m = Q/2\pi$, Q being the discharge of the sink, we have: $y \to Q/2q_0$. The line $y = Q/2q_0$ is an asymptote of the streamline passing through the stagnation point. Similarly, for $y \leqslant 0$, the equation of the streamline is:

$$-\,m\pi = q_0 y + m\theta \tag{7.8.42}$$

and as $\theta \to 0$, $y \to -\,Q/2q_0$. Thus $y = \pm\, y_0 = \pm\, Q/2q_0$ are the two asymptotes of the dividing streamline (or *water divide*) as no fluid crosses this curve. All the flow ($q_0 \cdot 2 \cdot Q/2q_0 = Q$) between these two lines reaches the sink. The equation of the dividing streamline may also be written as:

$$y/x = \pm \tan(q_0 y/m) \begin{cases} + & \text{for } y > 0 \\ - & \text{for } y < 0. \end{cases} \tag{7.8.43}$$

7.8.4 Conformal Mapping

In paragraph 7.8.2 and in the examples in paragraph 7.8.3, we have considered the relationship $\zeta = f(z)$ between the complex number $z = x + iy$ and the complex potential $\zeta = \Phi + i\Psi$. Geometrically, z represents a point in the xy plane or *z-plane* (or the physical plane, where the flow actually takes place). With Φ and Ψ as coordinates, we may visualize another plane—*the ζ-plane* (or the complex potential plane). In this way, the functional relationship $\zeta = f(z)$ sets up a correspondence between points, curves and region in one plane, and their images in another plane. We speak of such a correspondence as a *mapping* (or a transformation) between the two planes. From the discussion in paragraphs 7.8.1 and 7.8.2, it follows that every function of a complex variable maps the z-plane onto the $\zeta = f(z)$ plane. In particular we are interested in the cases where $f(z)$ is analytic, i.e., $J \equiv \partial(\Phi, \Psi)/\partial(x, y) \neq 0$, where J is defined by (7.8.10). We can then solve for x and y as single-valued functions of Φ and Ψ.

Let $\zeta = f(z)$ be analytic at some point $z = z_0$ and $d\zeta/dz = f'(z) \neq 0$. Then to any curve C passing through z_0 (fig. 7.8.11) in the z-plane, there corresponds a curve C' in the ζ-plane passing through $\zeta_0 = f(z_0)$.

The derivative $f'(z) = d\zeta/dz$ is a complex number, say $A \exp(i\beta)$, with:

$$\beta = \arg f'(z) = \lim_{\Delta z \to 0} \left(\arg \frac{\Delta\zeta}{\Delta z} \right); \qquad A = |f'(z)| = \lim_{\Delta z \to 0} \left| \frac{\Delta\zeta}{\Delta z} \right| \tag{7.8.44}$$

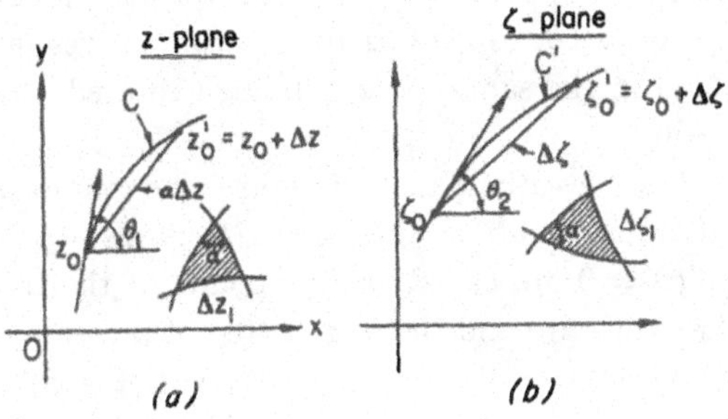

FIG. 7.8.11. Conformal mapping.

In the limit, as $\Delta z \to 0$, the argument of Δz approaches the angle θ_1. Similarly, as $\Delta\zeta$ is the image of Δz, its argument approaches the angle θ_2 as $\Delta\zeta \to 0$. Hence:

$$\beta = \arg f'(z) = \theta_2 - \theta_1; \qquad \theta_2 = \beta + \theta_1. \tag{7.8.45}$$

This means that where $f(z)$ is analytic, in the transformation from the z-plane to the ζ-plane, the directed tangent to a curve at a point z_0 is rotated through an angle $\beta = \arg f'(z)$. Since $f'(z)$ has only one value at that point, any two curves intersecting at an angle α at that point will both be rotated by the same angle β (and in the same direction) so that the angle between their images in the ζ plane will remain α.

As $\Delta z \to 0$, $A = |f'(z)|$ represents a *local magnification factor*; infinitesimal lengths in the z-plane are magnified by the factor A in the transformation to the ζ-plane. Since $|f'(z)|$ is independent of the direction of the chord Δz, it follows that all curves passing through the considered point z_0 are mapped onto curves passing through ζ_0, all of which are magnified by the same ratio $|f'(z)|$. As $\Delta z \to \varepsilon > 0$, $A \sim |f'(z)|$ is only an approximation, there is some distortion in the length $\Delta\zeta$, the degree of distortion depending on the magnitude of ε. Thus, large figures in the z-plane may map onto shapes in the ζ-plane that bear no resemblance to the original shapes. This is also due to the fact that $f'(z)$ varies from point to point in the z-plane. Smaller figures (such as the infinitesimal triangles of fig. 7.8.**11**) are mapped onto *similar*

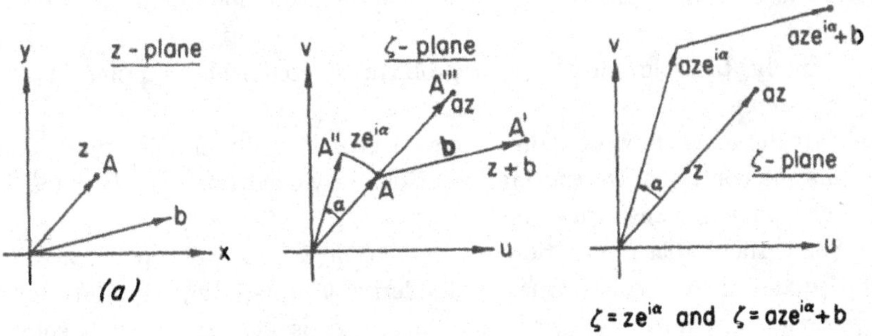

FIG. 7.8.12. Conformal mappings by $\zeta = z + b$, $\zeta = az$, $\zeta = z\exp(i\alpha)$ and $\zeta = az\exp(i\alpha) + b$.

closed figures in the ζ-plane with a certain magnification. It should be emphasized that the angles formed by corresponding intersecting curves in the two planes are preserved exactly (and in the same sense, although rotated) even for large figures, except where $f'(z) = 0$.

A *mapping* of the type described above is said to be *conformal*. If $\zeta = \zeta(z)$ represents a conformal transformation, then $z = z(\zeta)$ is also conformal.

Points at which $f'(z) = 0$ are called critical points of the transformation; there the magnification factor is zero and the angles are not preserved conformally. Also, at a critical point, the inverse function $z = z(\zeta)$ is not analytic.

The main advantage of conformal mapping in solving flow problems is that solutions of the Laplace equation, $\Phi = \Phi(x, y)$ and $\Psi = \Psi(x, y)$, remain solutions when subjected to conformal transformation. Let us consider a conformal mapping from the $z(=x + iy)$-plane to a $g(=u + iv)$-plane.

It can easily be shown that:

$$\partial^2\Phi/\partial x^2 + \partial^2\Phi/\partial y^2 = |g'(z)|^2(\partial^2\Phi/\partial u^2 + \partial^2\Phi/\partial v^2); \qquad |g'(z)|^2 = (\partial u/\partial x)^2 + (\partial u/\partial y)^2. \tag{7.8.46}$$

As long as $g'(z) \neq 0$, and since $\nabla^2_{x,y}\Phi = 0$, we have

$$\partial^2\Phi/\partial u^2 + \partial^2\Phi/\partial v^2 = 0. \tag{7.8.47}$$

A similar result may be derived for Ψ.

It can also be shown that:

$$\nabla_{u,v}\Phi = \left(\frac{\partial\Phi}{\partial u}\,\mathbf{1u} + \frac{\partial\Phi}{\partial v}\,\mathbf{1v}\right)|g'(z)| \tag{7.8.48}$$

where **1u** and **1v** are unit vectors in the u and v directions.

Unless the geometry of the boundaries of the flow domain in the z-plane is rather simple, it is very difficult to solve Laplace's equation for $\Phi = \Phi(x, y)$ or $\Psi = \Psi(x, y)$. However, having shown above that for any $g(u + iv)$-plane, onto which the z-plane is conformally mapped, Laplace's equation (in the uv coordinates) is still satisfied by Φ and Ψ, we may attempt to solve for Φ and Ψ in the g-plane, and then carry back the solution by the inverse transformation to the original z-plane. The main advantage lies in the fact that in the uv-plane, the boundaries of the flow domain, transformed from the z-plane by the same mapping function, often take on a simpler form.

Accordingly, the solution of a flow problem by conformal mapping involves the following steps.

(a) Map the given flow domain R onto a geometrically simpler flow domain R' by means of a conformal transformation. Boundary conditions (for Φ and Ψ) remain unchanged on the transformed boundaries.

(b) Solve the Laplace equation for Φ (and/or Ψ) in the transformed domain.

(c) Use the inverse transformation to derive $\Phi = \Phi(x, y)$, $\Psi = \Psi(x, y)$.

Sometimes a sequence of conformal mappings is needed before a solution for Φ and/or Ψ can be obtained.

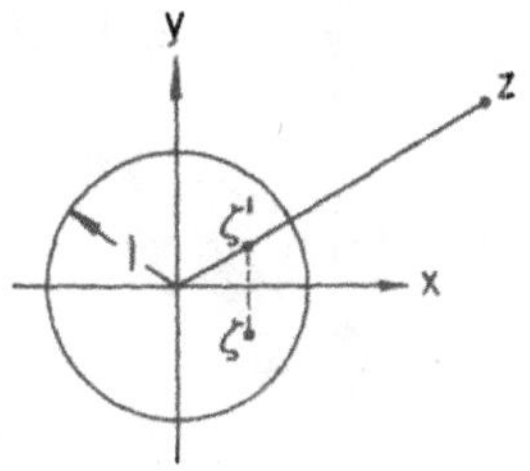

FIG. 7.8.13. Mapping by $\zeta = 1/z$.

Some useful conformal transformations are listed below:

(i) $$\zeta = z + b \tag{7.8.49}$$

this is *pure translation* ($A \to A'$ in fig. 7.8.12); each point in the ζ-plane is obtained by translating its corresponding point in the z-plane through the vector $\mathbf{b}$;

(ii) $$\zeta = z \exp(i\alpha) \tag{7.8.50}$$

this is *pure rotation* through an angle α (counterclockwise); point A'' in figure 7.8.12 is the image of A by (7.8.50);

(iii) $$\zeta = az, \qquad a > 0 \text{ and real} \tag{7.8.51}$$

this is the *linear amplification* ($A \to A'''$ in fig. 7.8.12b); the mapping consists of multiplying the vector z by the factor a. Figure 7.8.12c shows the combined transformation $\zeta = az \exp(i\alpha) + b$.

In this and in the previous transformation, straight lines and circles in the z-plane remain such in the ζ-plane. Circles passing through the origin in the z-plane become straight lines in the ζ-plane.

(iv) $$\zeta = 1/z \quad \text{or} \quad \zeta = (1/r) \exp(-i\theta). \tag{7.8.52}$$

This is the *inversion and reflection transformation.* The second transformation in (7.8.52) shows that (7.8.52) is an inversion with respect to the unit circle centered at the origin (fig. 7.8.13). This means that the point ζ' lies on the radius drawn through point z, with its distance $|\zeta|$ from the origin satisfying $|\zeta'| \cdot |z| = 1$. This inversion is followed by a reflection $\zeta = \bar{\zeta}'$ in the real axis (fig. 7.8.13). Thus, points outside the unit circle are mapped onto points inside the circle, and vice versa. Circles passing through the origin $z = 0$ transform into straight lines in the ζ-plane. Lines in the z-plane transform into circles through the origin $\zeta = 0$, unless they pass through $z = 0$, in which case their images are lines through the origin $\zeta = 0$. This transformation is most important in the hodograph method (sec. 8.3) in view of the fact that the free surface maps onto a circle in the hodograph plane.

(v) $$\zeta = \frac{z - z_0}{z - \bar{z}_0} \exp(i\theta) \tag{7.8.53}$$

where $\zeta = 0$ is the image of z_0, and $0 \leqslant \theta \leqslant 2\pi$, $\mathrm{Im}(z_0) > 0$, maps the upper half of the z-plane onto the unit circle $|\zeta| \leqslant 1$ (fig. 7.8.14).

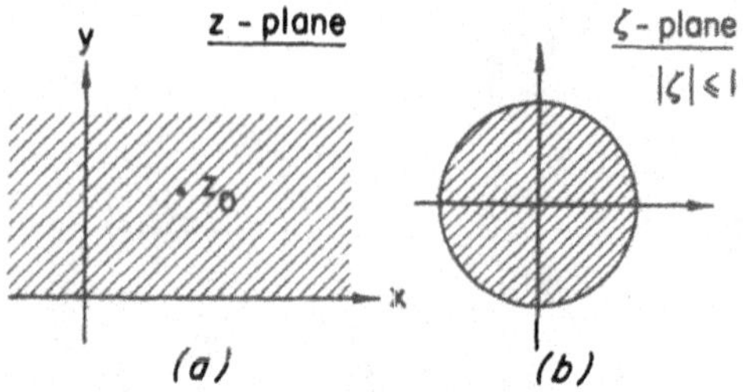

FIG. 7.8.14. Mapping by $\zeta = \exp(i\theta)\,(z - z_0)/(z - \bar{z}_0)$.

Other transformations of interest are shown in figures 7.8.15 through 7.8.25.

Several examples are given below of solutions to flow problems by applying a conformal mapping to the given flow.

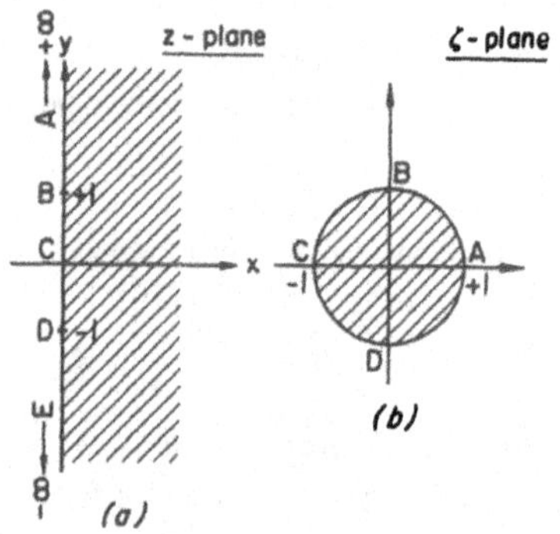

FIG. 7.8.15. Mapping by $\zeta = (z - 1)/(z + 1)$.

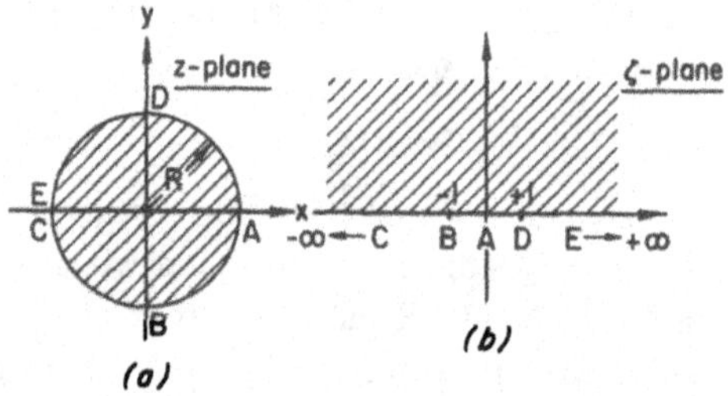

FIG. 7.8.16. Mapping by $\zeta = i(R - z)/(R + z)$.

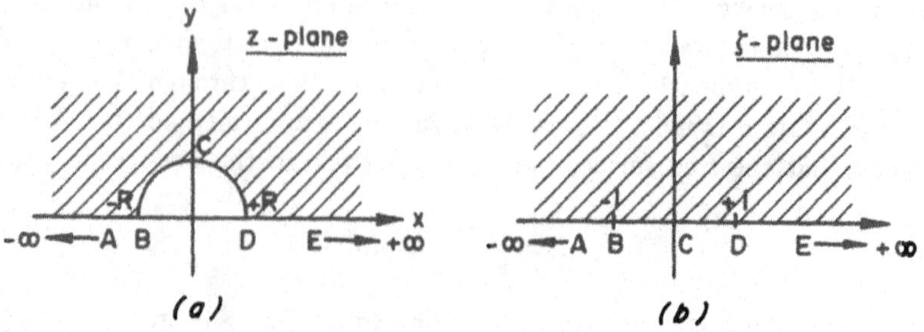

FIG. 7.8.17. Mapping by $\zeta = (z/R + R/z)/2$.

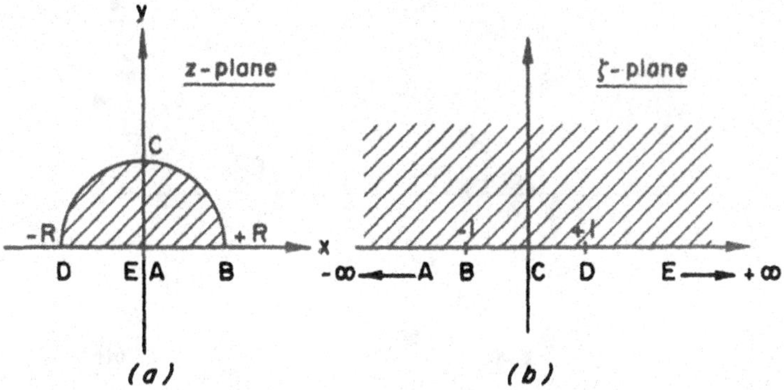

FIG. 7.8.18. Mapping by $\zeta = -(z/R + R/z)/2$.

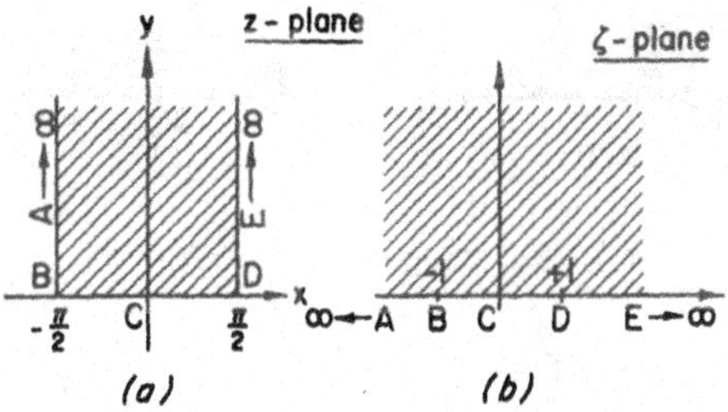

FIG. 7.8.19. Mapping by $\zeta = \sin z$.

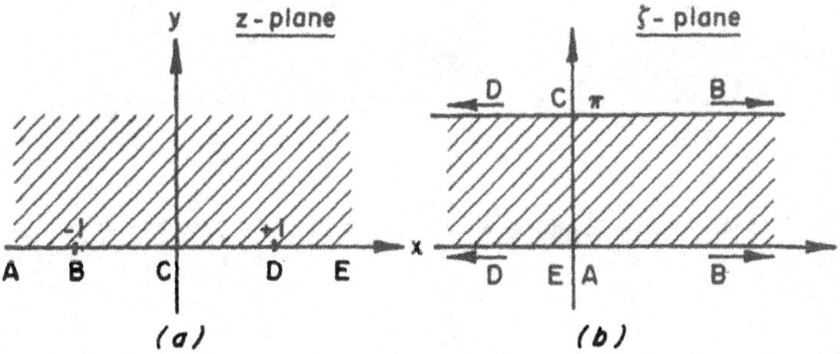

FIG. 7.8.20. Mapping by $\zeta = \ln[(z - 1)/(z + 1)]$.

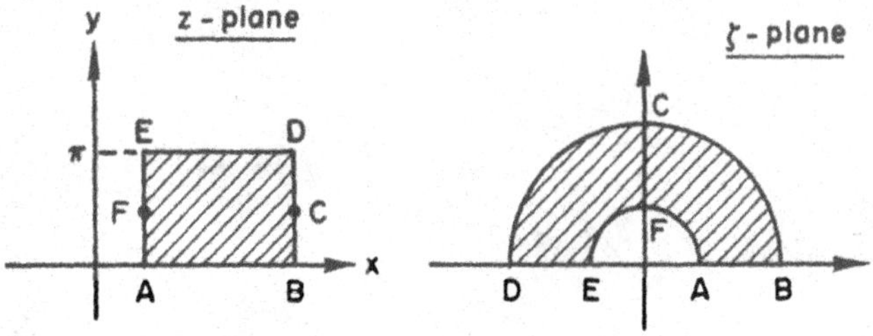

FIG. 7.8.21. Mapping by $\zeta = \exp z$.

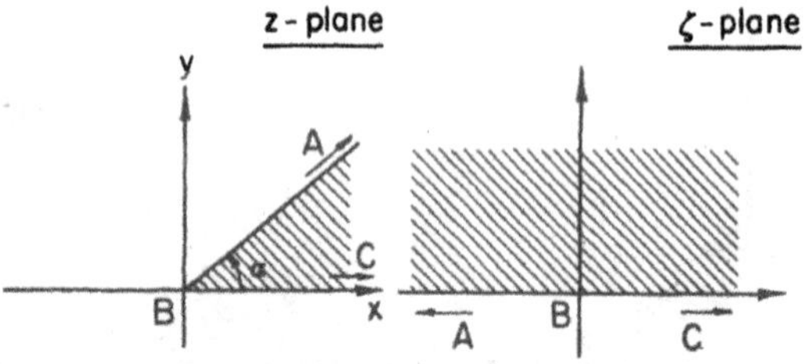

FIG. 7.8.22. Mapping by $\zeta = z^{\pi/\alpha}$.

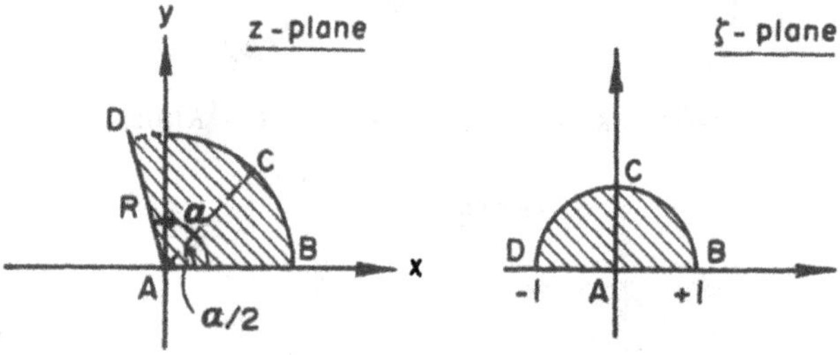

FIG. 7.8.23. Mapping by $\zeta = (z/R)^{\pi/2}$.

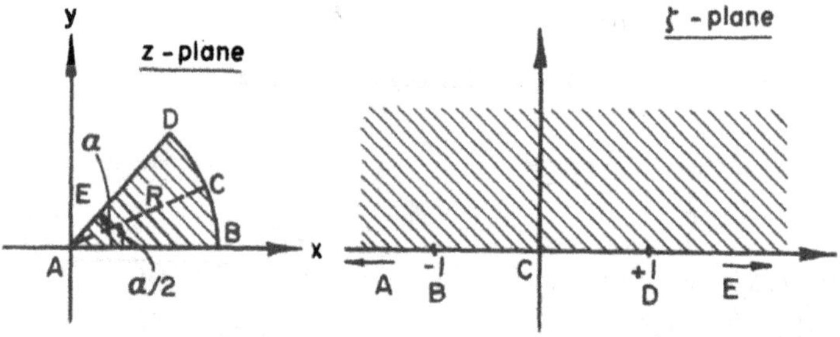

FIG. 7.8.24. Mapping by $\zeta = -\frac{1}{2}[(z/R)^{\pi/2} + (R/z)^{\pi/2}]$.

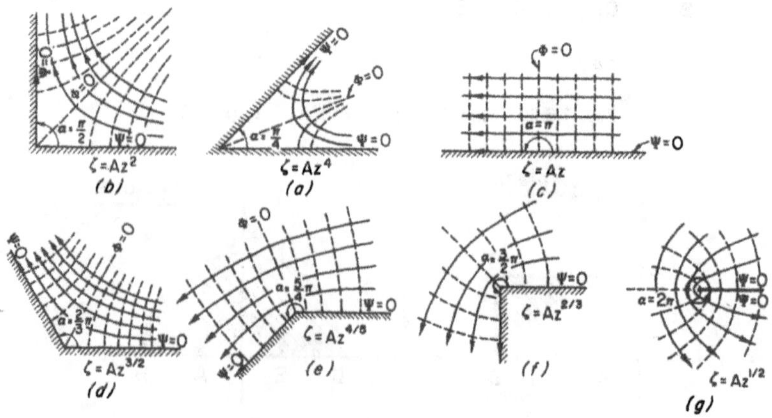

FIG. 7.8.25. Flows inside and around corners.

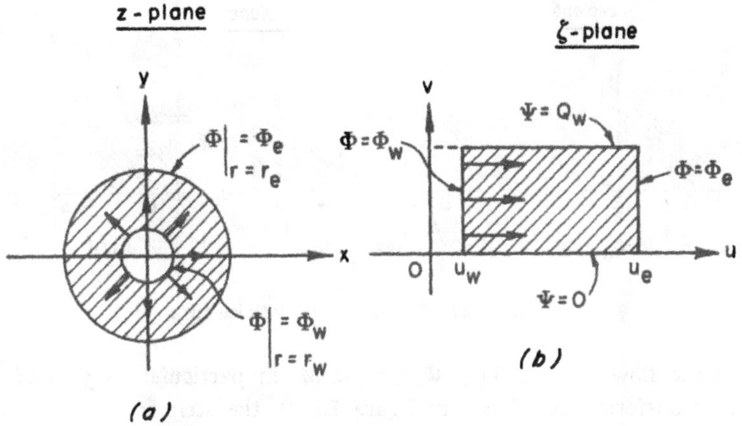

FIG. 7.8.26. Solution of a radial flow by conformal mapping.

Example 1. Consider the problem of steady plane radial flow between two concentric equipotential boundaries: $\Phi = \Phi_w$ at $r = r_w$ and $\Phi = \Phi_e$ at $r = r_e$ (fig. 7.8.26). By the logarithmic transformation:

$$\zeta = (Q_w/2\pi) \ln z = u + iv \tag{7.8.54}$$

which is the inverse of the transformation shown in figure 7.8.21; this region is mapped onto a rectangle in the ζ-plane. In this plane, we must determine $\Phi = \Phi(u, v)$ such that $\partial^2\Phi/\partial u^2 + \partial^2\Phi/\partial v^2 = 0$ in the rectangle. The boundary conditions are shown in figure 7.8.26b. The relationships between the coordinates in the two planes are:

$$u = \frac{Q_w}{2\pi} \ln r; \qquad v = \frac{Q_w}{2\pi} \theta; \qquad u_w = \frac{Q_w}{2\pi} \ln r_w; \qquad u_e = \frac{Q_w}{2\pi} \ln r_e; \quad \text{etc.}$$

The solution in the ζ-plane is straightforward:

$$\Phi = A + Bu \tag{7.8.55}$$

where A and B are constants to be determined from the boundary conditions. We obtain:

$$\Phi = \Phi|_{u_w} + \frac{\Phi|_{u_e} - \Phi|_{u_w}}{u_e - u_w} (u - u_w). \tag{7.8.56}$$

By returning to the z-plane, this becomes:

$$\Phi = \Phi_w + \frac{\Phi_e - \Phi_w}{\ln(r_e/r_w)} \ln(r/r_w) \tag{7.8.57}$$

which is the solution of the problem in terms of r (or x and y).

Example 2. Consider the steady flow in the semi-infinite strip $-\pi/2 < x < \pi/2$, $y > 0$, shown in figure 7.8.27a. The solution $\Phi = \Phi(x, y)$ is to be bounded at all

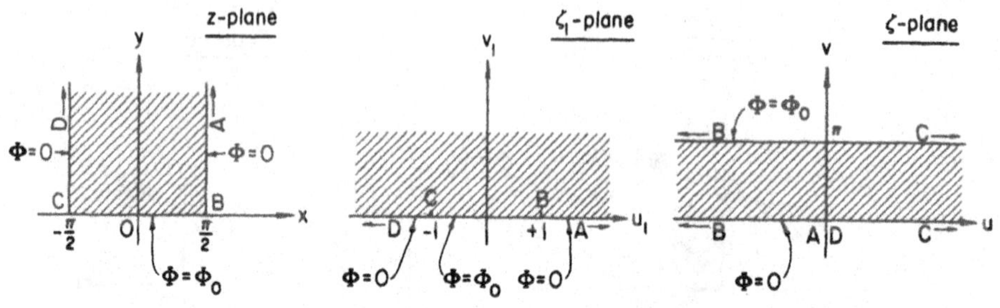

FIG. 7.8.27. Flow in a semi-infinite strip.

points of the flow domain, i.e., $|\Phi(x, y)| < M$, in particular as $y \to \infty$.

By the transformation shown in figure 7.8.19, the strip in question is mapped onto the upper half of a ζ_1-plane (fig. 7.8.27b). Then the transformation:

$$\zeta = \ln[(\zeta_1 - 1)/(\zeta_1 + 1)] \tag{7.8.58}$$

shown in figure 7.8.20 is employed to map the upper half of the ζ_1-plane onto an infinite strip between the lines $v = 0$ and $v = \pi$ in the ζ-plane (fig. 7.8.27*c*).

The solution $\Phi = \Phi(u, v)$ of $\nabla^2\Phi = 0$ in the ζ-plane, which satisfies the boundary conditions shown in figure 7.8.27c, is:

$$\Phi = \Phi_0 v/\pi. \tag{7.8.59}$$

From (7.8.58) we have:

$$v = \tan^{-1}[2v_1/(u^2{}_1 + v^2{}_1 - 1)] \tag{7.8.60}$$

where the $\tan^{-1}$ function is in the range 0 to π. Hence (7.8.59) becomes:

$$\Phi = (\Phi_0/\pi) \tan^{-1}[2v_1/(u^2{}_1 + v^2{}_1 - 1)]. \tag{7.8.61}$$

Equation (7.8.61) is a solution for the flow shown in figure 7.8.27b. In order to obtain the solution in the given z-plane, we note that:

$$\zeta_1 = \sin z; \qquad u_1 = \sin x \cosh y; \qquad v_1 = \cos x \sinh y.$$

Hence, equation (7.8.61) becomes:

$$\Phi = \frac{\Phi_0}{\pi} \tan^{-1}\left[\frac{2 \cos x \sinh y}{\sin^2 x \cosh^2 y + \cos^2 x \sinh^2 y - 1}\right]. \tag{7.8.62}$$

This can be reduced to:

$$\Phi = \frac{2\Phi_0}{\pi} \tan^{-1}\left(\frac{\cos x}{\sinh y}\right) \tag{7.8.63}$$

where the $\tan^{-1}$ function has the range 0 to $\pi/2$, its argument being nonnegative. Hence, $|\Phi(x, y)| \leqslant \Phi_0$.

Example 3. Consider the flow in the quadrant shown in figure 7.8.28a. By using the transformation:

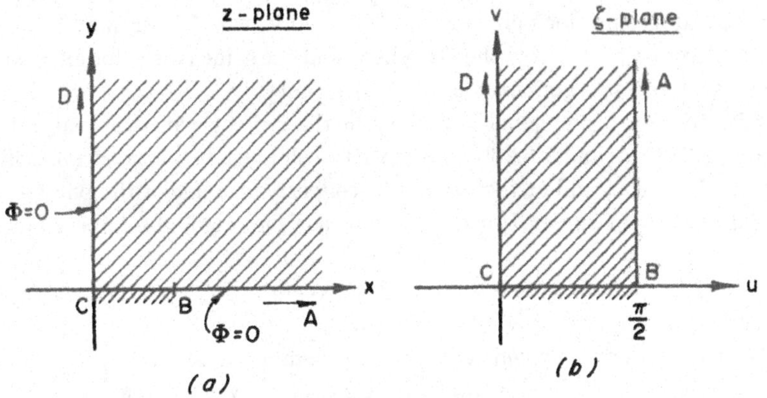

FIG. 7.8.28. Flow in a quadrant with a partly impervious boundary.

$$z = \sin \zeta \tag{7.8.64}$$

we map the quadrant $x \geqslant 0$, $y \geqslant 0$ onto the semi-infinite strip $0 \leqslant u \leqslant \pi/2$, $v \geqslant 0$ (fig. 7.8.28b). The solution of the flow in this strip is obviously:

$$\Phi = 2\Phi_0 u/\pi. \tag{7.8.65}$$

In order to return to the z-plane, we note that from (7.8.64) we have:

$$x = \sin u \cosh v, \qquad y = \cos u \sinh v; \qquad \frac{x^2}{\sin^2 u} - \frac{y^2}{\cos^2 u} = 1.$$

When this is solved for u and the result inserted in (7.8.65), we obtain

$$\Phi = \frac{2\Phi_0}{\pi} \sin^{-1} \tfrac{1}{2}\{[(x+1)^2 + y^2]^{1/2} - [(x-1)^2 + y^2]^{1/2}\}. \tag{7.8.66}$$

In many cases a simple transformation from the z- to the ζ-plane cannot be obtained. When the flow domain in the z-plane has the form of a polygon, the Schwarz–Christoffel transformation discussed in paragraph 7.8.5 may be used.

7.8.5 The Schwarz–Christoffel Transformation

The *Schwarz–Christoffel transformation* maps the *interior* of a *polygonal boundary* onto the upper half of the ζ-plane. We have seen above (par. 7.8.4) that it is sometimes simpler to solve the flow problem in this half plane than in the given z-plane. If the transformation is a conformal one, the boundary conditions, originally prescribed along the polygonal boundary, are specified in the ζ-plane on the real ($v = 0$) axis.

We deal here with a *simply connected polygon* with the following properties:

(a) it is possible to go from any assigned point of the boundary to any other assigned point of the boundary by following a path that never leaves the boundary; the boundary is said to be *connected*;

(b) the boundary divides the plane into two regions: *interior* to the boundary and *exterior* to it; interior points are such that any two of them may be connected by a path that never intersects the boundary, and the same is true for exterior points. Boundaries of the polygon may extend to infinity.

In order to map a polygonal boundary in the z-plane onto the real axis of the ζ-plane (figs. 7.8.29a and 7.8.29b), we must translate each vertex of the polygon to a point on the u-axis, and at the same time convert the angle of the polygon at that vertex into a straight angle. It can be shown that the required mapping function is:

$$z = K \int [(\zeta - u_1)^{(\alpha_1/\pi)-1}(\zeta - u_2)^{(\alpha_2/\pi)-1} \cdots (\zeta - u_n)^{(\alpha_n/\pi)-1}] \, d\zeta + C \quad (7.8.67)$$

where K and C are complex constants, and the symbols α_i and u_i are shown in figure 7.8.29. By rewriting the constant K as $k \exp(i\beta)$, we see that it comprises an arbitrary magnification factor k and a rotation β. By considering the segment $u > u_n$, we recognize that β is the angle through which the infinite segment to the right of $u = u_n$ is rotated when it is mapped (from a finite or infinite segment). By writing $C = C_1 + iC_2$, we see that the additive constant C represents an arbitrary translation of the polygon, without rotation or distortion, by the vector $C_1 + iC_2$.

It can be shown (Nehari 1952) that the numbers $u_1, u_2, \ldots, u_n$ and the complex constants K and C can always be chosen so that any given polygon in the z-plane is made to correspond point by point to the real axis ($v = 0$) in the ζ-plane. Furthermore, the mapping can be shown to establish a one-to-one correspondence between points in the interior of the polygon in the z-plane and points in the upper half of the ζ-plane.

It can also be shown (Wylie 1951) that in mapping a polygon of n sides, three of the values $u_1, u_2, \ldots, u_n$ can be assigned arbitrarily, following which $n - 3$ remain to be determined.

In many problems, a vertex of a polygon, usually at infinity, corresponds to $\zeta = \infty$. In this case $dz/d\zeta$ contains one less term than usual, and hence one less parameter. Then, only two of the $n - 1$ images u_i ($i = 1, \ldots, n$) can be specified arbitrarily.

Since ζ is analytic everywhere except at the points $u_1, u_2, \ldots, u_n$, the transformation is conformal. To avoid the difficulty arising from the point $\zeta = u_i$ as z passes through the vertices z_i, the images ζ_i in the ζ-plane pass around indented semicircles

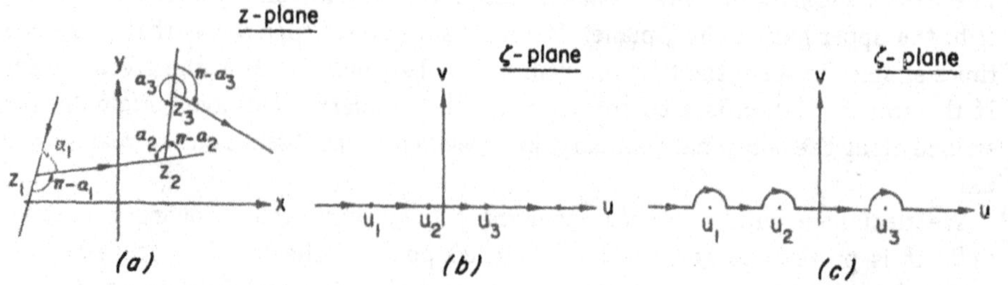

FIG. 7.8.29. The Schwarz-Christoffel transformation.

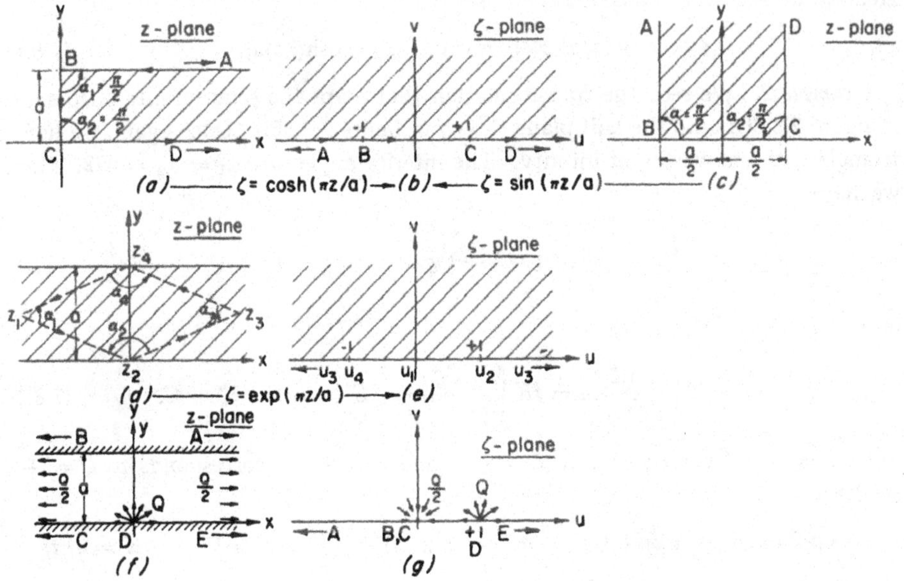

FIG. 7.8.30. Examples of Schwarz-Christoffel transformations.

at the points u_i (fig. 7.8.29c). The radius of these semicircles may be made as small as we wish by adjusting $|K|$.

The transformation (7.8.67) is called the *Schwarz–Christoffel transformation* in honor of the German mathematician H. A. Schwarz (1843–1921) and the Swiss mathematician E. B. Christoffel (1829–1900), who discovered it independently (1869 and 1867, respectively).

Many illustrations of the Schwarz–Christoffel transformation may be found in the literature. Several such applications are considered here.

Example 1. We seek the transformation that maps the semi-infinite strip in figure 7.8.30a onto the half plane $v > 0$ (fig. 7.8.30b).

Because we have a vertex at infinity, only two points (B and C) on the u-axis could be chosen arbitrarily. Hence we have:

$$\frac{dz}{d\zeta} = K(\zeta+1)^{(\pi/2)/\pi-1}(\zeta-1)^{(\pi/2)/\pi-1} = K(\zeta+1)^{-1/2}(\zeta-1)^{-1/2} = K/(\zeta^2-1)^{1/2}$$

or:

$$z = K\int\frac{d\zeta}{\zeta^2-1} = K\cosh^{-1}\zeta + C. \tag{7.8.68}$$

Since $z = 0$ corresponds to $\zeta = 1$, we have $C = 0$, and since $z = ia$ corresponds to $\zeta = -1$, we have:

$$ia = K\cosh^{-1}(-1) = K(i\pi) \quad \text{or} \quad K = a/\pi.$$

Hence, the required transformation is:

$$z = (a/\pi)\cosh^{-1}\zeta \quad \text{or} \quad \zeta = \cosh(\pi z/a). \tag{7.8.69}$$

Example 2. We seek the transformation that maps the semi-infinite strip shown in figure 7.8.30c onto the half plane shown in figure 7.8.30b. Once again, we have a triangle with one vertex at infinity. The interior angles are $\alpha_1 = \alpha_2 = \pi/2$. Hence we have:

$$\frac{dz}{d\zeta} = K(\zeta + 1)^{(\pi/2)/\pi - 1}(\zeta - 1)^{(\pi/2)/\pi - 1}\, d\zeta + C$$

or:

$$z = K\int \frac{d\zeta}{\sqrt{\zeta^2 - 1}} = iK\int \frac{d\zeta}{\sqrt{1 - \zeta^2}} + C = iK\sin^{-1}\zeta + C. \tag{7.8.70}$$

Since $z = -a/2$ corresponds to $\zeta = -1$ and $z = +a/2$ corresponds to $\zeta = +1$, we have:

$$-a/2 = iK(-\pi/2) + C; \qquad a/2 = iK(\pi/2) + C; \qquad C = 0, \qquad iK = a/\pi,$$

and the required transformation is:

$$z = (a/\pi)\sin^{-1}\zeta; \qquad \zeta = \sin(\pi z/a). \tag{7.8.71}$$

Example 3. The infinite strip $0 < y < a$ may be considered as the limiting case of a rhombus with vertices at $z_1, z_2 = 0$; $z_3, z_4 = ia$, as the points z_1 and z_3 are moved to infinity to the left and to the right, respectively (fig. 7.8.30d). In the limit, the interior angles α become: $\alpha_1 = 0$; $\alpha_2 = \pi$; $\alpha_3 = 0$; $\alpha_4 = \pi$.

Leaving u_4 to be determined, and choosing $u_1 = 0$, $u_2 = 1$ and $u_3 = \infty$ (and hence omitted from the expression for $dz/d\zeta$), we have:

$$\frac{dz}{d\zeta} = K(\zeta - 0)^{(0/\pi)-1}(\zeta - 1)^{(\pi/\pi)-1}(\zeta - u_4)^{(\pi/\pi)-1} = \frac{K}{\zeta}$$

or:

$$z = K\int \frac{d\zeta}{\zeta} = K\ln\zeta + C. \tag{7.8.72}$$

Since $\zeta = 1$ when $z = 0$, it follows that $C = 0$. Because the point z lies on the real axis when $\zeta = x$, $x > 0$, the constant K must be real. The image of the point $z = ia$ is $\zeta = u_4$, where u_4 is a negative number; therefore:

$$ia = K\ln u_4 = K\ln|u_4| + K\pi i.$$

By equating real and imaginary parts, we see that $|u_4| = 1$ and $K = a/\pi$. Hence, the transformation that maps the infinite strip onto the half plane is:

$$z = (a/\pi)\ln\zeta; \qquad \zeta = \exp(\pi z/a). \tag{7.8.73}$$

From this it follows that $u_4 = -1$. Also $z_3(\infty, 0) \to \zeta_3(\infty, 0)$; $z_3(\infty, a) \to \zeta_3(-\infty, 0)$; $z_1(-\infty, 0)$ and $z_1(-\infty, a) \to \zeta_1(0, 0)$.

Example 4. Consider a source of strength Q (dim L^2T^{-1}) placed at the origin of the z-plane (fig. 7.8.30f). The transformation (7.8.73) maps the strip onto the upper half of the ζ-plane. As in other boundary conditions, since the mapping is conformal, a source or a sink at a point in the z-plane corresponds to an equal source or sink at the image of that point in the ζ-plane. Hence, we have a source of strength $Q/2$ at (0, 0) of the ζ-plane, and a source of strength Q at (1, 0) of the same plane (fig. 7.8.30g). This gives rise to the complex potential (par. 7.8.3):

$$f(z) = \Phi + i\Psi = \frac{Q}{2\pi} \ln \zeta - \frac{Q}{\pi} \ln(\zeta - 1). \tag{7.8.74}$$

Note that here ζ denotes the plane with coordinates u, v and not the complex potential $\Phi + i\Psi$. Inserting ζ from (7.8.73), we obtain:

$$\Phi + i\Psi = -\frac{Q}{\pi}\ln\frac{\zeta - 1}{\zeta^{1/2}} = -\frac{Q}{\pi}\ln(\zeta^{1/2} - \zeta^{-1/2}) = -\frac{Q}{\pi}\ln[\exp(\pi z/2a) - \exp(-\pi z/2a)]$$

or:

$$f(z) = \Phi + i\Psi = -\frac{Q}{\pi} \ln\left(2 \sinh\frac{\pi z}{2a}\right). \tag{7.8.75}$$

Polubarinova-Kochina (1952, 1962), Aravin and Numerov (1953) and Harr (1962) give many examples in connection with confined flow under impervious structures in which this transformation is useful.

In many cases, the integral in the Schwarz–Christoffel transformation cannot be evaluated in terms of a finite number of elementary functions. In such cases the transformation may still be useful, although the solution of the problem by using it may become quite complicated.

7.8.6 Fictitious Flow in the $\tilde{\omega}$-Plane

The function $\zeta = \zeta(z)$ maps the z-plane conformally into the complex-potential ζ-plane. We may consider another mapping, namely that of the z-plane onto the conjugate complex velocity $\tilde{\omega} = q_x - iq_y$-plane. Since $q_x = q_x(x, y)$ and $q_y = q_y(x, y)$, we may express $\tilde{\omega}$ as an analytic function of the complex variable $z = x + iy$ by the relationship:

$$\tilde{\omega} = g(z), \qquad z = G(\tilde{\omega}). \tag{7.8.76}$$

Let us assume that this mapping gives a one-to-one correspondence between points of a flow domain R in the z-plane and a region R' of the $\tilde{\omega}$-plane (coordinates $u = q_x$ and $v = -q_y$). Then the equipotential lines Φ = const, and the streamlines Ψ = const, corresponding to a flow in R, will be mapped onto corresponding families of curves in the region R'. We have:

$$\frac{d\zeta}{d\tilde{\omega}} = \frac{d\zeta}{dz}\frac{dz}{d\tilde{\omega}} = \frac{d\zeta/dz}{d\tilde{\omega}/dz} = \frac{\zeta'(z)}{g'(z)}. \tag{7.8.77}$$

Since $\zeta(z)$ and $g(z)$ are analytic, it follows that ζ is an analytic function of $\tilde{\omega}$ except at points in the $\tilde{\omega}$-plane that correspond to points in the z-plane where $g(z) = 0$. Hence, in the plane $\tilde{\omega}$, with its rectangular coordinates u and v, the real and imaginary parts of $\zeta[G(\tilde{\omega})]$ satisfy the Laplace equation in rectangular coordinates. Hence, we may regard the configuration (Φ, Ψ) in the $\tilde{\omega}$-plane as representing a new, fictitious two-dimensional flow pattern of an incompressible fluid plane. The complex velocity of this fictitious flow is:

$$\tilde{\omega}^* = q^*_u - iq^*_v = -d\zeta/d\tilde{\omega} \tag{7.8.78}$$

or:

$$q^*_u - iq^*_v = -\frac{d\zeta}{dz}\frac{dz}{d\tilde{\omega}} = \frac{\zeta'(z)}{g'(z)} = \frac{q_x - iq_y}{g'(z)}. \tag{7.8.79}$$

From (7.8.79) it follows that:

$$(q^{*2}_u + q^{*2}_v)^{1/2} = (q_x^2 + q_y^2)^{1/2}/|g'(z)| = q/|g'(z)|. \tag{7.8.80}$$

This means that the absolute specific discharge at a point $\tilde{\omega}_0$ in the $\tilde{\omega}$-plane is obtained by dividing the absolute specific discharge q at the corresponding point z_0 in the z-plane by $|g'(z)|$. Equation (7.8.79) also gives the relationship between the two specific discharge vectors.

The fictitious flow is used by Polubarinova-Kochina (1952, 1962) in the numerical solution of the hodograph plane (par. 8.3.6).

7.9 Numerical Methods

In many cases, the partial differential equations governing flow through porous media cannot be solved by exact analytical methods. In certain cases, although a general analytic solution can be obtained, say in the form of an infinite series, it is very difficult and tedious to apply it to a specific problem. This is especially so when the boundaries of the flow domain have an irregular shape. In chapter 11, models and analogs are described that are used as tools for solving such problems. In the present chapter, some numerical and graphic techniques are outlined for the same purpose. A comparison between the two methods of solution is given in section 11.1.

Only basic principles of the numerical methods, especially those applicable to the partial differential equations governing flow through porous media, are given here. For details, the reader is referred to a large number of texts available on this subject (e.g., Shaw 1953; Richtmeyer 1957; Forsythe and Wasow 1960; Thom and Apelt 1961; Todd 1962; McCracken and Dorn 1964; Smith 1965; Rosenberg 1969). Paragraph 7.9.4 gives a graphic method for solving the finite difference equation in one-dimensional flow problems.

7.9.1 Method of Finite Differences

In most numerical methods of solving differential equations, the first step is to replace the latter by algebraic *finite difference equations*. These are relationships

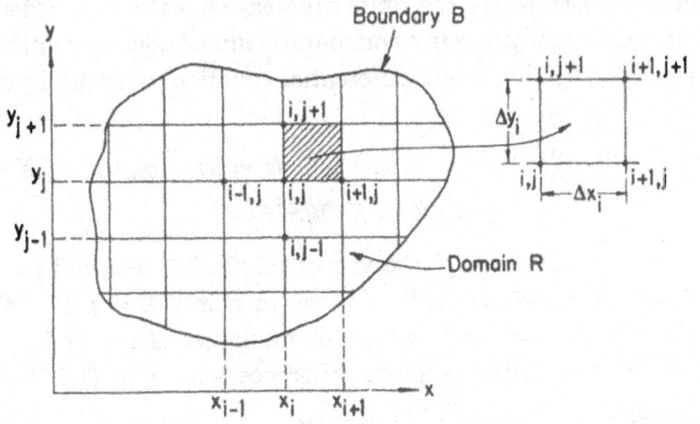

FIG. 7.9.1. A grid for a numerical solution.

among values of the dependent variable (say, φ) at neighboring points of the x, y, z, t space. The numerical solution of the series of simultaneous equations thus obtained gives the values of the dependent variables at a predetermined number of discrete points (*grid points*) throughout the domain investigated.

For the sake of simplicity, our discussion will be limited mainly to two-dimensional flows.

In order to solve a problem in the xy plane, the flow domain R enclosed by a boundary B is divided by a mesh of grid lines as shown in figure 7.9.1. The distance between grid lines need not be constant throughout the flow domain. Yet, unless a specific need for a variable grid spacing arises, the spacing, often with $\Delta x_i = \Delta y_i$ is maintained constant. The grid lines then form a *network of squares*. For time-dependent problems, time is divided into increments Δt. In general all time increments are chosen equal.

The differential equation is replaced by a *finite difference equation* written in terms of the values of the dependent variable at the grid points. The solution of the difference equation, or the set of difference equations, is carried out numerically, usually by means of a high speed digital computer. Denoting the exact solution of the differential equation by S, the exact solution of the difference equations by D and the numerical solution of the difference equation by N, we call $|S - D|$ the *truncation error* and $|D - N|$ the *numerical, or the round-off error*. The condition for convergence of the solution is that $|S - D| \to 0$ everywhere in the solution domain. The condition for *stability* is that everywhere in the solution domain $|D - N| \to 0$. The problem is to find N such that over the whole region of interest $|S - N|$ is smaller than some error criterion. As $(S - N) = (S - D) + (D - N)$, the total error is made up of the truncation error and the round-off error. The truncation error is due to the arbitrary form selected for the finite difference equation, and is often the larger part of the total error.

The value of the unknown function, φ, at point (i, j) at time $t = \sum_1^n \Delta t$, as computed from the finite difference equation, is denoted by $\varphi_{i,j}^n$. To illustrate some of the forms that may be selected for the finite difference equation, consider the problem of one-dimensional unsteady flow in a horizontal confined aquifer of uniform thickness. The flow is governed by the partial differential equation, written in nondimensional variables:

$$\partial H/\partial T = \partial^2 H/\partial X^2; \qquad 0 < X < 1; \qquad H = (\varphi - \varphi_0)/L, \qquad X = x/L,$$
$$T = K\,Dt/SL^2. \tag{7.9.1}$$

Let the X axis be divided into J intervals of length ΔX each, and the time interval T be divided into N increments ΔT. The set of points $X = j\,\Delta X$ and $T = n\,\Delta T$ in the XT-plane is the *net* (*grid, lattice*) of mesh size $\Delta X \times \Delta T$.

One possible way of writing a finite difference analog of (7.9.1) is:

$$\frac{H_j^{n+1} - H_j^n}{\Delta T} = \frac{H_{j+1}^n - 2H_j^n + H_{j-1}^n}{(\Delta X)^2}; \qquad \begin{array}{l}(j = 1, 2, \ldots, J),\\ (n = 0, 1, \ldots, N).\end{array} \tag{7.9.2}$$

Equation (7.9.2) is based on the approximate expressions:

$$\left(\frac{\partial H}{\partial t}\right)_j^n = \frac{H_j^{n+1} - H_j^n}{\Delta T}; \qquad \left(\frac{\partial^2 H}{\partial X^2}\right)_j^n = \frac{H_{j+1}^n - 2H_j^n + H_{j-1}^n}{(\Delta X)^2} \tag{7.9.3}$$

derived by applying Taylor's expansion theorem to the terms in (7.9.1).

An estimate of the error (in 7.9.2), as in other finite difference equations, can be obtained by considering the terms neglected in the Taylor expansion. Boundary conditions are stated as specified values H_0^n and H_j^n $(n = 0, 1, \ldots, N)$; initial conditions are given as $H_j^0 = f(j\,\Delta X)$, $(j = 0, 1, \ldots, J)$.

The scheme employed in writing (7.9.2) is called *explicit* because the equation at each point j includes only one unknown, namely H_j^{n+1}. The solution at the new time level $n + 1$ is computed, one point at a time, from three known values at time level n. The explicit solution begins with the initial condition H_j^0 $(j = 1, \ldots, J)$, and is propagated in time by solving for H^{n+1} from H^n. This scheme is also referred to as a *forward-in-time scheme* (fig. 7.9.2a).

By making ΔX and ΔT sufficiently small, the solution of the difference equation becomes a good approximation of the given differential equation. A finer mesh will result in a smaller error. Yet, the magnitude of ΔX and ΔT cannot be chosen arbitrarily. It can be shown (Richtmeyer 1957) that in order to get a stable solution of (7.9.2), i.e., a solution in which the round-off error dampens out rather than increases as the solution progresses, one must choose ΔX and ΔT such that:

$$\Delta T/(\Delta X)^2 \leqslant \tfrac{1}{2}. \tag{7.9.4}$$

This constraint makes the solution of (7.9.2) time consuming.

This stability constraint is avoided if (7.9.1) is represented by:

$$\frac{H_j^{n+1} - H_j^n}{\Delta T} = \frac{H_{j+1}^{n+1} - 2H_j^{n+1} + H_{j-1}^{n+1}}{(\Delta X)^2}. \tag{7.9.5}$$

This is an *implicit scheme* (also called a *backward-in-time scheme*); it is shown in

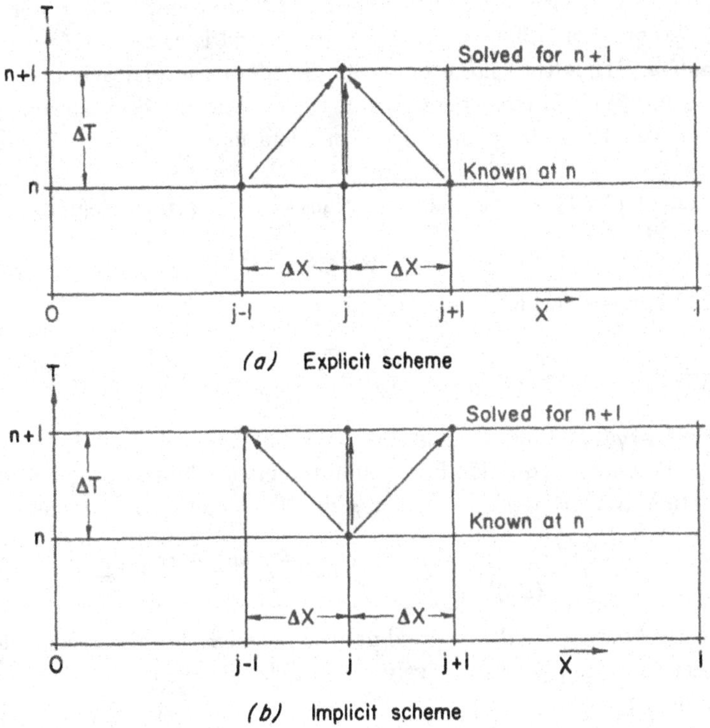

FIG. 7.9.2. Schemes for the finite difference equations.

figure 7.9.2b. To solve the equations one needs the boundary conditions at $j = 1$ and $j = J$ for all time levels.

There are three unknowns in (7.9.5) for each j. The equation is written for all js and the resulting set of equations is solved *simultaneously*. The implicit scheme is clearly superior in efficiency, although at every time step one must solve a set of simultaneous equations. The matrix of coefficients for the solution is tridiagonal, i.e., it has nonzero elements only on the main diagonal and the two adjacent diagonals. The solution of such a set is obtained by successive eliminations, and there exists a very efficient method, called the *Thomas algorithm*, to perform it (Shamir and Harleman 1966). For every time step the implicit scheme requires more computations than the explicit scheme, but because of the larger time steps that may be used a complete solution is less time consuming. The advantage of (7.9.5), however, is that it is unconditionally stable, whereas (7.9.2) has the advantage that each unknown H_j^{n+1} is determined from the known values H_j^n by solving a single equation. An implicit scheme is much more efficient, but usually requires a rather complicated program as compared to the relatively simple one needed for the explicit scheme.

A better finite difference analog of (7.9.1) is:

$$\frac{H_j^{n+1} - H_j^n}{\Delta T} = \varepsilon \frac{H_{j+1}^{n+1} - 2H_j^{n+1} + H_{j-1}^{n+1}}{(\Delta X)^2} + (1 - \varepsilon) \frac{H_{j+1}^n - 2H_j^n + H_{j-1}^n}{(\Delta X)^2} \tag{7.9.6}$$

where $0 < \varepsilon < 1$. With $\varepsilon = 0$ this reduces to (7.9.2), and with $\varepsilon = 1$ to (7.9.5). With $\varepsilon = \frac{1}{2}$, equation (7.9.6) is centered in time, and is called the *Crank–Nicholson approximation*. Thus the choice of the finite difference analog is quite arbitrary. In forming the finite difference analogs of the derivatives, one can also use points farther removed from the point under consideration and possibly achieve better accuracy.

If instead of (7.9.1) we have an equation with a variable coefficient:

$$\partial\varphi/\partial t = D(x)\, \partial^2\varphi/\partial x^2; \qquad D = D(x) \tag{7.9.7}$$

the finite difference formula takes the form:

$$\frac{1}{12}\frac{\varphi_{j+1}^{n+1} - \varphi_j^{n}}{D_{j+1}\,\Delta t} + \frac{5}{6}\frac{\varphi_j^{n+1} - \varphi_j^{n}}{D_j\,\Delta t} + \frac{1}{12}\frac{\varphi_{j-1}^{n+1} - \varphi_j^{n}}{D_{j-1}\,\Delta t} = \frac{(\delta^2\varphi)_j^{n+1} + (\delta^2\varphi)_j^{n}}{2(\Delta x)^2} \tag{7.9.8}$$

where $(\delta^2\varphi)_j^n = \varphi_{j+1}^n - 2\varphi_j^n + \varphi_{j-1}^n$ and $D_j = D(j\,\Delta x)$.

To solve the Laplace equation $\nabla^2\varphi = 0$ in the plane, the following finite difference approximation is written for a regular grid ($\Delta x =$ const, $\Delta y =$ const):

$$\frac{\varphi_{i+1,j} - 2\varphi_{i,j} + \varphi_{i-1,j}}{(\Delta x)^2} + \frac{\varphi_{i,j+1} - 2\varphi_{i,j} + \varphi_{i,j-1}}{(\Delta y)^2} = 0 \tag{7.9.9}$$

where the two terms are the approximate expressions for the second derivatives with respect to x and y, respectively; $x = i\,\Delta x$, $y = j\,\Delta y$. For $\Delta x = \Delta y = \Delta$, (7.9.9) becomes:

$$\varphi_{i,j} = \tfrac{1}{4}(\varphi_{i-1,j} + \varphi_{i+1,j} + \varphi_{i,j-1} + \varphi_{i,j+1}). \tag{7.9.10}$$

Thus the value of the function at any grid point is the average of its values at the immediately neighboring points. Equation (7.9.9) is *a five-point Laplace operator*, as it involves a "*molecule*" (or "star") of five points. A greater number of points may be used to express the Laplace operator.

The simple algebraic equation (7.9.9) or (7.9.10), which must be satisfied at every interior point of the flow domain, serves also as the basis for the relaxation method discussed in the following paragraph. In these equations, as in all other elliptic partial differential equations, stability and convergence are assured.

In three-dimensional flow ($x = i\,\Delta x$, $y = j\,\Delta y$, $z = k\,\Delta z$) we obtain:

$$\begin{aligned}\nabla^2\varphi \cong (\varphi_{i+1,j,k} + \varphi_{i-1,j,k} - 2\varphi_{i,j,k})/(\Delta x)^2 + (\varphi_{i,j+1,k} + \varphi_{i,j-1,k}\\ - 2\varphi_{i,j,k})/(\Delta y)^2 + (\varphi_{i,j,k+1} + \varphi_{i,j,k-1} - 2\varphi_{i,j,k})/(\Delta z)^2 = 0.\end{aligned} \tag{7.9.11}$$

Or, with $\Delta x = \Delta y = \Delta z = \Delta$:

$$\varphi_{i,j,k} = (\varphi_{i+1,j,k} + \varphi_{i-1,j,k} + \varphi_{i,j+1,k} + \varphi_{i,j-1,k} + \varphi_{i,j,k+1} + \varphi_{i,j,k-1})/6. \tag{7.9.12}$$

In cylindrical coordinates, the finite difference equation representing the Laplacian takes the form:

$$\nabla^2\varphi \cong \frac{1}{r_{i,j,k}}\left[\frac{2r_{i,j,k} + \Delta r}{2(\Delta r)^2}(\varphi_{i+1,j,k} - \varphi_{i,j,k}) + \frac{2r_{i,j,k} - \Delta r}{2(\Delta r)^2}(\varphi_{i-1,j,k}\right.$$

$$- \varphi_{i,j,k}) + \frac{\varphi_{i,j+1,k} - \varphi_{i,j,k}}{r_{i,j,k}(\Delta\theta)^2} + \frac{\varphi_{i,j-1,k} - \varphi_{i,j,k}}{r_{i,j,k}(\Delta\theta)^2}$$

$$+ \frac{r_{i,j,k}}{(\Delta z)^2}(\varphi_{i,j,k+1} - \varphi_{i,j,k}) + \frac{r_{i,j,k}}{(\Delta z)^2}(\varphi_{i,j,k-1} - \varphi_{i,j,k})\Bigg] = 0 \qquad (7.9.13)$$

where $r = i\,\Delta r$, $\theta = j\,\Delta\theta$, $z = k\,\Delta z$.

The finite difference approximation of the Poisson equation for the case $\Delta x = \Delta y = \Delta$ is:

$$\varphi_{i,j} = \tfrac{1}{4}(\varphi_{i+1,j} + \varphi_{i-1,j} + \varphi_{i,j+1} + \varphi_{i,j-1} + \Delta^2 F_{i,j}). \qquad (7.9.14)$$

An implicit finite difference approximation of the diffusion equation (6.4.7) is:

$$(H_{i,j}^{n+1} - H_{i,j}^{n})/\Delta T = (H_{i+1,j}^{n+1} + H_{i-1,j}^{n+1} - 2H_{i,j}^{n+1})/(\Delta X)^2 + (H_{i,j+1}^{n+1} + H_{i,j-1}^{n+1} - 2H_{i,j}^{n+1})/(\Delta Y)^2. \qquad (7.9.15)$$

The corresponding explicit form is:

$$(H_{i,j}^{n+1} - H_{i,j}^{n})/\Delta T = (H_{i,j+1}^{n} + H_{i,j-1}^{n} - 2H_{i,j}^{n})/(\Delta X)^2 + (H_{i,j+1}^{n} + H_{i,j-1}^{n} - 2H_{i,j}^{n})/(\Delta Y)^2 \qquad (7.9.16)$$

with a stability criterion:

$$\Delta T/(\Delta X)^2 + \Delta T/(\Delta Y)^2 \leqslant \tfrac{1}{2} \qquad (7.9.17)$$

that limits the size of the time step that can be used. This often renders the rather simple scheme impractical because of the long computer time required.

The implicit scheme (7.9.15), on the other hand, is unconditionally stable, but results in a set of simultaneous equations, each of which has five unknowns. Peaceman and Rachford (1955) proposed an *alternating direction procedure* that is more efficient than either the explicit or implicit schemes, when the problem is in more than one dimension. Bjordamnen and Coats (1967) and Carter (1967) also discuss this method. For the solution of the heat conduction equation in the plane, this scheme contains two equations at each time step:

$$\frac{H_{i,j}^{n+1/2} - H_{i,j}^{n}}{(\Delta T/2)} = \frac{H_{i+1,j}^{n+1/2} + H_{i-1,j}^{n+1/2} - 2H_{i,j}^{n+1/2}}{(\Delta X)^2} + \frac{H_{i,j+1}^{n} + H_{i,j-1}^{n} - 2H_{i,j}^{n}}{(\Delta Y)^2} \qquad (7.9.18)$$

$$\frac{H_{i,j}^{n+1} - H_{i,j}^{n+1/2}}{(\Delta T)/2} = \frac{H_{i+1,j}^{n+1/2} + H_{i-1,j}^{n+1/2} - 2H_{i,j}^{n+1/2}}{(\Delta X)^2} + \frac{H_{i,j+1}^{n+1} + H_{i,j-1}^{n+1} - 2H_{i,j}^{n+1}}{(\Delta Y)^2}. \qquad (7.9.19)$$

Only one of the two terms on the right-hand side of each of these equations includes unknowns, while the other is at the old time. Equation (7.9.18) results in a tridiagonal set of equations for $H^{n+1/2}$ and (7.9.19) in a tridiagonal set for H^{n+1}. Although each equation separately is subject to a stability criterion, the combination of the two equations is unconditionally stable.

The finite difference method of solution can be used also for solving nonlinear equations such as those encountered in phreatic flows (chap. 8). For example, the explicit scheme of (8.2.5) for a homogeneous aquifer (K = const) is:

$$\frac{(h^2)^{n+1}_{i+1,j} + (h^2)^{n+1}_{i-1,j} - 2(h^2)^{n+1}_{i,j}}{(\Delta x)^2} + \frac{(h^2)^{n+1}_{i,j+1} + (h^2)^{n+1}_{i,j-1} - 2(h^2)^{n+1}_{i,j}}{(\Delta y)^2}$$

$$= \frac{n_e}{K}\frac{h^{n+1}_{i,j} - h^n_{i,j}}{\Delta t} - \frac{N^{n+1/2}_{i,j}}{K\,\Delta x\,\Delta y}. \tag{7.9.20}$$

The stability criterion imposes a limitation even more severe than in the linear case.

It is possible to make the numerical solution easier by linearizing (8.2.5), replacing $h\,\partial h/\partial x$ by $h^n\,\partial h^{n+1}/\partial x$ and $h\,\partial h/\partial y$ by $h^n\,\partial h^{n+1}/\partial y$. A general scheme for solving (8.2.5) with $K \neq$ const is, then (Shamir 1967):

$$\begin{aligned}
&\varepsilon_x\{\Delta_{x+1/2}[KH^n(H^{n+1})] + \Delta_{x-1/2}[KH^n(H^{n+1})]\}\\
&+ (1-\varepsilon_x)\{\Delta_{x+1/2}[KH^n(H^n)] + \Delta_{x-1/2}[KH^n(H^n)]\}\\
&+ \varepsilon_y\{\Delta_{y+1/2}[KH^n(H^{n+1})] + \Delta_{y+1/2}[KH^n(H^{n+1})]\}\\
&+ (1-\varepsilon_y)\{\Delta_{y+1/2}[KH^n(H^n)] + \Delta_{y-1/2}[KH^n(H^n)]\}\\
&= (n_e)_{i,j}(H^{n+1}_{i,j} - H^n_{i,j})/\Delta t - N^{n+1/2}_{i,j}/\Delta x\,\Delta y
\end{aligned} \tag{7.9.21}$$

where, for example:

$$\Delta_{x+1/2}[KH^n(H^{n+1})] = K_{i+1/2,j}H^n_{i+1/2,j}(H^{n+1}_{i+1,j} - H^{n+1}_{i,j})/(\Delta x)^2$$

$$\Delta_{x-1/2}[KH^n(H^n)] = K_{i-1/2,j}H^n_{i-1/2,j}(H^n_{i-1,j} - H^n_{i,j})/(\Delta x)^2$$

and the values at the middle between grid points are obtained from a linear interpolation

$$K_{i+1/2,j} = (K_{i+1,j} + K_{i,j})/2; \qquad H^n_{i+1/2,j} = (H^n_{i+1,j} + H^n_{i,j})/2.$$

Equation (7.9.21) is explicit when $\varepsilon_x = \varepsilon_y = 0$, and fully implicit when $\varepsilon_x = \varepsilon_y = 1.0$; $\varepsilon = \frac{1}{2}$ yields a centered (in time) equation. An alternating direction scheme is obtained from equation (7.9.21) by setting $\varepsilon_x = 1.0$ and $\varepsilon_y = 0$, and then $\varepsilon_x = 0$ and $\varepsilon_y = 1.0$ at alternating steps.

Other equations governing flow through porous medium phenomena can be handled in a similar manner. For example, Peaceman and Rachford (1962), Shamir and Harleman (1966) and Reddell and Sunada (1970) discuss numerical solution by the method of finite differences of the dispersion equation (chap. 10); Watson (1967) deals with the problem of unsaturated flow (sec. 9.4). Shamir and Dagan (1970) present a numerical solution for the movement of a salt water–fresh water interface in a coastal aquifer. Douglas et al. (1959), Quon et al. (1967) and Carter (1967) discuss the numerical solution of multiphase (e.g., oil–water) flow problems encountered in reservoir engineering (chap. 9). The two last cases involve parameters, such as relative permeability and capillary pressure, that are functions of saturation and that are obtained experimentally. Therefore, they are generally given graphically, or in the form of tables, but not as analytic expressions.

All the equations mentioned above are solved subject to conditions specified on the boundary B of the flow domain R (fig. 7.9.1). The Dirichlet condition of specified value of the dependent variable along the boundary, and the Neumann condition of specified value of the derivative, are the more common conditions encountered in such problems. These and other conditions are discussed in detail in section 7.1.

In certain cases it is possible to fill the entire domain R with a uniformly spaced network such that the boundary of R intersects the network only at mesh points. This case is shown as boundary B_1 in figure 7.9.3a. For each interior nodal point, a finite difference molecule approximating the governing partial differential equation can be applied. When the values of the dependent variable, say φ, are specified at nodal points on the boundary, these values are incorporated in the finite difference expressions for the nodes near the boundaries.

Sometimes the boundary of the flow domain R takes the irregular form B_2 shown in figure 7.9.3b. Two procedures are available for handling such cases. In the first one, the partial differential equation considered is approximated at every point internal to the boundary B_2. At a point such as O in figure 7.9.3b, the regular molecule is modified by taking into account both the boundary conditions and the nonstandard spacing 01 and 02 (fig. 7.9.3b). In the second procedure the considered equation is approximated only at those interior points for which the regular molecule applies without any modification (i.e., to node points inside the dashed line B_3). This leaves an outer ring of points (between B_2 and B_3) at which, instead of requiring the φ-values to satisfy the considered equation, we require that these values satisfy finite difference approximations of the boundary conditions. The nonstandard spacings of the boundary are incorporated in these latter requirements.

For example, the finite difference approximation of the Laplacian according to the first procedure, with prescribed φ along the boundary (i.e., φ_1 and φ_2 are known), becomes:

$$\nabla^2\varphi = \frac{2}{\Delta^2}\left[\frac{\varphi_1}{a(1+a)} + \frac{\varphi_2}{b(1+b)} + \frac{\varphi_3}{1+a} + \frac{\varphi_4}{1+b} - \frac{(a+b)\varphi_0}{ab}\right] + 0(\Delta) \quad (7.9.22)$$

or:

$$\varphi_0 = \frac{ab}{a+b}\left[\frac{\varphi_1}{a(1+a)} + \frac{\varphi_2}{b(1+b)} + \frac{\varphi_3}{1+a} + \frac{\varphi_4}{1+b}\right] \quad (7.9.23)$$

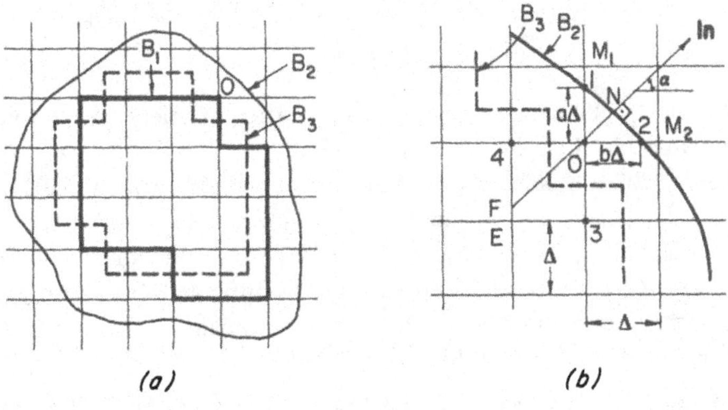

FIG. 7.9.3. Uniform grid applied to an irregular domain.

According to the second procedure we obtain φ_0 by using a linear interpolation from point 2 to 4 or from 1 to 3:

$$\varphi_0 = \frac{b}{1+b}\varphi_4 + \frac{1}{1+b}\varphi_2 \quad \text{or} \quad \varphi_0 = \frac{a}{1+a}\varphi_1 + \frac{1}{1+a}\varphi_3 \tag{7.9.24}$$

or by taking the average of the two interpolations. When $(\partial\varphi/\partial n)_N$ is given (i.e., a boundary condition of the second kind) we approximate it by $(\varphi_0 - \varphi_F)/\overline{OF}$, obtaining φ_F by a linear interpolation between points 4 and E.

For more details on the treatment of boundary conditions see the texts on numerical methods listed at the beginning of this section.

7.9.2 The Method of Finite Elements

Whereas the finite difference method is based on a finite difference analog of the partial differential equation, the *finite element technique* uses a functional associated with the partial differential equation itself. A *functional* (see any text on calculus of variations, e.g., Gelfand and Fomin 1963) means a correspondence that assigns a definite (real) number to each function (or curve) belonging to some class. It is a kind of function where the dependent variable is itself a function.

Consider (Shamir 1967), for example, the following integral in two independent variables x, y and two dependent variables $U_1 = U_1(x, y)$ and $U_2 = U_2(x, y)$:

$$I = \iint_{(R)} F(x, y, U_1, U_2, \partial U_1/\partial x, \partial U_1/\partial y, \partial U_2/\partial x, \partial U_2/\partial y)\, dx\, dy. \tag{7.9.25}$$

The integral is to be made *stationary*, that is, we seek $U_1(x, y)$ and $U_2(x, y)$, which will make I a minimum (we shall use "min" throughout, although the same applies to "max" as well). This is done by requiring that the *variation* (or *differential*) of I vanishes, i.e., $\delta I = 0$.

It can be shown (e.g., Gelfand and Fomin 1963; Hildebrand 1962) that this requirement holds *only if* the following partial differential equations are satisfied:

$$\frac{\partial}{\partial x}\left(\frac{\partial F}{\partial U_{1x}}\right) + \frac{\partial}{\partial y}\left(\frac{\partial F}{\partial U_{1y}}\right) - \frac{\partial F}{\partial U_1} = 0; \qquad \frac{\partial}{\partial x}\left(\frac{\partial F}{\partial U_{2x}}\right) + \frac{\partial}{\partial y}\left(\frac{\partial F}{\partial U_{2y}}\right) - \frac{\partial F}{\partial U_2} = 0 \tag{7.9.26}$$

where (only here!) subscripts x and y denote differentiation with respect to that independent variable (i.e., $U_{1x} \equiv \partial U_1/\partial x$, etc.).

Equations (7.9.26) are called the *Euler equations* associated with (7.9.25). In these equations U_1, U_{1x}, U_{1y}, U_2, U_{2x} and U_{2y}, in addition to x and y, are treated as independent variables. These equations are the *necessary* conditions for I to be stationary. *Sufficiency* conditions are often very difficult to establish analytically, and one resorts to considerations of the original problem to show that a minimum (or maximum) of I was indeed found.

The finite element technique is based on the solution of the variational problem in its original form, i.e., equation (7.9.25). Once the differential equations describing the problem have been formulated, we seek the functional for which they are the Euler equations. Then, instead of solving the differential equation, we can solve the minimization problem directly. The function to be minimized is an integral over the region of interest (an area integral for the two-dimensional case).

Next, the solution field is divided into elements. The shape, size and distribution of the elements is arbitrary. It is assumed that the value of the dependent variable varies *linearly* over each element. This means that the value of the dependent variable(s) at any point within the element is determined uniquely by the values of the variable(s) at the nodes of the element and the position of the point under consideration inside the element.

The contribution of each element to the integral given by equation (7.9.25) can be expressed in terms of the values of the dependent variables at the nodes of the element and its geometry. By differentiating this expression with respect to the dependent variable at each node, and adding up the resulting equation for all the elements in the field, a set of simultaneous equations is obtained in which the unknowns are the values of the dependent variables at the nodes, and the coefficients are functions of the coordinates of the nodes. The right-hand side is zero, as we are seeking a stationary point.

Boundary conditions are transposed from conditions along *sides* of the element to conditions at *nodes* of the element. For example, a given flux along a side is divided (in some proportional way) into two discrete fluxes at the ends of this side. A given value of the dependent variable is represented by its values at the nodes (and assumed to vary linearly between them).

In summary, the finite element technique uses the following procedure.

(a) For the partial differential equation that governs the considered flow, derive the associated variational problem.

(b) Divide the field into elements.

(c) Formulate the variational functional within the element.

(d) Take derivatives with respect to the dependent variable(s) at all nodes of the element.

(e) "Assemble" the equations for all elements.

(f) Express the boundary condition in terms of nodal values.

(g) Incorporate the boundary condition into the equations and solve.

(h) The shape and size of the elements is arbitrary. Different shapes (triangles, rectangles, etc.) can be used simultaneously. Smaller elements can be chosen in regions where there are rapid variations in the properties of the material or in the values of the dependent variables.

As the dependent variables are assumed to vary linearly along the side of each element, they are continuous from one element to the next. Their derivatives, however, may have a discontinuity along the side.

Shamir (1967) applied the finite element method to steady flow in an inhomogeneous anisotropic aquifer and to one-dimensional hydrodynamic dispersion (chap. 10) in a homogeneous isotropic medium. Javandel and Witherspoon (1968) applied the finite element method to transient flow in porous media. Volker (1969) applied the method to nonlinear flow in porous media. Neumann and Witherspoon (1970) applied the method to a steady seepage with a free surface.

7.9.3 Relaxation Methods

The relaxation method is a process for obtaining steadily improved approximations of the solution of the simultaneous algebraic difference equations. The method is applicable to steady-state problems described by the Laplace or the Poisson equations. In the steady-state problems, the boundary conditions are specified on all the boundaries. In the nonsteady-state problems, the initial distribution of the dependent variable is given, but there is no condition to be satisfied by the dependent variable at some later time. This is equivalent to saying that there is a free boundary with respect to the independent variable time. Sometimes we assume that the steady-state solution describes approximately the required solution of a nonsteady-state problem, e.g., when t is large; this provides the missing condition.

Following the procedure already described in paragraph 7.9.1, the first step is to replace the continuous investigated flow domain by a rectangular or a square grid. The differential equation is replaced by the corresponding difference equations. For the Laplace equation, for example, these are given by (7.9.10). Let us define a residual R_0 corresponding to point 0:

$$R_0 = \sum_{i=1}^{4} \varphi_i - 4\varphi_0 \tag{7.9.27}$$

which represents the amount by which the equation is in error at that point. If all the values of φ are correct, R_0 will be zero everywhere. In the relaxation method one begins by guessing the values of φ at all grid points; in general the initial residuals do not vanish everywhere. Although one is free to start from an arbitrary set of values, the closer they are to the correct values, the fewer computations are involved. The relaxation process consists in adjusting the values of φ, so that eventually all residuals approach zero, leaving a residual that corresponds to a required accuracy. The liquidation of residuals is achieved by a "relaxation pattern." The unit relaxation pattern is the same for every interior grid point: a change of $+1$ in the value of φ at any point of the field will change the residual there by -4, and the residuals at each of the four neighboring points by $+1$. If the error is $+4$ at that particular point, an additional increment of $\Delta\varphi = 1$ will remove this error there, but will introduce errors of $+1$ at the four surrounding points. Repeating this procedure at different grid points will gradually spread the residuals and reduce their value. When constant boundary values are specified, the relaxation of points near the boundary does not pass on residuals to the fixed boundary points. Special techniques are described in the literature by which the above basic operation can be accelerated. For example, only the largest existing residual is relaxed at each step.

With the notation of figure 7.9.4 (after Thom et al. 1961), let us introduce the sums:

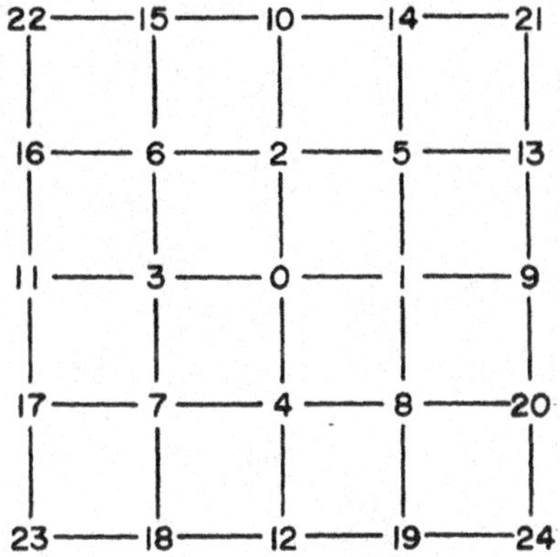

FIG. 7.9.4. Notation for the sums S_i.

$$S_1 = \varphi_1 + \varphi_2 + \varphi_3 + \varphi_4$$

$$S_2 = \varphi_5 + \varphi_6 + \varphi_7 + \varphi_8$$

$$S_3 = \varphi_9 + \varphi_{10} + \varphi_{11} + \varphi_{12}$$

$$S_4 = \varphi_{13} + \varphi_{14} + \varphi_{15} + \varphi_{16} + \varphi_{17} + \varphi_{18} + \varphi_{19} + \varphi_{20}$$

$$S_5 = \varphi_{21} + \varphi_{22} + \varphi_{23} + \varphi_{24}. \tag{7.9.28}$$

The Laplace equation can then be replaced by one of the following expressions:

$$4\varphi_0 = S_1$$

$$20\varphi_0 = 4S_1 + S_2$$

$$12\varphi_0 = 2S_1 + S_2$$

$$476\varphi_0 = 46S_3 + 32S_4 + 9S_5$$

$$100\varphi_0 = 10S_3 + 7S_4 + S_5. \tag{7.9.29}$$

These expressions may also be derived by the Taylor expansion (as in par. 7.9.1) with certain terms eliminated by summation and other terms neglected as higher power terms of the increments Δx and Δy. These "large molecule formulas" are better approximations of the Laplace differential equation and speed up the relaxation process. Equations of still larger molecules are also available.

Because the increments are much smaller than the values of φ themselves, working with the former is advisable, as this reduces the number of computations required.

For the two-dimensional Poisson equation $\nabla^2\varphi = g(x, y)$, we obtain for $g(x, y) =$ const $= G$ and $h = \Delta x = \Delta y$:

$$4\varphi_0 = S_1 - h^2G$$

$$20\varphi_0 = 4S_1 + S_2 - 6h^2G$$

$$100\varphi_0 = 10S_3 + 7S_4 + S_5 - 118h^2G. \tag{7.9.30}$$

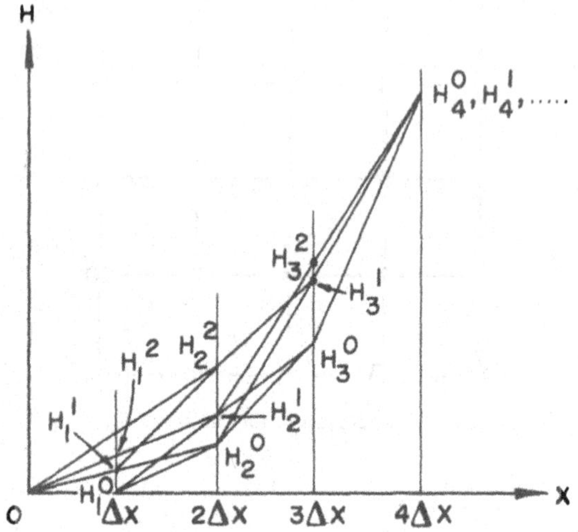

FIG. 7.9.5. Schmidt's graphical method for the solution of (7.9.1).

When $g(x, y)$ is not constant, but is known at every grid point, one proceeds similarly, aiming to reduce the residual R_0 to zero, where:

$$R_0 = S_1 - 4\varphi_0 + h^2 g(x, y). \tag{7.9.31}$$

The treatment of boundary condition is similar to that described in paragraph 7.9.1.

7.9.4 Schmidt's Graphic Method

Several graphic solution methods of the one-dimensional partial differential equations are mentioned in the literature. Their main advantage lies in their simplicity. Schmidt's (1924) method is described here as an example of such a method.

We consider a problem described by (7.9.1), written in its finite difference form (7.9.2). If ΔT and ΔX are chosen such that $\Delta T/(\Delta X)^2 = \frac{1}{2}$, equation (7.9.2) becomes:

$$H_j^{n+1} = (H_{j+1}^n + H_{j-1}^n)/2. \tag{7.9.32}$$

This means that the value of H at point X at time $(T + \Delta T)$ is equal to the arithmetic mean of H at the neighboring points $(X + \Delta X)$ and $(X - \Delta X)$ at time T. Figure 7.9.5 describes Schmidt's graphic method of solution, which is based on (7.9.32). The boundary conditions in the example of figure 7.9.5 are: $H(0, T) = 0$ and $H(4\,\Delta X,\ T) = H_4^0 =$ const. The initial conditions (at $T = 0$) are given by $H(0, 0) = 0$, $H(\Delta X, 0) = 0$, $H(2\,\Delta X, 0) = H_2^0$, $H(3\,\Delta X, 0) = H_3^0$ and $H(4\,\Delta X,\ 0) = H_4^0 =$ const. Shown are points H_1^2, H_2^2, H_3^2.

Figure 7.9.6 gives a graphic solution by Schmidt's method of the equation:

$$T\,\partial^2 h/\partial x^2 + N = n_e\,\partial h/\partial t; \qquad x > 0 \tag{7.9.33}$$

which describes unsteady, one-dimensional flow in an unconfined aquifer with a

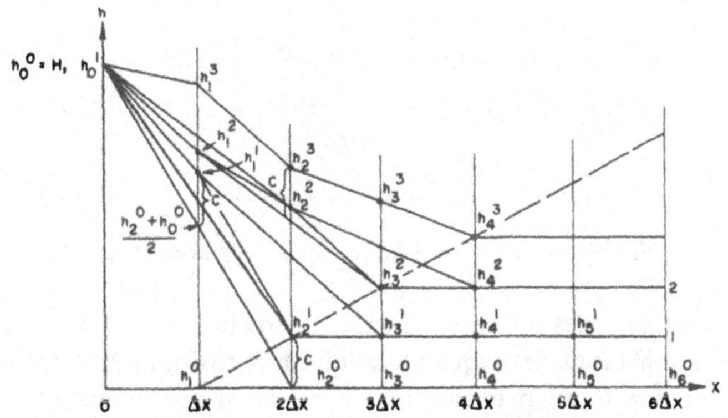

FIG. 7.9.6. Schmidt's graphical method for the solution of (7.9.33); after Thomas (1960).

constant rate of accretion N at the surface. This is the linearized equation based on the assumption that changes in water table elevation are relatively small. The finite difference equation of (7.9.33) is:

$$T(h_{j+1}^n - 2h_j^n + h_{j-1}^n)/(\Delta x)^2 + N = n_e(h_j^{n+1} - h_j^n)/\Delta t$$

or:

$$h_{j+1}^n - 2h_j^n + h_{j-1}^n + N(\Delta x)^2/T = [n_e(\Delta x)^2/T\,\Delta t](h_j^{n+1} - h_j^n). \tag{7.9.34}$$

If we choose Δx and Δt such that $N(\Delta x)^2/T = 2C$, and $n_e(\Delta x)^2/T\,\Delta t = 2$, equation (7.9.34) becomes:

$$(h_{j+1}^n + h_{j-1}^n)/2 + C = h_j^{n+1}. \tag{7.9.35}$$

The graphic solution of (7.9.35) is shown in figure 7.9.6 (after Thomas 1960). The initial conditions are $h(x, 0) = 0$. At $t = 0$ the water table elevation at $x = 0$ is suddenly raised to $h = H =$ constant, where it remains for $t > 0$.

Thomas (1960) gives examples of graphic solutions by Schmidt's method for several additional equations describing flow of ground water in aquifers. He also describes the solution of equations describing radial flow. When in (7.9.33) the parameter (n_e/T) varies with the distance x, one can modify the x-scale according to:

$$x' = x[(n_e/T)/(n_e/T)_0]^{1/2} \tag{7.9.36}$$

where x' is the modified scale and $(n_e/T)_0$ is some reference value of n_e/T. Once the problem is solved in terms of x', the solution can easily be rewritten in terms of x.

7.10 Flow Nets by Graphic Methods

In *two-dimensional flow problems*, the flow net discussed in detail in section 6.6 can be obtained by graphic methods without actually solving (analytically or numerically) the Laplace equation. In fact, the derived flow net is the solution sought. Only isotropic media need be considered here, as anisotropic flow domains can be trans-

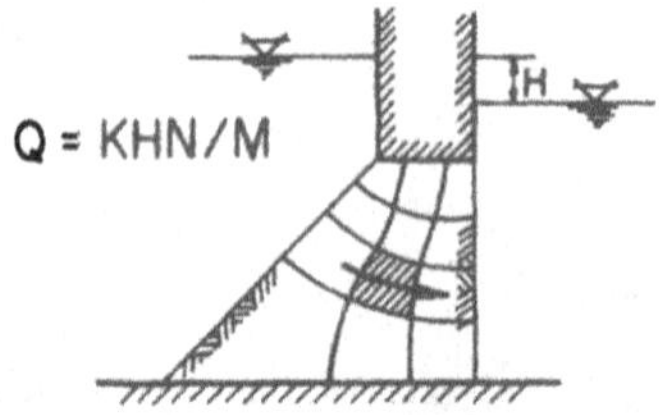

FIG. 7.10.1. Flow-net made of squares ($N = 4$, $M = 3$).

formed into isotropic ones by the method described in section 7.4. In most graphic methods, the flow net is made of squares, as it is easier to draw squares than rectangles. In many cases, the solution is carried out by free-hand drawing.

Most graphic methods are quick, and give the investigator a valuable understanding of the flow pattern. Long practice is required, however, to become proficient in flow-net sketching. Several methods that include explicit sets of instructions for drawing flow nets are available in the literature.

In engineering practice, the most common procedure for obtaining flow nets graphically is the so-called *Forchheimer method* (Forchheimer 1930). This is a trial and error method based on drawing small squares. The squares are drawn here (as in all other methods) so that the boundary conditions are satisfied. The term "square" is used here and below to denote a curvilinear square whose sides intersect each other orthogonally and whose line segments joining centers of opposite sides are of equal length. The flow net is drawn with an integer number (N) of streamtubes through each of which the discharge is Q/N, Q being the total discharge (per unit width). The number of equipotential drops M need not be an integer. Figure 7.10.1 shows an example of such a flow net. When a phreatic surface (streamline in steady flow) constitutes part of the boundary, the boundary condition $\varphi = z$ on it is also employed. The presence of a seepage face is also taken into account.

Casagrande (1937) gives a number of useful hints for sketching flow nets. (a) Use a small number of streamtubes (e.g., four to five) in the first attempt. (b) Do not adjust details before the flow net as whole is approximately correct. (c) Avoid sharp transitions between curved and straight segments of streamlines and equipotentials. (d) The squares along each tube should vary gradually in size.

Taylor (1948) suggests a procedure that he calls the "procedure of explicit trials." The procedure is shown in figure 7.10.2.

Eck (1944) claims that the small squares are not sufficiently defined. He defines them as four enveloping mutually orthogonal curves circumscribing a common central circle (fig. 7.10.3). A much simpler checking method is the orthogonality of the diagonal shown in figure 7.10.4 (Kozeny 1953).

Irmay (in Bear et al. 1968), following Breitenoder (1942), describes another method—the isogradient–isocline method.

Difficulties in the construction of flow nets occur at *singular points* where the specific discharge is either zero (stagnation point) or infinite (*cavitation point*). At such points the requirement of orthogonality of streamlines and equipotential lines

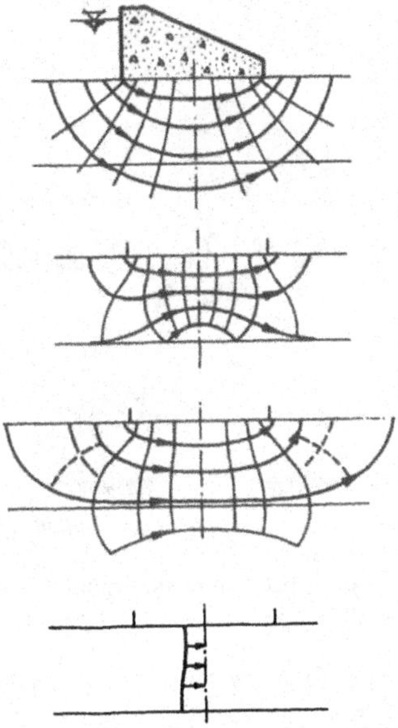

Fig. 7.10.2. Flow-net sketching by Taylor's procedure of explicit trials (Taylor, 1948).

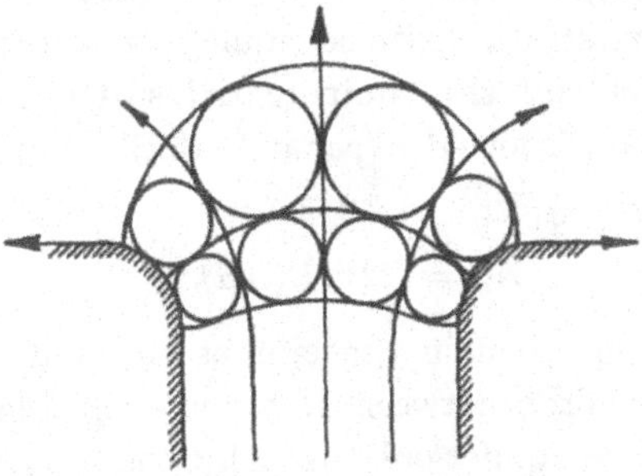

Fig. 7.10.3. Squares and circles (Eck, 1944).

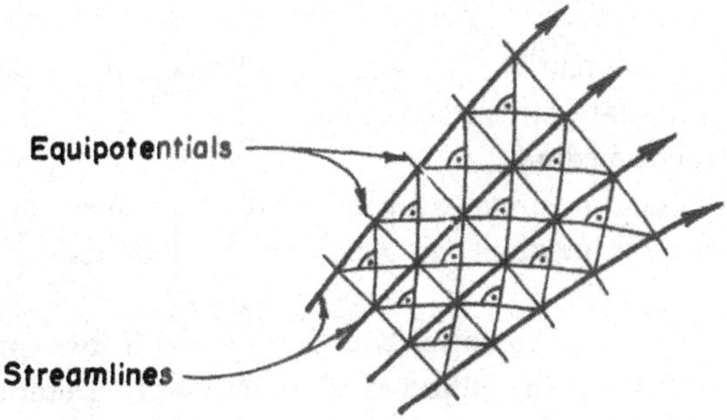

Fig. 7.10.4. Checking orthogonality of diagonals (Kozeny, 1953).

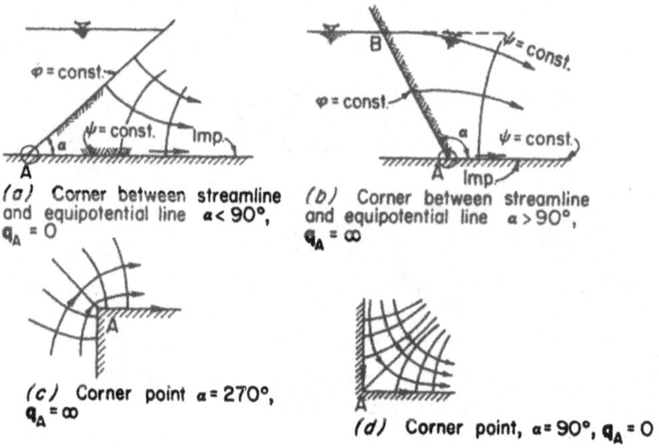

FIG. 7.10.5. Examples of singular points.

breaks down. Streamlines, which as a rule do not intersect each other, meet at singular points. Figure 7.10.5 gives several examples of singular points.

Finally, when the $\varphi - \psi$ flow net is drawn in meridianal planes of radially (or axisymmetrically) converging or diverging flows, we must take into account the change in the area through which the flow takes place as we approach the well. When a well fully penetrates a confined aquifer of constant thickness, the flow is everywhere horizontal and the equipotential surfaces are vertical concentric cylinders. Then the rate of flow ΔQ (per unit thickness) between two meridianal planes is given by:

$$\Delta Q = K2\pi r B(\Delta H/\Delta r). \tag{7.10.1}$$

This means that $r/\Delta r$ must remain constant as the well is approached if $\Delta\varphi =$ constant is the potential drop between adjacent equipotential lines.

If the well penetrates the aquifer only partially, the situation is more complicated and no rule exists for the spacing between equipotential lines.

Exercises

7.1 Express the equipotential boundary conditions in two-dimensional flow in terms of an appropriate stream function:

(a) for an isotropic medium;
(b) for an anisotropic medium, x, y principal directions; and
(c) for an anisotropic medium, x, y not principal directions.

7.2 A segment of an equipotential boundary $\varphi = 0$ in two-dimensional flow in the xy-plane is described by the ellipse $x^2/a^2 + y^2/b^2 = 1$. Determine the direction of $\mathbf{q}$ at points along the boundary when (a) the medium is isotropic and (b) when the medium is anisotropic.

7.3 Write the boundary conditions in terms of φ along an impervious boundary in an anisotropic medium. The geometry of the boundary is given by $x^2 + y^2 + z^2 = R^2 = \text{const.}$

7.4 Derive the boundary condition for a steady phreatic surface without accretion starting from the statement that this boundary is a streamline $\Psi = \text{const.}$

7.5 Derive the boundary condition for a steady phreatic surface without accretion in 3-dimensional flow where x, y, z are not principal directions.

7.6 Prove (7.1.51) for points B, C, D and E shown in fig. 7.1.6a.

7.7 Prove (7.1.67).

7.8 Determine the angle of refraction, β'', of a streamline upon passage through an interface between two isotropic media (K', K'') when the angle of incidence is $\beta' = 30°$ (fig. 7.1.12)
(a) when $K' = 10K'$, (b) when $K' = 0.1K''$.

7.9 A straight line boundary separates two anisotropic media. In both media, the principal directions are parallel and normal to this line. Develop the law of refraction for streamlines and equipotentials.

7.10 Give a complete mathematical statement of the two-dimensional steady flow in an isotropic medium shown in (a) fig. 7.1.4 (symmetric case), (b) fig. 7.1.6a, (c) fig. 7.1.6b.

7.11 Give a complete mathematical statement of the problem of unsteady radially converging flow to a fully penetrating well of radius r_w pumping at a constant rate Q_w from an infinite confined homogeneous aquifer (T, S). Initially the piezometric surface is everywhere horizontal ($\varphi = \varphi_0$).

7.12 Repeat Ex. 7.11 for a flowing well, where the head at the well for $t \geqslant 0$ is maintained constant at some $\varphi = \varphi_1$ ($< \varphi_0$) above the ground surface.

7.13 Repeat Ex. 7.11 but the well's screen extends only from the middle of the aquifer to its bottom.

7.14 A vertical two-dimensional flow domain is made up of two regions (K_1, K_2) as shown in fig. 7.E.1. Give a complete mathematical statement of the steady flow problem.

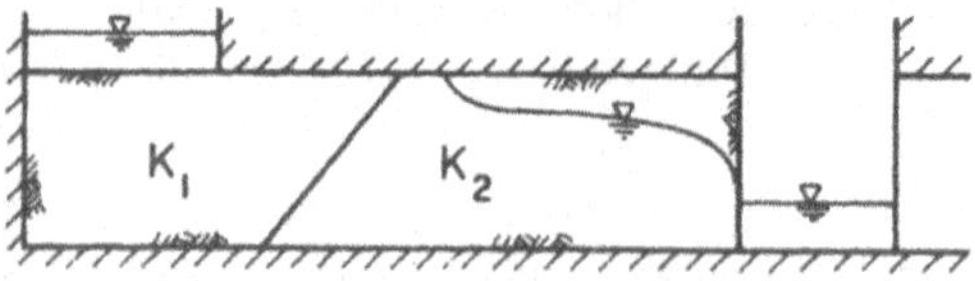

Vertical cross-section

FIG. 7.E.1.

7.15 Give a complete mathematical statement of unsteady flow to a well in an infinite phreatic homogeneous, isotropic aquifer where the water table is initially at an elevation h_0 above the horizontal impervious bottom, when (a) a constant rate of pumping Q_w is maintained at the well, (b) a constant water level h_w is maintained at the well.

7.16 What is the shape of the hodograph of (a) a phreatic surface and (b) a straight line seepage face, when the principal directions are x', y'; x' makes an angle α with the $+x$ axis.

7.17 Use (7.3.11) to obtain the range of values for q_x and q_y when $q_z = 0$ and $\partial\varphi/\partial t > 0$.

7.18 Draw the hodograph representation of the boundaries of the flow domain in the triangular dam shown in fig. 7.E.2.

7.19 Repeat Ex. 7.18 for (a) $\theta < \pi/2$, (b) $\theta > \pi/2$.

7.20 Draw the hodograph representation of the boundaries of the flow domains shown in fig. 7.E.3.

7.21 The relationship (7.4.12) was derived for a homogeneous domain. Show that the same relationship also holds for an inhomogeneous domain (i.e., k_{xp}, k_{yp} and k_{zp} are functions of x, y, z). Discuss the consequences.

7.22 Develop relationships similar to (7.4.22) and (7.4.23), but for a two-dimensional field. Use the resulting relationships to prove (7.4.26).

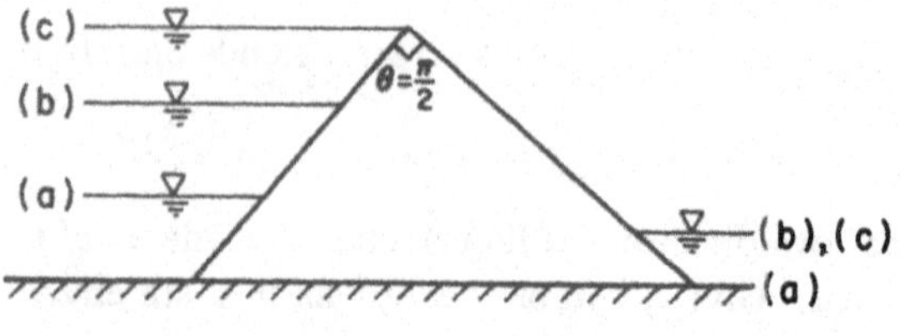

FIG. 7.E.2.

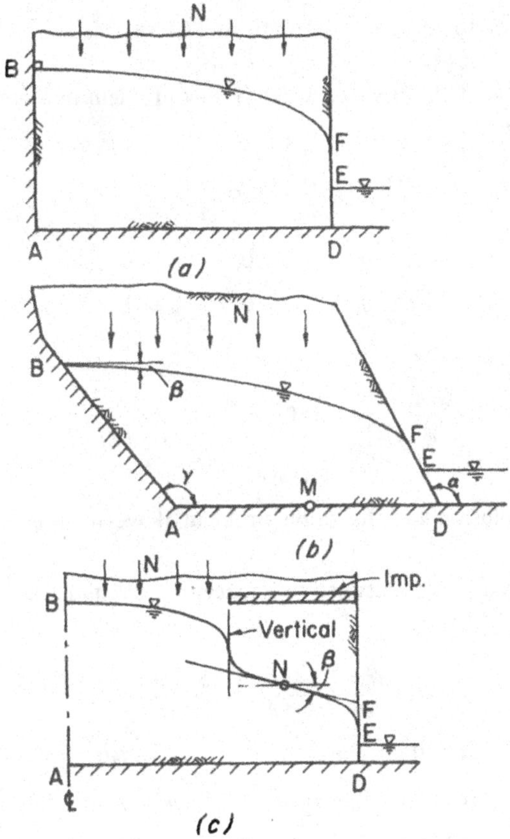

FIG. 7.E.3.

7.23 Determine the scaling conditions for problem 3 of par. 7.4.2 if we choose arbitrarily z_r, n_r (i.e., n_e) and K_e.

7.24 The approximate equation of (7.4.36) for small values of u is

$$s = \frac{Q_w}{4\pi T} \ln \frac{2.25\, Tt}{Sr^2}$$

What form will this equation take in an anisotropic aquifer ($T_x \neq T_y$)? What is the shape of the equipotentials and of the streamlines in this case.

7.25 Let the solution of a certain steady two-dimensional flow problem in an isotropic leaky aquifer be

$$Q = \frac{4KH}{\pi} \sum_{n=1,3,5,\ldots} \frac{1}{n} \left\{ \text{cotgh} \left[\frac{n\pi D}{2B} \left(1 + \frac{\sigma' K}{D} \right) \right] \right\}^{-1}$$

where H and D are lengths in the vertical (y) direction, and B is a length in the horizontal (x) direction. What will the corresponding solution be for an anisotropic domain, with $K_x \neq K_y$?

7.26 Use (7.4.26) to obtain (7.1.19) from (7.1.18).

7.27 The solution of the one-dimensional flow problem in a semi-infinite domain $x > 0$ described by

$$T \frac{\partial^2 \varphi}{\partial x^2} = S \frac{\partial \varphi}{\partial t}$$

$$t \leqslant 0, \qquad x \geqslant 0, \qquad \varphi = 0$$

$$t > 0, \qquad x = 0, \qquad \varphi = 1$$

is

$$\varphi = \frac{2}{\pi^{1/2}} \int_{x/2(Tt/S)^{1/2}}^{\infty} \exp(-\alpha^2)\, d\alpha.$$

What will be the solution for the same problem if we have $\varphi = f(t)$ at $x = 0$?

7.28 Unsteady radial flow between two concentric circular boundaries is described by

$$\frac{\partial^2 \varphi}{\partial r^2} + \frac{1}{r}\frac{\partial \varphi}{\partial r} = \frac{S}{T}\frac{\partial \varphi}{\partial t}$$

$$t > 0, \qquad r = r_1, \qquad \varphi = f_1(t)$$

$$t > 0, \qquad r = r_2, \qquad \varphi = f_2(t)$$

$$t \leqslant 0, \qquad r_1 < r < r_2, \qquad \varphi = g(r).$$

Use the principle of superposition to break this problem into three subproblems the solutions of which are φ_1, φ_1 and φ_3 and write (in principle) the solution to the given problem.

7.29 One-dimensional unsteady flow in a bounded domain $0 < x < L$ is described by

$$\frac{\partial^2 \varphi}{\partial x^2} = \frac{S}{T}\frac{\partial \varphi}{\partial t}$$

$$\text{at } x = 0, \qquad t > 0, \qquad \varphi = A$$

$$\text{at } x = L, \qquad t > 0, \qquad \varphi = B$$

$$0 < x < L, \qquad t = 0, \qquad \varphi = f(x).$$

Show how to solve this problem by superposition of a case of steady flow and a case where the ends are maintained at $\varphi = 0$.

7.30 An artificial recharge pond has the form of a square 100 m × 100 m with a horizontal bottom and vertical sides. The ground water table is at a very large depth.

Assuming shallow water in the pond (e.g., < 3 m deep) and a sandy clay with $K = 2$ m/d and $n = 0.20$, determine:

(a) The volume of water infiltrating through the pond's bottom during 5 days if the water level in the pond is maintained at a constant $H = 1$ m above the bottom.

(b) How long will it take for the pond to become empty if initially the water in the pond is 3 m deep?

7.31 Determine $\varphi = \varphi(r)$ in steady radial flow to a fully penetrating well of constant discharge in a homogeneous confined aquifer. At $r = r_w$, $\varphi = \varphi_w$; at $r = R$, $\varphi = \varphi_0$.

7.32 Repeat Ex. 7.31 for flow obeying the nonlinear law: $-\partial\varphi/\partial r = Wq_r + bq_r^2$, where W and b are constants.

7.33 A horizontal sand column of length L, cross-section A and hydraulic conductivity K connects two vertical circular reservoirs of cross sectional areas B_1 (at $x = 0$) and B_2 (at $x = L$). Determine the difference in elevations $h = h(t)$ of the water levels in the two reservoirs, if the initial difference $h(0) = H_0$.

7.34 Denote the water level in the reservoir connected to $x = 0$ by h_0 and in the reservoir connected at $x = L$ by h_L. Initially the reservoir connected at $x = 0$ is filled with fluid 1 (γ_1, μ_1) while the column and the other reservoir are filled with fluid 2 (γ_2, μ_2). Assuming that a sharp front with no capillary pressure across it separates the two fluids at every instant in the column and that flow takes place from $x = 0$ to $x = L$, determine the location of the front when h_0 and h_L are maintained constant.

7.35 An infinite confined aquifer is intersected by a straight line stream which may be considered as an equipotential boundary. The equation of the stream is $y = x$. A well of constant discharge is located at point $(d, 0)$. The aquifer is anisotropic with $T_x \neq T_y$. Develop the equation of unsteady flow to this well.

7.36 Draw images for a single well located at the center of a rectangle, three sides of which are impervious, while the fourth is an equipotential.

7.37 What is the flow described by the complex potential ζ:

(a) $$\zeta = iq_0 z$$

(b) $$\zeta = -q_0 z \exp(-i\alpha) + \frac{Q_w}{2\pi} \ln \frac{z + d}{z - d}$$

(c) $$\zeta = \ln(z^2 - d^2)$$

(d) $$\zeta = x^2 - y^2 - 2ixy.$$

7.38 What is the complex potential describing the flow in a homogeneous infinite strip bounded by $y = \pm a$ when a point source is located at the origin.

7.39 Verify the transformations:
(a) $\zeta = \sin z$ shown in fig. 7.8.19 (what is the transformation $\zeta = \sinh z$?)
(b) $\zeta = z^{\pi/\alpha}$ shown in fig. 7.8.22.

7.40 Use the Schwarz-Christoffel transformation to map the rectangle with vortices at $(\pm a/2, 0)$, $(\pm a/2, b)$ on the upper half of the ζ plane.

7.41 Use the method of relaxation to solve for φ in a square with three sides at $\varphi = 10$ m and the fourth side at $\varphi = 50$ m. Use four nodal points.

7.42 Derive the complex potential ζ for an infinite array of alternating sources and sinks spaced a distance d apart. (Place a sink at the origin.)

7.43 With $\zeta = \zeta(x, y, t)$ expressing the elevation of points on a moving phreatic surface, show that for points on this surface we have

$$\frac{n}{K}\frac{\partial \zeta}{\partial t} = \frac{\partial \varphi}{\partial x}\frac{\partial \zeta}{\partial x} + \frac{\partial \varphi}{\partial y}\frac{\partial \zeta}{\partial y} - \frac{\partial \varphi}{\partial z}.$$

What is the equivalent expression for a nonisotropic medium?

CHAPTER 8

Unconfined Flow and the Dupuit Approximation

In this chapter, we consider flows bounded above by a phreatic surface. Such flows occur in phreatic aquifers (par. 1.1.3) encountered in ground water hydrology. The thickness of the capillary fringe (par. 9.4.2) above the phreatic surface is assumed to be much smaller than that of the saturated domain below the phreatic surface.

The boundary condition on a phreatic surface was given in paragraphs 7.1.3 through 7.1.6. The nonlinearity of this boundary condition, together with the fact that the location of this boundary is a priori unknown and is, in fact, part of the required solution, makes an exact analytical solution of a flow problem with such a boundary most difficult, if not practically impossible, in all but a very limited number of cases. Some two-dimensional steady flow problems may be solved by the hodograph method (sec. 8.3). Numerical methods (sec. 7.9) are often employed. A way to circumvent some of the difficulties is to derive analytically approximate solutions based on a linearization of the boundary conditions and/or the nonlinear continuity equations describing unconfined flows. Some of these methods are discussed in the present chapter.

8.1 The Dupuit Approximation

The *Dupuit approximation* is among the most powerful tools for treating unconfined flows. In fact, it is the only simple tool available to most engineers and hydrologists for solving such problems.

8.1.1 The Dupuit Assumptions

Dupuit (1863) developed a theory based on a number of simplifying assumptions resulting from the observation that in most ground water flows the slope of the phreatic surface is very small. In steady two-dimensional unconfined flow without accretion in the vertical xz plane, the phreatic surface is a streamline. At every point P along it, the specific discharge q_s (fig. 8.1.1a) is given by Darcy's law:

$$q_s = -K\, d\varphi/ds = -K\, dz/ds = -K \sin\theta. \qquad (8.1.1)$$

As θ is very small, Dupuit suggested that $\sin\theta$ be replaced by the slope $\tan\theta = dh/dx$. The assumption of a small θ is equivalent to assuming that *equipotential surfaces are vertical* (i.e., $\varphi = \varphi(x)$ is independent of z) and the *flow essentially horizontal*, or to assuming that we have a *hydrostatic pressure distribution*. Thus the Dupuit assumptions lead to the specific discharge expressed by:

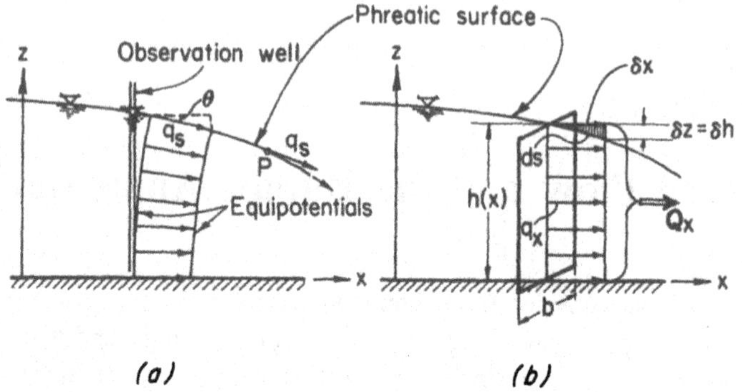

FIG. 8.1.1. The Dupuit assumptions.

$$q_x = - K\, dh/dx; \qquad h = h(x) \tag{8.1.2}$$

and to the total discharge through any vertical surface of width b (fig. 8.1.1b):

$$Q_x = - Kbh(x)\, dh/dx. \tag{8.1.3}$$

It should be emphasized that all these assumptions may be considered as good approximations in regions where θ is indeed small and the flow essentially horizontal.

The important advantage gained by employing the Dupuit assumptions is that the number of independent variables of the original problem (x, z) has been reduced by one; in (8.1.3) z does not appear as an independent variable. We have here an extension of what is known as the *hydraulic approach* to fluid flows. In hydraulic flows, also called *one-dimensional flows*, we neglect in a nonuniform flow variations or changes in velocity, pressure, etc., transverse to the main flow direction. In figure 8.1.1 this is the x direction. At every cross-section perpendicular to the flow direction, conditions are expressed in terms of average values of velocity, density and other properties over the cross-section. The whole flow is considered as a *single streamtube*. This is the engineering approach that is most useful in treating pipe flow, flow in open channels, etc. Here we consider a single streamtube bounded by two streamlines: the phreatic surface and the impervious bottom; the cross-sections of interest are vertical. Thus, the usual dependent variable, the piezometric head $\varphi = \varphi(x, z)$, is replaced by another variable $h(x)$. Also, since at a point on the free surface $p = 0$ and $\varphi = h$, we assume that the vertical cross-sections are equipotential surfaces on which $\varphi = h =$ constant.

The Dupuit assumptions actually amount to neglecting the vertical flow component $q_z = - K\, \partial\varphi/\partial z$. The value of q_z varies from $q_z = 0$ along the horizontal impervious boundary, to $q_z = - K\, \partial\varphi/\partial z = - K \sin^2\theta$ along the phreatic surface.

Let us obtain an estimate of the error introduced in determining $h = h(x)$ by the Dupuit assumptions. From the exact expression for the discharge per unit width (q'_x) through a phreatic aquifer, we obtain:

$$q'_x = -K_x \int_0^{h(x)} [\partial\varphi(x, z)/\partial x]\, dz = -K_x \frac{\partial}{\partial x}\left[\int_0^{h(x)} \varphi(x, z)\, dz - h^2/2\right]$$

$$= -K_x\, \partial\varphi'/\partial x; \qquad \varphi' = h\bar{\varphi} - h^2/2; \qquad h\bar{\varphi} = \int_0^{h(x)} \varphi(x, z)\, dz \tag{8.1.4}$$

since, by Leibnitz' rule of integration,

$$\frac{\partial}{\partial x}\left[\int_0^{h(x)} \varphi(x, z)\, dz\right] = \int_0^{h(x)} (\partial\varphi/\partial x)\, dz + \varphi(x, h)\, \partial h/\partial x.$$

According to the Dupuit assumptions:

$$q'_x = -K_x h\, \partial h/\partial x = -K_x\, \partial(h^2/2)/\partial x \tag{8.1.5}$$

so that actually we have replaced φ' by $h^2/2$ in the Dupuit assumptions.

Integrating by parts the expression for φ' in (8.1.4), we obtain:

$$\varphi' = \int_0^{h(x)} \varphi(x, z)\, dz - h^2/2 = z\varphi(x, z)\Big|_0^{h(x)} - \int_0^{h(x)} z(\partial\varphi/\partial z)\, dz - h^2(x)/2$$

$$= \frac{h^2(x)}{2}\left[1 + \frac{2}{K_z h^2(x)} \int_0^{h(x)} z q_z(x, z)\, dz\right] \tag{8.1.6}$$

where $q_z = q_z(z) = -K_z\, \partial\varphi/\partial z$, $q_z(h) < q_z < 0$. Since along the steady phreatic surface $\varphi = h$, we have:

$$d\varphi/dx \equiv dh/dx = (\partial\varphi/\partial x)|_{z=h} + (\partial\varphi/\partial z)|_{z=h}\, dh/dx = -(q_x/K_x) - (q_z/K_z)\, dh/dx$$

where q_x and q_z are evaluated at $z = h$. Since $q_z/q_x = dh/dx$, we obtain:

$$q_z|_{z=h} = -K_x(dh/dx)^2/[1 + (K_x/K_z)(dh/dx)^2]. \tag{8.1.7}$$

Hence, the range for the integral term in (8.1.6) is:

$$0 > \int_0^{h(x)} z q_z(x, z)\, dz > -\frac{K_x i^2}{1 + (K_x/K_z)i^2} \cdot \frac{h^2}{2}; \qquad i \equiv dh/dx$$

and therefore:

$$0 < \frac{h^2/2 - \varphi'}{h^2/2} < \frac{(K_x/K_z)i^2}{1 + (K_x/K_z)i^2}. \tag{8.1.8}$$

The error in replacing φ' in the exact expression (8.1.4) by $h^2/2$ in the Dupuit expression (8.1.5) is small as long as $(K_x/K_z)i^2 \ll 1$, where i gives the slope of the phreatic surface.

The Girinskii potential Φ^g presented in (5.8.4) is an extension of the definition of $\varphi'(x)$ to a stratified aquifer, with $K = K(z)$.

Kirkham (1967) indicates paradoxes that are inherent in the Dupuit assumptions (and the *Dupuit–Forchheimer theory* (par. 8.2.1) based on these assumptions) as, for example, water leaving the flow domain through the sloping phreatic surface (fig. 8.1.1b) because of the horizontal flow assumptions. He suggests a slab-slot model (fig. 8.1.3) obtained by cutting a large number of vertical, parallel, infinitely permeable, equally spaced, infinitesimally thin slots into a porous medium. Between adjacent slots we have the slab-column of soil. Together, the slabs and the slots produce a fictitious soil that follows exactly and without paradoxes the Dupuit assumptions, and hence leads to the Dupuit–Forchheimer drainage theory in two dimensions, discussed in paragraph 8.2.1.

Although essentially a one-dimensional approach, the Dupuit approximation, as described above, is also extended to cases where both q_x and q_y are nonzero, yet q_z is sufficiently small to be neglected. As in the one-dimensional case considered above, the assumption is that equipotential surfaces are vertical, the flow is horizontal (i.e., in xy plane) and the slope of the water table at some point on it represents the constant hydraulic gradient along a vertical line passing through this point. This is expressed for an isotropic medium as:

$$q_x = -K\,\partial h/\partial x; \qquad q_y = -K\,\partial h/\partial y; \qquad h = h(x, y) \tag{8.1.9}$$

where both q_x and q_y are constant along vertical lines as they are independent of z.

Although originally applied to steady phreatic surface flows without accretion, the Dupuit approximation has subsequently been applied to nonsteady flows where $h = h(x, y, t)$ (Boussinesq 1903, 1904) and to flows where accretion takes place (sec. 8.2). The observation well in figure 8.1.1a indicates the error introduced.

The Dupuit simplification cannot be applied to regions where the vertical flow component cannot be neglected. Such flow conditions occur as a seepage face is approached (fig. 8.1.2a) or at a crest in a phreatic surface with accretion (fig. 8.1.2b). Another example is the region close to the impervious vertical boundary of figure 8.1.2a. It is obvious that the assumption of vertical equipotentials fails at, and in the vicinity of, such a boundary. Only at distances $x > \sim 2h_0$ have we equipotentials that may be approximated by vertical lines or surfaces. It is important to note

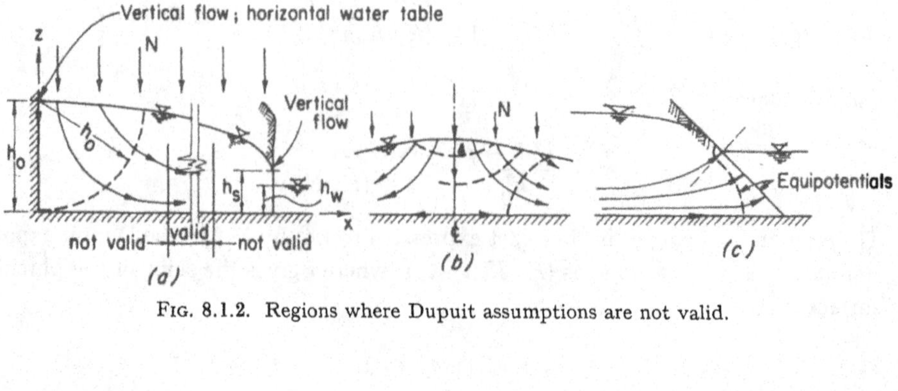

FIG. 8.1.2. Regions where Dupuit assumptions are not valid.

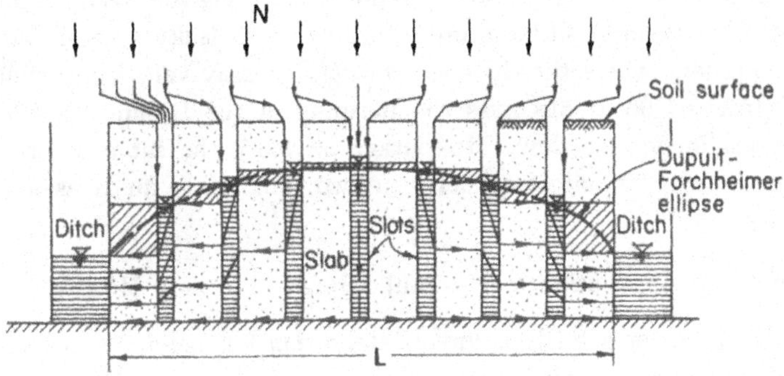

FIG. 8.1.3. Kirkham's slot-slab model for a rational explanation of the Dupuit-Forchheimer drainage theory.

here that in cases with accretion it is not sufficient to observe that the slope of the phreatic surface is very small (or sometimes even zero) in order to apply the Dupuit assumptions. One must verify that the vertical flow component may indeed be neglected before applying the Dupuit assumptions. Another case to which the Dupuit assumptions should be applied with care is that of unsteady flow in a decaying phreatic surface mound. In this case no accretion takes place, yet around the crest the flow is strictly vertically downward (fig. 8.1.2b). At some distance from the crest, say $1.5 \div 2$ times the value of h at the crest, the approximation of instantaneous vertical equipotentials may again be valid. One should note that all figures here, as in fact most figures describing aquifer flows, are highly distorted.

From the discussion above, one may conclude that the Dupuit assumptions, and the hydraulic approach, are applicable wherever the lengths of interest in the direction of flow are much larger (say $> 1.5 \div 2$) than the thickness of the saturated layer. We shall refer to such flows as *shallow flows*. The method of small perturbations may be used to show that the Boussinesq continuity equations (sec. 8.2), which are based on the Dupuit assumptions, may in fact be considered as a first approximation of the exact equations (Dagan 1964). The small parameter in the series development appearing in this method is $(D/L)^2$, where D is a characteristic thickness and L is a characteristic horizontal length. Dagan (1968) derived a Dupuit solution of steady flow toward wells by matched asymptotic expansions.

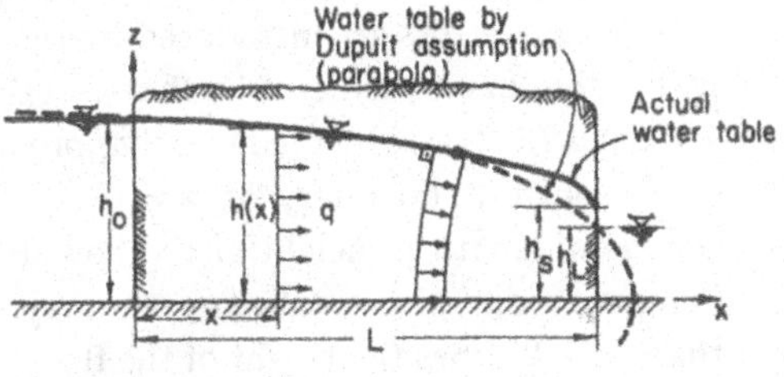

FIG. 8.1.4. Steady unconfined flow between two reservoirs.

The assumptions of essentially horizontal flow (neglecting vertical flow components) and vertical equipotential surfaces are also applied to leaky aquifers and to flow in confined aquifers, where the thickness $b = b(x, y)$ may vary from point to point (sec. 6.4). Irmay (1966a) discusses the meaning of the Dupuit and Pavlovskii's approximations in aquifer flow. In paragraph 9.5.7 the same assumptions are applied to flow with an interface between two immiscible fluids as a portion of the boundary.

8.1.2 Examples of Application to Hydraulic Steady Flows in Homogeneous Media

Equation (8.1.3) can be solved by direct integration for one-dimensional flows and for radially converging or diverging flows.

Consider the flow per unit width between the two reservoirs (vertical faces) shown in figure 8.1.4. Following the Dupuit assumptions, the total discharge (per unit width) through a vertical cross-section is:

$$Q = qh(x) = -Kh(x)\,dh/dx = \text{constant}. \tag{8.1.10}$$

Note that Q in (8.1.10) is the same as q'_x in (8.1.4). Integrating between $x = 0$ and any distance x, we obtain:

$$Q\,dx = -Kh(x)\,dh; \qquad Q\int_{x'=0}^{x} dx' = -K\int_{h'=h_0}^{h} h'\,dh'$$

$$Q = -K(h_0^2 - h^2)/2x; \qquad h^2 = h_0^2 - 2Qx/K. \tag{8.1.11}$$

With $h = h_L$ at $x = L$, we obtain:

$$Q = K(h_0^2 - h_L^2)/2L; \qquad h_L^2 = h_0^2 - 2QL/K \tag{8.1.12}$$

known as the Dupuit–Forchheimer discharge formula.

Equation (8.1.11) describes a parabolic water table $h = h(x)$ that passes through $h = h_0$ at $x = 0$. At $x = 0$ and $X = L$ it has the slopes:

$$dh/dx|_{x=L} = -Q/Kh_L; \qquad dh/dx|_{x=0} = -Q/Kh_0. \tag{8.1.13}$$

This means that at $x = 0$ there is an inconsistency, as the vertical boundary at $x = 0$ is an equipotential surface and the water table there should have a horizontal tangent. We also have an inconsistency at the downstream boundary $x = L$. The seepage surface between h_L and h_s is absent in the solution, and the water table (which is a streamline) makes an angle with the vertical equipotential boundary that is different from 90°. All these errors are introduced by the Dupuit approximation. The discrepancy between the curves predicted by the exact theory of the phreatic surface boundary (pars. 7.1.3 and 7.3.2) and by the Dupuit approximation is negligible except in the vicinity of the outflow boundary at $x = L$. As the flow becomes more shallow, i.e., as the ratio of length to height of the flow domain increases, the region of appreciable discrepancy diminishes. Again, one may use the simple rule that at distances larger than 1.5÷2 times the height of the flow domain, the solution based on the Dupuit assumptions is sufficiently accurate for practical purposes.

Consider the case of flow to a single trench. This is similar to the case illustrated by figure 8.1.4, except that the boundary at $x = 0$ does not exist. By integrating (8.1.10) from some arbitrary x to $x = L$ where $h = h_L$, we obtain:

$$h^2 = h_L{}^2 + 2Q(x_L - x)/K. \tag{8.1.14}$$

As $x \to -\infty$, $h \to \infty$, which is obviously impossible. We therefore conclude that steady flow in an infinite aquifer without a source of one kind or another is not possible.

The exact discharge per unit width (i.e., without using the Dupuit assumptions) through a vertical cross-section of height $h(x)$ in a homogeneous porous medium is given by:

$$Q = \int_0^{h(x)} q_x \, dz = -K \int_0^{h(x)} \frac{\partial \varphi(x, z)}{\partial x} dz = -K \frac{\partial}{\partial x}\left[\int_0^{h(x)} \varphi(x, z)\, dz - \frac{h^2}{2}\right] = -K \frac{\partial \varphi'}{\partial x} \tag{8.1.15}$$

where φ' is defined by (8.1.4).

Integrating with respect to x, we obtain (Charni 1951):

$$(Q/K)x = -\int_0^{h(x)} \varphi(x, z)\, dz + h^2/2 + C. \tag{8.1.16}$$

At $x = 0$, $\varphi = h(x) = h_0$, hence:

$$C = \int_0^{h_0} \varphi(0, z)\, dz - h_0{}^2/2 = h_0{}^2/2; \qquad (Q/K)x = -\int_0^{h(x)} \varphi(x, z)\, dz + h^2/2 + h_0{}^2/2 \tag{8.1.17}$$

valid for every x. At $x = L$: $h = h_s$ and $\varphi(L, z) = h_L$ for $0 < z \leqslant h_L$; $\varphi = z$, for $h_L \leqslant z \leqslant h_s$. Hence:

$$(Q/K)L = -\int_0^{h_s} \varphi(x, z)\, dz + \frac{h_s{}^2}{2} + \frac{h_0{}^2}{2} = -\int_0^{h_L} h_L\, dz - \int_{h_L}^{h_s} z\, dz + \frac{h_s{}^2}{2} + \frac{h_0{}^2}{2} = \frac{h_0{}^2}{2} - \frac{h_L{}^2}{2}$$

or:

$$Q = \frac{K}{2L}(h_0{}^2 - h_L{}^2) \tag{8.1.18}$$

which is the same as (8.1.12). An interesting conclusion of this result is that although the shape of the water table $h = h(x)$, as given by (8.1.11), is only an approximate one (e.g., the seepage face is neglected), the discharge given by (8.1.18) is an *exact* expression.

In a similar way it can be shown that the Dupuit–Forchheimer formula for radially converging flow to a well in a phreatic aquifer:

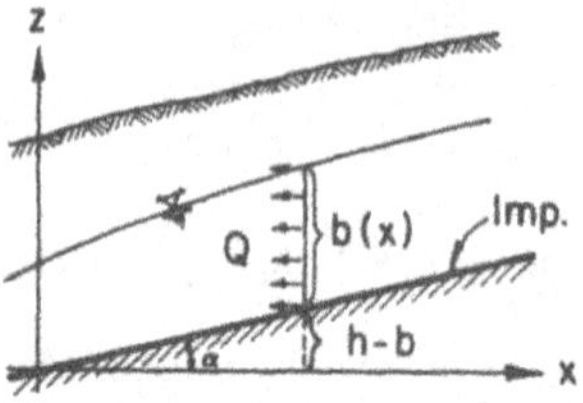

FIG. 8.1.5. Unconfined flow on an inclined bottom.

$$Q_w = \pi K(h_0^2 - h_w^2)/\ln(r_w/R) \tag{8.1.19}$$

obtained by employing the Dupuit assumptions with $h = h_w$ at $r = r_w$ and $h = h_0$ at $r = R$, is the exact expression for Q_w.

It appears that when the flow domain is bounded (in the horizontal direction) by vertical equipotential boundaries that extend to the impervious bottom, the error inherent in the Dupuit approximation is just balanced by the error introduced in the boundary condition by ignoring the seepage surface.

It is of interest to note that (8.1.18) may also be written as:

$$Q = K\frac{h_0 + h_L}{2} \cdot \frac{h_0 - h_L}{L} = K\bar{h}\frac{\Delta h}{L}; \qquad \bar{h} = \frac{h_0 + h_L}{2}; \qquad \Delta h = h_0 - h_L$$

i.e., as if we had a confined flow of thickness $\bar{h}$ caused by the driving head $\Delta h/L$.

A solution by direct integration can also be obtained for the case of unconfined flow on a sloping flat impervious bottom that makes an angle α with the horizontal x axis (fig. 8.1.5). Using $b = b(x)$ to denote the thickness of the saturated flow, $h = h(x)$ to denote the height of the phreatic surface above a horizontal datum level, and assuming a small slope with essentially horizontal flow, we obtain:

$$Q = Kb(x)\, dh/dx. \tag{8.1.20}$$

With $h - b = x \tan \alpha$, we obtain from (8.1.20):

$$Q = K(h - x \tan \alpha)\, dh/dx; \qquad dx/dh + (K/Q)x \tan \alpha = Kh/Q. \tag{8.1.21}$$

This is a linear differential equation whose general solution is:

$$x = A \exp\{-(K/Q)h \tan \alpha\} + (Kh \tan \alpha - Q)/K \tan^2\alpha \tag{8.1.22}$$

or:

$$h - x \tan \alpha = B \exp(-Kh \tan \alpha/Q) \tag{8.1.23}$$

where B is a constant. For $\tan \alpha = dh/dx$, $B = 0$. This is the case of uniform flow with a plane water table parallel everywhere to the impervious bottom.

Several cases of $B \neq 0$, $\tan \alpha > 0$ or $\tan \alpha < 0$ are presented by Polubarinova-Kochina (1952, 1962).

Dupuit (1863) and Pavlovsky (1931) derive flow equations in terms of the coordinate s along the bottom (fig. 8.1.6), using the Dupuit assumptions together with the assumption that along the vertical, which is an equipotential $\varphi = h$, the specific

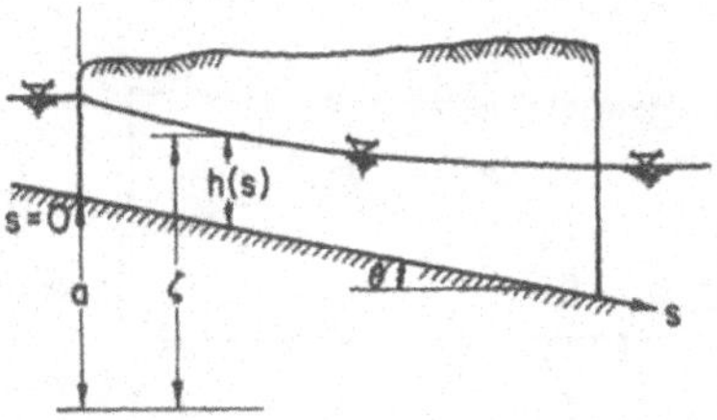

FIG. 8.1.6. Flow with an inclined bottom between two reservoirs.

discharge is constant, parallel to the bottom, and expressed by:

$$q_s = -K\, d\zeta/ds. \tag{8.1.24}$$

This is somewhat different from the Dupuit assumptions as stated above; however, the approximation is better as the direction of s is closer to that of the average streamline. The total discharge is given by:

$$Q = -Kh(s)\, d\zeta/ds; \qquad h(s) = \zeta - (a - i^*s), \qquad i^* = \sin\theta \tag{8.1.25}$$

and therefore:

$$Q = -Kh(dh/ds - i^*). \tag{8.1.26}$$

Note that $i^* > 0$ means a downward bottom slope. For a horizontal bed, $i^* = 0$. The analysis for the various types of water table shapes (see Aravin and Numerov 1953, 1965) is obtained by integrating (8.1.26).

(a) For $i^* > 0$ (downward bottom slope), we obtain:

$$i^* = \frac{h}{H^*}\left(i^* - \frac{dh}{ds}\right) = \eta(i^* - H^*\, d\eta/ds); \qquad H^* = Q/Ki^*; \qquad \eta = h/H^*$$

$$(i^*/H^*)\, ds = [\eta/(\eta - 1)]\, d\eta = d\eta + d\eta/(\eta - 1) \tag{8.1.27}$$

which, upon integration between (s_1, η_1) and (s_2, η_2), yields:

$$(i^*/H^*)(s_2 - s_1) = \eta_2 - \eta_1 + \ln[(\eta_2 - 1)/(\eta_1 - 1)]. \tag{8.1.28}$$

(b) For $i^* < 0$ (rising bottom), we obtain:

$$(i^{*\prime}/H^{*\prime})(s_2 - s_1) = \eta'_1 - \eta'_2 + \ln[(1 + \eta'_2)/(1 + \eta'_1)] \tag{8.1.29}$$

where $i^{*\prime} = |i^*|$; $\eta' = h/H^{*\prime}$; $H^{*\prime} = Q/Ki^{*\prime}$. Figure 8.1.6 gives an example of a water table obtained from (8.1.28).

8.1.3 Unconfined Flow in an Aquifer with Horizontal Stratification

Consider the case of two horizontal layers (K', K'') connecting two reservoirs (fig. 8.1.7).

Case 1, $a < h_{L1} < h_s$. The flow is made up of a sum of two parts, each corresponding to one of the layers. Using the Dupuit assumptions, we obtain:

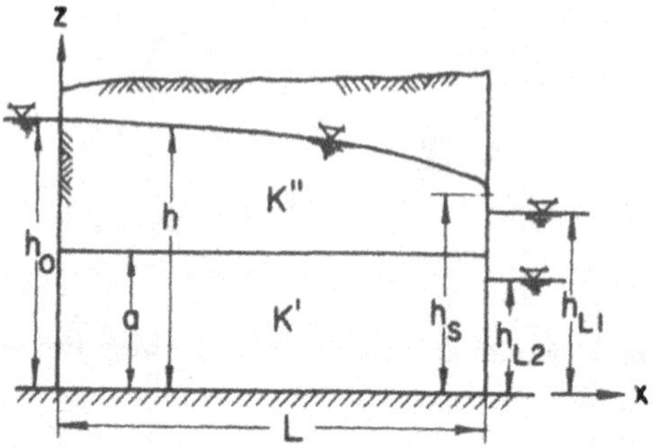

FIG. 8.1.7. Unconfined flow in a horizontally stratified aquifer.

$$Q = -K'a\,dh/dx - K''(h-a)\,dh/dx. \tag{8.1.30}$$

It is of interest to note at this point that (8.1.30) can also be obtained directly from $Q = -\partial\Phi^g/\partial x$, where Φ^g is Girinskii's potential (par. 5.8.4).

Upon integration, with $x = 0$, $h = h_0$; $x = L$, $h = h_{L1}$ (thus neglecting the presence of a seepage face), we obtain:

$$QL = -K'a(h_{L1} - h_0) - K''(h_{L1}^2 - h_0^2)/2 + K''a(h_{L1} - h_0)$$

or:

$$Q = \frac{K''}{2L}(h_0 - h_{L1})\left[h_0 + h_{L1} - 2a + 2\frac{K'}{K''}a\right]. \tag{8.1.31}$$

We could also integrate to any distance x and obtain an approximate shape for the phreatic surface.

Following the procedure outlined in paragraph 8.1.2 of examining the discharge in an exact manner, we write:

$$Q = -K'\int_0^a \frac{\partial\varphi(x,z)}{\partial x}\,dz - K''\int_a^h \frac{\partial\varphi(x,z)}{\partial x}\,dz$$

$$= -K'\frac{d}{dx}\int_0^a \varphi(x,z)\,dz - K''\left[\frac{d}{dx}\int_a^h \varphi(x,z)\,dz - \varphi(x,h)\frac{dh}{dx}\right].$$

By integration:

$$Qx = -K'\int_0^a \varphi(x,z)\,dz - K''\int_a^h \varphi(x,z)\,dz + \frac{K''}{2}h^2 + C.$$

At $x = 0$, $0 \leqslant z \leqslant h_0$: $\varphi(x,z) = h_0$. Hence:

$$C = K'h_0a + K''h_0(h_0 - a) - \frac{K''}{2}h_0^2 = ah_0(K' - K'') + K''\frac{h_0^2}{2}.$$

Therefore:

$$Qx = -K' \int_0^a \varphi(x, z)\, dz - K'' \int_a^h \varphi(x, z)\, dz + \frac{K''}{2} h^2 + ah_0(K' - K'') + K'' \frac{h_0^2}{2}.$$

Now, at $x = L$:

$$\varphi(x, z) = h_{L1}, \quad 0 \leqslant z \leqslant h_{L1}; \qquad \varphi(x, z) = z, \quad h_{L1} < z \leqslant h_s. \tag{8.1.32}$$

Hence:

$$QL = -K' \int_0^a h_{L1}\, dz - K'' \int_a^{h_{L1}} h_{L1}\, dz - K'' \int_{h_{L1}}^{h_s} z\, dz + \frac{K''}{2} h_s^2 + ah_0(K' - K'') + K'' \frac{h_0^2}{2}$$

leading to (8.1.31), so that again (8.1.31) is an exact expression for Q.

Case 2, $h_{L2} < a < h_s$. In this case, we may still write (8.1.30), and integrate between $x = 0$ and any $x = L_1$, to obtain the shape of the phreatic surface between them. However, as $x \to L$, the Dupuit approximation is no longer valid.

In the exact determination of Q in case 1, we may replace (8.1.32) by:

$$x = L: \quad \varphi(x, z) = h_{L2}, \quad 0 \leqslant z \leqslant h_{L2}; \qquad \varphi(x, z) = z, \quad h_{L2} \leqslant z \leqslant h_s.$$

This leads to:

$$Q = \frac{K''}{2L} (h_0 - a)^2 + \frac{K'}{L} a \left(h_0 - \frac{h_{L2} + a}{2} \right) + \frac{K'}{L} h_{L2} \frac{(a - h_{L2})}{2}$$

or:

$$Q = \frac{K''}{2L} (h_0 - a)^2 + \frac{K'}{L} a(h_0 - a) + \frac{K'}{L} \frac{a^2 - h_{L2}^2}{2}. \tag{8.1.33}$$

In (8.1.33), one may think of the first term as representing the flow in the upper layer, the second term as representing the flow in the lower layer, and the third term as a correction that vanishes when $a = h_{L2}$. Outmans (1964) analyzes a general case of horizontal stratification with layers of different hydraulic conductivity. His basic equation is Charni's (1951) equation (8.1.16), written as a sum along a vertical line.

Equation (8.1.31) may also be written as:

$$Q = \frac{K''}{2L} (h_0 - h_{L1}) \left[h_0 + h_{L1} - 2a + \frac{2K'}{K''} a \right]$$

$$= K'a \frac{h_0 - h_{L1}}{L} + \frac{K''}{2L} [(h_0 - a)^2 - (h_{L1} - a)^2].$$

As in (8.1.30), this shows again that when the Dupuit assumptions are employed, the total flow is obtained as the sum of a confined-type flow in the lower layer and a phreatic flow in the upper one. Now consider phreatic flow in a layer of hydraulic

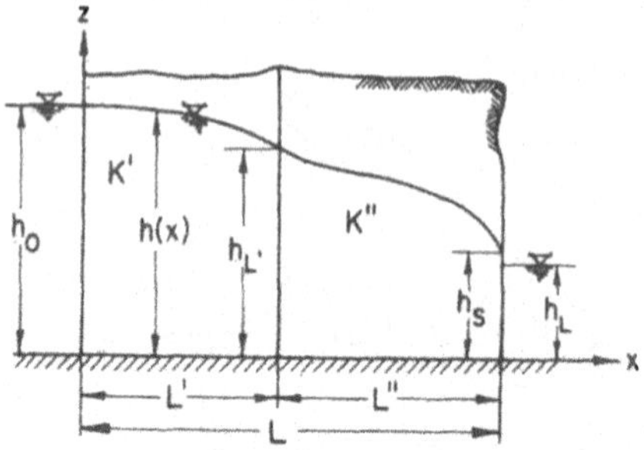

FIG. 8.1.8. Unconfined flow through a vertically stratified aquifer.

conductivity K'. By inserting $K' = K''$ in the last equation, we may conclude that the discharge between the two reservoirs of figure 8.1.7 can always be obtained as a sum of a confined flow of an arbitrary thickness a and a phreatic flow above it. This conclusion remains valid when we have a real separation by some thin impervious layer, creating a confined layer of thickness a underlying a layer with phreatic flow. Although the head distribution is different in the two layers, the resulting equation for discharge remains unchanged. Note that:

$$Q = \frac{K}{2L}(h_0^2 - h_L^2) = Ka\left(\frac{h_0 - h_L}{L}\right) + \frac{K}{2L}[(h_0 - a)^2 - (h_L - a)^2].$$

8.1.4 Unconfined Flow in an Aquifer with Vertical Strata

Consider the flow shown in figure 8.1.8. Using the Dupuit assumptions, we obtain by (8.1.11), for the region $0 \leqslant x \leqslant L'$:

$$h^2 = h_0^2 - \frac{2Qx}{K'}; \qquad h_{L'}^2 = h_0^2 - \frac{2QL'}{K'}; \qquad 0 \leqslant x \leqslant L' \tag{8.1.34}$$

where $h_{L'}$ is as yet unknown. For the right-hand side we have, again by (8.1.11):

$$h^2 = h_{L'}^2 - \frac{2Q(x - L')}{K''}; \qquad h_L^2 = h_{L'}^2 - \frac{2Q(L - L')}{K''};$$

$$h_{L'}^2 = h_L^2 + \frac{2QL''}{K''}; \qquad L' \leqslant x \leqslant L. \tag{8.1.35}$$

Hence:

$$Q = \frac{h_0^2 - h_L^2}{2[(L''/K'') + (L'/K')]}; \qquad h^2 = h_0^2 - \frac{(h_0^2 - h_L^2)}{K'[(L''/K'') + (L'/K')]}x; \qquad 0 < x < L' \tag{8.1.36}$$

$$h^2 = h_L^2 + \frac{h_0^2 - h_L^2}{K''[(L''/K'') + (L'/K')]}(x - L), \qquad L' < x < L. \tag{8.1.37}$$

It is interesting to note in (8.1.36) how the two-layer resistances are combined in series to form the total resistance of the stratified aquifer.

Following the procedure of determining the discharge in an exact manner, we start from (8.1.4) written for the K'-region:

$$Q = K' \frac{\partial}{\partial x}\left[\frac{h^2}{2} - \int_0^{h(x)} \varphi(x, z)\, dz\right]; \qquad \frac{Q}{K'} x = \tfrac{1}{2}h^2 - \int_0^{h(x)} \varphi(x, z)\, dz + C.$$

At $x = 0$, $\varphi(0, z) = h_0$, hence $C = -h_0^2/2$. Therefore we have:

$$\frac{Qx}{K'} = \frac{h_0^2}{2} + \frac{h^2}{2} - \int_0^h \varphi(x, z)\, dz, \qquad 0 < x \leqslant L'. \tag{8.1.38}$$

We now write (8.1.4) for the K'' region and integrate it for $x = L$ with: $\varphi(x, z) = h_L$ for $0 \leqslant z \leqslant h_L$ and $\varphi(x, z) = z$ for $h_L \leqslant z \leqslant h_s$. We obtain:

$$\frac{Q}{K''} = \frac{\partial}{\partial x}\left[\left(\frac{h^2}{2}\right) - \int_0^h \varphi(x, z)\, dz\right]; \qquad \frac{Q}{K''} x = \tfrac{1}{2}h^2 - \int_0^h \varphi(x, z)\, dz + C; \qquad L' < x < L \tag{8.1.39}$$

$$\frac{QL}{K''} = \tfrac{1}{2}h_s^2 - \int_0^{h_L} h_L\, dz - \int_{h_L}^{h_s} z\, dz + C = -\frac{h_L^2}{2} + C; \qquad C = \frac{QL}{K''} + \frac{h_L^2}{2}$$

$$\frac{Qx}{K''} = \frac{h^2}{2} - \int_0^h \varphi(x, z)\, dz + \frac{QL}{K''} + \frac{h_L^2}{2}, \qquad L' \leqslant x \leqslant L. \tag{8.1.40}$$

At $x = L'$ the upper limit of the integral should be identical in both (8.1.38) and (8.1.40). Hence, eliminating the integral from the two equations, we obtain:

$$Q\left(\frac{L'}{K'} + \frac{L''}{K''}\right) = \frac{h_0^2}{2} - \frac{h_L^2}{2}; \qquad Q = \frac{h_0^2 - h_L^2}{2[(L'/K') + (L''/K'')]}. \tag{8.1.41}$$

Again, the expression for Q derived by the Dupuit approximation is identical to the one based on an exact integration.

The discussion above can easily be extended to a larger number of strata. Outmans (1964) shows that for n strata, we obtain:

$$Q = (h_0^2 - h_L^2)/2 \sum_{i=1}^{n} (L_i/K_i). \tag{8.1.42}$$

8.1.5 Unconfined Flow in a Two-Dimensional Inhomogeneous Medium

The discussion in paragraphs 8.1.3 and 8.1.4 may be combined by considering the general case of flow with an impervious horizontal bottom and $K = K(x, z)$.

Since by Dupuit's assumptions $\partial\varphi/\partial x$ $(= \partial h/\partial x)$ is independent of z, we have:

$$Q = -\int_{z=0}^{z=h(x)} K(x, z') \frac{\partial \varphi}{\partial x} dz' = -\frac{\partial h}{\partial x} \int_{z=0}^{z=h(x)} K(x, z')\, dz' \tag{8.1.43}$$

Since $h = h(x)$, $\partial h/\partial x = dh/dx$, we obtain:

$$Q\, dx = -\left[\int_{z=0}^{z=h(x)} K(x, z')\, dz'\right] dh = -h\bar{K}(x)\, dh; \quad \bar{K}(x) = [1/h(x)] \int_0^{h(x)} K(x, z')\, dz'. \tag{8.1.44}$$

With $h(0) = H_0$ and $H(L) = H_L$, and $Q =$ constant, equation (8.1.44) may be integrated to yield:

$$Q \int_{x=0}^{L} \frac{1}{\bar{K}(x)} dx = -\frac{h^2}{2}\Big|_{H_0}^{H_L} = \frac{H_0^2 - H_L^2}{2}. \tag{8.1.45}$$

Thus, by comparing (8.1.45) with (8.1.18) for a homogeneous medium, it is possible to define an *equivalent hydraulic conductivity* of the saturated block of soil between $x = 0$ and $x = L$, by:

$$\frac{1}{K_{eq}} = \frac{1}{L} \int_0^L \frac{1}{\bar{K}(x)} dx = \frac{1}{L} \int_0^L \left[h(x) \Big/ \int_0^{h(x)} K(x, z')\, dz'\right] dx. \tag{8.1.46}$$

8.2 Continuity Equations Based on the Dupuit Approximation

Equations (8.1.9), which express the Dupuit assumptions, contain three unknowns: $h = h(x, y, t)$, $q_x = q_x(x, y, t)$ and $q_y = q_y(x, y, t)$, for which only two equations are available. As usual, in order to obtain a solution, we must introduce another equation—the continuity equation.

In the following paragraphs we shall derive the *continuity equation for phreatic flow*, assuming that the Dupuit assumptions are valid.

8.2.1 The Continuity Equation

Consider the fluid balance for the control box bounded by vertical surfaces at $x - \delta x/2$, $x + \delta x/2$, $y - \delta y/2$ and $y + \delta y/2$ as shown in figure 8.2.1. From below, the box is bounded by the horizontal impervious bottom of the aquifer, while from above it is bounded by the phreatic surface. Following the procedure described in paragraph 6.2.1, with q' the specific discharge through the entire saturated thickness (per unit width of aquifer; dim L^2/T), we obtain the excess mass inflow over outflow in the x and y directions during a period of time δt:

$$\delta y\,[(\rho q'_x)|_{x-\delta x/2, y} - (\rho q'_x)|_{x+\delta x/2, y}]\, \delta t + \delta x\,[(\rho q'_y)|_{x, y-\delta y/2} - (\rho q'_y)|_{x, y+\delta y/2}]\, \delta t.$$

Only inflow from accretion takes place in the z-direction:

$$\rho N(x, y, t)\, \delta x\, \delta y\, \delta t$$

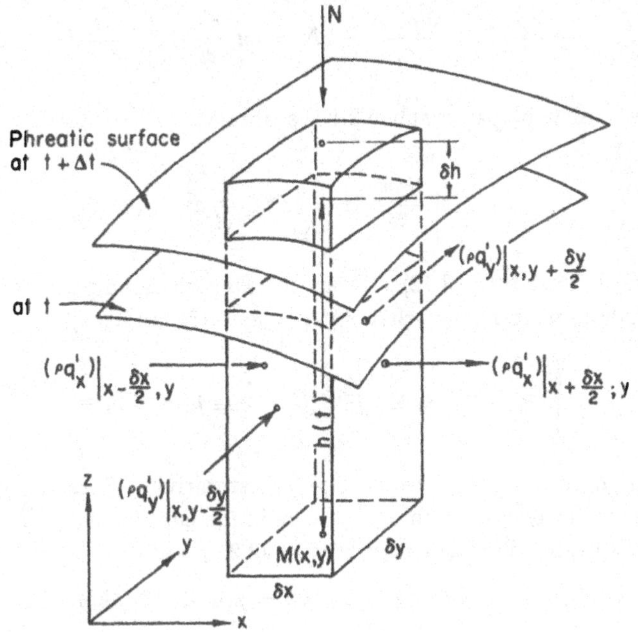

FIG. 8.2.1. Nomenclature for the continuity equation for phreatic flow with a horizontal impervious base.

where $N > 0$ (positive downward) means accretion; $N < 0$ means evaporation and/or transpiration from the phreatic surface.

The excess of inflow over outflow goes into storage in the control box in two ways. First, the phreatic surface rises by a vertical distance δh (measured at the center of the box) so that a volume of porous medium, $\delta x \, \delta y \, \delta h$, becomes saturated. The volume of water absorbed by this volume of porous medium depends on the initial moisture in the porous medium. Using the concept of specific yield (S_y), or the equivalent one of effective porosity (n_e, see pars. 7.1.4 and 9.4.3), we obtain for the mass stored in this way:

$$\rho n_e \, \delta x \, \delta y [h|_{t+\Delta t} - h|_t].$$

Second, the pressure everywhere in the water-saturated box of volume $h \, \delta x \, \delta y$ rises. Following the discussion on elastic storativity due to the compressibility of soil and water (secs. 6.2 and 6.3), we obtain for this part of the storage:

$$\int_0^h [(\rho n)|_{t+\Delta t} - (\rho n)|_t] \, \delta x \, \delta y \, dz.$$

Combining the expressions above and passing to the limit as δx, δy, $\delta t \to 0$, we obtain:

$$-\frac{\partial(\rho q'_x)}{\partial x} - \frac{\partial(\rho q'_y)}{\partial y} + \rho N = \rho n_e \frac{\partial h}{\partial t} + \int_0^h \frac{\partial(\rho n)}{\partial t} dz. \qquad (8.2.1)$$

From the discussion in paragraph 6.3.1 we obtain:

$$\int_0^h \frac{\partial(\rho n)}{\partial t} dz = \int_0^h S^*_{0p} \frac{\partial p}{\partial t} dz$$

where S^*_{0p} is defined by (6.3.12). From the Dupuit assumptions of hydrostatic pressure distribution, it follows that $\partial p/\partial t = \rho g\, \partial h/\partial t$, and:

$$\int_0^h [\partial(\rho n)/\partial t]\, dz = \rho g S^*_{0p} h\, \partial h/\partial t = h S^*_{0\varphi^*}\, \partial h/\partial t \qquad (8.2.2)$$

where $hS^*_{0\varphi^*}/\rho$ may be replaced by the aquifer's storativity, S, as defined by (6.4.2), i.e., volume of water released from the storage in the entire column of height, h, per unit decline of head. We may now rewrite (8.2.1) as:

$$-\partial(\rho q'_x)/\partial x - \partial(\rho q'_y)/\partial y + \rho N = \rho(n_e + S^*_{0\varphi^*} h/\rho)\, \partial h/\partial t. \qquad (8.2.3)$$

For all practical purposes, $n_e \gg S^*_{0\varphi^*}h/\rho$, and ρ is virtually constant. Hence (8.2.3) becomes:

$$-\partial q'_x/\partial x - \partial q'_y/\partial y + N = n_e\, \partial h/\partial t \qquad (8.2.4)$$

or, introducing Dupuit's assumptions for q'_x and q'_y in an isotropic aquifer, with $K = K(x, y)$:

$$\frac{\partial}{\partial x}\left(Kh \frac{\partial h}{\partial x}\right) + \frac{\partial}{\partial y}\left(Kh \frac{\partial h}{\partial y}\right) + N = n_e \frac{\partial h}{\partial t}$$

$$\frac{\partial}{\partial x}\left[\frac{K}{2}\frac{\partial(h^2)}{\partial x}\right] + \frac{\partial}{\partial y}\left[\frac{K}{2}\frac{\partial(h^2)}{\partial y}\right] + N = n_e \frac{\partial h}{\partial t}. \qquad (8.2.5)$$

When $K = K(x, y, z)$, we must replace K in (8.2.5) with

$$\bar{K} = \frac{1}{h}\int_0^h K(x, y, z')\, dz'.$$

Sometimes the aquifer's bottom is not horizontal. Then, let $h(x, y, t)$ and $\eta(x, y)$ denote the elevation of the phreatic surface and of the aquifer's bottom, respectively, above some datum level.

If $\nabla\eta$ is sufficiently small that Dupuit's assumption of horizontal flow in the aquifer is still applicable, we have $q'_x = -K(h - \eta)\, \partial h/\partial x$, $q'_y = -K(h - \eta)\, \partial h/\partial y$ and (8.2.5) becomes:

$$\frac{\partial}{\partial x}[K(h - \eta)\, \partial h/\partial x] + \frac{\partial}{\partial y}[K(h - \eta)\, \partial h/\partial y] + N = n_e\, \partial h/\partial t. \qquad (8.2.6)$$

Another way of writing (8.2.6) is by introducing the actual thickness of the flow domain $b(x, y, t) = h(x, y, t) - \eta(x, y)$. For a homogeneous isotropic aquifer, we obtain:

$$\frac{\partial}{\partial x}\left(b\frac{\partial b}{\partial x}\right) + \frac{\partial}{\partial y}\left(b\frac{\partial b}{\partial y}\right) + \frac{\partial}{\partial x}\left(b\frac{\partial \eta}{\partial x}\right) + \frac{\partial}{\partial y}\left(b\frac{\partial \eta}{\partial y}\right) + \frac{N}{K} = \frac{n_e}{K}\frac{\partial b}{\partial t} \tag{8.2.7}$$

where $\nabla\eta$ represents the slope of the impervious base. For grad $\eta = I_x \mathbf{1x} + I_y \mathbf{1y}$, with $I_x = \text{const}$ and $I_y = \text{const}$, we obtain:

$$\nabla \cdot (b\,\nabla b) + I_x\,\partial b/\partial x + I_y\,\partial b/\partial y + N/K = (n_e/K)\,\partial b/\partial t. \tag{8.2.8}$$

By the transformation

$$x' = x + I_x tK/n_e, \qquad y' = y + I_y tK/n_e, \qquad t' = t$$

(8.2.8) becomes:

$$\frac{\partial}{\partial x'}(b\,\partial b/\partial x') + \frac{\partial}{\partial y'}(b\,\partial b/\partial y') + N/K = (n_e/K)\,\partial b/\partial t' \tag{8.2.9}$$

which now has a form similar to (8.2.5).

Equations (8.2.5) and (8.2.6) are known as *Boussinesq's equations for unsteady flow* in a phreatic aquifer with accretion.

For a homogeneous aquifer ($K = \text{const}$), equations (8.2.5) and (8.2.6) become, for a horizontal bottom:

$$\frac{K}{2}\left[\frac{\partial^2(h^2)}{\partial x^2} + \frac{\partial^2(h^2)}{\partial y^2}\right] + N \equiv \frac{K}{2}\nabla^2{}_{xy}h^2 + N = n_e\frac{\partial h}{\partial t} \tag{8.2.10}$$

and for a nonhorizontal bottom:

$$K\left[\frac{\partial}{\partial x}\left(h\frac{\partial h}{\partial x}\right) + \frac{\partial}{\partial y}\left(h\frac{\partial h}{\partial y}\right) - \frac{\partial}{\partial x}\left(\eta\frac{\partial h}{\partial x}\right) - \frac{\partial}{\partial y}\left(\eta\frac{\partial h}{\partial y}\right)\right] + N = n_e\frac{\partial h}{\partial t}. \tag{8.2.11}$$

For *steady flow* with accretion we obtain for a horizontal bottom aquifer:

$$\frac{K}{2}\left[\frac{\partial^2(h^2)}{\partial x^2} + \frac{\partial^2(h^2)}{\partial y^2}\right] + N = 0; \qquad \frac{K}{2}\nabla^2 h^2 + N = 0. \tag{8.2.12}$$

This is a *Poisson equation* in h^2.

In the absence of accretion (8.2.12) becomes:

$$\partial^2(h^2)/\partial x^2 + \partial^2(h^2)/\partial y^2 = \nabla^2{}_{xy}(h^2) = 0 \tag{8.2.13}$$

which is the Laplace equation for h^2. Equation (8.2.10) is known as *Forchheimer's equation*. Sometimes (8.2.9) is also called Forchheimer's equation.

If the medium is anisotropic (in this case it means when $K_x \neq K_y$), we must introduce in (8.2.4):

$$q'_x = -K_x h\,\partial h/\partial x \quad \text{and} \quad q'_y = -K_y h\,\partial h/\partial y.$$

Subsequent equations will be modified accordingly.

The most interesting feature of Boussinesq's equation (8.2.5) is that it is a *nonlinear equation* for the unknown variable $h = h(x, y, t)$. This is in spite of the fact that we have employed the Dupuit assumptions. This fact makes it necessary to introduce the various linearization techniques discussed in section 8.4. However, Forchheimer's equation (8.2.13) is linear in h^2. Actually, this is nothing new as for this case Girinskii's potential (par. 5.8.4) is nothing but $\Phi^g = Kh^2/2$, so that (8.2.12) and (8.2.13) become:

$$\nabla^2\Phi^g + N = 0 \quad \text{and} \quad \nabla^2\Phi^g = 0. \tag{8.2.14}$$

The discussion presented in paragraph 6.4.4 can be extended to show that the Boussinesq equation (8.2.5) can be obtained from the general continuity equation (6.4.15), where $\mathbf{q}$ is the specific discharge, by averaging over the vertical line. The starting point is the average equation (6.4.22) in which we replace b with $h - \eta$, ζ_2 with h, ζ_1 with η, $N'' = N$, $N' = 0$, $\bar{\varphi} = \varphi(x, y, \zeta_2) = \varphi(x, y, \zeta_1) = h$ (because of the Dupuit assumptions). We obtain:

$$\frac{\partial}{\partial x}\left[(h-\eta)\frac{\partial h}{\partial x}\right] + \frac{\partial}{\partial y}\left[(h-\eta)\frac{\partial h}{\partial y}\right] + \frac{N}{K} = \frac{n_e}{K}\frac{\partial h}{\partial t} + \frac{S_s}{K}(h-\eta)\frac{\partial h}{\partial t}. \tag{8.2.15}$$

With $n_e \gg S_s h$, equation (8.2.15) becomes identical to (8.2.6) written for a homogeneous aquifer.

In section 8.4 it is shown that the Boussinesq equation is the first approximation of the exact solution.

8.2.2 Boundary and Initial Conditions

In order to solve a partial differential equation for a particular case of interest, we must specify the boundary conditions with respect to h that exist along boundaries of the flow domain. In an unsteady flow problem, we must also specify the initial conditions throughout the flow domain.

A detailed discussion on boundary and initial conditions in three-dimensional flows was given in section 7.1. However, here we are dealing with plane flow; the vertical coordinate is deleted. Obviously, this is an approximation that gives good results in cases where the Dupuit assumptions are valid. The phreatic surface no longer appears as a boundary of the flow domain. This is the major advantage of introducing the Dupuit assumptions. Accordingly, all boundary and initial conditions must be specified only in terms of the x, y (or r, θ) coordinates. Also, the accretion, instead of being part of the (phreatic) boundary condition, is introduced as a distributed source function $N = N(x, y, t)$ or $N = N(r, \theta, t)$.

For example, the boundary condition (7.1.6) of prescribed potential will become:

$$h = h(x, y) \quad \text{or} \quad h = h(x, y, t) \quad \text{on } C$$

where C is the boundary of the flow domain given as $F = F(x, y)$. When the actual boundary is not a vertical surface, we must replace it with an equivalent fictitious vertical boundary and use it to solve the problem. For a boundary of prescribed flux, we specify:

$$q'_n = q'_n(x, y) \quad \text{or} \quad q'_n = q'_n(x, y, t) \quad \text{on } C$$

where $q' = q'(x, y)$ is the discharge through the entire thickness of the aquifer, per unit (horizontal) width of aquifer.

For initial conditions we must specify $h = h(x, y)$ throughout the entire flow domain at some initial time.

8.2.3 Some Solutions of Forchheimer's Equation

Two examples of solutions obtained by direct integration are given below.

As a first example, consider the two-dimensional phreatic flow with accretion shown in figure 8.2.2. The flow is described by (8.2.12) written as:

$$(K/2)\,\partial(h^2)/\partial x^2 + N = 0; \qquad h = h(x). \tag{8.2.16}$$

By integrating twice, we obtain:

$$h^2 = -(N/K)x^2 + ax + b \tag{8.2.17}$$

where a and b are integration constants to be determined by the boundary conditions. For $N > 0$, equation (8.2.17) describes an ellipse. For $N < 0$, it describes an hyperbola.

For the boundary conditions $x = 0, h = h_0$; $x = L, h = h_L$, and $N > 0$, we obtain

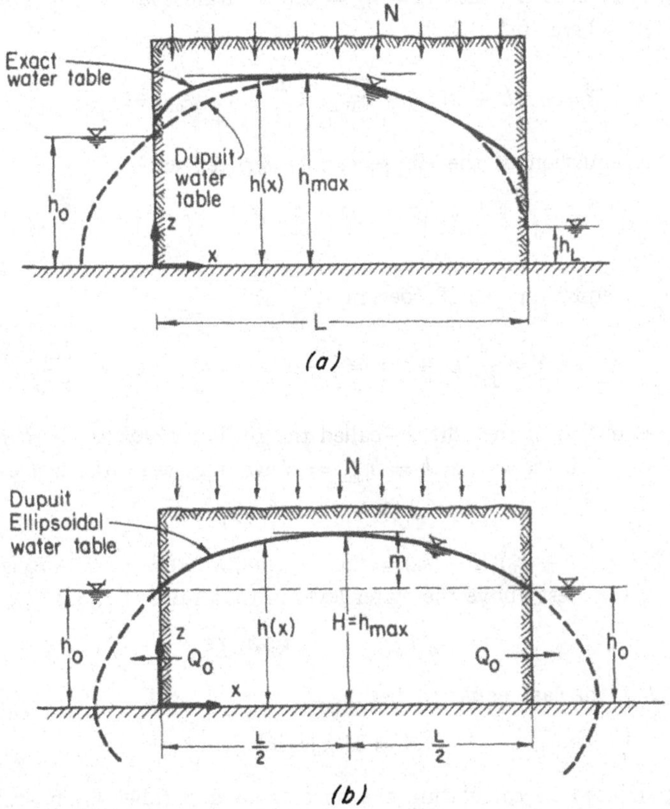

FIG. 8.2.2. Phreatic flow with accretion.

the equation for the shape of the phreatic surface (fig. 8.2.2a):

$$h = \left[h_0^2 - \frac{h_0^2 - h_L^2}{L} x + \frac{N}{K}(L - x)x\right]^{1/2} \tag{8.2.18}$$

The discharges Q_0 and Q_L toward the outlets at $x = 0$ and $x = L$, respectively, are:

$$\frac{Q_0}{Q_L} = \frac{K}{2L}(h_0^2 - h_L^2) \mp \frac{NL}{2}. \tag{8.2.19}$$

Somewhere in the region $0 < x < L$ the ellipsoidal phreatic surface reaches its highest elevation. This point is the *water divide*. There $dh/dx = 0$ and hence (by the Dupuit assumptions) $q'_x = 0$. Denoting x of the water divide by x_{WD}, we obtain from (8.2.18):

$$x_{WD} = \frac{L}{2} - \frac{K}{N}\frac{h_0^2 - h_L^2}{2L}. \tag{8.2.20}$$

From (8.2.20) it follows that under certain conditions h_0 and h_L, $x_{WD} = 0$. We may also have $x_{WD} < 0$ which means that there is no water divide and water enters the aquifer from the upstream reservoir.

Of special interest is the case $h_0 = h_L = 0$ (i.e., drains on the impervious bottom of the aquifer) where (fig. 8.2.2b):

$$h^2 = \frac{N}{K}(L - x)x; \qquad x_{WD} = \frac{L}{2}; \qquad h^2_{max} = \frac{NL^2}{4K} \tag{8.2.21}$$

Therefore, the equation of the ellipse can be written as:

$$\left(\frac{h}{h_{max}}\right)^2 + \left(\frac{X - L/2}{L/2}\right)^2 = 1. \tag{8.2.22}$$

When $h_0 = h_L$, equation (8.2.18) becomes:

$$h^2 - h_0^2 = \frac{N}{K}(L - x)x = \frac{2Q_0}{LK}(L - x)x; \qquad Q_0 = \frac{NL}{2} \tag{8.2.23}$$

which is the equation of the ellipse—called the *Dupuit–Forchheimer ellipse*—shown in figure 8.2.2b. With $x = L/2$, $h = h_{max} = H$ we may write the last equation as:

$$N = 4K(H^2 - h_0^2)/L^2. \tag{8.2.24}$$

Since $H^2 - h_0^2 = (H + h_0)(H - h_0) = (2h_0 + m)m$, where $m = H - h_0$ is the height of the water table crest above the water level in the drains (fig. 8.2.2b), we obtain:

$$N = (8Kh_0m + 4Km^2)/L^2. \tag{8.2.25}$$

When $m \ll 2h_0$, the last equation becomes:

$$N = 8Kh_0\, m/L^2 \tag{8.2.26}$$

which is often used in calculating the spacing in a parallel drain system where $h_0 \ll L$. This equation is equivalent to neglecting the flow above h_0.

The case $N < 0$ can be analyzed in a similar way. Depending on the values of h_0, L, K and N, the resulting hyperbola may have a minimum point (in the region $0 < x < L$) fed from both reservoirs, or the lowest point of the water table may be at the downstream reservoir so that only the upper reservoir is feeding water to the aquifer.

As a second example, consider the steady symmetrical flow to a well in a phreatic aquifer with accretion. The flow is described by (8.2.12), which, when rewritten in radial coordinates, becomes:

$$\frac{d}{dr}\left(rh\frac{dh}{dr}\right) + \frac{rN}{K} = 0. \tag{8.2.27}$$

After integration we obtain:

$$h^2 = -r^2N/2K + a \ln r + b. \tag{8.2.28}$$

With $r = r_w$, $h = h_w$ and $r = R$, $h = h_R$, equation (8.2.28) becomes:

$$h^2 - h_w^2 = -\frac{(r^2 - r_w^2)N}{2K} + \frac{(h_R^2 - h_w^2)2K + (R^2 - r_w^2)N}{2K \ln(R/r_w)} \ln\frac{r}{r_w}. \tag{8.2.29}$$

The discharge rate $Q(r)$ at some distance r is:

$$Q(r) = 2\pi Krh\frac{dh}{dr} = \pi Kr\frac{d(h^2)}{dr} = -\pi r^2N + \pi\frac{(h_R^2 - h_w^2)2K + (R^2 - r_w^2)N}{2\ln(R/r_w)}. \tag{8.2.30}$$

Since $R \gg r_w$ (8.2.30) reduces to:

$$Q(r) = -\pi r^2N + \pi\frac{(h_R^2 - h_w^2)2K + R^2N}{2\ln(R/r_w)}. \tag{8.2.31}$$

Let the water divide $Q(r) = 0$ occur at $r = R^*$. Then $Q(r_w) = \pi(R^{*2} - r_w^2)N$ and we obtain:

$$h^2 - h_w^2 = \frac{NR^{*2}}{K}\ln\frac{r}{r_w} - \frac{N}{2K}(r^2 - r_w^2). \tag{8.2.32}$$

8.2.4 Some Solutions of Boussinesq's Equation

Because of the nonlinearity of Boussinesq's equation (8.2.5), only a small number of exact solutions are known to date. Polubarinova-Kochina (1952, 1962), Aravin and Numerov (1953, 1965) and Irmay (in Bear, Zaslavsky and Irmay 1968) review some of these solutions.

Boussinesq (1904) was the first to present an exact solution of (8.2.5) for one-dimensional flow in the x direction, by the method of separation of variables (sec. 7.8). He thus solved:

$$\partial(h\,\partial h/\partial x)/\partial x = (n_e/K)\,\partial h/\partial t. \tag{8.2.33}$$

Assuming that the resulting $h = h(x, t)$ can be written as $h(x, t) = X(x)T(t)$ and substituting this expression into (8.2.33), he obtained:

$$\frac{1}{X}\frac{d^2(X^2)}{d^2x} = \frac{2n_e}{T^2K}\frac{dT(t)}{dt}. \tag{8.2.34}$$

As the left-hand side of (8.2.34) depends only on x, whereas the right-hand side depends only on t, equation (8.2.34) is valid only if both sides are equal to some constant independent of x and t, usually denoted by $-\lambda^2$. We obtain:

$$\frac{dT}{dt} + \frac{\lambda^2 K}{2n_e}T^2 = 0 \tag{8.2.35}$$

$$\frac{d^2(X^2)}{dx^2} + \lambda^2 X = 0 \tag{8.2.36}$$

which are integrable ordinary differential equations. The solution of (8.2.35) is given by:

$$T(t) = 1/(A + \lambda^2 Kt/2n_e) \tag{8.2.37}$$

where A is a constant of integration. The solution of (8.2.36) is given by Boussinesq in the form:

$$X(x) = BF\left(\frac{\lambda x}{\alpha\sqrt{3B}} + \frac{C}{\alpha}\right); \qquad \alpha = \int_0^1 \frac{\tau\, d\tau}{\sqrt{1-\tau^3}} = \frac{\sqrt{\pi}\,\Gamma(\frac{2}{3})}{3\Gamma(\frac{7}{6})} \approx 0.862 \tag{8.2.38}$$

where B and C are constants of integration, and $F(s)$ is an auxiliary function defined by:

$$s = \frac{1}{\alpha}\int_0^F \frac{\tau\, d\tau}{\sqrt{1-\tau^3}}; \qquad s \leqslant 0, \qquad F \leqslant 1. \tag{8.2.39}$$

Also, from the definition of $F(s)$ we have

$$\frac{dF}{ds} = \frac{1}{ds/dF} = \frac{\alpha\sqrt{1-F^3}}{F}, \quad \text{with} \quad \left.\frac{dF}{ds}\right|_{s=1} = 0.$$

Leibenzon (1947) gives for F the approximate expression $F \approx (1.321 - 0.142\,s - 0.179\,s^2)\sqrt{s}$ with a relative error of about 5%. A table of $F(s)$ is given by Aravin and Numerov (1953, 1965). By combining the results for $X(x)$ and $T(t)$, Boussinesq obtained:

$$h = BF\left(\frac{\lambda x}{\alpha\sqrt{3B}} + \frac{C}{\alpha}\right)\Big/(A + \lambda^2 Kt/2n_e). \tag{8.2.40}$$

Consider now the flow shown in figure 8.2.3. The impervious vertical boundary at $x = L$ may also be considered as a line of symmetry. The initial and boundary conditions for this case are:

I.C.: $\quad h(x, 0) = h_0(x)$

B.C.: $\quad h(0, t) = h_0(0) = 0$

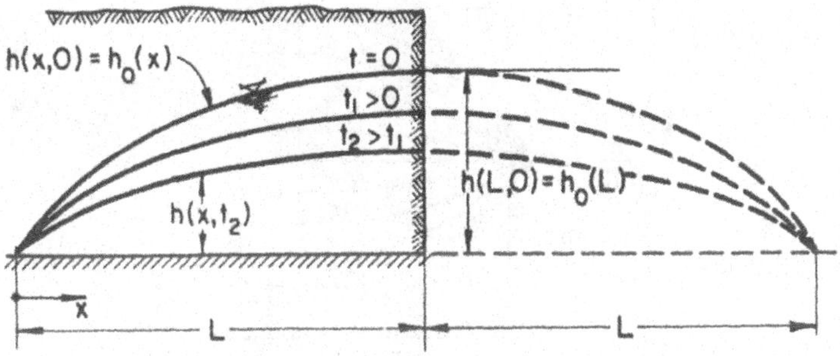

FIG. 8.2.3. Declining phreatic surface between parallel drains.

$$\partial h/\partial x|_{x=L} = \partial h_0/\partial x|_{x=L} = 0$$

(i.e., impervious boundary in view of the Dupuit assumption of horizontal flow only). Actually, at $x = L$ the Dupuit assumptions are not valid.

We also have:

$$h_0(x) = \frac{B}{A} F\left(\frac{\lambda x}{\alpha\sqrt{3B}} + \frac{C}{\alpha}\right); \qquad h(x, t) = \frac{h_0(x)}{1 + \lambda^2 Kt/2n_e A} \tag{8.2.41}$$

therefore:

$$0 = \frac{B}{A} F\left(\frac{C}{\alpha}\right), \qquad C = 0$$

$$\left.\frac{\partial h}{\partial x}\right|_{x=L} = \frac{B}{A}\frac{\alpha}{F}\sqrt{1 - F^3}\frac{\lambda}{\alpha\sqrt{3B}} = 0, \qquad F|_{x=L} = 1, \qquad \frac{\lambda L}{\alpha\sqrt{3B}} = 1$$

$$h_0(L) = B/A; \qquad h_0(x) = h_0(L) F\left(\frac{x}{L}\right). \tag{8.2.42}$$

Hence:

$$h(x, t) = h_0(x)/[1 + \beta K h_0(L)/n_e L^2 t]; \qquad \beta = \tfrac{3}{2}\alpha^2 \approx 1.12. \tag{8.2.43}$$

Thus, given the initial water table $h_0(x)$, we can use (8.2.43) to determine $h = h(x, t)$.

A second exact solution of (8.2.33) was obtained by Boussinesq in the following way. Let us introduce a new variable $\xi = \sqrt{n_e/2Kt}\,x$ in (8.2.33) so that the unknown is now $h = h(\xi)$. Since:

$$\frac{\partial h}{\partial x} = \frac{dh}{\partial \xi}\frac{\partial \xi}{\partial x} = \sqrt{\frac{n_e}{2Kt}}\frac{dh}{d\xi}; \qquad \frac{\partial h}{\partial t} = \frac{dh}{d\xi}\frac{\partial \xi}{\partial t} = -\sqrt{\frac{n_e}{2Kt}}\frac{x}{2t}\frac{dh}{d\xi},$$

we obtain from (8.2.32):

$$\frac{d}{d\xi}\left(h\frac{dh}{d\xi}\right) + \frac{dh}{d\xi} = 0. \tag{8.2.44}$$

The new variable ξ is a modified form of the Boltzmann transformation: $\xi = x/t^{1/2}$.

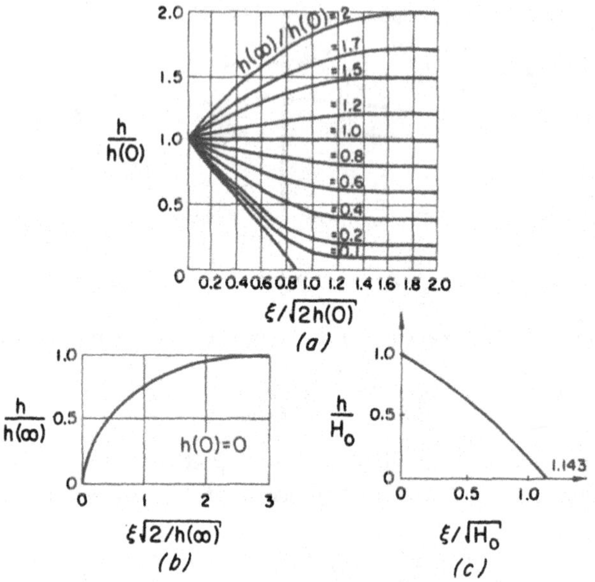

FIG. 8.2.4. Graphical representation of the solutions of (8.2.43) (after Polubarinova-Kochina 1949, 1952).

Polubarinova-Kochina (1949) presents the solution of (8.2.44) for the case $h(0) = h_1$; $h(\infty) = h_2$ in the graphic form shown in figure 8.2.4a. For the special case $h_0 = 0$, Polubarinova-Kochina's solution (1949) is given in figure 8.2.4b.

Detailed solutions of (8.2.44) are given by Polubarinova-Kochina (1952, 1962; see also Aravin and Numerov 1953, 1965) for several additional cases.

Irmay (chap. 8 in Bear, Zaslavsky and Irmay 1968), following Polubarinova-Kochina (1952) and Pirverdian (1960), describes the extension of Boussinesq's method of introducing the function $h(\xi)$ of the single variable ξ. According to this extension, the following new variables are introduced:

$$\eta = \theta^{\nu} f(\alpha); \qquad \alpha = \xi'/\theta^m \tag{8.2.45}$$

where the dimensionless variables η, θ and ξ' are defined by:

$$\eta = h/H; \qquad \theta = tKH/n_e L^2; \qquad \xi' = x/L \tag{8.2.46}$$

H and L being characteristic depth and length of the flow domain, e.g., $H = h(L, 0)$. With the dimensionless variables defined by (8.2.46), equation (8.2.33) becomes:

$$\frac{\partial}{\partial \xi'}\left(\eta \frac{\partial \eta}{\partial \xi'}\right) = \frac{\partial \eta}{\partial \theta}. \tag{8.2.47}$$

By introducing the transformations (8.2.45) into (8.2.47), we obtain the ordinary differential equation *in one independent variable*:

$$\frac{d}{d\alpha}\left(f \frac{df}{d\alpha}\right) + m\alpha \frac{df}{d\alpha} - (2m-1)f = 0; \qquad \nu = 2m - 1. \tag{8.2.48}$$

The special case $m = \frac{1}{2}$, $\nu = 0$, known as the Boltzmann transformation (par. 6.4.5), gives:

$$\frac{d}{d\alpha}\left(f\frac{df}{d\alpha}\right) + \frac{\alpha}{2}\frac{df}{d\alpha} = 0 \tag{8.2.49}$$

which is identical to (8.2.44) since $\xi = \alpha\sqrt{H/2}$. Another special case is $m = 0$, $\nu = -1$. This gives:

$$\frac{d}{d\alpha}\left(f\frac{df}{d\alpha}\right) + f = 0; \qquad \frac{d^2(f^2)}{d\alpha^2} + 2f = 0 \tag{8.2.50}$$

which is similar to (8.2.36). The usefulness of these and similar transformations is limited by the fact that the initial and boundary conditions must also be expressible in terms of α only.

A third case of interest (Irmay, in Bear et al. 1968) is $m = \frac{1}{3}$, $\nu = -\frac{1}{3}$. This gives:

$$\frac{d}{d\alpha}\left(f\frac{df}{d\alpha}\right) + \tfrac{1}{3}\alpha\frac{df}{d\alpha} + \tfrac{1}{3}f = 0; \qquad \frac{d}{d\alpha}\left(f\frac{df}{d\alpha} + \frac{\alpha f}{3}\right) = 0; \qquad f = C\alpha - \frac{\alpha^2}{6} \tag{8.2.51}$$

with a solution:

$$\eta = C/\theta^{1/3} - \xi^2/6\theta. \tag{8.2.52}$$

This solution describes a spreading of a ground water mound (in a vertical two-dimensional plane, fig. 8.2.5). Its base (the points $\eta = 0$) increases with time as:

$$\eta = 0, \qquad \xi|_{\eta=0} = l(t)/L = \sqrt{6C\theta^{1/3}}; \qquad l(t)/L = \sqrt{6C}(tKH/n_eL^2)^{1/3}. \tag{8.2.53}$$

This means that $l(t)$ increases as $t^{1/3}$.

At $x = 0$ (i.e., $\xi = 0$), the mound's height is:

$$\eta|_{\xi=0} = h|_{x=0}/H = C/\theta^{1/3}. \tag{8.2.54}$$

Several other cases (in the two-dimensional vertical plane) were studied by Irmay (chap. 8 of Bear et al. 1968).

The case of *radial flow without accretion* is described by:

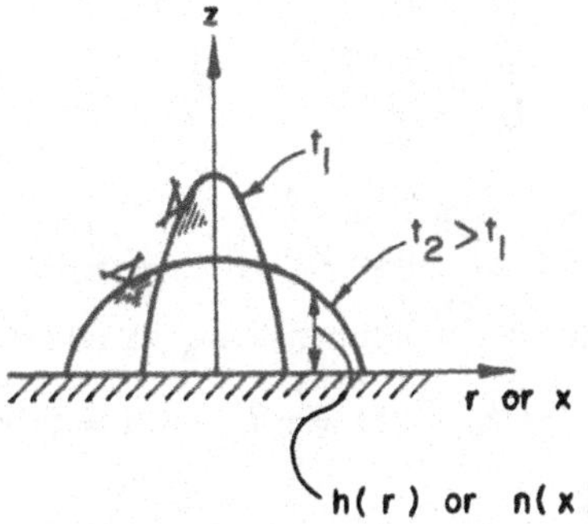

FIG. 8.2.5. Spreading of a ground water mound.

$$\frac{1}{2}\left(\frac{\partial^2 h^2}{\partial r^2} + \frac{1}{r}\frac{\partial h^2}{\partial r}\right) = \frac{n_e}{K}\frac{\partial h}{\partial t}. \tag{8.2.55}$$

As in the two-dimensional case considered above for the solution of (8.2.33), the method of separation of variables leads to:

$$h = R(r)T(t)$$

with:

$$dT/dt - (\lambda^2 K/n_e)T^2 = 0 \tag{8.2.56}$$

$$d(rR\,dR/dr)/dr - \lambda^2 rR = 0. \tag{8.2.57}$$

A special solution of (8.2.57) is $R = r^2$ with $\lambda^2 = 8$. The solution of (8.2.56) is $T = 1/(C_1 - \lambda^2 Kt/n_e) = 1/(C_1 - 8Kt/n_e)$. Hence, the corresponding solution of (8.2.55) is:

$$h = r^2/(C_1 - 8Kt/n_e) \tag{8.2.58}$$

where C_1 (dim L) is a constant of integration. This solution describes the decay of a mound (fig. 8.2.5) that has the shape of a paraboloid of revolution. At $t = 0$, $h \equiv h(r, 0) = r^2/C_1$.

Another axisymmetrical solution of (8.2.55) that describes a decaying mound of constant volume in the form of a paraboloid of revolution is:

$$h = (n_e/Kt)(Crt^{1/2} - r^2/8). \tag{8.2.59}$$

The base of this mound ($r|_{h=0}$ in fig. 8.2.5) increases with time as:

$$r(t) = (8C_2)^{1/2}t^{1/4}. \tag{8.2.60}$$

Galin, Karpycheva and Shkirich (1960) solve the problem of lenticular spreading of a ground water mound (fig. 8.2.5) in a somewhat different way. Again, the problem is described by (8.2.33) for the one-dimensional case and by (8.2.55) for the radial flow case. Galin et al. (1960) also require that the solution be such that the volume of water included in the spreading mound remains constant (say, U_0). For the one-dimensional case, for example, this is expressed by the condition:

$$\int_{-l(t)}^{+l(t)} h(x, t)\,dx = U_0 = \text{const.}$$

As initial conditions they use:

$$t = 0, \qquad h \equiv h(x, 0) = h_0(x); \qquad -l_0 < x < +l_0.$$

By introducing the following nondimensional variables:

$$u \equiv u(\xi, \tau) = h/h_0; \qquad \xi = x/l_0; \qquad \tau = (Kh_0/2n_e l_0^2)t;$$

Galin et al. (1960) obtain:

$$\partial u/\partial \tau = \partial^2 u^2/\partial \xi^2 \tag{8.2.61}$$

$$\tau = 0; \quad u(\xi, 0) = u_0(\xi); \quad -1 < \xi < +1$$
$$u(\xi, 0) = 0; \quad \xi < -1; \quad \xi > +1.$$

They seek a solution in the form of a parabola that may be described by:

$$u = \beta(\tau)\,[l(\tau) - \xi^2] \tag{8.2.62}$$

and which satisfies:

$$\text{for} \quad \tau = 0, \quad U_0(\xi) = 1 - \xi^2, \quad -1 < \xi < +1.$$

In view of the requirement that the volume of water in the mound remain constant, we have:

$$2\int_0^1 (1 - \xi^2)\, d\xi = \tfrac{4}{3} = 2\int_0^{l(\tau)} \beta(\tau)\,[l^2(\tau) - \xi^2]\, d\xi = \tfrac{4}{3}\beta(\tau) l^3(\tau).$$

Hence, $\beta(\tau) = 1/l^3(\tau)$. Therefore:

$$u(\xi, \tau) = \frac{1}{l^3(\tau)}\,[l^2(\tau) - \xi^2]. \tag{8.2.63}$$

Inserting (8.2.63) into the partial differential equation (8.2.61) for u leads to:

$$\frac{\partial u}{\partial \tau} - \frac{\partial^2 u^2}{\partial \xi^2} = \left[4 - l^2(\tau)\frac{dl(\tau)}{d\tau}\right]\left[\frac{1}{l^4(\tau)} - \frac{3}{l^6(\tau)}\xi^2\right] = 0$$

or:

$$4 - l^2(\tau)\, dl(\tau)/d\tau = 0; \quad l(\tau) = (C + 12\tau)^{1/3}.$$

Because $l(\tau) = 1$ for $\tau = 0$, we obtain $C = 1$. Hence the solution is:

$$u(\xi, \tau) = [(1 + 12\tau)^{2/3} - \xi^2]/(1 + 12\tau) \tag{8.2.64}$$

or, in a dimensional form:

$$h(x, t) = \frac{h_0}{1 + (6Kh_0/n_e l_0^2)t}\left[\left(1 + \frac{6Kh_0}{n_e l_0^2}t\right)^{2/3} - \left(\frac{x}{l_0}\right)^2\right]. \tag{8.2.65}$$

The width of the spreading mound is:

$$2l = 2l_0(1 + 6Kh_0 t/n_e l_0^2)^{1/3}. \tag{8.2.66}$$

Following a similar procedure for the case of an axially symmetric mound, Galin et al. (1960) obtain:

$$h(r, t) = \frac{h_0}{1 + (8Kh_0/n_e r_0^2)t}\left[\left(1 + \frac{8Kh_0}{n_e r_0^2}t\right)^{1/2} - \left(\frac{r}{r_0}\right)^2\right] \tag{8.2.67}$$

where at $\tau = 0$, $r = r_0$ gives the extent of the mound. At $\tau > 0$, we have

$$r(t)|_{h=0} = r_0(1 + 8Kh_0 t/n_e r_0^2)^{1/4}. \tag{8.2.68}$$

The results (8.2.67) and (8.2.68) should be compared with (8.2.59) and (8.2.60).

Galin et al. present experimental verification of their results.

8.3 The Hodograph Method

The hodograph method is applicable to steady two-dimensional flows in the vertical plane and is actually the only method by means of which an exact analytical solution can be derived. As in confined flows, when the geometry of the boundaries becomes complicated, even this method fails to yield an analytic solution.

This method is given here, and not in chapter 7, because of its special applicability to *steady free surface flows*. In section 9.6 this method is mentioned also as a tool for solving problems involving a steady abrupt interface between two immiscible fluids. In addition to the hodograph method, several other methods related to the hodograph method are included in the present section for the sake of completeness.

The hodograph plane and the procedure for mapping any two-dimensional flow domain (especially those involving a free surface) onto it are described in section 7.3. Here we shall describe the actual procedure of solving flow problems and present several examples to demonstrate it. We shall assume that the medium is homogeneous and isotropic.

Some of the concepts and terms used in this section are briefly reviewed in section 7.8, dealing with methods of solution based on the theory of functions.

Only cases with a one-to-one correspondence between the z and the ω (or $\tilde{\omega}$) plane are considered in this section. An example of a case whose hodograph representation takes the form of a *Riemann surface* is given in section 9.6.

8.3.1 The Functions ω and $\tilde{\omega}$

Consider the functions (par. 7.8.2):

$$\omega = q_x + iq_y; \qquad \tilde{\omega} = q_x - iq_y \tag{8.3.1}$$

where $q_x = q_x(x, y)$ and $q_y = q_y(x, y)$ are the components of the specific discharge vector $\mathbf{q}$ in the directions x and y (directed vertically upward), respectively, and $\tilde{\omega}$ is the *complex conjugate* of ω. ω is called the *complex specific discharge*; $\tilde{\omega}$ is the *conjugate complex specific discharge* (par. 7.8.2). In the same way as the complex number $z = x + iy$ describes a point P with coordinates x, y in the *physical xy plane*, the number ω describes a point (q_x, q_y) in the *hodograph plane* (sec. 7.3), and the number $\tilde{\omega}$ describes a point whose coordinates are $(q_x, -q_y)$ in a plane whose axes are q_x and $-q_y$. Geometrically the conjugate $\tilde{\omega}$ of ω is the reflection of ω in the axis of reals (fig. 8.3.1). The $\tilde{\omega}$-plane is sometimes called the *inverse hodograph plane*.

The complex function $\tilde{\omega}$ is analytic. Expressing $\tilde{\omega}$ in terms of Φ, we have:

$$\tilde{\omega} = \omega_1 - i\omega_2 = q_x - iq_y = -\partial\Phi/\partial x + i\,\partial\Phi/\partial y. \tag{8.3.2}$$

It can easily be verified that $\tilde{\omega}$ is analytic since its real ($\omega_1 \equiv q_x$) and imaginary ($\omega_2 = -q_y$) parts, together with their first derivatives, are continuous and single valued (we assume so) and satisfy the Cauchy–Riemann conditions:

$$\partial\omega_1/\partial x = \partial\omega_2/\partial y; \qquad \partial\omega_1/\partial y = -\partial\omega_2/\partial x. \tag{8.3.3}$$

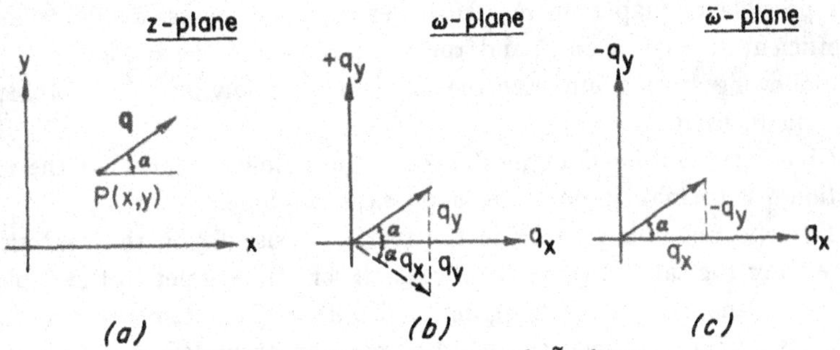

FIG. 8.3.1. Definition of z, ω, and $\tilde{\omega}$ planes.

Indeed, both conditions lead to the Laplace equation $\nabla^2\Phi = 0$, which is valid for this type of flow. However, a similar examination of ω reveals that *it is not an analytic function.*

We may now relate ω to the complex potential:

$$\zeta = f(z) = \Phi(x, y) + i\Psi(x, y) \tag{8.3.4}$$

defined in paragraph 7.8.2. The function $\zeta = f(z)$ is analytic. For a single-valued function that is analytic at a point, the derivative at that point is unique, i.e., independent of the direction dz along which the derivative is taken. In paragraph 7.8.2 we have shown that:

$$d\zeta/dz = \partial\zeta/\partial x|_{iy=\text{constant}} = \partial\Phi/\partial x + i\,\partial\Psi/\partial x$$

$$d\zeta/dz = \partial\zeta/\partial(iy)|_{x=\text{constant}} = \partial\Phi/\partial(iy) + i\,\partial\Psi/\partial(iy) = -\,i\,\partial\Phi/\partial y + \partial\Psi/\partial y. \tag{8.3.5}$$

Hence:

$$-\,d\zeta/dz = (q_x - iq_y) = \tilde{\omega}; \qquad |d\zeta/dz| = (q_x^2 + q_y^2)^{1/2} = |\mathbf{q}| = q. \tag{8.3.6}$$

We also have:

$$-\,dz/d\zeta = -\,1/(d\zeta/dz) = 1/(q_x - iq_y) = (q_x + iq_y)/q^2 = \omega/q^2. \tag{8.3.7}$$

In polar coordinates we may write:

$$-\,dz/d\zeta = 1/(q_x - iq_y) = (\exp i\alpha)/q; \qquad \alpha = \tan^{-1}(q_y/q_x).$$

8.3.2 The Hodograph Method

From (8.3.5) through (8.3.7) it follows that:

$$\tilde{\omega}(\zeta) = -\,d\zeta/dz; \qquad dz = -\,d\zeta/\tilde{\omega}(\zeta) \tag{8.3.8}$$

$$z = -\int \frac{d\zeta}{\tilde{\omega}(\zeta)} + \text{const} \tag{8.3.9}$$

which relates ζ (and hence also Φ and Ψ) to points z in the xy physical plane. We thus see that knowledge of the relationship between ζ and $\tilde{\omega}$ is essential to the application of this method. This knowledge can be obtained by mapping the z-plane on both the $\tilde{\omega}$- and the ζ-planes, and then mapping ζ on $\tilde{\omega}$. We have seen that it is *always* possible to map z on $\tilde{\omega}$. However, in many cases, available information is insufficient to map z on ζ and the method cannot be applied.

The following steps summarize the procedure for solving a flow problem by the hodograph method.

(a) Given the contour C of the boundary of a flow domain, use the rules given in section 7.3 to map it onto the hodograph ω-plane.

(b) By reflecting this contour in the real (q_x) axis, obtain the contour Γ in the $\tilde{\omega}$-plane. For the sake of simplicity, we shall sometimes refer to the $\tilde{\omega}$-plane as the hodograph plane. Since the mapping of C onto Γ (i.e., from the z-plane onto the $\tilde{\omega}$-plane) is a *conformal* one, the sense of rotation around C and around Γ should be the same.

(c) Map the boundary C on the ζ-plane (the complex potential plane). The mapped region in the ζ-plane will usually take the form of an infinite strip or a rectangle.

(d) Determine a conformal mapping relationship $\tilde{\omega} = \tilde{\omega}(\zeta)$ that maps the $\tilde{\omega}$-plane on the ζ-plane. Sometimes this cannot be derived directly and an intermediate plane, or several such planes, is required. For example, we may map the flow domain in the $\tilde{\omega}$-plane onto the upper half of some $\xi = \mu + i\nu$ plane by $\tilde{\omega} = \tilde{\omega}(\xi)$; we also map the flow domain represented in the ζ-plane on the same upper half of the ξ-plane by $\zeta = \zeta(\xi)$. Finally, we establish the required relationship $\tilde{\omega} = \tilde{\omega}(\zeta)$.

(e) Using (8.3.9), we integrate:

$$z = -\int_{\zeta=\zeta_0}^{\zeta} \frac{d\zeta'}{\tilde{\omega}(\zeta')} = x + iy; \qquad \zeta_0 = \Phi_0 + i\Psi_0; \qquad \zeta = \Phi + i\Psi. \tag{8.3.10}$$

By comparing real and imaginary parts separately, we obtain the parametric equations:

$$x = x(\Phi, \Psi), \qquad y = y(\Phi, \Psi) \tag{8.3.11}$$

from which the solution is obtained in the form:

$$\Phi = \Phi(x, y), \qquad \Psi = \Psi(x, y). \tag{8.3.12}$$

A transformation that is most useful in the application of the hodograph method is the *inversion transformation*. To understand its significance, consider the circle C of radius ρ and the point $P(x, y)$ of figure 8.3.2. The inversion transformation maps point P onto point P', located on the same ray from the origin (also center of circle), in such a way that:

$$\overline{OP} \cdot \overline{OP'} = \rho^2. \tag{8.3.13}$$

Point P' is said to be the image of P in the circle C, and vice versa. We then have:

$$x' = \rho^2 x/(x^2 + y^2); \qquad y' = \rho^2 y/(x^2 + y^2) \tag{8.3.14}$$

from which we get:

$$x' + iy' = \rho^2(x + iy)/(x^2 + y^2) = \rho^2/(x - iy) \tag{8.3.15}$$

or, with $z = x + iy$, $\bar{z} = x - iy$, $z' = x' + iy'$, $\bar{z}' = x' - iy'$:

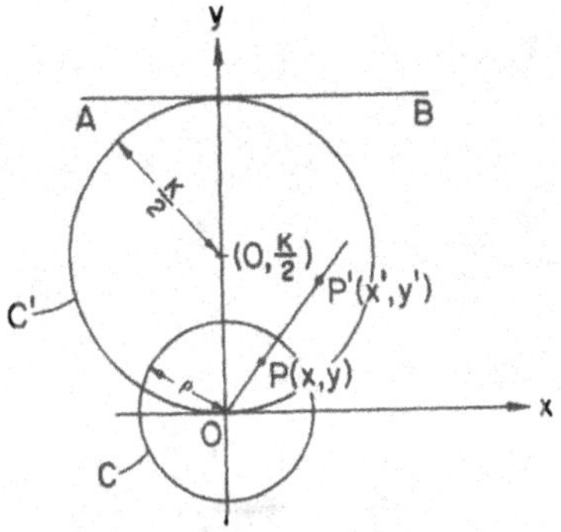

FIG. 8.3.2. The inversion transformation.

$$z' = \rho^2/\bar{z}. \tag{8.3.16}$$

Another transformation:

$$z' = \rho^2/z \tag{8.3.17}$$

is a combination of the inversion with a reflection in the real (x) axis of the z-plane. By applying the inversion (8.3.16) with respect to the circle, C, to points on it, they map onto themselves. The center of the circle maps onto a point at infinity (i.e., into the circle of infinite radius), while any circle not centered at 0, and any straight line map onto a circle or a straight line.

Of special interest is the mapping of a circle passing through the origin (C' in fig. 8.3.2). By the inversion transformation (8.3.16) we obtain:

$$x^2 + (y - K/2)^2 = (K/2)^2; \qquad x = \rho^2 x'/(x'^2 - y'^2), \qquad y = \rho^2 y'/(x'^2 + y'^2). \tag{8.3.18}$$

Several examples are given in the following paragraph to demonstrate the application of this method to the solution of problems of flow through porous media. Additional examples are given in section 9.6 in connection with the interface in coastal aquifers. Many examples can be found in the literature (e.g., Muskat 1937; Mkhitaryan 1947; Polubarinova-Kochina 1952, 1962; Aravin and Numerov 1953, 1965; Harr 1962).

8.3.3 Examples Without a Seepage Face

Example 1, confined flow under an impervious dam (Polubarinova-Kochina 1952, 1962). The physical plane is shown in figure 8.3.3a. AB is an equipotential $\Phi = + KH/2$; BCE is an impervious boundary (i.e., $\Psi = 0$); EF is an equipotential boundary $\Phi = - KH/2$. Because of symmetry, it is sufficient to consider only the region $ABCDA$ ($x \leqslant 0, y \leqslant 0$). Figures 8.3.3b and 8.3.3c show the mapping of the flow domain $ABCDA$ onto the $\tilde{\omega}$-plane and onto the ζ-plane, respectively. Figure 8.3.3b is obtained by using the rules of section 7.3 and then reflecting the result in the q_x axis. All three flow domains are indicated by shading.

Our next step is to find the relationship that maps the $\tilde{\omega}$-plane onto the ζ-plane. This can be done by using the Schwarz–Christoffel transformation (Churchill 1948

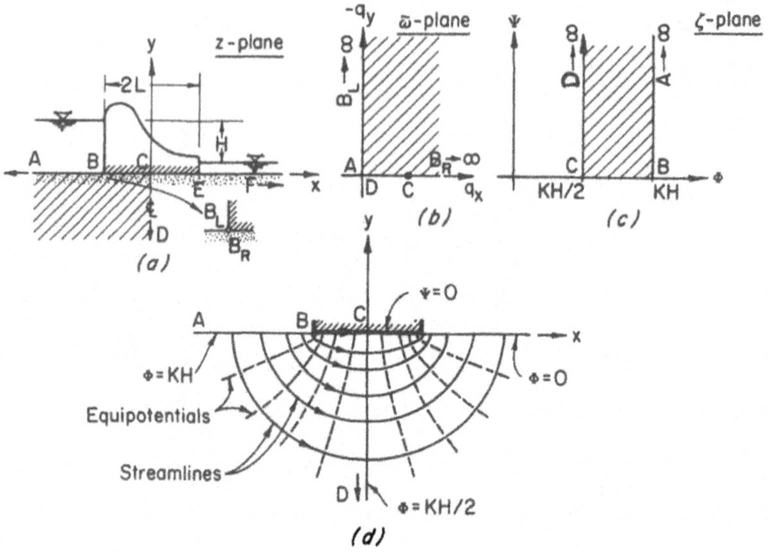

FIG. 8.3.3. z, $\tilde{\omega}$ and ζ planes for flow under an impervious dam on a pervious formation of infinite depth.

and sec. 7.8) and an auxiliary ξ-plane. Both the $\tilde{\omega}$- and the ζ-planes are mapped onto the upper half of ξ-plane. The relationship between $\tilde{\omega}$ and ζ is obtained by eliminating ξ. The result of this procedure is:

$$\tilde{\omega} = C \operatorname{cosec}(\pi\zeta/KH) \tag{8.3.19}$$

where C is a constant. Using (8.3.10), we obtain:

$$z = -\int \frac{d\zeta'}{C \operatorname{cosec}(\pi\zeta'/KH)} + \text{const}$$

and:

$$z = C_1(KH/\pi)\cos(\pi\zeta/KH) + C_2 \tag{8.3.20}$$

where C_1 and C_2 are constants. To determine these constants we introduce the boundary conditions:

at B: $\quad \zeta = \Phi + i\Psi = KH; \qquad z = x + iy = -L$

at C: $\quad \zeta = \Phi + i\Psi = KH/2; \qquad z = 0.$

This leads $C_2 = 0$, $C_1 = \pi L/KH$, and (8.3.20) becomes:

$$z = L\cos(\pi\zeta/KH) \tag{8.3.21}$$

or:

$$\frac{x}{L} + i\frac{y}{L} = \cos\frac{\pi}{KH}(\Phi + i\Psi) = \cos\frac{\pi\Phi}{KH}\cosh\frac{\pi\Psi}{KH} - i\sin\frac{\pi\Phi}{KH}\sinh\frac{\pi\Psi}{KH}.$$

By separating real and imaginary parts, we obtain:

$$\frac{(x/L)}{\cosh(\pi\Psi/KH)} = \cos\frac{\pi\Phi}{KH}; \qquad \frac{(y/L)}{\sinh(\pi\Psi/KH)} = -\sin\frac{\pi\Phi}{KH}.$$

From these we obtain:

$$\frac{(x/L)^2}{\cosh^2(\pi\Psi/KH)} + \frac{(y/L)^2}{\sinh^2(\pi\Psi/KH)} = 1; \qquad \frac{(x/L)^2}{\cos^2(\pi\Phi/KH)^2} - \frac{(y/L)^2}{\sin^2(\pi\Phi/KH)} = 1. \tag{8.3.22}$$

The first of equations (8.3.22) describes a family of confocal ellipses—the streamlines—each for a different constant value of Ψ. The second equation in (8.3.22) describes a family of confocal hyperbolas—the equipotentials. Both families are shown in figure 8.3.3d.

This is the required solution. From it one may determine $\mathbf{q}$ by solving first for $\Phi = \Phi(x, y)$ and then using Darcy's law, or by recalling that $\tilde{\omega} = q_x - iq_y = -d\zeta/dz$.

Example 2, a triangular dam on an impervious bottom without tail water. Figure 8.3.4a shows the z-, ω- and $\tilde{\omega}$-planes (Harr 1962). BC is a seepage face. Since the triangular flow domain ABC is similar in the z-plane and in the $\tilde{\omega}$-plane, we have:

$$\tilde{\omega} = C_1 z \tag{8.3.23}$$

where C_1 is constant. For point B:

$$z = H + iH; \qquad \tilde{\omega} = (K + iK)/2 = q_x - iq_y \tag{8.3.24}$$

where K is the dam's hydraulic conductivity. Hence, $C_1 = K/2H$ and:

$$\tilde{\omega} = Kz/2H = -d\zeta/dz. \tag{8.3.25}$$

By integration:

$$\zeta = -Kz^2/4H + C_2.$$

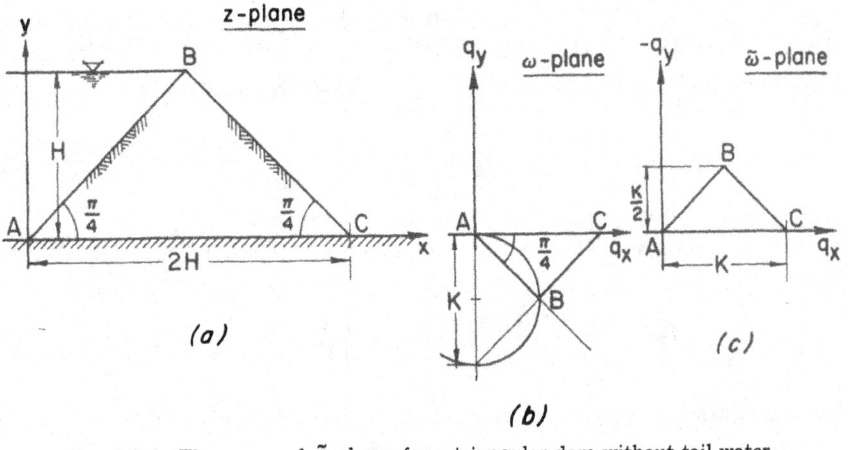

FIG. 8.3.4. The z, ω, and $\tilde{\omega}$ planes for a triangular dam without tail water.

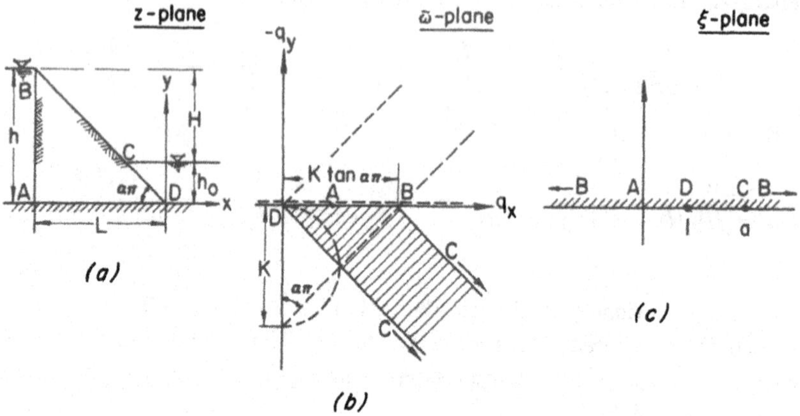

FIG. 8.3.5. A triangular dam with tail water.

For point B: $z = H + iH$, $\Phi = KH$, $\Psi = 0$. Hence, $C_2 = KH(2+i)/2$ and:

$$\zeta = -Kz^2/4H + KH(2+i)/2. \tag{8.3.26}$$

By separating real from imaginary parts, we obtain:

$$\zeta = \Phi + i\Psi = -K(x^2 + 2ixy - y^2)/4H + KH(2+i)/2$$

$$\Phi = K(4H^2 - x^2 + y^2)/4H; \qquad \Psi = K(H^2 - xy)/2H. \tag{8.3.27}$$

Along the bottom $\Psi = Q = KH/2$, and Φ varies according to $\Phi = K(4H^2 - x^2)/4H$. From (8.3.27) it follows that the equipotentials are given by $x^2 - y^2 = \text{const}$ and the streamlines by $xy = \text{const}$ (i.e., hyperbolas normal to the upstream face).

Example 3, a triangular dam with vertical upstream face and with tail water (Harr 1962). This example is given to illustrate the use of an auxiliary plane. Figure 8.3.5a shows the physical z-plane. Figure 8.3.5b shows the $\tilde{\omega}$-plane; the dashed lines indicate that boundaries in the ω-plane are reflected in the q_x-axis, so that the shaded area in the flow domain $ABCDA$ is mapped onto the $\tilde{\omega}$-plane. Selecting an auxiliary ξ-plane, we use the Schwarz–Christoffel transformation (par. 7.8.5) to map the z-plane onto this plane. From (7.8.67) and figure 8.3.5c we have:

$$z = M_1 \int_1^{\xi} \frac{d\xi'}{(\xi' - 1)^{1-\alpha}\xi'^{1/2}} \tag{8.3.28}$$

where M_1 is a constant. For B: $x = -h \operatorname{cotg} \alpha\pi$, $y = h$; $\xi = +\infty$. We have:

$$M_1 = \frac{h(i - \operatorname{cotg} \alpha\pi)}{J}; \qquad J = \int_1^{\infty} (\xi - 1)^{\alpha-1}\xi^{-1/2}\, d\xi. \tag{8.3.29}$$

By substituting $\xi = 1/p$ we note that J is the *beta function* $B(\frac{1}{2} - \alpha, \alpha)$ (Wylie 1951):

$$B(m, n) = \int_0^1 \xi^{m-1}(1-\xi)^{n-1}\,d\xi, \qquad m, n > 0$$

which may also be expressed in terms of *gamma functions* Γ:

$$B(m, n) = \frac{\Gamma(m)\Gamma(n)}{\Gamma(m+n)}; \qquad \Gamma(m) = \int_0^\infty s^{m-1}\exp(-s)\,ds. \tag{8.3.30}$$

Hence we have:

$$J = [\Gamma(\tfrac{1}{2}-\alpha)\Gamma(\alpha)]/\Gamma(\tfrac{1}{2}) \tag{8.3.31}$$

and J and M_1 may be determined for any given value of α. For example, for point C we have:

$$M_1 = \frac{h_0(1-\operatorname{cotg}\alpha\pi)}{J'}, \qquad J' = \int_1^a (\xi-1)^{\alpha-1}\xi^{-1/2}\,d\xi. \tag{8.3.32}$$

Hence, we obtain:

$$J' = J - \int_0^{1/a} (1-p)^{\alpha-1}p^{-\alpha-1/2}\,dp = J - B_{1/a}(\tfrac{1}{2}-\alpha, \alpha) \tag{8.3.33}$$

where $B_{1/a}(\frac{1}{2}-\alpha, \alpha)$ is the *incomplete beta function* (Abramowitz and Stegun 1965).

From (8.3.29), (8.3.32) and (8.3.33), we obtain:

$$\frac{h_0}{h} = 1 - \frac{B_{1/a}(\frac{1}{2}-\alpha, \alpha)}{B(\frac{1}{2}-\alpha, \alpha)}. \tag{8.3.34}$$

From this, by using tables of beta functions (e.g., Abramowitz and Stegun 1965) we may derive a for any given pair of α and h_0/h.

Next we map the $\tilde{\omega}$-plane of figure 8.3.5b on the upper half of the ξ-plane. Again, the Schwarz–Christoffel transformation is employed:

$$\tilde{\omega} = M_2 \int_1^\xi (\xi'-1)^{-1/2-\alpha}(\xi'-a)^{-1}\,d\xi' = -\,d\zeta/dz \tag{8.3.35}$$

where M_2 is a constant. For B: $\xi = -\infty$, $\tilde{\omega} = K\tan\alpha\pi$. Hence:

$$M_2 = iK\tan\alpha\pi \cdot \exp(i\pi\alpha)/J''$$

$$J'' = \int_1^{-\infty} (1-\xi)^{-1/2-\alpha}(\xi-a)^{-1}\,d\xi = (a-1)^{-1/2-\alpha}\Gamma(\tfrac{1}{2}-\alpha)\Gamma(\tfrac{1}{2}+\alpha). \tag{8.3.36}$$

We thus have both z and $\tilde{\omega}$ (i.e., $-\,d\zeta/dz$) related to ξ. In order to find the relationship between them, we multiply the expressions for $d\zeta/dz$ in (8.3.35) by $dz/d\xi$ derived

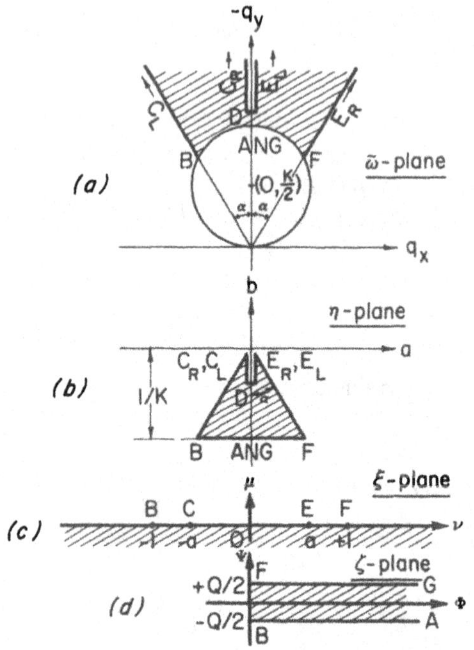

FIG. 8.3.6. Flow from a trapezoidal ditch.

from (8.3.28) and obtain:

$$\zeta - \zeta_0 = \Delta\zeta = -M_1M_2 \int_{\xi_0}^{\xi} \left[\xi'^{-1/2}(\xi' - 1)^{\alpha-1} \int_1^{\xi} (\tau - 1)^{-1/2-\alpha}(\tau - a)^{-1}\, d\tau \right] d\xi'. \tag{8.3.37}$$

From (8.3.37) and (8.3.28) we have a relationship between ζ and z in a parametric form. In principle, this is the required solution.

Example 4, seepage from a trapezoidal ditch (Polubarinova-Kochina 1952, 1962). This example is presented as an illustration of the use of the inversion transformation presented in paragraph 8.3.2.

Figure 7.3.13 shows the z- and ω-planes of this problem. Figure 8.3.6a shows the $\tilde{\omega}$-plane. Figure 8.3.6b shows the inversion of this plane:

$$\eta = 1/\tilde{\omega} = 1/(q_x - iq_y) \tag{8.3.38}$$

by the transformation (8.3.17) with $\rho^2 = 1$, and the origin as center:

$$x' = x/(x^2 + y^2); \quad y' = -y/(x^2 + y^2), \quad x = x'/(x'^2 + y'^2), \quad y = y'/(x'^2 + y'^2).$$

The circle: $x^2 + (y - K/2)^2 = (K/2)^2$ becomes: $y' = -1/K$; the straight line $y = x\tan(90° - \alpha)$ becomes: $y' = x'\tan(90° - \alpha)$, etc.

We now map the flow domain in the η-plane onto the lower half of the auxiliary ξ-plane by:

$$\eta = M \int_0^{\xi} \frac{\xi' \, d\xi'}{(1 - \xi'^2)^{1/2+\alpha/\pi}(a^2 - \xi'^2)^{1-\alpha/\pi}} + N = M\Omega(\xi) + N. \tag{8.3.39}$$

Introducing:

$$J' = \Omega(a) = \int_0^{a} \frac{\xi' \, d\xi'}{(1 - \xi'^2)^{1/2+\alpha/\pi}(a^2 - \xi'^2)^{1-\alpha/\pi}}$$

$$J'' = \int_a^{1} \frac{\xi' \, d\xi'}{(1 - \xi'^2)^{1/2+\alpha/\pi}(\xi'^2 - a^2)^{1-\alpha/\pi}} \tag{8.3.40}$$

and substituting $x = (\xi'^2 - a^2)/(1 - a^2)$ into the expression for J'', we obtain:

$$J'' = \frac{1}{2(1 - a^2)^{1/2}} \int_0^1 x^{\alpha/\pi - 1}(1 - x)^{-1/2-\alpha/\pi} \, dx = \frac{\Gamma(\alpha/\pi)\Gamma(\frac{1}{2} - \alpha/\pi)}{2(1 - a^2)^{1/2}\pi^{1/2}}. \tag{8.3.41}$$

To eliminate the constant N from (8.3.39) we note that E: $\xi = a$ and $\eta = 0$. Hence $N = -MJ'$. To evaluate the constant M we note that at F: $\xi = 1$ and $q_x = K \sin\alpha \cos\alpha$, $-q_y = K \cos^2\alpha$; $\bar{\omega} = q_x - iq_y = K \cos\alpha \sin\alpha + iK \cos^2\alpha = Ki \cos\alpha \exp(-i\alpha)$, $\eta = \exp(i\alpha)/iK \cos\alpha = (\tan\alpha - i)/K$. Then (8.3.39) becomes:

$$\frac{1}{Ki \cos\alpha} \exp(i\alpha) = M[\Omega(1) - J'].$$

Noting that $\Omega(1) = J' - \exp(i\alpha) J''$, we obtain $M = i/KJ'' \cos\alpha$, and:

$$\eta = \frac{i}{KJ'' \cos\alpha} [\Omega(\xi) - J']. \tag{8.3.42}$$

Figure 8.3.6d shows the ζ-plane. The flow domain has the form of an infinite half strip. It is mapped on the lower half of the ξ-plane by:

$$\zeta = (iQ/\pi) \sin^{-1}\xi \tag{8.3.43}$$

where Q is the discharge rate through the bottom of the ditch. We therefore have from (8.3.8):

$$\frac{dz}{d\xi} = \frac{1}{\bar{\omega}} \frac{d\zeta}{d\xi} = \eta \frac{d\zeta}{d\xi} = \frac{-Q}{\pi KJ''(\cos\alpha)\sqrt{1 - \xi^2}} [\Omega(\xi) - J'] \tag{8.3.44}$$

which, after integration with respect to ξ between any two points 1 and 2, yields:

$$z_2 - z_1 = -\frac{Q}{\pi KJ'' \cos\alpha} \left[\int_{\xi_1}^{\xi_2} \frac{\Omega(\xi) - J'}{\sqrt{1 - \xi^2}} \, d\xi \right]. \tag{8.3.45}$$

Detailed integrations are given by Vedernikov (1934), Polubarinova-Kochina (1952, 1962) and Harr (1962). For point E, the integration limits are: $z_1 = 0$, $z_2 = B_1/2$,

$\xi_1 = 0$, $\xi_2 = a$. For the segment EF: $z_F = B/2 + ih$; $\xi_F = 1$, $z_E = B_1/2$, $\xi_E = a$. Along the phreatic surface FG: $z_G = L - i\infty$, $\xi_G = +\infty$; for any intermediate point along the phreatic surface we introduce $z_1 = x_1 + iy_1$, separate (8.3.45) into its real and imaginary parts, and derive the parametric equations $x = x(\xi)$, $y = y(\xi)$ describing the shape of the phreatic surface.

Example 5, problems with a seepage face. Consider the case of the earth dam shown in figure 8.3.7a. Figure 8.3.7b shows the hodograph $\tilde{\omega}$-plane. Figure 8.3.7c shows the ζ-plane, the knowledge of which is essential to any solution by the hodograph method as described above. However, it immediately becomes obvious that the exact position of points E and F and the shape of the curve connecting them (i.e., the representation of the seepage face) in the ζ-plane *are unknown.*

Hence, although it is *always* possible to map the z-plane onto the $\tilde{\omega}$-plane, whenever a seepage face is present, it is, in general, impossible to map the flow domain onto the $\zeta = \Phi + i\Psi$-plane. The only cases where this mapping is possible are those with a horizontal seepage (or outflow) face that is also an equipotential surface (fig. 8.3.8a). In example 2, we avoided the use of the ζ-plane since we could map $\tilde{\omega}$ directly on z. In example 3, we avoided the use of the unknown ζ-plane by introducing the auxiliary ξ-plane, relating both $\tilde{\omega}$ ($= -d\zeta/dz$) and z to ξ. This was possible because of the simple form of the flow domain in the z-plane (i.e., no phreatic surface).

Several other approaches, using various mapping functions, are reviewed in the following paragraphs.

8.3.4 Hamel's Mapping Function

This procedure, first introduced by Hamel (1934), is described in detail by Muskat

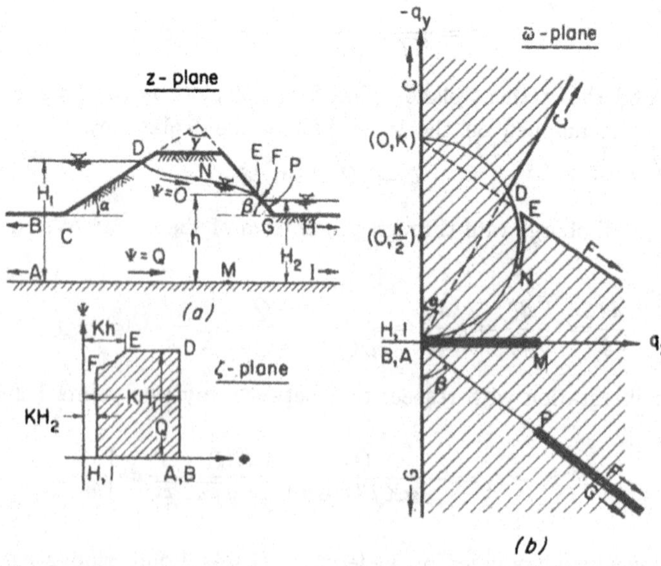

FIG. 8.3.7. The ζ plane for an earth dam with a seepage face.

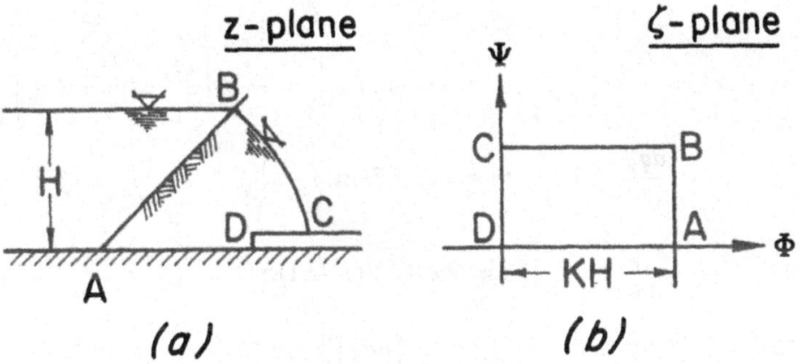

FIG. 8.3.8. A horizontal filter.

(1937) for the particular case of a dam with vertical faces given below as an example.

We already know that it is always possible to map the z-plane on the $\omega = q_x + iq_y$ or $\tilde{\omega} = q_x - iq_y$ planes. Let us define another function:

$$T = \tau + i\theta = -\ln[-f''(z)] = -\ln[-d^2\zeta/dz^2]. \tag{8.3.46}$$

Then:

$$-\frac{d^2\zeta}{dz^2} = -\frac{d}{dz}\left(\frac{d\zeta}{dz}\right) = -\frac{d[d\zeta/dz]/dt}{dz/dt} = \frac{\dot{q}_x - i\dot{q}_y}{n(q_x + iq_y)} \tag{8.3.47}$$

where $\dot{q}_x$ and $\dot{q}_y$ are rates of change of q_x and q_y with time t, and n is porosity. Since $-d\zeta/dz = \tilde{\omega}$, we have:

$$-d^2\zeta/dz^2 = d(q_x - iq_y)/d(x + iy). \tag{8.3.48}$$

Let us define:

$$\dot{q}_x - i\dot{q}_y = nA\exp(-i\beta); \qquad q_x + iq_y = q\exp(i\alpha) \tag{8.3.49}$$

where $q^2 = q_x^2 + q_y^2$; $(nA)^2 = \dot{q}_x^2 + \dot{q}_y^2$; and α and β are the moduli. Hence:

$$-d^2\zeta/dz^2 = (A/q)\exp[-i(\alpha + \beta)] \tag{8.3.50}$$

$$\tau + i\theta = -\ln(A/q)\exp[-i(\alpha + \beta)] = -\ln(A/q) + i(\alpha + \beta). \tag{8.3.51}$$

By comparing real and imaginary parts, we have:

$$\tau = -\ln(A/q), \qquad \theta = \alpha + \beta \quad \text{for} \quad A/q > 0$$
$$\theta = \alpha + \beta - \pi \quad \text{for} \quad A/q < 0. \tag{8.3.52}$$

We now examine how the various types of boundaries in the z-plane are mapped onto the T-plane.

(a) *A streamline* or *an impervious boundary* which makes a constant angle with the $+x$ axis. Along this boundary:

$$dy/dx = q_y/q_x = \tan\alpha; \qquad q_y = q_x \tan\alpha, \qquad dy = (\tan\alpha)\,dx$$

$$-\frac{d^2\zeta}{dz^2} = \frac{d(q_x - iq_y)}{d(x + iy)} = \frac{dq_x}{dx}\cdot\frac{(1 - i\tan\alpha)}{(1 + i\tan\alpha)} = \frac{dq_x}{dx}(\cos 2\alpha - i\sin 2\alpha). \tag{8.3.53}$$

Therefore:

$$\tau + i\theta = -\ln\left[\frac{dq_x}{dx}(\cos 2\alpha - i\sin 2\alpha)\right] = -\ln\left[\frac{dq_x}{dx}\exp(-2i\alpha)\right]. \quad (8.3.54)$$

For $\quad \dfrac{dq_x}{dx} > 0, \qquad \theta = 2\alpha + 2m\pi. \qquad (8.3.55)$

For $\quad \dfrac{dq_x}{dx} < 0, \qquad \theta = 2\alpha + (2m-1)\pi$

$$\tau = -\ln\left[\frac{dq_x}{dx}\right]. \quad (8.3.56)$$

(b) For an *equipotential boundary*, following the same procedure as in (a) above, we have:

$$dy/dx = -q_x/q_y = \tan\alpha$$

$$-\frac{d^2\zeta}{dz^2} = \frac{d(q_x - iq_y)}{d(x+iy)} = \frac{dq_x}{dy}\cdot\frac{(1 + i\,\mathrm{cotg}\,\alpha)}{(\mathrm{cotg}\,\alpha + i)} = -\frac{dq_y}{dx}(\tan 2\alpha + i\cos 2\alpha). \quad (8.3.57)$$

For $\quad dq_x/dy > 0, \qquad \theta = -\tan^{-1}(\cos 2\alpha/\sin 2\alpha) + 2m\pi.$

For $\quad dq_x/dy < 0, \qquad \theta = -\tan^{-1}(\cos 2\alpha/\sin 2\alpha) + (2m-1)\pi. \qquad (8.3.58)$

For example, for $\alpha = \pi/2$ (i.e., a vertical equipotential), $dq_x/dy > 0$, $\theta = \pi/2 + 2m\pi$; for $\alpha = 0$ (a horizontal equipotential), $dq_x/dy = -dq_y/dx > 0$, $\theta = -\pi/2 + 2m\pi$. τ may be obtained by the procedure described in (a) above.

(c) For a *seepage face*, making a constant angle α with the $+x$ axis, we have (par. 7.1.8), with $\beta_{sx} = \alpha$; $\beta_{sz} = \pi/2 - \alpha$:

$$-q_x\cos\alpha - q_y\sin\alpha = K\sin\alpha. \quad (8.3.59)$$

Hence:

$$-dq_x\cos\alpha = dq_y\sin\alpha$$

and:

$$-\frac{d^2\zeta}{dz^2} = \frac{dq_x}{dy}\frac{(1 + i\,\mathrm{cotg}\,\alpha)}{(\mathrm{cotg}\,\alpha + i)} \quad (8.3.60)$$

which yields for τ and θ the same expressions as for an equipotential surface.

(d) For a *steady free surface*, we may start from

$$q_x^2 + q_y^2 + Kq_y = 0; \qquad dq_x/dq_y = -(2q_y + K)/2q_x. \quad (8.3.61)$$

Since $dy/dx = q_y/q_x$, we obtain:

$$-\frac{d^2\zeta}{dz^2} = \frac{dq_y}{dx}\frac{\{-[(2q_y+K)/2q_x] - i\}}{[1 + i(q_y/q_x)]} = -\frac{dq_y}{dx}\frac{(2q_xq_y + Kq_x/2) + i(q_x^2 - q_y^2 - Kq_y/2)}{-Kq_y}.$$

(8.3.62)

Since along the phreatic surface: $q_x = -K \sin\alpha\cos\alpha$, $q_y = -K\sin^2\alpha$, we obtain from (8.3.62):

$$-\frac{d^2\zeta}{dz^2} = \frac{dq_y}{dx}\frac{\exp(-3i\alpha)}{2\sin\alpha}. \tag{8.3.63}$$

For $dq_y/dx > 0, \quad \theta = 3\alpha + 2m\pi$

For $dq_y/dx < 0, \quad \theta = 3\alpha + (2m-1)\pi.$

With these expressions for θ along the boundaries, the solution procedure is as follows.

Having mapped the flow domain in the z-plane on the hodograph, $\tilde{\omega}$-plane, we use the expressions derived above to compute and record the values of θ corresponding to varions portions of the boundary. Then the $\tilde{\omega}$-plane is mapped on a new ξ-plane ($\xi = \nu + i\mu$). All the boundaries in the $\tilde{\omega}$-plane are mapped onto only the real axis of the ξ-plane.

Now we use Poisson formula, which gives the value of $T = \tau + i\theta$ for each point in the ξ-plane from the known values of θ along the real axis of this plane:

$$-iT = \theta - i\tau = -i\tau_0 + \frac{1}{\pi i}\int_{-\infty}^{+\infty} \frac{\theta(\nu)\cdot(\xi\nu+1)}{(\nu-\xi)(1+\nu^2)}\,d\nu \tag{8.3.64}$$

where τ_0 is an arbitrary constant. From this integral we obtain an expression $T = F_1(\xi)$. Since $T = -\ln[-d^2\xi(z)/dz^2] = -\ln(d\tilde{\omega}/dz)$, and with the transformation of $\tilde{\omega}$ on the ξ-plane, $\xi = F_2(\tilde{\omega})$, we have:

$$-\ln(d\tilde{\omega}/dz) = F_2(\tilde{\omega}). \tag{8.3.65}$$

This leads to:

$$d\tilde{\omega}/dz = \exp\{-F_2(\tilde{\omega})\}; \qquad dz = d\tilde{\omega}/\exp\{-F_2(\tilde{\omega})\}$$

$$z = \int \frac{d\tilde{\omega}}{\exp\{-F_2(\tilde{\omega})\}} + C_1 = F_3(\tilde{\omega}) \tag{8.3.66}$$

$$\tilde{\omega} = F_3^{-1}(z) = -\frac{d\zeta}{dz}; \qquad \zeta = -\int F_3^{-1}(z)\,dz + C_2. \tag{8.3.67}$$

From (8.3.66) we obtain $q_x = q_x(x, y)$, $q_y = q_y(x, y)$, and also the equation for the phreatic surface. From (8.3.67), we obtain $\Phi = \Phi(x, y)$ and $\Psi = \Psi(x, y)$.

Let us illustrate Hamel's method by the example of an earth dam with vertical upstream and downstream surfaces (fig. 8.3.9a). The use of Hamel's method is dictated by the presence of the seepage face BC, the shape of which in the ζ-plane is unknown. The flow domain in the z-plane is mapped onto the $\tilde{\omega}$-plane (fig. 8.3.9b).

By the procedure described above, we determine the value of θ for each portion of the boundary in the $\tilde{\omega}$-plane:

along BC (seepage face): $\quad dq_x/dy < 0, \quad \theta = \pi/2 - \pi = -\pi/2$

along CDE (phreatic surface): $\quad dq_y/dx < 0, \quad \theta = 3\alpha - \pi$

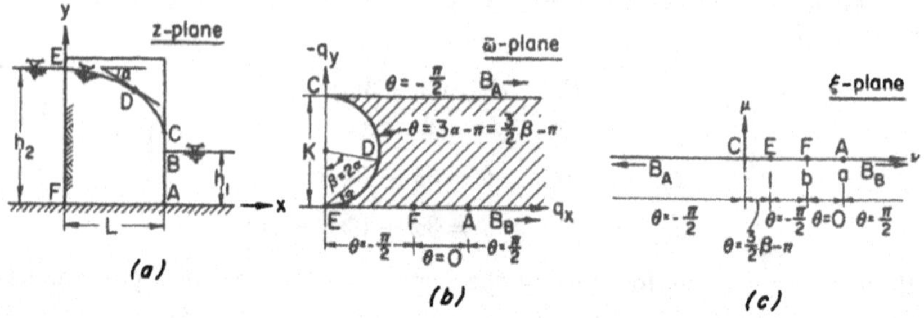

FIG. 8.3.9. The z, $\tilde{\omega}$ and ξ planes for an earth dam with vertical faces (Muskat, 1937).

along EF (equipotential): $dq_x/dy < 0,\quad \alpha = \pi/2,\quad \theta = -\pi/2$

along FA (streamline): $dq_x/dx > 0,\quad \alpha = 0,\quad \theta = 0$

along AB (equipotential): $dq_x/dy > 0,\quad \alpha = \pi/2,\quad \theta = \pi/2.$

The $\tilde{\omega}$-plane is mapped on the ξ-plane by the modular elliptic function:

$$\xi = \lambda(i\tilde{\omega}/K + 1). \tag{8.3.68}$$

Several typical values of this function are: $\lambda(0) = 0$, $\lambda(1) = 1$. $\lambda(\infty) = -\infty$, $\lambda(1 + i\infty) = +\infty$. Muskat (1937) discusses in detail some properties of this function. Hence:

	B_B	A	F	E	C	B_A
$\tilde{\omega}$	$+\infty, 0$	$q_x(A), 0$	$q_x(F), 0$	$0, 0$	$0, K$	∞, K
ξ	$+\infty, 0$	$+a, 0$	$+b, 0$	$1, 0$	$0, 0$	$-\infty, 0$

The constants a and b (fig. 8.3.9c) are to be determined from the boundary conditions. The semicircle CDE is mapped onto the segment EC. The transformation is performed by using the equation:

$$\tilde{\omega} - Ki/2 = (K/2)\exp\{i(\beta - \pi/2)\}. \tag{8.3.69}$$

For the circle in the $\tilde{\omega}$-plane, we obtain:

$$\xi = \lambda[1 + \exp(i\beta)]/2. \tag{8.3.70}$$

The ξ-plane is shown in figure 8.3.9c. Using Poisson's formula (8.3.64), we obtain by integrating over the various segments:

$$\theta - i\tau = -\tau_0 + \frac{1}{\pi i}\left[-\frac{3}{2}\int_0^1 \frac{\beta(\nu)\,d\nu}{\nu - \xi} + \frac{\pi}{2}\ln\frac{\xi - 1}{(a - \xi)(\xi - b)\xi}\right] \tag{8.3.71}$$

where τ_0 is a constant. The first term in the square brackets of (8.3.71) gives the contribution of the segment CE. The second one is the contribution of all other segments of the ν-axis. Following the procedure leading to (8.3.66) and (8.3.67), we obtain:

$$z = \int \frac{d\tilde{\omega}}{\exp[-F_2(\tilde{\omega})]} + C = C \int \left[\frac{1-\xi}{\xi(\xi-a)(\xi-b)}\right]^{1/2} \exp\left[-\frac{3}{2\pi}\int_0^1 \frac{\beta(\nu)\, d\nu}{\nu-\xi}\right] d\tilde{\omega}. \tag{8.3.72}$$

In principle, this is the solution of the problem as it gives z as a function of $\tilde{\omega} = q_x - iq_y$, and hence $q_x = q_x(x, y)$ and $q_y = q_y(x, y)$. Muskat (1937) gives details of this analysis.

Thus, Hamel's approach provides a tool for solving problems involving a seepage face.

8.3.5 Zhukovski's and Other Mapping Functions

This is another approach (Polubarinova-Kochina 1952, 1962; Harr 1962) that is also applicable to cases with a seepage face.

Using the definition of piezometric head, we define a function θ_1:

$$\theta_1 = p/\gamma = \Phi - Ky; \qquad \Phi = K\varphi. \tag{8.3.73}$$

The function θ_1 is harmonic in x and y (i.e., $\nabla^2\theta_1 = 0$ when flow in the vertical xy plane is considered), as are Φ and Ψ. A conjugate function of θ_1 is θ_2, defined by:

$$\theta_2 = \Psi + Kx. \tag{8.3.74}$$

With θ_1 as the real part and θ_2 as the imaginary part, a complex function θ is now derived:

$$\theta = \theta_1 + i\theta_2 = (\Phi - Ky) + i(\Psi + Kx) = (\Phi + i\Psi) + iK(x+iy) = \zeta + iKz \tag{8.3.75}$$

where $\zeta = \Phi + i\Psi$ is the complex specific discharge potential, and $z = x + iy$. In some publications the specific discharge potential is defined by $\Phi' = -K\varphi$, then $\theta'_1 = \Phi' + Ky$, $\theta'_2 = \Psi' - Kx$, $\theta' = \zeta' - iKz$, $\zeta' = \Phi' + i\Psi'$.

Along both a free surface and a seepage face, $\Phi = Ky$; hence in either case we have $\theta_1 = 0$. Hence, by mapping the flow domain in the z-plane onto the θ-plane where θ_1 (parallel to x) and θ_2 (parallel to y) are the coordinate axes, both the free surface and the seepage face will be mapped on the ordinate axis. When the boundary of saturation (par. 7.1.7) is being considered, we have $p = -p_c$ along it. Hence $\theta_1 = -p_c/\gamma = \text{const}$, and the boundary is mapped onto a line parallel to the ordinate. Again, as in the case of the hodograph mapping (par. 8.3.2), the a priori unknown geometry in the physical, z-plane, becomes a known one through this transformation. In fact, it is because of this advantage that Zhukovski (1923) introduced the θ function that is called *Zhukovski's function*. It is also called the *complex pressure* (*or pressure head*) (Bear, Zaslavsky and Irmay 1968). The θ-plane is also called *Zhukovski's θ-plane*. Aravin and Numerov (1953, 1965) refer to the method of solution that employs Zhukovski's function as the *Vedernikov–Pavlovski method*.

Now,

$$\frac{d\theta}{dz} = \frac{d(\zeta + iKz)}{dz} = \frac{d\zeta}{dz} + iK = -\tilde{\omega} + iK. \tag{8.3.76}$$

Hence:

$$z = \int \frac{d\theta}{-\tilde{\omega} + iK}. \tag{8.3.77}$$

Once we know the mapping function of θ on $\tilde{\omega}$ (either directly or through intermediate ξ-planes), equation (8.3.77) may be solved to yield a relationship between θ and z. From this, by comparing real and imaginary parts, we obtain $\Phi = \Phi(x, y)$ and $\Psi = \Psi(x, y)$.

As in the hodograph method described in paragraphs 8.3.2 and 8.3.3, in many cases we do not know the shape of the boundary of the flow domain in the θ-plane, especially when we have equipotential (inflow and outflow) boundaries.

Some authors use Zhukovski's function defined as $\theta = -i(\zeta + iKz) = Kz - i\zeta$, or $G = z - i\zeta/K$ (Aravin and Numerov 1953, 1965). In this case the seepage face and the phreatic surface are mapped on the real axis of the θ-plane.

An example involving both a seepage face and a phreatic surface is the case of *a trapezoidal drainage ditch with a very shallow water table* (Vedernikov 1939; Polubarinova-Kochina 1952, 1962). Figure 8.3.10a shows the z-plane. Figure 8.3.10b

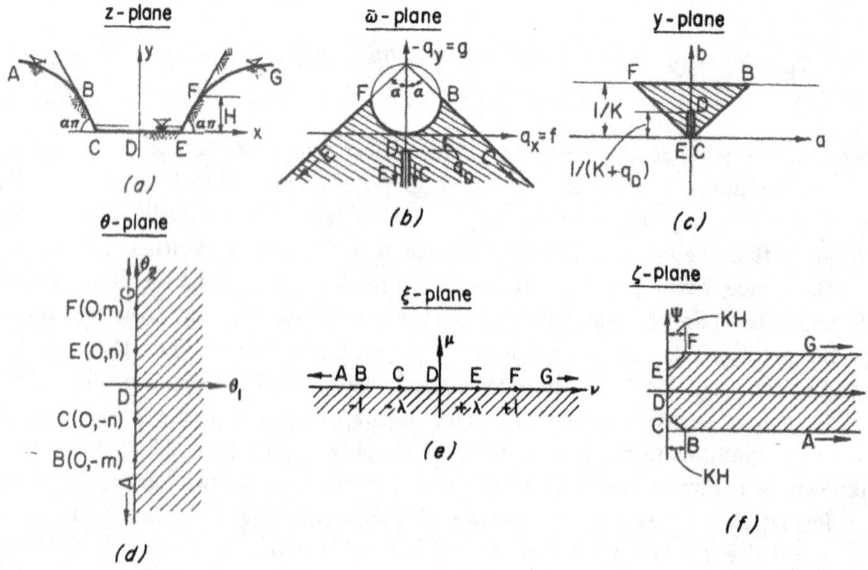

FIG. 8.3.10. A trapezoidal drainage ditch.

shows the $\tilde{\omega}$-plane. Since the entire boundary of the flow is composed of a phreatic surface, a horizontal equipotential line and a seepage face, we have on it $\Phi - Ky = 0$, and it is most convenient to use Zhukovski's function (8.3.75). Figure 8.3.10c shows the $1/\tilde{\omega}$-plane (par. 8.3.2) where the transformation (8.3.17) is performed with respect to a circle (radius unity) centered at point $M(0, iK)$ in the $\tilde{\omega}$-plane. The transformation in this case becomes:

$$\eta - \eta_0 = 1/(\widetilde{\omega - \omega_0}) = 1/(\tilde{\omega} - iK).$$

By choosing $\eta_0 = 0$, we obtain

$$\eta = 1/(\tilde{\omega} - iK). \tag{8.3.78}$$

Let $\eta = a + ib$ and $\tilde{\omega} = f + ig$. Then, from (8.3.78) it follows that:

$$a + ib = \frac{1}{f + i(g - K)} = \frac{f - i(g - K)}{f^2 + (g - K)^2};$$

$$a = \frac{f}{f^2 + (g - K)^2}; \qquad b = -\frac{g - K}{f^2 + (g - K)^2}$$

or

$$f + i(g - K) = \frac{1}{a + ib} = \frac{a - ib}{a^2 + b^2}; \qquad f = \frac{a}{a^2 + b^2}; \qquad g = K - \frac{b}{a^2 + b^2}$$

and the circle $f^2 + (g - K/2)^2 = (K/2)^2$ becomes $b = 1/K$, and line $g = K + f \tan \beta$ becomes $b = -a \tan \beta = a \tan(-\beta)$. These relationships lead to the transformation shown in figure 8.3.10c.

We note that from (8.3.76) it follows that:

$$dz/d\theta = -1/(\tilde{\omega} - iK) = -\eta. \tag{8.3.79}$$

We now map the η-plane on the lower half of a ξ-plane by using the Schwarz–Christoffel transformation:

$$\eta = M \int_0^\xi \frac{\xi' \, d\xi'}{(1 - \xi'^2)^{\alpha+1/2}(\lambda^2 - \xi'^2)^{1-\alpha}} + N \tag{8.3.80}$$

where λ is defined in figure 8.3.10e. Polubarinova-Kochina (1952, 1962) gives the details of the analysis leading to the constants M and N. The result of this analysis is:

$$\eta = -\frac{dz}{d\theta} = -\frac{2\sqrt{1 - \lambda^2}\, iA}{K} \int_0^\xi \frac{\xi' \, d\xi'}{(1 - \xi'^2)^{\alpha+1/2}(\lambda^2 - \xi'^2)^{1-\alpha}} + \frac{i}{q_D + K} \tag{8.3.81}$$

where $q_D = q_y$ at D.

Our next step is to map the θ-plane shown on figure 8.3.10d onto the same lower half of the ξ-plane. We obtain:

$$\theta = im\xi. \tag{8.3.82}$$

At F, $\theta = im = \zeta_F + iKz_F = KH + iQ/2 + iK(x_F + iH) = i(Q/2 + Kx_F)$. Hence, $m = Q/2 + Kx_F$ (fig. 8.3.10f). Hence $d\theta = im\, d\xi = i(Q/2 + Kx_F)\, d\xi$. We now insert this expression in (8.3.81) to obtain an expression, in the form of an integral along ξ, relating z to ξ and hence to θ, to ζ, etc. A detailed analysis is given by Polubarinova-Kochina (1952, 1962).

8.3.6 A Graphic Solution of the Hodograph Plane

We have seen above that we can always map the flow domain onto the $\tilde{\omega} = q_x - iq_y$ plane. Our problem is to determine $\zeta = \zeta(\tilde{\omega})$ or $\tilde{\omega} = \tilde{\omega}(\zeta)$ in order to apply the hodograph method. Along the boundary of a confined flow domain, we know the values of Φ and Ψ. Along a phreatic surface or a seepage face both Φ and Ψ may vary, but we know the relationships governing these variations. For example:

(a) along a seepage face and a phreatic surface without accretion, $\Phi = Ky$;

(b) along a phreatic surface with accretion $\Psi + Nx = \text{const.}$

The mapping $\zeta = \zeta(\tilde{\omega})$ is a conformal one (both $\zeta = \Phi + i\Psi = f(z)$ and $\tilde{\omega} = q_x - iq_y = g(z)$ are analytic functions of z (paragraph 7.8.4)). This means that the two orthogonal families of curves, $\Phi = \text{const}$ and $\Psi = \text{const}$, in the ζ-plane (fig. 8.3.11a) are mapped onto two orthogonal families of curves in the $\tilde{\omega}$-plane (fig. 8.3.11b); they correspond to $\Phi = \text{const}$ and $\Psi = \text{const}$. We thus have (par. 7.8.4)

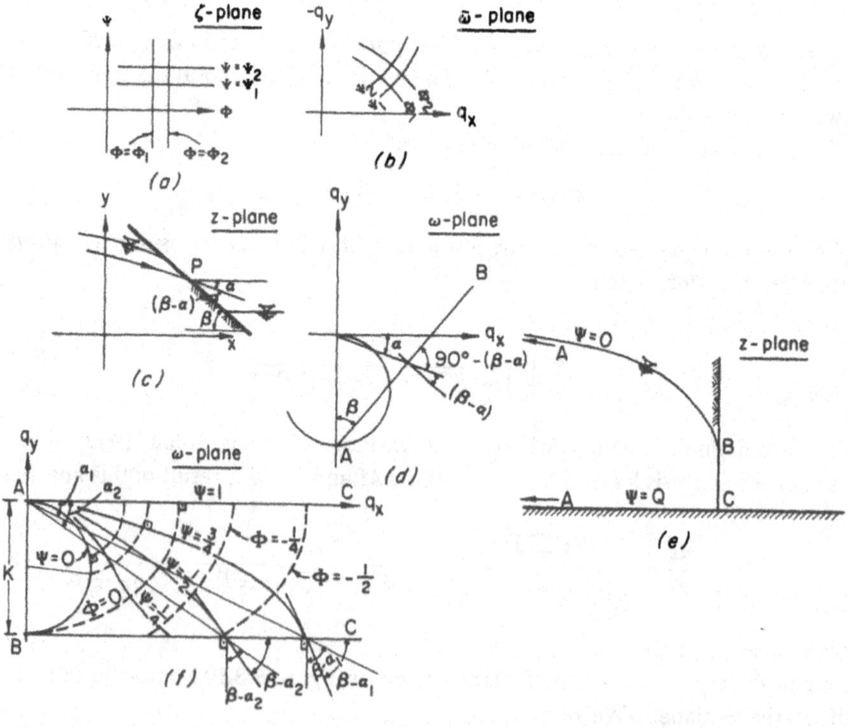

FIG. 8.3.11. A graphical solution of the hodograph plane.

a fictitious flow in the $\tilde{\omega}$-plane with:

$$\partial^2\Phi/\partial q_x^2 + \partial^2\Phi/\partial q_y^2 = 0, \qquad \partial^2\Psi/\partial q_x^2 + \partial^2\Psi/\partial q_y^2 = 0. \tag{8.3.83}$$

With the given boundary conditions on Φ and Ψ in the $\tilde{\omega}$-plane, we may now solve for $\Phi = \Phi(q_x, -q_y)$, $\Psi = \Psi(q_x, -q_y)$ by any of the known methods (numerical, graphic, analytical or by means of an analog). Whenever a seepage face is present, we use the condition $\Phi = Ky$ on it in a trial and error procedure.

Another interesting observation is the following. Figures 8.3.11c and d show a point P on the seepage face in the z- and ω-planes. Since the mapping is a conformal one, the angle between the streamline and the seepage face at a point P on the latter is preserved (actually, since we use the ω-plane, rather than the $\tilde{\omega}$-plane, its sense is reversed). Hence, for every point P on the line representing the seepage face in the ω-plane, we determine α and from it the angle β-α that the streamline at P makes with the seepage face. This value serves as a guide in the graphic determination of the Φ-Ψ orthogonal network. An example (Polubarinova-Kochina 1952, 1962) is given in figure 8.3.11e. The flow is from infinity (point A with $\Phi = \infty$) to the seepage face BC. We set $\Phi = 0$ arbitrarily at point B.

Once the Φ-Ψ network has been derived in the ω- (or $\tilde{\omega}$-) plane by a graphic or an analog technique, we must transform the curves from this plane back to z-plane. To do this we note that:

$$\tilde{\omega} = -\partial\zeta/dz = q_x - iq_y = q\exp(-i\alpha); \qquad z = -\int \frac{d\zeta}{\tilde{\omega}(\zeta)} + \text{const}$$

where $q = |\mathbf{q}|$ and the argument α is the angle between $\mathbf{q}$ and $+x$. Also, by (8.3.7):

$$1/\tilde{\omega} = \omega/q^2 = \exp(i\alpha)/q.$$

Hence, the expression:

$$d\zeta/\tilde{\omega}(\zeta) = (1/q)\exp(i\alpha)\, d(\Phi + i\Psi)$$

may be developed into its real and imaginary parts:

$$z = x + iy = -\int \frac{1}{q}(\cos\alpha\, d\Phi - \sin\alpha\, d\Psi) - i\int \frac{1}{q}(\sin\alpha\, d\Phi + \cos\alpha\, d\Psi). \tag{8.3.84}$$

Along a streamline, $d\Psi = 0$, and therefore the streamline in the z-plane may be derived from:

$$x = -\int_{\Phi_0}^{\Phi} \frac{\cos\alpha}{q}\, d\Phi; \qquad y = -\int_{\Phi_0}^{\Phi} \frac{\sin\alpha}{q}\, d\Phi \tag{8.3.85}$$

along an equipotential line, $d\Phi = 0$. Its equation in the z-plane is, therefore, given by:

$$x = \int_{\Psi_0}^{\Psi} \frac{\sin\alpha}{q}\, d\Psi; \qquad y = -\int_{\Psi_0}^{\Psi} \frac{\cos\alpha}{q}\, d\Psi \tag{8.3.86}$$

where Φ_0, Ψ_0 are arbitrary values of Φ and Ψ. The integration of (8.3.85) and (8.3.86) can be performed graphically from the curves $\Phi = \text{const}$ and $\Psi = \text{const}$ in the ω-plane (fig. 8.3.11f). Integrating (8.3.85) along $\Psi = 0$ (i.e., the half circle AB) gives the shape of the phreatic surface. The condition $\Phi = Ky$ (with $\Phi = 0$ at B) may be used to examine the accuracy of the Φ distribution along the phreatic surface.

The advantage of graphic solutions in the hodograph plane over a graphic solution directly in the z-plane is that in the former the geometry of the seepage face and the phreatic surface is known.

8.4 Linearization Techniques and Solutions

Although the nonlinearity introduced by the presence of a free-surface boundary condition may be removed by introducing the Dupuit approximation, the resulting partial differential equation describing the flow is still a nonlinear one. Several examples of solving this equation—the Boussinesq equation—in an exact manner were described in section 8.2. Another approach to the solution of such flow problems is to linearize either the partial differential equation describing the phreatic surface boundary condition, or the Boussinesq equation. In both cases the problem becomes a linear one, the solution of which can be obtained by standard mathematical methods. Several linearization techniques are described in the following paragraphs.

8.4.1 First Method of Linearization of the Boussinesq Equation

This simplest method has been extensively employed in solving phreatic flow problems in the fields of ground water hydrology and agricultural drainage.

When the depth (over a horizontal impervious bottom) of the saturated zone $h = h(x, y, t)$ varies only slightly, i.e., $h = \bar{h} + h'$; $h' \ll \bar{h}$, where $\bar{h}$ is the average thickness and h' is the deviation of h from $\bar{h}$, equation (8.2.10) becomes:

$$K\bar{h}[\partial^2 h/\partial x^2 + \partial^2 h/\partial y^2] + N = n_e\, \partial h/\partial t. \tag{8.4.1}$$

In a similar way (8.2.11) for a sloping impervious bottom becomes:

$$K\bar{h}\left(\frac{\partial^2 h}{\partial x^2} + \frac{\partial^2 h}{\partial y^2}\right) - K\left[\frac{\partial}{\partial x}\left(\eta\frac{\partial h}{\partial x}\right) + \frac{\partial}{\partial y}\left(\eta\frac{\partial h}{\partial y}\right)\right] + N = n_e\frac{\partial h}{\partial t}. \tag{8.4.2}$$

For the case of one-dimensional flow in the xz plane (8.4.1) and (8.4.2) reduce to:

$$K\bar{h}\frac{\partial^2 h}{\partial x^2} + N = n_e\frac{\partial h}{\partial t} \tag{8.4.3}$$

$$K(\bar{h} - \eta)\frac{\partial^2 h}{\partial x^2} - K\frac{\partial h}{\partial x}\frac{\partial \eta}{\partial x} + N = n_e\frac{\partial h}{\partial t}. \tag{8.4.4}$$

One should recall that $\bar{h}$ is the elevation of the water table above some horizontal reference level (say the horizontal bottom of the aquifer, whenever this bottom is impervious), whereas when the impervious bottom is not horizontal, $b = h - \eta$ is the saturated thickness of the flow domain. The main feature of equations (8.4.1) through (8.4.4) is that they are *linear in h*. Equations (8.2.7) through (8.2.9) may be linearized with respect to b in a similar manner.

The most common case is that of flow without accretion in the xz vertical plane with a horizontal impervious bottom. The linearized partial differential equation, whose solution in this case is $h = h(x, t)$, is derived from (8.4.1):

$$K\bar{h}\frac{\partial^2 h}{\partial x^2} = n_e \frac{\partial h}{\partial t} \quad \text{or} \quad \frac{\partial^2 h}{\partial x^2} = \frac{n_e}{\bar{T}}\frac{\partial h}{\partial t}; \qquad a^2 \frac{\partial^2 h}{\partial x^2} = \frac{\partial h}{\partial t}; \qquad a^2 = \frac{\bar{T}}{n_e} \tag{8.4.5}$$

where $\bar{T} = K\bar{h}$ plays the role of *aquifer transmissivity*, while n_e is the effective porosity (or specific yield) of the aquifer. Equation (8.4.5) is a parabolic, second-order, linear partial differential equation known as the *one-dimensional heat conduction, or diffusion, equation*. A vast number of solutions for various boundary conditions is available in works dealing with heat conduction (e.g., Carslaw and Jaeger 1959), molecular diffusion (e.g., Crank 1956), and many other fields of physics where this equation is often encountered.

Some of the more important elementary solutions of (8.4.5) are:

(a) *the source solution* (Carslaw and Jaeger 1959); it can easily be shown that:

$$h = t^{1/2} \exp(-n_e x^2/4\bar{T}t) \tag{8.4.6}$$

satisfies (8.4.5). The main properties of this solution are:

$$h \to 0 \quad \text{as} \quad t \to 0 \quad \text{for fixed} \quad x \neq 0$$

$$h \mapsto \infty \quad \text{as} \quad t \to 0 \quad \text{at} \quad x = 0$$

$$\int_{-\infty}^{+\infty} n_e h \, dx = 2n_e(\pi\bar{T}/n_e)^{1/2} \quad \text{for all} \quad t > 0.$$

Thus, this solution corresponds to the case where the quantity of water $2n_e(\pi\bar{T}/n_e)^{1/2}$ per unit length is instantaneously released at $x = 0$ at time $t = 0$. Since (8.4.5) is linear, so that the principle of superposition (sec. 7.5) is applicable, solutions of (8.4.5) may be obtained by integrating (8.4.6) with respect to either x or t;

(b) *the error function solution* (Carslaw and Jaeger 1959) follows from the discussion in (a) above. Equation (8.4.5) is also satisfied by:

$$\int_0^x t^{-1/2} \exp(-n_e x^2/4\bar{T}t)\, dx = 2(\bar{T}/n_e)^{1/2} \int_0^{x/2\sqrt{\bar{T}t/n_e}} \exp(-\xi^2)\, d\xi. \tag{8.4.7}$$

With the definition of the *error function* $\Phi(\alpha)$ as:

$$\Phi(\alpha) \equiv \operatorname{erf} \alpha = \frac{2}{\sqrt{\pi}} \int_{\xi=0}^{\alpha} \exp(-\xi^2)\, d\xi \tag{8.4.8}$$

the error function solution becomes:

$$h = A \operatorname{erf}(x/2\sqrt{\bar{T}t/n_e}) = A \operatorname{erf}(n_e x^2/4\bar{T}t)^{1/2} = A \operatorname{erf} \alpha \tag{8.4.9}$$

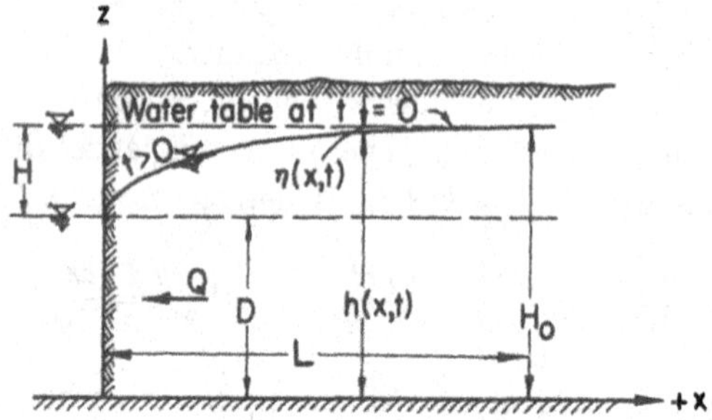

FIG. 8.4.1. Return flow from bank storage into a ditch.

where $\alpha = (n_e\, x^2/4\, \tilde{T}t)^{1/2}$ and A is a constant determined in each case by the boundary conditions.

(c) A third elementary solution presented by Carslaw and Jaeger (1959) is a solution of the form $t^m f(\sqrt{n_e\, x^2/4\, \tilde{T}t})$, where $f(\alpha)$ satisfies the differential equation:

$$(d^2f/d\alpha^2) + 2\alpha(df/d\alpha) - 4mf = 0; \qquad \alpha = (n_e\, x^2/4\, \tilde{T}t)^{1/2}. \tag{8.4.10}$$

Several examples of solutions of (8.4.5) satisfying specified initial and boundary conditions are given below.

Example 1. Consider the case shown in figure 8.4.1, where at $t = 0$ the water level in the stream is suddenly lowered by a distance H. Assuming $H \ll D$, we may define the average transmissivity by $\tilde{T} = KH_0$. For the boundary and initial conditions:

$$t \leqslant 0, \quad x > 0: \quad h = H_0 \quad (\text{or} \quad \eta = 0); \qquad t > 0, \quad x = 0: \quad h = D \quad (\eta = H_0 - D)$$

the solution of (8.4.5) satisfying these conditions is:

$$\eta(x, t) = H_0 - h = (H_0 - D)[1 - \operatorname{erf} \alpha] = (H_0 - D) \operatorname{erfc} \alpha$$

$$h = D + (H_0 - D) \operatorname{erf} \alpha; \qquad \alpha = (n_e\, x^2/4\, \tilde{T}t)^{1/2}. \tag{8.4.11}$$

The outflow rate at $x = 0$ is given by:

$$Q = 2(H_0 - D)KH_0/\sqrt{4\pi\tilde{T}t/n_e}. \tag{8.4.12}$$

The volume of water drained during time t from the aquifer into the channel is given by:

$$U(t) = \int_0^t Q\, dt = (H_0 - D)n_e(4\tilde{T}t/\pi n_e)^{1/2}. \tag{8.4.13}$$

If in figure 8.4.1 we set $D = 0$, so that the water level in the channel drops instantaneously from H_0 to 0, equation (8.4.11) becomes:

$$h(x, t) = H_0 \operatorname{erf} \alpha. \tag{8.4.14}$$

One should recall that we are solving the linearized equation based on the Dupuit

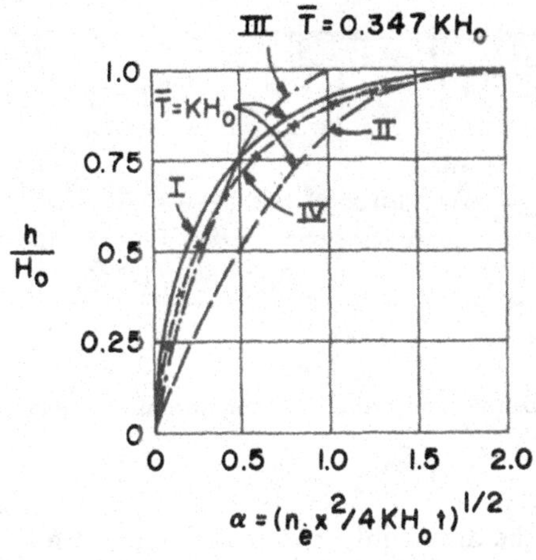

FIG. 8.4.2. Comparison between exact and approximate solutions for an instantaneous drop of water in a channel (after Aravin and Numerov, 1953, 1965).

assumptions, which means that the solution (8.4.14) deviates from the exact one in the region $h \ll H_0$. This is shown in figure 8.4.2 (after Aravin and Numerov 1953, 1965) where curve I gives the exact solution of (8.4.3) with $N = 0$, and curve II gives the approximate solution (8.4.14) (where $\bar{T} = KH_0$). Charni (1951) showed that the discrepancy between the two curves may be reduced by a proper choice of the "average" transmissivity $\bar{T}$ (actually of $\bar{h}$). He suggested that $\bar{T}$ be determined by the condition that the rate of discharge at the boundary $x = 0$ from the aquifer into the channel be the same in the exact and in the approximate solution. For the approximate solution we have:

$$Q|_{x=0} = -\bar{T}\frac{\partial h}{\partial x} = -\bar{T}\frac{\partial}{\partial x}\left[H_0 \operatorname{erf}\left\{\left(\frac{n_e x^2}{4\bar{T}t}\right)^{1/2}\right\}\right]\Bigg|_{x=0} = H_0\sqrt{\frac{n_e\bar{T}}{\pi t}}. \qquad (8.4.15)$$

Comparing this expression with the corresponding one obtained from the exact solution, Charni suggests $\bar{T} = 0.347KH_0$ (or $\bar{h} = 0.347H_0$). Inserting this average transmissivity in (8.4.14), he obtained the improved result shown graphically as curve III of figure 8.4.2.

The problem of choosing an average value $\bar{h}$ is also encountered in investigating drainage problems that are typical shallow flow problems. Usually the drainage ditches in a parallel drain system do not reach the impervious bottom of the aquifer (as is shown in fig. 8.2.2b). When the parallel drain system is made of tile drains, the convergence of the streamlines toward the outlets is more pronounced, so that the assumption of horizontal flow in the vicinity of the drains fails. Moreover, in the region where the flow converges, because of the reduction in the cross-section, the resistance to the flow increases.

Hooghoudt (1937, 1940; see Wesseling 1962) overcomes this difficulty by dividing

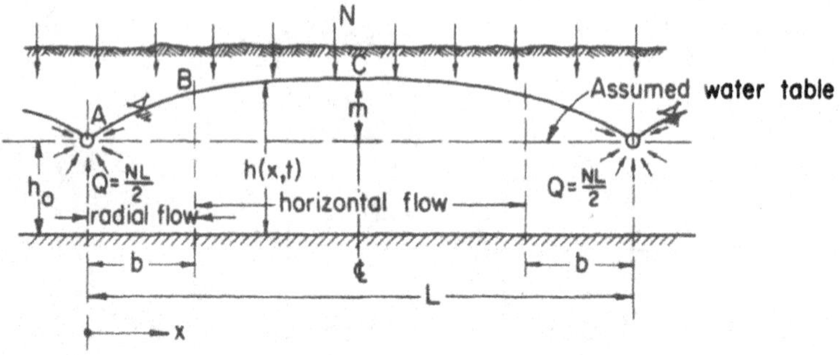

FIG. 8.4.3. Combined horizontal and radial flow for an impermeable bottom at shallow depth (Hooghoudt's method).

the flow field between the drains into two parts: one close to the drains where the flow is practically radial toward the drains (which then behave like horizontal wells), and one where the flow is essentially horizontal (fig. 8.4.3). For steady flow, he uses the method of images (with respect both to the horizontal impervious bottom and to the water table taken as the horizontal impervious boundary passing through the drains) and determines the distance $x = b$ from the drain to the point where the equipotentials are vertical for all practical purposes. His analysis leads to:

$$b = h_0/\sqrt{2}. \tag{8.4.16}$$

He then determines $\Delta\varphi_{BC}$ separately, assuming essentially horizontal flow, and $\Delta\varphi_{BA}$, assuming radial flow.

For the combined horizontal and radial flow with a finite h_0 he obtains:

$$m \equiv \varphi_C - \varphi_A = \frac{NL}{K}\left\{\frac{1}{\pi}\ln\frac{h_0}{\sqrt{2}\,r_0} + \frac{1}{\pi}\ln\frac{\sin(\frac{1}{2}\pi\eta/\sqrt{2})\,[\cosh(4\pi\eta) - \cos(\pi\eta\sqrt{2})\,]^{1/2}}{\pi\eta\sinh(2\pi\eta)} + \frac{(L - h_0\sqrt{2})^2}{8h_0 L}\right\} \tag{8.4.17}$$

where $\eta = h_0/L$. This can be written compactly as:

$$m = \varphi_C - \varphi_A = \frac{NL}{K}F(r_0, L, h_0) \tag{8.4.18}$$

where F is a function defined by (8.4.17). Comparing (8.4.18) with (8.2.26) in which m is replaced by $\Delta\varphi = \varphi_C - \varphi_A$ we obtain:

$$N = K\,\Delta\varphi/LF = 8Kh_0\,\Delta\varphi/L^2. \tag{8.4.19}$$

This means that taking into account the convergence of the flow in the vicinity of the drains is equivalent to replacing the actual depth h_0 by an equivalent one h'_0 expressed by $h'_0 = L/8F$. The equivalent depth h'_0 should also replace h_0 in (8.2.25).

Wesseling (1964) indicates that Hooghoudt's table of equivalent depths differs by less than 5% from a solution for steady flow to drains obtained by Kirkham (1958) by more rigorous mathematical procedures:

$$\varphi_C - \varphi_A = \frac{NL}{K} F^*(r_0/L, h_0/L)$$

where:

$$F^* = \frac{1}{\pi}\left\{\ln\frac{L}{\pi r_0} + \sum_{n=1}^{\infty}\left[\frac{1}{n}\left(\cos\frac{2\pi n r_0}{L} - \cos n\pi\right)\left(\operatorname{cotgh}\frac{2\pi n h_0}{L} - 1\right)\right]\right\}. \quad (8.4.20)$$

Graphs given by Toksöz and Kirkham (1961) facilitate the use of (8.4.20). A review of several other steady-state theories for drainage is presented by Kirkham (1966).

Example 2. Edelman (1947) solves the flow shown in figure 8.4.1 by solving (8.4.5), or its equivalent form, in terms of the drawdown $\eta = H_0 - h$:

$$\partial^2\eta/\partial x^2 = (n_e/KH_0)\,\partial\eta/\partial t; \qquad T = KH_0 \quad (8.4.21)$$

with the initial conditions: $t \leqslant 0, 0 \leqslant x < \infty, \eta = 0$, for various cases of the boundary condition at $x = 0$. For example:

(a) instantaneous drawdown at $x = 0$: $t \geqslant 0$, $x = 0$, $\eta = \eta_0 = H_0 - D$. This is the same as example 1 above.

(b) The water in the channel at $x = 0$ drops gradually so that the discharge rate from the aquifer into the channel is maintained at a constant value of Q_0: $t \geqslant 0$, $x = 0$, $Q = Q_0$. The solution is:

$$\eta = \frac{Q_0(Tt/n_e)^{1/2}}{T}\,[2\alpha\operatorname{erf}\alpha - 2\pi^{-1/2}\exp(-\alpha^2)] \quad (8.4.22)$$

where α is defined by (8.4.11).

(c) The rate of discharge from the aquifer into the reservoir varies linearly with time:

$$t \geqslant 0, \qquad x = 0, \qquad Q = bt.$$

The solution is:

$$\eta = \tfrac{2}{3}(bn_e/T^2)(Tt/n_e)^{3/2} - [(2\alpha^3 + 3\alpha)(1 - \operatorname{erf}(\alpha)) - 2(\alpha^2 + 1)\,\pi^{-1/2}\exp(-\alpha^2)]. \quad (8.4.23)$$

Edelman (1947) presents these solutions in tabular form.

Solutions of drainage problems have been presented by many investigators; among them we may mention Dumm (1954, 1960, 1962), Glover (e.g., 1953, 1960, 1966), Schilfgaarde (1957, 1963), Brooks (1961) and Hamad (1962). A detailed summary is given by Schilfgaarde (1970).

In the presence of accretion, $N = N(x, t)$, the partial differential equation to be solved for $h = h(x, t)$ is (8.4.3). For the infinite domain ($-\infty < x < +\infty$), the solution of (8.4.3) has the form:

$$h(x, t) = \frac{1}{2a\sqrt{\pi t}} \int_{-\infty}^{+\infty} h_0(s) \exp\left\{- \frac{(x-s)^2}{4a^2t}\right\} ds + \int_0^t \frac{d\tau}{2a\sqrt{\pi(t-\tau)}} \int_{-\infty}^{+\infty} N(s, \tau)$$
$$\cdot \exp\left\{- \frac{(x-s)^2}{4a^2(t-\tau)}\right\} ds \tag{8.4.24}$$

where $h(x, 0) \equiv h_0(x)$ gives the initial phreatic surface elevations (at $t = 0$) and $a^2 = K\bar{h}/n_e$. The first term on the right-hand side of (8.4.24) represents the response to the initial conditions, whereas the second term represents the influence of the distributed source function $N(x, t)$.

When $N = 0$, the second term on the right-hand side of (8.4.24) vanishes. Then since:

$$\int_0^\infty \exp(-a^2x^2) \cos 2bx \, dx = \left(\frac{\sqrt{\pi}}{2a}\right) \exp\left(-\frac{b^2}{a^2}\right),$$

and therefore:

$$\int_0^\infty \exp(-a^2\alpha t) \cos \alpha(x' - x) \, d\alpha = \sqrt{\frac{\pi}{4a^2t}} \exp\left\{-\frac{(x - x')^2}{4a^2t}\right\},$$

the first term on the right-hand side of (8.4.24) may also be written as:

$$\frac{1}{\pi} \int_{-\infty}^{+\infty} ds \int_0^\infty h_0(s) \cos[\alpha(s - x)] \exp\{-a^2\alpha t\} \, d\alpha,$$

a form which would be suggested by the Fourier integral (Carslaw and Jaeger 1959).

An example of practical interest is the case of the spreading or decay of a *ground water mound* without accretion (i.e., $N = 0$). Initially the mound has a constant elevation h_1 in the region $-L \leqslant x \leqslant +L$ (fig. 8.4.4a):

$$t = 0, \qquad -L \leqslant x \leqslant +L, \qquad h = h_1 = \text{const}$$
$$|x| > L, \qquad h = h_0 < h_1.$$

The solution for the infinite domain $-\infty < x < +\infty$ is:

$$h - h_0 = \frac{h_1 - h_0}{2}\left\{\operatorname{erf}\frac{L - x}{2a\sqrt{t}} + \operatorname{erf}\frac{L + x}{2a\sqrt{t}}\right\}. \tag{8.4.25}$$

Figure 8.4.4b gives (8.4.25) in a graphic form.

Carslaw and Jaeger (1959), section 2.2 (c), present a solution for the decay of a mound that initially has the form:

$$t = 0, \quad |x| > L, \quad h = 0; \quad 0 < x < L, \quad h = h_0(L - x)/L; \quad -L < x < 0,$$
$$h = h_0(L - x)/L.$$

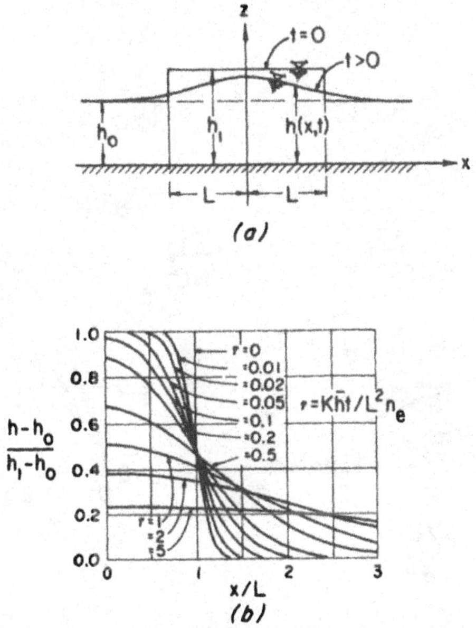

FIG. 8.4.4. Decay of a ground water mound.

Another case of practical interest is the build-up of a mound by accretion $N =$ const $\neq 0$ over the region $|x| < L$ (with $N = 0$ for $|x| > L$); the initial conditions are $h = h_0 =$ const in the entire infinite domain. We then obtain from (8.4.24):

$$h - h_0 = \frac{N}{2n_e} \int_0^t \left[\Phi\left(\frac{L-x}{2a\sqrt{t-\tau}}\right) + \Phi\left(\frac{L+x}{2a\sqrt{t-\tau}}\right) \right] d\tau; \qquad \Phi(\alpha) \equiv \operatorname{erf} \alpha. \quad (8.4.26)$$

By integrating (8.4.26), and recalling that $\Phi(-x) = -\Phi(x)$, we obtain an expression for h in the regions $|x| < L$ and $|x| > L$. Polubarinova-Kochina (1952, 1962) presents the expressions resulting from these integrations. Figure 8.4.5 shows these results in graphic form.

Polubarinova-Kochina (1952, 1962) also discusses the build-up (by $N \neq 0$) and decay (with $N = 0$) of a ground water mound in the presence of evapotranspiration (negative accretion). Denoting evapotranspiration by E (positive when upward), the Dupuit–Forchheimer theory leads to the equation:

$$K\bar{h} \frac{\partial^2 h}{\partial x^2} + N - E = n_e \frac{\partial h}{\partial t}. \qquad (8.4.27)$$

In many cases $E = E(h)$, i.e., evapotranspiration is a function of the depth of the water table below the soil surface. Polubarinova-Kochina presents a solution of (8.4.27) for the case $E = A + Bh$, where A and B are constant coefficients (with A depending on the depth of impervious bottom below the soil surface).

When the equation to be solved is (8.4.3), one may simplify it by introducing the

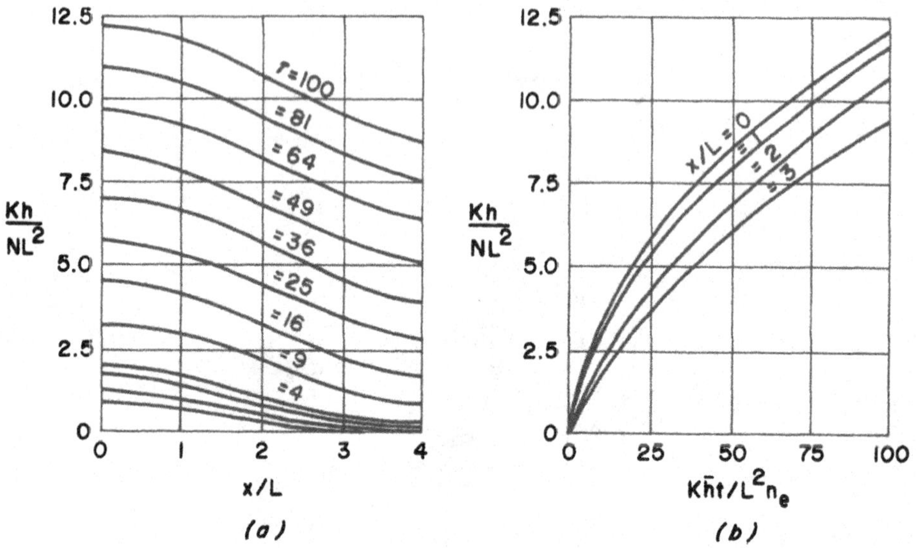

FIG. 8.4.5. Build-up of a ground water mound by accretion in the region $|x| < L$.

transformation:

$$h' = h + (N/2K\bar{h})x^2. \tag{8.4.28}$$

Equation (8.4.3) then becomes the heat conduction equation (8.4.5) in h'.

Solutions of a large number of problems of phreatic flows are given by Carslaw and Jaeger (1959), as many heat conduction problems can easily be "translated" into ones of phreatic flows described by (8.4.3) and (8.4.5). One should recall that (8.4.5) is identical to (6.4.7) describing (without any approximation) flow in a homogeneous confined aquifer of constant thickness.

For the two-dimensional flow without accretion (described by (8.4.1) with $N = 0$ and $a^2 = K\bar{h}/n_e$) in the infinite plane, Carslaw and Jaeger (1959) present a solution satisfying the initial conditions: $t = 0$: $|x| < L_1$, $|y| < L_2$, $h = h_1$, with $h = h_0$ everywhere else, in the form:

$$h - h_0 = \frac{h_1 - h_0}{4}\left\{\operatorname{erf}\frac{L_1 - x}{2a\sqrt{t}} + \operatorname{erf}\frac{L_1 + x}{2a\sqrt{t}}\right\}\left\{\operatorname{erf}\frac{L_2 - y}{2a\sqrt{t}} + \operatorname{erf}\frac{L_2 + y}{2a\sqrt{t}}\right\}. \tag{8.4.29}$$

Chapter 5 in Carslaw and Jaeger (1959) deals with flow in a rectangle.

Polubarinova-Kochina (1952, 1962) presents the solution of (8.4.1) without accretion ($N = 0$) for the flow in the semi-infinite plane $y \geqslant 0$ with:

$$t \leqslant 0, \qquad h(x, y, 0) = f(x, y) \quad \text{and} \quad h(x, 0, t) = g(x, t):$$

$$h(x, y, t) = \frac{1}{4\pi a^2 t}\int_{-\infty}^{+\infty} dx' \int_0^{\infty} f(x', y')\left[\exp\left\{-\frac{(x - x')^2 + (y - y')^2}{4a^2 t}\right\}\right.$$

$$- \exp\left\{-\frac{(x-x')^2+(y+y')^2}{4a^2t}\right\}\Bigg]dy' + \frac{y}{4a^2\pi}\int_0^t dt' \int_{-\infty}^{+\infty} \frac{g(x',t')}{(t-t')^2}$$

$$\cdot \exp\left\{-\frac{(x-x')^2+y^2}{4a^2(t-t')}\right\}dx'. \tag{8.4.30}$$

She then presents a solution for the case:

$$f(x,y) = h_0; \qquad g(x,t) = \begin{matrix} h_1 & \text{for} & x<0 \\ h_2 & \text{for} & x>0. \end{matrix}$$

For radial flow cases (8.4.1) becomes:

$$K\bar{h}\left(\frac{\partial^2 h}{\partial r^2} + \frac{1}{r}\frac{\partial h}{\partial r}\right) + N(r,t) = n_e\frac{\partial h}{\partial t}. \tag{8.4.31}$$

Solutions are also given by Carslaw and Jaeger (1959, chap. 7).

8.4.2 The Second Method of Linearization of the Boussinesq Equation

Equation (8.2.10) may be rewritten as:

$$\frac{1}{2}\left(\frac{\partial^2 h^2}{\partial x^2} + \frac{\partial^2 h^2}{\partial y^2}\right) + \frac{N}{K} = \frac{n_e}{2Kh}\frac{\partial h^2}{\partial t}. \tag{8.4.32}$$

If variations in h are such that Kh/n_e may be considered a constant, say $K\bar{h}/n_e$, where $\bar{h}$ is some average (constant) value of h, then (8.4.32) *is linear in* h^2. Obviously, the flow problem itself is linear in h^2 only if its initial and boundary conditions are also linear in h^2.

Another form of (8.4.32) is:

$$a^2\left(\frac{\partial^2 u}{\partial x^2} + \frac{\partial^2 u}{\partial y^2}\right) + N^* = \frac{\partial u}{\partial t}; \qquad u = \frac{h^2}{2}; \qquad N^* = \frac{N\bar{h}}{n_e}, \qquad a^2 = \frac{K\bar{h}}{n_e} \tag{8.4.33}$$

where Φ^g of (5.8.30) is actually Ku.

For the one-dimensional case, equation (8.4.33) becomes:

$$a^2\,\partial^2 u/\partial x^2 + N^* = \partial u/\partial t \tag{8.4.34}$$

which for $N = 0$ becomes the heat conduction equation:

$$a^2\,\partial^2 u/\partial x^2 = \partial u/\partial t. \tag{8.4.35}$$

In order to solve any of (8.4.33) through (8.4.35) we must specify the boundary and the initial conditions in terms of u.

As a simple example, consider the case of phreatic flow in the half plane $x \geqslant 0$ to a channel along $x = 0$. Initially we have steady flow with a parabolic phreatic surface. Then, at $t = 0$, the water level in the channel is instantaneously lowered from h_1 to $h_2 < h_1$ (Verigin 1949; Aravin and Numerov 1953, 1965):

$$t = 0, \qquad h^2 - h_1{}^2 = (2Q_0/K)x \tag{8.4.36}$$

where Q_0 is positive toward the channel. The initial and boundary conditions are:

$$x \geqslant 0, \quad t \leqslant 0, \quad u = h_1{}^2/2 + (Q_0/K)x$$
$$x = 0, \quad t > 0, \quad u = h_2{}^2/2. \tag{8.4.37}$$

Using the solution presented by Carslaw and Jaeger (1959, sec. 2.4(c)), we obtain as a solution of (8.4.35) satisfying (8.4.37):

$$u = h^2/2 = h^2{}_2/2 + (h_1{}^2/2 - h_2{}^2/2)\text{erf}\{x/2a\sqrt{t}\} + (Q_0/K)x \tag{8.4.38}$$

As $t \to 0$, the third term on the right-hand side of (8.4.38) vanishes and we obtain steady flow toward the channel with a parabolic water table. From (8.4.38) it follows that the rate of seepage into the channel at $x = 0$ is:

$$q'|_{x=0} = K\, \partial u/\partial x = Q_0 + [(h_2{}^2 - h_1{}^2)/2]Kn_e/\pi\bar{h}t. \tag{8.4.39}$$

In the more general case, with initial condition $g(x)$ and boundary conditions $g_1(t)$, we may apply the principle of superposition (par. 7.5.3). Letting $u = u_1 + u_2$, such that:

$$u_1(x, 0) = g(x), \qquad u_1(0, t) = 0; \qquad u_2(x, 0) = 0, \qquad u_2(0, t) = g_1(t)$$

we obtain:

$$u = \frac{1}{2a\sqrt{\pi t}} \int_0^\infty g(x') \left[\exp\left\{ -\frac{(x - x')^2}{4a^2t} \right\} - \exp\left\{ -\frac{(x + x')^2}{4a^2t} \right\} \right] dx'$$
$$+ \frac{2}{\sqrt{\pi}} \int_{x/2a\sqrt{t}}^\infty g_1\left(t - \frac{x'^2}{4a^2t'^2}\right) \cdot \exp(-t'^2)\, dt'. \tag{8.4.40}$$

For the special case $h(0, t) \equiv 0$ (and hence $g_1(t) = 0$), and $h(x, 0) = H_0$ (i.e., an initially horizontal water table at $h = H_0$ dropping at $x = 0$ to 0), we obtain:

$$u = (1/2)H_0{}^2\Phi(x/2a\sqrt{t}); \qquad h(x, t) = H_0[\Phi(x/2a\sqrt{t})]^{1/2}. \tag{8.4.41}$$

This problem was considered in paragraph 8.4.1 (compare (8.4.41) with (8.4.14)). Curve IV in figure 8.4.2 shows (8.4.41). It follows that the solution (8.4.40) obtained by the second method of linearization is closer to the exact solution (curve I) than the solution (8.4.14) obtained by the first method of linearization.

Marino (1967), following Hantush (1963, 1967), employs this method of linearization to solve the problem of the rise and decay of a ground water mound below a spreading ground (fig. 8.4.6). The problem is stated mathematically in the following way.

Determine $h_1 = h_1(x, t)$ in the region $0 < x < L$ and $h_2 = h_2(x, t)$ in the region $L < x < \infty$, such that:

for Region I *for Region II*

$$\frac{K\bar{h}}{n_e}\frac{\partial^2 h_1{}^2}{\partial x^2} + \frac{2\bar{h}}{n_e}N(t) = \frac{\partial h_1{}^2}{\partial t} \qquad \frac{K\bar{h}}{n_e}\frac{\partial^2 h_2{}^2}{\partial x^2} = \frac{\partial h_2{}^2}{\partial t}$$

$$N(t) \begin{array}{ll} = 1 & \text{for} \quad 0 < t < t_0 \\ = 0 & \text{for} \quad t > t_0 \end{array} \qquad h_2(x, 0) = h_i$$

$$K h_1 \frac{\partial h_1}{\partial x}\bigg|_{x=0, t \geqslant 0} = 0 \qquad h_2(\infty, t) = h_i$$

$$h_1(x, 0) = h_i \qquad h_1(L, t) = h_2(L, t)$$

and:

$$K h_1(\partial h_1/\partial x)|_{x=L, t\geqslant 0} = K h_2(\partial h_2/\partial x)|_{x=L, t\geqslant 0}$$

where symbols are explained in figure 8.4.6, and $\bar{h}$ is a weighted mean of the depth of saturation during the period of flow. Marino solves this set of equations for h_1 and h_2 using the Laplace transform, and compares the solutions with experimental results obtained by means of a Hele-Shaw analog (sec. 11.4). He concludes that the experimental results are in close agreement with those obtained analytically when: (a) $N \leqslant 0.2K$; (b) $(h - h_i) \leqslant 0.5h_i$. He also concludes, on the evidence of this comparison, that the average hydraulic head in a vertical section of a flow system is indeed approximately equal to the height of the water table whenever the rise of the water table is equal to or less than 50% of the initial depth. When this condition prevails, the relative deviations between observed and calculated values of h are less than 6%. The maximum relative deviation is 12.2% when $(h - h_i)/h_i >$ 20.

Hantush (1963, 1967), who first presented the analytical solution, also gives results for spreading basins in the form of a rectangle and a circle.

8.4.3 The Third Method of Linearization of the Boussinesq Equation

In this method, proposed by Charni (1951), Boussinesq's equation for one-dimensional flow is rewritten as:

$$\frac{K}{n_e}\frac{\partial}{\partial x}\left(f\,\frac{\partial h}{\partial x}\right) = \frac{\partial f}{\partial x} \tag{8.4.42}$$

where $f = f(h)$ is some function of h to be determined below. A new function

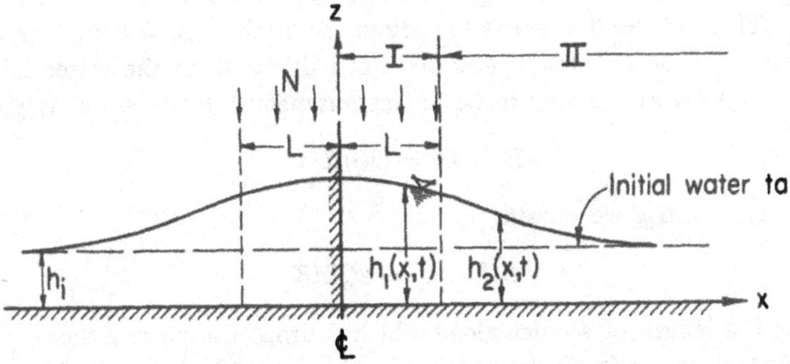

FIG. 8.4.6. Build-up and decay of ground water mound under a spreading pond (after Marino, 1967).

$$u = \int f(h)\, dh \tag{8.4.43}$$

is then introduced such that:

$$\frac{\partial u}{\partial x} = f\frac{\partial h}{\partial x}; \qquad \frac{\partial^2 u}{\partial x^2} = \frac{\partial}{\partial x}\left(f\frac{\partial h}{\partial x}\right); \qquad \frac{\partial u}{\partial t} = \frac{du}{df}\frac{\partial f}{\partial t}.$$

With these transformations (8.4.42) becomes:

$$\frac{K}{n_e}\frac{du}{df}\frac{\partial^2 u}{\partial x^2} = \frac{\partial u}{\partial t}. \tag{8.4.44}$$

By choosing f so that $df/du = 1/A =$ const, equation (8.4.44) becomes the (linear) homogeneous heat conduction equation considered in paragraph 8.4.1. From (8.4.43) and (8.4.44) it follows that:

$$df/f = A\, dh; \qquad f = B\exp(Ah) \tag{8.4.45}$$

where A and B are constants. Thus, by using (8.4.45) to define the function f, equation (8.4.42) reduces to the homogeneous heat conduction equation. The constants A and B are chosen such that $f \equiv h$ at the boundaries. For example, if at $x = 0$, $h = h_1$ and at $x = L$, $h = h_2$, we obtain $h_1 = B\exp(Bh_1)$; $h_2 = B\exp(Ah_2)$; $A = [\ln(h_1/h_2)]/(h_1 - h_2)$, $B = h_2\exp(-Ah_2)$.

8.4.4 The Method of Successive Steady States

According to this method (Lembke 1887; Weber 1928; Polubarinova-Kochina 1952, 1962; Aravin and Numerov 1953, 1965), sometimes called the *quasi-steady state method*, we assume that at every instant of time, the phreatic surface has the shape of a steady phreatic surface and that the unsteady process may be regarded as a sequence of steady states. A partial justification of this assumption stems from the observation that temporal variations of flow characteristics are much smaller than spatial ones.

Following Aravin and Numerov (1953, 1965), consider the case of unsteady, one-dimensional flow to a ditch (fig. 8.4.1). Initially, the water table is horizontal at $h = H_0$. Then, at $t = 0$, it starts to vary in the ditch $h|_{x=0} = D(t)$. The Dupuit–Forchheimer solution for steady flow toward a ditch where the water table is at an elevation D (assuming this to be an instantaneous steady state) is given by:

$$h^2 - D^2 = (2Q/K)x. \tag{8.4.46}$$

With $x = L$, $h = H_0$, we obtain:

$$H_0^2 - D^2 = 2QL/K \tag{8.4.47}$$

where L is the length of aquifer along which h drops during the time t.

The volume of soil U (> 0) drained during this interval of time is obtained from (8.4.46) and (8.4.47):

$$U(t) = n_e\left[H_0L - \int_0^L h(x)\,dx\right] = n_eH_0L - n_e\int_0^L \left(D^2 + \frac{2Q}{K}x\right)^{1/2} dx$$

$$= n_eL\left[H_0 - \frac{2}{3}\frac{H_0^3 - D^3}{H_0^2 - D^2}\right] = \frac{n_eK}{6Q}(H_0 - D)^2(H_0 + 2D). \tag{8.4.48}$$

This volume must be equal to:

$$U(t) = \int_0^t Q(t')\,dt'. \tag{8.4.49}$$

For the general case of $D = D(t)$, by differentiating (8.4.49), we obtain:

$$\frac{Kn_e}{6}\,d\left[\frac{(H_0 - D)^2(H_0 + 2D)}{Q}\right] = Q\,dt. \tag{8.4.50}$$

Multiplying both sides of (8.4.50) by $(H_0 - D)^2(H_0 + 2D)3Kn_e/Q$ and integrating, we obtain:

$$\left[\frac{(H_0 - D)^2(H_0 + 2D)}{2Q/K}\right]^2 = \frac{3K}{n_e}\int_0^t (H_0 - D)^2(H_0 + 2D)\,dt + C \tag{8.4.51}$$

where C is a constant of integration. In view of (8.4.47), equation (8.4.51) becomes:

$$\left[\frac{(H_0 - D)(H_0 + 2D)L}{(H_0 + D)}\right]^2 = \frac{3K}{n_e}\int_0^t (H_0 - D)^2(H_0 + 2D)\,dt + C. \tag{8.4.52}$$

From $L|_{t=0} = 0$ it follows that $C = 0$. Equation (8.4.52) may be rewritten as:

$$L = \sqrt{K\,Dt/n_e}\,F \tag{8.4.53}$$

where F is a known function of time defined by (8.4.52). Once we obtain the influenced range L from (8.4.53), we may insert the result in (8.4.47) to derive Q and then obtain $h(x)$ from:

$$(h^2 - D^2)/(H^2{}_0 - D^2) = x/L. \tag{8.4.54}$$

For the special case $D = \text{const}$, F is also a constant. When $H_0 = 0$ and $D = \text{const} > 0$, we obtain:

$$L = \sqrt{3KDt/2n_e} \approx 1.23\sqrt{KDt/n_e}. \tag{8.4.55}$$

As a second example, consider a case of a phreatic surface with accretion. Figure 8.4.7 shows an initially horizontal water table and a drain that opens at $t = 0$. Assuming instantaneous steady flow, we have at time t,

$$(h^2 - h_0^2)/(H_0^2 - h_0^2) = x/L; \qquad Q = K(H_0^2 - h_0^2)/2L. \tag{8.4.56}$$

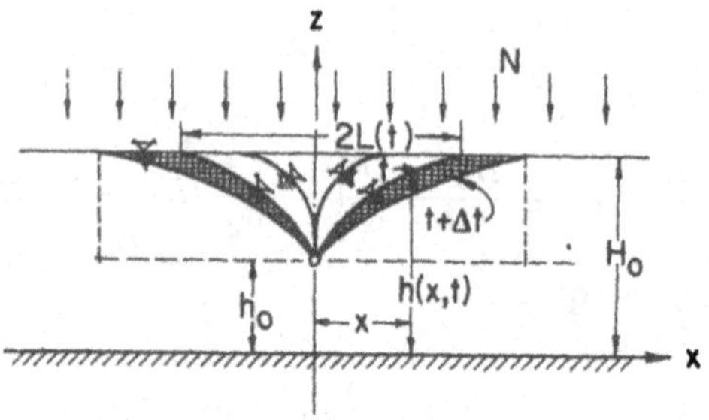

FIG. 8.4.7. Successive steady phreatic surfaces.

During the interval dt, the volume of water withdrawn from the aquifer is $Q\,dt$. This volume is equal to the sum of the volume of water drained from the soil body between the two phreatic surfaces at t and $t + dt$ (equal to $n_e(H_0 - h_0)\,dL/3$), and the accretion $NL\,dt$. Hence:

$$Q\,dt \equiv K\frac{H_0^2 - h_0^2}{2L}\,dt = n_e\frac{(H_0 - h_0)}{3}\,dL + NL\,dt \tag{8.4.57}$$

or:

$$\frac{2L\,dL}{K(H_0^2 - h_0^2) - 2NL^2} = \frac{3\,dt}{n_e(H_0 - h_0)}.$$

Integrating, with $L(0) = 0$, yields:

$$L = L(t) = \left[\frac{K(H_0^2 - h_0^2)}{2N}\left(1 - \exp\left\{-\frac{6\,Nt}{n_e(H_0 - h_0)}\right\}\right)\right]^{1/2} \tag{8.4.58}$$

$Q(t)$ is obtained from (8.4.56). As $t \to \infty$, $L \to L_\infty$ (region influenced by the drain):

$$L_\infty = [K(H_0^2 - h_0^2)/2N]^{1/2}. \tag{8.4.59}$$

To obtain $Q(t)$, we insert $L(t)$ of (8.4.78) into Q. When $N = 0$, (8.4.57) leads to:

$$L = 1.732[K(H_0 + h_0)\,t/n_e]^{1/2}. \tag{8.4.60}$$

8.4.5 The Method of Small Perturbations

In paragraphs 8.4.1 through 8.4.4, methods and techniques were presented aimed at the linearization of the Boussinesq equation, which is an approximation based on the Dupuit assumptions. The Dupuit assumption itself was introduced because of the nonlinearity of the phreatic surface boundary condition. The linearization techniques introduced in the present paragraph will be applied both to the nonlinear Boussinesq equation and to the nonlinear phreatic surface boundary condition in the exact statement of the unconfined flow problem. Some of these techniques are also employed in paragraph 9.5.6 in connection with the fresh-water–salt-water interface in a coastal aquifer.

The exact statement of a problem of phreatic flow in a homogeneous isotropic domain is the following (fig. 8.4.8).

Determine $\varphi(x, y, z, t)$ in the flow domain so that φ satisfies: (a) the partial differential equation $\nabla^2\varphi = 0$ at all points of the domain, (b) the phreatic surface boundary condition (7.1.46) rewritten as:

$$n_e \frac{\partial \varphi}{\partial t} = K\left[\left(\frac{\partial \varphi}{\partial x}\right)^2 + \left(\frac{\partial \varphi}{\partial y}\right)^2 + \left(\frac{\partial \varphi}{\partial z}\right)^2\right] - \frac{\partial \varphi}{\partial z}(K + N) + N \quad \text{on } z = \zeta \qquad (8.4.61)$$

and (c) appropriate boundary conditions on all other boundaries of the flow domain. Since the treatment of these conditions is rather simple and straightforward, they will not be mentioned in the analysis that follows. In unsteady flow problems, initial conditions $\varphi = \varphi(x, y, z, 0)$ must also be specified. As on the phreatic surface $p = 0$, and hence $\varphi = \zeta$, the initial conditions (at $t = 0$) along the phreatic surface may be specified as $\varphi = \zeta = f(x, y)$. However, the position of the phreatic boundary is unknown; it is part of the required solution. Once $\varphi = \varphi(x, y, z, t)$ is known, elevations ζ of the phreatic surface are determined from the condition $p = 0$. Hence:

$$\zeta = \varphi|_{z=\zeta}; \qquad \zeta = \zeta(x, y, t). \qquad (8.4.62)$$

Following the method employed in the *theory of waves of small amplitude* (see, for example, Finkelstein 1957; Stoker 1957; or Wehausen and Laitone 1960), and the works of Polubarinova-Kochina (1952, 1962) and Dagan (1964) in flow through porous media, we shall assume that both φ and ζ can be developed in a power series of a small parameter ε:

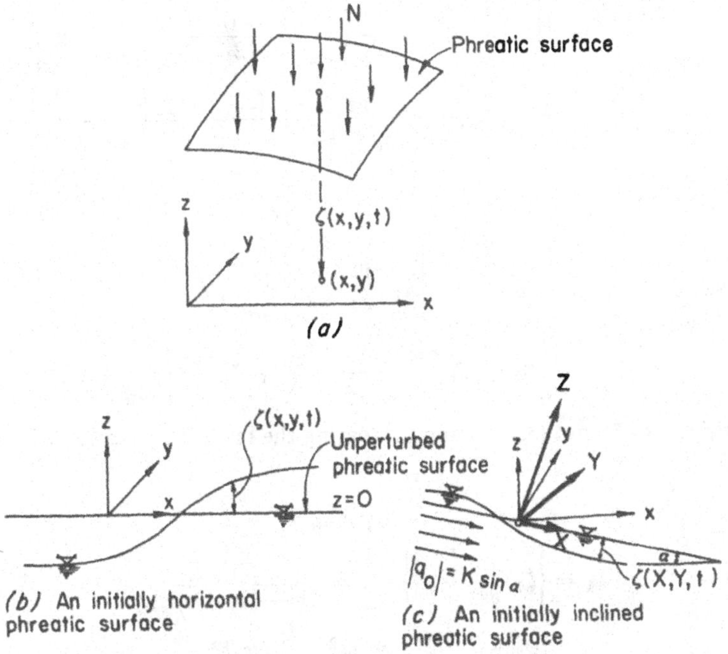

FIG. 8.4.8. Nomenclature for the method of small perturbations.

$$\varphi(x, y, z, t) = \varphi_0(x, y, z) + \varepsilon\varphi_1(x, y, z, t) + \varepsilon^2\varphi_2(x, y, z, t) + \cdots$$
$$\zeta(x, y, t) = \zeta_0(x, y) + \varepsilon\zeta_1(x, y, t) + \varepsilon^2\zeta_2(x, y, t) + \cdots \tag{8.4.63}$$

where ε, the nature of which is not specified, is a small dimensionless parameter that has the nature of a small perturbation. Setting $\varepsilon = 0$ means that there is no perturbation; thus φ_0 and ζ_0 are the nonperturbed, steady-state values of φ and ζ.

By substituting (8.4.63) into the continuity equation, we obtain:

$$\nabla^2\varphi \equiv \frac{\partial^2\varphi}{\partial x^2} + \frac{\partial^2\varphi}{\partial y^2} + \frac{\partial^2\varphi}{\partial z^2} = \left(\frac{\partial^2\varphi_0}{\partial x^2} + \frac{\partial^2\varphi_0}{\partial y^2} + \frac{\partial^2\varphi_0}{\partial z^2}\right)$$
$$+ \varepsilon\left(\frac{\partial^2\varphi_1}{\partial x^2} + \frac{\partial^2\varphi_1}{\partial y^2} + \frac{\partial^2\varphi_1}{\partial z^2}\right) + \varepsilon^2\left(\frac{\partial^2\varphi_2}{\partial x^2} + \frac{\partial^2\varphi_2}{\partial y^2} + \frac{\partial^2\varphi_2}{\partial z^2}\right) + \cdots = 0. \tag{8.4.64}$$

Since ε is a constant (8.4.64) is valid only if:

$$\nabla^2\varphi_0 = 0; \qquad \nabla^2\varphi_1 = 0; \qquad \nabla^2\varphi_2 = 0, \text{ etc., for higher order terms.} \tag{8.4.65}$$

Thus, all φ_is are harmonic functions.

Next we insert (8.4.63) into the phreatic surface boundary conditions (8.4.61) and (8.4.62) to be satisfied on $z = \zeta$. First, the value of φ and its derivatives on the phreatic surface $z = \zeta$ are derived by expanding the φ_is in a Taylor series around the surface $z = \zeta_0$:

$$\varphi(x, y, z, t)\big|_{z=\zeta} = \varphi_0\big|_{z=\zeta_0} + \varepsilon\zeta_1 \frac{\partial\varphi_0}{\partial z}\bigg|_{z=\zeta_0} + \frac{(\varepsilon\zeta_1)^2}{2}\frac{\partial^2\varphi_0}{\partial z^2}\bigg|_{z=\zeta_0} + \cdots$$
$$+ \varepsilon\left[\varphi_1\big|_{z=\zeta_0} + \varepsilon\zeta_1\frac{\partial\varphi_1}{\partial z}\bigg|_{z=\zeta_0} + \frac{(\varepsilon\zeta_1)^2}{2}\frac{\partial^2\varphi_1}{\partial z^2}\bigg|_{z=\zeta_0} + \cdots\right]$$
$$+ \varepsilon^2\left[\varphi_2\big|_{z=\zeta_0} + \varepsilon\zeta_1\frac{\partial\varphi_2}{\partial z}\bigg|_{z=\zeta_0} + \frac{(\varepsilon\zeta_1)^2}{2}\frac{\partial^2\varphi_2}{\partial z^2}\bigg|_{z=\zeta_0} + \cdots\right] + \cdots \tag{8.4.66}$$

or:

$$\varphi(x, y, z, t)\big|_{z=\zeta} = \varphi_0\big|_{z=\zeta_0} + \varepsilon\left[\zeta_1\frac{\partial\varphi_0}{\partial z} + \varphi_1\right]\bigg|_{z=\zeta_0}$$
$$+ \varepsilon^2\left[\varphi_2 + \zeta_2\frac{\partial\varphi_0}{\partial z} + \frac{\zeta_1{}^2}{2}\frac{\partial^2\varphi_0}{\partial z^2} + \zeta_1\frac{\partial\varphi_1}{\partial z}\right]\bigg|_{z=\zeta_0} + 0(\varepsilon^3). \tag{8.4.67}$$

By inserting (8.4.67) into (8.4.62) and comparing the resulting equation with (8.4.63), we obtain:

$$\zeta_0 = \varphi_0\big|_{z=\zeta_0} \tag{8.4.68}$$

$$\zeta_1 = \left\{\zeta_1\frac{\partial\varphi_0}{\partial z} + \varphi_1\right\}\bigg|_{z=\zeta_0} = \frac{\varphi_1\big|_{z=\zeta_0}}{[1 - (\partial\varphi_1/\partial z)\big|_{z=\zeta_0}]}$$
$$\zeta_2 = \frac{\varphi_2 + (\zeta_1{}^2/2)\,\partial^2\varphi_0/\partial z^2 + \zeta_1\partial\varphi_1/\partial z}{1 - \partial\varphi_0/\partial z}\bigg|_{z=\zeta_0} \tag{8.4.69}$$

etc., for higher order terms. By inserting (8.4.68) and (8.4.69) into (8.4.67), we obtain an expression for $\varphi = \varphi(x, y, z, t)$ on the free surface $z = \zeta$ in terms of φ_0, φ_1, φ_2, etc., and their derivatives evaluated on $z = \zeta_0$.

Next we expand all other time and space derivatives appearing in (8.4.61) around $z = \zeta_0$. For example:

$$\left.\frac{\partial\varphi}{\partial t}\right|_{z=\zeta} = \varepsilon \left.\frac{\partial\varphi_1}{\partial t}\right|_{z=\zeta_0} + \varepsilon^2 \left.\frac{\partial\varphi_2}{\partial t}\right|_{z=\zeta_0} + \varepsilon^3 \left.\frac{\partial\varphi_3}{\partial t}\right|_{z=\zeta_0}$$

$$+ \varepsilon\,(\zeta_1 + \varepsilon^2\zeta_2 + \cdots)\left.\left(\varepsilon\frac{\partial^2\varphi_1}{\partial z\,\partial t} + \varepsilon^2\frac{\partial^2\varphi_2}{\partial z\,\partial t} + \varepsilon^3\frac{\partial^2\varphi_3}{\partial z\,\partial t} + \cdots\right)\right|_{z=\zeta_0} + \cdots$$

$$= \varepsilon\left.\frac{\partial\varphi_1}{\partial t}\right|_{z=\zeta_0} + \varepsilon^2\left.\left\{\frac{\partial\varphi_2}{\partial t} + \zeta_1\frac{\partial^2\varphi_1}{\partial z\,\partial t}\right\}\right|_{z=\zeta_0} + 0(\varepsilon^3) \tag{8.4.70}$$

where ζ_1 may be expressed by (8.4.69). Similarly:

$$\left.\frac{\partial\varphi}{\partial x}\right|_{z=\zeta} = \left.\frac{\partial\varphi_0}{\partial x}\right|_{z=\zeta_0} + \varepsilon\left.\left\{\frac{\partial\varphi_1}{\partial x} + \zeta_1\frac{\partial^2\varphi_0}{\partial x\,\partial z}\right\}\right|_{z=\zeta_0}$$

$$+ \varepsilon^2\left.\left\{\frac{\partial\varphi_2}{\partial x} + \zeta_1\frac{\partial^2\varphi_1}{\partial x\,\partial z} + \zeta_2\frac{\partial^2\varphi_0}{\partial x\,\partial z}\right\}\right|_{z=\zeta_0} + 0(\varepsilon^3). \tag{8.4.71}$$

In a similar way we develop all the other derivatives and insert them into (8.4.61). We obtain:

$$n_e\left\{\varepsilon\frac{\partial\varphi_0}{\partial t} + \varepsilon^2\left(\frac{\partial\varphi_2}{\partial t} + \zeta_1\frac{\partial^2\varphi_1}{\partial t\,\partial z}\right)\right\}$$

$$= K\left\{\left[\frac{\partial\varphi_0}{\partial x} + \varepsilon\left(\frac{\partial\varphi_1}{\partial x} + \zeta_1\frac{\partial^2\varphi_0}{\partial x\,\partial z}\right) + 0(\varepsilon^2)\right]^2 + \left[\frac{\partial\varphi_0}{\partial y} + \varepsilon\left(\frac{\partial\varphi_1}{\partial y} + \zeta_1\frac{\partial^2\varphi_0}{\partial y\,\partial z}\right) + 0(\varepsilon^2)\right]^2\right.$$

$$\left. + \left[\frac{\partial\varphi_0}{\partial z} + \varepsilon\left(\frac{\partial\varphi_1}{\partial z} + \zeta_1\frac{\partial^2\varphi_0}{\partial z^2}\right) + 0(\varepsilon^2)\right]^2\right\}$$

$$- (K+N)\left[\frac{\partial\varphi_0}{\partial z} + \varepsilon\left(\frac{\partial\varphi_1}{\partial z} + \zeta_1\frac{\partial^2\varphi_0}{\partial z^2}\right) + 0(\varepsilon^2)\right] + N = 0 \tag{8.4.72}$$

where all terms are evaluated on the surface $z = \zeta_0$.

By (a) neglecting terms of $0(\varepsilon^3)$, (b) inserting the appropriate expressions for ζ_1 and ζ_2 from (8.4.69) in (8.4.72), and (c) separating terms having as coefficients ε to the same power, we obtain the following set of conditions:

$$\varepsilon^0\colon\ K\left[\left(\frac{\partial\varphi_0}{\partial x}\right)^2 + \left(\frac{\partial\varphi_0}{\partial y}\right)^2 + \left(\frac{\partial\varphi_0}{\partial z}\right)^2\right] - (K+N)\frac{\partial\varphi_0}{\partial z} + N = 0 \text{ on } z = \zeta_0 \tag{8.4.73}$$

which is immediately recognized as the condition on a steady phreatic surface with accretion.

$$\varepsilon^1:\quad n_e\frac{\partial\varphi_1}{\partial t}=2K\left[\frac{\partial\varphi_0}{\partial x}\frac{\partial\varphi_1}{\partial x}+\frac{\partial\varphi_0}{\partial y}\frac{\partial\varphi_1}{\partial y}+\frac{\partial\varphi_0}{\partial z}\frac{\partial\varphi_1}{\partial z}\right.$$
$$\left.+\zeta_1\left(\frac{\partial^2\varphi_0}{\partial x\,\partial z}\frac{\partial\varphi_0}{\partial x}+\frac{\partial^2\varphi_0}{\partial y\,\partial z}\frac{\partial\varphi_0}{\partial z}+\frac{\partial^2\varphi_0}{\partial z^2}\frac{\partial\varphi_0}{\partial z}\right)\right]$$
$$-(K+N)\left[\frac{\partial\varphi_1}{\partial z}+\zeta_1\frac{\partial^2\varphi_0}{\partial z^2}\right]\quad\text{on } z=\zeta_0 \qquad (8.4.74)$$

or, in the more compact form:

$$n_e\frac{\partial\varphi_1}{\partial t}=2K\left[\nabla\varphi_0\cdot\nabla\varphi_1+\tfrac{1}{2}\zeta_1\frac{\partial}{\partial z}(\nabla\varphi_0)^2\right]-(K+N)\left(\frac{\partial\varphi_1}{\partial z}+\zeta_1\frac{\partial^2\varphi_0}{\partial z^2}\right). \qquad (8.4.75)$$

For ε^2, already in the compact form, we obtain:

$$\varepsilon^2:\quad n_e\left(\frac{\partial\varphi_2}{\partial t}+\zeta_1\frac{\partial^2\varphi_1}{\partial z\,\partial t}\right)=K\left[(\nabla\varphi_1)^2+\zeta_1^2\left(\frac{\partial}{\partial z}\nabla\varphi_0\right)^2+2\zeta_1\frac{\partial}{\partial z}(\nabla\varphi_0\cdot\nabla\varphi_1)\right.$$
$$\left.+2\nabla\varphi_0\cdot\nabla\varphi_2+\zeta_2\frac{\partial}{\partial z}(\nabla\varphi_0)^2\right]$$
$$-(K+N)\left[\frac{\partial\varphi_2}{\partial z}+\zeta_1\frac{\partial^2\varphi_1}{\partial z^2}+\zeta_2\frac{\partial^2\varphi_0}{\partial z^2}\right]\quad\text{on } z=\zeta_0. \qquad (8.4.76)$$

Obviously, this procedure could be continued for higher order terms. The same technique is also applied to all other boundary conditions of the flow domain.

Regrouping these equations for the φ_is and ζ_is, we obtain:

$$\begin{aligned}&\nabla^2\varphi_0=0\\ &K(\nabla\varphi_0)^2-(K+N)\,\partial\varphi_0/\partial z+N=0 &&\text{on } z=\zeta_0\\ &\zeta_0=\varphi_0 &&\text{on } z=\zeta_0\end{aligned} \qquad (8.4.77)$$

$$\begin{aligned}&\nabla^2\varphi_1=0\\ &n_e\frac{\partial\varphi_1}{\partial t}=2K[\nabla\varphi_0\cdot\nabla\varphi_1+\tfrac{1}{2}\zeta_1\,\partial(\nabla\varphi_0)^2/\partial z]\\ &\qquad-(K+N)[\partial\varphi_1/\partial z+\zeta_1\,\partial^2\varphi_0/\partial z^2] &&\text{on } z=\zeta_0\\ &\zeta_1=\varphi_1/(1-\partial\varphi_0/\partial z) &&\text{on } z=\zeta_0\end{aligned} \qquad (8.4.78)$$

$$\begin{aligned}&\nabla^2\varphi_2=0\\ &n_e(\partial\varphi_2/\partial t+\zeta_1\,\partial^2\varphi_1/\partial z\,\partial t)\\ &=K\left[(\nabla\varphi_1)^2+\zeta_1^2\left(\frac{\partial}{\partial z}\nabla\varphi_0\right)^2+2\zeta_1\frac{\partial}{\partial z}(\nabla\varphi_0\cdot\nabla\varphi_1)+2\nabla\varphi_0\cdot\nabla\varphi_2+\zeta_2\frac{\partial}{\partial z}(\nabla\varphi_0)^2\right]\\ &\quad-(K+N)[\partial\varphi_2/\partial z+\zeta_1\,\partial^2\varphi_1/\partial z^2+\zeta_2\,\partial^2\varphi_0/\partial z^2] &&\text{on } z=\zeta_0\\ &\zeta_2=[(\zeta_1^2/2)\,\partial^2\varphi_0/\partial z^2+\zeta_1\,\partial\varphi_1/\partial z+\varphi_2]/(1-\partial\varphi_0/\partial z) &&\text{on } z=\zeta_0\end{aligned} \qquad (8.4.79)$$

etc., for higher order terms.

Thus, the original problem of solving for φ and ζ has been replaced by a series of subproblems of solving first for φ_0, ζ_0, then for φ_1, ζ_1 (using the results of φ_0, ζ_0), etc.

It follows from (8.4.77) that no advantage is gained with respect to φ_0, which describes the unperturbed steady state, as the problem remains a *nonlinear* one on account of the boundary condition on the free surface, the only advantage being that it is now specified on the known surface $z = \zeta_0$. However, (8.4.78) and (8.4.79) reveal certain farreaching simplifications. For, once the steady-state solution for φ_0 and ζ_0 is obtained (or is given as known initial conditions), the phreatic surface conditions for φ_1 and φ_2 (and higher order terms, whenever such terms are being considered) are *linear*. Moreover, these conditions are now specified on the a priori *known* boundary $z = \zeta_0$ (instead of the originally unknown one, $z = \zeta$).

In spite of the linearity of the boundary conditions with respect to φ_1 and φ_2, the equations are still rather complicated. Two special simple cases of initial steady flow lead to relatively simple flow problems.

Case 1, an initially horizontal water table with no accretion, $N = 0$ (i.e., initially the water under the phreatic surface is at rest (fig. 8.4.8b)). In this case $\varphi_0 \equiv 0$ and $\zeta_0 = 0$, and we obtain for φ_1 and φ_2:

$$\begin{aligned} &\nabla^2\varphi_1 = 0, && z \leqslant 0 \\ &n_e \frac{\partial \varphi_1}{\partial t} = -K \frac{\partial \varphi_1}{\partial z} && \text{on } z = 0 \\ &\zeta_1 = \varphi_1 \end{aligned} \tag{8.4.80}$$

$$\begin{aligned} &\nabla^2\varphi_2 = 0, && z \leqslant 0 \\ &n_e \frac{\partial \varphi_2}{\partial t} + K \frac{\partial \varphi_2}{\partial z} = \varphi_1 \frac{\partial}{\partial z}\left(\frac{\partial \varphi_1}{\partial z} - n_e \frac{\partial \varphi_1}{\partial t}\right) + K(\nabla \varphi_1)^2 && \text{on } z = 0 \\ &\zeta_2 = \varphi_1 \frac{\partial \varphi_1}{\partial z} + \varphi_2 && \text{on } z = 0. \end{aligned} \tag{8.4.81}$$

Case 2, uniform flow with an inclined plane phreatic surface without accretion. In this case (fig. 8.4.8c) it is convenient to employ an inclined coordinate system X, Y, Z:

$$X = x \cos\alpha - z \sin\alpha; \qquad Y = y; \qquad Z = x \sin\alpha + z \cos\alpha. \tag{8.4.82}$$

We obtain for the different approximations:

$$\begin{aligned} &\varphi_0 = -q_0 X/K = -X \sin\alpha \\ &\zeta_0 = 0 \end{aligned} \tag{8.4.83}$$

$$\nabla^2\varphi_1 = 0, \qquad Z \leqslant 0$$

$$\left.\begin{aligned} &n_e \frac{\partial \varphi_1}{\partial t} = -K\left[\sin\alpha \frac{\partial \varphi_1}{\partial X} + \cos\alpha \frac{\partial \varphi_1}{\partial Z}\right] \\ &\zeta_1 = \frac{\varphi_1}{\cos\alpha} \end{aligned}\right\} \quad \text{on } Z = 0 \tag{8.4.84}$$

$$\nabla^2\varphi_2 = 0, \qquad Z \leqslant 0$$

$$n_e \frac{\partial\varphi_2}{\partial t} + K\left[\sin\alpha\frac{\partial\varphi_2}{\partial X} + \cos\alpha\frac{\partial\varphi_2}{\partial Z}\right]$$

$$= K\left[\left(\frac{\partial\varphi_1}{\partial X}\right)^2 + \left(\frac{\partial\varphi_1}{\partial Y}\right)^2 + \left(\frac{\partial\varphi_1}{\partial Z}\right)^2\right] + n_e\zeta_1\frac{\partial}{\partial t}\left(\sin\alpha\frac{\partial\varphi_1}{\partial X} - \cos\alpha\frac{\partial\varphi_1}{\partial Z}\right)$$

$$+ K\zeta_1\left(\sin^2\alpha\frac{\partial^2\varphi_1}{\partial X^2} - \cos^2\alpha\frac{\partial^2\varphi_1}{\partial Z^2}\right) \qquad \text{on} \quad Z = 0$$

$$\zeta_2 = \frac{\zeta_1(-\sin\alpha(\partial\varphi_1/\partial X) + \cos\alpha(\partial\varphi_1/\partial Z)) + \varphi_2}{\cos\alpha} \qquad \text{on} \quad Z = 0. \qquad (8.4.85)$$

In the presence of accretion ($N \neq 0$), or when the initial condition does not correspond to a steady one, we start the linearization procedure described above from (8.4.63) with $\varphi_0 = 0$ and $\zeta_0 = 0$, and assume that N is small, of the order of ε, so that we may write $N = \varepsilon N_1$. This will lead to the following linearized statement of the phreatic flow problem in terms of the first and second approximations:

$$\varphi|_{z=\zeta} = \varphi|_{z=0} + (\varepsilon\zeta_1 + \varepsilon^2\zeta_2 + \cdots)\left.\frac{\partial\varphi}{\partial z}\right|_{z=0} + \cdots$$

$$= (\varepsilon\varphi_1 + \varepsilon^2\varphi_2 + \cdots)|_{z=0} + (\varepsilon\zeta_1 + \varepsilon^2\zeta_2 + \cdots)\left.\frac{\partial(\varepsilon\varphi_1 + \varepsilon^2\varphi_2 + \cdots)}{\partial z}\right|_{z=0} + \cdots$$

$$= \varepsilon\varphi_1 + \varepsilon^2\left(\varphi_2 + \zeta_1\frac{\partial\varphi_1}{\partial z}\right) + 0(\varepsilon^3) \qquad \text{on} \quad z = 0$$

$$\frac{\partial\varphi}{\partial t} = \varepsilon\frac{\partial\varphi_1}{\partial t} + \varepsilon^2\left(\frac{\partial\varphi_2}{\partial t} + \zeta_1\frac{\partial^2\varphi_1}{\partial z\,\partial t}\right) + 0(\varepsilon^3) \qquad \text{on} \quad z = 0$$

$$\frac{\partial\varphi}{\partial x} = \varepsilon\frac{\partial\varphi_1}{\partial x} + \varepsilon^2\left(\frac{\partial\varphi_2}{\partial x} + \zeta_1\frac{\partial^2\varphi_1}{\partial x\,\partial z}\right) + 0(\varepsilon^3) \qquad \text{on} \quad z = 0.$$

Finally we obtain, neglecting terms of $0(\varepsilon^3)$ or higher:

$$n_e\left[\frac{\partial\varphi_1}{\partial t} + \varepsilon^2\left(\frac{\partial\varphi_2}{\partial t} + \zeta_1\frac{\partial^2\varphi_1}{\partial z\,\partial t}\right)\right]$$

$$= K\left[\left(\varepsilon\frac{\partial\varphi_1}{\partial x} + \varepsilon^2\frac{\partial\varphi_2}{\partial x} + \varepsilon^2\zeta_1\frac{\partial^2\varphi_1}{\partial x\,\partial z}\right)^2 + \left(\varepsilon\frac{\partial\varphi_1}{\partial y} + \varepsilon^2\frac{\partial\varphi_2}{\partial y} + \varepsilon^2\zeta_1\frac{\partial^2\varphi_1}{\partial y\,\partial z}\right)^2\right.$$

$$\left. + \left(\varepsilon\frac{\partial\varphi_1}{\partial z} + \varepsilon^2\frac{\partial\varphi_2}{\partial z} + \varepsilon^2\zeta_1\frac{\partial^2\varphi_1}{\partial z^2}\right)^2\right]$$

$$- (K + \varepsilon N_1)\left(\varepsilon\frac{\partial\varphi_1}{\partial z} + \varepsilon^2\frac{\partial\varphi_2}{\partial z} + \varepsilon^2\zeta_1\frac{\partial^2\varphi_1}{\partial z^2}\right) + \varepsilon N_1.$$

Hence:

$$\nabla^2\varphi_1 = 0, \qquad z < 0$$

$$n_e \frac{\partial \varphi_1}{\partial t} + K\frac{\partial \varphi_1}{\partial z} = N_1 \qquad \text{on} \quad z = 0$$

$$\zeta_1 = \varphi_1 \qquad \text{on} \quad z = 0 \tag{8.4.86}$$

$$\nabla^2\varphi_2 = 0, \qquad z < 0$$

$$n_e \frac{\partial \varphi_2}{\partial t} + K\frac{\partial \varphi_2}{\partial z} = K\left[\left(\frac{\partial \varphi_1}{\partial x}\right)^2 + \left(\frac{\partial \varphi_1}{\partial y}\right)^2 + \left(\frac{\partial \varphi_1}{\partial z}\right)^2\right]$$

$$- K\zeta_1 \frac{\partial^2 \varphi_1}{\partial z^2} - n_e\zeta_1 \frac{\partial^2 \varphi_1}{\partial z\, \partial t} - N_1 \frac{\partial \varphi_1}{\partial z} \qquad \text{on} \quad z = 0$$

$$\zeta_2 = \varphi_2 + \zeta_1 \frac{\partial \varphi_1}{\partial z} \qquad \text{on} \quad z = 0. \tag{8.4.87}$$

For the initial condition (at $t = 0$) given in the form $\varphi(x, y, z, 0)|_{z=\zeta} = \zeta(x, y, 0) = f(x, y)$, on $z = 0$, we obtain:

$$\varphi_1(x, y, 0, 0) = \zeta_1(x, y, 0) = f(x, y)/\varepsilon$$

$$\zeta_2(x, y, 0) = 0.$$

In view of the last equation of (8.4.87), the last condition may be replaced by:

$$\varphi_2(x, y, 0, 0) = - \varphi_1(x, y, 0, 0)\, \partial\varphi_1(x, y, 0, 0)/\partial z. \tag{8.4.88}$$

Thus the first approximation is equivalent to neglecting the quadratic terms in (8.4.61) and assuming $N \ll K$. With $N = 0$, the first approximation (8.4.80) was formulated and employed by Polubarinova-Kochina (1952, 1962) for solving some of the nonlinear equations presented in section 8.2. It was also used by others, including Boulton (1954) and Beliacova (1955), by Kirkham (1958) for steady flow, by Galin (1959) and by Dagan (1964a and b) for solving different problems of phreatic surface flow in porous media. Kirkham (1958) suggests that this approximation is equivalent to replacing the region above $z = 0$ by a medium with horizontal conductivities K_x and $K_y = 0$, while the vertical conductivity is $K_z = \infty$. This approximation may be satisfactory when the deviations of the actual phreatic surface elevations from those of the assumed one ($z = 0$) are small relative to a characteristic horizontal length of the flow domain.

From the phreatic surface boundary condition for φ_1 and φ_2 in (8.4.78) through (8.4.81) written for $N = 0$ and $\zeta_0 = 0$, it follows that this condition has the general form:

$$n_e \frac{\partial \varphi_1}{\partial t} + K\frac{\partial \varphi_i}{\partial z} = F\left(\varphi_1, \varphi_2, \ldots, \varphi_{i-1}, \frac{\partial \varphi_1}{\partial x}, \ldots, \frac{\partial^{i-1}\varphi_{i-1}}{\partial z^{i-1}}\right) \qquad \text{on} \quad z = 0. \tag{8.4.89}$$

Thus, the problem of solving for the ith approximation φ_i, $i = 1, 2, \ldots$ in a flow domain D with boundaries S and S_0 may be stated as:

determine $\varphi' = \varphi'(x', y', z', t')$ satisfying:

$$\nabla^2\varphi' = 0 \qquad \text{in } D$$

$$\frac{\partial\varphi'(x', y', 0, t')}{\partial t'} + \frac{\partial\varphi'(x', y', 0, t')}{\partial z'} = F'(x', y', t') \qquad \text{on } S_0$$

$$\varphi' + \beta\frac{\partial\varphi'}{\partial n'} = 0 \qquad \text{on } S$$

$$\varphi'(x', y', 0, 0) = f'(x', y') \qquad \text{on } S_0 \text{ at } t = 0 \qquad (8.4.90)$$

where S_0 represents the free surface boundary, S represents all the other boundaries on which we have assumed the boundary condition $\varphi + \beta\, \partial\varphi/\partial n = 0$ and we have introduced the dimensionless variables: $\varphi' = \varphi/L$, $x' = x/L$, $y' = y/L$, $z' = z/L$ and $t' = Kt/n_e L$, with L some characteristic length of the flow domain. In (8.4.90), the subscript i was omitted. The functions $F'(x', y', t')$ and $f'(x', y')$ are known functions that depend for each term φ_i of the expansion of φ of the previous terms of the series, on singularities and on the initial distribution of φ.

Dagan (1966) solves (8.4.90) by introducing a Green function similar to the one used by Finkelstein (1957) in solving wave problems. He defines the Green function $G(M, M_0, \tau) \equiv G(x, y, z, x_0, y_0, z_0, \tau)$ by (fig. 8.4.9):

$$G(M, M_0, \tau) = \frac{1}{r_{MM_0}} + H(M, M_0, \tau) \text{ in } D \qquad (8.4.91)$$

where $H(M, M_0, \tau)$ is harmonic in D,

$$G(M, M_0, \tau) + \beta(M_0)\, \partial G(M, M_0, \tau)/\partial n_{M_0} = 0 \qquad \text{for } M_0 \text{ on } S$$

$$\partial G(M, P_0, \tau)/\partial\tau + \partial G(M, P_0, \tau)/\partial z_{P_0} = 0 \qquad \text{on } S_0$$

$$G(M, P_0, 0) = 0 \qquad \text{on } S_0. \qquad (8.4.92)$$

He then shows that the solution of (8.4.90) is obtained from:

$$\varphi(M, t) = \frac{1}{4\pi}\iint_{(S_0)}\int_0^t \frac{\partial G(M, P_0, \tau)}{\partial\tau} F'(P_0, t-\tau)\, d\tau\, dS_0$$

$$+ \frac{1}{4\pi}\iint_{(S_0)} \frac{\partial G(M, P_0, t)}{\partial\tau} f'(P_0)\, dS_{P_0}. \qquad (8.4.93)$$

Dagan (1966) also derives a corresponding Green function for two-dimensional flows and presents a detailed discussion of these Green functions. He also presents (1967) solutions of the decay of a phreatic surface mound in two and three dimensions by using Green functions.

8.4.6 The Shallow Flow Approximation (Dagan 1964)

A phreatic *shallow flow* domain is one where the characteristic horizontal length L (e.g., distance between drains) is much larger than its characteristic vertical dimension

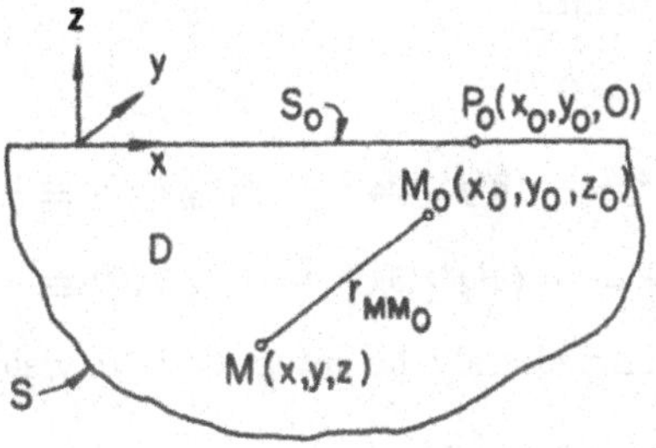

FIG. 8.4.9. The flow domain in the linearized approximation (after Dagan, 1967).

D (e.g., depth of impervious bottom below drains, or thickness of saturated flow domain at the inflow boundary).

The procedure described below is similar to the one employed by Friedrichs (1948) in the shallow-water wave theory. For the sake of simplicity we shall assume that the aquifer (fig. 8.4.10) has a horizontal impervious bottom. We shall employ the following dimensionless variables:

$$x' = x/L; \quad y' = y/L; \quad z' = z/D; \quad t' = (KD/n_e L^2)t; \quad N' = N/K;$$
$$\varphi' = \varphi/D; \quad \zeta' = \zeta/D \quad \text{where} \quad D \ll L.$$

By expanding φ' and ζ' in a power series in the small parameter $\sigma = D^2/L^2$, we obtain:

$$\varphi' = \varphi'_0(x', y', z', t') + \sigma\varphi'_1(x', y', z', t') + \sigma^2\varphi'_2(x', y', z', t') + \cdots \tag{8.4.94}$$
$$\zeta' = \zeta'_0(x', y', t') + \sigma\zeta'_1(x', y', t') + \sigma^2\zeta'_2(x', y', t') + \cdots . \tag{8.4.95}$$

As in the method of small perturbations (par. 8.4.5), we shall seek the equations satisfied by φ'_0, φ'_1, etc., by introducing (8.4.94) and (8.4.95) into the exact statement of the flow problem:

$$\nabla^2\varphi' \equiv \partial^2\varphi'/\partial x'^2 + \partial^2\varphi'/\partial y'^2 + \partial^2\varphi'/\partial z'^2 = 0 \tag{8.4.96}$$

into (8.4.61) and into other boundary and initial conditions. From (8.4.96), we obtain:

$$\nabla^2\varphi' \equiv \sigma(\nabla^2_{x'y'}\varphi'_0 + \sigma\nabla^2_{x'y'}\varphi'_1 + \sigma^2\nabla^2_{x'y'}\varphi'_2 + \cdots) + \frac{\partial^2}{\partial z'^2}(\varphi'_0 + \sigma\varphi'_1 + \sigma^2\varphi'_2 + \cdots)$$
$$= \frac{\partial^2\varphi'_0}{\partial z'^2} + \sigma\left(\frac{\partial^2\varphi'_1}{\partial z'^2} + \nabla^2_{x'y'}\varphi'_0\right) + \sigma^2\left(\frac{\partial^2\varphi'_2}{\partial z'^2} + \nabla^2_{x'y'}\varphi'_1\right) + 0(\sigma^3) = 0 \tag{8.4.97}$$

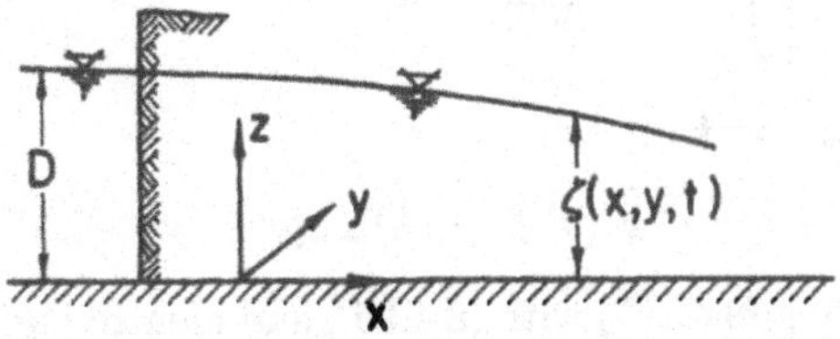

FIG. 8.4.10. Nomenclature for the shallow flow approximation.

which can be separated into:

$$\partial^2\varphi'_0/\partial z'^2 = 0 \tag{8.4.98}$$

$$\partial^2\varphi'_1/\partial z'^2 = -(\partial^2\varphi'_0/\partial x'^2 + \partial^2\varphi'_0/\partial y'^2) \equiv -\nabla^2_{x'y'}\varphi'_0 \tag{8.4.99}$$

$$\partial^2\varphi'_2/\partial z'^2 = -(\partial^2\varphi'_1/\partial x'^2 + \partial^2\varphi'_1/\partial y'^2) \equiv -\nabla^2_{x'y'}\varphi'_1. \tag{8.4.100}$$

These equations may immediately be integrated to yield:

$$\varphi'_0 = C_{01}z' + C_{00} \tag{8.4.101}$$

$$\varphi'_1 = -\int dz' \int \nabla^2_{x'y'}\varphi'_0\, dz' + C_{11}z' + C_{10} \tag{8.4.102}$$

$$\varphi'_2 = -\int dz' \int \nabla^2_{x'y'}\varphi'_1\, dz' + C_{21}z' + C_{20}, \quad \text{etc.,} \tag{8.4.103}$$

where $C_{ij}(x', y', t')$ are integration functions to be determined from the boundary conditions of the problem. For example, using the impervious boundary condition $\partial\varphi'/\partial z' = 0$ on $z = 0$, we obtain from (8.4.94):

$$\partial\varphi'_0/\partial z' = 0; \qquad \partial\varphi'_1/\partial z' = 0; \qquad \partial\varphi'_2/\partial z' = 0 \qquad \text{on } z = 0. \tag{8.4.104}$$

Hence: $C_{01} = 0$ and φ'_0 is independent of z' (i.e., $\partial\varphi'_0/\partial z' = 0$ everywhere). Hence φ'_0 describes a horizontal flow with a hydrostatic pressure distribution. It is thus equivalent to the Dupuit approximation.

Similarly, from (8.4.104) it follows that:

$$C_{11} = 0; \qquad \frac{\partial\varphi'_1}{\partial z'} = -\left(\frac{\partial^2\varphi'_0}{\partial x'^2} + \frac{\partial^2\varphi'_0}{\partial y'^2}\right) z'$$

$$C_{21} = 0; \qquad \frac{\partial\varphi'_2}{\partial z'} = -\left(\frac{\partial^2\varphi'_1}{\partial x'^2} + \frac{\partial^2\varphi'_1}{\partial y'^2}\right) z'. \tag{8.4.105}$$

On the phreatic surface $z' = \zeta'$, φ' may be expanded as follows:

$$\varphi'|_{z'=\zeta'} = \varphi'|_{z'=\zeta'_0} + (\sigma\zeta'_1 + \sigma^2\zeta'_2 + \cdots)\left.\frac{\partial\varphi'}{\partial z'}\right|_{z'=\zeta'_0} + \cdots$$

$$= \varphi'_0|_{z'=\zeta'_0} + \sigma\left.\left(\varphi'_1 + \zeta'_1\frac{\partial\varphi'_0}{\partial z'}\right)\right|_{z'=\zeta'_0} + \sigma^2\left(\varphi'_2 + \zeta'_1\frac{\partial\varphi'_1}{\partial z'} + \tfrac{1}{2}\zeta'^2_1\frac{\partial^2\varphi'_0}{\partial z'^2}\right.$$

$$\left.\left. + \zeta'_2\frac{\partial\varphi'_0}{\partial z'}\right)\right|_{z'=\zeta'_0} + 0(\sigma^3). \tag{8.4.106}$$

Also, on the phreatic surface:

$$\varphi'(x', y', z', t')|_{z'=\zeta'} = \zeta'. \tag{8.4.107}$$

Hence, by comparing (8.4.107) with (8.4.95), and identifying terms of equal powers of σ, we obtain:

$$\zeta'_0 = \varphi'_0 \qquad \text{on} \quad z' = \zeta'_0$$

$$\zeta'_1 = \varphi'_1 + \zeta'_1 \frac{\partial \varphi'_0}{\partial z'} \qquad \text{on} \quad z' = \zeta'_0$$

$$\zeta'_2 = \varphi'_2 + \zeta'_2 \frac{\partial \varphi'_0}{\partial z'} + \tfrac{1}{2}\zeta'^2_1 \frac{\partial^2 \varphi'_0}{\partial z'^2} + \zeta'_1 \frac{\partial \varphi'_1}{\partial z'} \qquad \text{on} \quad z' = \zeta'_0 \qquad (8.4.108)$$

etc. Rewriting (8.4.108) as expressions for φ'_0, φ'_1, φ'_2, and comparing these expressions with (8.4.101) through (8.4.103) written for $z' = \zeta'_0$, we obtain in view of (8.4.104):

$$\varphi_0' = \zeta_0'; \; C_{10} = \zeta'_1 + \frac{\zeta'^2_0}{2} \nabla^2 \varphi'_0$$

$$\varphi'_1 = \tfrac{1}{2}(\zeta'^2_0 - z'^2)\, \nabla^2 \zeta'_0 + \zeta'_1$$

$$\nabla^2 \varphi'_1 = \tfrac{1}{2} \nabla^2(\zeta'^2_0 \nabla^2 \zeta'_0) - \tfrac{1}{2} z'^2 \nabla^2 \nabla^2 \zeta'_0 + \nabla^2 \zeta'_1$$

$$\varphi'_2 = [\tfrac{1}{2} \nabla^2(\zeta'_0 \nabla^2 \zeta'_0)] \frac{z'^2}{2} - \frac{1}{24} z'^4 \nabla^2 \nabla^2 \zeta'_0 + \frac{z'^2}{2} \nabla^2 \zeta'_1 + C_{20} \qquad (8.4.109)$$

$$C_{20} = \zeta'_2 + \zeta'_0 \zeta'_1 \nabla^2 \zeta'_0 - \tfrac{1}{2} \nabla^2(\zeta'_0 \nabla^2 \zeta'_0) \zeta'^2_0/2$$

$$+ \frac{1}{24} \zeta'^4_0 \nabla^2 \nabla^2 \zeta'_0 - \frac{\zeta'^2_0}{2} \nabla^2 \zeta'_1$$

$$\varphi'_2 = \zeta'_2 + \zeta'_0 \zeta'_1 \nabla^2 \zeta'_0 + \tfrac{1}{4}(z'^2 - \zeta'^2_0)\, \nabla^2(\zeta'_0 \nabla^2 \zeta'_0)$$

$$- \frac{1}{24}(z'^4 - \zeta'^4_0)\, \nabla^2 \nabla^2 \zeta'_0 + \tfrac{1}{2}(z'^2 - \zeta'^2_0)\, \nabla^2 \zeta'_1. \qquad (8.4.110)$$

In (8.4.109) and (8.4.110),

$$\nabla^2(\;) \equiv \nabla^2_{x'y'}(\;) = \partial^2(\;)/\partial x'^2 + \partial^2(\;)/\partial y'^2.$$

Our next step is to insert the expansions of φ' and its derivatives on the phreatic surface into the phreatic surface boundary condition (8.4.61) rewritten in terms of the primed dimensionless variables:

$$\sigma \frac{\partial \varphi'}{\partial t'} = \sigma \left(\frac{\partial \varphi'}{\partial x'}\right)^2 + \sigma \left(\frac{\partial \varphi'}{\partial y'}\right)^2 + \left(\frac{\partial \varphi'}{\partial z'}\right)^2 - (1 + N') \frac{\partial \varphi'}{\partial z} + N'. \qquad (8.4.111)$$

By inserting (8.4.106) into (8.4.111) with $N' = \sigma N'_1$ and $\partial \varphi'_0/\partial z' \equiv 0$, we obtain, by identifying terms of equal order of σ:

$$\left.\begin{aligned}
\sigma:&\quad \frac{\partial \varphi'_0}{\partial t'} = \left(\frac{\partial \varphi'_0}{\partial x'}\right)^2 + \left(\frac{\partial \varphi'_0}{\partial y'}\right)^2 - \frac{\partial \varphi'_1}{\partial z'} + N'_1 \\
\sigma^2:&\quad \frac{\partial \varphi'_1}{\partial t'} = \left(\frac{\partial \varphi'_1}{\partial z'}\right)^2 + 2\left[\frac{\partial \varphi'_0}{\partial x'}\frac{\partial \varphi'_1}{\partial x'} + \frac{\partial \varphi'_0}{\partial y'}\frac{\partial \varphi'_1}{\partial y'}\right] - \frac{\partial \varphi'_2}{\partial z'} \\
&\quad - \zeta'_1 \frac{\partial^2 \varphi'_1}{\partial z'^2} - N'_1 \frac{\partial \varphi'_1}{\partial z'}
\end{aligned}\right\} \quad \text{on} \quad z' = \zeta'_0 \qquad (8.4.112)$$

etc. In order to obtain equations in terms of ζ'_0, ζ'_1, etc., alone, we insert (8.4.109) and (8.4.110) into (8.4.112). We obtain on $z' = \zeta'_0$:

$$\frac{\partial \zeta'_0}{\partial t'} = \left(\frac{\partial \zeta'_0}{\partial x'}\right)^2 + \left(\frac{\partial \zeta'_0}{\partial y'}\right)^2 + \zeta'_0 \nabla^2_{x'y'}\zeta'_0 + N'_1 = \frac{\partial}{\partial x'}\left(\zeta'_0 \frac{\partial \zeta'_0}{\partial x'}\right) + \frac{\partial}{\partial y'}\left(\zeta'_0 \frac{\partial \zeta'_0}{\partial y'}\right) + N'_1 \tag{8.4.113}$$

etc., for higher order terms. Once ζ'_0, ζ'_1, etc., are known, equation (8.4.108) is used to determine φ'_0, φ'_1, φ'_2, etc., and hence also φ'.

Of special interest is (8.4.113), which, when rewritten in the original dimensional variables, becomes:

$$n_e \frac{\partial \zeta_0}{\partial t} = K\left[\frac{\partial}{\partial x}\left(\zeta_0 \frac{\partial \zeta_0}{\partial x}\right) + \frac{\partial}{\partial y}\left(\zeta_0 \frac{\partial \zeta_0}{\partial y}\right)\right] + N \tag{8.4.114}$$

which is identical to the Boussinesq equation (8.2.5) for a homogeneous medium. Thus it appears (Dagan 1967) that h appearing in the nonlinear Boussinesq equation is equivalent to the first-order approximation of the actual φ. The second-order approximation φ_1 satisfies a *linear* nonhomogeneous parabolic equation with variable coefficients.

Finally, it may be of interest to analyze the case where the flow is a shallow one with $D \ll L$, and at the same time the deviations from some steady state are small (par. 8.4.5). This case is shown in figure 8.4.11. We shall assume that ζ_0 appearing in (8.4.114) and the corresponding φ_0 may be expressed, following the method of small perturbations (par. 8.4.5), as:

$$\begin{aligned} \zeta_0(x, y, t) &= \zeta_{00}(x, y) + \varepsilon\zeta_{01}(x, y, t) + \cdots \\ \varphi_0(x, y, t) &= \varphi_{00}(x, y) + \varepsilon\varphi_{01}(x, y, t) + \cdots \end{aligned} \tag{8.4.115}$$

where ζ_{00} and φ_{00} represent the average steady state, and ζ_{01}, φ_{01}, etc., represent nonsteady perturbations. Inserting (8.4.115) into (8.4.114), we obtain for the terms in ε:

$$\frac{\partial}{\partial x}\left[\zeta_{00} \frac{\partial \zeta_{00}}{\partial x}\right] + \frac{\partial}{\partial y}\left[\zeta_{00} \frac{\partial \zeta_{00}}{\partial y}\right] = 0 \tag{8.4.116}$$

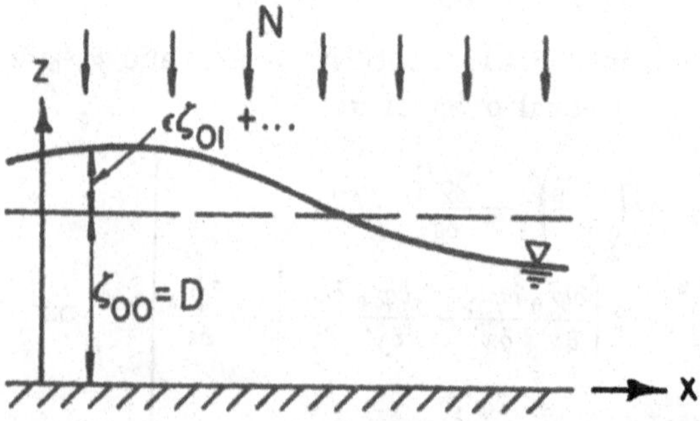

FIG. 8.4.11. Small phreatic surface perturbations in shallow flow.

$$\frac{n_e}{K}\frac{\partial\zeta_{01}}{\partial t} = \frac{\partial}{\partial x}\left[\zeta_{00}\frac{\partial\zeta_{01}}{\partial x}\right] + \frac{\partial}{\partial y}\left[\zeta_{00}\frac{\partial\zeta_{01}}{\partial y}\right] + \frac{\partial}{\partial x}\left[\zeta_{01}\frac{\partial\zeta_{00}}{\partial x}\right] + \frac{\partial}{\partial y}\left[\zeta_{01}\frac{\partial\zeta_{00}}{\partial x}\right] + N_1 \tag{8.4.117}$$

$$\varphi_{00} = \zeta_{00}, \qquad \varphi_{01} = \zeta_{01} \tag{8.4.118}$$

etc., where $N = \varepsilon N_1$. Once the steady-state solution is known, or is obtained from a solution of (8.4.116), equation (8.4.117) becomes a linear equation in ζ_{01}.

For the special case of an initially horizontal phreatic surface and $N = 0$, $\zeta_{00} \equiv D$ and (8.4.117) becomes:

$$n_e\, \partial\zeta_{01}/\partial t = KD\, \nabla^2_{xy}\zeta_{01} \tag{8.4.119}$$

which is the linear parabolic heat conduction equation.

Exercises

8.1 Show that q_z on a phreatic surface can be approximated by $q_z = -K(q'/Kh)^2$, where q' is the flow through the entire height of the unconfined flow as expressed by the Dupuit assumptions.

8.2 Cauchy's theorem states that along the boundary C of a flow domain in the xz-plane

$$\int_{(c)} \Phi\, dz + \Psi\, dx = 0.$$

Use this theorem to derive (8.1.12).

8.3 Derive the shape of the water table in steady radially converging flow to a well of radius r_w, pumping at a constant rate Q_w from an unconfined aquifer. At $r = R$ the water table elevation is maintained at $h = H_0$. If at $r = r_w$, h is maintained at $h = h_w$, show that the shape of the water table is independent of Q_w and K.

8.4 Show that although the expression for the discharge in Ex. 8.3 is based on the Dupuit assumptions, it nevertheless gives the exact discharge to the well.

8.5 Let the impervious bottom of fig. 8.1.3 be a plane rising from left to right at an angle θ (with the horizontal x). Determine the rate of flow Q towards the reservoir on the right.

8.6 Let the impervious bottom of fig. 8.1.6 be made of two parts: a length L_1 with a slope θ and a horizontal length L_2. Determine the rate of discharge between the two reservoirs and the water table elevation at $s = L_1$.

8.7 The water levels in two observation wells in a stratified phreatic aquifer are + 97.2 m and + 98.6 m. The distance between the wells is 400 m. The aquifer's bottom is at an elevation of + 4.8 m. The aquifer's hydraulic conductivity up to an elevation of + 20.0 m is $K_1 = 2.2$ m/d. Above this elevation, its hydraulic conductivity is $K_2 = 16.2$ m/d. Estimate the aquifer's discharge.

8.8 Let the block of soil between the two reservoirs of fig. 8.1.6 be stratified vertically: $0 < x < 6$ m, $K_1 = 0.2$ m/d; 6 m $< x <$ 8 m, $K_2 = 0.02$ m/d and 8 m $< x < L(= 14$ m), $K_3 = 0.2$ m/d. Estimate the rate of flow for $h_0 = 5$ m and $h_L = 2$ m.

8.9 A phreatic aquifer in the shape of an infinite strip of width L is located between two parallel streams. The aquifer has a horizontal bottom. Accretion N is added to the ground water flowing in the strip. The (constant) water levels in the two streams are at H_1 and H_2 above the horizontal bottom.

(a) Sketch the phreatic surface and the flow net.

(b) Determine the rates of discharge into each of the two streams for $K = 6$ m/d, $L = 2855$ m, $H_2 = 27.4$ m, $H_1 = 18.8$ m, $N = 820$ mm/year.

(c) What is the elevation of the highest point of the water table?

8.10 Use the Dupuit assumption to obtain $h = h(x)$ for the flow domain shown in fig. 8.1.2a. Sketch in the flow net and discuss weak points in the derived solution.

8.11 In the partially confined partially phreatic flow shown in fig. 8.E.1, determine b and the rate of flow Q.

8.12 Derive the partial differential equation describing unsteady radially converging flow to a well of constant discharge in a phreatic aquifer in the presence of accretion N.

8.13 Given a block of soil between two parallel fully penetrating drainage ditches with vertical walls (fig. 8.2.2a). The ditches are located at a distance of $L = 120$ m apart and the water levels in them is maintained constant at heights of $h_0 = 5$ m and $h_L = 10$ m above the impervious horizontal bottom. Because of the slope of the ground surface, the natural replenishment varies according to $N = Ax + B$; $A = 0.003\ yr^{-1}$, $B = 100\ mm/yr$. The flow is steady.

(a) Sketch the phreatic surface and the flow-net.

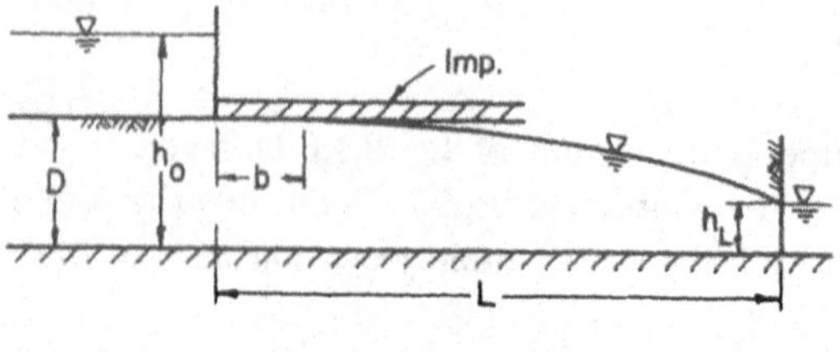

FIG. 8.E.1.

(b) Give an exact statement of the flow problem assuming a homogeneous isotropic medium.

(c) Determine the rates of flow to each channel for a homogeneous isotropic medium.

(d) Determine the shape of the phreatic surface for the case $K = cz + D$.

8.14 Linearize (8.2.9) by assuming $b = \bar{b} + b'$, $b' \ll \bar{b}$.

8.15 Use the hodograph method to solve the problem of steady flow to a drain on the bottom of an aquifer of finite depth.

8.16 Solve (8.4.5) to obtain a solution of the following one-dimensional phreatic flows:

(a) The domain is $-\infty < x < +\infty$. At $t = 0$, $h = f(x)$.
(b) The domain is $x \geqslant 0$; at $t = 0$, $h = g(x)$. At $x = 0, t > 0$, $h = h_0 = \text{const.}$

8.17 What is the partial differential equation (based on the Dupuit assumption) describing flow with accretion in a leaky phreatic aquifer? Assume that the semi-pervious layer is underlain by a very pervious formation so that the piezometric head in it is maintained practically constant. Then

(a) linearize the resulting equation by the first method of linearization (par. 8.4.1),

(b) linearize the same equation by using the second method of linearization (par. 8.4.2).

8.18 Use the method of successive steady states to study the unsteady flow between the two reservoirs of fig. 8.1.3. Initially steady flow takes place. Then h_L is maintained constant while $h_0 = h_0(t)$. Determine $h_0 = h_0(t)$ and $Q = Q(t)$.

CHAPTER 9

Flow of Immiscible Fluids

Some basic concepts, techniques and methods of investigation related to problems involving the flow of two fluids, and especially two liquids, through a porous medium domain are presented in this chapter. Such problems are often encountered in ground water hydrology and oil reservoir engineering. The discussion here is based on the theory of the flow of a homogeneous fluid presented in the previous chapters.

This is by no means a complete or comprehensive treatise on the subject. For additional information, the reader is referred to the works cited and to the vast literature existing on the subject. Of special interest are the works of Muskat (1937), Scheidegger (1960), Marle (1965) and Oroveanu (1966). Morel Seytoux (1969) and Bear (1970) also present summaries on flow of immiscible fluids.

9.1 Introduction

9.1.1 Types of Two-Fluid Flows

Two types of flow are possible when two or more fluids in motion occupy a porous medium domain.

Type 1, miscible displacement. In this case the two fluids are completely soluble in each other. The interfacial tension between the two fluids is zero, and the two fluids dissolve in each other. *A distinct fluid–fluid interface does not exist.* Often, especially in ground water hydrology, this type of flow is referred to as *hydrodynamic dispersion* (chap. 10).

Type 2, immiscible displacement. In this case we have a simultaneous flow of two or more immiscible fluids or phases (e.g., oil, water and gas) in the porous medium domain. The interfacial tension between the two fluids is nonzero, and a distinct fluid–fluid interface separates the fluids within each pore. A capillary pressure difference exists across the interface at each point on it. This type of flow is discussed in sections 9.3 and 9.4.

9.1.2 The Abrupt Interface Approximation

Actually, since the fluids are either miscible or immiscible, an abrupt interface between them in the macroscopic sense, i.e., a continuous surface completely separating the two fluids, cannot exist. In miscible displacement, even if the two fluids (e.g., fresh water and salt water) are initially separated by an abrupt interface (a situation which is, in fact, possible only under laboratory conditions), a *transition zone* due to hydrodynamic dispersion is immediately created. Across the zone, the composition of the fluid varies from that of one fluid to that of the other fluid. Owing to capillary phenomena, the same is true for immiscible fluids.

Yet, in many cases of practical interest, the transition zone in the case of miscible fluids is narrow relative to the size of the flow domain, or the displacement of immiscible fluids is almost complete, such that for all practical purposes a fictitious *abrupt interface* may be assumed to separate the two fluids. On each side of this imaginary interface we have only a single phase (or nearly so). Although this is clearly an approximate approach, whenever justified it appreciably simplifies the treatment of the two-fluid problem on hand. This type of flow is treated in sections 9.5 through 9.8.

9.1.3 Occurrence

The various types of two-fluid flows in porous media mentioned above occur in many engineering problems.

In practically all oil reservoirs, the simultaneous flow of oil, water and gas, is encountered, especially in the process of production. Of special interest is the *water drive* mechanism in which water partially or completely displaces oil and gas by intruding into the oil-producing region along its boundaries. Simultaneous flow also occurs in *secondary recovery operations* where gas, air or water is injected into a reservoir in order to increase its productivity after it has reached a state of substantially complete depletion of the initial energy required for oil expulsion. Of special interest in this respect is the case of *water flooding* where water is injected into the oil reservoir.

In oil reservoir engineering we also encounter flow of the miscible displacement type in the form of a secondary recovery process called *solvent drive*. In this process, the injected fluid is miscible with the displaced fluid. A mixed zone is created between the two fluids, behind which we have a region occupied only by the displacing fluid. Liquified propane or butane gases are examples of injected fluids in this case.

In ground water hydrology, where we deal essentially with water, we encounter only displacements of the first type, namely miscible displacements. The best known case is that of sea water intrusion into coastal aquifers (sec. 9.7). Sea water and fresh ground water are miscible fluids, and a transition zone develops between them. Across this zone the density of the water varies gradually from that of fresh water to that of sea water.

Another case where a transition zone develops as a result of hydrodynamic dispersion occurs when water of one quality is introduced into an aquifer containing water of another quality by various artificial recharge methods (surface spreading techniques or injection through wells). The different quality may serve in this case as a tracer by means of which the fluids and their mixture are recognized. Of special interest in this respect are the cases of sewage or radioactive waste disposal into aquifers, sometimes into deep formations.

Under certain conditions, the transition zone in these cases is narrow, say with respect to the dimensions of the injected water body or with respect to those of the aquifer, so that the abrupt interface approach may be implemented.

Unsaturated flow is a term often used by hydrologists and soil scientists to describe the flow of water and water vapor through a soil where the void space is partly filled with air. Essentially, this is a case of a simultaneous flow of two immiscible fluids, water and air, except that often the assumption is made that the air is practically immobile. This type of flow is discussed in section 9.4. The case of flow with change of phase, resulting from temperature or pressure variations, is not treated in this book.

9.2 Interfacial Tension and Capillary Pressure

A brief review of some concepts underlying the theory of simultaneous flow of immiscible fluids in porous media is given in this section.

9.2.1 Saturation and Fluid Content

When the void space of the porous medium is filled by two or more immiscible fluids (liquids or gases), the *saturation* (or *degree of saturation*) at a point, with respect to a particular fluid, is defined as the fraction of the void volume of the porous medium occupied by that particular fluid within an REV (par. 4.5.4) around the considered point.

$$S_\alpha = \frac{\text{volume of fluid } \alpha \text{ within an REV}}{\text{volume of voids within an REV}}; \qquad \sum_{(\alpha)} S_\alpha = 1. \tag{9.2.1}$$

In the theory of unsaturated flow (sec. 9.4), the (volumetric) *water content* (or moisture content) is defined as:

$$c = \frac{\text{volume of water within an REV}}{\text{bulk volume of REV}}. \tag{9.2.2}$$

This definition can be extended to any fluid α such that the (volumetric) fluid α content c_α is defined by:

$$c_\alpha = \frac{\text{volume of fluid } \alpha \text{ within an REV}}{\text{bulk volume of REV}} = nS_\alpha; \qquad \sum_{(\alpha)} c_\alpha = n \tag{9.2.3}$$

where n is the porosity of the porous medium.

In soil mechanics (Taylor 1948), *water content* (or moisture content) on a weight basis is sometimes defined as the ratio of weight of water to weight of solids within a sample.

9.2.2 Interfacial Tension and Wettability (Scheidegger 1960; Adamson 1967)

When a liquid is in contact with another substance (another liquid immiscible with the first, a gas or a solid), there is a free *interfacial energy* present between them. The interfacial energy arises from the difference between the inward attraction of the molecules in the interior of each phase and those at the surface of contact. Since a surface possessing free energy contracts if it can do so, the free interfacial energy

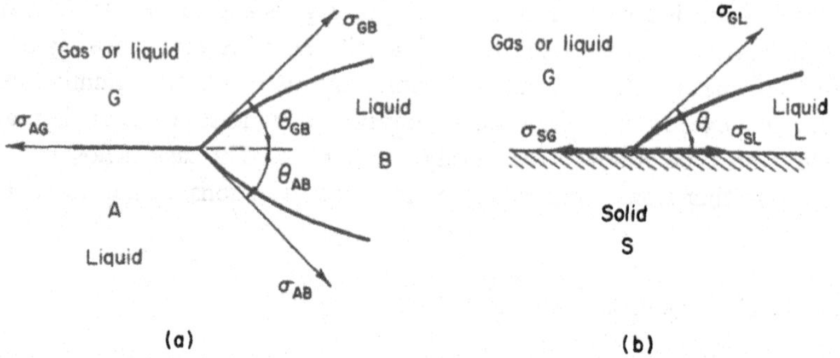

FIG. 9.2.1. Interfacial tensions.

manifests itself as *interfacial tension.* Thus the interfacial tension σ_{ik}, which is a constant for any pair of substances i and k, is defined as the amount of work that must be performed in order to separate a unit area of substance i from substance k. The interfacial tension σ_i between a substance i and its own vapor is called *surface tension.* The interfacial tension σ_{ik} between substances i and k, and the corresponding surface tensions, are related to each other by *Dupré's formula*:

$$W_{ik} = \sigma_i + \sigma_k - \sigma_{ik} \tag{9.2.4}$$

where W_{ik} is the amount of work required to separate a unit area of an interface between substances i and k into two substance-vapor interfaces i and k. The surface tension σ_i and the interfacial tension σ_{ik} are temperature dependent. Hence parameters related to σ_{ik} (e.g., capillary pressure) are also temperature dependent. Common units of σ_{ik} are dyne/cm or erg/cm^2 (dim F/L).

Figure 9.2.1a shows two immiscible liquids and a third fluid (liquid or gas) in contact with each other. Equilibrium requires that:

$$\sigma_{AG} = \sigma_{AB} \cos \theta_{AB} + \sigma_{GB} \cos \theta_{GB}. \tag{9.2.5}$$

Equation (9.2.5) can be satisfied if $\sigma_{AG} < (\sigma_{AB} + \sigma_{GB})$, and a lens of liquid B will be formed. If $\sigma_{AG} > (\sigma_{AB} + \sigma_{GB})$, equilibrium is not possible and liquid B will spread out between A and G.

Figure 9.2.1b shows two immiscible liquids (or a liquid and a gas) in contact with a solid surface. In this figure, θ denotes the angle between the *interface* and the surface SL. By convention, θ $(0 < \theta < 180°)$ is measured through the denser fluid. Equilibrium requires that:

$$\sigma_{GL} \cos \theta = \sigma_{SG} - \sigma_{SL}; \qquad \cos \theta = (\sigma_{SG} - \sigma_{SL})/\sigma_{GL}. \tag{9.2.6}$$

Equation (9.2.6), called *Young's equation,* states that $\cos \theta$ is defined by the ratio of the energy released in forming a unit area of interface between a solid S and a liquid L to the energy required for forming a unit interface between a fluid G and the liquid L. Sometimes a factor is introduced to account for roughness in the solid's surface. Bikerman (1958) questions (9.2.6) as it does not consider the force

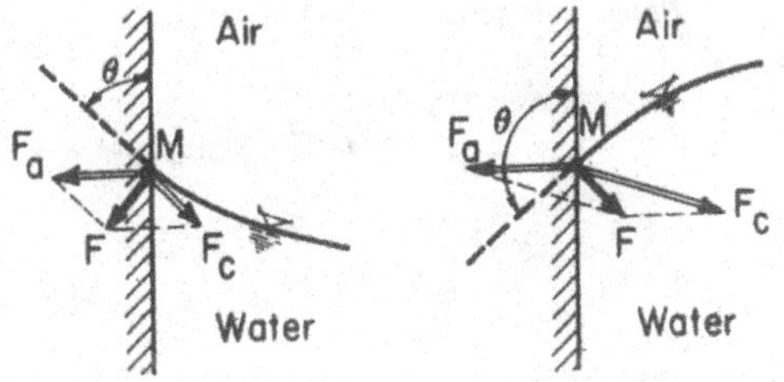

(a) (b)

FIG. 9.2.2. Contact angle between a water-air interface and a solid. (a) Water wetting the solid. (b) Water nonwetting the solid.

components normal to the surface.

From (9.2.6) it follows that no equilibrium is possible if $(\sigma_{SG} - \sigma_{SL})/\sigma_{GL} > 1$. In this case, liquid L will spread indefinitely over the solid. This leads to the concept of *wettability* of a solid by a liquid.

The angle θ in (9.2.6) is called the *contact angle*. The product $\sigma_{GL} \cos \theta$ is called the *adhesion tension*; it determines which fluid will *preferentially wet* the solid, i.e., adhere to it and spread over it.

When $\theta < 90°$, the fluid (e.g., L in fig. 9.2.1b) is said to wet the solid and is called the *wetting fluid*. When $\theta > 90°$, the fluid (G in fig. 9.2.1b) is called a *nonwetting fluid*. A zero adhesion tension indicates that both fluids have an equal affinity for the surface. In any system similar to that shown in figure 9.2.1b, it is possible to have either a fluid-L-wet or a fluid-G-wet surface, depending on the chemical composition of the fluids and the solid. Wettability has, therefore, only a relative meaning. Figure 9.2.2 shows the contact angle between a water–air interface and a solid.

Interfacial tension and wettability may be different when a fluid–fluid interface (e.g., an air–liquid interface) is advancing or receding on a solid surface. This phenomenon is called *hysteresis* (see below).

In an oil reservoir the fluids are usually oil and water and we speak of an *oil-wet* or a *water-wet rock*. Most reservoirs are assumed water wet. Some rocks seem to be partially wet with respect to more than one fluid. Figure 9.2.3 shows the fluid distribution in an oil-wet and a water-wet rock.

With this concept of wettability, we may distinguish three types of fluid saturation between the limits 0% and 100% (Pirson 1958). Figures 9.2.3a through 9.2.3c show a water-wet sand. Figures 9.2.3d through 9.2.3f show an oil-wet sand. At a very low water saturation (fig. 9.2.3a), water forms rings called *pendular rings* around the grain contact points. At this low water saturation the rings are isolated and do not form a continuous water phase, except for a very thin film of water of nearly molecular thickness on the grains' surfaces. Practically no pressure can be transmitted from one ring to the next within the water phase. As the wetting phase saturation increases, the pendular rings expand until a continuous wetting fluid phase is formed. The saturation at which this occurs is called *equilibrium saturation to the wetting*

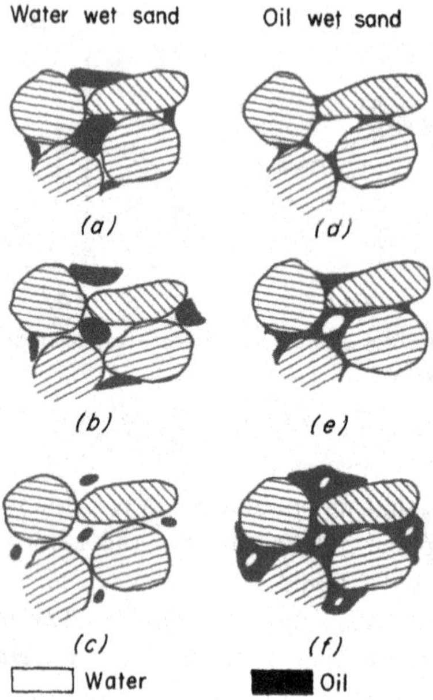

FIG. 9.2.3. Possible fluid saturation states.

phase. Above this critical saturation, the saturation is called *funicular* and the flow of the wetting phase is possible. As the saturation of the wetting fluid increases, a situation develops in which the nonwetting fluid is no longer a continuous phase; it breaks into individual droplets lodged in the larger pores (fig. 9.2.3c). The nonwetting phase is then said to be in a state of *insular saturation.* A globule of this type can move only if a pressure difference sufficient to squeeze it through capillary restrictions is applied across it in the wetting phase. Similar considerations may be applied to the oil-wet sand (figs. 9.2.3d,e,f).

A stage not mentioned above is the *adsorbed stage* where the pore space is largely filled by air while water (at very low saturation) is present within the pore space on adsorption sites of the solid as a continuous or a discontinuous film of one or more molecular layers (Stallman 1964). Sometimes these films provide the continuity of the wetting phase (water) in the pendular stage.

9.2.3 Capillary Pressure

When two immiscible fluids are in contact in the interstices of a porous medium, a discontinuity in pressure exists across the interface separating them. Its magnitude depends on the interface curvature at the point. Here "point" is the microscopic point inside the void space. The difference in pressure (p_c) is called *capillary pressure*:

$$p_c = p_{nw} - p_w \tag{9.2.7}$$

where p_{nw} is the pressure in the nonwetting phase and p_w is the pressure in the

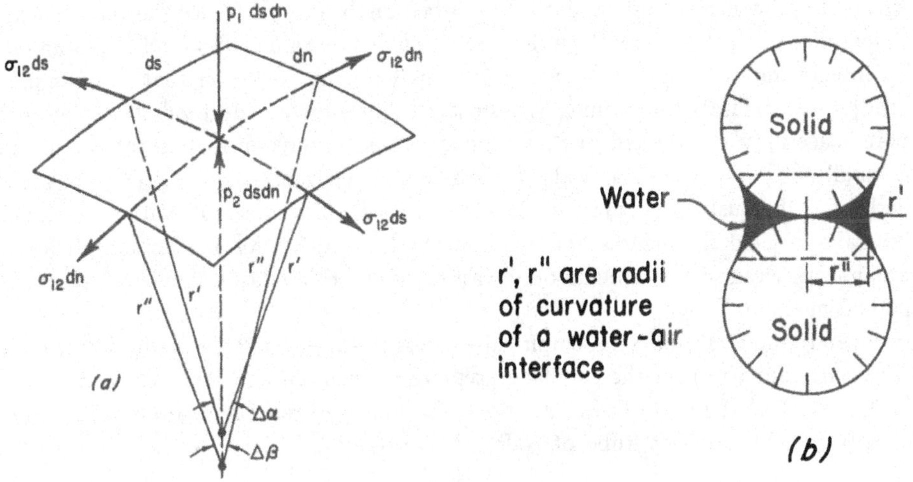

FIG. 9.2.4. Equilibrium at a curved interface between two immiscible fluids.

wetting phase. It may be determined by considering equilibrium on an elementary area around a point on the interface (fig. 9.2.4). The local interface within a pore is curved and has two principal radii of curvature, r' and r'', in two orthogonal planes. By considering the change in direction of the forces acting on opposite sides, we obtain:

$$\Delta p = p_c = p_2 - p_1 = \sigma_{12}(1/r' + 1/r'') = 2\sigma_{12}/r^* \tag{9.2.8}$$

where r^* is the mean radius of curvature ($2/r^* = 1/r' + 1/r''$). Equation (9.2.8) is known as the *Laplace equation for capillary pressure.* The capillary pressure is thus a measure of the tendency of a porous medium to suck in the wetting fluid phase or to repel the nonwetting phase. Figure 9.2.4b shows r' and r'' for a water–air interface in a pendular ring. When $r' = r'' = r^*$ (9.2.8) becomes:

$$p_c = 2\sigma_{12}/r^*. \tag{9.2.9}$$

In soil science, the negative of the capillary pressure (expressed as pressure head) is often called *suction* or *tension.*

In an actual porous medium, all terms in equations (9.2.7) through (9.2.9) have the meaning of a *statistical average taken over the void space in the vicinity of a considered point in the porous medium.* Because of the dependency of p_c locally (i.e., within each pore) on σ_{12} and r^* (which is of the order of magnitude of the pore (or grain) size), it depends on the geometry of the void space (e.g., expressed as a pore-size distribution), on the nature of solids and liquids (e.g., in terms of the contact angle θ) and on the degree of saturation (say, S_w of the wetting fluid that determines the volume of fluid accumulating as pendular rings at the points of contact between grains). In natural porous media, the geometry of the void space is extremely irregular and complex and cannot be described analytically. Hence the geometrical shape of the interface cannot be defined through the values of r' and r'' obtained by integrating over the whole interface, satisfying the condition that a minimum energy be consumed in forming the surface. Instead, an idealized model of the pore space may be adopted (e.g., a capillary tube, spheres of a constant radius or a bundle of parallel circular rods) for which the relationship $p_c = p_c(S_w)$ may be derived. Obviously this approach can only indicate the effects of various factors, but fails to yield the macroscopic relationship $p_c = p_c(S_w)$ for any actual medium. Laboratory experiments are the only way to derive this relationship for any given porous medium.

Several empirical and semi-empirical expressions are available in the literature, which attempt to relate the capillary pressure to medium and fluid properties and to S_w. Most of them are based on one of the idealized porous media models. For example, for a capillary tube of radius r, we obtain:

$$p_c = (2\sigma/r) \cos \theta. \tag{9.2.10}$$

One should recall that (arbitrarily) r in (9.2.10) is taken as negative. For other shapes of the cross-section, $2/r$ is replaced by $1/r^*$, where r^* is some equivalent radius equal to the ratio of volume to surface of the capillary. It is sometimes also called the *hydraulic radius of the capillary*. Leverett (1941) suggests a semi-empirical approach (using dimensional analysis), showing that the function:

$$J = J(S_w) = (p_c/\sigma)\sqrt{k/n}; \qquad p_c = p_c(S_w) \tag{9.2.11}$$

called the *J-Leverett function*, reduces to a common curve when plotted for several unconsolidated sands (fig. 9.2.5). In (9.2.11) k is the medium's permeability, n is its porosity and the ratio $(k/n)^{1/2}$ may be interpreted as some mean pore diameter.

Some authors suggest:

$$J(S_w) = (p_c/\sigma \cos \theta)\sqrt{k/n}. \tag{9.2.12}$$

Other authors (e.g., Rose and Bruce 1949) show (fig. 9.2.6) that different curves $J(S_w)$ characterize different formations.

An example of a consistent system of units to be used in defining the J-function is: for p_c (or for $Q_w g h_c$): gr/cm sec², for k: cm² and for σ: dyne/cm = gr/sec².

Brooks and Corey (1964) show that when the experimental data $p_c = p_c(S_e)$, where $S_e = (S_w - S_{w0})/(1 - S_{w0})$ is the *effective, or reduced, saturation*, and S_{w0} is the *irreducible wetting fluid saturation* (fig. 9.2.7a and b), are drawn on a log-log paper, a straight line is obtained except for S_e close to unity. They suggest that the negative slope of this curve λ (*pore-size distribution*) and the intercept p_b (value of p_c obtained by extending the straight line to $S_e = 1$) be used as constants characterizing the medium. This approach is extremely useful, e.g., when choosing sand for a sand box model, as the entire function $p_c = p_c(S_e)$ is represented by two constants. p_b

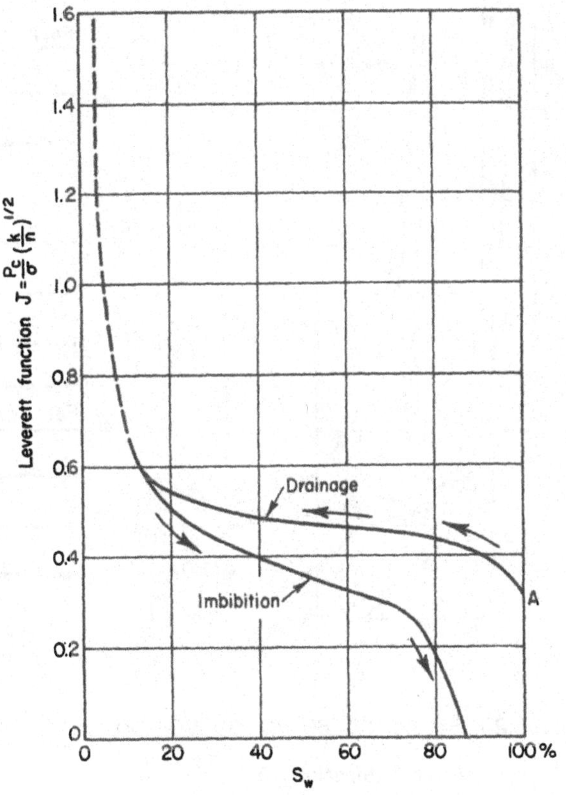

FIG. 9.2.5. Typical Leverett functions for sand (Leverett, 1941).

is called the *bubbling pressure* or the *threshold pressure* (see discussion at end of par. 9.2.4).

Capillary pressure is subject to *hysteresis* as the angle of contact θ is a function of the direction of the displacement; θ may have different values if equilibrium is approached by advancing or receding over a surface. For example (fig. 9.2.8), an advancing angle of contact occurs when water has a tendency to advance through oil, and a receding angle of contact when oil has a tendency to advance through water. Figure 9.2.9 shows this phenomenon for an air–water interface. This effect is sometimes called the *rain drop effect* (fig. 9.2.9c). Figure 9.2.8 shows the difference in contact angle in a static wetting fluid–nonwetting fluid interface and the one that occurs when the interface is displaced. This character of contact angle is the reason for the difference in capillary pressure curves $p_c = p_c(S_w)$ derived from a static experiment and those derived under dynamic conditions. Another mechanism causing hysteresis is the geometry of the void space with many bottlenecks. This is called the *ink-bottle effect*.

Because of the hysteresis phenomenon, the relationships between the J-function (or the capillary pressure) and S_w are not unique (par. 9.2.4) and we cannot determine

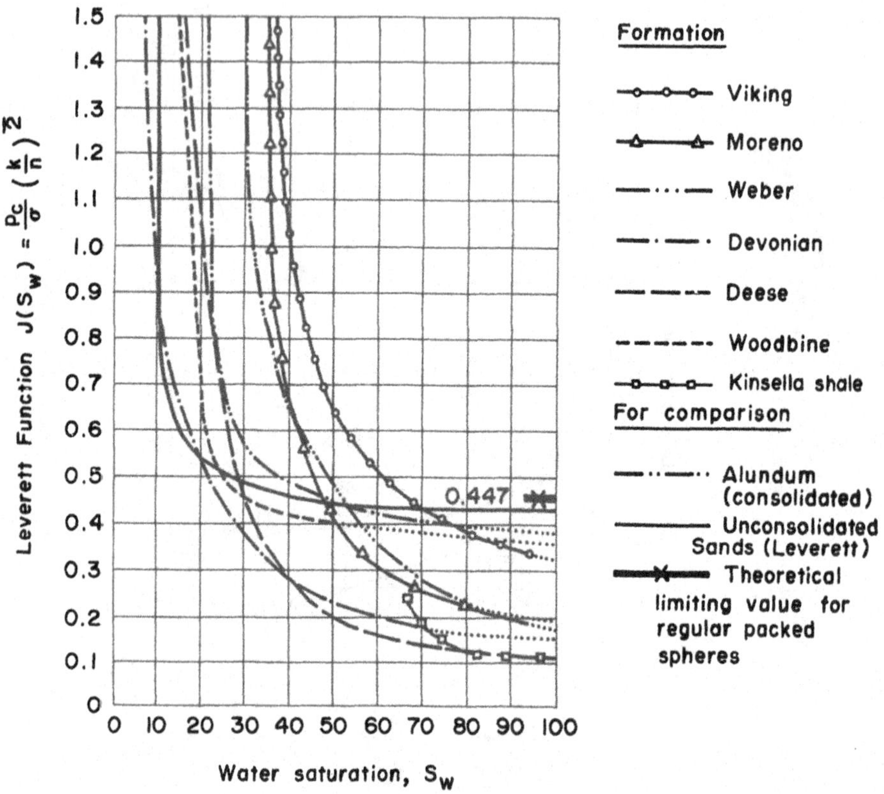

FIG. 9.2.6. Leverett function for various formations (Rose and Bruce, 1949).

p_c from a knowledge of S_w, or vice versa, without knowing the past wetting–drying history of the sample considered.

Finally, the capillary pressure–saturation relationship discussed above determines the saturation distribution across the transition zone between different hydrocarbons in a reservoir (fig. 1.1.7). Across the water–oil transition zone the water saturation increases with depth so that at the base of this transition zone the pore space is completely filled with water and hence the capillary pressure is zero. At any height h above the plane of zero capillary pressure, we have the capillary pressure p_c:

$$p_c(z) = (\gamma_w - \gamma_{nw})h \tag{9.2.13}$$

where γ_w and γ_{nw} are the specific weights of the heavier (here wetting) and the lighter (here nonwetting) fluids, respectively. To obtain (9.2.13) consider two points in the transition zone, with elevations z_1, z_2 and corresponding capillary pressures $p_{c1} = p_{nw1} - p_{w1}$ and $p_{c2} = p_{nw2} - p_{w2}$. At equilibrium, both fluids are stagnant, hence:

$$\varphi_{nw1} = \varphi_{nw2}; \qquad \varphi_{w1} = \varphi_{w2}.$$

Therefore:

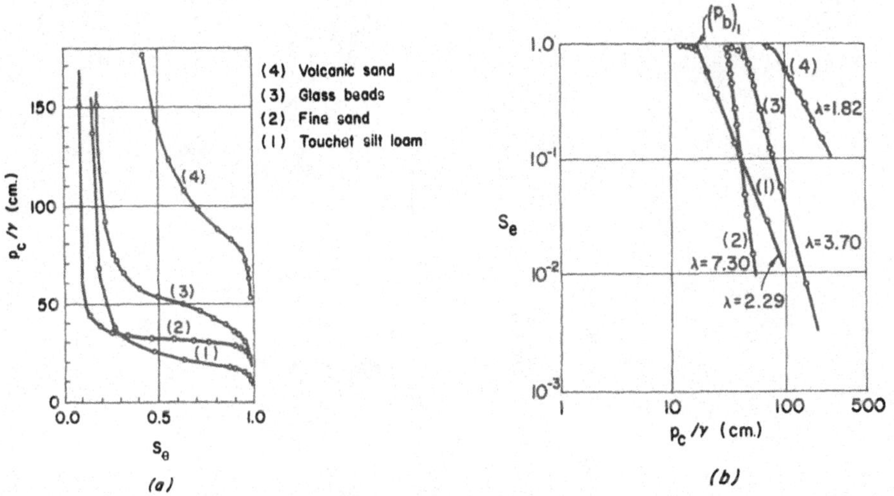

FIG. 9.2.7. Capillary pressure head as a function of effective saturation for porous materials of various pore-size distributions (Brooks and Corey, 1964).

$$z_1 + \frac{p_{nw1}}{\gamma_{nw}} = z_2 + \frac{p_{nw2}}{\gamma_{nw}}; \qquad z_1 + \frac{p_{w1}}{\gamma_w} = z_2 + \frac{p_{w2}}{\gamma_w}$$

$$z_1 - z_2 = \frac{1}{\gamma_w - \gamma_{nw}} (p_{c1} - p_{c2}). \tag{9.2.14}$$

If $\gamma_w > \gamma_{nw}$ and $p_{c1} > p_{c2}$, then $z_1 > z_2$. This also means $S_2 > S_1$. If, as a special case, $p_{c2} = 0$, i.e., point z is at the free water surface, we obtain (9.2.13). If from laboratory tests on a core taken from the reservoir we obtain the relationship $p_c = p_c(S_w)$ between capillary pressure and wetting fluid saturation, it is possible to use it together with (9.2.13) in order to determine the saturation distribution above the plane of $S_w = 1$ (*called free water level* when the wetting fluid is water). The elevation $S_{oil} = 1$ in the region of transition from oil to gas is called *free oil level.*

9.2.4 Drainage and Imbibition

Because of the capillary hysteresis phenomena described above, different capillary pressure curves $p_c = p_c(S_w)$ may be obtained, depending on whether a sample is

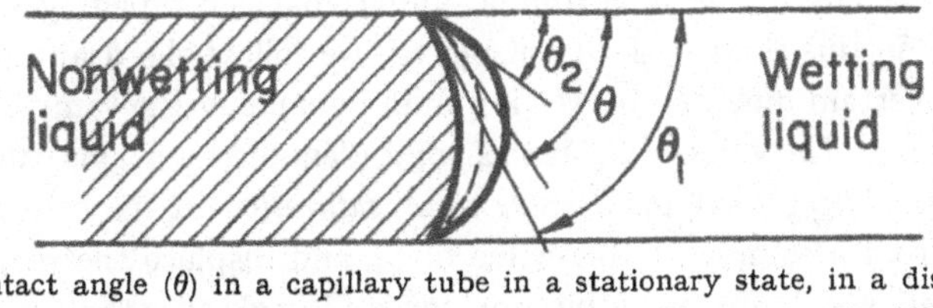

FIG. 9.2.8. Contact angle (θ) in a capillary tube in a stationary state, in a displacement of a nonwetting liquid by a wetting one (θ_1) and in a displacement of a wetting liquid by a nonwetting one (θ_2).

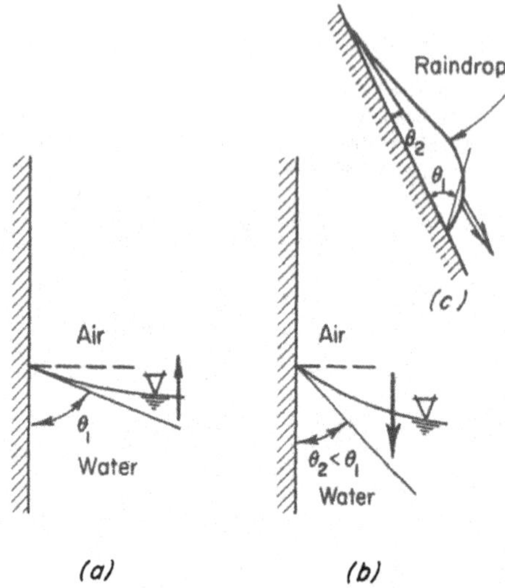

FIG. 9.2.9. Hysteresis in contact angle (rain drop effect).

initially saturated with a wetting or with a nonwetting fluid. In each case, the fluid initially saturating the sample is slowly displaced by the other fluid.

When a sample is initially saturated with a wetting fluid, the process is called *drainage*, and the curve $p_c = p_c(S_w)$ is called the *drainage curve*. The process by which a wetting fluid displaces a nonwetting fluid that initially saturates a porous medium sample, by capillary forces alone, is called *imbibition*. For example, if a sample is completely saturated with a nonwetting fluid, and some wetting fluid is introduced on its surface (or the sample is immersed in a wetting fluid), the wetting fluid will tend to flow in *spontaneously* along the solid walls of the pores, displacing the nonwetting fluid. The wetting fluid is said to displace the nonwetting one by imbibition. Imbibition is thus displacement due only to capillary forces. In a vertical displacement, equilibrium is reached when the wetting fluid has accumulated in those pores that permit the greatest curvatures of the fluid–fluid interfaces, i.e., in the smallest pores. Under such conditions capillary forces equal those of gravity.

The curve describing the relationship $p_c = p_c(S_w)$ during imbibition (i.e., rise in S_w) is called the *imbibition curve*. Figure 9.2.10 shows a typical capillary pressure–wetting fluid saturation relationship including the effect of hysteresis (kerosene and water in a sandstone).

Following these curves, we start from a sample that is completely saturated by a wetting fluid. As this fluid is slowly displaced, we follow the drainage curve. We observe that a certain quantity of wetting fluid remains in the sample even at high capillary pressures. The value of S_w at this point (denoted by S_{w0}) is called *irreducible saturation of the wetting fluid* (or *connate water saturation* in the case of water). If we now start from a sample at this value of S_w and displace the nonwetting fluid by a wetting one, we obtain the imbibition curve. We observe that at zero capillary pressure there remains a certain amount of the nonwetting fluid; this is the *residual*

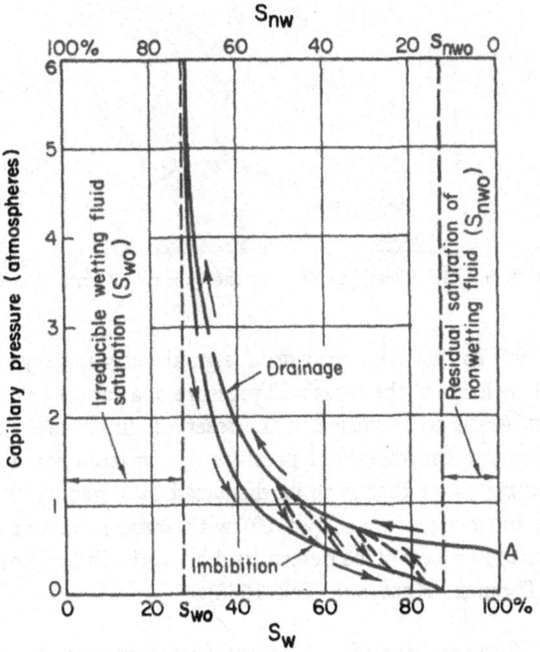

FIG. 9.2.10. Typical capillary pressure—wetting fluid saturation curves illustrating hysteresis.

saturation of the nonwetting fluid (denoted by S_{nw0}). In unsaturated flows, S_{nw0} indicates the amount of entrapped air that is the nonwetting fluid. At $S_{nw} \leqslant S_{nw0}$, the nonwetting fluid ceases to flow.

It is also possible to start the imbibition process from any point on the drainage curve, or to start the drainage process from any point on the imbibition curve (the dashed lines, called *drying scanning curve* and *wetting scanning curve,* shown in fig. 9.2.10). In this way the capillary pressure depends not only on the saturation at a certain instant, but also on the history of the particular sample under consideration. For a given capillary pressure, a higher saturation is obtained when the sample is being drained than when imbibition takes place. As long as the porous medium remains stable, the hysteresis loop can be repeatedly retraced.

In most fluid flow problems, capillary hysteresis is not a serious problem, because the flow regime (drainage or imbibition) usually dictates which curves should be used.

Point A on the drainage (or desaturation) curve in figures 9.2.5 and 9.2.10 shows that if a sample is initially saturated by a wetting fluid (i.e., $S_w = 1$), a certain pressure must be reached in the nonwetting fluid before the latter can begin to penetrate the sample, displacing the wetting fluid contained in it (fig. 9.2.11). In other words, a certain capillary pressure must be built up at the interface between the two fluids before drainage of the wetting fluid starts. The minimum pressure needed to initiate this displacement is called the *threshold pressure* (or *bubbling*

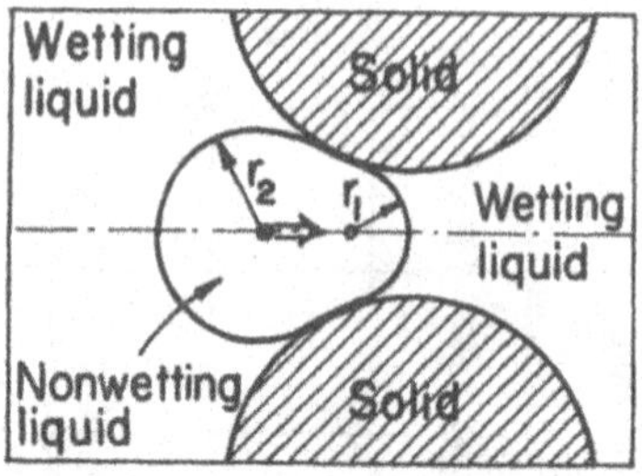

FIG. 9.2.11. Sketch explaining the concept of entry value.

pressure, p_b of fig. 9.2.7b; or *nonwetting fluid* (e.g., air or gas) *entry value*). In reservoir engineering the knowledge of the threshold pressure of a caprock saturated with water is important when gas is to be stored in the reservoir underneath the caprock. The simplest expression for the threshold pressure p_{ce}, written for a capillary tube of radius r saturated by water that is to be displaced by a gas, is (9.2.10). For actual porous media we must replace r in (9.2.10) with some mean (or equivalent) grain diameter r^*. Because r^* cannot be determined analytically, p_{ct} must be determined experimentally (Thomas, Katz and Tek 1968).

9.2.5 Saturation Discontinuity at a Medium Discontinuity

Let a layer of coarse sand (1) overlie a stratum of tight consolidated sandstone (2). Their capillary pressure curves are shown in figure 9.2.12. The two media are assumed to be water wet. Let a two-phase fluid mixture (e.g., oil and water or air and water) be present (under static or dynamic conditions) on both sides of the surface of contact. Given the wetting fluid saturation S_{w1} in medium 1, we wish to determine the saturation S_{w2} in medium 2. If saturations are assumed to be funicular in both media, we must have continuity of pressures across the boundary. Hence, at the

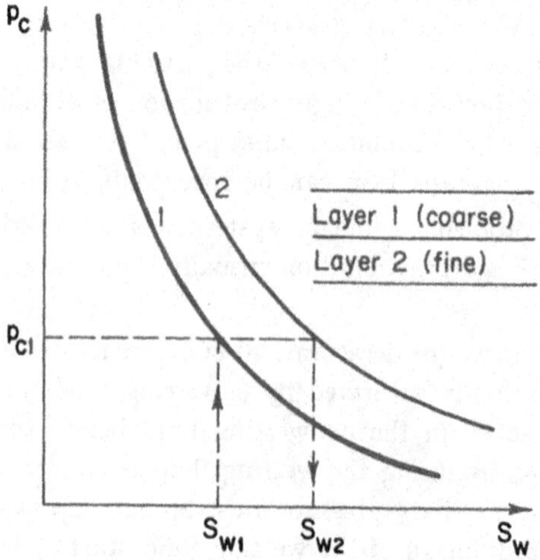

FIG. 9.2.12. Fluid saturation discontinuity between layers of different $p_c(S_w)$ characteristics.

surface of contact:

$$p_{w1} = p_{w2}; \qquad p_{nw1} = p_{nw2}; \qquad p_{c1} = p_{c2}.$$

The horizontal line in figure 9.2.12 connects the two equal values of p_c on both curves. We use it to determine S_{w2}. This means that a discontinuity in S_w must exist at a boundary of this kind.

The phenomenon described above also occurs when a porous medium overlies a liquid continuum (porosity = 100%) that serves as an external boundary of the flow domain. When in this case a wetting fluid displaces the nonwetting one toward this boundary, no wetting fluid will cross the exit face before $S_w = 1 - S_{nw0}$ occurs there. This phenomenon is known as *end effect* or *capillary end effect.*

At the beginning of paragraph 9.3.1 we have an example of steady, one-dimensional flow of two fluids in a column of length L, assuming that the saturations S_w and S_{nw}, and hence the effective permeabilities, remain uniform along the column. However, observations show that assumptions of uniform saturations along the column are good approximations only for sufficiently high values of q_w and q_{nw}. They fail for low specific discharge values. The reason for this is the end effect defined above. When a wetting fluid displaces a nonwetting one from the column initially saturated by the latter, the wetting fluid will not flow out until S_w at the outflow face ($x = L$) has been built up to some critical value. Initially only nonwetting fluid is present at $x = L$ and therefore we have $p_c = 0$. Consequently, no wetting fluid will appear outside the outflow face until p_c is also reduced to zero just inside the outflow face. As S_w increases during the experiment, the imbibition curve of figure 9.2.10 is applicable. According to this curve $p_c = 0$ at $S_{nw} = S_{nw0}$ (or $S_w = 1 - S_{nw0}$). However, at this critical value of nonwetting fluid saturation, the nonwetting fluid forms a discontinuous phase and ceases to flow. From this discussion it follows that as long as $S_w < 1 - S_{nw0}$ at the outflow face, only nonwetting fluid can leave the column through its outflow face. However, at $S_{nw} = S_{nw0}$, the permeability k_{rnw} of the nonwetting fluid becomes zero. This means that an infinite pressure gradient must exist in the nonwetting fluid if this fluid is to flow at a saturation of S_{nw0}.

The end effects are of special importance and should never be overlooked when a relative permeability test is performed under steady flow conditions. In such experiments, especially with relatively short cores, the end effects tend to concentrate excessive wetting fluid saturation near the outflow face when the major flow component is the nonwetting fluid, thus reducing the calculated nonwetting fluid effective permeability. End effects also occur at the circumference of wells in producing oil reservoirs and in fractured reservoirs.

9.2.6 Laboratory Measurement of Capillary Pressure

It is obviously impossible to use (9.2.10) for the determination of $p_c(S_w)$ of natural, or randomly ordered, porous media because of the complicated shapes of the pore openings of the medium. Even in simple media (e.g., spheres) it is difficult to find

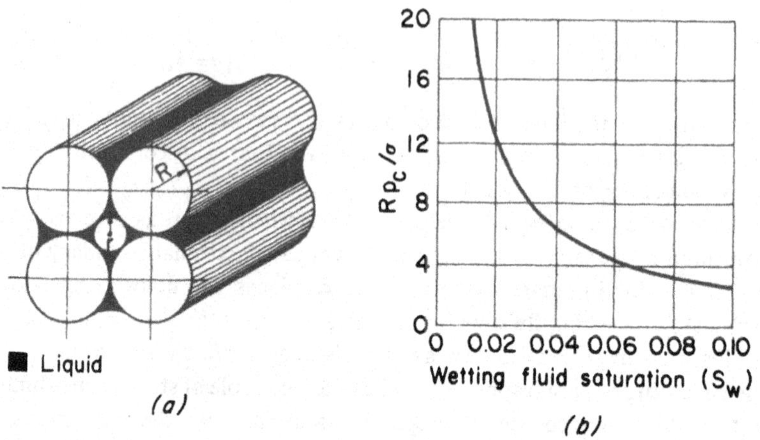

FIG. 9.2.13. Capillary pressure in a cubic packing of circular rods with liquid-air interfaces (after Collins, 1961).

the relationship between the mean interfacial radius r^* and the saturation. For example, consider the case of a porous medium in the form of circular rods of radius R in a cubic packing (fig. 9.2.13a; Collins 1961). The porosity of this medium is $n = (4R^2 - \pi R^2)/4R^2 = 1 - \pi/4$. If the radius of the interfacial meniscii is r, then the saturation of the liquid is given by:

$$S = \frac{4}{3\pi}\left[(\rho^2 + 2\rho)^{1/2} - \cos^{-1}\frac{1}{1+\rho} - \rho^2 \sin^{-1}\frac{1}{1+\rho}\right]; \qquad \rho = \frac{r}{R}. \quad (9.2.15)$$

Since by (9.2.8) $p_c = \sigma_{12}/r$ (as $r = r'$ and $r'' = \infty$) it is possible to use (9.2.15) for the computation of $p_c = p_c(S)$. The result which holds until the adjacent interfaces make contact is shown in figure 9.2.13b. Smith (1933) derives an equation for spheres of uniform size.

There are a number of laboratory methods for measuring the relationship $p_c = p_c(S_w)$ of porous media. They may be divided into two main groups of methods:

(a) *displacement methods*, based on the establishment of *successive states* of *hydrostatic equilibrium*, and

(b) *dynamic methods*, based on the establishment of *successive states of steady flow* of a pair composed of a wetting and a nonwetting fluid.

Of these two groups, the group of displacement methods is the more commonly used one. Several of these methods are described briefly below.

The *porous diaphragm method*, or the *Welge restored state method*, or *desaturation method*, is based on the drainage of a sample initially saturated with a wetting fluid. A schematic diagram of the apparatus is shown in figure 9.2.14. The sample or core saturated by the wetting fluid is placed inside the chamber, which is filled with a nonwetting fluid. The lower surface of the tested sample is placed on a permeable membrane or diaphragm made of fritted glass, porcelain, cellophane or other porous materials. The basic requirement is that the membrane will have a uniform pore-size

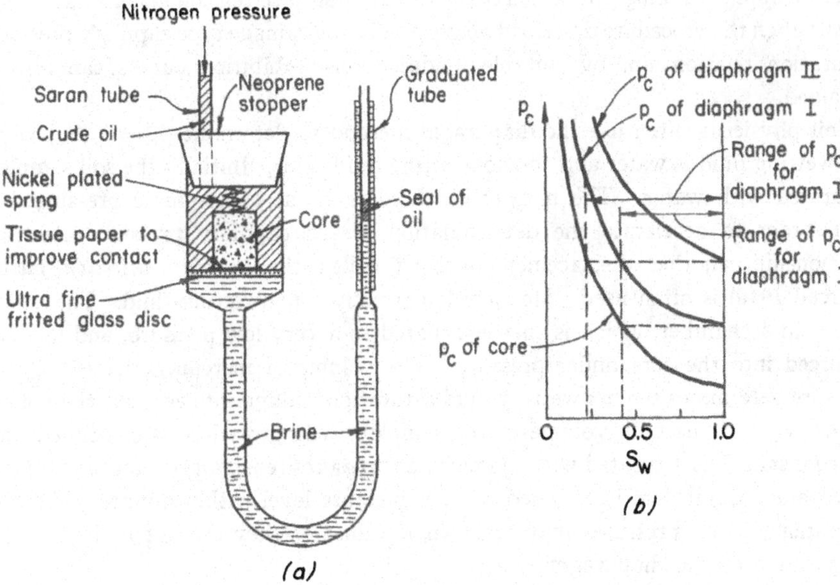

FIG. 9.2.14. Schematic diagram of a porous diaphragm device for capillary pressure determination (Welge and Bruce, 1947).

distribution containing pores of such size that the selected displacing fluid will not penetrate the membrane when the pressures applied to the displacing fluid are below some predetermined value (threshold pressure; par. 9.2.4) (fig. 9.2.14b). The membrane is initially saturated with the fluid to be displaced.

Wetting fluid, forced out of the sample by increasing the pressure in the nonwetting fluid, or sometimes by applying a suction by vacuum to the displaced fluid, drains into a graduated tube extending from the membrane. This permits the determination of the volume of displaced wetting fluid and hence the saturation of the tested sample. The displacement process is carried out in steps by increasing in small increments the pressure applied to the nonwetting fluid. At each pressure level the tested sample is allowed to approach a state of *static equilibrium*, a process which is usually time consuming (e.g., 10–40 days for a complete $p_c = p_c(S_w)$ curve), owing to the vanishing pressure differentials causing flow as the sample approaches equilibrium at each pressure level. When equilibrium is attained the capillary pressure is determined from the difference in pressure between the nonwetting (displacing) and the wetting (displaced) fluids, while the saturation of the sample is determined from the amount of displaced fluid. Other methods for determining sample saturation, such as electrical measurements (Martin et al. 1938) and X rays (Boyer, Morgan and Muscal 1947; Morgan, McDowell and Doty 1950, Laird and Putnam 1951; Norel 1964; Vachaud 1968), are sometimes used.

The diaphragm method is often used in reservoir engineering as it permits the use of actual reservoir fluids (e.g., water and oil). When air or mercury is used, a correction must be applied, taking into account the different surface tension and contact angle.

Although the procedure described above leads to a drainage type capillary pressure–saturation relationship, by suitable modifications imbibition curves can also be obtained.

Soil physicists often use the diaphragm method to determine $p_c = p_c(S_w)$ where the wetting fluid is water and the nonwetting fluid is air. Initially the soil sample is saturated with water. The nonwetting fluid is air at atmospheric pressure.

In order to accelerate the determination of the capillary pressure–saturation relationship, another displacement method called the *mercury injection method* (Purcell 1949) is often used. Mercury is normally a nonwetting fluid. The core is placed in a chamber, which is then evacuated to a very low pressure, and mercury is forced into the core under pressure. The volume of mercury injected at each pressure determines the nonwetting fluid saturation (taking into account the volume of the cell and the bulk volume of the sample). The procedure is continued until the core sample is saturated with mercury, or pressures reach a predetermined value. In general, equilibrium is attained at each pressure level within minutes. Another advantage is that pressures may reach high values as they are not limited by the entry value of the diaphragm.

A third static method for determining $p_c = p_c(S_w)$ is the *centrifuge method*. The high acceleration in the centrifuge increases the field of force acting on the fluids in the sample. This is equivalent to subjecting the tested sample to an increased gravitational force. When the centrifuge is turned on and maintained at a certain speed until no more liquid drains out of the sample, the volume of liquid is read in the small burette attached to the centrifuge cups. This volume determines the saturation of the sample corresponding to the rotating speed. The latter is related to p_c by:

$$p_c(r_1) = (r_2^2 - r_1^2)\omega^2 \Delta\rho/2 \qquad (9.2.16)$$

where ω is the angular velocity of the centrifuge and r_1 and r_2 are the radii of rotation to the inner and the outer faces of the sample, respectively. In general, an advantage of this method is that the capillary pressure–saturation relationship is obtained within a very short time (e.g., hours).

Amyx et al. (1960) compare several $p_c = p_c(S_w)$ curves obtained by the various methods discussed above.

Of the dynamic methods, the most commonly used is the one reported by Brown (1951). The apparatus (based on Hassler's principle) shown in figure 9.2.15, is the same apparatus used for relative permeability determinations (par. 9.3.7). Its main feature is the way it controls the capillary pressure at both ends of the sample. This is accomplished by placing the tested core between two membranes, or porous plates, which are permeable only to the wetting fluid. These permit maintenance of a uniform saturation throughout the length of the core even at low flow rates. The membranes permit the passage of the wetting phase (a liquid), but not the nonwetting phase (a gas). They also permit separate pressure measurements in each of the two phases. Initially the tested sample is saturated by the wetting

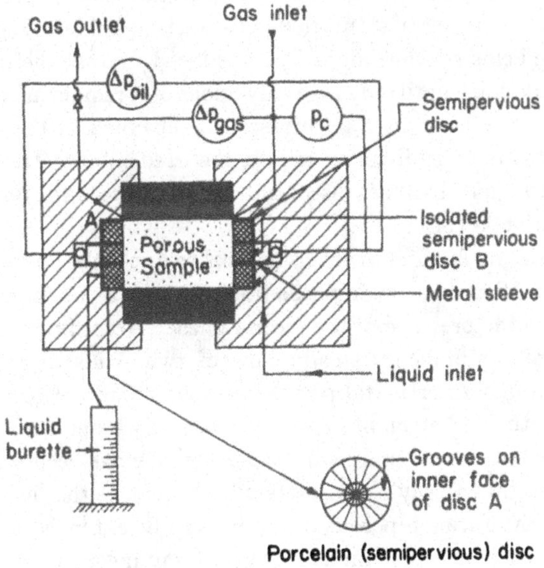

FIG. 9.2.15. Hassler's apparatus for relative permeability determination (after Osoba et al., 1951).

phase. It is then placed in the apparatus between the two membranes (with tissue paper to ensure good contact). Each membrane is divided by a metal sleeve into an inner disc, B, and an outer ring, A.

The liquid is introduced through ring A, while gas is introduced through radial grooves on the inner face of this ring. The pressure in the wetting phase is measured through the disc B. The capillary pressure is equal to the pressure difference between the nonwetting phase and the wetting one at the inflow face. Both phases pass through the sample at rates such that the pressure drop in both across the tested sample is the same, and hence p_c will be the same at both ends of the sample. When equilibrium is reached, the sample is removed and its saturation is determined by weighing.

9.3 Simultaneous Flow of Two Immiscible Fluids

Following the discussion on the continuum approach (sec. 1.3), when the simultaneous movement of two or more fluids is being considered, each of the fluids is regarded as a continuum completely filling the flow domain (at a fluid content that is a function of the space coordinates and of time). The various continua occupy the entire flow domain simultaneously.

9.3.1 The Basic Motion Equations

Many investigators conclude from experiments that when two immiscible fluids flow simultaneously through a porous medium each fluid establishes its own tortuous paths, which form very stable channels. They assume that a unique set of channels corresponds to every degree of saturation. If a wetting fluid (at S_w) and a nonwetting fluid (at S_{nw}) are being considered, as S_{nw} is reduced, the channels of the nonwetting fluid tend to break down until only isolated regions of it remain at residual nonwetting fluid saturation. Similarly, as S_w decreases, the channels of the wetting fluid tend to break down and become discontinuous at the irreducible wetting fluid saturation. When any of these fluids becomes discontinuous throughout the flow domain, no flow of that fluid can take place.

With these concepts in mind, it seems natural to apply the concept of permeability established for the flow of a single-phase fluid through a porous medium, modifying its value owing to the presence of the second phase. Accordingly, one may visualize an experiment of simultaneous steady flow of two fluids (denoted by subscripts $\alpha = 1$ and 2) through a horizontal porous medium column of constant cross-section A and finite length L. Let pumps force constant discharge rates Q_1 of fluid 1 and Q_2 of fluid 2 through this column. Once a steady flow has been established for both fluids, we assume that Darcy's law, originally describing the flow of a single-phase fluid completely saturating a porous medium, may be extended to describe the flow of each of the two immiscible fluids flowing simultaneously through the column:

$$q_1 = (k_1/\mu_1)\,\Delta p_1/L; \qquad q_2 = (k_2/\mu_2)\,\Delta p_2/L \tag{9.3.1}$$

where $q_\alpha = Q_\alpha/A_\alpha$, $\alpha = 1, 2$ and Δp_α is the pressure drop in the αth fluid.

In (9.3.1) k_1 and k_2 are called *effective permeabilities* of the medium to fluids 1 and 2, respectively. Obviously these depend on the structure of the porous medium involved—specifically, on the permeability k of the medium to a single-phase fluid completely saturating it—and on the respective saturations.

Often the ratios:

$$k_{r1} = k_1/k, \qquad k_{r2} = k_2/k \tag{9.3.2}$$

are used, called *relative permeability to fluid* 1 *and to fluid* 2, respectively.

Numerous experiments (Scheidegger 1960, p. 217) seem more or less to verify the validity of (9.3.1), with the relative permeabilities depending on the nature of the porous medium, on the preferential wettability within the couple of flowing fluids and on the saturations, but independent of the viscosities of the fluids and their specific discharges. However, one should recall that there is no way to derive $k_{r\alpha}$ ($\alpha = 1, 2$) except through idealized medium models or through experiments such as the one described schematically above. In fact, the effective permeabilities are *defined* by (9.3.1) or similar equations, which already presuppose the dependence of k_α on k and S_α alone. Additional comments on relative permeability are given in paragraph 9.3.2.

Equations (9.3.1), as a *working assumption* describing the simultaneous flow of immiscible fluids with sufficient accuracy for all practical purposes, can be generalized to include three-dimensional flows. For a single-phase fluid (compressible or incompressible) completely saturating an isotropic porous medium, the motion equation is:

$$\mathbf{q} = -(k/\mu)(\operatorname{grad} p - \rho \mathbf{g}) = -(k/\mu)(\operatorname{grad} p + \rho g \mathbf{1z}). \tag{9.3.3}$$

For an anisotropic medium, using the double index summation convention, it is:

$$q_j = -(k_{ij}/\mu)(\partial p/\partial x_i + \rho g\, \partial z/\partial x_i); \qquad j, i = 1, 2, 3. \tag{9.3.4}$$

A heuristic extension of (9.3.4) to the simultaneous flow of two immiscible fluids (denoted by subscripts 1 and 2) leads to the following equations for an anisotropic porous medium:

$$\begin{aligned} q_{i1} &= -\frac{k_{ij1}}{\mu_1}\left(\frac{\partial p_1}{\partial x_j} + \rho_1 g \frac{\partial z}{\partial x_j}\right) = -k_{ij}\frac{k_{r1}}{\mu_1}\left(\frac{\partial p_1}{\partial x_j} + \rho_1 g \frac{\partial z}{\partial x_j}\right) \\ q_{i2} &= -\frac{k_{ij2}}{\mu_2}\left(\frac{\partial p_2}{\partial x_j} + \rho_2 g \frac{\partial z}{\partial x_j}\right) = -k_{ij}\frac{k_{r2}}{\mu_2}\left(\frac{\partial p_2}{\partial x_j} + \rho_2 g \frac{\partial z}{\partial x_j}\right) \end{aligned} \tag{9.3.5}$$

where $\underline{\mathbf{k}}_1$ and $\underline{\mathbf{k}}_2$ are the two effective permeabilities. In this form, but for an isotropic medium, the equations are suggested by Wyckoff and Botset (1936) and Muskat et al. (1937). It is assumed that k_{r1} and k_{r2} are independent of direction, although this problem warrants further research. Another problem that arises in this connection is how the various k_{ij}s vary with saturation. It is possible to rewrite (9.3.5), for $\rho_1, \rho_2 =$ const., in terms of a piezometric head defined separately for each fluid:

$$\varphi_\alpha = z + p_\alpha/\rho_\alpha g; \qquad \alpha = 1, 2. \tag{9.3.6}$$

For a compressible fluid, φ_α must be replaced by φ^*_α defined by (5.9.1).

One should recall that for a given point (physical point) in the flow domain, p_1 and p_2 are pressures in the phase 1 and in the phase 2 fluids, respectively, each representing an average over the volume of fluid of that phase included in the REV for which the point considered is centroid.

9.3.2 Relative Permeability

Underlying the extension of the motion equation of a single fluid to the simultaneous flow of two or more fluids is the concept of *relative permeability*. At first sight, it seems natural to assume that when the flow of one of the fluids at a point is being considered, since part of the pore space in the vicinity of that point is occupied by another fluid, the permeability of the porous medium would be reduced with respect to the fluid considered. This means that the relative permeability depends only on the saturation. Figure 9.3.1 shows typical relative permeability curves for a pair of fluids: a wetting fluid (subscript w) and a nonwetting fluid (subscript nw). In figure 9.3.2 the wetting fluid is water (subscript w) and the nonwetting fluid is gas (subscript g). The figure also indicates some differences between a consolidated sand and an unconsolidated one. Both fluids are in motion only for a wetting fluid saturation S_w which is above S_{w0} (irreducible wetting fluid saturation) and below $1 - S_{nw0}$, where S_{nw0} is the residual nonwetting fluid saturation. At S_{nw0}, k_{rw} is usually much less than 1 whereas k_{rnw} at S_{w0} approaches the value of 1. Point A in figure 9.3.1, where $S_w = 1 - S_{nw0}$, is also called *equilibrium saturation*.

The rapid decline of k_{rw} indicates that the larger pores are occupied first by the

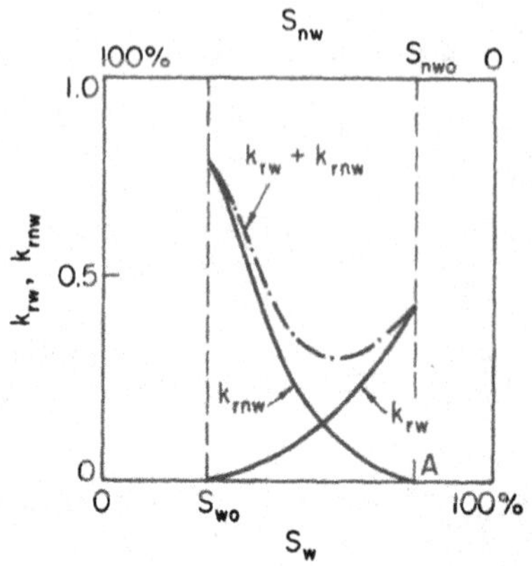

FIG. 9.3.1. Typical relative permeability curves (e.g., wetting fluid = water, nonwetting fluid = oil).

nonwetting phase. As S_{nw} increases, the average pore size saturated by the wetting fluid becomes progressively smaller. This is demonstrated by a rapid rise in k_{rnw}. In other words, above S_{nw0} the nonwetting fluid occupies larger pores than does the wetting one.

Many researchers have attempted to evaluate the influence of various fluid and flow parameters on relative permeability. One observation is that since one of the fluids preferentially wets the solid and adheres to its surfaces, the nonwetting fluid should be surrounded everywhere by the wetting fluid. One may think of a Poiseuille-type concentric flow in a tube as a model of such a flow in a porous medium. This means that the picture of each fluid establishing its own channels of flow through the medium is rather questionable. Moreover, since wettability is subject to hysteresis,

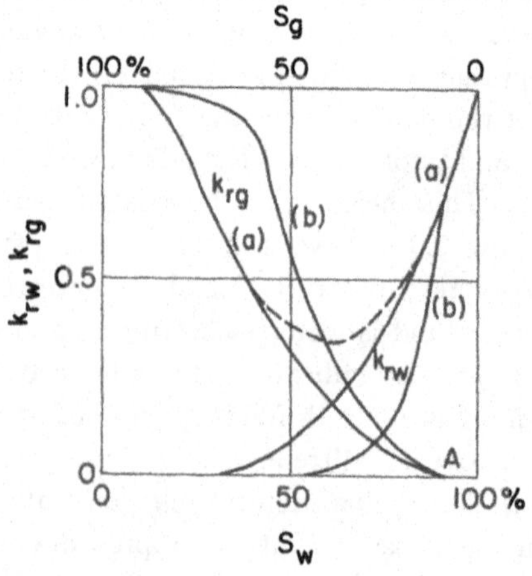

FIG. 9.3.2. Typical relative permeability to gas and water. (a) Unconsolidated sand. (b) Consolidated sand (Botset, 1940).

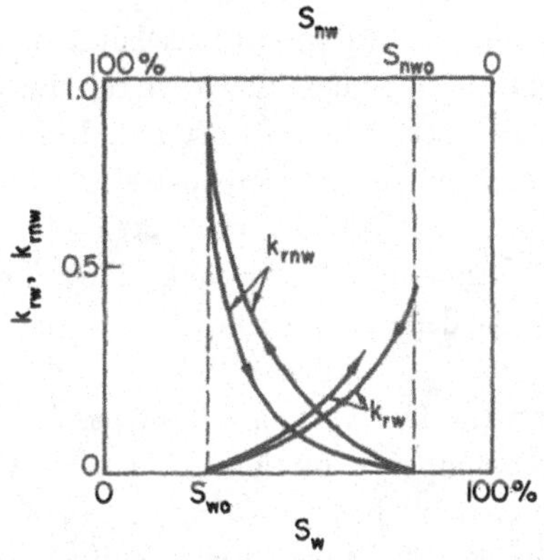

FIG. 9.3.3. Effect of hysteresis on relative permeability.

relative permeability should also be affected by this phenomenon, i.e., it should depend on the manner in which the two fluids are distributed within the pore space and on the saturation history of the sample. Figure **9.3.3** shows relative permeabilities as obtained for different saturation histories of oil and water.

The effect of pressure gradients on relative permeability has been investigated and reported by many authors. However, it seems that no definite conclusion has been reached. From available data, it seems (Muskat 1937) that for oil–water mixtures, the permeabilities, at least to the nonwetting phase, are higher at higher pressure gradients. Muskat (1937) concludes that while physical considerations suggest that the equilibrium permeability and saturation values should not be entirely independent of the pressure gradient, neglecting such effects is warranted at least as a first approximation.

Another factor, the effect of which on relative permeability has been studied and reported by many authors, is the interference that one fluid exerts on the other within the pore space as a result of the *difference in viscosity* between them. This effect is demonstrated by the fact that the sum $k_{rw} + k_{rnw}$ (dashed line in figs. 9.3.1 and 9.3.2) is less than unity, i.e., $(k_w + k_{nw}) < k$. Another example that seems to indicate that the viscosity ratio μ_w/μ_{nw} should affect k_{rw} and k_{rnw} is that the relative permeability to oil of a reservoir rock containing small amounts of connate water in pendular saturation might be greater than unity, an observation that contradicts the assumption that k_{rw} or k_{rnw} in (9.3.5) is independent of the fluid's viscosity. In the case of oil and water, Russell and Charles (1959) state that the presence of a thin layer of water on the solid surfaces, in analogy to a thin layer of water on the internal surface of a pipe, acts as a "lubricant," thus reducing resistance to the flow. All of these phenomena indicate that a transfer of viscous forces takes place across the fluid–fluid interface within the pore space. Yuster (1953), who also discusses the flow of fluid mixtures through porous media, suggests that a *finite velocity* must be taken as the boundary condition at the fluid–fluid interface. By considering a Poiseuille-type concentric flow in a circular tube (as a model representing the simultaneous flow of two immiscible fluids in a porous medium), he concludes that the viscosity ratio μ_w/μ_{nw} affects the specific discharge of the two fluids, and hence the extension of Darcy's law to two-phase flow with relative permeability independent of the viscosity ratio is not permitted. Scott and Rose (1953) choose to assume that Yuster's capillary tube model overemphasizes the magnitude of the ratio of fluid–fluid interfacial area to fluid–solid surface area. By replacing portions of the fluid–fluid interface by a thin impervious wall, they are able to demonstrate a corresponding decrease in the dependence of relative permeabilities upon the viscosity ratio.

Rose (1960) starts from the fact that one fluid passing over another contiguous immiscible fluid will always impart a motion to the second fluid. One of his models representing the porous medium has the geometry of a wide crack with parallel surfaces; the other model is a crack model with eddy pockets containing one fluid circulating in a vortex-like motion. By solving for the flow in these models, caused by applying a driving force to one or to both fluids, Rose can show the magnitude of the "lubrication effect" and concludes that the volumetric throughput of a particular fluid that only partially fills the pore space is sometimes greater than its throughput when it completely fills the pore space. This condition of increased relative permeability is met when the saturation of the fluid considered approaches unity, as long as the contiguous fluid has a lower viscosity. He also shows the effect of the viscosity ratio on relative permeability to be a second-order effect, thus explaining why most experimental data fail to show this dependence.

From the discussion above it seems that, although in principle relative permeability, as defined by extending the equation of motion to the simultaneous flow of two immiscible fluids, depends on several other factors as well as upon saturation, available experimental evidence indicates that this formal extension and the concept of relative permeability that depends only on saturation *is a good approximation for all practical purposes.*

Many references on relative permeability are given by Pirson (1958, p. 92), Odeh (1959), Scheidegger (1960, p. 217), Rose (1960) and Amyx et al. (1960).

Scheidegger (1960), Pirson (1958), and Amyx et al. (1960) list several methods for determining the effective and the relative permeability of a given sample. In some methods, a capillary equilibrium exists at all times among the various phases. In others, the fluids are introduced into the core at the same pressure and possibly seek equilibrium at a point away from the inlet. In practically all these methods a small core sample is mounted either in a plexiglass or a pressurized rubber sleeve. The two phases (oil and gas, oil and water, or gas and water) are introduced simultaneously at the inlet end through different piping systems. Most tests are of the desaturation type where the sample is initially at $S_w = 100\%$. The two fluids are introduced steadily at a predetermined ratio. After a certain time a steady-state equilibrium is established and the same ratio is measured at the outflow end. Under these conditions, the saturation of the test section is determined (by electrodes that determine core resistivity, or by a volumetric balance of the two fluids) and permeability is calculated. The test is then repeated with an increased injection ratio.

Instead of measuring relative permeability directly, by determining the two rates of flow (of the two fluids through a given core) and the saturation, some authors develop indirect methods based on the relationship between saturation and capillary pressure (fig. 9.2.10) or on the pore-size distribution. Such methods are usually based on some conceptual model of the porous medium for which the rate of flow can be determined in an exact manner. However, in all these conceptual model approaches, one can never circumvent the need to perform experiments in order to derive constants and coefficients introduced during the analytical development.

As an example, consider the following development presented by Wyllie and Gardner (1958). The development is based on a capillary model (par. 5.10.1) combined with a random interconnection of pores.

Consider a bundle of capillary tubes whose radii r ($r_1 < r < r_2$) have a distribution function $\alpha(r)$. Because at $S = S_{w0}$ the wetting phase is immobile, the discussion can be simplified by assuming that the wetting fluid at $S \leqslant S_{w0}$ is part of the solid structure, so that the irreducible saturation becomes zero. This leads to the introduction (Corey 1957) of the definition of *effective saturation* $S_e = (S_w - S_{w0})/(1 - S_{w0})$, sometimes called ***reduced saturation*** (par. 9.2.3), which is actually the wetting fluid saturation of the modified medium (taking into account the modified porosity). To make the bundle of capillary tubes further resemble an actual porous medium with interconnected pores, the bundle is assumed to be cut into a large number of thin slices. Then the short pieces of tubes in each slice are rearranged randomly and the slices reassembled.

At a certain value of capillary pressure p_c (assuming that it is the same for S_e as for S_w), the nonwetting fluid (say, gas) occupies all pores larger than a size r' defined by (9.2.9):

$$r' = 2\sigma/p_c. \tag{9.3.7}$$

The wetting phase reduced saturation S_e is related to r' by:

$$S_e = \int_{r_1}^{r'} \pi r^2 \alpha(r)\, dr \bigg/ \int_{r_1}^{r_2} \pi r^2 \alpha(r)\, dr \tag{9.3.8}$$

which means that $\alpha(r)$ can, in principle, be determined from the curve $p_c = p_c(S_e)$.

In any slice of area A of the model, the area nS_eA is occupied by the wetting fluid (say, water) in pores of radii between r_1 and r'. An equal area is occupied by the wetting fluid in neighboring slices. However, parts of these areas are not connected because of the random distribution (in the model) of the pores in each slice. Considering a point on the interface between two neighboring slices, the probability that it lies in the wetting fluid in each slice is nS_e, and hence, the probability that it lies in both slices simultaneously (making interconnected pores) is $(nS_e)^2$. Using similar reasoning, Wyllie and Gardner (1958) show that since the probability of a wetting-fluid-filled pore in a slice is nS_e, the area common to a *single* pore of cross-sectional area πr^2 in one slice, and *all* the wetting-fluid-filled pores in a neighboring slice, is therefore $\pi r^2 n S_e$. Thus the passage of water takes place from an area πr^2 to a constricted area $\pi r^2 n S_e$. One may visualize the constricted area as a pore of smaller radius r'' such that $\lambda\pi r''^2 = \pi r^2 n S_e$, or

$$r'' = (nS_e/\lambda)^{1/2} r \tag{9.3.9}$$

where $\lambda\,(\geqslant 1)$ is a numerical coefficient that reflects the manner in which the available interconnected total pore area is distributed; it also depends on $\alpha(r)$.

Poiseuille's law is now used (as in par. 5.10.1) to describe the flow through the capillary model:

$$Q_p = -(\pi r''^4 \gamma\beta/8\mu)J = -(\pi r^4 n^2 S_e^2 \gamma\beta/8\mu\lambda^2)J. \tag{9.3.10}$$

Assuming that both β and λ are independent of $\alpha(r)$, the specific discharge $q = Q/A$ is given by:

$$q = -\frac{\pi n^3 S_e^2 \beta\gamma}{8\mu\lambda^2} J \int_{r_1}^{r'} r^4 \alpha(r)\, dr \Big/ \int_{r_1}^{r_2} \pi r^2 \alpha(r)\, dr; \qquad nA = \int_{r_1}^{r_2} \pi r^2 \alpha(r)\, dr. \tag{9.3.11}$$

Now, by differentiating (9.3.8) with respect to r' and replacing r' by r we obtain:

$$dS_e = r^2 \alpha(r)\, dr \Big/ \int_{r_1}^{r_2} r^2 \alpha(r)\, dr. \tag{9.3.12}$$

Eliminating r from (9.3.11) by using (9.3.7) and (9.3.12), we obtain

$$q = -\frac{n^3 S_e^2 \beta\gamma}{8\mu\lambda^2} J \int_{r}^{r'} r^2 \left(\int_{r_1}^{r_2} r^2 \alpha(r)\, dr \right) dS_e \Big/ \int_{r_1}^{r_2} \pi r^2 \alpha(r)\, dr = -\frac{n^3 S_e^2 \beta\gamma\sigma^2}{2\mu\lambda^2} J \int_0^{S_e} \frac{dS_e'}{p_c^2(S_e')}. \tag{9.3.13}$$

Hence:

$$q = \frac{k_w \gamma}{\mu} J; \qquad k_w = \frac{n^3 S_e^2 \beta\sigma^2}{2\lambda^2} \int_0^{S_e} \frac{dS_e'}{p_c^2(S_e')}; \qquad k = \frac{n^3 \beta\sigma^2}{2\lambda^2} \int_0^1 \frac{dS_e'}{p_c^2(S_e')}. \tag{9.3.14}$$

Or, in terms of S_w:

$$k_{rw} = S_e^2 \int_{S_{w0}}^{S_w} \frac{dS_w'}{p_c^2(S_w')} \Big/ \int_{S_{w0}}^{1} \frac{dS_w}{p_c^2(S_w)}, \qquad k = \frac{n^3 \beta\sigma^2}{2\lambda^2} (1 - S_{w0})^{-1} \int_{S_{w0}}^{1} \frac{dS_w}{p_c^2(S_w)} \tag{9.3.15}$$

where it is assumed that β/λ^2 is not a sensitive function of S_w. Similarly, for the nonwetting phase:

$$k_{rnw} = [(1 - S_e)^2] \int_{S_e}^{1} \frac{dS_e'}{p_c^2(S_e')} \Big/ \int_0^1 \frac{dS_e'}{p_c^2(S_e')}. \tag{9.3.16}$$

Equations (9.3.15) and (9.3.16), known also as *Burdine's equations*, are also derived by Burdine (1953) by employing the hydraulic radius theory (par. 5.10.2). A detailed description of this theory is given by Brooks and Corey (1964).

One should note here that a weakness of (9.3.16) stems from the fact that, whereas the concept of relative permeability is a dynamic one, the relationship $p_c = p_c(S_e)$ is derived from laboratory experiments carried out under stationary (i.e., no flow) conditions.

Corey (1954) finds that for a large number of consolidated porous rocks the ratios of the integrals on the right-hand side of (9.3.15) and (9.3.16) are approximately equal to S_e^2 and $(1 - S_e)^2$, respectively. He therefore proposes, as a convenient approximation, the equations:

$$k_{rw} \propto S_e^4; \qquad k_{rnw} \propto (1 - S_e)^2(1 - S_e^2). \tag{9.3.17}$$

These are convenient equations as the only parameter needed for their use is S_{w0}. Brooks and Corey (1964) discuss in detail the validity of (9.3.17), and conclude that although these equations are often used by petroleum engineers because of their simplicity, their validity is rather limited, as they imply that:

$$S_e = (C/p_c)^2, \qquad p_c \geqslant C \tag{9.3.18}$$

where C is some constant. For most porous media this last equation is not strictly correct.

Actually, for a medium having a completely uniform pore-size distribution, the value of the exponent in Corey's equation (9.3.17) would be 3.0. Averjanov (1950, as given by Polubarinova-Kochina 1952, 1962) and Irmay (1954) suggest similar equations with values of 3.5 and 3.0, respectively, as exponents. The experimental verification of their equations is probably due to the fact that they use rather uniform unconsolidated sands; consolidated rocks are often much less uniform. Topp and Miller (1965) obtain the exponent 3 for glass beads of uniform size.

To extend the range of validity of (9.3.18), Brooks and Corey (1964) suggest the following equations on the basis of a large number of drainage experiments with various types of media:

$$S_e = (p_b/p_c)^{\lambda} \qquad \text{for} \quad p_c \geqslant p_b \tag{9.3.19}$$

where λ, called ***pore-size distribution index***, is a number that characterizes the pore-size distribution, and p_b, referred to as ***bubbling pressure*** by Brooks and Corey (1964), is approximately the minimum value of p_c on the drainage cycle at which a continuous nonwetting phase exists in the pore space (fig. 9.2.7).

With (9.3.19), equations (9.3.15) and (9.3.16) become:

$$k_{rw} = (S_e)^{(2+3\lambda)/\lambda} = (p_b/p_c)^{(2+3\lambda)}; \qquad p_c \geqslant p_b \tag{9.3.20}$$

$$k_{rnw} = (1 - S_e)^2(1 - S_e^{(2+\lambda)/\lambda}) = [1 - (p_b/p_c)^{\lambda}]^2[1 - (p_b/p_c)^{2+\lambda}]. \tag{9.3.21}$$

In this way a theoretical relationship, supported by experimental evidence, is obtained among the variables p_c, S_e, k_{rw} and k_{rnw}. The characteristic constants λ and p_b must be determined experimentally. Brooks and Corey, although limiting to isotropic media the validity of the results shown above, find them applicable to a wide range of pore-size distributions.

The literature contains many other formulas for relative permeability, based on an analysis of a porous medium model of one type or another. Scheidegger (1960) reviews several such theories. Among them he describes in detail the work of Fatt and Dykstru (1951), who use a capillary model, Rose (1949), who extends Kozeny's model (par. 5.10.3) to multiple-phase flow, and Rapoport and Leas (1951), who employ an approach similar to that of Rose.

9.3.3 Mass Conservation in Multiphase Flow

In the absence of sources and sinks, the mass conservation equation (4.2.10) for a substance whose density is g_α and average velocity is $\mathbf{V}_{G_\alpha}$ is:

$$\partial g_\alpha/\partial t + \operatorname{div}(g_\alpha \mathbf{V}_{G_\alpha}) = 0.$$

When this equation is applied to two phases participating simultaneously in a flow through a porous medium, a separate mass conservation equation must be written for each of them. To obtain the conservation equation for a phase α, we insert in it:

$$g_\alpha = nS_\alpha \rho_\alpha = \text{mass of fluid } \alpha \text{ per unit volume of porous medium}$$

$$n = \text{porosity of porous medium}$$

$$g_\alpha \mathbf{V}_{G_\alpha} = nS_\alpha \rho_\alpha \mathbf{V}_\alpha = \rho_\alpha \mathbf{q}_\alpha = \text{mass flux (mass per unit area per unit time) of fluid } \alpha.$$

Then:

$$\partial(nS_\alpha \rho_\alpha)/\partial t + \operatorname{div}(\rho_\alpha \mathbf{q}_\alpha) = 0. \tag{9.3.22}$$

In two-phase flow, $\alpha = 1, 2$. In three-phase flow (e.g., water, oil and gas), $\alpha = 1, 2, 3$.

For a nondeformable medium, $n = \text{const.}$ Hence:

$$n\, \partial(S_\alpha \rho_\alpha)/\partial t + \operatorname{div}(\rho_\alpha \mathbf{q}_\alpha) = 0. \tag{9.3.23}$$

For a deformable medium, the ideas presented in section 6.3 are applicable, with or without possible movement of the solid particles.

For a homogeneous incompressible fluid in a nondeformable medium, equation (9.3.23) becomes:

$$n\, \partial S_\alpha/\partial t + \operatorname{div} \mathbf{q}_\alpha = 0. \tag{9.3.24}$$

These equations are the basis for a mathematical statement of the multiphase flow problem.

9.3.4 Statement of the Multiphase Flow Problem

In (9.3.24), the specific discharge $\mathbf{q}_\alpha$ of each phase is expressed as a function of the pressure, saturation, viscosity and density of that phase (par. 9.3.1). For example, in two-phase flow there are 15 dependent variables:

$$p_\alpha, p_c, (q_i)_\alpha, \mu_\alpha, \rho_\alpha, S_\alpha; \qquad i = 1, 2, 3; \qquad \alpha = 1, 2.$$

Hence, we need 15 equations to derive a complete solution—i.e., a complete description of the flow (velocity, pressure and saturation distribution within the flow domain and their time variations) and the fluids (i.e., density and viscosity as functions of time and position). These are: (a) the six equations of motion included in (9.3.5); (b) equation (9.2.1), i.e., $\sum S_\alpha = 1$; (c) two equations of state for density, $\rho_\alpha = \rho_\alpha(p_\alpha)$; (d) two equations of state for viscosity, $\mu_\alpha = \mu_\alpha(p_\alpha)$; (e) two equations of continuity, one for each phase; (f) equation (9.2.7) relating the difference in pressures to p_c, and (g) capillary pressure–saturation relationship $p_c = p_c(S_w)$.

It is possible to eliminate the six dependent variables $(q_i)_\alpha$ by inserting (9.3.5) into the continuity equations. For a flow of two homogeneous incompressible fluids in an anisotropic medium, the problem reduces to the solution of:

$$n\frac{\partial S_1}{\partial t} - \frac{\partial}{\partial x_i}\left[k_{ij}\frac{k_{r1}}{\mu_1}\left(\frac{\partial p_1}{\partial x_j} + \rho_1 g\frac{\partial z}{\partial x_j}\right)\right] = 0$$

$$n\frac{\partial S_2}{\partial t} - \frac{\partial}{\partial x_i}\left[k_{ij}\frac{k_{r2}}{\mu_2}\left(\frac{\partial p_2}{\partial x_j} + \rho_2 g\frac{\partial z}{\partial x_j}\right)\right] = 0$$

$$S_1 + S_2 = 1$$

$$p_2 - p_1 = p_c(S_1) \qquad (9.3.25)$$

for the four unknowns S_1, S_2, p_1, p_2. For this case of homogeneous incompressible fluids, the piezometric heads $\varphi_1 = z + p_1/\rho_1 g$ and $\varphi_2 = z + p_2/\rho_2 g$ may be introduced. If the fluids (or one of them) are compressible, the appropriate continuity equations must be used (par. 9.3.6). Note the comment about k_{r1}, k_{r2} following (9.3.5).

Since we have here an initial value problem, a solution can be obtained only if the first two equations of (9.3.25) are supplemented by appropriate boundary and initial conditions (sec. 7.1). One should note that boundary conditions should be specified for each of the fluids participating in the flow.

As an example, consider steady, one-dimensional flow in a horizontal homogeneous column of length L of two incompressible fluids. The steady flow is obtained by starting from a medium at irreducible (connate) water saturation, and introducing a mixture of the fluids until an equilibrium is reached and maintained thereafter. The four equations of (9.3.25) simplify to:

$$\partial q_w/\partial x = 0; \qquad q_w = -(k_w/\mu_w)\,\partial p_w/\partial x$$

$$\partial q_{nw}/\partial x = 0; \qquad q_{nw} = -(k_{nw}/\mu_{nw})\,\partial p_{nw}/\partial x$$

$$p_{nw} - p_w = p_c$$

$$S_w + S_{nw} = 1. \qquad (9.3.26)$$

Assuming that at equilibrium S_w and S_{nw} are uniform within the medium (reasonable at high flow rates), k_w, k_{nw} and p_c, which depend on S_w and S_{nw}, also remain so (i.e., independent of x). Hence (9.3.26) becomes:

$$q_w = (k_w/\mu_w)\,\Delta p_w/L; \qquad q_{nw} = (k_{nw}/\mu_{nw})\,\Delta p_{nw}/L \qquad (9.3.27)$$

where Δp_w and Δp_{nw} are the respective pressure differences across L. Since p_c is constant, we have $\Delta p_w = \Delta p_{nw}$.

At low flow rates, saturations are not uniform. Then, we obtain from (9.3.26):

$$q_w = -(k_w/\mu_w)(\partial p_w/\partial x) \quad \text{and} \quad q_{nw} = -k_{nw}/\mu_{nw}[(\partial p_w/\partial x) + (\partial p_c/\partial S_w)(\partial S_w/\partial x)] \tag{9.3.28}$$

where $\partial p_c/\partial S_w$ must be determined from the curve $p_c = p_c(S_w)$.

For three-dimensional flow in a homogeneous isotropic medium, we obtain from (9.3.15):

$$\mu_1 \mathbf{q}_1/kk_{r1} = -\nabla p_1 - \rho_1 g \nabla z; \qquad \mu_2 \mathbf{q}_2/kk_{r2} = -\nabla p_2 - \rho_2 g \nabla z$$
$$p_c = p_2 - p_1. \tag{9.3.29}$$

With the total specific discharge $\mathbf{q} = \mathbf{q}_1 + \mathbf{q}_2$, and $\Delta\rho = \rho_1 - \rho_2$, we obtain from (9.3.29):

$$\mathbf{q}_1 = [\mathbf{q} + (kk_{r2}/\mu_2)(\nabla p_c - g\,\Delta\rho\,\nabla z)]/(1 + k_{r2}\mu_1/k_{r1}\mu_2). \tag{9.3.30}$$

When both fluids are incompressible, the continuity equation for fluid 1 in (9.3.25) simplifies to:

$$n\,\partial S_1/\partial t + \operatorname{div} \mathbf{q}_1 = 0. \tag{9.3.31}$$

By substituting (9.3.26) in (9.3.31) and recalling that $k_{r1} = k_{r1}(S_1)$, $k_{r2} = k_{r2}(S_2)$, $p_c = p_c(S_1)$, and $n\,\partial(S_1 + S_2)/\partial t = 0 = \operatorname{div}(\mathbf{q}_1 + \mathbf{q}_2) = \operatorname{div} \mathbf{q}$, we obtain:

$$n\,\partial S_1/\partial t + \operatorname{div}\{[\mathbf{q} + (kk_{r2}/\mu_2)(\nabla p_c - g\,\Delta\rho\,\nabla z)]/(1 + k_{r2}\mu_1/k_{r1}\mu_2)\}$$
$$= n\frac{\partial S_1}{\partial t} + \frac{\partial H}{\partial S_1}\mathbf{q}\cdot\operatorname{grad} S_1 + \frac{k}{\mu_2}\operatorname{div}\left(Hk_{r2}\frac{\partial p_c}{\partial S_1}\operatorname{grad} S_1\right) - \frac{k\,\Delta\rho g}{\mu_2}\frac{d}{dS_1}(Hk_{r2})\frac{\partial S_1}{\partial z} = 0 \tag{9.3.32}$$

where $H = 1/(1 + k_{r2}\mu_1/k_{r1}\mu_2)$ and z is positive upward.

It should be noted at this point that generally the experiments leading to the relationships $p_c = p_c(S_1)$ and $k_{r1} = k_{r1}(S_1)$, $k_{r2} = k_{r2}(S_2)$ are carried out under laboratory conditions, usually of no flow for $p_c = p_c(S_1)$ and steady flow for $k_{r\alpha} = k_{r\alpha}(S_\alpha)$, which may differ appreciably from the unsteady field flows described by (9.3.32).

In general, an analytic solution of (9.3.25) or (9.3.32) is not possible owing to the nonlinearity of the continuity equations resulting from the dependence of relative permeability on saturation, and hence also on pressure. Another difficulty stems from the fact that the medium properties are given, in general, in fhe form of curves that summarize laboratory experiments performed on samples. Therefore, in most cases of practical interest, solutions are obtained numerically with the aid of high speed digital computers.

9.3.5 The Buckley–Leverett Equations

Buckley and Leverett (1942) solve (9.3.26) for one-dimensional flow, neglecting gravity, capillarity and liquid compressibility. Their solutions are called the *Buckley–Leverett assumptions.*

Consider an inclined homogeneous reservoir of constant thickness b (fig. 9.3.4a).

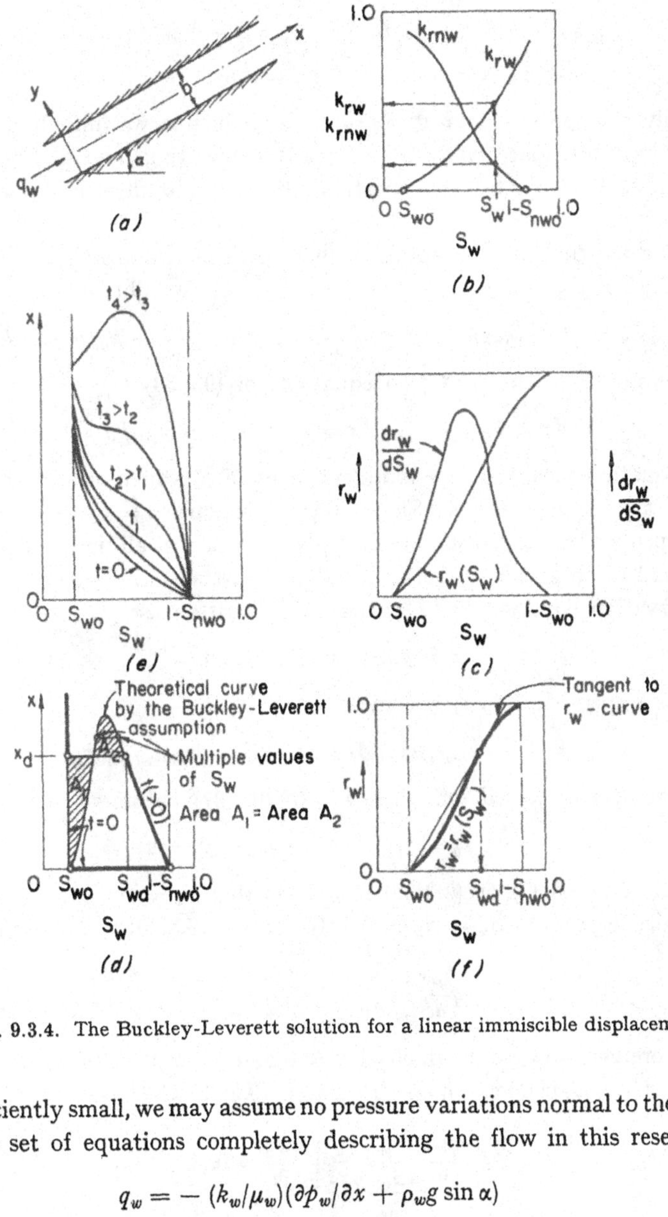

FIG. 9.3.4. The Buckley-Leverett solution for a linear immiscible displacement.

If b is sufficiently small, we may assume no pressure variations normal to the reservoir's axis. The set of equations completely describing the flow in this reservoir is:

$$
\begin{aligned}
q_w &= -\,(k_w/\mu_w)(\partial p_w/\partial x + \rho_w g \sin\alpha)\\
q_{nw} &= -\,(k_{nw}/\mu_{nw})(\partial p_{nw}/\partial x + \rho_{nw} g \sin\alpha)\\
n\,\partial S_w/\partial t &+ \partial q_w/\partial x = 0\\
n\,\partial S_{nw}/\partial t &+ \partial q_{nw}/\partial x = 0\\
S_w &+ S_{nw} = 1\\
p_c &= p_{nw} - p_w.
\end{aligned}
\qquad (9.3.33)
$$

The initial and boundary conditions of the problem are:

$$t \leqslant 0, \qquad 0 \leqslant x \leqslant L, \qquad S_{nw} = 1 - S_{w0}$$

$$t > 0, \qquad x = 0, \qquad q_w = \text{const}; \qquad q_{nw} = 0,$$

i.e., initially the formation is at $S_w = S_{w0}$ (irreducible wetting fluid saturation), the remaining pore space being filled by the nonwetting fluid at $S_{nw} = 1 - S_{w0}$. The wetting fluid is injected at a constant rate at $x = 0$, displacing the nonwetting fluid.

By adding the first two equations of (9.3.33) and introducing the new variables $r_w = q_w/q$ and $r_{nw} = q_{nw}/q = 1 - r_w$, $q = q_w + q_{nw}$, we obtain:

$$r_w = q_w/q = [1 + (k_{nw}/q\mu_{nw})(\partial p_c/\partial x - \Delta\rho g \sin \alpha)]/(1 + k_{nw}\mu_w/\mu_{nw}k_w). \qquad (9.3.34)$$

We also obtain from the next two equations of (9.3.33):

$$\partial q/dx = \partial(q_w + q_{nw})/\partial x = -n\, \partial(S_w + S_{nw})/\partial t = 0. \qquad (9.3.35)$$

The last equation means that q is independent of x, which should be obvious as the liquids are incompressible. Since at the boundary $x = 0$, we have $q = q_w =$ const, independent of time, it follows that q is a constant at all times. This conclusion remains valid when at $x = 0$, both q_w and q_{nw} exist and are constant. Hence, the first conservation equation in (9.3.33) may be written as:

$$q\, \partial r_w/\partial x + n\, \partial S_w/\partial t = 0. \qquad (9.3.36)$$

When p_c is neglected (9.3.34) becomes:

$$r_w = [1 - (k_{nw}/q\mu_{nw})\, \Delta\rho g \sin \alpha]/(1 + k_{nw}\mu_w/\mu_{nw}k_w). \qquad (9.3.37)$$

If gravity is also neglected (or when $\alpha = 0$) (9.3.37) becomes:

$$r_w = 1/(1 + k_{nw}\mu_w/k_w\mu_{nw}). \qquad (9.3.38)$$

From (9.3.34), (9.3.37) or (9.3.38) it follows that $r_w = r_w(S_w)$, as the effective permeabilities depend on $S_w = S_w(x, t)$ only. Hence (9.3.36) may be rewritten as:

$$q \frac{dr_w}{dS_w} \frac{\partial S_w}{\partial x} + n \frac{\partial S_w}{\partial t} = 0. \qquad (9.3.39)$$

In the more general case in which at $x = 0$ and $q = q(t)$, it follows from (9.3.35) that $q = q(t)$ is independent of x. Equation (9.3.39) then becomes:

$$\frac{q(t)}{n} \frac{dr_w}{dS_w} \frac{\partial S_w}{\partial x} + \frac{\partial S_w}{\partial t} = 0. \qquad (9.3.40)$$

The continuity equation for S_{nw} also reduces to (9.3.40).

This is a quasilinear first-order partial differential equation for the single dependent variable $S_w = S_w(x, t)$, because dr_w/dS_w is a function of S_w. It can be treated by numerical methods or by the *method of characteristics.*

The total derivative of $S_w(x, t)$ with respect to time is:

$$dS_w/dt = (\partial S_w/\partial x)\, dx/dt + \partial S_w/\partial t. \qquad (9.3.41)$$

If $x = x(t)$ is chosen to coincide with an advancing surface of fixed S_w, then on such surface we have:

$$dS_w/dt = 0 \tag{9.3.42}$$

and the velocity W at which the front of a given S_w is advancing is given by (9.4.41) in which we insert $dS_w/dt = 0$:

$$W = dx/dt|_{\text{for given } S_w} = -(\partial S_w/\partial t)/(\partial S_w/\partial x). \tag{9.3.43}$$

By combining (9.3.40) with (9.3.43), we obtain:

$$W = \left.\frac{dx}{dt}\right|_{\text{for given } S_w} = \frac{q(t)}{n}\frac{dr_w}{dS_w}. \tag{9.3.44}$$

Thus the partial differential equation (9.3.36) has been replaced by the two ordinary differential equations (9.3.42) and (9.3.44)—called the *Buckley–Leverett equations*. By integrating (9.3.44) with respect to time we obtain:

$$x|_{S_w}(t) - x|_{S_w}(0) = \frac{U(t) - U(0)}{nA}\left(\frac{dr_w}{dS_w}\right), \qquad U(t) = \int_0^t q(t)A\,dt \tag{9.3.45}$$

where $x|_{S_w}(t)$ and $x|_{S_w}(0)$ are the coordinates x of a plane at which a specified saturation S_w exists at times t and zero, respectively. $U(t)$ and $U(0)$ are the cumulative total volumes passed through the system at times t and zero, respectively.

Equation (9.3.45) gives the saturation distribution $S_w = S_w(x, t)$, given the distribution of S_w at $t = 0$. From (9.3.34) it follows that r_w appearing in (9.3.45) is a function of k_w and k_{nw}, which in turn are functions of S_w only. Hence, once r_w, and from it dr_w/dS_w, are evaluated for every value of S_w, the saturation distribution at any time t can be determined from (9.3.45) and from the known values of S_w at $t = 0$.

Since both $k_w = k_w(S_w)$ and $k_{nw} = k_{nw}(S_{nw})$ are usually given only in a graphic form (fig. 9.3.4b), the function $r_w = r_w(S_w)$, and from it dr_w/dS_w, are also determined graphically. From figure 9.3.4b we determine both k_{rw} and k_{rnw} for every value S'_w of S_w. From this information and (9.3.37) we plot r_w versus S_w and then dr_w/ds_w versus S_w (fig. 9.3.4c). Figure 9.3.4d shows $x = x(S_w)$ as derived from (9.3.44) for some value of t (> 0) for the case where at $x = 0$, $q =$ const. At $x = 0$, $S_{nw} = S_{nw0}$, $S_w = 1 - S_{nw0}$.

When $r_w = r_w(S_w)$ is determined by (9.3.37), the resulting curve may either have an inflection point or not. In the latter case, the resulting curve $x = x(S_w)$ for some value $t > 0$ will have a bell shape as shown in figure 9.3.4d. This means that two values of S_w correspond to the same value of x. Obviously such multiple values are physically impossible. This situation results because we have neglected the capillary pressure in determining W. The time at which such multiple values occur depends on the initial saturation distribution. In figure 9.3.4e, the phenomenon is observed only at $t > t_3$. When initially $S = S_{w0}$ at all points $x > 0$, the bell-shaped curve with multiple points occurs at all times observed.

The multiple values of S_w can be eliminated by introducing a discontinuity in saturation at some point $x = x_d$ (fig. 9.4.4d). The method of obtaining a solution using this device was suggested by Welge (1952) and improved later by Johnson, Bossler and Naumann (1959). Marle (1965) reviews these solutions and discusses them in detail because of their importance in connection with laboratory techniques for the determination of relative permeability. He also gives solutions that take gravity and capillary pressure into account.

Detailed analyses by various investigators lead to the conclusion that for high flow rates, the Buckley–Leverett solution (i.e., neglecting capillary effects) gives a relatively good approximation of the actual saturation distribution. At low flow rates the difference becomes large. In both cases the main difference is in the zone of the advancing front where saturation gradients (and hence capillary pressures) are of importance.

9.3.6 Simultaneous Flow of a Liquid and a Gas

In this case, which is of interest in reservoir engineering (in *gas drives*), we shall use oil (subscript o) and gas (subscript g) as examples for the liquid and gas considered. To generalize the discussion, and mainly to show the difficulties involved, we shall assume that the oil is compressible, with a *formation volume factor* for oil β_o:

$$\beta_o = U_o(p, T)/U_{os} \tag{9.3.46}$$

where U_{os} is the volume of oil at standard conditions and $U_o(p, T)$ is the volume of oil at the pressure p and the temperature T in the reservoir. We shall also assume that *change of phase is possible*.

The gas solubility in oil, s, is defined by:

$$s(p, T) = U_{gs}/U_{os} \tag{9.3.47}$$

where U_{gs} and U_{os} denote volumes of gas and oil, respectively, at standard conditions. These are the volumes obtained when a sample of oil, initially at the reservoir's pressure and temperature, is brought to atmospheric pressure and standard temperature.

With these definitions, the motion equation for oil is:

$$\mathbf{q}_o = -(kk_{ro}/\mu_o)(\operatorname{grad} p_o - \rho_o \mathbf{g}). \tag{9.3.48}$$

As gas is transported, both in the gas phase and in the liquid phase, its mass flux $\mathbf{J}_g$ is given by:

$$\begin{aligned} \mathbf{J}_g &= \rho_g \mathbf{q}_g + \frac{s\rho_{gs}}{\beta_o}\mathbf{q}_o \\ &= -\rho_g \frac{kk_{rg}}{\mu_g}(\operatorname{grad} p_g - \rho_g \mathbf{g}) - \frac{s\rho_{gs}}{\beta_o}\frac{kk_{ro}}{\mu_o}(\operatorname{grad} p_o - \rho_o \mathbf{g}) \end{aligned} \tag{9.3.49}$$

where: $\rho_g \mathbf{q}_g$ is the specific mass discharge of free gas and $U_{gs}\rho_{gs}/U_o = s\rho_{gs}/\beta_o$ is the mass of gas dissolved in, and carried by, the oil per unit volume of oil in the reservoir. In the absence of dissolved gas, or if we consider the flow of free gas only, we introduce

$s = 0$ in (9.3.49).

The mass conservation equations written in terms of density of oil and gas at standard conditions are for oil:

$$\operatorname{div}(\rho_{os}\mathbf{q}_o/\beta_o) = -n\frac{\partial}{\partial t}(\rho_{os}S_o/\beta_o) \tag{9.3.50}$$

where $\rho_o = \rho_{os}/\beta_o$ is the oil density under reservoir conditions, and for gas:

$$\operatorname{div}[\rho_g\mathbf{q}_g + (s\rho_{gs}/\beta_o)\mathbf{q}_o] = -n\frac{\partial}{\partial t}(\rho_g S_g + s\rho_{gs}S_o/\beta_o) \tag{9.3.51}$$

where $n\rho_g S_g$ is the mass of free gas, and $ns\rho_{gs}S_o/\beta_o$ is the mass of dissolved gas per unit volume of the reservoir. We also have:

$$S_o + S_g = 1; \qquad p_g - p_o = p_c(S_o, T) \tag{9.3.52}$$

and equations of state for ρ_o, ρ_g, μ_o and μ_g. Altogether we have eight equations for the eight independent variables p_o, p_g, S_o, S_g, ρ_o, ρ_g, μ_o, μ_g; $\mathbf{q}$ is expressed in terms of p_o and ρ_o by (9.3.48). In principle a solution is possible. Examples of solutions are given by Oroveanu (1966).

9.3.7 Laboratory Determination of Relative Permeability

Analytical expressions for relative permeability such as (9.3.15) and (9.3.16) or (9.3.20) and (9.3.21), derived from porous medium models, may be used for calculating k_{rw} and k_{rnw}. These expressions, and others not mentioned here, involve the capillary pressure–saturation relationship that must be determined experimentally. Another possibility is to determine k_{rw} and k_{rnw} directly by laboratory experiments.

Two types of laboratory experiments are used for the determination of relative permeability: steady flow experiments and unsteady flow (or displacement) experiments. The most commonly used methods are based on steady flow. While many such methods are described in the literature, they are all based on essentially the same procedure. The tested sample, or core, is mounted either in a plexiglass tube or in a pressurized rubber sleeve, and a steady flow of the two fluids (two liquids, or a liquid and a gas) is established through it. The two fluids used in the experiment are introduced simultaneously at the inflow end, at a certain ratio, through separate piping systems. A steady flow, i.e., inflow equals outflow, is reached within 2–40 hours, depending on the sample's permeability and the method used. At this stage the pressures in the two fluids at either end of the sample, the rates of flow and the saturation are determined. Using equations such as (9.3.1), the relative permeability corresponding to the saturation established during the experiment can be calculated. The injected ratio is then increased, removing more of the wetting phase, until once more a steady flow is established. The process is repeated until a complete relative permeability curve is obtained. Tests starting from $S_w = 100\%$ are called *desaturation tests*. In a *resaturation process* the tested sample is initially at $S_{nw} = 100\%$ and declines to $S_w = 100\%$; the injection ratios start with a high rate of flow of the non-wetting phase. The results obtained, using desaturation and resaturation processes, exhibit a hysteresis effect of the type discussed in paragraph 9.2.3.

Sometimes, to avoid end effects at the ends of the tested core (see below and par. 9.2.5), only the central part of the core is considered as a tested sample; pressure differences and saturations are measured only for this portion of the core. In this case, fluid saturations are usually determined by inserting electrodes at both ends of the tested section and measuring electrical resistance of the core. X rays or γ rays from a radioactive source are also sometimes used (Norel 1964). In other cases, saturations are determined by weighing the sample at every stage of steady flow, or by maintaining a volumetric balance of all fluids injected and produced from the sample.

The pressures in the fluids are measured by using appropriate porous plates (membranes). These are wetted (or are artificially wet) and saturated by the fluid, the pressure of which is to be measured through them. Through these pervious plates the pressures are conveyed to pressure gauges.

The most commonly used steady flow methods for relative permeability determination are the *Hassler method*, the *Hafford method* and the *dispersed feed method*. The various methods differ from each other mainly in the manner in which the two fluids are introduced into the cores and adjustment is made for end effects.

General discussions and review of these methods (and others not mentioned here) are presented by Brownscombe et al. (1950), Rose (1951, 1951a), Osoba et al. (1951) and Richardson et al. (1952).

Unsteady flow techniques were developed by Welge (1952) and Johnson, Bossler and Naumann (1959), among others. Whereas Welge only derived the ratio k_{rw}/k_{rnw}, the method introduced by Johnson, Bossler and Naumann permits the determination of the relative permeabilities k_{rw} and k_{rnw} separately. In both developments, the effects of capillarity and gravity are neglected. End effects are minimized by using high flow rates.

9.4 Unsaturated Flow

Unsaturated (air–water) flow (i.e., the flow of water at saturation less than 100%, with stagnant air filling that portion of the void space not occupied by water) is nothing but a special case of simultaneous flow of two immiscible fluids, where the nonwetting fluid (air) is *assumed* to be stagnant. Whereas the subject of immiscible fluids has been extensively treated by reservoir engineers, unsaturated flow has been developed by soil physicists and hydrologists as a separate subject, with its own terminology, because of its importance in soil sciences and in investigations of irrigation, drainage of agricultural lands, infiltration, etc. A vast amount of literature about unsaturated flow exists in books and professional journals. Among many summaries and reviews on this subject we may mention Wesseling (1961), Remson and Randolph (1962), Klute (1967), Zaslavsky (in Bear, Zaslavsky and Irmay 1968), Stallman (1967), Childs (1967), Swartzendruber (1969) and Philip (1970). We may also mention the proceedings of two symposia: the IASH symposium on water in the unsaturated zone, held in Wageningen in 1966, and the ICID sympo-

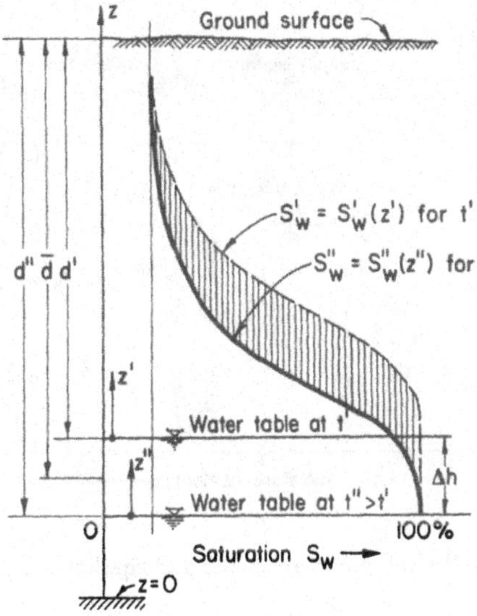

FIG. 9.4.1. Typical moisture distributions above the water table.

sium on soil water, held in Prague in 1967.

Because the basic principles underlying the simultaneous flow of two immiscible fluids in porous media have already been presented in sections 9.2 and 9.3, the present section presents mainly supplementary comments, and an attempt to unify, to some extent, the treatment of the two topics.

In the water–air system considered here, *water is the wetting phase* while *air is the nonwetting one.* The problem of water movement in the vapor phase is treated only very briefly.

Because only the water is *assumed* to be in motion, it is often more convenient to employ the concept of water content c, defined by (9.2.2), instead of the degree of saturation (S), defined by (9.2.1). Except when dealing with phenomena at the microscopic level (i.e., inside the pore), all values of dependent variables and parameters have the meaning of an average taken over an REV around considered points in space.

9.4.1 Capillary Pressure and Retention Curve

Most of the discussion presented in section 9.2 is directly applicable to unsaturated (air–water) flow with some slight modifications in the nomenclature.

Figure 9.4.1 shows some typical profiles of moisture distributions above the water table (see par. 1.1.2). Figure 9.4.2 gives some of the terms used for levels of water (or moisture) content in unsaturated flow studies. Obviously, the word "point," used to define such moisture levels as "wilting point," actually indicates a range rather than a well defined point.

In a drainage process, as water saturation is reduced and air enters the pore

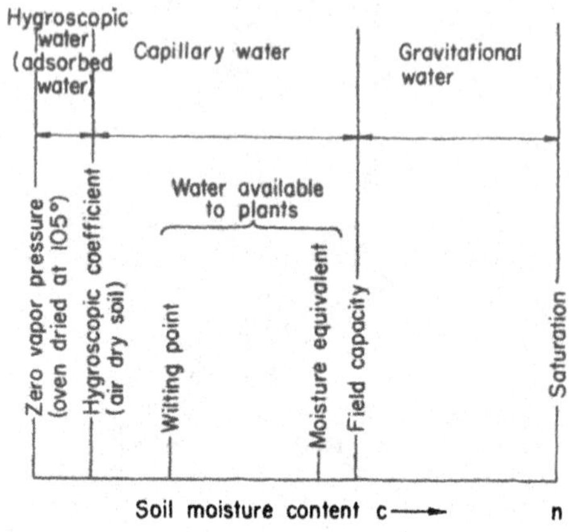

FIG. 9.4.2. Soil moisture classes and equilibrium points.

space, air–water interfaces are formed in the pores. Figures 9.2.3a through 9.2.3c, with the darkened areas representing air rather than oil, may be used to describe the water distribution in the pore space during drainage; the passage is from *funicular water* (fig. 9.2.3c) to *pendular water* (fig. 9.2.3a).

The discussion (and figures) on the interfacial tension in paragraph 9.2.2 is applicable to unsaturated flow. We may use figure 9.2.1b with G denoting air and L denoting water.

Figure 9.2.4a describes an element of the air–water interface, with subscript 2 denoting the air (nonwetting fluid) and subscript 1 denoting the water (wetting fluid). Across such an interface we have a pressure difference called *capillary pressure*. If the air pressure is taken as a datum for measuring the pressure in the wetting fluid, i.e., $p_{nw} = 0$, we have from (9.2.7):

$$p_c = p_{nw} - p_w = -p_w \tag{9.4.1}$$

where p_w denotes the pressure in the water immediately adjacent to interface. However, in the unsaturated soil, many such (microscopic) interfaces exist in the vicinity of every point, so that when (9.4.1) is averaged to give a macroscopic relationship in a porous medium domain, p_w will denote the (average) pressure in the water while its negative value gives the capillary pressure. As p_w itself is negative (i.e., less than atmospheric), p_c is positive. Sometimes p_c is referred to as *suction* or *tension*.

Sometimes, instead of working with the concept of capillary pressure, the concept of a *capillary pressure head* h_c (dim L) is used:

$$h_c = p_c/\gamma_w \equiv -p_w/\gamma_w;\ p_{\text{air}} = 0 \tag{9.4.2}$$

where γ_w is the specific weight of the water. Equations (9.2.8) and (9.2.10) relate the capillary pressure to the mean radius of curvature, to the air–water interfacial tension and to the contact angle θ.

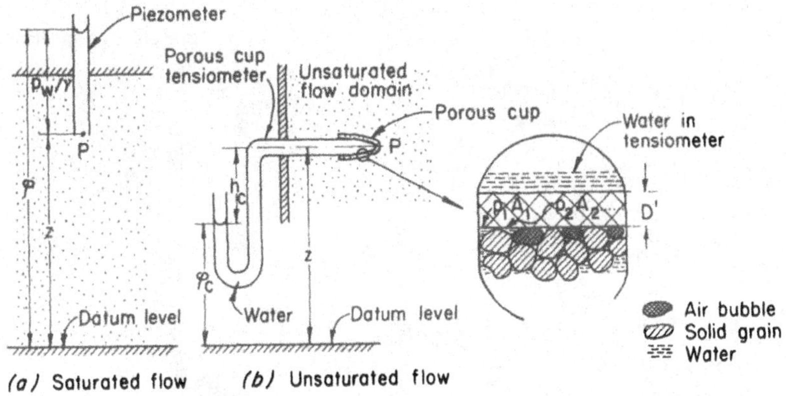

FIG. 9.4.3. Definition diagrams for φ and φ_c.

Another term often used by soil scientists is the pF defined by:

$$pF = \log_{10} h_c; \qquad h_c \text{ in cm} \tag{9.4.3}$$

as the logarithmic scale facilitates the presentation of a wide range of h_c in a single diagram.

As in saturated flow of water, we may define also in unsaturated flow a piezometric head $\varphi = z + p_w/\gamma_w$ at every point of the flow domain. Often the term *capillary head* (symbol φ_c) is used to denote the piezometric head in unsaturated flow (fig. 9.4.3) with $p_{\text{air}} = 0$:

$$\varphi_c = z + p_w/\gamma_w = z - p_c/\gamma_w = z - h_c; \qquad p_c > 0. \tag{9.4.4}$$

Many authors use the symbol ψ for the negative of the pressure head so that $\psi \equiv -p_w/\gamma_w > 0$.

The device used for measuring the capillary head φ_c in an unsaturated zone is the *tensiometer*. The tensiometer (a name introduced by Richards and Gardner 1936) is terminated in the soil by a water-wet porous cup (or in some cases by a porous plate) that is permeable to water, but impermeable to air when the pores of the cup, or the plate, are filled with water. The porous membrane (cup or plate) is necessary to establish hydraulic contact between the water in the tensiometer and in the soil. Because in unsaturated flow the pressure is always negative ($p_w < 0$), $h_c > 0$, and the water level in the open end of the manometer (fig. 9.4.3b) will always stand below the point P where the capillary head is being determined. Sometimes h_c is very large and heavier liquids (usually mercury) are used in the manometer or the tensiometer (fig. 9.4.4).

The capillary pressure varies with the mean curvature of the (microscopic) meniscii within the pores. As this curvature varies with the saturation, the capillary pressure is also a function of saturation. The relationship $p_c = p_c(S_w)$ must be determined experimentally. The *diaphragm* (or *porous plate*) *method* is the main method employed by soil scientists, although some of the other methods described in paragraph 9.2.6

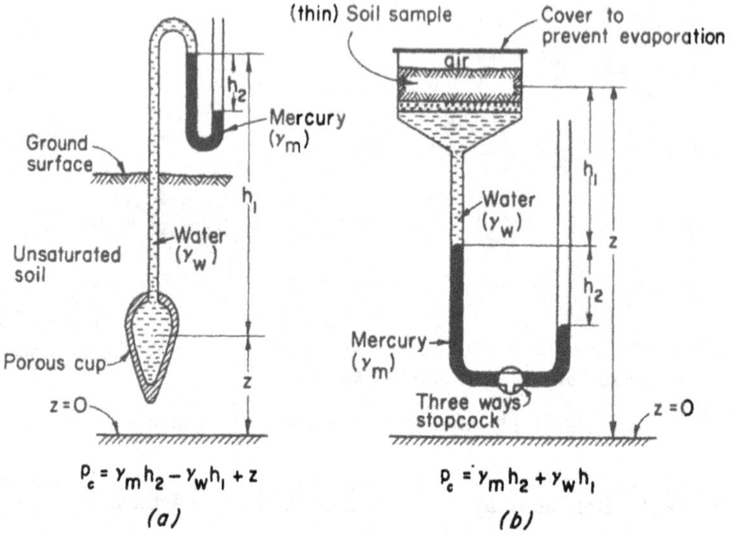

FIG. 9.4.4. The use of mercury tensiometer manometers. (a) Measurement of capillary pressure in the field, (b) the Haines apparatus for determining $p_c = p_c(S_w)$.

(e.g., the centrifuge method) are also applicable. Figure 9.4.4b shows the Haines apparatus for determining $p_c = p_c(S_w)$. The procedure is explained in paragraph 9.2.6. Sometimes the air chamber above the soil sample is connected to a regulated pressure source. In such an apparatus (called *pressure cell apparatus*), the pressure is limited only by the *bubbling pressure* (also called *air entry value* or *threshold pressure*) of the porous plate (or by the mechanical strength of the pressure cell). The bubbling pressure is the minimum pressure difference across the wetted porous plate, or membrane, required to force air through the membrane. It depends on the size of the membrane pores. Membrane materials are available with bubbling pressures of as high as $25 \div 50$ atmospheres.

Zaslavsky (in Bear, Zaslavsky and Irmay 1968) shows how the tensiometer gives the average pressure in the water phase in the soil. It is important to note that sometimes equilibrium (i.e., no flow) is reached only after a prolonged time, depending on the resistance (D'/K') of the membrane to the flow.

Figures 9.2.7, 9.2.10 and 9.4.5 show typical examples of $p_c = p_c(S_w)$ curves. Note the different abscissas and ordinates used in the various curves. In soil science, these curves are called *retention curves*, as they show how water is retained in the soil by capillary forces against gravity. Some authors refer to the drainage retention curve as a *desorption curve* and to the imbibition curve as a *sorption curve*. Point A in figure 9.4.5a is the *critical capillary head* h_{cc}. If we start from a saturated sample, say in the apparatus shown in figure 9.4.4a and produce a capillary head h_c, no water will leave the sample (i.e., no air will penetrate the sample) until the critical capillary head is reached. When expressed in terms of pressure, point A of figure 9.4.5a is the *bubbling pressure*. As the value of h_c is increased, the initial small

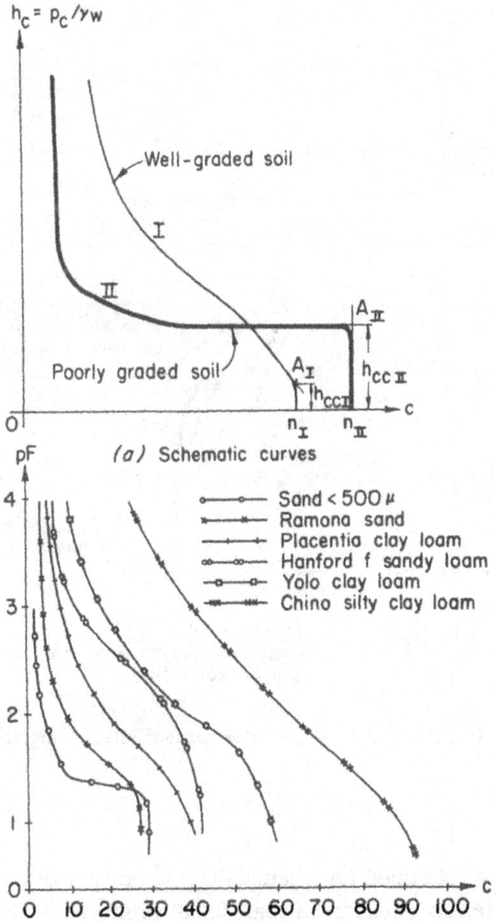

(a) Schematic curves

(b) Curves obtained during desaturation (from Richards and Weaver, 1944)

FIG. 9.4.5. Typical retention curves in soils.

reduction in c is associated with the retreat of the air–water meniscii into the pores at the external surface of the sample. Then, at the critical value h_{cc} the larger pores begin to drain.

The shape of the retention curve, and hence also the threshold pressure, depends on the pore-size distribution and pore shapes of the porous medium (fig. 9.4.5). It also depends on the initial water distribution. Brooks and Corey (1964), following Burdine (1953), prefer to draw the retention curve as a relationship between h_c and the *effective saturation* (or *reduced saturation*) S_e defined by:

$$S_e = (S - S_{w0})/(1 - S_{w0}) \tag{9.4.5}$$

where S_{w0} is the irreducible wetting fluid saturation (par. 9.2.3). Figure 9.2.7b shows

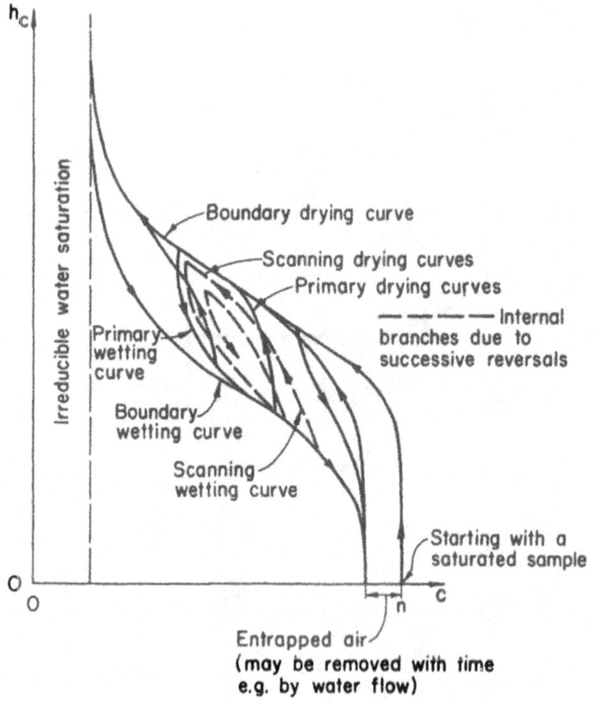

FIG. 9.4.6. Hysteresis in the capillary head-water content relationship for a coarse material in stable packing (after Klute, 1967).

several such curves.

The retention curve obtained for a sample is different for drainage and for imbibition because of hysteresis phenomena (par. 9.2.3) produced by hysteresis in contact angle, geometrical hysteresis effect (e.g., the ink bottle effect) of entrapped air upon rewetting and compaction or consolidation that changes the total volume of the sample as the latter is drained. In the absence of the last two factors, the drainage and imbibition curves form a closed loop (figs. 9.2.10 and 9.4.6). In a fine-grained material, the effect indicated in figure 9.4.6 as caused by entrapped air is caused by consolidation or shrinkage. Thus, whenever the relationship $p_c = p_c(S_w)$ is employed, past wetting–drying history may affect the results.

Finally, at an interface between two media, the discussion in paragraph 9.2.5 is applicable. As p_c must be the same on both sides of the interface between the two media (each having its own retention curve), the water content will be different on both sides of the interface (fig. 9.2.12).

9.4.2 The Capillary Fringe

The relationship (9.2.14) for equilibrium conditions is applicable to the zone of aeration in the soil. However, as in this case $\gamma_{nw} = \gamma_{air} \ll \gamma_w = \gamma_{water}$, we have:

$$z_1 - z_2 = (p_{c1} - p_{c2})/\gamma_w. \tag{9.4.6}$$

If $z_2 = 0$ is chosen as a point on the phreatic surface $p_{c2} = 0$ and we have:

$$z_1 = p_{c1}(S_w)/\gamma_w$$

or, in general:

$$z = z(S_w) = p_c(S_w)/\gamma_w. \tag{9.4.7}$$

Equation (9.4.7) gives the moisture distribution above the phreatic surface. As z increases upward, so does p_c. This means that S_w decreases with height above the phreatic surface. Figure 9.4.1 shows two curves $S_w = S_w(z)$, each representing an equilibrium corresponding to a stationary phreatic surface. Comparing these figures with the retention curves (e.g., fig. 9.4.5), it becomes obvious that these are actually the same curves. This is also a direct consequence of a comparison between (9.4.7) and (9.4.2).

In many cases of interest, immediately above the water table ($p = 0$) we have a zone that is saturated with water, or nearly so, because a certain suction must be reached before any substantial reduction in water content can be produced. Then, above this zone, there is a marked drop in the water content with a relatively small rise in the capillary pressure. This zone contains most of the water present in the zone of aeration. From figures 9.4.5a and 9.4.5b it is clear that this statement better describes the situation for poorly graded or coarse-textured soil (sand, gravel, etc.), but is also valid for fine-textured or well graded soils when the water table is sufficiently deep below the ground surface. As this phenomenon is analogous to the rise in a capillary tube where the water rises to a certain height above the free water surface, with a fully saturated tube below the miniscus and zero saturation above it, the nearly saturated zone above the phreatic surface, when it occurs, is called the *capillary fringe*, or *capillary rise* (fig. 1.1.2). Thus, in figure 9.4.5, h_{cc} is the capillary rise for the poorly graded soil.

The capillary fringe is thus an *approximate* practical concept that is very useful and greatly simplifies the treatment of phreatic flows when we wish to take into account the fact that a certain saturated (or nearly so) zone is present above the phreatic surface.

The following table gives some ranges of capillary rises (Silin–Bekchurin 1958).

Table 9.4.1

Material	Capillary Rise [cm]
Coarse sand	2–5
Sand	12–35
Fine sand	35–70
Silt	70–150
Clay	> 200–400

Often relationships such as:

$$h_{cc} = \frac{2.2}{d_H}\left(\frac{1-n}{n}\right)^{3/2} \tag{9.4.8}$$

for sands (Mavis and Tsui 1939) are used where h_{cc} is in inches, n is porosity, and d_H (in inches) is the harmonic mean grain diameter. Another equation for sand is:

$$h_{cc} = \frac{0.45}{d_{10}}\frac{1-n}{n} \tag{9.4.9}$$

where both d_{10} (effective particle diameter) and h_{cc} are in cm (Polubarinova-Kochina 1952, 1962). When the water is not pure, changes in surface tension σ and the contact angle θ should be taken into account.

Figure 9.4.7 shows the pressure distribution below the phreatic surface and in the capillary fringe above it. Often we use the symbol p_c to denote $p_{cc} = \gamma_w h_{cc}$. Similarly, h_c is used to denote the critical capillary head of h_{cc}.

A capillary fringe also exists immediately adjacent to any free surface bounding a saturated body of water, and not just above a phreatic surface. However, except for the capillary fringe above a horizontal water table, the water in the capillary fringe is not stagnant. As a water table or a free surface moves, the capillary fringe moves with it. If the water table rises fast, the movement of the capillary fringe may lag behind, sometimes considerably. The same is true for a declining water table. The latter may drop fast while the soil above it slowly drains until a new capillary fringe is established.

As an example, consider the case where at $t = 0$ a dry column of soil is brought in contact with a water table. Imbibition will start immediately. At every instant, the pressure just below the rising surface that bounds the saturated zone is: $p = -\gamma_w h_{cc}$. When this surface is at a distance ζ above the water table, the pressure gradient is $\partial p/\partial z = -(\gamma_w h_{cc}/\zeta)$ = constant with elevation z because of continuity. Assuming that we have complete saturation (e.g., no entrapped air) below this surface, we obtain:

$$q_z = -\frac{k}{\mu}\left(\frac{\partial p}{\partial z} + \gamma_w\right) = \frac{k\gamma_w}{\mu}\left(\frac{h_{cc}}{\zeta} - 1\right). \tag{9.4.10}$$

By continuity:

$$q_z = n\frac{d\zeta}{dt}.$$

Hence:

$$n\frac{d\zeta}{dt} = \frac{k\gamma_w}{\mu}\left(\frac{h_{cc}}{\zeta} - 1\right) = \frac{k\gamma_w}{\mu}\left(\frac{h_{cc}-\zeta}{\zeta}\right)$$

which, after integration, yields the capillary rise as a function of time:

$$t = \frac{nh_{cc}\mu}{k\gamma_w}\left[\ln\left(\frac{h_{cc}}{h_{cc}-\zeta}\right) - \frac{\zeta}{h_{cc}}\right]. \tag{9.4.11}$$

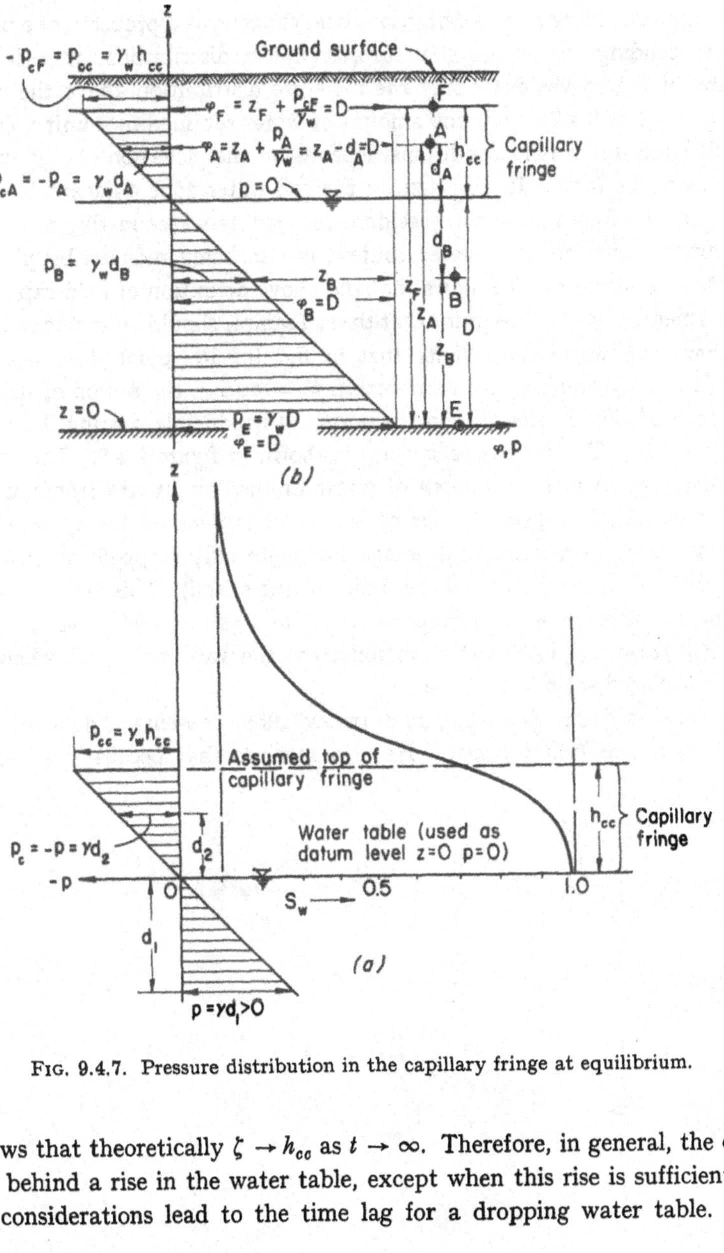

FIG. 9.4.7. Pressure distribution in the capillary fringe at equilibrium.

This shows that theoretically $\zeta \to h_{cc}$ as $t \to \infty$. Therefore, in general, the capillary rise lags behind a rise in the water table, except when this rise is sufficiently slow. Similar considerations lead to the time lag for a dropping water table.

9.4.3 Field Capacity and Specific Yield

Field capacity is usually defined as that value of water content remaining in a unit volume of soil after downward gravity drainage has ceased, or materially done so, say, after a period of rain or excess irrigation. A difficulty inherent in this definition is that no quantitative specification of what is meant by "materially ceased" is given. Gardner (1967) discusses this point.

Although, according to this definition, field capacity is a property of a unit volume of soil (depending on the soil structure, grain-size distribution, etc.), it is obvious from any of the curves describing the moisture distribution above the water table (e.g., fig. 9.4.2 or 9.4.7a) that the amount of water retained in a unit volume of soil at equilibrium under field conditions depends on the elevation of this unit volume above the water table. In addition, in the soil water zone adjacent to the ground surface (fig. 1.1.2), equilibrium is seldom reached as water in this zone constantly moves up or down and the water content is also being reduced by plant uptake. From these observations it follows that the above definition of field capacity should be supplemented by the constraint that the soil sample should be at a point sufficiently high above the water table (note that in fig. 9.2.10, point A is not the water table). Thus, returning to the relationship $p_c = p_c(S_w)$, the notion of field capacity of unsaturated flow is identical to the notion of irreducible wetting fluid saturation in figure 9.2.10. The field capacity (c_0) is shown in figure 9.4.8. The complement of the field capacity, i.e., volume of water drained by gravity from a unit volume of saturated soil, is called *effective porosity* and is denoted by n_e ($= n - c_0$).

At any point, once gravity drainage has materially stopped, a certain amount of moisture is retained in the soil (per unit volume of soil). This amount can be used to define the *retained moisture content* $c_r(z, t)$ and the *aerated porosity* $n_a(z, t)$ (fig. 9.4.8). Both c_r and n_a vary with elevation above the water table and, when the water table is displaced, also with time.

Zaslavsky (in Bear, Zaslavsky and Irmay 1968) presents a detailed discussion on the concept of *field capacity*. He also suggests that because true equilibrium

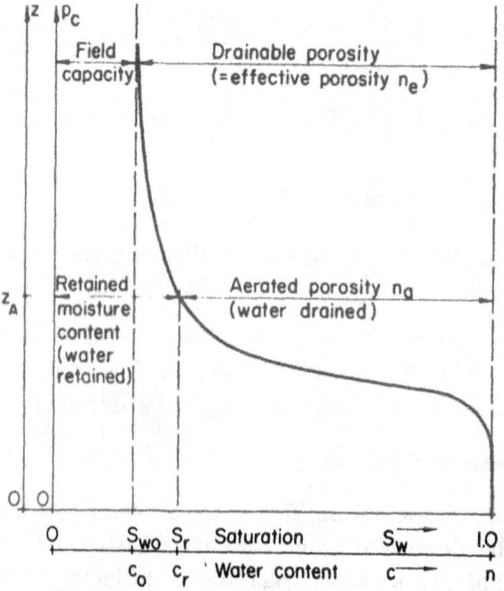

FIG. 9.4.8. Field capacity and effective porosity.

conditions (with stagnant water) are seldom reached under actual field conditions, one should consider field capacity as a dynamic state rather than as a state of equilibrium. He gives a detailed summary of the dependence of the field capacity on various parameters (e.g., permeability, soil structure and texture, temperature and barometric pressure, etc.). Hausenberg and Zaslavsky (1963) suggest a field capacity range located on both sides of the inflection point of the retention curve. Richards (1950) suggests defining field capacity as the water content corresponding to a suction of 0.1 to 0.5 atmospheres.

Specific yield is another unsaturated flow concept often employed in investigations of drainage of agricultural lands and in ground water hydrology. It is defined as the *average* amount of water per unit volume of soil drained from a soil column extending from the water table to the ground surface, per unit lowering of the water table. The corresponding amount of water retained in the soil against gravity when the water table is lowered is called *specific retention.* When expressed in terms of moisture content, we obtain for every instant:

$$c_y + c_r = n \tag{9.4.12}$$

where c_y is specific yield and c_r is specific retention. By dividing (9.4.12) by porosity, n, we obtain the same relationship in terms of saturation:

$$S_y + S_r = 1. \tag{9.4.13}$$

Thus, specific retention is a field concept, obtained by averaging what actually happens in natural soil in the zone of aeration when the water table is lowered.

When the water table is instantaneously lowered (or lowered relatively fast), say by a vertical distance Δh as a result of drainage (fig. 9.4.1), the changes in the moisture distribution lag behind and reach a new equilibrium (or practically so) only after a certain time interval that depends on the type of soil. A time lag will also take place when infiltration causes the water table to rise.

The total amount of water ΔW_d drained out of a column of soil of unit cross-sectional area as the water table at t' moves to a new position at t'' is (fig. 9.4.1):

$$\Delta W_d = n\,\Delta h + \int_{z'=0}^{d'} nS'_w(z', t')\,dz' - \int_{z''=0}^{d''} nS''_w(z'', t'')\,dz''. \tag{9.4.14}$$

The volume of water drained out of the column per unit decline of the water table, or specific yield corresponding to the average depth $\bar{d} = (d' + d'')/2$, is:

$$S_y(d', d'') = \frac{\Delta W_d}{n\,\Delta h} = 1 + \frac{1}{\Delta h}\int_{z'=0}^{d'} S'_w(z', t')\,dz' - \frac{1}{\Delta h}\int_{z''=0}^{d''} S''_w(z'', t'')\,dz''. \tag{9.4.15}$$

For a *homogeneous isotropic soil,* the two curves $S'_w = S'_w(z', t')$ and $S''_w = S''_w(z'', t'')$ are identical in shape. If, in addition, both water table positions are sufficiently deep below the ground surface, the two curves will merge at $S_w = c_0/n$.

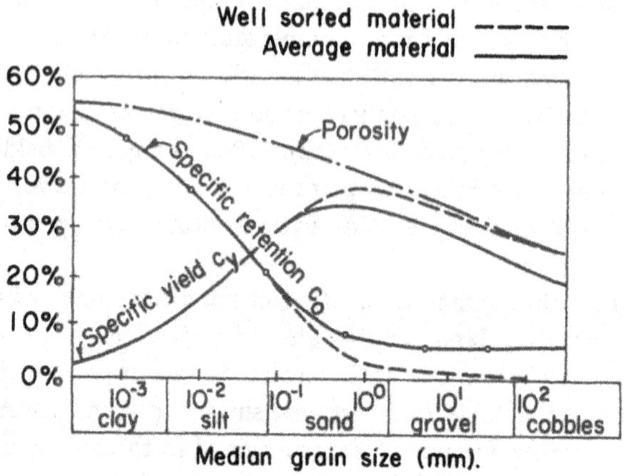

FIG. 9.4.9. Relationship between water storage properties and median grain size of alluvium from large valleys. (From Conkling et al. 1934, as modified by Davis and De Wiest, 1966).

Hence the total amount of water drained from the column is:

$$\Delta W_d = \Delta h(n - c_0). \tag{9.4.16}$$

Accordingly, the specific yield (in terms of saturation) is:

$$S_{y\infty} \equiv S_y|_{d\to\infty} = \frac{\Delta W_d}{n\,\Delta h} = \frac{n - c_0}{n} = 1 - \frac{c_0}{n} = 1 - S_{w0} \tag{9.4.17}$$

and in terms of water content:

$$c_{y\infty} \equiv c_y|_{d\to\infty} = n - c_0. \tag{9.4.18}$$

The corresponding changes in the amounts of water retained in the column, per unit decline of head, are:

$$S_r = 1 - S_{y\infty} = S_{w0}; \qquad c_{r\infty} = n - c_{y\infty} = c_0. \tag{9.4.19}$$

It is thus apparent that for a homogeneous isotropic soil and very deep water table, the specific retention is identical to the field capacity or to the irreducible water saturation, and the effective porosity is identical to the specific yield. For such conditions, figure 9.4.9 shows the relationship between specific yield and specific retention for various soils.

However, when the soil is inhomogeneous (e.g., composed of layers) or when the water table is a shallow one, the moisture distribution curves corresponding to the two water table positions are no longer parallel, and the identities above are no longer valid; we must distinguish between field capacity and specific retention.

When the time lag is also taken into account, as it takes time for drainage to be completed, we obtain a specific yield that is time dependent and that approaches asymptotically the values corresponding to the depths considered (fig. 9.4.10). The term *drainable porosity* is sometimes used to denote the instantaneous specific yield.

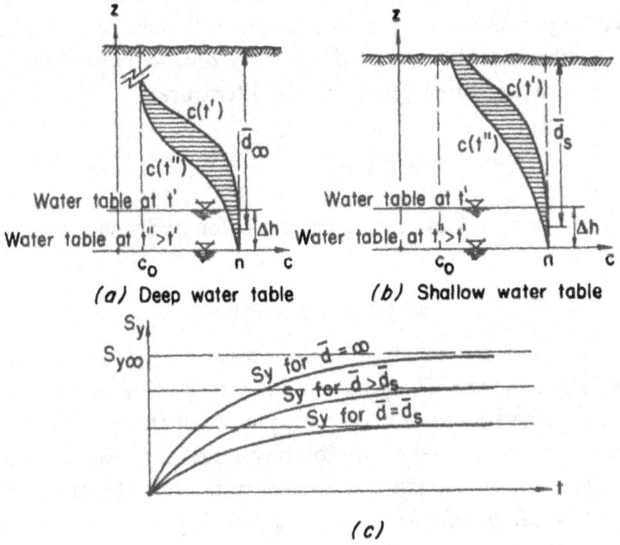

FIG. 9.4.10. Effect of depth and time on specific yield.

When the water table is rising or falling slowly, the changes in moisture distribution have sufficient time to adjust continuously and the time lag vanishes.

9.4.4 The Motion Equation

The entire discussion presented in paragraph 9.3.1 is applicable, except that here we assume that air (the nonwetting fluid) is (a) stagnant (i.e., $q_{nw} = 0$) and (b) at a constant pressure, usually taken as zero (i.e., $p_{nw} = 0$).

Thus, following many investigators, among whom we may mention Buckingham (1907), Richards (1931), Childs (1936), Childs and Collis-George (1950), we assume that the Darcy law, or the three-dimensional motion equation, is applicable to unsaturated flow in isotropic media in the form:

$$\mathbf{q} = -\frac{k(c)}{\mu} \nabla(p + \gamma z) = -\frac{k(c)\gamma}{\mu} \nabla\left(\frac{p}{\gamma} + z\right) \tag{9.4.20}$$

where $\mathbf{q}$ is the (volumetric) flux (volume of water per unit area per unit time), p is the pressure in the water phase and the permeability k is a function of the moisture content. One should recall that the average velocity, $\mathbf{V}$, is related to $\mathbf{q}$ by $\mathbf{q} = c\mathbf{V}$. In deformable media, $\mathbf{q}$ is with respect to the moving solid grains. As here $p_{nw} \equiv p_{\text{air}} = 0$, we have $p_c = -p_w \equiv -p$ and $p_c = p_c(c)$, so that (9.4.20) may be written as:

$$\mathbf{q} = -\frac{k(c)}{\mu} \nabla(-p_c + \gamma z) = -\frac{k(p_c)}{\mu} \nabla(-p_c + \gamma z). \tag{9.4.21}$$

The permeability in (9.4.21) is a *function* of the water content c, and not a constant, as in saturated flow. Following the discussion in section 9.3.1 it may be called *effective permeability*. As usual, $K(c) = k(c)\rho g/\mu$ will then be the *effective hydraulic conductivity*. Sometimes $K(c)$ is called *capillary conductivity*. Two other forms of the motion equation are often given in the literature:

$$\mathbf{q} = -K(c)\,\nabla\varphi = -\frac{K(c)}{\gamma}\nabla p - K(c)\mathbf{1z} \tag{9.4.22}$$

where $\varphi = z + p/\gamma = z - \psi$ is the piezometric (or hydraulic) head, defined as φ_c in (9.4.4), and:

$$\mathbf{q} = K(c)\,\nabla\psi - K(c)\mathbf{1z} \tag{9.4.23}$$

where $\psi = -p/\gamma > 0$ is the negative of the pressure head. In our notation, in unsaturated flow, ψ is positive. The pressure head itself ($-\psi = p/\gamma$; pressure energy per unit weight of water) expresses the energy state of the water in the unsaturated soil. It is therefore often referred to (by soil physicists) as *moisture potential, capillary potential* (p_c/γ), etc. When p/γ is expressed in units head of water, we may identify $-g\psi\,(p/\rho)$ with the *differential specific Gibbs function of soil water*. The liquid and vapor phases are then related to each other by:

$$h_r = \exp\{-g\psi/RT^0\} \tag{9.4.24}$$

where h_r is relative humidity, R is the gas constant for water vapor and T^0 is the absolute temperature.

For horizontal flows, the term $K(c)\mathbf{1z}$ in (9.4.22) and (9.4.23) vanishes. It is important to recall that the relationship $p_c = p_c(c)$ is not a unique one, and that because of hysteresis, the history of wetting and drying may play an important role in the analysis of flow problems.

Nevertheless, assuming for a moment that the relationship $p_c(c)$ is a unique one (as when dealing with a drainage problem only), we have $p_c = p_c(c)$, or $\psi = p_c/\gamma = \psi(c)$. Hence, equation (9.4.23) becomes:

$$\mathbf{q} = K(c)\frac{d\psi}{dc}\nabla c - K(c)\mathbf{1z} = \left[\frac{K(c)}{dc/d\psi}\right]\nabla c - K(c)\mathbf{1z}. \tag{9.4.25}$$

Klute (1952) called the group $D(c) = -K(c)\,d\psi/dc = -K/[dc/d\psi] = [K/\gamma]\,dc/dp$ the *diffusivity coefficient* or the *capillary diffusivity* (dim L^2T^{-1}). Then (9.4.25) becomes:

$$\mathbf{q} = -D(c)\,\nabla c - K(c)\mathbf{1z} \tag{9.4.26}$$

and for horizontal, two-dimensional flow in the xy plane:

$$\mathbf{q} = -D(c)\,\nabla c. \tag{9.4.27}$$

Sometimes (not here) the definitions $C = dc/dp$ and $D = -K/(dc/dp)$ are introduced. The similarity between (9.4.27) and Fick's law of diffusion explains why the term diffusivity was introduced here. Zaslavsky (in Bear, Zaslavsky and Irmay 1968) commented that the use of the term diffusivity (in the mathematical sense), may be misleading here because of the second term on the right-hand side of (9.2.26) and because $\mathbf{q}$ denotes here a volumetric flux and not a mass flux. He suggested for $D(c)$, the term *capillary response*, as it expresses the response of the system (in the form of flow) to the moisture gradient. One should note that $D(c)$ is made up of the ratio $K(c)/[-dc/d\psi]$, i.e., a ratio between the conductivity $K(c)$ and $-dc/d\psi = C(c)$, which may be interpreted as a *water capacity* (see comment after (9.4.45)). This observation improves the analogy to other diffusivities (used in the mathematical sense) such as thermal diffusivity, electric diffusivity, aquifer diffusivity (T/S, par. 6.4.2), etc., which have a similar structure. The minus sign is introduced because c increases as ψ decreases.

As we shall see below, the dependence of D on c, or of K on ψ, introduces a *nonlinearity* into the continuity equation. In nonhorizontal flows, the gravity term in the motion equation also makes the continuity equation a rather difficult one for exact solution by analytical methods.

As in paragraph 9.3.1, we shall assume (as very little work has been published to date on this subject) that the motion equation of unsaturated flow in isotropic media can be extended to anisotropic media by assuming, in analogy to permeability of saturated (single-phase) flow, that $k(c)$ is a second-rank symmetrical tensor with components $k_{ij}(c)$.

Although there is sufficient experimental evidence to justify the use of the above-mentioned equations (in the same range of Reynold's number for which Darcy's law is valid), there is also experimental evidence in the literature that at very low hydraulic gradients (which serve here as the driving force), the flow deviates from Darcy's law. Curves such as *I* (fig. 9.4.11), indicating a flow for every $\nabla\varphi > 0$, or *II*, indicating a *threshold gradient* that must be attained before any flow can be initiated, are given in the literature. Klute (1967) attributes these deviations from Darcy's law to electro-osmotic effects, particle movement, quasicrystalline water structure, etc. Bolt and Groenevelt (1969) indicate coupling phenomena (par. 4.4.3)

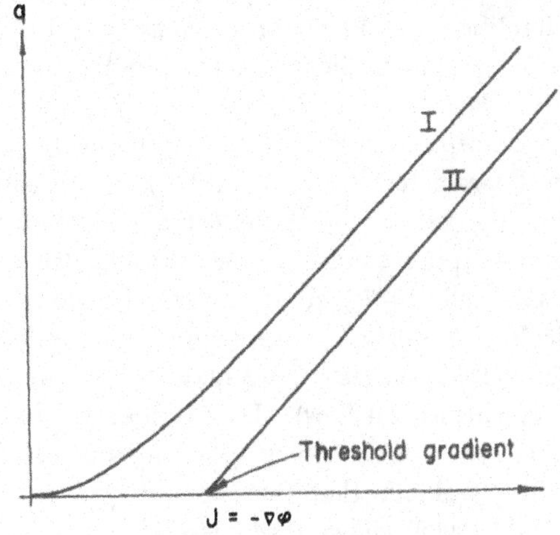

FIG. 9.4.11. Deviation from Darcy's law (after Klute, 1967).

as a possible cause of non-Darcian behavior of water in soil. Additional comments on this point are given in paragraph 5.3.2.

Up to this point, only isothermal flow of salt-free water has been considered. This has enabled us to write the relationship $p_c = p_c(c)$. As was mentioned in paragraph 9.2.3, temperature variations that affect interfacial tension σ also affect capillary pressure, so that for *nonisothermal flow* of salt-free water we must consider $p_c = p_c(c, T^0)$, $\psi = \psi(c, T^0)$, etc. By introducing these relationships in (9.4.25), we obtain:

$$\mathbf{q} = (K\partial\psi/\partial c)\,\nabla c - K\nabla z + (K\partial\psi/\partial T^0)\,\nabla T^0 \tag{9.4.28}$$

where the third term introduces the effect of temperature gradient on the water (i.e., the liquid phase) flux, in addition to the terms indicating the effects of moisture content gradient and of gravity. The effect of solute concentration C (leading to $p_c = p_c(c, T^0, C)$ since $\sigma = \sigma(c, T^0, C)$) can be introduced in a similar way.

We may introduce another diffusivity:

$$D_c = -K\partial\psi/\partial c \quad \text{and} \quad D_T = -K\partial\psi/\partial T^0$$

so that (9.4.28) may be rewritten as:

$$\mathbf{q} = -D_c(c, T^0, C)\,\nabla c - D_T(c, T^0, C)\,\nabla T^0 - K(c, T^0, C)\nabla z. \tag{9.4.29}$$

In the case of an anisotropic medium, D_c and D_T are second-rank symmetrical tensors: $\underline{\mathbf{D}}_c$ and $\underline{\mathbf{D}}_T$. Philip and de Vries (1957) suggest for $\partial\psi/\partial T^0$ the expression $(\psi/\sigma)\, d\sigma/dT^0 = A\psi$ (neglecting effects of solute concentration), where in the range of 10°C to 20°C, $A = -2.09 \times 10^{-3}$ °C^{-1}. This expression is derived from (9.2.9), which shows a proportionality between p_c and σ (with $2/r^*$ expressing some function of the moisture content). One should note that this is not temperature induced flow, in the sense of a coupled process as discussed in paragraph 4.4.3, nor is it the movement in the vapor phase, as discussed in paragraph 9.4.9 below.

Stallman (1964) simplifies (9.4.29) for one-dimensional flow in the $\mathbf{1s}$ direction, assuming that both the thermal gradient and the capillary pressure gradient are in the direction $\mathbf{1s}$ of the flow.

In introducing the concept of a soil diffusivity (actually only a mathematical device, as the basic soil parameters are $K(c)$ and $C(c)$), it was assumed that $\psi = \psi(c)$ was a unique relationship, except for the effect of hysteresis. However, several authors question this assumption. For example, Swartzendruber (1968), in a discussion of a paper by Vachaud (1967), questions the correctness of using the static $\psi(c)$ function to calculate ψ and $\nabla\psi$ corresponding to any transient c.

Gardner and Gardner (1951) suggest that ψ may also depend on the water-content gradient $\partial c/\partial x$ (in one-dimensional flow). Davidson et al. (1966) report data from which they conclude that the equilibrium $c(\psi)$ depends upon the applied increments $\Delta\psi$, thus implying the possibility that the transiently dynamic $\psi(c)$ differs from the static one. Vachaud (1968) shows that there is very little difference between the diffusivity obtained from a static $\psi(c)$ and that obtained from a dynamic $\psi(c)$.

9.4.5 Relative Permeability of Unsaturated Soils

The permeability $k(c)$ employed in the previous paragraph is nothing but the *effective permeability* defined in paragraph 9.3.1. For some reason, soil physicists, who have extended Darcy's law to unsaturated flow, have preferred the use of effective permeability $k(c)$ to that of relative permeability, which is commonly used in the equations of motion when two or more fluids move simultaneously. Nevertheless, since relative permeability is nothing but the ratio of effective permeability to permeability at saturation, we may apply the entire discussion in paragraph 9.3.2 to relative permeability of unsaturated soil. Obviously, in unsaturated flow, we have only relative permeability k_{rw} to the wetting fluid (water).

Figure 9.4.12 shows the variations of relative permeability with saturation, according to experiments by Wyckoff and Botset (1936). As the saturation decreases, the large pores drain first so that the flow takes place through the smaller pores. This causes both a reduction in the cross-sectional area available for the flow, and an increase in tortuosity of the flow paths. The combined effect causes a rather rapid reduction in the hydraulic conductivity as the moisture content decreases. As the water films become thinner, certain phenomena at the solid–water interface come into play, causing a further reduction in hydraulic conductivity. Among such factors we may mention an increase of viscosity of the water in close proximity to the solid surfaces. Point A indicates the irreducible water (wetting fluid) saturation S_{w0}. The corresponding irreducible water content will be denoted by $c_0 = nS_{w0}$. At point A (somewhere between 10 and 15% in fig. 9.4.12), the water phase becomes discontinuous, existing only in the very small pores or as isolated wedges. Several authors suggest relationships between the permeability (k), conductivity (K) and saturation S ($\equiv S_w$) or the water content c. Childs and Collis-George (1950) assume:

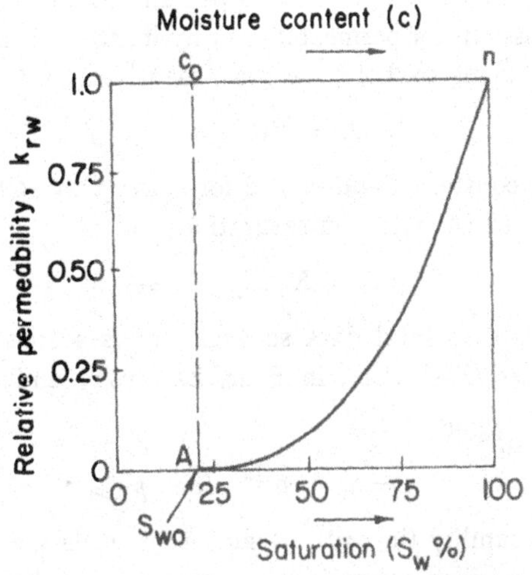

FIG. 9.4.12. Relative permeability of unsaturated sand according to experiments by Wyckoff and Botset (1936) and theoretical analysis by Irmay (1954).

$$K = Bc^3/M^2 \tag{9.4.30}$$

where M is the specific surface area of the soil phase and B is a constant.

Irmay (1954) derives a similar relationship assuming that resistance to flow offered by the solid matrix is proportional to the solid–liquid interfacial area. The hydraulic conductivity K then becomes proportional to the hydraulic radius, i.e., to the volume of voids divided by the wetted area:

$$R = U_v/A_s = nU_b/M(1-n)U_b = n/M(1-n), \qquad A_s = MU_s \tag{9.4.31}$$

(e.g., for a cubic arrangement of spheres of diameter d, the specific surface is $M = 6/d$), where subscripts v, b, and s denote voids, bulk and solids, respectively. This model leads to the cubic parabolas:

$$K = K_0\left(\frac{c-c_0}{n-c_0}\right)^3 = K_0\left(\frac{S-S_{w0}}{1-S_{w0}}\right)^3; \qquad \left(\frac{K}{K_0}\right)^{1/3} = \frac{c-c_0}{n-c_0}. \tag{9.4.32}$$

The experimental curve of figure 9.4.12 fits such a cubic parabola. Experiments by several authors with soils of uniform grain size (e.g., Hausenberg and Zaslavsky 1963; Topp and Miller 1965), seem to agree with the relationship in (9.4.32).

Corey (1957) finds that for many consolidated rocks a relative permeability proportional to $S_w{}^4$ describes unsaturated flow. Irmay's model, like most other models, fails to consider an additional resistance that might be introduced at the (microscopic) air–water interfaces present in the pores.

Hydraulic conductivity may also be presented as a function of the capillary pressure head ψ. However, the relationship $K(\psi)$ shows much hysteresis, probably because of the hysteresis in the function $\psi(c)$ discussed in paragraph 9.4.1. The effect of hysteresis is shown in figure 9.4.13. The figure also shows how, upon rewetting to zero pressure head, the relative hydraulic conductivity is less than that obtained at true saturation owing to the air entrapped in the pores.

Among other expressions for permeability or hydraulic conductivity of unsaturated flow suggested by various authors, we may mention:

Gardner (1958):
$$K = a/(b + \psi^m) \tag{9.4.33}$$

where a, b and m are constants, and $m \approx 2$ for heavy clay soil and $m \approx 4$ for sand; K_0 is the hydraulic conductivity at saturation;

Gardner (1958):
$$K = K_0 \exp(-a\psi) \tag{9.4.34}$$

which does not fit experimental data so well, but is sometimes more convenient for analytical purposes (Zaslavsky, in Bear, Zaslavsky and Irmay 1968);

Brooks and Corey (1966):
$$k = k_0 \quad \text{for} \quad p_c \leqslant p_b$$
$$k = k_0(p_b/p_c)^n \quad \text{for} \quad p_c \geqslant p_b \tag{9.4.35}$$

where k_0 is the permeability (in cm^2) at saturation, p_b is the bubbling pressure (in dynes/cm^2) and the exponent n is an index of the pore-size distribution of the porous medium. Additional information on (9.4.35) and experimental verifications are given by Laliberte, Corey and Brooks (1966).

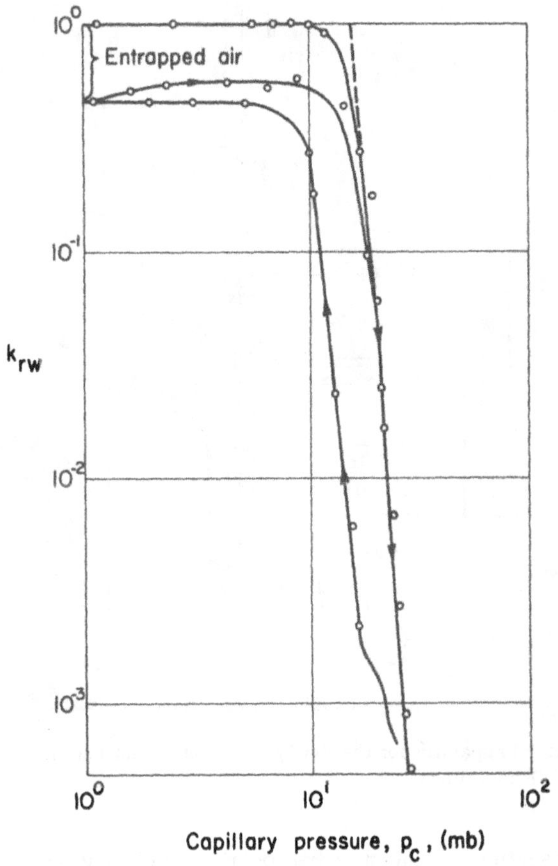

FIG. 9.4.13. Relative hydraulic conductivity as a function of capillary pressure for Loveland fine sand (Klute, 1967 as adapted from Brooks and Corey, 1964).

Finally, let us comment on laboratory determination of permeability or hydraulic conductivity of unsaturated porous media, as a supplement to the discussion in paragraph 9.3.7. Typical equipment is described by Anat et al. (1965) and by Laliberte, Corey and Brooks (1966). Another typical arrangement, described by Klute (1965), is shown in figure 9.4.14. Most methods are based on the solution of the one-dimensional continuity equation for steady and unsteady flow, with or without the gravity term.

A typical permeameter for steady flow consists of an impervious cylinder, in which the studied porous medium sample is placed between two porous plates. The air entry value of these plates should be larger than the suctions to be used during the test. By means of constant level reservoirs, constant suctions are maintained at both ends of the sample. When the suction at one end is increased over the one at the other end, flow takes place. Some of the difficulties introduced by the presence of *end effects* were discussed in paragraph 9.3.7. To circumvent these difficulties, the

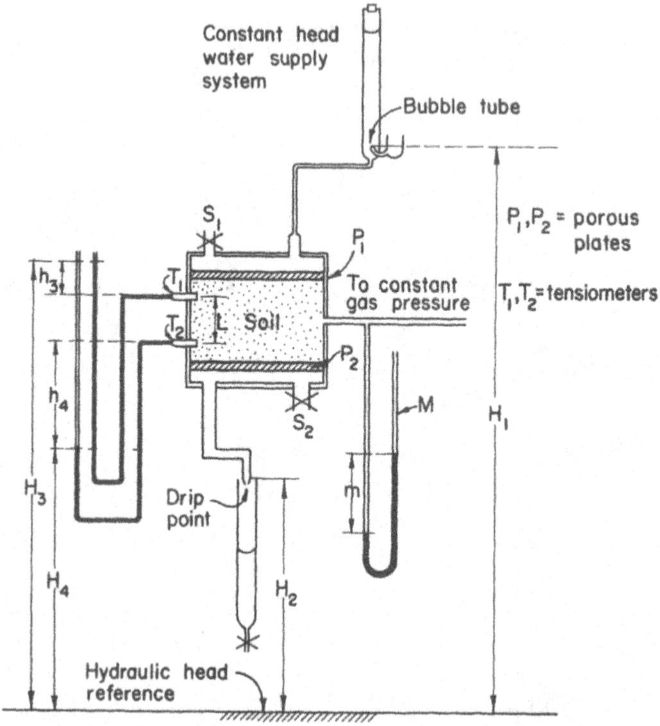

FIG. 9.4.14. Diagram of apparatus for the steady-state method for determining the conductivity of unsaturated soil (Klute, 1965).

measurements of suction are taken at two points sufficiently removed from the two ends that there is no need to assume either a constant hydraulic conductivity or a constant head loss along the entire sample (although saturation varies).

In the typical arrangement employed by Klute (1965), suction is measured in two tensiometers at a distance L apart. From Darcy's law for the case of vertical flow shown in figure 9.4.14 we obtain:

$$K(\bar{h}) = \frac{q}{(H_3 - H_4)/L}; \qquad \bar{h} = \rho_1 \frac{m}{\rho} + \frac{h_3 + h_4}{2} \tag{9.4.36}$$

where q is the (constant) specific discharge through the sample, $\bar{h}$ is the mean pressure head at which K is determined ($h = -\psi$), m is the gauge pressure in the cell, expressed as a column of fluid of length m, and ρ_1 and ρ are the densities of the manometer fluid and of the water, respectively. The degree of saturation of the sample is changed and controlled by subjecting the sample to a controlled gas-phase pressure that is greater than atmospheric by an amount proportional to the height of the fluid column m.

By starting the measurement at or near saturation and setting up a series of steady-state flows, a set of paired values K and $\bar{h}$ can be obtained on the drainage curve. The hysteresis behavior of $K(\psi)$ can be obtained by manipulating the apparatus through any desired set of drainage and rewetting cycles.

The *outflow method*, originally suggested by Richards (1949) and later refined and modified by others (e.g., Gardner 1956; Miller and Elrick 1958; Rijtema 1959; Kunze and Kirkham 1962; Klute 1965), has been used to obtain estimates of the hydraulic conductivity and diffusivity functions. The method is based on the measurement of the volume of water outflow as a function of time from a sample placed in a pressure cell. The slab of soil is placed on a porous plate or membrane (similar to the arrangements in figs. 9.2.14 and 9.2.15 for determining the capillary pressure curve) and is brought to equilibrium with a certain gas-phase pressure in the cell. By increasing the gas (e.g., air) pressure in the cell by a small increment, water is forced out through the membrane. The volume of water draining out of the sample is measured as a function of time. The procedure is then repeated by further increasing the gas pressure.

These data are then introduced into the solution of the unsteady, one-dimensional, continuity equation in which the effect of gravity is neglected.

Bruce and Klute (1956) also describe a method for derivation of the diffusivity function from analysis of the moisture-content distribution at a given time in an effectively semi-infinite horizontal column. Zaslavsky and Ravina (1965) describe a method that they call the *moisture moment method*, which circumvents the need to use average values, as is done in the outflow method described above. Instead, they express the moisture distribution along a horizontal permeameter by measuring changes in the first moment of the moisture-content distribution.

Determination of hydraulic conductivity and diffusivity may, in principle, be made in any flow system, whether steady or transient, for which an analytic solution is available. If we wish to avoid the use of average values (e.g., of moisture content or pressure-head gradient) taken over the entire length of the sample, we must measure point values of moisture content and of pressure head and their variations in time during an experiment. Pressures are measured by pressure transducer tensiometers (Watson 1965; Klute and Peters 1966). Moisture content may be measured by a gamma-ray attenuation system (e.g., Vachaud 1968a; Davidson et al. 1963), by a neutron scattering device (e.g., McHenry 1963) or by a gamma-radiation or X-ray absorption method (Norel 1963, 1964, 1966).

9.4.6 The Continuity Equation

Assuming no sources or sinks of moisture within the unsaturated flow domain, we may start from the mass conservation equation:

$$\partial(\rho nS)/\partial t + \operatorname{div}(\rho \mathbf{q}) = 0; \qquad \partial(\rho c)/\partial t + \operatorname{div}(\rho \mathbf{q}) = 0 \tag{9.4.37}$$

where symbols S and $\mathbf{q}$ refer to the water. As water is assumed incompressible in the range of pressures considered in unsaturated flow, equation (9.4.37) becomes:

$$\partial(nS)/\partial t + \operatorname{div} \mathbf{q} = 0; \qquad \partial c/\partial t + \operatorname{div} \mathbf{q} = 0. \tag{9.4.38}$$

For a *nondeformable* medium, $n =$ const, and (9.4.38) reduces to:

$$n\,\partial S/\partial t + \operatorname{div}\mathbf{q} = 0. \qquad (9.4.39)$$

In a deformable, or consolidating medium, we must take into account variations in n, and also the fact that $\mathbf{q}$, as defined by the equations of motion (par. 9.4.4), is with respect to the moving grains. By introducing these two factors into the continuity equation and following a procedure similar to that presented in section 6.3, the partial differential equations describing flow and consolidation may be derived (Philip 1969).

By combining the continuity equation (9.4.38) with any of the equations of motion in paragraph 9.4.4, we obtain various forms of the continuity equation for a homogeneous isotropic medium:

$$\partial c/\partial t - \operatorname{div}[K(c)\operatorname{grad}\varphi] = 0; \qquad \varphi = z - \psi$$

$$\partial c/\partial t + \operatorname{grad} K(c)\cdot\operatorname{grad}\psi + K(c)\,\nabla^2\psi - \partial K(c)/\partial z = 0. \qquad (9.4.40)$$

Since $c = c(\psi)$, $K = K(\psi)$, $\partial c/\partial t = (\partial\psi/\partial t)(dc/d\psi) = -\,C(c)\,\partial\psi/\partial t$ and $\operatorname{grad} K(c) = [dK(c)/d\psi]\operatorname{grad}\psi$ and we assume that $c = c(\psi)$, $K = K(c)$ and $K = K(\psi)$ are known functions, we obtain:

$$\frac{\partial\psi}{\partial t}\frac{dc}{d\psi} + \operatorname{div}[K(c)\operatorname{grad}\psi] - \frac{\partial K(c)}{\partial z} = 0 \qquad (9.4.41)$$

or:

$$\partial\psi/\partial t = A(\psi)(\nabla\psi)^2 + D(\psi)\,\nabla^2\psi - A(\psi)\,\partial\psi/\partial z$$

$$D(\psi) = -\frac{K(c)}{dc/d\psi} = \frac{K(c)}{C(c)}; \qquad A(\psi) = -\frac{dK(c)/d\psi}{dc/d\psi}$$

$$\frac{A(\psi)}{D(\psi)} = \frac{dK(c)/d\psi}{K(c)} = \gamma\frac{dK(c)/dp_c}{K(c)} = \gamma d[\ln K(c)]/dp_c. \qquad (9.4.42)$$

In terms of p_c $(=\gamma\psi)$ or p $(=-p_c)$, we obtain:

$$\partial p_c/\partial t = \frac{1}{\gamma}A(\psi)\cdot(\nabla p_c)^2 + D(\psi)\,\nabla^2 p_c - A(\psi)\,\partial p_c/\partial z \qquad (9.4.43)$$

$$\partial p_c/\partial t = -\frac{1}{\gamma}A(\psi)\,\nabla p_c\cdot\nabla(\gamma z - p_c) - D(\psi)\,\nabla^2(\gamma z - p_c) \qquad (9.4.44)$$

$$\partial p/\partial t = \frac{1}{\gamma}A(\psi)\cdot(\nabla p)^2 - D(\psi)\,\nabla^2 p + A(\psi)\,\partial p/\partial z. \qquad (9.4.45)$$

In the equations above, $C(c) = -\,\gamma(dc/dp_c) \equiv \gamma(dc/dp) \equiv -\,dc/d\psi$ has the meaning of *water capacity* or specific water capacity, or *specific water storativity* of an unsaturated medium, in analogy to the thermal capacity appearing in the heat conduction equation. Because of hysteresis, the water capacity $C(c)$ is a single-valued function only if everywhere in the flow domain we have either drainage or imbibition. Otherwise it depends upon the particular scanning curve (fig. 9.4.6) that describes the process that the porous medium is undergoing. Describing the sequence of wetting and drying by $c_1, c_2, c_3, \ldots$, we have:

$$p = p(c, c_1, c_2, c_3, \ldots)$$

so that:

$$dp = \frac{\partial p}{\partial c} dc + \sum_{r=1}^{n} \frac{\partial p}{\partial c_r} dc_r. \tag{9.4.46}$$

One should note that to take into account the history of drying and wetting as expressed by (9.4.45), equation (9.4.45) should be used to define all pressure gradients in the motion and continuity equations of unsaturated flow (Childs 1967). When solving a problem by numerical methods using digital computers, the double-valuedness of $c = c(p_c)$ can be taken into account by first checking the type of process, and then picking up the value of water capacity from the appropriate curve.

An important advantage of using the continuity equation in the form (9.4.45), in which the pressure $p = p(x, y, z, t)$ is the dependent variable, is that it is also valid for the saturated region below the water table where $K(c) = K = \text{const}$ and $c = n$. This enables one to treat a flow domain that is partly saturated (in the zone of saturation) and partly unsaturated (in the zone of aeration) as one continuous system having a single dependent variable. One should note that although it is assumed here that the water is an incompressible fluid, $\partial p/\partial t \neq 0$ because of the initial and boundary conditions of the problem.

Another form of the continuity equation is based on (9.4.26):

$$\partial c/\partial t = \operatorname{div}[D(c) \operatorname{grad} c] + \partial K(c)/\partial z \tag{9.4.47}$$

$$\partial c/\partial t = [dD(c)/dc] \cdot (\nabla c)^2 + D(c)\, \nabla^2 c + [dK(c)/dc]\, \partial c/\partial z. \tag{9.4.48}$$

It is always assumed that $K(c)$, $C(c)$, $D(c)$, etc., appearing in the various forms of the continuity equation, are known functions of the water content. In homogeneous soils (not in stratified, or layered soils), these functions are also functions of position and time: $K = K[c(x, y, z, t)]$, $C = C[c(x, y, z, t)]$, etc. All the partial differential equations presented above, as well as those presented in the remaining part of this paragraph, which are various forms of the continuity equation for unsaturated flow, are *nonlinear equations*, as conductivity and diffusivity are functions of the dependent variable (c, ψ, etc.). As will be discussed below, this fact dictates the methods of solution applicable to these equations. Sometimes the equations may be linearized by applying a certain transformation.

Whenever the gravity term, i.e., the last term on the right-hand side of the continuity equations given above, can be neglected, as in horizontal, or approximately horizontal, flow, the continuity equation has the mathematical form of the *nonlinear heat conduction equation*. Under such conditions, there is some advantage in using the form (9.4.47) without the gravity term, because of some simple expressions that have been suggested for $D(c)$. For example, Gardner and Mayhugh (1958) suggest for the intermediate range of water contents the exponential expression:

$$D = D_0 \exp[\beta(c - c_0)] \tag{9.4.49}$$

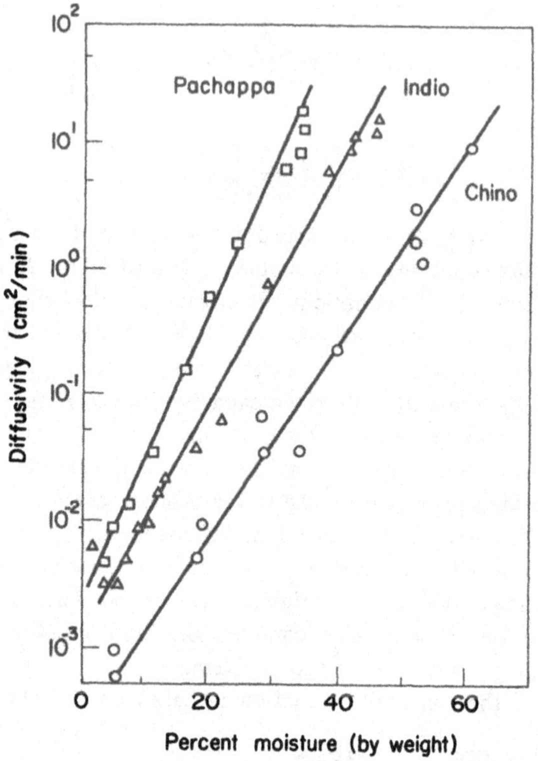

FIG. 9.4.15. Diffusivity-water content data for three soils (from Gardner, 1958).

where D_0 is the diffusivity corresponding to $c = c_0$, and β and c_0 are empirical parameters. Figure 9.4.15 shows data of $D(c)$ for three soils that seem to justify (9.4.49). Zaslavsky (in Bear, Zaslavsky and Irmay 1968), following Bruce and Klute (1956) and Hausenberg and Zaslavsky (1963), emphasized the fact that (9.4.49) fails in the range near saturation, owing to the presence of a marked inflection point on the $\psi = \psi(c)$ curve in some granular or well structured soils; D may have a maximum and a minimum somewhere between field capacity and complete saturation, and a range where it almost does not vary with c.

When the gravity term cannot be neglected, there is no advantage in using the form (9.4.47). On the other hand, there is some advantage in using the forms including $K(c)$, especially where hysteresis is to be taken into account, as in general $K(c)$ is practically hysteresis free (Klute 1967).

Sometimes, as in paragraph 9.3.5, we wish to follow the advance of a front of a specified water content, which is also a front of a specified p_c (in view of the relationship $p_c = p_c(c)$) or specified ψ as $\psi = \psi(c)$. In the latter case our dependent variable in one-dimensional flow is $x = x(\psi, t)$ rather than $\psi = \psi(x, t)$, which is a solution of (9.4.42) written for one dimension only, i.e., of:

$$\partial\psi/\partial t = A(\psi)(\partial\psi/\partial x)^2 + D(\psi)\, \partial^2\psi/\partial x^2. \tag{9.4.50}$$

In order to transform (9.4.50) where $\psi = \psi(x, t)$ into an equation with the dependent variable x and independent variables ψ and t, we employ the following relationships (King and Corey 1962):

$$\frac{\partial\psi}{\partial x}\frac{\partial x}{\partial t}\frac{\partial t}{\partial\psi} = -1 \quad \text{or} \quad \frac{\partial x}{\partial t} = -\frac{\partial\psi}{\partial t}\frac{\partial x}{\partial\psi}$$

and:

$$\frac{\partial^2\psi}{\partial x^2} = -\left(\frac{\partial^2 x}{\partial\psi^2}\right)\bigg/\left(\frac{\partial x}{\partial\psi}\right)^3. \tag{9.4.51}$$

Both relationships are obtained from the general solution of (9.4.50), which can be written in the form $F(\psi, x, t) = 0$. This expression may be considered as expressing implicitly either $\psi = \psi(x, t)$, $x = x(\psi, t)$.

By substituting these relationships in (9.4.50) we obtain:

$$\frac{\partial x}{\partial t} = -A(\psi)\frac{1}{\partial x/\partial\psi} + D(\psi)\frac{\partial^2 x/\partial\psi^2}{(\partial x/\partial\psi)^2} \tag{9.4.52}$$

the solution of which gives $x = x(\psi, t)$. Following the discussion on the Boltzmann transformation presented in paragraph 6.4.5 (King and Corey 1962), we may apply the method of separation of variables to (9.4.52) in order to determine $x = x(\psi, t)$. Assuming a solution in the form:

$$x = \Psi(\psi)T(t) \tag{9.4.53}$$

where Ψ is a function of ψ only and T is a function of t only, equation (9.4.52) is replaced by the two ordinary differential equations:

$$T(t)\frac{dT(t)}{dt} = k \tag{9.4.54}$$

$$\frac{1}{\Psi}\left[D(\psi)\frac{d^2\Psi/d\psi^2}{(d\Psi/d\psi)^2} - A(\psi)\frac{1}{d\Psi/d\psi}\right] = k. \tag{9.4.55}$$

By integrating (9.4.54) we obtain:

$$T(t) = \sqrt{2kt + C_1}. \tag{9.4.56}$$

Thus the solution of (9.4.52) is:

$$x = \sqrt{2kt + c_1}\Psi(\psi) \tag{9.4.57}$$

where $\Psi(\psi)$ is obtained by solving (9.4.55).

The function $\Psi(\psi)$ is a single-valued function of ψ only. Given a value of $\psi = \psi_0$, it has one and only one value $\Psi(\psi_0)$, so that (9.4.57) becomes:

$$x|_{\psi=\psi_0} = \Psi(\psi_0)\sqrt{2kt + C_1} \tag{9.4.58}$$

where $C_1 = 0$ for $x = 0$ at $t = 0$. This gives the rate of advance of a plane of constant capillary head. Because of the relationship between ψ and c, this also gives the rate of advance of a front of a given water content. Equation (9.4.58) with $C_1 = 0$ is identical to the *Boltzmann transformation* which in this case has the form:

$$\lambda(\psi) = xt^{-1/2} \tag{9.4.59}$$

provided $c = c(\psi)$, i.e., provided the water content c is a single-valued function of the capillary head only (King and Corey 1962).

Thus, applying the Boltzmann transformation (9.4.59) to (9.4.50), we obtain the ordinary differential equation:

$$D(\psi)\, d^2\psi/d\lambda^2 + A(\psi) \cdot (d\psi/d\lambda)^2 + (\lambda/2)\, d\psi/d\lambda = 0. \tag{9.4.60}$$

In a similar way, this transformation can also be applied to other forms of the one-dimensional continuity equations. Equation (9.4.60) can also be rewritten as:

$$\frac{d}{d\lambda}\left[D(c)\frac{dc}{d\lambda}\right] + \frac{\lambda}{2}\frac{dc}{d\lambda} = 0. \tag{9.4.61}$$

As mentioned in paragraph 6.4.5, the Boltzmann transformation is useful whenever the initial and the boundary conditions are also expressible in terms of λ. Thus, if initially $(t = 0)$ $c \neq$ const for $x > 0$, we cannot express c as $c(\lambda)$. On the other hand, the transformation is applicable to the region $x > 0$ with constant initial moisture content for $x > 0$ and constant moisture content $x = 0$ for $t > 0$.

In axially-symmetric flow in a horizontal plane, the Boltzmann transformation is:

$$\mu = rt^{-1/2}. \tag{9.4.62}$$

The Boltzmann transformation in the form $\nu = Rt^{-1/2}$ is also applicable to spherically-symmetric flow in which the gravity term is neglected. This approximation is justified only during the first stages of the unsaturated flow process.

As with the solution of other flow problems, the solution of the partial differential equations of unsaturated flow requires the specification of initial and boundary conditions. The latter are mathematical statements of the space and time distributions of ψ (or p or c, depending on the partial differential equation to be solved) or of the water fluxes on the external boundaries of the flow domain. However, unlike the case in saturated flow, the statement of water content distribution alone is not sufficient because $K(c)$, or $\psi(c)$, is subject to hysteresis. It is also necessary to state whether a drying or a wetting process is taking place along the boundary. An impervious boundary is a special case of a boundary along which the flux is specified; it is a boundary along which the component of the flux normal to the boundary is zero.

When the flow domain is made up of regions of different media, the discussion in paragraph 9.2.5 is applicable. Each region that is homogeneous in itself possesses its own properties $K(c)$, $C(c)$, etc. The conditions along the boundaries between the various media are that pressure (or pressure head, capillary pressure, etc.) and the normal component of the flux must both be continuous across the boundary. From figure 9.2.12 it follows that continuity in pressure means a discontinuity in water content.

Of special interest are several types of boundaries.

Type 1, a soil surface that serves as the upper boundary of a flow domain. The soil surface may be exposed to atmospheric pressure; however, the pressure in the water phase at this boundary is less than atmospheric and water cannot leave the flow domain unless its pressure rises above atmospheric (see discussion on end effect in par. 9.2.5). In paragraph 7.1.9 this type of boundary—referred to as a capillary exposed surface—was presented as the upper boundary of an assumed saturated capillary fringe bounding a saturated flow domain. When water ponds above a soil surface, the pressure along the bottom of the pond (which in this case is the upper boundary of the flow domain) depends on the depth of ponded water.

Type 2, a steady or unsteady phreatic surface that serves as a lower boundary of an unsaturated flow domain. In the capillary fringe approximation, the upper boundary of the capillary fringe is the lower boundary of the unsaturated flow domain. Along such boundaries we specify either $c = n$ (or $S_w = 1$) or the pressure $p = 0$ (or $-p = p_c$ at the upper boundary of the capillary fringe).

When moisture is introduced into a dry soil, or into a soil at the irreducible water content c_0, the moisture advances behind the wetting front (line or surface). This advancing wetting front serves as a boundary to the region behind the front where unsteady unsaturated flow takes place. This is analogous to the case of an advancing free surface, except that there we assume that behind the advancing front we have saturated flow. As in dry soil (or soil at $c = c_0$), the permeability is zero; in order to have a finite flux, we must have an infinite ∇c (or $\nabla \psi$, ∇p, etc.), i.e., a step-like change in saturation (and hence in pressure, capillary head, etc.) at the advancing front.

As the front is an iso-c surface (and hence also iso-p, iso-ψ, etc.) we may use the discussion in paragraph 7.1.4. For example, we may start from (7.1.34) rewritten as:

$$\frac{\partial \psi}{\partial t} + \frac{1}{c - c_0} \mathbf{q} \cdot \nabla \psi = 0. \tag{9.4.63}$$

Since

$$\frac{\partial \psi}{\partial t} = \frac{\partial c}{\partial t} \frac{d\psi}{dc}; \qquad \nabla \psi = \frac{d\psi}{dc} \nabla c,$$

we obtain:

$$\frac{\partial c}{\partial t} + \frac{1}{c - c_0} \mathbf{q} \cdot \nabla c = 0. \tag{9.4.64}$$

Expressing $\mathbf{q}$ by (9.4.26), we obtain the boundary condition at the advancing front in the form:

$$\frac{\partial c}{\partial t} - \frac{1}{c - c_0} \left[D(c)(\nabla c)^2 + K(c) \frac{\partial c}{\partial z} \right] = 0. \tag{9.4.65}$$

Appropriate boundary conditions must also be specified for the ordinary continuity differential equation that is obtained after applying the Boltzmann transformation to the partial differential equation of continuity. For example, for a semi-infinite domain ($x > 0$), the boundary conditions for (9.4.61) may take the form:

$$\begin{aligned} &(t=0, \quad x>0, \quad c=0), \qquad &&\lambda=\infty, \quad c=0 \\ &(t>0, \quad x=0, \quad c=c_0), \qquad &&\lambda=0, \quad c=c_0 \\ &(t>0, \quad x=\infty, \ c=0), \qquad &&\lambda=\infty, \quad c=0. \end{aligned} \tag{9.4.66}$$

In addition to the Boltzmann transformation, which transforms the partial differential equation of one-dimensional horizontal flow into an ordinary one, another transformation (Kirchhoff 1894, Crank 1956; Carslaw and Jaeger 1959) is often employed when the coefficient in the equation is a function of the dependent variable. To demonstrate this transformation we start from (9.4.26) and (9.4.47):

$$\mathbf{q} = -D(c)\,\nabla c - K(c)\mathbf{1z} \tag{9.4.67}$$

$$\partial c/\partial t = \operatorname{div}[D(c)\operatorname{grad} c] + \partial K(c)/\partial z \tag{9.4.68}$$

and introduce a new variable F such that:

$$D_0\,\nabla F = D(c)\,\nabla c \tag{9.4.69}$$

which is equivalent to:

$$D_0 F = \int_{c'=c_0}^{c} D(c')\,dc', \qquad D_0\,dF = D(c)\,dc \tag{9.4.70}$$

where D_0 is any value $D(c)$ and $F(c_0) = 0$. From (9.4.69) it follows that F is essentially a potential whose gradient is proportional to the flux. It is called the *diffusivity potential*. By inserting (9.4.69) into (9.4.68) and (9.4.67) we obtain:

$$\frac{D_0}{D(F)}\frac{\partial F}{\partial t} = D_0\nabla^2 F + \frac{dK(F)}{dF}\frac{\partial F}{\partial z} \tag{9.4.71}$$

$$\mathbf{q} = -D_0\,\nabla F - K(F)\mathbf{1z}. \tag{9.4.72}$$

For steady flow, $\partial F/\partial t = 0$ and (9.4.71) reduces to:

$$D_0\,\nabla^2 F + [dK(F)/dF]\,\partial F/\partial z = 0. \tag{9.4.73}$$

In the absence of the gravity term (i.e., $\partial F/\partial z = 0$) we can see clearly the simplification introduced by using the new variable F. For example (9.4.71) reduces to the Laplace equation in F. Obviously, for this transformation to be of any use, boundary and initial conditions must be expressible in terms of F. Owing to hysteresis, F is not a single-valued function of c.

A stream function Ψ can also be defined for unsaturated two-dimensional flow in a manner similar to its definition in saturated flow (par. 6.5.2). Thus, the determination of a network consisting of equipotentials $\varphi =$ const and/or streamlines $\Psi =$

const constitutes a complete solution of an unsaturated flow problem. The following conditions exist:

$$q_x = -\frac{\partial \Psi}{\partial y} = -K\frac{\partial \varphi}{\partial x}; \qquad q_y = \frac{\partial \Psi}{\partial x} = -K\frac{\partial \varphi}{\partial y} \tag{9.4.74}$$

where we recall that K is a function of the pressure (or capillary head). For a homogeneous isotropic medium, we may replace the effective hydraulic conductivity K by $K = K_0 K_r$, where K_0 is constant hydraulic conductivity at saturation and K_r is relative hydraulic conductivity. We may then define a potential function $\Phi = K_0 \varphi$ so that (9.4.74) will become:

$$\frac{\partial \Psi}{\partial y} = K_r \frac{\partial \Phi}{\partial x} (= -q_x); \qquad \frac{\partial \Psi}{\partial x} = K_r \frac{\partial \Phi}{\partial y} (= q_y). \tag{9.4.75}$$

The families of curves $\Phi = \text{const}$ and $\Psi = \text{const}$ are orthogonal to each other.

As in the case of saturated flow (par. 6.5.6), for certain boundary conditions it is more convenient to solve two-dimensional saturated flow problems with Φ and Ψ as dependent variables, rather than x and y. To do that, we first employ the rules of change of variables of implicit functions and obtain from (9.4.75):

$$\frac{\partial y}{\partial \Psi} = \frac{1}{K_r}\frac{\partial x}{\partial \Phi}; \qquad \frac{\partial y}{\partial \Phi} = -K_r \frac{\partial x}{\partial \Psi}. \tag{9.4.76}$$

One should note that, as in saturated flow (par. 6.5.2), from (9.4.74) we obtain by differentiation div $\mathbf{q} = 0$. It follows from (9.4.37) that the stream function Ψ in this case is valid only when $\partial c/\partial t = 0$, i.e., in steady flow.

From (9.4.76) we obtain:

$$\frac{\partial}{\partial \Phi}\left(\frac{1}{K_r}\frac{\partial x}{\partial \Phi}\right) + \frac{\partial}{\partial \Psi}\left(K_r \frac{\partial x}{\partial \Psi}\right) = 0. \tag{9.4.77}$$

In a similar way, equations with y as the dependent variable may be derived.

9.4.7 Methods of Solution and Examples

As in the general case of immiscible displacement, the major difficulty in solving any problem of unsaturated flow stems from the fact that the various medium coefficients involved (e.g., $k = k(S)$, $D = D(c)$, etc.) are available only in the form of experimental results and not in the form of analytical expressions. Moreover, all these expressions are subject to hysteresis, although the one least susceptible to hysteresis is $K(c)$, whereas $K(\psi)$ exhibits the strongest influence of hysteresis. These considerations virtually preclude exact analytical solutions of unsaturated flow problems, leaving possible only numerical and graphic methods. In addition, because all these coefficients are functions of the dependent variable (whether c, p or ψ), the partial differential equations are nonlinear, a fact that further complicates the problem.

Irmay (1966) presents a large number of analytic solutions to the nonlinear diffusion equation with a gravity term, describing unsaturated flow. Solutions are also presented by Klute (1952), Gardner (1958), Philip (1955, 1957a,b,c,d,e,f,g, 1958a,b, 1960a,b, 1968a), Youngs (1957) and Brutsaert (1968), among many others. To obtain analytic solutions, all these authors introduce analytic expressions of one form or another for $K(c)$ or $D(c)$.

Several methods of solutions and examples are presented below, demonstrating these difficulties. In all cases the medium is nondeformable.

Case 1, steady linear flow in the direction **1s**, *making an angle* α *with the* $+\ x$ *axis* (Irmay 1966; Zaslavsky, in Bear, Zaslavsky and Irmay 1968). We may start from any of the motion equations and corresponding continuity equations written for the direction **1s** alone. For example, from (9.4.23) written in the form:

$$q_s = K(\psi) \cdot [d\psi/ds - \sin\alpha]; \qquad \sin\alpha = dz/ds \tag{9.4.78}$$

we obtain:

$$\int_{s=s_1(\psi_1)}^{s=s_2(\psi_2)} ds = s_2 - s_1 = \int_{\psi_1}^{\psi_2} \frac{d\psi}{q_s/K(\psi) + \sin\alpha}. \tag{9.4.79}$$

If $K = K(\psi)$ is known (from experiments or as an assumed analytical expression) the integration in (9.4.79) can be performed numerically for various constant values of q_s. From (9.4.22) it follows for $(s_2 - s_1) > 0$, $q_s \geqslant 0$, $\varphi_2 \leqslant \varphi_1$, i.e.,

$$(z_2 - \psi_2) \leqslant z_1 - \psi_1 \quad \text{or} \quad (z_2 - z_1) \leqslant \psi_2 - \psi_1.$$

As $(z_2 - z_1) = (s_2 - s_1)\sin\alpha$, we obtain in general for $\sin\alpha \neq 0$:

$$(s_2 - s_1) \leqslant \frac{\psi_2 - \psi_1}{\sin\alpha}. \tag{9.4.80}$$

For the case $q_s/K(\psi) \ll \sin\alpha$, we obtain from (9.4.79) $z_2 - z_1 = \psi_2 - \psi_1$, i.e., $\varphi_2 \cong \varphi_1$; the moisture distribution approaches that of hydrostatic equilibrium.

As $(s_2 - s_1) \to \infty$, $[q_s/K(\psi) + \sin\alpha] \to 0$. Under these conditions $q_s \to -K(\psi)\sin\alpha$.

For *horizontal flow* with $s_1 = x_1$, $\psi = \psi_1$; $s_2 = x_2$, $\sin\alpha = 0$, equation (9.4.79) becomes:

$$q_x(x_2 - x_1) = \int_{\psi_1}^{\psi_2} K(\psi)\, d\psi = \int_{\psi_1}^{\psi_2} \left[K(\psi)\frac{d\psi}{dc}\right] dc = -\int_{c_1}^{c_2} D(c)\, dc \tag{9.4.81}$$

which can be integrated if $D = D(c)$ is known. From (9.4.81) we also have:

$$q_x = D_0 \frac{F(c_1) - F(c_2)}{x_2 - x_1} \tag{9.4.82}$$

where $F = F(c)$ is the known function of the water content defined by (9.4.69). For $x_2 > x_1$, we must have $F(c_1) > F(c_2)$, i.e., $c_1 > c_2$. If in a horizontal column of length L we have a water boundary with $c = c_1 = n$ at $x = x_1 = 0$ and an outlet into air at $x = x_2 = L$, where $c = c_0$, we obtain:

$$q = \frac{D_0}{L} [F(n) - F(c_0)]. \tag{9.4.83}$$

From (9.4.81) it follows that for $q_x > 0$, $x_2 - x_1 > 0$ and $\psi_2 > \psi_1$ (i.e., $p_2 < p_1$, $c_2 < c_1$), as $(x_2 - x_1)$ increases, maintaining ψ_2 and ψ_1 unchanged, q decreases.

For *vertically upward* flow, (e.g., movement from the water table toward the soil surface), $\sin \alpha = 1$ and $s \to z$. We obtain from (9.4.79):

$$z_2 - z_1 = \int_{\psi_1}^{\psi_2} \frac{d\psi}{q_z/K(\psi) + 1}; \qquad q_s \equiv q_z > 0; \qquad z_2 > z_1 > 0. \tag{9.4.84}$$

From (9.4.26) we obtain:

$$0 < q_z = -D(c)\frac{dc}{dz} - K(c); \qquad (z_2 - z_1)q_z = -\int_{c_1}^{c_2} \frac{D(c)\, dc}{1 + K(c)/q_z}. \tag{9.4.85}$$

When the downward flow is toward a stationary water table where $c = c_{sat}$, $z = 0$, $p = 0$, we obtain from (9.4.26):

$$q_{-z} = D\, dc/dz + K; \qquad D\, dc/dz = q_{-z} - K; \qquad dz = D\, dc/(q_{-z} - K)$$

$$z = \int_0^z dz' = \int_{c_{sat}}^{c} \frac{D(c')\, dc'}{q_{-z} - K(c')} \tag{9.4.86}$$

where q_{-z} is positive downward. From (9.4.86) it follows that the nature of the relationships $K(c)$ and $D(c)$ will cause z to tend rapidly to infinity as the moisture content approaches the value at which $K = q_{-z}$. This means that beyond a certain height above the water table, the moisture profile tends asymptotically to a uniform moisture content at which K equals the rate at which water is infiltrating into the soil at the ground surface. This is also obvious directly from the motion equation in (9.4.86): $q_{-z} = K$ for $dc/dz = 0$.

Figure 9.4.16 shows schematic curves of $\psi = \psi(z)$ and $c = c(z)$ obtained by integrating (9.4.84) and (9.4.86). The exact shape depends on the chosen functions $K = K(\psi)$ or $K = K(c)$. For $q_z = 0$ we obtain a hydrostatic distribution with $\psi = z$. For downward flow q_{-z} is negative.

Case 2, steady downward flow (infiltration) through a layered soil (Zaslavsky 1964a,b, and in Bear, Zaslavsky and Irmay 1968). Figure 9.4.17 shows water ponded above a two-layered soil ($K_2 > K_1$). In saturated flow (sec. 5.8) through the two-layered soil, the specific (downward) discharge is, by (5.8.7):

$$q = \bar{K}\frac{\Delta\varphi}{L} = (H_0 + D_1 + d_2)/(D_1/K_1 + d_2/K_2). \tag{9.4.87}$$

The loss of head $\Delta\varphi$ in the lower layer is qd_2/K_2. In order to obtain a zone of zero or negative pressure we must have:

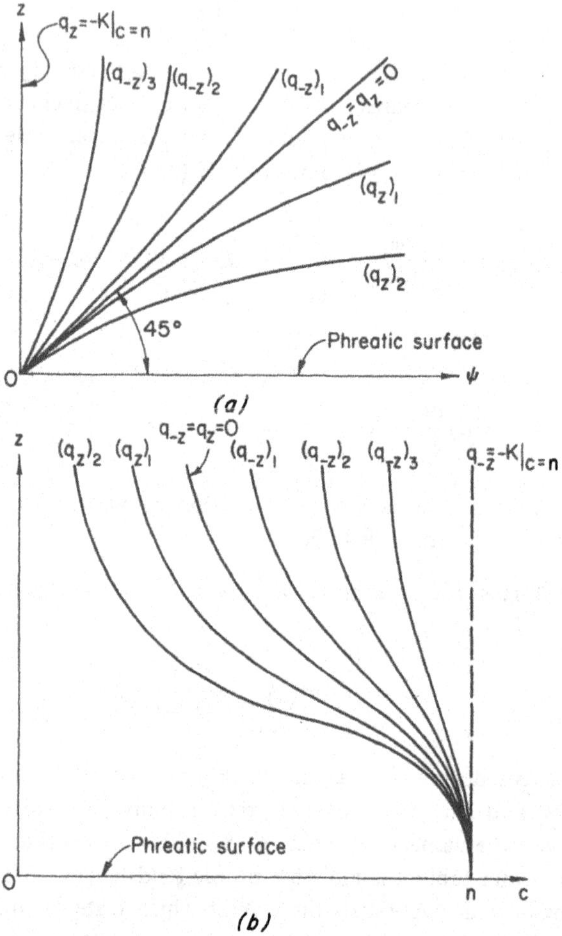

FIG. 9.4.16. Schematic capillary pressure head (a) and water content (b) distributions in steady vertically downward (q_{-z}) and upward (q_z) unsaturated flows.

$$qd_2/K_2 \leqslant d_2 \quad \text{or} \quad q \leqslant K_2. \tag{9.4.88}$$

When combined with (9.4.87), equation (9.4.88) becomes:

$$H_0 \leqslant D_1(K_2/K_1 - 1) \tag{9.4.89}$$

where the equal sign corresponds to zero (atmospheric) pressure throughout d_2. When this situation occurs and air cannot enter, flow may remain saturated under small negative pressures (fig. 9.4.17a). However, as the negative pressure increases and the *air entry value* (par. 9.4.1) is exceeded (by the difference between atmospheric pressure and the pressure in the soil), air will enter and the flow in the lower layer will become unsaturated. Actually, unsaturated flow will take place between points A and B of figure 9.4.17b. As we have a passage from a positive pressure at the top

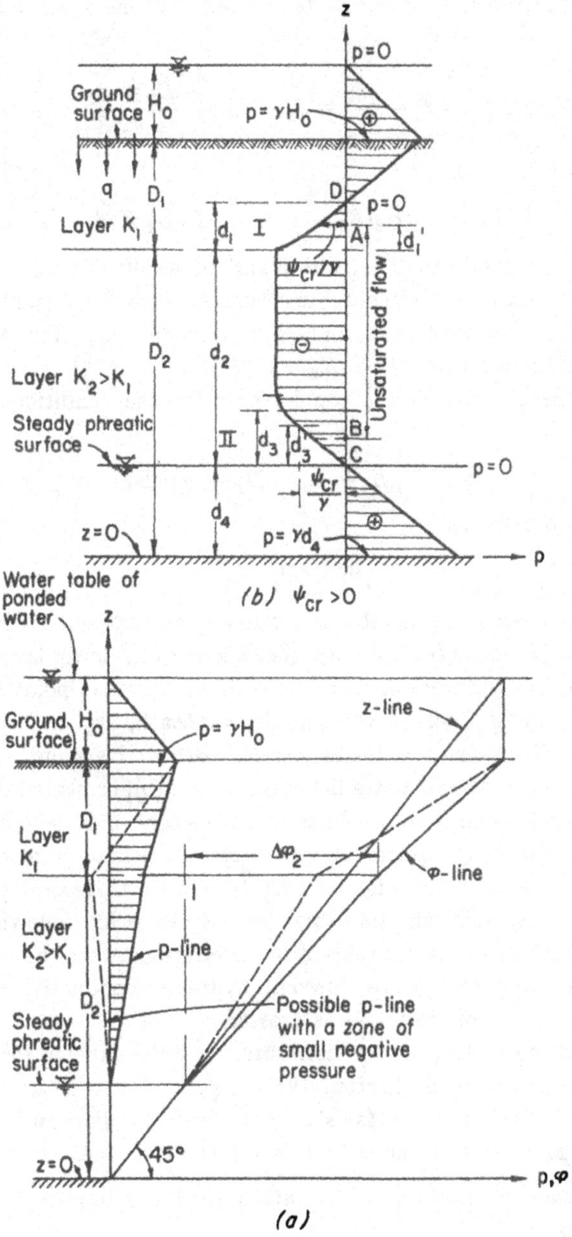

FIG. 9.4.17. Pressure distributions in steady saturated (a) and unsaturated (b) flow in a layered soil (after Zaslavsky, 1964a,b).

of the upper layer, to a negative pressure at the interface between the two layers, a zone of unsaturated flow with negative pressure will develop in the lower part of the upper, less pervious, layer (fig. 9.4.17a). We may modify condition (9.4.89) by requiring that the air entry value (expressed in terms of $\psi_{cr} = p_{cc}/\gamma_w$) will be exceeded:

$$d_2 - \Delta\varphi_2 \equiv d_2 - \frac{qd_2}{K_2} \equiv d_2 - \frac{H_0 + D_1 + d_2}{(D_1/K_1 + d_2/K_2)} \cdot \frac{d_2}{K_2} > \psi_{cr}$$

leading to the condition:

$$H_0 < D_1(K_2/K_1 - 1) - \psi_{cr}(K_2D_1/K_1d_2 + 1). \tag{9.4.90}$$

For example, for dune sand, ψ_{cr} was found to be 40–60 cm.

Under the conditions leading to unsaturated flow, two transition zones exist (fig. 9.4.17b): I of thickness d_1 and II of thickness d_3. Throughout the region of unsaturated flow $K_1 = K_1(\psi)$, $K_2 = K_2(\psi)$.

The discharge q_1 through the top layer under the conditions shown in figure 9.4.17b is:

$$q_1 = K_1(H_0 + \psi_{cr} + D_1 - d_1)/(D_1 - d'_1) \tag{9.4.91}$$

and for $\psi_{cr} = 0$ $(d'_1 \to d_1)$:

$$q_1 = K_1(H_0 + D_1 - d_1)/(D_1 - d_1). \tag{9.4.92}$$

When (a) d_2 is very large, (b) the flow is steady and the soil is wetted from the top by infiltration, and (c) a point X exists somewhere in the lower layer where the water content is less than saturation, there must be at least one point along the column between C and D (fig. 9.4.17b) where $d\psi/dz = 0$ (Zaslavsky, in Bear, Zaslavsky and Irmay 1968). There $d\varphi/dz = 1$ and $q = K$. Below this point we must either be near a wetting front, which contradicts assumption (b), or near the phreatic surface, which contradicts assumption (a). Thus we have a zone of downward flow at constant pressure and $\nabla\varphi = 1$, except in the region of thickness d_3 immediately above the phreatic surface, as shown in figure 9.4.17b. Above this point q cannot increase or diminish because of assumption (b). One should recall, following the discussion in paragraph 9.2.5, that on both sides of the interface between the two layers we have different saturations, and hence different hydraulic conductivities; therefore $\nabla\varphi$ changes abruptly as this interface is crossed.

Figure 9.4.18 shows the pressure distribution in a five-layered soil profile assuming zero critical capillary head (Partom 1967).

Figure 9.4.19 (Zaslavsky 1964a) shows the pressure distribution at the passage from a more permeable medium to a less permeable one.

Case 3, unsteady movement in a horizontal column $x \geqslant 0$. The continuity equation for this case is:

$$\frac{\partial c}{\partial t} = \frac{\partial}{\partial x}\left[D(c)\,\frac{\partial c}{\partial x}\right]. \tag{9.4.93}$$

When the column is initially at a moisture content c_i, and subsequently has the plane $x = 0$ maintained at $c = c_n$, we have:

$$t \leqslant 0, \quad x > 0, \quad c = c_i; \qquad t > 0, \quad x = 0, \quad c = c_n. \tag{9.4.94}$$

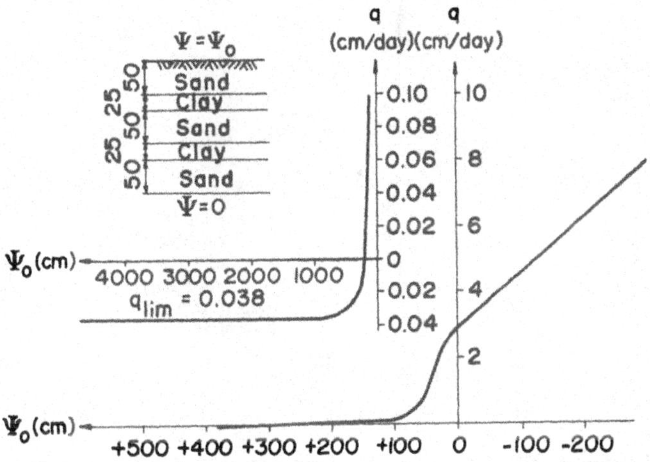

(a) Specific discharge q as a function of the suction head Ψ_0

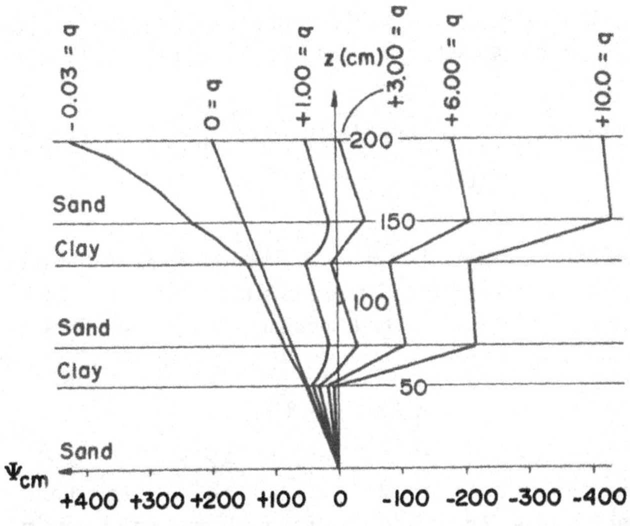

(b) Suction head profiles for different values of the specific discharge q (cm/day)

FIG. 9.4.18. Pressure distribution in steady unsaturated flow with zero critical capillary head, in a five layered soil (Partom, 1967, Zaslavsky, 1964a).

For $c_n > c_i$, equations (9.4.93) and (9.4.94) describe the wetting of the column, either under tension, or for $c_n = c_{\text{saturation}} = n$ (water content at zero suction), by free water being introduced at $x = 0$. For $c_i > c_n$ the same equations describe the removal of water by the application of a constant suction at $x = 0$.

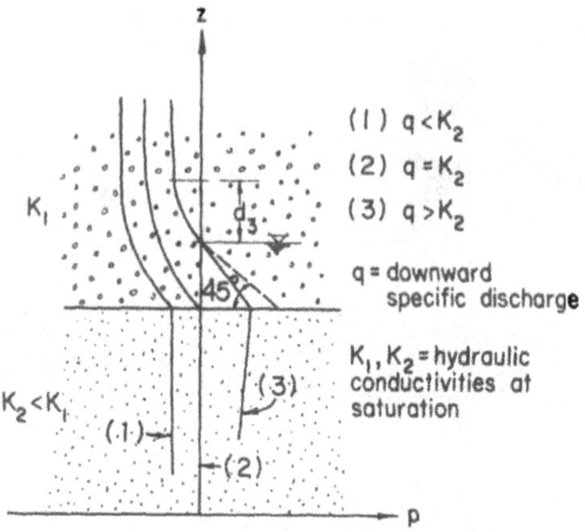

FIG. 9.4.19. Pressure distribution at the passage from a more pervious layer to a less pervious one (Zaslavsky, 1964a).

Klute (1952) uses a numerical method suggested by Crank and Henry to solve this problem. Crank (1956, p. 166), following the works of Hiroshi Fujita, presents formal solutions of (9.4.93) satisfying (9.4.94) with $c_i = 0$, for three specific expressions of $D = D(c)$:

$$\text{(a)}\quad D = D_0/(1 - \alpha c); \qquad \text{(b)}\quad D = D_0/(1 - \alpha c)^2; \qquad \text{(c)}\quad D = D_0/(1 + 2ac + bc^2) \tag{9.4.95}$$

where α, a and b are constants.

Another method of solution in which convergence is rapid is suggested by Philip (1955). When the Boltzmann transformation (1894, pars. 6.4.5 and 9.4.6) is applied to (9.4.93), we obtain (9.4.61). By integrating (9.4.61) we obtain:

$$\int_{c_i}^{c_n} \lambda\, dc = -2D\, dc/d\lambda \tag{9.4.96}$$

subject to the condition $c = c_n$ for $\lambda = 0$. Philip (1955) then obtains a solution of (9.4.96) by means of a forward integration with one initial condition determined by trial and improved by iteration. The solution as $c \to c_n$ is $x \to \infty$.

Philip (1960a,b) presents a very general method of obtaining exact solutions of (9.4.93), subject to (9.4.94), written in the form:

$$t \leqslant 0, \quad x > 0, \quad c = 0; \qquad t > 0, \quad x = 0, \quad c = 1. \tag{9.4.97}$$

It embraces all possible exact solutions, including as a special case those of Fujita and those implicit in Philip's (1955) work.

Brutsaert (1968a), commenting that none of the functions $\lambda = \lambda(c)$ and the corresponding $D = D(c)$ seem to be directly applicable to infiltration in soils, adapts one such solution to the problem of horizontal unsaturated flow with the boundary condition (9.4.94) with $c_n = n$. This solution is:

$$\lambda = 1 - c_*^m; \qquad m > 0; \qquad c_* = (c - c_i)/(n - c_i) \tag{9.4.98}$$

corresponding to the diffusivity:

$$D = mc_*^m [1 - c_*^m/(1 + m)]/2. \tag{9.4.99}$$

In these equations c_* is the *normalized moisture content.* When c_i is the irreducible moisture content $c_0 = nS_0 \equiv nS_{wo}$ (par. 9.2.3) c_* becomes $c_e = (c - c_0)/(n - c_0)$, which is the *effective, or reduced moisture content* (corresponding to the effective saturation $S_e = c_e/n$ defined in par. 9.2.3). With this expression, Brutsaert solves (9.4.40) for horizontal flow.

Yeh and Franzini (1968) also solve the problem of unsteady movement in a horizontal column. In their case, moisture is moving into a semi-infinite horizontal column of homogeneous soil initially at a moisture content c_0, under the influence of a certain applied pressure at one end.

Power series solutions of (9.4.93) for exponential and linear diffusivity functions are presented by Scott et al. (1962).

Case 4, unsteady movement in the vertical direction. This is the case of downward *infiltration* or upward evaporation, say, from a water table. The governing equation is:

$$\frac{\partial c}{\partial t} = \frac{\partial}{\partial z}\left[D(c)\frac{\partial c}{\partial z}\right] + \frac{\partial K(c)}{\partial z} \tag{9.4.100}$$

(with $+ z$ vertically upward). Boundary conditions depend on the particular problem being studied. For example:

$$t \leqslant 0, \quad z > 0, \quad c = c_i; \qquad t > 0, \quad z = 0, \quad c = c_n. \tag{9.4.101}$$

For $c_n > c_i$, the conditions (9.4.101) describe "capillary rise" with water being supplied at the base of the column, either under constant tension or, when $c_n = n$, as free water available in excess (a water table boundary). For $c_n < c_i$ the phenomenon described is drainage of the column $z > 0$ caused by applying a constant suction at its base. Another example of boundary condition is for the semi-infinite column $z \leqslant 0$:

$$t \leqslant 0, \quad z < 0, \quad c = c_i; \qquad t > 0, \quad z = 0, \quad c = c_n \tag{9.4.102}$$

which for $c_n > c_i$ describes *infiltration* (entry of water into a soil profile either directly from rainfall or from ponded water) into a vertical column. When at the surface $z = 0$ we have $c_n = n$ (i.e., saturation) we have the condition of free water being available in excess. The case of $c_i > c_n$ represents the removal of water from the column by applying a constant suction at the surface (case of evaporation). The phenomenon of infiltration has been the subject of many mathematical and experimental investigations (e.g., Miller and Richard 1952; Bruce and Klute 1956; Gardner and Mayhugh 1958; Hanks and Bower 1962; Nielsen and Bigger 1962; Rawlins and Gardner 1963; Ferguson et al. 1963; Whisler and Klute 1966 and

Ali Ibrahim and Brutsaert 1968). Philip, in a series of papers (1957–1958) applies a series method to the solution of this problem. He employs a power series in $t^{1/2}$:

$$z' = \lambda(c)t^{1/2} + \chi(c)t + \psi(c)t^{3/2} + \omega(c)t^2 + \cdots + f_m(c)t^{m/2} \qquad (9.4.103)$$

where the coefficients $|\lambda| \gg |\chi| \gg |\psi| \cdots$ are solutions of a series of ordinary differential equations that can be solved rapidly by simple numerical methods (Philip 1955).

From the theory developed by Philip it is also possible to determine cumulative infiltration and the rate of infiltration, and to study the effect of infiltration and initial moisture on them.

Rubin (1966) solves the problem of infiltration from ponded water by numerical analysis. He assumes that vertical infiltration of rainwater and its subsequent downward movement is governed by:

$$\frac{\partial c}{\partial t} = \frac{\partial}{\partial z}\left[K(H)\frac{\partial H}{\partial z}\right] + \frac{\partial K(H)}{\partial z}, \qquad z \leqslant 0 \qquad (9.4.104)$$

where $H = p/\gamma$ is the pressure head; or, with z' replacing $-z$:

$$\frac{\partial c}{\partial t} = \frac{\partial}{\partial z'}\left[K(H)\frac{\partial H}{\partial z'}\right] - \frac{\partial K(H)}{\partial z'}, \qquad z' \geqslant 0. \qquad (9.4.105)$$

Initial and boundary conditions are:

$z' \geqslant 0,\ t = 0,\ H = H_i$ (initial soil moisture pressure head)

$z' = \infty,\ t \geqslant 0,\ H = H_i$

$z' = 0,\ t_p \geqslant t > 0,\ R = -K(H)\dfrac{\partial}{\partial z'}(H - z') = -K(H)\dfrac{\partial H}{\partial z'} + K(H)$

$z' = 0,\ t > t_p,\ H = H_u$ (nonnegative pressure head during rainpond infiltration)

(9.4.106)

where R is infiltration rate, H_i is taken so low that, for all practical purposes, $K(H_i) = 0$, and t_p is the occurrence time of incipient ponding. Rubin also takes $H_u = 0$.

Equation (9.4.105) and the boundary conditions include two dependent variables c and H. In order to change the system into one containing only a single variable, Rubin, following Crank (1956) and Carslaw and Jaeger (1959), introduces a new variable $s(H)$:

$$s = s(H) = \frac{1}{M}\int_{H_{\max}}^{H} K(H')\,dH'; \qquad M = \int_{H_{\max}}^{H_i} K(H')\,dH' \qquad (9.4.107)$$

where $H_{\max}$ and H_i, respectively, are upper and lower bounds of H in the porous medium under consideration. Some results of Rubin's analysis are shown in figure 9.4.20. Liakopoulos (1966) also presents a theoretical approach to the solution of the one-dimensional infiltration problem. He presents solutions obtained by numerical solution of the partial differential equation (9.4.45).

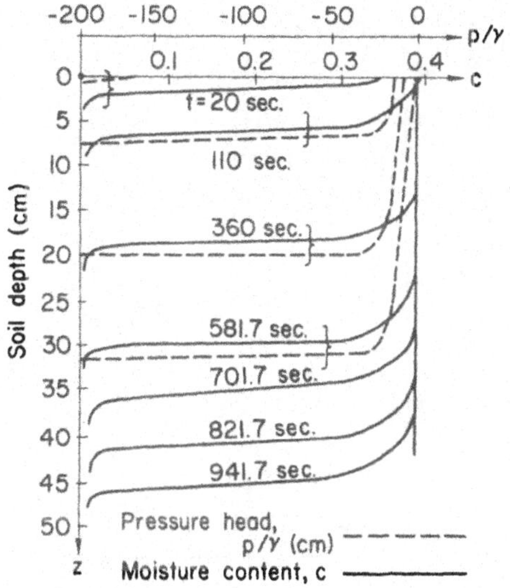

FIG. 9.4.20. Soil moisture content and pressure head profiles during rainpond infiltration. Soil is initially air-dry Rehovot sand (Rubin, 1966).

Some additional comments on infiltration and redistribution of moisture in a vertical column are given in paragraph 9.4.8.

The problem of unsaturated, two-dimensional flow in the vertical *xz*-plane has been studied by various investigators, mainly in connection with artificial recharge operations by means of infiltration basins or ponds. Among others, we may mention the works of Nelson (1962), Reisenauer (1962) and Bouwer (1964).

Philip (1966) treats the problem of absorption and infiltration in two- and three-dimensional flow systems with radial symmetry. He presents solutions for absorption (i.e., without gravity effects) and infiltration from circular and spherical cavities.

9.4.8 Additional Comments on Infiltration and Redistribution of Moisture

When water is applied in excess at the top of a soil column, the water will enter into the soil at a rate depending on the moisture content in the column and on the physical properties of the soil. Figure 9.4.21 shows how water advances downward in a soil column. In the first stages the moisture profile gradually changes, but later it maintains almost a constant form moving downward. In a column of uniform soil that is initially at a constant moisture content, the downward movement of this profile is at a constant velocity.

In their classical works on infiltration, Bodman and Colman (1943, and Colman and Bodman 1944) distinguish a number of zones during infiltration in a soil column (fig. 9.4.21):

(a) *a saturated zone* reaching a depth of approximately 1.5 cm;

(b) *a transition zone,* extending to a depth of about 5 cm, in which moisture content decreases rapidly with depth;

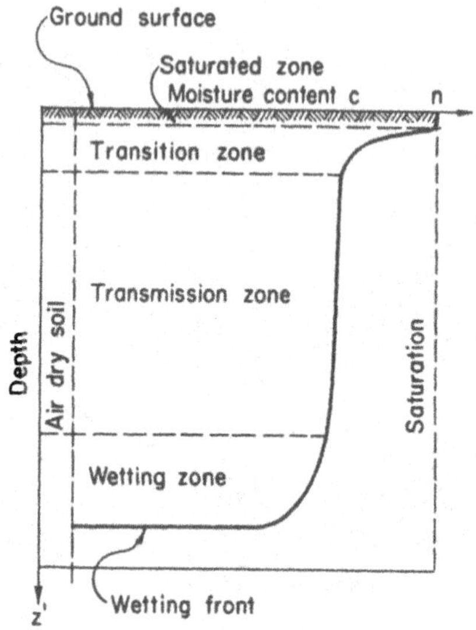

FIG. 9.4.21. Moisture zones during infiltration (Bodman and Colman, 1943).

(c) *the transmission zone* which increases in length as infiltration proceeds: throughout this zone c is approximately constant;

(d) *the wetting zone* with, again, a rather rapid drop in c;

(e) *the wetting front* (or *wet front*), which is a zone of a very steep moisture gradient. It represents the visible limit of moisture penetration into the soil.

Evidently, water is supplied to the wetting front from the saturated zone across the transmission zone in which the moisture content, and hence the hydraulic conductivity, may be assumed to be constant. Making use of this assumption, Van Duin (1955, 1956) derived an equation for the advance of the wetting front. Ignoring the water depth over the soil surface, and assuming that flow is caused only by capillary forces and gravity, the advance of a wetting front is given by:

$$\frac{dz'}{dt} = \frac{K(c)}{\Delta c} \frac{z' + \psi}{z'} \tag{9.4.108}$$

where z' is depth of the wetting front and Δc ($c_{\text{final}} - c_{\text{initial}}$) is the storage capacity of the transmission zone. By integrating (9.4.108) with $z' = 0$ for $t = 0$, and assuming the suction ψ at the wetting front to be constant, we obtain:

$$t = \frac{\Delta c}{K(c)} \left[z' - \psi \ln \frac{z' + \psi}{\psi} \right]. \tag{9.4.109}$$

Assuming that c_{final} through the profile is constant, so that Δc is also constant and the cumulative infiltration is $I = z' \Delta c$, we obtain from (9.4.109):

$$I = Kt \bigg/ \left(1 - \frac{\psi}{z'} \ln \frac{z' + \psi}{\psi} \right). \tag{9.4.110}$$

Equation (9.4.109) holds also for $z' \gg \psi$. Hence, after a certain period of infiltration (which will be shorter the higher the initial moisture content of the soil) the infiltration rate $i = dI/dt$ will become constant.

Various investigators suggest other formulas for the description of infiltration. A review of some such formulas is given by Philip (1957g). For example, from (9.4.109), Van Duin (1955) suggests:

$$z = a(Kt/\Delta c)^{0.55}; \qquad 0 \leqslant z \leqslant 0.5\psi; \qquad i = a\Delta c(Kt/\Delta c)^{0.55} \tag{9.4.111}$$

where a is a constant depending on the physical properties of the soil.

Another example of an infiltration formula is (Philip 1957a):

$$i = At^{1/2} + Bt \tag{9.4.112}$$

where A is a function of the initial and saturated moisture contents, and B is a parameter related to the analysis developed in Philip's infiltration theory.

Horton (1939) suggests:

$$i = i(t) = i_f + (i_i - i_f)\exp(-At) \tag{9.4.113}$$

where i_i and i_f are the initial and final infiltration rates and A is a constant. This equation predicts an asymptotic approach to a constant infiltration rate i for large t.

If water is applied to the soil surface at a rate less than the ultimate rate of infiltration for the ponded water case, then infiltration will proceed indefinitely with no water being ponded on the soil surface. If, however, the applied flux exceeds this limit, ponding will occur after a finite time.

In continuous infiltration, we have wetting only everywhere along the column. It is therefore possible to obtain the diffusivity D from the appropriate (i.e., wetting) retention curve and the relative permeability curve. The same is true for the case of drainage only; in this case the drainage curve is to be used for determining D. However if infiltration stops at a certain period, there follows a period of redistribution of the soil moisture along the vertical column owing to gravity. During this redistribution, drainage will occur along part of the column, while wetting, or absorption, will take place along another part. Hence hysteresis phenomena play an important role in this case.

Several authors solve the problem of redistribution of soil moisture analytically. Among them we may mention Youngs (1958a,b), Staple (1966) and Rubin (1967).

9.4.9 Comments on Vapor Movement in Unsaturated Flow

At positive temperatures ($T^0 > 0°C$) the void space of a porous medium is occupied by liquid (or liquids), vapors and inert gases. At negative temperatures it may contain solidified liquids (e.g., ice), subcooled liquids, vapors and inert gases. In the present paragraph we shall present some brief comments on the vapor movement which at very low (water) saturations (low unsaturated hydraulic conductivities) may constitute a significant part of the total fluid flow.

A vapor in equilibrium with its liquid is said to be *saturated*, and the equilibrium pressure at which this occurs is called the *vapor pressure*. Vapor pressure is affected by several factors, such as temperature, pressure in the liquid phase and pressure in the gas phase (Stallman 1964).

The pressure of a vapor in equilibrium with its liquid depends largely upon the temperature. This dependence is described by the Clausius–Clapeyron equation. If it is assumed that the specific volume of the liquid v_w is negligible with respect to the specific volume of the vapor v_v, that the heat of vaporization of the liquid at constant temperature, λ, is constant over a wide range of temperatures, and that the vapor satisfies the equation of state of an ideal gas, the dependence of the vapor pressure p_v on the temperature T^0 may be written in the form (Fermi 1937):

$$p_v = C \exp(-\lambda M/RT^0) \tag{9.4.114}$$

where C is a constant, M is the molecular weight of the vapor, T^0 is the absolute temperature and R is the gas constant (4.62×10^6 ergs/degree gram for water vapor). Thus the vapor pressure increases with temperature. In general, since vapor pressure is rather sensitive to temperature changes, relatively small temperature gradients may produce appreciable vapor pressure gradients.

Edlefsen and Anderson (1943) suggest that p_v depends on the pressure in the liquid phase, p, according to the equation:

$$(dp_v/dp)|_{T^0=\text{const}} = v_w/v_v = \rho_v. \tag{9.4.115}$$

The fraction v_w/v_v is generally very small; hence the effect of pressure change in the liquid phase on vapor pressure change is very small. Edlefson and Anderson also discuss the influence of concentration of a dissolved solute in the liquid (solvent) on p_v.

Finally, vapor pressure over a concave water meniscus is less than the vapor pressure over a free flat water surface. Thus, a vapor pressure deficiency exists over a meniscus (say in a capillary tube or pore in a porous medium) when compared with the vapor-pressure over a free flat water surface. The relationship between the vapor pressure and the radius of curvature of the meniscus is described by the Kelvin equation (Thompson 1871, see detailed development by Stallman 1964):

$$\frac{2\sigma}{r^*} = -\frac{\rho_w p_{v0}}{\rho_{v0}} \ln \frac{p_v}{p_{v0}} + (p_v - p_{v0}) \tag{9.4.116}$$

where the nomenclature is the same as that used in (9.2.9), ρ_w and ρ_{v0} are the densities of the water and of the vapor (at $p = p_{v0}$), respectively. Usually $(p_v - p_{v0})$ is very small, so that Kelvin's equation becomes:

$$\frac{2\sigma}{r^*} = -\frac{\rho_w}{\rho_{v0}} p_{v0} \ln \frac{p_v}{p_{v0}}. \tag{9.4.117}$$

In (9.4.117) r^* is taken as positive if the meniscus is concave to the vapor. Hence it follows that for $r^* > 0$ the pressure over the meniscus is less than over a flat surface.

Stallman (1964) also shows that vapor pressure varies with elevation.

With this information we can now consider the movement of vapor in the soil interstices. Vapor transmission may be by ordinary molecular diffusion through the gas (usually air) present in the pore space, or by convection with the gas, should the latter be moving. As stated at the beginning of the present section, in unsaturated flow we usually assume that the air is stagnant, so that we have only the diffusion-type motion. The movement of the water vapor in soils under isothermal and essentially salt-free conditions was studied experimentally by Hanks (1958), Rose (1963a,b) and Jackson (1964a,b,c), among others. A detailed theoretical discussion is presented by Philip and de Vries (1957).

If the gas phase is immobile, the rate of vapor mass flux $\mathbf{J}_v$ (say, in gr/sec/cm² of porous medium) follows the simple Fickian concept of mass transport by molecular diffusion in a stagnant ambient fluid (par. 4.1.2):

$$\mathbf{J}_v = - n_a \underline{\mathbf{D}}^*_{dv} \nabla \rho_v \tag{9.4.118}$$

where $n_a \underline{\mathbf{D}}^*_{dv}$ is the coefficient of diffusion of water vapor into the soil (porous medium), $n_a = n - c$ is the air-filled porosity and ρ_v is the density of the water vapor (say, in gr/cm³). Following the discussion in paragraph 10.4.4, the coefficient $\underline{\mathbf{D}}^*_{dv}$ (components $(D^*_{dv})_{ij}$), a second-rank symmetrical tensor, is related to the medium's tortuosity $\underline{\mathbf{T}}^*$ (components T^*_{ij}) and to the coefficient of molecular diffusion of water vapor in a fluid continuum D_{dv} by:

$$n_a \underline{\mathbf{D}}^*_{dv} = n_a \underline{\mathbf{T}}^* D_{dv}; \qquad n_a (D^*_{dv})_{ij} = n_a T^*_{ij} D_{dv} \tag{9.4.119}$$

(where the symbol of average over the medium tortuosity is omitted). Sometimes $\underline{\mathbf{D}}^*_{dv}$ is referred to as the molecular diffusion in the porous medium. In this case the flux is related to a unit area of air-filled voids.

Some authors suggest replacing n_a in (9.4.119) with some function $f(n_a)$ of the aerated porosity. Philip and de Vries (1957) also multiply the right-hand side of (9.4.119) by a "mass flow factor," ν, introduced to allow for the mass flow of vapor arising from the difference in boundary conditions governing the air and vapor components of the diffusion system. This factor is close to unity at normal soil temperatures.

By combining (9.4.119) with (9.4.24), where $h_r = \rho_v/\rho_{v0}$ (ratio of density of water vapor to density of saturated water vapor), we obtain the mass flux:

$$\mathbf{J}_v = n_a \underline{\mathbf{D}}^*_{dv} \frac{\rho_v g}{R T^0} \nabla \psi. \tag{9.4.120}$$

The volumetric flux is $\mathbf{q}_v = \mathbf{J}_v/\rho_w$. In general, for a saturation $S_w > 15\%$, the vapor movement under isothermal conditions for these materials is negligible.

De Vries and Kruger (1966) discuss the value of the diffusion coefficient (D_{dv}) of water vapor in air.

Philip and de Vries (1957) show that the relative humidity h_r depends only on c and is practically independent of T^0, whereas $\rho_{0v} = \rho_{0v}(T^0)$. Hence:

$$\nabla \rho_v = h_r \nabla \rho_{v0} + \rho_{v0} \nabla h_r = h \frac{d\rho_{v0}}{dT^0} \nabla T^0 + \rho_{v0} \frac{dh_r}{dc} \nabla c \tag{9.4.121}$$

$$\mathbf{J}_v = -n_a \underline{\mathbf{D}}^*_{dv} \left\{ \left(h_r \frac{d\rho_{v0}}{dT^0} \right) \nabla T^0 - \frac{\rho_w g}{RT^0} \frac{\partial \psi}{\partial c} \nabla c \right\} \tag{9.4.122}$$

which may be written as:

$$\mathbf{q}_v = \mathbf{J}_v / \rho_w = -\underline{\mathbf{D}}_{Tv} \nabla T^0 - \underline{\mathbf{D}}_{cv} \nabla c \tag{9.4.123}$$

where ρ_w = density of liquid water. Thus the vapor flux, like the liquid water flux in nonisothermal flow (par. 9.4.4), has been separated into two components, one due to temperature gradient and another due to moisture gradient. Philip and de Vries (1957) analyze and evaluate in detail the *isothermal vapor diffusivity* $\underline{\mathbf{D}}_{cv}$ and the *thermal vapor diffusivity* $\underline{\mathbf{D}}_{Tv}$. They also present a detailed discussion of the effect of the interactions of vapor and liquid phases. Actually the introduction of the coefficient $\underline{\mathbf{D}}_{cv}$ is questionable since it is not uniquely defined because of the hysteresis in $\partial\psi/\partial c$.

By combining (9.4.29) for flux of the liquid phase with (9.4.123) for the flux of the vapor phase we obtain the total volumetric flux $\mathbf{q}_{\text{total}}$ (dim L^2T^{-1}):

$$\mathbf{q}_{\text{total}} = \mathbf{q} + \mathbf{q}_v = -\underline{\mathbf{D}}_{T\,\text{total}} \nabla T^0 - \underline{\mathbf{D}}_{c\,\text{total}} \nabla c - \underline{\mathbf{K}} \nabla z \tag{9.4.124}$$

where:

$$\underline{\mathbf{D}}_{T\,\text{total}} = \underline{\mathbf{D}}_T + \underline{\mathbf{D}}_{Tv} = \textit{thermal moisture diffusivity}$$

$$\underline{\mathbf{D}}_{c\,\text{total}} = \underline{\mathbf{D}}_c + \underline{\mathbf{D}}_{cv} = \textit{isothermal moisture diffusivity.}$$

Philip and de Vries discuss the relative magnitude of D_T, D_{Tv}, D_c and D_{cv} for fine- and coarse-textured soils.

A detailed analysis of nonisothermal flow of water liquid and vapor is given by Stallman (1964). He also considers the possible exchange of mass between the liquid phase and the gas phase caused by condensation of vapor from the gas phase, and evaporation of the liquid phase.

As should be readily apparent by now, the motion equation (e.g., in the form (9.4.124) in itself does not provide a complete description of the flow problem. It contains five variables (q_1, q_2, q_3, T, c) for the solution of which only three equations (embodied in 9.4.124) are provided. We need two more equations. These are the equations of conservation of moisture and of thermal energy. For isothermal conditions the equation of conservation of thermal energy is not needed, as ∇T^0 does not appear in the equation of motion.

The equation of moisture conservation takes the form:

$$\partial c/\partial t + \text{div}(\mathbf{q}_{\text{total}}) = 0. \tag{9.4.125}$$

The equation of conservation of thermal energy takes the form (Philip and de Vries 1957):

$$C_T \, \partial T^0/\partial t = \text{div}(\lambda \nabla T^0) - L \, \text{div}(\underline{\mathbf{D}}_{cv} \nabla c) \tag{9.4.126}$$

where C_T (in cal cm^{-3} °C^{-1}) is the volumetric heat capacity of the soil, λ (in cal sec^{-1}

cm^{-1}) is the thermal conductivity (including the thermal distillation effect), L is the latent heat of vaporization of water, and the second term on the right-hand side represents distillation effects induced by the moisture gradient. Equations (9.4.125) and (9.4.126) must be solved simultaneously for T^0 and c.

The general principles of heat and mass transfer are described in section 10.7. Only those aspects that are relevant to the problem of vapor transfer in unsaturated flow are presented here. In addition to the works of Philip and de Vries (1957), Jackson (1964a,b,c), Jackson et al. (1965) and Stallman (1964), mentioned previously, one can also find information on nonisothermal flow of water (liquid and moisture or vapor) in the works of Taylor and Cary (1960, 1964) and Cary and Taylor (1967), who treated this subject from the point of view of irreversible thermodynamics, and in the work of Luikov and Mikhailov (1961).

9.5 Immiscible Displacement with an Abrupt Interface

9.5.1 The Abrupt Interface Approximation

In many cases of practical interest, the transition zone between two immiscible or miscible liquids flowing simultaneously in a porous medium domain is narrow relative to the dimensions of the regions occupied by each liquid alone (or mainly by it). Under such conditions one may assume that each liquid is confined to a well defined portion of the flow domain, with an *abrupt interface* separating the two domains. An approximation of this type is often used in studying fluid injection programs for the recovery of oil from an oil reservoir. It is also used in investigations concerning the interface between fresh water and salt water in coastal aquifers (sec. 9.7), whenever the transition zone between these two *miscible liquids* is relatively narrow.

With this approximation in mind, let R denote some porous medium region occupied by two liquids, 1 and 2 (fig. 9.5.1). Let B_1 and B_2 denote the portions of the external boundary of R in contact with liquids 1 and 2, respectively. If the fluids are incompressible, a piezometric head can be defined within each subregion:

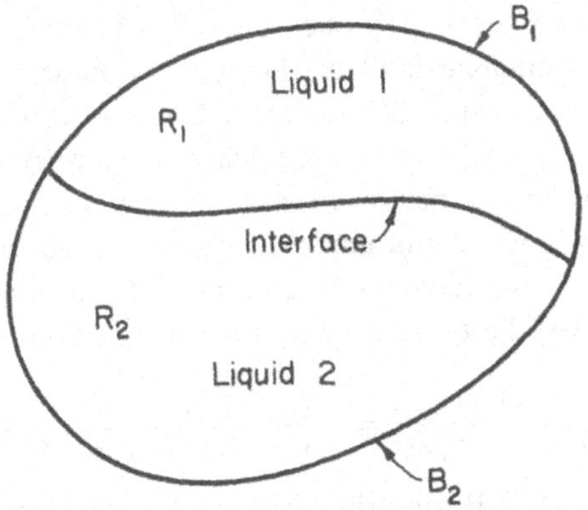

FIG. 9.5.1. An abrupt interface separating two immiscible liquids.

$$\varphi_1 = \varphi_1(x, y, z, t) = z + p/\gamma_1 \quad \text{in} \quad R_1$$
$$\varphi_2 = \varphi_2(x, y, z, t) = z + p/\gamma_2 \quad \text{in} \quad R_2. \tag{9.5.1}$$

The flow of each fluid within its own subregion is governed by gradients of its own piezometric head. When $\rho = \rho(p)$, we use Hubbert's potential (sec. 5.9). To solve a problem involving the flow of these two liquids means to determine the piezometric heads φ_1 in R_1 and φ_2 in R_2 (fig. 9.5.1) such that φ_1 and φ_2 each satisfies: (a) an appropriate continuity equation in its respective subregion, (b) initial conditions (in unsteady flow) in its respective subregion, and (c) boundary conditions on its respective portion of the external boundary. In addition, φ_1 and φ_2 must satisfy (simultaneously) boundary conditions on their common boundary—the interface.

For example, if the two liquids are incompressible and the medium is homogeneous, isotropic and nondeformable, the continuity equation to be satisfied by each liquid is Laplace's equation, and we have:

$$\nabla^2\varphi_1 = 0 \text{ in } R_1; \qquad \nabla^2\varphi_2 = 0 \text{ in } R_2. \tag{9.5.2}$$

The boundary conditions on B_1 and B_2 may be any of those encountered in a flow of a single fluid (sec. 7.1). The boundary conditions on the interface are derived in paragraph 9.5.2.

Assuming thus that the two fluids are completely separated by an abrupt interface, the velocity $\mathbf{V}$ of the interface at every point on it is related to the specific discharge $\mathbf{q}$ at that point by:

$$\mathbf{V} = \mathbf{q}/n$$

where n is porosity. This relationship can be written twice, once for each of the two fluids. However, continuity considerations require that the component of the velocity normal to the interface be the same in both regions:

$$V_n = V_{n1} = V_{n2}; \qquad q_n = q_{n1} = q_{n2}; \qquad V_n = \mathbf{V} \cdot \mathbf{1n}; \qquad q_n = \mathbf{q} \cdot \mathbf{1n} \tag{9.5.3}$$

where $\mathbf{1n}$ is the unit vector in the direction of the normal to the interface.

Sometimes, although the interface approximation is employed for two immiscible fluids, the separation is not a complete one. Instead, we have in each region a certain residual saturation of immobile fluid of the type essentially occupying the other region. In the case of an assumed interface between a wetting fluid (e.g., water) and a nonwetting one (e.g., oil) in a reservoir, where the former is injected to displace the latter, we actually have a narrow transition zone, so that on one side of the advancing interface we have oil and immobile water at irreducible water saturation S_{w0}, and on the other side we have water and immobile oil at residual oil saturation S_{nw0} (fig. 9.3.1). Under these conditions, the velocity (V_n) of the interface is:

$$V_n = \frac{(q_{nw})_n}{n(1 - S_{w0} - S_{nw0})} = \frac{(q_w)_n}{n(1 - S_{w0} - S_{nw0})}; \qquad q_w = q_{nw} \tag{9.5.4}$$

where $n(1 - S_{w0} - S_{nw0})$ indicates the portion of a cross-section through which flow can take place.

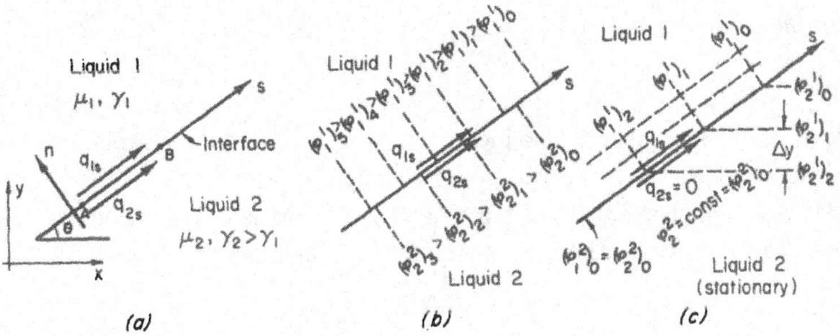

FIG. 9.5.2. Dynamic equilibrium conditions at a stationary interface.

On each side of a moving or stationary interface, only one fluid is present; its motion is governed by the gradient of its own piezometric head.

9.5.2 Piezometric Heads and Dynamic Equilibrium Conditions at a Stationary Interface

Let AB represent a small segment of a stationary interface completely separating two incompressible fluids ($\alpha = 1, 2$) in two-dimensional flow in a vertical xy-plane (fig. 9.5.2a).

The flow in each region is governed by Darcy's law:

$$\mathbf{q}_\alpha = -K_\alpha \nabla\varphi_\alpha = -(k\gamma_\alpha/\mu_\alpha)\nabla(y + p/\gamma_\alpha) \tag{9.5.5}$$

where K_α is the isotropic hydraulic conductivity of the α fluid. Since each fluid is confined to its own subregion and the interface is stationary, the interface acts as an impervious boundary to each of the two subregions; in each subregion, only components of the specific discharge that are tangent to the interface are possible. Neglecting capillary pressure at the interface, continuity of pressure requires that the pressure be the same when a point on the interface is approached from both sides. Hence, the component of the pressure gradient tangential to the interface must also be the same on both sides of the interface.

With the nomenclature of figure 9.5.2a, the tangential flow in the direction s in both regions is given by:

$$q_{1s} = -\frac{k\gamma_1}{\mu_1}\frac{\partial\varphi_1}{\partial s} = -\frac{k}{\mu_1}\left(\frac{\partial p}{\partial s} + \gamma_1\frac{\partial y}{\partial s}\right)$$

$$q_{2s} = -\frac{k\gamma_2}{\mu_2}\frac{\partial\varphi_2}{\partial s} = -\frac{k}{\mu_2}\left(\frac{\partial p}{\partial s} + \gamma_2\frac{\partial y}{\partial s}\right). \tag{9.5.6}$$

By eliminating $\partial p/\partial s$ from both equations of (9.5.6), we obtain:

$$\partial y/\partial s = \sin\theta = (q_{1s}\mu_1 - q_{2s}\mu_2)/k(\gamma_2 - \gamma_1) \tag{9.5.7}$$

where θ is the angle that the interface makes with the $+x$ axis. Following Hubbert (1940), several conclusions may be drawn from (9.5.7).

Conclusion 1. For immobile liquid 2, $q_{2s} = 0$, and (9.5.7) becomes:

$$\partial y/\partial s = \sin\theta = q_{1s}\mu_1/k(\gamma_2 - \gamma_1), \qquad q_{1s} = \frac{k\gamma_1}{\mu_1}\frac{\gamma_2 - \gamma_1}{\gamma_1}\frac{\partial y}{\partial s} = K'_1\frac{\partial y}{\partial s} \tag{9.5.8}$$

where $K'_1 = K_1(\gamma_2 - \gamma_1)/\gamma_1$. This means that the interface rises in the direction of the flow. As q_{1s} increases, θ also increases. This occurs, for example, in a stationary coastal interface (sec. 9.7) as the coast is approached.

For the sake of comparison, let a *phreatic surface* serve as an upper boundary of a fluid 1 flow domain in the vertical xy-plane. On this surface we have, by (8.1.1):

$$q_s = -K_1\,\partial\varphi_1/\partial s = -K_1\,\partial y/\partial s = -K_1 \sin\theta \tag{9.5.9}$$

where θ is the angle that the tangent to the phreatic surface makes with the $+x$ axis. By comparing (9.5.8) with (9.5.9), one observes the analogy that exists between a stationary phreatic surface and a stationary interface. The former drops in the direction of the flow, reaching at most an angle of $\theta = \pi/2$; then $q_s = -K$, i.e., directed downward. The latter rises in the direction of the flow, reaching at most an angle of $\theta = \pi/2$; then $q_s = K'_1$ directed vertically upward. Also $(\partial\varphi_1/\partial s)_{\max} = -(\gamma_2 - \gamma_1)/\gamma_1$, which means that the component along the interface of the piezometric head gradient cannot exceed a certain finite critical value. The analogy, shown here for a simple case of a stationary interface and free surface is, obviously, more general and exists also in the case of unsteady three-dimensional flow (par. 9.5.3). It is of interest as sometimes it is more convenient to study a problem involving an interface by investigating the corresponding phreatic surface problem (e.g., by means of a Hele-Shaw analog (sec. 11.4)).

Conclusion 2. If $q_{1s} = q_{2s} = q_s$, equation (9.5.7) becomes:

$$\sin\theta = q_s(\mu_1 - \mu_2)/k(\gamma_2 - \gamma_1). \tag{9.5.10}$$

If, also, $\mu_1 = \mu_2$, $\sin\theta = 0$, i.e., the interface is horizontal. This is a quasistationary situation. We also have $\sin\theta = 0$ when $q_s = 0$ (stationary horizontal interface), or when $q_{1s}\mu_1 = q_{2s}\mu_2$.

Conclusion 3. When $q_{1s} = 0$, $\sin\theta = -q_{2s}\mu_2/k(\gamma_2 - \gamma_1)$. The interface will tilt downward in the direction of the flow. At most we may have $\theta = -\pi/2$. Then $(\partial\varphi_2/\partial s)_{\max} = -(\gamma_2 - \gamma_1)/\gamma_2$.

Conclusion 4. When both liquids are in motion, θ will attain some intermediate value and the difference between the components along the interface of their piezometric head gradient cannot exceed a certain finite critical value. If this critical value is exceeded (or when the critical values in conclusions 1 and 3 above are exceeded), a stationary interface cannot exist. The interface will start moving until a new equilibrium is reached.

Let $\varphi^\beta{}_\alpha$ denote the piezometric head in a region occupied by liquid α measured by the liquid level in an observation well filled with liquid β:

$$\varphi^\beta{}_\alpha = y + p/\gamma_\beta, \qquad \alpha, \beta = 1, 2. \tag{9.5.11}$$

At any point within the flow domain we have the relationship between elevation and potentials in the forms:

$$y = [\gamma_2/(\gamma_2 - \gamma_1)]\varphi^2{}_2 - [\gamma_1/(\gamma_2 - \gamma_1)]\varphi^1{}_2 \tag{9.5.12}$$

or:

$$y = [\gamma_2/(\gamma_2 - \gamma_1)]\varphi^2{}_1 - [\gamma_1/(\gamma_2 - \gamma_1)]\varphi^1{}_1. \tag{9.5.13}$$

The elevation of points along the interface (neglecting capillary pressure) is obtained by eliminating p from the two equations in (9.5.1):

$$y = [\gamma_2/(\gamma_2 - \gamma_1)]\varphi^2{}_2 - [\gamma_1/(\gamma_2 - \gamma_1)]\varphi^1{}_1. \tag{9.5.14}$$

If liquid 1 is flowing while liquid 2 (with $\gamma_2 > \gamma_1$) is stationary, the entire liquid 2 region is at a constant $\varphi^2{}_2$ potential, say $(\varphi^2{}_2)_0$. However, if in this region we measure potentials by means of liquid 1 (i.e., $\varphi^1{}_2$), equipotentials will be horizontal lines (figs. 9.5.2c and 9.5.3). This follows from (9.5.12) with $\varphi^2{}_2 = (\varphi^2{}_2)_0 = \text{const}$. The difference in elevation between two adjacent such surfaces is:

$$\Delta y = -[\gamma_1/(\gamma_2 - \gamma_1)]\Delta\varphi^1{}_2. \tag{9.5.15}$$

Surfaces of $\varphi^2{}_1 = \text{const}$ are parallel to the interface (fig. 9.5.2c).

Similar consideration can be applied to the case $q_{1s} = 0$, $q_{2s} \neq 0$ (fig. 9.5.2b).

Lusczynski (1961), in connection with density variations across a coastal interface, defines several types of head, taking into account the presence of a transition zone.

Type 1, point water head φ_{ip}. At a point i (occupied by fluid i) in ground water of variable density, this is the water level (referred to a given datum level) in an observation well filled sufficiently with fluid i (of density ρ_i) to balance the pressure p_i at that point:

$$\varphi_{ip} = z_i + p_i/\rho_i g \tag{9.5.16}$$

where z_i is the vertical coordinate of the considered point.

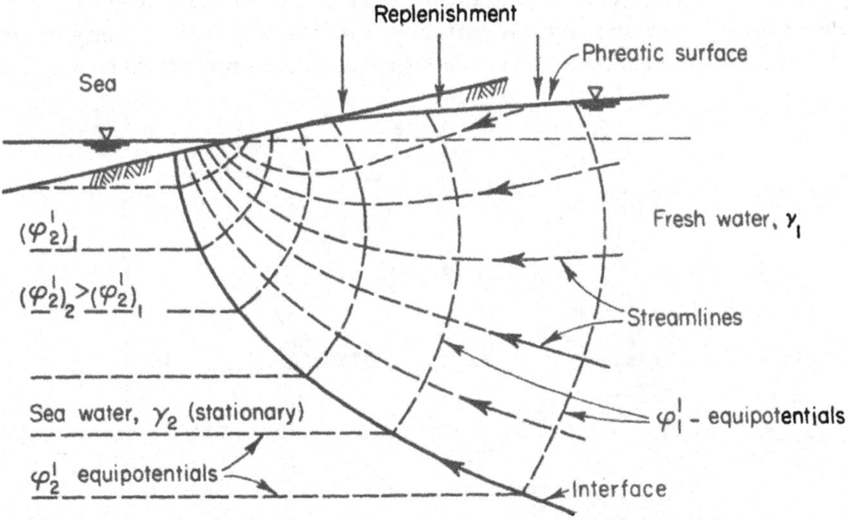

FIG. 9.5.3. Equipotentials near the coast.

Type 2, fresh water head φ_{if}. This is obtained when the observation well is filled with fresh water (of density ρ_f) to a level sufficient to balance the pressure p_i at point i:

$$\varphi_{if} = z_i + p_i/\rho_f g. \tag{9.5.17}$$

Type 3, environmental water head φ_{in}. This is obtained when the observation well is filled with water of variable density, such that the variations of density are identical to those encountered along a vertical line from point i to the phreatic surface:

$$\varphi_{in} = z_i + p_i/\rho_a g; \qquad \rho_a = \frac{1}{\varphi_{in} - z_i} \int_{z_i}^{\varphi_{in}} \rho(x, y, z')\, dz' \tag{9.5.18}$$

where ρ_a is the average water density between point i and the free surface above i. Lusczynski (1961) relates the three types of head to each other and determines the specific discharge components in an anisotropic medium in terms of the head gradients:

$$q_x = -(k_x \rho_f g/\mu_i)\, \partial\varphi_{if}/\partial x; \qquad q_y = -(k_y \rho_f g/\mu_i)\, \partial\varphi_{if}/\partial y; \qquad q_z = -(k_z \rho_f g/\mu_i)\, \partial\varphi_{in}/\partial z \tag{9.5.19}$$

where x, y, z are also the principal directions of permeability.

9.5.3 The Boundary Conditions Along an Interface

In the present paragraph we consider boundary conditions that φ_1 and φ_2 must satisfy on the interface separating the two liquids.

The interface (stationary or moving) is a *material* (*or fluid*) *surface* that is always composed of the same fluid particles. This means that at each point on it the component V_n of the velocity $\mathbf{V}$, normal to the interface in each of the subregions, is equal to the velocity of the interface. A portion of the interface in the vicinity of point P on it is shown in figure 9.5.4. The porous medium is assumed to be homogeneous

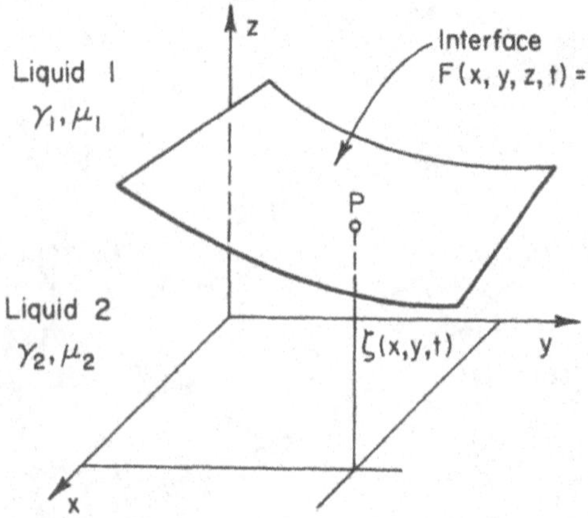

FIG. 9.5.4. An abrupt interface between immiscible liquids.

with permeability k (or k_x, k_y, k_z in the case of anisotropy with x, y, z as principal directions) and porosity n. The interface surface can be represented by an equation whose general form is:

$$F(x, y, z, t) = 0. \tag{9.5.20}$$

As this interface moves, because it is a material surface, we have by (4.1.30):

$$dF/dt = \partial F/\partial t + (\mathbf{V} \cdot \nabla F) = \partial F/\partial t + (\mathbf{q} \cdot \nabla F)/n = 0 \tag{9.5.21}$$

where dF/dt is the *substantial derivative* (par. 4.1.4; see also discussion on phreatic surface in par. 7.1.3). Applying (9.5.21) to each of the two liquids, we obtain:

$$\partial F/\partial t + (\mathbf{q}_1 \cdot \nabla F)/n = 0; \qquad \partial F/\partial t + (\mathbf{q}_2 \cdot \nabla F)/n = 0. \tag{9.5.22}$$

When necessary, $\mathbf{V}$ can be related to $\mathbf{q}$ by (9.5.4); then $\mathbf{q}_1$ and $\mathbf{q}_2$ must be expressed through the appropriate effective permeabilities: at S_{w0} for the wetting fluid, and at S_{nw0} for the nonwetting one.

Now, since the elevation of a point on the interface is given by $\zeta(x, y, t)$, i.e.: $z = \zeta(x, y, t)$ on the interface, equation (9.5.20) for the interface becomes:

$$F(x, y, z, t) = z - \zeta(x, y, t) = 0. \tag{9.5.23}$$

This expression can now be inserted in (9.5.22), leading, in a homogeneous isotropic medium, to the boundary conditions on the interface:

$$-n\, \partial\zeta/\partial t - \nabla\Phi_1 \cdot \nabla(z - \zeta) = 0; \qquad -n\, \partial\zeta/\partial t - \nabla\Phi_2 \cdot \nabla(z - \zeta) = 0 \tag{9.5.24}$$

where $\Phi_1 = K_1\varphi_1$ and $\Phi_2 = K_2\varphi_2$. However, since we are interested in a boundary condition in terms of φ_1 and φ_2 or Φ_1 and Φ_2 alone, we insert in (9.5.24) the value of the elevation ζ as defined for points on the interface by (9.5.14):

$$\zeta = (\gamma_2/\Delta\gamma)\varphi_2 - (\gamma_1/\Delta\gamma)\varphi_1; \qquad \Delta\gamma = \gamma_2 - \gamma_1 > 0. \tag{9.5.25}$$

When $p_2 - p_1 = p_c$, we add $-p_c/\Delta\gamma$ on the right-hand side of (9.5.25). Thus we have for F (on the interface) either $F(x, y, z, t) = z - \zeta(x, y, t) = 0$ or $F(x, y, z, t) = z - (\gamma_2/\Delta\gamma)\varphi_2 + (\gamma_1/\Delta\gamma)\varphi_1 = 0$. The boundary conditions for (9.5.24) along the interface then become:

$$\alpha_1 n \frac{\partial\Phi_1}{\partial t} - \alpha_2 n \frac{\partial\Phi_2}{\partial t} - \alpha_1 \left[\left(\frac{\partial\Phi_1}{\partial x}\right)^2 + \left(\frac{\partial\Phi_1}{\partial y}\right)^2 + \left(\frac{\partial\Phi_1}{\partial z}\right)^2\right]$$
$$+ \alpha_2 \left[\frac{\partial\Phi_1}{\partial x}\frac{\partial\Phi_2}{\partial x} + \frac{\partial\Phi_1}{\partial y}\frac{\partial\Phi_2}{\partial y} + \frac{\partial\Phi_1}{\partial z}\frac{\partial\Phi_2}{\partial z}\right] - \frac{\partial\Phi_1}{\partial z} = 0 \quad \text{on } \zeta = z$$

$$\alpha_1 n \frac{\partial\Phi_1}{\partial t} - \alpha_2 n \frac{\partial\Phi_2}{\partial t} + \alpha_2 \left[\left(\frac{\partial\Phi_2}{\partial x}\right)^2 + \left(\frac{\partial\Phi_2}{\partial y}\right)^2 + \left(\frac{\partial\Phi_2}{\partial z}\right)^2\right]$$
$$- \alpha_1 \left[\frac{\partial\Phi_1}{\partial x}\frac{\partial\Phi_2}{\partial x} + \frac{\partial\Phi_1}{\partial y}\frac{\partial\Phi_2}{\partial y} + \frac{\partial\Phi_1}{\partial z}\frac{\partial\Phi_2}{\partial z}\right] - \frac{\partial\Phi_2}{\partial z} = 0 \quad \text{on } \zeta = z \tag{9.5.26}$$

or in the compact form:

$$n\left(\alpha_1\frac{\partial\Phi_1}{\partial t}-\alpha_2\frac{\partial\Phi_2}{\partial t}\right)-\alpha_1(\nabla\Phi_1)^2+\alpha_2(\nabla\Phi_1\cdot\nabla\Phi_2)-\partial\Phi_1/\partial z=0\quad\text{on }\zeta=z$$

$$n\left(\alpha_1\frac{\partial\Phi_1}{\partial t}-\alpha_2\frac{\partial\Phi_2}{\partial t}\right)+\alpha_2(\nabla\Phi_2)^2-\alpha_1(\nabla\Phi_1\cdot\nabla\Phi_2)-\partial\Phi_2/\partial z=0\quad\text{on }\zeta=z$$

(9.5.27)

where $\alpha_i=\mu_i/k\Delta\gamma=\gamma_i/K_i\Delta\gamma$, $i=1,2$; K_i is the hydraulic conductivity with respect to the ith fluid. The specific discharge potential Φ is introduced because the medium is homogeneous and isotropic.

It is interesting to note that for $\Phi_1\equiv 0$, $\gamma_1=0$, and $\gamma_2=\gamma$ (9.5.27) becomes the boundary condition on a phreatic surface (i.e., air–water interface, par. 7.1.4).

Another case of special interest occurs when fluid 2 is stationary ($\Phi_2\equiv 0$). Then (9.5.27) becomes a steady boundary condition:

$$\alpha_1(\nabla\Phi_1)^2+\partial\Phi_1/\partial z=0\quad\text{on }\zeta=z. \tag{9.5.28}$$

Thus the problem stated in paragraph 9.5.1 is one of determining the two harmonic functions φ_1 and φ_2 (or Φ_1 and Φ_2), each in its respective subregion, satisfying (9.5.2), boundary conditions on the external boundaries and (9.5.27) on the interface separating the two liquids. Once φ_1 and φ_2 are known, equation (9.5.25) is used to determine the shape of the interface $\zeta=\zeta(x,y,t)$ or $F(x,y,z,t)=0$.

An exact analytical solution of a moving interface by present available mathematical tools is most complicated because of the nonlinearity of (9.5.27) and because the latter is given on an a priori unknown surface $z=\zeta$. Some approximate methods of solution are given in section 9.7. The shape of a stationary interface in two-dimensional flow in a vertical plane, when one of the fluids is stationary, can be derived by the hodograph method discussed in section 9.6.

If the porous medium is anisotropic, with x, y and z as principal directions, equation (9.5.26) becomes:

$$n\frac{\gamma_1}{\Delta\gamma}\frac{\partial\varphi_1}{\partial t}-n\frac{\gamma_2}{\Delta\gamma}\frac{\partial\varphi_2}{\partial t}-\frac{\gamma_1}{\Delta\gamma}\cdot\frac{\gamma_1}{\mu_1}\left[k_x\left(\frac{\partial\varphi_1}{\partial x}\right)^2+k_y\left(\frac{\partial\varphi_1}{\partial y}\right)^2+k_z\left(\frac{\partial\varphi_1}{\partial z}\right)^2\right]$$
$$+\frac{\gamma_2}{\Delta\gamma}\frac{\gamma_1}{\mu_1}\left[k_x\frac{\partial\varphi_1}{\partial x}\frac{\partial\varphi_2}{\partial x}+k_y\frac{\partial\varphi_1}{\partial y}\frac{\partial\varphi_2}{\partial y}+k_z\frac{\partial\varphi_1}{\partial z}\frac{\partial\varphi_2}{\partial z}\right]-\frac{k_z\gamma_1}{\mu_1}\frac{\partial\varphi_1}{\partial z}=0$$

$$n\frac{\gamma_1}{\Delta\gamma}\frac{\partial\varphi_1}{\partial t}-n\frac{\gamma_2}{\Delta\gamma}\frac{\partial\varphi_2}{\partial t}+\frac{\gamma_2}{\Delta\gamma}\cdot\frac{\gamma_2}{\mu_2}\left[k_x\left(\frac{\partial\varphi_2}{\partial x}\right)^2+k_y\left(\frac{\partial\varphi_2}{\partial y}\right)^2+k_z\left(\frac{\partial\varphi_2}{\partial z}\right)^2\right]$$
$$-\frac{\gamma_1}{\Delta\gamma}\cdot\frac{\gamma_2}{\mu_2}\left[k_x\frac{\partial\varphi_1}{\partial x}\frac{\partial\varphi_2}{\partial x}+k_y\frac{\partial\varphi_1}{\partial y}\frac{\partial\varphi_2}{\partial y}+k_z\frac{\partial\varphi_1}{\partial z}\frac{\partial\varphi_2}{\partial z}\right]-\frac{k_z\gamma_2}{\mu_2}\frac{\partial\varphi_2}{\partial z}=0\quad\text{on }\zeta=z.$$

(9.5.29)

9.5.4 Horizontal Interface Displacement

Some simple cases of interface displacement will be considered here. In all cases the medium is assumed to be isotropic.

Case 1, linear displacement. Consider horizontal flow in the direction $+x$ of two incompressible fluids in a homogeneous porous medium of length L. The two fluids are separated by an interface in the form of a plane whose normal is in the direction $+x$. As this interface advances, let fluid 1 (wetting fluid, e.g., water) displace fluid 2 (nonwetting fluid, e.g., oil). At some instant, t, the interface is at a distance ξ from the origin $x = 0$. Hence the interface equation (9.5.20) becomes:

$$F(x, t) \equiv x - \xi(t) = 0. \tag{9.5.30}$$

Since the flow is horizontal, the problem may be formulated in terms of the pressure as follows: Determine p_w and p_{nw} such that:

$$\begin{aligned} \partial^2 p_w/\partial x^2 &= 0, \qquad 0 < x \leqslant \xi(t) \\ \partial^2 p_{nw}/\partial x^2 &= 0, \qquad \xi(t) \leqslant x < L \end{aligned} \tag{9.5.31}$$

$$p_w = p_{nw} \text{ (neglecting } p_c) \text{ on } \xi = \xi(t) \tag{9.5.32}$$

$$\frac{k_{w0}}{\mu_w}\frac{\partial p_w}{\partial x} = \frac{k_{nw0}}{\mu_{nw}}\frac{\partial p_{nw}}{\partial x} = -q_x \quad \text{on } \xi = \xi(t) \tag{9.5.33}$$

where k_{w0} is the permeability to the wetting fluid (water) at residual nonwetting (oil) saturation, and k_{nw0} is the permeability to the nonwetting fluid at irreducible wetting fluid saturation. External boundary conditions are:

$$x = 0, \quad p_w = p'; \qquad x = L, \quad p_{nw} = p''. \tag{9.5.34}$$

By integrating (9.5.31) and using conditions (9.5.32) through (9.5.34), we obtain:

$$p_w = -\frac{p' - p''}{ML + (1 - M)\xi}x + p'; \qquad p_{nw} = \frac{p' - p''}{ML + (1 - M)\xi}M(L - x) + p'' \tag{9.5.35}$$

where $M = (k_{w0}/\mu_w)/(k_{nw0}/\mu_{nw})$ is called the *mobility ratio*. Hence:

$$q_x = \frac{k_{w0}}{\mu_w}\cdot\frac{p' - p''}{ML + (1 - M)\xi}\cdot \tag{9.5.36}$$

Since $\partial F/\partial t = -\partial\xi/\partial t$ and $\mathbf{V}\cdot\nabla F = V_x\,\partial F/\partial x = V_x$, we obtain:

$$-\frac{\partial\xi}{\partial t} + \frac{k_{w0}}{\mu_w n(1 - S_{w0} - S_{nw0})}\frac{p' - p''}{ML + (1 - M)\xi} = 0. \tag{9.5.37}$$

Replacing $\partial\xi/\partial t$ by $d\xi/dt$ and integrating, we obtain:

$$t = \frac{\mu_w n(1 - S_{w0} - S_{nw0})L^2}{k_{w0}(p' - p'')}\left[M\left(\frac{\xi}{L}\right) + \tfrac{1}{2}(1 - M)\left(\frac{\xi}{L}\right)^2\right] \tag{9.5.38}$$

for the time required for the front to advance from $x = 0$ to some distance $x = \xi$. From (9.5.38) it follows that the front either accelerates or decelerates, depending on whether $M > 1$ or $M < 1$. In the special case of $M = 1$, the front moves at a constant velocity. Figure 9.5.5 shows (9.5.38) in a graphic form (Muskat 1937).

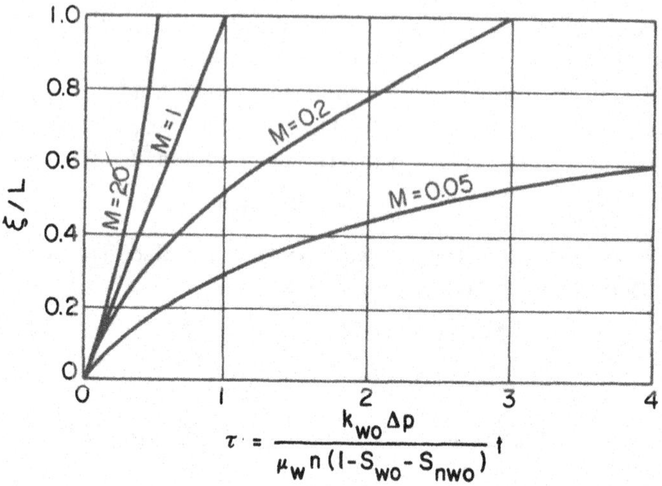

FIG. 9.5.5. Linear interface displacement (Muskat, 1937).

Case 2, radial encroachment. Muskat (1937) and Scheidegger (1960) present the solution for radial encroachment where the interface separating the displacing water from the displaced oil has the form of a circle in the horizontal plane advancing from $r = r''$ to $r = r' < r''$. The boundary conditions are $p = p''$ at $r = r''$, $p = p'$ at $r = r'$. The interface equation is $F \equiv r - \rho(t) = 0$ and the conditions to be satisfied on it are $p_w = p_{nw}$ and $(k_{wo}/\mu_w)\ \partial p_w/\partial r = (k_{nwo}/\mu_{nw}) \partial p_{nw}/\partial r$. The continuity equation for radial flow of incompressible fluids in a homogeneous isotropic medium is

$$\frac{\partial}{\partial r}\left(r\frac{k_{wo}}{\mu_w}\frac{\partial p_w}{\partial r}\right) = 0; \qquad \frac{\partial}{\partial r}\left(r\frac{k_{nwo}}{\mu_{nw}}\frac{\partial p_{nw}}{\partial r}\right) = 0. \tag{9.5.39}$$

The solution is obtained following the same procedure as that described above for the case of linear displacement. For $k_{wo} = k_{nwo} = k$, $S_{wo} = S_{nwo} = 0$, and p'' and p' independent of time, the solution is:

$$\frac{4k(p'' - p')}{\mu_w r''^2 n}t = \left(\frac{\rho}{r''}\right)^2\left[\ln\left(\frac{\rho}{r''}\right)^2 - M\ln\left(\frac{\rho}{r'}\right)^2 + M - 1\right] - M\ln\left(\frac{r'}{r''}\right)^2 - (M - 1). \tag{9.5.40}$$

Other cases given in the literature deal with water drives and water flooding techniques for secondary recovery in oil reservoirs. Scheidegger (1960) lists some of these works. In most cases, solutions are obtained by numerical or analog methods.

Case 3, interface displacement in two-dimensional horizontal flow. Equation (9.5.21), the solution of which gives $F(x, y, z, t)$, can be written, using the summation convention for subscripts i, j, k only, in the compact form:

$$\frac{\partial F}{\partial t} - \frac{K_\alpha}{n}\frac{\partial \varphi_\alpha}{\partial x_i}\frac{\partial F}{\partial x_i} = 0, \qquad \alpha = 1, 2 \tag{9.5.41}$$

where $x_1 \equiv x$, $x_2 \equiv y$, $x_3 \equiv z$, and α denotes either the wetting or the nonwetting fluid. Following Muskat (1937), let us introduce new orthogonal curvilinear coordinates:

$$\xi_1 = \xi_1(x, y, z); \qquad \xi_2 = \xi_2(x, y, z); \qquad \xi_3 = \xi_3(x, y, z)$$

so that:

$$\frac{\partial \xi_i}{\partial x_j}\frac{\partial \xi_k}{\partial x_j} = 0 \quad \text{and} \quad \frac{\partial \varphi_\alpha}{\partial x_i}\frac{\partial F}{\partial x_i} = \frac{\partial \varphi_\alpha}{\partial \xi_j}\frac{\partial F}{\partial \xi_j}\left[\sum_{(i)}\left(\frac{\partial \xi_j}{\partial x_i}\right)^2\right].$$

If surfaces $\xi_1 = \text{const}$ are made to coincide with equipotential surfaces $\varphi_\alpha = \text{const}$, then:

$$\frac{\partial \varphi_\alpha}{\partial x_i}\frac{\partial F}{\partial x_i} = \frac{\partial F}{\partial \xi_1}\left(\frac{\partial \xi_1}{\partial x_i}\frac{\partial \xi_1}{\partial x_i}\right) \quad \text{or} \quad \nabla \varphi_\alpha \cdot \nabla F = |\nabla \xi_1|^2 \frac{\partial F}{\partial \xi_1}. \tag{9.5.42}$$

Hence (9.5.41) can be written as:

$$\frac{\partial F}{\partial t} - \frac{K_\alpha}{n}\left(\frac{\partial \xi_1}{\partial x_i}\frac{\partial \xi_1}{\partial x_i}\right)\frac{\partial F}{\partial \xi_1} = 0 \quad \text{or} \quad \frac{\partial F}{\partial t} - \frac{K_\alpha}{n}|\nabla \xi_1|^2 \frac{\partial F}{\partial \xi_1} = 0 \tag{9.5.43}$$

where $|\nabla \xi_1|^2$ must be expressed as a function of ξ_1, ξ_2 and ξ_3. Equation (9.5.43) can be integrated to yield:

$$F(\xi_1, \xi_2, t) = t + \frac{n}{K_\alpha}\int \frac{d\xi_1}{|\nabla \xi_1|^2} + f(\xi_2, \xi_3) = \text{const} \tag{9.5.44}$$

where f is an arbitrary function to be chosen so that F assumes its known initial form at $t = 0$. For two-dimensional plane flow, neglecting the difference in density and viscosity between the two fluids, it is convenient to choose the family of streamlines $\psi = \text{const}$, which is everywhere orthogonal to the curves $\varphi = \text{const}$, as coordinate ξ_2. Essentially, we then have the case of steady flow of a homogeneous fluid, but with some fluid surface (say made of marked fluid particles) advancing in the flow domain. One should recognize this possibility of steady flow (i.e., velocity and pressure remain everywhere unchanged with time), yet with a moving interface. Then equation (9.5.44) becomes:

$$F(\varphi, \psi, t) = t + \frac{n}{k}\int \frac{d\varphi}{|\nabla \varphi|^2} + f(\psi) = \text{const} \tag{9.5.45}$$

in which the integral is to be evaluated along a streamline $\psi = \text{const}$.

As an example, Muskat (1937) considers the flow pattern produced by a single pumping well at a distance d from a line drive which at $t = 0$ coincides with the x axis. The x axis is an equipotential boundary ($\varphi = \varphi_0$) and remains so as the interface advances toward the well located at $x = 0$, $y = d$. Since the flow is steady, we have for this case:

$$\varphi_0 - \varphi = \frac{-Q}{4\pi K}\ln\frac{x^2 + (y-d)^2}{x^2 + (y+d)^2}; \qquad \psi = -\frac{Q}{2\pi K}\tan^{-1}\frac{2\,dx}{x^2 + y^2 - d^2} \tag{9.5.46}$$

where Q is the flux per unit thickness into the well. The denominator in the integral of (9.5.45) may then be shown to be:

$$|\nabla\varphi|^2 = \frac{Q^2}{4\pi^2 d^2 K^2}(\cosh\xi_1 + \cos\xi_2)^2; \qquad \xi_1 = \frac{2\pi K(\varphi_0 - \varphi)}{Q}; \qquad \xi_2 = -\frac{2\pi K\psi}{Q}.$$

The shape at any time t of the interface composed of fluid particles leaving the line drive at $t = 0$ is given by:

$$F(\xi_1, \xi_2, t) = t - n\frac{2\pi d^2}{Q}\int\frac{d\xi_1}{(\cosh\xi_1 + \cos\xi_2)^2} + f(\xi_2) = C = \text{const.}$$

The initial conditions require that $f(\xi_2) = C$. Hence:

$$t = \frac{2\,d^2 n}{Q\sin^2\xi_2}\left[\frac{\sinh\xi_1}{\cosh\xi_1 + \cos\xi_2} - 2\operatorname{cotg}\xi_2\tan^{-1}\left(\tanh\frac{\xi_1}{2}\tan\frac{\xi_2}{2}\right)\right]. \tag{9.5.47}$$

Figure 9.5.6a shows (9.5.47) plotted with some streamlines. The advancing interface will reach the pumping well at $t = 2\pi n d^2/3Q$ (or: $\tau = \frac{1}{3}$). The total area (A) swept out by the advancing interface during that time is $A = Qt/n = 2\pi d^2/3$ or two-thirds the area of a circle of radius d.

Muskat (1937) also solves the problem of a pumping well and an injection well of equal strength lying along the y axis at $(0, \pm d)$ in an infinite homogeneous flow domain. The interface, composed of fluid particles leaving the injection well at $t = 0$, is advancing from the injecting well toward the pumping one. For this case, the potential distribution and the stream function are also given by (9.5.46). The constant time curves (Muskat 1937) are:

$$t = \frac{2\pi n d^2}{Q\sin^2\xi_2}\left[\frac{\sinh\xi_1}{\cosh\xi_1 + \cos\xi_2} - 2\operatorname{cotg}\xi_2\tan^{-1}\left(\tanh\frac{\xi_1}{2}\tan\frac{\xi_2}{2}\right)\right.$$
$$\left. - \frac{\sinh\xi^0{}_1}{\cosh\xi^0{}_1 + \cos\xi_2} + 2\operatorname{cotg}\xi_2\tan^{-1}\left(\tanh\frac{\xi^0{}_1}{2}\tan\frac{\xi_2}{2}\right)\right] \tag{9.5.48}$$

where $\xi_1 = \xi^0{}_1$ is the equipotential defining the injection well at $(0, -d)$. The advancing interface is shown in figure 9.5.6b. In this case the injected fluid will reach the pumping well at $t = 4\pi n d^2/3Q$ ($\tau = \frac{2}{3}$).

A somewhat different approach is suggested by Shchelkachev and Lapuc (in Oroveanu 1966, p. 152). Using the complex potential (par. 7.8.2):

$$\zeta = f(z) = \varphi + i\psi; \qquad z = x + iy$$

where:

$$\frac{df(z)}{dz} = \frac{\partial\varphi}{\partial x} + i\frac{\partial\psi}{\partial x} = -\frac{1}{K}(q_x - iq_y) = -\frac{n}{K}\frac{d\bar{z}}{dt}; \qquad \bar{z} = x - iy. \tag{9.5.49}$$

From (9.5.49), we obtain:

$$t = -\frac{n}{K}\int_{(C)}\frac{d\bar{z}}{df(z)/dz} \tag{9.5.50}$$

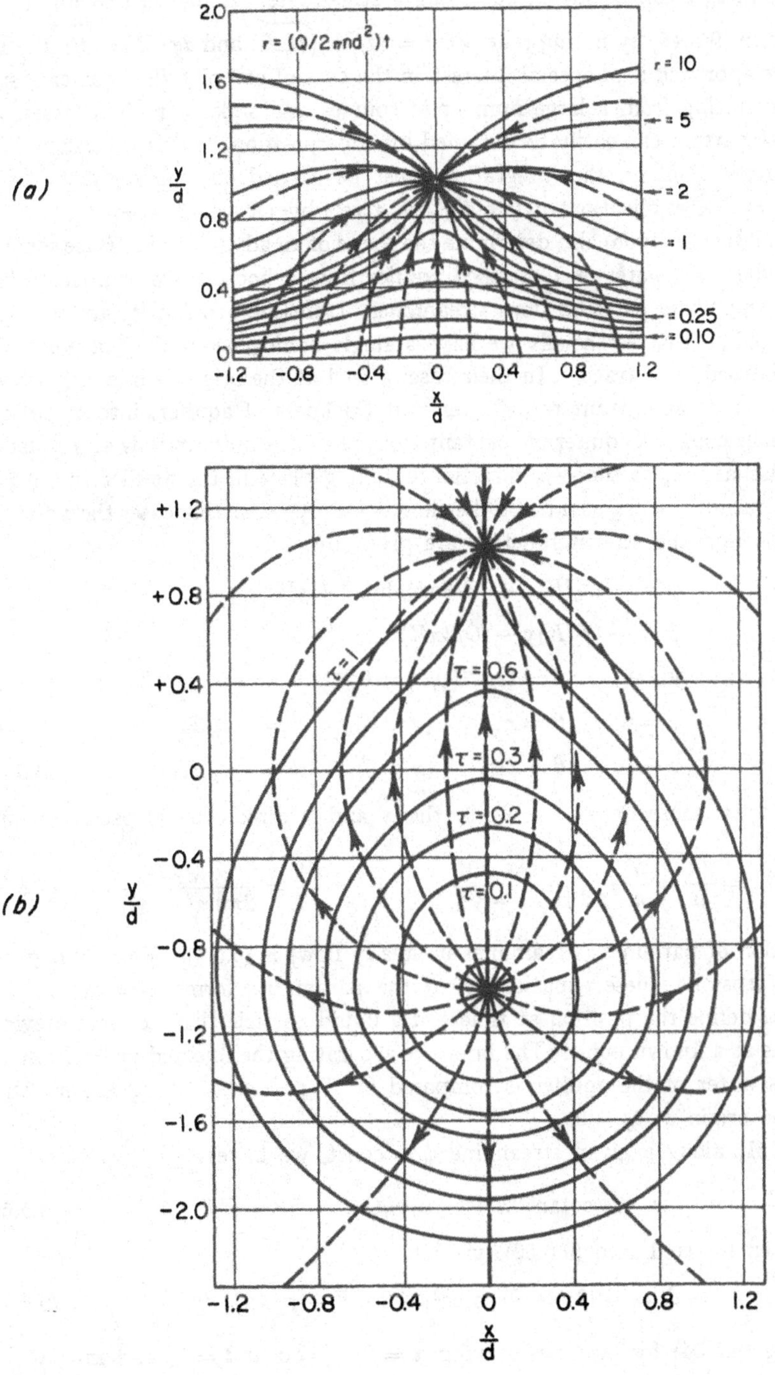

FIG. 9.5.6. Advancing interfaces: (a) Advance of a line drive. (b) From an injection well to a pumping well (Muskat, 1937).

where the integration is carried out along a streamline. Equation (9.5.50) can be derived from (9.5.45) by noting that $|\nabla\varphi|^2 = (df/dz)\overline{(df/dz)}$ and $d\varphi/d\bar{z} = \overline{(df/dz)}$. The use of this approach is of special interest in the case of complex flow patterns such as those resulting from a large number of sources and sinks. In these cases, the $\varphi - \psi$ flow pattern can easily be obtained by superposition. However, except for a small number of cases, the integration must be carried out numerically. Some examples of the application of (9.5.50) are given by Oroveanu (1966).

Bear and Jacobs (1965) also determine the shape of an advancing interface separating the indigenous water of a confined aquifer from a body of water injected into it by following water particles along streamlines. Differences in density and viscosity are neglected so that essentially we have a single homogeneous fluid, a portion of which is labeled by a tracer. In their case, a well at the origin is injecting tracer-labeled water at a constant rate Q (per unit thickness of aquifer) into an infinite homogeneous confined aquifer of constant thickness b. A uniform flow at a constant specific discharge q_0 in the $+\,x$ direction is taking place in the aquifer (fig. 9.5.7). Elastic storativity is neglected and the flow is steady. For this case, the potential distribution and the stream function are given by:

$$\begin{aligned} &\varphi = -(q_0/K)x + (Q/4\pi K)\ln(x^2+y^2)\\ &\psi = -(q_0/K)y - (Q/2\pi K)\tan^{-1}(y/x)\\ &y=0, \quad \theta = \tan^{-1}(y/x) = 0, \quad \psi = 0\\ &y=0, \quad \theta = \pi, \quad (x<0), \quad \psi = -Q/2K\\ &y=0, \quad \theta = 2\pi, \quad (x>0), \quad \psi = -Q/K. \end{aligned} \tag{9.5.51}$$

The velocity components V_x, V_y in the x and y directions, respectively, are:

$$V_x = \frac{dx}{dt} = \frac{q_0}{n} + \frac{Qx}{2\pi n(x^2+y^2)}; \qquad V_y = \frac{dy}{dt} = \frac{Qy}{2\pi n(x^2+y^2)}. \tag{9.5.52}$$

The equation of pathlines (streamlines in steady flow) is (6.5.1). When integrated in two dimensions, these equations yield the parametric forms $x = x(t, t_0)$, $y = y(t, t_0)$ that define the position at time t of a liquid particle that at some previous time t_0 was at a known point. The interface separating the injected water from the indigenous water of the aquifer is composed of all particles leaving the injecting well at the origin at $t_0 = 0$.

By (9.5.51) along a given streamline $\psi = \text{const}$, we have:

$$x = y\,\text{cotan}\,[2\pi(\Psi + q_0 y)/Q]; \qquad \Psi = K\psi \tag{9.5.53}$$

which, when inserted into (9.5.52), yields:

$$dt = -(2\pi n y/Q)\,\text{cosec}^2[-2\pi(\Psi + q_0 y)/Q]\,dy. \tag{9.5.54}$$

Integrating (9.5.54) for particles leaving $x = 0$, $y = 0$ at $t = 0$, and introducing the dimensionless variables:

$$\xi = (2\pi q_0/Q)x; \qquad \eta = (2\pi q_0/Q)y; \qquad \tau = (2\pi q_0^2/nQ)t \tag{9.5.55}$$

yields:

$$\tau = \xi + \ln[\sin\theta/\sin(\eta + \theta)]; \qquad \theta = \tan^{-1}(y/x). \tag{9.5.56}$$

Equation (9.5.56) defines the time required for the front to reach the point (ξ, η). The equation of the front at $\tau = \tau_0$ is:

$$[\cos\eta + (\xi/\eta)\sin\eta]\exp(-\xi) = \exp(-\tau_0) = \text{const.} \tag{9.5.57}$$

The shapes of the advancing front are shown in figure 9.5.7. As $\tau \to \infty$ we have $\xi = -\eta/\tan\eta$, which is the limiting front position (*water divide*). The front does not advance beyond $\eta = \pm\pi$. For $\tau \to \infty$ and $\eta = 0$, we obtain $\xi = -1$ or $\xi = +\infty$. The stagnation point $V_x = 0$, $V_y = 0$ is, by (9.5.52) at $\xi = -1$, $\eta = 0$. The front along the ξ axis moves according to:

$$\tau = \xi - \ln(1 + \xi). \tag{9.5.58}$$

When this method is applied to more complicated flow patterns, numerical methods of integration must be used.

9.5.5 Interface Displacement in the Vertical Plane

From the discussion in paragraph 9.5.3 it follows that an exact analytical solution of the shape and position of a moving interface is beyond our capabilities at the present. Several approximate approaches are presented throughout the present chapter.

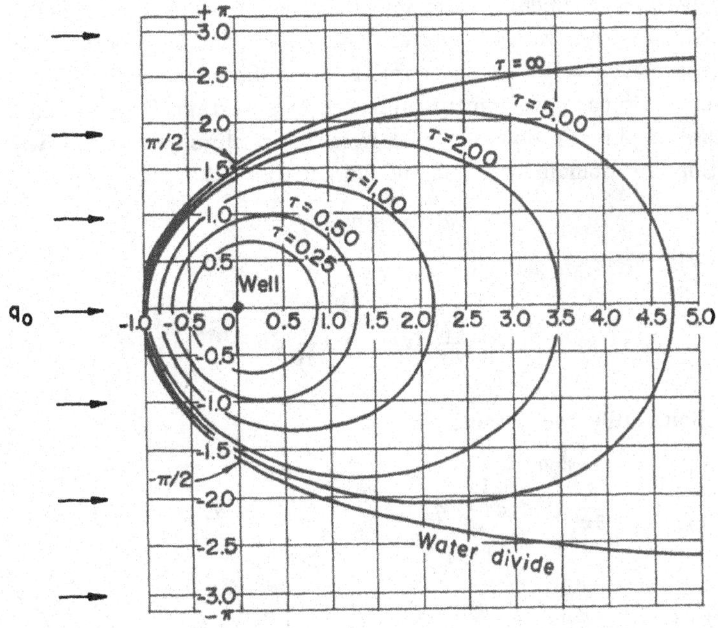

FIG. 9.5.7. An advancing interface in the case of a well injecting water into an aquifer with uniform flow (Bear and Jacobs, 1965).

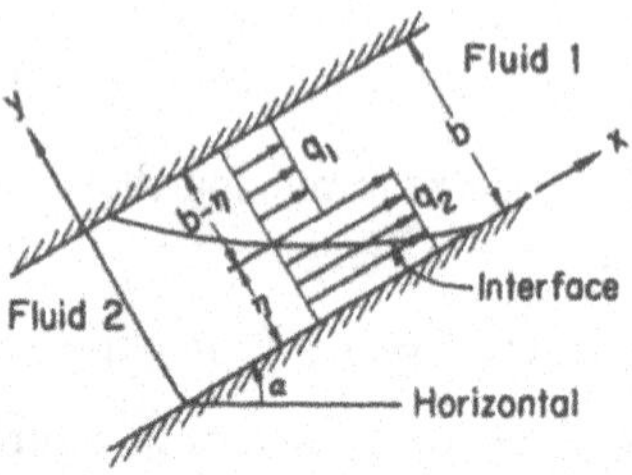

FIG. 9.5.8. Displacement in an inclined reservoir.

Consider the case of an inclined homogeneous and isotropic reservoir of constant thickness b in which fluid 2 (ρ_2, μ_2) is displacing fluid 1 ($\rho_1 < \rho_2$, μ_1), the displacement taking place from left to right (fig. 9.5.8). A simple approach to the case of an inclined homogeneous isotropic reservoir in which supercritical flow of two incompressible fluids takes place will now be considered as an example.

We shall assume that supercritical flow exists (Marle 1965) with an interface that is long with respect to the thickness b of the reservoir and that, in both fluids, flow components in the y direction, i.e., normal to impervious reservoir boundaries, are relatively small and may be neglected. In a way, this is equivalent to the Dupuit assumptions of essentially horizontal flow employed in ground water flows with a phreatic surface (chap. 8). We also assume that within each fluid, the specific discharge (q_x) varies only with x. Obviously, with these assumptions, the continuity equation div $\mathbf{q} = 0$, which is valid for every region separately, cannot rigorously be satisfied. Instead, we shall derive a global continuity equation for each fluid by averaging div $\mathbf{q} = 0$ over the normal to the flow direction.

Following a similar development presented in paragraph 6.4.4 leading to (8.2.15) for the case of phreatic flow, we start from the continuity equation written for a point within the domain occupied by fluid 2 (fig. 9.5.8):

$$\partial q_x/\partial x + \partial q_y/\partial y = 0 \tag{9.5.59}$$

and integrate it:

$$\int_0^{\eta(x,t)} \frac{\partial q_x}{\partial x}\, dy + \int_0^{\eta(x,t)} \frac{\partial q_y}{\partial y}\, dy = 0.$$

Using Leibnitz' rule, we obtain:

$$\frac{\partial}{\partial x}\int_0^{\eta(x,t)} q_x\, dy - q_x\big|_{y=\eta}\frac{\partial \eta}{\partial x} + q_y\big|_{y=\eta} - q_y\big|_{y=0} = 0.$$

Since

$$q_y\big|_{y=0} = 0; \qquad \int_0^{\eta(x,t)} q_x\, dy = Q_{x2}; \qquad q_y\big|_{y=\eta} = n\, d\eta/dt; \qquad q_x\big|_{y=\eta} = n\, dx/dt\big|_{y=\eta}$$

and $d\eta/dt = \partial\eta/\partial t + (\partial\eta/\partial x)\, dx/dt|_{y=\eta}$, we obtain:

$$\partial Q_{x2}/\partial x + n\, \partial\eta/\partial t = 0 \tag{9.5.60}$$

as the continuity equation for fluid 2. In a similar way, we obtain for fluid 1:

$$\partial Q_{x1}/\partial x + n\, \partial(b - \eta)/\partial t = 0. \tag{9.5.61}$$

In both equations n may be replaced by some effective porosity in the case of incomplete separation. Equations (9.5.60) and (9.5.61) are the *shallow depth* (or the *Boussinesq*) approximate equations of paragraph 8.2.3.

Based on the Dupuit assumptions of $\varphi_\alpha = \varphi_\alpha(x, t), \alpha = 1, 2$, stated above, we have:

$$Q_{x1} = q_1(b - \eta) = -\frac{k_1\rho_1 g}{\mu_1}\frac{\partial\varphi_1}{\partial x}(b - \eta); \qquad Q_{x2} = q_2\eta = -\frac{k_2\rho_2 g}{\mu_2}\frac{\partial\varphi_2}{\partial x}\eta \tag{9.5.62}$$

where k_1 and k_2 have the same, or different, values, depending on whether we have complete or incomplete separation. Also, at the interface, equality of pressure leads to:

$$\rho_2\varphi_2 - \rho_1\varphi_1 = (\rho_2 - \rho_1)(x \sin\alpha + y \cos\alpha) \quad \text{at} \quad y = \eta$$

or:

$$\rho_2\, \partial\varphi_2/\partial x - \rho_1\, \partial\varphi_1/\partial x = (\rho_2 - \rho_1)\, [\sin\alpha + (\partial\eta/\partial x) \cos\alpha]. \tag{9.5.63}$$

By adding (9.5.60) and (9.5.61), we have $\partial(Q_{x1} + Q_{x2})/\partial x \equiv \partial Q/\partial x = 0$ or $Q = Q(t)$, independent of x.

By combining (9.5.60) and (9.5.63), we obtain:

$$\frac{k_2}{\mu_2}\frac{\partial}{\partial x}\left\{\eta\,\frac{Q - g(\rho_2 - \rho_1)\,[\sin\alpha + (\partial\eta/\partial x)\cos\alpha]k_1/\mu_1\,(b - \eta)}{\eta k_2/\mu_2 + (b - \eta)k_1/\mu_1}\right\} + n\frac{\partial\eta}{\partial t} = 0. \tag{9.5.64}$$

For the horizontal reservoir, we set $\alpha = 0$ in the equations above (see par. 9.7.4).

Equation (9.5.64) is a second-order partial differential equation in the dependent variable η. Since Q (or q) is a function of t only, and the fluids are assumed incompressible, its value in the region $0 \leqslant x \leqslant L$ at any instant depends only on its value at $x = L$ or $x = 0$. Its solution yields the shape of the advancing interface: $\eta = \eta(x, t)$. Unfortunately the equation is nonlinear, so that an exact analytical solution even in this approximate form is not possible without further simplifications (see for example par. 9.7.4). However, as in the application of the Dupuit assumptions (sec. 8.1), the advantage of using these equations is that the dependent variable η is a function of two independent variables x and t, whereas the original dependent variable φ is a function of the three independent variables x, y, t.

The analysis may easily be extended to three dimensions with the elevation of points on the interface expressed by $\zeta = \zeta(x, y, t)$. The resulting equations will be:

$$\frac{\partial Q_{x2}}{\partial x} + \frac{\partial Q_{y2}}{\partial y} + n\frac{\partial\zeta}{\partial t} = 0; \qquad \frac{\partial Q_{x1}}{\partial x} + \frac{\partial Q_{y1}}{\partial y} + n\frac{\partial(b - \zeta)}{\partial t} = 0 \tag{9.5.65}$$

where $Q_{x2} = \zeta q_{x2}$, $Q_{y2} = \zeta q_{y2}$, $Q_{x1} = (b - \zeta)q_{x1}$, $Q_{y1} = (b - \zeta)q_{y1}$. By neglecting $\partial^2\eta/\partial x^2$ we actually assume that the interface has the form of an elongated tongue with a very small curvature (theoretically a straight line). Dietz (1953) presents an approximate solution that is based on this assumption. Oroveanu (1966, pp. 191–212) describes several approximate solution methods, and Le Fur and Sourieau (1963) and Marle (1965) give detailed analyses and discussions of (9.5.65).

If the two fluids are compressible, the appropriate continuity and motion equations should be used. These may also be averaged over a line normal to the direction of the flow in order to obtain global equations.

A different approach to the solution of interface problems, taking into account density and viscosity differences, is suggested by Pilatovski (1958, 1961), de Josselin de Jong (1960) and Jacquard (1962). They seek a single harmonic function in the regions occupied by the two fluids, such that it satisfies all external boundary conditions except along the interface itself, this being taken into account by a singularity property of the function. Pilatovski and Jacquard use the *single-layer potential* in order to preserve the jump in the normal derivative of the potential perpendicular to the interface. De Josselin de Jong replaces the two fluids with a single hypothetical fluid, and places vortices along the interface to account for the change in properties of the actual fluids. The magnitude of vorticity is chosen so that the values of the specific discharge in the hypothetical fluid are everywhere identical to those in the actual fluids. The flow of the hypothetical fluid is determined by the theory of potential flow, using the transformed boundary conditions and the singularities at the vortex points. The magnitude of the vortices for fluids of equal viscosity but different density is proportional to the horizontal component of the density gradient. If the initial position of the interface is known, its subsequent positions may be obtained by first using potential theory to determine the velocity distribution at a certain instant, then displacing the interface according to the velocities at various points along it. This leads to a step-wise method for determining the movement of the interface. Problems with a gradual change in density within a transition zone may be handled in a similar way. Differences in viscosity are accounted for by varying the vorticity according to the discharge which is a priori unknown and is part of the required solution. The analysis of this type of problem is much more complicated. In principle this is an analytic approach, yet the various integrations involved must be carried out numerically or by means of analogs. Additional remarks on numerical methods for determining the position of the interface are given in the following section.

9.5.6 Numerical and Graphic Methods

Numerical or analog methods may yield results when direct analytical solutions of the exact or the approximate equations fail to do so. In general, the solution itself is carried out on a high speed digital computer. Some comments on numerical methods are given in section 7.9. However, because of the nonlinearity of the boundary conditions on the interface, and owing to the fact that the location of this boundary, even in steady flow with a stationary interface, is a priori unknown, it is most difficult to solve interface problems even by numerical techniques and computers.

Under certain simplifying assumptions, a second numerical approach is possible. This approach is based on the fact that the interface is a material surface, and hence at every instant its motion (in magnitude and direction) is determined by the velocity of fluid particles along it. Hence, if this velocity can be determined for a sufficient number of particles (points) along the interface, the new position of the interface can be obtained by considering the displacement of each particle during a small time interval Δt. The displacement of each particle is then given by $\mathbf{\Delta s} = \mathbf{V}\,\Delta t$.

This method of solution becomes especially simple when the mobility ratio (ratio of (k/μ) of displacing fluid behind the front to (k/μ) of the displaced fluid ahead of the front) is equal to unity, or assumed to be approximately so, and when gravity effects are neglected, i.e., in the horizontal interface displacement (par. 9.5.4). Under such conditions, the flow of the two fluids would occur in the same manner as if the two fluids were a single homogeneous fluid. The moving interface is then a collection of labeled particles of the same fluid.

In steady flow it is possible, for example, to use potential theory in order to determine the flow pattern, i.e., the equipotentials and the streamlines. In complicated geometries, numerical or analog methods (e.g., the electric analog of the electrolytic tank type) are used to determine the φ-ψ flow net. Once the streamlines in the field of flow have been established, it is possible to determine the movement of the interface by displacing fluid particles belonging to it along streamlines. Thus, if ds denotes an element of length along a streamline, the time required for a fluid particle belonging to the interface to travel a length $s(t)$ along the streamline is given by:

$$t = \int_0^{s(t)} \frac{ds}{V_s(s)} = -\frac{n}{K}\int_0^{s(t)} \frac{ds}{\partial\varphi/\partial s} \tag{9.5.66}$$

where V_s is the velocity component in the direction tangent to the streamline. When the potential distribution is obtained numerically or by means of an analog, the computation of t by (9.5.66) is carried out numerically:

$$t = \sum_{(i)} (\Delta t)_i = (n/K) \sum_{(i)} [(\Delta s)_i^2/(\Delta\varphi)_i] \tag{9.5.67}$$

where $(\Delta s)_i$ is an increment of length (along the streamline) along which the potential drop is $(\Delta\varphi)_i$.

Of special interest is the case of a number of injection and pumping wells operating simultaneously in a given aquifer or oil reservoir. In this case, assuming that the mobility ratio is unity, it is required to determine the history of the advance of the injected fluid front. It is also required to determine the instant of ***breakthrough***, i.e., the time when the injected fluid reaches the pumping well, and the ratio between the two fluids in the pumped fluid after that instant. Except for simple well arrangements and steady flow (where φ and ψ do not vary with time), an analytical solution is practically impossible.

As an illustration of a simple numerical solution, consider the following example. Let $A_{i,k}$, $i = 1, 2, \ldots, N$, denote N points (fluid particles), each with coordinates

$(x_{i,k}, y_{i,k})$ along a moving interface (in the xy horizontal plane), at time $t = k\,\Delta t$ from the onset of injection. Let points B_j, $j = 1, 2, \ldots, M$, with coordinates (x_j, y_j) denote the location of wells in the field. The discharge of each well is $Q^k{}_j$ (positive or negative), during Δt from $k\,\Delta t$ to $(k+1)\,\Delta t$. All particles $A_{i,k}$ will move during Δt to their new positions $A_{i,k+1}$, thus defining the interface at time $t = (k+1)\,\Delta t$. Denoting the distance from $A_{i,k}$ to B_j by $r_{i,k,j}$, the velocity at $A_{i,k}$ caused by well B_j is given by:

$$V_{i,k,j} = \frac{\alpha Q^k{}_j}{r_{i,k,j}}; \qquad \alpha = 1/2\pi nb; \qquad (r_{i,k,j})^2 = (x_{i,k} - x_j)^2 + (y_{i,k} - y_j)^2 \tag{9.5.68}$$

where n and b are, respectively, the porosity and the constant thickness of the confined formation. Summing components in the x and y directions of the elementary displacements of the fluid particles, as caused by the simultaneous operation of all the wells, we obtain:

$$x_{i,k+1} - x_{i,k} = \Delta t \cdot \alpha \sum_{j=1}^{M} Q^k{}_j (x_{i,k} - x_j)/(r_{i,k,j})^2$$

$$y_{i,k+1} - y_{i,k} = \Delta t \cdot \alpha \sum_{j=1}^{M} Q^k{}_j (y_{i,k} - y_j)/(r_{i,k,j})^2. \tag{9.5.69}$$

In writing (9.5.69), variations of velocity during Δt have been neglected. By repeating this procedure for a sufficient number of points and for successive time intervals, often on a digital computer, the history of the interface movement is obtained. In such computations it is always convenient to assume that at time $t = \Delta t$ the front in the vicinity of each injection well has the shape of a circle of radius $r_0 = (2\alpha Q\,\Delta t)^{1/2}$.

If $Q^k{}_j$ varies with time, the particles $A_{i,k}$ are displaced approximately along pathlines, rather than along streamlines. As Δt is made smaller, a better approximation is obtained. The main difference between this approach and the one described before is that, in the latter case, particles are displaced *exactly* along known streamlines, whereas here the moving particles trace streamlines (or pathlines, if $Q^k{}_j$ varies with time) with an accumulating error. As Δt is reduced, the error becomes smaller. However, the advantage of this approach is that one need not solve for the φ–ψ distribution in the entire field at the end of each time interval.

The procedure described by (9.5.70) can also be carried out graphically by summing elementary displacement vectors. Figures 9.5.9 and 9.5.10 show the graphic method and an example of its application.

9.5.7 Approximate Solutions Based on Linearization

As shown in paragraph 9.5.3, the interface equations (9.5.27) are nonlinear partial differential equations. In order to permit an analytical solution, linearized approximations of these equations have been proposed by various authors. Some of these equations are described below (see also sec. 8.4).

Because the case of two fluids separated by an interface is more general than that of a single fluid bounded by a phreatic surface, some of the linearization methods

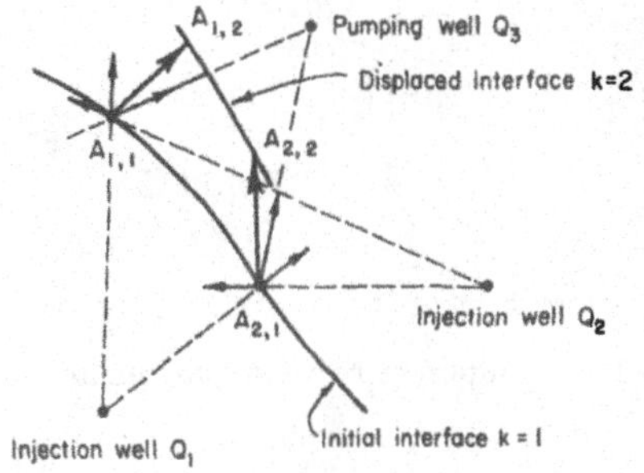

FIG. 9.5.9. Graphical method for determining successive interface positions.

are introduced here in detail in addition to their presentation in chapter 8. There they were used for the special case of one moving fluid underlying a second stagnant one, the pressure being assumed constant throughout the second fluid (air above the phreatic surface).

Example 1, the small perturbation approximation (par. 8.4.5; Bear and Dagan 1964; Dagan 1964 and Dagan 1964a,b,c; Dagan and Bear 1968). Figure 9.5.11 shows an abrupt interface completely separating two immiscible, incompressible fluids in a homogeneous confined aquifer. As the porous medium is homogeneous and isotropic, use is made of the specific discharge potential $\Phi = K\varphi$.

With the nomenclature of figure 9.5.11, let us assume that the specific discharge potential Φ in each subregion may be expressed as a sum of a power series of a small parameter:

$$\Phi_i(x, y, z, t) = \Phi^0{}_i(x, y, z) + \varepsilon\Phi'{}_i(x, y, z, t) + \varepsilon^2\Phi''{}_i(x, y, z, t) + \cdots, \quad i = 1, 2 \quad (9.5.70)$$

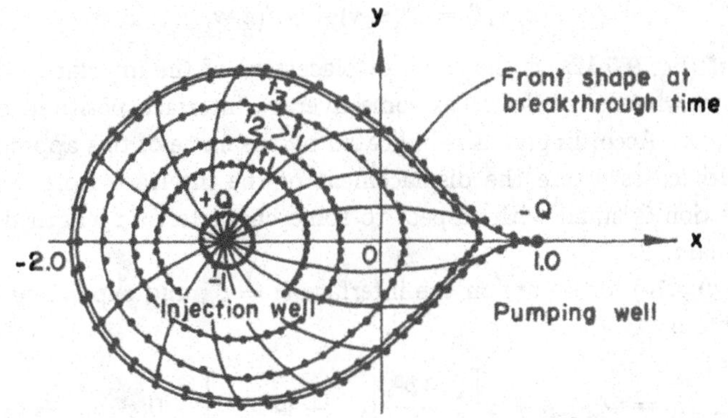

FIG. 9.5.10. Interface positions for an injection well and a pumping well of equal strength.

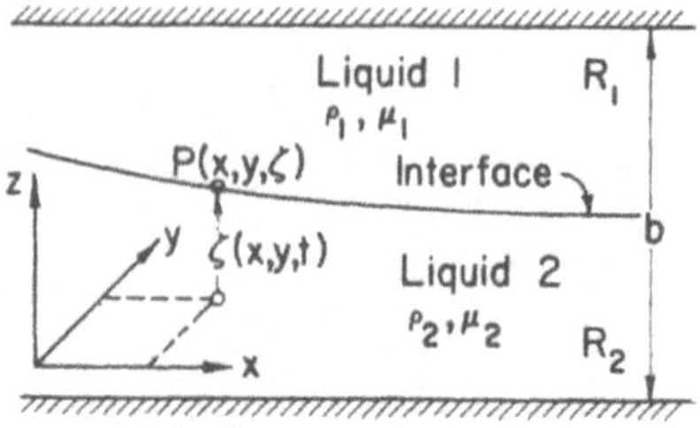

FIG. 9.5.11. An interface between two immiscible liquids.

where $\Phi^0{}_1$ and $\Phi^0{}_2$ are steady-state specific discharge potential distributions describing an average flow pattern or an initial steady one. The remaining terms of the series in (9.5.70) describe (in each region) deviations from the steady, or average, Φ-distribution, with ε having the character of a *perturbation*. An assumption, commonly used in the solution of similar problems in physics and mathematics, is that $\varepsilon \ll 1$. This assumption is obviously valid when the deviations of the unsteady flow from the steady one are indeed small.

Although the method to be described below (often used in the theory of waves of small amplitude) is also applicable to second- (Dagan 1964) and higher order linearizations, it will be described and employed here only to derive a first-order linearized solution. This means that only the first two terms on the right-hand side of (9.5.70) will be used.

By inserting (9.5.70) into (9.5.2), and recalling that by the definition of $\Phi^0{}_i$:

$$\nabla^2\Phi^0{}_1 = 0 \text{ in } R_1; \qquad \nabla^2\Phi^0{}_2 = 0 \text{ in } R_2 \tag{9.5.71}$$

we obtain:

$$\nabla^2\Phi'_1 = 0 \text{ in } R_1; \qquad \nabla^2\Phi'_2 = 0 \text{ in } R_2. \tag{9.5.72}$$

In a similar manner we assume that the elevation ζ of points along the interface may be expanded in a power series, the first two terms of which are:

$$\zeta(x, y, t) = \zeta^0(x, y) + \varepsilon\zeta'(x, y, t). \tag{9.5.73}$$

In (9.5.73), $\varepsilon\zeta'$(fig. 9.5.12a) is the small displacement of the interface caused by the unsteady flow, whereas ζ^0 describes some average interface position, or an initial steady-state one. Accordingly, as is indicated by the name of this approach, we deal here only with cases where the displacement of the interface from some average or initial position is small with respect to some characteristic length dimension of the flow domain.

Rewriting (9.5.70) for points on the interface ($z = \zeta$), and expanding in a Taylor series, we obtain:

$$\Phi_i\big|_{z=\zeta=\zeta^0+\varepsilon\zeta'} = \Phi^0{}_i\big|_{z=\zeta^0} + \varepsilon\left[\zeta'\,\frac{\partial\Phi^0{}_i}{\partial z}\bigg|_{z=\zeta^0} + \Phi'_i\big|_{z=\zeta^0}\right] + 0(\varepsilon^2); \qquad i = 1, 2.$$

Similar expressions may be derived for:

$$\Phi_1|_{z=\zeta}; \qquad \Phi_2|_{z=\zeta}; \qquad \partial\Phi_1/\partial x|_{z=\zeta}; \qquad \partial\Phi_2/\partial x|_{z=\zeta}$$

and for all other expressions appearing in (9.5.26). For example:

$$\partial\Phi_1/\partial x|_{z=\zeta} = \partial\Phi^0{}_1/\partial x|_{z=\zeta^0} + \varepsilon(\partial\Phi'_1/\partial x + \zeta'\, \partial^2\Phi^0{}_1/\partial x\, \partial z)|_{z=\zeta^0} + 0(\varepsilon^2). \quad (9.5.74)$$

In this way, all terms are expressed *for points on the average* (or initial) *interface with elevations* $\zeta^0(x, y)$.

By introducing these expressions into the boundary conditions of the interface (9.5.25) and (9.5.26), and collecting terms of the same power of ε, we obtain the following relationships for ζ', Φ'_1 and Φ'_2:

$$\zeta' = (\alpha_2\Phi'_2 - \alpha_1\Phi'_1)/(1 - \alpha_2\, \partial\Phi^0{}_2/\partial z + \alpha_1\, \partial\Phi^0{}_1/\partial z) \quad \text{on } z = \zeta^0 \qquad (9.5.75)$$

$$\begin{aligned} n(\alpha_1\, \partial\Phi'_1/\partial t - \alpha_2\, \partial\Phi'_2/\partial t) &- 2\alpha_1[\nabla\Phi^0{}_1 \cdot \nabla\Phi'_1 + \zeta'\, \nabla\Phi^0{}_1 \cdot \nabla(\partial\Phi^0{}_1/\partial z)] \\ &+ \alpha_2[\nabla\Phi^0{}_1 \cdot \nabla\Phi'_2 + \zeta'\, \nabla\Phi^0{}_1 \cdot \nabla(\partial\Phi^0{}_2/\partial z) + \nabla\Phi^0{}_2 \cdot \nabla\Phi'_1 \\ &+ \zeta'\, \nabla\Phi^0{}_2 \cdot \nabla(\partial\Phi^0{}_1/\partial z)] - (\partial\Phi'_1/\partial z + \zeta'\, \partial^2\Phi^0{}_1/\partial z^2) = 0 \quad \text{on } z = \zeta^0. \\ n(\alpha_2\, \partial\Phi'_2/\partial t - \alpha_1\, \partial\Phi'_1/\partial t) &- 2\alpha_2[\nabla\Phi^0{}_2 \cdot \nabla\Phi'_2 + \zeta'\, \nabla\Phi^0{}_2 \cdot \nabla(\partial\Phi^0{}_2/\partial z)] \\ &+ \alpha_1[\nabla\Phi^0{}_2 \cdot \nabla\Phi'_1 + \zeta'\, \nabla\Phi^0{}_2 \cdot \nabla(\partial\Phi^0{}_1/\partial z) + \nabla\Phi^0{}_1 \cdot \nabla\Phi'_2 \\ &+ \zeta'\, \nabla\Phi^0{}_1 \cdot \nabla(\partial\Phi^0{}_2/\partial z)] + (\partial\Phi'_2/\partial z + \zeta'\, \partial^2\Phi^0{}_2/\partial z^2) = 0, \quad \text{on } z = \zeta^0. \end{aligned} \qquad (9.5.76)$$

The advantages of replacing (9.5.25) and (9.5.26) by (9.5.75) and (9.5.76) are that

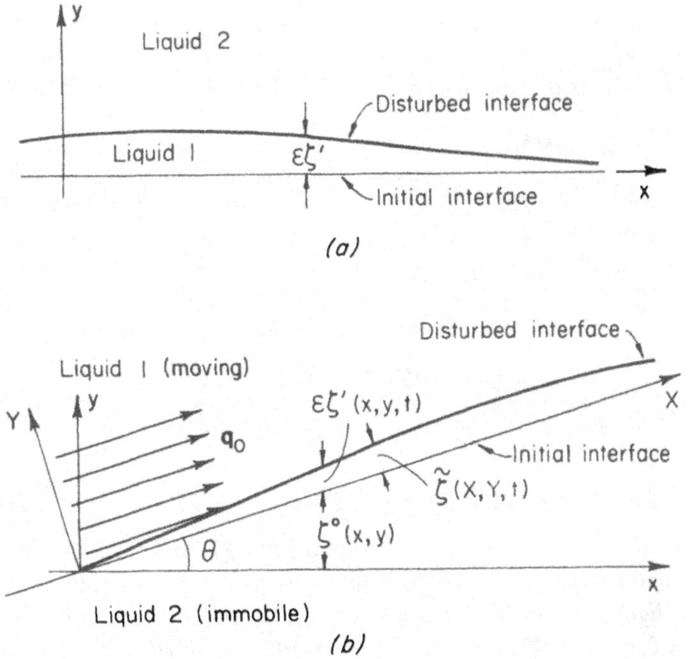

FIG. 9.5.12. Initial and disturbed interfaces in (a) horizontal and (b) inclined flows.

the latter are written for the fixed (in time) interface $z = \zeta^0$ and that the boundary conditions (9.5.76) are linear in Φ'_1 and Φ'_2.

For the steady-state solution we must solve (9.5.71) with the boundary conditions:

$$\left.\begin{aligned} \zeta^0 &= \alpha_2\Phi^0{}_2 - \alpha_1\Phi^0{}_1 \\ \alpha_1(\nabla\Phi^0{}_1)^2 - \alpha_2(\nabla\Phi^0{}_1\cdot\nabla\Phi^0{}_2) + \partial\Phi^0{}_1/\partial z &= 0 \\ \alpha_2(\nabla\Phi^0{}_2)^2 - \alpha_1(\nabla\Phi^0{}_1\cdot\nabla\Phi^0{}_2) - \partial\Phi^0{}_2/\partial z &= 0 \end{aligned}\right\} \quad \text{on} \quad z = \zeta^0. \tag{9.5.77}$$

Thus the problem has been transformed into one of first solving the steady flow problem (9.5.71) for $\Phi^0{}_1(x, y, z)$ satisfying (9.5.77) with a fixed interface of elevation $\zeta^0(x, y)$, and then solving (9.5.72) for ζ', Φ'_1 and Φ'_2 satisfying (9.5.75) and (9.5.76). No advantage has been gained with respect to the steady flow problem; the boundary conditions (9.5.77) in this case are still nonlinear. They are the same as (9.5.27), written for steady flow. However, in certain cases the initial steady state solution is known.

The following two cases are of special interest.

(a) *Initially, uniform flow* (q_0) *takes place in the upper* (γ_1) *liquid, while the lower liquid* (γ_2) *is stagnant.*

In this case (fig. 9.5.12b), it is convenient to introduce a rotated Cartesian coordinate system X, Y, Z, such that the plane $Z = 0$ coincides with the initial planar interface:

$$X = x\cos\theta + z\sin\theta; \qquad Y = y; \qquad Z = -x\sin\theta + z\cos\theta. \tag{9.5.78}$$

Then (fig. 9.5.12a), for the steady state:

$$\Phi^0{}_1 = -q_0X, \qquad \Phi^0{}_2 \equiv 0, \qquad \tilde{\zeta}^0 = 0, \qquad \tilde{\zeta} = \varepsilon\tilde{\zeta}' \tag{9.5.79}$$

where $\tilde{\zeta} = \tilde{\zeta}^0 + \tilde{\zeta}'$ is the interface elevation in the X, Y, Z system. Also, from (9.5.7):

$$q_0 = \partial\Phi^0{}_1/\partial x; \qquad \partial z/\partial x = \sin\theta = \alpha_1 q_0 = \mu_1 q_0/k\,\Delta\gamma. \tag{9.5.80}$$

By transforming (9.5.76) from the x, y, z into the X, Y, Z system and expressing $\Phi^0{}_1$, $\Phi^0{}_2$, and $\sin\theta$ by (9.5.78) through (9.5.79), the following conditions along the interface $Z = \tilde{\zeta}^0 = 0$ are obtained:

$$\begin{aligned} &n(\alpha_1\,\partial\Phi'_1/\partial t - \alpha_2\,\partial\Phi'_2/\partial t) + (\sin\theta)\,\partial\Phi'_1/\partial x - (\cos\theta)\,\partial\Phi'_1/\partial z \\ &\qquad - (\mu_2/\mu_1)(\sin\theta)\,\partial\Phi'_2/\partial x = 0 \\ &n(\alpha_1\,\partial\Phi'_1/\partial t - \alpha_2\,\partial\Phi'_2/\partial t) + (\cos\theta)\,\partial\Phi'_2/\partial z = 0 \end{aligned} \tag{9.5.81}$$

$$\tilde{\zeta}' = (\alpha_2\Phi'_2 - \alpha_1\Phi'_1)/\cos\theta. \tag{9.5.82}$$

Equations (9.5.72) in X, Y, Z, together with (9.5.81) and (9.5.82) and appropriate boundary conditions on the external boundaries, should yield results for Φ'_1 and Φ'_2.

(b) *Both liquids are initially at rest (fig. 9.5.12a).*

Here $\theta = 0$, $q_0 = 0$, $\Phi^0{}_1 \equiv 0$, $\zeta^0 = 0$, $\Phi_1 \equiv \varepsilon\Phi'_1$, $\Phi_2 \equiv \varepsilon\Phi'_2$. Hence the conditions along the horizontal interface $z = 0$, are:

$$\left.\begin{aligned} n(\alpha_1\,\partial\Phi'_1/\partial t - \alpha_2\,\partial\Phi'_2/\partial t) - \partial\Phi'_1/\partial z = 0 \\ n(\alpha_1\,\partial\Phi'_1/\partial t - \alpha_2\,\partial\Phi'_2/\partial t) - \partial\Phi'_2/\partial z = 0 \\ \zeta' = \alpha_2\Phi'_2 - \alpha_1\Phi'_1 \end{aligned}\right\} \quad \text{on} \quad z = 0. \tag{9.5.83}$$

Equations (9.5.83) have also been derived by Kidder (1956) and Cailleau et al. (1963).

It may be interesting to visualize the physical interpretation of (9.5.83). Let the horizontal interface be divided into a large number of squares, each of area A in the $z = 0$ plane (fig. 9.5.13). A closed tube of cross-sectional area $a = nA$ connects the two points: N'' immediately below and N' immediately above this interface. It is assumed that the average interface is impervious and that the flow of fluid displaced by the moving interface is possible only through this connecting tube that resists the flow through it. The interface in the connecting tube indicates the position of the moving interface.

An example of the application of this method is given in paragraph 9.7.5. Several examples are given by Bear and Dagan (1964) and Dagan and Bear (1968).

Example 2, combining the shallow flow and the small perturbation approximations (Bear and Dagan 1964). This is a linearization of equations (9.5.65) that are based on the Dupuit assumptions. It is assumed that the horizontal aquifer (fig. 9.5.11) is relatively thin ($b \ll L$, where L is some characteristic horizontal length) and that the deviations of the interface from its initial position are small with respect to b. Starting from (9.5.65), with $q_{x1} = -\,\partial\Phi_1/\partial x$, $q_{y1} = -\,\partial\Phi_1/\partial y$, etc., and assuming that ζ, q_{xi}, q_{yi} and Φ_i ($i = 1, 2$) may be expressed as a sum of a steady-state term and a deviation, we obtain:

$$\begin{aligned} \zeta(x,y,t) &= \zeta^0(x,y) + \varepsilon\zeta'(x,y,t) \\ q_{xi}(x,y,t) &= q^0{}_{xi}(x,y) + \varepsilon q'_{xi}(x,y,t), \qquad i = 1, 2 \\ q_{yi}(x,y,t) &= q^0{}_{yi}(x,y) + \varepsilon q'_{yi}(x,y,t), \qquad i = 1, 2 \\ \Phi_i(x,y,t) &= \Phi^0{}_i(x,y) + \varepsilon\Phi'_i(x,y,t), \qquad i = 1, 2. \end{aligned} \tag{9.5.84}$$

By inserting (9.5.84) into (9.5.65), and neglecting terms of $0(\varepsilon^2)$, the following equations are obtained for the steady-state part:

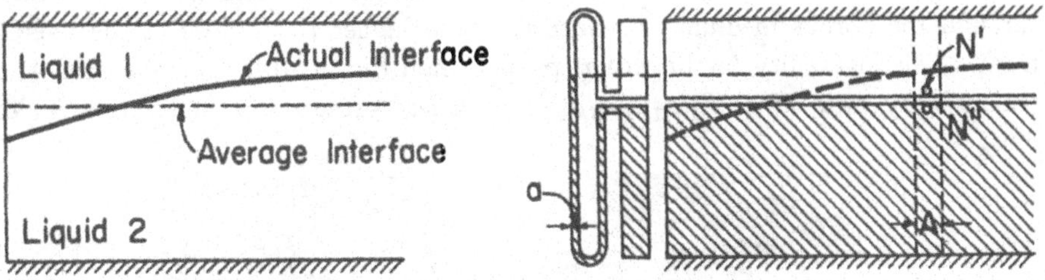

(a) Actual conditions *(b)* The Small Perturbation Approximation

FIG. 9.5.13. Physical interpretation of (9.5.83).

$$\frac{\partial}{\partial x}[(b-\zeta^0)q^0_{x1}] + \frac{\partial}{\partial y}[(b-\zeta^0)q^0_{y1}] = 0$$

$$\frac{\partial}{\partial x}[\zeta^0 q^0_{x2}] + \frac{\partial}{\partial y}[\zeta^0 q^0_{y2}] = 0. \tag{9.5.85}$$

For the unsteady deviations, the equations are:

$$n\frac{\partial \zeta'}{\partial t} = \frac{\partial}{\partial x}[(b-\zeta^0)q'_{x1}] + \frac{\partial}{\partial y}[(b-\zeta^0)q'_{y1}] - \frac{\partial}{\partial x}(\zeta' q^0_{x1}) - \frac{\partial}{\partial y}(\zeta' q^0_{y1})$$

$$-n\frac{\partial \zeta'}{\partial t} = \frac{\partial}{\partial x}[\zeta^0 q'_{x2}] + \frac{\partial}{\partial y}[\zeta^0 q'_{y2}] + \frac{\partial}{\partial x}[\zeta' q^0_{x2}] + \frac{\partial}{\partial y}[\zeta' q^0_{y2}]. \tag{9.5.86}$$

For the simple case where initially both liquids are at rest, $q^0_{xi} = q^0_{yi} = 0$. Then $\zeta^0 = \text{const}$, and (9.5.86) becomes:

$$n\frac{\partial \zeta'}{\partial t} = (b-\zeta^0)\left[\frac{\partial q'_{x1}}{\partial x} + \frac{\partial q'_{y1}}{\partial y}\right]$$

$$-n\frac{\partial \zeta'}{\partial t} = \zeta^0\left[\frac{\partial q'_{x2}}{\partial x} + \frac{\partial q'_{y2}}{\partial y}\right]. \tag{9.5.87}$$

Since $q'_{xi} = -\partial\Phi'_i/\partial x$, $q'_{yi} = -\partial\Phi'_i/\partial y$ and $\zeta' = \alpha_2\Phi'_2 - \alpha_1\Phi'_1$, equation (9.5.87) can be rewritten in terms of ζ' only:

$$n\left(\frac{\alpha_1}{b-\zeta^0} + \frac{\alpha_2}{\zeta^0}\right)\frac{\partial \zeta'}{\partial t} = \frac{\partial^2 \zeta'}{\partial x^2} + \frac{\partial^2 \zeta'}{\partial y^2}. \tag{9.5.88}$$

Equation (9.5.88) is a *linear* partial differential equation of the parabolic type. It is analogous to the approximate equations describing flow with a phreatic surface (sec. 8.4). The resistivity $1/K\bar{h}$ in (8.4.5) is replaced in (9.5.88) by $[\alpha_1/(b-\zeta^0) + \alpha_2/\zeta^0]$.

9.5.8 Interface Stability

Up to this point, it has been assumed that the abrupt interface between two immiscible fluids, which is a macroscopic approximate concept, is a regular surface. However, under certain conditions, this approximation fails and, instead of the displacement of the interface as a whole in a regular form, protuberances occur that may advance through the porous medium at velocities much higher than those of the average front. This instability (or fingering) phenomenon (fig. 9.5.14) always occurs when the mobility ratio $M = (k_1/\mu_1)/(k_2/\mu_2) > 1$, where subscripts 1 and 2 refer to the

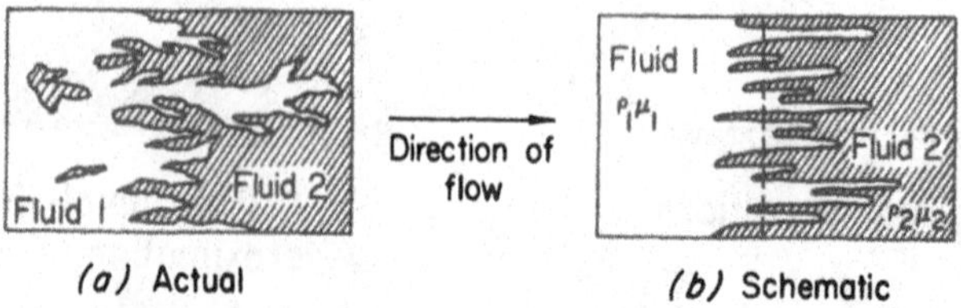

FIG. 9.5.14. Instability at a moving interface.

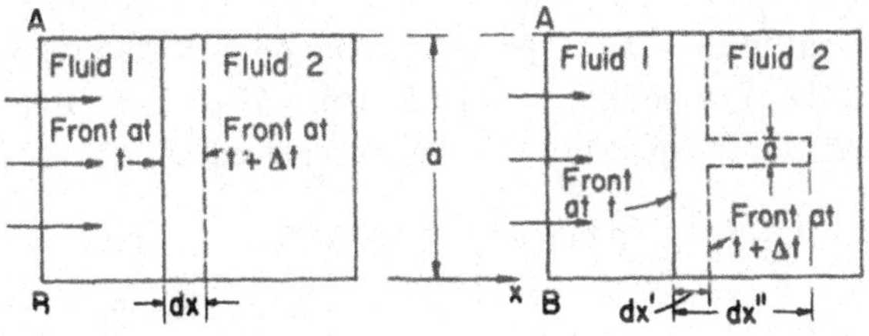

(a) Stable displacement *(b)* Model of finger

FIG. 9.5.15. Analysis of fingering (Scheidegger, 1960b).

displacing and displaced fluids, respectively. The phenomenon of instability, which is initiated by the local *microscopic inhomogeneity* of the medium, is most important in oil reservoirs where it reduces sweep efficiencies. There, when oil is displaced by water, k_1 is the permeability to water at residual oil saturation, k_2 is the permeability to oil at connate water saturation.

Because the frontal instability occurs at a microscopic level, its analysis in an exact manner is not possible. The random nature of the microscopic porous structure suggests an approach that involves a statistical description of the porous medium. Scheidegger (1960a) introduces the effect of heterogeneities contained in the porous medium by introducing a random perturbation term into the differential equation of motion for each spectral component of the fingers. By doing so he arrives at an equation describing the growth in time of each spectral component of the fingers. He shows that under given external conditions, fingering should be independent of the velocity of the front's displacement.

A different approach is to study simplified displacement models that are assumed to be representative of the fingering phenomenon.

Scheidegger (1960b) uses a model of an abrupt front in horizontal flow in order to analyze the phenomenon of fingering. Let figure 9.5.15a represent the horizontal displacement during a time interval dt, assuming a sharp front and a stable displacement obeying Darcy's law. As the front moves at an average velocity V, the displacement dx and the displaced volume U are given by:

$$dx = V\,dt; \qquad U = na\,dx. \tag{9.5.89}$$

Also:

$$-\mathbf{V} = (k_1/\mu_1 n)\operatorname{grad} p_1 = (k_2/\mu_2 n)\operatorname{grad} p_2 \tag{9.5.90}$$

where n is porosity and p_1 and p_2 are the pressures in the two fluids.

Let a tiny region on the front be (locally) more permeable than the surrounding region. Then, as the front advances, it will move more rapidly through this more permeable region, giving rise to a small "bump," or "finger." Figure 9.5.15b shows how such a "finger" develops. In this case, the displacement caused by the same volume U entering through AB is:

$$U = n(a'\,dx' + a''\,dx''); \qquad a'' + a' = a. \tag{9.5.91}$$

Such a finger can be maintained stable if there is no tendency for it either to spread out or to retreat and disappear. This is possible only if the pressure gradient in it is the same as in the surrounding liquid. Because $M \neq 1$, we must have $V_1 \neq V_2$. With grad $p_1 =$ grad p_2, equation (9.5.90) is replaced by:

$$n(\mu_1/k_1)V_1 = n(\mu_2/k_2)V_2.$$

With $V_1 = dx'/dt$, $V_2 = dx''/dt$ and $M = (k_1/\mu_1)/(k_2/\mu_2)$, we obtain:

$$dx'/dx'' = M. \tag{9.5.92}$$

According to Scheidegger, a finger will develop only if by its formation it consumes less energy than the corresponding stable displacement of the whole front. In the first case (fig. 9.5.15a), the work W^A required to move fluid 1 into the space of width dx and fluid 2 out of this space is:

$$W^A = a(dx)^2 \operatorname{grad} p_1 + a(dx)^2 \operatorname{grad} p_2 = na(dx)^2[\mu_1/k_1 + \mu_2/k_2](dx/dt). \tag{9.5.93}$$

In the second case, figure 9.5.15b, the work required is:

$$W^B = n[(\mu_1/k_1)a'(dx')^2 + 2(\mu_2/k_2)a''(dx'')^2 + (\mu_2/k_2)a'(dx')^2](dx'/dt). \tag{9.5.94}$$

From $W^B \leqslant W^A$ Scheidegger shows that fingering (i.e., $dx'' > dx'$) will occur whenever $M > 1$.

A somewhat different approach is presented by Marle (1965). In order to introduce the effect of density differences as well, he studies a vertical displacement of a horizontal front (fig. 9.5.16). In the stable case (without a finger), the specific discharges q_1 of fluid 1 and q_2 of fluid 2 are given by:

$$q_1 = -(k_1/\mu_1)(\partial p_1/\partial x + \rho_1 g); \qquad q_2 = -(k_2/\mu_2)(\partial p_2/\partial x + \rho_2 g). \tag{9.5.95}$$

The front moves at a velocity $v_F = q_1/n = q_2/n$, where n stands for either the porosity (for complete separation) or for $n(1 - S_{w0} - S_{nw0})$ in the case of incomplete separation; k_1 and k_2 are effective permeabilities. When a finger is formed, let V_D, q_D be the velocity and the specific discharge of its edge (point D in fig. 9.5.16). Within the finger:

$$q_D = -(k_1/\mu_1)(\partial p_2/\partial x + \rho_1 g) = nV_D. \tag{9.5.96}$$

In (9.5.96), $\partial p_2/\partial x$ is used because the finger is completely surrounded by fluid 2, which imposes its pressure on the fluid within the finger: $V_D > V_F$ means that

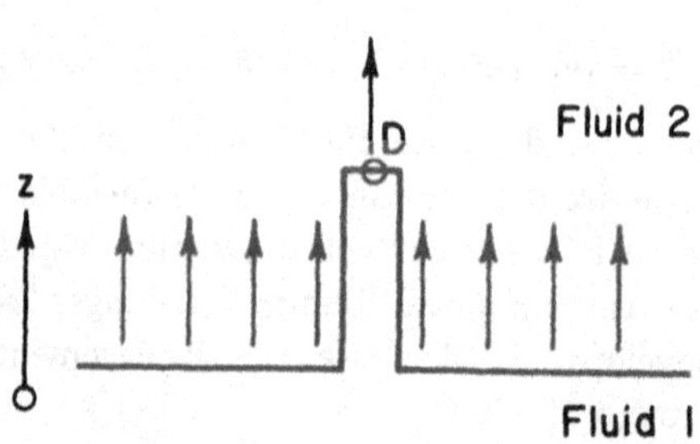

FIG. 9.5.16. A finger in the vertical direction

the finger tends to grow, and vice versa. From (9.5.95) and (9.5.96) it follows that:

$$V_D - V_F = -(k_1/n\mu_1)(\partial p_2/\partial x - \partial p_1/\partial x) \tag{9.5.97}$$

or, using (9.5.97):

$$V_D - V_F = -(k_1/n\mu_1)[nV_F(\mu_1/k_1 - \mu_2/k_2) + g(\rho_1 - \rho_2)]. \tag{9.5.98}$$

The last equation may also be written in the form:

$$V_D - V_F = (M-1)(V_F - V_C); \quad V_C = -k_1 g(\rho_1 - \rho_2)/n\mu_1(1-M). \tag{9.5.99}$$

Let $V_F > 0$ denote vertically upward flow; $g > 0$. Then:

(a) if $\mu_2/k_2 < \mu_1/k_1$ (i.e., $M < 1$) and $\rho_1 > \rho_2$; $V_C < 0$ and, always, $V_D - V_F < 0$, the front is stable;

(b) if $\mu_2/k_2 < \mu_1/k_1$ (i.e., $M < 1$) and $\rho_1 < \rho_2$: always $V_C > 0$, but there are two possibilities: if $V_F > V_C$ the front is stable; if $V_F < V_C$, the front is unstable;

(c) if $\mu_2/k_2 > \mu_1/k_1$ (i.e., $M > 1$) and $\rho_1 > \rho_2$, the displacement is stable if $V_F < V_C$, and unstable if $V_F > V_C$;

(d) if $\mu_2/k_2 > \mu_1/k_1$ (i.e., $M > 1$) and $\rho_1 < \rho_2$, we cannot have $V_F < V_C$ because $V_C < 0$, and the front is always unstable. The velocity V_C is a certain critical velocity that determines the stability of the front.

Jacquard and Seguier (1962) also investigate this problem, reaching similar conclusions.

Another force that may affect the stability of a front is the capillary force. Only approximate theories (e.g., Chouke, Van Meurs and Van de Pool (1959)) exist on this subject. The capillary forces act mainly in the direction perpendicular to the fingers and tend to equalize saturations. Therefore, these forces always tend to stabilize the displacement front.

9.6 Determining the Steady Interface by the Hodograph Method

The hodograph method for solving two-dimensional single liquid flow problems (in the vertical xy-plane), where a phreatic surface forms part of the external boundary of the flow domain, is described in detail in section 8.3. Here this method is applied to flow problems where an abrupt interface (par. 9.5.1) separates a stagnant liquid from a moving one in steady two-dimensional flow in a vertical plane. In this case the flow domain is that occupied by the moving liquid; the interface forms a portion of the external boundary of this domain.

The various types of boundaries and boundary conditions have been discussed in section 7.1, and the boundary conditions along an interface in paragraph 9.5.2. Another type of boundary, which does not appear when a single liquid is considered, is the boundary between the flow domain in the porous medium and an adjacent body of liquid continuum (e.g., a reservoir, sea, lake, etc.) containing the same liquid as that which is stationary in the porous medium on the other side of the interface. The boundary condition existing in such a case is treated in paragraph 9.6.1.

For the sake of simplicity, the medium is assumed to be homogeneous and isotropic, so that the specific discharge potential Φ can be employed. The effect of anisotropy may be introduced by using the analysis presented in section 7.4.

9.6.1 Boundary Conditions

In addition to the boundary conditions discussed in section 7.1, we should also consider the following.

Type 1, boundary condition along a stationary abrupt interface between a moving liquid (γ_1) *and a stationary liquid* ($\gamma_2 > \gamma_1$). This boundary condition for an isotropic medium is given by (9.5.28). When written for two-dimensional flow in the vertical xy-plane, it takes the form:

$$\alpha_1[(\partial\Phi_1/\partial x)^2 + (\partial\Phi_1/\partial y)^2] + \partial\Phi_1/\partial y = 0 \tag{9.6.1}$$

or, in terms of specific discharge components:

$$q_x^2 + q_y^2 - K'_1 q_y = 0; \qquad q_x^2 + (q_y - K'_1/2)^2 = (K'_1/2)^2. \tag{9.6.2}$$

When the upper liquid (γ_1) is stationary, while the lower liquid ($\gamma_2 > \gamma_1$) is flowing, we have from (9.5.27):

$$\alpha_2[(\partial\Phi_2/\partial x)^2 + (\partial\Phi_2/\partial y)^2] - \partial\Phi_2/\partial y = 0 \tag{9.6.3}$$

or:

$$q_x^2 + q_y^2 + K'_2 q_y = 0; \qquad q_x^2 + (q_y + K'_2/2)^2 = (K'_2/2)^2 \tag{9.6.4}$$

as the boundary condition along the interface.

Type 2, boundary condition along an outflow face of liquid 1 into a reservoir of liquid 2 ($\gamma_2 > \gamma_1$). The portion of the bottom of the sea through which fresh water is drained from the aquifer into the sea is an example of this type of boundary. The boundary condition along a surface of this type is $(\varphi^2{}_2 \equiv)\varphi_2 = \text{const.}$ However, since all boundary conditions should be expressed in terms of $(\varphi^1{}_1 \equiv)\varphi_1$ (of the flowing liquid 1), we must use (9.5.14) in order to express the boundary condition in terms of φ_1 only. We obtain:

$$\varphi_1 = \varphi_2(\gamma_2/\gamma_1) - y(\Delta\gamma/\gamma_1); \qquad \Delta\gamma = \gamma_2 - \gamma_1 \tag{9.6.5}$$

where $\varphi_2 = \text{const.}$ For example, by setting $\varphi_2 = 0$ we obtain:

$$\varphi_1 = -y\,\Delta\gamma/\gamma_1 \tag{9.6.6}$$

which is not an equipotential boundary in terms of φ_1. We may also write the boundary conditions as:

$$\nabla\varphi_2 \times \mathbf{1n} = 0 \tag{9.6.7}$$

which in view of (9.6.5) may be written as:

$$[\nabla\varphi_1 + (\Delta\gamma/\gamma_1)\mathbf{1y}] \times \mathbf{1n} = 0$$

$$(\partial\varphi_1/\partial x)\cos\alpha_{ny} - (\partial\varphi_1/\partial y + \Delta\gamma/\gamma_1)\cos\alpha_{nx} = 0$$

or:

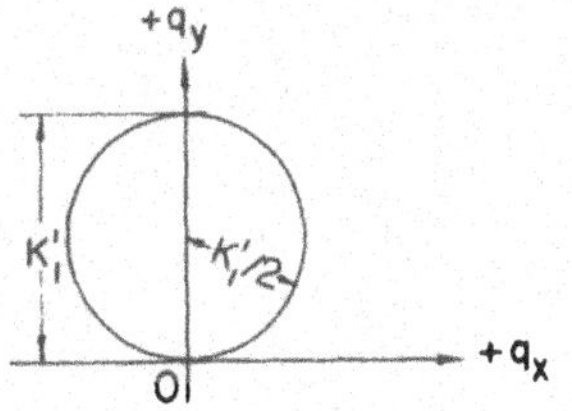

FIG. 9.6.1. Hodograph representation of interface (upper liquid is moving).

$$q_y = q_x \tan(\alpha_{sx} - 90°) + K'_1; \qquad K'_1 = K_1 \Delta\gamma/\gamma_1. \tag{9.6.8}$$

Thus both types of boundary conditions are expressed in terms of the specific discharge components q_x and q_y of the flowing liquid.

It is most interesting to note the striking similarity of (9.6.2) and (9.6.8) to (7.3.10) and (7.3.7), respectively. The main difference is that K in the case of the phreatic surface and the seepage face is replaced in the case of a moving liquid 1 by $-K'_1$. When the lower liquid 2 is moving, while the overlying liquid 1 is stationary, the same similarity exists with K replaced by K'_2.

9.6.2 Description of Boundaries in the Hodograph Plane

In addition to the boundaries considered in section 7.3, the boundaries discussed in paragraph 9.6.1 are described in the hodograph $q_x + iq_y$ plane (section 7.3) in the following way.

Case 1, the stationary interface between a moving liquid 1 overlying a stationary liquid 2. From (9.6.2) it follows that this boundary is described in the hodograph plane by a circle of radius $K'_1/2$ centered at $q_x = 0$, $q_y = K'_1/2$ (fig. 9.6.1).

If the lower liquid is moving, it follows from (9.6.4) that the hodograph representation of the interface is a circle of radius $K'_2/2$ centered at $(0, -K'_2/2)$.

Case 2, the outflow corresponding to case 1. From (9.6.8) it follows that when the outflow face is a straight line making an angle α_{sx} with $+x$ axis, its hodograph representation is a straight line passing through $(0, K'_1)$ and making an angle $\alpha_{sx} - 90°$ with the $+q_x$ axis (fig. 9.6.2).

Two examples of hodograph mappings of an interface in a coastal aquifer under various conditions are shown in figure 9.6.3. In both cases the lower liquid is stationary. It is interesting to note that at point A, the tangent to the interface is *vertical*. One should also note that $K'_1 \ll K$ (not properly scaled in the figures).

9.6.3 Examples

Several examples are given below for the solution of flow problems where an interface forms part of the boundary of the flow domain. The solution is obtained by the hodograph method described in detail in section 8.3.

Example 1. Consider (Bear and Dagan 1964) the interface in an infinite coastal confined aquifer of constant thickness D (fig. 9.6.4a) in which fresh water is discharged

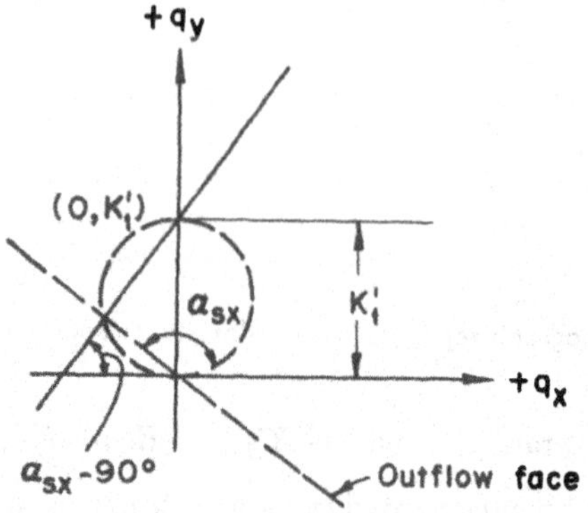

FIG. 9.6.2. Hodograph representation of outflow face.

to the sea through the horizontal outflow face AB above a stationary interface BE. Figures 9.6.4b and 9.6.4c show the flow domain $FGABEF$ mapped onto the $\tilde{\omega}$- and ζ-planes, respectively. Figure 9.6.4d shows the flow domain in the $\tilde{\omega}'$-plane, where $\tilde{\omega}' = K'/\tilde{\omega}$. Note that at point A we set (arbitrarily) $\Phi_1 = K\varphi_1 = 0$. At point E, $y = -D$, $p = \gamma_2 D$, $\Phi_2 = 0$; $\Phi_1 = K'D = \Phi_0$.

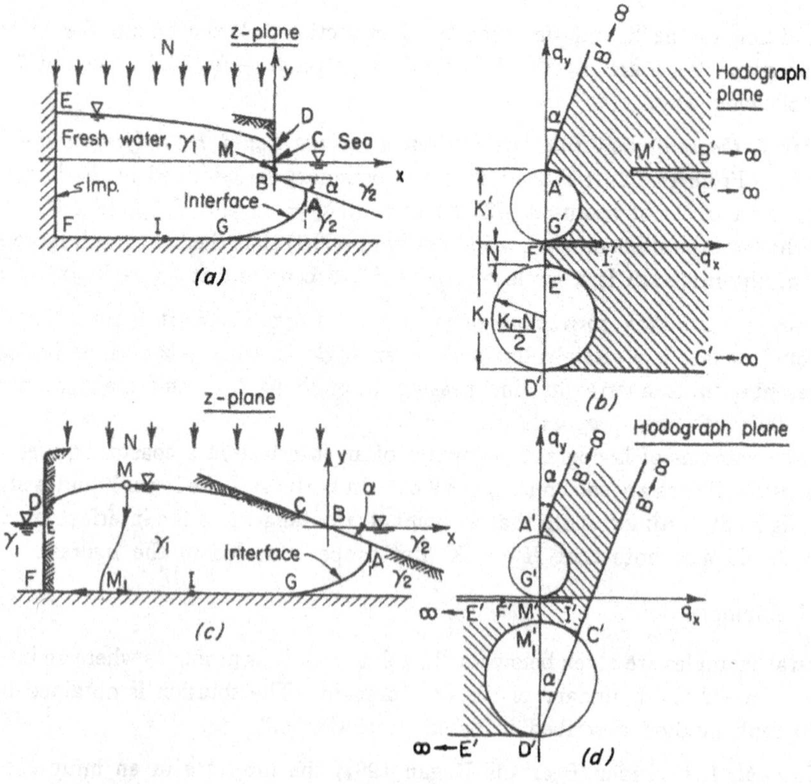

FIG. 9.6.3. Examples of hodograph representation of boundaries.

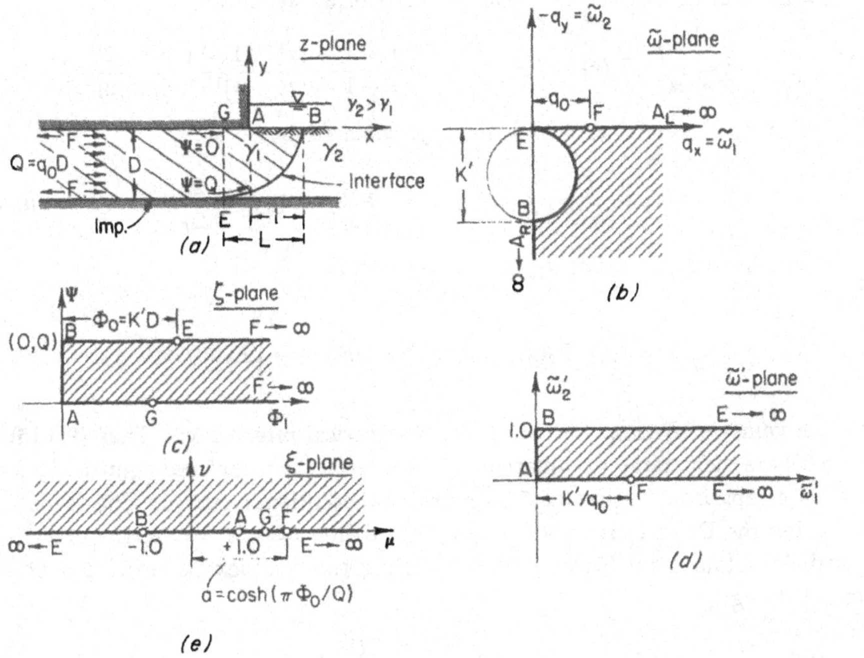

FIG. 9.6.4. Determining the interface in a confined aquifer by the hodograph method.

The flow region in both the $\tilde{\omega}$- and the $\tilde{\omega}'$-planes is now mapped onto the upper half of the ξ-plane by:

$$\xi = \cosh \pi\tilde{\omega}' = \cosh(\pi K'/\tilde{\omega}) \tag{9.6.9}$$

$$\zeta = \frac{Q}{\pi} \cosh^{-1} \frac{a\xi - 1}{a - \xi} \tag{9.6.10}$$

where $a = \cosh(\pi K'D/Q) = \pi K'/q_c$. With (8.3.8), (8.3.9), (9.6.9) and (9.6.10), we obtain:

$$dz = -\frac{d\zeta/d\xi}{\tilde{\omega}(\xi)} d\xi = -\frac{Q(a^2-1)^{1/2}\cosh^{-1}\xi}{\pi^2 K'(a-\xi)(\xi^2-1)^{1/2}} d\xi. \tag{9.6.11}$$

To derive the parametric equations of the interface in the z-plane, we integrate (9.6.11) along the segment BE. In the ξ-plane this means integration in the range $-\infty < \xi < -1$. Rewriting (9.6.11) with $\eta = \cosh^{-1}(-\xi)$, we obtain:

$$dz = \frac{Q}{\pi^2 K'} \frac{(a^2-1)^{1/2}}{a + \cosh\eta} (\eta + i\pi)\, d\eta. \tag{9.6.12}$$

Then, by separating real from imaginary parts, we obtain:

$$dx = \frac{Q}{\pi^2 K'} \frac{(a^2-1)^{1/2}}{a+\cosh\eta} \eta\, d\eta; \qquad dy = \frac{Q}{\pi K'} \frac{(a^2-1)^{1/2}}{a+\cosh\eta} d\eta. \tag{9.6.13}$$

By integrating the second equation in (9.6.13) we obtain:

$$y = \frac{Q}{\pi K'} F_1(\eta); \qquad F_1(\eta) = \ln \frac{a + 1 + (a^2 - 1)^{1/2} \tanh(\eta/2)}{a + 1 - (a^2 - 1)^{1/2} \tanh(\eta/2)}. \tag{9.6.14}$$

For large values of a, y may be approximated by:

$$y = \frac{Q}{\pi K'} [\eta - F_3(\eta)], \qquad F_3(\eta) = \ln \frac{e^\eta + 2a}{1 + 2a}. \tag{9.6.15}$$

From the first equation in (9.6.13) we obtain:

$$x = \frac{Q}{\pi^2 K'} F_2(\eta); \qquad F_2(\eta) = \eta F_1(\eta) - \int_0^\eta F_1(\eta')\, d\eta'. \tag{9.6.16}$$

The values of F_2 may be obtained by numerical integration. Thus (9.6.15) and (9.6.16) are the parametric equations of the interface in a confined aquifer (fig. 9.6.5).

For comparison, figure 9.6.5 includes two additional curves: one is obtained by using the Dupuit assumptions (sec. 9.7) with $y = 0$ at $x = 0$, the other is the parabola obtained by Glover (1959) by using the complex potential $\zeta = \Phi + i\Psi$ in satisfying:

$$z = \zeta^2 K'/2Q \tag{9.6.17}$$

for an aquifer of infinite depth. By separating real from imaginary parts he obtains:

$$\varphi = [x + (x^2 + y^2)^{1/2}]^{1/2} (Q/K')^{1/2} = \Phi/K$$
$$\psi = [-x + (x^2 + y^2)^{1/2}]^{1/2} (Q/K')^{1/2} = \Psi/K. \tag{9.6.18}$$

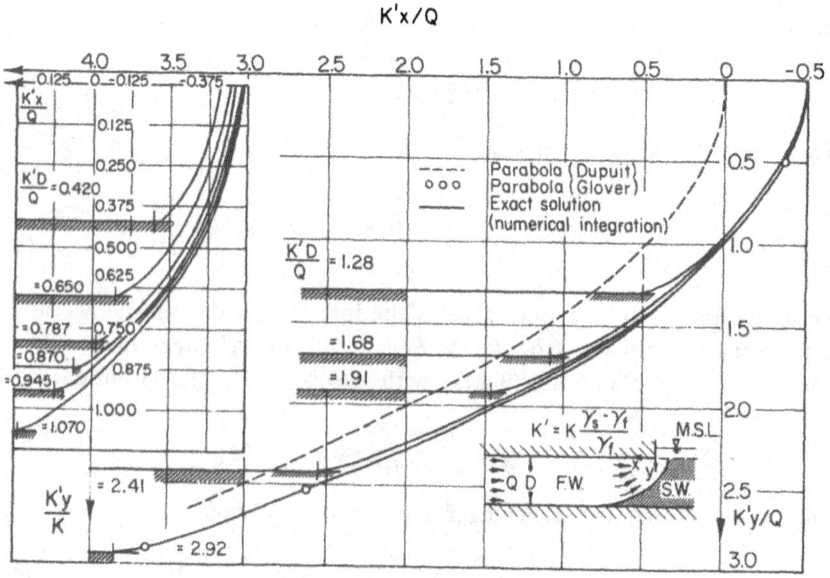

FIG. 9.6.5. Shapes of an interface in a confined aquifer.

The corresponding interface is then described by the parabola:

$$y^2 = (2Q/K')x + (Q/K')^2 \tag{9.6.19}$$

with:

$$y = 0, \qquad x = x_0 = -Q/2K'.$$

Two additional conclusions may be drawn from the exact solution derived above. The first concerns the length of the outflow face (AB in fig. 9.6.4a):

$$l \equiv \overline{AB} = \int_{-1}^{+1} \frac{Q(a^2-1)^{1/2}}{\pi^2 K'} \frac{\cosh^{-1}\xi}{(a-\xi)(\xi^2-1)^{1/2}} \, d\xi \tag{9.6.20}$$

which for sufficiently large values of $K'D/Q$ ($K'D/Q > 1.3$) becomes:

$$l \equiv \overline{AB} = Q/2K'. \tag{9.6.21}$$

This result agrees with Glover's solution (9.6.19) for an aquifer of infinite depth.

The second conclusion concerns the piezometric head $\varphi(G)$ at point G located on the upper confining layer directly above the interface toe E. This point will be of special interest in section 9.7. From $q_x = -\partial\Phi/\partial x$, and with $q_y = 0$ along AG, we obtain from (9.6.9):

$$\tilde{\omega}' = q_x - iq_y = q_x = \frac{\pi K'}{\cosh^{-1}\xi} = -\frac{\partial\Phi}{\partial x}$$

$$\Phi(G) = K\varphi(G) - \int_0^{L-l} (\pi K'/\cosh^{-1}\xi)\, dx = [Q(a^2-1)^{1/2}/\pi] \int_0^{\beta_G} \{1/(a-\cosh\beta)\}\, d\beta \tag{9.6.22}$$

where $\beta = \cosh^{-1}\xi$ and β_G is obtained from:

$$\overline{AG} = \int_0^{L-l} dx = [Q(a^2-1)\pi^2] \int_0^{\beta_G} \{\beta/(a-\cosh\beta)\}\, d\beta. \tag{9.6.23}$$

Example 2, upconing toward a sink (drain) in an aquifer of infinite extent in all directions (Bear and Dagan 1964). Consider a point, sink A, of constant strength, Q, located at some distance above an interface between fresh water and salt water (fig. 9.6.6a). When upconing takes place, this distance is denoted by b. Figure 9.6.6b shows the $\tilde{\omega}$-plane of the left half of the flow domain in the z-plane.

An inflection point B, where the velocity makes an angle β ($0 < \beta < \pi/2$) with the horizontal axis, exists somewhere along the interface. The case $\beta = \pi/2$ (fig. 9.6.6d) corresponds to the limit case for which salt water is still at rest.

The case $\beta = \pi/2$ is called *critical situation* in the following discussion. The corresponding discharge is called the critical discharge Q_{cr}. The hodograph representation of half the flow region in the critical situation is shown in figure 9.6.6e. Figure 9.6.6c shows the ζ-plane for the general case of upconing.

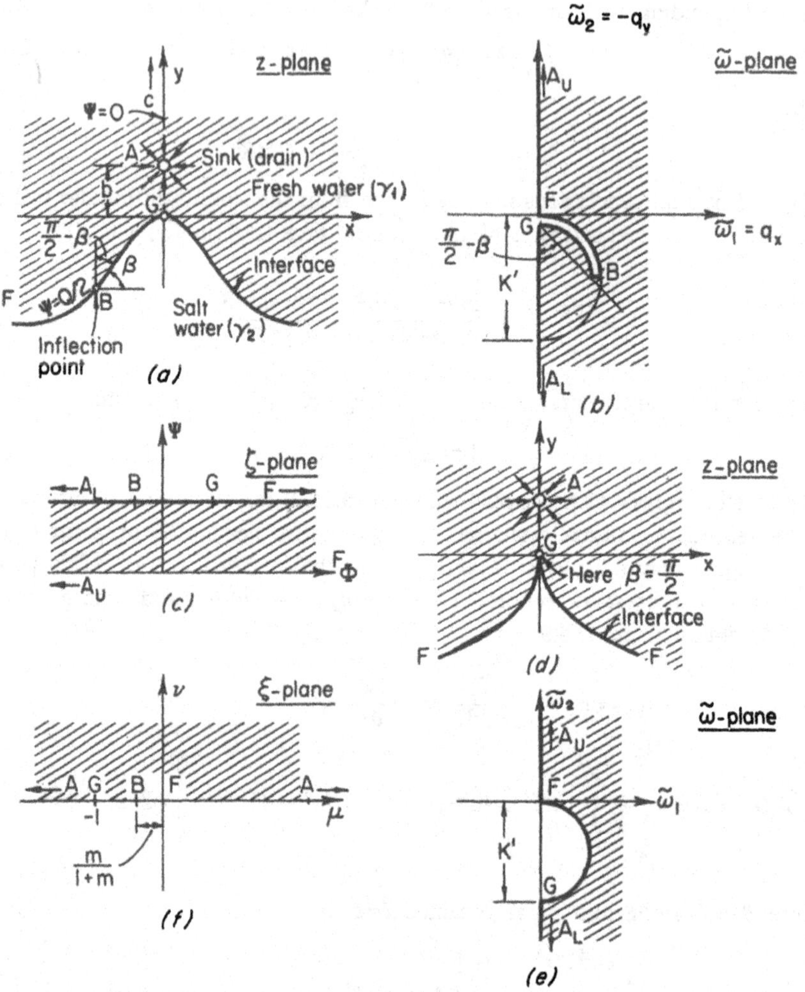

Fig. 9.6.6. Interface upconing towards a sink in an infinite flow domain (Bear and Dagan, 1964).

The flow region in the $\tilde{\omega}$-plane is now mapped on the upper half of the $\xi = \mu + i\nu$ plane (fig. 9.6.6f), by:

$$\tilde{\omega} = -iK'\left[1 - \tfrac{1}{2}\operatorname{cotg}\beta\left\{\left(\frac{m(1+\xi)}{\xi}\right)^{1/2} - \left(\frac{\xi}{m(1+\xi)}\right)^{1/2}\right\}\right]^{-1}$$

$$m = \tan^2(\pi/2 + \beta/2). \tag{9.6.24}$$

The strip in the ζ-plane is mapped on the upper half of the ξ-plane by:

$$\zeta = \frac{Q}{2\pi}\ln \xi. \tag{9.6.25}$$

By (8.3.9) we have:

$$\frac{dz}{d\xi} = \frac{Q}{2\pi i K'}\frac{1}{\xi}\left\{1 - \tfrac{1}{2}\operatorname{cotg}\beta\cdot\left[\left(\frac{m(1+\xi)^{1/2}}{\xi}\right) - \left(\frac{\xi}{m(1+\xi)}\right)^{1/2}\right]\right\} \tag{9.6.26}$$

$$z = \frac{Q}{2\pi i K'}\left[\ln\frac{\xi}{[\xi^{1/2} + (1+\xi)^{1/2}]^{1/2}} + m^{1/2}\left(\frac{1+\xi}{\xi}\right)^{1/2}\operatorname{cotg}\beta\right] + C. \tag{9.6.27}$$

A similar result was also obtained by Polubarinova-Kochina (1952, 1962). The constant C may be taken as zero if the origin is located at the peak G.

The shape of the interface (FBG in fig. 9.6.6a) is obtained from (9.6.27) for $-1 \leqslant \xi \leqslant 0$ on the real axis. Let $\xi = -\eta^2/(1+\eta^2)$; $0 \leqslant \eta \leqslant \infty$. Then the parametric equations of the interface take the form:

$$\frac{xK'\pi}{Q} = -\tan^{-1}\eta - \frac{m^{1/2}\operatorname{cotg}\beta}{2}\frac{1}{\eta} + \frac{\pi}{2} \tag{9.6.28}$$

$$\frac{yK'\pi}{Q} = -\tfrac{1}{2}\ln\frac{\eta^2}{1+\eta^2}. \tag{9.6.29}$$

The distance b from the top of the upconed interface (fig. 9.6.6a) to the sink is obtained by integrating (9.6.26) between $\xi = -\infty$ and $\xi = -1$:

$$\frac{\pi K' b}{Q} = \tfrac{1}{2}(\ln 4 - m^{1/2}\operatorname{cotg}\beta). \tag{9.6.30}$$

At the critical situation $\beta = \pi/2$, and

$$K' b_{cr}/Q = 0.097. \tag{9.6.31}$$

Figure 9.6.7 shows the shapes of the upconed interface for various values of β. To determine the value of β in a specific case, we must know one point of the interface in the xy-plane and determine β from (9.6.28) and (9.6.29). The value of Q_{cr} may be derived in a similar manner.

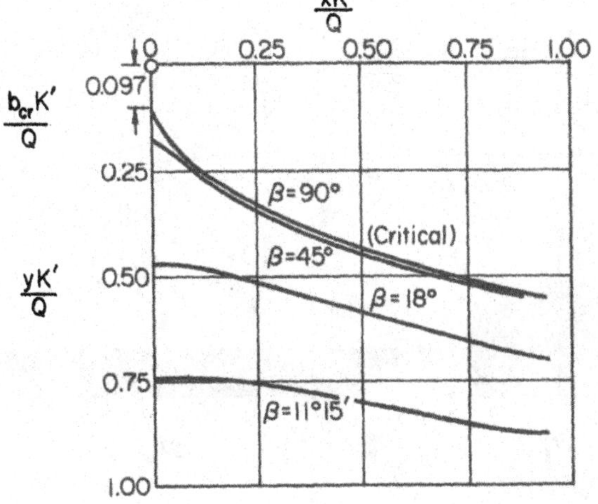

FIG. 9.6.7. Shapes of interface below an operating sink (Bear and Dagan, 1964).

Example 3. This example (De Josselin de Jong 1964, 1965) is introduced to demonstrate a solution in which the hodograph plane takes the form of a *Riemann surface.* This case is shown in figure 9.6.8a. The drain intercepts only part (Q_1) of the fresh water discharge, whereas the remaining part, $Q_2 = Q - Q_1$, flows to the sea bottom. The physical z-plane is shown on figure 9.6.8a. The interface (DM_2M_1F in fig. 9.6.8a) has two inflection points: M_1 and M_2.

Figure 9.6.8b shows the hodograph $\tilde{\omega}$-plane of this flow. In this plane, the fresh water flow domain is mapped onto the region $BA_1A_2SG_1G_2FM_1M_2DB$, which has the form of a Riemann surface (Churchill 1948). The boundaries A_1A_2 and G_1G_2

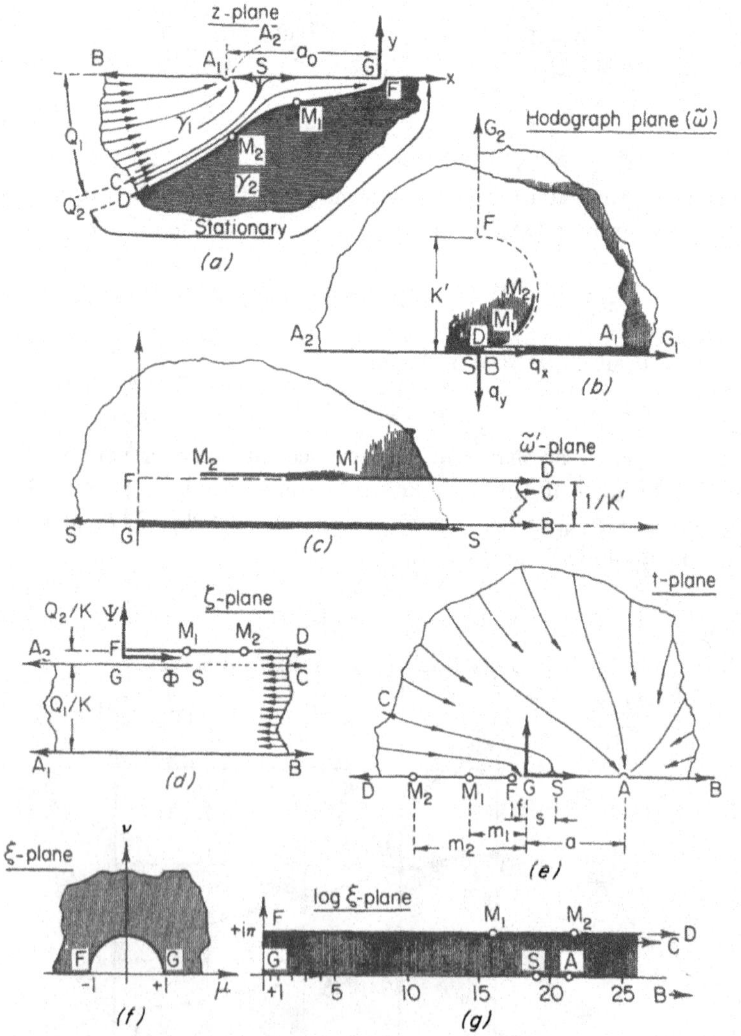

FIG. 9.6.8. Hodograph solution for a sink operating above an interface (de Josselin de Jong, 1964, 1965).

are circles of infinite radii. The segment FM_1M_2 is a slot in the lower (shaded) sheet, and there is a narrow "bridge" between the points S and M_1 that connects the region FG_2G_1 to A_2. From A_2, over the large circle, to A_1, and farther toward B, there is an overlap encircling the slot at M_2. This overlap terminates at the beak that contains B and D. Thus the flow domain in the $\tilde{\omega}$-plane has the form of a double sheet. When the $\tilde{\omega}$-plane is cut out of paper, by gluing two sheets together, it is possible to make a model that is simply connected.

Figure 9.6.8d shows the complex potential ζ-plane. The discharge rates Q_1 and Q_2, through the drain and through the outflow face, FG, respectively, are considered as known values. The location of the stagnation point S is as yet unknown. It is to be determined in the solution process.

To determine the shape of the interface in the z-plane, we map the flow domain in both the ζ- and the $\tilde{\omega}$-planes onto the upper half of the t-plane (fig. 9.6.8e). Then (8.3.9) is applicable.

To map ζ on t, we use the Schwarz–Christoffel transformation which leads to (de Josselin de Jong 1964, 1965):

$$\zeta(t) = \alpha_2[\ln \xi - (a-s)a^{-1/2}(a+f)^{-1/2}\mathscr{L}(t, f, a)] + \beta_2 \tag{9.6.32}$$

where:

$$\xi = [t^{1/2} + (t+f)^{1/2}]^2/f; \qquad \mathscr{L}(t, f, a) = 2\ln[t^{1/2}(a+f)^{1/2} + (t+f)^{1/2}a^{1/2}] - \ln f(a-t)$$

and the constants α_2, β_2, a, f, s must be determined from the boundary conditions.

The $\tilde{\omega}'$-plane is determined from the $\tilde{\omega}$-plane by the inverse transformation $\tilde{\omega}' = 1/\tilde{\omega}$. Then $\tilde{\omega}'$ is mapped onto the upper half of the ξ-plane by:

$$\tilde{\omega}' = \frac{1}{\pi K'}\left[\ln \xi - \frac{(a-s)t^{1/2}(t+f)^{1/2}}{(t-s)a^{1/2}(a+f)^{1/2}} \ln \xi_a\right] \tag{9.6.33}$$

where $\ln \xi_a = (m_1 + s)(m_2 + s)a^{1/2}(a+f)^{1/2}/s(s+f)(a-s)$, and the various constants are determined from the boundary conditions.

Finally, these expressions are inserted in (8.3.10) from which the shape of the interface is derived in a parametric form (for details see de Josselin de Jong 1964, 1965).

9.7 The Interface in a Coastal Aquifer

9.7.1 Occurrence

Coastal aquifers have, in general, a hydraulic gradient toward the sea that is the recipient of their excess fresh water (replenishment minus pumpage). Owing to the presence of sea water in the aquifer formation under the sea bottom, a zone of contact is formed between the lighter fresh water (γ_f) flowing to the sea, and the denser underlying sea water ($\gamma_s > \gamma_f$). Typical cross-sections with interfaces are shown in figure 9.7.1. In all cases there exists a body of sea water (often in the form of a wedge) underneath the fresh water. One should note that these figures are highly distorted (as are most figures describing flow in aquifers) and not drawn

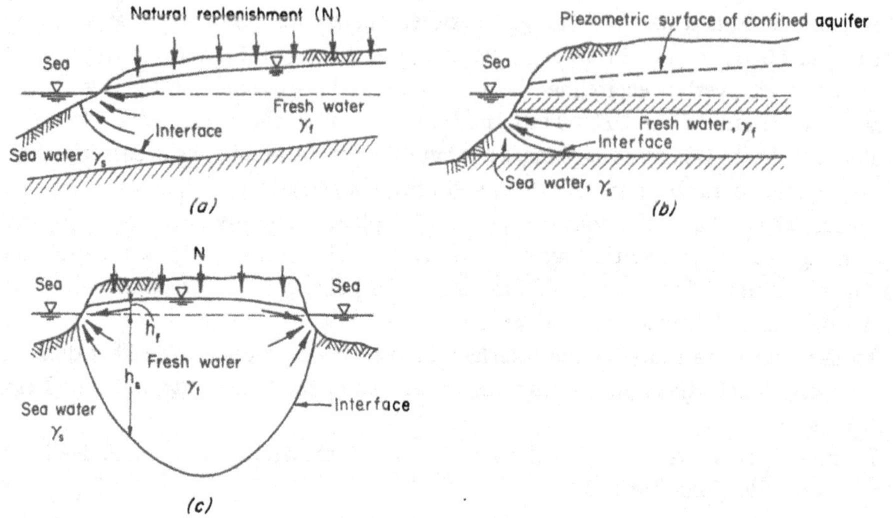

FIG. 9.7.1. Interfaces in coastal aquifers (highly distorted figures).

to scale. In addition, the vertical scales above and below sea level differ.

Fresh water and sea water are actually miscible fluids and therefore the zone of contact between them takes the form of a transition zone caused by ***hydrodynamic dispersion*** (chap. 10). Across this zone the density of the mixed water varies from that of fresh water to that of sea water. However, under certain conditions, the width of this zone is relatively small (e.g., when compared with the thickness of the aquifer), so that the abrupt interface approximation introduced in paragraph 9.5.1 is applicable. For example, observations (Jacobs and Schmorak 1960; Schmorak 1967) along the coast of Israel indicate that indeed this assumption of an abrupt interface is justified. On the other hand, Cooper (1959) describes a case where the transition zone is very wide so that the interface approximation is no longer valid. In what follows, we shall assume that an abrupt interface separates the regions occupied by the two fluids.

Under natural undisturbed conditions in a coastal aquifer, a state of equilibrium is maintained, with a stationary interface and a fresh water flow to the sea above it. At every point along this interface, the slope is determined by (9.5.8). The continuous change of slope (fig. 9.7.1) results from the fact that as the sea is approached, q_{1s} increases. By pumping from the coastal aquifer in excess of replenishment, the water table (or the piezometric surface) in the vicinity of the coast is lowered to the extent that the piezometric head in the fresh water body becomes less than in the adjacent sea water wedge, and the interface starts to advance inland. This phenomenon is called ***sea water intrusion*** (or ***encroachment***). As the interface advances, the transition zone widens; nevertheless, we shall assume that the abrupt interface approximation remains valid. When the advancing interface reaches inland pumping wells, the latter become contaminated. Extensive research is being carried out

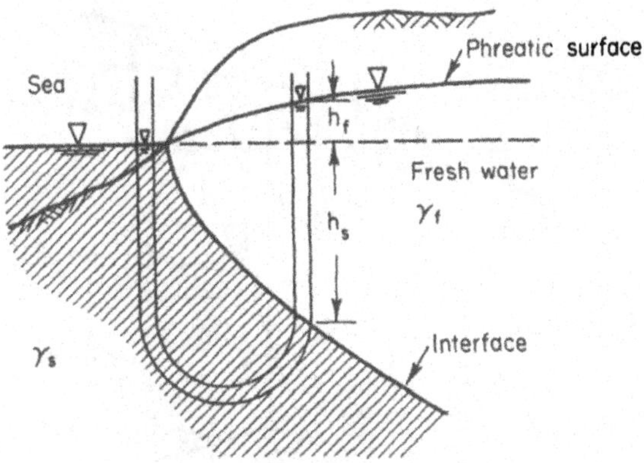

FIG. 9.7.2. Ghyben-Herzberg interface model.

in many parts of the world with the objectives of understanding the mechanism of sea water intrusion and learning to control it in order to improve the yield of coastal aquifers. Some elements of these investigations are reviewed here.

Actually, most of the theoretical background and solution approaches have already been presented in section 9.5 (especially pars. 9.5.3 and 9.5.5) and need not be repeated here. The difficulties inherent in an exact analytical solution should by now be familiar to the reader. Some approximate approaches often employed by ground water hydrologists (in addition to those already discussed in par. 9.5.5) are given below.

A tool that is relatively simple, yet most useful for solving complicated interface management problems in coastal aquifers, is the vertical Hele–Shaw analog described in section 11.4.

Unless otherwise specified, it will be assumed in the present section that the flow is two-dimensional in a vertical plane.

9.7.2 The Ghyben–Herzberg Approximation

Beginning with Badon–Ghyben (1888) and Herzberg (1901), investigations of the coastal interface have been aimed at determining the relationship between its shape and position, and the various hydrological components of a ground water balance in the region near the coast. Figure 9.7.2 shows the idealized Ghyben–Herzberg model of an interface in a coastal phreatic aquifer. Essentially, Ghyben and Herzberg assume static equilibrium and a hydrostatic pressure distribution in the fresh water region with stationary sea water. Instead, we may assume dynamic equilibrium, but with horizontal flow in the fresh water region. This means that equipotentials are vertical lines or surfaces (*identical to the Dupuit assumption*). With the nomenclature of figure 9.7.2, we have under these conditions:

$$h_s = [\gamma_f/(\gamma_s - \gamma_f)]h_f \equiv \delta h_f; \qquad \delta = \gamma_f/(\gamma_s - \gamma_f). \tag{9.7.1}$$

For example, for $\gamma_s = 1.025$ gr/cm^3, $\gamma_f = 1.000$ gr/cm^3, $\delta = 40$ and $h_s = 40h_f$, i.e., at any distance from the sea, the depth of the interface below sea level is 40 times

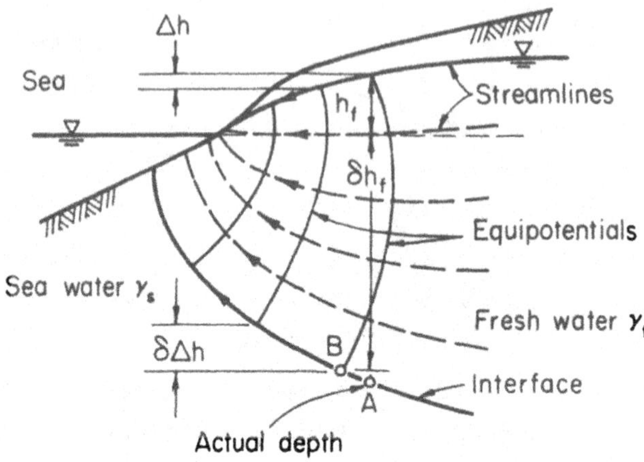

FIG. 9.7.3. Actual flow pattern near the coast.

the height of the fresh water table above it. Obviously, as the sea is approached, the assumption of quasistatic equilibrium with horizontal flow is no longer valid, because vertical flow components can no longer be neglected. Moreover, in figure 9.7.2 no outlet is left for the fresh water flow to the sea. Figure 9.7.3 shows the actual steady flow conditions near the coast. Point A on the interface indicates the actual depth of the interface at that distance from the coast. Point B corresponds to the Ghyben–Herzberg approximation. From (9.7.1) it follows that point A is at a depth of δh_f below sea level. If we set $\varphi^2{}_2 = 0$ (i.e., immobile sea water) in (9.5.14), we see that the difference between the Ghyben–Herzberg approximation (9.7.1) and the exact expression (9.5.14) stems from the difference between h_f and $\varphi^1{}_1$ (i.e., between the assumed vertical equipotential with $\varphi^1{}_1 = h_f$ and the actual curved one). In a confined aquifer, h_s in (9.7.1) is the depth of a point on the interface below sea level, whereas h_f is the (fresh water) piezometric head.

From (9.5.14) it follows that when the interface is in motion, $\varphi^2{}_2\ (\equiv \varphi_2)$ also affects the shape of the interface.

Bear and Dagan (1962a) investigate the validity of the Ghyben–Herzberg relationship. They find that in a confined horizontal aquifer of constant thickness D (fig. 9.6.4a), the approximation is good, within an error of 5%, for determining the depth of the interface toe (point E in fig. 9.6.4a), and hence also the length of the intruding sea water wedge, provided $\pi K'D/Q > 8$, where Q is the fresh water discharge to the sea. Figure 9.7.4 shows this relationship; the exact solution is obtained by the hodograph method (par. 9.6.3, example 1). The piezometric head φ_G is given by (9.6.22).

As the coast is approached, the depth of the interface is more than that given by the Ghyben–Herzberg relationship (fig. 9.7.3). In the case of a phreatic aquifer, a seepage face is also present above sea level. One should also note that for a downward sloping flat sea bottom it follows from the discussion in paragraph 9.6 that the interface terminates at the sea bottom with a tangent in the vertical direction.

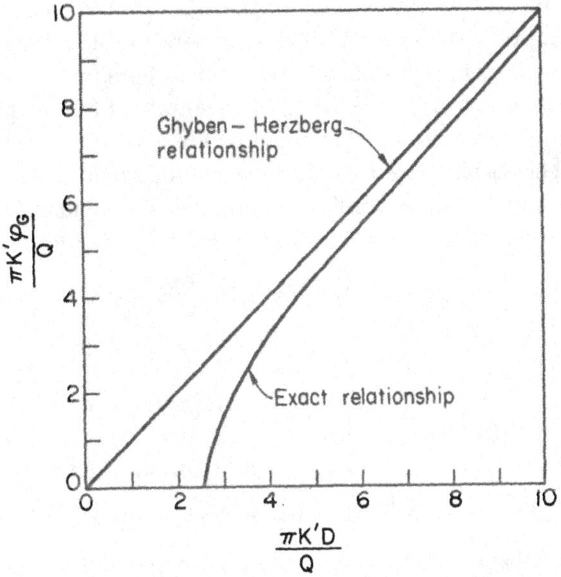

FIG. 9.7.4. Piezometric head above the interface toe in a confined aquifer.

9.7.3 Determining the Shape of a Stationary Interface by the Dupuit–Ghyben–Herzberg Approximation (Santing 1957)

Glover (1959) uses the similarity between (9.5.8) and (9.5.9), and between the boundary conditions on a phreatic surface and those on a stationary interface, to determine the shape of the latter in an infinitely thick confined aquifer, using the method of complex variables (fig. 9.7.5). The results have already been presented (par. 9.6.3, example 1).

Although the hodograph method provides solutions for some relatively simple geometries, for most practical cases (e.g., when the aquifer is a phreatic one, etc.) this approach is too complicated to be considered as a tool for engineering purposes.

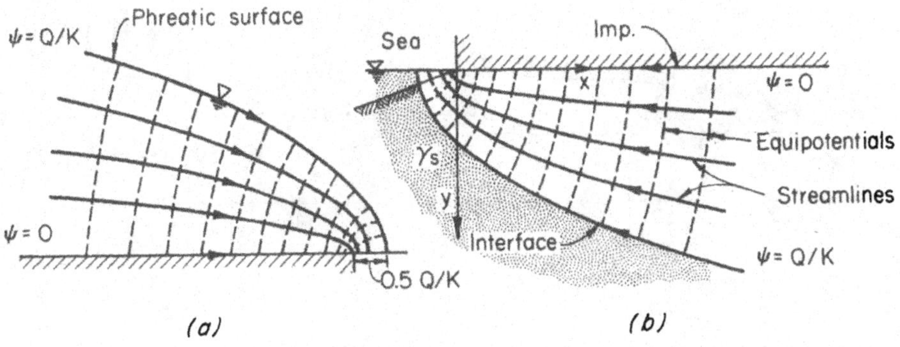

FIG. 9.7.5. The flow net near the coast (Glover, 1959).

A more suitable tool is the combination of the Ghyben–Herzberg relationship with the Dupuit assumption of horizontal flow considered in detail in sections 8.1 and 8.2. The method is demonstrated by means of some simple examples.

Consider the stationary interface in the horizontal confined aquifer of constant thickness shown in figure 9.7.6a. Let the origin $x = 0$ be located at the interface toe (point G). The seaward fresh water flow at this point is Q_0. Using Dupuit's assumption of horizontal flow of fresh water (and vertical equipotentials), continuity leads to:

$$Q_0 = \text{const} = -K_f h(x)\, \partial\varphi(x)/\partial x; \qquad K \equiv K_f = k\gamma_f/\mu_f, \qquad \varphi \equiv \varphi_f. \quad (9.7.2)$$

Since by (9.7.1) $h_s = d + h(x) = \delta\varphi$; $\delta\varphi_0 = d + D$, equation (9.7.2) becomes:

$$Q_0 = -\frac{Kh}{\delta}\frac{dh(x)}{dx} \quad \text{or} \quad Q_0 = -K[\delta\varphi(x) - d]\frac{d\varphi(x)}{dx}. \quad (9.7.3)$$

By integrating, with $x = 0$, $\varphi = \varphi_0$ (or $h = D$), we obtain:

$$Q_0 x = K[D^2 - h^2(x)]/2\delta; \qquad Q_0 x = K\,\delta(\varphi^2{}_0 - \varphi^2)/2 - K\,d(\varphi_0 - \varphi) \quad (9.7.4)$$

which shows that the interface has the form of a parabola.

At $x = L$ we set $h = 0$, $\varphi = d/\delta$. Then, with $\varphi_0 = (D + d)/\delta$ we obtain:

$$Q_0 L = \frac{K\varphi_0}{2}(\delta\varphi_0 - 2d) + \frac{K}{2\delta}d^2 = \frac{K}{2\delta}D^2. \quad (9.7.5)$$

The relationship among the length L of sea water intrusion, the discharge Q_0 to the sea and the piezometric head φ_0 above the toe is clearly expressed by this equation.

Figure 9.7.6b shows a phreatic aquifer with uniform accretion N (say, natural replenishment from precipitation). Again, assuming that the flow in the aquifer is essentially horizontal and $h(x) = \delta\varphi(x)$, continuity leads to:

$$Q_0 + Nx = -K(h + \varphi)\,\partial\varphi/\partial x = -K(1 + \delta)\varphi\,\partial\varphi/\partial x. \quad (9.7.6)$$

Integrating from $x = 0$, $\varphi = \varphi_0$, $h = D$, leads to:

$$\varphi^2{}_0 - \varphi^2 = (2Q_0 x + Nx^2)/K(1 + \delta). \quad (9.7.7)$$

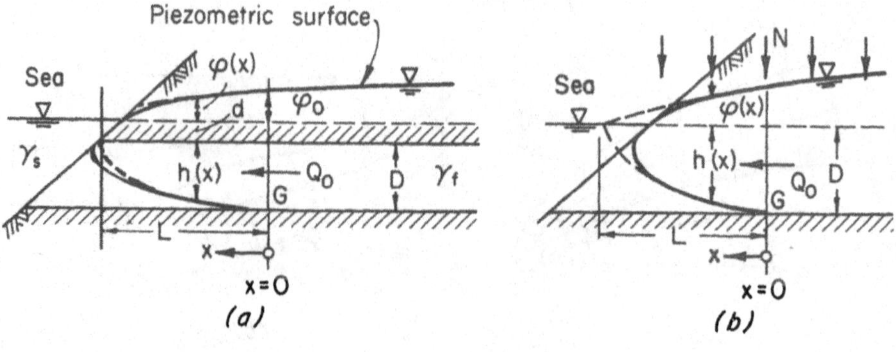

FIG. 9.7.6. The shape of the interface by the Dupuit-Ghyben-Herzberg approximation.

At $x = L$, $\varphi = 0$, hence:

$$\varphi^2{}_0 = \frac{2Q_0L + NL^2}{K(1+\delta)} \quad \text{or} \quad Q_0 = \frac{KD^2}{2L}\frac{(1+\delta)}{\delta^2} - \frac{NL}{2}; \qquad \varphi_0 = \frac{D}{\delta}. \tag{9.7.8}$$

For $N = 0$:

$$\varphi^2{}_0 = 2Q_0L/K(1+\delta) = D^2/\delta^2. \tag{9.7.9}$$

Again, the interface has a parabolic shape. In figures 9.7.6a and 9.7.6b, the dashed lines describe the Dupuit approximation. Several additional cases are solved by Santing (1957a).

The interesting outcome of (9.7.8) and (9.7.9), from the practical point of view, is the relationship between Q_0 and L. As Q_0 increases, L decreases, and vice versa. This indicates that the length of sea water intrusion is a decision variable. By controlling Q_0, one can control the extent of sea water intrusion. The decision on the length of the intruding sea water wedge, or equivalently how much fresh water should be drained to the sea, is part of the coastal aquifer management problem. The equations derived above provide the tools that aid in making these decisions.

Equation (9.7.9) relates φ_0 (piezometric head above the toe) to L. By controlling φ_0 (say, by means of artificial recharge), the water table may be lowered both landward and seaward of the toe, without causing any additional sea water intrusion. Landward of the toe, water levels may fluctuate as a result of some optimal management scheme. If pumpage is taking place seaward of the toe, the interface there will rise and may contaminate wells if their screened portion is not at a sufficient distance above it.

9.7.4 Approximate Solution for the Moving Interface

Up to this point, only a steady coastal interface has been considered. However, any change in the fresh water flow regime is followed by a transition period during which the flow in the aquifer is unsteady and the interface is moving slowly toward its new equilibrium position. The exact mathematical statement of the problem was presented in paragraph 9.5.3, where it was also emphasized that an exact analytical solution is not yet possible. Some approximate approaches were presented in paragraphs 9.5.3 and 9.5.6.

The Dupuit approximation employed in the previous paragraph for the two-dimensional steady flow of fresh water in a vertical plane may also be employed for unsteady flow with a moving interface. Actually, this was done in paragraph 9.5.5 where the flow of two fluids in an inclined confined reservoir was considered. Using the same procedure as that leading to (9.5.60) through (9.5.64), we may generalize the discussion to a phreatic aquifer with accretion $N = N(x, t)$ (fig. 9.7.7). In addition we shall assume that the aquifer's bottom is not necessarily horizontal, so that $D = D(x)$, and we shall use the symbol n_e (9.7.9) to denote phreatic storativity. The depth of the interface below sea level is denoted by $\eta = \eta(x, t)$. The permeability of the aquifer may also vary with location so that $k = k(x)$.

With the nomenclature of figure 9.7.7, the continuity equations for the region

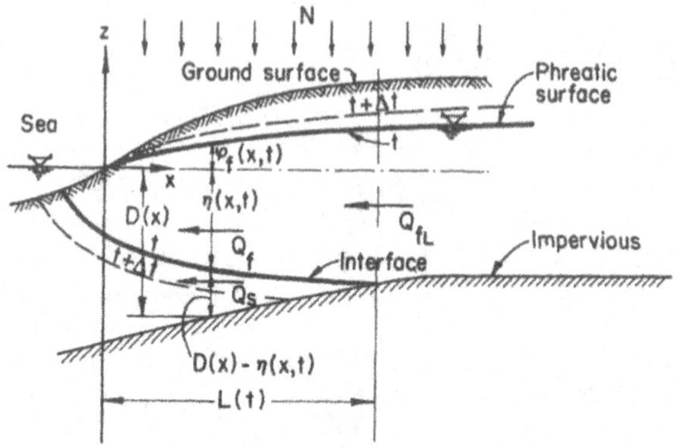

FIG. 9.7.7. Nomenclature for a moving interface in a phreatic aquifer.

$x \leqslant L$ are, for fresh water:

$$n_e \, \partial\varphi_f/\partial t + n \, \partial\eta/\partial t - \partial Q_f/\partial x - N = 0 \tag{9.7.10}$$

and for salt water:

$$n \, \partial\eta/\partial t + \partial Q_s/\partial x = 0 \tag{9.7.11}$$

where Q_f and Q_s are the total fresh water and salt water discharges, respectively (taken positive toward the sea), and $\varphi_f = z + p/\gamma_f$ is the fresh water potential. Applying the Dupuit assumptions to each fluid separately we may now express Q_f and Q_s by:

$$Q_f = K_f \, (\eta + \varphi_f) \, \partial\varphi_f/\partial x; \qquad Q_s = K_s(D - \eta) \, \partial\varphi_s/\partial x \tag{9.7.12}$$

where $\varphi_s = z + p/\gamma_s$ is the salt water potential (in the region occupied by salt water), $K_f = k\gamma_f/\mu_f$ and $K_s = k\gamma_s/\mu_s$. Inserting (9.7.12) into (9.7.10) and (9.7.11), we obtain:

$$n_e \frac{\partial\varphi_f}{\partial t} + n \frac{\partial\eta}{\partial t} - \frac{\partial}{\partial x}\left[K_f(\eta + \varphi_f) \frac{\partial\varphi_f}{\partial x}\right] - N = 0 \tag{9.7.13}$$

$$n \frac{\partial\eta}{\partial t} + \frac{\partial}{\partial x}\left[K_s(D - \eta) \frac{\partial\varphi_s}{\partial x}\right] = 0. \tag{9.7.14}$$

We have two equations here in the three independent variables η and φ_f and φ_s. However, the latter are related to each other by (9.5.25), which, in this case, has the form:

$$-\eta = (\gamma_s/\Delta\gamma)\varphi_s - (\gamma_f/\Delta\gamma)\varphi_f \quad \text{or} \quad \eta = (\gamma_f/\Delta\gamma)\varphi_f - (\gamma_s/\Delta\gamma)\varphi_s. \tag{9.7.15}$$

With (9.7.15), equation (9.7.14) may be written as:

$$n \frac{\partial\eta}{\partial t} + \frac{\partial}{\partial x}\left[K_s(D - \eta) \frac{\partial}{\partial x}\left(\frac{\gamma_f}{\gamma_s}\varphi_f - \eta \frac{\Delta\gamma}{\gamma_s}\right)\right] = 0. \tag{9.7.16}$$

Equations (9.7.13) and (9.7.16) may now be solved to yield $\varphi_f = \varphi_f(x, t)$ and

$\eta = \eta(x, t)$ in the region $0 \leqslant x \leqslant L$. Recalling that the Dupuit assumptions underlie the development of these equations, the boundary conditions are:

$$x = 0; \qquad \varphi_f = 0, \qquad \eta = 0$$
$$x = L; \qquad \eta = D \quad \text{and} \quad Q_{fL} = K_f(D + \varphi_f)\, \partial\varphi_f/\partial x$$

where Q_{fL} is known as a function of t, say, from solving for the flow in the region $x > L$.

In a confined aquifer, equations (9.7.13) and (9.7.16) become:

$$n\, \partial\eta/\partial t - \partial Q_f/\partial x = 0; \qquad n\, \partial(D - \eta)/\partial t - \partial Q_s/\partial x = 0 \tag{9.7.17}$$

with:

$$Q_f = K_f \eta\, \partial\varphi_f/\partial x; \qquad Q_s = K_s(D - \eta)\, \partial\varphi_s/\partial x. \tag{9.7.18}$$

The initial conditions for η and φ_f require a knowledge of the shape and position of both the interface and the phreatic surface at $t = 0$ in the region $0 \leqslant x \leqslant L$. By differentiating (9.7.15) with respect to x and introducing $Q_0 = Q_f + Q_s$, we obtain:

$$\frac{\partial\eta}{\partial x} = \frac{Q_f}{K'_f \eta} + \frac{Q_f}{K'_s(D - \eta)} - \frac{Q_0}{K'_s(D - \eta)} \tag{9.7.19}$$

where $K'_f = K_f\, \Delta\gamma/\gamma_f$ and $K'_s = K_s\, \Delta\gamma/\gamma_s$.

By combining (9.7.17) and (9.7.19), we obtain:

$$n\frac{\partial\eta}{\partial t} - \frac{\partial}{\partial x}\left[\frac{Q_0/K'_s(D - \eta) + \partial\eta/\partial x}{1/K'_f\eta + 1/K'_s(D - \eta)}\right] = 0. \tag{9.7.20}$$

This is a nonlinear second-order partial differential equation from which $\eta = \eta(x, t)$ may be derived, given Q_0, boundary conditions and initial conditions. The main difficulty stems from the fact that in addition to the nonlinearity, the boundary condition is a moving one as $L = L(t)$.

A similar equation was derived in paragraph 9.5.5, where the reader was referred to Le Fur and Sourieau for a detailed solution and analysis.

In a coastal aquifer, $\mu_s \approx \mu_f$ so that $K'_f \cong K'_s$ and (9.7.20) becomes:

$$n\frac{\partial\eta}{\partial t} - \frac{\partial}{\partial x}\left[\frac{Q_0\eta}{D} + \frac{K'_f\eta(D - \eta)}{D}\frac{\partial\eta}{\partial x}\right] = 0. \tag{9.7.21}$$

This equation is identical to (9.5.64) with $b \rightarrow D$, $k_2\mu_1/k_1\mu_2 \rightarrow 1$, $\alpha = 0$, $Q_0 = -bq$ and $K'_2 \rightarrow K'_f$. At $x = L$, $Q_s = 0$ and therefore $Q_{fL} = Q_f|_{x=L} \equiv Q_L$ is the fresh water discharge entering the region above the interface through the cross-section at $x = L$.

It is possible to extend this development to two-dimensional flow in the horizontal xy-plane, with $\eta = \eta(x, y)$. In this case we may have $k = k(x, y)$ and $N = N(x, y)$, $D = D(x, y)$. Boundary and initial conditions must be specified accordingly.

Although by employing the Dupuit assumptions the problem of the moving interface has been stated in a form simpler than the exact one presented in paragraph

9.5.3, the resulting equations are still nonlinear and cannot be solved by exact analytical methods. Their solution may be achieved by various numerical techniques (e.g., finite difference methods) using high speed digital computers (e.g., Shamir and Dagan 1970).

Bear and Dagan (1964b) present approximate solutions for an interface moving in a coastal aquifer. Figure 9.7.8 gives the nomenclature for this solution. Initially, the seaward fresh water discharge $Q^0{}_{fL}$ is equal to the discharge to the sea $Q^0{}_{f0}$; both discharges are positive in the direction of the sea. At $t = t_0$, the discharge $Q^0{}_{fL}$ above the toe is suddenly reduced to a new constant value Q'_{fL} such that $0 < Q'_{fL} < Q^0{}_{fL}$, and the interface starts moving landward until a new equilibrium is reached. During the transition period, the fresh water outflow to sea at $x = 0$ is reduced from $Q^0{}_{f0}$ to Q'_{fL}. When the new discharge above the toe is negative ($Q'_{fL} < 0$, i.e., flow above the toe is in a landward direction), no equilibrium is reached, and the interface will continue to advance landward. In this case the fresh water discharge at $x = 0$ gradually decreases from $Q^0{}_{f0} > 0$, reaches zero and becomes negative. For $Q_{f0} \leqslant 0$ sea water will penetrate into the aquifer.

Instead of attempting to solve (9.7.21), Bear and Dagan (1964b) follow the approximate quasisteady (or successive steady states) approach suggested by Polubarinova-Kochina (1952, 1962, see par. 8.4.4). Following this approach $Q_f(x, t)$ varies from Q'_{fL} at $x = L$, $\eta = D$, to $Q_{f0}(t)$ at $x = 0$, $\eta \approx 0$, according to some predetermined rule. The fresh water seaward discharge $Q_{f0}(t) \equiv Q_f(0, t)$ is regarded as an unknown variable of the problem. Continuity considerations applied to the volume of salt water $U_s(t)$ lead to:

$$n\, dU_s(t)/dt = -Q'_s(0, t) = Q_{f0}(t) - Q'_{fL}; \qquad U_s(t) = \int_{\eta=0}^{D} x(\eta, t)\, d\eta. \quad (9.7.22)$$

For the two variables of the problem: $\eta = \eta(x, t)$ and $Q_{f0}(t)$, we now have the following two equations. From (9.7.19) with Q_{fL} replaced by Q'_{fL} and $K'_f \approx K'_s \approx K'$,

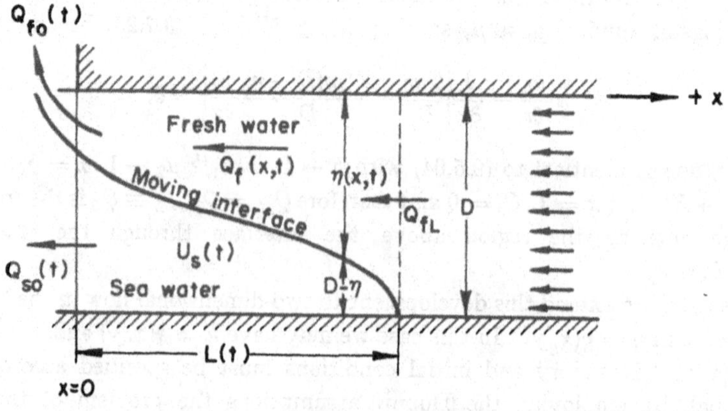

FIG. 9.7.8. Nomenclature for a moving interface in a confined aquifer.

we obtain:

$$\frac{\partial \eta}{\partial x} = \frac{Q_f D}{K'\eta(D-\eta)} - \frac{Q'_{fL}}{K'(D-\eta)}. \tag{9.7.23}$$

From (9.7.22) we obtain:

$$n \frac{d}{dt} \int_0^D x(\eta, t)\, d\eta = Q_{f0}(t) - Q'_{fL}. \tag{9.7.24}$$

Up to this point, the only assumptions underlying (9.7.23) and (9.7.24) are shallow flow and $K'_s \approx K'_f$. Q_f is still unknown. The proposed method may be summarized by the following three steps.

(a) An assumption is made regarding the variations in $Q_f(x, t)$ as a function of Q'_{fL}, $Q_{f0}(t)$ and $\eta(x, t)$, i.e., $Q_f = Q_f(\eta, Q_{f0}, Q'_{fL})$. The assumed distribution should satisfy the following conditions: at $x = 0$, $\eta = 0$, $Q_f = Q_{f0}(t)$; at $x = L$, $\eta = D$, $Q_f = Q'_{fL}$.

(b) With Q_f of (a), equation (9.7.23) is integrated to yield $\eta = \eta[x, Q_{f0}(t)]$.

(c) η of (b) is inserted in (9.7.24), which is then solved for $Q_{f0}(t)$.

The accuracy of the solution depends entirely on the accuracy of the assumed variations of Q_f.

Two approximations are considered by Bear and Dagan.

Case 1. The first approximation is a linear variation of Q_f with η (fig. 9.7.9a):

$$Q_f(\eta, t) = Q'_{fL} + [Q_{f0}(t) - Q'_{fL}](D - \eta)/D = Q_{f0}(t) - [Q_{f0}(t) - Q'_{fL}]\eta/D. \tag{9.7.25}$$

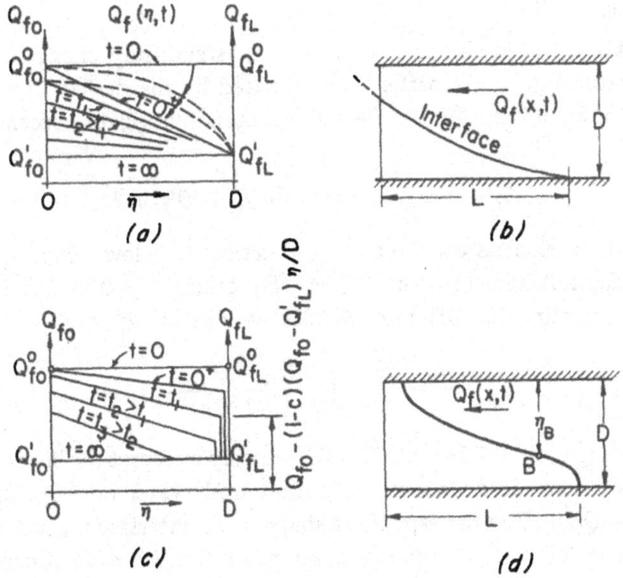

FIG. 9.7.9. First and second approximation for $Q_f = Q_f(\eta, t)$ and resulting interfaces.

By inserting (9.7.25) in (9.7.23), we obtain:

$$\partial\eta/\partial x = Q_{f0}(t)/K'\eta. \tag{9.7.26}$$

By introducing $Q_s = Q_0 - Q_f = 0$ and $Q_f = Q_{f0}$ in (9.7.19), equation (9.7.26) is obtained again. Hence, one may interpret (9.7.26) as describing *an instantaneous steady state* (at some time t) with stationary salt water and with $Q_{f0} = Q_f =$ constant.

Integrating (9.7.26) with $x = 0$, $\eta = 0$, yields:

$$\eta^2(x, t) = [2Q_{f0}(t)/K']x. \tag{9.7.27}$$

Equation (9.7.27) describes interfaces that have the shape of a parabola (fig. 9.7.9b). By inserting (9.7.27) into (9.7.24), we obtain:

$$\frac{6\,dt}{nK'D^3} = \frac{dQ_{f0}}{(Q_{f0})^2(Q'_{fL} - Q_{f0})}. \tag{9.7.28}$$

Integrating (9.7.28), with: $t = 0$, $Q_{f0} = Q^0{}_{f0}$, $Q'_{fL} \neq 0$, gives:

$$\frac{6t}{nK'D^3} = \frac{1}{(Q'_{fL})^2}\ln\frac{Q_{f0}}{Q^0{}_{f0}} - \frac{1}{Q'_{fL}}\left(\frac{1}{Q_{f0}} - \frac{1}{Q^0{}_{f0}}\right) - \frac{1}{(Q'_{fL})^2}\ln\frac{Q'_{fL} - Q_{f0}}{Q'_{fL} - Q^0{}_{f0}}. \tag{9.7.29}$$

For the special case of $Q'_{fL} = 0$, we obtain from (9.7.28):

$$Q_{f0} = [12t/nK'D^3 + 1/(Q^0{}_{f0})^2]^{-1/2}. \tag{9.7.30}$$

Equations (9.7.29) and (9.7.30) describe the temporal variations of the fresh water discharge to the sea when at $t = 0$ the seaward fresh water flow above the toe is changed from $Q^0{}_{fL}$ to Q'_{fL}. Experiments (Bear and Dagan 1964) seem to indicate that (9.7.29) and (9.7.30) are valid only for $Q'_{fL} > Q^0{}_{fL}$ (i.e., a seaward movement of the toe, fig. 9.7.9b).

Case 2. When the discharge rate at $x = L$ is suddenly changed from $Q^0{}_{fL}$ to Q'_{fL}, the variations in Q_f are better approximated by the dashed lines than by the straight lines of fig. 9.7.9a. Hence, Bear and Dagan investigate a second approximation:

$$Q_f(\eta, t) = Q_{f0} - (1 - c)(Q_{f0} - Q'_{fL})\eta/D \tag{9.7.31}$$

where c is a dimensionless constant to be determined below. This approximation is shown by the full lines in figure 9.7.9c. By taking $c = 0$ in (9.7.31) we obtain (9.7.25). By inserting (9.7.31) into (9.7.23) we obtain:

$$\frac{\partial\eta}{\partial x} = \frac{Q_{f0}}{K'\eta} + \frac{c(Q_{f0} - Q'_{fL})}{K'(D - \eta)}. \tag{9.7.32}$$

Again, by comparing (9.7.19) with (9.7.32) we may interpret the latter as describing instantaneously a steady flow with a constant fresh water flow Q_{f0} and salt water flow of $c(Q_{f0} - Q'_{fL})$. The corresponding shape of the interface is given in fig. 9.7.9d. From (9.7.32) it follows that an inflection point ($\partial^2\eta/\partial x^2 = 0$) occurs at:

$$\eta_B = D/\{1 + [c(1 - Q'_{fL}/Q_{f0})]^{1/2}\}. \tag{9.7.33}$$

Experiments in a Hele–Shaw analog of cases where the interface moves landward as a result of the change from $Q^0{}_{fL}$ to Q'_{fL} show that an inflection point indeed occurs at $\eta_B > 0.75D$ when $Q'_{fL} = 0$. This corresponds approximately to $c \cong 0.1$. As small variations in the value of c do not appreciably affect the final result, Bear and Dagan (1964b) integrate (9.7.32) with $c = 0.1$ and give the shape of the moving interface in the form of $x = x(\eta, Q_{f0}, Q'_{fL})$.

From (9.7.24), an expression is derived from $Q_{f0} = Q_{f0}(t)$. Results are given in graphic form (since numerical integration is used) both for $Q'_{fL} > 0$ and $Q'_{fL} < 0$; in both cases $Q'_{fL} < Q^0{}_{fL}$. For the special case of $Q'_{fL} = 0$, Bear and Dagan obtain:

$$Q_{f0}/Q^0{}_{fL} = \{1/[14.3t(Q^0{}_{fL})^2/nK'D^3 + 1]\}^{1/2}. \qquad (9.7.34)$$

According to Bear and Dagan (1964b), experiments indicate that (a) the second approximation is applicable only to cases of landward interface movement caused by a reduction in fresh water seaward flow, (b) the first approximation is applicable, with sufficient accuracy for practical purposes, only to the case of a seaward movement of the interface, and (c) both approximations are valid only as long as Q_{f0} remains sufficiently large.

For $Q'_{Lf} = 0$, i.e., stopping all seaward fresh water flow above the toe, it is possible to obtain the movement of the toe of the interface by determining the value of x at $\eta = D$. The result for both approximations is:

$$x|_{\eta=D} = (mK'Dt/n + x_0{}^2)^{1/2} \qquad (9.7.35)$$

where x_0 is the value of L at $t = 0$. For the first approximation, $m = 3$; for the second approximation $m = 1.75$. By comparison, Rumer and Harleman (1963) obtain $m = 1$ as part of their approximate solution for the moving interface.

The discussion in the present paragraph, although yielding several possible approximations for the problem of the moving interface in a coastal aquifer, seems to indicate that better solutions may be derived either by using the Hele–Shaw analog (sec. 11.4), which is an excellent tool even for rather complex problems, or by using a high speed digital computer for solving the nonlinear equations (Shamir and Dagan 1970).

9.7.5 Interface Upconing

Under certain circumstances, as a result of sea water intrusion, wells may pump from the fresh water region above an interface. As a result of the pumping, and the potential distribution established in the two fluids, the interface below a well (or a drain in two-dimensional flow) will rise, at first slowly and then faster, toward the pumping well. Under certain conditions, a new equilibrium may be reached before the upconed interface reaches the pumping well. Otherwise, the rising interface will eventually reach the pumping well in a cusp-like form. When pumping stops, the upconed interface (in the form of a mound) undergoes decay toward the initial steady-state shape of the interface. When, at the same time, seaward flow of fresh water takes place above the interface, the upconed interface is displaced seaward as it rises and decays. As was explained earlier, a transition zone, rather than an abrupt interface, separates the two miscible fluids. This transition zone becomes wider as the interface moves. Nevertheless, the abrupt interface approximation will be used in the present paragraph.

In oil reservoirs, the same phenomenon, known as *water coning* (e.g., Muskat 1937; Muskat 1949; and Pirson 1958), occurs when oil is withdrawn from a well located above an oil–water interface.

To date, only approximate analytical solutions are known, in addition to analog and numerical ones. Two such approximations are described below. For the case of a stationary upconed interface in the vertical plane, the hodograph method is applicable. An example is given in section 9.6 (example 2).

Example 1, Muskat's approximation. Muskat's approximation deals with steady flow and *relatively small interface rises* (say, with respect to the distance from the initial interface to the sink). Under these conditions, Muskat assumes that the lower (heavier) liquid is stationary, so that the interface behaves as an impervious boundary. He also assumes that the shape of local upconing of the interface beneath the sink (well or drain) does not appreciably affect the potential distribution in the region occupied by the lighter liquid, obtained by assuming that the upconing does not exist. With these assumptions, he derives the shape of the upconed interface from the potential distribution on some hypothetical horizontal plane introduced as an impervious boundary to the flow domain. Usually, the initial horizontal interface is taken as that boundary.

In figure 9.7.10, a point sink withdraws liquid 1 at a rate Q from a confined homogeneous and isotropic aquifer of thickness D. From Muskat's assumption it follows that:

$$\Delta\zeta = \zeta_M - \zeta_N = \frac{(\Phi_N - \Phi_M)\mu_1}{k(\gamma_2 - \gamma_1)} \simeq \frac{(\Phi_{N'} - \Phi_{M'})\mu_1}{k(\gamma_2 - \gamma_1)} = \frac{\varphi_{N'} - \varphi_{M'}}{\Delta\gamma/\gamma_1} \tag{9.7.36}$$

where $\varphi_{M'}$ and $\varphi_{N'}$ are liquid 1 piezometric heads at points M' and N' on the xy-plane. Actually, this is also the assumption that underlies the linearized approximation based on small perturbations (par. 9.5.7). In fact, the analytic considerations given there may serve as a justification for the Muskat approximation.

In (9.7.36), the steady-state potential distribution $\varphi(x, y)$ on the hypothetical impervious boundary may be obtained analytically (for simple cases), numerically (mainly by using digital computers), or by means of an electric analog. Bear and

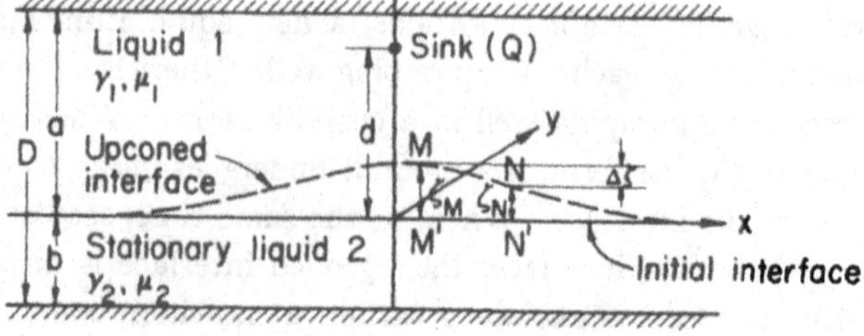

FIG. 9.7.10. Upconing by Muskat's approach.

Dagan (1966) use an electric analog of the electrolytic tank type to determine $\varphi(x, y)$, and from it the shape of the upconed interface.

Example 2, unsteady upconing by the method of small perturbations. This approximation (par. 9.5.7) is used by Bear and Dagan (1964a) and Dagan and Bear (1968) to determine the shape of the upconed interface below a sink pumping in two- and three-dimensional flows.

As an example, consider the case of a sink in the form of a drain located at a distance d above an initially horizontal interface. The flow is two-dimensional in the xz-plane (fig. 9.7.10). The notation of paragraph 9.5.7 is used.

Mathematically stated, the problem is to solve the Laplace equations for φ_1 and φ_2 in regions 1 and 2, respectively. Because the two fluids are initially stationary, $\varphi^0{}_1$, $\varphi^0{}_2$ and ζ^0 are identically zero, and we have only to determine the unsteady parts φ'_1, φ'_2 and ζ'. For the sake of simplicity, we shall henceforth denote $\varepsilon\varphi'_1$, $\varepsilon\varphi'_2$ and $\varepsilon\zeta'$ by φ_1, φ_2 and ζ. The piezometric heads φ_1 and φ_2 must satisfy the following boundary conditions:

$$t \geqslant 0; \quad -\infty < x < \infty: \quad z = a, \quad \partial\Phi_1/\partial z = 0; \quad z = -b, \quad \partial\Phi_2/\partial z = 0 \tag{9.7.37}$$

and the drain's discharge:

$$t \geqslant 0: \qquad \lim_{r \to 0} r(\partial\Phi_1/\partial r) = Q/2\pi; \qquad r^2 = x^2 + (z-d)^2. \tag{9.7.38}$$

Along the interface we have from (9.5.83):

$$\begin{aligned} &\beta_1\, \partial\Phi_1/\partial t - \beta_2\, \partial\Phi_2/\partial t - \partial\Phi_1/\partial z = 0 \\ &\beta_1\, \partial\Phi_1/\partial t - \beta_2\, \partial\Phi_2/\partial t - \partial\Phi_2/\partial z = 0 \end{aligned} \tag{9.7.39}$$

where: $\beta_i = n\alpha_i$; $i = 1, 2$. The elevations of the interface are given by $\zeta = \alpha_2\Phi_2 - \alpha_1\Phi_1$. Hence the initial condition is:

$$t = 0, \quad z = 0, \quad \zeta = \alpha_2\Phi_2 - \alpha_1\Phi_1 = 0. \tag{9.7.40}$$

At infinity, the interface remains undisturbed. Hence we have there:

$$z = 0, \quad x = \pm\infty, \quad \zeta = \alpha_2\Phi_2 - \alpha_1\Phi_1 = 0. \tag{9.7.41}$$

Because of the singularity introduced by the drain, Φ_1 is decomposed into two parts, Φ_{11} and Φ_{12}, both satisfying the Laplace equation:

$$\Phi_1(x, z, t) = \Phi_{11}(x, z) + \Phi_{12}(x, z, t) \tag{9.7.42}$$

and such that Φ_{11} represents the singular character of Φ_1 at $x = 0$, $z = d$, whereas Φ_{12} is an unsteady-state solution that is regular everywhere in region 1. Hence:

$$\Phi_{11} = (Q/4\pi) \ln[x^2 + (z-d)^2] - (Q/4\pi) \ln[x^2 + (z+d)^2]. \tag{9.7.43}$$

The boundary conditions for Φ_{12} and Φ_2 are:

$$z = a, \quad \frac{\partial\Phi_{12}}{\partial z} = \frac{\partial\Phi_1}{\partial z} - \frac{\partial\Phi_{11}}{\partial z} = \frac{Q}{2\pi}\left[\frac{a-d}{x^2 + (a-d)^2} - \frac{a+d}{x^2 + (a+d)^2}\right] \tag{9.7.44}$$

$$z = -b; \qquad \partial\Phi_2/\partial z = 0 \tag{9.7.45}$$

$$z=0,\quad \Phi_{11}=0: \qquad \beta_1\,\partial\Phi_{12}/\partial t-\beta_2\,\partial\Phi_2/\partial t-\partial\Phi_{12}/\partial z=-Qd/\pi(x^2+d^2)$$

$$\beta_1\,\partial\Phi_{12}/\partial t-\beta_2\,\partial\Phi_2/\partial t-\partial\Phi_2/\partial z=0. \qquad (9.7.46)$$

The initial conditions become:

$$t=0,\quad z=0,\quad \zeta=\alpha_2\Phi_2-\alpha_1\Phi_{12}=0. \qquad (9.7.47)$$

Since Φ_{12} and Φ_2 are everywhere regular, we assume that they can be represented by Fourier integral transforms permitting separation of variables:

$$\Phi_{12}=\frac{1}{\pi}\int_0^\infty[A_1(\lambda,t)\exp(-\lambda z)+A_2(\lambda,t)\exp(\lambda z)]\cos\lambda x\,d\lambda \qquad (9.7.48)$$

$$\Phi_2=\frac{1}{\pi}\int_0^\infty A_3(\lambda,t)\cosh\lambda(z+b)\cos\lambda x\,d\lambda. \qquad (9.7.49)$$

In this form both Φ_{12} and Φ_2 satisfy the Laplace equation; A_1, A_2 and A_3 are functions to be determined by the boundary and the initial conditions (9.7.44), (9.7.46) and (9.7.47). The boundary condition (9.7.45) is satisfied by (9.7.49).

Using tables, the right-hand side of (9.7.44) and the first equation of (9.7.46) are now represented in the form of a Fourier integral similar to (9.7.48):

$$\frac{Q}{2\pi}\left[\frac{a-d}{x^2+(a-d)^2}-\frac{a+d}{x^2+(a+d)^2}\right]$$

$$=\frac{Q}{2\pi}\int_0^\infty\{\exp[-\lambda(a-d)]-\exp[-\lambda(a+d)]\}\cos\lambda x\,d\lambda \qquad (9.7.50)$$

$$\frac{Q}{2\pi}\frac{d}{x^2+d^2}=\frac{Q}{\pi}\int_0^\infty\exp(-\lambda d)\cos\lambda x\,d\lambda. \qquad (9.7.51)$$

By inserting (9.7.48) through (9.7.51) into (9.7.44), (9.7.46) and (9.7.47), three equations in A_1, A_2, A_3 and λ are derived. Also for ζ:

$$\zeta=\frac{1}{\pi}\int_0^\infty[\alpha_2A_3\cos\lambda b-\beta_1(A_1+A_2)]\cos\lambda x\,d\lambda. \qquad (9.7.52)$$

Introducing:

$$\xi=\alpha_2A_3\cos\lambda b-\beta_1(A_1+A_2) \qquad (9.7.53)$$

we observe, in view of (9.7.52), that ξ is the inverse transform of ζ, such that:

$$\zeta=\frac{1}{\pi}\int_0^\infty\xi(\lambda,t)\cos\lambda x\,d\lambda. \qquad (9.7.54)$$

After some algebraic manipulations, Bear and Dagan (1964a) obtain from (9.7.53) a partial differential equation for ζ. By solving this equation and inserting the result in (9.7.54), they obtain:

$$\zeta = \frac{\alpha_1 Q}{\pi} \int_0^\infty \frac{1}{\lambda} \frac{\cosh \lambda(a-d)}{\sinh \lambda a} \left\{1 - \exp\left(\frac{-\lambda}{\beta_1 \operatorname{cotgh} \lambda a + \beta_2 \operatorname{cotgh} \lambda b}\right) t\right\} \cos \lambda x \, d\lambda. \quad (9.7.55)$$

With the dimensionless variables: $L = \lambda a$, $Z = \zeta/\alpha_1 Q$; $T = t/\beta_1 a$, $B = b/a$, $X = x/a$; $D = d/a$, and $M = \mu_2/\mu_1$, equation (9.7.55) becomes:

$$Z = \frac{1}{\pi} \int_0^\infty \frac{\cosh L(1-D)}{L \sinh L} \{1 - \exp[-LT/(\operatorname{cotgh} L + M \operatorname{cotgh} LB)]\} \cos LX \, dL. \quad (9.7.56)$$

Because of symmetry, the crest of the upconing interface is at $X = 0$. There, (9.7.56) becomes:

$$Z_{crest} = \frac{1}{\pi} \int_0^\infty \frac{\cosh L(1-D)}{L \sinh L} \{1 - \exp[-LT/(\operatorname{cotgh} L + M \operatorname{cotgh} LB)]\} \, dL. \quad (9.7.57)$$

For a very thick aquifer, $a \to \infty$, $B \to \infty$, equations (9.7.56) and (9.7.57) become:

$$Z = \frac{1}{2\pi} \ln \frac{(X/D)^2 + (T'+1)^2}{(X/D)^2 + 1}; \qquad T' = t(\beta_1 + \beta_2)d \quad (9.7.58)$$

$$Z_{crest} = \frac{1}{\pi} \ln(T'+1). \quad (9.7.59)$$

Equation (9.7.59) is also derived by Pilatovski (1958).

Figure 9.7.11 shows ζ and ζ_{crest} for an infinitely thick aquifer in a graphic form. The results are presented for the case of an anisotropic aquifer ($k_x \neq k_z$) using the transformation given in section 7.4.

Bear and Dagan (1964a) and Dagan and Bear (1968) also apply this method to cases where the initial interface makes an angle α with the $+x$ axis. Based on sand box model experiments they conclude that the approximation solutions are valid up to displacements that, at the crest of the upconing interface, reach one-third of the distance from the initial interface to the sink.

9.7.6 The Dupuit–Ghyben–Herzberg Approximation for an Unsteady Interface in a Thick Aquifer

In a thick aquifer we may neglect the flow in the salt water zone. This is, for example, the case shown in figure 9.7.12 where we have a fresh water lens floating on a large salt water body. The lens, created, for example, by artificial recharge, is bounded only by a phreatic surface and an interface. For this case (9.7.10) is valid. When extended to flow in the horizontal plane, equation (9.7.10) with $n \approx n_e$ becomes, in terms of h_f ($\equiv \varphi_f$ in (9.7.10)):

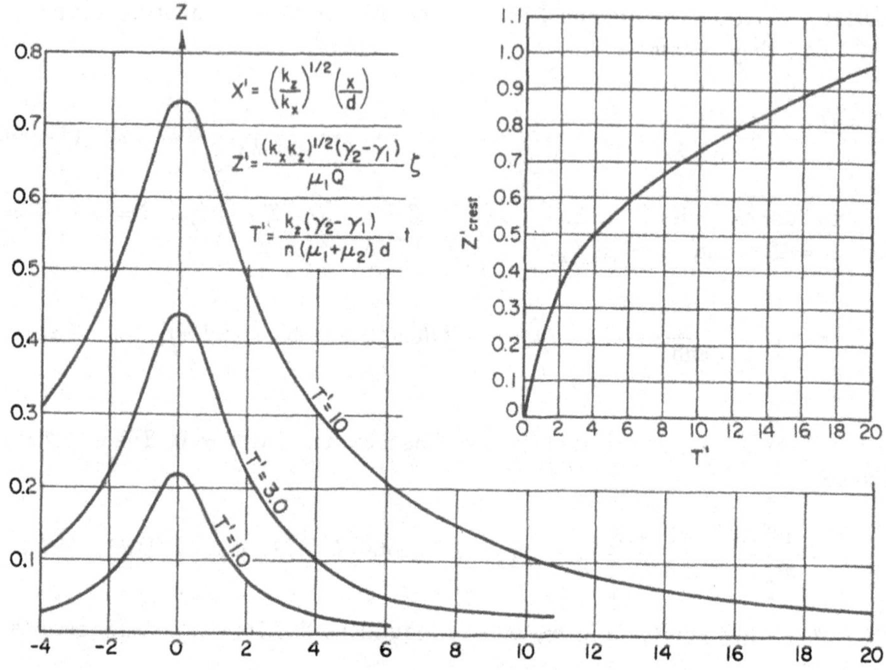

FIG. 9.7.11. The shape of the interface below a sink in two-dimensional flow.

$$\nabla^2 h_f^2 + 2(\nabla h_f)^2 + 2N/(1 + \delta)K_f - (n_e/K_f h_f)\, \partial h_f^2/\partial t = 0 \qquad (9.7.60)$$

or in terms of h_s ($\equiv \eta$ in (9.7.10)):

$$\nabla^2 h_s + 2(\nabla h_s)^2 + 2N\delta^2/(1 + \delta)K_f - (n_e\delta/K_f h_s)\, \partial h_s^2/\partial t = 0. \qquad (9.7.61)$$

The same equation can be obtained from (6.4.22) or by averaging div $\mathbf{q} = 0$ over a vertical line from $-h_s$ to $+h_f$, as was done to obtain (9.5.60).

Equations (9.7.60) and (9.7.61) are nonlinear partial differential equations. Following the discussion in section 8.4, we may apply to them various linearization tech-

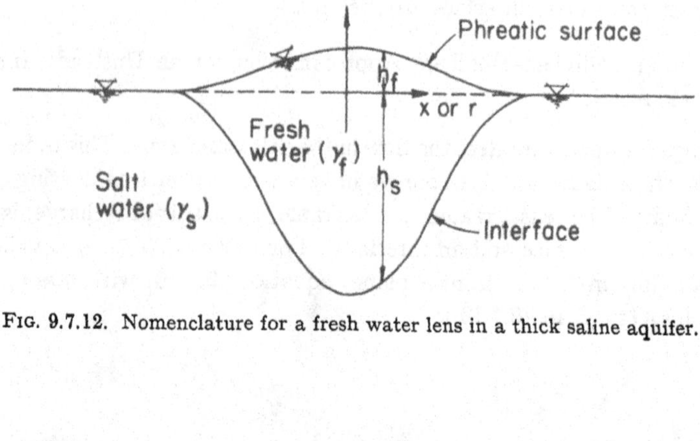

FIG. 9.7.12. Nomenclature for a fresh water lens in a thick saline aquifer.

niques. For example, we may apply to them the second method of linearization described in paragraph 8.4.2, by replacing the variable coefficient $n_e/K_f h_f$ by an assumed constant one $n_e/K'_f \bar{h}_f$, where $\bar{h}_f$ is the average value of h_f over the region of interest and during the period of flow. By doing so, (9.7.60) becomes a linear equation in the dependent variable h_f^2. Similar considerations may also be applied to (9.7.61) leading to a linear equation in h_s^2.

Hantush (1968) employs this approach to derive solutions of the linearized equation for a fresh water lens created by artificially recharging a thick saline aquifer through recharging areas of various shapes.

Exercises

9.1 Assume (a) a sharp interface between oil and water in an anticlinal oil reservoir, (fig. 9.E.1), (b) no capillary pressure at the interface, and (c) a caprock saturated with water in the reservoir. Determine the relationship between the highest possible thickness of the oil zone (h_{max}) and the caprock's entry pressure.

9.2 Derive a relationship between the capillary pressure across the interface shown in fig. 9.E.2 and the height h of the interface.

9.3 (a) A test for $p_c = p_c(S_w)$ is carried out in the laboratory with gas and water. How can the results be modified to yield $p_c = p_c(S_w)$ for reservoir conditions where we have oil and water?

(b) Calculate the height above the free water level in a reservoir from which a core was taken if a laboratory test with water and gas gave the following results:

$$p_c \text{ at the lab.} = 18 \text{ psi}; \qquad S_w = 0.35$$

$$\sigma_{wg} = 72 \text{ dynes/cm}; \qquad \sigma_{wo} = 24 \text{ dynes/cm}$$

$$\rho_o = 53 \text{ lb/ft}^3; \qquad \rho_w = 68 \text{ lb/ft}^3.$$

9.4 Determine the distance of the advancing front in each of the two tubes shown in fig. 9.E.3, assuming no pressure drop in the air phase. Show under what conditions air may be entrapped in one of the tubes.

9.5 Fatt and Dykstra (1951) suggest the following expression for relative

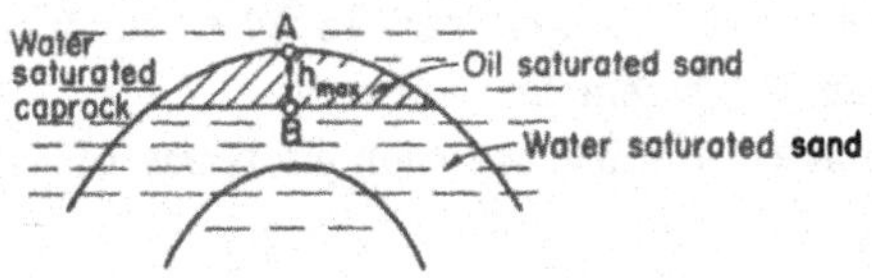

FIG. 9.E.1.

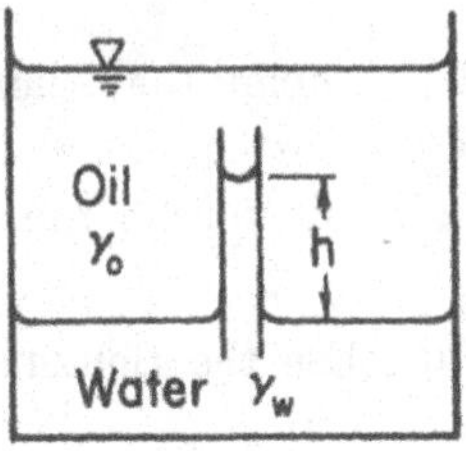

FIG. 9.E.2

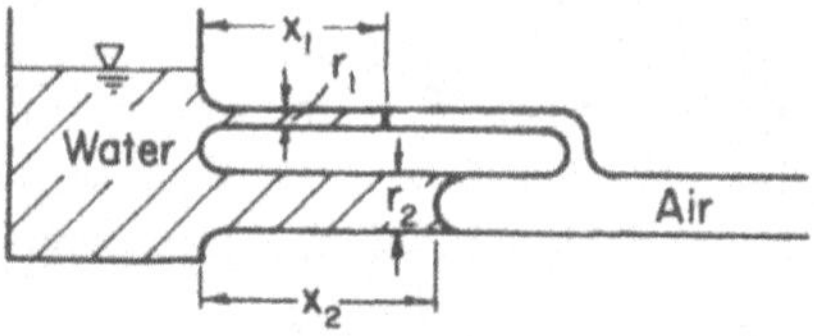

FIG. 9.E.3.

permeability to the wetting fluid:

$$k_{rw} = \frac{\int_0^{S_w} dS_w/p_c^3(S_w)}{\int_0^1 dS_w/p_c^3(S_w)}.$$

Use this expression together with the data on $p_c = p_c(S_w)$ of fig. 9.2.10 (drainage) to obtain the curve $k_{rw} = k_{rw}(S_w)$.

9.6 Use the Welge method of solution in order to solve the Buckley-Leverett equations describing one-dimensional flow in an oil reservoir. Data:

1. $k_{rw}(S_w)$ and $k_{rnw}(S_w)$ are given in fig. 9.3.1.
2. The reservoir's thickness and width are 20 m and 1320 m, respectively.
3. The daily production of the wells causing the movement in the reservoir in 5050 m³/d (at reservoir conditions).
4. The reservoir's porosity is $n = 18\%$.
5. $\mu_w = 0.6$ cp, $\mu_0 = 1.4$ cp (reservoir conditions).

Determine the saturation distribution along the reservoir at $t = 60$ days and 200 days.

9.7 Show that

$$R^2|_{S_w}(t) - R^2|_{S_w}(0) = \frac{U(t) - U(0)}{\pi D n} \frac{dr_w}{dS_w}$$

is the relationship corresponding to (9.3.45) in radially diverging flow in a horizontal reservoir of constant thickness D; R denotes radial distance.

9.8 Write the equation for one-dimensional horizontal unsaturated flow (a) in terms of c, (b) in terms of ψ, and (c) in terms of p.

9.9 Employ the technique described in sec. 7.4 and par. 12.2.7 to show that a front of constant ψ advances such that $xt^{-1/2}$ remains constant.
(Hint: start from the one-dimensional continuity equation in terms of ψ.)

9.10 Use the Boltzmann transformation $\rho = r/\sqrt{t}$ to reduce the equation

$$\frac{1}{r}\frac{\partial}{\partial r}\left(rD\frac{\partial c}{\partial r}\right) = \frac{\partial c}{\partial t}$$

describing horizontal radially symmetrical flow to $d(\rho D\, dc/d\rho)/d\rho + \rho^2(dc/d\rho)/2 = 0$.

9.11 Determine the distribution of tension (ψ) in a vertical column of soil in steady flow with a constant supply of water q_{-z} at the ground surface. Data: $K = K_0 \exp(-a\psi)$; $K_0 = 10^{-2}$ cm/sec; $a = 10^{-2}$ (cm H_2O)$^{-1}$. The water table is at a depth $L = 500$ cm. $q_{-z} = 0$, 10^{-3}, and 10^{-2} cm/sec.

9.12 Water from a pond in which the water level is maintained constant (H above ground surface) infiltrates into a dry soil. Determine the downward advance of a wetting front (assumed abrupt), taking into account a constant capillary pressure at the front.

9.13 An abrupt interface separates between two layers of water in a vertical saturated sand column of length $L = 150$ cm. The lower part of the column (50 cm) is filled with salt water ($\gamma_s = 1.03$), while the upper part (100 cm) is filled with fresh water. Draw the potential distribution along the column:

(a) in terms of a fresh water potential $\varphi_f = z + p/\gamma_f$;
(b) in terms of a salt water potential $\gamma_s = z + p/\gamma_s$;
(c) in each part of the column according to its own potential.

9.14 Prove (9.5.19).

9.15 Give an exact mathematical statement of the problem of flow shown in fig. 9.6.3c, assuming that the elevation of point E varies with time.

9.16 Given a horizontal homogeneous sand column (n, k) saturated with oil (γ_0, μ_0) connected at $x = 0$ to a water tank with a constant free surface at elevation h_w above the column's axis and at $x = L$ to an oil tank with a constant free surface at elevation h_0 above the column's axis. Assuming an abrupt oil-water interface and complete displacement, how long will it take for a complete displacement of the water from the column.

9.17 Develop expressions similar to (7.1.72) for the refraction of the interface upon passing a plane separating two media of different permeabilities.

9.18 Draw the hodograph representation of the flow domain shown in fig. 9.6.3a, assuming that EF is an equipotential boundary (with a corresponding change in the shape of the phreatic surface).

9.19 Describe in the hodograph the flow domain shown in fig. 9.E.4.

9.20 A U-tube is filled with sand saturated with fresh water and salt water, with a sharp interface separating the two liquids. If at $t = 0$, fresh water is added to

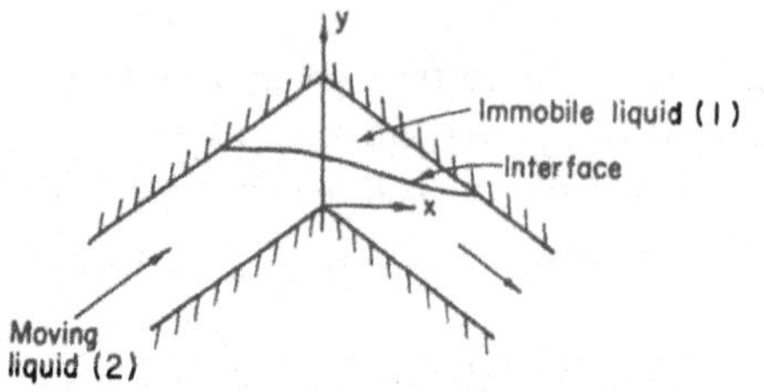

FIG. 9.E.4.

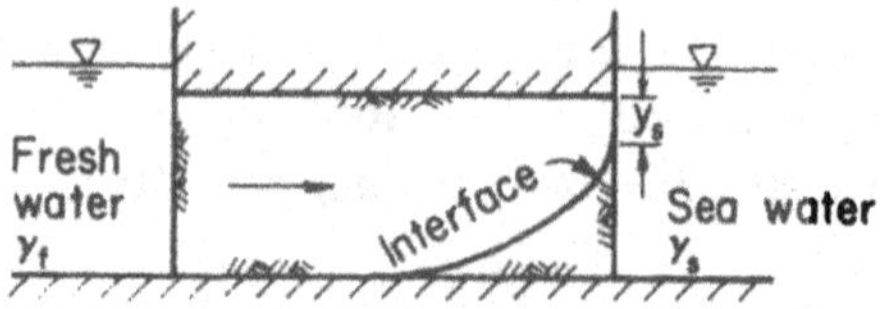

FIG. 9.E.5.

one side of the tube raising instantaneously the fresh water level by Δh, describe the movement of the interface for $t > 0$. Where will the interface be as $t \to \infty$?

9.21 Draw the flow net (streamlines and equipotentials) in the fresh water region above the interface shown in fig. 9.7.1a. Assume steady flow and a homogeneous isotropic aquifer.

9.22 Describe the domain of flow shown in fig. 9.E.5 in the hodograph plane.

9.23 Derive (9.7.60) (a) by averaging div $\mathbf{q} = 0$ over the vertical, and (b) from (6.4.22) assuming that the Dupuit assumptions are applicable to the fresh water region. Use the nomenclature of fig. 9.7.12.

CHAPTER 10

Hydrodynamic Dispersion

Actually, the basic equation describing hydrodynamic dispersion has already been developed in paragraph 4.6.2 by considering the basic mass conservation equation for a species in a solution. Using a conceptual porous medium model and an averaging procedure carried out over a representative elementary volume (REV) of the medium, a macroscopic description of this phenomenon was derived. We shall return now to reconsider in detail the variation in density and/or concentration of a species in a solution in an attempt to obtain a better insight into the mechanisms that govern these variations and the (macroscopic) medium and flow characteristics that determine them.

10.1 Definition of Hydrodynamic Dispersion

Consider saturated flow through a porous medium, and let a portion of the flow domain contain a certain mass of solute. This solute will be referred to as a *tracer*. The tracer, which is a labeled portion of the same liquid, may be identified by its density, color, electrical conductivity, etc. Experience shows that as flow takes place the tracer gradually spreads and occupies an ever-increasing portion of the flow domain, beyond the region it is expected to occupy according to the average flow alone (fig. 10.1.1). This spreading phenomenon is called ***hydrodynamic dispersion***

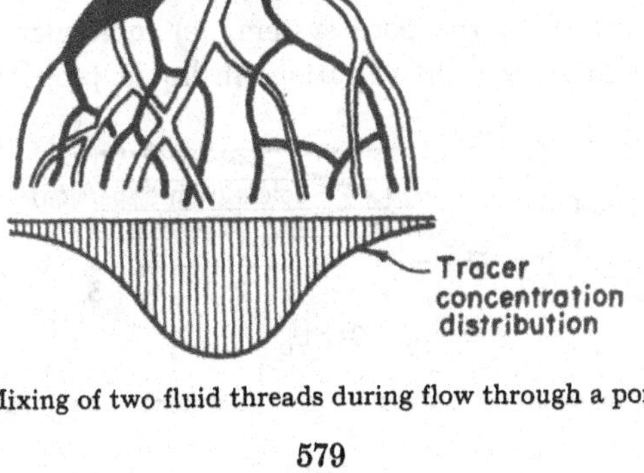

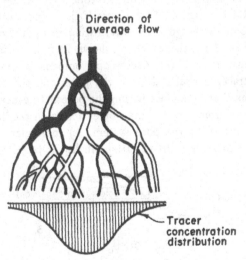

FIG. 10.1.1. Mixing of two fluid threads during flow through a porous medium.

(*dispersion, miscible displacement*) in a porous medium. It is a nonsteady, irreversible process (in the sense that the initial tracer distribution cannot be obtained by reversing the flow) in which the tracer mass mixes with the nonlabeled portion of the liquid. If initially the tracer-labeled liquid occupies a separate region, with an abrupt interface separating it from the unlabeled liquid, this interface does not remain an abrupt one, the location of which may be determined by the average velocity expressed by Darcy's law. Instead, an ever-widening *transition zone* is created, across which the tracer concentration varies from that of the tracer liquid to that of the unmarked liquid.

One of the earliest observations of these phenomena is reported by Slichter (1905), who uses an electrolyte as a tracer in studying the movement of ground water. Slichter observes that at an observation well downstream of the injection point, the tracer's concentration increases gradually, and that even in a uniform (average) flow field the tracer advances in the direction of the flow in a pear-like shape that becomes longer and wider as it advances.

The dispersion phenomenon may be demonstrated by a simple experiment. Consider steady flow in a cylindrical column of homogeneous sand, saturated with water. At a certain instant ($t = 0$), a tracer-marked water (e.g., water with NaCl at a low concentration so that the effect of density variations on the flow pattern is negligible) starts to displace the original unlabeled water in the column. Let the tracer concentration $C = C(t)$ at the end of the column be measured and presented in a graphic form, called a *breakthrough curve*, as a relationship between the *relative tracer distribution* ε and time, or volume of effluent:

$$\varepsilon(U) = [C(t) - C_0]/(C_1 - C_0) \tag{10.1.1}$$

where C_0, C_1 are the tracer concentrations of the original and of the displacing water, respectively, and U is the volume of effluent.

In the absence of dispersion, the breakthrough curve should have taken the form of the dashed line in figure 10.1.2, where U_0 is the pore volume of the column, and Q is the constant discharge. Actually, owing to hydrodynamic dispersion, it will take the form of the S-shaped curve shown in full line in figure 10.1.2.

Hydrodynamic dispersion is the macroscopic outcome of the actual movements of the individual tracer particles through the pores and the various physical and chemical phenomena that take place within the pores. In general, such movements and phenomena result from: (a) external forces acting on the liquid, (b) the microscopic intricate geometry of the pore system, (c) molecular diffusion caused by tracer concentration gradients, (d) variations in liquid properties, such as density

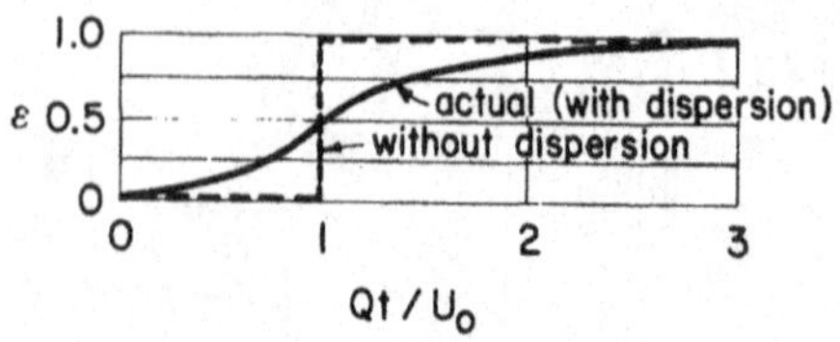

FIG. 10.1.2. Breakthrough curves in one-dimensional flow in a sand column.

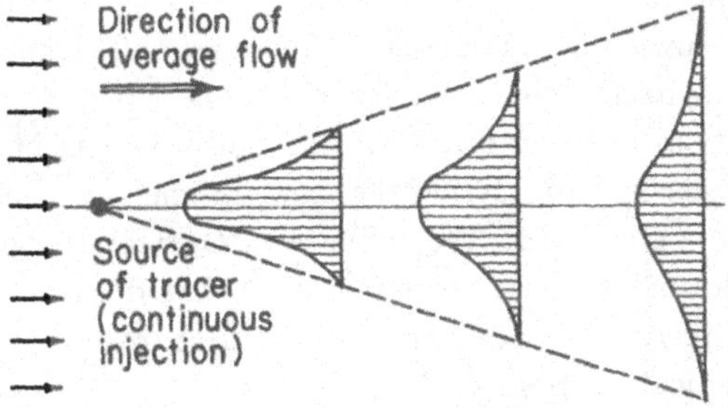

FIG. 10.1.3. Variations in tracer concentration (Danel, 1952).

and viscosity, that affect the flow pattern, (e) changes of the tracer's concentration due to chemical and physical processes within the liquid phase, and (f) interactions between the liquid and the solid phases.

There are two basic transport phenomena involved: *convection* and *molecular diffusion.* The complicated system of interconnected passages comprising the microstructure of the medium causes a continuous subdivision of the tracer's mass into finer offshoots. Variations in local velocity, both in magnitude and direction, along the tortuous flow paths and between adjacent flow paths as a result of the velocity distribution within each pore, cause any initial tracer mass within the flow domain to spread and occupy an ever-increasing volume of the porous medium (fig. 10.1.3). The two basic elements in this kind of mixing (often called *mechanical dispersion* or *convective diffusion*) are, therefore, *flow* and the *presence of the pore system* through which flow takes place. In addition to inhomogeneity on this microscopic scale (presence of pores, grains, etc.), we may also have inhomogeneity on a macroscopic scale due to variations in permeability from one portion of the flow domain to the next. This inhomogeneity also contributes to the mechanical dispersion of the tracer.

In general, we may have a convective mass transport in both a laminar flow regime, where the liquid moves along definite paths that may be averaged to yield streamlines, and a turbulent flow regime where the turbulence may cause yet an additional mixing. In what follows, we shall focus our attention only on flow of the first type.

An additional mass transport phenomenon, which occurs simultaneously with mechanical dispersion, is that caused by *molecular diffusion* resulting from variations in tracer concentration within the liquid phase. Actually, the separation between the two processes is artificial, as hydrodynamic dispersion includes *both* processes in an inseparable form. However, molecular diffusion alone does take place also in the absence of motion (both in a porous medium and in a liquid continuum). Because molecular diffusion depends on time, its effects on the overall dispersion will be more significant at low flow velocities. The relationship between the two phenomena is considered below.

The interaction between the solid surface of the porous matrix and the liquid may take several forms: adsorption of tracer particles on the solid surface, deposition, solution, ion exchange, etc. All these phenomena cause changes in the concentration of the tracer in the flowing liquid. Radioactive decay and chemical reactions within the liquid also cause tracer concentration changes.

In general, variations in tracer concentration cause changes in the liquid's density and viscosity. These, in turn, affect the flow regime (i.e., velocity distribution) that depends on these properties. We define an *ideal tracer* as one that is inert with respect to its liquid and solid surroundings and that does not affect the liquid's properties. At relatively low concentrations, the ideal tracer approximation is sufficient for most practical purposes.

10.2 Occurrence of Dispersion Phenomena

Hydrodynamic dispersion phenomena occur in many problems of ground water flow, in chemical engineering processes, in oil reservoir engineering, etc. In ground water flow we encounter it in: (a) the transition zone between salt water and fresh water in coastal aquifers; (b) artificial recharge operations where water of one quality is introduced into aquifers containing water of a different quality (mixing of water due to hydrodynamic dispersion is among the various objectives of artificial replenishment); (c) secondary recovery techniques in oil reservoirs, where the injected fluid dissolves the reservoir's oil; (d) radioactive and reclaimed sewage waste disposal into aquifers; (e) the use of tracers, such as dyes, electrolytes and radioactive isotopes, in hydrology, petroleum engineering and other scientific and engineering research projects; (f) the use of reactors packed with granular material in the chemical industry; and (g) the movement of fertilizers in the soil and the leaching of salts from the soil in agriculture.

10.3 Review of Some Hydrodynamic Dispersion Theories

Although a complete theory leading to the partial differential equations describing hydrodynamic dispersion was presented in chapter 4, it is of interest to review several other approaches. Most of them lead to the same macroscopic description. Nevertheless, some yield a better insight into the various mechanisms and parameters than others.

Two approaches are commonly employed. In the first one the porous medium is replaced by a fictitious, greatly simplified, model in which the mixing that occurs can be analyzed by *exact mathematical methods*. A single capillary tube, a bundle of capillaries, an array of cells, etc., are examples of such models (e.g., Taylor 1953; Danel 1952; Bear and Todd 1960). Models of this kind are also often considered in the flow of a single liquid (sec. 5.10). Because of their simplicity, regular models are amenable to exact mathematical analysis. Yet as a model becomes simpler, fewer factors affecting dispersion, and especially transverse dispersion, can be taken into account. The second approach, more commonly used currently, is to construct a statistical (conceptual) model of the microscopic motion of the marked fluid particles and to average these motions in order to obtain a macroscopic description of them. In either approach, one must start from the spreading that takes place in an elementary channel. Several examples of the two approaches are considered in the following paragraphs.

It must be emphasized that in both approaches—the simplified porous medium model and the statistical model—the final test is the experiment. An experiment is also the *only* way to determine the various coefficients that appear in the macroscopic equations derived from these models. There is no way to obtain them from the mathematical analysis itself, although some models may be considered more refined as they relate gross coefficients to more elementary medium properties, which still must be determined experimentally.

10.3.1 Capillary Tube and Cell Models

One of the simplest porous medium models is a bundle of capillary tubes. Taylor (1953) investigates the displacement of a liquid ($C = 0$) by another liquid ($C = C_0$) miscible with the first in a straight capillary tube of radius R. Both liquids have the same density and viscosity. In the absence of molecular diffusion, and because of the parabolic velocity distribution across the tube, a group of marked fluid particles, initially on a plane perpendicular to the tube's axis, will lie at a later time on a paraboloidal surface. This means that the initial tracer mass is continuously spreading along the tube as flow continues.

For any cross-section and time, we may define an average concentration $\bar{C}(x, t)$ as the ratio of the area occupied by the displacing fluid to the total area πR^2 of the tube's cross-section. The velocity distribution across the tube is $V(r) = 2\bar{V}(1 - r^2/R^2)$, where r is measured from the tube's axis, $\bar{V} = Q/\pi R^2$ is the average velocity, and Q is the discharge through the tube. With the nomenclature of fig. 10.3.1, continuity requires that:

$$\pi R^2 \bar{V}\bar{C}(x, t) = C_0 \int_0^{r^*} 2\pi r V(r)\, dr; \qquad r^*(x) = R(1 - x/2\bar{V}t)^{1/2}. \qquad (10.3.1)$$

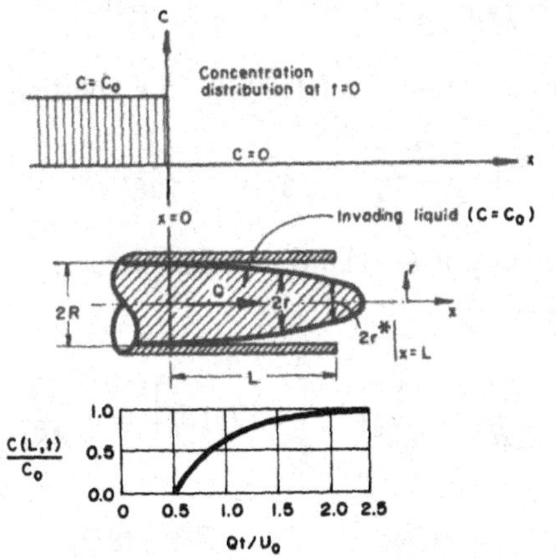

FIG. 10.3.1. Breakthrough curve in a capillary tube without molecular diffusion.

Hence, for $Qt/U_0 \geqslant \frac{1}{2}$, where $U_0 = \pi R^2 L$, we obtain:

$$\bar{C}(x, t)/C_0 = 1 - x^2/4\bar{V}^2t^2; \qquad \bar{C}(L, t)/C_0 = 1 - U_0^2/4Q^2t^2. \tag{10.3.2}$$

If molecular diffusion along the advancing interface is also considered, one obtains an equation for $C = C(x, r, t)$ in the form:

$$\frac{\partial C}{\partial t} = D_d\left(\frac{\partial^2 C}{\partial r^2} + \frac{1}{r}\frac{\partial C}{\partial r} + \frac{\partial^2 C}{\partial x^2}\right) - 2\bar{V}\left(1 - \frac{r^2}{R^2}\right)\frac{\partial C}{\partial x} \tag{10.3.3}$$

where D_d is the coefficient of molecular diffusion (assumed independent of C). Equation (10.3.3) may be simplified by deleting the second derivative $\partial^2 C/\partial x^2$, because in most cases, radial diffusion predominates over longitudinal diffusion. Rewritten in a moving coordinate system (10.3.3) then becomes:

$$\partial^2 C/\partial \rho^2 + (1/\rho)\, \partial C/\partial \rho = (R^2/D_d)\, \partial C/\partial t + 2(\bar{V}R^2/D_d)(\tfrac{1}{2} - \rho^2)\, \partial C/\partial \xi \tag{10.3.4}$$

where $\rho = r/R$ and $\xi = x - \bar{V}t$. Boundary conditions at the wall are: $\rho = 1$, $\partial C/\partial \rho = 0$. Also, at $\rho = 0$, C is finite. Initial conditions are: $t = 0$, $C = 0$ for $x > 0$, $C = C_0$ for $x \leqslant 0$.

Taylor (1953, 1954) obtains approximate solutions corresponding to two extreme cases:

(a) $2L/\bar{V} \ll R^2/14.4D_d$: axial convection dominates over radial diffusion;

(b) $2L/\bar{V} \gg R^2/14.4D_d$: radial diffusion dominates, i.e., the time required for radial concentration differences to be appreciably reduced by radial diffusion is short relative to the time required for longitudinal convection to cause appreciable radial concentration variations. This assumption is approximately applicable at relatively low velocities.

The solution of case (a) is given by (10.3.2).

As a first approximation for the solution of case (b), let the concentration gradient $(\partial C/\partial \xi)$ along the tube's axis be taken as constant. Then the solution of (10.3.4) is:

$$C = C^0 + \frac{R^2\bar{V}}{4D_d}\frac{\partial \bar{C}}{\partial \xi}(\rho^2 - \tfrac{1}{2}\rho^4); \qquad \frac{\partial C}{\partial t} = 0; \qquad C^0 \equiv C|_{\rho=0}. \tag{10.3.5}$$

With the average concentration $\bar{C}$ of (10.3.1), the solution (10.3.5) becomes:

$$C = \bar{C} + \frac{R^2\bar{V}}{4D_d}\cdot\frac{\partial \bar{C}}{\partial \xi}(-\tfrac{1}{3} + \rho^2 - \tfrac{1}{2}\rho^4); \qquad \frac{\partial C}{\partial t} = 0 \tag{10.3.6}$$

where $\partial \bar{C}/\partial \xi$ is independent of ξ. The tracer flux (mass of tracer per unit time per unit cross-sectional area) is:

$$J_c = \frac{Q_c}{\pi R^2} = \frac{1}{\pi R^2}\left[2\pi \int_0^1 CV(\rho)\, d\rho\right] = -\frac{R^2\bar{V}^2}{48D_d}\frac{\partial \bar{C}}{\partial \xi} \tag{10.3.7}$$

$$J_c = -D'\, \partial \bar{C}/\partial \xi; \qquad D' = R^2\bar{V}^2/48D_d. \tag{10.3.8}$$

This means that the tracer is dispersed by both longitudinal convection and radial molecular diffusion relative to a plane moving with velocity $\bar{V}$, exactly as though it were being diffused by a process that obeys *Fick's law* of molecular diffusion, but with a diffusion coefficient $D' = (R^2/48D_d)\bar{V}^2$. One should notice here the dependence of D' on $\bar{V}^2$. The condition that longitudinal molecular diffusion is negligible with respect to the longitudinal dispersion expressed by D' leading to the above results, is (Taylor 1953, 1954):

$$(48)^{1/2}(L/R) < L\bar{V}/D_d < 4(L/R)^2.$$

Conservation of mass during this diffusion-like process may be expressed by:

$$\partial\bar{C}/\partial t = D' \, \partial^2\bar{C}/\partial\xi^2 \quad \text{or} \quad \partial\bar{C}/\partial t = D' \, \partial^2\bar{C}/\partial x^2 - \bar{V} \, \partial\bar{C}/\partial x. \tag{10.3.9}$$

For the initial conditions $x < 0$, $\bar{C} = C_0$ and $x \geqslant 0$, $\bar{C} = 0$ (10.3.9) yields (von Rosenberg 1956):

$$\bar{C}/C_0 = \tfrac{1}{2}\left[1 \pm \operatorname{erf}\frac{x - \bar{V}t}{2\sqrt{D't}}\right] \qquad \begin{matrix} + & \text{for} & x - \bar{V}t < 0 \\ - & \text{for} & x + \bar{V}t > 0 \end{matrix} \tag{10.3.10}$$

where $\operatorname{erf} z = (2/\pi) \int_0^z \exp(-\alpha^2)\, d\alpha$ (fig. 10.3.2). The point of $\bar{C}/C_0 = 50\%$ moves with the fluid's average velocity. From (10.3.10) it also follows that at any instant, the width of the transition zone, i.e., the length over which most of the concentration change (say, from 90% to 10% of total concentration change) takes place, is directly proportional to the square root of the velocity and inversely proportional to the square root of D'. At any average velocity, the length of the transition zone increases as the square root of the average distance traversed. If L denotes the length of the zone along which $\bar{C}$ varies from 0.9 C_0 to 0.1 C_0, we obtain:

$$L = 0.52R\bar{V}(t/D_d)^{1/2} \tag{10.3.11}$$

which shows that the breakthrough curve flattens as $\bar{V}$ increases, and becomes steeper, i.e., tends to a *piston-like* flow, when D_d increases. Van Deamter et al. (1955) obtain a more generalized solution of (10.3.3). Danel (1952) and others extend the single capillary tube model to a bundle of capillary tubes of different diameters. The resulting dispersion is shown to depend both on the parabolic velocity distribution within each tube and on the distribution of average velocities in the tubes.

Aris (1956) extends Taylor's analysis to straight tubes of any cross-section. He shows that the results described above lead to an *effective diffusion coefficient* of the form:

$$D' = D_d + \delta R^2\bar{V}^2/D_d; \qquad D'/D_d = 1 + (\delta' R\bar{V}/D_d)^2; \qquad \delta' = \delta^{1/2} \tag{10.3.12}$$

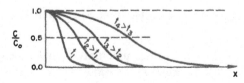

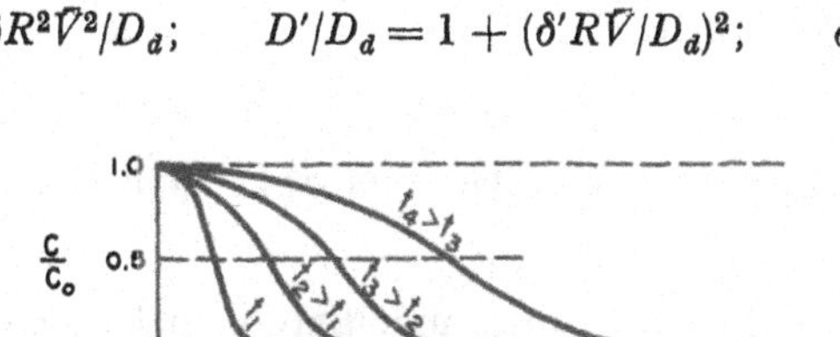

FIG. 10.3.2. Typical concentration distribution during miscible displacement.

where δ (restricted to a narrow range; e.g., for an elliptical cross-section $5/12 < 48\delta < 1$) is a coefficient depending on the shape of the capillary tube's cross-section. Aris shows that the dispersion causes an average concentration distribution in the form of a normal (Gauss) distribution around the center of gravity, which is displaced at the mean velocity of flow. The coefficient D' of (10.3.12) is obtained from the variance of this distribution. D' is usually called the coefficient of dispersion (dimensions L^2T^{-1}). Blackwell's (1957) experiments verify (10.3.12).

As another example of a regular type of model, consider the simplified model used by Bear (1960) in studies of one-dimensional dispersion. His model consists of an array of small cells with interconnecting short channels. It is assumed that when a liquid with a certain tracer concentration enters a cell occupied by a liquid of a different concentration, it displaces part of it, while the liquids remaining in the cell *immediately* mix to form a new homogeneous liquid. We may refer to such a cell as a *perfect mixer*. To obtain perfect mixing, the true movement of the tracer particles (through molecular diffusion or turbulence) must be much faster than the average liquid flow. Let U denote the volume of a perfect mixer cell filled with liquid, and $C^{(i)}$, C and $C^{(0)}$ be the tracer concentrations in the liquid entering the container, dwelling in it and leaving it, respectively. The variation in tracer concentration in a cell is then given by:

$$\partial C/\partial t = (Q/U)(C^{(i)} - C) \quad \text{or} \quad \partial C/\partial t = (C^{(i)} - C)/\tau \tag{10.3.13}$$

where $\tau = U/Q$ is the average *residence time* of the liquid in the cell and Q is the discharge. For $C^{(i)} = 0$ and $C|_{t=0} = C_0$, we obtain:

$$\partial C/\partial t = -C/\tau; \qquad C = C_0 \exp(-t/\tau). \tag{10.3.14}$$

Wentworth (1948) was probably the first to suggest the perfect mixing model in an array of cells as an explanation for the creation of a transition zone at a moving interface between fresh water and salt water in a coastal aquifer. One may visualize the pores in a porous medium as such cells. Aris and Amundson (1957) also present an analysis of dispersion as a process of mixing in cells.

In the analysis suggested by Bear (1960), the dispersion phenomenon may be regarded as a combination of two processes: (a) complete mixing in the elementary cells, and (b) translation at the average flow velocity from one cell to the next through the connecting channels. A tracer balance for the jth cell of an array, leads to:

$$\partial C^{(0)}{}_j(t)/\partial t + C^{(0)}{}_j(t)/\tau = C^{(i)}{}_j(t)/\tau. \tag{10.3.15}$$

The solution of (10.3.15) is:

$$C^{(0)}{}_j(t) = \exp(-t/\tau)\left[\int_0^t \frac{1}{\tau}\exp(t/\tau)\, C^{(i)}{}_j(t)\, dt + \text{const}\right]. \tag{10.3.16}$$

For an instantaneous inflow that causes an initial concentration of C_0 in the first cell, $C^{(i)}{}_j(t) = \bar{C}\,\delta(t)$, where $\delta(t)$ is the *Dirac function*, and $\bar{C} = C_0\tau$ may be interpreted as the tracer quantity per unit discharge. We obtain:

$$C^{(0)}{}_j(t) = C_0(t/\tau)^{j-1}\,[\exp(-t/\tau)]/(j-1)!. \qquad (10.3.17)$$

This is the *Poisson probability distribution* $P(\xi = N) = \lambda^N \exp(-\lambda)/N!$, with $N = j - 1$ and $\lambda = t/\tau$. For a constant inflow of concentration C_0 in the first cell, and with $\tau = \text{const}$ for all cells, we obtain:

$$C^{(0)}{}_j = C_0\left[1 - \exp(-t/\tau)\sum_{i=1}^{j-1}\frac{1}{i!}\left(\frac{t}{i}\right)^i\right]. \qquad (10.3.18)$$

If we add translation through the cells, with $j\,\Delta t$ the time of translation through j cells, so that of the total time t, $t - j\,\Delta t$ is left for the mixing, and $j\,\Delta t \approx N\,\Delta t$, we obtain:

$$C^{(0)}{}_j(t) = \begin{cases} C_0\dfrac{[(t-N\,\Delta t)/\tau]\exp\{-(t-N\,\Delta t)/\tau\}}{N!} & \text{for} \quad t > N\,\Delta t \\ 0 & \text{for} \quad t < N\,\Delta t. \end{cases} \qquad (10.3.19)$$

By the *central limit theorem* of statistics, as N becomes large, equation (10.3.19) may be approximated by the *normal distribution*:

$$\frac{C^0{}_j(t)}{C_0} = \frac{1}{(2\pi N)^{1/2}}\exp\left\{-\frac{(T-N)^2}{2N}\right\} \qquad (10.3.20)$$

or:

$$\frac{C(x,t)}{C_0} = \frac{1}{(4\pi D't)^{1/2}}\exp\left\{-\frac{(x-\bar{x})^2}{4D't}\right\} \qquad (10.3.21)$$

where $T = (t - N\,\Delta t)/\tau$; $\bar{x} = \bar{V}t$; $\bar{V} = \Delta l/(\tau + \Delta t)$; Δl is the distance between centers of cells; $D' = \frac{1}{2}(\Delta l)^2\tau^2/(\tau + \Delta t)^3 = \frac{1}{2}[(\Delta l')^2/\Delta l]\bar{V} = a_l\bar{V}$; $\Delta l'$ is the length of a mixing cell $(\tau\bar{V})$; $a_l = \frac{1}{2}[(\Delta l')^2/\Delta l]$ is a medium property that may be called the *medium's longitudinal dispersivity*, to be considered in detail below. One should note that here D' is proportional to $\bar{V}$ (average flow velocity) whereas in the capillary tube model we had $D' \sim \bar{V}^2$ (par. 10.3.1). Thus we see that even a model of a rather simple form may lead to results applicable to actual porous media. Obviously, before reaching the conclusion that (10.3.21) is indeed applicable to actual porous media, one must perform experiments to verify the validity of this equation and to determine D' experimentally. One must also check the dependence of D' on $\bar{V}$. However, an approach of this kind (as with the capillary tube model) fails to yield the transverse dispersion observed in flow through porous media.

Among other dispersion theories based on regular models one may mention those of Danckwerts (1953), Turner (1957), Perkins and Johnston (1963), Deans (1963) and Coats and Smith (1964), who directed their attention to the presence of *dead-end pores*.

10.3.2 Statistical Models

The basic idea underlying the statistical approach is that it is impossible to give an exact mathematical description of the motion of a single tracer particle, a description that seemingly should be necessary for any forecast of spreading or dispersion of a cloud of tracer particles. One should recall at this point that the path of a single tracer particle, which is an ion or a molecule, should not be identified with the path of a liquid particle, which is an instantaneous ensemble of a large number of molecules, the identity of which changes as a result of molecular diffusion. One may visualize the path of a tracer particle as the vectorial result of two motions: one along the pathline of a liquid particle, and another from one pathline to an adjacent one as a molecular diffusion process. The nature of both motions, the first determined by the intricate internal geometry of the medium, and the second by the random character of molecular diffusion, prevents any deterministic prediction of paths of tracer particles.

The basic postulate of the statistical approach is that, although it is impossible to predict the exact path of an individual tracer particle, owing to the large number of unknown factors governing its motion, one may employ the rules of probability theory to predict the spatial distribution at any later time of a cloud of many tracer particles that were initially at a close proximity, and that move under the same average conditions. In this approach, one considers the random motion of an anonymous tracer particle that represents an ensemble of many such particles, the paths of which have common statistical characteristics. Obviously, stating that the motion of a tracer particle is a random one is only our way of describing its path, which is completely defined by the laws of physics. The probability distribution of the location of a single particle may be interpreted as the spatial distribution of the relative concentration of a cloud of tracer particles originating from the neighborhood of a certain point at a certain time, and moving under the same average conditions.

This idea conforms to the postulates of probability theory of *random phenomena*. A random phenomenon is one that, when observed in repeated experiments performed under *apparently* the same conditions, yields a different result each time. These random variations are caused by secondary uncontrolled factors that change from one experiment to the next. Nevertheless, certain trends, or regularities, are observed when a large number of similar experiments is performed. Hence, although the result of each of a series of similar experiments cannot be predicted, the overall behavior, or the mean results, may still be derived.

The conditions for a successful statistical approach are: (a) that we have a porous medium *completely disordered* at the microscopic level; (b) that the application of statistical considerations is justified; and (c) that we may actually be able to measure the parameters that appear in the resulting macroscopic description.

The progress of a liquid particle through a porous medium follows an exact, predetermined path described by the Navier–Stokes equations, and by appropriate boundary conditions. Because the boundary conditions are unknown, it has been suggested (Scheidegger 1954, 1958) that instead of considering a specific porous medium, one should investigate a whole *ensemble* of porous media assumed to be macroscopically identical, i.e., having identical macroscopic properties. Each possible path of a liquid particle moving through a porous medium takes place at an ever-changing velocity through a system of interconnected pores considered as a particular sample of this ensemble. The entire porous medium domain may be considered as an assembly of a large number of such samples that differ from each other. Obviously, several particles may move through the same sample along different pathlines. Each pathline may be considered as a vector sum of elementary displacements.

Instead of attempting to determine the velocity at a certain time of a definite liquid particle at a certain point in space, one may study the assembly of all possible velocities (or elementary displacements) associated with many particles, occupying at that time a small volume around the considered point. With this approach, the single-valued velocity of a definite particle at a certain point is replaced by a large number of possible velocities of a large number of anonymous particles. A *probability distribution* may be assigned to this assembly of velocities and referred to this point.

In the case of a homogeneous fluid in spatially averaged uniform flow through a homogeneous medium, the random character of the velocity distribution results only from the nature of the porous matrix. Over a domain including many points, the velocity becomes a stationary *random-vector function* of time and space. At any given instant it is the function of space only.

As a particle moves through a homogeneous medium, it encounters all the conditions present in the many samples of the flow domain; hence, the *ergodic hypothesis* of interchanging time averages with ensemble averages is applicable. By the *law of large numbers* (e.g., Feller 1958), after a sufficiently long time, the time-averaged velocity characteristics of a single particle may replace averages taken over the assembly of many particles that move under the same flow conditions.

The problem of the spreading of a cloud of labeled particles is thus reduced in the statistical approach to the problem of the random motion of a single anonymous tracer particle through an ensemble of random porous media.

According to Scheidegger (1954, 1958), the application of the statistical approach requires: (a) an assumption about the statistical (average) properties of the porous medium (the ensemble), (b) an assumption on the microdynamics of the flow, i.e., on the relationships among the forces, the liquid properties and the resulting velocity during each small time step (in general, flow is assumed laminar); and (c) a choice of the type of statistics to be employed, i.e., the probability of occurrence of events during small time intervals within the chosen ensemble. This may take the form of correlation functions between velocities at different points or different times, or joint-probability densities of the local velocity components of the particle as functions of time and space or a probability of an elementary particle displacement. The chosen correlation function determines the type of dispersion equation derived.

Before describing several typical works based on the statistical approach, certain important general comments seem appropriate. As the total particle travel time becomes much larger than the time interval during which its successive (local) velocities are still correlated, the total displacement may be considered as a sum of a large number of elementary displacements statistically independent of one another. Then, the probability distribution of the particle's total displacement tends to the normal (*Gauss*) *distribution*. This is based on the *central limit theorem* of probability, which states that no matter what the probability distribution is for the elementary displacements (or for the velocity during these displacements), as the number of steps increases, the probability distribution of the total displacement tends to normal. In view of the ergodic principle, this distribution also represents the spatial distribution of displacements of a cloud of initially close particles.

In order to express these concepts in a mathematical form, let $\bar{\mathbf{V}}$ denote the average velocity vector (components $\bar{V}_i$) in a field of uniform flow of a homogeneous fluid through a homogeneous porous medium of infinite area. At any instant t, the component $V_i(t)$ in the direction i of the local velocity vector $\mathbf{V}$ of a marked fluid particle (ideal tracer), may be expressed as:

$$V_i(t) = \bar{V}_i + \mathring{V}_i(t) \tag{10.3.22}$$

where $\mathring{V}_i(t)$ is the deviation of the particle's instantaneous velocity from the average one. Equation (10.3.22) states that the movement of a particle may be considered as a superposition of two motions: an average motion at a velocity $\bar{V}_i$, which obeys some macroscopic, nonrandom law (e.g., Darcy's law), and a random velocity $\mathring{V}_i(t)$. By definition:

$$x_i(t) = \int_0^t V_i(\theta)\, d\theta;$$

$$\bar{x}_i(t) = \int_0^t \bar{V}_i(\theta)\, d\theta = \int_0^t d\theta \int_{-\infty}^{+\infty} V_i(\theta) f(V_i)\, dV_i = \int_{-\infty}^{+\infty} x_i(t) f(x_i)\, dx_i$$

where $f(V_i)$ and $f(x_i)$ are the probability distributions of V_i and x_i, respectively. Here and in what follows, unless otherwise specified, no summation is performed on i; further; a bar above a symbol denotes a probability average. By the ergodic property mentioned above, for sufficiently large times, the particle's time-averaged velocity $\bar{V}_i(t)$ approaches $\bar{V}_i$, and therefore $\overline{\mathring{V}_i(t)} = 0$.

To simplify the discussion on the random motion, let the movement be referred to a coordinate system $\mathring{x}_1, \mathring{x}_2, \mathring{x}_3$ that moves at a velocity $\bar{\mathbf{V}}$:

$$\mathring{x}_i(t) = x_i(t) - \bar{x}_i(t); \qquad i = 1, 2, 3. \tag{10.3.23}$$

A particle starting from the origin at $t = 0$, will reach a point $\mathring{x}_i(t)$ at some later time t such that:

$$d\mathring{x}_i = \mathring{V}_i(t)\, dt; \qquad \mathring{x}_i = \int_0^t \mathring{V}_i(\theta)\, d\theta. \tag{10.3.24}$$

The spread of the particle's possible displacements around its average displacement $\bar{x}(t)$ at that time is described by the *matrix of covariances* (correlation moments) $\overline{\mathring{x}_i(t)\mathring{x}_j(t)}$. By definition:

$$\mathrm{Cov}(x_i, x_j) \equiv \overline{\mathring{x}_i \mathring{x}_j} = \int_{-\infty}^{+\infty}\int_{-\infty}^{+\infty} (x_i - \bar{x}_i)(x_j - \bar{x}_j) f(x_i, x_j)\, dx_i\, dx_j = \overline{x_i x_j} - \bar{x}_i \bar{x}_j \quad (10.3.25)$$

where $f(x_i, x_j)$ is the *joint probability density* of x_i and x_j; for $i = j$, the covariances become the variances σ^2 of x_i: $\mu_{ii} \equiv \sigma^2(x_i) = \mathrm{Var}(x_i) = \overline{(\mathring{x}_i)^2}$, where σ is the *standard deviation* of x_i.

Now, $V_i(t)$ and $V_j(t)$—components in the i and j directions of the velocity vector $\mathbf{V}(t)$—are two *stationary random functions* of time. A random function $V_j(t)$ is called *stationary* if its *correlation coefficient* $r(t, t')$ is independent of the arguments t, t', but depends only on the interval $(t' - t)$ between them. In general:

$$r_{ij}(t, t') = \overline{\mathring{V}_i(t)\mathring{V}_j(t')}/\sigma[V_i(t)]\sigma[V_j(t')]. \quad (10.3.26)$$

In addition we require that the average, the variance and the other moments remain constant in time, i.e., $r_{ij}(t, t') = r_{ij}(t, t + \tau) = r_{ij}(\tau)$.

Consider now the derivative:

$$\frac{d}{dt}[\mathring{x}_i(t)\mathring{x}_j(t)] = \frac{d\mathring{x}_i(t)}{dt}\mathring{x}_j(t) + \mathring{x}_i(t)\frac{d\mathring{x}_j(t)}{dt}$$

$$= \int_0^t \mathring{V}_i(t)\mathring{V}_j(\theta)\, d\theta + \int_0^t \mathring{V}_i(\theta)\mathring{V}_j(t)\, d\theta. \quad (10.3.27)$$

By averaging (10.3.27), we obtain:

$$\overline{\frac{d}{dt}[\mathring{x}_i(t)\mathring{x}_j(t)]} = 2r_{ij}(0)\overline{\mathring{V}_i(t)\mathring{V}_j(t)} \int_0^t r_{ij}(\tau)\, d\tau \quad (10.3.28)$$

where $r_{ij}(\tau)$ is the *coefficient of correlation* between the velocity V_i at t and V_j at $(t + \tau)$. From (10.3.28):

$$\overline{\mathring{x}_i(t)\mathring{x}_j(t)} = 2r_{ij}(0)\overline{\mathring{V}_i(t)\mathring{V}_j(t)} \int_0^t\int_0^\theta r_{ij}(\tau)\, d\tau\, d\theta$$

$$= 2r_{ij}(0)\overline{\mathring{V}_i(t)\mathring{V}_j(t)} \int_0^t (t - \tau) r_{ij}(\tau)\, d\tau. \quad (10.3.29)$$

Because $V_i(t)$ and $V_j(t)$ are stationary random functions, $\overline{\mathring{V}_i(t)\mathring{V}_j(t)}$ is a constant, independent of t. If t_0 is the time during which velocities are still effectively correlated (*autocorrelation time*):

$$t_0 = \int_0^\infty r_{ij}(\tau)\, d\tau. \quad (10.3.30)$$

We may investigate two cases.

Case 1, $t \gg t_0$. Then (10.3.29) becomes:

$$\overline{\mathring{x}_i \mathring{x}_j} \approx 2r_{ij}(0)\overline{\mathring{V}_i(t)\mathring{V}_j(t)}\, t_0 t. \tag{10.3.31}$$

Introducing the notation D_{ij}, which will later be defined as the *coefficient of mechanical dispersion*, we may write:

$$D_{ij} = \overline{\mathring{x}_i \mathring{x}_j}/2t = r_{ij}(0)\overline{\mathring{V}_i \mathring{V}_j}\, t_0. \tag{10.3.32}$$

For $i = j$, $r_{ij}(0) = 1$, and:

$$D_{ii} = \overline{\mathring{x}_i^2}/2t = \overline{\mathring{V}_i^2}\, t_0; \qquad \sigma^2(x_i) = 2D_{ii}t. \tag{10.3.33}$$

Case 2, $t \ll t_0$. One obtains (Scheidegger 1960):

$$\overline{\mathring{x}_i \mathring{x}_j} = r_{ij}(0)\overline{\mathring{V}_i \mathring{V}_j}\, t^2; \qquad D_{ij} = \overline{\mathring{x}_i \mathring{x}_j}/2t = \tfrac{1}{2} r_{ij}(0)\overline{\mathring{V}_i \mathring{V}_j}\, t$$

$$D_{ii} = \overline{\mathring{x}_i^2}/2t = (\overline{\mathring{V}_i^2}/2)t; \qquad \sigma^2(x_i) = \overline{\mathring{V}_i^2}\, t^2 \tag{10.3.34}$$

which expresses the fact that for very short-time intervals there is no random process and every particle progresses with its velocity. One should note that (10.3.31) and (10.3.32) do not prove that the spreading of a cloud of marked particles is given by solutions of the diffusion equation with the effective diffusivity D_{ij}. This is so only if it is shown separately that the probability distribution of x_i is normal. If each elementary displacement $d\mathring{x}_i(t)$ is a random variable, the total deviation of the particle's position from its average position tends to a normal distribution only if $t \gg t_0$. As stated above, this is a consequence of the central limit theorem. The general form of the normal density function is (Chandrasekhar 1943):

$$\phi(x_1, x_2, x_3, t) = \frac{1}{(2\pi)^{3/2}|\mu|^{1/2}} \exp\left[-\frac{1}{2}\frac{\mu'_{ij}}{|\mu|}(x_i - \bar{x}_i)(x_j - \bar{x}_j)\right] \tag{10.3.35}$$

where the summation convention is employed; $|\mu|$ is the determinant of the correlation matrix $[\mu_{ij}]$:

$$[\mu_{ij}] = \begin{bmatrix} \mu_{11} & \mu_{12} & \cdots & \mu_{1n} \\ \mu_{21} & & \cdots & \\ \mu_{n1} & & \cdots & \mu_{nn} \end{bmatrix} \tag{10.3.36}$$

μ_{ij} is the correlation moment defined by (10.3.25) and μ'_{ij} is the cofactor of μ_{ij} in the determinant $|\mu_{ij}|$.

In (10.3.35), ϕ gives the relative tracer-mass concentration in percents of total mass per unit volume of the medium at the considered point; it is to be regarded as an average over a small volume of space around the point. The tracer concentration C (mass of tracer per unit volume of solution) is related to the relative *concentration* ϕ by:

$$C = \phi m/n \tag{10.3.37}$$

where n is the medium's porosity and m is the total tracer mass.

The exponent $(\mu'_{ij}/|\mu_{ij}|)\mathring{x}_i\mathring{x}_j$ in (10.3.35) is a quadratic form that describes a family of constant ϕ-surfaces at time t in the (x_1, x_2, x_3) coordinate system. It is a family of confocal ellipsoids centered at $(\bar{x}_1, \bar{x}_2, \bar{x}_3)$. It is easily verified that (10.3.35) may be written in the form:

$$\phi(x_1, x_2, x_3, t) = \phi_1(x_1, t)\phi_2(x_2, t)\phi_3(x_3, t) \tag{10.3.38}$$

where $\phi_i(x_i, t)$, $i = 1, 2, 3$, is the *normal density function* of the single random variable x_i at time t. Thus, in the one-dimensional case, the density function $\phi(x, t)$ is:

$$\phi(x, t) = [(2\pi)^{1/2}\sigma]^{-1} \exp\{-(x - \bar{x})^2/2\sigma^2\}; \qquad \sigma = (2Dt)^{1/2}; \qquad D \equiv D_{ii}. \tag{10.3.39}$$

We have shown here how (10.3.35) can be derived from certain general assumptions. However, these results do not show the relationships between σ_i or μ_{ij} and statistically averaged medium and flow parameters. This information may be obtained by choosing and analyzing the microdynamics of the flow within each elementary time interval.

As an illustration of the principles outlined above, let us review briefly several works employing the statistical approach. Most of them study dispersion in uniform steady flow through a homogeneous isotropic medium.

The simplest statistical model of dispersion is the one-dimensional *random walk*. In this model, a particle performs displacements along a straight line in the form of a series of steps of equal length, each step being taken either in the forward or in the backward direction with equal probability $\frac{1}{2}$. After N such steps the particle may be at any of the points: $-N, -N+1, \ldots, -1, 0, +1, \ldots, N-1, N$. Assuming that each step is equally likely to be taken in either direction, independently of all previous displacements, the probability of any sequence of N steps is $(\frac{1}{2})^N$. Hence, it can be shown that the probability $P(M, N)$, that the particle will arrive at a point M after N displacements, is:

$$P(M, N) = \frac{N!}{((N+M)/2)!((N-M)/2)!}(\tfrac{1}{2})^N$$

which is the *Bernoulli distribution*. The root mean square displacement is $(N)^{1/2}$. As $N \to \infty$ and $M \ll N$:

$$P(M, N) = (2/\pi N)^{1/2} \exp(-M^2/2N). \tag{10.3.40}$$

If l is the length of each step such that $x = Ml$, the probability element $P(x)\,\Delta x$ that the particle is likely to be in the interval $(x, x + \Delta x)$ after N displacements with $\Delta x > l$, is:

$$P(x, N)\,\Delta x = P(M, N) \cdot (\Delta x/2l) = (1/2\pi N l^2)^{1/2} \exp(-x^2/2Nl^2).$$

If the particle suffers n displacements per unit time (i.e., its velocity is $u = nl$), the probability that it will find itself between x and $x + \Delta x$ at time t is:

$$P(x, t)\,\Delta x = \left[\frac{1}{2(\pi D t)^{1/2}} \exp(-x^2/4Dt)\right] \Delta x; \qquad D = (\tfrac{1}{2})nl^2 = (l/2)u. \tag{10.3.41}$$

Compare D here with D' in (10.3.21). This random walk approach is the basis for molecular diffusion theories, Brownian motion, etc., where l has the meaning of mean free path of a particle.

Chandrasekhar (1943) presents a discussion on the general, three-dimensional random walk, using Markov's method.

He considers a particle that performs a sequence of elementary steps, each represented by a displacement vector $\mathbf{r}_i$: $\mathbf{r}_1, \mathbf{r}_2, \ldots$. The displacements are independent of each other, and he determines the position of the particle after n displacements. The probability of a step $\mathbf{r}_i$ being in the range $(\mathbf{r}_i, \mathbf{r}_i + d\mathbf{r}_i)$ is given by:

$$\tau_i(\mathbf{r}_i)\, d\mathbf{r}_i = \tau_i(x_i, y_i, z_i)\, dx_i\, dy_i\, dz_i$$

where x_i, y_i, z_i are the components of $\mathbf{r}_i$ and τ_i is the joint probability distribution of x_i, y_i, z_i. The location of the particle after n steps is defined by the radius vector $\mathbf{R}(X, Y, Z)$:

$$\mathbf{R}(X, Y, Z) = \sum_{i=1}^{n} \mathbf{r}_i; \qquad X = \sum_{i=1}^{n} x_i; \qquad Y = \sum_{i=1}^{n} y_i; \qquad Z = \sum_{i=1}^{n} z_i.$$

Chandrasekhar shows that for $n \gg 1$, the probability $W(\mathbf{R})$ that $\mathbf{R}$ will be in the range $(\mathbf{R}, \mathbf{R} + d\mathbf{R})$ is:

$$W(\mathbf{R})\, d\mathbf{R} = \frac{1}{(2\pi)^{3/2}\sigma_X\sigma_Y\sigma_Z} \exp\left\{-\frac{(X-\bar{X})^2}{2\sigma_X^2} - \frac{(Y-\bar{Y})^2}{2\sigma_Y^2} - \frac{(Z-\bar{Z})^2}{2\sigma_Z^2}\right\}$$

where $\bar{X} = (1/n)\sum_{i=1}^{n} x_i$, etc. This expression describes a normal distribution in a three-dimensional space. If the particles undergo, on the average, $\bar{n}$ displacements per unit time, $n = \bar{n}t$, we may define the average velocity of the particle by: $V_x = \bar{n}\bar{x} = n\bar{x}/t = \bar{X}/t$; $V_y = \bar{n}\bar{y} = \bar{Y}/t$, $V_z = \bar{n}\bar{z} = \bar{Z}/t$. Then the dispersion coefficients become:

$$D_{11} = n\overline{x^2}/2t = \sigma_X^2/2t; \qquad D_{22} = n\overline{y^2}/2t = \sigma_Y^2/2t; \qquad D_{33} = \sigma_Z^2/2t$$

and we obtain for $W(\mathbf{R})$:

$$W(\mathbf{R}, t)\, dR = \frac{dX\, dY\, dZ}{(4\pi t)^{3/2}(D_{11}D_{22}D_{33})^{1/2}} \exp\left\{-\frac{(X - V_x t)^2}{4D_{11}t} - \frac{(Y - V_y t)^2}{4D_{22}t} - \frac{(Z - V_z t)^2}{4D_{33}t}\right\}. \tag{10.3.42}$$

If, instead of considering a single particle, we consider a large number of particles with no interference, all starting from the same origin $\mathbf{R} = 0$ at $t = 0$, $W(\mathbf{R}, t)\, d\mathbf{R}$ may be interpreted as that portion of the total number of particles that will be found at time t in the range $(\mathbf{R}, \mathbf{R} + d\mathbf{R})$.

Equation (10.3.42) is an elementary solution of the partial differential equation:

$$\frac{\partial W}{\partial t} = D_{11}\frac{\partial^2 W}{\partial X^2} + D_{22}\frac{\partial^2 W}{\partial Y^2} + D_{33}\frac{\partial^2 W}{\partial Z^2} - V_x\frac{\partial W}{\partial X} - V_y\frac{\partial W}{\partial Y} - V_z\frac{\partial W}{\partial Z}.$$

In a coordinate system X_1, X_2, X_3, which does not coincide with the principal axes of dispersion, this last equation, using matrix notation, becomes:

$$\frac{\partial W}{\partial t} = D_{ik}\frac{\partial^2 W}{\partial X_i\,\partial X_k} - V_i\frac{\partial W}{\partial X_i}.$$

We shall return to these equations at a later stage.

Scheidegger (1954), who also employed the theory of random walk, extending it to three dimensions, based his statistical analysis on the following assumptions.

(a) The homogeneous isotropic porous medium constitutes an ensemble of a large number of systems, or samples, having identical macroscopic characteristics.

(b) The fluid is a homogeneous continuum, each particle of which has a path of its own.

(c) The motion of a particle through a specific porous medium (i.e., a specific sample) is made up of a sequence of straight elementary displacements of equal duration. The direction and length of each displacement take on random values, with no correlation among displacements. Under such conditions, a sufficiently long path of a particle through a specific system incorporates all the conditions encountered in the ensemble of systems. Hence we may apply the *ergodic property*, which enables us to interpret a temporal average along a single path as a spatial average over a large ensemble of paths.

(d) Laminar laws of flow govern the movement in each step, and external forces are uniform and steady.

With these assumptions, and neglecting molecular diffusion, the time interval $(0, t)$ is divided into N equal intervals $\Delta t = t/N$, in each of which the particle undergoes a displacement vector $\boldsymbol{\xi}$ with Cartesian components (ξ, η, ζ). In Scheidegger's treatment, the liquid is homogeneous and a particle is an elementary volume of liquid that remains undivided as it moves through the porous medium. The probability density of a displacement $\boldsymbol{\xi}$ within Δt is denoted by $p(\boldsymbol{\xi})$. By definition:

$$\iiint_{-\infty}^{+\infty} p(\xi, \eta, \zeta)\, d\xi\, d\eta\, d\zeta = 1. \tag{10.3.43}$$

For a sufficiently large number N, Scheidegger obtains the normal (Gaussian) distribution:

$$p(x, y, z, t) = (4\pi Dt)^{-3/2} \exp\{-[(x-\bar{x})^2 + (y-\bar{y})^2 + (z-\bar{z})^2]/4Dt\} \tag{10.3.44}$$

where $D = N\sigma^2/2t = \sigma^2/2\Delta t$, $\bar{x} = V_x t$, $\bar{y} = V_y t$ and $\bar{z} = V_z t$; V_x, V_y and V_z are the components of the uniform velocity field.

In view of the ergodic property, equation (10.3.44) describes the spatial tracer concentration at time t of a large number of particles that started from the vicinity of the same origin at $t = 0$, and traveled along independent paths under similar statistical conditions. Such an ensemble of particles is normally distributed around its center, which travels with the average velocity of the flow.

Scheidegger (1954) obtains his solution by assuming that there is no correlation between the ith and the $(i-1)$st step, and that all steps are of equal duration.

He also assumes that because the medium is isotropic, the probability is the same in all directions. This leads to a normal distribution and to a D in (10.3.44) that is the same in all directions. It will be shown below that this results from an erroneous assumption of isotropic spreading. The spreading is not isotropic, that is, it is different in the direction of the mean flow (longitudinal dispersion) and perpendicular to it (transverse dispersion), although the medium is isotropic.

Danckwerts (1953), who studies the efficiency of chemical reactors, also employs the random walk model (in one dimension). However, he introduces into his model the *residence time* of an elementary step, i.e., the time during which a fluid particle passes through an elementary channel or through a certain length of the medium. In his model, and in other *random residence time models*, the flow takes place through a sequence of cells in each of which a complete mixing takes place. Unlike the deterministic cell theories described above, here there is a certain probability that a fluid particle will leave the cell, the probability being independent of the length of time the particle stays in the cell. The time of passage of a fluid particle through a sequence of cells simulating the porous medium, which is the sum of the time intervals spent in the individual cells of the sequence, is a random function that tends to normal as the number of cells becomes large. As in (10.3.21), the variance of the distribution is proportional to the number of cells. The longitudinal dispersion is then described by the diffusion equation with an effective diffusivity proportional to the product of the average velocity and the length of a cell.

Scheidegger's (1954) work was probably the first attempt to treat hydrodynamic dispersion in a three-dimensional space. However, he does not take molecular diffusion into account and as a result, his coefficient of dispersion is a scalar with no distinction between the longitudinal and the transversal directions. Before proceeding to describe other studies in which this distinction is recognized and molecular diffusion is introduced, let us examine in more detail the mechanism that causes transversal dispersion, i.e., dispersion in a direction perpendicular to the average flow. An example of such spreading is given in figure 10.1.1. Discussions of transversal dispersion are also given by Simpson (1962, 1969).

As an example, consider a field of steady uniform average flow in a horizontal plane. In laminar flow, streamlines coincide with pathlines, and once a liquid particle joins a streamline, it will follow the same path as the particle preceding it along the same streamline. The exact (microscopic) flow pattern is unknown, yet, as explained above, it is possible to study the statistical characteristics common to a large number of labeled liquid particles starting from the same point. The velocity of flow along a streamline fluctuates around the mean flow direction so that its amplitude and the nature of the fluctuations remain approximately constant in time. Hence, molecular diffusion is the only mechanism that can cause continued widening (in a direction perpendicular to the flow) of the zone occupied by the tracer as the flow proceeds. This mechanism causes an exchange of tracer particles between adjacent streamlines. This in itself is not sufficient. The second necessary condition for the existence of lateral spreading, as observed in experiments, is the presence of junctions at which intensive mixing takes place among fluid streams entering from several channels. Only through such branching can a tracer be transported to points at appreciable distances from the average flow axis. Lateral transport through molecular diffusion within a single channel is limited by the dimensions of the channel. The number of particles transported by the diffusion process from one streamline to the next depends on the length of time during which the particles remain within each passage, i.e., on the velocity along the streamlines. It is molecular diffusion that makes hydrodynamic dispersion an irreversible process.

De Josselin de Jong (1958, 1958a), who also used the statistical approach, was probably the first to express in an analytical way the fact that longitudinal dispersion is greater than transversal dispersion. In Scheidegger's (1954) random walk model, each unit step $r_j(x_j, y_j, z_j)$ is characterized by the lack of relationships among its components x_j, y_j, z_j (in a Cartesian coordinate system), and the lack of preference with respect to the main flow direction. In addition, it is assumed that each step has the same duration. Therefore, after a time interval $T \gg \Delta t$, the number of steps performed by each particle is $N = T/\Delta t$.

In de Josselin de Jong's (1958) work, this number is not a constant. He visualizes a porous medium model as composed of a network of interconnected straight channels of equal length, oriented at random, but uniformly distributed in all directions, in which an average uniform flow takes place in the z-direction (fig. 10.3.3). In this model, the pressure gradient in each channel is proportional to $\cos\theta$, where θ is the angle between the direction s of the channel and of the uniform flow. Because of radial diffusion in the narrow channels, one assumes that tracer particles move with a mean velocity within each channel. With these assumptions, the residence time t_j in the jth channel is:

$$t_j = l/V_0 \cos\theta$$

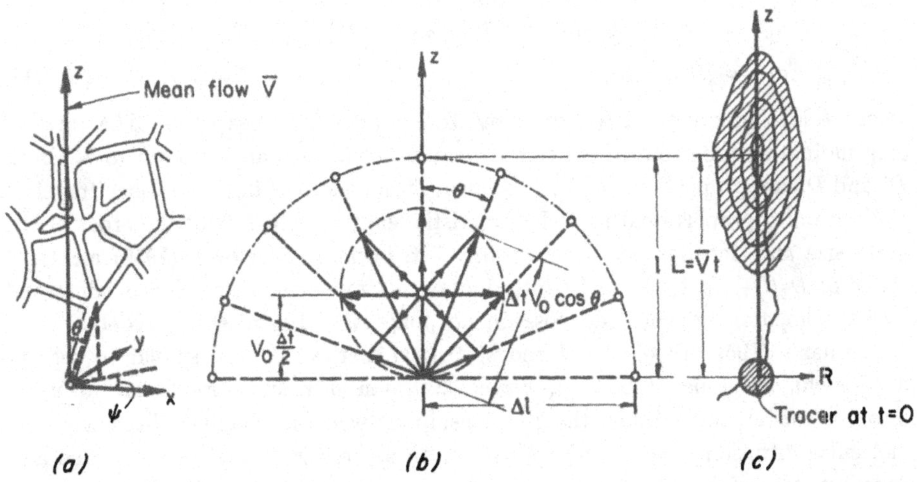

FIG. 10.3.3. Path of particle and final distributions of dispersed fluid (de Josselin de Jong, 1958).

where V_0 is the velocity in a channel oriented in the z-direction and l is the constant length of an elementary channel. The components of the jth displacement are:

$$x_j = l \sin \theta_j \cos \psi_j; \qquad y_j = l \sin \theta_j \sin \psi_j; \qquad z_j = l \cos \theta_j \qquad (10.3.45)$$

where θ and ψ are the angles in a spherical coordinate system (fig. 10.3.3).

The probability $p(\theta, \psi)$ that a tracer particle will choose a direction defined by the angular interval $(\theta, \theta + \Delta\theta)$ and $(\psi, \psi + \Delta\psi)$ when arriving at a junction, is assumed to be proportional to the ratio between the discharge $Q(\theta, \psi)$ in that direction and the total discharge Q:

$$p(\theta, \psi) = Q(\theta, \psi)/Q = (1/\pi) \cos \theta \sin \theta. \qquad (10.3.46)$$

The above assumptions are different from those assumed in the random walk theories described above. If the velocity in a channel is $V = V_0 \cos \theta$ (fig. 10.3.3b), then during $\Delta t < l/V$, a particle moves a distance $\Delta t\, V_0 \cos \theta$ in each channel, and all particles entering a junction simultaneously will be distributed on the circumference of a sphere centered along the z-axis at $\Delta t V_0/2$. The difference between the present approach and that of ordinary random walk lies in the fact that here all channels have the same length, but the velocity in them varies with their direction. Therefore, particles reach the end of their respective channels at different times and are confronted with new "choices." Each particle, therefore, undergoes a different number of displacements in a given time interval.

Using Markov's method, de Josselin de Jong obtains the probability of a particle arriving at a given point at time t after a large number of displacements. The result is a normal distribution in three dimensions.

Figure 10.3.3c shows surfaces of equal concentration resulting from a point injection. These are ellipsoids of revolution centered at a point that travels with the mean velocity $\bar{V} = V_0/3$. The standard deviations σ_L, σ_T of the distribution in the longitudinal and transverse directions, respectively, are:

$$\sigma_L^2 = (l/3)(\lambda + 0.173)\bar{V}t = (l/3)(\lambda + 0.173)L; \qquad D_L = \sigma_L^2/2t$$
$$\sigma_T^2 = \tfrac{3}{8} l\bar{V}t = \tfrac{3}{8} lL; \qquad D_T = \sigma_T^2/2t \qquad (10.3.47)$$

where λ is a function related to L by $3L/l = (\lambda - 2.077) \exp(2\lambda)$. The ratio of longitudinal to transverse dispersion generally lies in the range from 5 to 7. Both D_L and D_T are proportional to $\bar{V}$. However, D_L increases with the distance traveled. Both are also proportional to a characteristic length l, which is of the order of the grain size and depends on the grain-size distribution. Blackwell (1959) mentions ratios of D_L/D_T up to 24. The introduction of transverse dispersion is supported by experimental evidence (de Josselin de Jong 1958, 1958a; Bear 1961b).

Saffman's (1959, 1960) model and statistical method is very similar to that of de Josselin de Jong (1958). However, Saffman introduces molecular diffusion into his model and studies the relationship between mechanical dispersion and molecular diffusion. His model consists of an assembly of randomly oriented and distributed straight circular tubes, equivalent to a statistically homogeneous, isotropic porous medium. The tubes, of equal length and cross-section, are connected to one another at the ends, and several tubes may start or end at these end-points (junctions). The dimensions of these tubes (or pores) are comparable to the sizes of the pores in the real porous medium. In this model, the path of a fluid particle may be regarded as a random one due to variations in direction and velocity.

Saffman introduces molecular diffusion in determining the residence time of a particle in an elementary tube. He assumes, as did de Josselin de Jong (1958), that the velocity of a liquid particle in a tube (of radius R) depends on the angle θ between the tube's axis and the direction of the average flow. However, his time of residence also depends on the distance of the particle from the tube's axis:

$$t = l/V = l/2\langle V\rangle(1 - r^2/R^2) = l/6\bar{V}(1 - r^2/R^2)\cos\theta \qquad (10.3.48)$$

where $\bar{V}$ is average flow velocity and $\langle V\rangle$ is the average velocity in the tube. A tracer particle also participates in the motion due to molecular diffusion, in addition to the flow of the liquid. This motion causes the tracer particle to move *across* the tube from one streamline to the next, as well as *along* each streamline relative to the ambient liquid. The difficulty in adding these two motions stems from the fact that the convective motion is described by the Lagrange approach of following a particle, while diffusion is described by Euler's approach of observing particles at a fixed place. Saffman, therefore, suggests an approximate approach, distinguishing among four cases.

Case 1, $l/\langle V\rangle \ll \tau_1$. Time of passage of an elementary mass through the tube is much shorter than the time τ_1 required for this mass to diffuse effectively across the tube. By analogy with molecular diffusion, Saffman assumes: $\tau_1 = (R/2)^2/2D_d$. In this case Saffman assumes τ = residence time $= l/\langle V\rangle$.

Case 2, $l/\langle V\rangle \ll \tau_1$, *but* $r/R \sim 1$ (i.e., the tracer particle enters the tube close to the wall). Since here the velocity is very small, there is sufficient time for its diffusion to adjacent streamlines, where the velocity is higher. This is equivalent to $\tau = l/\langle V\rangle + \tau_1$.

Case 3, $\tau_1 < l/\langle V\rangle < \tau_0$. Where τ_0 is the time required for effective longitudinal diffusion to take place: $\tau_0 = l^2/2D_d$. For this case Taylor (1953) shows that the tracer mass spreads with respect to the average motion with a spreading coefficient $R^2\langle V\rangle^2/48D_d$. In a very narrow tube ($R \ll l$) we may then assume $\tau = l/\langle V\rangle$. Otherwise ($R < l$): $\tau = l/\langle V\rangle + \tau_1$.

Case 4, $\tau_0 \ll l/\langle V\rangle$. Then $\tau = \tau_0$.

In this model, the displacement of a single particle after N independent steps in a random variable with components parallel to the x, y, z coordinate system:

$$Z_N = \sum_{j=1}^{N} l_j \cos\theta_j; \qquad X_N = \sum_{j=1}^{N} l_j \sin\theta_j \cos\psi_j; \qquad Y_N = \sum_{j=1}^{N} l_j \sin\psi_j \sin\theta_j$$

where the average flow is in the z direction, θ and ψ are shown on figure 10.3.3a. The time required for the N steps is the random variable:

$$T_N = \sum_{j=1}^{N} t_j.$$

Saffman calculates the probability that a particle will choose a streamline between r and $r + dr$ in a tube of direction between θ and $\theta + d\theta$, ψ and $\psi + d\psi$ by:

$$dP = (4/\pi) \sin\theta \cos\theta \cdot (r/R^2) [1 - (r^2/R^2)] \, d\theta \, d\psi \, dr \qquad (10.3.49)$$

which differs from the probability function employed by de Josselin de Jong in that it includes the velocity distribution within the tube. Saffman then determines the location of the particle and the other statistical parameters for $l_j = l = \text{const}$ after a large number of steps N.

Assuming that for a long time, $\bar{N} = \frac{3}{2}\bar{V}t/l = \frac{3}{2}(L/l)$, Saffman obtains: $\bar{x} = L = \bar{V}t$, $\bar{y} = \bar{z} = 0$, and shows that the displacement components perpendicular to the average flow are normally distributed around the mean with a variance independent of molecular diffusion:

$$\text{Var}\, y(t) = \text{Var}\, z(t) = \tfrac{3}{8}\bar{V}lt = (\tfrac{3}{8}l)L. \qquad (10.3.50)$$

In the direction of the flow $(x - \bar{V}t)/t$ is approximately normally distributed when $Pe \equiv l\bar{V}/D_d \gg 1$. The expression $Pe = l\bar{V}/D_d$, called the *Peclet number*, defines the ratio between rate of transport by convection to the rate of transport by molecular diffusion. Here Pe is related to the length l.

Based on his analysis, Saffman then defines coefficients of effective longitudinal and transverse diffusivities D_L, and D_T, which depend on the average Darcy velocity ($\bar{V}$), on a pore (or channel) length (l), on a pore radius (R) (related to the intrinsic permeability), on the molecular diffusivity (D_d) and on the time (t) from the initial instant:

$$D_L = \frac{1}{2t}\overline{(x - \bar{V}t)^2} = \tfrac{1}{2}\bar{V}lS^2; \qquad S^2 = \ln\frac{3\bar{V}\tau_0}{l} - \tfrac{1}{12}; \qquad \tau_0 = l^2/2D_d \quad (10.3.51)$$

for

$$\bar{V}\tau_0/l \ll [N \ln(3\bar{V}\tau_0/l)]^{1/2}$$

$$D_T = \frac{1}{2t}\overline{y^2} = \frac{3}{16}\bar{V}l$$

where N is the number of steps that should be sufficiently great that total displacement of the individual tracer particles will become statistically independent.

Saffman's second work (1960) is devoted to the case where molecular diffusion and macroscopic mixing are of the same order of magnitude, i.e., $\bar{V}l/D_d = 0(1)$. Again the model consists of a random network of capillaries. The Lagrangian correlation function $\overline{\mathring{V}_i \mathring{V}_j}$ is introduced, and longitudinal (D_L) and lateral (D_T) dispersion coefficients are derived for all values of $\bar{V}l/D_d$ less than some large value. It is assumed that $l/\bar{V} \gg R^2/8D_d$, i.e., the time scale for molecular diffusion to smooth out concentration variations across the cross-section of the capillary tube of radius

R, is small compared to the mean time spent by a particle in a capillary. Saffman then adapts the analysis of Taylor (1953, 1954) and Aris (1956), who show that in this case the concentration is approximately uniform over the cross-secton and that the tracer disperses in a manner relative to an axis moving at a velocity $\langle V \rangle$, according to the diffusivity given by (10.3.12). This means that the dispersion in the network of capillaries is the same as if the velocity in each capillary were uniform throughout it, and the material is subject to a diffusion-like process along the capillary with a diffusivity D'. Based on these assumptions, Saffman considers the velocity of a tracer particle at any instant as a sum:

$$V = \langle V \rangle + V_{D'}; \qquad \langle V \rangle = 3\bar{V}\cos\theta \tag{10.3.52}$$

where $\langle V \rangle$ is the mean velocity of motion in the capillary and $V_{D'}$ is a random velocity due to the diffusion process with diffusivity D' expressed by (10.3.12). He then calculates the covariance of velocity components in the direction of the average (Darcy) velocity $\bar{V}$ and perpendicular to it. Integrating these covariances for a large t, he obtains for longitudinal dispersion:

$$D_L = \frac{1}{3}D_d + \frac{3}{80}\frac{R^2\bar{V}^2}{D_d} + \frac{l^2\bar{V}^2}{4}\int_0^1 (3\alpha^2-1)^2\frac{M\coth M - 1}{D'M^2}\,d\alpha$$

$$M = \frac{3}{2}\frac{\bar{V}l\alpha}{D'}; \qquad D' = D_d + \frac{3}{16}\alpha^2\left(\frac{R^2\bar{V}^2}{D_d}\right); \qquad \alpha = \cos\theta. \tag{10.3.53}$$

The first term in D_L is the contribution of molecular diffusion in the longitudinal direction. The second term expresses the contribution of dispersion in a single capillary as derived by Taylor (1953). The third term expresses the resulting convective dispersion.

If the liquid is at rest ($\bar{V} = 0$) we obtain $D_L = D_d/3$. Saffman comments that laboratory experiments in granular material gave $D_L \sim (\frac{2}{3})D_d$, which indicates that a granular material cannot be represented as a random network of capillaries. He therefore suggests replacing $D_d/3$ by some empirical value $m D_d$. For l he uses the value of average grain size, while for R ($\leqslant l/5$) he suggests $R = (24\,k/n)^{1/2}$, where k is medium permeability and n is porosity.

For transverse dispersion he obtains:

$$D_T \approx m\,D_d + \frac{1}{80}\frac{R^2\bar{V}^2}{D_d} + \frac{9\bar{V}^2l^2}{8}\int_0^1 \alpha^2(1-\alpha^2)\frac{M\coth M - 1}{D'M^2}\,d\alpha. \tag{10.3.54}$$

For example:

(a) $Pe \ll 1$, $\quad D_L/D_d \sim m + Pe^2/15$; $\quad D_T/D_d = m + Pe^2/40$; $\quad \frac{1}{3} \leqslant m \leqslant \frac{2}{3}$

(b) $1 \ll Pe \ll 8(R/l)^2$

$D_L/D_d \sim (Pe/6)\ln(3Pe/2) - (17Pe/72) - (R/l)^2Pe^2/48 + (m + 4/9) + 0(1/Pe)$

$D_T/D_d \sim 3Pe/10 + (R/l)^2Pe^2/40 + (m - \frac{1}{3}) + 0(1/Pe)$.

For $Pe < 1$, these results are similar to those obtained by Beran (1955), who obtains, for $Pe < 1$, $D = D_d + (\gamma/D_d)d^2V^2$, where γ is a constant depending on the grain-size distribution and d is a characteristic grain dimension.

Hiby (1959; (in Saffman 1960) and 1962) obtains for glass beads of diameter l:

$$D_L/D_d = 0.67 + 0.65Pe/(1 + 6.7Pe^{-1/2}) \tag{10.3.55}$$

which agrees with Saffman's results for $R/l = \frac{1}{8}$ in the wide range $10^{-2} \leqslant Pe \leqslant 10^2$. However, for transversal dispersion, Saffman's values are higher than Hiby's. On the other hand, experimental results obtained by Blackwell, Rayne and Terry (1959) yield:

$$D_L/D_d = 8.8Pe^{1.17} \quad \text{for} \quad Pe > 0.5. \tag{10.3.56}$$

The works of de Josselin de Jong and Saffman, reviewed above, clearly indicate the existence of transversal dispersion. In discussing the general form of the partial differential equation describing hydrodynamic dispersion, Scheidegger (1965) states that the differential equation of dispersion may be obtained by purely statistical considerations of a random process. An example of such a process is the *Markov process* (e.g., de Josselin de Jong 1958), defined as a stochastic process in which the prediction of the probability distribution of a random variable at time t, given its distribution at some time $t_0 < t$, depends only on that given distribution at time t_0. Any additional information regarding this random variable at times t prior to t_0 does not affect the resulting probability distribution for $t > t_0$.

The equation describing such a process is the *Kolmogorov equation*, which has the form of the diffusion equation.

Let $\xi(t)$ be an ensemble of random variables depending on the parameter t (time), and let $F(t, x; \tau, y)$ describe the probability that at some time τ, the random variable $\xi(t)$ will be smaller than y, given that at some earlier time $t < \tau$ we had $\xi(t) = x$:

$$F(t, x; \tau, y) \equiv P[\xi(\tau) < y | \xi(t) = x] \tag{10.3.57}$$

$f(t, x; \tau, y) = \partial F(t, x; \tau, y)/\partial y$; it satisfies:

$$\int_{-\infty}^{y} f(t, x; \tau, \eta)\, d\eta = F(t, x; \tau, y); \qquad \int_{-\infty}^{+\infty} f(t, x; \tau, \eta)\, d\eta = 1.$$

We shall also assume that the random process is continuous so that:

$$\lim_{\Delta t \to 0} \frac{1}{\Delta t} \int_{|y-x| \geqslant \varepsilon} F(t - \Delta t, x; t, \eta)\, d\eta = 0$$

for any positive constant ε.

With $\partial F/\partial x$, $\partial^2 F/\partial x^2$ continuous for all t, x, y and $\tau > t$, F satisfies the *first Kolmogorov equation*:

$$\frac{\partial F}{\partial t} = -a(t, x)\frac{\partial F}{\partial x} - \frac{b(t, x)}{2}\frac{\partial^2 F}{\partial x^2} \tag{10.3.58}$$

where we have:

$$\lim_{\Delta t \to 0} \frac{1}{\Delta t} \int_{|y-x|<\varepsilon} (y-x)F(t-\Delta t, x; t, y)\, dy = a(t, x)$$

$$\lim_{\Delta t \to 0} \frac{1}{\Delta t} \int_{|y-x|<\varepsilon} (y-x)^2 F(t-\Delta t, x; t, y)\, dy = b(t, x)$$

and the convergence is uniform with respect to x.

If, in addition to the conditions above, a density $f(t, x; \tau, y) = \partial F/\partial y$ exists and:

$$\partial F/\partial \tau,\ \partial[a(\tau, y) f(t, x; \tau, y)]/\partial y \quad \text{and} \quad \partial^2[b(\tau, y) f(t, x; \tau, y)]/\partial y^2$$

exist and are continuous, f will satisfy *the second (or forward) Kolmogorov equation*:

$$\frac{\partial f}{\partial \tau} = -\frac{\partial}{\partial y}[a(\tau, y) f] + \frac{1}{2}\frac{\partial^2}{\partial y^2}[b(\tau, y) f]. \tag{10.3.59}$$

To obtain a physical interpretation of a and b one should notice that:

$$\int (y-x)\, F(t-\Delta t, x; t, y)\, dy = \overline{[\xi(t) - \xi(t-\Delta t)]} = \overline{\xi(t)} - \overline{\xi(t-\Delta t)} = \overline{\Delta\xi}$$

$$\int (y-x)^2\, F(t-\Delta t, x; t, y)\, dy = \overline{[\xi(t) - \xi(t-\Delta t)]^2} = \overline{(\Delta\xi)^2}$$

from which we have:

$$a(t, x) = \overline{d\xi}/dt; \qquad b(t, x) = \lim_{\Delta t \to 0} [\overline{(\Delta\xi)^2}/\Delta t]. \tag{10.3.60}$$

This means that a represents the average rate of change of $\xi(t)$, whereas b represents the average of the square of the change per unit time. If $\xi(t)$ represents the coordinates of a point moving under the influence of random factors, $a(t, x)$ may be interpreted as its velocity at time t and point x, whereas b denotes the mean of the squares of incremental displacements per unit time (rate of growth of variances). In each case the average is taken over all possible displacements during Δt.

It is thus shown that if the dispersion process is regarded as a Markov process, the average tracer concentration is described by the diffusion equation obtained from the second Kolmogorov equation.

10.3.3 Spatial Averaging

The spatial averaging approach, which is another statistical approach, was explained and presented in detail in sections 4.5 through 4.7, where it was used to derive the fundamental conservation equations for a porous medium. The hydrodynamic dispersion equation (equation of mass conservation of a species) was one of these equations. Each parameter or variable appearing in these (averaged) equations has the meaning of an average of the local values of the considered parameter or variable, taken over a representative elementary volume (REV) around the point to which the average value is assigned.

As a simple example of this approach, let us start from (4.3.7) written for a point in the fluid continuum within the pore space:

$$\partial\rho_\alpha/\partial t = -\operatorname{div}(\rho_\alpha \mathbf{V}^* + \mathbf{J}^*_\alpha) + I_\alpha. \tag{10.3.61}$$

In the absence of chemical reactions resulting in the production of α-species, $I_\alpha = 0$. The diffusive flux $\mathbf{J}^*_\alpha$ is expressed by (4.1.12):

$$\mathbf{J}^*_\alpha = -\rho D_{\alpha\beta} \operatorname{grad}(\rho_\alpha/\rho) = -D_{\alpha\beta}[\operatorname{grad}\rho_\alpha - (\rho_\alpha/\rho)\operatorname{grad}\rho]. \tag{10.3.62}$$

For the case of a dilute system with $\rho \approx$ const, or when grad(ln ρ) $\ll$ grad(ln ρ_α), we obtain:

$$\mathbf{J}^*_\alpha \approx -D_{\alpha\beta}\operatorname{grad}\rho_\alpha. \tag{10.3.63}$$

By combining (10.3.61) written for $I_\alpha = 0$, with (10.3.63), we obtain:

$$\partial\rho_\alpha/\partial t + \operatorname{div}(\rho_\alpha \mathbf{V}^* - D_{\alpha\beta}\operatorname{grad}\rho_\alpha) = 0. \tag{10.3.64}$$

We now express every velocity and density as a sum of an average value (taken over an REV) and a deviation:

$$\mathbf{V}^* = \bar{\mathbf{V}}^* + \mathring{\mathbf{V}}^* \quad (\text{i.e.,}\quad V^*_i = \bar{V}^*_i + \mathring{V}^*_i,\quad i = 1, 2, 3); \qquad \rho_\alpha = \bar{\rho}_\alpha + \mathring{\rho}_\alpha \tag{10.3.65}$$

where $\bar{\mathbf{V}}^*$ is the average mass velocity in the porous medium, and the average $\bar{a}$ of a property a is defined as:

$$\bar{a} = \frac{1}{(\varDelta U_0)_v}\int_{(\varDelta U_0)_v} a\, dU_v; \qquad \bar{\mathring{a}} \equiv 0. \tag{10.3.66}$$

By inserting (10.3.65) into (10.3.64) we obtain:

$$\partial(\bar{\rho}_\alpha + \mathring{\rho}_\alpha)/\partial t + \operatorname{div}[(\bar{\rho}_\alpha + \mathring{\rho}_\alpha)(\bar{\mathbf{V}}^* + \mathring{\mathbf{V}}^*) - D_{\alpha\beta}\operatorname{grad}(\bar{\rho}_\alpha + \mathring{\rho}_\alpha)] = 0 \tag{10.3.67}$$

or:

$$\partial\bar{\rho}_\alpha/\partial t + \partial\mathring{\rho}_\alpha/\partial t + \operatorname{div}\bar{\rho}_\alpha\bar{\mathbf{V}}^* + \operatorname{div}(\bar{\rho}_\alpha\mathring{\mathbf{V}}^* + \mathring{\rho}_\alpha\bar{\mathbf{V}}^*) + \operatorname{div}\mathring{\rho}_\alpha\mathring{\mathbf{V}}^*$$
$$- \operatorname{div}(D_{\alpha\beta}\operatorname{grad}\bar{\rho}_\alpha) - \operatorname{div}(D_{\alpha\beta}\operatorname{grad}\mathring{\rho}_\alpha) = 0. \tag{10.3.68}$$

We now multiply each term of (10.3.68) by dU_v, integrate over $(\varDelta U_0)_v$ and divide the result by $(\varDelta U_0)_v$. In other words, we average (10.3.68) term by term over the REV. Assuming (and this assumption is valid subject to certain constraints) that the average of a gradient or a divergence is equal to the gradient or the divergence of the average, respectively, and noting that $\bar{\mathring{\rho}}_\alpha = 0$, $\overline{\mathring{\rho}_\alpha\bar{\mathbf{V}}^*} = 0$, $\overline{\bar{\rho}_\alpha\mathring{\mathbf{V}}^*} = 0$, we obtain:

$$\partial\bar{\rho}_\alpha/\partial t + \operatorname{div}(\bar{\rho}_\alpha\bar{V}_\alpha) + \operatorname{div}(\overline{-D_{\alpha\beta}}\operatorname{grad}\bar{\rho}_\alpha) + \operatorname{div}(\overline{\mathring{\rho}_\alpha\mathring{\mathbf{V}}^*}) = 0. \tag{10.3.69}$$

To obtain (10.3.69) we have assumed that no correlation exists between grad $\bar{\rho}_\alpha$ and $D_{\alpha\beta}$; $\bar{D}_{\alpha\beta}$ has the meaning of average molecular diffusivity in the porous medium.

The term $\overline{\mathring{\rho}_\alpha\mathring{\mathbf{V}}^*}(\equiv \overline{\rho_\alpha\mathring{\mathbf{V}}^*})$, as explained in detail in paragraph 4.6.2, has the meaning of the average *dispersive flux* resulting from velocity fluctuations. We now assume that this dispersive flux, in analogy to diffusive fluxes in general, may be expressed as:

$$\overline{\rho_\alpha \overset{\circ}{\mathbf{V}}{}^*} = -\underline{\mathbf{D}} \operatorname{grad} \bar{\rho}_\alpha \quad \left(\text{or} \quad \overline{\rho_\alpha \overset{\circ}{V}{}^*_i} = -D_{ij}\, \partial\bar{\rho}_\alpha/\partial x_j\right) \qquad (10.3.70)$$

where $\mathbf{D}$ (second rank tensor with components D_{ij}) is the coefficient of mechanical dispersion. Equation (10.3.69) then becomes:

$$\partial\bar{\rho}_\alpha/\partial t = \frac{\partial}{\partial x_i}\left(D'_{ij}\, \partial\bar{\rho}_\alpha/\partial x_j\right) - \partial(\bar{\rho}_\alpha \bar{V}^*_i)/\partial x_i; \qquad D'_{ij} = D_{ij} + (\bar{D}_{\alpha\beta})_{ij} \quad (10.3.71)$$

which is similar to the dispersion equation derived in paragraph 4.6.2; D'_{ij} is the coefficient of hydrodynamic dispersion. The shortcoming of this one-step averaging as compared with the two-step averaging performed in paragraph 4.6.2 is that we have no insight into the structure of D'_{ij} and its dependence on medium and flow characteristics.

An equation similar to (10.3.71) is derived by the same procedure of spatial averaging by Harleman and Rumer (1962).

Whitaker (1967) also volume averages the fundamental transport equation in a general porous medium. His dispersion equation takes the form:

$$\frac{\partial\bar{C}}{\partial t} = \frac{\partial}{\partial x_j}\left[D_{\alpha\beta}\left(\frac{\partial\bar{C}}{\partial x_j} + R\tau_j\right)\right]$$

$$+ \frac{\partial}{\partial x_j}\left[\left(A^I{}_{jik}\bar{V}_i + A^{II}{}_{jilk}\bar{V}_i\bar{V}_l + A^{III}{}_{jilk}\bar{V}_i\frac{\partial\bar{C}}{\partial x_l}\right)\frac{\partial\bar{C}}{\partial x_k}\right] - \bar{V}_i\frac{\partial\bar{C}}{\partial x_i} \qquad (10.3.72)$$

where the tensors A^I, A^{II} and A^{III} are symmetrical and are functions of the transport properties of the fluid; R is a coefficient and $\boldsymbol{\tau}$ is a *tortuosity vector* which includes the sinuousness of the pores as well as their expansions and contractions. When flow is such that molecular diffusion is negligible compared to dispersion, tortuosity does not enter the problem.

In general, Whitaker's expression for the dispersive flux vector $(\overline{C\overset{\circ}{\mathbf{V}}})$ is different from that suggested by (4.6.24) or (10.3.70). His dispersion equations (10.3.72) is, therefore, different from (4.6.27) or (10.3.71).

10.4 Parameters of Dispersion

10.4.1 The Coefficients of Mechanical Dispersion and of Hydrodynamic Dispersion

The coefficient of mechanical dispersion $\underline{\mathbf{D}}$ appearing in the dispersion equation has been analyzed by various investigators. It depends on the flow pattern, e.g., through the velocities, on the Peclet number, and on some basic medium characteristics. Many investigators consider the sum $\mathbf{D}' = (\underline{\mathbf{D}} + \underline{\mathbf{D}}_d{}^*)$ as the coefficient of dispersion (or rather, of hydrodynamic dispersion), which depends on velocity, on molecular diffusion and on the medium characteristics. Equation (10.3.12) may serve as an example. Obviously, works dealing with essentially one-dimensional flow cannot recognize the tensorial nature of $\underline{\mathbf{D}}'$ or of $\underline{\mathbf{D}}$ (when molecular diffusion is neglected).

Taylor (1953) in his one-dimensional analysis obtains D proportional to $\bar{V}^2$. Bear and Todd (1960) in their one-dimensional analysis suggest $D = a_I\bar{V}$, i.e., D is proportional to $\bar{V}$, with a_I being some characteristic medium length. In the analysis of variances given in section 10.3, the dispersion, or the spread of particles around their average displacement, is described by the matrix of covariances $\overline{\mathring{x}_i(t)\mathring{x}_j(t)}$. There it is shown that these covariances may be related to the coefficient of dispersion D_{ij} for large times in the form $D_{ij} = r_{ij}(0)\overline{\mathring{V}_i\mathring{V}_i}t_0$, where t_0 is the time during which velocities are still effectively correlated. If, following Nikolaevskij (1959), we let $r_{ij}(0)t_0 = L/\bar{V}$, where $\bar{V}$ is the absolute value of the average velocity and L is a characteristic medium length (e.g., length of a channel in a porous medium model), we obtain:

$$D_{ij} = (\overline{\mathring{V}_i\mathring{V}_j/\bar{V}})L, \tag{10.4.1}$$

i.e., $\underline{\mathbf{D}}$ is proportional to the first power of the velocity.

Scheidegger (1957) summarizes his analysis on the two possible relationships between D and $\bar{V}$ according to the role played by molecular diffusion: (a) $D \sim a'\bar{V}^2$, where a', a constant of the porous medium alone (*dynamic dispersivity*), is derived by a dynamic procedure applicable when there is enough time in each flow channel for appreciable mixing to take place by molecular transverse diffusion; (b) $D \sim a''\bar{V}$, where a'', another constant of the porous medium (*geometric dispersivity*), is derived by a geometric procedure applicable where there is no appreciable molecular transverse diffusion from one streamline into another.

Thus, in all models in which the combined effect of a velocity distribution across a channel and transverse molecular diffusion are considered (e.g., Taylor 1953), the coefficient of dispersion is proportional to $\bar{V}^2$. Where only the mean motion in a channel is considered, while mixing occurs at junctions connecting different channels, disregarding molecular diffusion one obtains $D \sim \bar{V}$. Of course, intermediate cases, where D is proportional to some power of the velocity between 1 and 2, lie between these two extremes.

Blackwell (1959), and many others, study the relationship between the coefficient of dispersion, velocity and molecular diffusion. Pfannkuch (1963) summarizes (fig. 10.4.1) 175 experimental values obtained in one-dimensional flow by Rifai, Kaufman and Todd (1956), Day (1956), Ebach and White (1958), Carberry and Bretton (1958), Blackwell (1959), Blackwell and Rayne (1959), Raimondi et al. (1959) and Brigham and Reed (1959). By dimensional analysis one can show that in general the *coefficient of hydrodynamic dispersion* $\underline{\mathbf{D}}' = (\underline{\mathbf{D}} + \underline{\mathbf{D}}^*{}_d)$, which includes the effects of both mechanical (or convective) dispersion and molecular diffusion, is a function of the *Peclet number* of molecular diffusion, $Pe \equiv Vd/D_d = (q/n)d/D_d$, where d is the mean grain size or any other characteristic medium length, and D_d replaces the coefficient of thermal conductivity in the Peclet number defined in heat conduction studies.

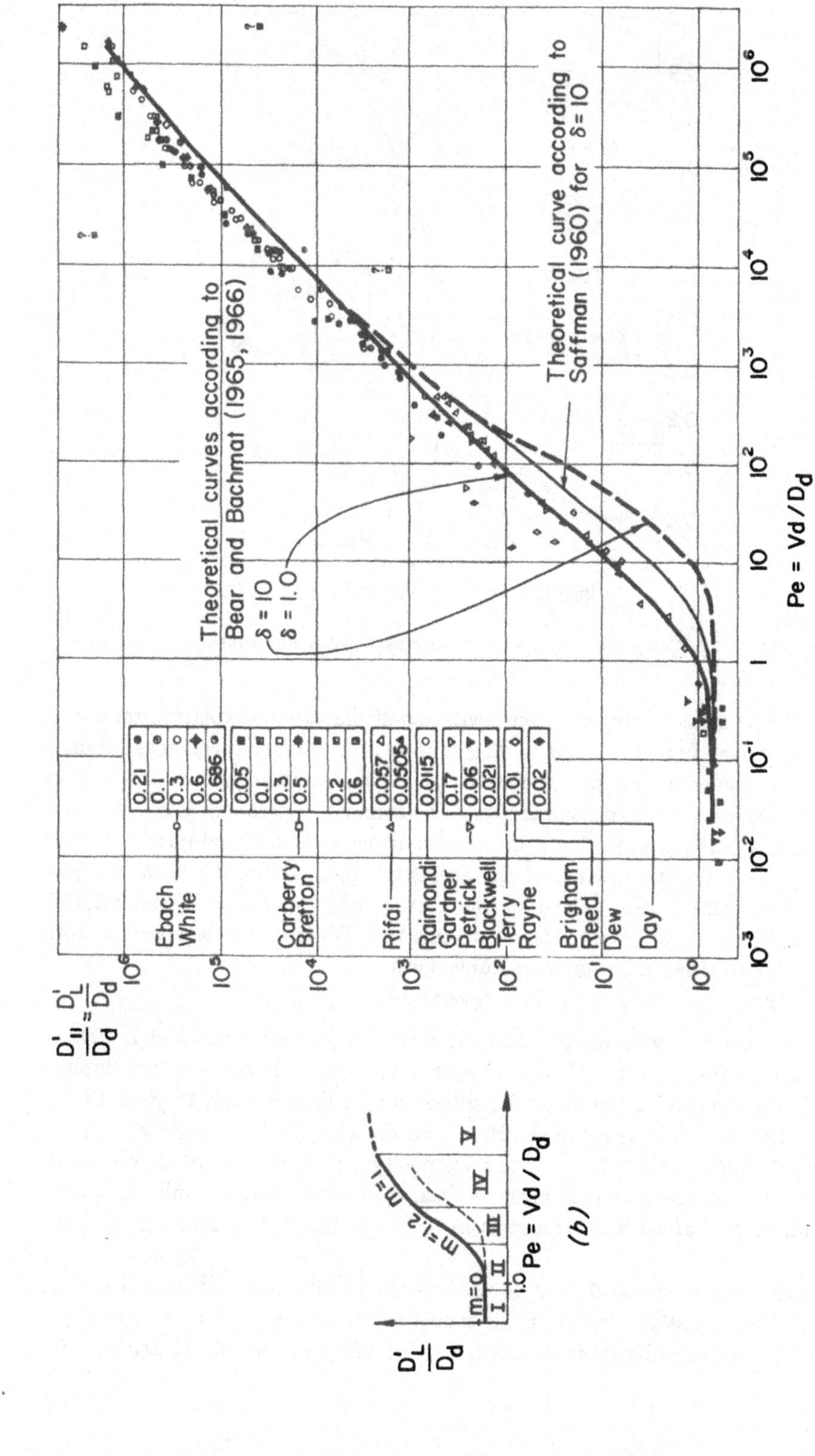

FIG. 10.4.1. Relationship between molecular diffusion and convective dispersion (after Pfannkuch, 1963; Saffman, 1960).

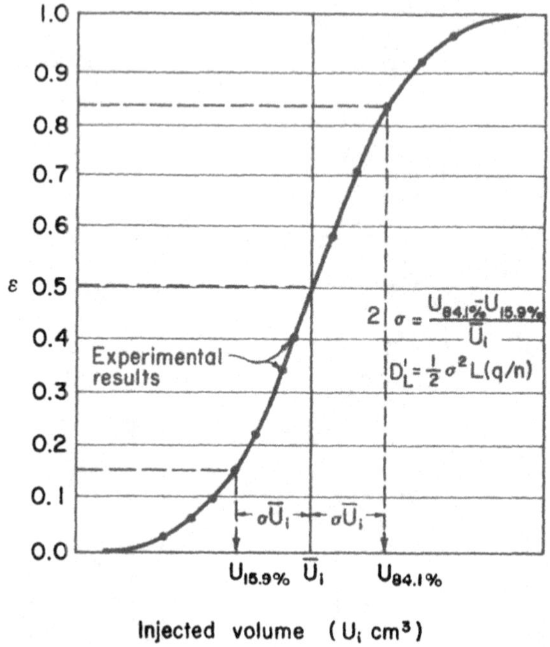

FIG. 10.4.2. Computing D'_L from a one-dimensional dispersion experiment.

Figure 10.4.1 gives a schematic representation of the various experimental results mentioned above. In it, D'_L is the coefficient of hydrodynamic dispersion as obtained from the experimental results. The subscript L denotes "longitudinal" here, since we are dealing with one-dimensional flow. The common experiment is one in which a fluid at constant concentration is introduced at one end of a sand-packed column of length L. The concentration of the effluent is recorded in the form of figure 10.4.2. From this figure, and from the corresponding one-dimensional solution (sec. 10.6), we obtain $D'_L = (\frac{1}{2})\sigma^2 L\, q/n$. In this way, D'_L involves the effects of both dispersion and diffusion in an inseparable form.

Figure 10.4.1 may be divided into several zones.

Zone I. This is a zone where molecular diffusion predominates, and $D'_L/D_d = f(Pe) = \text{const.}$ Here $a_I\, q/n \ll T^*D_d$. The effect of a porous medium is only through T^* (< 1), which modifies the molecular diffusion in a fluid domain to yield $D^*_d = T^*D_d$ for fluid in a porous medium. Here, and elsewhere in the present and in the following sections, T^* is the medium's tortuosity (sec. 4.8). To be in this zone, the time of travel through a pore must be equal to or longer than t_0 required for molecular diffusion to affect mass transport in the longitudinal direction: $t > t_0$; $t_0 \propto d^2/2D_d$.

An experiment with $q = 0$ may be used to obtain D^*_d and from this the value of T^*. Saffman (1959) obtains in his model for this case $D'_L = T^*D_d = (\frac{1}{3})D_d$. However, for an actual porous medium, a value of approximately $(\frac{2}{3})D_d$ (e.g., for loose sand) seems better.

Zone II. This corresponds to a Peclet number of diffusion between 0.4 and 5 (approximately). In this zone the effect of molecular diffusion is of the same order of magnitude as that of mechanical dispersion and the sum of both should be considered.

Zone III. Here the main spreading is caused by mechanical dispersion combined with transversal molecular diffusion. The two mechanisms are no longer additive, as they interfere with each other. In this zone: $t_1 < t \ll t_0$; $t_1 = R^2/8D_d$. Experimental results for this range yield:

$$D'_L/D_d = \alpha(Pe)^m; \qquad \alpha \approx 0.5; \qquad 1 < m < 1.2. \tag{10.4.2}$$

In this zone transversal molecular diffusion tends to reduce longitudinal spreading. Obviously, theories on dispersion in which molecular diffusion is neglected, or in which molecular diffusion is considered as an additive term, fail to explain the behavior in this region.

Zone IV. This is a region of dominant mechanical dispersion (as long as we stay in the range of validity of Darcy's law). The effects of molecular diffusion are negligible. In the diagram we obtain a straight line at 45°:

$$D'_L/D_d = \beta Pe; \qquad \beta \approx 1.8. \tag{10.4.3}$$

Zone V. This is another zone of pure mechanical dispersion, but the effects of inertia and turbulence may no longer be neglected. Their role is equivalent to the role of transversal molecular diffusion in zone III. The slope of the curve in this zone tends to be less than one.

Fewer experiments are available on the coefficient of transversal dispersion. Denoting the coefficient of transverse hydrodynamic dispersion by D'_T, one should expect the relationship between D'_T/D_d and Pe to be similar to that of D'_L/D_d.

In earlier studies (e.g., Scheidegger 1954), dispersion was taken as isotropic, i.e., D is considered a scalar. However, even in isotropic porous media we observe longitudinal (i.e., in the direction of the flow) and transversal (i.e., in the direction perpendicular to the flow) dispersion that differ from each other. When generalized to anisotropic media, the problem becomes even more complicated. It was shown in paragraph 4.6.2 that the coefficient of mechanical dispersion D_{ij} is a second-rank symmetrical tensor that depends on the characteristics of the medium and on the flow. For the sake of simplicity let us consider here only the range where $D \sim V$.

Bear (1960, 1961a) observes, in determining the tracer distribution resulting from a point injection in a uniform field of flow, that the components V_i of the velocity vector $\mathbf{V}$, or L_i of the displacement $\mathbf{L}$, play no role in the analysis. He suggests that this is probably due to the nature of the porous medium with its interconnected channels distributed at random in all directions, and to the fact that the true local velocity in the channels (neglecting molecular diffusion) causes the dispersion, and not the average velocity $\mathbf{V}$ over the entire cross-section and its components V_i (say, in a Cartesian coordinate system x_i). Since in an isotropic medium the dispersion

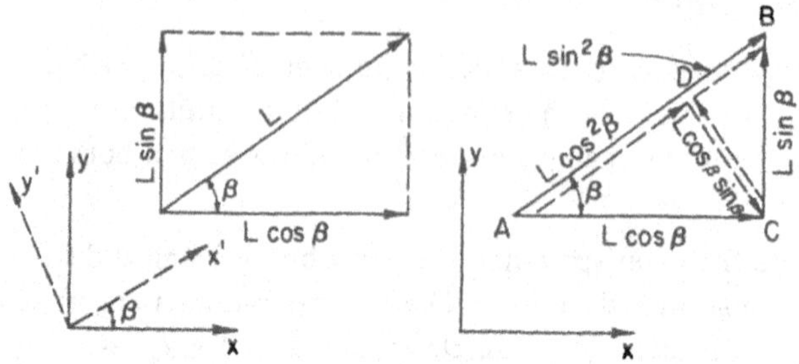

FIG. 10.4.3. The resolution of the displacement L.

in the direction of the mean displacement **L** and perpendicular to it (fig. 10.4.3a) can be obtained from the longitudinal and the transversal coefficients of dispersion: $D_L t = a_I L$ and $D_T t = a_{II} L$, where a_I and a_{II} are medium constants, but not from the components L_i of **L** in the x_i system (x, y in the two-dimensional case of fig. 10.4.3), Bear suggests resolving the components L_i further, and taking their projections L_{ij} on the direction of the flow and perpendicular to it. Thus, displacement is defined as a second-rank tensor which in two dimensions takes the form:

$$[L_{ij}] = \begin{bmatrix} L\cos^2\beta & L\sin\beta\cos\beta \\ L\sin\beta\cos\beta & L\sin^2\beta \end{bmatrix} = L\begin{bmatrix} \cos^2\beta & \sin\beta\cos\beta \\ \sin\beta\cos\beta & \sin^2\beta \end{bmatrix}. \qquad (10.4.4)$$

In a three-dimensional space, $i, j = 1, 2, 3$. Then, since in our case the covariance $(\sigma^2)_{kl}$, which defines D_{kl}, e.g., in (10.3.33) extended to $l \neq k$, transforms as a second-rank tensor, Bear states as a working hypothesis that:

$$(\sigma^2)_{kl} = 2a_{ijkl}L_{ij} = 2D_{kl}t \qquad (10.4.5)$$

where a_{ijkl}—a fourth-rank tensor—is the *medium's (geometrical) dispersivity*. In general, a_{ijkl} $(i, j, k, l = 1, 2, 3)$ contains 81 components, of which all but those with four equal indices, or with two pairs of equal indices, are equal to zero.

Equation (10.4.4) may be rewritten for a field of uniform flow, with $L = Vt$, in the form:

$$[L_{ij}] = \frac{L}{V^2}\begin{bmatrix} (V\cos\beta)^2 & V\sin\beta\cdot V\cos\beta \\ V\sin\beta\cdot V\cos\beta & (V\sin\beta)^2 \end{bmatrix} = \left[\frac{V_iV_j}{V}\right]t. \qquad (10.4.6)$$

Hence, we obtain from (10.4.5) and (10.4.6):

$$D_{kl} = a_{ijkl}V_iV_j/V \qquad (10.4.7)$$

where the summation convention is employed.

The general form of (10.4.7) has also been obtained by Nikolaevskij (1959), starting from an analogy with the statistical theory of turbulence. He states that for an isotropic medium, the point tensor of dispersion must be a tensor of even rank (Batchelor 1953). He thus obtains (10.4.7) as the simplest form that satisfies this requirement. De Josselin de Jong and Bossen (1961) obtained (10.4.7) by extending Bear's (1960, 1961) analysis to uniform flow in an isotropic medium. They also show that for isotropic media the tensor a_{ijkl} has only two distinct components.

Poreh (1965) also investigates the tensor form of D_{ij} and its relationship to the components V_i. From macroscopic symmetry considerations, following the work of Robertson (1940), he obtains the following general forms for an isotropic medium:

$$D_{ij} = a\delta_{ij} + bV_iV_j \tag{10.4.8}$$

where a and b are considered continuous functions of V^2. Expanding D_{ij} in a power series he obtains:

$$D_{ij} = a_0\delta_{ij} + a_1V^2\delta_{ij} + b_0V_iV_j + 0(V^4) \tag{10.4.9}$$

where a_0, a_1, b_0 are constants depending on the properties of the medium, of the fluid and of the tracer; δ_{ij} is the Kronecker delta.

For an anisotropic medium with an axis of symmetry in the direction $\mathbf{1\lambda}$, Poreh obtains:

$$D_{ij} = a'\delta_{ij} + b'V_iV_j + c'\lambda_i\lambda_j + d'(V_i\lambda_j + V_j\lambda_i) \tag{10.4.10}$$

where a', b', etc., are functions of V^2 and of $V\cos(\mathbf{V}, \mathbf{1\lambda})$.

Harleman and Rumer (1962) suggest for D_{ij}:

$$D_{ij} = a_{ijkl}(V_kV_l/V)^{n_{ijkl}} \tag{10.4.11}$$

where n_{ijkl} has a form similar to a_{ijkl}.

10.4.2 The Medium's Dispersivity

For an isotropic medium, Bear (1960, 1961a) relates the medium's dispersivity a_{ijkl} (a fourth-rank symmetrical tensor) to the *two* constants: a_I = *longitudinal dispersivity of the medium*, and a_{II} = *transversal dispersivity of the medium*. For two dimensions:

$$\begin{aligned} &a_{1111} = a_{2222} = a_I; \qquad a_{1122} = a_{2211} = a_{II};\\ &a_{1112} = a_{1121} = a_{1211} = a_{1222} = a_{2111} = a_{2122}\\ &\quad = a_{2212} = a_{2221} = 0;\\ &a_{1212} = a_{1221} = a_{2112} = a_{2121} = \tfrac{1}{2}(a_I - a_{II}). \end{aligned} \tag{10.4.12}$$

In three dimensions, we must add similar terms, but with subscript 3 replacing 2.

Scheidegger (1961) studies the symmetry properties of a_{ijkl}, which in a three-dimensional space includes 81 components, and extends the analysis above to the general case of an anisotropic medium. He shows that because of certain symmetry properties: $a_{ijkm} = a_{ijmk}$; $a_{ijkm} = a_{jikm}$; $a_{ijkm} = a_{jimk}$; $a_{ijkm} = a_{kmij}$. Actually, Scheidegger (1961) leaves the last equality as questionable, while Bear (1961a) assumes that this symmetry does exist in isotropic media. Even in an anisotropic medium only 36 terms are nonzero. However, the number of constants in specific types of anisotropy is not given. In an isotropic medium, 21 terms are nonzero.

Scheidegger (1961) writes (10.4.12) (for an isotropic medium) in the general form:

$$a_{ijkm} = a_{II}\delta_{ij}\delta_{km} + \frac{a_I - a_{II}}{2}(\delta_{ik}\delta_{jm} + \delta_{im}\delta_{jk}). \tag{10.4.13}$$

In a Cartesian coordinate system x_i, the tensor a_{ijkm} may be presented in a symbolic form by a 6×6 matrix with two indices $a_{\alpha\beta}$; in this matrix, α represents the indices ij while β represents km, as follows:

α or β	1	2	3	4	5	6
ij or km	11	22	33	12	13	23

For an isotropic medium we therefore obtain from (10.4.13):

$$[a_{\alpha\beta}] = \begin{bmatrix} a_I & a_{II} & a_{II} & 0 & 0 & 0 \\ a_{II} & a_I & a_{II} & 0 & 0 & 0 \\ a_{II} & a_{II} & a_I & 0 & 0 & 0 \\ 0 & 0 & 0 & (a_I - a_{II})/2 & 0 & 0 \\ 0 & 0 & 0 & 0 & (a_I - a_{II})/2 & 0 \\ 0 & 0 & 0 & 0 & 0 & (a_I - a_{II})/2 \end{bmatrix}. \qquad (10.4.14)$$

By inserting (10.4.13) in (10.4.7) we obtain:

$$D_{ij} = a_{II} V \delta_{ij} + (a_I - a_{II}) V_i V_j / V. \qquad (10.4.15)$$

Whitaker (1967, see also discussions by Greenkorn and Kessler (1969) and by Patel and Greenkorn (1970)) suggests that A^{III}_{jikl} of (10.3.72) may be assumed negligible. Then (10.3.72) becomes the usual dispersion equation:

$$\partial \bar{C}/\partial t = \partial (D'_{jk}\, \partial \bar{C}/\partial x_k)/\partial x_j - \bar{V}_i\, \partial \bar{C}/\partial x_i$$

where the coefficient of hydrodynamic dispersion D'_{jk} is defined by:

$$D'_{jk} = D_{\alpha\beta}(\delta_{jk} + RB^I{}_{jk}) + A^I{}_{jik}\bar{V}_i + A^{II}{}_{jilk}\bar{V}_i\bar{V}_l.$$

In this expression $B^I{}_{jk}$ is a tensor expressing the medium's tortuosity. At very low velocities the last two terms on the right-hand side become negligible and D'_{jk} becomes the molecular diffusivity of the porous medium. At higher velocity, the diffusion term is less important and we obtain:

$$D'_{jk} = A^I{}_{jik}\bar{V}_i + A^{II}{}_{jilk}\bar{V}_i\bar{V}_l.$$

Whitaker (1967) gives detailed expressions for $A^I{}_{jik}$ and $A^{II}{}_{jilk}$.

The main difference between Whitaker's expression for D'_{jk} and the expression for D'_{jk} obtained by Bear (1960, 1961a), Nikolaevskij (1959) and de Josselin de Jong and Bosson (1961), is that it incorporates a third-order tensor as well as a fourth order. Nikolaevskij (1959) eliminated A^I since he was considering only an isotropic medium. Patel and Greenkorn (1970) suggest that Whitaker's expression for D'_{jk} as presented above is the correct one for anisotropic media, and that for isotropic media, only the fourth-rank tensor is to be retained. However, Whitaker's (1967) analysis predicts that for uniform flow in an isotropic medium, the dispersivity tensor includes only one distinct component as the longitudinal dispersivity of the medium is exactly three times the value of the transversal one. Note that in the last three equations the average velocity is denoted by $\bar{\mathbf{V}}$, whereas in the other equations of this section it is denoted by $\mathbf{V}$.

Let us choose the Cartesian coordinate system x_i such that one of its axes, say x_1, coincides with the direction of the average velocity V. Under such conditions we obtain from (10.4.15):

$$D_{11} = a_I V, \qquad D_{22} = a_{II} V \quad \text{and} \quad D_{ij} = 0 \quad \text{for} \quad i \neq j.$$

Then D_{ij} is given by:

$$[D_{ij}] = \begin{bmatrix} a_I V & 0 & 0 \\ 0 & a_{II} V & 0 \\ 0 & 0 & a_{II} V \end{bmatrix}. \tag{10.4.16}$$

The axes of the coordinate system in which D_{ij} is expressed by (10.4.16) are called the *principal axes of dispersion.*

Bear and Bachmat (1965, 1966) show that in the more general case the coefficient of dispersion D_{ij} is related to a_{ijkm} by (10.4.15) rather than by (10.4.7) where $f(Pe, \delta) = 1$. Equations (10.4.15) and (10.4.16) should be modified accordingly.

Bachmat and Bear (1964) discuss the general form of D_{ij} (given in a Cartesian coordinate system by (10.4.15)) of an isotropic medium in any coordinate system. They show that the principal axes of D_{ij} are such that one of them coincides at every point with the direction of the average velocity $\mathbf{V}$, and the other two (in a three-dimensional space) are perpendicular to the first. In general, for an isotropic medium we have:

$$D_{ij} = \lambda g_{ij} V + 2\mu V_i V_j / V \tag{10.4.17}$$

$$\lambda = a_{II}; \qquad \mu = (a_I - a_{II})/2; \qquad V \equiv |\mathbf{V}|.$$

Here, the covariant components V_i and V_j of the velocity vector V and the covariant components D_{ij} of the coefficient of dispersion are in terms of the considered coordinate system; $g_{ij} = (\partial x^p / \partial y^i)(\partial x^p / \partial y^j)$ is a covariant symmetrical tensor of the second rank. It is called the *fundamental tensor of the Riemannian space.* In Cartesian coordinates g_{ij} becomes δ_{ij}.

Bear and Bachmat (1966) make an attempt (through the analysis of their porous medium model described in secs. 4.5 and 4.6) to relate the dispersivity a_{ijkm} to basic medium characteristics by using physical considerations.

In D_{ij} defined by (4.6.25) we have two parts: the correlation $\overline{\mathring{V}'_i \mathring{V}'_j}$, and $\overline{(\text{div}\, \mathbf{V}_\alpha)^{-1}}$. The latter expresses the relative rate of growth of the volume occupied by an elementary mass of solute as it moves. This growth occurs mainly during the passage through the medium's model junctions, where liquid entering from various channels is mixed, say by molecular diffusion. The rate of growth should, accordingly, be proportional to the number of channels traversed by tracer particles per unit time. If $\bar{L}$ is the average length of a channel, and $(\bar{V}_\alpha)_s$ is the average velocity of a particle in a channel of direction $\mathbf{1s}$, then $(\bar{V}_\alpha)_s/\bar{L}$ expresses the number of junctions traversed per unit time, while $\overline{L}/(\overline{V}_\alpha)_s$ expresses the average time of passage between two adjacent junctions. A similar expression may be suggested for the relative expansion in the direction normal to the flow. Bear and Bachmat (1966) suggest $(\overline{V}_\alpha)_a/\bar{a}$, where $\bar{a}$ is a characteristic length of the channel's cross-section and $\overline{(V_\alpha)_a}$ is the average tracer velocity across the channel due to concentration gradients. Thus they come up with:

$$\overline{(\text{div}\,\mathbf{V}_\alpha)^{-1}} \approx \overline{(\text{div}\,\mathbf{V}_\alpha)}^{-1} \propto [2\overline{(V_\alpha)_a}/\bar{a} + \overline{(V_\alpha)_s}/\overline{L}]^{-1}. \tag{10.4.18}$$

Taking $\bar{a}^2/2D_d$ as representing the time required for concentrations to equalize over a channel's cross-section so that $\overline{(V_\alpha)_a}/\bar{a} \approx 2D_d/\bar{a}$, and $\overline{(V_\alpha)_s} = \overline{(V_\alpha)_s - V^*} + \overline{V^*} = 2D_d/\overline{L} + \overline{V}^*$, Bear and Bachmat (1966) end up with:

$$\overline{(\text{div}\,\mathbf{V}_\alpha)^{-1}} \propto \frac{\overline{L}}{\overline{V}^*}\frac{Pe}{Pe + 2 + 4\delta^2} \equiv \frac{\overline{L}}{\overline{V}^*} f(Pe, \delta) \tag{10.4.19}$$

where $Pe = \overline{L}\,\overline{V}^*/D_d$, and $\delta = \overline{L}/\bar{a}$ (i.e., characterizes the shape of the channel), and $f(Pe, \delta)$ is defined by (10.4.19). Obviously other estimates of the relative rate of growth and the effect of the interplay between molecular diffusion and convection on it will lead to different forms of the function $f(Pe, \delta)$ (fig. 10.4.1).

The moment of correlation $\overline{\mathring{V}'_i\mathring{V}'_j}$ between the velocity components can be obtained by expressing $\mathring{V}'_i$, V'_i, $\mathring{V}'_j$ and V'_j by (4.7.11) and (4.7.15), in which the local derivative term is neglected. Introducing certain simplifying assumptions, Bear and Bachmat (1966) obtain:

$$\overline{\mathring{V}'_i\mathring{V}'_j} \approx \overline{B\mathring{T}^*_{ik}B\mathring{T}^*_{jl}}\,(\overline{BT^*_{km}})^{-1}\,(\overline{BT^*_{ln}})^{-1}\,\overline{V^*_m}\,\overline{V^*_n}. \tag{10.4.20}$$

Hence:

$$D_{ij} = \overline{B\mathring{T}^*_{ik}B\mathring{T}^*_{jl}}(\overline{BT^*_{km}})^{-1}(\overline{BT^*_{ln}})^{-1}\frac{\overline{L}\,\overline{V^*_m}\,\overline{V^*_n}}{\overline{V^*}} f(Pe, \delta) = a_{ijmn}\frac{\overline{V^*_m}\,\overline{V^*_n}}{\overline{V^*}} f(Pe, \delta) \tag{10.4.21}$$

where:

$$a_{ijmn} = [\overline{B\mathring{T}^*_{ik}B\mathring{T}^*_{jl}}/\overline{BT^*_{km}}\,\overline{BT^*_{ln}}]\overline{L} \tag{10.4.22}$$

is the *medium's dispersivity* (dim L).

From (10.4.21) it follows that the coefficient of mechanical dispersion D_{ij} may be expressed as a product of three factors: the medium's dispersivity, a second-rank tensor composed of the velocity components, and a nondimensional function depending on δ and Pe.

If we use the principal axes of D_{ij} as a Cartesian coordinate system, e.g., in the case of uniform flow with $\overline{\mathbf{V}^*} = \text{const}$, equation (10.4.21) becomes:

$$[D_{ij}] = \begin{bmatrix} a_I\overline{V^*} & 0 & 0 \\ 0 & a_{II}\overline{V^*} & 0 \\ 0 & 0 & a_{II}\overline{V^*} \end{bmatrix} f(Pe, \delta). \tag{10.4.23}$$

The considerations above are extended to a *general* (nonorthogonal) *curvilinear coordinate* system (Bear and Bachmat 1965, 1966).

To conclude this discussion on the parameters of dispersion, the following nomenclature is suggested:

$D'_{ij} = D_{ij} + (D^*_d)_{ij}$ = coefficient of hydrodynamic dispersion
D_{ij} = coefficient of mechanical (or convective) dispersion
$(D^*_d)_{ij} = T^*_{ij}D_d$ = coefficient of molecular diffusion in a porous medium
a_{ijkl} = dispersivity of the porous medium
a_I = longitudinal dispersivity of an isotropic porous medium
a_{II} = transversal dispersivity of an isotropic porous medium
T^*_{ij} = tortuosity, or tensor of directions, of a porous medium

where typical components of the various tensors are indicated.

10.4.3 Dispersivity–Permeability Relationship

Whereas the dispersivity of a medium determines the spreading of a definite fluid portion, the permeability determines the average flux of a fluid through the porous medium. Both are medium geometric characteristics. Although both depend on the fundamental medium property BT^*_{ij}, they are two distinct characteristics. The permeability k_{ij} depends on the average of BT^*_{ij} (par. 4.8.1):

$$k_{ij} = n\overline{BT^*_{ij}}. \tag{10.4.24}$$

If we choose a coordinate system that coincides with the principal directions of k_{ij}, we obtain:

$$k_{ij} = k\delta_{ij}, \quad \text{where} \quad k_{ij} = 0 \quad \text{for} \quad i \neq j.$$

In general, $\overline{BT^*_{ij}}$ (*medium's average directional conductance*) at a point may be expressed as:

$$\overline{BT^*_{ij}} = \bar{B}\overline{T^*_{ij}} + \overline{\mathring{B}\mathring{T}^*_{ij}} = \bar{B}\overline{T^*_{ij}} + \overline{\mathring{B}\mathring{\alpha}_i}\alpha_j + \overline{\mathring{B}\mathring{\alpha}_j}\alpha_i. \tag{10.4.25}$$

It can be shown that for an isotropic medium, $\overline{\mathring{B}\mathring{T}^*_{ij}} = 0$. Then:

$$k_{ij} = n\bar{B}\bar{T}^*\delta_{ij}. \tag{10.4.26}$$

The dispersivity a_{ijmn} depends on the variance of BT^*_{ij}. Through this basic medium characteristic, one may find an intrinsic relationship between permeability and dispersivity.

Let us consider the particular case of an *isotropic medium* (with respect to k_{ij}). Then, $k = n\bar{B}\overline{T^*} = (1/3)n\bar{B}\overline{T^*_{kk}}$, and the dispersivity given by (10.4.21) becomes:

$$a_{ijmn} = (n^2\bar{L}/k^2)\overline{B\mathring{T}^*_{im}B\mathring{T}^*_{jn}} \tag{10.4.27}$$

from which:

$$a_I = a_{1111} = (n^2\bar{L}/k^2)\overline{(B\mathring{T}^*_{11})^2}; \qquad a_{II} = a_{2211} = (n^2\bar{L}/k^2)\overline{(B\mathring{T}^*_{21})^2}.$$

If we also assume that $B = \text{const} = \bar{B}$, we obtain:

$$a_I = (n^2\bar{L}\overline{B^2}/k^2)\overline{(\mathring{T}_{11})^2} = \overline{(\mathring{T}_{11}/\bar{T})^2}\bar{L}. \tag{10.4.28}$$

As an example, let us consider the special case of a model consisting of straight circular channels of constant radius R and a uniform distribution of directions:

$$B = \frac{r^2 - R^2}{4}; \quad \bar{B} = R^2/8; \quad \frac{\overline{(\mathring{B})^2}}{\bar{B}^2} = \tfrac{1}{3}; \quad \overline{T_{ij}} = \int_{(\Omega)} T_{ij}(\Omega) f(\Omega)\, d\Omega; \quad \int_{(\Omega)} f(\Omega)\, d\Omega = 1 \tag{10.4.29}$$

where $d\Omega$ is a solid angle (fig. 10.4.4). Then:

$$d\Omega = \sin\theta\, d\theta\, d\psi; \qquad f(\Omega) = 1/2\pi \quad \text{(isotropic medium)}$$

$$T_{11} = \cos^2\theta; \qquad \overline{(\mathring{T}_{11})^2} = \overline{(T_{11})^2} - (\bar{T}_{11})^2$$

$$\bar{T}_{11} = \frac{1}{2\pi}\int_0^{2\pi}\int_0^{\pi/2} \cos^2\theta \sin\theta\, d\theta\, d\psi = \tfrac{1}{3}$$

$$\overline{T_{11}^{\,2}} = \frac{1}{2\pi}\int_0^{2\pi}\int_0^{\pi/2} \cos^4\theta \sin\theta\, d\theta\, d\psi = \tfrac{1}{5}$$

$$\overline{(\mathring{T}_{11})^2} = 4/5, \text{ etc.}$$

Assuming no correlation between B and T, we obtain:

$$a_{1111} \equiv a_I = \frac{\overline{(B\mathring{T}_{11})^2}}{(\bar{B}\bar{T}_{11})^2}\bar{L} = \left[\overline{\left(\frac{\mathring{B}}{\bar{B}}\right)^2} + \overline{\left(\frac{\mathring{T}_{11}}{\bar{T}_{11}}\right)^2} + \overline{\left(\frac{\mathring{B}}{\bar{B}}\right)^2\left(\frac{\mathring{T}_{11}}{\bar{T}_{11}}\right)^2}\right]\bar{L}$$

$$= \left[\frac{1}{3} + \frac{4}{5} + \frac{1}{3}\cdot\frac{4}{5}\right]\bar{L} = \frac{7}{5}\bar{L}, \quad \text{etc.}$$

and:

$$k = \left(\frac{7}{2880}\right)^{1/2} \frac{nR^2}{(a_I/\bar{L})^{1/2}} = \frac{1}{24} nR^2 = \tfrac{1}{3} n\bar{B}. \tag{10.4.30}$$

This expression relates the permeability of the medium to the longitudinal dispersivity, to porosity, to some characteristic pore cross-section and to some characteristic pore length. Harleman et al. (1963) obtain by experiments expressions relating permeability to dispersivity.

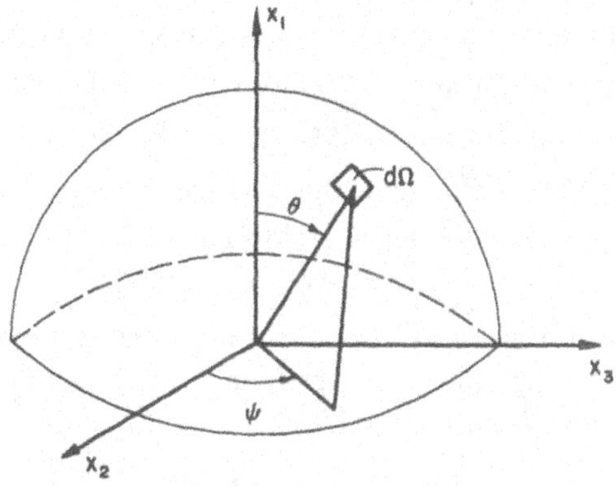

FIG. 10.4.4. Definition sketch for the determination of T_{ij}.

10.5 The Governing Equations and Boundary Conditions

In the present section, the symbol C will be used to denote the concentration of a tracer, i.e., mass of tracer per unit volume of solution ($C \equiv \rho_\alpha$). The term tracer will be used to denote any species of interest in a solution.

In order to determine the tracer-concentration distribution $C = C(x, y, z, t)$, one must solve the partial differential equation that expresses the tracer mass conservation in the flow domain considered, subject to certain boundary and initial conditions. These equations and conditions are discussed in the present section.

Unless otherwise specified, we shall denote the *average velocity* vector by the symbol $\mathbf{V} = \mathbf{q}/n$. For a homogeneous fluid ($\rho =$ const), we know from paragraph 4.1.1 that the mass average velocity $\mathbf{V}^*$ is equal to the volume average velocity $\mathbf{V}'$, and therefore $\mathbf{V}$ stands for either one of them. For a nonhomogeneous fluid (i.e., $\rho \neq$ const), equation (4.1.16) is valid:

$$\mathbf{V}^* - \mathbf{V}' = -(D_{\alpha\beta}/\rho)\,\nabla\rho,$$

i.e., $\mathbf{V}' \neq \mathbf{V}^*$. However, assuming that the contribution of the solute's molecular diffusion to the total mass flux is negligible, i.e., $|D_{\alpha\beta}\,\nabla\rho| \ll |\rho\mathbf{V}^*|$, we have $\mathbf{V}^* \simeq \mathbf{V}' \equiv \mathbf{V}$. We shall consider only incompressible fluid.

10.5.1 The Partial Differential Equation in Cartesian Coordinates

In paragraph 4.6.2 it was shown that for an incompressible fluid, the tracer mass conservation is expressed by the partial differential equation in Cartesian coordinates:

$$\partial C/\partial t = \operatorname{div}(\underline{\mathbf{D}}' \cdot \operatorname{grad} C) - \mathbf{V} \cdot \operatorname{grad} C; \; n = \text{const} \qquad (10.5.1)$$

or, in indicial notation:

$$\partial C/\partial t = \partial(D'_{ij}\,\partial C/\partial x_j)/\partial x_i - V_i\,\partial C/\partial x_i. \qquad (10.5.2)$$

When $\operatorname{div} \mathbf{V} \neq 0$, the last terms in (10.5.1) and (10.5.2) become $\operatorname{div}(\mathbf{V}C)$ and $\partial(V_iC)/\partial x_i$, respectively. Equations (10.5.1) and (10.5.2) are obtained from (4.6.27). The coefficient of dispersion $\underline{\mathbf{D}}'$ appearing in (10.5.1) and (10.5.2) is the sum of the coefficient of mechanical dispersion ($\underline{\mathbf{D}}$) and the coefficient of molecular diffusion in porous media ($D_d\underline{\mathbf{T}}^*$). The concentration C, the velocity $\mathbf{V}$ and the tortuosity $\mathbf{T}^*$ appearing in (10.5.1) and (10.5.2) represent the average values $\bar{C}$ (actually $\bar{\rho}_\alpha$), $\bar{\mathbf{V}}'$ and $\bar{\underline{\mathbf{T}}}^*$, respectively, taken over an REV of the medium.

Two possible extensions of (10.5.2) will be considered.

Extension 1. The tracer being considered may undergo *radioactive decay* as it is being transported through the porous medium. The change in tracer concentration resulting from this decay is expressed by:

$$\partial C/\partial t = -\lambda C \tag{10.5.3}$$

where λ is the *decay constant of the tracer* (equal to the reciprocal of the tracer's mean lifetime). The dispersion equation (10.5.2) for n = const then becomes:

$$\partial C/\partial t = \partial(D'_{ij}\,\partial C/\partial x_j)/\partial x_i - V_i\,\partial C/\partial x_i - \lambda C. \tag{10.5.4}$$

Extension 2. The tracer considered may undergo chemical reactions and/or adsorption to the surface of the porous matrix. Adsorption here is the tracer mass transfer (in the form of solute ions) from the liquid to the surfaces of the solid phase, controlled by the chemical properties of the solute and the solid matrix, and by the solute concentration in the liquid phase.

Many *isotherms*, or *adsorption equations*, that relate C in the liquid phase to the tracer concentration F (mass of tracer per unit volume of solid phase) on the solid surface, under equilibrium or nonequilibrium conditions, are available in the literature (e.g., Lapidus and Amondson 1952; Amondson 1960; Bondarev and Nikolaevskij 1962). For example, we may use the nonequilibrium relationship:

$$\partial F/\partial t = \beta(C - F/a_2) = aC - bF;\ \beta,\ a_2,\ a,\ b = \text{constants}. \tag{10.5.5}$$

In the presence of adsorption and radioactive decay the concentration equation (10.5.2) becomes:

$$\frac{\partial}{\partial t}\left(C + \frac{1-n}{n}F\right) = \frac{\partial}{\partial x_i}\left(D'_{ij}\frac{\partial C}{\partial x_j}\right) - V_i\frac{\partial C}{\partial x_i} - \lambda\left(C + \frac{1-n}{n}F\right). \tag{10.5.6}$$

In principle, the two equations (10.5.5) and (10.5.6) yield (subject to appropriate initial and boundary conditions on both F and C) the desired solutions.

If we express F in (10.5.6) by the equilibrium isotherm, $F = a_1C$, we obtain:

$$\frac{\partial C}{\partial t}\left(1 + \frac{1-n}{n}a_1\right) = \frac{\partial}{\partial x_i}\left(D'_{ij}\frac{\partial C}{\partial x_j}\right) - V\frac{\partial C}{\partial x_i} - \lambda C\left(1 + \frac{1-n}{n}a_1\right) \tag{10.5.7}$$

where a_1 is a constant, or:

$$R_d\frac{\partial C}{\partial t} = \frac{\partial}{\partial x_i}\left(D'_{ij}\frac{\partial C}{\partial x_j}\right) - V_i\frac{\partial C}{\partial x_i} - \lambda R_d C \tag{10.5.8}$$

where R_d is a *retardation factor*; it retards the appearance of a breakthrough curve. By writing (10.5.8) for n = const. in the form:

$$\frac{\partial C}{\partial t} = \frac{\partial}{\partial x_i}\left(\frac{D'_{ij}}{R_d} \cdot \frac{\partial C}{\partial x_j}\right) - \frac{V_i}{R_d}\frac{\partial C}{\partial x_i} - \lambda C \tag{10.5.9}$$

we may conclude that adsorption produces an effect that is equivalent to a reduction in the average velocity (recalling the relationship between D_{ij} and $\mathbf{V}$) and in the coefficient of molecular diffusion in the porous medium.

If we introduce moving coordinates:

$$x' = x - V_x t, \qquad y' = y - V_y t, \qquad z' = z - V_z t; \qquad t' = t \tag{10.5.10}$$

(10.5.2) becomes:

$$\frac{\partial C}{\partial t'} = \frac{\partial}{\partial x'_i}\left(D'_{ij}\frac{\partial C}{\partial x'_j}\right). \tag{10.5.11}$$

Under certain conditions, a state of equilibrium may be reached with no further change in the tracer concentration distribution. In this case, $C = C(x, y, z)$ and we must insert $\partial C/\partial t = 0$ in the partial differential equation corresponding to the considered problem.

10.5.2 The Partial Differential Equation in Curvilinear Coordinates

In a general curvilinear coordinate system y^i, (10.5.2) may be written in the form:

$$\frac{\partial C}{\partial t} = \nabla_i\left[D'^{ij}\frac{\partial C}{\partial y^j}\right] - V^i \nabla_i C; \; n = \text{const} \tag{10.5.12}$$

where ∇_i is the divergence symbol. In an *isotropic medium*, with D'^{ij} expressed in curvilinear coordinates, equation (10.5.12) becomes:

$$\frac{\partial C}{\partial t} = \frac{1}{\sqrt{g}}\frac{\partial}{\partial y^i}\left\{\sqrt{g}\, f(Pe, \delta)\left[(a_I - a_{II})\frac{V^i V^j}{V} + a_{II} g^{ij} V\right]\frac{\partial C}{\partial y^j}\right.$$
$$\left. + \sqrt{g}\, D_d T^* g^{ij}\frac{\partial C}{\partial y^j}\right\} - V^i\frac{\partial C}{\partial y^i} \tag{10.5.13}$$

where:

$$g^{ij} = (\partial y^i/\partial x)(\partial y^j/\partial x) + (\partial y^i/\partial y)(\partial y^j/\partial y) + (\partial y^i/\partial z)(\partial y^j/\partial z).$$

In an orthogonal system, we have:

$$g^{ij} = \begin{cases} (h^i)^2 = 1/(h_i)^2 & \text{for } i = j \\ 0 & \text{for } i \neq j \end{cases} \tag{10.5.14}$$

where the h's are scale factors, and (10.5.13) becomes:

$$\frac{\partial C}{\partial t} = \frac{1}{h_1 h_2 h_3}\frac{\partial}{\partial y^i}\left\{\frac{h_1 h_2 h_3}{(h_i)^2}\left(f(Pe, \delta)\left[(a_I - a_{II})\frac{V_i V_k}{V}\frac{1}{(h_k)^2}\frac{\partial C}{\partial y^k} + a_{II} V\frac{\partial C}{\partial y^i}\right]\right.\right.$$

$$+ D_d T^* \frac{\partial C}{\partial y^i}\bigg)\bigg\} - \frac{V_i}{(h_i)^2}\frac{\partial C}{\partial y^i}; \qquad T^{*ij} = T^* g^{ij} \tag{10.5.15}$$

where the summation is performed over both i and k, except in $(h_k)^2$ and $(h_i)^2$, and f is defined by (10.4.19). In (10.5.15), and everywhere in this section, V is the absolute value of the average velocity vector $\mathbf{V}$. Equation (10.5.15) is the general equation of hydrodynamic dispersion in an isotropic homogeneous medium in laminar flow.

As explained in section 10.4, the principal axes of D^{ij} in an isotropic medium are three orthogonal axes of which one is parallel to the average velocity vector. Taking the principal axes of the dispersion as axes of a coordinate system y^i, we have: $V_1 \neq 0$, $V_2 = V_3 = 0$, and with $V_i = Vh_i \cos(\mathbf{V}, \mathbf{1y}^i)$ (10.5.15) becomes:

$$\frac{\partial C}{\partial t} = \frac{1}{h_1 h_2 h_3}\bigg\{\frac{\partial}{\partial y^1}\bigg[\frac{h_2 h_3}{h_1}(a_I f(Pe, \delta)V + D_d T^*)\frac{\partial C}{\partial y^1}\bigg]$$
$$+ \frac{\partial}{\partial y^2}\bigg[\frac{h_3 h_1}{h_2}(a_{II} f(Pe, \delta)V + D_d T^*)\frac{\partial C}{\partial y^2}\bigg]$$
$$+ \frac{\partial}{\partial y^3}\bigg[\frac{h_1 h_2}{h_3}(a_{II} f(Pe, \delta)V + D_d T^*)\frac{\partial C}{\partial y^3}\bigg]\bigg\} - \frac{V}{h_1}\cos(\mathbf{V}, \mathbf{1y}^1)\frac{\partial C}{\partial y^1}. \tag{10.5.16}$$

Equation (10.5.16) may also be written in terms of lengths along coordinate lines. If ds^i is an element of length along the coordinate line y^i, we have:

$$ds^i = h_i\, dy^i \qquad \text{(no summation on } i).$$

Then $(1/h_i)\, \partial C/\partial y^i = \partial C/\partial s^i$, and (10.5.16) becomes:

$$\frac{\partial C}{\partial t} = \frac{1}{h_2 h_3}\frac{\partial}{\partial s^1}\bigg[h_2 h_3 (a_I f(Pe, \delta)V + D_d T^*)\frac{\partial C}{\partial s^1}\bigg] + \frac{1}{h_1 h_3}\frac{\partial}{\partial s^2}\bigg[h_3 h_1 (a_{II} f(Pe, \delta)V + D_d T^*)\frac{\partial C}{\partial s^2}\bigg]$$
$$+ \frac{1}{h_1 h_2}\frac{\partial}{\partial s^3}\bigg[h_1 h_2 (a_{II} f(Pe, \delta)V + D_d T^*)\frac{\partial C}{\partial s^3}\bigg] - V\cos(\mathbf{V}, \mathbf{1s}^1)\frac{\partial C}{\partial s^1}. \tag{10.5.17}$$

The family of streamlines may serve as the family of coordinate lines y^1 in (10.5.16) and (10.5.17). If in *two-dimensional flow* in a homogeneous isotropic medium we use the family of streamlines $\Psi' =$ const as y^1, and the family of equipotential lines $\Phi' =$ const as y^2, where $\Phi' = -\Phi/n = -K\varphi/n$ is the *velocity potential*, we obtain:

$$y^1 = \Phi' = \Phi'(x, y); \qquad s^1{}_{\Phi'} = s_{\Phi'}; \qquad dy^1/ds^1 = d\Phi'/ds_{\Phi'} = |\text{grad}\, \Phi'| = V = 1/h_1$$
$$y^2 = \Psi' = \Psi'(x, y); \qquad s^2{}_{\Psi'} = s_{\Psi'}; \qquad dy^2/ds^2 = d\Psi'/ds_{\Psi'} = |\text{grad}\, \Psi'| = 1/h_2.$$

Since $\partial\Phi'/\partial x = \partial\Psi'/\partial y$ and $\partial\Phi'/\partial y = -\partial\Psi'/\partial x$, we also have $|\text{grad}\, \Phi'| = |\text{grad}\, \Psi'|$ and $h_1 = h_2 = 1/V$.

With these results, and with $h_3 = 1$, for plane flow, we obtain from (10.5.16):

$$\frac{\partial C}{\partial t} = V^2\bigg\{\frac{\partial}{\partial \Phi'}\bigg[(fVa_I + D_d T^*)\frac{\partial C}{\partial \Phi'}\bigg] + \frac{\partial}{\partial \Psi'}\bigg[(fVa_{II} + D_d T^*)\frac{\partial C}{\partial \Psi'}\bigg] - V^2\cos(\mathbf{V}, \mathbf{1\Phi}')\frac{\partial C}{\partial \Phi'} \tag{10.5.18}$$

and from (10.5.17):

$$\frac{\partial C}{\partial t} = V\left\{\frac{\partial}{\partial s_{\Phi'}}\left[\left(\frac{Va_I}{V+A}+\frac{D_dT^*}{V}\right)\frac{\partial C}{\partial s_{\Phi'}}\right]+\frac{\partial}{\partial s_{\Psi'}}\left[\left(\frac{Va_{II}}{V+A}+\frac{D_dT^*}{V}\right)\frac{\partial C}{\partial s_{\Psi'}}\right]\right.$$
$$\left.-\cos(\mathbf{V}, \mathbf{1s}_{\Phi'})\frac{\partial C}{\partial s_{\Phi'}}\right\} \qquad (10.5.19)$$

where $A = (2+4\delta^2)D_d/L = \text{const}$. The $\Phi' - \Psi'$ coordinates in this case are the *natural coordinates* of plane flow.

In the special case of uniform flow in the direction $\mathbf{1x}$ in the xy-plane: $ds_{\Psi'} = dy$, $ds_{\Phi'} = dx$; $\cos(\mathbf{V}, \mathbf{1s}_{\Phi'}) = 1$. Then the dispersion equation becomes:

$$\frac{\partial C}{\partial t} = \left(\frac{V^2a_I}{V+A}+D_dT^*\right)\frac{\partial^2 C}{\partial x^2}+\left(\frac{V^2a_{II}}{V+A}+D_dT\right)\frac{\partial^2 C}{\partial y^2}-V\frac{\partial C}{\partial x}. \qquad (10.5.20)$$

When $A \ll V$ (10.5.20) becomes:

$$\frac{\partial C}{\partial t} = (Va_I + D_dT^*)\,\partial^2C/\partial x^2 + (Va_{II}+D_dT^*)\,\partial^2C/\partial y^2 - V\,\partial C/\partial x. \qquad (10.5.21)$$

In uniform flow in the direction $\mathbf{1x}$ in the xyz space, the dispersion equation corresponding to (10.5.21) is:

$$\partial C/\partial t = (Va_I + D_dT^*)\,\partial^2C/\partial x^2 + (Va_{II}+D_dT^*)(\partial^2C/\partial y^2 + \partial^2C/\partial z^2) - V\,\partial C/\partial x. \qquad (10.5.22)$$

In the symmetrical (or one-dimensional) case of flow in the direction $\mathbf{1x}$, $\partial C/\partial y = 0$, and (10.5.20) becomes:

$$\partial C/\partial t = [V^2a_I/(V+A) + D_dT^*]\,\partial^2C/\partial x^2 - V\,\partial C/\partial x. \qquad (10.5.23)$$

If $D_dT^* \gg V^2a_I/(V+A)$, we have:

$$\partial C/\partial t = [V^2a_I/(V+A)]\,\partial^2C/\partial x^2 - V\,\partial C/\partial x. \qquad (10.5.24)$$

If, also, $V \gg A$, we obtain:

$$\partial C/\partial t = a_IV\,\partial^2C/\partial x^2 - V\,\partial C/\partial x \qquad (10.5.25)$$

which is commonly used as the one-dimensional dispersion equation. For a compressible fluid, the last term in (10.5.25) should be replaced by $\partial(CV)/\partial x$.

A special case of interest is that of radial convergent or divergent plane flow into a sink or from a source at the origin. In the polar coordinate system (r, θ), $ds_{\Phi'} = dr$, $ds_{\Psi'} = r\,d\theta$; $V = G/r$ where G is a constant. We obtain:

$$\frac{\partial C}{\partial t} = V\left\{\frac{\partial}{\partial r}\left[\left(\frac{Va_I}{V+A}+\frac{D_dT^*}{V}\right)\frac{\partial C}{\partial r}\right]+\frac{1}{r}\frac{\partial}{\partial \theta}\left[\left(\frac{Va_{II}}{V+A}+\frac{D_dT^*}{V}\right)\frac{1}{r}\frac{\partial C}{\partial \theta}\right]\pm\frac{\partial C}{\partial r}\right\} \qquad (10.5.26)$$

where $V = Q/2\pi Bnr$, and the plus and minus signs correspond to radially convergent and divergent flows, respectively. If we have axisymmetric conditions, we set

$\partial C/\partial\theta = 0$ in (10.5.26). Because V varies with r, the relationship between $a_I V$ and $D_d T^*$ may also depend on r: we may have $a_I V \gg D_d T^*$ near the origin, but $a_I V \ll D_d T^*$ for larger r.

For $V \gg A$ and $a_I V \gg D_d T^*$: we obtain for the symmetrical case of $\partial C/\partial\theta = 0$:

$$\partial C/\partial t = a_I V\, \partial^2 C/\partial r^2 \pm V\, \partial C/\partial r. \tag{10.5.27}$$

Equation (10.5.27) is used to describe hydrodynamic dispersion in radially converging or diverging plane flow. Difficulties in solution are encountered because V is proportional to $1/r$.

10.5.3 Initial and Boundary Conditions

As in any other boundary value problem described by a partial differential equation, the solution $C = C(x, y, z, t)$ of the partial differential equation describing hydrodynamic dispersion must also satisfy specified initial and boundary conditions. For the sake of simplicity, the discussion will refer to (10.5.2). We shall assume that inside the flow domain, C is a continuous function of time and space, and that its first derivative with respect to time and its first two derivatives with respect to space coordinates are also continuous.

Discussion 1, initial conditions. As an initial condition we must specify the concentration distribution at some initial time $t = 0$ at all points of the flow domain:

$$C(\mathbf{x}, 0) = f_1(\mathbf{x}); \qquad \mathbf{x} \equiv x_1, x_2, x_3 \tag{10.5.28}$$

where f_1 is a known function of $\mathbf{x}$. If $f_1(\mathbf{x})$ is discontinuous at certain points, curves or surfaces, we require that this discontinuity will disappear after an infinitesimally short time and that within that limit, as $t \to 0$, the derived concentration distribution will approach the initial one at all points where initially C was continuous.

As a special case of interest, we may mention an initial condition describing an instantaneous injection of a slug, or pulse, of a specified size:

$$C(\mathbf{x}, t_0) = (M/n)\,\delta(\mathbf{x} - \mathbf{x}_0) \tag{10.5.29}$$

where δ is the *Dirac delta function*, M is the tracer's total mass injected at $t = t_0$ at $\mathbf{x} = \mathbf{x}_0$, and n is the medium's porosity.

Discussion 2, boundary conditions. The conditions along the boundaries of a flow domain throughout which hydrodynamic dispersion occurs depend on the type of medium and fluid present in the region just outside these boundaries. All boundary conditions are based on the requirement that for any point of the boundary surface, the tracer's mass flux $\mathbf{J}$ (mass per unit area of boundary surface) normal to the boundary must be equal on both sides of a stationary boundary.

Let superscript a denote the porous medium domain inside the boundaries, and superscript b denote the domain external to the boundaries. The requirement that the mass fluxes normal to the boundary be equal on both sides of the latter is expressed by:

$$\mathbf{J}^a \cdot \mathbf{1\nu} = \mathbf{J}^b \cdot \mathbf{1\nu}; \qquad J^a{}_\nu = J^b{}_\nu \tag{10.5.30}$$

where $\mathbf{1v}$ (components ν_i) denotes a unit vector in the direction of the outward normal to the boundary surface. Several types of boundary conditions may occur.

(a) The external domain b is also a porous medium with porosity n^b. From (10.5.30), we have:

$$n^a\nu_i(-D'_{ij}\,\partial C/\partial x_j + V_iC)^a = n^b\nu_i(-D'_{ij}\,\partial C/\partial x_j + V_iC)^b \qquad (10.5.31)$$

where, for a homogeneous medium $D'_{ij} = D_{ij} + D_dT^*_{ij}$ and:

$$n\,D_{ij} = a_{ijmn}(q_mq_n/q)f^*(Pe^*, \delta); \qquad nf^* = f(Pe^*, \delta); \qquad Pe^* = nPe = qL/D_d. \qquad (10.5.32)$$

If both media are homogeneous and isotropic, the condition (10.5.31) becomes:

$$\left\{f^*\left[(a_I - a_{II})\nu_i\frac{q_iq_j}{q} + a_{II}q\nu_j\right]\frac{\partial C}{\partial x_j} + nD_dT^*\nu_j\frac{\partial C}{\partial x_j} - \nu_iq_iC\right\}^a$$
$$= \left\{f^*\left[(a_I - a_{II})\nu_i\frac{q_iq_j}{q} + a_{II}q\nu_j\right]\frac{\partial C}{\partial x_j} + nD_dT^*\nu_j\frac{\partial C}{\partial x_j} - \nu_iq_iC\right\}^b. \qquad (10.5.33)$$

For the one-dimensional case (10.5.33) reduces to:

$$\left(f^*a_Iq\,\frac{\partial C}{\partial x} + nD_dT^*\,\frac{\partial C}{\partial x} - qC\right)^a = \left(f^*a_Iq\,\frac{\partial C}{\partial x} + nD_dT^*\,\frac{\partial C}{\partial x} - qC\right)^b. \qquad (10.5.34)$$

Obviously in this case we have a second condition on the boundary, namely $C^a = C^b$.

(b) The external domain is impervious both to diffusion and to mass flow. Then:

$$[(f^*a_{II}q + nD_dT^*)\nu_j\,\partial C/\partial x_j]^a = 0. \qquad (10.5.35)$$

(c) The external medium is a liquid continuum (e.g., a sea, lake, river, reservoir) in which the tracer's concentration is specified, independent of the tracer's mass flux through the domain's boundary. If the tracer's concentration along the external side of the boundary is denoted by $C(\mathbf{x}', t)$, where $\mathbf{x}'$ denotes points along the boundary, the flux through the boundary is given by $J^b{}_\nu = C(\mathbf{x}', t)\nu_iq_i - [D_d\nu_i\,\partial C/\partial x_i]^a$. Then the boundary condition becomes:

$$[C(\mathbf{x}', t) - C]\nu_iq_i = [D_d\nu_i(\partial C/\partial x_i)]^b_{\mathbf{x}=\mathbf{x}'} - [n\nu_i(D_{ij} + D_dT^*_{ij})\,\partial C/\partial x_j]^a, \qquad (10.5.36)$$

where, in general, $[\partial C/\partial x_i]^b_{\mathbf{x}=\mathbf{x}'} = 0$. From (10.5.36) it follows that, in general, the tracer concentration on both sides of the boundary are not the same.

For the one-dimensional case in a homogeneous isotropic medium (10.5.36) reduces to:

$$[C(x', t) - C]q = [D_d\,\partial C/\partial x]^b - [(nD_dT^* + f^*a_Iq)\,\partial C/\partial x]^a. \qquad (10.5.37)$$

When the *specific discharge vector* $\mathbf{q}$ *in* (10.5.36) *is normal to the boundary*, equation (10.5.36) for a homogeneous isotropic medium becomes:

$$[C(x', t) - C]q = (D_d - nD_dT^* - f^*a_Iq)\,\partial C/\partial s_\nu \qquad (10.5.38)$$

where s_ν is the distance measured along the normal.

When the tracer concentrations on both sides of the boundary become equal, we obtain the boundary condition:

$$C(\mathbf{x}', t) = C \quad \text{and} \quad \partial C/\partial s_\nu = 0. \tag{10.5.39}$$

This last type of boundary condition is possible after a sufficiently long time if $C(\mathbf{x}', t)$ remains constant. In general, for short times, equation (10.5.39) is not possible.

Dankwerts (1953) comments with respect to $\partial C/\partial x$ at an outflow end ($x = L$) that $\partial C/\partial x|_{x=L} < 0$, which means that $C|_{x=L} > C|_{x<L}$ (i.e., $C_L > C$) is not possible, $\partial C/\partial x|_{x=L} > 0$ is also impossible as it produces a minimum in C inside the flow domain. Hence we must have $\partial C/\partial x = 0$ at $x = L$.

(d) The external domain (outside an outflow boundary) is a vacuum (or a gas continuum). In this case the tracer concentration is equal on both sides of the boundary, i.e.:

$$C(\mathbf{x}', t) = C. \tag{10.5.40}$$

As there is no molecular diffusion outside the porous medium domain, we obtain from (10.5.36):

$$n\nu_i(D_{ij} + D_d T^*_{ij})\, \partial C/\partial x_j = 0. \tag{10.5.41}$$

(e) For an infinite domain without supply or withdrawal of tracer at infinity, we require that as $\mathbf{x} \to \infty$, C will have a finite fixed value, say C_0:

$$\lim_{\mathbf{x}\to\infty} C = C_0. \tag{10.5.42}$$

10.5.4 Solving the Boundary Value Problems

The tracer concentration distribution $C = C(x, y, z, t)$ within a specified flow domain is obtained by solving the appropriate partial differential equation of hydrodynamic dispersion (pars. 10.5.1 and 10.5.2) subject to a specified initial tracer concentration distribution and to boundary conditions with respect to the tracer's concentration (par. 10.5.3). If adsorption also takes place, we must also solve for $F = F(x, y, z, t)$ using an adsorption equation as a second equation. Appropriate boundary and initial conditions must also be specified with respect to F. We shall consider only isothermal flow.

Two cases may be distinguished.

Case 1, homogeneous liquid. In a homogeneous liquid (ρ = const, μ = const) the concentration distribution does not affect the velocity distribution. Hence, the solution of a dispersion problem is made up of two *independent subproblems.* In the first one, the velocity distribution is determined (analytically, numerically or by models) for all points of the flow domain. The resulting velocity distribution is inserted in the dispersion equation, which is then solved to yield the concentration distribution in the flow domain.

Case 2, inhomogeneous liquid. In the case of inhomogeneous liquids, the two problems must be solved *simultaneously*, as the instantaneous average velocity distribution depends on the concentration distribution at that instant (through its effect on ρ and μ). Moreover, changes in the average velocity vector at a given point change the principal axes of dispersion at that point.

We must, therefore, determine seven dependent variables simultaneously: the concentration C, three velocity components V_i ($i = 1, 2, 3$), density ρ, viscosity μ, and pressure p. The medium parameters, permeability, k_{ij}, and dispersivity, a_{ijmn}, as well as the function $f(Pe, \delta)$, must also be known. In what follows we shall assume that (4.7.16) is the motion equation and that the fluid is inhomogeneous and incompressible. We shall emphasize the fact that ρ varies by distinguishing between $\mathbf{V}'$ and $\mathbf{V}^*$.

Under these assumptions, the equations at our disposal are:

(a) three motion equations:

$$V^*_i = -(k_{ij}/n\mu)(\partial p/\partial x_j + \rho g\, \partial z/\partial x_j) \tag{10.5.43}$$

(b) two equations of state $\rho = \rho(C, p)$ and $\mu = \mu(C, p)$. For the approximation of a dilute incompressible system we may assume:

$$\rho = \rho_0 + \alpha(C - C_0), \qquad \mu = \mu_0 + \beta(C - C_0) \tag{10.5.44}$$

where C_0 is a reference concentration and α and β are considered constants as a first approximation;

(c) the dispersion equation (equation of tracer mass conservation):

$$\frac{\partial C}{\partial t} = \frac{\partial}{\partial x_i}\left[(D_{ij} + D_d T^*_{ij})\frac{\partial C}{\partial x_j}\right] - V'_i \frac{\partial C}{\partial x_i}; \qquad D_{ij} = a_{ijmn}\frac{V^*_m V^*_n}{V^*} f(Pe, \delta) \tag{10.5.45}$$

(d) the equation of mass conservation of the incompressible fluid system:

$$\partial \rho/\partial t + \partial(\rho V^*_i)/\partial x_i = 0 \tag{10.5.46}$$

(e) the relationship between $\mathbf{V}^*$ and $\mathbf{V}'$:

$$V^*_i - V'_i = -D_d T^*_{ij}\, \partial \rho/\partial x_j. \tag{10.5.47}$$

In general this set of equations cannot be solved by analytic methods; numerical methods must be employed.

Even for a homogeneous fluid, the solution of the dispersion equation, except for a small number of simple one- and two-dimensional cases, is not an easy task, and usually solutions are obtained by numerical methods using digital computers. Examples of such solutions are given by Peaceman and Rachford (1962), Stone and Brian (1963), Garder, Peaceman and Rachford (1964), among others. Shamir and Harleman (1966) present an improved numerical method (using a digital computer) for solving the hydrodynamic dispersion equation for a homogeneous fluid (ρ = const, μ = const) in steady three-dimensional flow through porous media. They develop the method and test it for two-dimensional problems where the dispersion equation is (10.5.18). A special type of unsteady flow problem can also be solved by this method. Recently, Redell and Sunada (1970) presented a numerical solution for dispersion described by the set of equations given above.

10.5.5 The Use of Nondimensional Variables

It is sometimes convenient to introduce nondimensional variables and parameters in studying dispersion phenomena. When this is done, solutions (as well as experimental results) may be presented in a general form, applicable to any set of numerical values of the fluid, medium and flow parameters involved.

The choice of the nondimensional variables follows the procedure described in section 7.4 and paragraph 11.2.7. In what follows, subscripts p and e will denote the value of a variable or a parameter in the prototype and in the equivalent systems, respectively, while subscript r will denote the ratio between a value in the equivalent system and the corresponding value in the prototype system.

As an example, consider the one-dimensional dispersion equation:

$$R_d\, \partial C/\partial t = D'\, \partial^2 C/\partial x^2 - (q/n)\, \partial C/\partial x - \lambda C R_d. \tag{10.5.48}$$

Following the procedure explained in section 7.4 and paragraph 11.2.7, we obtain:

$$C_r(R_d)_r/t_r = D'_r C_r/x_r^2 = q_r C_r/n_r x_r = \lambda_r C_r (R_d)_r. \tag{10.5.49}$$

From these three equations we obtain:

$$(R_d)_r/t_r = D'_r/x_r^2; \qquad D'_r/x_r^2 = q_r/n_r x_r; \qquad q_r/n_r x_r = \lambda_r (R_d)_r. \tag{10.5.50}$$

Hence:

$$t_r D'_r/(R_d)_r x_r^2 = 1; \qquad x_r q_r/n_r D'_r = 1; \qquad (R_d)_r x_r \lambda_r n_r/q_r = 1 \tag{10.5.51}$$

or:

$$(q\, \Delta x/nD')_e = (q\, \Delta x/nD')_p; \qquad (q^2\, \Delta t/n^2 R_d D')_e = (q^2\, \Delta t/n^2 R_d D')_p. \tag{10.5.52}$$

In the special case of $D' = a_I q/n$, we obtain from (10.5.51) nondimensional length and time:

$$\xi = x/a_I, \qquad \tau = qt/na_I R_d. \tag{10.5.53}$$

The nondimensional distance ξ is thus the distance measured in units of longitudinal dispersivity a_I, while the nondimensional time may be interpreted as the average distance traveled by particles at the reduced velocity (q/nR_d), again measured in units of a_I. Since $\lambda = 1/T$, where T is the half-life time of a radioactive tracer, and referring to T/R_d as the effective half-life of the tracer in the presence of adsorption, we also have:

$$\tau = q(T/R_d)/na_I \tag{10.5.54}$$

which may be interpreted as the distance traveled by radioactive tracer particles during the effective half-life of the tracer, measured in units of a_I.

10.6 Some Solved Problems

Only a few typical examples of analytic solutions are given here. The objective is to show typical boundary and initial value problems. In all cases the tracer is an ideal one, the medium is homogeneous and isotropic, and we seek $C(\mathbf{x}, t)$. In the presence of adsorption, sometimes a solution can be obtained by introducing a retardation factor (paragraph 10.5.1) that modifies the average velocity.

10.6.1 One-dimensional Flow

Case 1, progress of a concentration front in an infinite column of porous medium. Both hydrodynamic dispersion and molecular diffusion are considered; radioactive decay and adsorption are neglected. The specific discharge is either $q =$ const, or $q = q(t)$ along the column. Initially, the column is saturated by two miscible liquids at different tracer concentrations with an abrupt interface (say, at $x = 0$) between them.

The partial differential equation governing the tracer distribution here is (10.5.25) with $A \to 0$:

$$\partial C/\partial t = D' \, \partial^2 C/\partial x^2 \pm (q/n) \, \partial C/\partial x, \qquad -\infty < x < +\infty \tag{10.6.1}$$

with $D' = a_I|q|/n + D_d T^*$, and the minus sign corresponds to flow in the $+x$ direction.

Initial conditions are:

$$t \leqslant 0, \qquad -\infty < x < 0, \qquad C = C_0$$
$$0 \leqslant x < +\infty, \qquad C = C_1. \tag{10.6.2}$$

Boundary conditions are:

$$t > 0, \qquad x = \pm\infty, \qquad \partial C/\partial x = 0$$
$$x = +\infty, \qquad C = C_1$$
$$x = -\infty, \qquad C = C_0. \tag{10.6.3}$$

Bear and Todd (1960) solve this problem by applying the Laplace transform to (10.6.1) through (10.6.3). The solution is:

$$\varepsilon(x, t) \equiv \frac{C(x, t) - C_0}{C_1 - C_0} = \tfrac{1}{2} \operatorname{erfc} \left\{ - \frac{x - \int_0^t [q(t)/n] \, dt}{2 [\int_0^t (a_I|q|/n + D_d T^*) \, dt]^{1/2}} \right\}. \tag{10.6.4}$$

For the special case of negligible diffusion (i.e., $a_I|q| \gg n D_d T^*$), we set $D_d = 0$ in (10.6.4).

For a constant q in the $\mathbf{1x}$ direction, equation (10.6.4) reduces to:

$$\varepsilon(x, t) \equiv \frac{C(x, t) - C_0}{C_1 - C_0} = \tfrac{1}{2} \operatorname{erfc} \left\{ - \frac{x - qt/n}{\sqrt{4D't}} \right\}. \tag{10.6.5}$$

From (10.6.4) it follows that the point $\varepsilon = 0.5$ travels with the mean flow, while the spreading expressed by σ^2 is proportional to the total path traveled. The solution (10.6.5) is shown in figure 10.6.1.

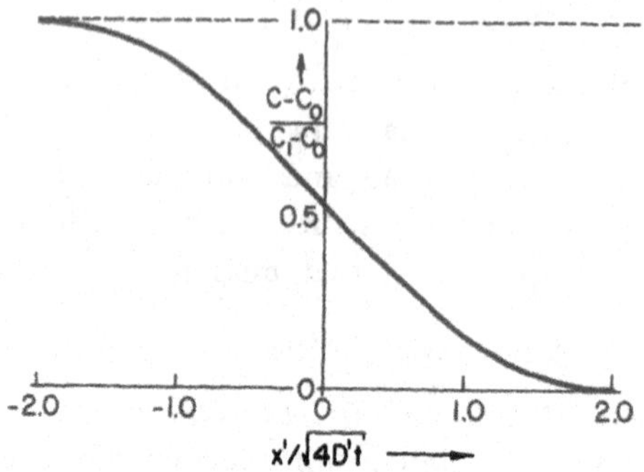

FIG. 10.6.1. Progress of a front.

In the presence of adsorption, the partial differential equation to be solved is (10.6.1) with $R_d\, \partial C/\partial T$ replacing $\partial C/\partial t$. The solution is obtained directly from (10.6.5) by replacing t with t/R_d.

Case 2, an infinite column of porous medium with steady flow q in the **1x** *direction.* At $t = 0$, a very thin slug of tracer-marked fluid is injected into the column at $x = 0$. As the slug moves downstream, the tracer's concentration is described by (10.6.1). For an observer moving with the average flow, this equation becomes (10.5.11):

$$\partial C/\partial t' = D'\, \partial^2 C/\partial x'^2 \tag{10.6.6}$$

where $t' = t$, $x' = x - (q/n)t$. This is the well known heat conduction equation.

The initial conditions here are in the form of a *Dirac delta function* $\delta(x)$:

$$C(x, 0) = (M/n)\, \delta(x) \tag{10.6.7}$$

where M is the total amount of tracer contained in the slug, and n is porosity. The Dirac distribution $\delta_m(x)$ is described by:

$$\begin{aligned} \delta_m(x) &= 1/m \quad \text{for} \quad 0 < x < m \\ \delta_m(x) &= 0 \qquad \text{elsewhere} \end{aligned} \tag{10.6.8}$$

where m is a positive small number. Then:

$$\delta(x) = \lim_{m\to 0} \delta_m(x).$$

Boundary conditions specified for $C(x', t')$ are:

$$\lim C(x', t') = 0, \quad |x'| \to \infty; \qquad \int_{-\infty}^{+\infty} C(x', t')\, dx' = M/n. \tag{10.6.9}$$

The solution of (10.6.6) for these initial and boundary conditions is (Crank 1956):

$$C(x, t) = \frac{M/n}{(4\pi D' t)^{1/2}} \exp\left[-\frac{x'^2}{4D't'}\right] = \frac{M/n}{\sqrt{2\pi}\,\sigma} \exp\left[-\frac{(x-\bar{x})^2}{\sqrt{2}\,\sigma}\right] \tag{10.6.10}$$

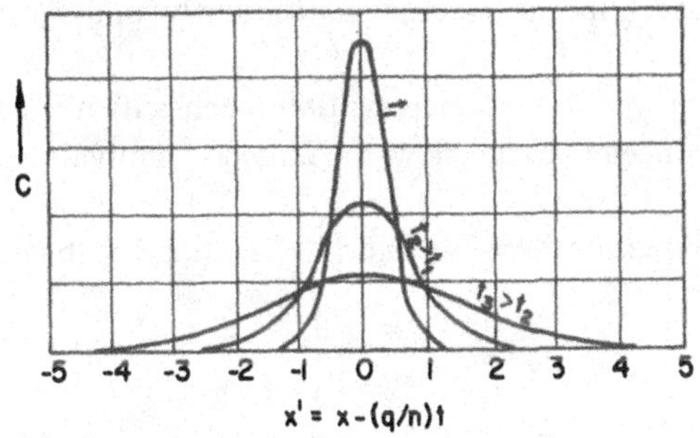

FIG. 10.6.2. Progress of a slug.

where $\bar{x} = (q/n)t$, $\sigma^2 = 2D't$. The shapes of the curves $C(x', t)$ are given in figure 10.6.2.

Case 3, an infinite column of porous medium with a constant continuous injection of a decaying tracer at $x = 0$. Along the column $q =$ const. The tracer's concentration distribution is described in this case by:

$$\partial C/\partial t = D' \, \partial^2 C/\partial x^2 - (q/n) \, \partial C/\partial x - \lambda C. \tag{10.6.11}$$

The elementary solution of (10.6.11) is:

$$C(x, t; t') = \frac{dM}{[4\pi D'(t - t')]^{1/2}} \exp\left\{- \frac{[x - q(t - t')/n]^2}{4D'(t - t')} - \lambda(t - t')\right\} \tag{10.6.12}$$

where $dM = C_0(q/n) \, dt'$ is the tracer mass injected during dt' at $x = 0$ and $t = t'$. To obtain the effect of a continuous injection, we must integrate (10.6.12). Denoting $\tau = t - t'$, we obtain the solution for a continuous injection:

$$C(x, t) = \frac{C_0 q/n}{(4\pi D')^{1/2}} \exp\left\{\frac{qx}{2D'n}\right\} \int_0^{\tau = t} \frac{1}{\sqrt{\tau}} \exp\left\{- \frac{a}{\tau} - b\tau\right\} d\tau \tag{10.6.13}$$

where $a = x^2/4D'$, $b = q^2/4D'n^2 + \lambda$. As $t \to \infty$ (10.6.13) reduces to:

$$C(x, \infty) = \frac{C_0}{\sqrt{1 + 4\lambda D'n^2/q^2}} \exp\left[\frac{qx}{2D'n} (1 - \sqrt{1 + 4\lambda D'n^2/q^2})\right]. \tag{10.6.14}$$

If in (10.6.13) we set $x = 0$, i.e., $a = 0$, we may integrate it and obtain:

$$C(0, t) = \frac{C_0}{\sqrt{1 + 4\lambda D'n^2/q^2}} \operatorname{erf}\left[\frac{q^2 t}{4D'n^2} + \lambda t\right] \tag{10.6.15}$$

i.e., $C(0, t) \neq C_0$. As $t \to \infty$, we obtain:

$$C(0, \infty) = C_0/\sqrt{1 + 4\lambda D'n^2/q^2}. \tag{10.6.16}$$

Case 4, semi-infinite column of porous medium, $x > 0$. The column is adjacent to a reservoir containing a tracer solution of constant concentration C_0. The flow in the column is maintained at a constant specific discharge q in the **1x** direction.

In addition, the tracer in the column continuously undergoes radioactive decay. Adsorption is absent.

We assume that at $x = 0$ the concentration reaches its ultimate value C_0 immediately upon commencement of flow. This is equivalent to assuming that $\lim_{t\to 0} \partial C/\partial x|_{x=0} = 0$.

The differential equation here is (10.6.11). Initial and boundary conditions are:

$$t \leqslant 0, \quad x \geqslant 0, \quad C = 0$$
$$t > 0, \quad x = 0, \quad C = C_0$$
$$t > 0, \quad x = \infty, \quad C = 0. \tag{10.6.17}$$

By applying the Laplace transform to (10.6.11) and (10.6.17), we obtain:

$$D'\, \partial^2 C^*/\partial x^2 - (q/n)\, \partial C^*/\partial x - (\lambda + p)C^* = 0, \qquad C^* = \int_0^\infty C \exp(-pt)\, dt$$

$$x = 0, \quad t > 0, \quad C^* = C_0/p. \tag{10.6.18}$$

The solution of (10.6.18) is:

$$C^*(p) = \frac{C_0}{p} \exp\left\{x\left[\frac{q}{2D'n} - \left(\frac{q^2}{4D'n^2} + \frac{\lambda + p}{D'}\right)^{1/2}\right]\right\} \tag{10.6.19}$$

which leads to the solution:

$$C(x, t) = C_0 \exp\left\{\frac{qx}{2D'n}\right\} \frac{1}{2\pi i} \int_{\tau - i\infty}^{\tau + i\infty} \frac{\exp(yt)}{y} \exp\left\{-\left(\beta^2 + \frac{y}{D'}\right)^{1/2}\right\} dy \tag{10.6.20}$$

$$\beta^2 = \frac{q^2}{4D'^2n^2} + \frac{\lambda}{D'}.$$

Integrating (10.6.20), we obtain (Grobner and Hofreiter 1949, 1950):

$$C(x, t) = \tfrac{1}{2} C_0 \exp(qx/2D'n) \left[\exp(-x\beta) \operatorname{erfc} \frac{x - \sqrt{(q/n)^2 + 4\lambda D'}\, t}{2(D't)^{1/2}} \right.$$
$$\left. + \exp(x\beta) \operatorname{erfc} \frac{x + \sqrt{(q/n)^2 + 4\lambda D'}\, t}{2(D't)^{1/2}}\right]. \tag{10.6.21}$$

When $\lambda = 0$ (10.6.21) becomes:

$$C(x, t) = \tfrac{1}{2} C_0 \left\{\operatorname{erfc} \frac{x - (q/n)t}{2(D't)^{1/2}} + \exp\left[\frac{xq}{D'n}\right] \operatorname{erfc} \frac{x + (q/n)t}{2(D't)^{1/2}}\right\} \tag{10.6.22}$$

derived also by Ogata and Banks (1961). Curves describing (10.6.22) are given in figure 10.6.3. According to them, the second term in (10.6.22) may be neglected when x/a_I is sufficiently large, a condition usually satisfied in practice at some distance from the inflow boundary (e.g., an error of 3% when $x/a_I > 500$).

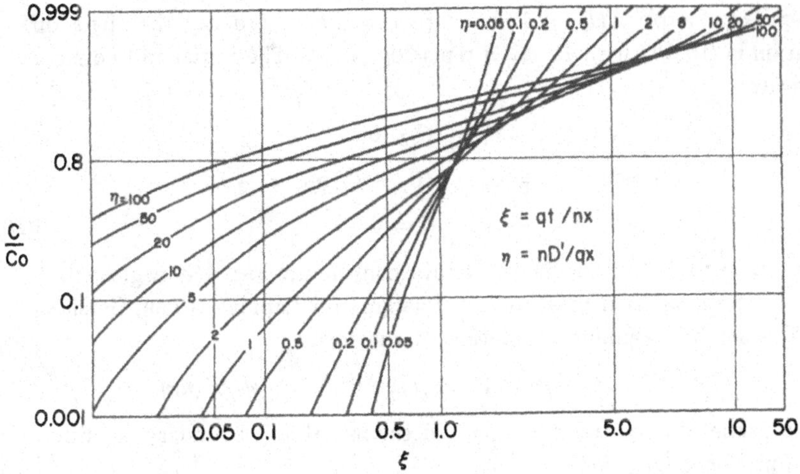

FIG. 10.6.3. Graphical representation of (10.6.22) (after Ogata and Banks, 1961).

Equation (10.6.22) also approximates (10.6.21) when molecular diffusion is neglected, and $4\lambda a_I n/q \ll 1$. Thus, the parameter $4\lambda a_I n/q$ is a criterion for the relative importance of radioactive decay of the tracer considered. It is a combination of the medium's parameters (a_I, n), the flow (q), and the tracer (λ).

When the second term of (10.6.21) may be neglected, we obtain:

$$C = \tfrac{1}{2}C_0 \exp\left\{\frac{xq}{2D'n}[1-(1+4\lambda D'n^2/q^2)^{1/2}]\right\} \cdot \operatorname{erfc}\left(\frac{x-(q/n)(1+4\lambda D'n^2/q^2)^{1/2}t}{2(D't)^{1/2}}\right). \tag{10.6.23}$$

As $t \to \infty$ (10.6.23) becomes:

$$C/C_0 = \exp[(xq/2D'n)(1-(1+4\lambda D'n^2/q^2)^{1/2})]. \tag{10.6.24}$$

This is also the steady-state solution of the dispersion equation.

The approximate form of (10.6.22) is:

$$C(x,t)/C_0 = \tfrac{1}{2}\operatorname{erfc}\{(x-\bar{x})/2(a_I\bar{x})^{1/2}\}; \qquad \bar{x} = qt/n. \tag{10.6.25}$$

If the variations of relative tracer concentration at x during an experiment in a column are recorded as a function of time, we obtain the breakthrough curve $\varepsilon = \varepsilon(t)$ shown in figure 10.4.2 where the injection volume $U = Qt$. The slope i of this curve at $x = \bar{x} = L$ where, according to (10.6.25), we have $C/C_0 = 50\%$, is given by:

$$i = (q/Ln)\sqrt{L/4\pi a_I}.$$

Hence:

$$a_I = (q/n)^2/4\pi Li^2. \tag{10.6.26}$$

This relationship can be used for the determination of the longitudinal dispersivity a_I in a column experiment.

Case 5, a semi-infinite column, but with a constant tracer mass flux at $x = 0$. In the presence of adsorption ($R_d \neq 1$), but without radioactive decay, the partial differential equation is (10.6.1) with $R_d\, \partial C/\partial t$ replacing $\partial C/\partial t$. The initial and boundary conditions are:

$$\begin{array}{lll} t \leqslant 0, & x > 0, & C = 0 \\ t > 0, & x = 0, & [C_0 - C]\, q/n = -D'\, \partial C/\partial x \\ t > 0, & x \to \infty, & C = 0 \end{array} \qquad (10.6.27)$$

where molecular diffusion in the liquid continuum $x < 0$ is neglected.

Following a solution presented by Bastian and Lapidus (1956), Gershon and Nir (1969) start by assuming a solution in the form:

$$C - C_0 = \omega(x, t) \exp\{(qx/2D'n)(1 - qt/2R_d nx)\}. \qquad (10.6.28)$$

The partial differential equation and the initial and boundary conditions for the new function $\omega(x, t)$ are:

$$R_d\, \partial\omega/\partial t = D'\, \partial^2\omega/\partial x^2$$

$$\begin{array}{lll} t \leqslant 0, & x > 0, & \omega = -C_0 \exp\{-qx/2D'n\} \\ t > 0, & x = 0, & -\partial\omega/\partial x + (q/2D'n)\omega = 0 \\ t > 0, & x \to \infty, & \omega = 0. \end{array} \qquad (10.6.29)$$

A solution of the system (10.6.29) is given by Carslaw and Jaeger (1959, sec. 14.2 III) for the case of heat conduction in a semi-infinite rod with radiation at $x = 0$. Gershon and Nir (1969) present this solution in the form:

$$\frac{C(x, t)}{C_0} = \tfrac{1}{2} \operatorname{erfc}\left(\frac{R_d x - qt/n}{2\sqrt{R_d D' t}}\right) - \tfrac{1}{2} \exp\left\{\frac{qx}{nD'}\right\} \operatorname{erfc}\left(\frac{R_d + qt/n}{2\sqrt{R_d D' t}}\right) \cdot \left(1 + \frac{R_d x + qt/n}{R_d D' n/q}\right)$$
$$+ \left(\frac{q^2 t}{\pi n^2 R_d D'}\right)^{1/2} \exp\left\{\frac{qx}{nD'} - \frac{(R_d x + qt/n)^2}{4R_d D' t}\right\}. \qquad (10.6.30)$$

They also present a steady-state solution for the same type of boundary condition, but with a radioactive tracer (otherwise steady flow cannot occur). The partial differential equation in this one-dimensional case is:

$$D' \frac{d^2C}{dx^2} - \frac{q}{n} \frac{dC}{dx} - \lambda R_d C = 0 \qquad (10.6.31)$$

whose general solution is:

$$C(x) = A_1 \exp\left\{\left(\frac{q}{2nD'} + \sqrt{\frac{q^2}{4n^2 D'^2} + \frac{R_d \lambda}{D'}}\right) x\right\}$$
$$+ A_2 \exp\left\{\left(\frac{q}{2nD'} - \sqrt{\frac{q^2}{4n^2 D'^2} + \frac{R_d \lambda}{D'}}\right) x\right\}$$

where A_1 and A_2 are integration constants to be determined by the boundary conditions. For the present case (of constant feed at $x = 0$), the solution is:

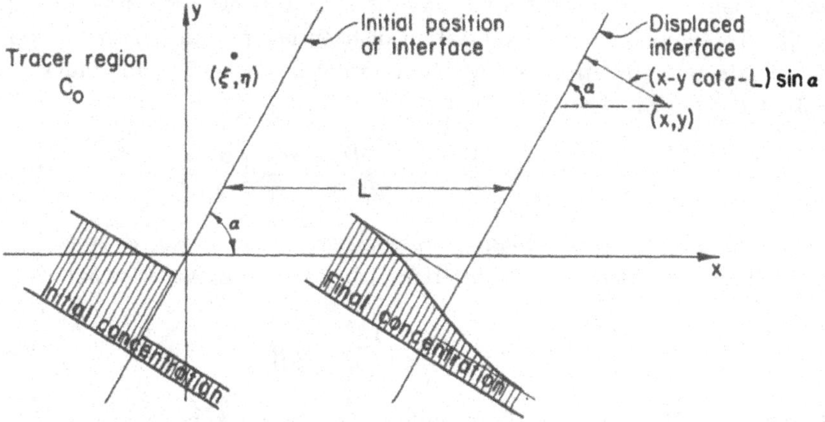

FIG. 10.6.4. Transition zone at a moving interface in uniform flow in a plane.

$$\frac{C(x)}{C_0} = \left[\frac{1}{2} + \sqrt{\frac{1}{4} + \frac{R_d \lambda D' n^2}{q^2}}\right]^{-1/2} \exp\left\{\frac{qx}{2D'n}\left(1 - \sqrt{1 + \frac{4R_d \lambda D' n^2}{q^2}}\right)\right\}. \quad (10.6.32)$$

Shamir and Harleman (1966) solve the case of dispersion in one-dimensional flow in a layered medium where the flow is normal to the layers.

10.6.2 Uniform Flow in a Plane

Case 6, a uniform flow at $q = const$ in the direction **1x** *in an infinite field.* Initially an abrupt straight line interface $y = x \tan \alpha$ (fig. 10.6.4) separates the tracer-labeled region at concentration C_0 from the region at $C = 0$.

The equation governing changes in $C(x, y, t)$ for this case is:

$$\frac{\partial C}{\partial t} = D' \frac{\partial^2 C}{\partial x^2} + D'' \frac{\partial^2 C}{\partial y^2} - \frac{q}{n}\frac{\partial C}{\partial x}; \qquad D' = a_I q/n + D_d T^*, \qquad D'' = a_{II} q/n + D_d T^* \quad (10.6.33)$$

where a_I and a_{II} are the longitudinal and the transversal dispersivities of the porous medium.

The distribution sought is obtained by integrating the elementary solution similar to (10.6.10) over the entire tracer-fluid region. If M is the mass of a point source injected at (ξ, η) at $t = 0$, the concentration distribution in the field at any later time is given by:

$$C(x, y, t) = \frac{M/n}{4\pi\sqrt{D'D''}\,t} \exp\left\{-\frac{(x - \xi - (q/n)t)^2}{4D't} - \frac{(y - \eta)^2}{4D''t}\right\} \quad (10.6.34)$$

where

$$\frac{M}{n} = \int_{-\infty}^{+\infty}\int_{-\infty}^{+\infty} C(x', y', t)\, dx'\, dy'.$$

Curves of equal tracer concentration have the form of ellipses centered at $(\xi + (q/n)t, \eta)$. To determine the concentration distribution resulting from the moving front, we integrate the effect of an infinite number of small point sources, each with $M = C_0 n \, d\xi \, d\eta$. The result is:

$$C(x, y, t) = \frac{C_0}{2} \operatorname{erfc} \left\{ \frac{[x - (q/n)t] \sin \alpha - y \cos \alpha}{[4(D' \sin^2\alpha + D'' \cos^2\alpha)t]^{1/2}} \right\} \tag{10.6.35}$$

which describes a *normal distribution perpendicular to the displaced front.*

When the motion is parallel to the interface, i.e., $\alpha = 0$, equation (10.6.35) becomes:

$$C(x, y, t) = \frac{C_0}{2} \operatorname{erfc} \left\{ \frac{-y}{(4D''t)^{1/2}} \right\}. \tag{10.6.36}$$

This corresponds to the case where uniform flow at $q = \text{const}$ is in the direction $\mathbf{1x}$, and initially $C = 0$ for $y \leqslant 0$ and $C = C_0$ for $y > 0$. It is interesting to note that $C(x, y)$ in this case is determined only by D''.

Case 7, a continuous injection at the origin into a uniform steady plane flow (in the xy plane). Much as in case 3, the solution is obtained by integrating the elementary solution of (10.6.34) over the period of injection. The effect of an instantaneous slug of mass $dM = C_0 Q \, dt$ is:

$$dC(x, y, t) = \frac{dM}{2\pi \sqrt{2D't} \sqrt{2D''t}} \exp \left\{ -\frac{(x - qt/n)^2}{4D't} - \frac{y^2}{4D''t} \right\}. \tag{10.6.37}$$

For a continuous injection, we obtain:

$$C = \frac{C_0 Q}{4\pi \sqrt{D'D''}} \int_{\theta=0}^{\theta=t} \frac{1}{(t-\theta)} \exp \left\{ -\left(\frac{x^2}{4D'} + \frac{y^2}{4D''} \right) \cdot \frac{1}{(t-\theta)} + \frac{2xq}{4D'n} - \frac{q^2}{4D'n^2} (t - \theta) \right\} d\theta. \tag{10.6.38}$$

For the steady concentration distribution, we obtain, by inserting $t = \infty$ in (10.6.38):

$$C(x, y) = \frac{C_0 Q}{2\pi \sqrt{D'D''}} \exp \left\{ \frac{qx}{2D'n} \right\} \cdot K_0 \left[\left(\frac{q^2}{4D'n^2} \left(\frac{x^2}{D'} + \frac{y^2}{D''} \right) \right)^{1/2} \right] \tag{10.6.39}$$

where K_0 is the modified Bessel function of second kind and zero order. This case can easily be extended to uniform flow in the *xyz* space.

10.6.3 Plane Radial Flow

Plane radially converging and diverging flows occur (as an approximation) in cases of artificial recharge of aquifers through wells, or injection into oil reservoirs, where the injected fluid differs from the indigenous fluid in the recharged formation, yet is miscible with it. In some cases the difference between the fluids is a natural one (e.g., difference in salinity); in others the difference is due to labeling the injected fluid with a tracer substance (radioactive tracer, dye, etc.).

Case 8, dispersion in radially symmetric diverging flow from a well, $R_d = 1$, $\lambda = 0$, *governed by* (10.5.26), *with* $\partial C/\partial\theta = 0$ *and* $A = 0$ (*for* $A \ll V$). When, in addition, $a_I V \gg D_d T^*$, the governing equation reduces to:

$$\frac{\partial C}{\partial t} = V a_I \frac{\partial^2 C}{\partial r^2} - V \frac{\partial C}{\partial r} \quad \text{or} \quad \frac{\partial C}{\partial \tau} = \frac{1}{\rho}\left(\frac{\partial^2 C}{\partial \rho^2} - \frac{\partial C}{\partial \rho}\right) \tag{10.6.40}$$

where

$$V = Q/2\pi r B n, \qquad \rho = r/a_I, \qquad \tau = Qt/2\pi B n a_I^2.$$

Because of the nonlinearity of these equations, resulting from the fact that $V = V(r)$, an exact analytical solution in a closed form is most difficult. Several approximate solutions suggested in the literature are discussed below.

For a well of radius r_w, injecting labeled fluid at a constant rate Q into a confined formation of porosity n and constant thickness B, the initial and boundary conditions are:

$$\begin{aligned} &t \leqslant 0, && r > r_w, && C = 0 \\ &t > 0, && r = r_w, && C = C_0 \\ &t > 0, && r = \infty, && \partial C/\partial r = 0. \end{aligned} \tag{10.6.41}$$

For the second equation in (10.6.40), they are:

$$\begin{aligned} &\tau \leqslant 0, && \rho > \rho_w, && C = 0 \\ &\tau > 0, && \rho = \rho_w, && C = 1 \\ &\tau > 0, && \rho = \infty, && \partial C/\partial \rho = 0 \end{aligned} \tag{10.6.42}$$

where C stands for C/C_0.

The Laplace transform of the second equation in (10.6.40) with the corresponding boundary conditions (10.6.42) is:

$$\frac{\partial^2 C^*}{\partial \rho^2} - \frac{\partial C^*}{\partial \rho} - \rho p C^* = 0; \quad C^*(\rho, p) = \int_0^\infty \exp(-p\tau)\, C(\rho, \tau)\, d\tau \tag{10.6.43}$$

$$C^*(\rho_w, p) = \int_0^\infty \exp(-p\tau)\, d\tau = \frac{1}{p}; \quad \left.\frac{\partial C^*}{\partial \rho}\right|_{(\infty, p)} = \int_0^\infty \exp(-p\tau) \frac{\partial C}{\partial \rho}(0, p) = 0. \tag{10.6.44}$$

By introducing the transformation

$$C^* = u \exp\{(z - p^{-2/3}/4)p^{-1/3}/2\},$$

equation (10.6.43) becomes:

$$u'' - uz = 0 \tag{10.6.45}$$

which is known as *Airy's equation*. A solution of (10.6.45) is given by Jahnke and

Emde (1945, p. 147). Applying this solution to (10.6.45), we obtain:

$$C^* = \exp(\rho/2)\, z^{1/2}[A_1 I_{1/3}(\tfrac{2}{3}z^{3/2}) + A_2 K_{1/3}(\tfrac{2}{3}z^{3/2})] \tag{10.6.46}$$

where A_1 and A_2 are constants and $I_{1/3}$ and $K_{1/3}$ are modified Bessel functions of first and second kinds, respectively, and of order $\frac{1}{3}$. Because of the asymptotic behavior of $I_\nu(x)$ and $K_\nu(x)$ as $x \to \infty$, only $A_1 \to 0$ will satisfy the given initial condition. Also, from the boundary condition at $\rho = \rho_w$:

$$A_2 = 1/[p \exp(\rho_w/2)\, z^{1/2} K_{1/3}(\tfrac{2}{3}z^{3/2})].$$

Hence:

$$C^* = \frac{1}{p} \exp\left\{\frac{\rho - \rho_w}{2}\right\} \cdot \left(\frac{p\rho + \frac{1}{4}}{p\rho_w + \frac{1}{4}}\right)^{1/2} \frac{K_{1/3}[\frac{2}{3}(p\rho + \frac{1}{4})^{3/2}/p]}{K_{1/3}[\frac{2}{3}(p\rho_w + \frac{1}{4})^{3/2}/p]}. \tag{10.6.47}$$

Finally, the inverse transform of (10.6.47) yields:

$$\frac{C}{C_0} = \sqrt{\frac{\rho}{\rho_w}} \exp\left\{\frac{\rho - \rho_w}{2}\right\} \frac{1}{2\pi i} \int_{\gamma - i\infty}^{\gamma + i\infty} \frac{\exp(\lambda\tau)}{\lambda} \left(\frac{\lambda + 1/4\rho}{\lambda + 1/4\rho_w}\right)^{1/2}$$

$$\cdot \frac{K_{1/3}[\frac{2}{3}\rho^{3/2}(\lambda + 1/4\rho)^{3/2}/\lambda]}{K_{1/3}[\frac{2}{3}\rho_w{}^{3/2}(\lambda + 1/4\rho_w)^{3/2}/\lambda]}\, d\lambda \tag{10.6.48}$$

where λ is the complex variable replacing p, and the path of integration is shown in figure 10.6.5.

Ogata (1958) gives the details of the integration in the complex plane. Following his work we obtain as a final solution of the radial dispersion problem:

$$\frac{C}{C_0} = 1 + \frac{2}{\pi} \exp\left\{\frac{r - r_w}{2a_I}\right\} \int_0^\infty \frac{\exp(-\nu^2 t)}{\nu} \left(\frac{\nu^2 r - G/4a_I}{\nu^2 r_w - G/4a_I}\right) M(\nu)\, d\nu \tag{10.6.49}$$

where

$$M(\nu) = \frac{J_{1/3}(\sigma) Y_{1/3}(\sigma') - Y_{1/3}(\sigma) J_{1/3}(\sigma')}{J_{1/3}{}^2(\sigma') + Y_{1/3}{}^2(\sigma')}; \qquad G = Q/2\pi Bn = Vr$$

$$\sigma = \frac{2}{3\sqrt{a_I G}} \frac{(\nu^2 r - G/4a_I)^{3/2}}{\nu^2}; \qquad \sigma' = \frac{2}{3\sqrt{a_I G}} \frac{(\nu^2 r_w - G/4a_I)^{3/2}}{\nu^2}.$$

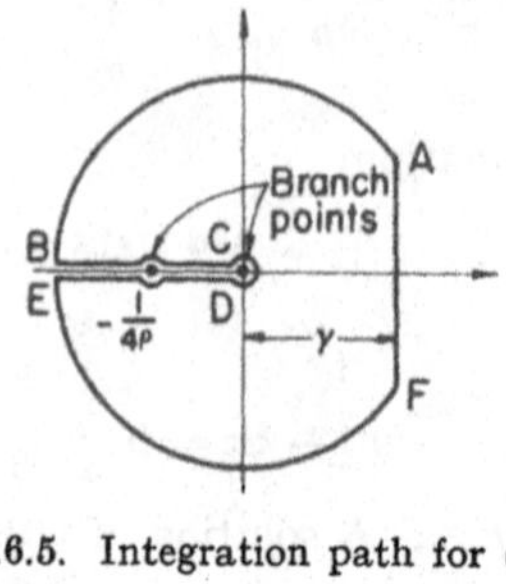

FIG. 10.6.5. Integration path for (10.6.48).

$J_{1/3}$ and $Y_{1/3}$ are Bessel functions of order $\frac{1}{3}$ of the first- and second-kinds, respectively. As a boundary condition at infinity Ogata stipulates $C(\infty, t) = 0$, rather than $\partial C/\partial t|_{x=0, t>0} = 0$.

Because of the difficulties inherent in obtaining numerically the values of the integral in (10.6.49), several authors suggest solving the radial dispersion equation (10.6.40) by numerical techniques. Ogata (1958), Hoopes and Harleman (1965, 1967), Shamir and Harleman (1966), among others, present such solutions.

Bondarev and Nikolaevskii (1962) obtain the following solution for the case $C(r, 0) = 0$ and $C(r_w, t) = C_0$:

$$\frac{C(r,t)}{C_0} = 1 - \frac{1 - \exp \rho + \rho \exp \sqrt{6\tau}}{1 - \exp \sqrt{6\tau} + \sqrt{6\tau} \exp \sqrt{6\tau}}$$

$$\tau = \frac{1}{a_L^2} \int_0^t \frac{Q(t)}{2\pi Bnr} dt, \qquad \rho = \frac{r}{a_1} \tag{10.6.50}$$

valid except during the first period of injection.

De Josselin de Jong (in Lau et al. 1959) suggests an approximate solution for dispersion in radially diverging flow from a well. His approach is based upon the assumptions that (a) the tracer is distributed in a nearly normal fashion, and that (b) the tracer distribution is a linear sum of two effects—one due to longitudinal dispersion and the other due to the divergence of streamlines. Hence, characterizing the tracer distribution by the standard deviation σ_r, he expresses the change $d\sigma_r$ as the sum:

$$d\sigma_r = d\sigma_1 + d\sigma_2 \tag{10.6.51}$$

where $d\sigma_1$ is caused by dispersion, and $d\sigma_2$ is caused by the diverging nature of the flow. From the discussion on dispersion in one-dimensional flow in paragraph 10.6.1, it follows that when molecular diffusion is neglected, σ_1 of the normal distribution is defined by:

$$\sigma_1 = \sqrt{2a_I\bar{r}} \tag{10.6.52}$$

where $\bar{r}$ is the average radius of the body of injected water. From (10.6.52) we obtain:

$$d\sigma_1/d\bar{r} = a_I/\sigma_1. \tag{10.6.53}$$

For a transition zone (across which C varies) whose width is characterized by σ_2, we obtain:

$$2\pi r\sigma_2 = \text{const}; \qquad d\sigma_2/d\bar{r} = \sigma_2/\bar{r}. \tag{10.6.54}$$

By inserting (10.6.53) and (10.6.54) into (10.6.51), and deleting the subscripts 1 and 2, as actually both $d\sigma_1$ and $d\sigma_2$ are increments of the same standard deviation σ, we obtain:

$$d\sigma/d\bar{r} = a_I/\sigma - \sigma/\bar{r}. \tag{10.6.55}$$

Solving the differential equation (10.6.55) with the boundary conditions $\sigma = 0$ at the well's radius $r = r_w$, $\sigma = \sigma_r$ at $r = \bar{r}$, yields:

$$\sigma_r^2 = \tfrac{2}{3}a_I(\bar{r}^3 - r^3_w)/\bar{r}^2; \quad \sigma_r = [\tfrac{2}{3}a_I(\bar{r}^3 - r^3_w)/\bar{r}^2]^{1/2}. \tag{10.6.56}$$

For $r_w \ll \bar{r}$, we obtain:

$$\sigma_r = [\tfrac{2}{3}a_I\bar{r}]^{1/2}. \tag{10.6.57}$$

By comparing $\sigma = \sigma_L$ for the linear case with σ_r for the radial case, we obtain:

$$\sigma_r/\sigma_L = 1/\sqrt{3} = 0.577. \tag{10.6.58}$$

With σ defined by (10.6.57), the tracer distribution according to de Josselin de Jong's approximation becomes:

$$\frac{C}{C_0} = \tfrac{1}{2}\,\text{erfc}\left\{\frac{r - \bar{r}}{\sqrt{2}\sigma}\right\} = \tfrac{1}{2}\,\text{erfc}\left\{\frac{r - \bar{r}}{\sqrt{\tfrac{4}{3}a_I\bar{r}}}\right\}. \tag{10.6.59}$$

Lau et al. (1959) present experiments that verify (10.6.59).

Raimondi et al. (1959) suggest an approximate solution based on the assumption that the influence of dispersion, expressed by the first term on the right-hand side of (10.6.40), becomes small in comparison to the local convective effect as the tracer moves away from the source. In other words, they assume that the influence of dispersion and diffusion on the tracer-concentration distribution as the tracer moves past any point becomes small as compared to the accumulated effect of dispersion and diffusion that has taken place up to that point. When the dispersion term in (10.6.40) is small, we obtain:

$$\frac{\partial C}{\partial t} \simeq -\frac{G}{r}\frac{\partial C}{\partial r}; \qquad \frac{\partial}{\partial r} \simeq -\frac{r}{G}\frac{\partial}{\partial t}; \qquad V = \frac{G}{r}. \tag{10.6.60}$$

Using (10.6.60) to express the dispersion term $a_I V\, \partial^2 C/\partial r^2$, equation (10.6.40) becomes the *approximate* dispersion equation:

$$\partial C/\partial t + (G/r)\, \partial C/\partial r = (a_I/G) r\, \partial^2 C/\partial t^2. \tag{10.6.61}$$

Raimondi et al. (1959) solve (10.6.61) for the case of a well continuously injecting a tracer at constant concentration C_0 at $r = 0$. Their solution can be written in the form:

$$\frac{C}{C_0} = \tfrac{1}{2}\,\text{erfc}\left\{\frac{r^2/2 - Gt}{\sqrt{\tfrac{4}{3}a_I\bar{r}^3}}\right\}. \tag{10.6.62}$$

The influence of molecular diffusion on the tracer distribution becomes important for r values such that $a_I q/n$ is comparable with $D_d T^*$. This influence can be accounted for using D'_{11} instead of $D_{11} = a_I V$ in (10.6.40). For this case, Raimondi et al. (1959) give the solution:

$$\frac{C}{C_0} = \tfrac{1}{2}\,\text{erfc}\left\{\frac{r^2/2 - Gt}{[\tfrac{4}{3}a_I\bar{r}^3 + (D_d T^*/G)\bar{r}^4]^{1/2}}\right\}. \tag{10.6.63}$$

Equation (10.6.63) is also obtained by Hoopes and Harleman (1965) by integrating the tracer concentration distribution resulting from an instantaneous tracer injection with respect to time.

It should be noted, however, that although (10.6.62) satisfies the boundary conditions of a constant rate of tracer supply and $C(\infty, t) = 0$ for $t > 0$, used by Raimondi et al., the initial condition $C(r, 0) = 0$ is not satisfied, as the approximate solution assumes a finite amount of tracer mass initially in the porous medium. The error introduced is large at the immediate vicinity of the injecting well. Hoopes and Harleman (1965) suggest a region of large errors within 10–20 particle diameters of the source. It should be noted that (10.6.63) also satisfies the boundary condition $C(0, t) = C_0$ for $t > 0$ with

$$\pi \bar{r}^2 nB = Qt; \qquad \bar{r}^2 = 2Gt, \qquad \bar{r} = \int_0^t (q/n)\, dt = (2Gt)^{1/2};$$

equation (10.6.62) becomes:

$$\frac{C}{C_0} = \tfrac{1}{2} \operatorname{erfc} \left\{ \frac{r^2/2 - \bar{r}^2/2}{\bar{r}\sqrt{\frac{4}{3}a_I\bar{r}}} \right\} = \tfrac{1}{2} \operatorname{erfc} \left\{ \frac{(r - \bar{r})(r + \bar{r})/2}{\bar{r}\sqrt{\frac{4}{3}a_I\bar{r}}} \right\}. \tag{10.6.64}$$

For $r + \bar{r} \simeq 2\bar{r}$, we obtain:

$$\frac{C}{C_0} = \tfrac{1}{2} \operatorname{erfc} \frac{r - \bar{r}}{\sqrt{\frac{4}{3}a_I\bar{r}}} \tag{10.6.65}$$

which is the same as (10.6.59) obtained by de Josselin de Jong.

Hoopes and Harleman (1965) carry out an extensive program of laboratory investigations in a sand box model in the form of 180° sector. On the basis of these experiments, they conclude that (10.6.63) may be considered a good approximation of dispersion in radially diverging flow. They also state that this equation approximates a numerical solution of (10.6.40) for distances larger than 20 particle diameters from the well.

Mercado and Bear (1965) made an attempt to extend de Josselin de Jong's approach to the case of radially converging flow (e.g., pumping from the well) immediately following an injection period. Their work also describes laboratory experiments and large-scale pumping and injection operations in the field.

The case of dispersion in steady flow between a well injecting tracer-labeled fluid and a nearby pumping well is analyzed in detail by Hoopes and Harleman (1965).

Using the Φ-Ψ coordinate system, the partial differential equation is (10.5.18) which, with $f = 1$ and $\cos(\mathbf{V}, \mathbf{1\Phi}') = 1$, becomes:

$$\frac{\partial C}{\partial t} = V^2 \left\{ \frac{\partial}{\partial \Phi'} \left[(Va_I + D_d T^*) \frac{\partial C}{\partial \Phi'} \right] + \frac{\partial}{\partial \Psi'} \left[(Va_{II} + D_d T^*) \frac{\partial C}{\partial \Psi'} \right] - \frac{\partial C}{\partial \Phi'} \right\} \tag{10.6.66}$$

where $\Phi' = \Phi/n$ and $\Psi' = \Psi/n$ are the velocity potential and its corresponding stream function, respectively; and Φ is the specific discharge potential.

The steady field of flow from an injection well of strength $Q/2\pi$ (per unit thickness of aquifer) at $(-d, 0)$ toward a pumping well of strength $Q/2\pi$ at $(+d, 0)$ is described by (7.8.24) and (7.8.25).

Two types of boundary conditions are possible at the injecting well: a boundary condition of a constant concentration C_0 and a boundary condition of a constant rate of tracer supply. Actually, from the discussion in paragraph 10.5.3, it follows that the appropriate boundary condition at the injection well $(r = r_w)$ is obtained from (10.5.36):

$$r = r_w, \qquad \Phi = \Phi_w = (Q/2\pi) \ln(r_w/2d)$$

$$[C_0 - C]\frac{Q}{2\pi B r_w} = -\left(nD_d T^* + f^* a_I \frac{Q}{2\pi B r_w}\right)\frac{Q}{2\pi B r_w}\frac{\partial C}{\partial \Phi}; \qquad C_0 \equiv C(r_w, t). \tag{10.6.67}$$

However, after a rather short while, we shall have the boundary condition $C = C_0$ at $r = r_w$.

For the pumping well, since C becomes equal on both sides of the boundary, we have:

$$r = r_w, \qquad \Phi = -(Q/2\pi) \ln(r_w/2d), \qquad \partial C/\partial\Phi = 0. \tag{10.6.68}$$

Hoopes and Harleman (1965, 1967) obtain solutions of (10.6.66) by assuming a product solution in the form:

$$C(\Phi', \Psi', t) = C_1(\Phi', t) C_2(\Psi', t) \tag{10.6.69}$$

where they show that $C_1(\Phi', t)$ represents the effects on the tracer concentration distribution of convection, dispersion and molecular diffusion along the streamlines, whereas $C_2(\Psi', t)$ represents the effects of (transversal) dispersion and molecular diffusion across the streamlines on the tracer concentration distribution.

Owing to mathematical difficulties inherent in solving (10.6.66) analytically, Hoopes and Harleman (1965) obtain approximate analytical solutions for four cases:

(a) no dispersion or diffusion along or transverse to streamlines; for this case (10.6.66) reduces to:

$$\partial C/\partial t + V^2\, \partial C/\partial\Phi' = 0 \tag{10.6.70}$$

(b) dispersion and diffusion along streamlines, but not across them; the equation in this case becomes:

$$\partial C/\partial t + V^2\, \partial C/\partial\Phi' = (a_I/V + D_d T^*/V^2)\, \partial^2 C/\partial t^2 \tag{10.6.71}$$

(similar to (10.6.61))

(c) tracer concentration distribution with dispersion and diffusion across streamlines; as longitudinal dispersion and diffusion are neglected, the equation to be solved is:

$$\partial C/\partial t + V^2\, \partial C/\partial\Phi' = V^2\, \partial[(a_{II} V + D_d\, T^*)\, \partial C/\partial\Psi']/\partial\Psi' \tag{10.6.72}$$

(d) combined influence of convection, dispersion and diffusion.

10.7 Heat and Mass Transfer

Up to this point, only mass transfer in a porous medium has been considered. The present section deals with a nonisothermal system in which both thermal energy (heat) and mass (of a solute) are transported in a porous medium. The theory of heat transfer in porous media is of interest because of its many applications. In ground water hydrology, it is of interest in connection with investigations of thermal springs, and more generally, in connection with ground water flow, taking into account the presence of the *geothermal gradient* (according to which the temperature increases with depth below the soil's surface at a rate of 1°C per 20–40 m, depending on the geographic location, type of rock, presence of thermal sources, volcanic activity, etc.). It is also of interest when relatively hot water from cooling installations is injected artificially into aquifers for the purpose of water conservation. Another field in which the theory of heat and mass transfer is of interest is reservoir engineering, especially in connection with thermal recovery processes. Processes involving heat and mass transfer in porous media are often encountered in the chemical industry.

Only saturated flow is treated in the present section. The problem of heat and fluid mass transfer in unsaturated flow in porous media is treated in detail by Luikov and Mikhailov (1961) and Luikov (1966). The problem of heat transfer by itself (in a fluid or a solid continuum) is assumed known to the reader (see, for example, Carslaw and Jaeger 1959).

10.7.1 Modes of Heat Transfer in a Porous Medium

In general, heat is transported in any of the following three modes: *conduction*, *convection* or *radiation*.

Mode 1, conduction. This is the mechanism of thermal energy exchange due to the exchange of kinetic energy among colliding molecules, yet without appreciable displacement of the latter. Heat or thermal energy transfer by conduction in a *fluid or a solid continuum* is described macroscopically by *Fourier's law* (1822), which expresses a proportionality between a heat flux ($\mathbf{J}_h$) and a temperature (T) gradient:

$$\mathbf{J}_h = -\lambda \operatorname{grad} T. \tag{10.7.1}$$

The coefficient of proportionality λ is the *thermal conductivity* of the substance through which the heat flux takes place. For example, for water at 32°F and 140°F, $\lambda = 0.377$ Btu/ft hr°F. For air at 32°F we have $\lambda = 0.0140$ Btu/ft hr°F. Dimensions of $\mathbf{J}_h$ are ML^0T^{-3}; its units in the c.g.s. system are cal cm^{-2} sec^{-1}, and in the English system Btu ft^{-2} hr^{-1}. Dimensions of λ are $MLT^{-3}\theta^{-1}$; its units in the c.g.s. system are cal/(cm^2 sec °C/cm) = cal/cm sec or in the English system Btu ft^{-1} hr^{-1} θ^{-1} (where 1 Btu ft^{-1} hr^{-1} °F^{-1} = 4.13×10^{-3} cal cm^{-1} sec^{-1} °C^{-1}). In a flowing fluid, $\mathbf{J}_h$ represents the flux of thermal energy (by conduction) relative to the local fluid motion. In an anisotropic medium, λ is a second-rank tensor.

In gases and liquids, thermal conductivity λ depends on temperature and on pressure. The thermal conductivity of gases at low density increases with increasing temperature, whereas the thermal conductivity of most liquids decreases with increasing temperature. Polar, or associated liquids, such as water, may exhibit a maximum in the curve $\lambda = \lambda(T)$. The thermal conductivity of gases also varies with pressure depending on whether the gas is mono- or polyatomic. In general, however, variations of λ with temperature are far more significant than its variation with pressure.

Mode 2, convection. This is the mechanism of thermal energy transport in fluids, associated with the actual mass movement of fluids from one region to another. As the fluid moves, it carries its own heat content with it. Convection imposed by external means is known as *forced convection*, while fluid motion caused by density difference due to temperature variations in the field of flow is called *free*, or *natural convection*.

Often we are interested in predicting the rate of heat transferred between a fluid and a solid bounding surface (e.g., fluid flowing either in a conduit or around a solid object). This is also the case encountered in flow through porous media where heat is transferred from the fluid to the solid and vice versa. Since the fluid adjacent to the solid surface is either stationary or in laminar motion, the mechanism of heat transfer at the solid–fluid interface is essentially one of conduction. In principle, such transfer could have been described by (10.7.1) written for this case in the form:

$$(J_h)_n = -\lambda_f\, \partial T_f/\partial n \tag{10.7.2}$$

where subscript f denotes fluid, $(J_h)_n$ is the component of the heat flux normal to the solid wall and $\partial T_f/\partial n$ is the temperature gradient in the fluid normal to the wall at the solid–fluid interface. If the details of the flow and heat transfer processes were known for any flow situation, the temperature distribution within the fluid could be determined and the rate of heat transfer through the solid–fluid interface calculated from (10.7.2). However, in many convection problems with surfaces having complicated geometries, the details of the latter, as well as the information regarding the velocity and temperature distributions, are unknown. Again, a porous medium is a typical example of such a case. In these cases the heat transfer information is largely empirical. It has been found convenient to define a *heat transfer coefficient* (or a *surface coefficient of heat transfer*):

$$h = (J_h)_n/(T_s - T_f) \tag{10.7.3}$$

where T_s is the solid surface temperature and T_f is some representative temperature of the fluid. Typical h values for liquids in free convection are 10–60 cal cm^{-2} hr^{-1} °C^{-1}.

Mode 3, radiation. This is essentially a mechanism of emission of electromagnetic waves that allows energy to be transported with the speed of light through regions of space that are devoid of any matter. When a solid body is heated, its surface emits radiation, called *thermal radiation*, primarily of a wavelength in the 0.1 to 10 micron range, the exact wavelength (or wavelengths) depending on the temperature and on the nature of the solid body.

In a porous medium whose void space is filled with a moving fluid, the three basic modes of transfer may manifest themselves in the six following ways (Lagarde 1965):

(a) heat transfer through the solid phase (considered as a continuum) by conduction;

(b) heat transfer through the fluid phase (considered as a continuum) by conduction;

(c) heat transfer through the fluid phase (considered as a continuum) by convection;

(d) heat transfer through the fluid phase by dispersion (i.e., *heat dispersion*) due to the presence of grains and an interconnected pore system (This mode of heat transfer is completely analogous to mass transfer by hydrodynamic dispersion. It is due to same reasons, namely the distribution of local velocities caused by the presence of grains and the thermal conduction of heat (which is analogous to the role played by molecular diffusion in the phenomenon of hydrodynamic dispersion). Heat dispersion tends to promote further spreading of the heat carried by the fluid.);

(e) heat transfer from the solid phase to the fluid one;

(f) heat transfer between solid grains by radiation, when the fluid is a gas.

Not all of these modes of heat transfer have equal significance in an actual case of flow through a porous medium. Their relative importance will be discussed in the following paragraphs.

10.7.2 Formulation of the Problem of Heat and Mass Transfer in a Fluid Continuum

Following the general approach to transport phenomena in porous media, we shall first present the mathematical formulation of the problem of heat and mass transfer in a fluid continuum filling the void space of the porous medium. Then, applying an averaging procedure of one kind or another to the equations comprising this formulation, the mathematical description of heat and mass transfer in a porous medium will be obtained.

For the sake of simplicity we shall assume that:

(a) the solid matrix is homogeneous, nondeformable and chemically inert with respect to the fluid;

(b) the fluid is single phase and Newtonian; its density (ρ_f) does not depend on pressure variations, but only on variations of solute concentration and temperature;

(c) the fluid is a binary system with a molecular diffusion coefficient D_d;

(d) the flow is in the laminar range;

(e) no chemical reactions take place among the fluid's species; and

(f) no heat sources or sinks exist in the fluid.

Under these conditions, the flow of heat and mass *in a fluid continuum* is described by the following set of equations (written in tensorial notation using the summation convention).

(a) The equation of mass conservation of the solute is (4.3.7) with $I_\alpha = 0$:

$$\frac{\partial \rho_\alpha}{\partial t} + \frac{\partial}{\partial x_i}(\rho_\alpha V^*_i) + \frac{\partial}{\partial x_i}(J^*_{\alpha i}) = 0; \qquad \mathbf{J}^*_\alpha = \rho_\alpha(\mathbf{V}_\alpha - \mathbf{V}^*). \qquad (10.7.4)$$

(b) The equation of mass conservation of the fluid, denoted by subscript f, is (4.3.14):

$$\frac{\partial \rho_f}{\partial t} + \frac{\partial}{\partial x_i}(\rho_f V^*_i) = 0. \tag{10.7.5}$$

(c) The conservation of linear momentum of the fluid is expressed by the Navier–Stokes equations:

$$\rho_f \frac{\partial V^*_i}{\partial t} + V^*_j \frac{\partial V^*_i}{\partial x_j} = -\frac{\partial p}{\partial x_i} + \mu \frac{\partial^2 V^*_i}{\partial x_j \partial x_j} - \rho g \frac{\partial z}{\partial x_i}. \tag{10.7.6}$$

(d) The conservation of energy of the fluid, neglecting the reversible rate of internal energy increase per unit volume by compression (p div $\mathbf{V}^*$), and introducing $P_{ij} \approx \mu(\partial V^*_i/\partial x_j + \partial V^*_j/\partial x_i)$, is expressed by:

$$\rho \frac{Du}{Dt} = -\frac{\partial}{\partial x_i}(J_{hi}) + \frac{\mu}{2}\left(\frac{\partial V^*_i}{\partial x_j} + \frac{\partial V^*_j}{\partial x_i}\right)\left(\frac{\partial V^*_i}{\partial x_j} + \frac{\partial V^*_j}{\partial x_i}\right) \tag{10.7.7}$$

where $D(\ )/Dt \equiv \partial(\ )/\partial t + V^*_i \, \partial(\ )/\partial x_i$ and μ is the internal energy per unit mass of fluid. In (10.7.7) energy transport by radiation is not taken into account.

In addition to these three conservation statements, we must also specify the constitutive assumptions for fluxes of heat and of mass.

(e) Fluxes of heat and mass as coupled processes taking into account the *Soret effect* and the *Dufour effect* (par. 4.4.3; de Groot and Mazur 1962):

$$J^*_{\alpha i} = -\rho_f D_d \, \partial\omega_\alpha/\partial x_i - L_S \, \partial T_f/\partial x_i \tag{10.7.8}$$

$$J_{hi} = -L_D \, \partial\omega_\alpha/\partial x_i - \lambda_f \, \partial T_f/\partial x_i \tag{10.7.9}$$

where $\omega_\alpha = \rho_\alpha/\rho$ and the coefficients L_S and L_D, represent the Soret and the Dufour effects.

Finally, we must add the equations of state, which under the conditions stated above take the form:

$$\rho_f = \rho_f(\rho_\alpha, T_f) = \rho_0 - \eta(T_f - T_{f0}) + \alpha(\rho_\alpha - \rho_{\alpha 0}) \tag{10.7.10}$$

where $\rho_f = \rho_0$ for $T_f = T_{f0}$ and $\rho_\alpha = \rho_{\alpha 0}$, and $u = u(\rho_\alpha, T_f) = c_f T_f$; c_f is the specific heat of the fluid at constant pressure.

Altogether we have eight equations for the eight dependent variables: ρ_f, ρ_α, p, $\mathbf{V}^*$, u, $\mathbf{J}_h$, $\mathbf{J}_\alpha$, T_f. In principle, given initial and boundary conditions, the problem can be solved.

10.7.3 Formulation of the Problem of Heat and Mass Transfer in a Porous Medium

The passage from the microscopic scale (inside the pore space) to the macroscopic one is carried out by an averaging procedure, often in conjunction with some conceptual model of the porous medium. The result of such procedure is a macroscopic mathematical description of the problem of heat and mass transfer in a porous medium in the form of a set of equations in the dependent variables ρ_α, ρ_f, μ, $\mathbf{V}^*$, T_f, T_s. Each of the fluid variables has the meaning of an average value, defined by (Dagan 1969):

$$x_{f,\text{average}} = \int_{(\Delta U_0)_v} x\, dU_v \Big/ \int_{(\Delta U_0)_v} dU_v. \tag{10.7.11}$$

A solid variable is defined by:

$$x_{s,\text{average}} = \int_{(\Delta U_0)_s} x\, dU_s \Big/ \int_{(\Delta U_0)_s} dU_s \tag{10.7.12}$$

where $(\Delta U_0)_v (= n\,\Delta U_0)$ is the volume of voids and $(\Delta U_0)_s = (1-n)\,\Delta U_0$ is the volume of solids within an REV of volume ΔU_0.

With these definitions, the fluxes $\mathbf{J}_\alpha$ and $\mathbf{J}_h$ are *per unit cross-sectional area of the fluid phase* only, and $\mathbf{J}_{hs}$ is *per unit cross-sectional area of the solid phase* only. We shall add an overscore over the symbol $\mathbf{J}$ (i.e., $\bar{\mathbf{J}}_\alpha = n\mathbf{J}_\alpha$; $\bar{\mathbf{J}}_h = n\mathbf{J}_h$; $\bar{\mathbf{J}}_{hs} = (1-n)\mathbf{J}_{hs}$) where we wish to emphasize that a flux is *with respect to a unit cross-sectional area of the porous medium.*

Using the same symbols to denote average values, we obtain the following average equations.

Equation 1, mass conservation of the solute (par. 4.6.2)

$$\frac{\partial \rho_\alpha}{\partial t} + \frac{\partial}{\partial x_i}(\rho_\alpha V^*_i) - \frac{\partial}{\partial x_i}\left(D_{ij}\frac{\partial \rho_\alpha}{\partial x_j}\right) + \frac{\partial}{\partial x_i}(J^*_{\alpha i}) = 0 \tag{10.7.13}$$

which is the equation of hydrodynamic dispersion, except that we have not expressed $\mathbf{J}^*_\alpha$ in terms of grad ρ_α. In (10.7.13) D_{ij} is the coefficient of mechanical hydrodynamic dispersion. The passage from (4.6.28) to (10.7.13) is based on the relationship:

$$\rho_\alpha \mathbf{V} = \rho_\alpha \mathbf{V}_\alpha + D_d T^* \nabla \rho_\alpha = \mathbf{J}^*_\alpha + \rho_\alpha \mathbf{V}^* + D_d T^* \nabla \rho_\alpha. \tag{10.7.14}$$

One should note that $\mathbf{J}^*_\alpha$ here is the flux per unit area of the fluid phase only (i.e., equivalent to average velocity and not to specific discharge).

Equation 2, mass conservation of the fluid (par. 4.6.3)

$$\frac{\partial \rho_f}{\partial t} + \frac{\partial}{\partial x_i}(\rho_f V^*_i) = 0 \tag{10.7.15}$$

where we have assumed that the mass flux carried by the average flow is much larger than that resulting from velocity fluctuations.

Equation 3, conservation of linear momentum of the fluid (equation of motion; sec. 4.7)

$$V^*_i + \frac{B\rho_f}{\mu}\frac{\partial V^*_i}{\partial t} = -\frac{k_{ij}}{n\mu}\left(\frac{\partial p}{\partial x_j} + \rho_f g \frac{\partial z}{\partial x_j}\right) \tag{10.7.16}$$

where k_{ij} is the medium's permeability tensor. As an approximation, we may replace in (10.7.16) $\mathbf{V}^*$ by $\mathbf{V}' = \mathbf{q}/n\ (\equiv \bar{\mathbf{q}}/n)$, where n is porosity.

Equation 4, conservation of energy of the fluid. By introducing $u = c_f T_f$ into

(10.7.7) and averaging the latter, we obtain:

$$\rho_f c_f \left(\frac{\partial T_f}{\partial t} + V^*_i \frac{\partial T_f}{\partial x_i} \right) = - \frac{\partial J_{hi}}{\partial x_i} - \frac{\partial}{\partial x_i} \left(E_{ij} \frac{\partial T_f}{\partial x_j} \right) + h_f(T_s - T_f) + \varepsilon \qquad (10.7.17)$$

where c_f is assumed constant; $- E_{ij}\, \partial T_f / \partial x_j$ ($\overline{T\mathring{V}^*_i}$ in analogy to $\overline{\rho_\alpha \mathring{V}^*_i} = - D_{ij} \partial \rho / \partial x_j$, both D_{ij} and E_{ij} are defined in terms of $\mathbf{V}^*$) expresses the *dispersive heat flux* (resulting from fluctuations in $\mathbf{V}^*$ and T); T_s is the solid's temperature, h is the heat transfer coefficient (discussed in detail by Green **1963**) and ε is the rate of energy dissipation (per unit volume of fluid) by viscous stresses.

Equations 5 and 6, conservation of energy of the solid matrix. With the internal energy of the solid u_s (per unit bulk volume of porous medium) defined by $u_s = c_s T_s$, we obtain the average heat conservation in the porous matrix in the form:

$$\rho_s c_s\, \partial T_s / \partial t = - \partial (J_{hsi}) / \partial x_i + h_s(T_f - T_s) \qquad (10.7.18)$$

$$J_{hsi} = - (\lambda_s)_{ij}\, \partial T_s / \partial x_j \qquad (10.7.19)$$

where ρ_s is the density of the solid, c_s is the heat capacity of the solid (assumed constant) and $\mathbf{J}_{hs}$ is the conductive heat flux in the solid (per unit area of solid); $\underline{\boldsymbol{\lambda}}_s$ is the *coefficient of thermal conductivity of the solid porous matrix* (a second-rank symmetrical tensor in an anisotropic medium).

Equations 7 and 8, the fluxes of heat and mass. By averaging (10.7.8) and (10.7.9) we obtain:

$$J_{\alpha i} = - \rho_f (D^*_d)_{ij}\, \partial \omega_\alpha / \partial x_j - (L_S)_{ij}\, \partial T_f / \partial x_j \qquad (10.7.20)$$

$$J_{hi} = - (L_D)_{ij}\, \partial \omega_\alpha / \partial x_j - (\lambda_f)_{ij}\, \partial T_f / \partial x_j \qquad (10.7.21)$$

where D^*_d is the coefficient of molecular diffusivity in a porous medium, λ_{ij} is the coefficient of thermal conductivity of the fluid saturating the void space of the porous medium domain, L_S and L_D are the averaged coefficients of the Soret and the Dufour effects, respectively. All four of these coefficients are microscopic transfer coefficients in a stationary fluid phase saturating the porous medium.

Equation 9, equation of state of the fluid's density

$$\rho_f = \rho_{0T} [1 - \eta(T_f - T_{f0})] + \alpha' (\rho_\alpha - \rho_{\alpha 0}). \qquad (10.7.22)$$

In all of these equations it is assumed that viscosity is a constant; otherwise we should add another equation of state $\mu = \mu(T_f, \rho_\alpha)$.

Altogether we have nine dependent variables: $\rho_f, \rho_\alpha, \mathbf{V}^*, p, T_f, T_s, \mathbf{J}_\alpha, \mathbf{J}_h, \mathbf{J}_{hs}$, for the determination of which we have nine equations.

In the case of a homogeneous fluid $\rho =$ const and $\rho_\alpha = 0$, we have only six equations.

In many cases of heat transfer in porous media, especially at low Reynolds number flows, we may neglect the difference in temperature between the fluid and the solid, i.e., $T_f = T_s \equiv T$. This corresponds to $h = \infty$. Houpert, Debouvrier and Ifly (**1965**) present a rough estimate of the time required for T_f to approach T_s (within

10%) in a porous medium saturated by water. For example, they estimate that for glass spheres of 1 mm diameter, this time equals approximately 1.3 seconds. For a fractured rock where the solid blocks have a thickness of 100 mm, this time is approximately 2 hours. Under such conditions, by introducing J_{hsi} and J_{hi} from (10.7.19) and (10.7.21), and noting that $nh_f = (1-n)h_s$, equations (10.7.17) and (10.7.18) reduce to:

$$[n\rho_f c_f + (1-n)\rho_s c_s]\frac{\partial T}{\partial t} + \rho_f c_f n V^*{}_i \frac{\partial T}{\partial x_i}$$

$$= \frac{\partial}{\partial x_i}\left\{[n(\lambda_f)_{ij} + (1-n)(\lambda_s)_{ij}]\frac{\partial T}{\partial x_j} + (L_D)_{ij}\frac{\partial \omega_\alpha}{\partial x_j}\right\} - \frac{\partial}{\partial x_i}\left(nE_{ij}\frac{\partial T}{\partial x_j}\right) + n\varepsilon. \quad (10.7.23)$$

For an isotropic medium $(\lambda_f)_{ij}$ and $(\lambda_s)_{ij}$ reduce to the scalars λ_f and λ_s. For a homogeneous fluid, the rate of energy dissipation ε appearing in (10.7.17) may be estimated from Darcy's law, following paragraph 4.4.3:

$$\varepsilon = \rho V_i^* \frac{\partial}{\partial x_i}\left(\frac{p}{\rho_f} + gz\right). \quad (10.7.24)$$

10.7.4 Comments on Some Heat and Mass Transfer Coefficients

A few comments on the various (macroscopic) heat and mass transfer coefficients appearing in the equations of paragraph 10.7.3 seem appropriate at this point.

Case 1, specific heat of solid (c_s) *and fluid* (c_f). The following table gives some typical values of specific heat (heat capacity) of some solids and fluids at constant pressure.

Table 10.7.1

Some Typical Values of Specific Heat

Material	T(°C)	C(kcal/kg °C)
Gases		
Air	50 (20 atm)	0.2480
Water vapor	100	0.4820
Oxygen	15	0.2178
Methane	15	0.5284
Metals		
Aluminum	20	0.214
Iron (cast)	20–100	0.1189
Copper	15–100	0.0931
Inorganic Compounds		
Calcium carbonate	0	0.203
Calcium sulfate	36	0.265

Table 10.7.1 (continued)
Some Typical Values of Specific Heat

Material	T(°C)	C(kcal/kg °C)
Liquid organic compounds		
Benzene (C_6H_6)	20	0.406
Ethyl alcohol	25	0.581
Dry rocks		
Basalt	20–100	0.20
Chalk	20–100	0.214
Clay	20–100	0.22
Granite	12–100	0.192
Quartz	12–100	0.188

From (10.7.23) we may derive an expression for the heat capacity of the saturated porous medium (effective heat capacity) c_e:

$$c_e = n\rho_f c_f + (1 - n)\rho_s c_s. \tag{10.7.25}$$

Somerton (1958) gives examples of c_e for rocks saturated with water at 32°C. For example, granular sandstone (small grains) made of 40% quartz with Elite clay, $c_e = 163.3$ cal/m³ °C.

In the case of two- or three-phase flows in a porous medium:

$$c_e = (1 - n)\rho_s c_s + S_w n\rho_w c_w + S_{nw} n\rho_{nw} c_{nw}. \tag{10.7.26}$$

Case 2, thermal conductivity (Lagarde 1965). From (10.7.23) we may derive an equivalent (or effective) heat capacity of a saturated porous medium λ_e:

$$\lambda_e = n\lambda_f + (1 - n)\lambda_s \tag{10.7.27}$$

where, for the sake of simplicity, we have considered an isotropic medium. In obtaining (10.7.23) by adding (10.7.17) and (10.7.18) we have implicitly assumed a simple *parallel conduction model*, i.e., a model in which conduction through the fluid phase and conduction through the solid phase occur separately but simultaneously, with no interchange of heat between the two media. Obviously, the actual situation is much more complicated as heat is interchanged continuously between the two phases. Another very simple model of conduction through the solid–fluid system is the *series conduction model*, which leads to an equivalent conductivity λ_e expressed by:

$$1/\lambda_e = (1 - n)/\lambda_s + n/\lambda_f. \tag{10.7.28}$$

Willye and Southwick (1954) propose:

$$\lambda_e = \varepsilon[\varepsilon_1\lambda_f + (1 - \varepsilon_1)\lambda_s] + (1 - \varepsilon)\frac{1}{\varepsilon_2/\lambda_f + (1 - \varepsilon_2)/\lambda_s} \tag{10.7.29}$$

where $n = \varepsilon\varepsilon_1 + (1 - \varepsilon)\varepsilon_2$, ε indicates the fraction of the total volume in which

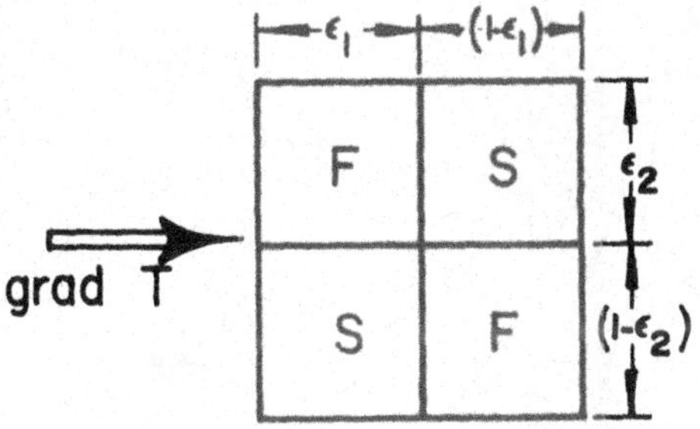

FIG. 10.7.1. Composite model of thermal conduction.

the conduction is essentially according to the parallel model, ε_1 is a parameter characterizing the absence of solid–solid contact, and ε_2 is a parameter characterizing the zone where both solid–solid and fluid–solid contacts exist.

Figure 10.7.1 shows another possible combination. Cardwell and Parsons (1945), in connection with averaging permeability of a porous medium, show that for the case illustrated by figure 10.7.1 with $\varepsilon_1 = \varepsilon_2 = \frac{1}{2}$, the average permeability will always lie between the arithmetic and harmonic mean permeabilities. Obviously this conclusion is valid for other types of conductivity (thermal, electric, etc.).

In addition to the difficulty of constructing a model simulating the passage of heat in the solid–fluid system comprising the porous medium, we have the difficulty of determining the value of λ_s itself, especially in natural media such as soils and rocks, or artificial ones, such as paper.

Other formulas for λ_e are suggested by Woodside and Messmer (1961) and by Asaad (1955).

The problem of deriving λ_e from a theoretical model becomes even more complicated in the case of two- (e.g., oil–water) or three-phase (e.g., gas–oil–water) flows in a porous medium.

Table 10.7.2

Material	Thermal Conductivity (cal/m sec °C)	Material	Thermal Conductivity (cal/m sec °C)
Silver	100	Water	0.11
Aluminum	50	NaCl (25%) solution	0.0138
Carbon	1.25	Benzene	0.038
Quartz	2	Petroleum	0.035
NaCl	0.8–1.5	Glass	0.12–0.26
Sandstone	0.9	Hydrogen	0.040
Limestone	0.5	Oxygen	0.006
Dolomite	0.4–1	Air	0.006
Clay	0.2–0.3	Methane	0.007

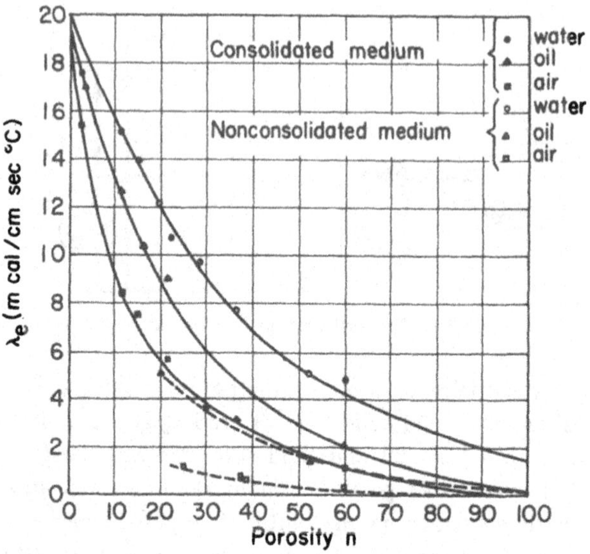

FIG. 10.7.2. Variation of λ_e with fluid and matrix properties (after Lagarde, 1965).

Some typical values of λ_s and λ_f (at ordinary temperatures) are given in the following table (Lagarde 1965).

Also, we may mention, as an example, a soil saturated by water at 32°C (Somerton, 1958). In granular, nonconsolidated sandstone, the clay mineral is Elite; 40% quartz; $n = 40\%$; $\lambda_e = 0.435$.

From the discussion above it becomes evident that the controlled experiment is the only means of obtaining λ_e.

Lagarde (1965) presents a summary of several experimental investigations for the determination of λ_e in connection with thermal recovery methods in oil reservoirs. In these investigations (e.g., Kunii and Smith 1960, 1961; Adivaraham, Kunii and Smith 1962), the effects of porosity, fluid and solid thermal conductivities, interstitial pressure, confinement pressure, etc., on λ_e are investigated. Figure 10.7.2b gives typical results of such experiments.

Case 3, the coefficient of thermal dispersion (E_{ij}). The *coefficient of thermal dispersion* E_{ij}, resulting from velocity and temperature fluctuations in the pore space, introduced in (10.7.17), is similar to the coefficient of mechanical hydrodynamic dispersion D_{ij}. Another analogy is between the effective molecular diffusivity defined by:

$$(D_h)_{ij} = (\lambda_e)_{ij}/[(1 - n)\rho_s c_s + n\rho_f c_f] \qquad (10.7.30)$$

and the molecular diffusivity in the porous medium $nT^*_{ij}D_d$. Both have the dimensions L^2T^{-1}.

As in hydrodynamic dispersion, it is possible to combine $(D_h)_{ij}$ and nE_{ij} into a single coefficient $nE'_{ij} = (D_h)_{ij} + nE_{ij}$ which is analogous to the coefficient of hydrodynamic dispersion $nD'_{ij} = n(D_{ij} + T^*_{ij}D_{\alpha\beta})$.

In the absence of fluid motion, E_{ij} (like D_{ij}) vanishes, and nE'_{ij} reduces to $(\lambda_e)_{ij}$.

However, two basic differences exist between the two dispersive phenomena. The coefficient of molecular diffusivity in dilute aqueous solutions in a porous medium is approximately equal to 5×10^{-6} cm²/sec. From figure 10.4.1 we have for $Pe = 1000$ (equivalent to $Re \approx 1$), $D_l/D_d = 1000$. On the other hand, the thermal diffusivity $D_h \approx 5 \times 10^{-4}$ cm²/sec, i.e., some 100 times larger than $T^*D_{\alpha\beta}$. This means that the effect of E_{ij} can be neglected with respect to that of D_h in a much wider range of velocities than in the case of molecular diffusion and hydrodynamic dispersion.

The second difference is that unlike mass transfer, heat is also transferred through the solid phase. This introduces a basic difference between E_{ij} and D_{ij}. As explained in the discussion of tortuosity in section 4.8, the concept of tortuosity is not applicable unless the transfer takes place only through the void space. Green (1963) concludes on the basis of available experimental data, that up to $Pe = 10^4$, E_{ij} and D_{ij} are practically identical, the influence of the passage of heat through the solid phase being insignificant. Moreover, the coefficient of longitudinal thermal dispersion is negligible as compared with the coefficient of thermal diffusivity at $Pe < 3000$, which corresponds approximately to the Darcian range $Re < 3$. In this range one may therefore assume $E_{ij} \approx 0$.

10.7.5 Simplifying the Macroscopic Heat and Mass Transfer Equations

Two possible simplifications of the set of equations describing heat and mass transfer have been mentioned in the previous paragraph. They are: $h = \infty$, i.e., $T_s = T_f (\equiv T)$, and $E_{ij} = 0$, i.e., dispersive heat flux much smaller than the conductive one. A few additional simplifications may be added (Dagan 1969).

(a) In most cases, the cross-coefficients $\underline{\mathbf{L}}_S$ and $\underline{\mathbf{L}}_D$, representing the Dufour and the Soret effects, are negligible in comparison to the direct coefficients, the ratio between the two being of the order of 10^{-3}. We may therefore assume $L_S \approx 0$ and $L_D \approx 0$ in the flux equations (10.7.20) and (10.7.21).

(b) Equation (10.7.15) may be rewritten as:

$$(1/\rho_f)\, \partial\rho_f/\partial t + (1/\rho_f) V^*_i\, \partial\rho_f/\partial x_i + \partial V^*_i/\partial x_i = 0. \tag{10.7.31}$$

In this equation the first two terms on the left-hand side are of *order of magnitude* $O(V_0\, \Delta\rho/\rho_0 L)$, where ρ_0, $\Delta\rho$, V_0 and L are characteristic fluid density, density difference, (average) velocity and length of flow domain, respectively. For cases where $\Delta\rho/\rho_0 \ll 1$, we may neglect these two terms with respect to the third one. This leads to the approximation:

$$\partial V^*_i/\partial x_i = 0. \tag{10.7.32}$$

(c) In (10.7.16), written for $k_{ij} = n\bar{B}\bar{T}_{ij}$ in the form:

$$\mu V^*_i/\bar{B} + \rho_f\, \partial V^*_i/\partial t = -\, \bar{T}_{ij}(\partial p/\partial x_j + \rho_f g\, \partial z/\partial x_j) \tag{10.7.33}$$

the inertial term $\rho_f\, \partial V^*_i/\partial t$ is of order $(\rho_0 V_0^2/L)$ while the frictional term $\mu V^*_i/\bar{B}$ is of order $O(\mu\, V_0/\bar{B})$. Since $\bar{B}$ is of order $O(d^2)$, where d is a characteristic pore (or grain) size, we have $\mu V^*_i/\bar{B} = O(\mu V_0/d^2)$. Hence, the order of magnitude of the ratio of the two terms is:

$$O\left(\frac{\rho_f\,\partial V^*_i/\partial t}{\mu V^*_i/B}\right)=O\left(\frac{\rho V_0^2/L}{\mu V_0/d^2}\right)=O\left(\frac{\rho V_0 d}{L}\frac{d}{L}\right)\equiv O\left(Re\,\frac{d}{L}\right)$$

where Re is the Reynolds number. Hence, for flow at low Re, and since $d/L \ll 1$, we may neglect the inertial term with respect to the frictional one in (10.7.16). The latter then reduces to:

$$V^*_i = -\frac{k_{ij}}{n\mu}\left(\frac{\partial p}{\partial x_j} + \rho_f g\,\frac{\partial z}{\partial x_j}\right). \tag{10.7.34}$$

(d) The dissipation term ε in (10.7.17) is by (10.7.24) of the order $O(J\rho_0{}^2V_0{}^2g^2/\mu)$, where J is the thermal equivalent of mechanical energy. All other terms in (10.7.17) are of the order $O(\rho_0 c_f\,\Delta T\,V_0/L)$. Hence the ratio of the dissipation terms to the other terms in (10.7.17) is of the order $O(J\rho_0 V_0 gL/\mu c_f\,\Delta T)$. In most cases of interest this term is exceedingly small and may be neglected with respect to the terms expressing changes in thermal energy due to heat transfer.

Summarizing the remarks above, we may use the following set of equations as a sufficiently approximate mathematical statement of the problem of heat and mass transfer in a porous medium:

$$\frac{\partial V^*_i}{\partial x_i} = 0 \tag{10.7.35}$$

$$V^*_i = -\frac{k_{ij}}{n\mu}\left(\frac{\partial p}{\partial x_j} + \rho_f g\,\frac{\partial z}{\partial x_j}\right) \tag{10.7.36}$$

$$\frac{\partial \rho_\alpha}{\partial t} = \frac{\partial}{\partial x_i}\left[(D_{ij} + D_d T^*_{ij})\,\frac{\partial \rho_\alpha}{\partial x_j}\right] - V'_i\,\frac{\partial \rho_\alpha}{\partial x_i} \tag{10.7.37}$$

$$\frac{\partial T}{\partial t} = \frac{(\lambda_e)_{ij}}{n\rho_f c_f + (1-n)\rho_s c_s}\,\frac{\partial^2 T}{\partial x_i x_j} - \frac{n\rho_f c_f}{n\rho_f c_f + (1-n)\rho_s c_s}\,V^*_i\,\frac{\partial T}{\partial x_i} \tag{10.7.38}$$

$$\rho_f = \rho_{0T}[1 - \eta(T - T_{f0})] + \alpha'(\rho_\alpha - \rho_{\alpha 0}) \tag{10.7.39}$$

$$\mu = \mu_{0T}\exp[-\,\delta(T - T_{f0})] + X\,\Delta\rho_\alpha. \tag{10.7.40}$$

We may add to these equations the approximation $\mathbf{V}^* \simeq \mathbf{V}' \equiv \mathbf{V}$, so that altogether we have six dependent variables: $\mathbf{V}$, ρ_f, μ, ρ_α, p, T, to be derived from the six equations (10.7.35) through (10.7.40). Needless to say, appropriate initial and boundary conditions must be added to complete the mathematical statement of the heat and mass transfer problem. As in the discussion of boundary conditions in section 7.1, here the first kind of boundary condition consists of a specification of the temperature distribution at the boundary of the considered domain, while the second kind of boundary condition involves a specification of the heat flux distribution over the boundary. The third kind of boundary condition is also of interest, it describes the amount of heat transferred in a unit time from a unit area of the investigated domain's boundary to the surrounding medium where the temperature is maintained constant, say, T_0:

$$(J_h)_n = h(T - T_0) = -\lambda\, \partial T/\partial n; \qquad h(T - T_0) + \lambda\, \partial T/\partial n = 0 \quad (10.7.41)$$

where h is the heat transfer coefficient (cal/cm² sec °C), and **1n** indicates the direction of the normal to the boundary. Other boundary conditions (e.g., between two domains of different properties) are possible.

As in the mass transfer problem discussed at the end of paragraph 10.5.4, equations (10.7.35) through (10.7.40) must be solved simultaneously. This is a most difficult undertaking, generally requiring the use of digital computers. Better known is the problem of heat transfer alone, i.e., when $\rho_\alpha =$ constant.

As an example, consider the problem of the rate of vertical ground water movement resulting from the earth's thermal profile (Bredehoeft and Papadopulos 1965).

The one-dimensional partial differential equation of heat transfer alone in a fully saturated homogeneous isotropic porous medium is obtained from (10.7.38):

$$c\rho(\partial T/\partial t) = \lambda_e(\partial^2 T/\partial z^2) - nc_f\rho_f V_z(\partial T/\partial z); \qquad 0 < z < L \qquad (10.7.42)$$

where $T = T(x, y, z, t)$, $\operatorname{div} \mathbf{V} = 0$, ρ_0, $c_f =$ density and specific heat of fluid, respectively, $c\rho \equiv n\rho_f c_f + (1 - n)\rho_s c_s$, and $\lambda_e =$ thermal conductivity of solid–fluid complex. If the flow is also steady, we obtain:

$$\partial^2 T/\partial z^2 - (nc_f\rho_f V_z/\lambda_e)\, \partial T/\partial z = 0; \qquad 0 < z < L. \qquad (10.7.43)$$

For:

$$z = -L, \qquad T = T_L; \qquad z = 0, \qquad T = T_0$$

the solution of (10.7.43) is:

$$\frac{T(z) - T_0}{T_L - T_0} = \frac{\exp(-\alpha z/L) - 1}{\exp \alpha - 1}; \qquad \alpha = \frac{nc_f\rho_f V_z L}{\lambda_e}. \qquad (10.7.44)$$

By plotting observed data on type-curves $(T(z) - T_0)/(T_1 - T_0) = f(\alpha, z/L)$, Bredehoeft and Papadopulos (1965) obtain an estimate of α, and from it an estimate of V_z.

10.7.6 Convective Currents and Instability

When in a stationary fluid continuum the density in some layer is greater than in an underlying one, a state of instability arises, as even a small disturbance may result in a completely changed regime produced by convective currents (i.e., currents that result from the tendency of the less dense fluid to rise). The difference in density may be either intrinsic (as when water condenses on the upper surface of an oil layer), or attributable to variations in salinity or temperature. In the latter cases, molecular diffusion and thermal conduction counteract the tendency for convective currents to arise. However, when some cause exists that tends to maintain the inequality in density, convective currents will appear in spite of the counteracting effect of diffusion and conduction. This is the case, for example, when a horizontal fluid layer has its upper surface kept at a constant temperature, while the lower boundary is kept at a constant higher temperature by heat supplied to it. The question of stability under these circumstances was studied, probably for the first time, by Lord Rayleigh (1916) in relation to experiments made by Bénard (1900). Among other authors mentioned in the literature on convective currents in bulk fluid masses are Jeffreys (1926, 1928) and Low (1929, 1930). From these and other works, it seems that instability will occur at some critical value of the expression $gH^3 \Delta\rho/\rho D_d \nu$, where H denotes the thickness of the fluid layer, $\Delta\rho/\rho$ denotes the fractional excess of density in the fluid at the top as compared with that of the fluid at the bottom surface, D_d is the molecular diffusivity and ν is the kinematic viscosity. When $\rho = \rho(T)$, i.e., in a thermally induced instability, a similar parameter, called the *Rayleigh number*, can also be expressed as $Ra = \eta \, \Delta T g L^3/D_h \nu_0$ where $\Delta T = T_1 - T_0$, η is the coefficient of linear thermal volume expansion, L is a representative length, D_h is thermal diffusivity of the liquid at T_0, and ν_0 is the kinematic viscosity at T_0. The critical value depends on the type of boundary conditions at the top and the bottom. When convection does occur, it will establish itself in cylindrical (columnar) cells that have the shape of irregular hexagons or pentagons whose axis is vertical. Pellow and Southwell (1940) review several works on this subject and examine the stability of a viscous liquid in a steady regime in which the temperature decreases uniformly between a lower horizontal heated surface and an upper horizontal cooled surface.

The same phenomenon of *convective currents*, arising from variations of density within a flow domain, may also occur in a fluid saturating a porous medium.

In nonisothermal flow of a homogeneous fluid, $\rho = \rho(T)$. In isothermal flow of an inhomogeneous fluid, $\rho = \rho(C)$. In the latter case, the convective currents (gravitational convection) are due to variations in solute concentration. Sometimes $\rho = \rho(T, C)$.

The equation of motion (10.7.34) may serve as a starting point for the discussion on convective currents. It may be rewritten in the form:

$$V^*_i = -\frac{k_{ij}\rho_0 g}{n\mu}\frac{\partial}{\partial x_j}\left(z + \frac{p}{\rho_0 g}\right) - \frac{k_{ij}g(\rho_f - \rho_0)}{n\mu}\frac{\partial z}{\partial x_j} \qquad (10.7.45)$$

where ρ_0 is some reference density. We may then interpret the motion as caused by two driving forces: one resulting from piezometric head differences, where the head $(z + p/\rho_0 g)$ is referred to a fictitious homogeneous fluid of density ρ_0, and the other resulting from a *buoyancy force*, directed vertically upward, acting on a fluid particle of density ρ_f imbedded in a fluid of density ρ_0. The orders of magnitude of the two terms on the right-hand side of (10.7.45) are $O(k\rho_0 g/n\mu)(\Delta\varphi_0/L)$ and $O(kg \, \Delta\rho/n\mu)$, respectively, where $\Delta\varphi_0$ and $\Delta\rho$ are characteristic head and density differences. The ratio between the two orders of magnitude is $O(\Delta\rho/\rho_0)/(\Delta\varphi_0/L)$. We may now distinguish between two cases, according to the relative magnitude of $\Delta\rho/\rho_0$ with respect to $\Delta\varphi_0/L$. When $\Delta\varphi_0/L$ is so small that $(\Delta\rho/\rho_0)/(\Delta\varphi/L) \gg 1$, the flow is determined mainly by the buoyancy force and the flow regime is one of *free convection*. When $(\Delta\rho/\rho_0)/(\Delta\varphi/L) \ll 1$, the flow is governed mainly by the external head gradients and the flow regime is one of *forced convection*. One should note that actually the criterion should be based on a comparison between the *vertical component of the gradient* and the vertically directed buoyancy force. A free convection flow regime may still exist when the horizontal components of the external driving force are large.

Obviously, this division into two different flow regimes is not a clear-cut one, and the definition of a flow regime as belonging to either of the two regimes is rather arbitrary, especially at intermediate values of $(\Delta\rho/\rho_0)/(\Delta\varphi_0/L)$. Moreover, we have used the characteristic quotients $\Delta\rho/\rho_0$ and $\Delta\varphi_0/L$ to characterize the entire flow domain. We could have based the discussion on local values of grad φ_0 and $\Delta\rho/\rho_0$, as the flow regime may be different in different portions of the flow domain depending on the local values of grad φ.

When $\rho = \rho(T)$ only, we may expand $\rho = \rho(T)$ in a Taylor series about some reference temperature T_0:

$$\rho = \rho|_{T=T_0} + \left.\frac{\partial\rho}{\partial T}\right|_{T=T_0} (T - T_0) + \cdots = \rho_0 - \rho_0\eta(T - T_0) + \cdots \tag{10.7.46}$$

where ρ_0 is the density at T_0 and η is the coefficient of volume expansion also evaluated at T_0. By comparing (10.7.46) with (10.7.39) we may identify: $\rho_{0T} = \rho_0$ and $T_{0f} = T_0$. By substituting the first two terms of (10.7.46) in the motion equation (10.7.36), we obtain:

$$V^*_i = -\frac{k_{ij}}{n\mu}\frac{\partial p}{\partial x_j} - \frac{k_{ij}g\rho_0[1 - \eta(T - T_0)]}{n\mu}\frac{\partial z}{\partial x_j}. \tag{10.7.47}$$

If there are no external forces (i.e., free convection), or if the conditions are such that the pressure distribution in the flow domain is hydrostatic (or approximately so), i.e., $\partial p/\partial x_j = -\rho_0 g\, \partial z/\partial x_j$, we obtain from (10.7.47):

$$V^*_i = -\frac{k_{ij}g\rho_0\eta(T - T_0)}{n\mu}\frac{\partial z}{\partial x_j}. \tag{10.7.48}$$

For an isotropic medium we replace k_{ij} by k. In vertical flow $V^*_i \equiv V_z$ and $\partial z/\partial x_j = 1$. The physical interpretation of (10.7.48), in which the variations in density due to variations in temperature are neglected, is that the viscous forces are just balanced by the buoyancy forces.

The problem of $\rho = \rho(C)$ can be treated in a similar manner.

One of the first works on convective currents in a porous medium saturated by a homogeneous incompressible fluid as a result of a vertical temperature gradient is that of Horton and Rogers, Jr. (1945). A series of papers follows, by Morrison, Rogers, Jr. and Horton 1949; Rogers, Jr. and Morrison 1950; Rogers, Jr. and Schilberg 1951 and Rogers, Jr. 1953. For the case of a horizontal layer bounded from above and below by a perfect heat-conducting medium, Horton and Rogers, Jr. (1945) show that the minimum temperature gradient for which convection can occur is approximately $4\pi^2 D_h\mu/k\rho_0 g\eta H^2$, where D_h is the thermal diffusivity, η is the coefficient of cubical expansion, ρ_0 is the density at zero temperature and H is the thickness of the layer. The value of this minimum temperature gradient exceeds the limiting gradient found by Rayleigh for a fluid continuum by a factor of $16H^2/27\pi^2 k\rho_0$.

Another early work is that of Lapwood (1948) who uses the *method of small disturbances* as a tool for studying the setting-up of convection currents in a layer of viscous fluid.

Lapwood (1948) considers a horizontal layer of saturated homogeneous isotropic porous medium $0 \leqslant z \leqslant H$. He describes the flow in this layer by a set of equations which is essentially the same as (10.7.35) through (10.7.40). Initially, a constant temperature is maintained across the layer. In order to investigate the stability of this equilibrium Lapwood assumes that small disturbances from this equilibrium occur. Denoting small disturbances of velocity, density, temperature and pressure by $\mathring{\mathbf{V}}^*$, $\mathring{\rho}$, $\mathring{T}$, and $\mathring{p}$ respectively, he obtains a set of equations from which these disturbances can be determined for various sets of boundary conditions.

For boundaries at $z = 0$ and $z = H$, which are impervious to flow and perfectly conducting (to heat), Lapwood determines the minimal G_1 of the initial uniform temperature gradient across the layer:

$$G_1 = 4\pi^2 D_h/K\eta H^2 \tag{10.7.49}$$

for which no convection takes place. He introduces a nondimensional coefficient:

$$Y = G_1 K\eta H^2/D_h = 4\pi^2 \tag{10.7.50}$$

where K is the hydraulic conductivity. Y may be called *Lapwood's convection parameter*. When G_1 is smaller than the critical value we have stability. When it is above this value, there is a possibility for convective currents. Such currents have cellular shape, with a temperature distribution and streamlines as shown in figure 10.7.3. They are also obtained from the assumption of steady flow.

One cannot predict from the first-order theory described above the size of the cells of convection. Other combinations of boundary conditions lead to similar results. The differences among the various cases are in the critical value of the dimensionless parameter $K\eta H^2 G_1/D_h$, which may differ from $4\pi^2$. Lapwood analyzes cases with various boundary conditions and derives the distribution of temperatures $T = T(z)$ in each case as a result of convective currents.

Rogers, Jr. (1953) presents theoretical considerations and experimental results on stability in a vertical column. His results indicate the occurrence of convective currents at values of the parameter $K\eta H^2 G_1/D_h$ much smaller than $4\pi^2$. He attributes these differences to changes in the liquid's viscosity and to the unstable regime created as a result of a nonlinear temperature distribution along the column. He incorporates these two factors in his analysis, in which he employs a variational method.

In a series of papers, Wooding (e.g., 1957, 1958, 1959, 1960, 1962, 1963, 1964) also investigates convection currents in saturated porous media.

He introduces the *convection parameter*:

$$Y = k\eta(T_1 - T_0)gH/D_h\nu_0; \qquad (T_1 - T_0)H = G_1 \tag{10.7.51}$$

(a) Isotherms $\overset{\circ}{T} = -Gz \sin (\pi x/\sqrt{2}H) \sin (\pi z/H) = $ const.

(b) Streamlines given by $\cos^2 \frac{\pi x}{\sqrt{2}H} \sin \frac{\pi z}{H} = $ const.

(c) Surface pattern $\dfrac{\cos (\pi y/\sqrt{2}H)}{\cos (\pi x/\sqrt{2}H)} = $ const.

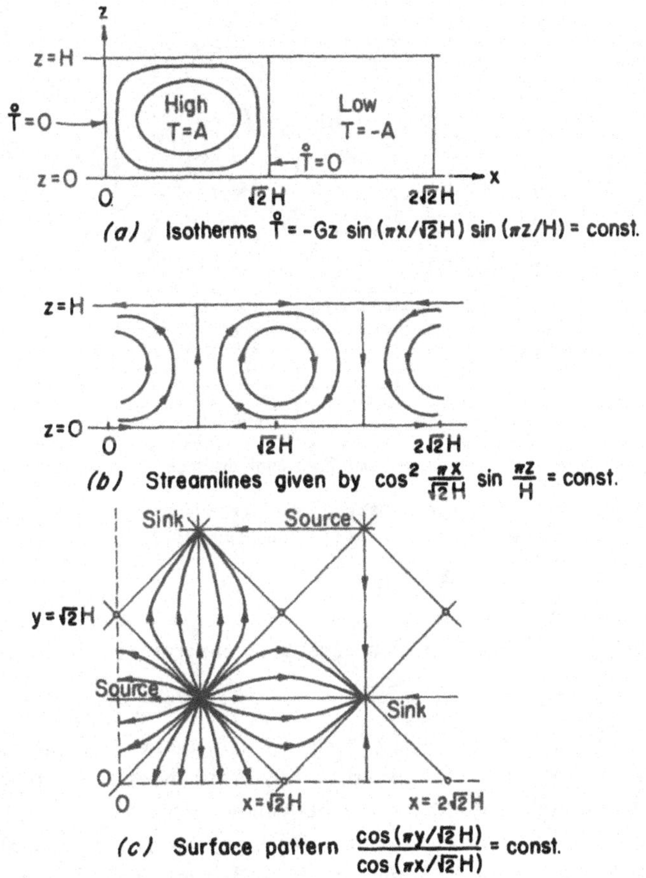

FIG. 10.7.3. Streamlines and isotherms of temperature deviations under conditions of instability (Lapwood, 1948).

which is identical to Lapwood's parameter defined in (10.7.50). This parameter may also be referred to as a modified Rayleigh number (Ra^*). $Y = 0$ corresponds to conduction only.

Donaldson's (1962) work is connected with the geothermal Wairakei region in New Zealand. In his paper he relates the volume flow rate of the convecting fluid and the temperature distribution in the ground to Wooding's convective parameter Y, for values of Y up to four times the critical value. He uses the usual perturbations approach combined with a numerical solution of the perturbation equations (Wooding 1957), assuming that viscosity and other liquid properties are constant and that only the density is temperature dependent. Figure 10.7.4 shows a typical result of Donaldson's studies.

In Wooding's, Donaldson's and some other investigations, free convection results from instability effects of temperature gradients only. Prats (1966), investigates the effect of horizontal flow on thermally induced convection. He considers saturated

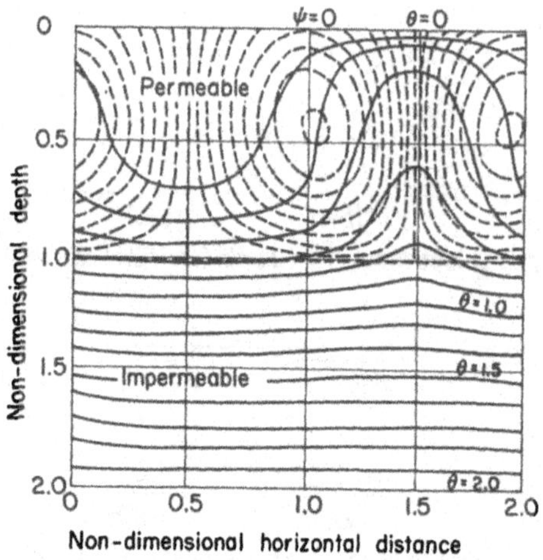

FIG. 10.7.4. Temperature and flow patterns for a two-layer system when the circulatory flow is combined with a recharge-discharge flow, $Y = 96$ (Donaldson, 1962).

flow in a horizontal layer of infinite area. The layer is bounded from above and below by impervious layers on which constant temperatures are maintained:

$$z = 0, \qquad T = T_0; \qquad z = H, \qquad T = T_1. \tag{10.7.52}$$

A horizontal uniform flow ($q_x = q_0$, $q_y = 0$, $q_z = 0$) in the $+ x$ direction initially takes place in this layer. Like Lapwood (1948), Prats also assumes that the fluid's density is constant unless it affects the buoyancy forces.

Prats follows the usual procedure of perturbing the system and allowing the convection effects to enter. He obtains equations that are analogous to those solved by Lapwood (1948) to obtain the disturbances due to free convection. The only difference is that his equations are written with respect to a frame of reference that moves at a velocity αq_0, which is the steady-state heat front velocity in the $+ x$ direction. Hence Lapwood's (1948) results presented above are immediately applicable by simply replacing x with $x - \alpha q_0 t$. Since the criterion for marginal stability does not depend on x it follows that it remains unaffected by the presence of horizontal currents. The critical free convection parameter (modified Rayleigh number for porous media) remains (10.7.50). The main difference, though, is that its results also refer to unsteady temperature and velocity distributions. Both become steady distributions from the point of view of an observer moving at a velocity αq_0. Figure 10.7.5 shows results obtained by Prats (1966).

Elder (1966) presents an experimental and numerical study on steady free convection in a homogeneous isotropic slab of porous medium when the system is heated from below. This boundary condition differs from the ones employed in the works described above. For the field equations he also uses the Boussinesq approximation employed by Wooding (1957). His equations include the stream function ψ and the temperature T as dependent variables.

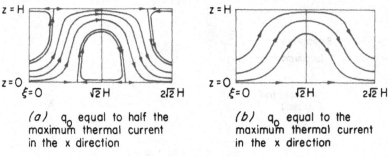

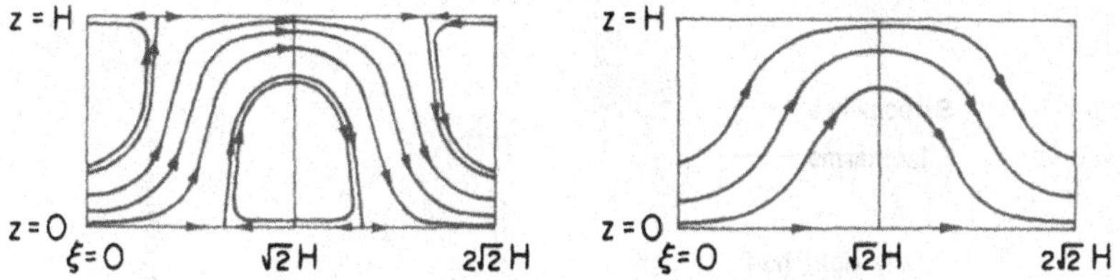

(a) q_0 equal to half the maximum thermal current in the x direction

(b) q_0 equal to the maximum thermal current in the x direction

FIG. 10.7.5. Streamlines for $q_0 \neq 0$ (Prats, 1966).

He also carries out experiments on a Hele–Shaw cell. He expresses the amount of heat transferred by the *Nusselt number*, which is a dimensionless thermal conductivity defined by:

$$Nu = Q_h/\lambda_e \Delta A\, \Delta T/H \tag{10.7.53}$$

where ΔA is the heated area, λ_e is the thermal conductivity of the saturated medium, and Q_h is the power transferred across the slab of thickness H and temperature difference ΔT. His experimental results, both for the packed column and for the Hele–Shaw cell, are presented as a relationship between the Nusselt and Rayleigh numbers (or convection parameter $= k\eta\, g\, \Delta T H/D_h \nu$). He also introduces the *discharge number* $B = gHk/D_h\nu$, which is a measure of the ratio of the imposed pressure forces.

From Elder's (1967) experiments it follows that at the critical Rayleigh number, when convective currents start, there is an increase in the heat discharge through the column or slab. This provides a means for determining the critical Rayleigh number. Elder's experiments give $Y \approx 40$, which is sufficiently close to Lapwood's $4\pi^2$. At very high Rayleigh numbers ($\sim 10^7$) viscous currents as well as thin boundary layers at the bottom and near the upper surface play a significant role. When the thickness of this layer is of the order of magnitude of the grain size, the equations based on averaging over a representative porous medium model are no longer valid.

Figure 10.7.6 shows a superposition of convective currents with flow resulting from forced discharge along a portion of the upper boundary and recharge along the vertical boundary.

Figure 10.7.7 shows calculated isotherms and stream functions for natural recharge; the Rayleigh number is $Y = 50$. Recharge occurs over the entire length of the upper boundary; the heated fluid rises as a fairly thin column even at this rather low Rayleigh number, and only about 30% of the fluid in the plume is circulated.

Up to this point we have considered only convective currents resulting from temperature-induced density differences. However, as mentioned at the beginning of this paragraph, ρ may also vary as a result of variations in a solute's concentration alone, or as a result of variations in both concentration and temperature. The problem of flow of a nonhomogeneous fluid where variations of ρ are due to variations

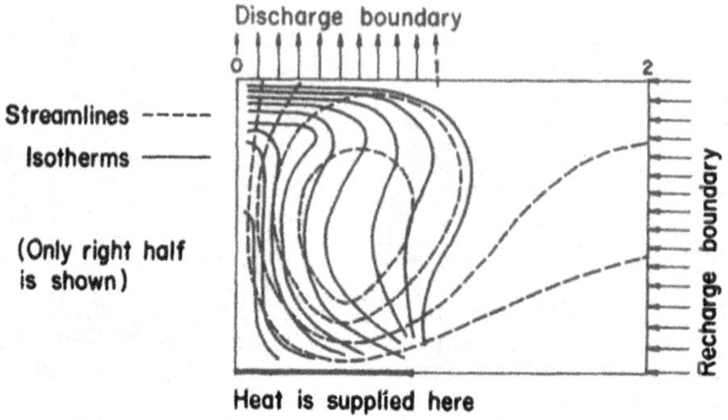

FIG. 10.7.6. Example of calculated distribution of temperature and stream function with forced discharge (Elder, 1967).

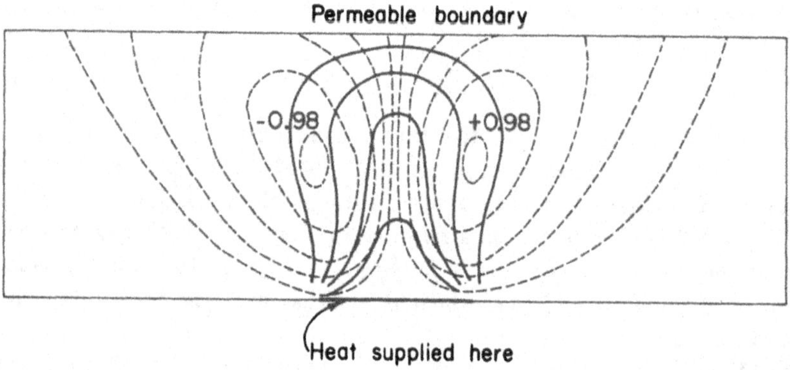

FIG. 10.7.7. Natural discharge, calculated isotherms and stream-functions for $Y = 5$ (Elder, 1967).

in solute concentration only is the subject matter of the present chapter on hydrodynamic dispersion. The system of equations formulating the mathematical statement of this problem was presented in section 10.5. The solution of this system of equations yields the distributions $C = C(x, y, z, t)$, hence, also $\rho = \rho(x, y, z, t)$. When initially we have two miscible fluids, with two distinct densities—say ρ_1 and $\rho_2 > \rho_1$—a *transition zone* between the fluids will develop as the motion progresses. However, it is not obvious that the motion will always be a stable one or that the rate of mixing described by the dispersion coefficient will be independent of the density difference. As an example for a questionable stability, one may think of the mixing of two miscible fluids (ρ_1, $\rho_2 > \rho_1$) in horizontal motion, one above the other, when the fluid of higher density overlies the fluid of lower density.

The problem of instability due to density differences is investigated by Wooding (1963), List (1965) and Nield (1968), among others.

10.7.7 Some Similitude Considerations

The main concepts of the theory of similarity are given in section 11.2 in connection with the scaling of models and analogs. Here we shall add comments on some dimensionless parameters that characterize heat and mass transfer in porous media.

As an example of how the modified inspectional analysis method (par. 11.2.7) is applied to problems of heat transfer, consider a simple problem of steady heat transfer in a two-dimensional homogeneous isotropic medium. The equation is:

$$D_h \nabla^2 T - V_x \, \partial T/\partial x - V_y \, \partial T/\partial y = 0. \tag{10.7.54}$$

As boundary conditions let us use:

$$\lambda \, \partial T/\partial y + h \, \Delta T = 0. \tag{10.7.55}$$

Following the procedure described in paragraph 11.2.7, we obtain from (10.7.54) and (10.7.55):

$$(D_h)_r T_r/x_r^2 = (V_x)_r T_r/x_r = (V_y)_r T_r/y_r \qquad \lambda_r T_r/y_r = h_r T_r. \tag{10.7.56}$$

Since the medium is isotropic, we obtain (e.g., by writing Fick's and Darcy's laws for the x and y directions, separately) $x_r = y_r \equiv L_r$ and $(V_x)_r = (V_y)_r \equiv V_r$. Hence, we obtain from (10.7.56):

$$\begin{aligned} L_r V_r/(D_h)_r = 1 \quad &\text{or} \quad L_m V_m/(D_h)_m = L_p V_p/(D_h)_p \\ h_r L_r/\lambda_r = 1 \quad &\text{or} \quad h_m L_m/\lambda_m = h_p L_p/\lambda_p \end{aligned} \tag{10.7.57}$$

as criteria for the similarity of the two systems. The subscripts m and p here denote two systems similar to one another.

The criterion VL/D_h in (10.7.57) is called the *Peclet number of heat transfer*:

$$Pe = VL/D_h. \tag{10.7.58}$$

In paragraph 10.4.1 we have defined a *Peclet number of molecular diffusion* with molecular diffusivity instead of thermal diffusivity and with the mean grain size as characteristic length. In general, any length dimension can be taken as the characteristic length L in the similarity parameter. It is essential, however, that for all systems the criterion be calculated according to the same characteristic length.

Another criterion for the parameter of similarity that can be derived from (10.7.57) is the *Nusselt number*:

$$Nu = hL/\lambda. \tag{10.7.59}$$

For similar phenomena, it is necessary to have the same values of the similarity parameters at all homologous points.

The Peclet number defined by (10.7.58) may be expressed as a product of the Reynolds number and the *Prandtl number* defined by $Pr = \nu/D_h$. Whereas the Reynolds number is a number characterizing dynamic similarity, the Prandtl number is a characteristic of the medium and the fluid.

The Prandtl number is a significant parameter in the study of systems in which energy and momentum transfer take place simultaneously. Physically it expresses the relative speed at which momentum and energy are propagated through the system.

The Peclet number for heat transfer may be interpreted as giving the ratio between heat transport by convection and heat transport by conduction, $(\rho c V \Delta T/L)/(\lambda \Delta T/L^2)$.

For problems of free convection in a fluid continuum the *Grashof number* (Gr) may be of importance:

$$Gr = \rho^2 g \eta L^3 \Delta T/\mu^2 \quad \text{or} \quad Gr = \rho g L^3 \Delta \rho/\mu^2. \tag{10.7.60}$$

Other dimensionless numbers appearing in the literature in connection with heat and mass transfer are:

$$\text{the } \textit{Lewis number}\text{:} \quad Le = \frac{D_h}{D_d} = \frac{\text{thermal diffusivity}}{\text{molecular diffusivity}} \tag{10.7.61}$$

which is of importance in systems subject to simultaneous heat and mass transfer,

$$\text{the } \textit{Schmidt number}\text{:} \quad Sc = \frac{\nu}{D_d} = \frac{\text{kinematic viscosity}}{\text{molecular diffusivity}} \tag{10.7.62}$$

which is of importance in isothermal systems undergoing simultaneous momentum and mass transfer and

$$\text{the dimensionless combination:} \quad \frac{Gr}{Re^2} = \frac{\rho \eta g \Delta T}{\rho V^2/L} = \frac{\text{buoyancy forces}}{\text{inertial forces}}; \quad Re = \frac{VL}{\nu}. \tag{10.7.63}$$

Most of these parameters are presented in the literature in connection with heat and mass transfer in a fluid continuum. However, in most cases their use can be extended to problems of heat and mass transfer in porous media by introducing slight modifications, such as qd/ν instead of VL/ν for Re.

The modified inspectional analysis described in paragraph 11.2.7 is an easy way to obtain these and other dimensionless parameters.

Exercises

10.1 Let a circular capillary tube be filled with a certain fluid at concentration C_1 (of some tracer) along $x < 0$ and at concentration C_2 along $x > 0$. Assuming an abrupt interface (i.e., neglecting molecular diffusion), determine the concentration distribution $\bar{C}(x, t)$ for $t > 0$, where $\bar{C}$ is the average tracer concentration across the tube after flow at an average velocity $\bar{V}$ starts at $t = 0$.

10.2 Fresh water at relative concentration $C/C_0 = 0$ is introduced into a sand column saturated with salt water at relative concentration $C/C_0 = 1.0$. As the salt water is displaced, the following measurements were made:

Distance along column (cm)	48.2	49.7	51.5	53.6	55.4	57.2	59.3	61.3	63.2	65.4	68.2	73.4
C/C_0 (%)	0.8	2.2	3.5	9.5	21.7	50.3	78.0	94.5	94.5	98.7	98.7	100

The average velocity of flow is 1.6 cm/min.

(a) How long after the initiation of flow were these readings taken?

(b) Determine the coefficient of dispersion D' and the dispersivity a_I.

10.3 Use (10.5.33) to derive the boundary conditions (a) when the external boundary b is impervious to fluid flow, but not to molecular diffusion, (b) when it is impervious to both fluid mass flow and diffusion.

10.4 (a) Write the boundary condition (10.5.36) for a homogeneous isotropic medium.

(b) Rewrite the condition in (a) for the case of one-dimensional flow where at $x = 0$ the column is connected to a large reservoir where $C = C_0 = \text{const.}$

10.5 Solve Case (2) of par. 10.6.1 for the case where the injected fluid is marked by a radioactive tracer.

10.6 Solve the one-dimensional dispersion equation in the presence of adsorption and decay for flow in an infinite column. The boundary and initial conditions are the same as in case (1) of par. 10.6.1.

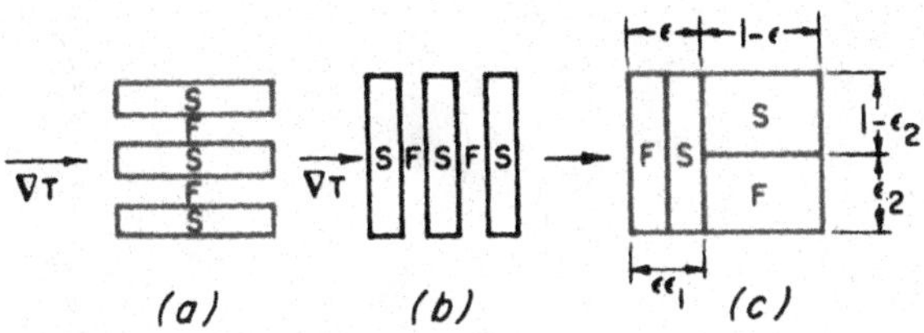

FIG. 10.E.1.

10.7 Develop expressions for an equivalent thermal conductivity λ_e for fluid solid combinations shown in fig. 10.E.1.

CHAPTER 11

Models and Analogs

Some elements of analytical and numerical methods for solving boundary and initial value problems of flow through porous media, often with the aid of a digital computer, were presented in chapter 7. Models and analogs are also tools for achieving the same goal, especially where a direct analytical solution is not possible because of the complexity of the model. The models and the analogs may therefore be regarded as *special purpose computers* each of which is designed to solve a particular problem. In most cases these are problems encountered in engineering practice. An attempt will be made to indicate the applications of each model and to emphasize the advantages and the limitations of each. In each particular case, advantages must be weighed against those of using numerical methods of solution by means of high speed digital computers.

In addition, the sand box model, and sometimes the Hele-Shaw analog, is used to observe phenomena as they actually occur in porous media in the course of developing basic theories and verifying theories based on these observations. Often these models and analogs are also used by scientists for verifying analytical or numerical results based on simplifying approximations.

11.1 General

To solve a problem of flow through a porous medium domain means to determine the *response*, usually in the form of head, or velocity, distribution, resulting from a given *excitation* in the form of boundary and initial conditions acting upon a fully specified *system*—the investigated flow domain, or field, within the porous medium. Specifications of the flow domain must include both the domain's boundaries and the medium's distributed characteristics (e.g., permeability) at all points within these boundaries. Excitations may act along the domain's external boundaries, at specified regions within the field, or they may be distributed in a continuous fashion throughout it. Only the macroscopic behavior of the system and the macroscopic medium characteristics are considered in the present chapter.

When an engineer or an applied scientist attacks a field problem, he usually has a specific, often limited, objective in mind. He is interested in specific answers, within a specified accuracy and usually at a minimum cost of labor and equipment. Although an analytical solution is, by far, the most satisfactory form of solution, and should by no means be overlooked by engineers, it frequently proves inadequate or impractical for engineering application. Most exact analytical techniques are effective only for very simple field geometries. In many cases it is difficult to interpret the results in a physical context, especially when the solution is in the form of an infinite series of certain mathematical functions, or in the form of integrals of complicated functions so that a digital computer is needed. Model and analog methods are an attempt to circumvent some of the shortcomings of a purely mathematical approach. One may consider the analog as a *single purpose computer* designed and constructed to solve a particular problem. This technique of reproducing the behavior of a phenomenon on a different and more convenient scale is known as *modeling*. Although each experiment solves a particular problem (i.e., for a given set of data), it is possible to employ the theory of similarity in order to generalize the result of an experiment, or a sequence of experiments, to cover a whole class of similar problems (each of which involves a different set of data).

The problem of the interface in a coast aquifer (sec. 9.7) may serve as an example of these considerations. Although the exact mathematical statement of a flow problem involving an interface between two fluids can easily be set up, an analytical solution, and sometimes even a numerical one, is practically impossible because of the nonlinearity of the problem. Yet, in many cases satisfactory solutions, i.e., predictions of interface movements in the aquifer, may be obtained by dimensionally scaled laboratory models or analogs in which the various active parameters are adjusted to have the same relative significance as in the actual problem investigated. In a properly scaled model or analog, the laboratory results faithfully duplicate the behavior of the system investigated (to the extent the latter is described by the equations upon which the model is based), but on a miniature dimensional scale and in a reduced time scale. It is also possible to employ a model or an analog in order to study the role of a single parameter by systematically varying its value in repeated experiments, maintaining all other parameters constant.

We have two systems: the investigated system (e.g., the aquifer), referred to as the *prototype* or *prototype system*, and the *analog system*. These two systems are said to be *analogous* if the characteristic equations describing their dynamic and kinematic behavior are similar in form. This is possible only if there is a *one-to-one correspondence* between elements belonging to the two systems. For every element in the prototype system, there must be present in the model (or analog) system an element having a similar excitation–response relationship. Furthermore, corresponding elements must be related to each other in a similar manner. Such a relationship between the two systems is often referred to as a *direct analogy*.

This is obviously true when the two systems involve the same physical phenomena, as in the case of a sand box model simulating the flow in an aquifer or an oil reservoir; both systems involve flow through a porous material. Generally, however, the physical dimensions of the two systems need not be the same. We shall use the term *model* (or scaled model) to denote an analog having the same dimensions as those of the prototype. In a model, every prototype (macroscopic) element is reproduced, different only in size. Analogs, on the other hand, are based on the analogy between systems belonging to entirely differential physical categories. In an analog, similarity is recognized by the following: (a) to each dependent variable and all its derivatives in the equations describing one system there corresponds a variable with corresponding derivatives in the equation of the second system, and (b) the independent variables and their derivatives are related to each other in the same manner in the two sets of equations. One should keep in mind that the reason underlying the existence of the analogy is not just the similarity of the equations—that is merely a clever device for recognizing it. Actually, the analogy stems from the fact that the characteristic equations in both systems (e.g., those describing conservation of mass and thermal energy, or Darcy's law and Ohm's law) represent the same principles of conservation and transport that govern physical phenomena. It is, therefore, possible to develop analogs without referring to the mathematical formulation. This approach is particularly advantageous where the mathematical expressions are excessively complicated or are unknown.

The most common equations that describe flow through porous media (e.g., ground water flow in aquifers or oil in reservoirs), and for which an analog solution is sought, are the Laplace equation (6.2.24), or its equivalent form (6.2.23) for anisotropic media, the diffusion type equations (6.2.32) or (6.4.7) for unsteady flow and the Poisson equation (8.2.12) for flow with accretion. As similar equations occur in heat flow, in electricity and in other branches of physics, one is led to use these phenomena as analogs to flow through porous media.

Analogs may be either of the *discrete* or the *continuous* type with respect to space variables. In discrete analogs, the behavior of the investigated system is defined only at specified points in space. In the continuous analog every prototype point is represented. In both types, time remains a continuous independent variable.

A special class of analogs includes mathematical rather than physical analogs. The investigated flow is first represented by a set of algebraic or differential equations. An assemblage of computing units, usually electrical, each capable of performing some specific mathematical operation (e.g., addition, multiplication, integration, etc.), is then interconnected to generate the required solution by performing the mathematical operations involved. Such systems are called ***indirect analogs, analog computers*** or ***differential analyzers***. They solve equations rather than simulate the behavior of the prototype. Analog computers are not considered in the present chapter (see, for example, Karplus 1958; Karplus and Soroka 1959; Volynskii and Bukhman 1960).

Obviously, no method of solution, whether analytical, numerical or analog, can circumvent the need for complete information regarding the field of flow, its parameters, boundaries, boundary conditions, etc. However, in many practical aquifer (or oil reservoir) problems, with complicated geology and boundary conditions, it is sufficient to base the initial construction of the analog on available data and on a rough estimate of the missing data, and to calibrate the analog by reproducing in it the known past history of the aquifer, e.g., pumpage and natural replenishment. The various analog components (boundaries, aquifer characteristics, etc.) are then adjusted until a satisfactory fit is obtained between the analog's response (e.g., head distribution) and the response actually observed in the aquifer. It is assumed that once the analog reproduces past history reliably, of course within the required range of accuracy, it may also be used to predict the aquifer's response to planned future operations.

Finally, a few remarks on the use of *digital computers* as compared with analogs in general, and with electrical analogs and analog computers in particular, in solving flow problems such as ground water flow in aquifers. In a digital computer, all independent variables (including time) are made discrete and the information is expressed for all points within the investigated domain and for all time intervals in terms of numbers having a specified number of significant figures. On the other hand, in most analogs the information is handled in a continuous fashion with respect to time. Even in the discrete electric analog (pars. 11.5.3 and 11.5.4) time remains a continuous parameter. Hence the accuracy (or number of meaningful significant figures in the solution) of the analog, or analog computer, is determined only by the accuracy of its components and measuring equipment. In a digital computer, the accuracy depends only on the number of significant figures carried in the computations, which, in turn, depends on the computer type, size, etc. Essentially the accuracy in this case is unlimited. Yet, one should bear in mind that accuracy, in both approaches, costs money. In most engineering applications, an accuracy of $\pm 1\%$ is sufficient. In many ground water forecasts the required accuracy is less than ± 10 or $\pm 20\%$, as often the accuracy of our knowledge of field data (aquifer permeability, storativity, boundaries, etc.) is even poorer than that. Under such conditions, an analog is frequently preferred. A second point to be considered is cost. In most simple cases, an analog is less expensive. For large regions, or unsteady three-dimensional problems, or in an inhomogeneous medium, the digital computer may be less expensive. All included, in many cases the choice between the two approaches is not a clear-cut one, and each case should be examined carefully to determine which tool is more appropriate.

Only the main types of analogs applicable to ground water flow problems and problems of flow in oil reservoirs are discussed in the following sections. The heat conduction analogy and the diffusion analogy have been omitted because of the technical difficulties inherent in their application.

In the general description of the analogs, an attempt has been made to convey to the reader the versatility of this tool as a means for solving field problems. The description is by no means complete; substantial additional information may be found in the literature, and much is left to the ingenuity of the investigator.

11.2 Scaling Principles and Procedure

11.2.1 The Two Systems

Two systems are being considered here: a *prototype system* (say, the aquifer) *and a model, or analog system.* In each system we have several variables and dimensional or dimensionless coefficients. Our objective in this section is to determine the relationships between the two groups of variables and coefficients, so as to be able to predict the behavior of the prototype from experiments performed on a model or analog that represents it.

Hydraulic models are often used in fluid mechanics when the equations of motion and the boundary conditions are too complex to permit a purely analytical or even a numerical prediction of prototype behavior. In the theory of flow through porous media, sand box models (sec. 11.3) are the equivalent of the hydraulic models. However, the discussion in the following paragraphs will be extended to include various types of analogs.

The conditions that a model of a phenomenon reproduces all aspects of behavior of the prototypes represented by it are known as conditions of *similitude.* In hydraulic models, as in the sand box model, we recognize the concepts of *geometric similarity, kinematic similarity* and *dynamic similarity.* There exist other types of similarity as well, all of which must be satisfied if we require a *complete similarity* to exist between the flow phenomena in the two systems.

11.2.2 Geometric Similarity

Geometric similarity implies that the ratios between all corresponding lengths in the two considered systems must be the same. Let $(\delta x)_p, (\delta y)_p, (\delta z)_p$ denote certain lengths in the prototype (designated by subscript p), and $(\delta x)_m, (\delta y)_m, (\delta z)_m$ be their corresponding lengths in the model (designated by subscript m). Their ratios are the *length scales*:

$$x_r = (\delta x)_m/(\delta x)_p; \qquad y_r = (\delta y)_m/(\delta y)_p; \qquad z_r = (\delta z)_m/(\delta z)_p \tag{11.2.1}$$

where subscript r denotes the ratio between a value in the model and the corresponding value in the prototype. Exact geometric similarity requires that:

$$x_r = y_r = z_r = l_r; \qquad l_r = (\delta l)_m/(\delta l)_p \tag{11.2.2}$$

where l denotes any length. Obviously, areas (A) and volumes (U) in the two systems are related to each other by:

$$A_r = l_r^2; \qquad U_r = l_r^3 \tag{11.2.3}$$

where A_r is the *area scale* and U_r is the *volume scale.* However, as will be shown below, often a *distorted model,* with $x_r \neq y_r \neq z_r$ (or $x_r = y_r \neq z_r$, etc.) is employed. In a distorted model, the length scale l_r depends on the direction in which this length is measured. If we consider a certain length δl in the direction of the unit vector $\mathbf{1}l$, then (Irmay 1964a):

$$\begin{aligned} l_r^2 &= [(\delta l)_m/(\delta l)_p]^2 = x_r^2(\alpha_{lx})^2{}_p + y_r^2(\alpha_{ly})^2{}_p + z_r^2(\alpha_{lz})^2{}_p \\ l_r &= x_r/(\alpha_{lx})_r = y_r/(\alpha_{ly})_r = z_r/(\alpha_{lz})_r \end{aligned} \tag{11.2.4}$$

where $\alpha_{lx} = \delta x/\delta l$, α_{ly} and α_{lz} are the direction cosines of $\mathbf{1}l$. Hence angles are not preserved in a distorted model.

In a distorted model, area scales also depend on the direction of the considered area:

$$(A_x)_r = y_r z_r; \qquad (A_y)_r = x_r z_r; \qquad (A_z)_r = x_r y_r \tag{11.2.5}$$

where A_{x_i} denotes a surface whose normal is in the direction of the unit vector $\mathbf{1x}_i$. The volume scale in a distorted model is given by:

$$U_r = x_r y_r z_r. \tag{11.2.6}$$

11.2.3 Kinematic Similarity

Kinematic similarity means similarity of the flow net composed of streamlines and equipotentials, i.e., the two flow nets are geometrically similar. In nonsteady flow, pathlines should also be geometrically similar. Obviously, since the boundaries of the flow domain will form some of the streamlines, or equipotentials, kinematically similar flows must also be geometrically similar. If we remain in the range of laminar flows, the converse of this statement is also true. Kinematic similarity, therefore, implies that the direction of the velocity remains unchanged and that the ratio between velocities (and accelerations) at all homologous points in the two systems is the same throughout the flow domain. This means that:

$$V_r = V_m/V_p = \text{const.} \tag{11.2.7}$$

Since $V = \delta l/\delta t$, we have:

$$V_r = l_r/t_r \tag{11.2.8}$$

and therefore, we must have $t_r = (\delta t)_m/(\delta t)_p = \text{const.}$ However, in a distorted model, since l_r varies with direction, V_r (as, in fact, any other vector property) will also vary with the direction of the velocity. We have:

$$(V_x)_r = x_r/t_r, \qquad (V_y)_r = y_r/t_r; \qquad (V_z)_r = z_r/t_r. \tag{11.2.9}$$

Hence, *it is impossible to maintain kinematic similarity in a distorted model.*

11.2.4 Dynamic Similarity

In *dynamically similar* systems, forces at homologous points and homologous times acting on homologous elements of fluid mass must be in the same ratio throughout the two systems. In addition, we therefore require geometric and kinematic similarity. Here forces are those of gravity, pressure, friction (or viscosity), inertia, elasticity and surface tension. The physical properties involved are density, viscosity, elasticity, etc. For example, expressing the force due to inertia by $f_i = \rho V^2 l^2$ and that due to viscosity by $f_v = \mu V l$, and requiring that their ratio remains constant at all homologous points of the model and the prototype, leads to:

$$\frac{(f_i)_m}{(f_i)_p} = \frac{(f_v)_m}{(f_v)_p} \quad \text{or} \quad \left(\frac{f_i}{f_v}\right)_m = \left(\frac{f_i}{f_v}\right)_p = \left(\frac{Vl\rho}{\mu}\right)_m = \left(\frac{Vl\rho}{\mu}\right)_p$$

$$(Re)_m = (Re)_p \tag{11.2.10}$$

where *Re* is the *Reynolds number* defined by $Re = Vl/\nu$, V being a characteristic velocity and l a characteristic length.

In hydraulic models we recognize other dimensionless numbers such as the *Froude number* ($Fr = V/\sqrt{gl}$), resulting from the constancy of the ratio of the forces of inertia and gravity (actually the square root of this ratio), the *Weber number* ($We = V\sqrt{l\rho/\sigma}$), equal to the ratio of inertial force to capillary force, etc. (see any textbook on fluid mechanics).

In most cases, however (and in particular in the sand box model), it is impossible to satisfy all requirements for dynamic similarity; some must be neglected, either because they do not apply or because the effects they represent are negligible with respect to other phenomena involved.

We may summarize the discussion on similarity, following Langhaar (1951), by considering any two functions $f_p(x_p, y_p, z_p, t_p)$ and $f_m(x_m, y_m, z_m, t_m)$ in the two systems, where (x_p, y_p, z_p) and (x_m, y_m, z_m) are the coordinates of homologous points and t_p and t_m are homologous times. The two systems are said to be similar if for homologous points and times $f_r = f_m/f_p$ remains constant throughout the considered system. We call f_r the *scale* (or *scale factor*) of f.

Although the discussion above relates to a true model, it can be extended to analogs as well. In general, two approaches to model scaling are usually employed: *dimensional analysis*, and *inspectional analysis*. Only some brief comments are presented on the former method because this subject is well covered in texts on fluid mechanics, and because we shall employ only the latter method in the following sections, as it is easily applicable to the analogs considered here.

11.2.5 Dimensional Analysis

Dimensional analysis is based on Fourier's (1822) *principle of homogeneity*, which states that any equation expressing a physical relationship between quantities must be *dimensionally homogeneous*, i.e., all terms in such an equation are similar in dimensions. All physical quantities are measured through comparison with units of one dimensional category or another. These categories, like the units themselves, are arbitrary. In the field of mechanics, however, all quantities are usually expressed as one or another combination of mass (M) (or force (F)), length (L) and time (T).

In general, the variables involved in a physical phenomenon are known, while the relationships among them are unknown. By a procedure of dimensional analysis the phenomenon considered may be formulated as relationships among a set of *dimensionless groups* of the variables. The immediate advantage of this procedure is that considerably less experimentation is required in order to establish relationships among the several variables involved. Instead of altering the value of each variable in the course of the experiments in order to study the relationship between any pair of variables, it is sufficient to study the relationships among the dimensionless groups, each of which consists of two or more variables. It is usually sufficient to vary the value of only one variable included in the dimensionless group in order to cause the group as a whole to vary over a required range. The Reynolds number (Re) defined by (11.2.10) is an example of a dimensionless group that involves four variables: V, l, ρ and μ.

Each dimensionless group is called a π (with no relation implied to the number 3.1416..). The number of dimensionless π's to be formed from a set of variables, known (or assumed) to be involved in a physical phenomenon, and their actual construction is given by *Buckingham's π theorem.*

According to this theorem (Buckingham 1915), when N quantities $Q_1, Q_2, \ldots, Q_N$ which can be expressed dimensionally by $f\,(< N)$ fundamental quantities (e.g., unit length, unit mass (or force), unit time), are related by a dimensionally homogeneous equation, the relationship among the N quantities can always be expressed in terms of exactly $N-r$ dimensionless products, or groups (i.e., πs), where $r\,(\leqslant f)$ is the rank of the *dimensional matrix* of the N quantities:

$$g(\pi_1, \pi_2, \ldots, \pi_{N-r}) = 0 \quad \text{or} \quad \pi_1 = g(\pi_2, \pi_3, \ldots, \pi_{N-r}). \qquad (11.2.11)$$

Actually, r is the number of independent fundamental quantities in a particular problem. For a proof of this problem, the reader is referred to texts on fluid mechanics (e.g., Li and Lam 1964). As an example of how the πs are constructed, consider the following example. It is assumed that the hydraulic conductivity K of a porous medium is related to some characteristic grain diameter d, to a dimensionless shape factor c, to the dynamic viscosity μ of the liquid and to its specific weight γ by:

$$K = cd^{k_1}\gamma^{k_2}\mu^{k_3} \qquad (11.2.12)$$

where k_1, k_2 and k_3 are exponents that must be determined. Expressing d, γ and μ by the fundamental dimensions F (force), L (length) and T (time), we obtain:

$$(L/T) = C(L)^{k_1}(F/L^3)^{k_2}(FT/L^2)^{k_3}.$$

By equating the exponents of F, L and T, we obtain:

$$1 = k_1 - 3k_2 - 2k_3$$

$$-1 = 0 + 0 + k_3$$

$$0 = 0 + k_2 + k_3.$$

Hence:

$$k_3 = -1, \qquad k_2 = 1, \qquad k_1 = 2, \quad \text{so that} \quad K = cd^2\gamma/\mu. \qquad (11.2.13)$$

In another form, this procedure will lead to the equivalent expression:

$$\pi = Kd^{-2}\gamma^{-1}\mu = \text{const.} \qquad (11.2.14)$$

If all the πs appearing in (11.2.11) are the same in model and in prototype, the form of the function g in the model will also be the same as in the prototype. *Scales* or *design criteria* are derived by equating the various πs in the model with their counterparts in the prototype. As the various dimensionless groups, known as the Reynolds number, the Weber number, the Froude number, etc. (par. 11.2.4), are actually π numbers, in order to achieve dynamic similarity we require equality among these numbers in the two systems. In a similar manner, dimensionless groups may be derived from the requirements of geometric and kinematic similarity. Obviously, only those dimensionless groups related to the problem at hand must be duplicated.

Let us see an example of how model scales are derived by requiring that the dimensionless groups be equal in model and prototype. Consider a case where in designing a model we must determine the following three items: a length scale $l_r = l_m/l_p$, a velocity scale $V_r = V_m/V_p$ and the fluid for the model (e.g., by determining μ_m and ρ_m, or equivalently, the scales $\mu_r = \mu_m/\mu_p$ and $\rho_r = \rho_m/\rho_p$, where μ_p and ρ_p are, for example, those of water). Assuming that viscous forces are predominant, we require equality of the Reynolds number in the model and in the prototype. Hence:

$$(Re)_m = (Re)_p; \qquad V_m l_m \rho_m/\mu_m = V_p l_p \rho_p/\mu_p \tag{11.2.15}$$

or:

$$V_r l_r \rho_r/\mu_r = 1 \tag{11.2.16}$$

which is a relationship among the scales for V, l, ρ and μ. For example, if we choose a certain fluid for the model (and this actually means that we choose ρ_r and μ_r) and a length scale, the velocity scale can be determined from (11.2.16). Using (11.2.8) we can then determine the time scale t_r. Obviously, the choice of fluid and the model size must be such that the model size, velocities, etc., are practical, and the influence of the other forces remains small in the model, as assumed.

If two or more forces are considered, the number of arbitrarily chosen parameters is reduced. In many cases, especially where three or more forces are considered, it is impossible to find a fluid to satisfy all conditions. In such cases, an exact, dynamically similar, model is impossible. Fortunately, in most practical cases only one force is predominant, and the model is designed according to the π representing this predominant force. The numbers that are thus neglected affect the accuracy of the data observed in the model and introduce certain errors (called *scale effects*).

The main advantage of the dimensional analysis method is that it is applicable to investigations in which the governing equations are unknown, yet the significant variables and parameters are known.

If, on the other hand, the differential equations governing the physical phenomena are known, we can use them to deduce the dimensionless groups and the similarity laws resulting from their duplication. This is done by the method of inspectional analysis described in the following paragraph.

11.2.6 Inspectional Analysis

In this paragraph, the discussion is extended to include analogs as well as models, although the term model will still be used to refer to both. We assume that all the aspects of the physical prototype system and the macroscopic phenomena occurring in it can be completely described by a known system of mathematical equations:

$$P(y_1, y_2, \ldots, y_m; x_1, x_2, \ldots, x_n; a_1, a_2, \ldots, a_k) = 0 \tag{11.2.17}$$

where x_j $(j = 1, 2, \ldots, n)$, are independent variables, y_i $(i = 1, 2, \ldots, m)$ are dependent variables and a_l $(l = 1, 2, \ldots, k)$ are dimensional parameters and coefficients.

The total number of variables and parameters is $N = m + n + k$. The system P may be composed of several types of equations, e.g., integral, differential, etc. The requirement of completeness means that the system of equations possesses a unique solution for the y_is in terms of the x_js and a_ls:

$$y_i = y_i(x_1, x_2, \ldots, x_n; a_1, a_2, \ldots, a_k). \tag{11.2.18}$$

Let M be another system of equations, analogous to P, completely defining the dynamic and kinematic behavior of the model system. This means that for every equation in P there is a corresponding equation in M and for each variable and operation in P, there is a corresponding variable and operation in M. The systems of equations may be written formally as:

$$P(y_1, y_2, \ldots, y_m; x_1, x_2, \ldots, x_n; a_1, a_2, \ldots, a_k) = 0$$

$$M(y'_1, y'_2, \ldots, y'_m; x'_1, x'_2, \ldots, x'_n; a'_1, a'_2, \ldots, a'_k) = 0 \tag{11.2.19}$$

where y'_i $(i = 1, 2, \ldots, m)$ are the dependent variables, x'_j $(j = 1, 2, \ldots, n)$ are the independent variables and a'_l $(l = 1, 2, \ldots, k)$ are constant parameters of the model system. Each of the equations involved must be dimensionally homogeneous, but it is not necessary that the physical dimensions of the corresponding variables in the two systems be the same.

As in paragraph 11.2.5, our objective now is to derive the $(N - r)$ dimensionless groups. However, in the inspectional analysis method, this objective is achieved by a simple set of transformations or change of variables appearing in the known equations, such that the dimensional variables in P and M become dimensionless ones. These transformations may generally be written as:

$$Y_i = Y_i(y_i; a_1, a_2, \ldots, a_k); \qquad i = 1, 2, \ldots, m$$

$$X_j = X_j(x_j; a_1, a_2, \ldots, a_k); \qquad j = 1, 2, \ldots, n \tag{11.2.20}$$

where X_j, Y_i are dimensionless variables. The systems P and M then become:

$$P^*(Y_1, Y_2, \ldots, Y_m; X_1, X_2, \ldots, X_n; A_1, A_2, \ldots, A_{k'}) = 0$$

$$M^*(Y'_1, Y'_2, \ldots, Y'_m; X'_1, X'_2, \ldots, X'_n; A'_1, A'_2, \ldots, A'_{k'}) = 0 \tag{11.2.21}$$

where the A'_ls $(1 \leqslant l \leqslant k')$ are now dimensionless parameters whose number, according to Buckingham's π theorem, is $k' = k - r$.

A simple way to convert the sets P and M into sets P^* and M^* including only dimensionless variables and groups is to divide each of the equations by some value of the same dimensions as those of the terms appearing in the equation characteristic of the considered system (e.g., characteristic length, velocity, time, etc).

The two systems P^* and M^* are identical in form. This means that the family of solutions of P^*:

$$Y_i = Y_i(X_1, X_2, \ldots, X_n; A_1, A_2, \ldots, A_{k'}) \tag{11.2.22}$$

corresponding to different sets of values of its dimensionless variables and parameters is identical to the family of solutions of M^* corresponding to different sets of values of its dimensionless variables and parameters. Let A_l and A'_l ($l = 1, \ldots, k'$) denote the dimensionless constant parameters (similarity groups) in the prototype and in the model systems, respectively. A solution of P^* will be identical to a solution of M^* if and only if $A_l = A'_l$ for all values of l. These are the required *scaling criteria*.

The use of inspectional analysis in the form described above or in a somewhat different form to be considered below, requires knowledge of the mathematical equations that describe the flow phenomena. It is obvious that only those features that are correctly described by the systems P and M are correctly represented in a model designed according to these procedures.

Approximate equations or empirical relations may also be used. In fact, the use of Darcy's law is also an approximation of the true flow within the interstices. On the other hand, an implicit knowledge of *all* pertinent variables and constant parameters involved is required when the method of dimensional analysis is employed. The result is a complete set of dimensionless groups whose physical meaning is, in general, clearer than that of the dimensionless groups A_l derived by the inspectional analysis method.

As an example for the application of the method of inspectional analysis, consider the case of *cold water drive* (Geertsma et al. 1956).

The problem in the horizontal xy-plane is described by the following set of equations:

$$\partial(\rho_0 q_{0x})/\partial x + \partial(\rho_0 q_{0y})/\partial y = -\,\partial(\rho_0 n S_0)/\partial t$$

$$\partial(\rho_w q_{wx})/\partial x + \partial(\rho_w q_{wy})/\partial y = -\,\partial(\rho_w n S_w)/\partial t \qquad (11.2.23)$$

$$q_{0x} = -\,(kk_{r0}/\mu_0)\,\partial p_0/\partial x; \qquad q_{0y} = -\,(kk_{r0}/\mu_0)\,\partial p_0/\partial y$$

$$q_{wx} = -\,(kk_{rw}/\mu_w)\,\partial p_w/\partial x; \qquad q_{wy} = -\,(kk_{rw}/\mu_w)\,\partial p_w/\partial y$$

$$p_0 = p' + b\sigma\cos\theta\sqrt{n/k}\,J(S_0); \qquad p_w = p' + (b-1)\sigma\cos\theta\sqrt{n/k}\,J(S_0)$$

where p' is the pressure that would prevail if the capillary pressure $(p_0 - p_w)$ were zero, b is a distribution factor for interfacial forces, $J(S_0)$ is the *Leverett function* of S_0 and all other symbols are defined in sections 9.2 and 9.3.

Let the following dimensionless independent variables be introduced: $X = x/l$, $Y = y/h$ and $T = t/n\tau$, where τ is some characteristic time, and l and h are characteristic lengths in the x and y directions, respectively, of the considered reservoir. With them, and assuming that both liquids are incompressible, the set of equations (11.2.23) becomes:

$$\partial(q_{0x}\tau/l)/\partial X + \partial(q_{0y}\tau/h)/\partial Y = -\,\partial S_0/\partial T$$

$$\partial(q_{wx}\tau/l)/\partial X + \partial(q_{wy}\tau/h)/\partial Y = -\,\partial S_w/\partial T = \partial S_0/\partial T$$

$$q_{0x}\tau/l = -\,k_{r0}(\mu_w/\mu_0)\,[\partial(kp'\tau/\mu_w l^2)/\partial X + (b\tau\sigma\cos\theta\sqrt{nk}/\mu_w l^2)(dJ/dS_0)(\partial S_0/\partial X)]$$

$$q_{0y}\tau/h = -\,k_{r0}(\mu_w/\mu_0)(l/h)^2[\partial(kp'\tau/\mu_w l^2)/\partial Y + (b\tau\sigma\cos\theta\sqrt{nk}/\mu_w l^2)(dJ/dS_0)(\partial S_0/\partial Y)]$$

(11.2.24)

and similar equations for $q_{wx}\tau/l$ and $q_{wy}\tau/h$.

Thus, the dimensionless groups appearing in the set of transformed equations above are:

(a) independent variables: X, Y, T;

(b) dependent variables: $\frac{q_{0x}\tau}{l}$; $\frac{q_{wx}\tau}{l}$; $\frac{q_{0y}\tau}{h}$; $\frac{q_{wy}\tau}{h}$; $\frac{pk\tau}{\mu_w l^2}$; S_0;

(c) similarity groups: $\frac{l}{h}$; $\frac{\mu_w}{\mu_0}$; $\frac{\tau\sigma\cos\theta\sqrt{nk}}{\mu_w l^2}$; $\frac{dJ}{dS_0}$; b; k_{r0}; k_{rw}.

Boundary conditions may add several more similarity groups.

11.2.7 Modified Inspectional Analysis

A somewhat different approach to the derivation of scales by inspectional analysis is also possible (Bear 1960a; Irmay 1964a; Luikov 1966). Once the two sets of homologous equations—for the prototype and for the model or analog—are established, we introduce ratios (scales) of model values to corresponding prototype values for all variables and constants involved. For example:

$$x_r = (\delta x)_m/(\delta x)_p; \qquad y_r = (\delta y)_m/(\delta y)_p; \qquad \varphi_r = (\delta\varphi)_m/(\delta\varphi)_p$$
$$K_{xr} = K_{xm}/K_{xp}; \qquad K_{yr} = K_{ym}/K_{yp}; \qquad K_{zr} = K_{zm}/K_{zp} \qquad (11.2.25)$$

where subscript m denotes model value, subscript p denotes prototype value and subscript r denotes the ratio between the two. These ratios are then inserted into one set of equations, e.g., the prototype equations, and the resulting set is compared with the unchanged one. By demanding that the two sets of equations become identical, scales are obtained.

As an example, consider the steady flow of a single incompressible fluid in a homogeneous anisotropic porous medium. The continuity equation describing the flow is:

$$K_{xp}\,\partial^2\varphi_p/\partial x_p{}^2 + K_{yp}\,\partial^2\varphi_p/\partial y_p{}^2 + K_{zp}\,\partial^2\varphi_p/\partial z_p{}^2 = 0. \qquad (11.2.26)$$

The same flow in an *isotropic* sand box model, where $K_{xm} = K_{ym} = K_{zm} = K_m$ is described by:

$$\partial^2\varphi_m/\partial x_m{}^2 + \partial^2\varphi_m/\partial y_m{}^2 + \partial^2\varphi_m/\partial z_m{}^2 = 0. \qquad (11.2.27)$$

Now, let the following ratios (or scales) be introduced:

$$x_r = (\delta x)_m/(\delta x)_p; \qquad y_r = (\delta y)_m/(\delta y)_p; \qquad z_r = (\delta z)_m/(\delta z)_p;$$
$$\varphi_r = (\delta\varphi)_m/(\delta\varphi)_p; \qquad K_{xr} = K_m/K_{xp}; \qquad K_{yr} = K_m/K_{yp}; \qquad K_{zr} = K_m/K_{zp}. \qquad (11.2.28)$$

With (11.2.28), (11.2.26) becomes:

$$K_m\cdot\frac{x_r{}^2}{K_{xr}\varphi_r}\frac{\partial^2\varphi_m}{\partial x_m{}^2} + K_m\frac{y_r{}^2}{K_{yr}\varphi_r}\frac{\partial^2\varphi_m}{\partial y_m{}^2} + K_m\frac{z_r{}^2}{K_{zr}\varphi_r}\frac{\partial^2\varphi_m}{\partial z_m{}^2} = 0. \qquad (11.2.29)$$

Equations (11.2.27) and (11.2.29) are identical (or proportional) if:

$$x_r^2/K_{xr} = y_r^2/K_{yr} = z_r^2/K_{zr}. \tag{11.2.30}$$

The two conditions included in (11.2.30) express the relationships among the scales x_r, y_r, z_r; K_{xr}, K_{yr}, K_{zr}. Additional (independent) equations will lead to more such relationships and/or more scales. If the total number of conditions, i.e., design criteria relating scales to each other, is N', and the total number of parameters (variables and constants) involved is N'' ($> N'$), scales of ($N'' - N'$) parameters may be chosen arbitrarily; the others will be determined by the N' conditions. The choice of the arbitrary parameters depends on the convenience of performing the model experiments of the problem on hand. Such local conditions may include size of model, available time, available medium and liquids, etc. In general, this modified form of inspectional analysis is simplest in application.

Comparing this procedure with the discussion presented in section 7.4, one can easily see that the procedure introduced there as a method for relating solutions in isotropic and anisotropic media is exactly identical to the one introduced here; the equivalent system mentioned there is the model (or analog) system considered here. The summary of procedure given at the end of paragraph 7.4.2 should therefore also be followed here. Many examples are given in the following sections.

As was mentioned above in the discussion of the method of dimensional analysis, a complete model similitude cannot always be achieved. One may be forced to neglect certain features that seem less important. A knowledge of the set of equations completely describing those flow features that are of interest is essential for application of this method.

It is always recommended to present experimental results (and also analytical results) in terms of dimensionless parameters. Such parameters are the πs derived in dimensional analysis. One can also use the relationships among scales to derive dimensionless parameters. For example, using (11.2.25) and Darcy's law, we have from $Q_r = K_{xr}\varphi_r b_r z_r / x_r$:

$$\left(\frac{Q\,\delta x}{K_x\,\delta\varphi b\,\delta z}\right)_m = \left(\frac{Q\,\delta x}{K_x\,\delta\varphi b\,\delta z}\right)_p. \tag{11.2.31}$$

Then, representing δx by some characteristic horizontal length L, δz by some characteristic vertical length D and $\delta\varphi$ by some characteristic difference φ_0, we obtain: $QL/K_x\varphi_0 bD$ as a *dimensionless discharge*. Similarly, from (11.4.9) or directly from Darcy's law $n\,dx/dt = -K_x\,\partial\varphi/\partial x$, we may derive a *dimensionless time*: $tK_x\varphi_0/nL^2$. In this way, dimensionless groups that must be identical in both the model and the prototype are determined.

When the experimental results are expressed in terms of these dimensionless groups, the resulting relationships, e.g., in a graphic form, can be extended to a wide range of cases of the same phenomenon, but with different values of the parameters involved. This introduces an important extension of the use of models, beyond their usefulness in solving specific problems.

11.3 The Sand Box Model

The sand box model (*granular* or *physical model*) is a reduced scale representation of the natural porous medium domain. It is a true model in the sense that both prototype and model involve flow through porous media.

11.3.1 Description

A sand box model consists of a rigid, watertight container, or box, filled with a porous matrix (sand, powdered or crushed glass or glass beads), one or more fluids, a supply system and measuring devices. The geometry of the box corresponds to that of the investigated flow domain. Sometimes symmetry properties (e.g., axial symmetry) of the investigated domain are such that it is sufficient to study in the model only part of the whole prototype field. The most common shapes are rectangular, radial and columnar. The sand column, i.e., a 2″–6″ pipe packed with sand, is the most common experimental tool for one-dimensional flow problems. Usually the box is made of a transparent material (e.g., lucite), especially when more than one liquid is present and a dye tracer is added to one liquid (or to both) in order to follow its movement at the walls. The box material and the porous matrix filling it must be stable and chemically inert with respect to the liquids in it. Heterogeneity due to permeability or porosity variations in the prototype may be simulated by varying the corresponding properties of the material used as a porous matrix in the model according to the scaling rules. The porous matrix in the model may be anisotropic. Piezometers and tensiometers, to measure piezometric heads and underpressures, may be inserted into the flow domain in the model.

A saturated *uniform packing*, with uniform sand characteristics and without entrapped air, is difficult to obtain. Sometimes this difficulty is overcome by pouring the sand (preferably in the form of glass beads) into the model, where a depth of a few centimeters of water is maintained above the sand already in the box. The box is vibrated during filling in order to obtain optimal packing. Whenever possible, *de-aired water* should be used. In prolonged tests it is also important to add disinfectants to the water (e.g., Formol) to prevent bacterial growth that causes clogging.

In order to eliminate wall effects, sand grains are often glued to the walls of the box. The wall effect is also reduced when the simulated porous domain is sufficiently large in the direction normal to the wall.

The *Christiansen filter* (or *transparent model*) is a special type of sand box model where, in addition to the transparent walls, the "sand" is made of a transparent material (often beads or crushed Pyrex glass). The liquid in the model is an oil or an aqueous solution having the same *refractive index* as that of the porous matrix. When the porous medium is saturated by the liquid, the entire solid–liquid system becomes completely transparent and the movement of an injected dye, or the movement of one of the liquids in a two-liquid system, can be observed inside the flow domain (Van Meurs 1957; de Josselin de Jong 1958a). Usually this type of model is used for two-dimensional flow domains where the width of the box is 2 to

3 cm. Bear (1961b) uses this type of model with a constant light source on one side of the model and a photoelectric cell on the other side to detect tracer-concentration changes resulting from hydrodynamic dispersion. The photocell reading depends on the light absorption of the liquid in the model.

In an ordinary two-dimensional sand box model, the same idea may be used, but with X rays or γ radiation replacing the ordinary light (Norel 1963, 1964). An absorbing component (e.g., $BaCl_2$, BaI_2 or KI) is added to one of the liquids. Another possibility is to use a γ-ray-emitting radioactive tracer.

When the liquids are immiscible (e.g., oil–water), methods based on electrical resistance often fail. The method of X-ray absorption (in two-dimensional models) is still applicable for saturation measurements.

The relative concentration of miscible liquids leaving the box (e.g., simulation of pumpage) may be determined by measuring the electrical conductivity of the mixture. When the liquids are immiscible, one may first separate them by separators, and then determine their ratio by measuring volumes of components.

The relative concentration inside the model in miscible displacement studies may be measured by means of small electric sondes. Such devices are usually made of two closely spaced electrodes, the spacing depending on the size of the sand grains. When an electric current passes through these electrodes from a constant current source, the voltage drop indicates the resistance of the saturated medium in the vicinity (or mainly in the vicinity) of the electrodes. Because of the unpredictable effect of the individual sand grains, each sonde must be calibrated *in situ* by running various standard concentrations through the model. In this way, each sonde indicates the average concentration of a certain small volume around it (Bear 1961b; Rumer 1962; Harleman and Rumer 1962; Bear and Dagan 1966a).

When the model is full and saturated, an experiment to determine the intrinsic permeability (k) of the sand-pack is performed. Usually this is done by letting one of the fluids flow in a pattern for which a known analytical description is available.

When the flow of two immiscible liquids is investigated in a model (for example, flow of oil and water in an oil reservoir), the model must simulate the wettability preference of the natural porous medium (e.g., water-wet or oil-wet). We may satisfy this requirement either by modifying the porous matrix by surface treatment or by a proper choice of solids and liquids. Clean silica grains are normally water-wet.

The model must simulate the proper boundary and initial conditions of the natural reservoir: fixed potentials, fixed supply and fixed extraction. This may be achieved by using inlets and outlets in the box walls connected to fixed-level reservoirs or to pumps.

In models simulating ground water aquifers, water is usually used. Sometimes, liquids of a higher viscosity are used in order to obtain a hydraulic conductivity in the model that will yield a suitable time scale. In simulating oil reservoirs, fluids with properties that vary widely from one reservoir to another, or even within the same reservoir if more than one fluid is involved, often must be simulated. The most important properties to be considered are: viscosity, density, interfacial tension and wettability. In choosing a liquid for the model, one must also consider corrosiveness, toxicity and inflammability. Temperature is the main factor affecting these properties. Desired fluid properties may sometimes be obtained by mixing fluids.

The sand box model is extensively used because of its special features that permit studies of phenomena related to the microscopic structure of the medium: unsaturated flow, miscible displacement, hydrodynamic dispersion, immiscible displacement, simultaneous flow of two or more liquids at different relative saturations, fingering, wettability, capillary pressure, etc. It is usually used to simulate flow under confined conditions, because of the difficulties caused by the capillary rise when the flow is a phreatic one. The capillary fringe in the sand box model is disproportionately larger than the corresponding capillary rise in the prototype, according to the model's vertical length scale. The flow in this zone may also introduce a significant error when the depth of the saturated region below the phreatic surface is relatively small.

In petroleum reservoir engineering, the sand box model is used extensively in investigations of water drives, gas drives, solvent drives and sweep efficiencies in oil fields. Examples are given by Muskat (1937), van Meurs (1957), Barnes (1962), Sourieau (1963), Watson et al. (1964) and Croissant (1964).

11.3.2 Scales

Unlike the cases of the various analogs discussed in the present chapter, there is no need here to prove the existence of an analogy between the sand box model and the prototype, as both involve flow through porous media. The various equations developed in chapters 5 and 6 for a single fluid, in chapter 9 for two immiscible fluids and in chapter 10 for miscible fluids are applicable to both the model and the prototype. In each case, the determination of scales, or design criteria, for the sand box model depends on the nature of the investigated problem (homogeneous fluid, several immiscible fluids, miscible displacement, etc.).

Several examples of scale determination are given below. The modified inspectional analysis method (par. 11.2.7) is employed.

Example 1, flow of single phase, slightly compressible fluids. An example for this case is *cold water drive* in oil reservoir engineering.

Let the equation governing the flow in a homogeneous anisotropic medium be:

$$\operatorname{div}[\rho(\underline{\mathbf{k}}\rho g/\mu)\cdot\operatorname{grad}\varphi^*] = n\beta\rho^2 g\,\partial\varphi^*/\partial t \tag{11.3.1}$$

where the directions x, y, z are principal directions of anisotropy and φ^* is defined by (5.9.1). By applying the scaling technique described in paragraph 11.2.7 to (11.3.1) and (5.9.2), written for an anisotropic medium; we obtain:

$$\frac{k_{xr}(\rho_r)^2\varphi^*_r}{x_r^2\mu_r} = \frac{k_{yr}(\rho_r)^2\varphi^*_r}{y_r^2\mu_r} = \frac{k_{zr}(\rho_r)^2\varphi^*_r}{z_r^2\mu_r} = \beta_r(\rho_r)^2 n_r\frac{\varphi^*_r}{t_r} \tag{11.3.2}$$

$$q_{xr} = \frac{k_{xr}\rho_r\varphi^*_r}{\mu_r x_r}; \qquad q_{yr} = \frac{k_{yr}\rho_r\varphi^*_r}{\mu_r y_r}; \qquad q_{zr} = \frac{k_{zr}\rho_r\varphi^*_r}{\mu_r z_r} \tag{11.3.3}$$

$$q_{xr} = \frac{n_r x_r}{t_r}; \qquad q_{yr} = \frac{n_r y_r}{t_r}; \qquad q_{zr} = \frac{n_r z_r}{t_r} \tag{11.3.4}$$

$$U_r = n_r x_r y_r z_r \tag{11.3.5}$$

where U denotes volume.

Altogether we have 16 independent variables: $x, y, z, k_x, k_y, k_z, q_x, q_y, q_z, n, \beta, \mu, \rho, \varphi^*, U, t$. These variables are related to each other by eight independent relationships. Hence, except for certain constraints, eight variables may be chosen arbitrarily. For example, we may choose $x_r, y_r, z_r, n_r, \mu_r, \beta_r, \rho_r, t_r$. Then q_{xr}, q_{yr}, q_{zr} are determined by (11.3.4) and U_r is determined from (11.3.5).

From (11.3.2):

$$k_{xm} = k_{xp}\beta_r n_r x_r^2 \mu_r/t_r; \qquad k_{ym} = k_{yp}\beta_r n_r y_r^2 \mu_r/t_r; \qquad k_{zm} = k_{zp}\beta_r n_r z_r^2 \mu_r/t_r.$$

From (11.3.3) and (11.3.4):

$$\varphi^*_r = \mu_r q_{xr} x_r/k_{xr}\rho_r = \mu_r n_r x_r^2/k_{xr} t_r \rho_r.$$

If we wish to measure pressure as a separate entity, we must add the condition:

$$\beta_r p_r = 1; \qquad \rho_r = \rho_{0r} \tag{11.3.6}$$

derived from (2.3.4). From the definition of φ^* it follows that:

$$\varphi^*_r = z_r = p_r/\rho_r. \tag{11.3.7}$$

Altogether we now have 18 independent variables and 11 independent relationships. We may choose seven variables arbitrarily. For example, choosing $x_r, y_r, z_r, n_r, \mu_r, \beta_r, t_r$, we obtain: $\varphi^*_r = z_r$; $p_r = 1/\beta_r$; $U_r = n_r x_r y_r z_r$; k_{xr}, k_{yr} and k_{zr} are determined from (11.3.2); q_{xr}, q_{yr} and q_{zr} are determined from (11.3.4).

From (11.3.3) and (11.3.6) we obtain:

$$\rho_r = \rho_{0r} = \mu_r x_r q_{xr}/k_{xr}\varphi_r = 1/\beta_r z_r.$$

If the model is isotropic, $k_{xm} = k_{ym} = k_{zm} = k_m$. This reduces the number of independent variables by two, namely from 16 to 18 (when pressure is not an independent variable). As the number of independent equations remains unchanged, only six independent variables may be chosen arbitrarily. For example, choosing $x_r, n_r, \mu_r, \beta_r, \rho_r, t_r$, we obtain:

$$y_r = x_r(k_{xp}/k_{yp})^{1/2}; \qquad z_r = x_r(k_{xp}/k_{zp})^{1/2}; \qquad k_m = \beta_r n_r x_r^2 \mu_r k_{xp}/t_r \tag{11.3.8}$$

q_{xr}, q_{yr} and q_{zr} are determined by (11.3.4); U_r is determined by (11.3.5) and φ^*_r is determined by (11.3.3):

$$\varphi^*_r = 1/\rho_r\beta_r. \tag{11.3.9}$$

One should note the restriction that no more than one length scale may be chosen arbitrarily.

Example 2, the simultaneous flow of two incompressible liquids. This type of flow is discussed in detail in section 9.5. The porous medium is assumed to be isotropic.

The flow is described by the set of equations (9.3.5) and (9.3.24), for $\alpha = 1$ and 2, and (9.2.12) where $\cos\theta$ may be replaced by some function $f(\theta)$.

By following the usual procedure (par. 11.2.7), we obtain from these equations the following relationships among the scales involved in this problem:

$$S_{1r} = S_{2r} = 1 \quad \text{or} \quad (S_1/S_2)_m = (S_1/S_2)_p; \qquad S_{1m} = S_{1p}; \qquad S_{2m} = S_{2p} \tag{11.3.10}$$

where the last equality is between the values of S_α at homologous points in the model and in the prototype. Also:

$$p_{1r} = p_{2r} = (\sigma_r/\sqrt{k_r/n_r})f_r J_r \tag{11.3.11}$$

where $f_r = f(\theta_m)/f(\theta_p)$ and $J_r = J(S_{1m})/J(S_{1p})$. Both f_r and J_r depend on the porous matrix and the fluids chosen to simulate the medium and the fluids of the reservoir model. However, f_r has a constant value for a given pair θ_m, θ_p, whereas J_r varies with S_{1m} or S_{1p}. We also have:

$$q_{1xr}/x_r = q_{1yr}/y_r = q_{1zr}/z_r = n_r S_{1r}/t_r \tag{11.3.12}$$

$$q_{2xr}/x_r = q_{2yr}/y_r = q_{2zr}/z_r = n_r S_{2r}/t_r \tag{11.3.13}$$

$$q_{\alpha xr} = k_r(k_{r\alpha})_r p_{\alpha r}/\mu_{\alpha r} x_r; \qquad q_{\alpha yr} = k_r(k_{r\alpha})_r p_{\alpha r}/\mu_{\alpha r} y_r; \qquad \alpha = 1, 2 \tag{11.3.14}$$

$$q_{\alpha zr} = k_r(k_{r\alpha})_r p_{\alpha r}/\mu_{\alpha r} z_r = k_r(k_{r\alpha})_r \rho_{\alpha r} z_r/\mu_{\alpha r}; \qquad \alpha = 1, 2 \tag{11.3.15}$$

where

$$(k_{r\alpha})_r = [k_{r\alpha}(S_1)]_m/[k_{r\alpha}(S_1)]_p.$$

Twenty-five parameters are involved here: x, y, z, S_1, S_2, p_1, p_2, q_{1x}, q_{1y}, q_{1z}, q_{2x}, q_{2y}, q_{2z}, σ, n, k, k_{r1}, k_{r2}, ρ_1, ρ_2, μ_1, μ_2, t, f, J. As we have 14 independent relationships among the scales, we may choose 11 of them arbitrarily. As $S_{1r} = S_{2r} = 1$ always, only nine may be chosen arbitrarily. As always, this choice is subject to certain constraints; for example, that only one of the three length scales may be chosen arbitrarily, or that either ρ_{1r} or ρ_{2r} may be so chosen. If we add the capillary pressure p_c as another parameter, we must also add one condition:

$$p_{2r} = p_{1r} = p_{cr}. \tag{11.3.16}$$

Again, nine parameters may be chosen arbitrarily.

As an example, let us choose x_r, n_r, t_r, k_r, ρ_{1r}, J_r, μ_{1r}, μ_{2r}, f_r (by choosing θ_m). Then:

$$y_r = z_r = x_r \quad (\text{or} \quad (\delta x/\delta y)_r = 1; \quad (\delta x/\delta z)_r = 1)$$

$$S_{1r} = S_{2r} = 1 \quad (\text{or} \quad (S_1/S_2)_r = 1)$$

$$p_{1r} = p_{2r} = p_{cr} = \rho_{1r} z_r = \rho_{1r} x_r \quad (\text{or} \quad (p_1/p_2)_r = 1)$$

$$\rho_{2r} = \rho_{1r} \equiv \rho_r \quad (\text{or} \quad (\rho_1/\rho_2)_r = 1)$$

$$q_{1xr} = q_{1yr} = q_{1zr} = q_{2xr} = q_{2yr} = q_{2zr} = n_r x_r/t_r$$

$$\sigma_r = \rho_r x_r \sqrt{k_r/n_r}/t_r J_r$$

$$(k_{r1})_r = \frac{n_r x_r \mu_{1r}}{t_r k_r \rho_{1r}}; \qquad (k_{r2})_r = \frac{n_r x_r \mu_{2r}}{t_r k_r \rho_{2r}}$$

from which it follows that:

$$[(k_{r1}/\mu_1)/(k_{r2}/\mu_2)]_r = 1.$$

A special complication arises from the fact that some of the parameters considered here, namely k_{r1}, k_{r2} and J (or $p_c(S_1)$) are functions of saturation. Moreover, their dependence on saturation is known only as an experimental curve depending on the nature of the porous medium. This imposes a severe restriction on the choice of the model's porous matrix. Unless the model's porous matrix is identical to that of the prototype, it is impossible to obtain constant scales for k_{r1} and k_{r2} independent of saturation. For the same reason, $J_r = [J(S_m)]/[J(S_p)] = \text{const}$ means that a certain proportionality must be maintained between the two curves in order to have a constant scale J_r. Because of these practically insurmountable difficulties, some authors use approximate scales, a procedure that is quite common in hydrodynamics.

This example, with some small modification, can also be used for scaling unsaturated flow (sec. 9.4).

Example 3, hydrodynamic dispersion (miscible displacement; chap. 10). The general case of hydrodynamic dispersion of an incompressible inhomogeneous fluid was summarized in paragraph 10.5.4. There it was shown that ten equations—(10.5.43) through (10.5.47)—are necessary to describe the problem. The dispersion equation itself for an isotropic medium can be written in the form

$$\frac{\partial C}{\partial t} = \frac{\partial}{\partial x_i}\left\{(a_I - a_{II})f(Pe, \delta)\frac{V^*_i V^*_j}{V^*}\frac{\partial C}{\partial x_j} + a_{II}f(Pe, \delta)V^*\frac{\partial C}{\partial x_i} + D_d T^*\frac{\partial C}{\partial x_i}\right\} - V'_i\frac{\partial C}{\partial x_i} \tag{11.3.17}$$

Following the procedure explained and employed in the present section, we obtain from (10.5.43) through (10.5.47) the following relationships among the variables V^*_i, V'_i, ρ, μ, p, C in an isotropic medium:

$$(V^*_i)_r = \frac{k_r p_r}{n_r \mu_r (x_i)_r} = \frac{k_r \rho_r z_r}{n_r \mu_r (x_i)_r} = (V'_i)_r \tag{11.3.18}$$

$$\rho_r = \rho_{0r} = \alpha_r C_r = \alpha_r (C_0)_r \tag{11.3.19}$$

$$\mu_r = \mu_{0r} = \beta_r C_r = \beta_r (C_0)_r \tag{11.3.20}$$

$$\frac{C_r}{t_r} = \frac{f_r(a_I)_r (V^*_i)_r (V^*_j)_r C_r}{(x_i)_r V^*_r (x_j)_r} = \frac{f_r(a_{II})_r (V^*_i)_r (V^*_j)_r C_r}{(x_i)_r V^*_r (x_j)_r}$$

$$= \frac{f_r(a_{II})_r V^*_r C_r}{(x_i)_r (x_i)_r} = \frac{(D_d)_r T^*_r C_r}{(x_i)_r (x_i)_r} = \frac{(V'_i)_r C_r}{(x_i)_r} \tag{11.3.21}$$

$$\frac{(V^*{}_i)_r}{(x_i)_r} = \frac{(V^*{}_j)_r}{(x_j)_r}; \qquad (V^*{}_i)_r = (V'{}_i)_r = \frac{(D_d)_r T_r \rho_r}{(x_i)_r}. \tag{11.3.22}$$

In all these equations, $i, j = 1, 2, 3$.

From (11.3.18) through (11.3.22), the following independent relationships may be attained:

$$1/t_r = (V^*{}_i)_r/(x_i)_r = (V^*{}_j)_r/(x_j)_r, \qquad i, j = 1, 2, 3 \tag{11.3.23}$$

$$(V^*{}_i)(x_i)_r = (V_j)^*(x_j)^* = (V_k)_r(x_k)_r \tag{11.3.24}$$

$$(x_i)_r = (x_j)_r = (x_k)_r \equiv x_r = z_r \quad \text{or:} \quad (x_1)_r = (x_2)_r = (x_3)_r \equiv x_r. \tag{11.3.25}$$

Hence, we also have:

$$(V^*{}_i)_r = (V^*{}_j)_r = (V^*{}_k)_r \equiv V_r \tag{11.3.26}$$

and we may replace $(x_i)_r$ and $(V^*{}_i)_r$, $i = 1, 2, 3$, by x_r and V_r, respectively, in (11.3.18) through (11.3.24).

From the definition of $f \equiv f(Pe, \delta)$, it follows that:

$$f = Pe/(Pe + A); \qquad fPe + fA = Pe$$

and hence:

$$f_r(Pe)_r = f_r A_r = (Pe)_r$$

$$f_r = 1, \qquad (Pe)_r = 1, \qquad A_r = 1 \quad (\text{i.e.,} \quad \delta_r = 1) \tag{11.3.27}$$

as was to be expected, since all of these parameters are dimensionless functions and should therefore be the same in the model as in the prototype.

With $(x_i)_r \equiv x_r$; $(V_i)_r \equiv V_r$ and $f_r = 1$, we now obtain:

(a), (b) $$t_r = x_r/V_r = k_r\rho_r/n_r T^*{}_r \mu_r \tag{11.3.28}$$

(c) $$p_r = \rho_r g_r x_r \tag{11.3.29}$$

(d) $$V_r = k_r \rho_r g_r/n_r \mu_r \tag{11.3.30}$$

(e), (f) $$\rho_r = (\rho_0)_r = \alpha_r (C_0)_r \tag{11.3.31}$$

(g), (h) $$\mu_r = (\mu_0)_r = \beta_r (C_0)_r \tag{11.3.32}$$

(i) $$C_r = (C_0)_r \tag{11.3.33}$$

(j) $$(a_I)_r = (a_{II})_r \equiv a_r \tag{11.3.34}$$

(k) $$a_r = x_r \tag{11.3.35}$$

(l) $$(D_d)_r (T^*)_r = V_r x_r. \tag{11.3.36}$$

Equation (11.3.28b) was obtained by using (4.7.12) rather than (10.5.43) as the motion equation.

Altogether we have here 12 independent relationships for the 18 scale factors: x_r, V_r, p_r, $(a_I)_r$, $(a_{II})_r$, k_r, n_r, $T^*{}_r$, ρ_r, $(\rho_0)_r$, μ_r, $(\mu_0)_r$, α_r, β_r, C_r, $(C_0)_r$, $(D_d)_r$. Six of these factors may therefore be chosen arbitrarily. A detailed example is given by Bachmat (1967a), who designed a model for the studies of the spreading of a solute in ground water flow.

By inserting (11.3.35) in (11.3.18) we obtain:

$$(Vk\rho/nT^*\mu a)_r = 1; \quad \text{or} \quad V_m k_m \rho_m / n_m T^*{}_m a_m = V_p k_p \rho_p / n_p T^*{}_p a_p \tag{11.3.37}$$

or, with $\nu = \mu/\rho$:

$$[V(k/nT^*a)/\nu]_r = 1. \tag{11.3.38}$$

In (11.3.38), k/nT^*a has the dimension of length. From the discussion in paragraph 10.4.3, $k = nBT^*$ so that $k/nT^*a = B/a$. This ratio characterizes the porous medium. Although its dimension is length, we must be careful, as B is proportional to the square of a length characterizing the cross-section of a channel, while a is proportional to a length $\tilde{L}$ of a channel of the porous medium.

Finally, from (11.3.36) we have:

$$(V\,\Delta x/D_d T^*)_r = 1 \quad \text{or} \quad (Va/D_d T^*)_r = 1. \tag{11.3.39}$$

In section 10.4, a Peclet number $Pe = VL/D_d$ was defined, where L is a characteristic channel length (or, in general, a length characterizing the microscopic structure of the porous medium). A comparison between this definition of Pe and (11.3.39) may lead (Bachmat 1967a) to a definition of another Peclet number:

$$Pe^* = Va/D_d T^* = qa/(nD_d T^*) \tag{11.3.40}$$

where a (either a_I or a_{II}) represents the characteristic length and $D_d T^*$ (or $nD_d T^*$) is the molecular diffusivity in a porous medium. Thus (11.3.39) requires the identity of Pe^* at homologous points in the model and in the prototype.

Although, seemingly, the above set of design criteria permits us to choose porous media, fluids, tracers, model dimensions, etc., for any model investigations, the simultaneous fulfillment of all design criteria given above imposes severe restrictions on such a choice. For example, we must choose a porous medium for the model such that n, k, a_I and a_{II} will satisfy the design criteria. In fact, we must also have $x_r = (a_I)_r = (a_{II})_r$, which is in itself a severe restriction. It is also most difficult to find an appropriate fluid. As in many other cases of model simulation, the solution to the problem is to ignore certain criteria, the effect of which is shown to be small in comparison to the others, and to achieve only partial similitude. Another possible approximation is to delete the molecular diffusion term in the dispersion equation. This will cause the requirement that Pe^* be an invariant to vanish. Several other approximations are mentioned by Bachmat (1967a).

Example 4, heat and mass transfer (sec. 10.7). The set of equations describing heat and mass transfer in a homogeneous isotropic medium is:

$$\partial V_i/\partial x_i = 0 \tag{11.3.41}$$

$$V_i = -(k/n\mu)(\partial p/\partial x_i + \rho g\, \partial z/\partial x_i) \tag{11.3.42}$$

$$\partial \rho_\alpha/\partial t = \partial(D'_{ij}\, \partial \rho_\alpha/\partial x_j)/\partial x_i - V'_i\, \partial \rho_\alpha/\partial x_i \tag{11.3.43}$$

(or in the equivalent form (11.3.17), with $C \equiv \rho_\alpha$)

$$\partial T/\partial t = D_h \nabla^2 T - (V_F)_i\, \partial T/\partial x_i \tag{11.3.44}$$

$$\rho = \rho_{0T}[1 - \eta(T - T_{f0})] + \alpha\rho_\alpha \qquad (11.3.45)$$

$$\mu = \mu_{0T}\exp[-\delta(T - T_{f0})] + \chi\rho_\alpha. \qquad (11.3.46)$$

These equations were obtained from (10.7.33) through (10.7.38) by introducing:

$$\rho \equiv \rho_f, \qquad \rho_{\alpha 0} = 0, \qquad D_h\,(\textit{thermal diffusivity}) = \lambda_e/[n\rho_f c_f + (1-n)\rho_s c_s];$$

$\mathbf{V}_F$ is the heat front velocity defined by

$$\mathbf{V}_F = \{n\rho_f c_f/[n\rho_f c_f + (1-n)\rho_s c_s]\}\mathbf{V}^*.$$

Following the usual procedure, we obtain the following relationships among the various scales:

$$x_r = y_r = z_r = L_r \qquad (11.3.47)$$

$$V_r = L_r/t_r = k_r p_r/n_r\mu_r L_r = k_r\rho_r/n_r\mu_r \quad (\text{taking } g_r \equiv 1) \qquad (11.3.48)$$

$$C_r/t_r = f_r(a_I)_r V_r C_r/L_r^2 = f_r(a_{II})_r V_r C_r/L_r^2 = (D^*_d)_r C_r/L_r^2 = V'_r C_r/L_r \qquad (11.3.49)$$

$$T_r/t_r = (D_h)_r T_r/L_r^2 = (V_F)_r T_r/L_r \qquad (11.3.50)$$

$$\rho_r = (\rho_{0T})_r = (\rho_{0T})_r \eta_r T_r = \alpha_r C_r \qquad (11.3.51)$$

$$\mu_r = (\mu_{0T})_r = \chi_r C_r \qquad (11.3.52)$$

$$\delta_r T_r = 1. \qquad (11.3.53)$$

The discussion on the difficulties inherent in the scaling of mass transfer (hydrodynamic dispersion) alone is also applicable here, except that here we have additional difficulties because of limitations in the choice of fluids. For example, for practical reasons we must choose the fluid so that:

$$\eta_r = 1; \qquad \alpha_r = 1; \qquad \chi_r = 1 \qquad (11.3.54)$$

(i.e., the same fluid in the model as in the prototype). It follows then that:

$$T_r = 1; \qquad \rho_r = (\rho_{0T})_r = 1; \qquad C_r = 1; \qquad \mu_r = 1. \qquad (11.3.55)$$

The condition $T_r = 1$ (i.e., the same temperatures in the model as in the prototype) is another serious constraint on the execution of laboratory work. With (11.3.53) and (11.3.54) we obtain:

$$V_r = L_r/t_r = k_r/n_r \qquad (11.3.56)$$

$$f_r = 1; \qquad (a_I)_r = (a_{II})_r \equiv a_r \qquad (11.3.57)$$

which means a severe restriction on the choice of porous medium for the model.

$$(D^*_d)_r = L_r^2/t_r = (D_q)_r \qquad (11.3.58)$$

$$V_r = (V_F)_r; \qquad \{n\rho_f c_f/[n\rho_f c_f + (1-n)\rho_s c_s]\}_r = 1. \qquad (11.3.59)$$

Dagan and Kahanovitz (1968) show a possibility that circumvents the restriction $T_r = 1$. They rewrite (11.3.42) in the form:

$$V_i = -\frac{k\rho_{0T}g}{\mu n}\frac{\partial\varphi}{\partial x_i} - \frac{k\,\Delta\rho g}{\mu}\frac{\partial z}{\partial x_i}; \qquad \rho = \rho_{0T} + \Delta\rho \qquad (11.3.60)$$

where the piezometric head φ is measured in terms of a fluid of density ρ_{0T}. Then the following relationships hold:

$$x_r = y_r = z_r = L_r; \qquad V_r = k_r(\Delta\rho)_r/\mu_r = L_r/t_r \tag{11.3.61}$$

$$(\rho_{0T})\varphi_r = (\Delta\rho)_r; \qquad \Delta\rho_r = (\rho_{0T})_r\eta_r T_r = \alpha_r C_r. \tag{11.3.62}$$

The relationships in (11.3.61) and (11.3.62) involve no approximation. As for viscosity, an approximation must be introduced, for example in the form $\mu_r = (\mu \text{ average})_m/(\mu \text{ average})_p$, because of the nonlinear variation of viscosity with temperature. A proper scaling of the various dispersion and diffusion coefficients (including thermal diffusion) requires further approximations.

11.4 The Viscous Flow Analogs

11.4.1 General

The *viscous flow analog* (*Hele–Shaw analog, parallel-plate analogy*) is a well known device for two-dimensional ground water investigations. It is based on the similarity between the differential equations governing saturated flow in a porous medium and those describing the flow of a viscous liquid in the narrow space between two parallel planes. The analog was first developed by Hele–Shaw (1897, 1898) for studying the potential flow patterns around variously shaped bodies. Zamarin (1931) was probably the first to apply this analog to the study of seepage through earth dams. Dachler (1936) applied it to ground water investigations. Since then it has been used extensively by many investigators for solving regional ground water flow problems, including such problems as artificial recharge, sea water intrusion, drainage, seepage through earth dams (e.g., Aravin 1938, 1941; Gunther 1940a,b; Dietz 1941; Kellog 1948; Santing 1951a,b, 1957; Todd 1954, 1955a,b; Kruysse and Bear 1956; Bear 1960a; Mikhailov 1956; Bear and Zaslavsky 1961; DeWiest 1962; Naor and Bear 1963; Sternberg and Scott 1963; and Columbus 1965). It has also been used for studies of oil production in reservoirs (e.g., Polubarinova-Kochina and Shkrich 1954; Bear, Jacobs and Braester 1966). Two types of viscous flow analogs will be described below: the vertical analog and the horizontal one. In fact, the two parallel plates may be placed at any angle with respect to the horizontal.

11.4.2 Description of the Vertical Hele–Shaw Analog

The analog consists of two parallel plates placed in a vertical position. At least one of the plates should be transparent (e.g., lucite). The plates, 6–30 mm thick, are kept at a fixed distance apart (usually 0.8 to 3.5 mm) by a network of screws and spacers (fig. 11.4.1). Fewer screws and spacers are needed if thicker (20–30 mm) plates are used. The size of the plates depends on the prototype dimensions and on the analog scales. A viscous liquid (or several immiscible, or virtually immiscible, liquids such as water, oil or glycerin) is allowed to flow in the narrow space between the plates. The flow domain between the plates is bounded partly by an impervious packing or spacer (fig. 11.4.1) simulating impervious boundaries of the flow domain

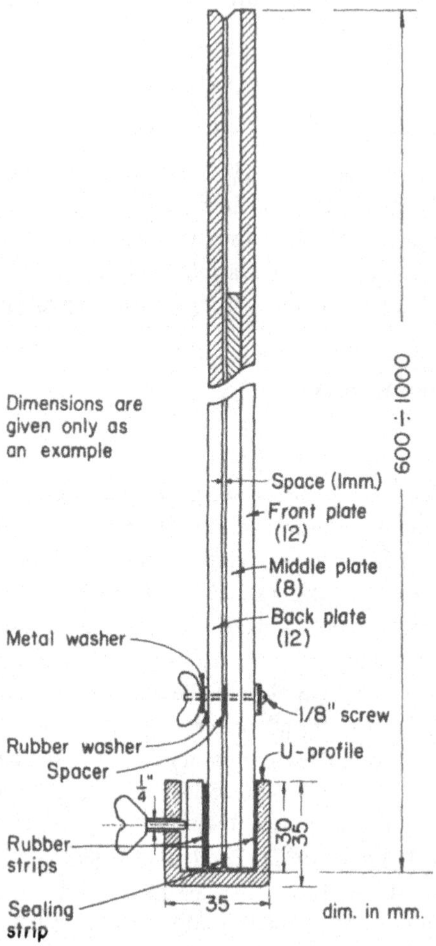

FIG. 11.4.1. Cross-section of a vertical Hele-Shaw analog.

in the prototype. Several typical analogs are shown in figure 11.4.2. Inhomogeneous hydraulic conductivity in the flow domain is simulated by varying the width of the interspace. This can be done by varying the thickness of the spacers (for a continuous variation of K), or by inserting into the space thin sheets of plastic with the geometrical shape of the region of reduced K, thereby leaving free a narrower space for the flow of the viscous liquid (fig. 11.4.2c). However, this is subject to severe restrictions (par. 11.4.4) that make the simulation of inhomogeneity, especially in the horizontal Hele–Shaw analog, almost impossible.

The ordinary analog is isotropic. By a proper scale distortion (sec. 11.2 and par. 11.4.4), an anisotropic flow domain may be simulated. By using a special kind of lucite with grooves for one of the plates, an analog with anisotropic permeability is obtained.

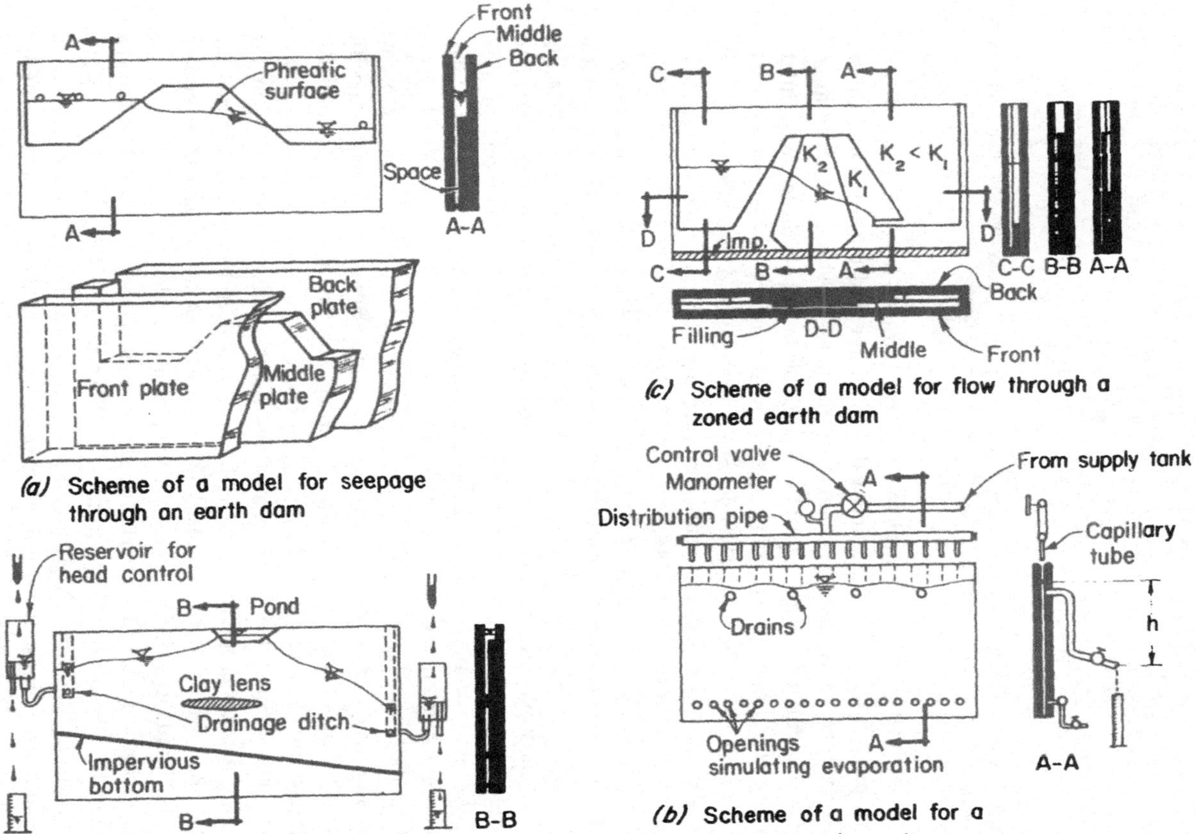

(a) Scheme of a model for seepage through an earth dam

(c) Scheme of a model for flow through a zoned earth dam

(b) Scheme of a model for flow from a spreading pond

(b) Scheme of a model for a parallel drain system

FIG. 11.4.2. Some typical vertical Hele-Shaw analogs (Bear and Zaslavsky, 1961).

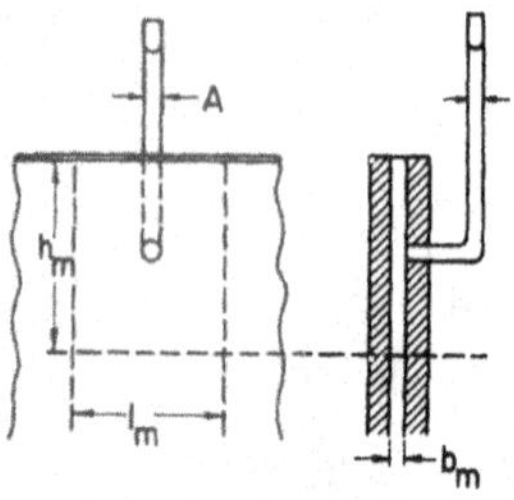

FIG. 11.4.3. Simulating specific storativity in a Hele-Shaw analog.

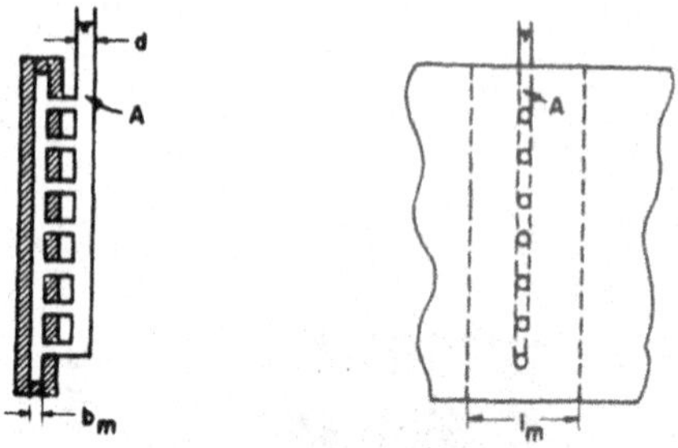

FIG. 11.4.4. Simulation of aquifer storativity in a vertical Hele-Shaw analog.

Liquid is introduced into the interspace, or withdrawn from it, through holes in the back plate fitted with pipes and valves. Inflow and outflow rates through these holes may be controlled by connecting them to fixed supply (or drainage) reservoirs, or to regulated pumps. Constant or variable head boundaries are simulated by connecting the proper points (curves, etc.) in the analog to control reservoirs. One should recall that the vertical Hele–Shaw represents a *two-dimensional flow domain*. Hence, every opening represents in the prototype a drain that is perpendicular to the investigated vertical cross-section.

Because, in the analog, pressures are small and the plates are rigid, aquifer storativity (sec. 6.4), resulting from soil and water compressibility, is simulated in the vertical analog by connecting vertical transparent pipes to openings in the back plate (fig. 11.4.3). In this way, the continuously distributed storativity is made discrete, each "storage pipe" representing the specific storage capacity of a certain area. The storage pipes serve at the same time as piezometers, indicating the piezometric head in the investigated flow domain.

With the notation of figure 11.4.3, the pipe of cross-sectional area A represents the specific storativity of a block of soil $b_m \times l_m \times h_m$. Hence, the model's specific storativity is $(S_s)_m = A/b_m \times l_m \times h_m$. Aquifer storativity can be simulated in asimilar manner (fig. 11.4.4). Then $S_m = A/b_m\, l_m$.

If two or more immiscible (or virtually immiscible) liquids are flowing in the interspace, the domain occupied by each of the liquids may be observed by adding a dye to the liquid (photo 11.1). In steady flow, streamlines may be traced by adding a dye to the liquid at various points through small nozzles.

Because viscosity plays an important role in analog scaling (par. 11.4.3), the analog should be placed in a temperature-controlled room. If this is not possible, the temperature must be measured at all inflow and outflow points during each experiment and scales must be recomputed according to the varying average temperature of the liquid in the analog.

11.4.3 Establishing the Analogy between Analog and Prototype

Consider flow between the two vertical parallel plates shown in figure 11.4.5a.

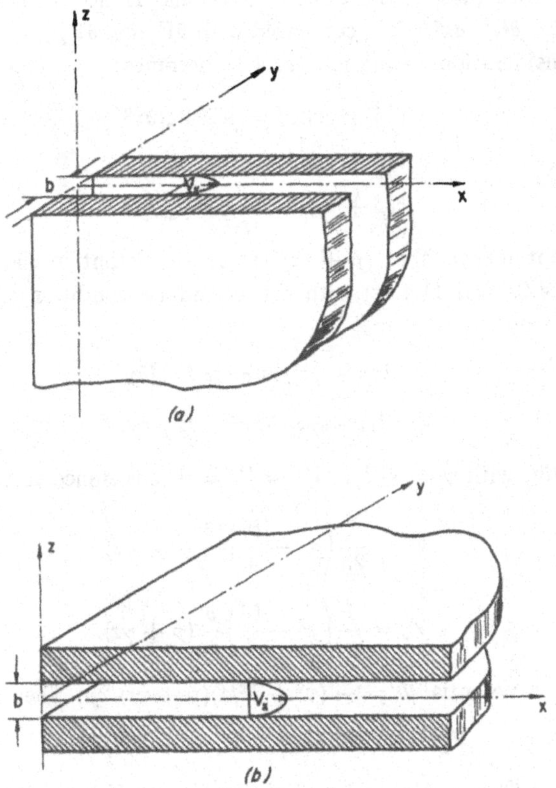

FIG. 11.4.5. Vertical and horizontal parallel plates.

The Navier–Stokes equations for a viscous incompressible fluid may be written in the form:

$$DV_x/Dt = f_x - \rho^{-1}\,\partial p/\partial x + \nu\,\nabla^2 V_x$$
$$DV_y/Dt = f_y - \rho^{-1}\,\partial p/\partial y + \nu\,\nabla^2 V_y$$
$$DV_z/Dt = f_z - \rho^{-1}\,\partial p/\partial z + \nu\,\nabla^2 V_z \qquad (11.4.1)$$

where $D(\ )/Dt$ represents the hydrodynamic derivative, V_x, V_y, V_z are the velocity components in the x, y, z directions, respectively, ν is the kinematic viscosity and f_x, f_y, f_z are components of the external force per unit mass acting on the liquid. For a liquid flowing in the narrow *vertical* space of width b, $V_y = 0$. In very slow motions (*creeping motions*) or in fluids with very high viscosities, the viscous forces are considerably greater than the inertial forces. If such a flow is assumed to take place between the plates, we may neglect the inertial terms, which constitute the left-hand side of (11.4.1). The only active body force is gravity, the potential of which is gz, so that $f_x = -\,\partial(gz)/\partial x = 0$, $f_y = -\,\partial(gz)/\partial y = 0$, $f_z = -\,\partial(gz)/\partial z = -\,g$. Since the liquid sticks to the plates, the velocity gradients in the y direction are much larger than those in the x and z directions. Hence, we may neglect $\partial V_x/\partial x$, $\partial^2 V_x/\partial x^2$, $\partial V_z/\partial x$, $\partial^2 V_z/\partial x^2$, when compared with $\partial V_x/\partial y$, $\partial V_z/\partial y$, $\partial^2 V_x/\partial y^2$, $\partial^2 V_z/\partial y^2$. With these considerations, equation (11.4.1) becomes:

$$\text{(a)} \quad \partial(p + \gamma z)/\partial x = \mu \, \partial^2 V_x/\partial y^2$$

$$\text{(b)} \quad \partial(p + \gamma z)/\partial y = 0$$

$$\text{(c)} \quad \partial(p + \gamma z)/\partial z = \mu \, \partial^2 V_z/\partial y^2. \qquad (11.4.2)$$

From (11.4.2b) it follows that $(p + \gamma z)$ remains constant in the y direction. By integrating (11.4.2a and 11.4.2c) with the boundary conditions: $y = 0$, $\partial V_x/\partial y = \partial V_z/\partial y = 0$, we obtain:

$$y \, \partial(p + \gamma z)/\partial x = \mu \, \partial V_x/\partial y$$

$$y \, \partial(p + \gamma z)/\partial z = \mu \, \partial V_z/\partial y. \qquad (11.4.3)$$

Integrating again, with $y = \pm b/2$, $V_x = V_z = 0$ (adherence to walls), we obtain:

$$V_x = \frac{1}{2\mu}\left(y^2 - \frac{b^2}{4}\right)\frac{\partial}{\partial x}(p + \gamma z)$$

$$V_z = \frac{1}{2\mu}\left(y^2 - \frac{b^2}{4}\right)\frac{\partial}{\partial z}(p + \gamma z). \qquad (11.4.4)$$

Let us define a potential $\Phi' = -(y^2 - b^2/4)(p + \gamma z)/2\mu$. Then (11.4.4) becomes:

$$V_x = -\,\partial\Phi'/\partial x; \qquad V_z = -\,\partial\Phi'/\partial z \qquad (11.4.5)$$

which means that Φ' is a *velocity potential* (for V_x and V_z) which depends on y. By integrating (11.4.4), we obtain the *specific discharge* in the interspace between the plates:

in the x direction: $$q_x = \frac{1}{b}\int_{-b/2}^{+b/2} V_x(y)\,dy = -\frac{b^2}{12}\frac{\gamma}{\mu}\frac{\partial}{\partial x}\left(z + \frac{p}{\gamma}\right)$$

in the z direction: $$q_z = \frac{1}{b}\int_{-b/2}^{b/2} V_z(y)\,dy = -\frac{b^2}{12}\frac{\gamma}{\mu}\frac{\partial}{\partial z}\left(z + \frac{p}{\gamma}\right). \qquad (11.4.6)$$

Introducing the *piezometric head*: $\varphi = z + p/\gamma$, and the hydraulic conductivity of the space between the plates, $K_m = \frac{1}{12}g(b^2/\nu)$, equation (11.4.6) becomes:

$$q_x = -K_m \, \partial\varphi/\partial x; \qquad q_z = -K_m \, \partial\varphi/\partial z. \qquad (11.4.7)$$

Hence, we have here a potential flow with respect to the specific discharge vector $\mathbf{q}$. Equations (11.4.7) may be rewritten in the form:

$$\mathbf{q} = -K_m \operatorname{grad} \varphi. \qquad (11.4.8)$$

The analogy between this expression and Darcy's law is obvious. Continuity will lead to $\nabla^2\varphi = 0$, which is analogous to the Laplace equation (6.2.24). It is also interesting to note the analogy between the expression for K_m in (11.4.7) and the expression for the hydraulic conductivity in (5.5.8). In both cases we have a term describing the capacity of the medium to transmit fluid (permeability, cd^2, $b^2/12$) and another term (g/ν) describing the liquid. The analogy to other continuity equations may be established in a similar manner.

In deriving the analogy above, it is assumed that we have *laminar flow* throughout the analog. Such flow may be characterized by a Reynolds number Re defined for the analog by $Re = qb/\nu$. Aravin and Numerov (1953, 1965) suggest $Re < 500$. Santing (1957) suggests $Re < 1000$. Instead of referring to these values of Re, one may observe streamlines in the analog and determine by using dyes whether or not the flow is laminar everywhere. Above these critical values, the flow becomes turbulent and the analogy fails.

11.4.4 Scales for the Vertical Analog

Once the analogy has been established by the equation of motion (11.4.6) and by the continuity or the mass conservation equation, scales may be derived by the inspectional method described in section 11.2, or by using the results of section 7.4. In all cases $K_m = b_m^2\, g/12\nu_m$, and we assume that x and z (in the anisotropic prototype) are the principal axes of permeability.

Example 1, flow of a single liquid in a confined domain without elastic storativity. We have:

$$x_r^2/K_{xr} = z_r^2/K_{zr} \quad \text{or} \quad x_r^2 K_{xp} = z_r^2 K_{zp}$$

$$Q_r = K_{xr}\varphi_r b_r z_r/x_r; \qquad t_r = n_r x_r^2/K_{xr}\varphi_r \tag{11.4.9}$$

for the eight scales: x_r, z_r, b_r, Q_r, t_r, n_r, φ_r, K_{xr}. Choosing $K_{xr} \equiv K_m/K_{xp}$ actually means choosing K_m or a combination of width b_m and a liquid viscosity ν_m. With the *three* equations of (11.4.9), this means that *five* scales may be chosen arbitrarily (only one of x_r and z_r). In the model $n_m = 1$, $b_m = b$ (representing some width b_p in the prototype).

If we wish to consider volume U of liquid, we must use the condition

$$U_r = Q_r t_r = n_r b_r x_r z_r. \tag{11.4.10}$$

Example 2, phreatic flow of a single liquid. In addition to the conditions of example 1, we have:

$$\varphi_r = z_r. \tag{11.4.11}$$

When accretion (N) takes place, we have:

$$N_r = Q_r/x_r b_r = n_r x_r z_r b_r/x_r b_r t_r = n_r z_r/t_r = n_r z_r K_{xr} z_r/n_r x_r^2 = K_{zr}. \tag{11.4.12}$$

Example 3, flow of two liquids separated by an interface. The liquids are assumed incompressible and the flow domain is homogeneous, anisotropic and nondeformable.

The flow (in the xz vertical plane) is governed by:

$$q^{(i)}_x = -(k_x\gamma^{(i)}/\mu^{(i)})\, \partial\varphi^{(i)}/\partial x;$$
$$q^{(i)}_z = -(k_z\gamma^{(i)}/\mu^{(i)})\, \partial\varphi^{(i)}/\partial z; \qquad i = 1, 2 \text{ represents the two liquids } (\gamma^{(2)} > \gamma^{(1)})$$
$$k_x\, \partial^2\varphi^{(i)}/\partial x^2 + k_z\, \partial^2\varphi^{(i)}/\partial z^2 = 0. \tag{11.4.13}$$

Neglecting capillary pressure, along the interface between the two liquids we have (par. 9.5.3):

$$\gamma_2\varphi_2 - \gamma_1\varphi_1 = (\gamma_2 - \gamma_1)z \tag{11.4.14}$$

and the two equations in (9.5.29). The average velocity is given by:

$$V_x = dx/dt = -(k_x\gamma^{(i)}/n\mu^{(i)})\, \partial\varphi^{(i)}/\partial x; \qquad i = 1, 2. \tag{11.4.15}$$

By writing these equations first for the prototype and then for the analog, and following the procedure outlined in sections 7.4 and 11.2, we obtain the following relationships among the various scales:

$$q^{(i)}_{xr} = k_{xr}\rho^{(i)}{}_r\varphi^{(i)}{}_r/\mu^{(i)}{}_r x_r; \qquad q^{(i)}_{zr} = k_{zr}\rho^{(i)}{}_r\varphi^{(i)}{}_r/\mu^{(i)}{}_r z_r; \qquad i = 1, 2 \tag{11.4.16}$$
$$x_r^2/k_{xr} = z_r^2/k_{zr} \tag{11.4.17}$$
$$\rho^{(1)}{}_r\varphi^{(1)}{}_r = \rho^{(2)}{}_r\varphi^{(2)}{}_r = (\Delta\rho)_r z_r \tag{11.4.18}$$
$$\rho^{(1)}{}_r = \rho^{(2)}{}_r \tag{11.4.19}$$
$$\mu^{(1)}{}_r = \mu^{(2)}{}_r \qquad (\text{or: } (\mu^{(1)}/\mu^{(2)})_r = 1) \tag{11.4.20}$$
$$t_r = n_r\mu^{(1)}{}_r x_r^2/k_{xr}\rho^{(1)}{}_r\varphi^{(1)}{}_r. \tag{11.4.21}$$

The ten independent equations (11.4.16) through (11.4.21) include the 14 independent variables: x_r, z_r, k_m, $q^{(1)}{}_{xr}$, $q^{(2)}{}_{xr}$, $q^{(1)}{}_{zr}$, $q^{(2)}{}_{zr}$, $\varphi^{(1)}{}_r$, $\varphi^{(2)}{}_r$, $\rho^{(1)}{}_r$, $\rho^{(2)}{}_r$, $\mu^{(1)}{}_r$, $\mu^{(2)}{}_r$, t_r (since $n_r = 1/n_p$). Therefore, four may be chosen arbitrarily. For example, if we choose: k_m, z_r, $\rho^{(1)}{}_r$ and $\mu^{(1)}{}_r$, we obtain:

$$q^{(1)}{}_{xr} = q^{(2)}{}_{xr} = k_{xr}(\rho^{(1)}{}_r/\mu^{(1)}{}_r)(k_{xp}/k_{zp})^{1/2} = (k_m{}^2/k_{xp}k_{zp})^{1/2}\rho^{(1)}{}_r/\mu^{(1)}{}_r$$
$$= (b^2/12)p^{(1)}{}_r/(k_{xp}k_{zp})^{1/2}\mu^{(1)}{}_r \tag{11.4.22}$$
$$q^{(1)}{}_{zr} = q^{(2)}{}_{zr} = k_{zr}\rho^{(1)}{}_r/\mu^{(1)}{}_r = (b_m{}^2/12k_{zp})\rho^{(1)}{}_r/\mu^{(1)}{}_r \tag{11.4.23}$$
$$x_r = z_r(k_{zp}/k_{xp})^{1/2} \tag{11.4.24}$$
$$\varphi^{(1)}{}_r = \varphi^{(2)}{}_r = z_r \tag{11.4.25}$$
$$\rho^{(1)}{}_r = \rho^{(2)}{}_r; \qquad \mu^{(1)}{}_r = \mu^{(2)}{}_r \tag{11.4.26}$$
$$t_r = (n_r\mu^{(1)}{}_r/k_{xr}\rho^{(1)}{}_r)z_r k_{zp}/k_{xp} = 12\mu^{(1)}{}_r k_{zp}z_r/b_m{}^2 n_p\rho^{(1)}{}_r. \tag{11.4.27}$$

One should recall that choosing k_m means choosing b_m.

It is possible to simulate in a vertical analog a nonhomogeneous medium, where the inhomogeneity is in the form of zones (e.g., a zoned dam) or layers, each of a constant permeability. For an isotropic prototype, we require that Q_r remain constant throughout the analog. This means $K_r b_r$ = const or $b_m{}^3/K_p$ = const in all portions of the analog.

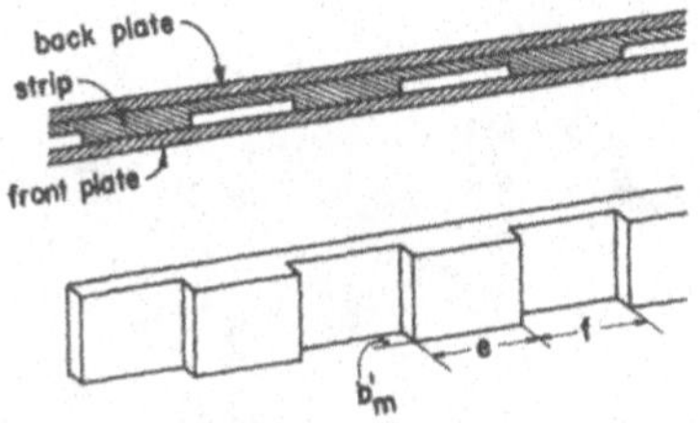

FIG. 11.4.6. Simulation of a thin semipervious layer on a vertical Hele-Shaw analog.

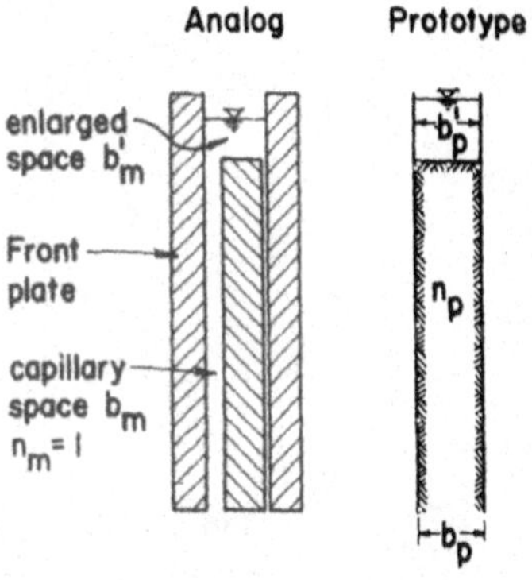

FIG. 11.4.7. Simulation of open liquid bodies in a vertical Hele-Shaw analog.

Although this condition seems at first to contradict the definition of K_m in (11.4.6), the reason for the third power is that b_m in the analog plays two roles: it represents the permeability of the analog (in the form b_m^2) and it represents the width of the investigated domain. When inhomogeneous media are being considered, this width remains unchanged while b_m varies in the analog. The apparent contradiction can be removed by considering a kind of transmissivity $T = Kb$ in both model and prototype. Then $T_m \propto b_m^3$.

Designating the layers in an anisotropic medium with $S_s \neq 0$ by a prime and a double prime, we have: $b'^3_m/k'_{xp} = b''^3_m/k''_{xp}$ and also $b'^3_m/k_{zp} = b''^3_m/k''_{zp}$, i.e., $k'_{xp}/k'_{zp} = k''_{xp}/k''_{zp}$. In addition, from the requirement of $t_r = \text{const}$ throughout the model, we have: $t_r = z_r^2 K_{zp}(S_s)_r/K_m = \text{const}$ or $(S_s)_r k_{zp}/b_m^2 = \text{const}$. By combining the two requirements $Q_r = \text{const}$ and $t_r = \text{const}$ (i.e., $U_r = Q_r t_r = \text{const}$, where U is volume of liquid), we obtain $(S_s)_r b_m = \text{const}$ throughout the analog. If the movement of fronts is also considered, the requirement $t_r = \text{const}$ leads to $n_p/b_m = \text{const}$. Altogether, in a layered analog, we may have to satisfy $b_m^3/k_{xp} = \text{const}$, $b_m^3/k_{zp} = \text{const}$, $(S_s)_r b_m = \text{const}$ and $n_p/b_m = \text{const}$. All these conditions can seldom be satisfied simultaneously. In steady flow, conditions involving time may be waived (in the absence of front movements), and the simulation of a nonhomogeneous aquifer becomes simpler.

When a relatively thin semipervious layer is present, treating it as a layer in a multilayered aquifer often leads to interspaces that are too narrow (e.g., 0.1 mm). One may then represent the semipervious layer by a strip that completely seals part of the interspace, leaving a wider interspace in the remaining part (fig. 11.4.6). We then simulate the resistivity b'/K' of this layer rather than b' and K' separately. Instead of calculating the resistivity of such a layer, it is better to determine it experimentally in the analog by measuring discharge through it under a controlled head.

Often, bodies of liquid adjacent to the flow domain in the porous medium must

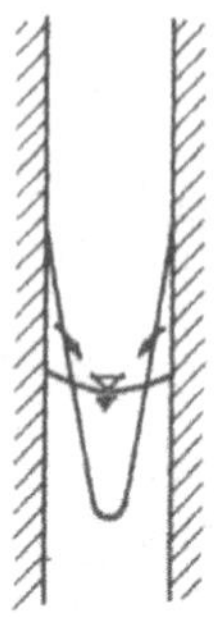

FIG. 11.4.8. Delayed storage in a vertical Hele-Shaw analog.

be simulated as boundary conditions. Rivers and lakes are examples. In these cases, the model is often made of three plates (figs. 11.4.1, 11.4.2a, 11.4.2b and 11.4.2c). The narrow space is maintained between two of them, while the third one, together with the middle one (which has the shape of the desired geometry of the open liquid boundary), increase the spacing up to the sum of the capillary space and the thickness of the middle plate (fig. 11.4.7). When the volume of liquid U leaving the open liquid body (say a reservoir or a ditch) and entering the interspace is important, the enlarged width must satisfy an additional condition:

$$b'_r = b_r/n_{\mathscr{p}}. \tag{11.4.28}$$

Because of the capillary rise of the liquid in the narrow analog interspace, the liquid level in it does not represent a phreatic surface ($p = 0$), but rather the upper boundary of saturation (par. 7.1.7). A capillary fringe of a certain height, depending on the width of the interspace, the liquid and the plates' material, exists between the two plates. The phreatic surface ($p = 0$) is obtained by deducting this height (which may vary with time in an unsteady flow problem) from the elevations of the liquid table in the analog.

As with any other analog, once scales have been chosen and the analog constructed, it must be calibrated. Of special importance is the determination of K_m, since in general it differs from the value computed from the nominal b_m. All discharge openings, the rainfall simulator, etc., must also be calibrated.

A rapid drawdown in the analog may introduce an error due to delayed liquid storage, especially when the viscosity is high, because of liquid sticking to the walls (fig. 11.4.8).

11.4.5 Recommended Applications of Vertical Analog

The main advantage of the vertical Hele-Shaw analog is its ability to simulate directly an abrupt interface between two liquids, and hence also a phreatic surface. In such studies, we enjoy the possibility of visual observation of the interface or the phreatic surface. A camera (still or movie) is often used to record the results. Streamlines or liquids of different kinds can be made visible by adding dyes. The analog is thus a special purpose computer for solving problems of steady and unsteady flows involving stationary or moving interfaces. Obviously, the use of this analog is restricted to problems of two-dimensional flows in the vertical plane. The flow domain, subject to certain restrictions, may be inhomogeneous and anisotropic.

Drainage problems (involving flows to ditches or tile drains) are typical of those problems that may be investigated by this tool, neglecting, of course, the presence of the capillary fringe or the unsaturated region above the phreatic surface. Another problem often studied by this analog (e.g., Santing 1951b; Naor and Bear 1963; Bear and Dagan 1964b) is that of the interface between fresh water and sea water in coastal aquifers (plate 11.1). The entire water balance of a coastal strip perpendicular to the coast (when the flow is everywhere toward the sea), may be simulated and investigated.

11.4.6 The Liquids

From the scaling rules (par. 11.4.4) it follows that viscosity and density are the most important properties of the liquids used in the analog. Of secondary importance are purity (except when it clogs outlets), corrosiveness, toxicity and inflammability. The most common liquids used are water (especially for steady flow problems) and various kinds of oils and glycerine. Sometimes a mixture of liquids is used to adjust liquid properties such as viscosity, density, interfacial tension and wettability.

11.4.7 The Horizontal Hele–Shaw Analog—Description and Scales

When the two plates are placed in a horizontal position (usually the upper plate is transparent), the interspace represents a horizontal confined aquifer or oil reservoir in which flow of a single fluid, or of two fluids, takes place (Kruysse and Bear 1956; DeWiest 1966). Variations in head in the vertical direction cannot be simulated in this analog. Essentially, the horizontal analog represents two-dimensional flow in the horizontal plane. Sometimes the analog is placed in an inclined position to represent an inclined reservoir. Figure 11.4.9 shows a horizontal Hele–Shaw analog for a water drive study in reservoir engineering. Bear and Zaslavsky (1962) used this model to study the movement of water bodies injected into aquifers in artificial replenishment operations through wells, where the injected water was labeled by a tracer.

In studies of interface problems, the horizontal analog can simulate only front movements; the shape of the interface in a vertical cross-section resulting from density differences cannot be simulated. Small variations in aquifer thickness can be simulated by varying the width b_m of the interspace, as long as the flow remains essentially two-dimensional.

Storativity of a confined aquifer or reservoir is represented in the analog by a network of a large number of short vertical pipes of diameter 10–30 mm, and length 10–15 cm (figs. 11.4.9 and 11.4.10), glued to the upper plate and connected to the interspace by holes in the upper plate. In this way, the continuously distributed storativity is made discrete: each "storage pipe" represents the storage capacity of the area (usually in the form of a square, e.g., 10 $\times$ 10 cm) around it. The pipes

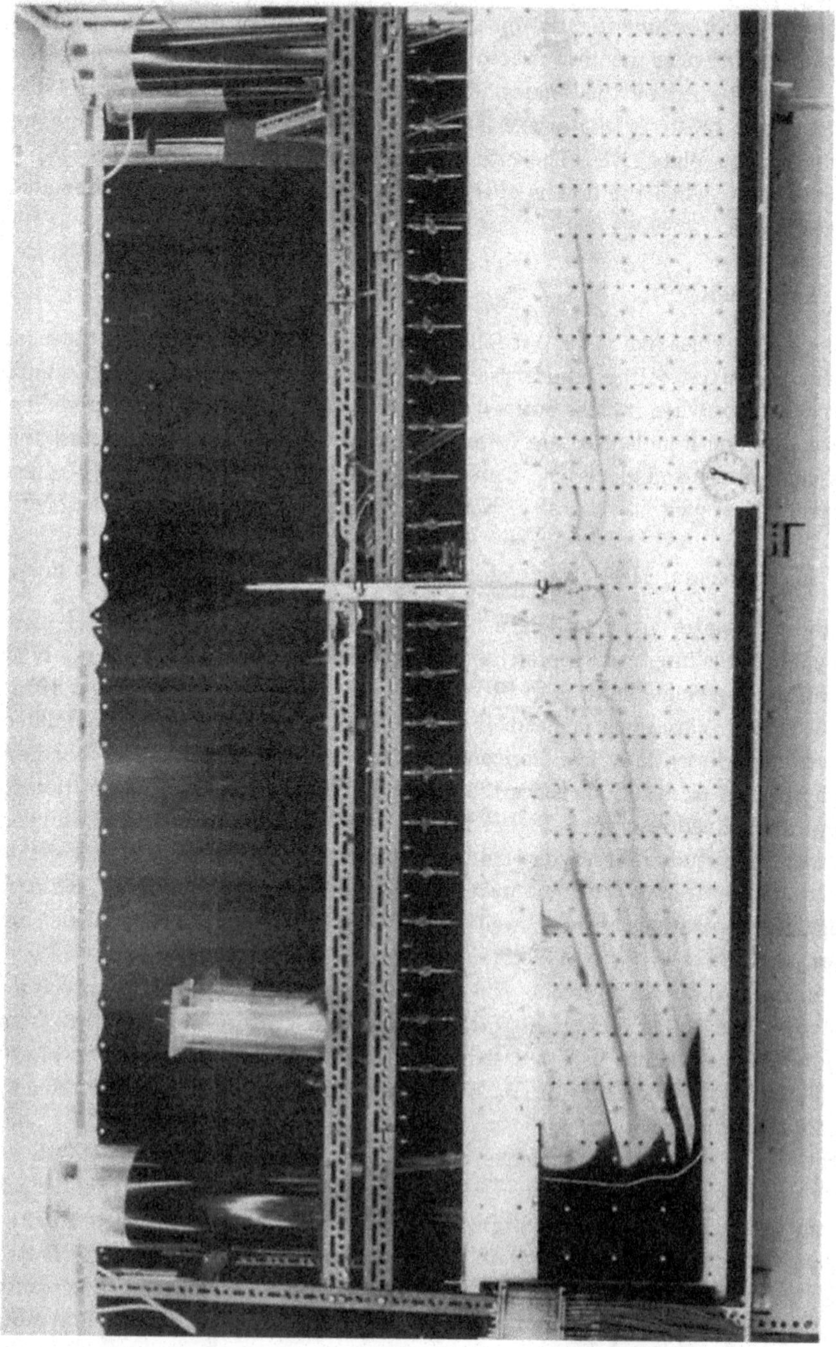

PLATE 11.1. A Hele-Shaw analog for investigations of sea water intrusion and artificial recharge along the Coastal Plane in Israel (Courtesy of Water Planning for Israel, Ltd.).

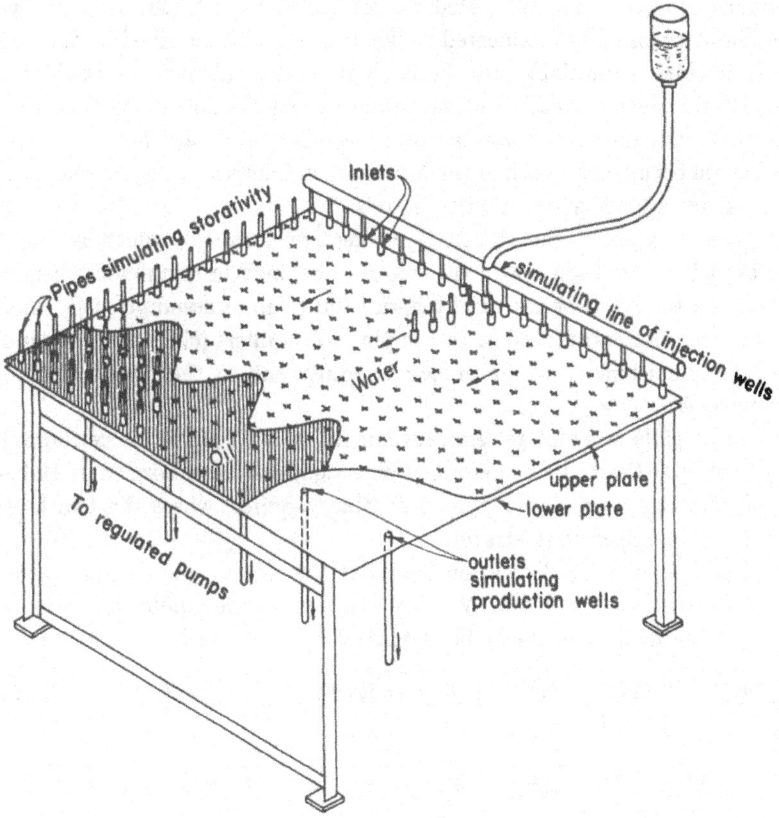

FIG. 11.4.9. Horizontal Hele-Shaw analog, for a water drive study (after Bear, Jacobs and Braester, 1966).

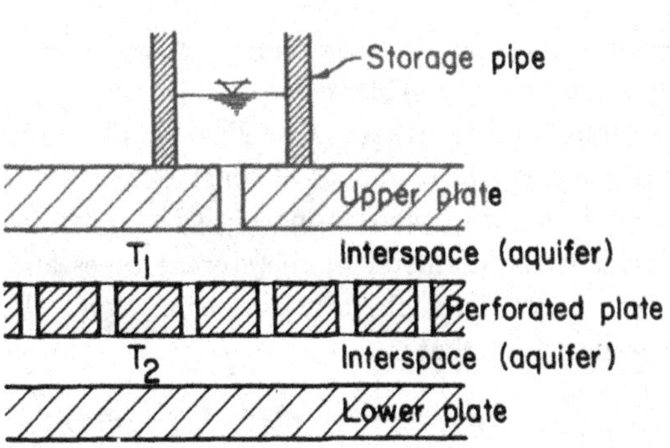

FIG. 11.4.10. Simulation of a semipervious layer in a horizontal Hele-Shaw analog.

are usually transparent so that the liquid level in them may easily be observed. They may thus serve as observation wells (piezometers). Pumping and recharge wells are simulated by holes in the lower plate fitted with valves and connected to regulated pumps or to small reservoirs where the head can be controlled.

Boundary conditions are simulated by controlled head tanks, as in the vertical analog. Similarly, all pipes connected to the interspace are fitted with short capillary tubes to increase appreciably the head required for introducing liquid into the analog. By this device, small head variations within the flow domain in the analog do not affect the initial calibrations of these inlets and outlets.

The domain occupied by each of the liquids in an immiscible displacement problem is observed by adding a dye to the liquids.

In a phreatic aquifer, the storativity (effective porosity), which is due to the storage capacity of the soil above the free surface, may be simulated whenever the flow may be assumed to be essentially horizontal (Dupuit assumptions). This is the approximation often introduced in flow in phreatic aquifers (chap. 8). The simulation of natural replenishment in the analog is equivalent to the linearized boundary condition $\partial\varphi/\partial t = N/n_e$.

It is also possible to simulate semipervious layers by a porous or perforated plate (fig. 11.4.10). DeWiest (1966) describes investigations by a horizontal Hele-Shaw analog of pumpage from a two-layered confined aquifer, where the two layers are separated by an impervious stratum.

As an example of scale derivation, consider the case of *ground water flow in an anisotropic aquifer with accretion* (N). Let the flow in the aquifer (phreatic aquifer with relatively small drawdown) be described by:

$$\partial(T_x\,\partial\varphi/\partial x)/\partial x + \partial(T_y\,\partial\varphi/\partial y)/\partial y + N = S\,\partial\varphi/\partial t; \qquad Q_x = -\,T_x\,\delta y\,\partial\varphi/\partial x. \quad (11.4.29)$$

This leads to:

$$T_{xr}\varphi_r/x_r^2 = T_{yr}\varphi_r/y_r^2 = N_r = S_r\varphi_r/t_r; \qquad Q_r = T_{xr}y_r\varphi_r/x_r. \quad (11.4.30)$$

For $T_x = T_y = T$, we have:

$$x_r = y_r; \qquad t_r = S_r x_r^2/T_r; \qquad N_r = S_r\varphi_r/t_r = T_r\varphi_r/x_r^2; \qquad Q_r = T_r\varphi_r. \quad (11.4.31)$$

These are *four* relationships for the *eight* scales: $x_r, y_r, \varphi_r, S_r, T_r, t_r, N_r, Q_r$. Four scales may, therefore, be chosen arbitrarily.

In a nonhomogeneous aquifer, where $T = T(x, y)$, the requirement $Q_r =$ const throughout the analog leads to $T_r = K_r \cdot b_r =$ const, or $b_m^3/K_p b_p =$ const. However, from the requirement $t_r =$ const, we obtain: $S_r/T_r =$ const, or $S_r T_p/b_m^3 =$ const, where $S_m = \pi d^2/4A$, d is the diameter of the storage pipes and A is the area represented by each such pipe. In a phreatic aquifer, $S \equiv n_e$. If, in addition, front movements are considered so that:

$$V_r = x_r/t_r = K_r\varphi_r/x_r n_r = T_r\varphi_r/x_r n_r b_r \quad (11.4.32)$$

we also have from $t_r =$ const, $T_r/n_r b_r = K_r/n_e =$ const, or $b_m^3 n_p/K_p =$ const. In a nonhomogeneous reservoir, it is seldom possible to satisfy all these conditions simultaneously.

Errors in a phreatic aquifer, resulting from the assumption of T independent of time (i.e., relatively small drawdowns), are especially large in the vicinity of sources or sinks (wells). Another error may result from the concentration of accretion, storativity, etc., at discrete points. This error is reduced by increasing the number of points.

11.4.8 Simulation of an Infinite Horizontal Aquifer

In many cases, because of the lack of additional information, aquifers and reservoirs are assumed infinite in area. When the flow in such aquifers is investigated by means of the horizontal Hele–Shaw analog (or the electrolytic tank analog, or the *RC*-network analog), large error may be introduced with respect to the exact analytical solution. One way to reduce this error is to simulate in the analog an area that is much larger than the actual area of interest. A more elegant way was suggested by Boothroyd et al. (1949) and used by de Jong (1962) in an electrolytic tank analog, and by De Wiest (1966) in a Hele–Shaw analog.

The basic idea is to transform the given infinite domain into a finite one by means of the *reciprocal or inversion transformation* (8.3.17), where ρ is any characteristic length of the given domain and $z = x + iy$, $z' = x' + iy'$ are the complex numbers defining the points (x, y) in the given domain and their corresponding points (x', y') in the transformed domain, respectively (fig. 11.4.11). This corresponds to the geometric inversion of a point z with respect to a circle of radius ρ, followed by a reflection in the (real) x-axis.

From (8.3.17) we obtain:

$$x' = \rho^2 x/(x^2 + y^2); \qquad y' = \rho^2 y/(x^2 + y^2) \tag{11.4.33}$$

so that lines $x = C_1 = \text{const}$ and $y = C_2 = \text{const}$ are mapped onto circles:

$$(x' - \rho^2/2C_1)^2 + y^2 = \rho^4/4C_1^2; \qquad x'^2 + (y' + \rho^2/2C_2)^2 = \rho^4/4C_2^2 \tag{11.4.34}$$

respectively (fig. 11.4.11). Points at $z = \infty$ are mapped onto the origin $z' = 0$ of the z'-plane.

The function $z' = f(z)$ in (8.3.17) is analytic. An harmonic function (sec. 7.8) of x and y transforms into a harmonic function of x' and y' under the change of variables $z = f(z')$, where f is an analytic function. Hence, the harmonic function $\varphi(x, y)$, satisfying $\partial^2\varphi/\partial x^2 + \partial^2\varphi/\partial y^2 = 0$, remains harmonic, i.e., $\varphi(x', y')$ satisfies

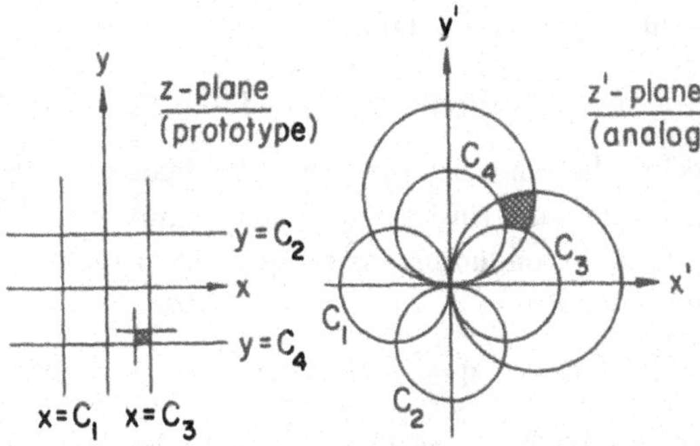

FIG. 11.4.11. Use of the inverse transformation $z' = \rho^2/z$ for studying an infinite flow domain ($z' = x' + iy'$, $z = x + iy$).

$\partial^2\varphi/\partial x'^2 + \partial^2\varphi/\partial y'^2 = 0$ under the change of variables arising from (8.3.17).

Let the flow in a confined aquifer be described by: $\varphi = \varphi(x, y)$ satisfying $\nabla^2_{xy}\varphi = (S/T)\, \partial\varphi/\partial t$. Then, by (8.3.17) and the transformation

$$(\nabla^2_{x'y'}\varphi)(dz'/dz)^2 = \nabla^2_{xy}\varphi; \qquad (dz'/dz)^2 = \rho^4/z^4, \qquad \varphi = \varphi(x', y')$$

will satisfy:

$$\nabla^2_{x'y'}\varphi = (S'/T')\, \partial\varphi/\partial t; \qquad T' \equiv T; \qquad S' = S(x^2 + y^2)^2/\rho^4 \qquad (11.4.35)$$

where T' and S' are the transmissivities and storativities of the transformed domain. Once the problem is solved in this domain in the form $\varphi = \varphi(x', y', t)$, the solution can be transformed back to yield $\varphi = \varphi(x, y, t)$.

Churchill (1948) shows that boundary conditions of $\varphi =$ const or $\partial\varphi/\partial n = 0$ remain unchanged in the $x'y'$-plane. Other boundary conditions change according to the transformation employed.

If A and A' are the areas represented by storage vessels of radii r and r' in the given z-plane and in the transformed z'-plane, respectively, then

$$S = \pi r^2/A; \qquad S' = \pi r'^2/A'.$$

However, from (11.4.34) it follows that $r'^2/r^2 = (A'/A)(dz/dz')^2 = 1$, or $r' = r$ (Churchill 1948). Hence, in the analog, beyond a certain region it is impossible to model the storativity. De Wiest (1966) shows a method of overcoming this difficulty.

11.5 Electric Analogs

Three types of analogs are considered in the present section: the continuous electric analog (electrolytic tank and conducting paper), the discrete electric analog (the resistance network and the resistance–capacitance network), and the ion motion analog. All three analogs are most powerful tools in studies involving flow through porous media (especially flow in aquifers). As explained in section 11.1, they should be considered as special purpose computers for the solution of the Laplace equation or the diffusion equation describing unsteady flow, and as such have their advantages and disadvantages when compared with other solution methods such as numerical methods by means of digital computers.

11.5.1 Description of the Electrolytic Tank and the Conducting Paper Analogs

The analogy is based on the similarity between the differential equations that govern the flow of a homogeneous fluid through a porous medium, and those governing the flow of electricity through conducting materials. In a porous medium, the flow obeys *Darcy's law*:

$$\mathbf{q} = -\underline{\mathbf{K}}\,\mathrm{grad}\,\varphi. \qquad (11.5.1)$$

The flow of an electric current through a conductor obeys Ohm's law:

$$\mathbf{i} = -\underline{\boldsymbol{\sigma}}\,\mathrm{grad}\,V \qquad (11.5.2)$$

where $\mathbf{i}$ is the electric current (vector) per unit area (ampere/cm^2), σ is the electric conductivity of the conducting medium (ohm^{-1} cm^{-1}) and V is the electric potential (volts).

The continuity equation for an incompressible fluid flowing through a rigid porous medium is:

$$\operatorname{div}(\underline{\mathbf{K}} \operatorname{grad} \varphi) = 0. \tag{11.5.3}$$

In steady flow of electricity in a conductor, the voltage V satisfies:

$$\operatorname{div}(\underline{\boldsymbol{\sigma}} \operatorname{grad} V) = 0. \tag{11.5.4}$$

A comparison between (11.5.1), (11.5.2) and (11.5.3), (11.5.4) leads to the conclusion that any problem of steady flow of an incompressible fluid having a potential $\varphi(x, y, z)$ may be simulated by the flow of an electric current in an analog. The analogy is between the following elements.

Fluid Flow through a Porous Medium	Flow of Electricity through a Conducting Medium
Porous medium	Conducting medium
Fluid potential φ	Electric potential V
Specific fluid discharge $\mathbf{q}$	Specific electric current $\mathbf{i}$
Hydraulic conductivity $\underline{\mathbf{K}}$	Electric conductivity $\underline{\boldsymbol{\sigma}}$

In addition, the geometry of the investigated flow domain and its boundary conditions must be simulated. Thus, the unknown potential $\varphi(x, y, z)$, or in a thin horizontal reservoir the pressure $p(x, y)$, in a specified domain with given boundary conditions is simulated by an electric potential $V(x, y, z)$ within an electrically conducting domain of a similar geometry and boundary conditions. Measurements of $V = V(x, y, z)$ in the analog provide a numerical solution of $\varphi(x, y, z)$ satisfying (11.5.3) and the given boundary conditions. If the investigated flow domain is anisotropic, we can either use a conductor with an anisotropic $\boldsymbol{\sigma}$ or employ the distortion approach suggested in section 7.4.

When solid conductors are employed, either alternating or direct currents may be used. If the conductor is an electrolyte solution (e.g., $CuSO_4$), an alternating current must be used, with a frequency of 50–400 Hz, in order to prevent polarization of the electrodes. The choice of the conducting material depends on the available instrumentation. The accuracy of the measurements depends on the relationship between the impedance of the analog and of the measuring instrument. In three-dimensional analogs, where potentials must be measured at internal points within the flow domain, electrolytic solutions are used (e.g. Debrine, 1970). In two-dimensional models, electrically conducting paper (see below) is often used.

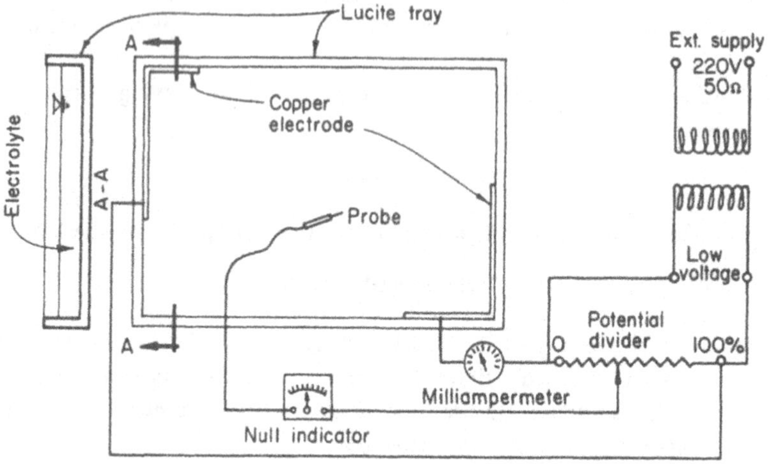

FIG. 11.5.1. Simplified circuit of electric analog (electrolytic tank).

The electrolytic tank analog consists of the following parts: (a) a watertight container having a geometry simulating, with or without distortion, that of the prototype flow domain; the geometry is determined by the analog's scales; (b) a conducting medium–an electrolyte—for example, tap water or a $CuSO_4$ solution of low conductivity placed in the container; (c) an electric circuit of low voltage to avoid electric shock.

Figures 11.5.1 and 11.5.2 show schematic diagrams of electric analogs of this type. The electric circuit consists of: (a) an AC power supply; a DC current may be used for solid conductors; (b) a potentiometer, or potential divider; (c) a null indicator; (d) a source of supply of fixed potentials for simulating boundary conditions of fixed potentials; (e) a source of supply of fixed currents, for simulating fixed current boundary conditions. Boundary conditions of fixed potentials are simulated in the analog by electrodes at the proper voltages. Impervious boundaries are simulated by insulators. If the flux through a portion of the boundary is specified as a boundary condition, a special source of electric currents is used, which can be regulated to supply a fixed current to an electrode, independent of the voltage at that point in the model. If φ varies along the boundary, the boundary in the analog is made of a large number of separate, small, closely spaced electrodes, each at the proper potential. The electrodes can be made of copper, brass or platinum. Details of the electric circuit are given in many books and publications (see list of references in Todd 1959 and De Wiest 1965). The reader is specially referred to: Karplus (1958), Karplus and Soroka (1959) and Volynskii and Bukhman (1965), who discuss the general aspects of analog simulation and analog methods.

When a two-dimensional flow domain (say, a horizontal aquifer of thickness D_p) consists of zones of different hydraulic conductivity K_p, or transmissivity T_p, electrolytes of different electric conductivity σ (e.g., by changing the electrolyte concentration) are used. We have the analogy between $T_p = D_pK_p$ and $T_m = D_m\sigma$.

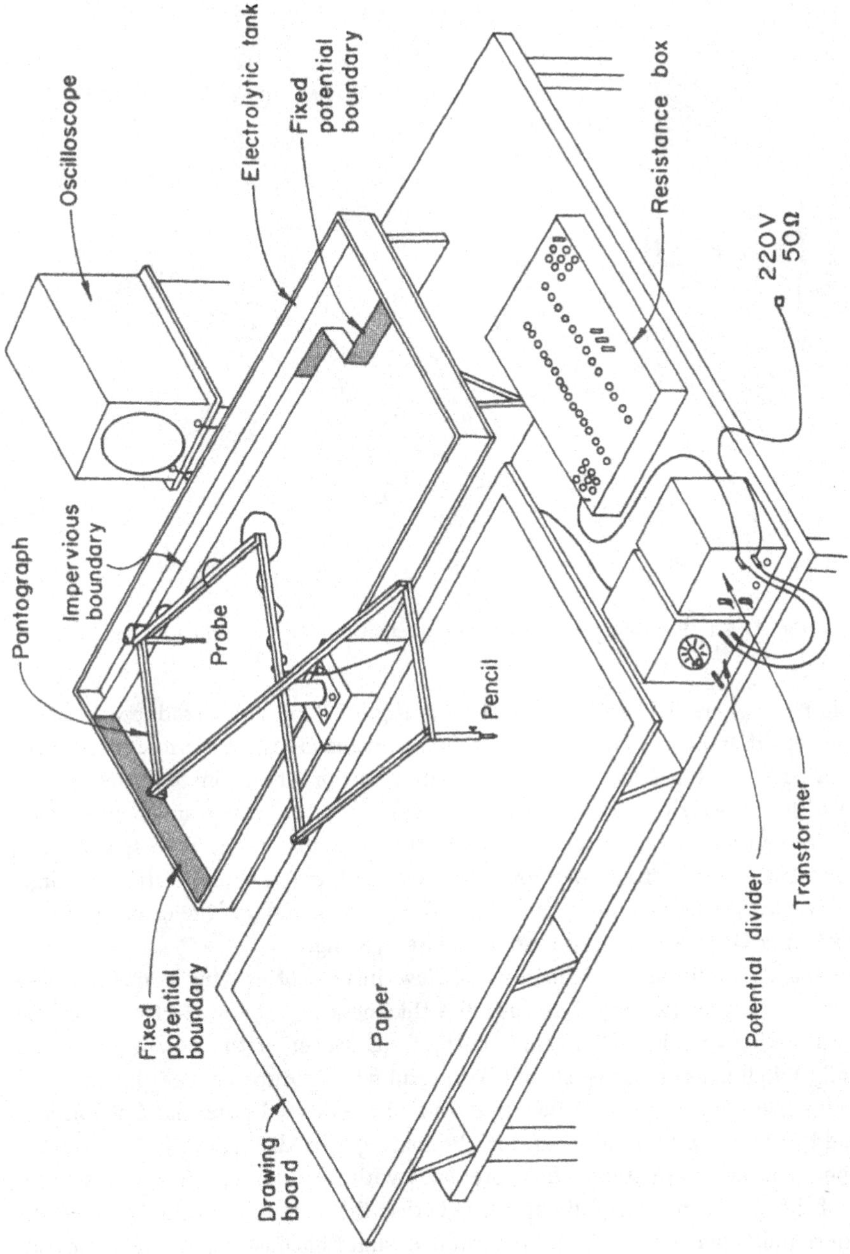

FIG. 11.5.2. Schematic view of the electric analog (Todd and Bear, 1959).

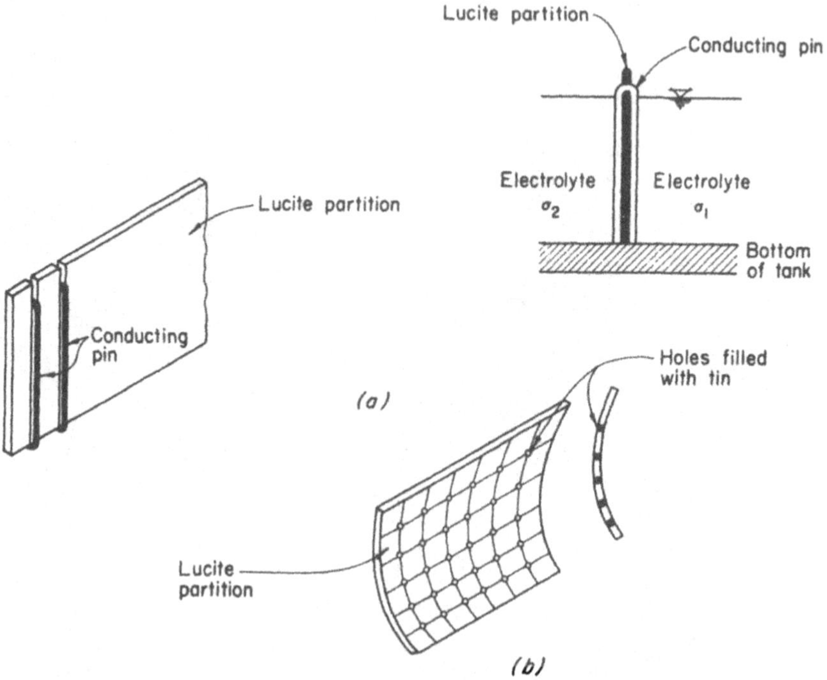

FIG. 11.5.3. Partition between zones of different electrolyte conductivity.

To avoid mixing, special insulating partitions (fig. 11.5.3a) are introduced between the zones of different σ. The pins or tin fillings transmit the electric current with no loss of potential from one side of the partition to the other. Small errors are due to the width of the strip and to the wires acting as an array of line sinks on one side and of line sources on the other. This last error depends on the wire spacing and diameter. The effect of streamline convergence and divergence at the arrays of sinks and sources diminishes rapidly with distance from the boundary, the error becoming negligible at a distance greater than the wire spacing.

In an analog simulating two-dimensional flow, instead of varying σ, one may use the same electrolytic solution, but vary the thickness D_m of the electrolyte in the tank by lowering or raising the tank's bottom. However, if the flow is to remain essentially twodimensional, one should beca reful when abrupt changes in thickness are involved, as they distort the flow pattern and the desired potential distribution. McDonald (1953) suggests methods for eliminating this distortion.

Of special interest are steady flow problems with an interface or a free surface as part of the boundary. The difficulty here is that this boundary of the flow domain is a priori unknown and must be determined simultaneously with the potential distribution. In solving such a problem by an electrolytic tank analog, it is necessary to adjust, in the analog, that portion of the boundary that represents the interface or the free surface until the boundary conditions on it are satisfied.

As an example, consider the steady two-dimensional flow through the earth dam shown in figure 11.5.4a. Along the (a priori unknown) phreatic surface AG, as well as along the seepage face GB, the boundary condition is $\varphi = z$ (sec. 7.1). In addition, the phreatic surface is a streamline. The location of point G is also unknown, except that it is on the vertical face of the dam. The electrolytic tank is a shallow tray similar to the one shown in figure 11.5.1. Initially, the tray is filled with clay up to some line above point B, but below the estimated phreatic surface. The remaining part is filled with the electrolyte solution. The electrode arrangement is shown in figure 11.5.4b. Then, electric potentials are measured along the estimated phreatic surface boundary, and the shape of the clay boundary is adjusted (by cutting away clay) until the condition $\varphi = z$ is satisfied at all points. This trial and error procedure is usually rapid. Once the shape of this boundary has been determined, equipotentials are traced inside the flow domain.

In two-dimensional flow problems, the shallow electrolytic tank may be replaced by a sheet made of a solid conductor. A sheet of insulating material (e.g., lucite) covered by a thin layer of a conducting paint can be used. A commonly used conducting sheet is the *Teledeltos conducting paper* manufactured by Western Union Telegraph Co. Either alternating or direct current may be used. The paper is made by adding carbon black (a conductor) to the paper pulp. The paper is then coated on one side with a lacquer (insulator) and a very thin layer of aluminum paint on the other side. It can be purchased in long rolls of varying widths. Sunshine Scientific Instrument Co. also manufactures a conducting paper. In Teledeltos paper (Type "L"), the resistance of a square sheet, taken between parallel sides, is approximately 3000 ohms. Electrode boundaries are readily obtained by applying a silver paint to the appropriate paper edges, or by attaching copper electrodes. The voltag source itself can be a simple dry cell, a storage battery, or a more elaborate electroni power supply. Conducting paper analogs are often used for solving stable-interfac

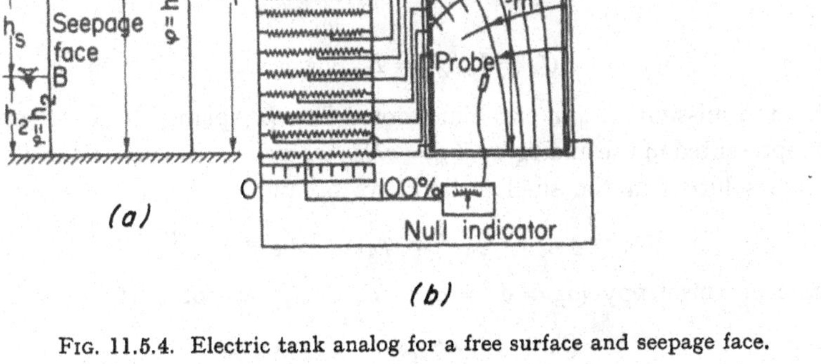

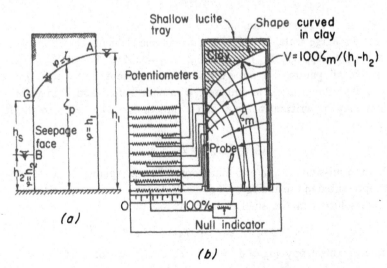

FIG. 11.5.4. Electric tank analog for a free surface and seepage face.

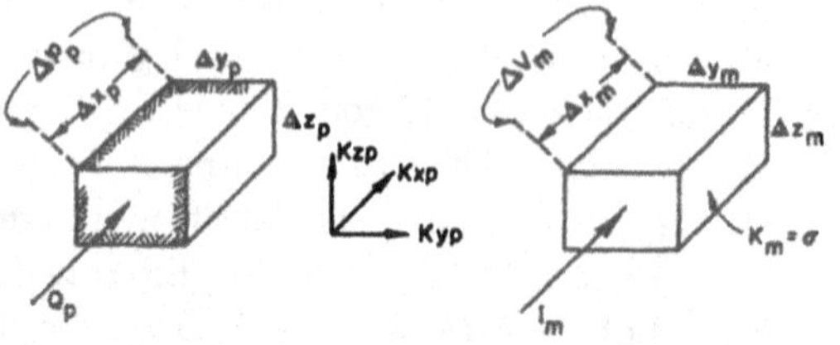

FIG. 11.5.5. The discharge scale for an electrolytic tank analog.

or free-surface problems, as the cutting of the paper in the trial and error procedure is very simple.

Generally, the accuracy of potential measurements is somewhat inferior to that obtained by using an electrolytic tank. A slight anisotropy often exists in the paper.

11.5.2 Scales for the Electrolytic Tank Analog

Scales are derived by the same inspectional analysis procedure described in section 7.4 and paragraph 11.2.7.

Let the dimensional quotient $\varphi_d = (\delta V)_m/(\delta\varphi)_p$ denote the ratio between a potential difference in the prototype and the corresponding voltage difference in the analog, and $Q_d = I_m/Q_p$ denote the ratio between the analog's current and the corresponding prototype's discharge. The subscript d (and not r as in the previous sections) was used here because the two variables are of a different physical nature. Thus, φ_d and Q_d are potential and discharge scales.

The length scales are derived from the continuity equations:

$$x_r^2/K_{xd} = y_r^2/K_{yd} = z_r^2/K_{zd} \tag{11.5.5}$$

where $K_m \equiv \sigma$, $K_{mx} \equiv \sigma_x$, etc.

For the discharge scale, consider the two-volume elements shown in figure 11.5.5. We have $Q_p = K_{xp}\,\Delta y_p\,\Delta z_p\,(\Delta\varphi)_p/(\Delta x)_p$; $I_m = \sigma\,\Delta y_m\,\Delta z_m\,(\Delta V)_m/(\Delta x)_m$; $K_d = \sigma/K_p$, where all parameters are expressed in consistent units, e.g., Q in m^3/day; K_{xp} in m/day; $\Delta\varphi$ in m; I in amperes; σ in ohm^{-1} cm^{-1} and V in volts. Similar equations may be written in the y and z directions. This leads to the following relationships:

$$Q_d = I_m/Q_p = K_{xd}y_rz_r\varphi_d/x_r. \tag{11.5.6}$$

If the transmissivity T_p in two-dimensional flow (xy-plane) in an isotropic prototype is represented in the analog by $T_m = \sigma D_m$, where D_m (2–3 cm) is the depth of the electrolyte solution in the shallow tank, we obtain:

$$x_r = y_r; \qquad Q_d = T_d\varphi_d; \qquad T_d = T_m/T_p. \tag{11.5.7}$$

In case of anisotropy ($\sigma_x = \sigma_y = \sigma_m$; $T_x \neq T_y$), we obtain:

$$Q_d = T_{xd}y_r\varphi_d/x_r = T_{yd}x_r\varphi_d/y_r. \tag{11.5.8}$$

An aquifer of varying thickness is simulated by filling the bottom of the tank with paraffin and carving it according to a configuration corresponding to the bottom of the aquifer. When the aquifer's thickness varies, but the variations relative to the average thickness are small, the flow is still considered twodimensional. Since varying the thickness of the electrolyte in the shallow tank is equivalent to varying the analog's transmissivity, we may use this procedure to simulate an inhomogeneous aquifer, where $T_p = T_p(x, y)$.

As in two-dimensional flow through porous media, we may define a *current function* W (analogous to ψ or Ψ, sec. 6.5) for the flow of electricity through an electric conductor, such that for an isotropic conductor we have:

$$\partial V/\partial x = (\partial W/\partial y)/\sigma_m; \qquad \partial V/\partial y = -(\partial W/\partial x)/\sigma_m \tag{11.5.9}$$

where $\sigma_m = \sigma_m(x, y)$. With (11.5.9), (11.5.4) written for two-dimensional flow in a nonhomogeneous conductor:

$$\frac{\partial}{\partial x}\left(\sigma_m \frac{\partial V}{\partial x}\right) + \frac{\partial}{\partial y}\left(\sigma_m \frac{\partial V}{\partial y}\right) = 0 \tag{11.5.10}$$

becomes:

$$\frac{\partial}{\partial x}\left(\frac{1}{\sigma_m} \frac{\partial W}{\partial x}\right) + \frac{\partial}{\partial y}\left(\frac{1}{\sigma_m} \frac{\partial W}{\partial y}\right) = 0. \tag{11.5.11}$$

If $\sigma_m = \text{const}$, both V and W satisfy the Laplace equation. Equations (11.5.9) through (11.5.11) are analogous to:

$$K_p\, \partial\varphi/\partial x = \partial\Psi/\partial y; \qquad K_p\, \partial\varphi/\partial y = -\partial\Psi/\partial x \tag{11.5.12}$$

$$\frac{\partial}{\partial x}\left(K_p \frac{\partial \varphi}{\partial x}\right) + \frac{\partial}{\partial y}\left(K_p \frac{\partial \varphi}{\partial y}\right) = 0 \tag{11.5.13}$$

$$\frac{\partial}{\partial x}\left(\frac{1}{K_p} \frac{\partial \psi}{\partial x}\right) + \frac{\partial}{\partial y}\left(\frac{1}{K_p} \frac{\partial \psi}{\partial y}\right) = 0. \tag{11.5.14}$$

If $K_p = \text{const}$, both φ and ψ satisfy the Laplace equation. There exist, therefore, two types of analogy:

type I, between φ and V: $\varphi = mV$; $\psi = nW$; $\sigma_m = (m/n)K_p$

type II, between φ and W: $\varphi = m'W$; $\psi = -n'V$; $\sigma_m = (n'/m')K_p$

where m, n, m', n' are the dimensional constant quotients that depend upon the system of units employed.

Experimentally, it is simpler to measure and plot the electric potential $V = V(x, y)$ than the current function $W = W(x, y)$. In the analog of type I, curves $V(x, y) = \text{const}$ give the equipotentials $\varphi = \text{const}$. In the analog of type II, they give the streamlines $\psi = \text{const}$. In isotropic media, these two families of curves form an orthogonal network. In a two-dimensional analog simulating flow between only two equipotential segments of the external boundary and the remaining portions

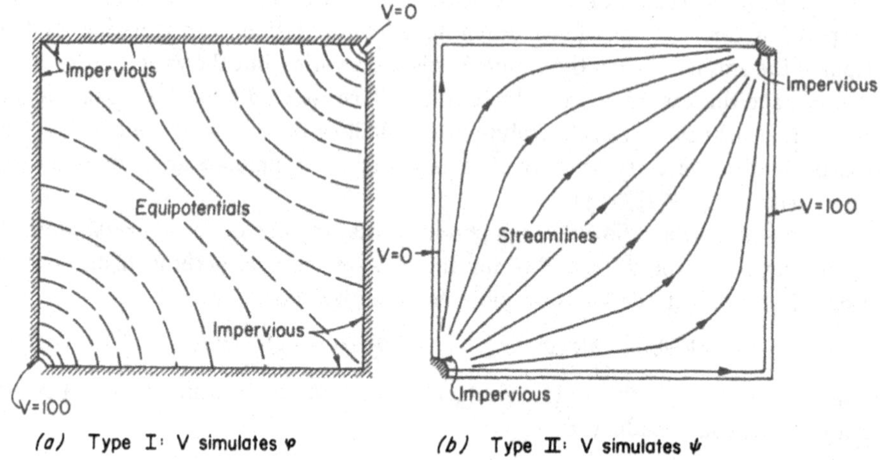

FIG. 11.5.6. The two types of the electrical analogy.

of the boundary being impervious (i.e., streamlines), we first map the equipotentials in the usual way. Then we interchange the conducting (electrodes) and the non-conducting portions of the boundary, so that the equipotential segments in the prototype are represented in the analog by streamlines (i.e., insulating boundaries), and trace streamlines by plotting lines $V = \text{const}$ in the analog (fig. 11.5.6). This procedure is not applicable when the flow is between more than two electrodes, or in three-dimensional flow, because stagnation points, or curves, which then occur along the impervious portions of the boundary in the prototype (i.e., on streamlines or stream surfaces), cannot be identified.

If the investigated flow domain consists of subdomains, each of which is anisotropic, the selection of model scales is governed not only by the laws of conversion of anisotropic media to isotropic ones (sec. 7.4), but also by the requirement that the layers have coincident adjoining boundaries after the transformation. When the boundary between two regions is an arbitrary curve, the boundaries of the transformed regions cannot be made to coincide with one another. However, since common points of the boundary can be identified in the transformed regions, each region is reproduced in the analog separately, and the boundaries are made of a large number of small electrodes connected to their counterparts by conductors of negligible resistance.

11.5.3 The Resistance Network Analog for Steady Flow

A basic difference exists between the discrete (or network) analogs, described in this section, and the continuous-type analogs described in the previous sections. In the latter, every point in the studied continuous field of flow has a homologous point in the analog, and the electrical parameters (e.g., conductivity, storativity) correspond directly to the distributed parameters of the prototype field, whereas in the former, electric circuit elements are concentrated in the network's node points to simulate the properties of portions of the continuous prototype field around

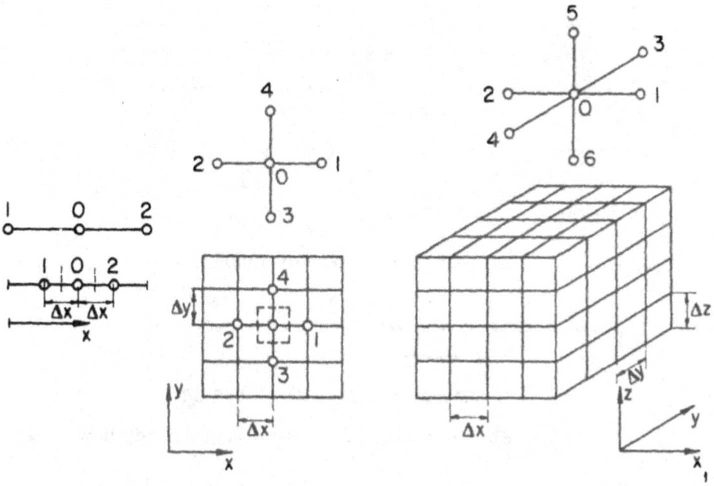

FIG. 11.5.7. One-, two- and three-dimensional discretized flow domains.

them. The unknown potentials, constituting the solution of the problem, are obtained only for those points that correspond to the nodes of the analog network. In general, it is permissible to make the continuous prototype field discrete in this fashion, provided the distance between adjacent nodes is sufficiently small. By reducing the spacing, the error resulting from this approximation may be kept sufficiently small to have a negligible effect upon the overall accuracy of the solution. However, the spacing in itself is not the only criterion for the error involved; the nature of the potential field being analyzed also affects the overall accuracy. One should recall that the second derivative, when approximated by a finite difference expression, involves an error—$\{[(\Delta x)^2/12]\, \partial^4\varphi/\partial x^4 + \cdots\}$. As $\Delta x \rightarrow 0$, the error is likewise reduced, although it depends also on the fourth and higher derivatives of φ. As the discrete electric analog is based on the finite-difference approximation of the equations to be solved, the errors involved in the discrete representation are the same as those occurring in this approximation.

In this paragraph, attention is focused on problems governed by equations such as:

$$\nabla^2\varphi = 0, \qquad K_x\, \partial^2\varphi/\partial x^2 + K_y\, \partial^2\varphi/\partial y^2 + K_z\, \partial^2\varphi/\partial z^2 = 0$$

$$K_x\, \partial^2\varphi/\partial x^2 + K_y\, \partial^2\varphi/\partial y^2 + K_z\, \partial^2\varphi/\partial z^2 = S_s\, \partial\varphi/\partial t$$

$$T_x\, \partial^2\varphi/\partial x^2 + T_y\, \partial^2\varphi/\partial y^2 = S\, \partial\varphi/\partial t$$

$$\partial^2\varphi/\partial x^2 + \partial^2\varphi/\partial y^2 = f(x, y)$$

$$\partial^2\varphi/\partial x^2 + \partial^2\varphi/\partial y^2 = f(x, y) + (S/T)\, \partial\varphi/\partial t.$$

The analogy may be established by two approaches: a mathematical approach, based on the finite difference approximation (par. 7.9.1), and a physical one. The latter will be adopted here.

Making the flow domain discrete (fig. 11.5.7) involves its replacement by an array,

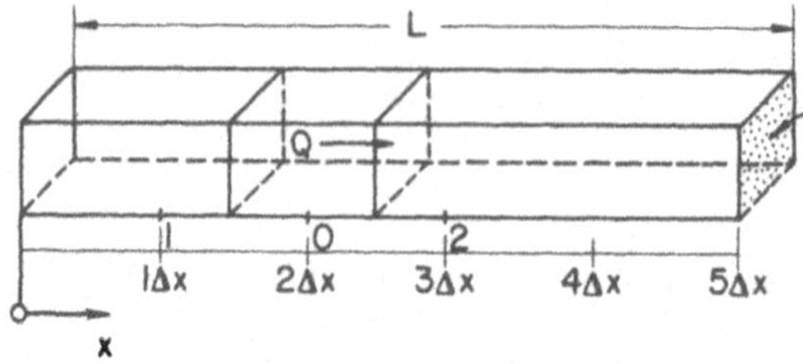

(a) Column of porous medium

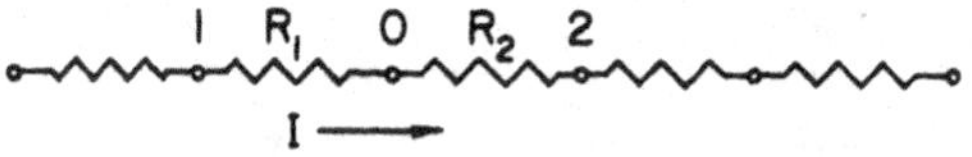

(b) The discretized equivalent system

FIG. 11.5.8. Electric analog for one-dimensional steady flow.

or a network, of lumped electric circuit elements. For the discharge through the soil block of figure 11.5.5a, we may write:

$$Q = K\,\Delta y\,\Delta z(\Delta\varphi/\Delta x) = \Delta\varphi/[\Delta x/K\,\Delta y\,\Delta z]$$

which is analogous to Ohm's law for the flow through an electric resistance $I = \Delta V/R$, where I is the current (amperes) and R is resistance (ohms). We thus see that the single electric resistance may simulate the resistance $R_p = \Delta x/K\,\Delta y\,\Delta z$ of the soil block. The analogy is between Q, $\Delta\varphi$ and R_p of the prototype, and I, ΔV and R of the analog.

For example, a steady one-dimensional flow through a sand column of cross-section A (fig. 11.5.8a) is simulated by an array of resistors (fig. 11.5.8b), each representing a length Δx of the sand column. Carrying out a water balance for the block of soil whose center is at 0, we obtain:

$$K_{01}A(\varphi_1 - \varphi_0)/\Delta x + K_{02}A(\varphi_2 - \varphi_0)/\Delta x = 0$$

$$(\varphi_1 - \varphi_0)/(\Delta x/K_{01}A) + (\varphi_2 - \varphi_0)/(\Delta x/K_{02}A) = 0 \qquad (11.5.15)$$

where K_{01} and K_{02} are the average hydraulic conductivities between nodes 0 and 1 and between 0 and 2, respectively. For a homogeneous medium, $K_{01} = K_{02}$, we obtain:

$$\varphi_1 - 2\varphi_0 + \varphi_2 = 0 \qquad (11.5.16)$$

which is also the finite difference approximation of $\partial^2\varphi/\partial x^2 = 0$.

For the array of resistors, by Kirchhoff's current law, $I =$ const requires:

$$(V_1 - V_0)/R_1 + (V_2 - V_0)/R_2 = 0 \qquad (11.5.17)$$

where V is the voltage at the nodes. If $R_1 = R_2$,

$$V_1 - 2V_0 + V_2 = 0. \qquad (11.5.18)$$

The analogy between φ and V, Q and I, and R and $\Delta x/KA$, is obvious.

Figure 11.5.9 shows the extension to a two-dimensional field of flow of transmissivity T. Instead of using the finite difference approximation of the continuity equation, we shall consider the continuity of flow through the rectangle $ABCD$. For the flow through BC, we have, by Darcy's law, approximately:

$$Q_{BC} = T_x \Delta y(\varphi_1 - \varphi_0)/\Delta x. \tag{11.5.19}$$

Similar expressions may be written for Q_{BA}, Q_{AD} and Q_{DC}. From $\Sigma Q = 0$, we obtain:

$$\frac{\varphi_1 + \varphi_2 - 2\varphi_0}{\Delta x/T_x \Delta y} + \frac{\varphi_3 + \varphi_4 - 2\varphi_0}{\Delta y/T_y \Delta x} = 0 \tag{11.5.20}$$

where the medium is assumed to be homogeneous, but anisotropic, with $T_x \neq T_y$. If $T_x = T_y$, and we choose a square network of sides $\Delta x = \Delta y$:

$$\varphi_1 + \varphi_2 + \varphi_3 + \varphi_4 - 4\varphi_0 = 0. \tag{11.5.21}$$

Similarly, for the arrangement of resistances of figure 11.5.9b, the total current flow into node 0 from the neighboring nodes is given by:

$$(V_1 - V_0)/R_1 + (V_2 - V_0)/R_2 + (V_3 - V_0)/R_3 + (V_4 - V_0)/R_4 = 0. \tag{11.5.22}$$

If $R_1 = R_2$, $R_3 = R_4$, we have:

$$(V_1 + V_2 - 2V_0)/R_1 + (V_3 + V_4 - 2V_0)/R_3 = 0 \tag{11.5.23}$$

which is analogous to (11.5.20). If $R_1 = R_2 = R_3 = R_4$, we have:

$$V_1 + V_2 + V_3 + V_4 - 4V_0 = 0 \tag{11.5.24}$$

which is analogous to (11.5.21). It is easy to construct an equation of continuity for an aquifer which is also inhomogeneous. In this case, as with (11.5.15), T_x and T_y should be replaced by their values at midpoints between the nodes. The extension to three-dimensional flow is obvious. A typical node in three-dimensional flow is shown in figure 11.5.10. In an anisotropic medium with principal directions x, y, z, the resistances in each direction are different. In an inhomogeneous medium, they

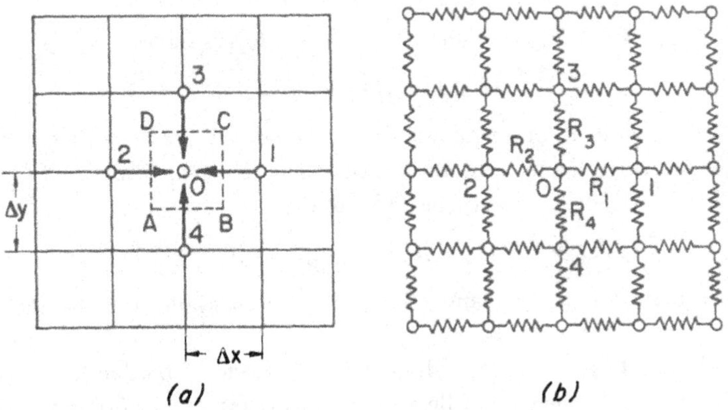

FIG. 11.5.9. An electric analog for two-dimensional steady flow.

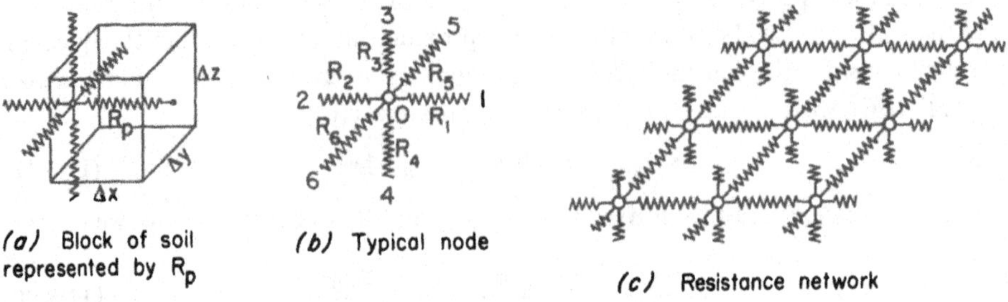

(a) Block of soil represented by R_p (b) Typical node (c) Resistance network

FIG. 11.5.10. An electric analog for three-dimensional steady flow.

also vary from node to node. We may define the hydraulic resistance R_p of a block of soil in a way similar to the definition of electrical resistance. For example (fig. 11.5.10a), from $Q = K_x \Delta y\, \Delta z\, (\Delta\varphi/\Delta x) = \Delta\varphi/(\Delta x/K_x \Delta y\, \Delta z) = \Delta\varphi/R_{xp}$, we have $R_{xp} = \Delta x/K_x \Delta y\, \Delta z$, which is analogous to R_1. This block of soil is also shown in figure 11.5.10a. Thus, each resistance depends both on the conductivity and on the chosen dimensions of the soil block. We often choose $\Delta x = \Delta y = \Delta z$.

The determination of scales is straightforward. Consider, for example, a two-dimensional flow described by (11.5.20) for the aquifer and by (11.5.23) for the analog. If the voltage drop ΔV corresponds to a head drop $\Delta\varphi$, then:

$$\frac{\Delta\varphi T_x \Delta y_p R_1}{\Delta x_p \Delta V} = \frac{\Delta\varphi T_y \Delta x_p R_3}{\Delta y_p \Delta V}; \qquad \frac{R_1}{1/T_x}\frac{\Delta y_p}{\Delta x_p} = \frac{R_3}{1/T_y}\frac{\Delta x_p}{\Delta y_p}. \tag{11.5.25}$$

If $\Delta x_p = \Delta y_p$:

$$R_1/(1/T_x) = R_3/(1/T_y). \tag{11.5.26}$$

For the discharge, if ΔQ corresponds to ΔI:

$$\frac{\Delta I}{\Delta Q} = \frac{\Delta V}{\Delta\varphi}\frac{1/T_x}{R_1}\frac{\Delta x_p}{\Delta y_p} = \frac{\Delta V}{\Delta\varphi}\frac{1/T_y}{R_3}\frac{\Delta y_p}{\Delta x_p}. \tag{11.5.27}$$

If $\Delta x_p = \Delta y_p$; $T_x = T_y = T$ and $R_1 = R_3 = R$, we have:

$$\Delta I/\Delta Q = (\Delta V/\Delta\varphi)(1/T)/R. \tag{11.5.28}$$

This means that if no other parameters are involved, we may choose two of the three ratios arbitrarily, and compute the third one from (11.5.28). With the dimensional quotient (scale) $\varphi_d = \Delta V/\Delta\varphi$, we obtain:

$$R_p = 1/T_p; \qquad R_d = R/R_p; \qquad Q_d = \varphi_d/R_d. \tag{11.5.29}$$

We must maintain φ_d, R_d and Q_d constant throughout the investigated flow domain.

The Laplacian $\nabla^2\varphi$, in polar, cylindrical or spherical coordinates, may also be made discrete, e.g., in the case of flow to wells or point sinks (Karplus 1958). R_p of a porous medium block between two nodes is always given by l/KA where l is the

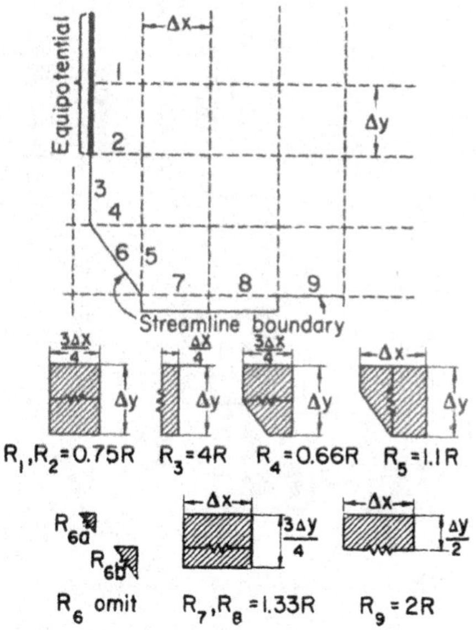

FIG. 11.5.11. An electric analog for a field with irregular boundaries (Karplus, 1958).

average length of the element in the direction of flow, K is the hydraulic conductivity in that direction, and A is the average cross-section perpendicular to the flow.

In many flow domains the boundaries do not coincide with the node points of a square or rectangle network. It is then possible to approximate that irregular boundary by a jagged line corresponding to the closest grid lines. By sufficiently reducing the net spacing, the error introduced by this approximation may be kept within acceptable limits. Special attention should be given to boundaries that are streamlines. These are represented by resistances corresponding to the areas they represent. A simple example (Karplus 1958) that requires no further explanation is shown in figure 11.5.11.

When the potential gradients are not uniform, a finer "mesh" may be employed in certain portions of the flow domain in order to obtain solutions of uniform accuracy or to achieve a higher degree of accuracy.

If we have two aquifers separated by a semipervious layer (K', b'), each aquifer is represented by its own resistance network, and we introduce between each node in the upper aquifer and the corresponding node in the lower aquifer a resistor corresponding to a resistance of $b'/K'a^2$, where a^2 is a horizontal area corresponding to each node in the aquifer.

It is also possible to combine a continuous analog with a discrete one. As an example, consider the two aquifers (fig. 11.5.12) separated by a semipervious layer (de Jong 1962). The two aquifers are simulated by two separate tanks. The semipervious layer is simulated by resistors connecting a square network of homologous points in the two aquifers. The problem of simulating an aquifer with accretion is discussed in the following paragraph.

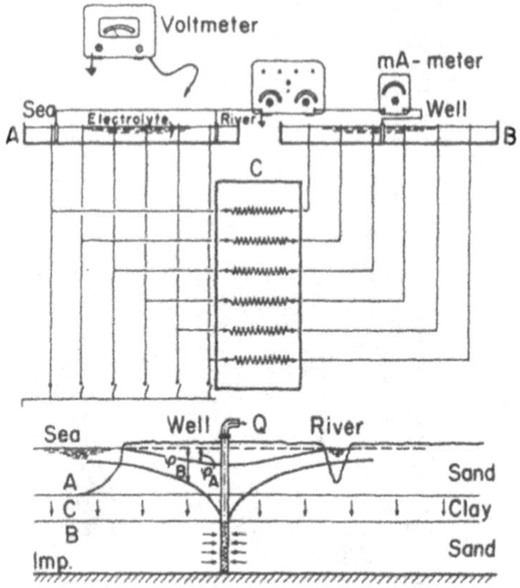

FIG. 11.5.12. An electric analog of two aquifers separated by an aquiclude (de Jong, 1962).

11.5.4 The Resistance–Capacitance Network for Unsteady Flow

In the analog of the diffusion equation (6.4.7), known as the *RC network* (resistance–capacitance network), capacitors are added to the nodes to simulate the storage capacity of the reservoir. A direct current is used. Equation (6.4.7), with a source term (to describe accretion) $N(x, t)$ added to its left-hand side, can be approximated for flow in a homogeneous isotropic aquifer by (fig. 11.5.13a and c):

$$T(\varphi_1 + \varphi_2 + \varphi_3 + \varphi_4 - 4\varphi_0) + a^2N = a^2S\ \partial\varphi/\partial t \qquad (11.5.30)$$

where we have taken $\Delta x = \Delta y = a$. For a typical node in the analog, we have:

$$(1/R)(V_1 + V_2 + V_3 + V_4 - 4V_0) + I = C\ \partial V/\partial t. \qquad (11.5.31)$$

The continuity equation (11.5.30) can also be written for $\Delta x \neq \Delta y$. Equations corresponding to inhomogeneous and anisotropic aquifers can also be written. In these difference-differential equations, obtained by considering the water balance of a square of area a^2 in the aquifer, and a balance of electric current at a node, we make space discrete, yet time remains a continuous variable. The analogy is between:

head:	φ (m)	— V (volt);	$\varphi_d = \Delta V/\Delta\varphi$
resistance:	$1/T$ (d/m^2)	— R (ohm);	$R_d = R/(1/T)$
storage:	a^2S (m^2)	— C (farad);	$C_d = C/a^2S$
time:	t (day)	— t (sec);	$t_r = t_m/t_p$
discharge:	Q (m^3/day)	— I (ampere);	$Q_d = I/Q$
volume:	U (m^3)	— U^* (coulomb);	$U_d = U/U^*$.

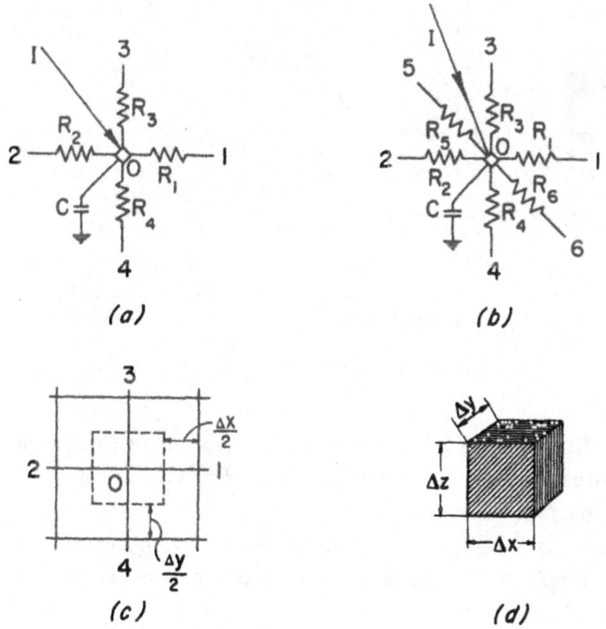

FIG. 11.5.13. *RC* Network for two- and three-dimensional domains.

Any other consistent system of units may also be used. Accordingly, the design criteria of the model are:

$$\varphi_d = Q_d R_d; \qquad Q_d = C_d \varphi_d / t_r; \qquad U_d = Q_d t_r. \tag{11.5.32}$$

A few comments should be added regarding the simulation of point sources and sinks in the analog. As an example, let us consider the simulation of a well in the aquifer, where the flow is described by (11.5.30). The common procedure is to simulate a well by introducing (or withdrawing) a regulated current into the appropriate node. When this is done, the potential (piezometric head) observed at a sufficient distance from the node is sufficiently accurate. In most cases, no serious error is encountered even at the nodes adjacent to the well. However, the potential observed at the well itself is erroneous as the analogy does not simulate correctly the resistance to the flow converging to the well (or diverging from it). This is illustrated in figure 11.5.14.

For R we have:

$$s = \frac{Q_{\text{well}}/4}{aT/a}; \qquad R = \frac{s}{Q_{\text{well}}/4} = \frac{1}{T}.$$

For R^* we have:

$$s = \frac{Q_{\text{well}}/4}{\pi T/2} \ln \frac{a}{r_w}; \qquad R^* = \frac{2}{\pi T} \ln \frac{a}{r_w} = \frac{2R}{\pi} \ln \frac{a}{r_w} \tag{11.5.33}$$

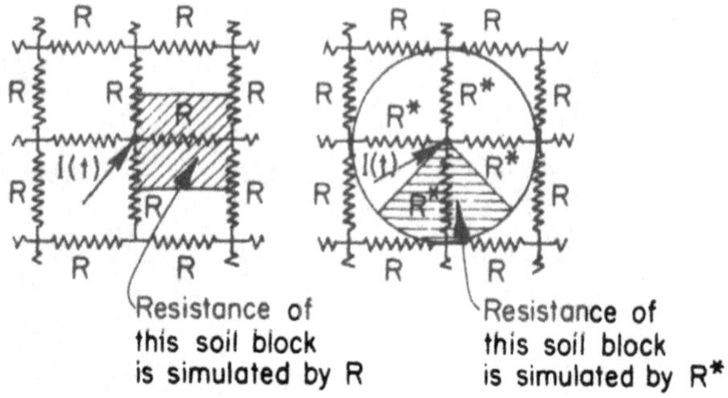

FIG. 11.5.14. Simulation of a well in an aquifer.

where s is the difference in head between the well and a nearby node. Thus, by replacing the resistances R near the well with R^*, the potential observed at the well itself will be more accurate.

For three-dimensional flow in a homogeneous isotropic confined domain with elastic storativity, we use the equation $K\,\nabla^2\varphi = S_s\,\partial\varphi/\partial t$, which may be approximated for $\Delta x = \Delta y = \Delta z = a$, by:

$$K(\varphi_1 + \varphi_2 + \varphi_3 + \varphi_4 + \varphi_5 + \varphi_6 - 6\varphi_0) = a^2 S_0\,\partial\varphi/\partial t. \tag{11.5.34}$$

For a typical node in an analog where $R_1 = R_2 = \cdots = R_6 = R$ (figs. 11.5.13b and 11.5.13d):

$$(1/R)(V_1 + V_2 + V_3 + V_4 + V_5 + V_6 - 6V_0) = C\,\partial V/\partial t. \tag{11.5.35}$$

In a homogeneous, but anisotropic domain ($K_x \neq K_y \neq K_z$), we have:

$$\frac{\varphi_1 + \varphi_2 - 2\varphi_0}{(\Delta x)^2/K_x} + \frac{\varphi_5 + \varphi_6 - 2\varphi_0}{(\Delta y)^2/K_y} + \frac{\varphi_3 + \varphi_4 - 2\varphi_0}{(\Delta z)^2/K_z} = S_s\frac{\partial\varphi}{\partial t} \tag{11.5.36}$$

where $R_1 = R_2$; $R_3 = R_4$; $R_5 = R_6$. The analogy is between S_s and C, a^2/K and the corresponding R. Design criteria may easily be derived:

$$\frac{\varphi_d K_{xd}}{x_r^2} = \frac{\varphi_d K_{yd}}{y_r^2} = \frac{\varphi_d K_{zd}}{z_r^2} = \frac{S_{sd}\varphi_d}{t_r}. \tag{11.5.37}$$

Often flow in a phreatic aquifer is considered. When variations in the water table elevations are small relative to the total thickness of the aquifer, the problem may be simplified by assuming that the aquifer is a confined one, thus neglecting the flow region above and below an average water table taken as the confining surface. However, in unsteady flow problems, the change in storage attributed to the fluctuating water table is an essential element of the problem. Hence, in three-dimensional flow with a phreatic surface, the storage may be simulated by appropriate capacitors connected to nodes representing the confining surface. In this case, the analogy is between C and $n_e a^2$, where a^2 is the area represented by a node, and n_e is the effective porosity (or specific yield) of the aquifer. We usually neglect $a^3 S_s$

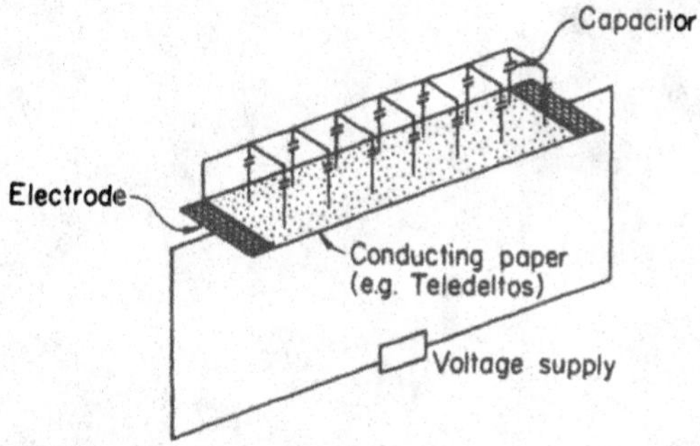

FIG. 11.5.15. A continuous electric analog with discretized capacitance (after Santing, 1963).

with respect to $a^2 n_e$. In two-dimensional flow, $n_e \gg S$, and the capacitors represent porosity storage only.

The ratio between the *RC* product of the analog network and the diffusivity of the reservoir, $a^2(S/T)$, determines the time scale of the analog. Two analog approaches are possible: the "long-time" analog, which yields a solution of a problem during 2–15 min, and the "short-time" or "repetitive" one, where the solution is obtained in less than 0.1 sec. In the first type, ordinary recorders can be used. In the latter type we must use cathode-ray oscilloscopes. This requires that the excitations be synchronized with the sweep frequency of the oscilloscope. When a storage oscilloscope is used, non-repetitive tests can be performed, with solutions obtained also in less than 0.1 sec. The "short-time" analog is the type more commonly used in ground water studies. Errors due to capacitor leakage are avoided. A portion of each cycle of the excitation is used to charge the circuit to its initial condition. Additional information can be found in numerous publications, e.g., Skibitzke (1960, 1961), Brown (1962), Bower (1960, 1962), Walton and Prickett (1963), Lagarde and Marrast (1963), and Walton (1970).

Bear and Schwarz (1966) give a detailed description of investigations of a regional ground water regime in an inhomogeneous aquifer. The investigations were carried out at the Hydrological Laboratory of Water Planning for Israel, Ltd. (Tel-Aviv, Israel). The analog was built by Elron Electronic Industries, Ltd. (Haifa, Israel). Plate 11.2 shows this analog.

A combination of a continuous analog and discrete storativity is also possible (fig. 11.5.15).

11.5.5 The Ion Motion Analog

The analogy is based on the fact that the velocity of ions in an electrolytic solution under the action of a DC voltage gradient is analogous to the average velocity of fluid particles under imposed potential gradients in a porous medium. Both electric and elastic storativities are neglected, and the flow is governed by $\nabla^2 V = 0$ and $\nabla^2 \varphi = 0$, respectively.

PLATE 11.2. An Electric Analog (*RC*) for investigations of the ground water regime in the coastal limestone aquifer in Israel (Courtesy of Water Planning for Israel, Ltd.).

The main advantage of the ion motion analog is that, in addition to the usual potential distribution, it permits a direct visual observation of the movement of an interface separating two immiscible fluids. This model is often used in oil reservoir engineering, mainly for two-dimensional displacement problems. It cannot be applied to ground water interface problems where gravity is involved.

Originally the ion-motion analog was limited to studies of the simple case of a homogeneous porous medium and two fluids of equal mobilities (Wyckoff et al., **1933**; Wyckoff and Botset **1934**; Muskat **1949**). Ramey (**1954**) uses an ion-motion analog for an infinite mobility ratio. Burton and Crawford (**1956**), and Bureau and Manasterski (**1963**) develop a new type of analog simulating a mobility ratio between the injected fluid and the displaced fluid of from 1 to 25, with stable fronts, permitting variations of the intrinsic permeability k of the reservoir. Gravity and storativity effects were not simulated.

The scaling is based on the similarity between Darcy's law and Ohm's law, governing the ion motion in an electrolytic solution.

Essentially, the analog consists of an electrolytic tank having the same geometry as the investigated flow domain. Positive and negative electrodes simulate inflow (such sources as injection wells, for example) and outflow boundaries (sinks, for example, pumping wells), respectively. Both two- and three-dimensional flow domains may be investigated.

A two-dimensional field may also be simulated by a shallow tank with a thin layer of electrolyte solution. When currents corresponding to the pumping and injection rates are introduced, ions will move from the sources to the sinks, simulating the motion of the injected fluid from the injection wells towards the production wells.

In order to obtain a sharp front in the analog, the effect of molecular diffusion of the ions should be made small compared to the velocity of the advancing front. This requirement is satisfied if the electrolyte is confined to a porous medium of low permeability. For two-dimensional fields, Wyckoff and Botset (**1934**) use a blotting paper as a porous matrix, and soak it with a diluted solution of potassium sulfate. Improved simulators are made of transparent containers with agar or gelatin (e.g., 1.5–2% by weight) as an inert porous matrix. The gelatin is made electrically conductive by adding an electrolyte to it.

Reservoirs with a variable thickness (or transmissivity) may be simulated by varying the thickness of the gelatin layer. A gelatin, or an agar solution containing 0.1 N zinc ammonium chloride, is often used to simulate the displaced reservoir fluid. This mixture, therefore, fills the tank initially. The displacing fluid is simulated by a mixture of gelatin and 0.1 N copper ammonium chloride. The latter mixture is placed in the vessels connected to injection wells, or to recharge boundaries in the analog. The invasion of the copper ammonium ions simulating the displacing fluid produces a blue coloration of the gelatin behind the advancing front. The advancing front is thus made visible (Karplus **1958**).

In early models of this type, OH^- ions simulated the injected fluid and a colorless solution containing phenolphthalein, as an indicator, made the moving ions visible

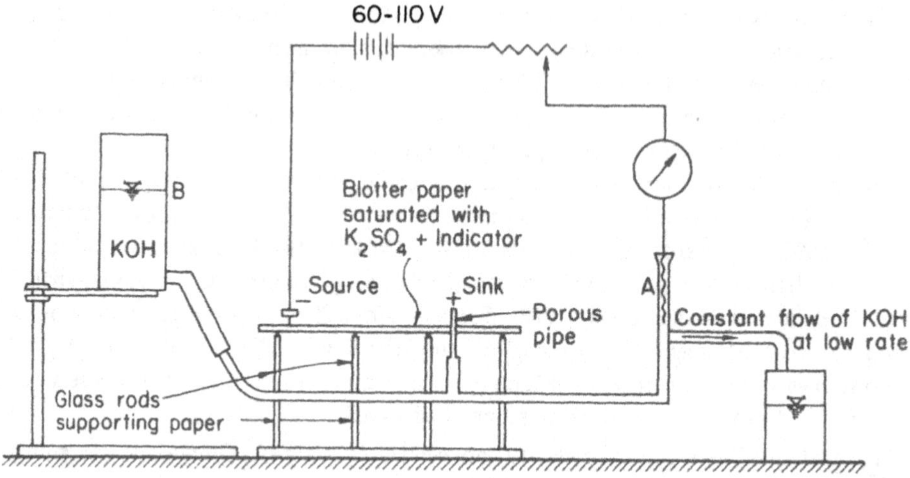

FIG. 11.5.16. A schematic diagram of the ion motion model (with blotting paper). OH ions are neutralized at the anode (after Wyckoff and Botset, 1934).

(phenolphthalein becomes red in the presence of OH^-), so that photographs of the moving fronts could be taken at regular time intervals. However, at the same time as OH^- ions move toward the positive electrodes, hydrogen ions move toward the negative electrodes. When both fronts meet, the progress of the indicator front stops. This can be avoided by neutralizing the H^+ upon formation. Figure 11.5.16 shows a method devised by Wyckoff and Botset (1934) for this purpose. The platinum electrode at A dips into an alkaline solution that is kept flowing at a low rate from the reservoir B into a beaker. In this way the anode is always alkaline and the H^+ ions are neutralized upon formation.

In the analogs simulating a mobility ratio larger than unity ($M > 1$), a mixture of gelatin and HCl of different concentrations is used. The lower concentration of HCl simulates the displaced fluid, while the higher HCl concentration simulates the displacing fluid.

11.6 The Membrane Analog

Consider a thin sheet of rubber stretched uniformly in all directions and clamped to a flat, plane frame (fig. 11.6.1). A tension σ' exists throughout the sheet. If the frame is distorted out of the plane in some arbitrary manner, the elevation of the membrane's surface will vary in a smooth manner from point to point to meet the new boundary conditions. If the elevations of the boundary (with respect to the original plane) do not vary too abruptly, the tension σ' is not appreciably affected by the distortion and may be considered constant. Under these conditions, equilibrium of the forces acting on an elementary area of the membrane (fig. 11.6.1) leads to the Laplace equation:

$$\partial^2\zeta/\partial x^2 + \partial^2\zeta/\partial y^2 = 0. \tag{11.6.1}$$

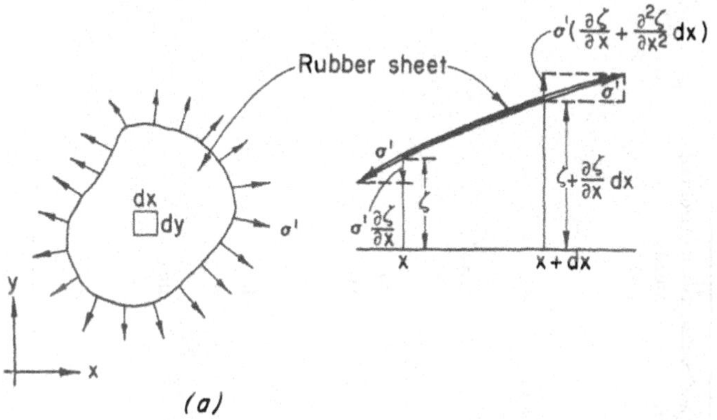

FIG. 11.6.1. The principle of the membrane analog.

When the stretched membrane is subjected to a transversal load (P) per unit area in the z direction, the equation expressing equilibrium for small slopes becomes the Poisson equation:

$$\partial^2\zeta/\partial x^2 + \partial^2\zeta/\partial y^2 = P/\sigma'. \tag{11.6.2}$$

The load is either the membrane's own distributed weight, or a specially applied air or water pressure. When so desired, the effect of the term P/σ' in (11.6.2) can be minimized by placing the membrane in a vertical position.

The analogy is based on the similarity between (11.6.1) and (11.6.2) and the corresponding equations that describe flow in aquifers: (11.6.1) is analogous to (6.2.24), while (11.6.2) is analogous to the approximate equation (8.4.3) with $\partial h/\partial t = 0$, describing steady two-dimensional flow in a horizontal homogeneous aquifer with uniform accretion at a constant rate N:

$$T(\partial^2\varphi/\partial x^2 + \partial^2\varphi/\partial y^2) + N = 0. \tag{11.6.3}$$

We make certain assumptions leading to (11.6.3). Similarly, in deriving (11.6.1) and (11.6.2) we have assumed that the variations in the elevations (or the deflections) ζ that represent the head $\varphi = \varphi(x, y)$, are relatively small, so that the tensile stress σ' (parallel to the initial membrane's plane), which represents the aquifer's transmissivity T, remains practically constant. At the same time, the shear stress perpendicular to the membrane represents the specific discharge per unit width of aquifer.

The membrane's edges are clamped on frames according to the head distribution along the boundaries. Hansen (1949, 1952) investigates a complicated multiple well problem by using an analog of this type (fig. 11.6.2). Piezometric heads at pumping or injecting wells are simulated by appropriate elevations of the membrane maintained at those points by means of frictionless pegs. Rates of flow are represented by the slopes.

The scales may be derived from (11.6.1) and (11.6.2) by the method described in sections 7.4 and 11.2. We obtain:

Table 11.7.1

Applicability of Models and Analogs

Feature	Sand Box Model	Hele–Shaw Analog		Electric Analogs			Membrane Analog
		Vertical	Horizontal	Electrolytic	*RC* Network	Ion Motion	
Dimensions of field	two or three	two	two	two or three	two or three	two (horizontal)	two (horizontal)
Steady or unsteady flow	both	both	both	steady	both	steady	steady
Simulation of elastic storage	yes, for two dimensions	yes	yes	yes, for two dimensions	yes	no	no
Simulation of capillary fringe and capillary pressure	yes	yes	no	no	no	no	no
Simulation of phreatic surface	yes[1]	yes[1]	no	yes[2]	no[3]	no	no
Simulation of anisotropic media[4]	yes	yes $k_x \neq k_z$	yes $k_x = k_y$	yes	yes	yes $k_x \neq k_y$	yes $k_x = k_y$
Simulation of medium inhomogeneity	yes	yes[5]	yes[5]	yes	yes	yes	no
Simulation of leaky formations	yes	yes	yes	yes[5]	yes	no	no
Simulation of accretion	yes	yes	yes	yes, for two dimensions	yes	no	yes
Flow of two liquids with an abrupt interface	approximately	yes	yes (no gravity)	no[6]	no[6]	yes (no gravity)	no
Hydrodynamic dispersion	yes	no	no	no	no	no	no
Simultaneous flow of two immiscible fluids	yes	no	no	no	no	no	no
Observation of streamlines and pathlines	yes, for two dimensions, near transparent walls for three dimensions	yes	yes	no	no	no	no

(1) Subject to restrictions because of the presence of a capillary fringe
(2) By trial and error for steady flow
(3) By trial and error for steady flow, or, as an approximation, for relatively small phreatic surface fluctuations
(4) By scale distortion in all cases, except for the *RC* network and sometimes the Hele–Shaw analog where the hydraulic conductivity of the analog can be made anisotropic
(5) With certain constraints
(6) For a stationary interface by trial and error

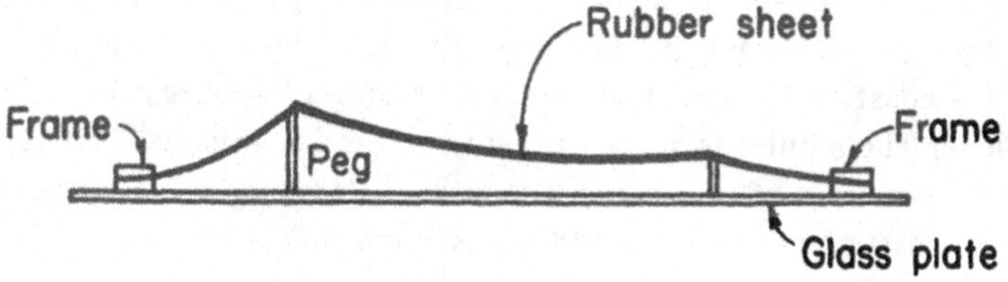

FIG. 11.6.2. The membrane analog.

$$x_r = y_r; \qquad \varphi_r = \delta\zeta/\delta\varphi; \qquad E_r \equiv (P/\sigma')/(N/T) = \varphi_r/x_r^2. \qquad (11.6.4)$$

De Josselin de Jong (1960) shows how a Moire pattern (Lightenberg 1955; Zienkiewicz and Holister 1965) provides a convenient procedure for obtaining contour lines of the deflected membrane. The main advantage of this device is that it yields a direct picture of the φ-ψ flow net. This procedure eliminates the need for point-to-point plotting of contour lines, as in the electrolytic tank analog (sec. 11.5).

The method is applicable mainly to cases of steady two-dimensional flow with a complicated boundary geometry, and with point sources and sinks (e.g., wells) within the flow domain.

11.7 Summary

Table 11.7.1 gives a summary of the applicability of the various types of models and analogs discussed in the present chapter.

Exercises

11.1 Determine the design criteria (scales) for unsaturated flow in a sand box model.

11.2 Design a Hele-Shaw analog for the drainage problem shown in fig. 8.2.2a: $L = 100$ m, $h_0 = 10$ m, $h_L = 5$ m, $N = 500$ mm/yr, Δt_p of interest is 1 month; the soil is anisotropic with $K_x = 27$ m/d, $K_y = 3$ m/d.

11.3 In the analog study of Ex. 11.2, the variable discharge to each of the ditches is plotted versus time. Suggest dimensionless discharge and time parameters.

11.4 Design a Hele-Shaw analog for steady flow through the dam shown in fig. 11.4.2. The dam's base is $L = 200$ m, its height is 30 m, the hydraulic conductivity $K_x = 5$ m/d, $K_y = 0.2$ m/d.

11.5 Determine the scales for an experiment of three-dimensional flow of a slightly compressible fluid (φ^*) in a homogeneous anisotropic porous medium.

11.6 Design a Hele-Shaw analog for investigating sea water intrusion in the coastal aquifer shown in fig. 9.7.1a (unsteady flow). Data: Depth of impervious bottom at the coast = 100 m below sea level. Depth of aquifer 8 km inland is 30 m below sea level (the aquifer terminates there with an impervious vertical boundary), $N = 300$ mm/year; Δt of interest is 6 months. Pumping takes place along a line 3 km from the coast at a rate of up to 300.000 m^3/month/km. The aquifer is anisotropic with $K_x = 64$ m/d, $K_z = 4$ m/d. Choose the necessary scales.

11.7 Determine the design criteria for a horizontal Hele-Shaw analog of the two leaky confined aquifers shown in fig. 11.4.10. Neglect the storativity of the semi-pervious layer and of the lower aquifer.

11.8 Design an electrolytic tank analog for studying the steady flow through the dam shown in fig. 8.1.7. Data: $L = 50$ m, $L' = L''$, $h_0 = 20$ m, $h_L = 5$ m, $K' = 2K'' = 0.1$ m/d (both media are isotropic). Describe the analog, its instrumentation and the experimental procedure.

11.9 Repeat Ex. 11.8, assuming that in both media the vertical permeability is 1/10 of the horizontal one.

11.10 The flow net shown in fig. 6.6.1 was obtained by means of an electrolytic tank analog. Explain the procedure employed to obtain both streamlines and equipotentials. Explain how to determine the dimensions of the electrolytic tank when

(a) the medium is homogeneous but anisotropic $K_x = 25K_y$,

(b) the medium is composed of two layers (the interface passing at some depth below the channel's bottom), each of which is anisotropic, but x and y are principal axes in both media, and

(c) when in the upper layer the principal axes are inclined.

11.11 Describe an electrolytic tank analog and an experimental procedure for solving the problem of upconing of an oil–water interface below a well in steady flow (fig. 9.7.10).

11.12 A homogeneous isotropic aquifer has the form of an infinite strip between two equipotential boundaries: $\varphi = 0$ along $y = 0$ and $\varphi = +10$ m along $y = 2000$ m. The aquifer has a transmissivity of 2000 m^2/d. Two wells are located in this aquifer: $Q_w = 200$ m^3/h at (200 m, 400 m) and $Q_w = 400$ m^3/h at (-1000 m, 1000 m). Design an electrolytic tank analog with tap water (conductivity 400 μmhos/cm) for determining the potential distribution in the aquifer. (Hints: Divide the aquifer into two parts one of which in the form of a square in the vicinity of the origin and employ the inverse transformation).

11.13 Given a homogeneous isotropic confined aquifer in the shape of a rectangle 4 km × 10 km. The two longer sides (parallel to x) are maintained at fixed potentials $\varphi = 0$ and $\varphi = +10$ m, while the other two boundaries are impervious. The aquifer is anisotropic with $T_x = 6000$ m²/d, $T_y = 1000$ m²/d. The aquifer's storativity is $S = 0.2$. A well located at the center of the area pumps a constant discharge of 500 m³/h during summer and rests during winter. Design an *RC* network analog which can be used to determine the potential distribution in the aquifer. Initially, steady flow without pumpage exists in the aquifer.

11.14 A boundary of a flow domain is represented by straight line segments connecting the following points: (0, 3) equipotential (0, 6) equipotential (4, 10) equipotential (9.5, 10) impervious (9.5, 0) equipotential (3.5, 0) impervious (3.5, 3) impervious (0, 3). Using the units of length in the x and y directions as sides of squares in a grid used for representing the flow domain by an *RC* network, determine the magnitude of the resistances and capacitors connected to grid points which are located on and adjacent to the boundaries. Use R_0 and C_0 to denote the corresponding values at interior grid points.

(Note: (a) Only resistances which are parallel to the grid lines are represented.

(b) $R = R_0\, a/b$ where a and b are average lengths of the area represented by the resistance in the direction of the latter and normal to it (why?).

(c) Capacitors are proportional to the areas represented by them.)

11.15 Determine scales and explain the structure (e.g., a typical node) of an *RC* network analog which can be used for studying the flow in a leaky confined aquifer described by (6.4.10).

11.16 Design an *RC* network for studying unsteady flow to a partially penetrating well in an anisotropic homogeneous confined aquifer of constant thickness; at $r = R$, $\varphi =$ const. Take into account both hydraulic conductivity and specific storativity of the aquifer.

11.17 (a) Show that at any instant in unsteady two-dimensional flow bounded in part by a phreatic surface, the downward velocity of a particle on the interface is given by

$$\frac{\partial \zeta}{\partial t} = \frac{K}{n}\left(\frac{\partial \varphi}{\partial x}\tan\theta - \frac{\partial \varphi}{\partial z}\right)$$

where θ is the slope angle of the phreatic surface ($\tan\theta = \partial\zeta/\partial x$). Modify this expression for anisotropic media.

(b) Since unsteady flow can be studied as a succession of steady states, explain how a resistance network analog can be employed, together with the above expression to determine the movement of a phreatic surface (e.g., decline of a ground water mound).

(Hint: The phreatic surface is a boundary on which the variation of φ is known).

Answers to Exercises

Chapter 4

4.1 $$\mathbf{V}'' = \sum_{\alpha=1}^{N} x_\alpha \mathbf{V}_\alpha$$

Chapter 5

5.1 $Q = 4.6\ \text{cm}^3/\text{sec}$; $q = 0.023\ \text{cm/sec}$; $V = 0.064\ \text{cm/sec}$

5.2 $K = 75\ \text{m/d}$

5.3 (a) $T = 32\ \text{m}^2/\text{h}$, (b) $J/\nu = \text{const.}$

5.5 (a) $Re = 2.7 \times 10^{-4} < 1$, (b) $V = 0.15\ \text{m/d}$

5.6 2670 psi

5.8 $|q| = 0.00945\ \text{K}$, $(\widehat{\mathbf{q}, \mathbf{1x}}) = 122°$

5.9 $(ax + by + c)(\partial^2\varphi/\partial x^2 + \partial^2\varphi/\partial y^2) + a\ \partial\varphi/\partial x + b\ \partial\varphi/\partial y = 0$

5.10 for $\nu = 0.01\ \text{cm}^2/\text{s}$, $Re > 10$ for $r < 0.053\ \text{m}$

5.12 $K_w = 0.746 \times 10^{-3}\ \text{cm/sec}$, $K_0 = 0.54 \times 10^{-5}\ \text{cm/sec}$

5.13 $q_x = -0.052\ \text{m/day}$, $q_y = 0.05\ \text{m/day}$

5.14 (a) $\mathbf{q} = 0.306\ \text{m/d}$; $\tan^{-1}(\widehat{\mathbf{q}, \mathbf{1x}}) = -18°45'$;

(b) Angles of $K_{x'x'}$ and $K_{y'y'}$ are 26°34′ and 116°34′ (clockwise with respect to the $+x$ axis)

5.15 $K_{x'x'} = 22.7$; $K_{y'y'} = 7.3$; $K_{x'y'} = -1.35$

5.16 (a) $K_J = 22.1\ \text{m/d}$, $\alpha_1 = 15°$;

(b) $K_q = 14.1\ \text{m/d}$, $\beta_1 = 76°$;

(c) $K_J \geqslant K_q$

5.17 $K_x = 20.8$, $K_y = 5.2$, $\theta = 25°5'$

5.18 (a) $K_x = 10$, $K_y = 2.5$, (b) $K_q = 6.3$, (c) $K_J = 3.98$,

(d) $K_{x'x'} = 8.13$, $K_{y'y'} = 4.37$, $K_{x'y'} = -3.25$

5.19 $$K = \frac{L}{[(1/A_1) + (1/A_2)]\,\text{at}} \ln \frac{h_0}{h}$$

5.20 $k = 395\ \text{md}$

5.22 $q = 0.0787\ \text{m/d}$, $\bar{K} = 39.4\ \text{m/d}$

5.23 $$p_B = -0.019 \text{ kg/cm}^2; \qquad p_B = 33 \text{ kg/cm}^2$$

5.24 $$Q = 1.27 \text{ m}^3\text{/d/m}$$

5.25 $$K_H = \frac{N_1K_1b_1 + N_2K_2b_2}{N_1b_1 + N_2b_2}$$

5.26 $$K_V = \frac{N_1b_1 + N_2b_2}{(N_1b_1/K_1) + (N_2b_2/K_2)}$$

5.28 $$Q = 2\pi D(\varphi_e - \varphi_w) \Big/ \left(\frac{1}{K_1}\ln\frac{r_1}{r_w} + \frac{1}{K_2}\ln\frac{r_e}{r_1}\right)$$

5.29 (a) $h_B = \dfrac{K_1 \sin\theta_1 + K_2 \sin\theta_2}{(k_2\cos\theta_2/L_2) + (K_1\cos\theta_1/L_1)}$, (b) $\dfrac{K_2}{K_1} \leqslant \dfrac{\cos\theta_1}{\cos\theta_2}$

5.30 $$\Phi^g = (K_1b_1 + K_2b_2)\varphi - \frac{K_1{b_1}^2}{2} - (2b_1 + b_2)\frac{K_2b_2}{2}$$

5.31 (a) $\Phi^g = K_1h^2/2$; (b) $\Phi^g = K_2h^2/2 + b_1(K_1 - K_2)(h - b_1/2)$

5.33 $$p^2(x) = {p_0}^2 - \frac{{p_0}^2 - {p_L}^2}{L}x$$

5.34 (a) $$M = \frac{\pi k h \rho_{at}}{\mu p_{at}} \frac{{p_e}^2 - {p_w}^2}{\ln r_e/r_w}$$

(b) $$p^2 = {p_w}^2 + \frac{{p_e}^2 - {p_w}^2}{\ln r_e/r_w}\ln\frac{r}{r_w}$$

5.35 $$q = 8.15 \text{ cm/s}$$

5.37 $$\sigma = 21 \text{ t/m}^2; \qquad \sigma' = 11 \text{ t/m}^2$$

5.38 (a) $\sigma = 19 \text{ t/m}^2$, $\sigma' = 11 \text{ t/m}^2$

(b) $\sigma = 23 \text{ t/m}^2$, $\sigma' = 11 \text{ t/m}^2$

5.39 (a) $\sigma = 23 \text{ t/m}^2$, $\sigma' = 13 \text{ t/m}^3$

(b) $\sigma = 23 \text{ t/m}^2$, $\sigma' = 9 \text{ t/m}^2$

5.40 q_z, (upward) $= 5.5$ m/d

Chapter 6

6.1 (a) $$r\frac{\partial^2\varphi}{\partial r^2} + \frac{\partial\varphi}{\partial r} = 0$$

(b) $$\frac{1}{r}\frac{\partial}{\partial r}\left(K_r r\frac{\partial\varphi}{\partial r}\right) + \frac{\partial}{\partial z}\left(K_z\frac{\partial\varphi}{\partial z}\right) = 0$$

6.3 $$\frac{k_{ij}}{\mu}\frac{\partial^2\varphi}{\partial x_i\,\partial x_j} = 0$$

6.4 $$\nabla^2 p = \frac{n\beta\mu}{k}\frac{\partial p}{\partial t}$$

6.9 $$BE = 55.2\%$$

6.11 $$\frac{\partial^2\varphi}{\partial r^2} + \frac{1}{r}\frac{\partial\varphi}{\partial r} + \frac{1}{r^2}\frac{\partial^2\varphi}{\partial\theta^2} = \frac{S}{T}\frac{\partial\varphi}{\partial t}$$

6.14 $$\partial^2 h^2/\partial r^2 + (1/r)\,\partial h^2/\partial r = (2n/K)\,\partial h/\partial t$$

6.18 $$(\mathrm{curl}\,\mathbf{q})_z \neq 0$$

6.19 $$\text{(a)}\quad \frac{1}{ax}\nabla^2\Psi - \frac{1}{ax^2}\frac{\partial\Psi}{\partial x} = 0$$

$$\text{(b)}\quad ay\frac{\partial^2\Psi}{\partial x^2} + by\frac{\partial^2\Psi}{\partial y^2} = 0$$

6.20 $$\text{(a)}\quad \mathbf{q} = 2b(\mathbf{1x} + \mathbf{1y}), \qquad \text{(b)}\quad \Phi = 2xyb + \text{const}$$

6.21 No, because $K\nabla^2\Psi - \nabla K\cdot\nabla\Psi \neq 0$

Chapter 7

7.1 (a) $(\partial\Psi/\partial y)/\cos\alpha_{nx} = -(\partial\Psi/\partial x)/\cos\alpha_{ny}$;

(b) $(\partial\Psi/\partial y)/K_x\cos\alpha_{nx} = -(\partial\Psi/\partial x)/K_y\cos\alpha_{ny}$

(c) $$(\partial\Psi/\partial y)\left(\frac{K_{yy}}{\cos\alpha_{nx}} + \frac{K_{xy}}{\cos\alpha_{ny}}\right) = -(\partial\Psi/\partial x)\left(\frac{K_{xx}}{\cos\alpha_{ny}} + \frac{K_{xy}}{\cos\alpha_{nx}}\right)$$

7.2 $$\tan\alpha = \frac{q_y}{q_x} = \frac{K_y}{K_x}\frac{a^2y}{b^2x}$$

7.3 $$2xK_x\frac{\partial\varphi}{\partial x} + 2yK_y\frac{\partial\varphi}{\partial y} + 2zK_z\frac{\partial\varphi}{\partial z} = 0$$

7.8 $$\text{(a)}\quad \beta'' = 3°18', \qquad \text{(b)}\quad \beta'' = 80°10'$$

7.24 $$s(r,t) = \frac{Q_w}{4\pi(T_xT_y)^{1/2}}\ln\frac{2.25t(T_xT_y)}{(x^2T_y + y^2T_x)S}$$

7.25 $$Q = \frac{4K_xH}{\pi}\sum_{n=1,3,5}^{\infty}\left\{\frac{1}{n}\,\mathrm{ctgh}\left[\frac{n\pi D}{2B}\left(\frac{K_x}{K_y}\right)^{1/2}\left(1 + \frac{\sigma' K_y}{D}\right)\right]\right\}^{-1}$$

7.27 $$\varphi = \frac{2}{\pi^{1/2}}\int_0^\infty f\left(t - \frac{x^2}{4(T/S)\alpha^2}\right)\exp(-\alpha^2)\,d\alpha$$

7.28 $$\varphi = \varphi_3(r,t) + \int_0^t f_1(\tau)\frac{\partial\varphi_1(r,t-\tau)}{\partial t} + f_2(\tau)\frac{\partial\varphi_2(r,t-\tau)}{\partial t}\,d\tau$$

7.30 (a) $0.108 \times 10^6\ \mathrm{m}^3$; (b) ~ 1.6 days

7.31 $$\varphi_0 - \varphi(r) = \frac{Q_w}{2\pi T}\ln\frac{R}{r}; \qquad \varphi(r) = \varphi_w + (\varphi_0 - \varphi_w)\frac{\ln(r/r_w)}{\ln(R/r_w)}$$

7.33 $$h(t) = H_0\exp\left\{-\frac{KA}{L}\left(\frac{1}{B_1} + \frac{1}{B_2}\right)t\right\}$$

7.34 $$\frac{\zeta^2}{2}(\mu_1 - \mu_2) + \mu_2 L\zeta = (h_0\gamma_1 + h_L\gamma_2)\frac{k}{n}t$$

7.41 $$\varphi_1 = 15, \quad \varphi_2 = 15, \quad \varphi_3 = 25, \quad \varphi_4 = 25$$

7.42 $$\zeta = -m \ln \tan(\pi z/2d)$$

Chapter 8

8.3 $$h^2 = H_0{}^2 - (H_0{}^2 - h_w{}^2)\frac{\ln R/r}{\ln R/r_w}$$

8.7 $$5.27 \text{ m}^3/\text{day}/\text{m}$$

8.8 $$Q = 0.066 \text{ m}^3/\text{day}/\text{m}$$

8.9 $$Q_2 = 2.85 \text{ m}^3/\text{d}/\text{m}, \quad Q_1 = 3.6 \text{ m}^3/\text{d}/\text{m}, \quad h_{\max} = 28.2 \text{ m}$$

8.10 $$h^2(x) = h_w{}^2 + \frac{N}{K}(L^2 - x^2)$$

8.11 $$b = \frac{KD}{Q}(h_0 - D), \quad Q = KD\left(h_0 - \frac{D^2 + h_L{}^2}{2D}\right)\Big/L$$

Chapter 9

9.1 $$h_{\max} = \frac{p_{cr}}{\gamma_w - \gamma_0}$$

9.2 $$h = p_c/(\gamma_w - \gamma_0)$$

9.3 (a) $$\frac{p_c, \text{ lab.}}{p_c, \text{ res.}} = \frac{\sigma_{wg}\cos\theta_{wg}}{\sigma_{w0}\cos\theta_{w0}},$$

(b) $h = 58$ ft

9.16 $$t = \frac{n}{k}\frac{L^2(\mu_w + \mu_0)}{2(h_0\gamma_0 + h_w\gamma_w)}$$

Chapter 10

10.3 (a) $$\left[(f^* a_{II} q + nD_d T^*) v_j \frac{\partial C}{\partial x_j}\right]_a = \left[(nD_d T^*) v_j \frac{\partial C}{\partial x_j}\right]_b$$

(b) $$\left[(f^* a_{II} q + nD_d T^*) v_j \frac{\partial C}{\partial y_j}\right]_a = 0$$

10.4 $$(C_0 - C)q = -(nD_d T^* + f^* a_I q)\,\partial C/\partial x$$

10.6 $$\frac{C(x,t) - C_0}{C_1 - C_0} = \tfrac{1}{2}\operatorname{erfc}\left[-\frac{xR_d - qt/n}{(4D'R_d t)^{1/2}}\right]$$

where $R_d = 1 + \dfrac{1-n}{n}a_1$ is defined in (11.5.7)

Bibliography

Abromowitz, M., and I. A. Stegun, *Handbook of Mathematical Functions*, Dover, New York, 1965.

Adamson, A. W., *Physical Chemistry of Surfaces* (1st ed., 1960), 2nd ed., Interscience, New York, 1967.

Adivaraham, P., D. Kunii, and J. M. Smith, Heat transfer in porous rocks with single phase fluid flowing, *J. Soc. Petrol. Eng.* No. 3, **2**, 290–296 (1962).

Ahmad, N., Physical Properties of Porous Medium Affecting Laminar and Turbulent Flow of Water, Ph. D. Thesis, Colorado State University, Fort Collins, Colo., 1967.

Ali Ibrahim, H., and W. Brutsaert, Intermittent infiltration into soils with hysteresis, *Proc. Amer. Soc. Chem. Eng.* No. HY1, **94**, 113–137 (1968).

Amyx, J. W., D. M. Bass Jr., and R. L. Whiting, *Petroleum Reservoir Engineering, Physical Properties*, McGraw-Hill, New York, 1960.

Anat, A., H. R. Duke, and A. T. Corey, Steady upward flow from water tables, Hydrology Paper No. 7, Colorado State University, Fort Collins, Colo., 1965.

Aravin, V. I., Basic problems in the experimental investigation of the flow of groundwater by means of a parallel plate model (in Russian), *Izv. NIIG* **23** (1938).

———, Experimental investigation of unsteady flow of groundwater (in Russian), *Trans. Sci. Res. Inst. Hydrotech. U.S.S.R.* **30**, 79–88 (1941).

———, and S. N. Numerov, *Theory of Motion of Liquids and Gases in Undeformable Porous Media* (in Russian), Gostekhnizdat, Moscow, 1953; English transl. by A. Moscona, Israel Program for Scientific Translation, Jerusalem, 1965.

Archie, G. E., The electrical resistivity log as an aid in determining some reservoir characteristics, *Trans. A.I.M.E.* **146**, 54–61 (1942).

Aris, R., On the dispersion of a solute in a fluid flowing through a tube, *Proc. Roy. Soc. A.* **235**, 67–77 (1956).

———, *Vectors, Tensors and the Basic Equations of Fluid Mechanics*, Prentice-Hall, Englewood Cliffs, N. J., 1962.

———, and N. R. Amundson, Some remarks on longitudinal mixing or diffusion in fixed beds, *J. Amer. Inst. Chem. Eng.* No. 2, **3**, 280–282 (1957).

Asaad, Y., A Study of Thermal Conductivity of Fluid Bearing Porous Rock, Ph. D. Thesis, University of California, 1955.

Athy, L. F., Density, porosity and compaction of sedimentary rocks, *Bull. Amer. Ass. Petrol. Geol.* No. 1, **14**, 1–24 (1930).

Averjanov, S. F., About permeability of subsurface soils in case of incomplete saturation, *Engineering Collection* (in Russian), **7** (1950); *see also*, Dependence of permeability of subsurface soils on their air content, *DAN* (*CCCP*) No. 2, **69** (1949).

Bachmat, Y., Basic transport coefficients as aquifer characteristics, *I.A.S.H. Symp. Hydrology of Fractured Rocks, Dubrovnik* **1**, 63–75 (1965).

———, On the similitude of dispersion phenomena in homogeneous and isotropic porous medium, *Water Resources Res.* No. 4, **3**, 1079–1083 (1967a).

———, and J. Bear, The general equations of hydrodynamic dispersion, *J. Geophys. Res.* No. 12, **69**, 2561–2567 (1964).

Badon-Ghyben, W., *Notes on the Probable Results of Well Drilling Near Amsterdam* (in Dutch), p. 21, Tijdschrift van het Koninklijk Inst. van Ing., The Hague, 1888.

Barnes, A. L., The use of a viscous slug to improve waterflood efficiency in a reservoir

partially invaded by bottom water, *J. Petrol. Tech.* **14**, 1147–1154 (1962).
Bastian, W. C., and L. Lapidus, Longitudinal diffusion in ion exchange and chromatographic columns, Finite columns, *J. Phys. Chem.* **60**, 816–817 (1956).
Batchelor, G. K., *Theory of Homogeneous Turbulence*, Cambridge University Press, London, 1953.
Bear, J., The Transition Zone Between Fresh and Salt Waters in Coastal Aquifers, Ph. D. Thesis, University of California, Berkeley, Calif., 1960.
———, Scales of viscous analog models for ground water studies, *Proc. Amer. Soc. Civil Eng. Hydraul. Div.* No. HY2, **86**, 11–23 (1960a).
———, On the tensor form of dispersion, *J. Geophys. Res.* No. 4, **66**, 1185–1197 (1961a).
———, Some experiments on dispersion, *J. Geophys. Res.* No. 8, **66**, 2455–2467 (1961b).
———, Two-liquid flows in porous media, in *Advances in Hydroscience* (V. T. Chow, ed.), Vol. 6, pp. 142–252, Academic Press, New York, 1970.
———, and Y. Bachmat, Hydrodynamic dispersion in non-uniform flow through porous media, taking into account density and viscosity differences (in Hebrew with English summary), Hydraulic Lab., Technion, Haifa, Israel, P.N. 4/66, 1966.
———, and Y. Bachmat, A generalized theory on hydrodynamic dispersion in porous media, *I.A.S.H. Symp. Artificial Recharge and Management of Aquifers, Haifa*, Israel, IASH, P.N. 72, 7–16 (1967).
———, and G. Dagan, The transition zone between fresh and salt waters in a coastal aquifer, Hydraulic Lab., Technion, Haifa, Israel; Prog. Rep. 1: The steady interface between two immiscible fluids in a two-dimensional field of flow, 1962a; Prog. Rep. 2: A steady flow to an array of wells above the interface, approximate solution for a moving interface, 1963; Prog. Rep. 3: The unsteady interface below a coastal collector, 1964a; Prog. Rep. 4: Increasing the yield of a coastal collector by means of special operation techniques, 1966; Prog. Rep. 5: The transition zone at the rising interface below the collector, 1966a.
———, and G. Dagan, The use of the hodograph method for ground water investigations, Dept. of Civ. Eng., Technion, Haifa, Israel, 1962.
———, and G. Dagan, Some exact solutions of interface problems by means of the hodograph method, *J. Geophys. Res.* No. 2, **69**, 1563–1572 (1964).
———, and G. Dagan, Moving interface in coastal aquifers, *Proc. Amer. Soc. Civil Eng. Hydraul. Div.* No. 4, **90**, 193–215 (1964b).
———, and G. Dagan, The relationship between solutions of flow problems in isotropic and anisotropic soils, *J. Hydrology* **3**, 88–96 (1965).
———, and M. Jacobs, On the movement of water bodies injected into aquifers, *J. Hydrology* No. 1, **3**, 37–57 (1965).
———, M. Jacobs, and C. Braester, The use of models and analoges in reservoir engineering, Hydraulic Lab., Technion, Haifa, Israel, P.N. 2/66, 1966.
———, and Y. Schwartz, Electric analog for regional groundwater studies, Water Planning for Israel, Ltd., Tel-Aviv, Israel, P.N. 609, 1966.
———, and D. K. Todd, The transition zone between fresh and salt waters in coastal aquifers, University of California, Water Resources Center Contrib. No. 29, 1960.
———, and D. Zaslavsky, Investigation on the drainage of the 108″ main conduit; stage I: The flow profile, Hydraulic Lab., Technion, Haifa, Israel, IASH, P.N. 1/61, (in Hebrew, English Abstract), 1961.
———, D. Zaslavsky, and S. Irmay, *Physical Principles of Water Percolation and Seepage*, UNESCO, Paris, 1968.
Belicova, V. K., Unsteady flow of groundwater towards an horizontal drain (in Russian), *Prik. Mat. Mek.* **19**, 234–239 (1955).
Benard, H., *Rev. Gen. Sci. Pur. Appl.* **2**, 1261, 1309 (1900).
Beran, N. J., Dispersion of Soluble Matter in Slowly Moving Fluids, Ph. D. Thesis, Harvard University, Cambridge, Mass., 1955.

Bikerman, J. J., *Surface Chemistry Theory and Application*, 2nd ed., Academic Press, New York, 1958.

Biot, M. A., General theory of three-dimensional consolidation, *J. Appl. Phys.* **12**, 155–164 (1941).

______, Theory of deformation of a porous viscoelastic anisotropic solid, *J. Appl. Phys.* No. 5, **27**, 459–467 (1956).

Bird, R. B., W. E. Stewart, and E. N. Lightfoot, *Transport Phenomena*, John Wiley, New York, 1960.

Blackwell, R. J., An investigation of miscible displacement processes in capillaries, paper presented at the Local Section Meeting of *Amer. Inst. Chem. Eng.*, Galveston, Texas, 1957.

______, Experiments on mixing by fluid flow in porous media, *Amer. Inst. Chem. Eng.* and *Soc. Petrol. Eng.* 52nd Annual Meeting, San Francisco, Preprint No. 29, 1959.

______, J. R. Rayne, and W. M. Terry, Factors influencing the efficiency of miscible displacement, *Trans. A.I.M.E.* **217**, 1–8 (1959).

Blake, F. C., The resistance of packing to fluid flow, *Trans. Amer. Inst. Chem. Eng.* **14**, 415–421 (1922).

Blick, E. F., Capillary orifice model for high speed flow through porous media I & EC, *Process Design and Development*, No. 1, **5**, 90–94 (1966).

Bodman, G. B., and E. A. Coleman, Moisture and energy conditions during downward entry of water into soils, *Soil Sci. Soc. Amer. Proc.* **8**, 116–122 (1943).

Bolt, G. H., and P. H. Groenevelt, Coupling phenomena as a possible cause for non-Darcian behavior of water in soil, *Bull. I.A.S.H.* No. 2, **14**, 17–26 (1969).

Boltzmann, L., Zur Integration der Diffusionsgleichung bei variablen Diffusions-koeffizienten, *Ann. Phys.* (*Leipzig*) **53**, 959–964 (1894).

Bondarev, E. A., and V. N. Nikolaevskij, Convective diffusion in porous media with the influence of adsorption phenomena, *The Acad. Sci. U.S.S.R.*, *J. Appl. Mech. Tech. Phys.* No. 5, **128–134** (1962).

Boothroyd, A. R., E. C. Cherry, and R. Maker, An electrolytic tank for the measurement of steady state response, transient response and allied problems of networks, *Proc. Inst. Elec. Eng.* London, No. 96, **1**, 163–177 (1949).

Botset, A. G., Flow of gas-liquid mixtures through consolidated sand, *Trans. A.I.M.E.* **136**, 91–108 (1940).

Boulton, N. S., The drawdown of the water table under non-steady conditions near a pumped well in an unconfined formation, *Proc. Brit. Inst. Civil Eng.* **3**, Part 3, 564–579 (1954).

Boussinesq, J., Recherches théoriques sur l'écoulement des nappes d'eau infiltrées dans le sol et sur débit de sources, *C.R.H. Acad. Sci.*, *J. Math. Pures Appl.* **10**, 5–78 (June, 1903), and 363–394 (Jan., 1904).

Bouwer, H., Unsaturated flow in ground water hydraulics, *Proc. Amer. Soc. Civil Eng.* No. HY5, **90**, 121–144 (1964).

Bower, H., A study of final infiltration rates from cylindrical infiltrometers and irrigation furrows with electrical resistance network, *Trans. Intern. Soc. of Soil Sci. 7th Congr.* **1**, Paper 6, Madison, Wisc. (1960).

______, Analyzing ground water mounds by resistance network, *Proc. Amer. Soc. Civil Eng. J. Irrig. and Drainage Div.* No. IR1, **88** 15–36 (1962).

Boyer, R. L., F. Morgan, and M. Muscat, A new method for measurement of oil saturation in cores, *A.I.M.E.* **170**, 15–29 (1947).

Bredehoeft, J. D., and I. S. Papadopulos, Rates of vertical groundwater movement estimated from the earth's thermal profile, *Water Resources Res.* No. 2, **1**, 325–328 (1965).

Breitenöder, M., *Ebene Grundwasserströmungen mit freier Oberfläche*, Springer, Berlin, 1942.

Brigham, W. E., P. W. Reed, and J. N. Dew, Experiments on mixing by fluid flow in porous media, *Amer. Inst. Chem. Eng.* and *Soc. Petrol. Eng. Joint Symp. Oil Recovery Methods, San Francisco* (1959).

Brooks, C. S., and W. R. Purcell, Surface area measurements on sedimentary rocks, *Trans. A.I.M.E.* **195**, 289–296 (1952).

Brooks, R. H., Unsteady flow of groundwater into drain tile, *Proc. Amer. Soc. Civil Eng.* No. IR2, **87**, 27–37 (1961).

———, and A. T. Corey, Hydraulic properties of porous media, Hydrology Papers, Colorado State University, Fort Collins, Colo., 1964.

———, and A. T. Corey, Properties of porous media affecting fluid flow, *Proc. Amer. Soc. Civil Eng.* No. IR2, **92**, 61–87 (1966).

Brown, H. W., Capillary pressure investigations, *Trans. A.I.M.E.* **192**, 67–74 (1951).

Brown, R. H., in Theory of Aquifer Tests, by J. G. Ferris, D. B. Knowles, R. H. Brown and R. W. Stallman, U.S. Geol. Survey, Water Supply Paper 1536-E, Washington D.C., 1962.

Brownscombe, E. R., R. L. Slobad, and B. H. Caudle, Laboratory determination of relative permeability, *Oil Gas J.* 66ff and 98ff (1950).

Bruce, R. R., and A. Klute, The measurement of soil moisture diffusivity, *Soil Sci. Soc. Amer. Proc.* **20**, 458–462 (1956).

Brutsaert, W., The adaptability of an exact solution to horizontal infiltration, *Water Resources Res.* No. 4, **4**, 785–789 (1968a).

———, and R. N. Weisman, Comparison of solutions of a nonlinear diffusion equation, *Water Resources Res.* No. 2, **6**, 642–644 (1970).

Buckingham, E., Studies in the movement of soil moisture, *U.S. Dept. Agr. Bur. Soils Bull.* 38, Washington D.C., 29–61 (1907).

———, Model experiments and the forms of empirical equations, *Trans. A.I.M.E.* **37** (1915).

Buckley, S. E., and M. C. Leverett, Mechanism of fluid displacement in sands, *Trans. A.I.M.E.* **146**, 107 (1942).

Burcik, E. J., *Properties of Petroleum Reservoir Fluids*, John Wiley, New York, 1961.

Burdine, N. T., Relative permeability calculations from pore-size distribution data, *Trans. A.I.M.E.* **198**, 71–77 (1953).

Bureau, M., and G. Manasterski, Modèle électrique pour l'étude du déplacement de l'huille d'un gisement par un fluide d'injection, *Rev. Inst. Fr. Petro. Ann. Combust. Liquids* **18**, 369–390 (Dec. 1963).

Burke, S. P., and W. B. Plummer, Gas flow through packed columns, *Ind. Eng. Chem.* **20**, 1196–1200 (1928).

Burton, M. B., Jr., and P. B. Crawford, Application of the gelatin model for studying mobility ratio effects, *Trans. A.I.M.E.* **207**, 333–337 (1956).

Cailleau, J., P. Chaumet, B. Jeanson, and N. Van Quy, Etude numérique et expérimentale de l'"edge coning", *Rev. Inst. Fr. Petrol.* **18**, 344–357 (1963).

Carberry, J. J., and R. H. Bretton, Axial dispersion of mass in flow through fixed beds, *J. Amer. Inst. Chem. Eng.* No. 3, **4**, 367–375 (1958).

Cardwell, W. J., Jr., and R. L. Parsons, Average permeabilities of heterogenous oil sands, *Trans. A.I.M.E.* **160**, 283–291 (1945).

Carman, P. C., Fluid flow through a granular bed, *Trans. Inst. Chem. Eng. London* **15**, 150–156 (1937).

———, Determination of the specific surface of powders I, *J. Soc. Chem. Indus.* **57**, 225–234 (1938).

———, *Flow of Gases through Porous Media*, Butterworths, London, 1956.

Carslaw, H. S., and J. C. Jaeger, *Conduction of Heat in Solids* (1st ed. 1946), 2nd ed., Oxford University Press, London, 1959.

Carter, R. D., Comparison of alternating direction explicit and implicit procedures in two-dimensional flow calculations, *J. Soc. Petrol. Eng.* No. 1, **7**, V–VI (1967).

Cary, J. W., and S. A. Taylor, The dynamics of soil water II. Temperature and solute effects, Monograph, No. 11 (R. M. Hagan, H. R. Haise, and T. W. Edminster, eds.), Chap. **13**, pp. 245–253, Amer. Soc. Agron., Madison, Wisc., 1967.

Casagrande, A., Seepage through dams, *J. New England Water Works Assoc.* **51**, 131–172 (1937); see also, *J. Boston Soc. Civil Eng.* 295–337 (1940).

Casimir, H. B. G., On Onsager's principle of microscopic reversibility, *Rev. Mod. Phys.* **17**, 343–350 (1945).

Chalky, J. W., J. Cornfield, and H. Park, A method of estimating volume–surface ratios, *Science* **110**, 295 (1949).

Chandrasekhar, S., Stochastic problems in physics and astronomy, *Rev. Mod. Phys.* **15**, 1–89 (1943).

Charni, I. A., A rigorous derivation of Dupuit's formula for unconfined seepage with seepage surface, *Dokl. Akad. Nauk U.S.S.R.*, No. 6, **79** (1951).

Chauveteau, G., and Cl. Thirriot, Régimes d'écoulement en milieu poreux et limite de la loi de Darcy, *La Houille Blanche*, No. 1, **22**, 1–8 (1967).

Childs, E. C., The transport of water through heavy clay soils I, III, *J. Agric. Sci.* **26**, 114–141, 527–545 (1936).

______, Soil moisture theory, in *Advances in Hydroscience* (Van T. Chow, Ed.), Vol. **4**, pp. 73–117, Academic Press, New York, 1967.

______, and N. Collis-George, The permeability of porous materials, *Proc. Roy. Soc. A* **201**, 392–405 (1950).

Chouke, R. L., P. Van Meurs, and C. Van der Poel, The instability of slow, immiscible, viscous liquid–liquid displacement in permeable media, *Petrol. Trans. A.I.M.E.* **216**, 188–194 (1959).

Christoffel, E. B., Sul problema delle temperature stazionarie e la rappresentazione di una data superficie, *Ann. Mat.* **1**, 89ff. (1867).

Churchill, R. V., *Introduction to Complex Variables and Applications*, McGraw-Hill, New York, 1948.

Coats, K., and B. D. Smith, Dead-end pore volume and dispersion in porous media, *J. Soc. Petrol. Eng.* 73–84 (March, 1964).

Collins, R. E., *Flow of Fluids Through Porous Materials*, Reinhold, New York, 1961.

Colman, E. A., and G. B. Bodman, Moisture and energy conditions during downward entry of water into moist and layered soils, *Proc. Soil Sci. Soc. Amer.* **9**, 3–11 (1944).

Columbus, N., Viscous model study of sea water intrusion in water table aquifers, *Water Resources Res.* No. 2, **1**, 313–323 (1965).

Cooper, H. H., Jr., A hypothesis concerning the dynamic balance of fresh water and salt water in a coastal aquifer, *J. Geophys. Res.* **64**, 461–467 (1959).

______, The equation of ground water flow in fixed and deforming coordinates, *J. Geophys. Res.* No. 20, **71**, 4785–4790 (1966).

Corey, A. T., The interrelation between gas and oil relative permeabilities, *Producer's Monthly* No. 1, **19**, 38–41 (1954).

______, Measurement of water and air permeability in unsaturated soils, *Proc. Soil Sci. Soc. Amer.* **21**, 7–10 (1957).

Cornell, D., and D. L. Katz, Flow of gases through consolidated porous media, *Ind. Eng. Chem.* **45**, 2145–2152 (1953).

Crank, J. *Mathematics of Diffusion*, Oxford University Press, New York and London, 1956.

Croissant, R., Étude expérimentale de l'écoulement diphasique dans un modèle rectangulaire de milieu poreux, Inst. Fr. Petrol., Ref. 10350, 1964.

Curie, P., *Oeuvres*, Gauthier-Villars, Paris, 1908.

Dachler, R., *Grundwasserströmung*, Springer, Vienna, 1936.

Dagan, G., The movement of the interface between two liquids in a porous medium with application to a coastal aquifer (in Hebrew), Ph. D. Thesis, Technion, Haifa, 1964.

———, Spacing drains by an approximate method, *Proc. Amer. Soc. Civil Eng. Irrigation and Drainage Div.* No. IR1, **90**, 41–56 (1964a).

———, Linearized solution of unsteady deep flow toward an array of horizontal drains, *J. Geophys. Res.* **69**, 3361–3369 (1964b).

———, Second-order linearized theory of free surface flow in porous media, *La Houille Blanche*, No. 8, 901–910 (1964c).

———, The solution of the linearized equations of free surface flow in porous media, *J. Mécanique* No. 2, **5**, 207–215 (1966).

———, Second-order theory of shallow free-surface flow in porous media, *Quart. J. Mech. Appl. Math.* Pt. 4, **20**, 517–526 (1967).

———, A derivation of Dupuit solution of steady flow toward wells by matched asymptotic expansions, *Water Resources Res.* No. 2, **4**, 403–412 (1968).

———, Some aspects of heat and mass transfer in porous media, *IAHR Symp. Fundamentals of Transport Phenomena in Porous Media*, Haifa, 1969.

———, and J. Bear, Solving the problem of local interface unconing in a coastal aquifer by the method of small perturbations, *J. IAHR* No. 1, **6**, 15–44 (1968).

———, and A. Kahanovitz, Mass and heat transfer in porous media (in Hebrew), A literature survey and preliminary experiments, Hydraulics Lab., Technion, Haifa, Israel P.N. 2/68, 1968.

Danel, P., The measurement of ground water flow, *Ankara Symp. Arid Zone Hydrology, Proc.* UNESCO, No. 2, 99–107 (1952).

Dankwerts, P. V., Continuous flow systems (distribution of residence times), *Chem. Eng. Sci.* No. 1, **2**, 1–13 (1953).

Darcy, H., *Les Fontaines Publiques de la Ville de Dijon*, Dalmont, Paris, 1856.

Davidson, J. M., J. W. Bigger, and D. R. Nielsen, Gamma radiation attenuation for measuring bulk density and transient water flow in porous materials, *J. Geophys. Res.* **68**, 4777–4783 (1963).

———, D. R. Nielsen, and J. W. Biggar, The dependence of soil water uptake and release upon the applied pressure increment, *Soil Sci. Soc. Amer. Proc.* **30**, 298–304 (1966).

Davis, S., and R. J. M. De Wiest, R., *Hydrogeology*, John Wiley, New York, 1966.

Davison, B. B., The flow of ground water through an earth filled cofferdam with vertical walls (in Russian), *Zap. Gosudars. Gidrologich. Inst.* **6**, 121 (1932).

Day, P. R., Dispersion of a moving salt-water boundary advancing through a saturated sand, *Trans. Amer. Geophys. Union* **37**, 595–601 (1956).

Deans, H. A., A mathematical model for dispersion in the direction of flow in porous media, *J. Soc. Petrol. Eng.* **49** (1963).

De Boodt, M. F., and D. Kirkham, Anisotropy and measurement of air permeability of soil clods, *Soil Sci.* **76**, 127–133 (1953).

Debrine, B. E., Electrolytical model study for collector wells under river beds, *Water Resources Res.* No. 3, **6**, 971–978 (1970).

De Groot, S. R., *Thermodynamics of Irreversible Processes*, North-Holland Publishing Co., Amsterdam, 1963.

———, and P. Mazur, *Non-Equilibrium Thermodynamics*, North-Holland Publishing Co., Amsterdam, 1962 (2nd printing 1963).

De Jong, J., Electric analog models for the solution of geohydrological problems (in Dutch), *Water* No. 4, **46** (1962).

De Josselin de Jong, G., Longitudinal and transverse diffusion in granular deposits, *Trans. Amer. Geophys. Union* **39**, 67–74 (1958).

———, Discussion on "longitudinal and transverse diffusion in granular deposits," *Trans. Amer. Geophys. Union* **39**, 1160–1161 (1958a).

———, Singularity distribution for the analysis of multiple fluid flow through porous media, *J. Geophys. Res.* **65**, 3739–3758 (1960).

———, Moire patterns of the membrane analogy for ground water movement applied to multiple fluid flow, *J. Geophys. Res.* **66**, 3625–3628 (1961).

———, *A Many Valued Hodograph in an Interface Problem*, Technische Hogeschool Delft, Afd., Weg en Waterbouwkunde, 1964.

———, A many valued hodograph in an interface problem, *Water Resources Res.* No. 4, **1**, 543–555 (1965).

———, The tensor character of the dispersion coefficient in anisotropic porous media, *1st IAHR Symp. Fundamentals of Transport Phenomena in Porous Media*, Haifa, 1969.

———, Generating functions in the theory of flow through porous media, in *Flow Through Porous Media* (R. J. M. de Wiest, Editor), pp. 377–400, Academic Press, New York, 1969a.

———, and M. J. Bossen, Discussion of paper by J. Bear, On the tensor form of dispersion, *J. Geophys. Res.* No. 10, **66**, 3623–3624 (1961).

———, E. H. De Leeuw, and A. Verruijt, *Consolidatie in Drie Dimensies* (in Dutch), Overdruk uit LGM, Mededlingen, 1963.

De Vries, D. A., and A. J. Kruger, On the value of the diffusion coefficient of water vapour in air, Colloques Inter. du Centre National de la Recherche Scientifique, Phénomènes de transport avec changement de phase dans les milieux poreux ou colloidaux, Paris, pp. 61–69, 1966.

De Wiest, R. J. M., Free surface flow in homogeneous porous medium, *Amer. Soc. Civil Eng. Trans.* **127**, 1045–1089 (1962).

———, *Geohydrology*, Wiley, New York, 1965.

———, Hydraulic model study of nonsteady flow to multiaquifer wells, *J. Geophys. Res.* No. 20, **71**, 4799–4810 (1966).

———, On the storage coefficient and the equations of groundwater flow, *J. Geophys. Res.* **71**, 1117–1122 (1966a).

——— (Ed.), *Flow Through Porous Media*, Academic Press, New York, 1969.

De Witte, A. J., A study of electric log interpretation methods in shaly formations, *Trans. A.I.M.E.* **14**, 103–110 (1955).

———, Saturation and porosity from electric logs in shaly sands, *Oil & Gas J.* No. 9, **55**, 89–93 (1957).

Dicker, D., and W. A. Sevian, Transient flow through porous mediums, *J. Geophys. Res.* No. 20, **70**, 5043–5053 (1965).

Dietz, D. N., Een modelproef ter bestudeering van niet-stationnare bewegingen van het grondwater (in Dutch), *Water* **25**, 185–188 (1941).

Donaldson, I. G., Temperature gradients in the upper layers of the earth's crust due to convective water flows, *J. Geophys. Res.* No. 9, **67**, 3449–3459 (1962).

Douglas, J., Jr., D. W. Peaceman, and H. H. Rachford, Jr., A method for calculating multidimensional immiscible displacement, *Trans. A.I.M.E.* **216**, 297–308 (1959).

Drake, L. C., and H. L. Ritter, Pore size distribution in porous materials, *Ind. Eng. Chem. Anal.* **17**, 782–787 (1945).

Dudgeon, C. R., An experimental study of the flow of water through coarse granular media, *La Houille Blanche* **7**, 785–801 (1966).

Duhamel, J. M. C., Mémoires sur la méthode générale relative au mouvement de la chaleur dans les corps solides plongés dans les milieux dont la temperature varie avec le temps, *J. Éc. Polytech. Paris* **14**, 20 (1833).

Dumm, L. D., Drain spacing formula, *Agr. Eng.* **35**, 726–730 (1954).

———, Drain spacing method used by the Bureau of Reclamation, Paper presented at *ARS-SCS Drainage Workshop*, Riverside, Calif., Feb. 1962.

———, and R. J. Winger, Subsurface drainage system design for irrigated area using transient flow concept, *Trans. Amer. Agr. Eng.* No. 2, **7**, 147–151 (1964).

Dupuit, J., *Études Théoriques et Pratiques sur le Mouvement des Eaux dans les Canaux Découverts et à Travers les Terrains Permeables*, 2nd ed., Dunod, Paris, 1863.

Ebach, E. A., and R. R. White, Mixing of fluids flowing through beds of packed solids, *J. Amer. Inst. Chem. Eng.* No. 2, **4**, 161–164 (1958).

Eck, B., *Technische Strömungslehre*, 2nd ed., Springer, Berlin, 1944.

Edelman, J. H., Over de berekening van grondwaterstomingen, Thesis, Delft, Netherlands (in Dutch), mimeographed, 1947.

Edlefsen, N. E., and A. B. C. Anderson, Thermodynamics of soil moisture, *Hilgardia* No. 2, **15**, 298 (1943).

Elder, J. W., Steady free convection in a porous medium heated from below, *J. Fluid Mech.* Part 1, **27**, 29–48 (1967).

Ergun, S., Fluid flow through packed columns, *Chem. Eng. Prog.* **48**, 89–94 (1952).

———, and A. A. Orning, Fluid flow through randomly packed columns and fluidized beds, *Ind. Eng. Chem.* **41**, 1179–1184 (1949).

Fair, G. M., and L. P. Hatch, Fundamental factors governing the streamline flow of water through sand, *J. Amer. Water Works Ass.* **25**, 1551–1565 (1933).

Fara, H. D., and A. E. Scheidegger, Statistical geometry of porous media, *J. Geophys. Res.* No. 10, **66**, 3279–3284 (1961).

Fatt, I., and H. Dykstra, Relative permeability studies, *Trans. A.I.M.E.* **192**, 249–256 (1951).

———, The network model of porous media I, II, and III, *Trans. A.I.M.E.* **207**, 144–159 (1956).

Feller, W., *An Introduction to Probability Theory and Its Applications*, 2nd ed., John Wiley, New York, 1958.

Ferrandon, J., Les Lois de l'écoulement de filtration, *Le Génie Civil* No. 2, **125**, 24–28 (1948).

Ferguson, H., and W. H. Gardner, Diffusion theory applied to water flow data using gamma ray absorption, *Soil Sci. Soc. Amer. Proc.* **27**, 243–245 (1963).

Fermi, E., *Thermodynamics*, Prentice-Hall, Englewood Cliffs, N. J., 1937.

Finkelstein, A. B., The initial value problem for transient water waves, *Comm. Pure Appl. Math.* **10**, 511–522 (1957).

Forchheimer, P., Wasserbewegung durch Boden, *Z. Ver. Deutsch. Ing.* **45**, 1782–1788 (1901).

———, *Hydraulik*, 3rd ed., Teubner, Leipzig, Berlin, 1930.

Forsythe, G. E., and W. R. Wasow, *Finite-Difference Methods for Partial Differential Equations*, John Wiley, New York, 1960.

Fourier, J. B. J., *Théorie Analytique de la Chaleur*, Paris, 1822; English transl. by Freeman, Cambridge 1878.

Friedrichs, K. O., On the derivation of the shallow water theory, *Comm. Pure Appl. Math.* No. 1, 109–134 (1948).

Galin, L. A., Unsteady filtration of ground water in the case of narrow ditch (in Russian), *Prikl. Mat. Mek. Moscow* **23**, 789–791 (1959).

———, Z. F. Karpycheva, and A. R. Shkirich, Lenticular spreading of underground water, *PMM* (in the English Translation) No. 3, **24**, 826–831 (1960).

Garabedian, P. R., *Partial Differential Equations*, John Wiley, New York, 1964.

Gardner, W. H., and W. Gardner, Flow of soil moisture in the unsaturated state, *Soil Sci. Soc. Amer. Proc.* **15**, 42–50 (1951).

Gardner, W. R., Calculation of capillary conductivity from pressure plate and flow data, *Soil Sci. Soc. Amer. Proc.* **20**, 317–320 (1956).

———, Some steady state solutions of the unsaturated moisture flow equation with application to evaporation from a water table, *Soil Sci.* No. 4, **85**, 228–232 (1958).

———, Water movement in the unsaturated soil profile, *ICID Soil Water Symp. Prague* 223–236 (1967).

———, and M. S. Mayhugh, Solutions and tests on the diffusion equation for the movement of water in soil, *Soil Sci. Soc. Amer. Proc.* **22**, 197–201 (1958).

Gassmann, F., *Viertelj., Naturforsch. Gesellsch. Zürich*, **96**, 1ff. (1951).

Geertsma, J., The effect of fluid pressure decline on volumetric changes of porous rocks, *Trans. A.I.M.E. Petrol.* **210**, 331–340 (1957).

———, G. A. Croes, and N. Schwartz, Theory of dimensionally scaled models of petroleum reservoirs, *Trans. A.I.M.E. Petrol.* **206** (1956).

Gelfand, I. M., and S. V. Fomin, *Calculus of Variations* (transl. from Russian by R. A. Silverman), Prentice-Hall, Englewood Cliffs, N. J., 1963.

Georghitza, St. I., On the plane steady flow of water through inhomogeneous porous media, *1st Symp. The Fundamentals of Transport Phenomena in Porous Media, IAHR*, Haifa, 1969.

Gershon, N., and A. Nir, Effects of boundary conditions of models on tracer distribution in flow through porous media, *Water Resources Res.* **5**, 830–839 (1969).

Girinskii, N. K., Generalization of some solutions for wells to more complicated natural conditions (in Russian), *Dokl. A.N. U.S.S.R.* No. **3**, **54** (1946).

Glover, R. E., Formulas for movement of ground water, Oahe Unit Missouri River Basin Project Bureau of Reclamation, Memorandum No. 657, Sect. D. 35–46 (1953).

———, The pattern of fresh water flow in a coastal aquifer, *J. Geophys. Res.* **64**, 439–475 (1959).

———, Studies of ground water movement, Technical Memorandum 657, Bureau of Reclamation, Denver, Colo., 1960.

———, Ground water movement, Engineering Monograph 31, Bureau of Reclamation, 1966.

Goldstein, S., *Modern Developments in Fluid Dynamics*, Oxford Clarendon Press, 1938.

Graton, L. C., and H. J. Fraser, Systematic packing of spheres with particular relation to porosity and permeability, *J. Geol.* No. 8, **43**, 785–909 (1935).

Green, D. W., Heat Transfer with Flowing Fluid Through Porous Media, Ph. D. Thesis, University of Oklahoma, 1963.

Greenkorn, R. A., and D. K. Kessler, Dispersion in heterogeneous nonuniform anisotropic porous media, *Flow Through Porous Media Symp. Ind. Eng. Chem.* No. 9, **61**, 14–32 (1969).

Gröbner, W., and N. Hofreiter, *Integraltafel*, Springer, Vienna, Vol. 1, 1949, Vol. 2, 1950.

Gunther, E., Lösung von Grundwasseraufgaben mit Hilfe der Strömung in dünnen Schichten, *Wasserkraft und Wasserwirtschaft* No. 3, **35**, 49–55 (1940a).

———, Untersuchung von Grundwasserströmungen durch analoge Strömungen zäher Flüssigkeiten, *Forschung auf dem Gebiete des Ingenieurwesens* **11**, 76–88 (1940b).

Hamel, G., Über Grundwasserströmung, *Z. Angew. Math. Mech.* **14**, 129 (1934).

Hammad, H. J., Depth and spacing of tile drain systems, *Proc. Amer. Soc. Civil Eng.* No. IR 1, **88**, 15–34 (1962).

Hanks, R. J., Water vapor transfer in dry soil, *Soil Sci. Soc. Amer. Proc.* **22**, 392–394 (1958).

———, and S. A. Bower, Numerical solution of the moisture flow equations for infiltration into layered soils, *Soil Sci. Soc. Amer. Proc.* **26**, 530–534 (1962).

Hansen, V. E., Evaluation of Unconfined Flow to Multiple Wells by Membrane Analogy, Ph.D. Thesis, Iowa City, State University, Iowa, 1949.

———, Complicated well problems solved by the membrane analogy, *Trans. Amer. Geophys. Union* **33**, 912–916 (1952).

Hantush, M. S., Hele-Shaw model study of the growth and decay of groundwater ridges, *J. Geophys. Res.* No. 4, **72**, 1195–1205 (1963).

———, Hydraulics of wells, in *Advances in Hydrosciences* (V. T. Chow, Ed.), Vol. 1, pp. 281–432, Academic Press, New York, 1964.

———, Growth and decay of groundwater mounds response to uniform percolation, *Water Resources Res.* No. 1, **3**, 227–234 (1967).

———, Unsteady movement of fresh water in thick unconfined saline aquifers, *Bull.*

I.A.S.H. No. 2, **XIII**, 40–60 (1968).

Hall, H. N., Compressibility of reservoir rocks, *Trans. A.I.M.E.* **198**, 309–316 (1953).

Happel, J., and H. Brenner, *Low Reynolds Number Hydrodynamics, with Special Application to Particulate Media*, Prentice-Hall, Englewood Cliffs, N. J., 1965.

Haring, R. E., and R. A. Greenkorn, A statistical model of a porous medium with non-uniform pores, *J. Amer. Inst. Chem. Eng.* No. 3, **16**, 477–483 (1970).

Harleman, D. R. E., P. F. Mehlhorn, and R. R. Rumer, Dispersion–permeability correlation in porous media, *J. Hydraul. Div. Amer. Soc. Civil Eng.* No. HY2, **89**, 67–85 (1963).

______, and R. R. Rumer, The dynamics of salt-water intrusion in porous media, Rep. No. 55, Hydrodyn. Lab., Dept. of Civil Engineering, Massachusetts Institute of Technology, Cambridge, Mass., 1962.

Harr, M. E., *Groundwater and Seepage*, McGraw-Hill, New York, 1962.

Hauzenberg, I., and D. Zaslavsky, The effect of size of water stable aggregates on field capacity (in Hebrew), Dept. of Civil Eng., Technion, Haifa, P.N. 35, 1963.

Hele-Shaw, H. S., Experiments on the nature of surface resistance in pipes and on ships, *Trans. Inst. Naval Architects* **39**, 145–156 (1897).

______, Experiments on the nature of surface resistance of water and streamline motion under certain experimental conditions, *Trans. Inst. Naval Architects* **40**, 21–46 (1898).

Henry, H. R., Salt water intrusion into fresh water aquifers, *J. Geophys. Res.* No. 11, **64**, 1911–1919 (1959).

Herzberg, A., Die Wasserversorgung einiger Nordseebäder (The water supply on parts of the North Sea Coast, in German), *J. Gasbeleucht. Wasserversorg.* **44**, 815–819, 842–844 (1901).

Hiby, J. W., Longitudinal and transverse mixing during single phase flow through granular beds, *Symp. Interaction between Fluids and Particles, Inst. Chem. Eng. London*, 312–325 (1962).

Hildebrand, F. B., *Advanced Calculus for Applications*, Prentice-Hall, Englewood Cliffs, N.J., 1962.

Hill, H. L., and J. D. Milburn, Effect of clay and water salinity of electrochemical behavior of reservoir rocks, *Trans. A.I.M.E.* **207**, 65–72 (1956).

Hooghoudt, S. B., Bijdragen tot de kennis van eenige natuurkundige grootheden van de grond, 6 Bepaling van de doorlatenheid in gronden van de tweede soort; theorie en toepassing van de kwantitative strooming van het water in ondiep gelegen grondlagen vooral in verband met ontwaterings – en infiltratievraagstuken, *Versl. Landbouwk. Ond.* **43**, 461–676 (1937).

______, Bijdragen tot de kennis van eenige natuurkundige grootheden van den grond, 7, algemeene beschouwing van het probleem van de detail ontwatering de infiltratie door middel van parallel loopende drains, greppels, slooten, en kanalen, *Versl. Landbouwk. Ond.* **46**, 515–707 (1940).

Hoopes, J. A., and D. R. F. Harleman, Waste water recharge and dispersion in porous media, *Techn. Rep.* No. 75, Hydrodynamic Lab., Massachusetts Institute of Technology, Cambridge, Mass., 1965.

______, and D. R. F. Harleman, Waste water recharge and dispersion in porous media, *Proc. Amer. Soc. Civil Eng.* No. HY5, **93**, 51–72 (1967).

Horton, C. W., and F. T. Rogers, Jr., Convection currents in a porous medium, *J. Appl. Phys.* **16**, 367–370 (1945).

Horton, R. E., Analysis of runoff-plot experiments with varying infiltration capacity, *Trans. Amer. Geophys. Union* **20**, 693–711 (1939).

Houpert, Debouvrier and Ifly, 1965.

Hubbert, M. K., The theory of ground water motion, *J. Geol.* **48**, 785–944 (1940).
Hubbert, M. K., Darcy law and the field equations of the flow of underground fluids, *Trans. Amer. Inst. Min. Metal. Eng.* **207**, 222–239 (1956).
Hutta, J. J., and J. C. Griffiths, Directional permeability of sandstones; a test technique, *Producers Monthly*, Nov. 26–34 and Oct. 24–31 (1955).
Iberall, A. S., Permeability of glass wool and other highly porous media, *J. Res. Nat. Bur. Stand.* **45**, 398–406 (1950).
Irmay, S., On the hydraulic conductivity of unsaturated soils, *Trans. Amer. Geophys. Union* **35**, 463–468 (1954).
———, Flow of liquid through cracked media, *Bull. Res. Council of Israel* No. 1, **5A**, 84 (1955).
———, Extension of Darcy law to unsaturated flow through porous media, *Proc. Symp. Darcy, Dijon, I.U.G.G., Int. Ass. Sci. Hydrol.* **41**, 57–66 (1956).
———, On the theoretical derivation of Darcy and Forchheimer formulas, *Trans. Amer. Geophys. Union* No. 4, **39**, 702–707 (1958); Discussion in *J. Geophys. Res.* No. 4, **64**, 486–487 (1959).
———, Réfraction d'un écoulement à la frontière separant deux milieux poreux anisotropes differents, *C.R.H. Acad. Sci.* **259**, 509–511 (1964).
———, La similitude hydrodynamique des modèles distordus, extrait du volume d'hommage au Professeur F. Campus 175–180 (1964a).
———, Seepage forces and the formation of piping in flow through porous media (in Hebrew), Dept. of Civil Eng., Technion, Haifa, 1964c.
———, Theoretical models of flow through porous media, *R.I.L.E.M. Symp. Transfer of Water in Porous Media, Paris* (1964d); *Bull. R.I.L.E.M.* **29**, 37–43 (Dec. 1965).
———, Solutions of the non-linear equation with a gravity term in hydrology, *I.A.S.H. Symp. Water in the Unsaturated Zone*, Wageningen, 1966.
———, On the meaning of the Dupuit and Pavlovskii's approximations in aquifer flow, Hydraulic Eng. Lab., Univ. of Calif., Berkeley, 1966a.
Jackson, R. D., Water vapor diffusion in relatively dry soil I. Theoretical considerations and sorption experiments, *Soil Sci. Soc. Amer. Proc.* **28**, 172–176 (1964a).
———, Water vapor diffusion in relatively dry soil II. Desorption experiments, *Soil Sci. Soc. Amer. Proc.* **28**, 464–466 (1964b).
———, Water vapor diffusion in relatively dry soil III. Steady state experiments, *Soil Sci. Soc. Amer. Proc.* **28**, 466–470 (1964c).
———, D. A. Rose, and H. L. Penman, Circulation of water in soil under a temperature gradient, *Nature* **205**, 314–316 (1965).
Jacob, C. E., On the flow of water in an elastic artesian aquifer, *Trans. Amer. Geophys. Union* **21**, 574–586 (1940).
———, in *Engineering Hydraulics* (H. Rouse, Ed.), pp. 321–386, John Wiley, New York, 1950.
Jacobs, M., and S. Schmorak, Sea water intrusion and interface determination along the coastal plane of Israel, State of Israel, Hydrological Service, Hydrological paper No. 6, 1960.
Jacquard, P., Calculs Numérique de Déplacement de Fronts, Rapport I.F.P., ref. 7844, 1962.
———, and P. Séguier, Mouvement de deux fluides en contact dans un milieu poreux, *J. Mech.* No. 4, **1**, 367–394 (1962).
Jahnke, E., and F. Emde, *Tables of Functions*, 4th ed., Dover, New York, 1945.
Javandel, I., and P. A. Witherspoon, Application of the finite element method to transient flow in porous media, *J. Soc. Petrol. Eng.* 241–252 (1968).
Jeffreys, H., *Phil. Mag.* **2**, 833–844 (1926).
———, *Proc. Roy. Soc. A* **118**, 195–208 (1928).
Johnson, E. F., D. P. Bossler, and V. O. Naumann, Calculation of relative permeability

from displacement experiments, *Trans. A.I.M.E. Petrol.* **216**, 370–372 (1959).

Johnson, W. E., and R. V. Hughes, Directional permeability measurements and their significance, *Producers Monthly* No. 1, **13**, 17–25 (November, 1948).

Jones, K. R., On the differential form of Darcy's law, *J. Geophys. Res.* No. 2, **67**, 731–732 (1962).

Karman, Th. U., and M. Biot, *Mathematical Methods in Engineering*, McGraw-Hill, New York, 1940.

Karplus, W. J., *Analog Simulation Solutions of Field Problems*, McGraw-Hill Series in Information Processing and Computers, McGraw-Hill, New York, 1958.

———, and W. W. Soroka, *Analog Methods—Computation and Simulation*, 2nd ed., McGraw-Hill, New York, 1959.

Katchalsky, A., and P. F. Curran, *Nonequilibrium Thermodynamics in Biophysics*, Harvard University Press, Cambridge, Mass., 1965.

Katz, D. L., D. Cornell, R. Kobayashi, F. H. Poettmann, J. A. Vary, J. R. Elenbaas, and C. F. Weinaug, *Handbook of Natural Gas Engineering*, McGraw-Hill, New York, 1959.

Kellog, F. H., Investigation of drainage rates affecting the stability of earth dams, *Trans. Amer. Soc. Civil Eng.* **113**, 1261–1309 (1948).

Kidder, R. E., Flow of immiscible fluids in porous media, An exact solution of a free boundary problem, *J. Appl. Phys.* No. 8, **27**, 867–869 (1956).

King, L. G., and A. T. Corey, A discussion of the diffusion equation, unpublished paper, Agr. Eng. Dept., Colorado State University, 1962.

Kirchhoff, G., *Vorlesungen über die Theorie der Wärme*, Barth, Leipzig, 1894.

Kirkham, C. E., Turbulent flow in porous media, Dept. of Civil Eng., University of Melbourne, 1967a.

Kirkham, D., Seepage of steady rainfall through soil into drains, *Trans. Amer. Geophys. Union* **39**, 892–908 (1958).

———, Steady state theories for drainage, *Proc. Amer. Soc. Civil Eng.* No. IR 1, **92**, 19–39 (1966).

———, Explanation of paradoxes in Dupuit–Forchheimer seepage theory, *Water Resources Res.* No. 2, **3**, 609–622 (1967).

Klinkenberg, L. J., The permeability of porous media to liquids and gases, *Amer. Petrol. Inst. Drilling Prod. Pract.* 200–213 (1941).

Klute, A., A numerical method for solving the flow equation for water in unsaturated materials, *Soil Sci.* **73**, 105–116 (1952).

———, Laboratory measurement of hydraulic conductivity of unsaturated soil, in *Methods of Soil Analysis* (C. A. Black, Ed.), Monograph 9, Chap. 16, Part 1, pp. 253–261, Amer. Soc. Agron., Madison, Wisc., 1965.

———, Notes on flow of water in unsaturated soils, presented at the M.I.T. Summer Session on Ground Water Hydrology and Flow Through Porous Media, 1967.

———, and D. B. Peters, Hydraulic and pressure head measurements with strain gauge pressure transducers, *I.A.S.H. Symp. Water in the Unsaturated Zone*, Wageningen, The Netherlands, 156–165 (1966).

Knudsen, M., *The Kinetic Theory of Gases*, Methuen, London, 1934; 3rd ed., 1950.

Kozeny, J., Über kapillare Leitung des Wassers im Boden, *Sitzungsber. Akad. Wiss. Wien* **136**, 271–306 (1927).

———, *Hydraulic*, Springer, Vienna, 1953.

Krumbein, W. C., and L. L. Sloss, *Stratigraphy and Sedimentation*, 1st ed. W. H. Freeman, San Francisco, Calif., 1951.

Kruysse, M. P. C., and J. Bear, Investigations of ground water movement and related problems in Zealand Flanders by means of a horizontal slit model (in Dutch), Government Inst. for Water Supply, The Hague, Netherlands, 1956.

Kunii, D., and J. M. Smith, Heat transfer characteristics of porous rocks, *Amer. Inst.*

Chem. Eng. No. 1, **6**, 71–78 (1960).

——, and J. M. Smith, Thermal conductivities of porous rocks filled with stagnant fluids, *J. Soc. Petrol. Eng.* **1**, 37–42 (1961).

Kunze, R. J., and D. Kirkham, Simplified accounting for membrane impedance in capillary conductivity determination, *Proc. Soil Sci. Soc. Amer.* **26**, 421–426 (1962).

Kutilek, M., Non-Darcian flow of water in soils (Laminar Region), *1st IAHR Symp. Fundamentals of Transport Phenomena in Porous Media*, Haifa, Israel, 1969.

Lagally, M., Ideale Flüssigkeiten (Ideal Fluids), Chapt. I, in *Mechanik der flüssigen und gasförmigen Körper* (R. Grammel, Ed.), Vol. 7, in *Handbuch der Physik* (H. Geiger and K. Scheel, Eds.), Springer, Berlin, 1927.

Lagarde, A., Considérations sur le transfert de chaleur en milieu poreux, *Rev. Inst. Fr. Petrol.* No. 2, **20**, 383–446 (1965).

Lagarde, N. M. A., and J. Marrast, Études des gisements sur l'analyseur électrique, *Inst. Fr. Petrol.*, ref. 8129 (January 1963).

Laird, A. D. K., and J. A. Putnam, Fluid saturation in porous media by *X*-ray technique, *Trans. A.I.M.E.* **192**, 275–283 (1951).

Laliberte, G. E., A. T. Corey, and R. H. Brooks, Properties of unsaturated porous media, Colorado State University, Hydrology Paper No. 17, 40 pp., 1966.

Landau, L. D., and E. M. Lifschitz, *Statistical Physics* (English transl. by E. Peierls and R. F. Peierls), Addison Wesley, Reading, Mass., 1958.

Langhaar, H. L., *Dimensional Analysis and Theory of Models,* John Wiley, New York, 1951.

Lapidus, L., and N. R. Amundson, Mathematics of adsorption in beds VI. The effect of longitudinal diffusion in ion exchange and chromatographic columns, *J. Phys. Chem.* **56**, 984–988 (1952).

Lapwood, E. R., Convection of a fluid in a porous medium, *Proc. Cambridge Phil. Soc.* **44**, 508–521 (1948).

Lau, L. K., W. J. Kaufman, and D. K. Todd, Dispersion of a water tracer in radial laminar flow through homogeneous porous media, Hydraulic Lab., University of California, Berkeley, 1959.

Le Fur, B., and P. Sourieau, Étude de l'écoulement diphasique dans une couche Inclinée et dans un modèle rectangulaire de milieu poreux, *Rev. Inst. Fr. Petrol.* **18**, 325 and 343 (1963).

Leibenzon, L. S., *The Flow of Natural Fluids and Gases in Porous Media* (in Russian), Gostekhizdat, Moscow, 1947.

Lembke, K. E., Groundwater flow and the theory of water collectors (in Russian), The Engineer, *J. of the Ministry of Communications* No. 2 (1886) No. 17–19 (1887).

Leverett, M. C., Capillary behavior in porous media, *Trans. A.I.M.E.* **142**, 341–358 (1941).

Li, W. H., and S. H. Lam, *Principles of Fluid Mechanics*, Addison Wesley, Reading, Mass., 1964.

Liakopoulos, A. C., Theoretical approach to the solution of the infiltration problem, *Bull. I.A.S.H.* No. 1, **11**, 69–110 (1966).

Lightfoot, E. N., and E. L. Cussler, Jr., Diffusion in liquids, *Selected Topics in Transport Phenomena* No. 58, **61**, 66–85 (1965).

Ligtenberg, F. K., The Moiré method—a new experimental method for the determination of moments in small slab models, *Proc. Soc. Expl. Stress Anal.* No. 2, **12** (1955).

Lindquist, E., On the flow of water through porous soil, Premier Congrès des Grands Barrages, Stockholm, **5**, 81–101 (1933).

List, G. J., The Stability and mixing of a density stratified horizontal flow in a saturated porous medium, W. M. Keck Lab. of Hydraulics & Water Resources, Rept. KH-R-11, Calif. Inst. of Tech. (same as Ph. D. Thesis), 1965.

Low, A. R., On the criterion for stability of a layer of viscous fluid heated from below, *Proc. Roy. Soc. A* **125**, 180–195 (1929).

———, *Proc. 3rd. Int. Congr. Appl. Mech.*, Stockholm, 1930.

Low, P. F., Physical chemistry of clay-water interaction, *Advan. in Agron.* **13**, 269–327 (1961).

Luikov, A. V., *Heat and Mass Transfer in Capillary Porous Bodies* (English transl. of original Russian book published by Academy of Sciences U.S.S.R., Minsk, 1961), Pergamon, New York, 1966.

———, and Y. A. Mikhailov, *Theory of Energy and Mass Transfer*, Prentice-Hall, Englewood Cliffs, N.J., 1961.

Lusczynski, N. S., Head and flow of ground water of variable density, *J. Geophys. Res.* No. 12, **66**, 4247–4256 (1961).

Luthin, J. N., *Drainage Engineering*, John Wiley, New York, 1966.

Maasland, M., Soil anisotropy and land drainage, Sec. II, in *Drainage of Agricultural Lands* (J. N. Luthin, Ed.), pp. 216–236, Amer. Soc. Agron., Madison, Wisc., 1957.

———, and D. Kirkham, Theory and measurement of anisotropic air permeability in soil, *Proc. Soil Sci. Soc. Amer.* No. 4, **19**, 395–400 (1955).

Marcus, H., and D. E. Evenson, Directional permeability in anisotropic porous media, Water Resources Center, Contribution No. 31, Hydrology Lab., University of California, Berkeley, 1961.

Marino, M. A., Hele-Shaw model study of the growth and decay of groundwater ridges, *J. Geophys. Res.* No. 4, **72**, 1195–1205 (1967).

Marle, C., Cours de Production, in *Les Écoulements Polyphasiques*, Vol. 4, Inst. Fr. Petrol. 1965.

Marshal, J. J., Permeability equations and their models, *Proc. Symp. Interaction Between Fluids and Particles*, European Federation of Chemical Engineers, London, 299–303 (1962).

Marthin, M., G. H. Murray, and W. J. Gillingham, Determination of the potential productivity of oil bearing formations by resistivity measurements, *Geophys.* **3**, 258ff. (1938).

Mavis, F. T., and T. P. Tsui, Percolation and capillary movements of water through sand prisms, Bull. 18, Univ. of Iowa, Studies in Eng., Iowa City, 1939.

Maxwell, J. C., *A Treatise on Electricity and Magnetism*, 2nd ed., Vol. I, Oxford, 1881.

McCracken, D. D., and W. S. Dorn, *Numerical Methods and Fortran Programming with Applications in Engineering and Science*, John Wiley, New York, 1964.

McDonald, D., The electrolytic analog in the design of high voltage power transformer, *Proc. Inst. Elect. Eng.* **100**, 145–166 (1953).

McHenry, J. R., Theory and application of neutron scattering in the measurement of soil moisture, *Soil Sci.* No. 5, **95**, 294–307 (1963).

Meinzer, O. E. (Ed.) *Hydrology*, Dover, New York, 1942.

Mercado, A., and J. Bear, Mixing of labeled water by injecting and pumping in the same well (in Hebrew), Water Planning for Israel, Tel Aviv, P.N. 511, 1965.

Mikhailov, G. K., On maximum gradients near drainage of earth dams (in Russian), *Trans. Acad. Sci. Div. Tech. Sci. U.S.S.R.* No. 2, 109–112 (1956).

Miller, E. E., and D. E. Elrick, Simplified accounting for membrane impedance in capillary conductivity determination, *Proc. Soil Sci. Soc. Amer.* **22**, 483–486 (1958).

Miller, R. D., and F. Richard, Hydraulic gradients during infiltration in soils, *Soil Sci. Soc. Amer. Proc.* **16**, 33–38 (1952).

Milne-Thomson, L. M., *Theoretical Hydrodynamics*, 3rd ed., Macmillan, New York, 1955; 4th ed., 1960.

Mkhitaryan, A. M., Seepage of water through earth dams on impervious basis (in Russian), *Izv. Akad. Nauk. Arm. S.S.R.* No. 5, 1947.

Monicard, R., Course de production, *Caractéristiques des roches Réservoirs Analyse des Carottes*, Vol. 1, Inst. Fr. Petrol. Tech., 1965.

Morel-Seytoux, H. J., Effect of Boundary Shape on Channel Seepage, Tech. Rep., No. 7,

Dept. of Civil Eng., Stanford University, 1961.

———, Introduction to flow of immiscible liquids in porous media, in *Flow through Porous Media* (R. J. M. de Wiest, Ed.), Chap. 11, pp. 456–516, Academic Press, New York, 1969.

Morgan, F., J. M. McDowell, E. C. Doty, Improvement in the X-ray saturation technique of studying fluid flow, *Trans. A.I.M.E.* **189**, 183–194 (1950).

Morrison, H. L., F. T. Rogers, Jr., and C. W. Horton, Convection currents in porous media II. Observations of conditions at the onset of convection, *J. Appl. Phys.* **20**, 1027–1029 (1949).

Morse, P. M., and H. Feshbach, *Methods of Theoretical Physics*, McGraw-Hill, New York, 1953.

Muskat, M., *The Flow of Homogeneous Fluids through Porous Media*, McGraw-Hill, New York, 1937; 2nd printing by Edwards, Ann Arbor, Mich., 1946.

———, *Physical Principles of Oil Production*, McGraw-Hill, New York, 1949.

———, R. D. Wycoff, H. G. Botset, and M. W. Meres, Flow of gas liquid mixtures through sands, *Trans. A.I.M.E. Petrol.* **123**, 69–96 (1937).

Naor, I., and J. Bear, Model investigations of coastal ground water interception (Caesarea Region), joint experimental coastal ground water collector project, Techn. Rep. No. 1, Tel Aviv, Water Planning for Israel Ltd., P.N. 262, 1963.

Nehari, Z., *Conformal Mapping*, McGraw-Hill, New York, 1952.

Nelson, R. W., Steady Darcian transport of fluids in heterogeneous, partially saturated porous media I. Mathematical and numerical formulation, General Electric, Hanford Atomic Products Operations, HW-72335, Pt. 1, 1962.

———, Stream functions for three-dimensional flow in heterogeneous porous media, *Int. Ass. Sci. Hydrol. Berkeley Symp.*, P.N. 64, 290–301 (1963).

Neuman, S. P., and P. A. Witherspoon, Finite element method of analyzing steady seepage with a free surface, *Water Resources Res.* No. 3, **6**, 889–897 (1970).

Nield, D. A., Onset of thermohaline convection in a porous medium, *Water Resources Res.* No. 3, **4**, 553–560 (1968).

Nielsen, D. R., J. W. Bigger, and J. M. Davidson, Experimental consideration of diffusion analysis in unsaturated flow problems, *Soil Sci. Soc. Amer. Proc.* **26**, 107–112 (1962).

Nikolaevskii, V. N., Convective diffusion in porous media, *J. Appl. Math. Mech.* (*P.M.M.*) No. 6, **23**, 1042–1050 (1959).

Norel, G., Mesure de saturation par absorption X, couples eau–air, huile–air, et eau–huile, *Inst. Fr. Petrol.*, Ref. 8613 (1963).

———, Absorption X des milieux poreux saturés, application à la mesure des saturation et des porosités, *Inst. Fr. Petrol.*, Ref. 10649 (1964).

———, Diffusion et fluorescence X en milieu poreux, *Inst. Fr. Petrol.*, Ref. 13568 (November 1966).

———, Analyse quantitative par absorption ou fluorescence d'un traceur stable. *Suppl. Bull. Inform. ATEN*, No. 65, 8–11 (1967).

Nutting, P. G., Physical analysis of oil sands, *Bull. Amer. Ass. Petr. Geol.* **14**, 1337–1349 (1930).

Odeh, A. S., Effect of viscosity ratio on relative permeability, *Trans. A.I.M.E.* **216**, 346–352 and discussion by C. F. Wienaug, pp. 352–353 (1959).

Ogata, A., Dispersion in Porous Media, Ph. D. Thesis, Northwestern University, Illinois, 1958.

———, and R. B. Banks, A solution of the differential equation of longitudinal dispersion in porous media, U.S. Geol. Survey, Professional Paper 411-A, 1961.

Onsager, L., *Phys. Rev.* **37**, 405–426; **38**, 2265–2279 (1931).

Oroveanu, J., *Flow of Multiphase Fluids through Porous Media* (in Romanian), Editura Academiei Republicii Socialiste Romania, Bucarest, 1966.

Oseen, C. W., Über die Stokessche Formel und über eine verwandte Aufgabe in der

Hydrodynamik, *Arkiv Math. Astro. Fys.* No. 29, **6** (1910).
Osoba, J. S., J. G. Richardson, J. K. Kerver, J. A. Hafford, and P. M. Blair, Laboratory measurements of relative permeability, *Trans. A.I.M.E.* **192**, 47–55 (1951).
Outmans, H. D., Dupuit's formula generalized for heterogeneous aquifers, *J. Geophys. Res.* No. 16, **69**, 3383–3386 (1964).
Parker, G. G., and V. T. Stringfield, Effects of earthquakes, trains, tides, winds and atmospheric pressure changes on water in geologic formations of Southern Florida, *Econ. Geology*, No. 5, **45**, 441–460 (1950).
Parsons, R. W., Permeability of idealized fractured rock, *J. Soc. Petrol. Eng.* **6**, 126–136 (1966).
Partom, I., Steady unsaturated vertical flow through multilayered soil, *Proc. of the 3rd Asian Regional Conf. on Soil Mechanics and Foundation Eng.* **I**, 165–167, Haifa, (1967).
Patel, R. D., and R. A. Greenkorn, On dispersion in laminar flow through porous media, *J. Amer. Inst. Chem. Eng.* No. 2, **16**, 332–334 (1970).
Peaceman, D. W., and H. H. Rachford, Jr., The numerical solution of parabolic and elliptic differential equations, *J. SIAM* No. 1, **3**, 28–41 (1955).
———, and H. H. Rachford, Jr., Numerical calculation of multidimensional miscible displacement, *J. Soc. Petrol. Eng.* No. 4, **2**, 327–339 (1962).
Pellew, A., and R. V. Southwell, On maintained convective motion in a fluid heated from below, *Proc. Roy. Soc. A* **176**, 312–343 (1940).
Perkins, T. K., and O. C. Johnston, A review of diffusion and dispersion in porous media, *J. Soc. Petrol. Eng.* **19**, 70–84 (1963).
Pfankuch, H. O., Contribution à l'étude des déplacement de fluides miscible dans un milieu poreux, *Rev. Inst. Fr. Petrol.* No. 2, **18**, 215–270 (1963).
———, On the correlation of electrical conductivity properties of porous systems with viscous flow transport coefficients, *Proc. IAHR 1st Int. Symp. Fundamentals of Transport Phenomena in Porous Media*, Haifa, 1969.
Philip, F. L., Physical chemistry of clay-water interaction, in *Advances in Agronomy* (A. G. Norman, Ed.), Vol. 13, pp. 269–327, Academic Press, New York, 1961.
Philip, J. R., Numerical solution of equations of the diffusion type with diffusivity-concentration dependent, *Trans. Faraday Soc.* **51**, 885–892 (1955).
———, Numerical solution of equations of the diffusion type with diffusivity concentration dependent II, *Australian J. Phys.* **10**, 29–42 (1957a).
———, The theory of infiltration 1. The infiltration equation and its solution, *Soil Sci.* **83**, 345–357 (1957b).
———, The theory of infiltration 2. The profile at infinity, *Soil Sci.* **83**, 435–448 (1957c).
———, The theory of infiltration 3. Moisture profile and relation to experiments, *Soil Sci.* **84**, 163–178 (1957d).
———, The theory of infiltration 4. Sorptivity and algebraic infiltration equations, *Soil Sci.* **84**, 257–264 (1957e).
———, The theory of infiltration 5. The influence of the initial moisture content, *Soil Sci.* **84**, 329–339 (1957f).
———, The physical principles of soil water movement during the irrigation cycle, *3rd Congr. Int. Comm. Irrig. Drainage* **8**, 125–154 (1957g).
———, The theory of infiltration 6: Effect of water depth over soil, *Soil Sci.* **85**, 278–286 (1958a).
———, The theory of infiltration 7, *Soil Sci.* **85**, 333–337 (1958b).
———, General method of exact solution of the concentration dependent diffusion equation, *Australian J. Phys.* **13**, 1–12 (1960a).
———, A very general class of exact solutions in concentration dependent diffusion, *Nature* **185**, 233 (1960b).
———, Absorption and infiltration in two- and three-dimensional systems, *I.A.S.H. Symp. Water in the Unsaturated Zone*, Wageningen, The Netherlands, 1966.

______, Hydrostatics and hydrodynamics in swelling media, *Proc. IAHR 1st Inter. Symp. Fundamentals of Transport Phenomena in Porous Media*, Haifa, 1969; *Water Resources Res.* **5**, 1070–1077 (1969).

______, Theory of Infiltration, in *Advances in Hydroscience* (V. T. Chow, Ed.), pp. 215–296, Academic Press, New York, 1969a.

______, Flow through porous media, *Ann. Rev. Fluid Mechan.* **2**, 177–204 (1970).

______, and D. A. de Vries, Moisture movement in porous materials under temperature gradients, *Trans. Amer. Geophys. Union* **38**, 222–232 (1957).

Piersol R. J., L. E. Workman, and M. C. Watson, Porosity, total liquid saturation and permeability of Illinois oil sands III, Geol. Survey, Report No. 67, 1940.

Pilatovski, V. P., On the problem of ground water encroachment to a drain in a layer of infinite thickness (in Russian), *Isv. Akad. Nauk S.S.S.R., Otd. Tech. Nauk.* **7**, 70–75 (1958).

______, On the displacement of the interface between two heavy fluids in a horizontal layer (in Russian), *Akad. Nauk. S.S.S.R., O.T.N. Mekh. Makhin.* **1**, 127–130 (1961).

Pirverdian, A. M., On self similar solutions in the theory of unsteady filtration of a gas in a porous medium (in Russian), *Prikl. Mat. Mekh. Moscow* **24**, 558–564 (1960).

Pirson, S. J., *Oil Reservoir Engineering*, McGraw-Hill, New York, 1958.

Polubarinova-Kochina, P. Ya., Unsteady seepage with an interface (in Russian), *Dokl. Nauk S.S.S.R., Moscow* **66**, 173–176 (1949).

______, *Theory of Ground Water Movement* (in Russian), Gostekhizdat, Moscow, 1952; English transl. by R. J. M. De Wiest, Princeton University Press, Princeton, N.J., 1962.

______, and A. R. Shkrich, On the problem of displacement of the oil-contour front (in Russian), *Trans. Acad. Sci. Div. Tech. Sci. U.S.S.R.* No. 11, 105–107 (1954).

Popovich, M., and C. Hering, *Fuels and Lubricants*, John Wiley, New York, 1959.

Poreh, M., The dispersivity tensor in isotropic and axisymmetric mediums, *J. Geophys. Res.* No. 16, **70**, 3909–3914 (1965).

Prager, W., *Introduction to Mechanics of Continua*, Ginn, Boston, Mass., 1961.

Prager, S., Viscous flow through porous media, *Phys. of Fluids* No. 12, **4**, 1477–1482 (1961a).

Prandtl, L., *Essentials of Fluid Dynamics*, Hafner, New York, 1952.

______, and O. G. Tietjens, *Fundamentals of Hydro- and Aeromechanics*, McGraw-Hill, New York, 1934.

Prašil, F., *Technische Hydrodynamik*, Springer, Berlin, 1913.

Prats, M., The effect of horizontal fluid flow on thermally induced convection currents in porous mediums, *J. Geophys. Res.* No. 20, **71**, 4835–4838 (1966).

Purcell, W. R., Capillary pressures—their measurement using mercury and the calculation of permeability therefrom, *Trans. A.I.M.E.* **186**, 39–46 (1949).

Quon, D., P. M. Dranchuk, S. R. Allada and P. K. Leung, Application of the alternating direction explicit procedure to two-dimensional natural gas reservoirs, *J. Soc. Petrol. Eng.* No. 2, **6**, 137–142 (1966).

Raats, P. A. C., Development of Equations Describing Transport of Mass and Momentum in Porous Media, with Special Reference to Soils, Ph. D. Thesis, University of Illinois, 1965.

Raimondi, P., G. H. F. Gardner, and C. B. Petrick, Effect of pore structure and molecular diffusion on the mixing of miscible liquids flowing in porous media, *Amer. Inst. Chem. Eng.* and *Soc. Petrol. Eng. Joint Symp. Oil Recovery Methods, San Francisco*, Preprint 43, 1959.

Ramey, H. J., The blotter type electrolytic model determination of areal sweeps in oil recovery by in-situ combustion, *Trans. A.I.M.E.* **201**, 119–123 (1954).

Rapoport, L. A., and W. J. Leas, Relative permeability to liquid in liquid-gas systems, *Trans. A.I.M.E.* **192**, 83–95 (1951).

Rawlins, S. L., and W. H. Gardner, A test of the validity of the diffusion equation for unsaturated flow of soil water, *Soil Sci. Soc. Amer. Proc.* **27**, 507–510 (1963).

Rayleigh, Lord, *Phil. Mag.* **32**, 529–546 (1916).

Reddell, D. L., and D. K. Sunada, Numerical simulation of dispersion in groundwater aquifers, Hydrology Paper No. 41, Colorado State University, June, 1970.

Reisenauer, A. E., Solution methods for multidimensional partially saturated flow in soils, General Electric, Hanford Atomic Products Operations, HW-SA-2854, 1962.

Remson, I., and J. R. Randolph, Review of some elements of soil moisture theory, U.S. Geol. Survey, Professional Paper 411-D, 1962.

Richards L. A., Capillary conduction of liquids through porous medium, *Physics* **1**, 318–333 (1931).

——, Methods of measuring moisture tension, *Soil Sci.* **68**, 95–112 (1949).

——, Experimental demonstration of the hydraulic criterion for zero water flow in unsaturated soil, *Trans. Int. Congr. Soil Sci. Amsterdam*, **1**, 67–68 (1950).

——, and W. Gardner, Tensiometers for measuring the capillary tension of soil water, *J. Amer. Soc. Agron.* **28**, 352–358 (1936).

Richardson, J. G., J. K. Kerver, J. A. Hafford, and J. S. Osoba, Laboratory determinations of relative permeability, *Trans. A.I.M.E.* **195**, 187–196 (1952).

Richtmeyer, R. D., *Difference Methods for Initial-Value Problems*, Interscience, New York, 1957.

Rideal, E., Introductory lecture, Proc. 10th Symp. Colston Res. Soc., *The Structure and Properties of Porous Materials* (D. H. Everett and F. S. Stone, Eds.), pp. 1–5, Butterworth, London, 1958.

Rifai, M. N. E., W. J. Kaufman, and D. K. Todd, Dispersion phenomena in laminar flow through porous media, Sanitary Engineering Rep. No. 3, I.E.R., Series 90, University of California, Berkeley, 1956.

Rijtema, P. E., Calculation of capillary conductivity from pressure plate outflow data with non-negligible membrane impedance, *Neth. J. Agr. Sci.* No. 3, **7**, 209–216 (1959).

Rizenkampf, B. K., Hydraulics of groundwater (in Russian) *Part 1*, No. 1, **1** (1938); *Part 2*, No. 2, **1**, State Univ. of Saratov, Series F.M.M., 1938.

Roberts, J. E., and J. M. de Souza, The Compressibility of sands, Paper presented at 61st Annual Meeting, Amer. Soc. for Testing and Materials, Boston, 1958.

Robertson, H. P., The invariant theory of isotropic turbulence, *Proc. Phil. Soc.* **36**, 209–223 (1940).

Robertson, J. M., *Hydrodynamics: in Theory and Application*, Prentice-Hall, Englewood Cliffs, N.J., 1965.

Rogers, F. T., Jr., Convection currents in porous media V. Variational form of the theory, *J. Appl. Phys.* **24**, 877–880 (1953).

——, and H. L. Morrison, Convection currents in porous media III. Extended theory of the critical gradient, *J. Appl. Phys.* **21**, 1170–1180 (1950).

——, and L. E. Schilberg, Observations in initial flow in a fluid obeying Darcy's law by radioactive tracer technique, *J. Appl. Phys.* **22**, 233–234 (1951).

Rose, D. A., Water movement in porous materials 1. Isothermal vapor transfer, *Brit. J. Appl. Phys.* **14**, 256–262 (1963a).

——, Water movement in porous materials 2. The separation of the components of water movement, *Brit. J. Appl. Phys.* **14**, 491–496 (1963b).

Rose, H. E., An investigation into the laws of flow of fluids through beds of granular material, *Proc. Inst. Mech. Eng.* **153**, 141–148 (1945).

Rose, W., Some problems of relative permeability measurement, *Proc. 3rd World Petrol. Congr. Sect. II* 446–459 (1951a).

——, The Muskat Model, Private Communication, 1966.

——, and W. A. Bruce, Evaluation of capillary characters in petroleum reservoir rock, *Trans. A.I.M.E.* **186**, 127–142 (1949).

———, Some problems of relative permeability measurement, paper presented at the Third World Petroleum Congress **2**, 446ff. (1951).

———, Fluid flow in petroleum reservoirs III. Effect of Fluid–Fluid Interfacial Boundary Condition, *Ill. Geol. Survey Circ.* 291 (1960).

Rosenberg, D. U. von, *Methods for the Numerical Solution of Partial Differential Equations*, Elsevier, New York, 1969.

Rothe, R., K. Pohlhausen, and F. Ollendorff, *Theory of Functions as Applied to Engineering Problems* (English transl. by A. Herzberg), Technology Press M.I.T., Cambridge, Mass., 1951.

Rouse, H. (Ed.), *Advanced Mechanics of Fluids*, John Wiley, New York, 1959.

Rubin, J., Numerical analysis of ponded rainfall infiltration, *I.A.S.H. Symp. Water in Unsaturated Zone*, Wageningen, The Netherlands, 440–451 (1966).

———, Numerical method for analyzing hysteresis affected post infiltration redistribution of soil moisture, *Soil Sci. Soc. Amer. Proc.* **31**, 13–20 (1967).

Rühl, W. von, and C. Schmid, Über das Verhältnis der vertikalen zur horizontalen absoluten Permeabilität von Sandsteinen, *Geolog. Jahrb.* **74**, 447–461 (1957).

Rumer, R. R., Longitudinal dispersion in steady and unsteady flow through porous media, *Proc. Amer. Soc. Civil Eng. J. Hydraulic Div.* No. HY4, **88**, 147–172 (1962).

———, Resistance to flow through porous media, in *Flow through Porous Media* (R. J. M. de Wiest, Ed.), Academic Press, New York, 1969.

———, and P. A. Drinker, Resistance to laminar flow through porous media, *Proc. Amer. Soc. Civil Eng.* No. HY5, **92**, 155–164 (1966).

———, and D. R. F. Harleman, Intruded salt water wedge in porous media, *Proc. Amer. Soc. Civil Eng. Hydrol. Div.* No. 6, **89**, 193–220 (1963).

Russell, T. W. F., and E. Charles, Effect of the less viscous liquid in the laminar flow of two immiscible liquids, *Canad. J. Chem. Eng.* **37**, 18–24 (1959).

Saffman, P. G., A theory of dispersion in a porous medium, *J. Fluid Mech.* No. 3, **6**, 321–349 (1959).

———, Dispersion due to molecular diffusion and macroscopic mixing in flow through a network of capillaries, *J. Fluid Mech.* No. 2, **7**, 194–208 (1960).

Santing, G., Infiltratie en modelonderzoek (in Dutch), *Water* No. 21, **35**, 234–238, No. 22, 243–246 (1951a).

———, Modèle pour l'étude des problèmes de l'écoulement simultané des eaux souterraines douces et salées, Assemblée Générale de Bruxelles, *Ass. Int. Hydrol. Scientifique* **2**, 184–193 (1951b).

———, A horizontal scale model, based on the viscous flow analogy, for studying groundwater flow in an aquifer having storage, *Proc. IUGG General Assembly, Toronto* 105–114 (1957).

———, Investigations of Ground Water in the Coastal Plain of Israel, Water Planning for Israel Ltd., P.N. 23, 1957a.

———, Modelonderzoek (in Dutch), in *Committee for Hydrological Research T.N.O.*, Proceedings of the Tech. Meeting 17, 23–47 (1963).

Scheidegger, A. E., Theoretical models of porous matter, *Producers Monthly* No. 10, **17**, 17–23 (1953).

———, Statistical hydrodynamics in porous media, *J. Appl. Phys.* No. 25, 994–1001 (1954).

———, On the theory of flow of miscible phases in porous media, *Proc. IUGG General Assembly, Toronto* **2**, 236–242 (1957).

———, Statistical approach to miscible displacement in porous media, *Bull. Canad. Inst. Min. Met.* 26–30 (1958).

———, *The Physics of Flow Through Porous Media*, 2nd ed., University of Toronto Press, Toronto, 1960.

———, Growth of instabilities in displacement fronts in porous media, *Phys. of Fluids*

No. 1, **3**, 94–104 (1960a).

______, On the stability of displacement fronts in porous media, *Canad. J. Phys.* **38**, 153–162 (1960b).

______, General theory of dispersion in porous media, *J. Geophys. Res.* **66**, 3273–3278 (1961).

Schilfgaarde, J. van, Design of tile drainage for falling water tables, *Proc. Amer. Soc. Civil Eng.* No. IR2, **82**, 1–11 (1963).

______, Theory of land drainage I. Approximate solutions to drainage flow problems, in *Drainage of Agriculture Lands* (J. N. Luthin, Ed.), pp. 79–112, *Amer. Soc. Agron.*, Madison, Wisc., 1957.

______, Theory of flow to drains, in *Advances in Hydroscience*, pp. 107–141, Academic Press, New York, 1970.

Schmidt, E., *Föppls Festschrift*, Springer, Berlin, 1924.

Schmorak, S., Salt water encroachment in the coastal plain of Israel, *I.A.S.H. Symp. Artificial Recharge and Management of Aquifers, Haifa*, 305–318 (1967).

Schneebeli, G., Expériences sur la limite de validité de la loi de Darcy et l'apparition de la turbulence dans un écoulement de filtration, *La Huille Blanche* No. 2, **10**, 141–149 (1955).

Schwartz, H. A., Über einige Abbildungsaufgaben, Crelle **70** (1869); *Gesammelte Abhandl.* **2**, 65ff. (1890).

Scott, E. J., R. J. Hanks, D. B. Peters, and A. Klute, Power series solutions of the one-dimensional flow equation for exponential and linear diffusivity functions, *USDA*, ARS 41-64, 1962.

Scott, P. H., and W. Rose, An explanation of the Yuster effect, *J. Petrol. Tech.* No. 1, **5**, 19–20 (1953).

Scott, R. F., *Principles of Soil Mechanics*, Addison-Wesley, Reading, Mass., 1963.

Shamir, U., The use of computers in ground water hydrology, Report No. 105, Hydrodynamics Lab., M.I.T., Cambridge, Mass. 1967.

______, and G. Dagan, Motion of the sea water interface in a coastal aquifer, Hydraulic Lab., Technion, Haifa, 1970.

______, and D. R. F. Harleman, Numerical and analytical solutions of the dispersion problems in homogeneous aquifers, Rept. No. 89, Hydrodynamics Lab., M.I.T., 1966.

Shaw, F. S., *Relaxation Methods*, Dover, New York, 1953.

Simpson, E. S., Transverse dispersion in liquid flow through porous media, U.S. Geol. Survey, Professional Paper 411-C, Washington, D.C., 1962.

______, Velocity and the longitudinal dispersion coefficient in flow through porous media, in *Flow through Porous Media* (R. J. M. de Wiest, Ed.), Chap. 5, pp. 201–214, Academic Press, New York, 1969.

Skempton, A. W., Effective stresses in soils, concrete and rocks, in *Conference on Pore Pressure and Suction in Soils*, pp. 4–16, Butterworths, London, 1961.

Skibitzke, H. E., The use of analog computers for studies in ground water hydrology, *J. Inst. Water Eng.* **17** (1960).

______, Electronic computers as an aid to the analysis of hydrologic problems, Publ. No. 52, *I.A.S.H.*, 1961.

Slichter, C. S., Field measurement of the rate of movement of underground waters, U.S. Geol. Survey, Water Supply Paper, No. 140, 1905.

Smith, G. O., *Numerical Solution of Partial Differential Equations*, Oxford University Press, New York/London, 1965.

Smith, W. O., Capillary flow through an ideal uniform soil, *Physics* **3**, 139–146 (1933).

Snow, D. T., A Parallel Plate Model of Fractured Permeable Media, Ph. D. Thesis, University of Calif., Berkeley, Calif., 1965.

Sokolnikoff, I. S., *Tensor Analysis*, 2nd ed., John Wiley, New York, 1964.

Somerton, H. W., Some thermal characteristics of porous rocks, *J. Petrol. Tech.* **10**,

Note 2008, 61–65 (1958).

Sourieau, P., Étude de déplacement de fluides dans les milieux poreux à l'aide de modèles physiques, Mémoires et travaux de la S.H.F., No. II, 181–192 (1963).

Spain, B., *Tensor Calculus*, Oliver & Boyd, London, 1960.

Spiegel, M. R., *Theory and Problems of Vector Analysis*, Schaum, New York, 1959.

Stallman, R. W., Multiphase fluid flow in porous media—A Review of Theories Pertinent to Hydrologic Studies, U.S. Geol. Survey, Prof. Papers No. 411-E, 51 pp., 1964.

______, Flow in the zone of aeration, in *Advances in Hydroscience* (Ven. T. Chow, Ed.), Vol. 4, pp. 151–195, Academic Press, New York, 1967.

Staple, W. J., Infiltration and redistribution of water in vertical columns of loam soil, *Soil Sci. Soc. Amer. Proc.* **30**, 553–558 (1966).

Sternberg, Y. M., and V. A. Scott, The Hele-Shaw model as a tool in ground water research, NWWA Conf., San Francisco, Sept., 1963.

Stoker, J. J., *Water Waves*, Interscience, New York, 1957.

Sullivan, R. R., and K. L. Hertel, The permeability methods for determining specific surface of fibers and powders, *Advances in Colloid Science*, Vol. 1, pp. 37–80, Interscience, New York, 1942.

Sunada, D. K., Turbulent flow through porous media, Water Resources Center, Contrib. No. 103, University of California, Berkeley, 1965.

Swartzendruber, D., Non-Darcy flow behavior in liquid-saturated porous media, *J. Geophys. Res.* No. 13, **67**, 5205–5213 (1962).

______, Comment on the paper: Determination of the hydraulic conductivity of unsaturated soils from an analysis of transient flow data by G. Vachaud (1967), *Water Resources Res.* No. 3, **4**, 659–660 (1968).

______, The flow of water in unsaturated soils, in *Flow through Porous Media* (R. J. M. de Wiest, Ed.), Chap. 6, pp. 215–292, Academic Press, New York, 1969.

Taylor, D. W., *Fundamentals of Soil Mechanics*, John Wiley, New York, 1948.

Taylor, G. I., Dispersion of soluble matter in solvent flowing slowly through a tube, *Proc. Roy. Soc. A*, No. 1137, **219**, 186–203 (1953).

______, The dispersion of matter in turbulent flow through a pipe, *Proc. Roy. Soc. A*. **223**, 446–468 (1954).

Taylor, S. A., and J. W. Cary, Analysis of simultaneous flow of water and heat or electricity with the thermodynamics of irreversible processes, *Proc. 7th Int. Congr. Soil Sci.*, 1960.

______, and J. W. Cary, Linear equations for the simultaneous flow of matter and energy, in a continuous soil system, *Soil Sci. Soc. Amer. Proc.* **28**, 167–172 (1964).

Terzaghi, K., Die Berechnung der Durchlässigkeitsziffer des Tones aus dem Verlauf der hydrodynamischen Spannungserscheinungen, *Sitz. Akad. Wiss.* Wien, Austria, **132**, 125–138 (1923).

______, *Erdbaumechanik auf bodenphysikalischer Grundlage*, Deuticke, Leipzig, 1925.

______, *Theoretical Soil Mechanics*, John Wiley, New York, 1943.

______, *From Theory to Practice in Soil Mechanics*, John Wiley, New York, 1960.

Thom, A., and L. J. Apelt, *Field Computations in Engineering and Physics*, Van Nostrand, New York, 1961.

Thomas, H. E., Ground water regions of the United States—Their storage facilities, Vol. 3, Inter- and Insular Affairs Comm., House of Representatives, 5 U.S. Congress, Washington, D.C., 1952.

Thomas, L. K., D. L. Katz, and M. R. Tek, Threshold pressure phenomena in porous media, *J. Soc. Petrol. Eng.* No. 2, **8**, 174–184 (1968).

Thomas, R. G., *Graphical Solution of Ground Water Flow Problems*, Stetson, Strauss and Dresselhaus, Los Angeles, Calif., 1960.

Thompson, W., On the equilibrium of vapor at a curved surface of liquid, *Philosophical Mag.* **42**, 448–452 (1871).

Todd, John, *Survey of Numerical Analysis*, McGraw-Hill, New York, 1962.

Todd, D. K., Unsteady flow in porous media by means of a Hele–Shaw viscous fluid model, *Trans. Amer. Geophys. Union* **35**, 905–916 (1954).

———, Flow in porous media studied in Hele–Shaw channel, *Civil Eng.* No. 2, **25**, 85 (1955a).

———, Laboratory research with ground water models, *Symp. Darcy*, Publ. 41, *Assoc. Int. Hydrol. Sci.* 109–206 (1955b).

———, *Ground Water Hydrology*, John Wiley, New York, 1959.

———, and J. Bear, River seepage investigations, Water Resources Center, Cont. No. 20, Hydraulic Lab., University of California, Berkeley, 1959.

Toksöz, S., and D. Kirkham, Graphic solution and interpretation of a new drain spacing formula, *J. Geophys. Res.* No. 2, **66**, 509–516 (1961).

Topp, G. C., and E. E. Miller, Hysteretic conductivity calculation versus measurement, 1964 meeting of Amer. Soc. Agron., *Agron. Abstr.* (1965).

Truesdell, C., Principles of continuum mechanics, Colloquium Lectures in Pure and Appl. Science, No. 5, Socony Mobil Co. Inc., 1961.

———, and R. A. Toupin, Classical field theories, in *Handbuch der Physik*, Vol. III/1, pp. 226–793, Springer, Berlin, 1960.

Tuinzaad, H., Influence of the atmospheric pressure on the head of artesian water and phreatic water, Assemblée Générale de Rome, *I.A.S.H.* **2**, 32–37 (1954).

Turner, G. A., The flow structure in packed beds, *Chem. Eng. Sci.* **7**, 156–165 (1957).

Vachaud, G., Determination of the hydraulic conductivity of unsaturated soils from an analysis of transient flow data, *Water Resources Res.* **3**, 697–705 (1967).

———, Reply to discussion of D. Swartzendruber on an earlier paper (Vachaud, 1967), *Water Resources Res.* No. 3, **4**, 661–664 (1968).

———, Contribution à l'étude des problèmes d'écoulement en milieux poreux non saturés, Thèse pour le grade de Docteur ès Sciences Physiques, Grenoble, France, 1968a.

Van Deemter, J. J., J. J. Brader, and H. A. Lauweir, Fluid displacement in capillaries, *Appl. Sci. Res. A.* No. 5, **5**, 374–388 (1955).

Van Duin, R. H. A., Tillage in relation to rainfall intensity and infiltration capacity of soils, *Neth. J. Agric. Sci.* No. 3, **3**, 182–191 (1955).

———, On the influence of tillage on conduction of heat, diffusion of air and infiltration of water in soil (in Dutch), Versl. Landbouwk, Onderz, The Hague, The Netherlands No. 62.7, 1956.

Van Meurs, P., The use of transparent three-dimensional models for studying the mechanisms of flow processes in oil reservoirs, *Trans. A.I.M.E. Petrol.* **210**, 295-301 (1957).

Van Quy, N., Étude numérique et experiméntale de l' "Edge Coning", *Rev. Inst. Fr. Petrol.* **18**, (1963).

Vedernikov, V. V., *Seepage from Channels* (in Russian), Gosstroÿzdat, Moscow, 1934; also in German: *Wasserkrafts-Wasserwirtsch.*, pp. 11–13, 1934.

———, Theory of filtration in soils and its application in problems of irrigation and drainage (in Russian), Bur. of Public Works, Moscow, U.S.S.R., 1939.

Verigin, N. N., On unsteady flow of soil waters in the neighbourhood of reservoirs (in Russian), *Dokl. A.N. S.S.S.R. Moscow* 1067–1070 (1949).

Verruijt, A., Elastic storage of aquifers, NSF sponsored Hydrology Institute at Princeton University, 1965; also in *Flow through Porous Media* (R. J. M. de Wiest, Ed.), pp. 331–376, Academic Press, New York, 1969.

Volynskii, B. A., and V. Ye. Bukhman, *Analogues for the Solution of Boundary Value Problems*, Pergamon, Elmsford, New York, 1965; English transl. of original Russian, published by Fizmatgiz, Moscow, 1960.

Von Engelhardt, W., and W. L. M. Tunn, The flow of fluids through sandstones, transl. by P. A. Witherspoon from *Heidelberger Beitr. Mineral Petrog.* **2**, 12–25, Ill. State Geol. Survey, Circular 194, 1955.

Von Mises, R., *Theorie der Wasserräder*, Leipzig, 1908.

Von Rosenberg, D. U., Mechanics of steady state single phase fluid displacement from porous media, *J. Amer. Inst. Chem. Eng.* **2**, 55–58 (1956).

Vreedenburgh, C. G. F., De parallelstroming door grond bestaande uit evenwijdige regelmatig afwissenlende lagen van verschillende dikte en doorlaatbaarheid, *Ingen. Ned. Indie* **8**, 111–113 (1937).

Walton, W. C., and T. A. Prickett, Hydrogeologic electric analog computers, *J. Amer. Soc. Civil Eng. Hydraulic Div.*, No. HY6, paper 3695, 67–91 (1963).

———, *Groundwater Resource Evaluation*, McGraw-Hill, New York, 1970.

Ward, J. C., Turbulent flow in porous media, *Proc. Amer. Soc. Civil Eng.* No. HY5, **90**, 1–12 (1964).

Watson, K. K., Some operating characteristics of a rapid response tensiometer system, *Water Resources Res.* **1**, 577–586 (1965).

———, Experimental and numerical study of column drainage, *Proc. Amer. Soc. Civil Eng. Hydraul. Div.* No. HY2, **93**, 1–15 (1967).

Watson, R. E., H. I. Silberberg, and B. H. Claudle, Model studies on the inverted nine-spot injection pattern, *J. Petrol. Tech.* **16**, 801–804 (1964).

Waxman, M. H., L. J. M. Smits, Electrical conductivities in oil bearing shaly sands, *J. Soc. Petrol. Eng.* No. 2, **8**, 107–122 (1968).

Weber, H., *Die Reichweite von Grundwasserabsenkungen mittels Rohrbrunnen*, Springer, Berlin, 1928.

Welge, H. J., and W. A. Bruce, The restored state method for determination of oil in place and connate water, *Drilling and Production Practices, Amer. Petrol. Inst.*, p. 166, 1947.

———, A simplified method for computing oil recovery by gas or water drive, *Trans. A.I.M.E.* **195**, 91–99 (1952).

Wentworth, C. K., Growth of the Ghyben-Herzberg transition zone under a rinsing hypothesis, *Trans. Amer. Geophys. Union* No. 1, **29**, 97–98 (1948).

Wesseling, J., Principles of unsaturated flow and their application to the penetration of moisture into the soil, *Tech. Bull. Inst. Land and Water Res.* No. 23, Wageningen, The Netherlands, 1961.

———, Some drainage problems and their solutions, Technion, Haifa, 1962.

———, A comparison of the steady state drain spacing formulas of Hooghoudt and Kirkham in connection with design practice, Tech. Bull. 34, Inst. for Land and Water Management Res., 1964.

Whisler, F. D., and A. Klute, Analysis of infiltration into stratified soil columns, *I.A.S.H. Symp. Water in the Unsaturated Zone, Wageningen*, 1966.

Whitaker, S., The equations of motion in porous media, *Chem. Eng. Sci.* **21**, 291–300 (1966).

———, Diffusion and dispersion in porous media, *J. Amer. Inst. Chem. Eng.* No. 3, **13**, 420–427 (1967).

White, A. M., *Trans. Amer. Inst. Chem. Eng.* **31**, 390ff. (1935).

Wilhelm, O., Classification of petroleum reservoirs, *Bull. Amer. Ass. Petrol. Geol.* **29**, 1537ff. (1945).

Williams, M., Estimation of interstitial water from the electric log, *Trans. A.I.M.E.* **189**, 295–308 (1950).

Winsauer, W. O., H. M. Shearin, P. H. Masson, and M. Williams, Resistivity of brine saturated sands in relation to pore geometry, *Bull. Amer. Ass. Petrol. Geol.* No. 2, **36**, 253–277 (1952).

Wooding, R. A., Steady state free thermal convection of liquid in a saturated porous medium, *J. Fluid. Mech.* **2**, 273–285 (1957).

———, An experiment on free thermal convection of water in saturated permeable material, *J. Fluid Mech.* **3**, 582–600 (1958).

———, The stability of a viscous liquid in a vertical tube containing porous material,

Proc. Roy. Soc. A. **252**, 120–134 (1959).
______, Rayleigh instability of a thermal boundary layer in flow through a porous medium, *J. Fluid Mech.* **9**, 183–192 (1960).
______, Free convection of fluid in a vertical tube filled with porous material, *J. Fluid Mech.* **13**, 129–144 (1962).
______, Convection in a saturated porous medium at large Rayleigh number or Peclet number, *J. Fluid Mech.* **15**, 527–544 (1963).
______, Mixing layer flows in a saturated porous medium, *J. Fluid Mech.* **19**, 103–112 (1964).
Woodside, W., and J. H. Messmer, Thermal conductivity of porous media, in (1) unconsolidated and (2) consolidated sands, *J. Appl. Phys.* **32**, 1688–1699 (1961).
Wright, D. E., Nonlinear flow through granular media, *Proc. Amer. Soc. Civil Eng. Hydraul. Div.* No. HY4, **94**, 851–872 (1968).
Wyckoff, R. D., and H. G. Botset, The flow of gas-liquid mixture through unconsolidated sands, *Physics* **7**, 325–345 (1936).
______, and H. G. Botset, An experimental study of motion of particles in systems of complex potential, *Physics* **5**, 265–275 (1934).
______, H. G. Botset, M. Muskat, and D. W. Reed, The measurement of the permeability of porous media for homogeneous fluids, *Rev. Sci. Instr.* **4**, 394–405 (1933).
Wylie, C. R., *Advanced Engineering Mathematics*, McGraw-Hill, New York, 1951.
Wyllie, M. R. J., and G. H. F. Gardner, The generalized Kozeny–Carman equation II. A novel approach to problems of fluid flow, *World Oil Prod. Sect.* 210–228 (April, 1958).
______, and W. D. Rose, Some theoretical considerations related to the quantitative evaluation of the physical characteristics of reservoir rock from electrical log data, *Trans. A.I.M.E.* **189**, 105–118 (1950).
______, and P. F. Southwick, An experimental investigation of the S.P. and resistivity phenomena in dirty sands, *Trans. A.I.M.E.* **201**, 43–56 (1954).
______, and M. B. Spangler, Application of electrical resistivity measurements to problems of fluid flow in porous media, *Bull. Amer. Ass. Petrol. Geol.* **36**, 359–403 (1952).
Yeh, William Wen-Cong, and J. B. Franzini, Moisture movement in a horizontal soil column under the influence of an applied pressure, *J. Geophys. Res.* No. 16, **73**, 5151–5157 (1968).
Yih, C. S., Stream functions in three-dimensional flows, *La Houille Blanche* **12**, 445–450 (1957).
Youngs, E. G., Moisture profiles during vertical infiltration, *Soil Sci.* No. 4, **84**, 283–290 (1957).
______, Redistribution of moisture in porous materials after infiltration I, *Soil Sci.* **86**, 117–125 (1958a).
______, Redistribution of moisture in porous materials after infiltration II, *Soil Sci.* **86**, 202–207 (1958b).
Yuster, S. T., Theoretical considerations of multiphase flow in idealized capillary systems, *Proc. 3rd World Petrol. Congr. The Hague* **2**, 437–445 (1953).
Zamarin, E. A., The flow of ground water under hydraulic structures (in Russian), Tashkent, 1931.
Zaslavsky, D., Flow model of water in saturated soil with non-uniform conductivity, Dept. of Civil Eng., Technion, Haifa, Publ. 15, 1962.
______, Theory of unsaturated flow into a non-uniform soil profile, *Soil Sci.* No. 6, **97**, 400–410 (1964a).
______, Saturation-unsaturation transition in infiltration to a non-uniform soil profile, paper presented at the *8th Congr. Int. Soc. Soil Sci. Bucharest* (1964b).
______, and I. Ravina, Measurement and evaluation of the hydraulic conductivity through moisture moment method, *Soil Sci.* **100**, 104–107 (1965).
Zhukovski, N. E., Seepage through dams, *Collected Works* **7**, 1923 (1949).
Zienkiewicz, O. C., and G. S. Holister, *Stress Analysis*, John Wiley, New York, 1965.

Index

A CATALOG OF SELECTED

DOVER BOOKS

IN SCIENCE AND MATHEMATICS

Astronomy

CHARIOTS FOR APOLLO: The NASA History of Manned Lunar Spacecraft to 1969, Courtney G. Brooks, James M. Grimwood, and Loyd S. Swenson, Jr. This illustrated history by a trio of experts is the definitive reference on the Apollo spacecraft and lunar modules. It traces the vehicles' design, development, and operation in space. More than 100 photographs and illustrations. 576pp. 6 3/4 x 9 1/4. 0-486-46756-2

EXPLORING THE MOON THROUGH BINOCULARS AND SMALL TELESCOPES, Ernest H. Cherrington, Jr. Informative, profusely illustrated guide to locating and identifying craters, rills, seas, mountains, other lunar features. Newly revised and updated with special section of new photos. Over 100 photos and diagrams. 240pp. 8 1/4 x 11. 0-486-24491-1

WHERE NO MAN HAS GONE BEFORE: A History of NASA's Apollo Lunar Expeditions, William David Compton. Introduction by Paul Dickson. This official NASA history traces behind-the-scenes conflicts and cooperation between scientists and engineers. The first half concerns preparations for the Moon landings, and the second half documents the flights that followed Apollo 11. 1989 edition. 432pp. 7 x 10. 0-486-47888-2

APOLLO EXPEDITIONS TO THE MOON: The NASA History, Edited by Edgar M. Cortright. Official NASA publication marks the 40th anniversary of the first lunar landing and features essays by project participants recalling engineering and administrative challenges. Accessible, jargon-free accounts, highlighted by numerous illustrations. 336pp. 8 3/8 x 10 7/8. 0-486-47175-6

ON MARS: Exploration of the Red Planet, 1958-1978--The NASA History, Edward Clinton Ezell and Linda Neuman Ezell. NASA's official history chronicles the start of our explorations of our planetary neighbor. It recounts cooperation among government, industry, and academia, and it features dozens of photos from Viking cameras. 560pp. 6 3/4 x 9 1/4. 0-486-46757-0

ARISTARCHUS OF SAMOS: The Ancient Copernicus, Sir Thomas Heath. Heath's history of astronomy ranges from Homer and Hesiod to Aristarchus and includes quotes from numerous thinkers, compilers, and scholasticists from Thales and Anaximander through Pythagoras, Plato, Aristotle, and Heraclides. 34 figures. 448pp. 5 3/8 x 8 1/2. 0-486-43886-4

AN INTRODUCTION TO CELESTIAL MECHANICS, Forest Ray Moulton. Classic text still unsurpassed in presentation of fundamental principles. Covers rectilinear motion, central forces, problems of two and three bodies, much more. Includes over 200 problems, some with answers. 437pp. 5 3/8 x 8 1/2. 0-486-64687-4

BEYOND THE ATMOSPHERE: Early Years of Space Science, Homer E. Newell. This exciting survey is the work of a top NASA administrator who chronicles technological advances, the relationship of space science to general science, and the space program's social, political, and economic contexts. 528pp. 6 3/4 x 9 1/4. 0-486-47464-X

STAR LORE: Myths, Legends, and Facts, William Tyler Olcott. Captivating retellings of the origins and histories of ancient star groups include Pegasus, Ursa Major, Pleiades, signs of the zodiac, and other constellations. "Classic." – *Sky & Telescope.* 58 illustrations. 544pp. 5 3/8 x 8 1/2. 0-486-43581-4

A COMPLETE MANUAL OF AMATEUR ASTRONOMY: Tools and Techniques for Astronomical Observations, P. Clay Sherrod with Thomas L. Koed. Concise, highly readable book discusses the selection, set-up, and maintenance of a telescope; amateur studies of the sun; lunar topography and occultations; and more. 124 figures. 26 halftones. 37 tables. 335pp. 6 1/2 x 9 1/4. 0-486-42820-6

Chemistry

MOLECULAR COLLISION THEORY, M. S. Child. This high-level monograph offers an analytical treatment of classical scattering by a central force, quantum scattering by a central force, elastic scattering phase shifts, and semi-classical elastic scattering. 1974 edition. 310pp. 5 3/8 x 8 1/2. 0-486-69437-2

HANDBOOK OF COMPUTATIONAL QUANTUM CHEMISTRY, David B. Cook. This comprehensive text provides upper-level undergraduates and graduate students with an accessible introduction to the implementation of quantum ideas in molecular modeling, exploring practical applications alongside theoretical explanations. 1998 edition. 832pp. 5 3/8 x 8 1/2. 0-486-44307-8

RADIOACTIVE SUBSTANCES, Marie Curie. The celebrated scientist's thesis, which directly preceded her 1903 Nobel Prize, discusses establishing atomic character of radioactivity; extraction from pitchblende of polonium and radium; isolation of pure radium chloride; more. 96pp. 5 3/8 x 8 1/2. 0-486-42550-9

CHEMICAL MAGIC, Leonard A. Ford. Classic guide provides intriguing entertainment while elucidating sound scientific principles, with more than 100 unusual stunts: cold fire, dust explosions, a nylon rope trick, a disappearing beaker, much more. 128pp. 5 3/8 x 8 1/2. 0-486-67628-5

ALCHEMY, E. J. Holmyard. Classic study by noted authority covers 2,000 years of alchemical history: religious, mystical overtones; apparatus; signs, symbols, and secret terms; advent of scientific method, much more. Illustrated. 320pp. 5 3/8 x 8 1/2. 0-486-26298-7

CHEMICAL KINETICS AND REACTION DYNAMICS, Paul L. Houston. This text teaches the principles underlying modern chemical kinetics in a clear, direct fashion, using several examples to enhance basic understanding. Solutions to selected problems. 2001 edition. 352pp. 8 3/8 x 11. 0-486-45334-0

PROBLEMS AND SOLUTIONS IN QUANTUM CHEMISTRY AND PHYSICS, Charles S. Johnson and Lee G. Pedersen. Unusually varied problems, with detailed solutions, cover of quantum mechanics, wave mechanics, angular momentum, molecular spectroscopy, scattering theory, more. 280 problems, plus 139 supplementary exercises. 430pp. 6 1/2 x 9 1/4. 0-486-65236-X

ELEMENTS OF CHEMISTRY, Antoine Lavoisier. Monumental classic by the founder of modern chemistry features first explicit statement of law of conservation of matter in chemical change, and more. Facsimile reprint of original (1790) Kerr translation. 539pp. 5 3/8 x 8 1/2. 0-486-64624-6

MAGNETISM AND TRANSITION METAL COMPLEXES, F. E. Mabbs and D. J. Machin. A detailed view of the calculation methods involved in the magnetic properties of transition metal complexes, this volume offers sufficient background for original work in the field. 1973 edition. 240pp. 5 3/8 x 8 1/2. 0-486-46284-6

GENERAL CHEMISTRY, Linus Pauling. Revised third edition of classic first-year text by Nobel laureate. Atomic and molecular structure, quantum mechanics, statistical mechanics, thermodynamics correlated with descriptive chemistry. Problems. 992pp. 5 3/8 x 8 1/2. 0-486-65622-5

ELECTROLYTE SOLUTIONS: Second Revised Edition, R. A. Robinson and R. H. Stokes. Classic text deals primarily with measurement, interpretation of conductance, chemical potential, and diffusion in electrolyte solutions. Detailed theoretical interpretations, plus extensive tables of thermodynamic and transport properties. 1970 edition. 590pp. 5 3/8 x 8 1/2. 0-486-42225-9

Engineering

FUNDAMENTALS OF ASTRODYNAMICS, Roger R. Bate, Donald D. Mueller, and Jerry E. White. Teaching text developed by U.S. Air Force Academy develops the basic two-body and n-body equations of motion; orbit determination; classical orbital elements, coordinate transformations; differential correction; more. 1971 edition. 455pp. 5 3/8 x 8 1/2. 0-486-60061-0

INTRODUCTION TO CONTINUUM MECHANICS FOR ENGINEERS: Revised Edition, Ray M. Bowen. This self-contained text introduces classical continuum models within a modern framework. Its numerous exercises illustrate the governing principles, linearizations, and other approximations that constitute classical continuum models. 2007 edition. 320pp. 6 1/8 x 9 1/4. 0-486-47460-7

ENGINEERING MECHANICS FOR STRUCTURES, Louis L. Bucciarelli. This text explores the mechanics of solids and statics as well as the strength of materials and elasticity theory. Its many design exercises encourage creative initiative and systems thinking. 2009 edition. 320pp. 6 1/8 x 9 1/4. 0-486-46855-0

FEEDBACK CONTROL THEORY, John C. Doyle, Bruce A. Francis and Allen R. Tannenbaum. This excellent introduction to feedback control system design offers a theoretical approach that captures the essential issues and can be applied to a wide range of practical problems. 1992 edition. 224pp. 6 1/2 x 9 1/4. 0-486-46933-6

THE FORCES OF MATTER, Michael Faraday. These lectures by a famous inventor offer an easy-to-understand introduction to the interactions of the universe's physical forces. Six essays explore gravitation, cohesion, chemical affinity, heat, magnetism, and electricity. 1993 edition. 96pp. 5 3/8 x 8 1/2. 0-486-47482-8

DYNAMICS, Lawrence E. Goodman and William H. Warner. Beginning engineering text introduces calculus of vectors, particle motion, dynamics of particle systems and plane rigid bodies, technical applications in plane motions, and more. Exercises and answers in every chapter. 619pp. 5 3/8 x 8 1/2. 0-486-42006-X

ADAPTIVE FILTERING PREDICTION AND CONTROL, Graham C. Goodwin and Kwai Sang Sin. This unified survey focuses on linear discrete-time systems and explores natural extensions to nonlinear systems. It emphasizes discrete-time systems, summarizing theoretical and practical aspects of a large class of adaptive algorithms. 1984 edition. 560pp. 6 1/2 x 9 1/4. 0-486-46932-8

INDUCTANCE CALCULATIONS, Frederick W. Grover. This authoritative reference enables the design of virtually every type of inductor. It features a single simple formula for each type of inductor, together with tables containing essential numerical factors. 1946 edition. 304pp. 5 3/8 x 8 1/2. 0-486-47440-2

THERMODYNAMICS: Foundations and Applications, Elias P. Gyftopoulos and Gian Paolo Beretta. Designed by two MIT professors, this authoritative text discusses basic concepts and applications in detail, emphasizing generality, definitions, and logical consistency. More than 300 solved problems cover realistic energy systems and processes. 800pp. 6 1/8 x 9 1/4. 0-486-43932-1

THE FINITE ELEMENT METHOD: Linear Static and Dynamic Finite Element Analysis, Thomas J. R. Hughes. Text for students without in-depth mathematical training, this text includes a comprehensive presentation and analysis of algorithms of time-dependent phenomena plus beam, plate, and shell theories. Solution guide available upon request. 672pp. 6 1/2 x 9 1/4. 0-486-41181-8

Mathematics–Logic and Problem Solving

PERPLEXING PUZZLES AND TANTALIZING TEASERS, Martin Gardner. Ninety-three riddles, mazes, illusions, tricky questions, word and picture puzzles, and other challenges offer hours of entertainment for youngsters. Filled with rib-tickling drawings. Solutions. 224pp. 5 3/8 x 8 1/2. 0-486-25637-5

MY BEST MATHEMATICAL AND LOGIC PUZZLES, Martin Gardner. The noted expert selects 70 of his favorite "short" puzzles. Includes The Returning Explorer, The Mutilated Chessboard, Scrambled Box Tops, and dozens more. Complete solutions included. 96pp. 5 3/8 x 8 1/2. 0-486-28152-3

THE LADY OR THE TIGER?: and Other Logic Puzzles, Raymond M. Smullyan. Created by a renowned puzzle master, these whimsically themed challenges involve paradoxes about probability, time, and change; metapuzzles; and self-referentiality. Nineteen chapters advance in difficulty from relatively simple to highly complex. 1982 edition. 240pp. 5 3/8 x 8 1/2. 0-486-47027-X

SATAN, CANTOR AND INFINITY: Mind-Boggling Puzzles, Raymond M. Smullyan. A renowned mathematician tells stories of knights and knaves in an entertaining look at the logical precepts behind infinity, probability, time, and change. Requires a strong background in mathematics. Complete solutions. 288pp. 5 3/8 x 8 1/2. 0-486-47036-9

THE RED BOOK OF MATHEMATICAL PROBLEMS, Kenneth S. Williams and Kenneth Hardy. Handy compilation of 100 practice problems, hints and solutions indispensable for students preparing for the William Lowell Putnam and other mathematical competitions. Preface to the First Edition. Sources. 1988 edition. 192pp. 5 3/8 x 8 1/2. 0-486-69415-1

KING ARTHUR IN SEARCH OF HIS DOG AND OTHER CURIOUS PUZZLES, Raymond M. Smullyan. This fanciful, original collection for readers of all ages features arithmetic puzzles, logic problems related to crime detection, and logic and arithmetic puzzles involving King Arthur and his Dogs of the Round Table. 160pp. 5 3/8 x 8 1/2. 0-486-47435-6

UNDECIDABLE THEORIES: Studies in Logic and the Foundation of Mathematics, Alfred Tarski in collaboration with Andrzej Mostowski and Raphael M. Robinson. This well-known book by the famed logician consists of three treatises: "A General Method in Proofs of Undecidability," "Undecidability and Essential Undecidability in Mathematics," and "Undecidability of the Elementary Theory of Groups." 1953 edition. 112pp. 5 3/8 x 8 1/2. 0-486-47703-7

LOGIC FOR MATHEMATICIANS, J. Barkley Rosser. Examination of essential topics and theorems assumes no background in logic. "Undoubtedly a major addition to the literature of mathematical logic." – *Bulletin of the American Mathematical Society.* 1978 edition. 592pp. 6 1/8 x 9 1/4. 0-486-46898-4

INTRODUCTION TO PROOF IN ABSTRACT MATHEMATICS, Andrew Wohlgemuth. This undergraduate text teaches students what constitutes an acceptable proof, and it develops their ability to do proofs of routine problems as well as those requiring creative insights. 1990 edition. 384pp. 6 1/2 x 9 1/4. 0-486-47854-8

FIRST COURSE IN MATHEMATICAL LOGIC, Patrick Suppes and Shirley Hill. Rigorous introduction is simple enough in presentation and context for wide range of students. Symbolizing sentences; logical inference; truth and validity; truth tables; terms, predicates, universal quantifiers; universal specification and laws of identity; more. 288pp. 5 3/8 x 8 1/2. 0-486-42259-3

Mathematics–Algebra and Calculus

VECTOR CALCULUS, Peter Baxandall and Hans Liebeck. This introductory text offers a rigorous, comprehensive treatment. Classical theorems of vector calculus are amply illustrated with figures, worked examples, physical applications, and exercises with hints and answers. 1986 edition. 560pp. 5 3/8 x 8 1/2. 0-486-46620-5

ADVANCED CALCULUS: An Introduction to Classical Analysis, Louis Brand. A course in analysis that focuses on the functions of a real variable, this text introduces the basic concepts in their simplest setting and illustrates its teachings with numerous examples, theorems, and proofs. 1955 edition. 592pp. 5 3/8 x 8 1/2. 0-486-44548-8

ADVANCED CALCULUS, Avner Friedman. Intended for students who have already completed a one-year course in elementary calculus, this two-part treatment advances from functions of one variable to those of several variables. Solutions. 1971 edition. 432pp. 5 3/8 x 8 1/2. 0-486-45795-8

METHODS OF MATHEMATICS APPLIED TO CALCULUS, PROBABILITY, AND STATISTICS, Richard W. Hamming. This 4-part treatment begins with algebra and analytic geometry and proceeds to an exploration of the calculus of algebraic functions and transcendental functions and applications. 1985 edition. Includes 310 figures and 18 tables. 880pp. 6 1/2 x 9 1/4. 0-486-43945-3

BASIC ALGEBRA I: Second Edition, Nathan Jacobson. A classic text and standard reference for a generation, this volume covers all undergraduate algebra topics, including groups, rings, modules, Galois theory, polynomials, linear algebra, and associative algebra. 1985 edition. 528pp. 6 1/8 x 9 1/4. 0-486-47189-6

BASIC ALGEBRA II: Second Edition, Nathan Jacobson. This classic text and standard reference comprises all subjects of a first-year graduate-level course, including in-depth coverage of groups and polynomials and extensive use of categories and functors. 1989 edition. 704pp. 6 1/8 x 9 1/4. 0-486-47187-X

CALCULUS: An Intuitive and Physical Approach (Second Edition), Morris Kline. Application-oriented introduction relates the subject as closely as possible to science with explorations of the derivative; differentiation and integration of the powers of x; theorems on differentiation, antidifferentiation; the chain rule; trigonometric functions; more. Examples. 1967 edition. 960pp. 6 1/2 x 9 1/4. 0-486-40453-6

ABSTRACT ALGEBRA AND SOLUTION BY RADICALS, John E. Maxfield and Margaret W. Maxfield. Accessible advanced undergraduate-level text starts with groups, rings, fields, and polynomials and advances to Galois theory, radicals and roots of unity, and solution by radicals. Numerous examples, illustrations, exercises, appendixes. 1971 edition. 224pp. 6 1/8 x 9 1/4. 0-486-47723-1

AN INTRODUCTION TO THE THEORY OF LINEAR SPACES, Georgi E. Shilov. Translated by Richard A. Silverman. Introductory treatment offers a clear exposition of algebra, geometry, and analysis as parts of an integrated whole rather than separate subjects. Numerous examples illustrate many different fields, and problems include hints or answers. 1961 edition. 320pp. 5 3/8 x 8 1/2. 0-486-63070-6

LINEAR ALGEBRA, Georgi E. Shilov. Covers determinants, linear spaces, systems of linear equations, linear functions of a vector argument, coordinate transformations, the canonical form of the matrix of a linear operator, bilinear and quadratic forms, and more. 387pp. 5 3/8 x 8 1/2. 0-486-63518-X

Mathematics-Probability and Statistics

BASIC PROBABILITY THEORY, Robert B. Ash. This text emphasizes the probabilistic way of thinking, rather than measure-theoretic concepts. Geared toward advanced undergraduates and graduate students, it features solutions to some of the problems. 1970 edition. 352pp. 5 3/8 x 8 1/2. 0-486-46628-0

PRINCIPLES OF STATISTICS, M. G. Bulmer. Concise description of classical statistics, from basic dice probabilities to modern regression analysis. Equal stress on theory and applications. Moderate difficulty; only basic calculus required. Includes problems with answers. 252pp. 5 5/8 x 8 1/4. 0-486-63760-3

OUTLINE OF BASIC STATISTICS: Dictionary and Formulas, John E. Freund and Frank J. Williams. Handy guide includes a 70-page outline of essential statistical formulas covering grouped and ungrouped data, finite populations, probability, and more, plus over 1,000 clear, concise definitions of statistical terms. 1966 edition. 208pp. 5 3/8 x 8 1/2. 0-486-47769-X

GOOD THINKING: The Foundations of Probability and Its Applications, Irving J. Good. This in-depth treatment of probability theory by a famous British statistician explores Keynesian principles and surveys such topics as Bayesian rationality, corroboration, hypothesis testing, and mathematical tools for induction and simplicity. 1983 edition. 352pp. 5 3/8 x 8 1/2. 0-486-47438-0

INTRODUCTION TO PROBABILITY THEORY WITH CONTEMPORARY APPLICATIONS, Lester L. Helms. Extensive discussions and clear examples, written in plain language, expose students to the rules and methods of probability. Exercises foster problem-solving skills, and all problems feature step-by-step solutions. 1997 edition. 368pp. 6 1/2 x 9 1/4. 0-486-47418-6

CHANCE, LUCK, AND STATISTICS, Horace C. Levinson. In simple, non-technical language, this volume explores the fundamentals governing chance and applies them to sports, government, and business. "Clear and lively ... remarkably accurate." – *Scientific Monthly.* 384pp. 5 3/8 x 8 1/2. 0-486-41997-5

FIFTY CHALLENGING PROBLEMS IN PROBABILITY WITH SOLUTIONS, Frederick Mosteller. Remarkable puzzlers, graded in difficulty, illustrate elementary and advanced aspects of probability. These problems were selected for originality, general interest, or because they demonstrate valuable techniques. Also includes detailed solutions. 88pp. 5 3/8 x 8 1/2. 0-486-65355-2

EXPERIMENTAL STATISTICS, Mary Gibbons Natrella. A handbook for those seeking engineering information and quantitative data for designing, developing, constructing, and testing equipment. Covers the planning of experiments, the analyzing of extreme-value data; and more. 1966 edition. Index. Includes 52 figures and 76 tables. 560pp. 8 3/8 x 11. 0-486-43937-2

STOCHASTIC MODELING: Analysis and Simulation, Barry L. Nelson. Coherent introduction to techniques also offers a guide to the mathematical, numerical, and simulation tools of systems analysis. Includes formulation of models, analysis, and interpretation of results. 1995 edition. 336pp. 6 1/8 x 9 1/4. 0-486-47770-3

INTRODUCTION TO BIOSTATISTICS: Second Edition, Robert R. Sokal and F. James Rohlf. Suitable for undergraduates with a minimal background in mathematics, this introduction ranges from descriptive statistics to fundamental distributions and the testing of hypotheses. Includes numerous worked-out problems and examples. 1987 edition. 384pp. 6 1/8 x 9 1/4. 0-486-46961-1

Mathematics–Geometry and Topology

PROBLEMS AND SOLUTIONS IN EUCLIDEAN GEOMETRY, M. N. Aref and William Wernick. Based on classical principles, this book is intended for a second course in Euclidean geometry and can be used as a refresher. More than 200 problems include hints and solutions. 1968 edition. 272pp. 5 3/8 x 8 1/2. 0-486-47720-7

TOPOLOGY OF 3-MANIFOLDS AND RELATED TOPICS, Edited by M. K. Fort, Jr. With a New Introduction by Daniel Silver. Summaries and full reports from a 1961 conference discuss decompositions and subsets of 3-space; n-manifolds; knot theory; the Poincaré conjecture; and periodic maps and isotopies. Familiarity with algebraic topology required. 1962 edition. 272pp. 6 1/8 x 9 1/4. 0-486-47753-3

POINT SET TOPOLOGY, Steven A. Gaal. Suitable for a complete course in topology, this text also functions as a self-contained treatment for independent study. Additional enrichment materials make it equally valuable as a reference. 1964 edition. 336pp. 5 3/8 x 8 1/2. 0-486-47222-1

INVITATION TO GEOMETRY, Z. A. Melzak. Intended for students of many different backgrounds with only a modest knowledge of mathematics, this text features self-contained chapters that can be adapted to several types of geometry courses. 1983 edition. 240pp. 5 3/8 x 8 1/2. 0-486-46626-4

TOPOLOGY AND GEOMETRY FOR PHYSICISTS, Charles Nash and Siddhartha Sen. Written by physicists for physics students, this text assumes no detailed background in topology or geometry. Topics include differential forms, homotopy, homology, cohomology, fiber bundles, connection and covariant derivatives, and Morse theory. 1983 edition. 320pp. 5 3/8 x 8 1/2. 0-486-47852-1

BEYOND GEOMETRY: Classic Papers from Riemann to Einstein, Edited with an Introduction and Notes by Peter Pesic. This is the only English-language collection of these 8 accessible essays. They trace seminal ideas about the foundations of geometry that led to Einstein's general theory of relativity. 224pp. 6 1/8 x 9 1/4. 0-486-45350-2

GEOMETRY FROM EUCLID TO KNOTS, Saul Stahl. This text provides a historical perspective on plane geometry and covers non-neutral Euclidean geometry, circles and regular polygons, projective geometry, symmetries, inversions, informal topology, and more. Includes 1,000 practice problems. Solutions available. 2003 edition. 480pp. 6 1/8 x 9 1/4. 0-486-47459-3

TOPOLOGICAL VECTOR SPACES, DISTRIBUTIONS AND KERNELS, François Trèves. Extending beyond the boundaries of Hilbert and Banach space theory, this text focuses on key aspects of functional analysis, particularly in regard to solving partial differential equations. 1967 edition. 592pp. 5 3/8 x 8 1/2. 0-486-45352-9

INTRODUCTION TO PROJECTIVE GEOMETRY, C. R. Wylie, Jr. This introductory volume offers strong reinforcement for its teachings, with detailed examples and numerous theorems, proofs, and exercises, plus complete answers to all odd-numbered end-of-chapter problems. 1970 edition. 576pp. 6 1/8 x 9 1/4. 0-486-46895-X

FOUNDATIONS OF GEOMETRY, C. R. Wylie, Jr. Geared toward students preparing to teach high school mathematics, this text explores the principles of Euclidean and non-Euclidean geometry and covers both generalities and specifics of the axiomatic method. 1964 edition. 352pp. 6 x 9. 0-486-47214-0

Mathematics–History

THE WORKS OF ARCHIMEDES, Archimedes. Translated by Sir Thomas Heath. Complete works of ancient geometer feature such topics as the famous problems of the ratio of the areas of a cylinder and an inscribed sphere; the properties of conoids, spheroids, and spirals; more. 326pp. 5 3/8 x 8 1/2. 0-486-42084-1

THE HISTORICAL ROOTS OF ELEMENTARY MATHEMATICS, Lucas N. H. Bunt, Phillip S. Jones, and Jack D. Bedient. Exciting, hands-on approach to understanding fundamental underpinnings of modern arithmetic, algebra, geometry and number systems examines their origins in early Egyptian, Babylonian, and Greek sources. 336pp. 5 3/8 x 8 1/2. 0-486-25563-8

THE THIRTEEN BOOKS OF EUCLID'S ELEMENTS, Euclid. Contains complete English text of all 13 books of the Elements plus critical apparatus analyzing each definition, postulate, and proposition in great detail. Covers textual and linguistic matters; mathematical analyses of Euclid's ideas; classical, medieval, Renaissance and modern commentators; refutations, supports, extrapolations, reinterpretations and historical notes. 995 figures. Total of 1,425pp. All books 5 3/8 x 8 1/2.
Vol. I: 443pp. 0-486-60088-2
Vol. II: 464pp. 0-486-60089-0
Vol. III: 546pp. 0-486-60090-4

A HISTORY OF GREEK MATHEMATICS, Sir Thomas Heath. This authoritative two-volume set that covers the essentials of mathematics and features every landmark innovation and every important figure, including Euclid, Apollonius, and others. 5 3/8 x 8 1/2. Vol. I: 461pp. 0-486-24073-8
Vol. II: 597pp. 0-486-24074-6

A MANUAL OF GREEK MATHEMATICS, Sir Thomas L. Heath. This concise but thorough history encompasses the enduring contributions of the ancient Greek mathematicians whose works form the basis of most modern mathematics. Discusses Pythagorean arithmetic, Plato, Euclid, more. 1931 edition. 576pp. 5 3/8 x 8 1/2.
0-486-43231-9

CHINESE MATHEMATICS IN THE THIRTEENTH CENTURY, Ulrich Libbrecht. An exploration of the 13th-century mathematician Ch'in, this fascinating book combines what is known of the mathematician's life with a history of his only extant work, the Shu-shu chiu-chang. 1973 edition. 592pp. 5 3/8 x 8 1/2.
0-486-44619-0

PHILOSOPHY OF MATHEMATICS AND DEDUCTIVE STRUCTURE IN EUCLID'S ELEMENTS, Ian Mueller. This text provides an understanding of the classical Greek conception of mathematics as expressed in Euclid's Elements. It focuses on philosophical, foundational, and logical questions and features helpful appendixes. 400pp. 6 1/2 x 9 1/4. 0-486-45300-6

BEYOND GEOMETRY: Classic Papers from Riemann to Einstein, Edited with an Introduction and Notes by Peter Pesic. This is the only English-language collection of these 8 accessible essays. They trace seminal ideas about the foundations of geometry that led to Einstein's general theory of relativity. 224pp. 6 1/8 x 9 1/4. 0-486-45350-2

HISTORY OF MATHEMATICS, David E. Smith. Two-volume history – from Egyptian papyri and medieval maps to modern graphs and diagrams. Non-technical chronological survey with thousands of biographical notes, critical evaluations, and contemporary opinions on over 1,100 mathematicians. 5 3/8 x 8 1/2.
Vol. I: 618pp. 0-486-20429-4
Vol. II: 736pp. 0-486-20430-8